CAMBRIDGE LIBRARY COLLECTION

Books of enduring scholarly value

Physical Sciences

From ancient times, humans have tried to understand the workings of the world around them. The roots of modern physical science go back to the very earliest mechanical devices such as levers and rollers, the mixing of paints and dyes, and the importance of the heavenly bodies in early religious observance and navigation. The physical sciences as we know them today began to emerge as independent academic subjects during the early modern period, in the work of Newton and other 'natural philosophers', and numerous sub-disciplines developed during the centuries that followed. This part of the Cambridge Library Collection is devoted to landmark publications in this area which will be of interest to historians of science concerned with individual scientists, particular discoveries, and advances in scientific method, or with the establishment and development of scientific institutions around the world.

Catalogue of Books and Papers Relating to Electricity, Magnetism, the Electric Telegraph, Etc.

In 1816, Sir Francis Ronalds (1788–1873) became the first physicist to demonstrate the possibility of an electric telegraph. Previously, the only telegraphs were semaphores – cumbersome signal towers capable of sending only two or three words per minute. However, his idea was dismissed by the Admiralty, where senior officials deemed any new telegraphs 'unnecessary'. Although his designs were soon to be superseded by those of the more successful Samuel Morse, Ronalds' devotion to telegraphy never waned; he spent much of his life collecting books on the subject. Upon his death, his collection was left to the Society of Telegraph Engineers, where it would become available to those most in need of it. Covering more than 13,000 titles, and including a short memoir of Ronalds, this book, first published in 1880, is a catalogue of that collection and other relevant works. It remains an invaluable resource for students in the history of science.

Catalogue of Books and Papers Relating to Electricity, Magnetism, the Electric Telegraph, Etc.

FRANCIS RONALDS
EDITED BY ALFRED J. FROST

CAMBRIDGE
UNIVERSITY PRESS

CAMBRIDGE UNIVERSITY PRESS

Cambridge, New York, Melbourne, Madrid, Cape Town,
Singapore, São Paolo, Delhi, Mexico City

Published in the United States of America by Cambridge University Press, New York

www.cambridge.org
Information on this title: www.cambridge.org/9781108052542

© in this compilation Cambridge University Press 2013

This edition first published 1880
This digitally printed version 2013

ISBN 978-1-108-05254-2 Paperback

PUBLISHED BY THE SOCIETY OF TELEGRAPH ENGINEERS.

CATALOGUE

OF BOOKS AND PAPERS RELATING TO

ELECTRICITY, MAGNETISM,

THE ELECTRIC TELEGRAPH, &c.

INCLUDING

THE RONALDS LIBRARY.

COMPILED BY

SIR FRANCIS RONALDS, F.R.S.

WITH A BIOGRAPHICAL MEMOIR.

EDITED BY

ALFRED J. FROST,

Acting-Librarian of the Society of Telegraph Engineers and Member of the Library Association of the United Kingdom.

LONDON : E. & F. N. SPON, 46, CHARING CROSS.

NEW YORK: 446, BROOME STREET.

—

1880.

UNWIN BROTHERS, PRINTERS, 109A, CANNON STREET, LONDON; AND CHILWORTH, SURREY.

PREFACE.

THE compilation of this Catalogue was commenced at an early date, and was continued up to the death of its compiler in 1873.

It contains over 13,000 entries, comprising not only the books, pamphlets, and other publications in the RONALDS LIBRARY, but also the titles of all other works on the subject of Electricity, Magnetism, &c. which came to the notice of its compiler.

The titles are arranged under the authors' names, and the anonymous entries have been placed, with a slight attempt at classification, under a separate heading at the end of letter A.

After Sir Francis Ronalds retired from the direction of Kew Observatory, he lived for many years abroad, chiefly in Italy, his principal occupation being the compilation of this Catalogue and the formation of his Library. He, consequently, had exceptional opportunities for obtaining the foreign works upon the subject, and the collection is extremely rich in such publications.

During this time it was one of his favourite pursuits to obtain as many of the Electrical works as possible which he required, which had belonged to those whose names have become eminent, and are identified with the subject. An instance of this may be given of the celebrated Volta's works, a large number of which he obtained from the Count Zanino Volta at great personal trouble and some expense. There are also in the Library many works which originally belonged to Arago, Faraday, Roget, and others.

Many of the most important works relating to Electricity, and, indeed, of most other branches of science, consist of papers contributed to the Transactions and Proceedings of scientific bodies, and to scientific periodicals. These papers have each been treated as separate works, and their titles are recorded under the author's name, a note being added giving the reference to the transactions or periodicals where they may be found.

A considerable number of the titles have been obtained from book-sellers' catalogues, and from notes and references in Electrical writings, and, where Sir Francis Ronalds has been unable to obtain the works, the source from which the title was obtained is in most cases given.

The date and place of birth, and, where possible, of the death of each author, have been added, the greater portion of these having been obtained from Poggendorff's "Handwörterbuch zur Geschichte der Exacten Wissenschaften," 2 vols. Roy. 8vo. Leipzig, 1863. This work was also carefully examined, and those parts extracted which relate to Electricity.

As the binding of the books in the Library has only just been completed, it was impossible, owing to the want of arrangement of the works, and to the large number of unbound pamphlets and papers, to refer to them to any extent for the correction of titles; besides which, there are between 3,000 and 4,000 titles of works entered in the Catalogue which are not in the Library.

The Catalogue is published without the addition of any new entries, such being the express desire of the donor of the Library, Mr. Samuel Carter. Great efforts were made to obtain his permission to make additions, and, as far as possible, to complete the Catalogue to the date of publication, but it is to be regretted that these efforts were unsuccessful. The Society has, however, at its disposal a large amount of material available for this purpose, and it is contemplated at a not very distant date to publish a Supplement, which will make the Catalogue more complete, at any rate as regards the more recent publications.

Great care has been taken to issue the Catalogue as free from error as possible, and it is hoped that it will be found as accurate as the nature of the work will permit.

There can be no doubt that it was the original intention of Sir Francis Ronalds to present his Library to the Royal Society (of which he was a Fellow), but previous to his death it was represented to him that it would add but little to the value of the Royal Society's collection, as they already possessed the greater number of the books. It was further represented that as his name was so much identified with the subject of Electrical Telegraphy, he would be conferring great benefit upon those specially interested in Electricity and the Electric Telegraph, by making the Library more available for their use than it could be in the Library of the Royal Society, in which place it would be, to a great extent, out of the reach of practical electricians and telegraph engineers. These arguments did not fully prevail with him, as the Society had only recently been founded, and could at that time scarcely be said to have been firmly established.

Sir Francis Ronalds became one of the earliest members of the Society, but, owing to his advanced age, he was unable to be present at any of its meetings. He died about two years after its foundation, but he had during this time some opportunity of seeing its rapid development, and no doubt this, coupled with the representations that had been made to him, considerably influenced him in the disposal of the Library. He died on the 8th August, 1873, finally bequeathing the Library to his brother-in-law, the late Mr. Samuel Carter, with a desire " that it should not be dispersed, but should be preserved in an entire state, and so as to be of as much use as possible to such persons as from time to time should be engaged in the pursuit of Electrical science and other cognate sciences."

Mr. Carter considered that he would be best carrying out the wishes of Sir Francis Ronalds by handing the Library over to the Society, which he did upon Trust, the following being the names of the Trustees : Samuel Carter, Esq., Dr. Edmund Ronalds, John Corrie Carter, Esq., John Martineau Fletcher, Esq., Sir William Thomson, F.R.S., D.C.L., Latimer Clark, Esq., M.I.C.E., Charles William Siemens, Esq., F.R.S., D.C.L., and Major (now Lieut.-Colonel) Frank Bolton.

There are several conditions upon which the Society retains the Library, the two principal being the publication of this Catalogue and the binding of the Library—the latter condition having just been carried out.

As it was the desire of Sir Francis Ronalds that the Library should be made as widely useful as possible, it is the intention of the Council of the Society, as soon as the arrangement of the books is completed, to open the Library to the Members of the Society, and under certain necessary restrictions to the public generally.

The thanks of the Society are due to the Commandeur E. D'Amico, Director-General of Italian Telegraphs, M. Alfred Niandet, of Paris, and Mr. R. Von Fischer-Treuenfeld, for their kindness in having undertaken the correction of the Italian, French, and German portions of the proofs respectively.

BIOGRAPHICAL MEMOIR

OF

SIR FRANCIS RONALDS, F.R.S.

BY

ALFRED J. FROST.

FRANCIS RONALDS was born in London, on the 21st February, 1788, and was the son of Francis Ronalds, Esq., a merchant of the same city. He was educated at a private school, at Cheshunt, his master being the Rev. E. Cogan.

The Ronalds family, who originally came from Scotland, appear to have settled at Brentford for some considerable time, for on reference to Faulkner's "Antiquities of Brentford and Chiswick," it will be found that there are no less than ten memorials in St. Lawrence Church, Brentford, erected between 1779 and 1833, to the memory of members of the family.

He early showed a taste for original experiment, and this, coupled with considerable mechanical and inventive ability, enabled him to construct with his own hands electrical and scientific apparatus of various kinds. He was also an excellent draughtsman, and evidence of this talent may be seen in most of his publications, the drawings of which are, in almost every case, his own.

At an early age he studied Electricity, and having (in 1814) made the acquaintance of M. De Luc—then engaged in a series of interesting electrical experiments—he was induced to turn his attention to the practical development of the science.

During 1814-15 he wrote several electrical papers, which were published in the *Philosophical Magazine ;* and a complete list of these and of his other writings will be found at pp. 438 and 439 of the Ronalds Catalogue.

The name of Ronalds will, however, continue to be best remembered from the fact that as early as 1816 he demonstrated by actual experiment the possibility of an electric telegraph, and showed that electricity could be practically used for conveying messages over long distances.

Being fully cognisant of the difficulties arising from defective insulation, and knowing that those who had previously experimented in the same direction had failed from this cause, he secured the success of his arrangements by adopting for his underground line a system of insulation vastly superior to any that had hitherto been employed.

Although the telegraph of Ronalds has been frequently described, and is now tolerably well known to telegraph engineers, a biographical notice of its inventor would be incomplete which did not include a short description of it, as well as an account of the reasons of its non-adoption, or at least of its not having been tried on a larger scale than the experimental line which he constructed would permit of.

The telegraph referred to was erected and worked in 1816, at Hammersmith, and is described and fully illustrated in a small and exceedingly interesting work which was published in 1823,* and now scarce : a second edition was published in 1871. The following is an extract from this work : "Upon a lawn or grassplot at Hammer-

---

* Description of an Electrical Telegraph, and of some other Electrical Apparatus. By Francis Ronalds. Demy 8vo., London, 1823.

smith* I erected two strong frames of wood at a distance of twenty
yards from each other, and each containing nineteen horizontal bars;
to each bar was attached thirty-seven hooks, and to the hooks were
applied as many silken cords, which supported a small iron wire (by
these means well insulated),which (making its inflections at the points
of support) composed in one continuous length a distance of rather
more than eight miles."

Having made many experiments with this over-head line, which
was the longest that had at that time been constructed, he gave his
attention to the most suitable line to be adopted for the telegraph
which he proposed. This he decided by proposing an underground
line, a length of which he constructed and laid down, and which he
describes as follows: "A trench was dug in the garden five
hundred and twenty-five feet in length, and four feet deep; in
this was laid a trough of wood two inches square, well lined on
the inside and out with pitch, and within this trough thick glass
tubes were placed, through which the wire ran." †

In order to prevent the tubes from breaking by the variation of
temperature, each length was laid a short distance from the next
length, and the joint made with soft wax. "The trough was then
covered with pieces of wood, screwed upon it whilst the pitch was hot;
they also were well covered with pitch, and the earth then thrown into
the trench again."

The method of conveying intelligence which he adopted may be
briefly described as follows:—At each end of the line was placed a
clock beating dead seconds, having upon its seconds arbor a light

---

* The house at Hammersmith where the line was erected, and where
the original experiments took place, is the first large house at the east
end of the Upper Mall.

† The first underground line laid by Cooke, between Euston and Cam-
den, in 1837, consisted of solid blocks of wood, in which were cut small
grooves for receiving the wires, and which were afterwards filled up by
wooden strips.

circular brass plate divided into twenty equal parts. Each division
was marked with the letters of the alphabet—leaving out the letters
J Q U W X and Z ;—there were also a series of figures from 1 to
10, and 10 preparatory signs. Before or over this disc was fixed
another brass plate, which had an aperture of such dimensions that
whilst the disc was carried round by motion of the clock only one of
the letters, figures, or preparatory signs upon it could be seen
through the aperture at the same time. In front of the clocks a
Canton's Electrometer of pith balls was suspended by an insulated
wire communicating with a cylindrical electrical machine of six
inches' diameter, which was connected with the line. These clocks
were adjusted (by a method described) to go synchronously.

Upon a current being sent from the electrical machine through the
line the pith balls diverged, and by noting the letter which appeared
through the aperture at the moment, a message could be spelt out—
or, as Ronalds says, " by the use of a telegraphic dictionary " (or code,
as it would now be called) " a word, or even a whole sentence, could
be conveyed by only two or three discharges."

The code, or " dictionary," proposed by Ronalds is described in the
work, and is accompanied by a plate which represents one leaf of it.

Having made a large number of experiments with these lines, and
having thoroughly proved the practicability of his invention, he decided
upon bringing it to the notice of the Government. This he did on
the 11th of July, 1816, in a letter addressed to Lord Melville, the
First Lord of the Admiralty, as follows :—-

" Mr. Ronalds presents his respectful compliments to Lord
Melville, and takes the liberty of soliciting his lordship's attention to
a mode of conveying telegraphic intelligence with great rapidity,
accuracy, and certainty, in all states of the atmosphere, either at
night or in the day, and at small expense, which has occurred to him
whilst pursuing some electrical experiments. Having been at some
pains to ascertain the *practicability* of the scheme, it appears to Mr.

Ronalds, and to a few gentlemen by whom it has been examined, to possess several important advantages over any species of telegraph hitherto invented, and he would be much gratified by an opportunity of demonstrating those advantages to Lord Melville by an experiment which he has no doubt would be deemed decisive, if it should be perfectly agreeable and consistent with his lordship's engagements to honour Mr. Ronalds with a call ; or he would be very happy to explain more particularly the nature of the contrivance if Lord Melville could conveniently oblige him by appointing an interview.

Upper Mall, Hammersmith, July 11, 1816."

Shortly afterwards Mr. Ronalds received the following in reply:—

"Admiralty, July 29th, 1816.

SIR,—I am desired by Lord Melville to acknowledge the receipt of your letter of the 11th instant. His Lordship has left town for some weeks, but he has requested me to see you on the subject of your discovery if you desire it.

I am, Sir,

Your most obedient, humble servant,

— Ronalds, Esq. R. N. HAY."

Mr. Ronalds replied that he did desire an interview, but before the nature of the invention had been made known except to Lord Henniker, Dr. Rees, Mr. Brande, and a few personal friends, he received the following laconic and curious communication from Mr. Barrow (afterwards Sir John Barrow), the Secretary of the Admiralty : —

" Mr. Barrow presents his compliments to Mr. Ronalds, and acquaints him with reference to his note of the 3rd instant, that telegraphs of any kind are now wholly unnecessary ; and that no other than the one now in use will be adopted.

Admiralty Office, 5th August, 1816."

Sir John Barrow was the author of the article " Telegraph," in the seventh edition of the Encyclopædia Britannica.

Ronalds was too far ahead of his time, and too purely a man of science to secure a hearing for his invention in these early days, and it was left to others to mature his ideas, and to establish the system which his prophetic eye had foreseen would one day transform the world.

In alluding to the above correspondence in a foot-note at page 24 of his work, Ronalds says: "I felt very little disappointment and not a shadow of resentment on the occasion, because everyone knows that telegraphs have long been great bores at the Admiralty. Should they *again* become *necessary*, however, perhaps electricity and electricians may be indulged by his Lordship and Mr. Barrow with an opportunity of *proving* what they are capable of in this way. I claim no indulgence for mere chimeras and chimera framers, and I hope to escape the fate of being ranked in that unenviable class."

It is probable, however, that he did feel some disappointment, for in the preface to his work he speaks of "taking leave of a science which once afforded him a favourite source of amusement," and adds that "he is compelled to bid a cordial adieu to electricity."

The only telegraph which had at that time been erected, and which is referred to in Mr. Barrow's communication, was the semaphore telegraph from Portsmouth to London. This telegraph was used for many years after the date now in question, and was the subject of a parliamentary return in 1843 (twenty-seven years afterwards), hereafter referred to. It was based upon the invention of Chappe, which had been previously used in France, and which was afterwards modified by Colonel Pasley and Sir Home Popham. It was worked by moveable arms, by means of which signals could be transmitted from station to station; but it could only be used in the daytime, and then only during fine weather. In foggy or misty weather it was more or less difficult and often impossible for one station to see the next, and signals could not be communicated; and it sometimes occurred that a message would be commenced in London or Portsmouth when the atmosphere was clear, and its transmission had to be delayed in consequence of some part of the route being enveloped in fog or mist.

In the parliamentary return referred to, bearing date the 2nd May, 1843, it is stated that during three years it was found impossible to use the semaphore from the Admiralty on no less than 323 days: the usual hours of working being from October 1st to February 28th, five hours per day; and from March 1st to September 30th, seven hours per day.

Judged by the light of our present requirements, it is almost impossible to realise the fact that the Government was contented with the means of conveying intelligence between their head-quarters and their principal naval station, then at their disposal, and which found any other telegraph "*wholly unnecessary.*"

The semaphore telegraph between London and Portsmouth is stated to have cost £2,000 per annum.

In the work before referred to, Ronalds foretells in a very striking manner the many public uses to which the telegraph was applicable. "Why," he says, "should not our Kings hold councils at Brighton with their Ministers in London? Why should not our Government govern at Portsmouth almost as promptly as in Downing Street? Why should our defaulters escape by default of our foggy climate? Let us have *electrical conversazione* offices communicating with each other all over the kingdom."

The phenomena of the "retardation of current," which so much occupied the attention of electricians and telegraph engineers during the early days of submarine telegraphy, was distinctly foreseen and described in this work.

At pp. 16—20 he provides for repairing the line and discovering faults; and in referring to the line being wilfully damaged, he says: "Should they still succeed in breaking the communication, hang them if you can catch them, damn them if you cannot, and mend it immediately in both cases."

The allusion to the escape of criminals is curious, seeing that a quarter of a century afterwards, one of the first practical uses of the telegraph of Cooke and Wheatstone (and which did more to popularise the telegraph than anything else) was its employment in effecting the capture of the Quaker Tawell, for the murder at Salthill.

The telegraph of Ronalds was worked by frictional electricity, and although in its crude form it might not have been at once capable of doing all that he promised, yet, had the Admiralty given the invention a fair trial in 1816, it might have been easily improved, and the whole history of the telegraph would in all probability have been changed; at any rate its powers would have been exercised, and the world might have profited by its advantages long before 1837.

During the time which elapsed between the erection of the experimental line in 1816 and the publication of the description of it in 1823, Ronalds, who was a philosopher as well as an inventor, went abroad, travelling through Europe to the East; this journey occupied between two and three years, and it is interesting to note that during this period he had (it would seem, from some of his letters written at the time) commenced the collecting of a portion of the electrical works which now form the fine library in the possession of the Society of Telegraph Engineers.

Besides the description of the telegraph there is included in the before-mentioned work " a description of an improved electrical machine; description of a new mode of electrical insulation, and of some experiments on Vesuvius; a new electrograph for registering the change of atmospheric electricity; a pendulum doubler; and an attempt to apply M. de Luc's electric column to the measurement of time."

Ronalds employed a local carpenter, named Eyles, to assist him in the erection of the telegraph and in laying the underground line. Eyles died in 1854, but his son, Silas Eyles, who was a young lad at the time, and who is now living at Barnes, distinctly remembers the line which his father assisted to erect. He gave the writer several interesting particulars with reference to some experiments which Ronalds performed before him. One of these experiments was to notice the time which elapsed between the connection of the electric machine with the line and the firing of the cannon at the other end— this he said he was unable to do, the cannon going off at the same moment the connection was made. He remembers Ronalds saying that if the line had been 500 miles long instead of eight the result would have been nearly the same.

The name of Ronalds had been long known to the few interested in the history of the electric telegraph, but public attention was more particularly drawn to his claims during the controversy between Cooke and Wheatstone as to the invention of their electric telegraph, a subject upon which Mr. Latimer Clark has thrown considerable light in his exhaustive Memoir of Sir William Fothergill Cooke.*

Prof. Wheatstone, in his " Reply "† to Mr. Cooke's pamphlet, says that " Mr. Cooke's principle (the principle of reciprocal communication) was no novelty, having formed part of previous inventions, . . especially having been developed completely and effectively by Mr. Ronalds in his telegraph, the description of which was published as early as 1823." In this " Reply " Prof. Wheatstone makes further references to the prior labours of Ronalds. See pp. 9, 17, 24, 64, and 66. Wheatstone, then a boy of about 15, was present at many of the principal experiments at Hammersmith.

It is a singular fact that Ronalds was well acquainted with Mr. Cooke, the father of the late Sir William Fothergill Cooke, as is shown by the following extract from Ronalds' diary, dated Dec. 14, 1855 :—

---

* Journal of the Society of Telegraph Engineers, Part xxviii. vol. viii. p. 361, 1st part, 1879.

† Reply to Mr. Cooke's Pamphlet, "The Electric Telegraph: Was it Invented by Professor Wheatstone?" Demy 8vo. London, 1855.

"At about the time of my having completed my electric telegraph, Mr. Cooke, then a surgeon at Brentford, well acquainted with
our family, and father of Mr. W. F. Cooke, called on us at the Upper
Mall, Hammersmith, on which occasion I explained to him minutely
and proved the perfect practicability for general use of my telegraph,
making him the recipient of many messages by it, from the room
over the coachhouse to the little toolhouse at the end of the garden,
and receiving messages in return from him. Mr. W. F. Cooke was a
very young boy at this time."

In a letter to Mr. Ronalds, dated 11th December, 1866, Mr.
(afterwards Sir) Wm. F. Cooke, writes :—

" Many years ago, when you were a young man and I was a boy,
my father, afterwards Dr. Cooke, of Durham, lived at Brentford.   I
think I recollect your living at Mrs. Hanin's cottage in the Butts. If
I am right it is a singular fact that two men so much associated with
the Electrical Telegraph should have been resident in the same small
town.   I have often thought what a fortunate thing it would have been
if I had known of your labours in 1837.   Letters of the alphabet,
three letters in a row, might have been distinguished by your clocks
by movement of a needle to the left, the middle row of the letters by
a flourish of the needle right and left, and the letters to the right by
the needle pointing in that direction."

Mr. Ronalds, in his reply to this letter dated 3rd January, 1867,
says:—

" It must have been your acquaintance with my cousin Dr. Henry
Ronalds to which you refer ; but I had the honour of knowing
your respected father, and well remember an obliging visit from him
when, in 1816, I resided at Hammersmith Upper Mall, on which
occasion I explained and exhibited to him the Electric Telegraph in
operation which I had completed.   I recollect also that he expressed
much satisfaction at what he saw.   You were good enough to say it
would have been a fortunate thing if, in 1837, you had known of my

previous labours. I much regret that you had not then seen my little work of 1823, in which they were fully detailed. Mr. Wheatstone, whom I had long known, was well acquainted with what I had done : it would seem, indeed, that two figures in the second plate of that pamphlet were closely imitated in your specification of 1840."

It should be added that Ronalds took out no patent, but was contented with publishing his invention to the world. The first patent for an electric telegraph was granted to Cooke and Wheatstone in 1837, for " Improvements in giving Signals and Sounding Alarums in distant places by means of electric currents transmitted through metallic circuits."

While there can be no doubt that Wheatstone was perfectly acquainted with the previous labours of Ronalds, the fact that Cooke was not so acquainted is to a great extent proved by the following letter from Mr. Cooke, written more than three years after the date of Cooke and Wheatstone's first patent, and which exists among the manuscript papers in the Ronalds Library :—

" 1, *Copthall Buildings*, 4*th December*, 1840.

My DEAR SIR,—I have been trying for some time to borrow an account of Mr. Ronalds' electric telegraph published by him.

Meeting you yesterday reminded me that you were acquainted with Mr. R., and I shall esteem it a favour if you will procure the loan of it for me as soon as you can.

The Parcel Delivery Co. would deposit it at the above address, or at 51, Notting Hill-square.

Yours very truly,

George Cooper, Esq.   WM. F. COOKE."

Mr. Cooper was known to Ronalds, and the above letter undoubtedly came into the latter's possession when the book referred to was borrowed for Mr. Cooke.

It only remains to be added that after a lapse of nearly fifty years

a portion of this telegraph was discovered in the garden at Hammer
smith. It is thus described by Mr. J. A. Peacock in a letter to Sir
Francis Ronalds of the 6th December, 1871, informing him that a
portion of his telegraph, together with a copy of his book, had been
placed in the Museum of the Pavilion at Brighton :—

"About five or six years ago I was in the garden (then rented by
a friend of mine) wherein this telegraph was laid down, when it was
dug for and found after a lapse of upwards of forty years, what was
then found and seen agreeing with the descriptions given in the book.
Several yards of copper-wire were found where the ground had not
been disturbed, by reason of a large rustic garden seat and alcove
having been over it; a glass tube, or the greater part of one, with the
copper wire in it (page 6*) ; and one of the joints with a short tube
(glass) (plate 2, fig. 1*) were also found : the copper-wire seemed to
be in perfect order. The wooden trough and pitch (page 6*) had
become consolidated with the earth, which was as hard as, and
formed an opening like that of a drain tile, or the run of a burrowing
animal."

This specimen, together with the original model of the dial made
by Ronalds in 1816, is now in the possession of Mr. Latimer Clark,
M.I.C.E.

Mr. Thomas Gibson, writing on the 1st April, 1870, says :—

"How well I remember when a schoolboy, fifty-five years ago,
seeing the clock apparatus in your little upper room over the stable,
connected with another at the bottom of the garden, of the meaning of
which I had but a very hazy apprehension ; also the lines of wire
stretched from frame to frame across the grassplot."

In 1825 Ronalds invented a perspective tracing instrument to facilitate drawing from nature, or from plans and elevations, a description
of which he published, in 1828, in a work entitled " Mechanical

---

* Description of an Electrical Telegraph, &c. By Francis Ronalds. 1823.

Perspective." During his travels he gave considerable attention to the curious Celtic antiquities, as at Stonehenge and other places, and he was enabled by means of his perspective tracing instrument, assisted by Dr. Blair, to procure exact perspective projections, taken from given noted stations, of the Celtic remains at Carnac, in Brittany. The result of these labours was published by Mr. Ronalds and Dr. Blair in 1836, entitled " Sketches at Carnac ; or, Notes concerning the present state of the Celtic Antiquities in that and some of the adjoining Communes." A description of this instrument was also published in a small quarto pamphlet, with illustrations.

Ronalds appears to have been an excellent workman, turning being his favourite pursuit. He has left behind several small specimens of his work, which show a high state of proficiency in the art. He contemplated the publication of a work upon this subject, and although a considerable portion was put into type, and several drawings for it were made, it was never published. There is no doubt he intended to make this work a most complete practical treatise on the subject, introducing many new contrivances and arrangements for delicate turning, and he also had the idea of adding an appendix of the names of tools, &c. in several languages, a part of this having been commenced.

At a meeting of the Council of the British Association for the Advancement of Science, held March 28th, 1842, Sir Roderick Murchison stated that he had reason to believe that the " building in Richmond Park formerly used as a Royal Observatory "* might, upon proper application being made to Her Majesty, be placed at the disposal of the British Association for scientific purposes ; the Royal Society, to whom the building had in the first instance been offered, having declined its acceptance. An application was afterwards made, and the Kew Observatory was, on the 26th May, 1842, placed at the

---

* This observatory was built for George III. about 1780, by Sir Wm. Chamoers.

disposal of the Association. Early in 1843 Ronalds was appointed its first Honorary Director and Superintendent, which post he held for nine years. During this period he made several full and exhaustive reports to the British Association, of which body he was for many years a Member of Council.

In the report read at the annual visitation of Greenich Observatory, June 1, 1844, the Astronomer Royal says :—

" In the autumn of last year I had an opportunity of examining the beautiful arrangements of the atmospheric electrometer at the Kew Observatory, which have been made under the superintendence of Professors Wheatstone* and Francis Ronalds, Esq., and it was impossible to see these without perceiving that considerable improvements might be made in our own by following the same plan, with such alterations as the difference of local circumstances rendered necessary."

In the appendix to Kaemtz's Meteorology,† Mr. C. V. Walker, F.R.S., says :—

" The electricity of the atmosphere is a branch of meteorology which has been cultivated on no settled and systematic manner until the present time. Indeed the observations that have been made at the Kew Observatory, a description of which it is our present purpose to give, constitutes so novel a feature in the science that Mr. Ronalds, whose whole time and talents appear to be enthusiastically devoted to perfecting them, says in the report submitted to the British Association at the meeting at York (1844), that they must be rather regarded as experimental and educational than otherwise. As the admirable arrangements at this Observatory will form a model for the guidance of others, we will give them in detail."

---

* Ronalds always denied, and has left his denial on record, that Wheatstone had any share whatever in the invention of the apparatus at Kew for registering atmospheric electricity.

† A Complete Course of Meteorology by L. F. Kaemtz, translated with notes and additions by C. V. Walker. London. 8vo. 1845.

On the 1st February, 1844, he was elected a Fellow of the Royal Society, and was admitted on the 29th, at the same time as Dr. Carpenter.

It is not generally known that Ronalds, during the time he occupied the position of Director of Kew Observatory, invented many important instruments, which have in no small measure advanced the science of meteorological and magnetical observation; his system of self-registration by photography* alone being sufficient to render his name famous.

In pursuance of proposals made to the British Association in 1840-1 and 1844,† Ronalds commenced in 1845, at Kew Observatory, a series of experiments on the photographic registration of all meteorological and terrestro-magnetical instruments. The first successful attempts were made in August, 1845, upon an atmospheric voltaic electrometer, a thermometer, a barometer, and a declination magnetometer. On the 24th September of the same year he obtained the first good specimen of registered atmospheric electricity, and on the 25th and frequently afterwards the apparatus which he had constructed enabled him to observe that under certain circumstances the electric potential increased with the light of the sun, but that such increased potential did not continue.‡

The Royal Society made a grant of £50 from the Wollaston Donation Fund to assist in the construction of the apparatus devised by Ronalds, for the self-registry of magnetical and meteorological instruments. In 1851 the Royal Society made a further grant of £100 to Ronalds for the purpose of an experimental trial of his self-registering instrument for six months. These experiments were commenced in April, and were being continued in January, 1852.

---

* Philosophical Transactions of the Royal Society for 1847, Part i. p. 111.

† *Vide* Report of British Association held at Southampton, 1846.

‡ *Vide* pp. 7 and 8, Epitome of the Electro-Meteorological and Magnetical Observations, Experiments, &c., made at Kew Observatory under the direction of Francis Ronalds, 8vo. pamph., Chiswick, 1848.

With reference to these grants it may be well to mention that Ronalds bequeathed a sum of £500 to the Royal Society, as set forth in the following extract from his will :—" I bequeath to the President, Council, and Fellows of the Royal Society the sum of five hundred pounds, in trust, for the augmentation of the Wollaston Fund, and in recognition of the advantages I derived, when Honorary Director of the Kew Observatory, from grants which I received from that Fund in aid of my not unsuccessful inventions of photographic instruments for the registration of terrestrial magnetism, atmospheric electricity, and other meteorological phenomena."

From a minute of the Council of the British Association at the meeting at Belfast, September 1, 1852, the following is extracted :—

" The experimental trial of Mr. Ronalds' magnetograph, which was in progress when the last report of the Council was made, has been completed, and detailed statements of the performance of each of the three instruments have been furnished by Messrs. Ronalds and Welsh, and are inserted in the Report of 1851. The Council have great pleasure in referring to these statements, as showing that Mr. Ronalds' adaptation of photography to record the magnetic variations is an effective and practically useful invention, supplying to those who may desire it the means of making and preserving a continuous registry of the phenomena."

During August, 1846, many of the magnetic photographs were submitted to a rigid comparison with the corrected readings of the Greenwich Observatory by Mr. Glaisher, the result being, as declared by him, " highly satisfactory."*

The self-registering barometer was in 1870, and probably now is, employed in the Meteorological Observatories instituted by the Board of Trade.

At a meeting of the Council of the British Association on the 8th

---

* Report of the Committee of the Kew Observatory, Aug. 9, 1849, signed on behalf of the Committee by Sir J. F. W. Herschel.

May, 1846, it was resolved " That the Kew Observatory be maintained in its present state of efficiency ;" because, after stating several other reasons, it appeared, both from the publications of the British Association and from the records of the establishment, that the systematic inquiry into the intricate subject of atmospheric electricity carried out by Mr. Ronalds had been productive of very material improvements in that subject, and had, in effect, furnished the model of the processes conducted at the Royal Observatory, and because those inquiries were still in progress.

The Toronto and Madrid Observatories were furnished with the self-registering instruments of Ronalds, the manufacture of them being carried out under his superintendence. Besides which he received application for advice and assistance from the Sardinian Government respecting meteorological instruments ; also from the Oxford Observatory for a barometrograph and thermometrograph ; and from the Astronomer at Trevandrum for electrical apparatus similar to that at Kew.*

Some of the original models of Ronalds' instruments are now in the South Kensington Museum.

The electrometers and electric-spark measurers used at the Greenwich Observatory in 1844, and for many years afterwards, were constructed from plans designed by Ronalds, and similar to those he had constructed for the Kew Observatory.

On the application of the Marquis of Northampton and Sir John Herschel, Her Majesty's Government granted to Ronalds a pecuniary recompense of £250 for his invention of photographic self-registering magnetical and meteorological apparatus. It will be remembered that Mr. Brooke also received a pecuniary recompense from the Government for his apparatus for similar purposes, which were invented about the same time, and which are fully described in

* See Minute of Council of British Association, November 29, 1852.

the Philosophical Transactions for 1846. As regards the priority of invention or the relative merits of the two systems, it may be observed that the matter had long before 1846 occupied the attention of Ronalds, and that we find him in 1840-41 and 1844 making proposals to the British Association on the subject, and that his first successful photographs were produced in August, 1845.

His apparatus for taking electrical observations was also adopted at the Royal Observatory at Greenwich.

In 1847 Ronalds made with Dr. Birt an experiment in which they sought to maintain a kite at an almost invariable given altitude, by the employment of three cords attached to it, and to points forming a triangle on the earth. This experiment was quite successful, and a short account of it was published in the Philosophical Magazine for September, 1847.

In 1852 Ronalds retired from the direction of Kew Observatory, where he had for nine years contributed so much to the advancement of electrical and magnetical observation, and his several reports to the British Association bear witness to the extent of his knowledge of the subject, and to the interest he felt in everything appertaining to it. On his retirement he received a grant from the Crown, on the recommendation of Lord Derby's Government, of a small pension of £75 per annum, for (as is stated in the warrant) his "eminent discoveries in electricity and meteorology."

After his retirement from Kew he lived for many years abroad, principally in Italy, where he continued the compilation of his Catalogue of Works on Electricity, Magnetism, and the Electric Telegraph, which, together with the formation of the "Ronalds Library," afforded him occupation during the remainder of his life.

There are in the library of the Society several interesting and curious unpublished MSS. upon various subjects by Ronalds, among which may be mentioned, the heads of a proposed History of

Electricity, Journal of a Tour in the Mediterranean, Egypt, Syria, Greece, &c. in 1819-20 ; Fires, Meteorology, Drawing, Surveying, &c.

A short biographical sketch of Ronalds (and others) was published by Wm. Walker,* to accompany the well-known engraving of " The Distinguished Men of Science of Great Britain, living in 1807-8, assembled at the Royal Institution," in which Ronalds' portrait appears ; and it is interesting to note that for a great number of years Ronalds was the only surviving member of the group.

In 1866 a letter was addressed to Lord Derby, modestly setting forth the early labours of Ronalds, but Mr. Barrington, in a letter to Mr. Samuel Carter, says that Lord Derby " declines re-opening the question of honours for the invention of the electric telegraph, in consequence of the length of time that has passed since the name of Ronalds has been connected with the invention."

At the Paris International Exhibition of 1867, the British Commissioners placed in one of the windows of the Machinery Gallery allotted to Great Britain a blind upon which was painted a diagram showing the instruments used by Ronalds in his telegraph of 1816.

In February, 1870, a memorial was addressed to the Right Hon. W. E. Gladstone on the subject of the invention of the electric telegraph, and on March 31, 1870, Ronalds received the honour of knighthood ; Her Majesty's intention of conferring this mark of distinction being conveyed to him in a letter from Mr. Gladstone as follows :—

" *Carlton House Terrace, S.W., March 28th*, 1870.

Sir,—I have great pleasure in announcing to you, that in acknowledgment of your early and remarkable labours in telegraphic

---

* Memoirs of the Distinguished Men of Science of Great Britain living n the year 1807-8, and Appendix, compiled and arranged by William Walker, un. 2nd edition. London, 1864.

investigations, Her Majesty will be pleased to confer upon you the honour of knighthood.

It is highly agreeable to me to make this communication, and
I have the honour to be, Sir,
Your very obedient and faithful servant,
W. E. GLADSTONE."

Sir William Fothergill Cooke, in a congratulatory letter to Sir F. Ronalds, says : "It is a singular fact that Brentford should have furnished two of the men who were most practical in their original views ; I might say *the only two men* who up to the year 1837 realised in their minds the electric telegraph as a *future fact.*

" I have before said that you were *before* the time. Had you taken the subject up again in 1835, it would have been all your own, I fully believe, or had we jointly taken it up in 1836, we should have divided the honours.

" May you long be spared to enjoy the ephemeral honour : which, however, so granted, links your name with one of the greatest adaptations of science the world can ever know ; next in usefulness to printing ; in its character, the most marvellous development of the Creator's powers vouchsafed to the practical service of man. To Him be the glory ! to us the deep and abiding satisfaction of being His instruments."

During the fierce controversy which raged between Cooke and Wheatstone as to their relative merit in the invention of *their* electric telegraph, the modest claims of Ronalds—the pioneer—were well-nigh forgotten ; and he had to thank his good constitution for having lived to see his early labour recognised. However valueless the honour of knighthood may have been to a man 83 years of age, it is still gratifying to know that that honour, though tardily given, was appreciated by one who has rendered such signal service to electrical science, and who must always stand as the first of English Telegraph Engineers.

In the *Illustrated London News* of the 30th April, 1870, will
be found a good portrait of Ronalds, together with a description
of his telegraph illustrated from the plates given in the work
of 1823.

There is a fine marble bust of Ronalds in the library of the Society of
Telegraph Engineers, by Mr. Edward Davis, presented to the Society
by their first president, Dr. C. W. Siemens, F.R.S.

There is also in the possession of Mrs. Samuel Carter a fine oil
painting of Ronalds by Mr. Hugh Carter.

Ronalds, who was never married, was a man of extremely sensitive
temperament and of very retiring manners, and this may in a great
measure account for his inventions never having been brought
prominently before the public. His whole life was devoted to science,
without any pecuniary reward, with the exception of the small Civil
List pension, his small patrimony being expended in his experiments,
and in the formation of the Ronalds Library.

He resided during the last ten years of his life at Battle, in Sussex,
in order to be near his sister, the wife of Mr. Samuel Carter. His
niece, Miss Julia Ronalds, lived with him up to the time of his
death, and devoted herself to his declining years, besides materially
assisting him during the latter part of his life in the compilation
of his Catalogue. He died on the 8th August, 1873, and was
buried at Battle.

THE
RONALDS CATALOGUE.

Works marked thus † are in the Library.
„   „   „   * are Meteorological, &c.

† **Abbadie,** A de. Sur le Tonnerre en Ethiopie. 4to. 158 pp.   *Paris,* 1858
   Mémoire sur un Météore (le Quobar) qu'il a aperçu sur différents points du
   globe : et Rapport sur le Météore.   M. Petit, Rapporteur. 8vo. 9 pp.
   (*Toulouse, Acad. 3me Série,* v. 303.)   *Toulouse,* 1849

† **Abeille,** J. L'electricité appliquée à la thérapeutique chirurgicale, et, en particulier, au traitement des accidents produits par les inhalations d'éther et de
   chloroforme. 8vo. 110 pp.   *Paris,* 1870

† **Abel,** F. A. Account of recent researches on the application of Electricity from
   different sources to the Explosion of Gunpowder. 8vo. 35 pp. (*Reprinted
   from the Quarterly Journal of the Chemical Soc.*)   *London*

   Abernethy, John. *Born* 1763 *at Abernethy ; died April* 20, 1831, *at Enfield.*
   (*Very celebrated Surgeon in London.*)
   Surgical and physical essays.   *London,* 1796
   Inquiry into the rationality of Mr. Hunter's theory of life. (*Zantedeschi,
   Trattato* ii. *Torpedo.*)

† **Abich,** H. Ueber einen bei Stawropol gefallenen Meteorstein. 8vo. 37 pp.
       Mosk (Moscow), 1860
†   Die Fulguriten im Andes und kleinen Ararat, nebst Bemerkungen über örtliche Einflüsse bei der Bildung elektrischer Gewitter. 8vo. 11 pp.
       Wien, 1869

   Abilgaard, Peter Christian. *Born December* 22, 1740, *at Kopenhagen ; died
   January* 21, 1801, *at Kopenhagen.*
   Tentamina electrica. (*Vide* Van Troostwyk and Krayenhoff, 1788.)

† **Abraham,** J. H. On new phenomena caused by the effect of Magnetic and Electric Influence, and suggestions for ascertaining the extent of the terrestrial
   magnetic atmosphere. 8vo. 6 pp. (*Phil. Mag.* i, 266.)   *London,* 1827
†   (On) Neutralizing the Magnetism of Watch-work. 8vo. 1 p. (*Phil. Mag.* i.
   470.)   *London,* 1827

† **Abria.** Anémomètre electro-magnétique. (*Vide* Moigno, 1852.)   *Bordeaux*
   Sur quelques propriétés physiologiques des courants d'induction. 8vo. 11 pp.
   (*Actes de l' Acad. de Bordeaux.*)   *Bordeaux,* 1842
†   Recherches galvanométriques sur les lois de l'induction des courants par les
   courants. 8vo. 11 pp. (*Actes de l' Acad. de Bordeaux.*)   *Bordeaux,* 1844
†   Sur la déclinaison, l'inclinaison, et l'intensité magnétique à Bordeaux, 8vo.
   8 pp. (*Actes de l' Acad. de Bordeaux.*)   *Bordeaux,* 1848

B

2   ABR—ACC

Abria—*continued.*

† Rapport sur un Discours de *Gavarret* le 14 Novembre, 1848.   8vo. 4 pp.
(*Actes de l' Acad. de Bordeaux.*)   *Bordeaux*, 1852

† Rapport sur l'appareil Fragneau destiné à prévenir les rencontres des trains de
chemins de fer.   8vo. 8 pp.   (*Actes de l' Acad. de Bordeaux.*)
                                                           Bordeaux, 1853

† Démonstration de plusieurs formules de *Gauss* relatives à l'action mutuelle de
deux aimants.   8vo. 16 pp.   (*Ext. des Mém. de la Faculté de Bordeaux.*)
                                                           Bordeaux, 1861

† Various Observations of Inclination and Declination.   *Bordeaux*

Academia de Madrid. Memorias de la Real Academia de Ciencias. 4to.
                                                          Madrid, 1850-51

Academia Moguntina Scientiarum. (*Vide* Erfurt.)

Academia Naturæ Curiosorum. (*Vide* Breslau.)

Académie de Paris. (*Vide* Paris Academy.)

Académie de Toulouse. Histoire et Mémoires. (*Vide* Toulouse Academy.)

Academy, Swedish. (*Vide* Schwedischen Akadem and Svenska Academien.)

Accademia degli Aspiranti Naturalisti di Napoli. Annali; Bullettini;
Esercitazioni.   (*Vide*, for each head, Naples, Accademia degli Aspiranti.)

† Atti dei Congressi degli *Scienziati Italiani* raccolti dall' Accademia. Con
Note dell' Editore Presidente della stessa (*i.e. Dr. Leonardo Dorotea.*) 4to.
6 vols.   *Napoli*, 1844-45

Accademia del Cimento. Saggi di naturali esperienze, descritte dal Segretario.
8vo. 2 vols.   *Milano*

† Saggj di naturali esperienze fatte nell' Accademia del Cimento, e descritte dal
Segretario. fol. 269 pp.   *Firenze*, 1666

† Saggj di naturali esperienze fatte nell' Accademia del Cimento, sotto la prote-
zione del Serenissimo Principe Leopoldo di Toscana, e descritte dal Segretario
di essa Accademia.   fol. 269 pp. (*reprint*)   *Firenze*, 1667

† Essays of natural experiments made in the Accademia del Cimento. (Englished
by Rd. Waller.) 4to. 19 plates. 180pp.   *London*, 1684

† Saggj di naturali esperienze, descritte dal Segretario. 2ª edizione. fol. 269 pp.
                                                           Firenze, 1691
Tentamina, &c. *Musschenbrock.* 4to. (*Vide* Musschenbrock, 1731.)   1731

† Saggi di naturali esperienze fatte nell' Accademia del Cimento, e descritte dal
Segretario di essa Accademia. With life of *Magalotti*, by Manni. 8vo.
                                                           Venezia, 1761
Saggi, with additions by Tozzetti. Aggrandimenti, Notizie, &c. 4to. (*Vide*
Tozzetti, 1780.)   1780

Saggi di naturali esperienze fatte nell' Accademia del Cimento. Terza edizione
Fiorentina : preceduta da notizie istoriche dell' Accademia stessa e seguitata
da alcune aggiunte (di Gazetti Antinori.) 4to.   *Firenze*, 1841
Aggiunte ai Saggi. (*Vide* Gazzeri, 1841.)   *Firenze*, 1841

† Intorno ad una nuova edizione dei Saggi. Relazione di Belli. 8vo. 14 pp.
(*Giorn. I. R. Instit. Lomb.* iii. 183.)   *Milano*, 1842

Accademia Italiana. Atti dell' Accademia Italiana. 4to. *Firenze*, 1808

Accademia Reale di Modena. (*Vide* Modena Academy).

Accademia di Napoli. Atti dalla fondazione sino al 1787 ; Atti della Sezione
Borbonica ; Rendiconto delle Adunanze, 1842 ; Memorie dal 1852 in avanti.
(*Vide*, for each head, Naples Academy.)

Accademia di Padova. Saggi ; Memorie ; Nuovi Saggi ; Rivista Periodica ;
Catalogo di Libri. (*Vide*, for each head, Padua Academy.)

Accademia Petropolitana. (*Vide* Petersburg Academy.)

Accademia Pontificia dei Nuovi Lincei. Atti dell'. 4to.   *Roma*

Accademia di Siena. (*Vide* Siena.)

Accademia Théodoro Palatina. (*Vide* Manheim.)

Accademia di Torino. (*Vide* Turin Academy.)

† **Acerbi** G. Notice of. *Born at Castelgoffredo May* 3, 1773 ; *died August* 25, 1847. Published in London in 1802, in English, Journey to the North Cape by Sweden, Finland, and Lapland (query anonymously). He established the Biblioteca Italiana in 1816. (*Giorn. I. R. Instit. Lomb.* xvi. 13.)

Achard, Franz Carl. *Born April* 28, 1753, *at* Berlin ; *died April* 20, 1821, *at Kunern.*

Abhandlung der Theoretischen und Practischen Lehre von der Electricität.

Von den Electrischen Erscheinungen, die durch Reiben des Quecksilbers auf verschiedene, vorzüglich harzige Körper hervorgebracht werden. (*Beschaft. der Berliner Gesellsch. Naturf.*) *Berlin,* 1775

Sur l'électricité de la Glace. (*Journ. litt. de Berlin,* 1776.) *Berlin,* 1776

Ueber die Kraft der Elektricität verglichen mit der Schwere. (*Beschaft. Naturf. Gesellsch. Berl.* i. 1775.) *Berlin,* 1775

Von den Elektrischen Erscheinungen durch Reiben des Quecksilbers. (*Beschaft. Naturf. Gesellsch. Berl.* ii. 1775.) *Berlin,* 1775

Ueber die Kraft der Elektricität, verglichen mit der Kraft der Schwere. 8vo. (*Reuss. Rep.* iv. 351.) *Berlin,* 1775

Expériences sur l'Electrophore, avec une théorie de cet instrument. 4to. (*Mem. de Berlin,* 1776, 122.) *Berlin,* 1776

Expériences sur la célérité avec laquelle les corps de différentes figures se chargent du fluide électrique ; et sur le rapport entre la quantité qu'ils en absorbent et la distance à laquelle ils sont d'un corps électrisé. 4to. (*Mem. de Berlin,* 1777, 25.) *Berlin,* 1777

Sur l'analogie qui se trouve entre la production et les effets de l'électricité et de la chaleur ; avec la description d'un instrument nouveau propre a mesurer la quantité de fluide électrique que peuvent conduire des corps de différentes natures placés dans les mêmes circonstances. 4to. (*Mem. de Berlin,* 1779, 27.) *Berlin,* 1779

† Dissertazione sulla differenza fra i corpi elettrici e i conduttori. 4to. 11 pp. (*Opus. Scelti,* iii. 313.) *Milano,* 1780

Expériences qui prouvent que les corps se chargent en raison de leur surface. 4to. (*Mem. de Berlin,* 1780, 47.) *Berlin,* 1780

Sammlung physik. und chemischer Abhandlungen. 8vo. *Berlin,* 1784

† Schriften (chymisch-physik). 8vo. *Berlin,* 1780

Vorlesungen über die Experimental-Physik. 8vo. *Berlin,* 1791

Kurze Anleitung, ländliche Gebäude gegen Gewitter-Schaden sicher zu stellen. 8vo. 78 pp. *Berlin,* 1798

Mémoire renfermant le récit de plusieurs expériences électriques faites dans différentes vues. 4to. (*Mem. de Berlin,* 1781, 9.) *Berlin,* 1781

† Articolo di Lettera al Landriani sull' Elettricità. 4to. (*Opus. Scelti,* v. 351.) *Milano,* 1782

† Relazione delle Sperienze dell' Achard, per decidere se i corpi ricevano, e disperdano l'elettricità in ragione delle superficie, o delle masse. 4to. (*Opus. Scelti,* vi. 199.) *Milano,* 1783

Observations sur l'électricité terrestre (*Mem. de Berlin,* 1786, 13.) *Belrin,* 1786-87

Sur l'irritation des nerfs et les contractions musculaires. 4to. (*Mem. de Berlin* 1790-91, 1, 14, 27.) *Berlin,* 1790-91

Ackermann, Johann Friedrich. *Born February* 3, 1726 ; *died June,* 1804, *at Kiel.*

On the Contact Theory. (*Salzb. Mediccchirurg.,* 1792, 287.) *Salzburg,* 1792

Versuch einer physischen Darstellung der Lebenskräfte organisirter Körper. 2 (or more) vols. 8vo. *Frankfurt,* 1798

Ackermann—*continued.*
Nachrichten von der sonderbaren Wirkung eines Wetterstrahles. 8vo. *Kiel,* 1772
Programma, quo morbus et sectio fulmine nuper adusti enarratur. 4to. *Kilæ,* 1774
Acqua. (*Vide* Dell' Acqua.)
Acta eruditorum Lipsiensia. (*Vide* Leipsig.)
Acta Helvetica. (*Vide* Basle.)
Acton, J., and **Lofft,** Capel. On the probability of Meteorolites being projected from the Moon. Galvanic and electrical experiments by Mr. Acton. 8vo. (*Phil. Mag.* li. 109.) *London,* 1818
Adams, Dudley. Electricity in the formation of nature. 8vo. 15 pp. *London,* 1820
A trifle or sketch, wherein the science and elements of Electricity are developed in a few books. 4to. *London,* 1823
Adams, George. *Born in* 1750 *at Southampton; died August* 14, 1795, *at Southampton.*
† An Essay on Electricity; to which is added an Essay on Magnetism. 1st edition, 8vo. 367 pp. *London,* 1784
† An Essay on Electricity and the mode of applying it to medical purposes; with an Essay on Magnetism. 2nd edition, corrected and enlarged. Frontispiece and 7 plates. 8vo. 476 pp. *London,* 1785
† Versuch über die Elektricität durch eine Menge methodisch-geordneter Experimenta erläutert; nebst einen Versuch über den Magnet. 6 plates, 8vo. 270 pp. *Leipzig,* 1785
† An Essay on Electricity; with an Essay on Magnetism. 3rd edition. Frontispiece, 7 plates, and 1 plate (belonging to the Supplement). 8vo. 468 pp. *London,* 1787
† Essay on Electricity; to which is now added a Letter to the Author from Mr. John Birch on the subject of Medical Electricity. 4th edition. 5 plates and frontispiece. 8vo. 588 pp. *London,* 1792
An Essay on Electricity; to which is added a Letter to the Author from Mr. John Birch on the subject of Medical Electricity. The 5th edition, with corrections and additions, by Wm. Jones. 8vo. 7 plates. 594 pp. *London,* 1799
Adamson, J. C. Notes on certain modes of measuring minute Intervals of Time. 4to. (*Trans. American Phil. Soc., New Series,* xi.) *Philadelphia,* 1857
Adanson, Michel. *Born April* 7, 1727, *at Aix; died August* 3, 1806, *at Paris.* (Poggendorff, i. 11, says that he first made known, in 1751, the Silurus electricus, and compared its shock with that of the Leyden phial. *Vide* Humboldt's Voyage au Senegal, pt. ii. Zoologie, p. 59. "Il a le mérite d'avoir reconnu le premier, en 1751, l'analogie des effets de ces poissons (le Gymnotus) avec ceux d'une bouteille de Leyde, mérite que l'on a faussement attribué à S'Gravesande et à Walsh."—Voyage au Sénégal about 1757. *Vide* Humboldt.)
† **Addenet,** R. F. Nouvelle théorie de l'Electricité et de la Nécessité des Forêts pour l'Agriculture et le Bien-être général. 8vo. 150 pp. *Paris,* 1843
† **Adley,** C. C. The story of the Telegraph in India. 94 pages and 1 map. 8vo. *London,* 1866
Admiralty Manual of Scientific Enquiry. *Vide* Herschell, Sir John.
Ditto. Deviation of the Compass. *Vide* Evans and Smith.
Adsigerius, Peter. (This is Peter Peregrinus. *Vide* Wenckebach.)
† **Æpinus,** Franz Ulrich Theodor. (*Born December* 13, 1724, *at Rostock; died August* 10, 1802, *at Dorpat.*)
Mémoires concernants quelques nouvelles expériences électriques remarquables. 4to. (*Mem. de Berlin,* p. 105.) *Berlin,* 1756
† Sermo Academ. de similitudine vis electricæ atque magneticæ. 4to. 32 pp. *Petropoli,* 1758
Descriptio ac explicatio novorum quorumdam experimentorum electricorum. 4to. 26 pp. (*Novi Comment. Acad. Petrop.* vii. 1761, 22 and 277.) *Petropoli,* 1758-59

Æpinus—*continued.*

† Tentamen theoriæ electricitatis et magnetismi. Accedunt Dissertationes duæ quarum prior phænomenon quoddam electricum, altera, magneticum, explicat. 7 plates. 4to. 390 pp. *Petropoli,* 1759

Rede über die Aehnlichkeit der magnetischen Kräfte. 8vo. *Leipzig,* 1760

† Recueil de différents Mémoires sur la Tourmaline. 5 plates. 8vo. 193 pp. *Petersburg,* 1762

† Additamentum ad Dissertationem de experimento magnetico, celeberr. Domini du Fay, continens nova experimenta magnetica detecta et explicata. 1 plate. 4to. 12 pp. (*Novi Comment. Acad. Petrop.* ix. 1864, p. 340.) *Petropoli,* 1762-63

Similitudinis effectuum vis magneticæ et electricæ : novum specimen. 4to. (*Novi Comment. Acad. Petrop.* x. 296. *Petropoli*

† Abhandlung von den Luft-Erscheinungen. 4to. 16 pp. *Petersburg,* 1763

† De electricitate barometrorum disquisitio. 4to. 22 pp. (*Novi Comment. Acad. Petrop.* xii. 1768, p. 303.) *Petropoli,* 1766-67

† Examen theoriæ magneticæ a celeberr. Tob. Mayero propositæ. 4to. 26 pp. (*Novi Comment. Acad. Petrop.* xii. 1768, p. 325.) *Petropoli,* 1766-67

Descriptio novi phænomeni electrici detecti in Chrysolitho sive Smaragdo Brasiliensi. 4to. 5 pp. (*Novi Comment. Acad. Petrop.* xii. 1768, 351.) *Petropoli,* 1766-67

Exposition de la Théorie de l'Electricité de M. Æpinus. *Paris,* 1787

Tentamen theoriæ Electricitatis et Magnetismi. 4to. *Petersburg and Leipzig,* 1787

Tentamen theoriæ Electricitatis et Magnetismi. 4to. *Petersburg,* 1791

Descriptio acuum magneticarum noviter inventarum, quæ vulgaribus præsantiores sunt, atque artificii, vires magnetum naturalium insigniter augendi. (*Acta Academiæ Moguntinæ,* ii. 255.)

Aernest, S. The Comet and Cometic Electricity. 8vo. *Edinburgh,* 1857

Affaidaties (Affaidatus). Physicæ et astronomicæ considerationes. (Contains " *De peouliari magnetis ad polum descensu, seu mavis de ipsius magnetis ad polum conversione, et De causa cur magnes ad se ferrum attrahat.*") *Venice,* 1549

Agobard. *Vide* St. Agobard.

Agmini. *Vide* Erckmann and Agmini.

Agricola, Jh. Theoria fulminum per Electricitatem illustrata. *Heidelberg,* 1771

Agut. *Vide* D'Agut.

Ahrens, J. E. W. Dissertatio Observationes et experimenta de qualitate et quantitate electricitatis corporis humani, in statu sano et morboso. *Kiel,* 1813

Aimé, Georges. *Born about 1813 ; died September 9, 1846, between Algiers and Médéah.*

† Mémoire sur le Magnétisme terrestre. (Relative to his observations made in 1841-2 in Algiers.) 8vo. 12 pp. (*Ext. Ann. Chim.* 3me série, tom. x.) *Paris*

† Recherches de Physique sur la Méditerranée. fol. *Paris,* 1845

Observations sur le Magnétisme terrestre. fol. 229 pp. *Paris,* 1846

Note sur un nouveau procédé d'aimanter. 4to. (*Vide* Dove, p. 260, and Pogg. i. 20.) *Paris*

De la masse du fluide électrique. 4to. (*Vide* Pogg. i. 20.) *Paris*

† **Airy,** Sir George Biddell. *Born July 27, 1801, at Alnwick.*
Account of experiments on Iron-built Ships, instituted for the purpose of discovering a correction for the Deviation of the Compass produced by the iron of the ships. 4to. 47 pp. (*Phil. Trans.* 1839, part 1.) *London,* 1839

† Description of the instruments and process used in the photographic self-registration of the magnetical and meteorological instruments at the Royal Observatory, Greenwich. 4to. 8 pp. 3 plates. (*Ext. from Introduc. to Greenwich Mag. and Met. Observations* for 1847.) *London,* 1849

Airy—*continued.*

† Results of Experiments on the Disturbance of the Compass in Iron-built Ships. Made at the desire of the Board of Admiralty. 4to. 16 pp.   *London,* 1840

On the equations applying to light under the action of Magnetism. 8vo. (*Phil. Mag.* xxviii.)   *London,* 1846

† Addresses to the individual Members of the Board of Visitors to the Royal Observatory, Greenwich, by the Astronomer-Royal, and Reports of the same read at the annual visitations.   *London*

† Results of the Magnetical and Meteorological Observations made at the Royal Observatory, Greenwich, for the years 1840 to 1850 inclusive. 10 vols.
   London

On the Deviations of the Compass in several ships. 4to.   *London,* 1855

† Description of the Galvanic Chronographic Apparatus of the Royal Observatory, Greenwich. Forming the Appendix to the Greenwich Observations, 1856. 4to. 8 pp.   *London,* 1857

† Results of the Magnetical and Meteorological Observations made at the Royal Observatory, Greenwich. 4to.   1870

† A Treatise on Magnetism. 8vo. 220 pp.   *London,* 1870

Vide also Christie and Airy.

Akademie der Wissenschaft zu München. Neue Abhandlungen, Denk-schriften, Bulletins. *Vide* Bavarian Academy.

Albanus, L. T., mit C. H. **Böse.** Materialien für Elektriker.   *Halle,* 1783-90

Albany Institute, Transactions of the. 3 vols. 8vo. (*Vide* Trübner's Bibl. Guide of 1859, p. 22.)   *Albany, New York,* 1831-55

Albert, C. L. De vi electrica in Amenorrhæa. 4to.   *Gottingen,* 1764

† **Albinus,** F. B. Specimen inaugurale de Meteoris ignitis. 4to. 79 pp.
   Lugduni Batavium, 1740

† **Albrecht,** T. Ueber die Bestimmung von Längen-Differenzen mit Hülfe des elektrischen Telegraphen. 4to. 83 pp.   *Leipzig,* 1869

† **Alden,** Timothy, Jun. Effects of Lightning on the House of Captain Manning in Portsmouth, New Hampshire ; in a letter to Dr. Eliot. 4to. 2 pp. (*Mem. Amer. Acad.* iii. p. 93.)   *Cambridge, U.S.,* 1809

An account of some Electrical Phenomena. 4to. (*Mem. Amer. Acad.* iii. 333.)
   Cambridge, U.S., 1815

Aldini, Giovanni. *Born April 16, 1762, at Bologna ; died January 16, 1834, at Milan.*

† Memoria intorno all' Elettricità animale diretta all' Amoretti. 4to. 12 pp. (*Opuscoli Scelti,* xvii. 231.)   *Milano,* 1794

† De animali electricitate Dissertationes duæ. 4to. 41 pp.   *Bologna,* 1794

† Sull' elettricità animale. Lettera al Moscati. 4to. 10 pp. (*Opuscoli Scelti,* xix. 217.) *Vide* also *Ann. di Chim. di Brugnatelli,* xiii. 135.   *Milano,* 1796

† Memoria intorno alcune elettriche esperienze al La Cépéde. 8vo. 37 pp. (*Ann. di Chim. di Brugnatelli,* xiv. 174.)   *Pavia,* 1797

† Lettera intorno all' Elettricità animale al Moscati. 8vo. 20 pp. (*Ann. di Chim. di Brugnatelli,* xiii. 135.)   *Pavia,* 1797

† Memoria sulla elettricità animale di Galvani allo Spallanzani. Aggiunte alcune elettriche sperienze di Aldini. 4to. 105 pp. and plates.   *Bologna,* 1797

† Memoria intorno ad alcune elettriche esperienze : al La Cépéde. 4to. 20 pp.
   Bologna, 1797

† Transunto d'una Memoria intorno ad alcune elettriche sperienze del Cittadino Aldini, al La Cépéde. 4to. 11 pp. (*Opuscoli Scelti,* xx. 73.)   *Milano,* 1798

† Transunto delle sperienze sul galvanismo dell' Aldini. 4to. 12 pp. 1 plate. (*Opuscoli Scelti,* xxi. 412.)   *Milano,* 1801

† Lettera al Vassalli sull' Elettricità. 8vo. 9 pp. (*Ann. di Chim. di Brugnatelli,* xix. 29.)   *Pavia,* 1802

Aldini—*continued.*

† Saggio di Esperienze sul Galvanismo. 9vo. 80 pp. (*Ann. di Chim. di Brugna-telli,* xix. 158.) *Pavia,* 1802

† Saggio di Esperienze sul Galvanismo. 3 parts, 8vo. *Bologna,* 1802

† Expériences faites par Aldini. 4to. (*Bulletin des Sciences de la Soc. Phil.* No. lxviii.) *Paris,* 1803

† An account of the late improvements in Galvanism, with Appendix. Experiments on the body of a malefactor. 4to. 221 pp. 4 plates. *London,* 1803

† Galvanic experiments on February 18, 1803, in Dr. Pearson's lecture-room. 8vo. 2 pp. (*Phil. Mag.* xv. 93.) *London,* 1803

† Précis des expériences galvaniques faites récemment à Londres et à Calais. 8vo. 48 pp. *Paris,* 1803

† Essai théorique et expérimental sur le Galvanisme, avec une série d'experiences faites en présence des commissaires de l'Institut National de France, et en divers amphithéatres anatomiques de Londres. 4to. 398 pp. *Paris,* 1804

† Essai théorique et expérimental sur le Galvanisme. 2 vols. 8vo. *Paris,* 1804

Theoretisch praktische Versuche über den Galvanismus. 2 vols. 8vo. *Leipzig,* 1804

Lettre de Milan au Professor Sue. 8vo. (*Vide* Sue, Hist. iv. 279.) *Paris, about* 1805

† Sul potere del solo Arco animale nelle contrazioni muscolari. Esperienze galvaniche fatte dal G. Aldini. 4to. 4 pp. 1 pl. (*Mem. Soc. Ital.* xiv. 329.) *Verona,* 1809

† General views on the application of Galvanism to medical purposes. 8vo. 96 pp. *London,* 1819

De animalis electricæ theoriæ ortu atque incrementis. Dissertatio. (*Poggendorff,* i. 27, *gives no date, place, or other indications.*)

Gabinetto Aldini di fisica e chimica applicata. Col ritratto dell' Aldini, e con litografia rappresentante il Gabinetto. 8vo. 11 pp. (*Tipografia Regia.*) *Bologna,* 1863

† **Alemanni,** Pietro. Account of some experiments upon the Decomposition of Water and the production of Muriatic Acid by the electrical pile, made at the Literary Society of Milan by M. Pierre Alemanni, Apothecary to his Majesty, and Member of that Society. 8vo. 5 pp. (*Phil. Mag.* xxvii. 339.) *London,* 1807

Alexander, W. Plan for an instantaneous telegraphic communication betwixt Edinburgh and London by means of electric or voltaic currents transmitted through metallic conductors under ground. 1837

† Description of a Model for an Electro-magnetic Telegraph. fol. in 4to. 3 pp. *Edinburgh,* 1840

† Plan and description of the original Electro-magnetic Telegraph, with prefatory note to the Royal Commissioners. 1 plate. 8vo. 30 pp. *London and Edinburgh,* 1851

Alibert. Eloges historiques de Galvani. (*Mém. de la Soc. Méd. d'Emulation de Paris,* vol. iv.) *Paris, about* 1802

Eloges historiques de Galvani. "Versione Italiana,' printed anonymously, but attributed to Alfonso Bonfiole Malvezzi. (*Vide* Gherardi Rapport, p. 19.) *Bologna,* 1802

† Eloges de Spallanzani, de Galvani, de Roussel, et de Bichat, suivis d'un Discours. 8vo. *Paris,* 1806

Nouveaux éléments de thérapeutique. (*1st Edition,* 1804 *and* 1805.) 1817

† **Aliés,** P. F. Mémoire sur le mécanisme secret et sensible de l'électricité naturelle. 8vo. 112 pp. *Montpellier,* 1780

† **Aliprando,** L. Diatriba de Electricitate ; habita in Gymnasio Regio Cremonensi, Anno 1777. 8vo. 92 pp. *Cremonæ*

† **Alison, R. E.** Narrative of an Excursion to the Summit of the Peak of Teneriffe on the 23d and 25th of February, 1829, including an Account of Phænomena attending a Waterspout at the Llanos de Gaspar, Teneriffe, in November 1826. 8vo. 8 pp. (*Phil. Mag.* viii. 23.)   *London,* 1830

Allamand, Jean Nicolas Sébastien. *Born September* 18, 1713, *at Lausanne ; died March* 2, 1787, *at Leyden.*)

† An account of some experiments lately made in Holland upon the fragility of unannealed Glass Vessels, communicated to the President. (*Phil. Trans.* xliii. 505.)   *London,* 1744-45

Wahrnehmung, &c.   *The Hague,* 1755

Bericht wie ein Mädchen, das eine gewisse Art vom Schlage getroffen hatte, vermittelst der Electricität wieder zurechte gebracht worden ist:—nebst einigen Stücken zur Geschichte solcher Personen welche durch die Electricität geheilt worden sind durch T. D. Gaubius. 8vo. (*Abhand. der Gesellsch. zu Haarlem, Th.* i. 264.)   *Haarlem,* 1757

On the Surinam Eel, with an account of many experiments, &c., by S'Gravesande. (*Trans. Haarlem Soc.*)   *Haarlem,* 1757

Allan, Thos. Electro-motive power.   *London,* 1852

Systems of Inland and Submarine Telegraphy. 8vo.   *London,* 1858
Systems of National Telegraphic Communication. 8vo.   *London,* 1859

† **Allard, L.** Electricité, Galvano-plastique. Tableau des effets de l'électricité et de ses applications à l'industrie. folio and 4to.   *Paris*

† **Allen, Z.** Philosophy of the Mechanics of Nature, &c. 8vo. 797 pp.
                    New York, 1852
Allen. (*Vide* Hare and Allen.)

Almanach des Lignes-Télégraphiques. 12mo.   *Paris,* 1855

Almeida. (*Vide* D'Almeida.)

† **Alquier, M. F.** Sur la distribution de l'électricité sur deux sphères conductrices mises en présence. Présentée à la Faculté des Sciences de Paris, le 30 Mars. 4to. 19 pp.   *Paris,* 1852

† **Althaus,** Freyhern von. Versuche über den Elektro-magnetismus ; nebst einer kurzen Prüfung der Theorie des Hern Ampère vom Freyhern Althaus ; mit einer Vorrede vom Hofrath Muncke. 8vo. 37 pp.   *Heidelburg,* 1821

† **Althaus, J.** Treatise on Medical Electricity, theoretical and practical, and its use in the treatment of Paralysis, &c. 8vo. 352 pp.   *London,* 1859

Die Electricität in der Medizin .8vo. (*Vide* Meyer, 1861, p. 43.) *Berlin,* 1860

† On Paralysis, Neuralgia, and other affections of the nervous system, and their successful treatment by Galvanisation and Faradisation. 3rd edition. sm. 8vo. 236 pp.   *London,* 1864

† A treatise on Medical Electricity, theoretical and practical, and its use in the treatment of Paralysis, Neuralgia, and other diseases, &c. 2nd edition. 8vo. 676 pp.   *London,* 1870

† On the medical use of Galvanization and Faradisation. New edition. 8vo. 69pp.
                    London, 1870

† **Ambrosoli, G.** Prime nozioni di fisica esposte. 8vo. 1189 pp. *Milano,* 1854-56

Ambschel, A. von. Translation in German of Herbert Dissertatio de vi electrica aquæ. (*Vide* Herbert, 1778.)   1778

American Academy of Arts and Sciences, Memoirs of the. Old series 4 vols. 4to.   *Boston,* 1785-1818

New series. 5 vols. 4to. (*From Trübner's Bibl. Guide,* 1859.)
                    Boston, 1833-55

† Proceedings of the, for 1846. Part of vols. i. to iv. 8vo. *Boston,* 1847

AME--AMO 9

American Association for the Advancement of Science. Proceedings of, 8vo. (*From Trübner's Bibl. Guide*, 1859, p. 19.)   1849-56

American Journal of Science. Conducted by Benjamin Silliman, B. Silliman, junior, and Dana. Published at New York and New Haven, U.S. (*From Trübner's Bibl. Guide*, 1859, pp. 489-90.)   1818-56

American National Institution. Bulletins of the proceedings of the. The 2nd Bulletin, 1842, contains Rogers's Explanation of his Land Telegraph, p. 154. 8vo.   *Washington*, 1841-46

† American Philosophical Society. Proceedings of, contained in ten incomplete volumes, and extending over the period of 1838-66. No. 27 has an account of the 100th Anniversary. 8vo.   *Philadelphia*

American Polytechnic Journal. (*From Kuhn of* 1866, p. 1091.)

American Annual of Scientific Discovery.

American, Scientific. (*Vide* Scientific American.)

Amici, G. B. Osservazioni sulla circolazione del succhio della chara. (*Pogg.* i. 35 *gives a French version, and Ann. Chim. Phys.* xiii. 1320.)   *Modena*, 1820

† Ammersin, Wendelino. Brevis relatio de electricitate propria lignorum. 12mo. 27 pp.   *Lucèrne*, 1753 *or* 1754

Kurze Nachricht von der eigenthumlichen Electricität des Holzes; aus dem Latin. 8vo. (*Vide* Jalabert, 1771, *which contains this translation as an appendix*.)   *Basel*, 1771

Amontons, G. On his Telegraph. 4to. (*In his Biography in the Mém. de Paris*, 1705.)   *Paris*, 1705

Amoretti, Carlo. *Born March* 13, 1741, *at Oneglia, near Genoa; died March* 25, 1816, *at Milan*.)

† Lettera al Soave su alcune sperienze elettriche. 4to. 5 pp. (*Opus. Scelti*, xix. 233.)   *Milano*, 1793

† Lettera al Fortis, su varj Individui che hanno la facoltà di sentire le sorgenti, le miniere, &c. 4to. 17 pp.   *Milano*, 1797

† Rabdomanzia, Lettera 1a. 4to. 34 pp. (*See below*, 1801-4.)   *Milano*, 1798

† Raguaglio ed analisi delle pietre cadute dal cielo, o almeno reputate tali. 4to.   *Milano*, 1803

† Breve storia del Galvanismo. 4to. 26 pp. (*Opus. Scelti*, xxii. 357.)   *Milano*, 1803

† Altre notizie sul Galvanismo o Voltaismo medico. 4to. 7 pp. (*Opus. Scelti*, xxii. 84.)   *Milano*, 1803

† Sugli Areoliti ossia pietre cadute dall' atmosfera. 4to. 4 pp. (*Nuova Scelta d' Opuscoli*, l. 48.)   *Milano*, 1804

† Rabdomanzia, Lettere 2a, 3a, 4a, 5a, 6a. 4to. (*Opus. Scelti*, xxi. 73, 170, 393.)   *Milano*, 1801-4

† Nuova Scelta d'Opuscoli interessanti sulle scienze e sulle arti tratti dagli Atti delle Accademie, &c., dalle opere Inglesi, Tedesche, &c. 2 vols. tom. i. 1804, ii. 1807, 4to.   *Milano*, 1804 *and* 1807

† (1.) Ragguaglio d'un Aerolite in Siberia. Translated from Gilbert's Annalen, 1807.

† (2.) Nota dell' Editore su Altri Aeroliti. 4to. 3 pp. (*Nuova Scelta d' Opus.* ii. 165 and 166.)   *Milano*, 1807

† Aeroliti caduti nel Dipartimento del Garda in Milano. 4to. 4 pp. (*Nuova Scelta d' Opus.* ii. 63.)   *Milano*, 1807

† Sui Turbini o Trombe di Terra osservati al Nord di Milano il giorno 6 Giugno, 1808. 4to. 8 pp. (*Nuova Scelta d' Opus.* ii. 302.)   *Milano*, 1807

Elogio dell' Amoretti. 4to. (*Ital. Soc. Mem.* xviii.)   *Modena*, 1820

Viaggio ai tre laghi colla vita dell' autore del G. Labus. 12mo. (*From Libri's Sale Cat.* 469.)   *Milano*, 1824

† **Amoretti,** Campi, Fromond e Soave. Scelta di Opuscoli interessanti tradotti da varie lingue. 36 vols. 12mo. (*Collated for Electricity, &c.,* and the articles on those subjects obtained by Sir Francis Ronalds in 1859.)

Milano, 1775-77

Sequel to the Scelta di Opuscoli. 22 vols. 4to.   *Milan,* 1801-3

Ampère, André Marie. *Born Jan.* 22, 1775, *at Lyons; died June* 10, 1836, *at Marseilles.*

† Account of his discoveries on the attraction and repulsion of two conductors, or portions of the same conductor joining the two extremities of the voltaic pile. 8vo. 2pp. (*Phil. Mag.* lvi. 308.)   *London,* 1820

† Analyse des Mémoires lus par M. Ampère à l'Académie les 18 et 25 Septembre, les 9 et 13 Octobre, 1820. Suite de l'Analyse. Mémoire lu le 6 Nov. 1820. 8vo. (*Ann. de Phy. de Bruxelles,* vii.)   *Paris*

Electro-magnetic Telegraph explained in a Memoir presented to the Academy, October 2, 1820. 8vo. (*An. de Phy. et. Chim.* xv. 72.) (*Vide* also *Anon. Electricity,* 1821.)   *Paris,* 1820

† Mémoires sur l'Action mutuelle de deux courans électriques, sur celle qui existe entre un courant électrique et un aimant ou le globe terrestre, et celle de deux aimants l'un sur l'autre. 8vo. 68 and 31 pp.   *Paris,* 1820-22

† Recueil d'Observations Electro-dynamiques contenant divers Mémoires, &c., relatifs à l'action mutuelle de deux courants électriques, &c. 8vo. (Scarce ; a new edition proposed.)   *Paris,* 1822

† Extrait d'une lettre au Professeur De la Rive sur des expériences electro-magnétiques. 8 pp. 1 plate. 8vo. (*Bibl. Univers.* xx.)   *Paris,* 1822

† Exposé méthodique des phénomènes electro-dynamiques et des lois de ces phénomènes. 8vo. 28 pp.   *Paris,* 1823

† Précis de la Théorie des phénomènes electro-dynamiques pour servir de supplément à son Recueil d'Observations électro-dynamiques et au Manuel d'électricité-dynamique de M. Demonferand. 8vo. 67 pp. 1 plate.   *Paris,* 1824

† Description d'un Appareil électro-dynamique. 1st edition. 8vo. 24 pp.   *Paris,* 1824

† Mémoire sur une nouvelle expérience Electro-dynamique, suivi d'une lettre à Gherardi. 8vo. 48 pp. (*Acad. de Paris,* 12 *Sept.,* 1825. *Ann. de Chim.,* 1825.)   *Paris,* 1825

† Précis d'un Mémoire sur l'Electro-dynamique. Extrait de la correspondance mathématique et physique des Pays Bas. 8vo.   *Gand,* 1825

† Précis d'un Mémoire sur l'Electro-dynamique. Lu à l'Acad. R. des Sciences, 21 Novembre, 1825. 8vo. 15 pp. 1 plate.   *Gand, n. d.*

Sur les piles sèches de M. Zamboni. 8vo. (*Ann. de Chim. et Phys.* xxix.)   *Paris,* 1825

Description d'un Appareil électro-dynamique. 2nd edition. 8vo. (Reprint.)   *Paris,* 1826

† Théorie des Phénomènes électro-dynamiques uniquement déduite de l'expérience. 4to. 226 pp. 2 plates.   *Paris,* 1826

Mémoire sur l'action mutuelle d'un Conducteur voltaïque et d'un Aimant. 8vo. 88 pp.   *Paris,* 1826

† Mémoire sur l'action mutuelle d'un Conducteur voltaïque et d'un Aimant. 4to. 88 pp. 3 plates.   *Bruxelles,* 1827

† Note sur l'action mutuelle d'un Aimant et d'un Conducteur voltaïque. 8vo. 29 pp. (*Ext. Ann. de Chim.* 1828.)   *Paris,* 1828

† Notice sur M. Ampère (par E. Littré). 8vo. 96 pp.   *Paris,* 1843

† **Ampère et Babinet.** Darstellung der neuen Entdeckungen über die Elektricität und den Magnetismus aus dem Französischen. 8vo.   *Leipzig,* 1822

Darstellung der neuesten Entdeckungen über Elektricität und Magnetismus. 8vo. 118 pp. 2 plates.   *Leipzig,* 1822

Ampère et Babinet—_continued._

† Exposé des nouvelles découvertes sur l'Electricité et Magnétisme de Oersted' Arago, Ampère, Davy, Biot, Erman, Schweiger, De la Rive, &c. (_Seyffer p. 377, mentions a German translation at Leipsig,_ 1822. _Ouerard says "imprimé d'abord dans le Supplément de la Chimie de Thompson."_) 8vo.
Paris, 1822

Vide also Hachette and Ampère.

Amsler, Jacob. _Born November 16, 1823, at Stalden bei Brugg._

† Zur Theorie der Vertheilung des Magnetismus im weichen Eisen. 4to. 26 pp.
Neuchatel, 1849

Amyot. Note historique sur le Télégraphe electrique. 4to. (_Comptes Rendus,_ vii. 80.)
Paris, 1838

Sur un procédé de Correspondance télégraphique au moyen de l'Electricité. 4to. (_Comptes Rendus_ v. 909, vii. 1162, xxvii. 271-293.) _Paris,_ 1838 and 1848

Andern. (_Vide_ Pfaff and Andern.)

† **Anderson,** Sir James. Existing and projected Telegraphic Routes to India considered. 8vo. 31 pp.
London, 1868

Anderson, Johann. _Born March 14, 1674, at Hamburg ; died March 3, 1743, at Hamburg._

Nachricht von Island, Grönland und der Strasse Davis, &c. 8vo. _Hamburg,_ 1746

Andria, Niccolo. _Born September 10, 1748, at Massafra, prov. Otranto ; died December 9, 1814, at Naples._

† Analyse de la série des circonstances particulières qui ont lieu dans les êtres organisés, et qui décident de leur existence. Du Siége et de la Faculté du Galvanisme et de la cause de l'excitabilité, traduite par A. Pitaro. 8vo.
Paris, 1805

Andrieux. Mémoire sur l'Application méthodique du Galvanisme au traitement des Maladies.
Paris, 1824

† De l'emploi du Galvanisme dans le traitement de la Gastrite chronique. Mémoire lu à l'Académie de Médecine en Février, 1833. 8vo. 16 pp.
Paris, 1835

Andry et Thouret. Observations et Recherches sur l'usage de l'Aimant en médecine, ou Mémoire sur le Magnétisme médicinal. 8vo. (_Reuss. Repert._ xii. 18.)
Paris, 1779

† Rapport sur les Aimants présentés par M. l'Abbé le Noble. (_Extrait des Régistres de la Société Royale de Médecine. Lu Mardi,_ 1 _Avril,_ 1783.) 4to.
1783

Anglade, J. G. Essai sur le Galvanisme appliqué à la Pathologie. 4to. (_Vide_ Sue Hist. iii. 73.)
Montpellier, 1803

Angus, Wm. Magnetism, its history. 18mo.
London, 1841

Annalen der Chemie, &c. Herausg. von Liebig, Wöhler, &c. A complete set. 8vo. vols. 1 to 116.
Heidelb. 1832-1860

Annalen der Telegraphie. _Vide_ Brix.

Annales de Chimie, ou Recueil de Mémoires concernant la Chimie, &c., par Guyton-Morveau, Lavoisier, Monge, Berthollet, Seguin ,Vauquelin, de Fourcroy, de Dietrich, Hassenfratz, et Adet; C. A. Prieur, Chaptal, Deyeux, Bouillon-Lagrange, Collet-Descollis, A. Laugier, Gay-Lussac, Thenard, &c. 8vo. 96 vols.
Paris, 1789-1815

† **Annales de l'Electricité Médicale.** Revue internationale de l'Electricité, de l'Electro-puncture, du Galvanisme, et du Magnétisme, appliquées à la Médecine et la Chirurgie; publiée par une Réunion de Médecins, sous la direction de Van Holsbeeck. 8vo. vols. i. iii. iv. and v.
Paris, Londres, et Bruxelles, 1861

Annales des Mines. 8vo.
Paris, 1816

Annales générales des Sciences Physiques. Par Bory de Saint Vincent, Drapiez, et Van Mons. 8vo. (*Cat. Roy. Soc.* 1839, p. 636.)

Bruxelles, 1819-21

Annales Télégraphiques. Publiées par un Comité de l'Administration des Lignes Télégraphiques. Plates. 8vo. *Paris*, 1858

† **Annali di Fisica, Chimica, &c.** *Vide* Majocchi. Fol. 28 vols. 1841-47

Annali delle Scienze del Regno Lombardo Veneto. 4to. (*Vide* Fusinieri Editor.) *Vicenza,Padua, &c.* 1831-45

Annali di Scienze Matematiche e Fisiche. 8vo. 8 vols. Plates. *Roma*, 1850-7

Annals of Philosophy, or Magazine of Chemistry and Mineralogy, by T. Thomson. 8vo. vols. i. to xvi. *London*, 1813-20

New Series, by R. Phillips. 8vo. vols. i. to x. *London*, 1821-26

Annuaire des Lignes Télégraphiques. 8vo. *Paris*, 1867

Annuaire des Sociétés savantes de la France et de l'Etranger. 8vo. *Paris*, 1846

Annuario Bibliografico Italiano. Pubblicato per cura del Ministero della Istruzione Pubblica. Anno I. 1863. 8vo. *Torino e Firenze*, 1866

Anonymous. (See end of Letter A, p. 16.)

Anselmo, Giorgio. Effets du Galvanisme sur des animaux noyés. (Continuation promised but not given.) 8vo. 18 pp. *Turin*, 1803

† Dello svolgimento dell' Elettricità nell' Economia animale e vegetale. Discorso accademico. 8vo. 42 pp. *Torino*

Anstruther, Sir Alexander. *Born in Fifeshire, Scotland ; died July* 1819 *at Isle de France.*

Essay on the nature, &c. of Heat, Electricity, and Light. 8vo. *London*, 1800

Antheaulme. Traité sur les Aimans artificiels. 4to. *Petersburg*, 1760

Mémoire sur les Aimants artificiels, qui a remporté le prix de l'Académie de St. Petersbourg. 12mo. *Paris*, 1761

Antinori. Notizie istoriche su due Saggi di Naturali esperienze fatte nell' Accademia del Cimento. *Firenze*, 1841

Patologia fisico-chimico-meccanico-animale. 16mo. 5 vols. *Firenze, n. d.* (*Vide* also Nobili and Antinori.)

Antonio, N. Bibliotheca Hispana Vetus. 2 vols. fol. *Matriti*, 1788

Bibliotheca Hispana Nova. 2 vols. fol. *Matriti*, 1783-88

Apelt, E. F. Die Theorie der Induction. 8vo. 204 pp. *Leipzig*, 1854

Arago, Dominique François Jean. *Born February,* 26, 1786, *at Estagel, near Perpignan ; died October* 2, 1853, *at Paris.*

Aimantation par l'action de l'Electricité ordinaire. (*Le Moniteur* 10 *Nov.* 1820.) *Paris*, 1820

(Découvert) que le Courant voltaique développe fortement la vertu magnétique dans des Lames de fer ou d'acier qui d'abord en etaient totalement privées. 8vo. (*An. de Chim. et Phys.* xv. 94.) *Paris*, 1820

† On Earthquakes and Magnetism. On 19th February, 1822, agitation of a needle caused, he thought, by an Earthquake. 8vo. (Account transmitted to the French Academy.) (*Phil. Mag.* lix. 233.) *London*, 1822

Sur les Déviations que les Métaux en mouvement font éprouver á l'aiguille aimantée. 4to. (*Ann. de Chim. et Phys.* xxxiii.) *Paris*, 1825

† Notices scientifiques. Sur la Grêle et des Paragrêles, &c. 12mo. *Paris*, 1827

† Notices scientifiques. Notice historique sur la Pile Voltaïque, &c. 12mo. *Paris*, 1833

† Notices scientifiques. Sur le Tonnerre. 12mo. 410 pp. *Paris*, 1837

† Quelle est la durée d'un éclair de la première ou de la seconde classe ? 12mo. 13 pp. *Paris*, 1837

Arago—*continued.*

Comment on peut constater l'avance ou le retard de deux phénomènes lumineux. (*From Moigno, who gives no indication as to date, place, &c. The quotation has reference to electrical sparks, &c.*)

Sämmtliche Werke mit Einleitung von Humboldt, herausgegeben von Hankel, Band iv. enthaltend über die Gewitter, Electricität, Magnetismus, und Nordlicht. 8vo. *Leipsig,* 1854

† Œuvres de François Arago publiées d'après son ordre sous la direction de M. J. A. Barral. Notices Scientifiques, Tome premier. 8vo. *Paris,* 1854

Principe des Télégraphes électriques. 8vo. *Paris,* 1854

† Meteorological Essays, with an Introduction by Humboldt, translated by Sabine. 8vo. *London,* 1855

† Œuvres publiées par Barral. Tome ii. des Notices Scientifiques. 8vo. *Paris,* 1855

Narrative of Freycinet's Voyage. 4to. (*Phil. Mag.* vol. lxi. p. 134.) 1823

Notice biographique de Volta. (*Vide* Volta.)

(*Vide* also Faria and Arago.)

† **Aratos.** Sternerscheinungen und Wetterzeichen, griech, über: und Erklärungen Voss. *Heidelburg,* 1824

Arbas. (*Vide* D'Arbas.)

† **Arcangeli,** Carlo. Sul magnetismo, sull' elettricismo, e sui raggi solari. Tre Lezioni. 8vo. *Firenze,* 1867

Archereau, H. A. Régulateur de la Lumière électrique. (*Du Moncel,* p. 217, *describes it (with figure), but gives no date, place, &c.*)

Archives des Découvertes et des Inventions nouvelles faites dans les sciences, les arts, &c., pendant 1808-30. 8vo. 23 vols. (*Cat. Roy. Soc.* 1839, 636.) *Paris,* 1809-31

Archives de l'Electricité. De la Rive, Editeur. (*Belonging to the Bibliothèque published at Geneva.*) (*Vide* La Rive, De)

† **Archives des Sciences Physiques.** (*Belonging to the Bibliothèque published at Geneva.*)

Archives für Wissenschaftliche Kunde von Russland. Herausgegeben von A. Erdman. 8vo. *Berlin,* 1841

Arden. (*Vide* Priestly, Versuche über Naturlehre.)

† **Arella,** Carnevale Antonio. Storia dell' Elettricità. 8vo. 2 vols. *Alessandria,* 1839

† **Arenstein,** J. Austria at the National Exhibition of 1862. 4to. *Vienna,* 1862

† **Argelander,** F. W. A. Verzeichniss von Nordlichtern beobachtet auf den Sternwarten zu Abo und Helsingfors in den Jahren 1823-37. 4to. 49 pp. *Helsingfors,* 1866

Armengaud. Génie industriel. (*From Kuhn,* 1866, *p.* 1360.)

Arnaud. (*Vide* Porna (or Poma) and Arnaud.)

Arnault. L'Institut. Journal des Académies et Sociétés scientifiques. (*Venet. Lomb. Inst.* v. 54.) *Paris,* 1835

† **Arndtsen,** Adam. Magnetiske undersögelser, anstillede med W. Weber's Diamagnetometer. 4to. 43 pp. 2 plates. *Christiania,* 1858

† Nickelens Ledningsmodstand. 4to. *Christiania,* 1858

Arnim, Ludvig Achim Von. *Born Jan.* 26, 1781, *at Berlin; died Jan.* 21, 1831, *at Wiepersdorff.*

† Versuch einer Theorie der electrischen Erscheinungen. 8vo. 146 pp. 1 plate. *Halle,* 1799

Arnott, Neil. Elements of Physics or Natural Philosophy, written for General Use in Non-technical Language. Comprising the new chapters on Electricity and Astronomy, with an outline of Popular Mathematics. The work complete in two Parts. (*The 5th Edition is dated* 1833.) *London*

14 ARN—AUB

† **Arnoux, E.** La Lettre Electrique. Nouveau service télégraphique. Le télégraphe électrique rendu populaire. 8vo. 106 pp. *Paris*, 1867

† **Arregoni.** Jasi meccanica, o trattato dei rimedii naturali meccanici. 8vo. 2 vols. *Lodi*, 1775.

Arrest. (*Vide* D'Arrest.)

Arrighetti, Niccolò. *Born November* 11, 1586, *at Florence*; *died June* 1, 1639, *at Florence.*
Epitome della teoria elettrica (in Latin.) *Sienna*, 1756

† **Arrot, A. R.** On some new cases of Voltaic Action, and on the construction of a Battery without the use of oxidizable metals. 8vo. 8 pp. *London*, 1842

Arschereau. (*Vide* Lemolt and Arschereau.)

Aschlund, A. Om Magnetens Forhold til seine Poler. *Kiobnhafn*, 1830

† **Asclepi,** Giuseppe. *Born April* 16, 1706, *at Macerata*; *died July* 21, 1776, *at Rome.*

† Osservazioni intorno alla Declinazione della Calamita fatte in Roma. 4to. 19 pp. (*Atti dell' Accad. di Siena*, ii. 107.) *Sienna*, 1763

Ash, Dr. Edward. On the action of Metals in contact with each other upon Water. Letter to Humboldt April 10, 1796. (This is a German Letter in Humboldt's Versuche, p. 472.) 1796

Asiatic Researches; or Transactions of the Society instituted in Bengal for inquiring into the History and Antiquities, the Arts, Sciences, and Literature of Asia. 4to. 12 vols. plates. *London*, 1799-1812. *Calcutta*, 1816

Asiatic Society, Transactions of the. 4to. *London*, 1827

Assalini, Ritter. Kurze Erläuterung des Zambonischen immerwährenden Elektromotors. *München*, 1816

Asson. Rapporto della commissione che ha fatto gli esperimenti sull' Elettropuntura, come mezzo congelante il sangue nelle arterie e sull' obliterazione delle vase (*sic.*) (*Gazette des Hôpitaux, No.* 48.) 1847

Astier, Charles Benoit. *Born March* 6, 1771, *at Mont-Dauphin, dep. Hautes Alpes.*
Notice sur les Paragrèles à pointes ; projet de paragrèles à flammes et expériences comparatives du pouvoir électrique des flammes et des pointes. 8vo. *Toulouse*,1829
Considérations sur les fonctions physiologiques des épines, et sur le rapport qu'elles paraissent avoir avec les méteores électriques. (*Annales de la Société linnéenne de Paris, November.*) 1825

Astronomical Expedition. The United States Naval Astronomical Expedition. Vol. vi. Magnetical and Meteorological Observations. Many plates and maps. 4 vols. 8vo. (Vols. iv. and v. not published.) *Washington*, 1855-56

Astronomical Society. Memoirs. 4to. (*Cat. Roy. Soc.* 1839.) *London*, 1822

† **Atkinson, H.** On Hypotheses proposed for explaining the origin of Meteoric Stones ; with remarks on Mr. Murray's letter on Aërolites inserted in the Philosophical Magazine. 8vo. 7 pp. (*Phil. Mag.* liv. 336.) *London*, 1819

Atkinson, E. (*Vide* Ganot, A.)

† **Atlantic Telegraph Company.** The Atlantic Telegraph. A History of preliminary experimental Proceedings, and a descriptive account of the undertaking ; published by order of the Directors of the Company. 8vo. 69 pp. 2 plates. *London, July*, 1857

† Chart of Soundings and Section of the Bottom of the Atlantic Ocean from Valentia, Ireland, to St. John's, Newfoundland. *London*, 1858

† **Aubé,** Ph. A. De l'Electricité soit de l'Ame universelle considerée dans ses forces motrices. 8vo. 56 pp. *Elbeuf*, 1852

Aubert. Elektrometische Flasche. 4to. (*Observations sur la Phys. Delametherie.*) *Paris*, 1789

His Elektrophorus. (*Journ. de l'Hist. Nat.*) 1787

Aubuisson. (*Vide* D'Aubuisson.)

"Auditor." (*Vide* Anon. Elect. 1808.)

Augustin, Friedrich Ludwig. *Born June 3, 1776, at Berlin.*

† Vom Galvanismus und dessen medicinischer Anwendung. 8vo. 64 pp. 1 plate.
Berlin, 1801

† Versuch einer Geschichte der galvanischen Elektricität und ihrer medicinischen Anwendung. 8vo. 284 pp. 1 plate. *Berlin,* 1803

† **Aurifabrum,** A. Succini historia. Ein kurtzer grundlicher Berichte woher der Agtstein oder Börnstein ursprunglich komme, das er kein Baumhartz sey sonder ein geschlecht des Bergwachs. Und wie man ihnen manigfaltiglich in artzneien möge gebrauchen. Durch Andream Aurifabrum. Vratislauienscm, Medicum. Sm. 4to. (*Scarce, curious, and esteemed.*) *Königsberg,* 1551

Auriol, L. J. d'. Manuel de la Correspondance secrète postale ou télégraphique. Guide pour composer et pour traduire les dépêches en chiffres. 8vo.
Paris, 1867

† **Austin** and **Clark.** Letter from Austin to Dr. Monro, junior, on the effect of Electricity in removing obstructions of the Menses. Remarks on the foregoing letter by the late Dr. D. Clark. Edinburgh, September 6, 1764. (*Essays and Observations,* iii. 116.) *Edinburgh,* 1764

Auzout, A. Observation made in Rome concerning the declination of the Magnet. (*Phil. Trans.,* &c., 1670, p. 1184.) 4to. *London,* 1670

Avanzamenti della Medicina e Fisica. Opera periodica che serve di seguito al Giornale Fisico Medico di L. Brugnatelli. 8vo. *Pavia,* 1796

† **Avelloni,** D. Lettera in cui si dimostra non esser necessario ricorrere al fuoco elettrico per ispiegare la formazione e gli effetti delle Meteore. 12mo.
Venezia, 1760

Avezac. (*Vide* D'Avezac.)

† **Avogadro,** A. Mémoire sur la construction d'un Voltimètre multiplicateur et sur son application. 4to. 40 pp. (*Mem. di Torino,* xxvii. 43.) *Torino,* 1823

† Mémoire sur les volumes atomiques et sur leur relation avec le rang que les corps occupent dans la série électro-chimique. 4to. 65 pp. (*Mem. di Torino,* serie 2, tom. viii. p. 129.) *Turin,* 1846

† Saggio di teoria matematica della Distribuzione dell' Elettricità sulla superficie dei corpi conduttori, &c. 4to. 29 pp. 1 plate. (*Ital. Soc. Mem.* xxiii. 156.)
Modena, 1844

(*Vide* also Botto et Avogadro.)

† **Ayrton,** W. E. On a quantitative Method of testing a "Telegraphic Earth." (*Journal of the Asiatic Society of Bengal.*) 1871

Azais, Pierre Hyacinthe. *Born March 1, 1766, at Corrèze.*

Théorie générale de l'Electricité, du Galvanisme, et du Magnétisme. 8vo. (*Phil. Mag.* xxviii. 182.) *London,* 1807

(*Vide* also Bapst and Azais.)

Azuni, D. A. Sull' Origine della Bussola Nautica. 8vo. *Venezia,* 1797

† Dissertation sur l'origine de la Boussole. 8vo. 133 pp. *Paris,* 1805

† Dissertation sur l'origine de la Boussole. 2nd edit. 8vo. 269 pp. *Paris,* 1809

Azzalini. Kurze Erläuterung des Zambonischen Electromotors. 8vo.
München, 1814

Kurze Erläuterung des Zambonischen immerwährenden Elektromotors. 4to.
München, 1816

16

ANONYMOUS·

Atlantic Telegraph.

† Origin and Progress of the Oceanic Electric Telegraph. 8vo. *London*, 1858

† The Atlantic Telegraph. Its History from the commencement of the undertaking, in 1854, to the sailing of the "Great Eastern" in 1866. 8vo. 16 pp.
London, 1866

Atlantic Telegraph, a history of experimental proceedings. 8vo. 1867

The Atlantic Telegraph. An authentic history of that great work ; with Biographies, Maps, Engravings, Portraits, &c. Dedicated to C. W. Field.
New York

Electricity, Electro-magnetism, Galvanism, &c.

Aufrichtige und unpartheyische Gedanken von der Electricität, worinnen der Herr Professor Bose in Wittenberg, gegen das so harte Urthiel der Hamburg unpartheyischen Correspondenz im 153 St. dieses Jahres vertheidiget wird. 4to.
Wittenberg, 1744

† Dissertation sur la cause de l'électricité des Corps et des Phénomènes qui en dépendent. Sujet proposé par l'Académie Royale des Sciences de Berlin pour le prix qui doit être distribué le 31 Mai, 1745. (*It is the Fourth Dissertation.*)
4to. 45 pp. *Berlin*, 1745

† Zweite Abhandlung von der Natur der Electricität, welche bei ber Königlichen Akademie der Wissenschaften in Berlin eingesendet und des Drucks würdig geschätzt worden. 4to. 192 pp. *Berlin*, 1745

† Dritte Abhandlung von den Eigenschaften, Wirkungen und Ursachen der Electricität, welche bei der Königlichen Akademie der Wissenschaften in Berlin eingesendet und wegen ihrer Ausarbeitung des Drucks würdig geschätz worden.
4to. 104 pp. (*Waitz's Collection of 4 Essays.*) *Berlin*, 1745

(F. W.) Schreiben an die Akad. zu Berlin, die Untersuchung von der Electricität, &c. betreffend. 4to. (*Cat. Royal Society.*) 1745

† Dissertation sur la cause de l'Electricité des Corps et des phénomènes qui en dépendent. Sujet proposé par l'Academie Royale des Sciences de Berlin pour le prix, le 31 Mai, 1745. 4to. 45 pp. (*Waitz's Collection of 4 Essays.*)
Berlin, 17—(?)

† A Philosophical Enquiry into the properties of Electricity. Letter to a friend.
8vo. 32 pp. *London*, 1746

Congetture fisiche intorno ai fenomeni della macchina elettrica. 8vo.
Roma, 1746

† Dell' Elettricismo, ossia delle forze elettriche dei Corpi. Aggiuntevi due Dissertazioni. 8vo. 391 pp. plates. (See also 1847.) *Venezia*, 1746

Mémoire sur l'Electricité. 37 pp. (*Vide* Boullanger.) *Paris*, 1746

Explication physique des effets de l'électricité * * * Professeur perpétuel de philosophie de l'Université d'Avignon. 12mo. 1 plate. 48 pp. *Avignon*, 1747

† Dell' elettricismo, ossia delle forze elettriche de' Corpi . . . aggiuntevi due Dissertazioni . . . 8vo. 364 pp. 1 plate. *Napoli*, 1747

† Recueil de Traités sur l'Electricité, traduits de l'Allemand et de l'Anglais. In 3 parts. 12mo.

1er Partie. Winckler. Essai sur la nature, les effets, et les causes de l'électricité, avec une description de deux nouvelles machines, traduit de l'Allemand.

2me Partie. Watson. Expériences et Observations pour servir à l'explication de la nature et des propriétés de l'électricité, proposées en trois lettres à la Société Royale de Londres. Traduits de l'Anglois d'après la 2me edition.

And Suites des expériences et observations, addressées a la Société Royale de Londres.

3mo Partie. Freke. Essai sur la cause de l'électricité. 2me édition, avec un supplément. Traduit de l'Anglois.

Electricity, &c.—*continued.*

Martin. Essai sur l'Electricité. Traduit de l'Anglois.

Each part is paged separately, and each work has a separate title-page. They are the works, printed separately, but united in this volume, under the above general title. At the end of the volume is the "Appiobation" enumerating them (as above.) *Paris,* 1748

† Cause et Mécanique de l'Electricité. 12mo. 214 pp. *Paris,* 1749

† Lettre à l'Abbé Nollet sur l'Electricité. 8vo. (*Vide* Boullanger.) *Londres,* 1749

† Lettera al Beccaria intorno al primo capo del suo Elettricismo artificiale. (Nollet?) Author's date is 3 Marzo, 1753. 1753

Electricitas. *Nuremberg,* 1758

Experiments on the instantaneous transmission of electric power through more than four Italian miles of wire. By "Filosofi presso il Monteiro." (*The Pyrotechnia,* vi. 944.) 1771 (?)

Lettre d'un Abbé * * * sur l'Electrophore. (*Vide* Jaquet.) *Vienne,* 1775

Electrical means of spreading information through immeasurable space, almost instantaneously. (*Le Mercure de France, Juin,* 1782, *No.* 23.) 1782

† (N***, l'Abbé.) Principes d'Electricité. Ouvrage traduit de l'Anglois par l'Abbe N*** auquel on a joint certaines notes interessantes et propres à confirmer les principes nouveaux de l'illustre Auteur (*i. e.* Lord Mahon). 8vo. 250 pp. 6 plates. (*Nollet died* in 1770). (*Vide* Lord Mahon, 1780.) *London,* 1781

† Dell' efficacità dei conduttori elettrici, Dubbj proposti ai Fisici moderni. 8vo. 1784

Materialien für Elektriker. 1er Lieferung. 8vo. (*Poggendorff,* i. 22, *says by Albanus.* (*Vide* Albanus.) *Halle,* 1788

Materialien für Elektriker. 2e Lieferung. *Halle,* 1790

(C. L.) Briefe über Electricität. 8vo. 1789

Revision der vorzüglichen Schwierigkeiten in der Lehre von der Elektricität. 8vo. (*Vide* Heinsius, iii. 373.) *Heidelberg,* 1789

† Intorno all' influenza dell' elettricità nel moto muscolare, &c. 8vo. 3 plates. (*Gior. Fis. Med.* i. 280. *Repetition of Galvani's experiments by Volta, Rezia, and Brugnatelli.*) *Pavia,* 1792

Lettre d'un ami au comte Prospero Balbo. (Sur les expériences de Galvani.) *Bibliotèque de Turin,* i. 261, 1792, *Mars.*) *Turin,* 1792

† Ragguaglio delle sperienze del Galvani sull' elettricità animale, estratto da una lettera al Conte Balbo. 8vo. 16 pp. (*Gior. Fisico-Medico di Brugnatelli,* ii. 94 *and* 97.) *Pavia,* 1792

Expériences galvaniques verifiées jusqu'à présent à l'école de Médecine au moyen de l'appareil imaginé par le Dr.Volta. 4to. (*Soc. Philomath, An.* 4, ii. 165.) *Paris,* 1796

† Esperienze galvaniche verificate alla Scuola di Medicina. 8vo. (*In Annali di Chimica di Brugnatelli,* xviii. 175. *Also in Magazin Enciclop.* No. 16.) *Pavia,* 1806

Introduzione all' Elettricità ; che contiene le nozioni esatte del fuoco elementare colla loro applicazione. 8vo. (*Opusc. Scelti* (*in* 4to.), xii. 23.) *Madrid*

† On the Electric Fluid. (*Phil. Mag.* xvi. 173.) *London,* 1803

† Galvanic experiments made by Carpue on the body of Michael Carney, from Notes taken by an eminent Physician. 8vo. 3 pp. (*Phil. Mag.* xviii. 90.) *London,* 1804

† (B. E.) On Galvanism, by a Correspondent. 8vo. 3 pp. (*Phil. Mag.* xviii. 170.) *London,* 1804

† Preliminary matter on Electricity, &c., in the volumes printed in 1805, 1809, 1816, and 1827, and in vol. i., printed in 1839, of the Second Series of the Turin Memoirs, (in all five articles). 4to. (Vassalli and others ?) *Torino,* 1805-39

† A new Electrical Phenomenon. 8vo. 1 p. (*Phil. Mag.* xxi. 162.) *London,* 1805

c

Electricity, &c.—*continued.*

† A new Fact in Galvanism. Communicated by a Correspondent. 8vo. 2 pp. (*Phil. Mag.* xxiv. 203.) *London,* 1806

† (R. F.) Sulla teoria del circuito galvanico. 8vo. 47 pp. (*Arago's copy.*) *Torino e Pisa, n. d.*

† On Electricity. Adhesion of two cylindrical conductors after a discharge, &c. (*Phil. Mag.* xxv. 72.) *London,* 1806

† (C. R.) On the partial fusion of Metals by the Electric Discharge. By a Correspondent. 3 pp. (*Phil. Mag.* xxiv. 73.) *London,* 1806

† (Auditor.) On Mr. Davy's Theory of Chemical Affinity, &c. 8vo. 4 pp. (*Phil. Mag.* xxxii. 370.) *London,* 1808-9

† Articles on Electricity, &c., in Rees's Encyclopædia. 4to. (*Vide* Cavallo and Cuthbertson, 1809.) *London, n. d.*

† (L.O.C.) On Galvanic Electricity. 8vo. 4 pp. (*Phil. Mag.* xlii. 476.) *London,* 1813

† (A. B.) Proposal for an improvement of the Galvanic Trough. 8vo. 2pp. 1 plate. (*Phil. Mag.* xliv. 15.) *London,* 1814

† On the Electric Fluid. 8vo. 1 p. (*Phil. Mag.* xlv. 218.) *London,* 1815

† (A. B.) On the Effects of Galvanism in Asthma and diseases of the respiratory organs. 8vo. 3 pp. (*Phil. Mag.* xlix. 419.) *London,* 1817

† Electrical Phenomena. (*Phil. Mag.* xlix. 390.) *London,* 1817

† (W.) On the supposed Repulsion of Electricity. (*Phil. Mag.* xlix. 208.) *London,* 1817

† Electrical Phenomena. (*Phil. Mag.* xlix. 314.) *London,* 1817

† Electrical Experiments. (*Phil. Mag.* lii. 376, 393, and 387.) *London,* 1818

† Note sur les expériences electro-magnetiques de Œrsted, Ampère, Arago, et Biot. 8vo. 26 pp. (*Arago's copy.*) (*Ext. des Annales des Mines,* 1820.) *Paris,* 1821

† (W. M. G.) On Electricity excited in paper. 8vo. 3 pp. (*Phil. Mag.* lxi. 330.) *London,* 1823

† Article Electricity in Encyclopædia Britannica (old edition), vol. iii. 239. 4to. 21 pp. *Edinburgh*

† Article Electricity in Encyclopædia Metropolitana. 4to. 132 pp. 5 plates. *London,* 1824-25

† Article Galvanism in Encyclopædia Metropolitana, vol. iv. 173 to 224. 4to. 52 pp. 1 plate. *London,* 1826

Navigation by means of Galvanism. (*Glasgow Mechanic's Magazine,* No. 23, October, 1826.) 1826

† Specimina Electrica. Habebitur exercitatio in Aula maxima Collegii Romani. 8vo. *Romæ,* 1826

† (E. W. B.) Remarks on one of the experiments from which Mr. Ritchie has inferred the inadequacy of the chemical theory of galvanism. 8vo. 5pp. (*Phil. Mag.* vii. 61.) *London,* 1830

† An attempt to simplify the Theories of Electricity and Light. 8vo. 48 pp. 1 plate. *Edinburgh,* 1834

(S. G.) Relazione ragionata sui fatti, &c., intorno alla singolare virtù dei pesci elettrici prima della scoperta del Galvanismo e della Pila Voltiana. 1838

Steam superseded : an Account of the Electro-magnetic Engine. 18mo. (*Vide* Taylor.) *London,* 1841

† Documents relatifs à l'emploi de l'Electricité pour mettre le feu aux fourneaux des mines et à la démolition des navires sous l'eau. 8vo. 88 pp. 1 plate. (*Extraits du Militaire Spectator Hollandais,* and from *The United Service Journal.*) *Paris,* 1841

† Straordinaria produzione di elettricità in una donna. Fol. 2 pp. (*L'Economista,* vol. i. dell anno 1844, p. 110.) *Milan,* 1844

† Dritte Abhandlung. Von den Eigenschaften und Ursachen der Electricität, welche bei der K. Acad. der Wissenschaften in Berlin eingesendet und . . . des Drucks würdig geschätzt werden. 4to. 102 pp. *Berlin,* 1845

Electricity, &c.—*continued.*

New Theory of Vegetable Physiology based on Electricity. 12mo.
London, 1847

Railway Appliances ; or the Rail, Steam, and Electricity. 12mo. *London,* 1848

Electro-Biologie, de wetenschappelijk verklaard. 8vo. (*Translation of Smee?*)
Amsterdam, 1852

Minas nuevas de guerra, y su aplicacion á la defensa como consecuencia de un reciente descubrimento para emplear la electricidad en la Voladuro de los Hornillos. 8vo. *Madrid,* 1854

† Part of Bulletin des Travaux de l'Académie pendant l'Année 1855. 8vo. (*Toulouse Acad.* 4to. Série v. 398.) At p. 398 : *Laroque propose comme sujet de prix : Recherches sur l'Electricité atmosphérique.* At p. 434 : *Petit, Notes sur Inclinaison et Declinaison.* *Toulouse,* 1855

Die geographische Verbreitung der elektrischen Fische. 1856

† Elektricitätslehre und Physik der Erde. (*Fortschritte der Physik im Jahre* 1856. *Dargestellt von der physikalischen Gesellschaft zu Berlin.* xii. *Jahrgang. Redegirt von A. Krönig.* 2 *Abtheilung.*) *Berlin,* 1856

† Electrical Disc and Experiments. By a positive Conductor. 8vo. 96 pp.
London, 1869

Vorträge, popular-naturwissenschaftliche über neuere Forschungen. 6te Heft. 8vo. 80 pp. (*Inhalt. Ueber Luft-elektricität, Nebel u. Höhenrauch. Nach Untersuchungen d. Verf.* v. *Ob-Lehr. Dr. F. Dellmann.*) *Kreuznach,* 1870

Haandbog i læren om elektricitet og magnetisme for 1ste ingenieur bataillon. Approberet af Krigsministeriet ved resolution of 18 Oktober, 1869. 8vo.
Kjobenhavn, 1870

Mémoire sur l'Eremophilus et l'Astroblepus, sur une nouvelle espèce de Pimelode, sur la Condor, une nouvelle espèce de Gymnote, sur l'Anguille électrique. 4to. (From *Wheldon's Cat.* No. 74, 1870.)

Notice sur un appareil électro-magnétique inventé par M. Clarke, Mécan. Anglais. 8vo. (*Zantedeschi?*) *Nancy, n.d.*

Electricity—History of.

Istoria dell' Elettricità. Firenze. (*Giornale dei Letterati d'Italia, Firenze,* tom. iv. part iv. art. 6 ; tom. v. part i. art. 2 ; part ii. art. 6.)
Firenze, 1747-48

† Histoire de l'Electricité, traduite de l'Anglais de Jh. Priestley, avec des notes critiques. 3 vols. 12mo. 9 plates. *Paris,* 1771

† Histoire générale et particulière de l'Electricité. (*Vide* Mangin.)

Electro-Metallurgy.

Nouveau Manuel complet de Galvanoplastie, &c. (From *Zantedeschi, Trattato* ii. 440.) (*Jacobi?*) *Paris,* 1843

. . . intorno all' Elettro-metallurgia Italiana quale fu dimostrata e svolta nella memoria letta alla Rl. Accad. . . . di Modena, 19 Dec. 1843, e nel relativo articolo portato dall' Appendice del Foglio periodico di Modena, 8 Feb. 1844. 8vo. (*Nuovi Ann. delle Scienze,* x. 435.) *Bologna,* 1844

Un Chapitre d'Electro-métallurgie. Fabrication des Plaques Argentées pour le daguerréotype. 12mo. 48 pp. *Paris,* 1852

† On the Application of Science to Electro-plate Manufactures. 8vo. (From the *Popular Science Review.*) (*Gore?*) *London, n.d.*

Manuale di galvano-plastica e della doratura chimica. Aggiuntovi un ricettario per l'argentatura galvanica sul vetro sulle norme di Figuier, Volta, Brugnatelli, ecc. 2 edizione. 16mo. 94 pp. *Trieste,* 1864

Medical Electricity.

† Lettere sopra l'elettricità principalmente per quanto spetta alla Medicina. 12mo. 81 pp. *Venezia,* 1747
Saggio d'esperienze sopra la medicina elettrica. *Venice,* 1745 ?

Medical Electricity—*continued.*

† Recueil sur l'électricité médicale ; dans lequel on a rassemblé les principales pièces publiées par divers savans, sur les moyens de guérir, en électrisant, les malades. 2nd edition. 2 vols. 12mo.

The authors in the 1st volume are—

1, Pivati ; 2, Veratti ; 3, Sauvages ; 4, Jalabert ; 5, De la Soue ; 6, Quelmalz ; 7, Zetzell ; 8, M . . .

Those in the 2nd volume are—

9, Bianchini ; 10, Deshais (vere *Sauvages*) ; 11, Dufay ; 12, Sauvages (lettre à Morand). *Paris,* 1763

† M . . . Réflexions sur les différents succés des tentatives de l'Electricité pour la guérison des Maladies. 12mo. 28 pp. (*Recueil sur l'Elect. Méd.*)
Paris, 1763

New Thoughts on Medical Electricity ; or, Attempt to discover the real uses of Electricity. (From Kuhn. Hist. ii. 319.) *Cambridge,* 1782

Sur les Tracteurs de Perkins. (*Mém. des Soc. Sav. et Let.* ii. 237.)

Recherches sur le Perkinisme, ou de l'influence des Tracteurs métalliques, inventés par le Dr. Perkins, sur certaines maladies. (*Annales de la Soc. de Médec. de Montpellier,* xxix. 274.) *Montpellier*

Observations sur l'Electricité, par M., Chirurgien de la Salpétrière. (A brochure.)

Consectaria Electro-medica. (*Vide* Bertholon, ii. p. 77.)

A Fortunate Cure. (*Guy's Hospital Reports,* vol. viii. part i. 1852, p. 108.)
London, 1852

Electric Telegraph.

† An Essay, tending to improve intelligible Signals, and to discover an universal Language. From an anonymous correspondent in France (probably the inventor of the Telegraph). Translated from the French. 4to. (*Trans. Amer. Phil. Soc.,* old series, vol. iv. p. 162.) *Philadelphia,* 1799

Rail, Electric Telegraph, and Electric Clock. 12mo. *London,* 1847

Modification of Keisser's Ideas (Telegraph). Letters with Tinfoil intervals. 1848. (*Forster's Bauzeitung,* 1848, p. 238.) 1848

An Article on the Electric Telegraph in the North British Review, vol. xxii. p. 543. (*Vide* W. F. Cooke in the "Reader" Correspondence, Mr. Ronalds and the Electric Telegraph.)

† Electro-magnetiske Telegrapher, en fortælling fra den nyere Tid. 8vo. (A Novel.) *Kjobenhavn,* 1851

The Telegraph Guide, containing general information for sending Telegrams to all parts of England, Scotland, Ireland, and the continents of Europe, Asia, Africa, and America ; with a Telegraphic Map of Europe. (*Supplement to the "Electrician."*) *London*

† Instruction pour les Télégraphistes de la Suisse, basée sur l'emploi du Système Steinheil pour l'établissement du Réseau et des Stations. 8vo. 7 plates. 128 pp. *Berne,* 1852

† Die elektrischen Telegraphen. Gemeinfassliche Erläuterungen über die Principien und Apparate der neueren Telegraphie. 8vo. 62 pp. *Stuttgart,* 1852

† Dienst-Anweisung für die telegraphische Correspondenz auf den Linien der Deutsch-Oesterreichischen Telegraphen-Vereins. 4to. 71 pp. *Berlin,* 1853

† Riassunto generale di Telegrafia elettrica. 8vo. *Milano,* 1854

Recueil de prescriptions relatives à l'usage des Télégraphes électriques en Suisse. (Pub. par la Direction du Téleg. Suisse.) 8vo. *Berne,* 1854

† Dienst-Anweisung für die telegraphische Correspondenz auf den Linien des Deutsch-Oestereichischen Telegraphen-Vereins. 4to. 80 pp. (Formula marked from A to Q.) *Berlin,* 1855

Le Père de la nouvelle Télégraphie. 4to. *Auch,* 1856

† Télégraphie privée. Tarifs pour la France et l'étranger à partir de Marseille. 8vo. *Marseille,* 1856

Electric Telegraph—*continued.*

† Cheap Telegraphs. 8vo. 12 pp. *London*

† Dienst-Instruction für die Telegraphen Verwaltung. 3 vols. 1118 pp. 4to.
Berlin, 1860

† Application de la Télégraphie électrique aux usages domestiques. 8vo. 42 pp.
1 plate. (*Extrait de la Revue générale de l'Architecture et des Travaux publics.*) *Paris*, 1862

Télégraphie domestique. Instructions sur la pose et l'entretien des Sonnettes électriques. 12mo. (*Vide* Breguet, 1865.) *Paris*, 1865

Sulla Telegrafia Italiana. 8vo. (From *L'Annuario Bibliografico*, an. i. del. 1863. p. 190.) *Torino*, 1866

† Dictionnaire pour la Correspondance télégraphique secrète ; précédé d'Instructions détaillées et suivi de la Convention télégraphique Internationale conclue à Paris, le 17 Mai, 1865. Par un Secrétaire de Légation. 12mo.
Paris and Strasburg, 1867

† Telegraphic Reform. The Post Office and the Electric Telegraph. 8vo. 46 pp.
(Reprinted from the *British Quarterly Review.*) *London*, 1867

Manuel de Correspondance télégraphique, par un Directeur du Télégraphe. 8vo·
(From *Reinwald's Catalogue* for 1868, p. 138.) *Paris*, 1868

† Special-Karte der Normal-Verbindungen der Telegraphen-Leitungen des Norddeutschen-Bundes, entworfen und gezeichnet im technischen Bureau der General Telegraphen Direction. Abgeschlossen den 1 Jan. 1868. Folio. 16 Chromolithographs. *Berlin*, 1868

† Notices of the Electro-magnetic Telegraph. (*Vide* Society for the Diffusion of Useful Knowledge (S.)

† Uebersicht der Stationen des Telegraphen Vereines, und des Grossherzogthums Luxemburg, nach Taxquadraten geordnet. 4to. 47 pp. *Berlin*, 1870

Magnetism.

Answer to some Magnetical Enquiries proposed. Pp. 423-24, year 1667, of these Transactions. 4to. (*Phil. Trans.* 1667, p. 478.) *London*, 1667

Several Observations of the respect of the Needle to a piece of Iron held perpendicularly, made by a Master of a Ship crossing the Æquinoctial Line. 4to. (*Phil. Trans.* 1685, p. 1213.) *London*, 1685

Recueil d'Expériences sur l'Aimant, avec la manière de les faire par rapport aux Aimants que l'on spécifie, et aux petites pièces et machines que l'on décrit. Sm. 8vo. 6 + 71 pp. (*Curious and rare.*) *Lyon*, 1686

† M.D.*** Traité de l'Aimant, divisé en deux parties. La première contenant les expériences, la seconde les raisons. 1st edit. 12mo. (*Vide* Dalance.)
Amsterdam, 1687

C*** J. A Paper about Magnetism, or concerning the changing and fixing the Polarisation of a piece of Iron. 4to. (*Phil. Trans. for* 1694, xviii. p. 258.)
London, 1695

† Lettres écrites sur le choix d'une hypothèse de l'Aimant. 12mo. 138 pp. *Lyon*, 1699

† Explication des propriétés de l'Aymant. 4to. 11 pp. 1727

† Traité de la Boussole (en Turc). 8vo. *Constantinople* 1144 = 1732

† Pièces sur l'Aimant. 4to. *Paris*, 1748

† Description des Courans Magnétiques. (*Basin ?*) *Strasbourg*, 1753

† A brief Theory of the North Magnetic Pole. 8vo. (*Note.* This is Lovet's Appendix to his *Philosoph. Essays.*) *Worcester*, 1766

Report on McCullagh's Compasses. *London*, 1778

Physica-magnetica. Variatio, declinationes. . . . (*Opusc. Act. Erudit. Lyps.* tom. i. p. 542.)

Betrachtungen über die Kraft des Magnets. (*Oekonom. Nachr. d. Gesellsch. in Schlesien*, iii. 220.)

Zusammenstellung der, in den letzten Jahren in dem Gebiete des Magnetismus gemachten Erfahrungen. (*Erdmann's Journ.* xlix. 1.)

Magnetism—*continued.*

Ency. Method. Pl. viii. Amusements de Physique. (*Magnetical Apparatus.*)

Ency. Method. Art. vi. 694. (*Artificial Magnetism.*)

† Trismi tonici cæt. per magnetem curatio. 4to. 22 pp. (*Volta's copy.*)

† Article Magnetism, in Encyclopædia Metropolitana. 4to. 113 pp. (*Vide* Barlow.)
London, n. d.

Remarques sur les Aimans artificiels de Basle (faites par Dietrich, bourgeois et artiste à Basle.) (*Acta Helv.* ii. 264.)

† Magnetism. Observations on the Dip and on the Intensity of the Magnetic Force. . . . By Hansteen. (*Phil. Mag.* lvii. 70.)   *London,* **1821**

† Observations on the Experiment of Mr. Murray, on the supposed relation between Caloric and Magnetism. "Experimenter." 8vo. (*Phil. Mag.*, lxi. 251.)
London, **1823**

Ency. Method. Physique. Art. Aimant.

Catalogue général des Livres composant les Bibliothèques du Département de la Marine, au mot Aimant. 8vo. 5 vols. (In the *Chambre de Commerce, Bordeaux.*)   *Paris*

An unnamed writer in the *Gazzetta di Venezia,* 25 Gennajo, 1839, No. 21, on "Magnetismo Transversale." (*Fusinieri, Riflessioni sopra il Magnetismo Transversale,* in his *Annali,* ix. 19.)   *Venezia,* **1839**

Terrestrial Magnetism. 8vo. (*Quarterly Review.*)   *London,* **1840**

(Exploration Scientifique de l'Algérie.) Observations sur le Magnétisme Terrestre. Recherches de physique sur la Méditerranée. 4to. 2 vols.   **1846**

R***. (*Rieu?*) Examen de quelques Questions de Physique. 8vo. 1 plate.
96 pp.   *Genève,* **1847**

Essai sur l'Influence que peut avoir la Construction des Corps sur leurs qualités magnétiques. 8vo. 26 pp.   *Genève,* **1852**

† Essai sur l'Influence de la Polarité moléculaire dans les Phénomènes électriques et magnétiques. 8vo. 1 plate. 57 pp.   *Genève,* **1853**

Essai sur le Mouvement du Fluide magnétique. 8vo. 21 pp. *Genève,* **1853**

De l'Action que parait exercer le Magnétisme Terrestre sur un Corps construit de couches égales juxta-posées. 8vo. 1 plate. 22 pp.   *Genève*

El Magnetismo. Consideraciones luminosas acerca de esta doctrina, por un doctor de la Facultad de Medicina de Paris. Traduccion Castellana. 8vo.
Madrid, **1855**

Walk through the Paris Exhibition of 1855. By F. R., &c. 12mo. (*Magnetical and Meteorological Inst.,* &c. By F. Ronalds.)   *London,* **1855**

Magnetic Variation.

(B . . . Dr.) Extract of a Letter concerning the present Declination of the Magnetic Needle and the Tydes at Bristol, 1668. 4to. (*Phil. Trans.* 1668, p. 726.)   *London,* **1668**

Abwechslung von der Magnet-Nadel, nebst einem Calendario magnetis declinantis et inclinantis. 8vo. (*Fragment, probably from a Scientific Journal printed in Germany about* 1746, pp. 311-347.)

† S——, Sir R. A Relation of the Effect of a Thunder-clap on the Compass of a Ship on the coast of New England. 3 pp. Also, a Letter concerning the former Relation. 2 pp. (*Phil. Trans. for* 1683, xiii. pp. 520-21.) *London,* **1683**

† Cartes de Déclination et Inclination de l'Aiguille Aimantée, rédigées d'après la Table des Observations Magnétiques faites après les Voyageurs depuis l'année 1775. (*Sic.*) (*Chambre de Commerce, Bordeaux.*)

† Declination of the Needle :

	At Toulouse in	1770	=	18° 59′
,,	,,	1780	=	20° 15′
,,	,,	1790	=	21° 5′
,,	,,	1804	=	21° 43′—(1st of Germinal, year 12.)

(*Phil. Mag.* xviii. 369. 1816.)   *London,* **1816**

Magnetic Variation—*continued.*

† Northern Expedition. 8vo. 6 pp. (*Phil. Mag.* lii. 225.) (*Contains statements as to extraordinary variations of declination and inclination of the magnetic needle.*) *London*, 1818

† Remarks on an important Error in a Table for computing Local Attraction, circulated by order of the Board of Longitude. 8vo. 3 pp. (*Phil. Mag.* lv. 357.) *London*, 1820

† Return of the Discovery Ships. 8vo. 6 pp. (*Phil. Mag.* lvi. 383.) *London*, 1820

† The Arctic Expedition. 8vo. 4 pp. 1 plate (*i.e.* map.) (*Phil. Mag.* lvi. 440.) *London*, 1820

† Tables of the Variation of the Magnetic Needle in different parts of the Globe. (*Edinb. Phil. Journal*, ix. p. 447, and x. p. 116.) *Edinburgh*, 1824

† Mittheilungen über ältere magnetische Declinations-Beobachtungen. 1es Heft. *Wien*, 1850

A Treatise on Terrestrial Magnetism. Containing an outline of the discoveries and theories connected therewith ; an inquiry as to whether the terrestrial sphere has four or only two magnetic poles ; on the probable causes of terrestrial magnetism ; on the irregularities observed in the secular variation, &c. *Edinburgh and London*, 1871

Meteorological Phenomena.

AURORA BOREALIS.

† (J. W.) An Account of the Aurora Borealis observed at Dublin on the 6th of February, 1720-21. 4to. 7 pp. (*Phil. Trans.* xxxi. for 1720-21, 180. *London*, 1721

Rerum Italicarum Scriptores, ab anno aeræ Christianæ 1000 ad 1600 ; quorum potissima pars nunc primum in lucem prodit ex Florentinarum Bibliothecarum codicibus. Tomus primus. Folio. (*Note.* Contains accounts of ancient auroræ, meteors, marvels, &c.) *Florentiæ*, 1748

† Account of an Aurora Borealis, from a Correspondent at Lancaster, Pennsylvania. 4to. 2 pp. (*Trans. Amer. Phil. Soc.* vol. i. old series, p. 338.) *Philadelphia*, 1769

† Article Aurora Borealis in Encyclopædia Britannica, 7th edition, vol. iv. p. 194. 4to. *London and Edinburgh*, 18—

† Aurora Borealis observed at Leeds, London, Derby, Paris, and in Germany, on the 8th February, 1817. 8vo. (*Phil. Mag.* xlix. pp. 155, 223, and 236.) "Herr Stark, at Augsburg, observed that Volta's electrometer and the magnetic needle were almost constantly in motion." Its increased variation was (at one time) 2° 7′ *London*, 1817

Note.—I know of but two other observations of strong electrical signs occasioned (seemingly) by an Aurora, viz., those of Canton, in the *Phil. Trans.* for 1753, p. 350 ; and of Morozzo (Count), in the *Mém. de Turin An.* 1784-5, tom. i. part ii. pp. 328 and 338. The last very extraordinary.— *Francis Ronalds.*

† Aurora Borealis in northern parts of England and south of Scotland and in Norfolk, &c., 17th October, 1819. 8vo. 3 pp. (*Phil. Mag.* liv. 388.) *London*, 1819

† Aurora Borealis of September 25, 1827. (*Phil. Mag.* ii. 395.) *London*, 1827

† Auroræ Boreales on 13th April, 1827, at Gosport (with falling stars, &c.), and on 17th February, 1827. 8vo. (*Phil. Mag.* i. 317.) *London*, 1827

† Aurora Borealis observed at Gosport on the 29th September, 1828. 8vo. (*Phil. Mag.* iv. 392.) *London*, 1828

† Aurora Borealis observed in London December 1, 1828. 8vo. (*Phil. Mag.* v. 77.) *London*, 1829

† Aurora Borealis 19th June, 1830. 8vo. (*Phil. Mag.* vii. 463.) *London*, 1830

† Auroræ Boreales, with meteors, on the 5th, 16th, and 17th October, 1830. 8vo. *London*, 1830

† Aurora Borealis, with meteors, on the 20th October, 1830. 8vo. (*Phil. Mag.* viii. 316.) *London*, 1830

24 ANONYMOUS.

Meteorological Phenomena—_continued._

† Aurora Borealis on the 7th November, 1830, observed at Gosport. 8vo. (*Phil. Mag.* viii. 392.) *London,* 1830

† Aurora Borealis, December 14, 1829. 8vo. (*Phil. Mag.* iv. 158.) *London,* 1830

† Auroræ Boreales, with meteors, on the 1st, 4th, and 7th November, 1830. 8vo. (*Phil. Mag.* ix. 79.) *London,* 1831

† Influence of the Aurora Borealis on the Magnetic Needle. 8vo. 1 p. (*Phil. Mag.* ix. 151.) *London,* 1831

† Aurora Borealis observed at Gosport on the 7th of January, 1831. 8vo. (*Phil. Mag.* ix. 233.) *London,* 1831

† Auroræ Boreales, with meteors, 19th and 20th April, 1831. (*Phil. Mag.* ix. 466.) *London,* 1831

† De Naturali Electricismo ejusque ad Aurorem Borealem applicatione. Dissertatio physicæ. 4to. pp. 11-30.

AEROLITES.

† (P. N. J.) On the production of Aërolites and other solids. 8vo. 3 pp. (*Phil. Mag.* xliv. 321.) *London,* 1814

† On Shooting Stars. 8vo. 2 pp. (*Phil. Mag.* liii. 201.) *London,* 1819

† Aërolite, fell 15th June, 1822, at Juvinas, a village in the arrondissement of l'Argentiers. 8vo. 1 p. A further account of it given by Firman. 8vo. (*Both in Phil. Mag.* lix. 235.) *London,* 1822

† Account of an Aërolite found near Croz (Finistère) on the 15th June, 1822. (*Phil. Mag.* lix. 67.) *London,* 1822

† Caduta d'un Aerolito. 8vo. (*Biblioteca Ital.* xcvii. 424.) *Milano,* 1840

† Bulletin des mois Juillet et Août. 8vo. (*Petit on Asteroids as the causes of High Temperature,* p. 436.) *Toulouse,* 1851

† Bulletin des mois de Juin, Juillet, et Août. 8vo. (*Petit on Asteroids,* p. 431.) *Toulouse,* 1853

METEORS, &c.

Histoire d'un Météore igné. 8vo. 6 pp. (*Mémoires de l'Académie de Dijon,* tom. i. xlii.) *Dijon,* 1769

† Account of an uncommonly luminous Meteor of 13th November, 1803. 8vo. 1 p. (*Phil. Mag.* xvii. 191.) (*Vide* Firmingham.) *London,* 1803

Des pierres Météoriques. 12mo. 30 pp. *Dresden,* 1804

Meteoric Phenomenon in Middle Quarters, in the parish of St. Elizabeth. 8vo. 2 pp. (*Phil Mag.* xx. 281.) (*This article appears under the heading of St. Iago de la Vega, Oct.* 8.) *London,* 1804

† Meteoric Stone in Russia, 8vo. 1 p. (*Phil. Mag.* xxviii. 373.) *London,* 1807

† On a Meteor of the 17th October, 1808. 8vo. 1 p. (*Phil. Mag.* xxxii. 95.) *London,* 1808

† Singular Sensations, or Shakes experienced on board the ship Favourite on July 14, 1810, in lat. 31° 56′, long. 39° 30′. 8vo. 1 p. (*Phil. Mag.* xxxvi. 395.) *London,* 1810

† Meteor (luminous) seen in Holland. 8vo. 2 pp. (*Phil. Mag.* xxxvi. 395.) (*Note.*—If this account is correct, it presents some very uncommon and interesting facts.) *London,* 1810

† Description of a splendid Meteor seen at Dublin on the 17th April, 1814. 8vo. 3 pp. (*Phil. Mag.* xliii. 303.) *London,* 1814

† Fall of Meteoric Stones in India. 8vo. 1 p. (*Phil. Mag.* xlvi. 155.) *London,* 1815

† A hard body, supposed to be of the nature of a Meteoric Stone, fell during a hard thunderstorm into a window at Glastonbury. 8vo. (*Phil. Mag.* xlviii. 235.) *London,* 1816

† A brilliant Meteor (described in Letter to the Editor, dated Ipswich, 1817). 8vo. 1 p. (*Phil. Mag.* l. 469.) *London,* 1817

Meteorological Phenomena—*continued.*

† Meteor at Aberdeen, &c. (*Phil. Mag.* liv. 75.) London, 1819

† Some Observations made at Clapham Common, Surrey, on the heavy storm that took place on the night of Sunday, 30th July, 1820. 8vo. 4 pp. "Philobius." (*Phil. Mag.* lvi. 120.) London, 1820

† A singular Meteoric Substance (black leaves like burnt paper, but harder) found in the Museum of Von Grotthus of Courland. According to the Ephemeris of the Leopold Acad., it fell, in great quantities, in Courland, on the 31st January, 1686. 8vo. (*Phil. Mag.* lvi. 157.) London, 1820

† Remarks upon Mosely's article on Solar Spots of 1816 in *Phil. Mag.* xlix. 182. 8vo. 3 pp. (*From New Monthly Mag. for January*, 1821. At p. 70, article on Magnetism and on Meteoric Stone; and p. 72, Chain Cables as Conductors.) London, 1821

† Meteoric Stone. An aërolite which fell near Kostritz, in Russia, analysed by Stromeyer. 8vo. (*Phil. Mag.* lvii. 70.) London, 1821

† Meteor at Brighton. 8vo. (*Phil. Mag.* lviii. 457.) London, 1821

† Meteors at Richmond, Virginia, &c. of extraordinary size, and at Rodez, in Aveyron. 8vo. 1 p. (*Phil. Mag.* lix. 399.) London, 1822

† Remarkable Meteor (seen at Kreil and at Copenhagen), (*Phil. Mag.* lxii. 238.) London, 1823

† Meteor and Earthquakes at Ragusa, in Dalmatia, on 20th August, 1823. 8vo. (*Phil. Mag.* lxii. 315.) London, 1823

† Account of a large Meteoric Stone . . . which fell near the village of Kadonah, in the district of Agra. 8vo. London, 1827

Rapport sur les travaux de l'Académie des Sciences, 1829. 8vo. (*Séances de l'Acad. de Bord.* 1829, pp. 43 and 59.) Bordeaux, 1829

Observations Météorologiques faites à Nijné-Taguilsk (Monts Oural). 4to. (*This is the Livraison for October* 1, 1839, *to December* 31, 1840. *It contains notices of Tonnerre, Orages, Aurors, Météores de Feu, &c. in the Remarks.*) Paris, 1842

Report (official) of the fall of a Thunderbolt, or Meteorolite, near Dover. 8vo. London, 1853

Boletin del Observatorio Fisico Meteorico de la Habana. Parte i. 4to. Habana, 1862

† Catalogue of the Meteorolites in the Museum of the Geological Survey of India, Calcutta. 4to. 9 pp. Calcutta, 1866

VARIOUS.

† A Description of the Phenomenon of March 6 last as it was seen on the ocean, near the coast of Spain. With an account of the return of the same sort of appearance on March 31 and April 1 and 2 following. 4to. 3 pp. (*Phil. Trans. for* 1714-15-16.) London, 1717

De cælesti quadam Flamma, An. 1743, xiii. Kal. Febr. 4to. (*Commentarii Bononienses*, tom ii. part i. p. 459.) Bononia

† Surprising Inundation of the Valley of St. John's, near Keswick, in Cumberland, on the 22d of August, 1749 ; in a letter from a young clergyman to his friend. Communicated by J. Lock. Read March 15, 1749. (*Phil. Trans.* xlvi. 362.) London, 1749-50

† Beschreibung eines Donnerschlages in Gothland. 8vo. (*K. Schwed. Akad.* xv. 80.) Hamburg and Leipzic, 1753

† Phenomena cælestia observata. 8vo. Romæ, 1754

† Osservazione del Globo luminoso apparso il di 11 Settembre, 1784. 4to. 1 p. (*Opus. Scelti.* vii. 284.) Milano, 1784

† A Fire-ball struck a house on the 4th July at East Norton. (*Phil. Mag.* xvi. 191.) London, 1803

† Description d'une Trombe de Terre ascendante. 4to. 5 pp. Turin, 1808

† Relazione di un Fenomeno osservato nel Porto di Napoli a' 24 Decembre, 1798. 4to. 7 pp. (*Atti dell' Accad. di Siena*, tom. ix. p. 30.) Siena, 1808

26 ANONYMOUS.

Meteorological Phenomena—_continued._

† Atmospherical Phenomena in the River St. Lawrence. 8vo. (*Phil. Mag.* xliv. 91.)
London, 1814

† Atmospheric Phenomenon, 11th September, 1814. Luminous arch near the City. 8vo. (*Phil. Mag.* xliv. 236.)
London, 1814

† Combination of the Electric Column, the Thermometer, Barometer, and Hygrometer in one instrument, for electro-atmospherical researches. 8vo. (*Phil. Mag.* xlix. 55.)
London, 1817

† Waterspout, on 27th June, 1817, observed from Kentish Town. 8vo. 3 pp. (*Phil. Mag.* l. 146.)
London, 1817

(T. H.) On the identity of Waterspouts and Whirlwinds. 8vo. (*Phil. Mag.* li. 103.)
London, 1818

† Destructive Waterspout. 8vo. 1 p. (*Phil. Mag.* lii. 68.)
London, 1818

† (E. F.) Atmospheric Phenomenon. 8vo. (*Phil. Mag.* lviii. 314.)
London, 1821

† Brilliant Phenomenon observed at sea in 1819. 8vo. 2 pp. (*Phil. Mag.* lix. 157.)
London, 1822

† Atmospheric Phenomena. An igneous meteor. Article dated Bamberg, Dec. 25, 1821. 8vo. (*Phil. Mag.* lix. 67.)
London, 1822

Violent Storms and Waterspouts in Belgium. 8vo. 2 pp. (*Phil. Mag.* lxii. 232.)
London, 1822

† On the Formation of Hail. From the *Gentleman's Mag.* (Electrical Theory). 8vo. (*Phil. Mag.* lix. 93.)
London, 1822

† (" Glosterian.") On the quantity of Rain collected in two rain-gauges placed at different heights, &c. 8vo. 5 pp. (*Phil. Mag.* lxi. 321.)
London, 1823

† Uncommon Electrical Phenomenon at Belfast on 11th March, 1849. 8vo. (*Phil. Mag.* xlix. 314.)
London, 1849

Cyclone at Masulipatam. Reprinted, with additions, from the *Madras Observer* of December 3, 1864. 14 pp.
Madras, 1864

Lightning, Lightning-conductors, &c.

Neueste Versuche zur Bestimmung der zweckmässigsten Form der Gewitterstangen. (*Deutsch Mus.*, Oct. 1778, pp. 351-62.)
1778

† Maniera pratica di fare li Conduttori . . . 4to. (*Printed by order of the Magistrato della Sanità di Venezia.*) (*Vide* Marzari.)
Venezia, 1787

Encyclopædia Method. Arts. Article, Paratonnerre.
1782

Einige gegen die Gewitterableiter gemachte Einwürfe beantwörtet. 8vo.
Frankfurt, 1790

† Dubbii sull' Efficacia dei Conduttori. 8vo. 122 pp. 1 plate. (*Vide* Bragadin.)
Venezia, 1795

Nachricht und Zeichnung von einer im Jahre 1778, am Schlossthurme, zu Dresden, angebrachten Ableitung. (*Schrift. d. Leipz. ökonomische Societät*, th. v. pp. 222-32.)

† Risposta dell' autore dei Dubbii sull' efficacia dei Conduttori, alla giunta al Giornale Astrometeorologico del Gr. Toaldo. (*Vide* Bragadin.)

† Plain Directions for safe Lightning Conductors for Lightning. 8vo. 49 pp. (*Note. Part of a work. Begins at p. 33.*)

Account of a mass of 7000 Bricks of a Wall displaced several feet by Lightning. 8vo. (*Manchester Memoirs*, ii. 2.)
Manchester, n.d.

Relazione del Turbine scoppiato in Venezia nel Giorno 16 Giugno, 1805. Data Venezia, 19 Giugno. 8vo. (*Da Rio Giornale*, ix. 266.)
Padova, 1805

† Instruction sur les Paratonnerres. Fol. 39 pp. (*Vide Comité des Fortifications.*)
Paris, 1808

(R. R.) Death by Lightning. Man killed at Colwall, near Ledbury, 1817. 8vo. (*Phil. Mag.* l. 315.)
London, 1817

† On Lightning Conductors of Straw.
1820

† On the Cure of a case of Paralysis by Lightning. 8vo. 2 pp. (*Phil. Mag.* lix. 287.)
London, 1822

Lightning, Lightning-conductors, &c.—*continued.*

Instruction sur les Paratonnerres. Fol. *Paris*, 1823

† Anleit. zur Verfertigung und Benutzung der Blitzableiter. 8vo. (*Vide* Gay Lussac, &c.) *Strasburg*, 1824

† Tubes formed by Lightning. 8vo. 1 p. (*Phil. Mag.* or *Annals*, iv. 228.) *London*, 1828

Memoir on Lightning Conductors. Reply to a Prize Question. Bordeaux, 1837. (*Vide* Bourges, Secretary of the Bordeaux Academy, Séance 1837, p. 83.) *Bordeaux*, 1837

On the Knowledge of the Ancients concerning Lightning Conductors. 8vo. (*Fraser's Magazine*, 1839?) *London*, 1839?

Sur l'Histoire du Paratonnerre. 1843. (*Le Portique*, 1re livraison, Jan. 1843, p. 51.) 1843

† (R. C.) (Richard or Richardot?) Nouveaux appareils contre les Dangers de la Foudre, ou les Paratonnerres popularisés. 8vo. 1 plate. 58 pp. *Paris*, 1846

† Part of Bulletin des Mois de Mars, Avril, Mai, Juin, Juillet, Août, 1854. 8vo. (*Toulouse Acad.* series 4, vol. iv. At p. 483, De Clos, *Effets de la Foudre sur un Paratonnerre.*) *Toulouse*, 1854

De la Construction des Paratonnerres. Quelques Réflexions sur le Rapport de la Commission de l'Académie des Sciences du 14 Janvier, 1867. 8vo. 29 pp. *Paris*, 1868

Thunderstorms.

† Beschreibung des erschrecklichen Ungewitters und grossen Wasserfluth, so am 29 Mai, 1613,Weimar und anderere Örte überfallen. 4to. 39 pp. *Leipzig*, 1613

Nachricht, glaubwürd. von 3 grossen Ungewittern oder Wettern am 8 Juli, 1663, zu Ullersdorff in Mähren. 663, 4. 1663

† Part of a Letter to D. G. Garden, giving an account of the Effects of a very extraordinary Thunder near Aberdeen. 4to. 3 pp. (*Phil. Trans. for* 1694, xviii.) *London*, 1695

† Riflessioni sopra . . Maffei . . Fulmini. 4to. 52 pp. (*Vide* Ferri.) *Vicenza*, 1748

† Difesa della . . . sentenza che i Fulmini discendono dalle nuvole . . . Riflessioni . . . Venezia, 1749. 4to. 184 pp. (*Vide* Costantini.) *Venezia*, 1749

Four (very short) Accounts (Italian), printed in 1769 in Brescia and in Venice, of a terrible "Fulmine" at Brescia in 1769. 12mo. *Brescia and Venice*, 1769

† Della maniera di preservare gli edificj dal Fulmine : Informazione al popolo, &c. 4to. 38 pp. (*Vide* Toaldo.) At p. 20 is inserted a translation of Saussure's Exposition, entitled *Manifesto ossia breve esposizione dell' utilità dei Conduttori elettrici.* *Venezia*, 1772

† Evénement produit par le Tonnerre le 27 Juin, 1772, à Paris. 12mo. 88 pp. *Paris*,1772

Verhaltungs Regeln bey nahen Donnerwettern, &c. . . . 8vo. 78 pp. Zweite und vermehrte Auflage. (*Vide* Lichtenberg) *Gotha*, 1775

† Della maniera di preservare . . dal Fulmine. 8vo. 22 pp. *Milano*, 1776

Gewitterbetrachtungen, Gebete, und Danksagung. 8vo. *Heilbronn*, 1780

† (M . .) Osservazioni sopra di un Fulmine, 1783. 4to. 10 pp. (*Opus. Scelti* vi. 347 ; *Tratta dal No.* 35 *del Giornale* . . . *dei Confini dell' Italia.*) *Milano*, 1783

† Nuovo metodo di costruire i Parafulmini praticato in Padova. 4to. 2 pp. (*Opuscoli Scelti*, vi. 380.) *Milano*, 1783

Dissertazione sopra i supposti Fulmini. 4to. (*Note. The figures represent ancient instruments of stone, mistaken, he says, by some authors for "Pietre di Fulmini.*") *Venezia*, 1784

Gewitterkatechismus. 8vo. *Augsburg and Wien*, 1797

Anzeigen der nothwend. Verhaltungsregeln bey nahen Gewittern. 8vo. *Gorlitz*, 1798

28 ANONYMOUS.

Thunderstorms—*continued.*

Onderrigtingen omtrent de wyse op welke menzich te gedragen heeft by het opkomen van Onweder, &c. 8vo. 2nd edition. 1 plate. 104 pp. (*Translated from the German.*) *Amsterdam*

De Fulminibus quibusdam (aus der Erde.) 4to. (*Commentarii Bononienses,* tom. ii. part i. pp. 460-64.)

Rathgeber bey Gewittern. 8vo. *Pirna,* 1807

Miscellaneous.

Curiöse Speculationes bey schlaflosen Nachten von einem Leibhaber der immergern speculirt. 8vo. *Chemnitz and Leipzig,* 1707

Schediasma physicum. 4to. *Vittemberg,* 1715

† Auszug aus verschiedenen eingelaufenen Berichten von seltsamen Knallen die man in Swenskeby gehört. 8vo. 3 pp. (*K. Schwed. Akad.* xiv. 326.) *Hamburg and Leipzig,* 1752

Nachricht von einer Erfindung einer Flinte aus welche man in 2 Minuten zehnmahl schiessen kann und wovon die Elektricität der Grund seyn soll. 8vo. *Stutgardt,* 1762

Schreiben eines Geistlichen. (*Vide* Jaquet.) *Wien,* 1776

Raccolta d' Opuscoli scientifici . . . di Autori Italiani. 4to. (From *Opusc. Scelti,* ii. 44.) *Ferrara,* 1779

Gedanken über die anziehenden Kräfte welche bei der chemischen Auflösung ze. (*Sic.*) 8vo. *Prag,* 1779

Table analytique, &c. Contents of the Dictionary of Sciences, Arts, &c., and its Supplement. 2 vols. folio. (*Opuscoli Scelti,* 4to. iii. 22.) *Paris,* 1780

Monatliche Früchte einer gelehrten Gesellschaft in Ungarn. 8vo. Pest. und Ofen. (*Kuhn N. Entdeck,* i. 27, who says that this book could scarcely be had, and that his account of it is derived from a notice in the *Allgemeine Deutschen Biblioth.* b. 74, st. 1, pp. 214-16. He refers to it on the subject of " the electric matter being a fire.") *Brachmonoth,* 1784

Geschichte einer ausserordentlichen Begebenheit. 8vo. *Frankfort,* 1785

Fisici dubbj dedicati agli amici del vero. 74 pp. 1791 ?

Comment. mortis historiam causa et signa sistens. *Goettingae,* 1794

Diario de los nuevos discumbrimientos, &c. 8vo. (*Opuscoli Scelti,* xviii. 24.) *Madrid,* 1793

† On the Opinions which have prevailed respecting the Nature of Alkalis and Earths. With Note by the Editor. 8vo. 5 pp. (*Phil. Mag.* xxxii. pp. 18-22.) *London,* 1808

† Extract of a short Article on Volcanoes, from a recent French Journal. 8vo. 1p. (*Phil. Mag.* xxxv. 235.) *London,* 1810

† Experiments on Capillary Syphons, with electrified and with heated Liquids. 8vo. (*Phil. Mag.* xlii. 202.) *London,* 1813

† A Review of some leading Points in the official Character and Proceedings of the late President of the Royal Society. 8vo. 14 pp. (*Phil. Mag.* lvi. pp. 161 and 241.) *London,* 1820

Dictionnaire des Découvertes, Inventions, &c., en France, 1789 à 1820. Par une Société de Gens de Lettres. 8vo. 17 vols. 1822-24

(R. N.) Notice respecting the real Inventor of the Steam Engine. 8vo. 2 pp. (*Phil. Mag.* lvii. 426.) Said to be Saml. Morland, 1632. *London,* 1821

Il propagatore dei Paragrandini convinto da sè stesso della loro inutilità ; ossia confutazione della difesa dei paragrandini di A, B, C, D. *Milano,* 1824

Notice sur les Travaux de la Classe des Sciences de l'Acad. de Toulouse. 8vo. (*Toulouse Acad.* 2ᵐᵉ série, i. 93.) *Toulouse,* 1827

Repertorium Bibliograph. 8vo. *London*

† (Z. L. V.) *i.e.,* Volta, Zanino, and Lo. Sopra alcuni Errori Biografici e Fisici del Biot. Sm. 8vo. 48 pp. (*Raccolta Pratica di Zanino, Volta, ed altri, Anno* ii.) (*Note. The errors referred to are those relative to* (*his father*) *A. Volta's writings, &c.*) *Como,* 1833

Miscellaneous—*continued.*

Lettere inedite di quaranta illustri Italiani, del secolo (Muratori, Volta, Tiraboschi, &c.) 8vo.   *Milano,* 1836

Une Masse de Grillons de 19 pieds, 10 pouces (Anglais) de puissance. (*Asiatic Journal for November,* 1838.)   1838

Account of Une Masse de Grillons de 19 pieds, 10 pouces (Anglais) de puissance. (*Asiatic Journal, November,* 1838.)   1838

Apparatus zur Entdeckung des Scheintodes im Grabe, erfunden von J. A. Meier. Nebst Bermerkungen eines prakt Arztes. 8vo.   *Berlin,* 1843

† Résumé des Travaux de l'Acad. pendant les Années 1839-41. 8vo. (*Toulouse Acad.* 2ᵐᵉ serie vi. pp. 4 and 6. References to *Pinaud, Fazeau, Hombre Firmas,* &c.)   *Toulouse,* 1843

† Bulletin du mois de Mai, 1850. 8vo. (At p. 248, *Leymerie on the Tourmaline.*)
                                                                    Toulouse, 1850

† Bulletin du mois de Mars. 8vo. (*Toulouse Acad.* 4ᵐᵉ série, i. 233. Notice of *Petit's* verbal communication on *Une Chute de Pluie par un Ciel parfaitement serein,* p. 234.)   *Toulouse,* 1851

† Bulletin des mois Avril et Mai. 8vo. (*Toulouse Acad.* 4ᵐᵉ série, i. 319-20. At p. 319, allusion to reply of *Petit* to *La Verrière ou Bolides.* At p. 320, *Lecture d'une Lettre par Petit . . . Chute de Pierre à Escal.*)
                                                                    Toulouse, 1851

Papers on Subjects connected with the Duties of the Corps of Royal Engineers (Blasting, &c.)   *London,* 1851

† Table Générale des 31 premiers Vols. des Comptes Rendus, Années 1835-50. 4to. 1018 pp.   *Paris,* 1853

† La Società di mutua Assicurazione contro la Grandine e gl' Incendi. Memoria al Parlamento Italiano. 4to.   *Milano,* 1861

Nouvelle Dissertation. (*Vide* Morin.)

† (B . . . R.) Insetti luminosi dell' America. 8vo. 2 pp. (*Bibliot. Ital.* lix. 424.)   *Milano,* 1830

B

B., A. (*Vide* Anon. Electricity.)   1814

B., W. A. (*Vide* Anon. Electricity.)   1830

B * * * , Dr. (*Vide* Anon. Magnetic Variation.)   1668

B * * * , R. (*Vide* Anon. Miscellaneous.)   1830

Baader, Franz Xavier von. (*Born March 27, 1765, at Munchen; died March 23, 1841, at Munchen.* Ueber den Blitz, als Vater des Lichts. . . . 8vo.
Nurnberg, 1815

† **Babinet,** Jacques. *Born March 5, 1794, at Lusignan, départ. Vienne; died October 21, 1872, Paris.* Exposé des nouvelles Découvertes . . . par Œrsted, &c. 8vo. (*Vide* Ampère et Babinet.)   *Paris,* 1822

Notice sur l'établissement des lignes Télégraphiques. 4to. (*Revue des Deux Mondes,* 1er *Semestre,* 1853.)   *Paris,* 1853

† Etudes et Lectures sur les sciences d'observation et leurs applications pratiques. Premier vol. 12mo. 221 pp.   *Paris,* 1855

† Etudes et Lectures sur les sciences d'observation et leurs applications pratiques. Sixième vol. 12mo. 261 pp.   *Paris,* 1860

Bacelli, Liberato Giovanni. *Born November 18, 1772, at Lucca; died October 21, 1835, at Bologna.* Lettere due sopra i fenomeni Elettro-magnetici. (*Venet. Lomb. Instit.* iv. 48.)   *Carpi,* 1820

† I fenomeni Elettro-magnetici a due leggi ridotti, con la loro cagione tolta dall' opinione Symmeriana. Ragionamento. 8vo. 86 pp. (*Vide* Barlocci, 1826, p. 280.)   *Modena,* 1821-28

† Risultati dell' esperienze fatte dal Bacelli intorno all' azione dei fili di metallo elettrici sugli aghi calamitati e non calamitati. Comunicati alla Rle. Accad. di Modena. (Dal *Messaggiere Modenese,* No. 38, il 12 Maggio, 1821.) Con note inedite del medesimo Professore. 8vo. (*Biblioteca Ital.* (*Stella*) xxiii. 76.)   *Milano,* 1821

Reference to his new Electrical Machine, described, with some hydraulic and optical apparatus, 19th July, 1834, at the University of Modena. (*Verona Poligrafo,* nuov. s. iii. 1834, p. 83.)   1834

Sul Magnetismo del rame e di altre sostanze. . . . (*Poggendorff,* i. 83.)

Bache, Alexander Dallas. *Born July 19, 1806, at Philadelphia.* On the diurnal Variation of the Horizontal Needle. 4to. (*Trans. Amer. Phil. Soc.* new series, v.)   *Philadelphia,* 1837

† Discussion of the Magnetic and Meteorological Observations made at the Girard College Observatory, Philadelphia, from 1840 to 1845. In Three Sections and Nine Parts. (*Smithsonian Contributions.*)   *Washington,* 1840-45

† Report (as Chairman) of the Compass Committee of the National Academy of Sciences. (*Vide* National Academy of Sciences.)   1864

† Observations of the Magnetic Intensity at twenty-one stations in Europe. Read March 6, 1840. 4to. (*Trans. Amer. Phil. Soc.,* new series, vii. 75.) (*Faraday's copy*)   *Philadelphia,* 1841

Notice of Experiments on Electricity developed by Magnetism. 8vo. (*Journal of the Franklin Inst.*)   *Philadelphia,* 1832

Note of the Effect upon the Magnetic Needle of the Aurora visible at Philadelphia 17th May, 1833. 8vo. (*Journal of the Franklin Inst. July,* 1835.)   1835

Observations on the Disturbance in the direction of the Magnetic Needle during the Aurora of 1833. 8vo. (*Journal of the Franklin Inst.*)   1833

Bache—*continued.*

† Observations upon the facts recently presented by Professor Olmsted in relation to Meteors seen on the 13th November, 1834. 8vo. (*Journal of the Franklin Inst.* xvii.) 1836

† 1. On Franklin. . . . Storms. 2. On Rain. 3. On the Discovery of the Non-conducting Power of Ice. 8vo. (*Journal of the Franklin Inst.* xvii.) 1836

† Notes and Diagrams illustrative of the directions of the forces acting at and near the surface of the earth in different parts of the Brunswick Tornado of June 19, 1835. 4to. 2 pp. (*Trans. Amer. Phil. Soc.* v. 407.) *Philadelphia,* 1837

Observations at the Magnetical and Meteorological Observatory at the Girard College, Philadelphia, made under the direction of A. D. Bache, LL.D. . . 3 vols. 8vo. (*Trübner's Bibl. Guide,* p. 195, says, Printed 1847.) *Washington.* 1840-45

† Records and Results of a Magnetic Survey of Pennsylvania and parts of adjacent States in 1840 and 1841, with some additional records and results of 1833-35, 1843, and 1862, and a Map. fol. 32 pp. *Washington,* 1863

† **Bache, A. D., and Courtenay, E. H.** Observations to determine the Magnetic Dip at Baltimore, Philadelphia, New York, West Point, Providence, Springfield, and Albany. 4to. Read November 7, 1834. (*Trans. Amer. Phil. Soc.* v. new series, p. 209.) *Philadelphia,* 1836

† On the relative horizontal Intensities of Terrestrial Magnetism at several places in the United States, with the investigation of corrections for temperature, and comparisons of the methods of oscillation, in full, aud in rarefied air. 4to. 31 pp. Read May 6, 1836. (*Trans. Amer. Phil. Soc.* v. 427.) *Philadelphia,* 1837

† **Bachhoffner, G. H.** A popular Treatise on Voltaic Electricity and Electromagnetism. 8vo. 35 pp. 1 plate. *London,* 1838

On the difference between the Effects of an Apparatus of Induction according to the placement of a cylinder of iron or a bundle of iron wires in the bobbin. (From Matteucci, 1854, p. 137.)

Vide also Sturgeon and Backhofner.

Bacon, Francis Lord. *Born January,* 22, 1561, *at London; died April* 9, 1626, *at Highgate.*
Physiological Remains. (*From his Works, edition of* 1803, vol. ii. p. 215.) *London,* 1648

Bacon. Chart of the Atlantic Telegraph. *London*

† **Baeblich, H.** Das Nordlicht nach den Resultaten der neusten Forschungen erklärt. *Berlin,* 1871

Bæckel. Thèse de l'Ozone. 4to. (*Leiber. Cat. of* 1865, p. 21.) 1856

Baehrens, J. C. F. Beschreibung einer neuen astronom-geometrischen Bussole. *Halle,* 1793, *und Zusätze dazu* 1794

Baer, W. Elektricität und Magnetismus. Die Gesetze und das Wirken dieser mächtigen Naturkräfte und ihre Bedeutung für das praktische Leben. 8vo. 324 pp. (*Zuchold's Cat.* 1863, p. 145.) *Leipzig,* 1863

Baffin, Wm. In 1612-1616 made voyages in the waters of N. W. America, in which he instituted observations of magnetic declination and inclination.

Bag. On Magnetism, or the doctrine of equilibrium. 12mo. *Detroit, U.S.,* 1845

Baggs, Isham. Electric Telegraphy by Steam. (Advert[t] in "*Times*," Dec. 11, 1857.)

† **Bagnolo, Giovanni Francesco Giuseppe.** *Born* 1709 *at Turin; died about* 1760. Lettera intorno all' Aurora Boreale, veduta la notte de' 16 Dec. 1737, e di alcune altre cose erudite . . . (*Callogere, Raccolta,* xx. 189. Original.) 12mo. 29 pp. 1 plate. *Venezia,* 1739

† **Bähr, J. K.** Ueber die Einwirkung der Reibungs-Elektricität auf das Pendel. 4to. 56 pp. *Dresden,* 1870

Baier, J. W. De Fulmine, fulgure, et tonitru hiemale. 1706

† **Baille, J.** L'électricité. 2nd edition. 12mo. 344 pp. (*Forms part of Bibliothèque des Merveilles par Charton.*)   *Paris*, 1869

Bailly, Charles François. *Born May 3, 1800, at Merlieux, near Laon.*

Bailly de Merlieux. Résumé complet de Météorologie . . . Météores . . . élec triques, &c. . . . 32mo.   *Paris*, 1830

Bailly, Jean Sylvain. *Born September,* 15, 1736, *at Paris ; died November* 12, 1793, *at Paris.* Rapport des Commissaires chargés par l'Académie des Sciences de l'examen du Magnétisme animal. 8vo.   *Paris*, 1784

† **Bain, A.** On Electro-magnetic Printing Telegraph and Clock. 8vo. (*Vide* Finlaison.)   *London*, 1843

A short History of the Electric Clocks, with explanations of their principles and mechanism, and instructions for their management and regulation. 8vo. 31 pp.
                                                            London, 1852

† Mémoire sur les Horloges, les Télégraphes, et les Lochs électriques. 4to. (*Comptes Rendus*, xxi. 885.)   *Paris*, 1845

M. Bain présente un Modèle de la Machine qui sert à former les signaux dans son télégraphe électrique. 4to. (*Comptes Rendus*, xxi. 885.)   *Paris*, 1845

M. Arago met sous les yeux de l'Acad. deux modèles d'horloges électriques inventés par M. Bain, et un loch, imaginé par le même ingénieur, loch qui donne la mesure, non seulement, comme les appareils ordinaires, de la vitesse du navire à un instant donné, mais de l'espace parcouru dans l'intervalle de deux observations, quelle qu'ait pu etre, pendant ce temps, la variation de la vitesse. 4to. (*Comptes Rendus*, xxi. 923.)

Note.—Moigno, p. 156, *says that the Memoir on the Log was missing, or suppressed in the Academy.*   *Paris*, 1845

The Petition of Alexander Bain. (*Vide* Moigno, 1852, p. 100.)   *London*, 1845

Sur quelques améliorations récemment introduites dans les Télégraphes Electriques. 4to. (*Comptes Rendus*, xxx. 478-525.)   *Paris*, 1850

† Electric Clocks. 4to. 3 pp.   *London*, 1852

Mr. Joseph Whitworth's Special Report, presented to the House February 6, 1854, chap. xi. p. 28, Electric Telegraph. (*The Report is headed New York Industrial Exhibition.*)   1854

Several Articles by Brewster concerning him. . . . (*North British Review.*)

† A Treatise on numerous applications of Electrical Science to the Useful Arts. Part i. 8vo. 35 pp.

Experiments (repeated) in Hyde Park. Moisture of the Earth or Water (natural body of) used as part of the circuit, &c. &c. 4to. (*Literary Gazette, June 4*, 1842.)   *London*, 1842

Vide also Finlaison, Wright and Bain, and Patents.

† **Bain and Wheatstone.** Controversy on the Electro-magnetic Telegraph Clock and the Electro-magnetic Printing Telegraph. 4to.
Note.— This is a collection of four Letters to the Editor of the *Literary Gazette*, inserted in the numbers for June 11 and 18 and August 6 and 20. They have been cut out, and the above title affixed in MS.
                                                            London, 1842

Bain (not Alex.) (*Vide* Reid and Bain.)

† **Bain, G.** Saggio sulla Variazione dell' Ago magnetico . . . Trad. dall' Inglese. 8vo. 148 pp.   *Venezia*

Bain, W. *Born* 1775, *at Culross, Perthshire ; died September* 11, 1853, *at Grange, Essex.* An Essay on the Variation of the Compass, showing how far it is influenced by a change in the direction of the ship's head, with an exposition of the dangers arising to navigators from not allowing for this change of variation. Interspersed with practical observations and remarks. 8vo. 140 pp. 1 chart.   *Edinburgh*, 1817

† **Bajon.** *Died* 1790. Descrizione di un Pesce che dà la scossa elettrica conosciuto a Cayenne sotto il nome d'Anguilla tremante. Traduzione. 12mo. (*Scelta d' Opuscoli.*) (*Is this translated from the Phil. Trans.* 1773, p. 481 ? *Vide* Humboldt, p. 50.) *Milano,* 1775 Mémoires pour servir à l'Histoire de Cayenne.

† **Baker,** H. A Letter from H. Baker to the President concerning several medical Experiments of Electricity. Read March 13, 1748. 4to. 6 pp. (*Phil. Trans.* xlv. 270.) *London,* 1748

† A Letter from H. Baker to the President, containing Abstracts of several Observations of Aurorae Boreales lately seen. 4to. (*Phil. Trans.* xlvi. 499.) *London,* 1749-50

† **Baker, Knight, Freeman, Fauquier, Miles, Martin, &c.** A collection of several Papers laid before the Royal Society concerning several Earthquakes felt in England and some neighbouring countries in the year 1750, 4to. (*Appendix to the Phil. Trans. for* 1750, vol. xlvi.) *London,* 1750

† **Baker.** Baker's patent Anti-Incrustator Apparatus for removing scale from Steam Boilers, and preventing its formation. 8vo. (*Principally Testimonials.*) *London,* 1868

Baker. (*Vide* Fenwick and Baker.)

Bakewell. Electricity : its History . . . and Applications. 8vo. (*From Advertisement.*) *London,* 1853

† **Bakewell,** F. C. A Manual of Electricity, practical and theoretical. Second edition, revised and enlarged. 8vo. 310 pp. *London and Glasgow,* 1857
A Manual of Electricity, theoretical and practical. Third edition, revised and enlarged. 8vo. *London,* 1859

† The Copying Telegraph, invented by Bakewell. 4to. *London,* 1850
Trial of the Autographic Electric Telegraph of Mr. Bakewell. 4to. (*Literary Gazette, Sept.* 23, 1847.) *London,* 1847

Balbi, Paolo Battista. *Born February* 17, 1693, *at Bologna ; died December* 7, 1772, *at Bologna.* Descriptio Ignei Globi per aera improviso excurrentis. 4to. (*Comment. Bononienses,* i. 285.) *Bonon.*

Balbo, Prospero, Conte. *Born July* 2, 1762, *at Chieri, Turin ; died March* 14, 1837, *at Turin.*

Note.—The great friend and patron of Beccaria ; sons and other relations still living in Turin (1860). Possibly some of Beccaria's valuable or curious MSS. left in his hands may be discoverable.—*F. R.*

Lettere di B. Franklin a G. Beccaria volgarizzate dal conte Prospero Balbo.

(*Note.*—This work begins at p. 145, and forms part of Eandi's Memorie Istoriche di Beccaria, 1783. There are six Letters—1st dated Londra, 29 May, 1766 ; 2nd, Lond. Settem. 1768 ; 3rd, Lond. 11 Agost. 1773 ; 4th, Lond. 25 May, 1774-5 ; 5th, Passy, 19 Nov. 1779 ; 6th, Passy, 19 Feb. 1781.)

Balcells, J. Lithologia Meteorica. 4to. *Barcelona,* 1854

Baldassini, F. Conte. Sopra gli Animali Microscopici considerati come causa unica della Fosforescenza del Mare, Osservazioni. 8vo. 34 pp. (*Poligrafo,* viii. 1835.) *Verona,* 1835

Baldi, L. Memoria sui Telegrafi Elettrici. *Pistoia,* 1851

† **Baldini,** Giovanni Francesco. *Born February* 4, 1677, *at Brescia ; died* . . . 1765, *at Tivoli bei Roma.*) Relazione dell' Aurora Boreale veduta in Roma li 16 Dec. 1737. 12mo. (*Calogera's Raccolta,* xvii. 47.) *Venezia,* 1738

Baldung. (*Vide* Waldung.)

† **Baldwin,** L. An Account of a very curious appearance of the Electric Fluid produced by raising a kite in the time of a thundershower ; in a Letter to J. Willard. 4to. (*Memoirs of American Acad.* i. part ii. 257, old series.) *Boston,* 1785

Baldwin—*continued.*

† Observations on Electricity, and an improved mode of constructing Lightning-rods ; in a Letter to Jos. Willard. 4to. (Letter dated January 25, 1797.) (*Memoirs of American Acad.* ii. part ii. 96, old series.) *Charlestown U.S.* 1804

† **Balestrini,** A. La Telegrafia Elettro-magnetica. 12mo. 48 pp. *Torino,* 1851

† **Balestrini,** P. A. Description du système de Télégraphie sous-marine de M. P. A. Balestrini, par M. P. de Branville. 8vo. 8 pp. (*Mémoires de la Soc. des Ingénieurs Civils.*) *Paris*

Ball. The Electric Telegraph applied to meteorological announcements. 8vo. (*Brit. Assoc. Report.* Meeting at Swansea in September, 1852.) *London,* 1852

Ballard. On the Magnetism of Drills. 4to. (*Phil. Trans.* 1698, p. 417.) *London,* 1698

Ballot. (*Vide* Buys-Ballot.)

Balthasar, Augustin. *Born September 23, 1632, at Anclam ; died November 20, 1688, at Greifswald.* (*Pogg.* i. 95.)

† **Bamberger,** B. Elektricität und Magnetismus, als Heilmittel. 8vo. *Berlin,* 1854

† **Bammacaro,** Niccolò. *Born . . . at Neapel ; died about 1778.* Tentamen de vi Electrica ejusque phænomenis. 8vo. 202 pp. *Neapoli,* 1748

† **Bancalari,** Michele Alberto. *Born 1805, at Chiavari.* Lettera al Belli sulla repulsione che l'Elettro-magnetismo esercita sulla fiamma e sul fumo. 4to. 1 p. (*Giorn. dell' I. R. Ist. Lomb.* nuova serie, i. 271.) *Milano,* 1847

Bancroft, Edward Nathaniel. Saggio sulla storia naturale della Guiana. Natural History of Guiana. (*From Prins,* p. 8, *on Electrical Fish.*) (*Humboldt, Voyage,* part ii. *Zoology,* p. 50, *says Essay on Guiana,* 1769, p. 191.)

† **Bangma,** O. S. Wiskundige verhandeling over het aardische Magnetismus. 4to. 75 pp. 1 plate. (*Lid van het Koninklijk Instituut den Wittenschappen Letterkunde en Schoone kunsten.*) *Amsterdam,* 1824

† **Banks,** Sir Joseph. *Born December 13, 1743 ; died June 19, 1820.* Biographical Memoir of the late Right Hon. Sir Joseph Banks, Bart. 8vo. (*Phil. Mag.* lvi. 40.) *London,* 1820

Bapst avec **Azais.** Explication et emploi du Magnétisme. 8vo. *Paris,* 1817

Bar, W. Elektricität und Magnetismus. (*Calvary Cat.* No. lxii. 1869, p. 2.) *Leipzig,* 1863

† **Barbancois,** Charles Hélion, Marquis de. *Born August 17, 1760, at Schloss Villegongis ; died March 17, 1822.* Lettre addressée à M. Delametherie, contenant un Essai sur le fluide électrique. 8vo. 70 pp. *Paris,* 1817

† **B** * * * **,** le Marquis de (*i. e.* Barbançois.) Deux Lettres écrites en 1819 à M. le Président de l'Acad. . . . La première relative à un système sur l'Electricité ; la seconde relative à un Tableau des Sciences. 8vo. 38 pp. and table. 1819

Barber. (*Vide* Morgan and Barber.)

† **Barberet,** Denis. *Born December 27, 1714, at Arnay le Duc, Burgundy ; died after 1776.*

Dissertation sur le rapport qui existe entre les phénomènes du Tonnerre et ceux de l'Electricité, qui a remporté le prix. 4to. 16 pp. *Bordeaux,* 1750

Discours qui a remporté le prix . . . de l'Acad. de Bordeaux en 1850.

Phénomènes du Tonnerre et le l'Electricité. 4to. (*Printed in or for the Mém. Acad. Dijon.*) *Dijon,* 1751

Sur la nature et la formation de la Grêle. 8vo. (*Mém de l'Acad. de Dijon,* tom. i. p. 1.) *Dijon,* 1769

† **Barbeu du Bourg.** Œuvres de Franklin, traduites de l'Anglais sur la 4ᵐᵉ édn. avec additions nouvelles, &c. 2 vols. 4to. 12 plates. *Paris,* 1773

Barbier, A. Recherches sur l'Electricité des Gaz et des Liquides. 8vo. 27 pp.
Bordeaux, 1858

† **Barbier de Tinan.** Mémoires sur les Conducteurs pour préserver les édifices de la foudre; par l'Abbé Jh. Toaldo; traduits de l'Italien avec des Notes et des Additions, par M. Barbier de Tinan. . . . 8vo. 241 pp. 3 plates.
Strasbourg, 1779

† (Nuove) Considerazioni sopra i conduttori del Sig. Barbier di Tinan. Traduz. dal Francese. 4to. 43 pp.

Note.—This is a printer's translation of Barbier's Considerations sur les Conducteurs en général, appended to his translation of Toaldo's Dei Conduttori per preservare gli edifizj da Fulmini. 4to. 1778, nuova edn.
Venezia, 1779

† Mémoire sur la manière d'armer d'un Conducteur la Cathédrale de Strasbourg et sa tour. 8vo. 34 pp. (*Volta's copy.*) *Strasbourg*, 1780

† **Barbieri, Ludovico Duca.** *Born June* 24, 1719, *at Vicenza ; died after* 1756. Discorso sopra la Generazione e Natura de' Fulmini. 12mo. (*Calogera's Rac-colta*, xxx. 291. Recitato nell' Accad. de Ricoviati in Padova, Maggio, 1741.) *Venezia*, 1744

† Lettera del Sig. Conte Barbieri al . . . Rev. Sig. D. Domenico S. Della Gene-razione e Natura de' Fulmini. 12mo. 68 pp. (*Calogera Raccolta d'Opuscoli*, tom. xli. p. 177.) *Venezia*, 1749

† **Barca, Alessandro.** *Born November* 26, 1741, *at Bergamo ; died June* 13, 1814, *at Bergamo.*

Conghiettura sulla Elettricità. 12mo. 6 pp. (*Scelta d'Opuscoli*, xvii. 87.)
Milano, 1776

Barca, G. M. Opere in Latino. *Messina*, 1756 (?)

Bardenfleth. Om Orkaner. . . . 4to. 70 pp. (*Kopenh. Acad.*) *Kopenhagen*

Bardenot, J. R. P. Les Recherches . . . de Bichat sur la Vie et la Mort réfutées. 8vo. *Paris*, 1824

† **Barham, H.** A Letter to the Publisher, giving a relation of a Fiery Meteor seen by him, in Jamaica, to strike into the earth ; with Remarks on the Weather, Earthquakes, &c., of that island. 4to. 2 pp. (*Phil. Trans. for* 1718, p. 837.) *London*, 1718

† **Barker, C.** On various Electrical Apparatus, &c. Letter to Editor, 30th August, 1839. (*Roget's copy.*) (*Annals of Electricity*, iv. 22.) *London*, 1839

† **Barker, F.** Dissertatio Physiologica inauguralis de inventu Galvani, vulgo, anima-lium electricitate. 8vo. 34 pp. (*Brugnatelli's copy.*) (*Giornale Fis. Med. di Brugnatelli, An.* ix. 1796, p. 58.) *Pavia*, 1796

† **Barker, Thos.** An account of an extraordinary Meteor seen in the county of Rutland, which resembled a waterspout. 4to. 2 pp. (*Phil. Trans.* xlvi. 248.)
London, 1749-50

† **Barker.** Mariner's Compass, with letter-press Instructions. Card-plate and. movable diagram. *London*, 1860

† **Barletti, Carlo.** *Died February*, 1800, *at Milan.*

Nuove Sperienze Elettriche, secondo la Teoria del Sig. Franklin e le produzioni del P. Beccaria. 8vo. 134 pp. *Milano*, 1771

† Physica specimina. (Electricitas.) 8vo. 184 pp. 1 plate. *Mediolani*, 1772

† Lettera al Volta sopra d'un nuovo Elettroforo. 12mo. 4 pp. (*Scelta d'Opus-coli*, xiv. 97.) *Milano*, 1776

† Dubbii e Pensieri sopra la Teoria degli Elettrici Fenomeni. (There is a Letter to Volta at p. 118.) 8vo. 136 pp. 1 plate. *Milano*, 1776

† Analisi di un nuovo fenomeno del Fulmine ed Osservazioni sopra gli usi medici della Elettricità. 4to. vi. 63 pp. 1 plate. *Pavia*, 1780

† Introduzione a nuovi principii della Teoria Elettrica dedotti dall' analisi de' fenomeni delle elettriche punte. Parte Prima. 4to. 54 pp. (*Mem. Soc. Ital.* tom. p. 1.) *Verona*, 1782

Barletti—*continued.*

† Introduzione a nuovi principii della Teoria Elettrica dedotti dall' analisi dei fenomeni delle elettriche punte. Parte Seconda. 4to. 122 pp. (*Mem. Soc. Ital..* tom. ii. p. 1.) *Verona*, 1784

† Fisica particolare e generale in Saggi . . . 4 volumi. Tomi i. ii. iii. di Calore, Meteorologia, Aerologia, Ottica. 8vo. *Pavia*, 1785-86

Descrizione de' fulmini di Porta Comasina, e del Duomo di Milano, e de' confronti loro coi principali effetti dei fulmini. 1785

Dei Conduttori del fulmine . . .

† Saggio analitico di alcune lucide Meteore. 4to. 38 pp. (Previously published in his Fisica particolare e generale, tom. ii.) (*Mem. Soc. Ital.* iii. 331.) *Verona*, 1786

† Della supposta eguaglianza di contraria Elettricità nelle due opposte facce del vetro, o di un strato resistente, per ispiegare la scarica o scossa della boccia di Leyden. 4to. 6 pp. (*Mem. Soc. Ital.* iv. 304.) *Verona*, 1788

† D'immutabile capacità, e necessaria contrarietà di eccesso e difetto di Elettricità negli opposti lati del vetro, o di altro strato resistente supposta da Franklin, per la spiegazione della carica e della scarica Elettrica nella boccia Leidense. 4to. *Verona*, 1794

† **Barlocci,** Saverio. *Born December 3, 1784, at Rome; died May 25, 1845, at Rome.* Congetture sull' origine de' Fuochi Vulcanici. 8vo. 10 pp. (*Giornale dell' Ital. Letteratura di Da Rio.* v. 38.) *Padova*, 1803

Esame comparativo di alcune ipotesi relative alla Elettricità atmosferica. 4to. (*Giornale Arcadia-Roma.*) *Roma*, 1818

† Saggio di Elettro Magnetismo dedotto dagli esperimenti instituiti nel gabinetto fisico della Università di Roma. 8vo. 74 pp. *Roma*, 1826

Sulla influenza della luce solare nella produzione dei fenomeni Elettrici e Magnetici. 8vo. 16 pp. *Roma*, 1829

Lezioni di Fisica sperimentale. 2 vols. 8vo. (*Bibliot. Ital.* lxxxv. 109, *and* lxxxvii. 414.) *Roma*, 1836-37

Fisica sperimentale (*Zantedeschi,* ii. 122.) *Roma*, 1837

Barlocci e Zantedeschi. Congetture sulla origine dell' Elettricità atmosferica (di Barlocci). Corredata di alcune note del . . . F. Zantedeschi. 8vo. 18 pp. *Verona*, 1831

Congetture . . . corredata di note del Zantedeschi . . . Continuazione e fine. 8vo. 14 pp. *Verona*, 1831

Barlow, Jas. (of New York). A new Theory accounting for the dip of the Magnetic Needle ; being an analysis of Terrestrial Magnetism. 8vo. *New York*, 1835

† **Barlow,** Peter. *Born October 13, 1776, at Norwich.* An Essay on Magnetic Attractions, particularly as respects the deviation of the Compass on shipboard, occasioned by the local influence of the guns, &c., with an easy practical method of observing the same in all parts of the world. 1st edition. 8vo. 145 pp. 1 plate. *London*, 1820

† A curious Electro-magnetic Experiment. 8vo. 2pp. (*Phil. Mag.* lx. 241.) *London*, 1822

† On the anomalous Magnetic Action of Hot Iron between the white and bloodred heat. 8vo. 8 pp. (*Phil. Mag.* lx. 343.) *London*, 1822

† An Account of some Electro-magnetic combinations for exhibiting Thermometric Phenomena, invented by Mr. James Marsh, of Woolwich ; with Experiments on the same. 8vo. 7 pp. 1 plate. (*Phil. Mag.* lxii. 321.) *London*, 1823

† An Essay on Magnetic Attractions, and on the Laws of Terrestrial and Electro Magnetism. 2d edition, much enlarged and improved. 8vo. 303 pp. *London*, 1823

† An Essay on Magnetic Attractions ; and on the Laws of Terrestrial and Electro Magnetism. With an Appendix, containing the results of Experiments made on shipboard from latitude 61° S. to latitude 80° N. 2nd edition. 8vo. 368 pp. *London*, 1824

Barlow, Peter—*continued*.

† Electro-Magnetism. 4to. 40 pp. and plates. (*Encyclopædia Metropolitana*, iv. 1-40.) *London*, 1824

† On the present situation of the Magnetic Lines of equal variation ; and their changes on the terrestrial surface. 4to. (*Phil. Trans.*) *London*, 1833

† Magnetism. (Article of the *Encyclopædia Metropolitana*.) Encyclopædia of British Arts, Manufactures, &c. 4to. *London*, 1855

† **Barlow,** W. H. On the spontaneous Electrical Currents observed in the wires of the Electric Telegraph. 4to. 12 pp. (*Phil. Trans.* part i. 1849.) *London*, 1849

Barlowe, William. *Born . . . in Pembrokeshire ; died* 1625 *in Easton, Winchester.* Magneticall Advertisements. 4to. 1613

Magneticall Advertisements ; or, divers Observations concerning the Loadstone. 4to. *London*, 1616

† Magneticall Advertisements ; or, Divers pertinent Observations and approved Experiments concerning the nature and properties of the Load-stone. Whereunto is annexed a briefe Discoverie of the idle Animadversions of Mark Ridley, Dr. in Physike, upon this Treatise entituled Magneticall Advertisements. 2d edition, sm. 4to. *London*, 1618

† **Barlowe,** W. and **Sturgeon.** Magnetical Advertisements ; or, diverse pertinent Observations and approved Experiments concerning the nature and property of the Loadstone. A new edition, with Notes by W. Sturgeon. *London*, 1843

Barnes, Robt. The use of Galvanism in Obstetric practice. 8vo. (*Lancet*, 1853, vol. ii. No. xxii. 456.) (*London*, 1853

Barneveld, William van. *Born January* 20, 1747, *at Hattem ; died June* 23, 1826, *at Hattem.*

Het onweesvuur in zyne en rigting en uitwerkels nagespoort mit de elektrische Stoffen vergeleken, etc. (*Poggendorff*, i. 104.) *Amsterdam*, 1780

Genecskundige Electriciteit, iii. stukken. (*From Guitard.*) *Amsterdam*, 1785

Medicinische Elektricität. 8vo. *Leipzig*, 1787

† **Baronio,** Giu. Saggio di naturali Osservazioni sulla Elettricità Voltiana, colla descrizione d'una nuova macchina a corona di persone e di un piliere tutto vegetabile. 8vo. 144 pp. *Milano*, 1806

† **Baronio,** Jos. Description of a Galvanic Pile formed of vegetable materials only. 8vo. 2 pp. (*Phil. Mag.* xxiii. 283.) *London*, 1806

Barrington, Daines. *Born* 1727 ; *died March* 11, 1800, *in London.*

Anglo-Saxon version of the Historiæ of Orosius, by Alfred the Great, with an English translation from the Anglo-Saxon, containing Accounts of Wars, Plagues, &c., Thunder and Lightnings, &c., from the Creation to A.D. 416. 8vo. *London*, 1773

Bartaloni, D. Lettera sopra il fulmine caduto nel dì 18 Ap. 1777, sulla spranga posta nella torre del palazzo pubblico della città di Siena. 8vo. (*There is also an English translation of the above.*) *Siena*, 1777

† Mem. sul conduttore Elettrico collocato nella torre della Piazza di Siena. 4to. 36 pp. 1 plate. (*Atti dell' Accad. di Siena*, vi. 253. 1781.) *Siena*, 1781

† Relazione sopra un supposto Fulmine caduto nella Cappella della Piazza di Siena il dì 7 Giugno dell' Anno 1784. 4to. 8 pp. (*Atti dell' Accad. di Siena*, vii. 61.) *Siena*, 1794

Barth, Johann Matthäus. *Died after* 1751.

De rariori quodam phænomeno cum terræ motu conjuncto, Ratisbonæ observato. 4to. (*Acta Acad. Nat. Curios.* iv. 491.) *Norimbergæ*

† De luce Barometrorum ut et aliis connexis argumentis, Epistola. 4to. 70 pp. *Lipsiæ*, 1716

Bartholinus, Thomas. *Born October* 20, 1616, *at Kopenhagen ; died December* 4, 1680, *at Hagested.* (*Poggendorff*, i. 109.)

Bartholinus—*continued.*

† De luce animalium, lib. iii. admirandis historiis rationibusque novis referti. small 8vo. 396 pp. *Lugd. Batav.* 1647

De luce hominum et brutorum, libri iii. novis rationibus et raris historiis secundum illustrati. 12mo. *Hafniæ,* 1659

De naturæ mirabilibus, questiones academicæ. 4to. *Hafniæ,* 1674

Barton, Sir John. Invented a floating Compass (schwimmenden Compass).

† **Barton,** B. S. On Meteoric Stones. Extr. from Valentia M. B. in the Epimenides Naturæ Curiosorum, Crystallus inter grandines e nubibus decidens, Anno 1724. 8vo. 2 pp. (*Phil. Mag.* xlvi. 154.) *London,* 1815

† **Bary,** Emile Louis François. *Born January* 13, 1799, *at Paris.*
Statique appliquée au Magnétisme. Note sur la manière de corriger . . . .
Boussoles. 8vo. 8 pp. 1 plate. (*Nouv. Ann. de Mathémat. Juin* 1844.)
Paris, 1844

Basevi, Emile. Sulla Conducibilità Elettrica del Vetro ridotto in fili o lamine.
Pisa, 1841

Basilius. (*Vide* Sinner-Basilius.)

Basin. (*Vide* Anon. Magnetism, 1753.)

Basle. Acta Helvetica Physico-mathematico-botanico-medica. 8 vols. 4to.
Basel, 1751-77
Nova Acta . . . 1 vol. 4to. 1787

Basse, F. H. Galvanische Versuche und Betrachtungen, die Leitung des galvanisch-elektrischen Fluidums, betreffend ; angestellt mit einer Voltaischen Säule aus 70 zweizölligen Metallplatten-Paaren von Kupfer und Zink. 8vo. (*Gilbert's Ann.* xiv. 26-37.) *Leipzig,* 1804

* **Battagia.** Delle Accademie Veneziane. Dissertazione storica. 8vo. 136 pp.
Venezia, 1826

† **Baudouin,** F. M. Observations sur le mode d'établissement des lignes Télégraphiques sous-marines. 8vo. 31 pp. *Paris,* 1858

Bauer, Fulgentius. *Born* 1731 ; *died March* 3, 1865, *at Vienna.*
Dissertatio experimentalis de Electricitatis theoria et usu, quam in Disputatione defendit. L. B. Pongraz. 4to. (*Pogg.* i. 115, *says* 1764.) *Viennæ,* 1762

† **Bauer, Marherr, und Kirchvogel.** Experimental Abhandlung von der Theorie und dem Nutzen der Electricität. Von F. Bauer. (Aus dem Latinis von L.D.)
Abhandlung von der Wirkung der Luft-Elektricität in dem menschlichen Körper.
Von P. A. Marherr. (Aus dem Latinis von L.D.)
Abhandlung von der Wirkung der Luft-Elektricität in dem menschlichen Körper.
Von A. B. Kirchvogel. (Aus dem Latinis von L.D.) 8vo.
Chur. and Lindau, 1770

Baumé, Antoine. *Born February* 26, 1728, *at Senlis ; died October* 15, 1804, *at Paris.* Chymie expérimentale, &c. 3 vols. 8vo. *Paris,* 1773

Baumer, Johan Wilhelm. *Born September* 10, 1719, *at Rehweiler, Frank. ; died August* 4, 1788, *at Giessen.*
Progr. de Electricitatis effectibus in corpore animali. 4to. *Erfurth,* 1755

Baumes. Réflexions sur le Galvanisme. (*Journal de Médecine, ou Annales de la Soc. de Méd. pratique de Montpellier, Germinal, an.* xi. No. iii.)
Montpellier, 1803

Baumgartner, Andreas. *Born November* 28, 1793, *at Friedberg, Bohm.*
Die Naturlehre, &c. *Wien,* 1823
Anfangsgründe der Naturlehre. 3 Auflagen. (*Poggendorff,* i. 118.)
Wien, 1837
Grundrisz der Naturlehre. 1 und 2 Lfg. 8vo. *Wien,* 1851-52
Anfangsgründe der Naturlehre. 4th edition, enlarged. 8vo. 388 pp. *Wien,* 1851
Anfangsgründe der Naturlehre. 5. vermehrte Auflage. 8vo. 192 pp. *Wien,* 1852

Baumgartner—*continued.*

Anfangsgründe der Naturlehre. 6to durchaus umgearbeitete und vermehrte Auflage. 8vo. 242 pp. *Wien,* 1855

4 Abhandlungen über Gewitter. 8vo. (*Wien Akad.*) *Wien*

Trattato di Fisica. (*Zantedeschi,* ii. 190.) *Padova*

Ueber die Wirkungen der natürlichen Elektricität auf elektromagnetische Telegraphen. 8vo. (*Sitzungsber. der Wien Akad. Klasse* i. 270.) *Wien,* 1849

† Weitere Versuche über den elektrischen Leitungswiderstand der Erde. 8vo. (*Sitzungsbuch der Königlichen Akad.* 1849, v. *und* vi. *Heft,* p. 28.) *Wien,* 1849

† Nachtrag zu meinem Aufsatz. Von der Umwandlung der Wärme in Elektricität. 8vo. 6 pp. (*Sitzungsberichte . . . der Kain. Akad. der Wissen.* xxxii.) *Wien*

Ueber die Leitkraft der Erde für Electricität. (*Sitzungsberichte der Wien Akad.* ii. 28.) *Wien*

Vide also Ettingshausen und Baumgartner.

†: **Baumgartner e Zambra.** Elementi di Fisica di A. Baumgartner. Traduzione e Continuazione di B. Zambra. 8vo. 296 and 189 pp. *Vienna,* 1858

† **Baumgartner,** A. Freih. Ueber Gewitter uberhaupt, Hagelwetter insbesondere. 8vo. 26 pp. (*Sitzungsberichte,* 1857, *der Kaiserl. Akad. der Wissenschaften,* xxiii.) *Wien,* 1857

† Von der Umwandlung der Wärme in Elektricität. Sitzung 13 Nov. 1856. 8vo. 12 pp. *Wien,* 1857

Baumhauer, Eduard Heinrich. *Born September* 18, 1820, *at Brüssel.*

De ortu lapidum Meteoricum. 8vo. *Trajecti,* 1844

Bavarian Academy. Abhundlungen der kurfürstl. Baierschen Akademie der Wissenschaften. 4to. *München,* 1764-76

Neue philosoph. Abhandlung. 4to. *Augsb. und Nurnb.* 1779-97

Denkschriften der Kön. Akad. der Wissenschaften zu München. 4to. 1808, 1832

Abhandlungen der kurfürst. Baierschen Akad. der Wissenschaften, historischen und philosophischen Inhalts. 10 Bde. 4to. *München,* 1763-76

Neue philosoph. Abhandlungen der Baierschen Akad. der Wissenschaften. 7 Bde. 4to. *München,* 1778-97

Denkschriften der königl. Akad. der Wissenschaften zu München. 4to. *München,* 1809

Idem für die Jahre 1811 und 1812. 4to. *München,* 1815

Idem für die Jahre 1813-17. 4to. *München,* 1816-20

Denkschriften . . . für die Jahre 1823 und 1824. 4to. *Sulzbach,* 1825

Note.—This collection seems to have been continued under the same title, with various intervals and modifications, down to the present time [1866] ; but the difficulty of obtaining correct dates, &c. from Heinsius is too great (or impossible.) Other authorities must be consulted.

† **Baxter,** H. F. An Experimental Enquiry, undertaken with a view of ascertaining whether any and what signs of current electricity are manifested during the organic process of secretion in living animals ; being an attempt to apply some of the discoveries of Faraday to physiology. 4to. (*Phil. Trans.* 1848, p. 243.) (*Faraday's copy.*) *London,* 1848

† **Bazin,** Gilles Augustin. *Born . . . in Paris ; died March,* 1754.

Description des Courants Magnétiques dessinés et gravés d'après la nature ; suivies de quelques observations sur l'Aimant, par * * * 4to. 52 pp. *Strasbourg,* 1753

† Supplément pour la Description des Courants Magnétiques. 4to. *Strasbourg,* 1754

† **Beardmore,** S. Terra-Voltaism. Remarks on the Application of a Terra-voltaic Couple to Submarine Telegraphs. 8vo. 51 pp. *London,* 1860

Beardmore—*continued.*

† The Globe Telegraph. An Essay on the Use of the Earth for the Transmission of Electric Signals. 8vo.   *London,* 1859

Beatson. On Shocks and Tones produced when the passage of the Current of the Spiral is renewed at very small intervals of time. pp. 242. (*Vide* Matteucci.)

† Beauchamp, Lord. An Account of the Fire-ball seen in the Air, and of the Explosion heard on December 11, 1741, by Lord B. 4to. 2 pp. (*Phil. Trans.* xli. 870-71.)   *London,* 1839-41
   (*Vide* also Fuller.)

† Beaufoy, Mark. Variation of the Magnetic Needle. Observations near Stanmore. Correction of Error. 8vo. (*Phil. Mag.* liii. 387.)   *London,* 1819

† Variation of the Magnetic Needle. 8vo. (*Phil. Mag.* lv. 394.) *London,* 1820

Beaume. (*Vide* La Beaume.)

† Beaumont-Brivazac. Electro-Magnetism, Animal. 8vo. 23 pp.   *Grenoble,* 1861

Beccari, Jacopo Bartolomeo. *Born July* 25, 1682, *at Bologna; died January* 18, 1767, *at Bologna.* Auroræ Borealis maximæ, anno 1726, ad xiv. Kal. Nov. descriptio. 4to. (*Comment. Bononienses,* tom. i. c. p. 288.) *Bononiæ,* 1731

† De quamplurimis phosphoris nunc primum detectis. Commentarius. 4to. 85 pp.
                                                           Bononiæ, 1744

† De quamplurimis phosphoris nunc primum detectis. Commentarius (primus). 4to. 44 pp. (*Comment. de Bonon. Scient. Instit.* tom. ii. parte ii. p. 136.)
                                                           Bononiæ, 1746

† De quamplurimis phosphoris nunc primum detectis. Commentarius. 4to. 22 pp. (*Comment. de Bonon. Scient. Instit.* tom. ii. parte iii. p. 498.)
                                                           Bononiæ, 1747

† Commentarii duo, de phosphoris naturalibus (ex actis Bononiensibus excerpti.) 8vo. 113 pp.   *Graesii,* 1768

Electricitas vindex experimentis atque observ. stabilita. 8vo.   *Grätz,* 1773

† De artificiali electricitate ex Benj. Franklini theoria, quam confirmavit, expolivit, auxitque. 4to. 15 pp.   *Torino, n. d.*

De electricitate communi. 4to. (*Comment. Bononienses,* iv. p. 88.)
                                                               Bononiæ

Beccari, J. B., et Bonzi. De vi, quam ipsa per se Lux habet, non colores modo, sed etiam texturam rerum salvis coloribus, immutandi. 4to. (*Comment. Bononienses,* iv. c. p. 74.)   *Bononiæ*

Beccaria, Giovanni Battista. *Born October* 3, 1716, *at Mondovi; died May* 27, 1781, *at Turin.*

† Dell' elettricismo artificiale e naturale. Libri due. 4to.   *Torino,* 1753

† Giunta al primo libro della Elettricità naturale. 4to.   *Torino,* 1753

† Lettre sur l'Electricité, addressée à Nollet. Traduite de l'Italien par De Lor. 12mo. (This is a translation of Beccaria's *Lettera al Nollet,* inserted at p. 144 of his *Dell' Elettricismo artificiale e naturale.* 4to. *Torino,* 1753.)
                                                                Paris, 1754

† Scientiarum Academicis Londinensibus atque Bononiensibus. (*Questa lettera fu, dallo stesso autore soppressa. Vide* Eandi Vita.*** di Beccaria, p. 154.)
                                                               Taurini, 1756

† Dell' Elettricismo. Lettere di G. Beccaria, dirette all' Ch. Sig. G. B. Beccari, coll' Appendice di un nuovo Fosforo descritto all' illust. Sig. Conte Ponte di Scarnafigi . . . Fol. 378 pp.   *Bologna,* 1758

Experiments in Electricity. 4to. (*Phil. Trans.* 1760, p. 514.) *London,* 1760

† (Lettera) a sua Altezza Reale il Sig. Duca di York. Sperienze, ed Osservazioni. 4to. 16 pp.
      (*Note.* On the Action of the Electric Spark and Lightning on Air ; on the Magnetic Direction given to Bodies by Lightning, &c.)   *Torino,* 1764

Beccaria—*continued.*

Novorum quorumdam in re electrica experimentorum specimen, quod Regiæ
Londinensi Societati mittebat die 26 Aprilis, 1766. T. B. B. . . . Fol.
(*Letto alla Soc. Reale di Londra*, il 4 Giugno, 1767, ed inserito nelle
Transaz. di quell' anno, lvii. p. 297.) *Taurini*, 1766

† Novorum quorumdam in re electrica experimentorum specimen quod Regiæ Lon-
dinensi Societati mittebat die 14 Jan. anno 1766. T. B. B. Fol. 4 pp. (*Phil.
Trans.* 1766. lvi. 105.) *Taurini*, 1766
A Paper on the Pekin Experiments. 4to. (*Phil. Trans.*) *London*, 1766 *or* 1767

† De electricitate vindice ad B. Franklinium . . . Epistola (date 20 February.)
Fol. *Taurini*, 1767
Note.—*The letter to which he refers, addressed to Franklin, was never
printed.*

† Experimenta atque Observationes quibus electricitas vindex late constituitur et
explicatur. 4to. 1 plate. 66 pp. *Taurini*, 1769

† De atmosphæra electrica. Ad Regiam Londinensem Societatem libellus. Fol.
7 pp. (*Phil. Trans.* 1770. lx. 277.) *Taurini*, 1769

† Elettricismo artificiale (al Duca di Chablis). 4to. 439 pp. and plates. *Torino*, 1772

† Della elettricità terrestre atmosferica a cielo sereno, Osservazioni. 4to. (Dedicate
a sua Alt. Rle. il Sig. Principe di Piemonte.) 54 pp. *Torino*, 1775

† A Treatise upon Artificial Electricity, in which are given Solutions of a number
of interesting Electric Phenomena hitherto unexplained. To which is added
an Essay on the mild and slow Electricity which prevails in the Atmosphere
during serene weather. (Translated from the original Italian of Father
Giambattista Beccaria.) 4to. 457 pp. 11 plates. *London*, 1776

† Articolo di Lettera . . . al Landriani sullo spezzamento de' vetri nell' atto della
scarica, e sopra un nuovo Elettrometro. 12mo. 7 pp. (Date, Torino, 25 Dec.
1775.) *Milano*, 1776

† Occhiale elettrico per ispiare la luce nella scossa della Torpedine. 12mo. 5 pp.
(*Scelta d' Opuscoli*, xix. 87.) *Milano*, 1776

† Lettera al Sig. Le Roy, sulle Stelle-cadenti. 12 mo. 12 pp. (*Says that a falling
star struck his kite. So also in his letter to Beccari*, p. 110, &c. *Scelta d'
Opuscoli*, xxi. 86.) *Milano*, 1776

† Articolo di lettera . . . intorno a due nuovi punti di analogia del Magnetismo
indotto dal Fulmine ne' Mattoni e nelle Pietre ferrigne. 12mo. 5 pp. (*Scelta
d' Opuscoli*, xxxii. 40.) *Milano*, 1777

† Lettera al Fromond sul cangiamento di colore prodotto dal fuoco. 4to. 5 pp.
(*Opuscoli Scelti*, ii. 378.) *Milano*, 1779

† Poscritta alla Lettera al Fromond. 4to. 4 pp. *Milano*, 1779

† Articolo d' altra Lettera. 4to. 1 pp. *Milano*, 1779

† De' Fiori elettrici. Lettera al Cavallo. 4to. 7 pp. (*Opuscoli Scelti*, iii. 243.)
Milano, 1780

† Lettera al . . . Priestley . . . intorno all' azione dal fuoco elettrico sulle calci
metalliche. 4to. 6 pp. (*Opuscoli Scelti*, iii. 377.) *Milano*, 1780

† Descrizione di un suo nuovo disegnatore del Fulmine (Ceraunografo). 8vo.
20 pp. 1780 (?)

† Nuovi sperimenti per confermare ed estendere la meccanica del fuoco elettrico.
4to. 19 pp. *Torino*, 1780

† Articolo di lettera all' Amoretti sulla luce delle Lagrime Britanniche. 4to. Dated
Torino, 16 Agosto, 1780. Original. (*Opuscoli Scelti*, iii. 284.) *Milano*, 1780

† *Vide* his Letter to Fromond, dated Torino, 27 Genna. 1754, describing his
Experiments, intended to prove that Electric Motions do not occur *in vacuo*.
4to. (Contained in Bianchi's *Elogio di Fromond*, p. 42.) *Cremona*, 1781

Elogio di Beccaria, di Tana. 8vo. (printed on vellum). *Torino*, 1781

42 BEC

Beccaria—*continued.*

Al Sig. Conte Cotti di Brusasco per la laurea . . . . Intorno alla naturalezza della cagione efficiente de' Temporali, e de' fenomeni compagni. 8vo.
Torino, 1781

† Memorie istoriche intorno agli studii del Padre Giambattista Beccaria (di Eandi, G. A.) 8vo. 116 pp *Torino,* 1783
(*Note.*—*The edited works are named in order of their dates. The unedited comprise very many works on* Thunder Conductors, St. Elmo Fire, Auroræ, Electricitas-vindex, &c.)

† Dell' elettricismo. Opere del Beccaria con molte note nuovamente illustrate. 2 vols. 4to. *Macerata,* 1793

A Letter or Dissertation to Count San Martino della Motta on Earthquakes (in which he refers to his works of 1753 and 1758.) 1793

Letter to the Princess Giuseppina di Carignano, on the Electricity of the Moon. 1793

Elogio del P. Beccaria (Anonymous). 8vo. (*From Opus. Scelti,* v. 19.) *Torino*

† Notizia storica di Giambattista Beccaria, scritta da A. M. Vassalli Eandi. 8vo. 22 pp. (*Lo Spettatore,* tom. v. p. 101.) *Milano,* 1816

(*Note.*—He left in the hands of Count Balbo fourteen Letters on Lightning, and one, "su i Baleni a caldo," *ready for printing;* also another, not quite completed, on Auroræ Boreales ; beside " un' infinità d'altri lavori non meno importanti, e degni tutti *nello stato in cui sono* d'essere donati al pubblico." These consist of Observations of the Effects of Lightning, Papers on Zodiacal Light, on Marine Waterspouts, on St. Elmo Fire, on the Effect of the Stroke of Lightning on the Tower at Siena, " sulla maniera di preservare gli edifizii dai Fulmini e specialmente i magazzini della polvere," and of Two Journals of Observations, one of Atmospheric Electricity, the other of Earthquakes and Atmospheric Electricity.)

(*N.B.* Sons and other relations of Count Balbo were recently living at Turin.)

† Portrait. Vercel del. Fiori sculp.

Becher, A. B. La aguja de las tormentas, ó sea Manual sobre huracanes para uso del navegante. Con la teoria de los huracanes puesta al alcance de todos. Puesto en castellano por D. Miguel Lobo. Con un apéndice que contiene el uso de la rosa transparente de Talco, tomada de la obra de Mr. Piddington. 8vo. 96 pp. *Barcelona,* 1856

Beck, Dominicus. *Born September* 27, 1732, *at Oepfingen b. Ulm; died February* 22, 1791, *at Salzburg.*

Beschreibung einer elektrischen Flinte. (*Pogg.* i. 125.) *Salzburg,* 1780

Fassliche Unterredung, Gebäude vor dem Einschlagen des Blitzes zu bewahren. 8vo. (*Heinsius,* i. 210.) *Salzburg,* 1786

† Kurzer Entwurf der Lehre von der Elektricität . . 8vo. 196 pp. *Salzburg,* 1787

Beckensteiner, C. Electrodoreur. 1841

† Etudes sur l'Electricité. Organes électriques chez l'Homme, ou corpuscules de Pacini, par les Drs. Heule et Kælliker. Précédées d'Observations sur l'Electricité Animale. 8vo. 1re livraison. 72 pp. *Paris and Lyon,* 1847

† Etudes sur l'Electricité. Nouvelle Méthode pour son emploi médical. 8vo. 340 pp. *Paris,* 1852

† Etudes sur l'Electricité. Traitement de l'Epilepsie par l'Electricité statique. Nouvelle Méthode. 8vo. 80 pp. 1 plate. Extrait du tom. iii. 1re livraison. *Paris, Lyon, Leipzig,* 1859

† Etudes sur l'Electricité. Nouvelle Méthode pour son emploi médical. Traitement de l'Epilepsie ; Traitement de la Chorée, ou Danse de St. Guy, &c. 3 vols. 8vo. 2nd edition. *Paris,* 1859-60

† Etudes sur l'Electricité. Traitement de la Chorée, ou Danse de St. Guy, par l'Electricité statique. Extrait du tom. iii. 8vo. 30 pp. 2me livraison. *Paris, Lyon, Leipzig,* 1860

Beckensteiner—*continued.*

† Etudes sur l'Electricité. Nouvelle Méthode pour son emploi médical. Tom. ii. 8vo. 320 pp. *Paris*, 1870

Becker, Christian August. *Born August 24, 1792, at Mühlhausen.*

Der mineralische Magnetismus . . . Heilkunst. 8vo. 202 pp. *Mülhausen*, 1829

† **Becket,** J. B. An Essay on Electricity, containing a Series of Experiments introductory to the study of that Science, in which are included some of the latest Discoveries, intended chiefly with a view of facilitating its application and extending its utility in Medical Purposes. 8vo. 151 pp. *Bristol*, 1773

Beckmann, Johann. *Born June 4, 1739, at Hoya; died February 3, 1811, at Göttingen.*

Physikal-ökonomische Bibliothek. 23 vols. 8vo. (jede bd. in 4 st.)
Gottemburg, 1770-1807

Beiträge zur Geschichte der Erfindungskunst. 8vo. 5 bde. zu 4 stck. (*Poggendorff*, i. 127, *says* 1784-1805.) *Leipzig*, 1780-1805

Geschichte der Erfindungen. 8vo. 2nd edit. (*From Exleben*, 489.) *Leipzig*, 1783

History of Inventions and Discoveries. 3 vols. 8vo. *London*

History of Inventions and Discoveries. Translated by Johnson. 4 vols. 8vo. 1814

Becquerel, Antoine Cesar. *Born March 8, 1788, at Chatillon-sur-Loing, département du Loiret; died January 18, 1878, Paris.*

Expériences sur le Développement de l'Electricité par la Pression. Lois du Développement. 8vo. (*Poggendorff says Ann. Chim.* 1823, vol. xxii.)
Paris, 1823

† Experiments on the Development of Electricity by Pressure. Laws of this Development. Continuation. 8vo. 12 and 8 pp. (*From Ann. de Chim.* xxii. 5. *Vide also Phil. Mag.* lxii. 204 and 263.) *London*, 1823

† Du Développement de l'Electricité par le Contact de deux Portions d'un même Métal . . 8vo. 20 pp. (*Lu à l'Acad.* 16 *Juin*, 1823 ; *Ann. de Chim.* xxiii. 1823.) *Paris*, 1823

† Des Effets Electriques qui se développent pendant diverses Actions Chimiques. 8vo. 13 pp. (*Lu à l'Acad.* 7 *Juin*, 1823 ; *Ann. de Chim.* xxiv. 1823.) *Paris*, 1823

Des Actions Magnétiques (ou . . . analogues) produites dans tous les Corps par l'Influence de Courants Electriques très énergiques. 8vo. 9 pp. (*Lu à l'Acad.* 1 *Mars*, 1824 ; *Ann. de Chim.* xxv. 1824.) *Paris*, 1824

† Mémoire sur les Décompositions Chimiques opérées avec des Forces Electriques à très-petite tension. 4to. 11 pp. (*Lu le 21 Août*, 1826 ; *Mém. de l'Acad.* xi. 33.) *Paris*, 18—

† Mémoire sur les Actions Magnétiques excitées dans tous les Corps par l'Influence d'Aimants très-énergiques. 4to. 13 pp. (*Lu le 17 Septembre*, 1827 ; *Mém. de l'Acad.* xi. 45.) *Paris*, 18—

Mémoire sur l'Electro-chimie et l'emploi de l'Electricité pour opérer des Combinaisons. 8vo. 40 pp. (*Ann. de Chim.* xli. 15.) *Paris*, 1829

Mémoire sur le Pouvoir Thermo-électrique des Métaux. 4to. 21 pp. (*Lu 3 Août*, 1829, *à l'Acad.; Paris Acad.* 1829.) *Paris*, 1829

Mémoire sur les Sulfures, Iodures, Bromures, &c., métalliques. 4to. 12 pp. (*Acad. Paris.*) *Paris*, 1829

Mémoire sur de Nouveaux Effets Electro-chimiques propres à produire des Combinaisons, et sur leur Application à la Cristallisation du Soufre et d'autres Substances. 4to. 15 pp. (*Lu à l'Acad.* 25 *Janv.* 1830 ; *Acad. Paris.*)
Paris, 1830 ?

Mémoire sur un Procédé Electro-chimique pour retirer le Manganèse et le Plomb des Dissolutions dans lesquelles ils se trouvent. 4to. 7 pp. (*Lu à l'Acad.* 3 *Mai*, 1830 ; *Acad. de Paris.*) *Paris*

† Decomposition of Sulphuret of Carbon by Electricity. 8vo. (*Phil. Mag. or Annals*, vii. 61 ; *Ann. de Chim. Sept.* 1829.) *London*, 1830

44 BEC

Becquerel, A. C.—*continued.*

† Considérations générales sur les Décompositions Electro-chimiques, et la Réduc
tion de l'Oxide de Fer, de la Zircone et de la Magnésie, à l'aide de Forces
Electriques peu considérables. 4to. 11 pp. (*Mém. de l'Acad.* tom. xii.
p. 581.) *Paris,* 1833

† Traité Expérimental de l'Electricité et du Magnétisme, et de leurs Rapports avec
les Phénomènes Naturels. 8vo. tom. i. and ii. *Paris,* 1834

Résultat d'un Voyage fait en Suisse, en Piémont, et en Italie, avec M. Brachet.
Expériences sur le Magnétisme Terrestre . . sur l'Electricité atmosphérique,
sur la Commotion Electrique de la Torpille, &c. 4to. (*Comptes Rendus,* i.
242.) *Paris,* 1835

† Traité Expérimental de l'Electricité et du Magnétisme. 1 vol. 8vo. tom. iii.
450 pp. *Paris,* 1835

† Traité Expérimental de l'Electricité et du Magnétisme. 1 vol. 8vo. tom. iv.
333 pp. *Paris,* 1836

Rapport sur divers Mémoires de M. Matteucci concernant l'Electricité Animale
en général, et particulièrement les Phénomènes de la Torpille. 4to. (*Comptes
Rendus,* v. 788.) *Paris,* 1837

† De quelques Propriétés nouvelles relatives au Pouvoir Phosphorescent de la
Lumière Electrique. 4to. 7 pp. (*Comptes Rendus,* viii. No. 7.) *Paris,* 1839

† Traité Expérimental de l'Electricité et du Magnétisme. 8vo. Tom. v. is in 2 parts.
First part, dated 1837, 316 pp. 2 plates ; second part, dated 1840, 288 pp.
5 plates. *Paris,* 1837-40

† Traité Expérimental de l'Electricité et du Magnétisme. Tom vi. in 2 parts.
First part, dated 1840, 440 pp. 5 plates ; second part, dated 1840, 547 pp.
18 plates. *Paris,* 1840

† Traité Expérimental de l'Electricité et du Magnétisme. Fol. Atlas pour la
2ᵉ partie du 5ᵉ vol. et le 6ᵉ et le 7ᵉ vol. *Paris,* 1840

† Traité de Physique considerée dans ses Rapports avec la Chimie et les Sciences
Naturelles. 8vo. tom. i. 544 pp. *Paris,* 1842

† Eléments d'Electro-chimie appliquée aux Sciences Naturelles et aux Arts. 8vo.
419 pp. *Paris,* 1843

† Mémoire sur l'Application Electro-chimique des Oxides et des Métaux sur des
Métaux. 8vo. 24 pp. (*Ann. de Chim.* 3ᵐᵉ serie, tom. viii. 1843.)
Paris, 1843

† Traité de Physique considerée dans ses Rapports avec la Chimie et les Sciences
Naturelles. Tom. ii. 651 pp. 7 plates. *Paris,* 1844

† Traité de Physique dans ses Rapports avec la Chimie et les Sciences Naturelles.
2 vols. 8vo. *Paris,* 1844

† Elemente der Electro-Chemie in ihrer Anwendung auf die Naturwissenschaften
und die Künste. Aus dem Französischen. 8vo. 488 pp. 3 plates.
Erfurth, 1845

† De la Polarité produite par les Décharges Electriques, et de son emploi pour la
détermination de la quantité d'Electricité ordinaire associée aux Parties con-
stituantes des Corps dans les Combinaisons. 4to. 24 pp. (*Comptes Rendus,*
1846, 1ᵉ serie, xxii. No. 10, p. 381.) *Paris,* 1846

Recherches sur les Causes de l'Electricité atmosphèrique et terrestre, et sur les
Effets Chimiques produits en vertu d'Actions lentes avec ou sans le Concours
de Forces Electriques. 4to. 145 pp. (*Extrait des Mémoires de l'Académie
des Sciences,* tome xxvii. 2ᵉ partie.) *Paris,* 1859

† Considérations générales sur la Théorie Electro-chimique. 4to. 6 pp. (*Comptes
Rendus,* 1849, 1ᵉ série, xxviii. No. 22, p. 658.) *Paris,* 1849

Recherches sur la Température de l'Air, au Nord, au Midi, loin et près des
Arbres. Suivi de : Note sur la Psychrométrie électrique ; nouveau Mémoire
sur la coloration électro-chimique et le Dépôt de Peroxyde de Fer sur des
lames de Fer et de Cuivre ; et Mémoire sur la Production électrique de la
Silice et de l'Alumine. 4to. 205 pp. *Paris,* 1863

Becquerel, A. C.—*continued.*

Le Dégagement de l'Electricité dans les Actions Chimiques. 8vo. *Paris,* 1854
† Sur la Conservation du Fer et de la Fonte dans l'Eau douce. 4to. 8 pp. (*Ext. des Comptes Rendus,* lix. 31 Oct. 1864.) *Paris,* 1864
† Eléments d'Electro-chimie appliquée aux Sciences Naturelles et aux Arts. 8vo. 626 pp. Deuxième édition, entièrement refondue. *Paris,* 1864

† **Becquerel et Brachet.** Lettre concernant des Expériences sur la Torpille faites en commun avec M. Brachet. 4to. (*Comptes Rendus,* iii. 135.) *Paris,* 1836

† **Becquerel, A. C., et Becquerel, Ed.** Eléments de Physique Terrestre et de Météorologie. 8vo. 706 pp. *Paris,* 1847
† Traité d'Electricité . . . Tome troisième, Magnétisme et Electro-magnétisme. 8vo. 412 pp. (*N.B. There are two plates, numbered* 6—*viz.* 6 *and* 6 *bis.*) *Paris,* 1856
† Traité d'Electricité et de Magnétisme et des Applications de ces Sciences à la Chimie, à la Physiologie, et aux Arts. Tome premier, Electricité. Principes généraux. 8vo. 456 pp. *Paris,* 1855
† Traité d'Electricité et de Magnétisme, et des Applications de ces Sciences à la Chimie, à la Physiologie, et aux Arts. Tome deuxième. Electro-chimie. 8vo. 475 pp. *Paris,* 1855
† Résumé de l'Histoire de l'Electricité et du Magnétisme, et des Applications de ces Sciences. 8vo. 300 pp. *Paris,* 1858
(*Note. Very imperfect lists of authors are annexed.*)

† **Becquerel, Biot, et Becquerel, Ed.** Mémoires sur la Phosphorescence par la Lumière Electrique. 4to. 27 pp. (*Archives du Muséum d'Histoire Naturelle,* tom. i.) *Paris,* 1839

Becquerel, Ad. Des Applications de l'Electricité à la Pathologie. 8vo. 52 pp. *Paris,* 1856
† Application de l'Electricité à la Thérapeutique. 1st edition. (?) *Paris,* 1857
† Traité des Applications de l'Electricité à la Thérapeutique Médicale et Chirurgicale. Deuxième édition. Revue, &c., et augmentée. 8vo. 550 pp. *Paris,* 1860

Becquerel, Edmond. *Born March* 24, 1820, *at Paris.* (*Vide* also Becquerel, A. C.)
† Des Effets Chimiques et Electriques produits sous l'Influence de la Lumière Solaire. Thèse présentée à la Faculté des Sciences, Paris, Août, 1840. 4to. 49 pp. *Paris,* 1840
Die Dagerotypie und Photographie in einer Sekunde. Aus dem Franz, nebst Bemerkungen über die Galvanotypie . . Uebersetzt von Oger. 12mo. *Achen,* 1841
† Des Lois du Dégagement de la Chaleur pendant le Passage des Courants Electriques à travers les Corps solides et liquides. 8vo. 52 pp. (*Ext. des Archives de l'Electricité de Genève.*) 1843
† Note sur l'Action du Magnétisme sur tous les Corps. 4to. 10 pp. (*Comptes Rendus,* 1846, 1e série, xxii. No. 23, p. 952.) *Paris,* 1846
† Recherches sur la Transmission de l'Electricité à travers des Gaz à des Températures élevées. Mémoire présenté à l'Acad. le 4 Juillet, 1853. (*Extrait des Ann. de Chim. et de Phys.* 3e série, xxxix.) 8vo. 48 pp. *Paris,* 18—
Recherches sur le Dégagement de l'Electricité dans les Piles Voltaïques. 1e partie, Force Electromotrice. 2e partie, Résistance à la Conductibilité. 3e partie, Puissance Chimique des Piles, Evaluation de leur Dépense. 8vo. 96 pp. (*Ext. du Cat. Lacroix a Miège,* p. 3. The 2nd and 3rd parts are probably also in the *Annales de Chimie.*) *Paris,* 1856
† Notice sur les Travaux Scientifiques de M. Ed. Becquerel. 4to. *Paris,* 1859
† Etudes sur la Conductibilité des Liquides dans les Tubes Capillaires, Rhéostat destiné à la Comparaison des grandes Résistances. 8vo. 24 pp. (*Ann. du Conservat. des Arts et Métiers,* Avril, 1861.) *Paris,* 1861

Becquerel, Edmond—*continued.*

† Effets Lumineux qui résultent de l'Action de la Lumière sur les Corps. Leçon faite le 19 Avril, 1861, à la Société Chimique de Paris. 8vo. 45 pp.
Paris, 1862

On the Transport of some Salts by Electrical Discharge. (*Comptes Rendus,* June, 1871.) (*From "Athenæum,"* Aug. 12, 1871.) *Paris,* 1871

† Mémoire sur la Conservation du Cuivre et du Fer dans la Mer. 4to. 8 pp. (*Ext. des Comptes Rendus,* lix. *Séance* 4 *Juillet,* 1864.) *Paris,* 1864

De l'Action du Magnétisme sur tous les Corps. 8vo. *Paris,* 1850

† Des Actions Electromotrices de l'Eau et des Liquides . . . sur les Métaux, &c., et des Effets Electriques qui ont lieu—1, dans le Contact de certaines Flammes et des Métaux ; 2, dans la Combustion. 8vo. 15 pp. (*Ann. de Chim. ?*)
Paris, 18—

† Exposé des Phénomènes Electriques qui précèdent et qui suivent les Actions Chimiques. 8vo. 35 pp. (*Ann. de Chim.?*) *Paris*

† Considérations générales sur les Changements qui s'opèrent dans l'état électrique des Corps . . . Seconde Partie. 4to. 34 pp. *Paris*

† Considérations générales sur les Changements qui s'opèrent dans l'état électrique des Corps . . Troisième Partie. 4to. 12 pp. (*Lu à l'Acad.* 27 Fev. 1832.) *Paris*

† Considérations générales sur les Applications des Sciences Physico-chimiques . . 4to. 12 pp. (*Comptes Rendus de l'Acad.* tom. xi. 16 Juillet, 1840.)
Paris, 1840

† Mémoire sur l'Application Electro-chimique des Oxides et des Métaux sur les Métaux . . . 4to. 16 pp. (*Comptes Rendus de l'Acad.* 3 Juillet, 1843.)
Paris, 1843

† Précipitation des Métaux par d'autres Métaux. 4to. 24 pp. (*Comptes Rendus de l'Acad.* tom. xviii. 18 Mars, 1844.) *Paris,* 1844

† Des Courants Electriques Terrestres. 4to. 18 pp. (*Comptes Rendus de l'Acad.* tom. xix. 18 Nov. 1844.) *Paris,* 1844

† Sur les Applications de l'Electro-chimie à l'Etude des Phénomènes de Décompositions et Recompositions terrestres. Premier mémoire. 4to. 28 pp. (*Comptes Rendus de l'Acad.* tom. xx. 26 Mai, 1845.) *Paris,* 1845

† De la Polarité par les Décharges Electriques, &c. 4to. 24 pp. (*Comptes Rendus de l'Acad.* tom. xxii. 9 Mars, 1846.) *Paris,* 1846

† Nouvelles Applications de l'Electricité à la Décomposition des Substances Minérales. 4to. 8 pp. (*Comptes Rendus de l'Acad.* tom. xxii. 13 Mai, 1846.)
Paris, 1846

† Mémoire sur les Circuits Electro-chimiques simples formés de Liquides. 4to. 12 pp. (*Comptes Rendus de l'Acad.* tom. xxiv. 29 Mars, 1847.)
Paris, 1847

† Sur les Fils très-fins de Platine et d'Acier ; et sur la Distribution du Magnétisme libre dans ces derniers. 8vo. 20 pp.

(*Vide* also Fremy and Becquerel.)

Beek, A. Van. Ueber das Magnetisiren des Stahls durch Maschinen-Elektricität. (*Gilb. Ann.* 1822. lxxii.) *Leipzig,* 1822

Neue Thermo-elektricität Versuche. 8vo. (*Gilb. Ann.* 1823. lxxiii.) *Leipzig,* 1823

† De l'Influence que le Fer des Vaisseaux exerce sur la Boussole, &c. . . Trad. du Hollandois, par Lipkin. (L . . .) 8vo. 71 pp. *Paris,* 1826

Ueber dauernde Einwirkungen, welche die Eigenschaften eines Metalls noch lange nach dessen Berührung mit einem anderen Metall erleiden. 8vo. (*Pogg. Ann.* xii. 1828.) *Leipzic,* 1828

Over de aanwending der Elektrische vonk, bij het doen van Microscopische nasporingen op snel bewegen ligchamen. (*Tijdschrift voor de wis-u. natur-kundige Wetenschapp,* &c. i. 1848.) 1848

(*Vide* also Moll and Van Beek.)

† **Beer, Aug.** Introduction à l'Electrostatique, à la Théorie du Magnétisme et à l'Electrodynamique. Ouvrage traduit et annoté par G. Vandermensbrugghe . . . d'après l'édition publiée après la mort de l'auteur, par M. Plücker. 8vo. 109 pp. *Paris,* 1868

† Einleitung in die Electrostatik. Die Lehre vom Magnetismus und die Electrodynamik. Nach dem Tode des Verfassers herausgegeben von Julius Plücker. 8vo. 418 pp. *Braunschweig,* 1865

† Electricität als Heilmittel gegen Nervenkrankheiten, Rheumatismen, Gelenksleiden, Drüsenanschwellungen, &c. 8vo. 52 pp. *Wien,* 1868

(*Vide* also Plücker and Beer.)

Beets. (*Vide* Du Bois-Reymond and Beets.)

† **Beetz, W.** von. Ueber die Leitungsfähigkeit füı Electricität, welche Isolatoren durch Temperaturerhöhung annehmen. 15 pp. (*Der Königl. Acad. der Wissensch. zu Berlin mitgetheilt.*)

† Ueber Magnetismus. Ein Vortrag gehalten im Wissenschaftlichen Verein am 13 März, 1852. 8vo. 28 pp. *Berlin,* 1852

† **Beggiato, F. S.** Elogio del Fusinieri. (*Vide* Fusinieri.)

Beguelin, Nicolas de. *Born June* 25, 1714, *at Courtlari by Bienne ; died Feb.* 3, 1789, *at Berlin.*

Extrait des Observations Météorologiques faites à Berlin pendant l'Année 1770. 4to. (*Nouv. Mém. de Berlin,* an. 1770, p. 75.) *Berlin,* 1770

† **Behn, F. D.** Das Nordlicht von 1770. (*Poggendorff,* i. 135.) *Lubec,* 1770

Beschreibung einiger merkwürdigen Nordlichter, &c. 8vo. 127 pp. *Lubeck,* 1783

Behr, Von. Ueber Elektrische Telegraphen. 4to. *Königsberg,* 1848

Vom Magnetismus und dessen Verhältniss zu den übrigen Naturkräften. 8vo. (*Unterhaltungen Königsberger, Naturwissenschaftliche.* Heft 2.) *Königsberg,* 1851

Behrends, T. G. B. Erfinder des nach ihm (und mehr nach Bohnenberger) benannten Elektrometers, beschreiben in dem Aufsatz. Das Merkwürdigste aus Versuchen über Elektricität. 8vo. (*Gilbert Ann.* 1806, xxiii.) *Leipzig,* 1806

On the Application of Galvanism to discover the reality of Death. Dissertatio qua demonstratur cor nervis carere. *Mogunt.* 1792

† **Behrmann, C.** Beobacht. über das Sternschnuppen angestellt auf den Sternwarten zu Göttingen und Münster, nebst daraus abgeleiteten Resultaten. 8vo. 49 pp. *Göttingen,* 1866

Beil. Die Anwendung elektromagnetischer Telegraphen für den Dienst der Eisenbahnen. 8vo. *Frankfurt,* 1845

† **Beinert, C. C.** Der Meteorit von Braunau am 14 Juli, 1847. Darstellung, Beschreibung und Analyse. 8vo. 51 pp. *Breslau,* 1848

Belavenetz, J. Captain. A Russian translation of the Admiralty Manual, with Notes. (*Evans and Smith's Admiralty Manual,* 2nd edition, 1863.) 116 pp. *Russia,* 1863

Beldani. Sulla forma, &c. (*Pila.*)

† **Belfield, Lefevre.** Des procédés Galvano plastiques. 8vo. 76 pp. (*Extr. de la Revue Scientifique et Industrielle.*) *Paris*

Belgian Academy. Catalogue des Livres de l'Académie de Belgique. 8vo. (*Gl. I. R. Inst. Lomb. nuo. s.* iii. 230.) *Bruxelles,* 1850

† **Belgrado, Giacomo.** *Born November* 16, 1704, *at Udine; died March* 26, 1780, *at Udine.*) I Fenomeni Elettrici con i corollari da loro dedotti, e con i fonti di ciò che rende malagevole la ricerca del principio Elettrico. All' Al. Rle. . . . Don Filippo Borbone, Duca di Parma, &c. 4to. 44 pp. *Parma,* 1749

48   BEL

Belgrado—*continued.*

De phialis vitreis ex minimi silicis casu dissilientibus.   4to.   (*Friedlander Cat.* 1869.)   *Patavii,* **1743**

† **Belknap,** Jeremy.   Extract of a Letter from Jeremy Belknap, containing Observations on the Aurora Borealis.   Read May 2, 1783.   4to.
Philadelphia, **1786**

Bell.   Animal Electricity and Magnetism.   8vo.   **1792**

Bella, Giovannantonio della.   (*Vide* Della Bella.)

† **Bellagatta,** A.   Trattamento sopra l'ignea apparenza . . . del 16 Dec. 1737. 12mo. 18 pp.   (*Callogera, Raccolta* xvii. 121.)   *Venezia,* **1738**

Bellani, Angelo.   *Born* 1776 *at Monza ; died* 1852 *at Milan.*

† Sulla produzione dell' acido muriatico ossigenato colla Pila . . . Lettera al Moscati.   4to. 18 pp.   (*Nuova Scelta d'Opuscoli,* i. 307.)   *Milano,* **1804**

† Osservazioni sopra varii argomenti . . . (Sulle pietre Meteoriche, p. 194.) 8vo. 10 pp.   (*Ann. di Chim. di Brugnatelli,* xxii. 189-194.)   *Pavia,* **1805**

† Nuove sperienze fisico chimiche . . . cogli elettromotori del Bellani . . . al Volta.   12mo. 86 pp.   *Milano,* **1806**

Difesa della Lettera . . . del . . . Volta al Marzari . . . 8vo. 29 pp.
Milano, **1823**

Su alcuni Fenomeni del vetro e del mercurio.   (*Memoranda received on January* 16, 1823, *by the Venetian-Lombardy Institution, but not printed in its Memoirs.*)

† Sopra la questione : se i cloruri metallici decompongono l'acqua.   8vo.   3 pp. (*Articolo 5 di Cenni su diversi Argomenti, Poligrafo di Verona di* 1832, x. 330.)   *Verona,* **1832**

† Dell' anteriorità di una scoperta attribuita a Davy.   Nota.   Applicazione del ferro alle lamine di Rame che foderano i bastimenti.   4to.   1 p.   (*Ann. del Reg. Lomb.-Veneto,* iv. 46.)   *Padova,* **1834**

† Sopra la questione : se i cloruri metallici decompongono l'acqua, ec.   Con Nota dell' Editore (*i. e.* Fusinieri.)   4to. 3pp.   (*An. del Reg. Lomb.-Veneto,* iv. 124.)   *Padova,* **1834**

† Sulla Grandine. Memoria.   4to.   52 pp.   (*Estratta dai fasci.* i. *e* ii. *del tomo* ii. *degli Opuscoli Matem. e Fisici.*)   *Milano,* **1834**

† Article headed, Ai Signori Direttori (on Falling Stars), with Notes by the Editors.   8vo. 8 pp.   (*Bibliot. Ital.* lxxxii. 470.)   *Milano,* **1836**

† Degli Aëroliti delle Pioggie o Nevi rosse e delle Nebbie, o esalazioni secche. Riflessioni.   8vo. 27 pp.   (*Estratta dalla Biblio. di Farmacia Chim. &c.*)
1836

† * Delle difficoltà che si oppongono allo stabilimento di Osservatorii meteorologici Discorso.   8vo.   (*Ital. Soc. Mem.* xxiii. 9.)   *Milano*

† Collettore di calorico, ossia nuovo strumento Meteorologico ; inventato e descritto da A. B.   4to. 9 pp. 1 plate.   (*Fusinieri's Annali,* vi. 200.)
Milano, **1886**

† Invito alla Riunione . . . di Torino . . . Sulla formazione della Grandine. (*Bibliot. Ital.* xcix. 65.)   8vo. 36 pp.   *Milano,* **1840**

Rettificazione . . . contro le Osservazioni del . . . Zantedeschi . . . nel bimestre iii. e iv. 1844, degli Annali delle Scienze del Regno Lombardo-Veneto.   8vo. (*Gior. I. R. Ist. Lomb.* xi. 209.)   **1845**

† Suono reso da un filo metallico teso in pien 'aria.   8vo.   (*Gior. I. R. Ist. Lombardo,* xv. 136.)   *Milano,* **1846**

Opuscoli diversi del . . . A. B. per la Maggiore parte estratti dagli Annali di Agricultura del Regno Lombardo-Veneto. 8vo. (*From Gior. I. R. Ist. Lombardo,* x. 370, 1845.)

Discorso sopra diversi argomenti Fisico-chimici.   (*V. Ext.* vi. 63.)

(*Vide* also Volta and Bellani.)

Bellavitis, Giusto. *Born* 1803, *at Bassano.* Lettera (di B. . . . He is the Anonimo matematico mentioned by Fusinieri) in risposta a quella che il . . . Fusinieri pubblicò negli Annali delle Scienze del Reg. Lomb.-Ven. Bimestre 2° del 1840, p. 91. 8vo. 13 pp. (*Verona Poligrafo,* iii. 1840, p. 220.) *Verona,* 1840 Sulle correnti Elettriche simultanee, ecc. Sperienze per verificare se vi possono esistere, in un medesimo conduttore, correnti Elettriche simultanee ed opposte. *Note.—Zantedeschi, in his paper, Ueber die Physikal-Studien, &c. der Italiener, im Jahre* 1858, p. 27, *gives a very long title, and says, in Atti dell' I. R. Inst. Veneto,* p. 113-147, 1858.

Bellemare. Système Controleur (télégraphe), présenté à l'Institut, Février, 1856.
Paris, 1856
Bellevue. (*Vide* Fleurian de Bellevue.)

Belli, Giuseppe. *Born November* 25, 1791, *at Calasca, Piémont; died June* 1, 1860, *at Pavia.*

Sulle induzioni Elettrostatiche. 8vo. (*Nuovo Cimento,* viiii 97, *anno* 1858.)
Torino, 1858

* Di un nuovo Psicrometro o Igrometro a raffreddamento per evaporazione. Memoria. 4vo. 33 pp. (*Nuovi Saggi dell' Accad. di Padova,* vi. 26.) *Padova,* 1847

† Corso elementare di Fisica Sperimentale. 3 vols. (the last one entirely on Static Electricity.) 8vo. 235 and vii. pp. 1 plate. *Milano,* 1830-31-38

† Di una nuova maniera di Macchina elettrica immaginata dal sig. Dott. Giuseppe Belli. 4to. 18 pp. 1 plate. (*Ann. del Reg. Lomb.-Veneto,* i. 111.) *Padova,* 1831

On the Action between two equal square Plates placed perpendicularly to a right line joining their centres. . . . 4to. *Milano,* 1832

On Repulsion, or rarefied Air. 4to. (*Opuscoli Matematici e Fisici,* tom. i. 376.)
Milano, 1832

† Riflessioni sulla Legge dell' attrazione molecolare. Memoria. 4to. 140 pp. 1 plate. (*Dei fascicoli,* 1, 2, 3, 4, *degli Opuscoli Matemat. e Fisici.*) *Milano,* 1833

† Sulla repulsione Elettrica nell' aria rarefatta. 4to. 2 pp. (*Fascicolo* 4° *degli Opuscoli Matematici e Fisici.*) *Milano,* 1833

† Sul dissiparsi, più facilmente . . . l'Elettricità negativa che non la positiva. Nota. 8vo. 5 pp. (*Bibliot. Ital.* lxxxi. 189.) *Milano,* 1836

† Sulla Elettricità negativa delle cascate. 8vo. 12 pp. 1 plate. (*Biblio. Ital.* lxxxiii. 32.) *Milano,* 1836

† Sulla dispersione delle due Elettricità. 2nd Mem. E sui residui delle scariche delle bocce di Leida. 8vo. 14 pp. (*Biblio. Ital.* lxxxv. 406-417.) *Milano,* 1837

† Sulla dispersione delle due Elettricità : sperienze. Continuazione (of 2nd Mem. called also 3rd Mem.) 8vo. 9 pp. 1 plate. (*Biblio. Ital.* lxxxvi. 276.)
Milano, 1837

† Dichiarazione intorno ad alcuni punti della scienza dell' Elettricità. 8vo. 2 pp. (*Biblio. Ital.* xc. 370.) *Milano,* 1838

† Di un nuovo apparecchio per le sperienze sull' origine dell' Elettricità voltiana. 8vo. 20 pp. 1 plate. *Milano,* 1840

† Della distribuzione dell' Elettrico nei corpi conduttori. Recitata 28 Otto. 1839. 4to. 99 pp. 1 plate. (*Mem. Soc. Ital.* xxii. 111.) *Modena,* 1841

† Considerazioni sulle trombe di terra e di mare. 8vo. 7 pp. (*Gior. I. R. Inst. Lomb.* xv. 232.) *Milano,* 1846

† Altre considerazioni sulle trombe di terra e di mare. 8vo. 13 pp. (*Gior. I. R. Inst. Lomb.* xvi. 72.) *Milano,* 1847

† Di una sperienza per imitare artificialmente le Trombe o vortici atmosferici. Comunicata 4 Marzo, 1847. 8vo. 9 pp. (*Gior. I. R. Inst. Lomb.* xvi. 307.)
Milano, 1847

Sulla possibilità di contrarie correnti Elettriche simultanee in uno stesso filo conduttore. 2nd Memoir. (From *Atti I. R. Inst. Lomb.* i. 76.) *Pisa,* 1857

† Sui mezzi di difendere i cronometri dal . . . Magnetismo terrestre. 4to. 2 pp. (*Atti dell' I. R. Inst. Lomb. Tornata* 4 *Giugno* 1857, vol. i. p. 9.)
Milano, 1858

E

Belli, Giuseppe—*continued.*
† Cenni sul Professore Giuseppe Belli dell' Ingegnere D. G. Cantoni. Small 8vo.
16 pp. (*Estr. dalla Perseveranza,* 19 *Giugno,* 1860.) *Milano,* 1860
(*Vide* also Kramer and Belli.)

Bellin, Jacques Nicolas. *Born* 1703, *at Paris; died November* 21, 1772, *at Paris.*
Carte des variations de la boussole et des vents . . . dans les mers les plus fréquentées. (*Young,* p. 441.) *Paris,* 1765

Bellingeri, Carlo Francesco. *Died May* 15, 1848, *at Turin.*
† In Electricitatem salivæ, muci, et puris simplicis et contagiosi experimenta.
4to. 32 pp. Lecta Jan. 1828. (*Mem. della R. Accad. delle Scienze di Torino.*)
Torino, 1828
Sperienze ed Osservazioni sul Galvanismo. 4to. (*Mem. de Turin,* xxiii. 143.)
Torino, 1818
† Sulla Elettricità del Sangue nelle Malattie. Saggio di esperimenti. 4to. 33 pp.
Presentato 30 Marzo, 1816 (*Mem. di Torino,* xxiv. 107.) *Torino,* 1820
† Sulla Elettricità dei Liquidi Minerali. Memoria presentata 23 Giugno, 1816.
4to. 19 pp. (*Mem. di Torino,* xxiv. 141.) *Torino,* 1820
† Memoria sull' Elettricità dell' Orina. 4to. 20 pp. Rimessa 24 Genn. 1819. (*Mem. di Torino,* xxiv. 459.) *Torino,* 1820
† Sulla proprietà Elettrica dei solidi animali. Memoria letta 12 Dec. 1819. 4to.
12 pp. (*Mem. di Torino,* xxv. 1.) *Torino,* 1820
† In electricitatem sanguinis, urinæ et bilis animalium experimenta habita. 4to.
24 pp. 1 plate. (*Mem. di Torino,* xxxi. 295.) *Torino,* 1827

† **Bellotti,** A. Dell' origine .e del progresso dell' Arte Telegrafica. 8vo. 40 pp.
2 plates. *Milano,* 1844

Beltrami, P. Nuova scoperta . . . per preservare le Campagne dalla Grandine. . .
8vo. 27 pp. *Milano,* 1823
† Risposta . . . alle critiche Osservazioni del . . . Majocchi sui Paragrandini e
Parafulmini, con Appendice di altra nuova probabile scoperta. . . . 8vo. 26 pp.
Milano, 1823
† Difesa dei Paragrandini. . . . 8vo. 176 pp. *Milano,* 1824
Paralogismi di . . . contro la difesa de' Paragrandini . . . e Manuale pratico
teorico. . . . 8vo. 171 pp. *Treviglio,* 1825
Li buoni effetti dei Paragrandini nell' An. 1825. 8vo. *Milano,* 1826
† Vendicazione dei Paragrandini. 12mo. 43 pp. *Lodi,* 1836

Bénard, M. Observations Ozonomètriques. 8vo. 19 pp. (From *Zuchold's Cat.* 1862,
p. 133.) *Le Havre,* 1862
† **Benedict,** M. Elektrotherapie. 8vo. 485 pp. *Wien,* 1868
Beneni, S. De Igne, Luce, et Fluido Electrico propositiones physicæ.
Florentiæ, 1790
† **Benet,** S. V. Colonel. Electro-Ballistic Machines and the Schultz Chronoscope.
4to. 47 pp. *New York,* 1866

Bennet, Abraham. *Born* 1750 ; *died May,* 1799.
Description of a new Electrometer. 4to. (*Phil Trans.* 1787, pp. 26-32.)
London, 1787
An Account of a Doubler of Electricity, or a Machine by which the least conceivable quantity of positive or negative electricity may be continually doubled,
till it becomes perceptible by common Electrometers or visible in sparks. 4to.
(*Phil. Trans.* 1787, p. 288.) *London,* 1787
† New Experiments on Electricity, wherein the causes of thunder and lightning, as
well as the constant state of positive or negative electricity in the air or clouds,
are explained ; with Experiments on Clouds of Powders and Vapours artificially diffused in the air. Also a Description of a Doubler of Electricity, and
of the most sensible Electrometer yet constructed, with other new Experiments and Discoveries in the Science, illustrated by explanatory plates. 8vo.
141 pp. *Derby,* 1789

Bennet, Abraham—*continued.*

† A new Suspension of the Magnetic Needle, intended for the discovery of minute quantities of magnetic attraction ; also an Air-vane of great sensibility ; with new Experiments on the Magnetism of Iron-filings and Brass. 4to. 18 pp. (*Phil. Trans.* 1792, p. 81.) *London,* 1792

† **Bennet, N.** Account of a Waterspout in Watuppa Pond, at Freetown, in a Letter to Jno. Davis (dated October 5, 1793.) 4to. (*Mem. Amer. Acad.* old series, ii. part ii. p. 70.) *Charlestown*

Benoit. (*Vide* Fournet and Benoit.)

Benzenberg, Johann Friedrich. *Born May 5, 1777, at Schöller, between Elberfeld and Düsseldorf ; died June 8, 1846, at Bilk.*

† Ueber Bestimmung der geographischen Lage durch Sternschnuppen. 8vo. *Hamburg,* 1802

† Die Sternschnuppen sind Steine aus den Mondvulkanen, die ein Durchmesser von 1, bis 5 Fuss haben, und welche bei 8,000 Fuss Geschwindigkeit in 1 Secunde nicht wieder auf den Mond zurückkommen und die dann mit Millionen um die Erde herumlaufen. 8vo. 80 pp. *Bonn,* 1834

† Die Sternschnuppen. 8vo. 375 pp. *Hamburg,* 1839

† **Benzenberg and Brandes.** Versuche die Entfernung, Geschwindekeit der Bahnen der Sternschuppen zu bestimmen. 8vo. 88 pp. *Hamburg,* 1800

Béraud, Laurent. *Born March 5, 1702, at Lyon ; died January 26, 1777, at Lyon.*

† Dissertation sur le rapport qui se trouve entre la cause des effets de l'Aimant et celle des Phénomènes de l'Electricité, qui a remporté le Prix de l'Académie (*Rare. Neither Æpinus nor Van Swinden could find a copy.*) 4to. 38 pp. *Bordeaux,* 1748

Mémoire sur cette question, Pourquoi les Corps Electriques par eux mêmes ne reçoivent ils pas l'Electricité par communication. 1749

Mémoire sur l'Electricité fulminante. (Bursting explosivity of Electrical Machine.) Lu 1750 (15 Avril), dans une Séance de la Soc. Royale de Lyon. *Lyon,* 1750

† Theoria Electricitatis. 4to. 48 pp. *Petersbourg,* 1755

La Physique des Corps Animés, 12mo. *Paris,* 1755

Eloge de Béraud, par Lefebvre. 12mo. *Lyon,* 1780

Béraud. (*Vide* Euler, Frisi, and Béraud.)

† **Berdoe, Marmaduke.** An Enquiry into the influence of Electric Fluid in the structure and formation of animated beings. 8vo. 183 pp. 4 plates. *Bath,* 1771

On the Electric Fluid. 8vo. 1773

† **Beretta, G.** Lettera all' Amorelli sul Tormalino del Monte di San Gottardo. 4to. 3 pp. (*Opuscoli Scelti,* viii. 404.) *Milano,* 1785

Bergen, Carl August von. *Born August 11, 1704, at Frankfurt on the Oder ; died October 7, 1759, at Frankfurt on the Oder.*

De Perseverantia virtutis Electricæ in Corpore electricitatis derivativæ. 4to. (*Nova Acta Acad. Nat. Curios.* ii. 158.) *Norimbergæ*

Observationes Meteorologicæ et Epidemicæ nec non electricæ. 1745

† **Berghaus, R.** Allgemeiner Erdmagnetischer Atlas. Consisting of 5 maps. Fol. *Gotha,* 1851

Bergman (or Bergmann), Torbern Olof. *Born March 9, 1735, at Katherinberg West Gothland ; died July 8, at Bad Medevi.*

† Anmerkungen vom stillen Wetterleuchten. 8vo. *Stockholm,* 1760

† A Letter concerning Electricity. 4to. (*Phil. Trans.* 1760, p. 907.) *London,* 1760

† Bemerkung von der Elektricität des Isländischen Krystales. 8vo. *Stockholm,* 1762

Bergman, Torbern Oloff—*continued.*

Observations on Auroræ Boreales in Sweden (1759). 4to.   (*Phil. Trans.*
1762, p. 479.)   *London,* 1762

Auroræ Boreales Annis 1759, 1760, 1761, et 1762, observatæ. 4to. (*Nova Acta*
Soc. Upsaliensis, tom. i. p. 118.)   *Upsala*

Observations in Electricity ; and on a Thunderstorm. 4to. (*Phil. Trans.* 1763,
p. 97.)   *London,* 1763

† Elektrische Versuche mit Seiden Bande von unterschiedlicher Farbe. 8vo.
Stockholm, 1763

† Some Experiments in Electricity. 4to. (*Phil. Trans.* 1764, p. 84.) *London,* 1764

Von der Höhe des Nordscheins. 8vo.   *Leipsic,* 1764

Rede von der Möglichkeit des Donners schädlichen Wirkungen vorzukommen.
4to.   *Stockholm,* 1764

† Elektrische Versuche, mit an einander geriebenenen Glas-Scheiben. 8vo.
Stockholm, 1765

† Commentarius de indole Electrica Turmalini. 4to.   *London,* 1766

† Zusatz zu der Abbandlung von der Höhe des Nordscheins. 8vo. 3 pp. (*K. Schwed.*
Akad. xxviii. 230.)   *Leipzig,* 1766

Experimenta Electrica, transitum commotionis per aquam illustrantia. 4to.
(*Nova Acta Soc. Upsaliensis,* vol. i. p. 111.)   *Upsala,* 1767

Von der Möglichkeit den schädlichen Wirkungen der Gewitter vorzubeugen.
(*Aus dem Schwedische Magazine,* i. 39.)   *Copenhagen,* 1768

† Zusatz zu Vorhergehenden, *i. e.* Wilcke, Bemerkungen bei einem den 30 May,
1769. . . . Donnerschlage. 8vo. 5 pp. (*K. Akad. Schwed. Abh.* xxxii. 128.)
Leipzig, 1770

† Braune Turmalinen nach ihrem Grundstoffe untersucht von I. R. 8vo. 14 pp.
(*Schwed. Akad.* xli. 199.)   *Leipzig,* 1779

† Precipitations Versuche mit Platina, Nickel, Kobalt, und Magnesium. 8vo·
12 pp. (*K. Schwed. Akad. Neue Abhand.* i. 269.)   *Leipzig,* 1780

† **Bergmann, L.** Die Telegraphie. 8vo. 80 pp.   *Leipzig,* 1853

Bergsma, P. A. Dissertatio Historico-physica de phosphorescentia per irradia-
tionem. 8vo.   *Ultrajecti ad Rhenum,* 1854

Bergstrasser, J. A. B. 5 Sendungen über sein Problem. 8vo.
Braunschweig, 1785-88

Erweiterung der Signal, &c. Schreiberei. (Not electrical.)   *Leipzig,* 1795

Ueber Signal-Ordre und Zeitschreiberei in die Ferne oder über Synthemato-
graphie und Telegraphie. 8vo. (Not electrical.)   *Frankfort,* 1795

Bérigny, Ad. Rapport sur les Observations Ozonomètriques faites avec le papier
Schönbein, autour de la caserne de Saint Cloud (Seine et Oise), et au milieu de
la cour à deux altitudes différentes, à sept heures et demie du soir pendant
trente et un jours, (du 6 Octobre au 5 Novembre, 1855 (inclusivement). 2 Mémoire. 8vo. 24 pp. (*Extrait des Mémoires de Médecine, de Chirurgie, et de
Pharmacie Militaires.*)   *Paris,* 1856

Recherches et Observations pratiques sur le Papier Ozonométrique. 8vo. 15 pp.
Paris, 1857

Vide also Deux-Ponts-Bérigny, E.A.

Berkel, A. Van. Reise nach Rio de Berbice (in der Sammlung, &c. aus dem
Holland.) 8vo. (*Gymnotus, Observations in* 1680 to 1689.)   *Memming,* 1789

† **Berkhout, J.** Dissertatio Physica Inauguralis, de fluxibus Thermo-electricis Acus
magneticæ ope, observatis. 8vo. 71 pp.   *Amstelodami,* 1843

Berlin. Gemeinnütz. Natur und Kunst Magazin.   *Berlin*

Berlin Academy. Abhandlungen, physikalische, der Königlichen Akademie der
Wissenschaften zu Berlin. 4to.   *Berlin*

BER

53

Berlinghieri, Vacca Leopold. *Born* 1732, *at Ponsacco, near Pisa ; died October* 6, 1812, *at Montefoscoli.*

Elementi di Fisica del Corpo umano in stato di salute. 4to. (*Giornale di Med. Pratica di Brera,* ix. 171.) *Pisa,* 1783

† Method of communicating Magnetism to a Bar of Iron without a magnet. 8vo. 1 p. (*Phil. Mag.* xxxv. 157.) *London,* 1810

Elogio di . . Vacca Berlinghieri scritto dal Tautini. F. 8vo. (*Giornale di Med. Pratica di Brera,* ix. 171 *and* (*continuation*) 298.) *Padova,* 1816

Bern. Mittheilungen der Naturforschung Gesellschaft in Bern. *Bern*

Bernard, Claude. Sur l'influence du nerf grand sympathique sur la chaleur animale. 4to. (*Comptes Rendus,* 29 *Mai*, 1852.) *Paris,* 1852

† **Bernard,** C. A. Die Functionen des elektrischen Fluidums, vorzüglich in Hinsicht des menschlichen Körpers im gesunden und kranken Zustande. Inaugural Dissertation. 8vo. 70 pp. *Wien,* 1838

† **Bernardi,** A. Sulla declinazione dell' Ago. 8vo. 4 pp. (*Biblio. Ital.* lxiv. 255.) *Milano,* 1831

Bernouilli, Daniel. *Born January* 29, 1700, *at Gröningen ; died March* 17, 1782, *at Basel.*

Mémoire sur la manière de construire les boussoles d'inclinaison pour faire, avec le plus de précision qu'il est possible, les observations d'inclinaison de l'aiguille aimantée, tant sur mer que sur terre, ce qui suppose des boussoles qui, étant mises dans un même lieu, donneront sensiblement la même inclinaison. 4to. (*Pièces de Prix de l'Académie de Paris,* v. *Mém.* 8.) *Paris,* 1743

Sur la cause physique de l'Aimant. 4to. (*Mém. de Paris,* 1746.) *Paris,* 1746

An Experiment on a Dog. (*Bonnefoy says, in Journal de Méd. Jan.* 1756, p. 56.) *Paris,* 1756

Le rapport de l'attraction de deux Aimants. (*Journal Helvétique, Nov.* 1758.) 1758

On the Magnets of Dietrich (civis Basiliensi). Description of, &c. (*Nouvelle Bibliothèque Germanique,* tom. xvi. part i. p. 225.)

Bernouilli, Daniel et Jean. Nouveaux Principes de Méchanique et de Physique tendant à expliquer la nature et les propriétés de l'Aimant. 4to. (*Pièces de Prix de l'Académie de Paris,* v. *Mém.* 12, p. 115.) *Paris,* 1746

Bernouilli, J. *Vide* Euler, &c.

Bernouilli, Johan, I. Nouvelle manière de rendre les Baromètres lumineux. (*Mém. de Paris,* 1700.) 1700

Nouveau Phosphore. (*Mém. de Paris,* 1701.) 1701

Expériences sur la lumière que rendent les Corps frottés dans l'obscurité. 4to. (*Mém. de Paris,* 1707, *Hist.* p. 1.) *Paris,* 1707

Dissertatio de Mercurio lucente. *Bâle,* 1719

Bernouilli, Johan (3rd). *Born November* 4, 1744, *at Basle, Switzerland ; died July* 13, 1807, *at Köpnick bei Berlin.*

Déclinaison de l'Aiguille aimantée observée à Berlin. 4to. (*Mém. de Berlin,* An. 1769, p. 490.) *Berlin,* 1769

Bernoulli, Cp. Ueber das Leuchten des Meers mit besonderer Hinsicht auf das Leuchten thierischer Körper. 8vo. *Göttingen,* 1803

† **Béron,** P. Atlas du Magnétisme Terrestre représentant l'aimantation de la terre par le soleil et l'aimantation du fer par la terre, avec un texte contenant l'explication de touts les faits magnétiques suivant les lois physiques. Appendice. Variations diurnes, annuelles, et séculaires des éléments magnétiques, &c. fol. 98 pp. 3 plates. *Paris,* 1860

Note.—The paging begins at p. 311. The three very large folding plates are not served in the volume ; they are numbered 8, 9, and 10. This is probably a part of his work, Grand Atlas Meteórologique (12 plates).

Réforme de la Physique par la découverte de la Stocchiométrie ; des équivalents Electriques. Par Pierre Béron. 1er livraison. Janvier et Février, 1860. 8vo. 136 pp. (*Réforme fondamentale des Sciences.*) *Paris,* 1860

Béron—*continued.*

Electrostatique, contenant l'explication des faits Electriques et Electrochimiques de toutes les sciences, suivi d'un Appendice sur la cause physique des maladies, &c. *Paris,* 1861

† Météorologie simplifiée et Télégraphie sans fils et sans cables. 8vo. 944 pp. *Paris,* 1863

† **Bertelli,** Barnabita e **Palagi.**. Sulla distribuzione delle correnti Elettriche nei conduttori. Esperienze. 8vo. 25 pp. 3 plates. *Bologna,* 1855

† Sopra Pietro Peregrino di Maricourt e la sua epistola de Magnete. Memoria prima. 4to. 32 pp. (*Estratto dal Bullettino di Bibliografia di Storia delle Scienze Matemat. e Fisiche,* tom. i. 1868.) *Roma,* 1868

† **Berthollet,** Claude Louis. *Born November* 9, 1748, *at Tulloire, Annecy, Savoy; died December* 6, 1822, *at Arcueil, Paris.*
Essai de Statique Chimique. *Paris,* 1803

† **Bertholon** de Saint Lazare. *Born* 1742, *at Lyon; died April* 21, 1800, *at Lyon.*
Rimedio per mal di denti. 12mo. 2 pp. (*Scelta d'Opuscoli,* xxl. 118.) *Milano,* 1776

Mémoire sur un nouveau moyen de se préserver contre la Foudre. 4to. *Montpellier,* 1777

Sur la cause Phosphorico-électrique des Aurorès Boréales. 4to. (*Journal de Phys.* 1778.) *Paris,* 1778

Mémoire sur cette question : Quelles sont les Maladies qui procèdent de la plus ou moins grande quantité du Fluide Electrique ? 8vo. *Lyon,* 1779

† De l'Electricité du corps humain dans l'état de santé et de maladie. Ouvrage couronné par l'Acad. de Lyon, dans lequel on traite de l'Electricité de l'Atmosphère, de son influence et de ses effets sur l'économie animale, &c. 8vo. 541 pp. *Paris,* 1780

Neue Elektrisirmaschine von Bertholon. (The Rubber in Motion.) 8vo, (*Lichtenbergs Mag.* i. 92.) *Gotha,* 1781

† Preisschrift über die Elektricität nach Medecinischen Gesichtspuncten betrachtet, nebst einem Anhange über die Gewitter und Gewitter Ableiter. 8vo. *Bern,* 1781

Sur l'Electricité, aérienne et artificielle. 4to. (*Journal de Phys.* 1782.) *Paris,* 1782

Lettre à M. de la Tourette, sur les Paratonnerres ascendants et descendants de la Ville de Lyon. (*Samml. zu Phys.* xix. *Mai* 1782, p. 382.) 1782

† De l'Electricité des Végétaux. Ouvrage dans lequel on traite de l'Electricité de l'Atmosphère sur les plantes, de ses effets sur l'économie des végétaux, de leurs vertus médico et nutrivo-électriques et principalement des moyens de pratique de l'appliquer utilement à l'agriculture, avec l'invention d'un électro-végétomètre. 8vo. 468 pp. *Lyon,* 1783

† Nouvelles Preuves de l'efficacité des Paratonnerres. 4to. 28 pp. 3 plates. *Montpellier,* 1783

† Estratto d'una Mem. sui Vegetabili che comunicano la scossa : e transunto d'una Mem. sulle Sabbie, &c., che trasmettono la scossa. 4to. (*Scelta d'Opuscoli, Nuova Ed.* iii. 80 and 389. *Originally printed in the* 12mo *edition in* 1777.) *Milano,* 1784

Ueber die Elektricität in Beziehung auf Pflanzen aus dem Französisch . . nebst Erfindung eines Elektrovegometers. 8vo. *Leigzig,* 1785

† De l'Electricité du Corps humain. Ouvrage couronné par l'Acad. de Lyon, dans lequel on traite de l'Electricité de l'Atmosphère, de son influence et de ses effets sur l'économie animale, des vertus médicales de l'Electricité, des découvertes modernes, et des différentes méthodes d'électrisation ; avec un grand nombre de Figures en taille-douce. 2nd edition, 2 vols. 8vo. vol. i. 522 pp. vol. ii. 518 pp. 6 plates. *Paris,* 1786

† De l'Electricité des Météores. Ouvrage dans lequel on traite de l'Electricité naturelle en général et des Météores en particulier. 2 vols. 8vo. *Paris,* 1787

Bertholon de Saint Lazare—*continued.*

Anwendung und Wirksamkeit der Elektricität zur Erhaltung und Wiederherstellung der Gesundheit des menschlichen Körpers. 2 Theile. 8vo.
Leipzig, 1788

Nouvelles expériences sur les effets de l'Electricité appliquée aux Végétaux. 4to. (*Journ. de Phys.* xxxv.)
Paris, 1789

Die Elektricität der Lufterscheinungen. 2 vols. 8vo.
Leignitz, 1792

* **Berthoud.** Eclaircissements sur l'invention, la théorie, la construction, etc. des machines pour la détermination des longitudes en mer.
Paris, 1773

* Histoire de la mesure du temps par les horloges.
Paris, 1802

Bertier (Berthier), J. E. Lettres sur l'Electricité. La physique des Comètes.
Paris, 1760

† **Bertin,** P. A. *Born February* 13, 1818, *at Besançon, dep. Doubs.*

Thèses ... sur les propriétés rotatives que prennent les corps transparents sous l'influence du Magnétisme, et Thèse ... sur les phénomènes chimiques qui accompagnent le passage de l'Electricité à travers les corps ... ou étincelles ... ou courants. Presentées à la Faculté des Sciences de Paris. 4to. 38 pp.
Paris, 1847

† Mémoire sur la rotation Electromagnetique des Liquides. 8vo. 26 pp. (*Extr. des Ann. de Chim.,* 3rd serie, vol. lv.)
Paris

† Sur la Polarisation circulaire magnétique. 8vo. (*Extr. des Ann. de Chim.* 3rd series, vol. xxiii.)
Paris, 1848

Discours sur les théories physiques en général et sur celles de l'Electricité en particulier. 8vo.
Strasbourg, 1859

Sur les phénomènes de Polarisation magnétique dans les verres trempés et dans les parallélépipèdes de Fresnel. 4to. (*Compt. Rendus,*1849,xxviii.) *Paris,*1849

Bertin, E. Quelques réflexions sur les poussières atmosphèriques, à propos d'un travail du Dr. de Vivenot sur un obscurcissement particulier du ciel en Sicile. 8vo. 20 pp.
Montpellier, 1867

Bertoncelli, B. Di un nuovo Barometro e di un nuovo Barometrografo. Memoria ... letta 23 Feb. 1854, all' Accademia. 8vo. (*Verona Accad. d'Agricul.* vol. xxx. p. 41, *on Hydrometrical Principles.*)
Verona, 1854

Bertrand, Joseph L. François. *Born March* 11, 1822, *at Paris.*

Sur quelques points de la théorie de l'Electricité. (*Liouville. Journ. de Mathém.* 1839, iv.)
1839

† Programme d'une Thèse sur la distribution de l'Electricité à la surface des corps. 4to. 7 pp.
Paris, 1839

† **Berrutti,** S. Elogio del ... Vassalli Eandi. 8vo.
1839

Esperienze sulla esistenza delle correnti Elettrico-fisiologiche negli animali a sangue caldo. Eseguite nel gabinetto di fisica della Reale Università dal Prof. S. Berrutti, in compagnia dei .. Botto, Girola, Bellingeri, Demarchi, e Malinverni. 8vo. 32 pp. (*Ital. Soc. Mem.* xxii. 30.)
Torino, 1840

Berzelius, Jöns Jacob, Freiherr von. *Born August* 29, 1779, *at Vafversunda Sorgärd im Stifte Linkoping, am Fusse des Omberg, zwichen dem Vettern und Takern See ; died August* 7, 1848, *at Stockholm.*

(*Poggendorff says that most of his works in the Academy, the Afh. Fisik, &c. are translated in the Ann. Chim. Phys. in der Zeitschrift von Gehlen, Schweigger, Gilbert, and Poggendorff, in which are also various smaller original treatises of his. He published more than 200 treatises and smaller articles.*)

† Afhandling om Galvanismen. sm. 8vo. 145 pp. (*Rare.*)
Stockholm, 1802

De Electricitatis Galvanicæ Apparatu Cel. Volta excitatæ in corpora organica, effectu.
Upsala, 1802

Berzelius—*continued.*

Försök med elektrisa Stapelus verkan på Salter och på nägra af deras baser (mit Hisinger). (*Afh. i Fisik, &c.* tom. i.) 1806

Elektroskopica försök med. fargade paper. (*Afh. i Fisik,* tom. i.) 1806

Elektriska Stapelns Theori. *Afh. i Fisik,* tom. ii.) 1807

Om Magnetnälens afvikelser. (*Afh. i Fisik,* tom. ii.) 1807

† * Proposed Improvement of the Hygrometer. Translated from Berzelius's *Phil. Journal,* 1808. 8vo. 4 pp. (*Phil. Mag.* xxxiii. 39.) *London,* 1809

Försök att. genom användandet af den Elektro-kemiska propotionerna, grundlaga ett. rent vetenskapligt Systems för Mineralogien. *Stockholm,* 1814

† An attempt to establish a pure Scientific System of Mineralogy by the application of the Electro-chemical theory and the chemical proportions. Translated from the Swedish original by John Black. 8vo. 144 pp. (*Brugnatelli's copy.*) *London,* 1814

Versuch durch Anwendung der elektrisch-chemischen Theorie, &c. ein System der Mineralogie zu begrunden. Aus dem Schwedischen von Gahlen. 8vo. *Nurnberg,* 1815

† Essai sur la Théorie des proportions chimiques et sur l'influence chimique de l'électricité. Traduit du Suédois. 8vo. 190 pp. *Paris,* 1819

Versuch über die Theorie der chemischen Proportionen und über die chemischen Wirkungen der Elektricität; nebst Tabellen. Nach dem schwedischen und franz. Original ausgearbeitet von R. A. Blode. 8vo. *Dresden,* 1820

Om det Magnetiska tillstandet hos de kroppar, som urladda elektriska stapeln. 8vo. (*Vetensk. Acad. Handl.* 1820.) *Stockholm,* 1820

† Jahrsbericht über die Fortschritte der physichen Wissenschaften. Aus dem Schwedischen von F. Whöler. 8vo. 1822 to 1851

Untersokn af. en Meteorsten. 8vo. (*Vetensk Acad. Handl.* 1828.) *Stockholm,* 1828

† Power of Metallic Rods or Wires to decompose water after their connection with the Galvanic pile is broken. 8vo. 2 pp. (*Phil. Mag.* viii. 229.) *London,* 1830

Om Meteorstenar. 8vo. (*Vetensk Acad. Handl.* 1834.) *Stockholm,* 1834

† Théorie des proportions chimiques et Table synoptique des poids atomiques des corps simples, &c. Revue, corrigée, et augmentée. 2d edition. 8vo. 476 pp. *Paris,* 1835

Rapport annuel sur le progrès de la Chimie. Traduit par Plantamour. 8 vols. 8vo. *Paris,* 1841-8

† Vollständiges Sach-und Namen-Register zum Jahr's-Bericht über die Fortschritte der physischen Wissenschaften, der Chemie und Mineralogie. 1-25 Jahrg. 8vo. 180 pp. *Tubingen,* 1847

Neues chemisches Mineralsystem nebst einer Zusammenstellung seiner älteren hierauf bezüglichen Arbeiten. 8vo. *Nurnberg,* 1847

Lehrbuch der Chemie. 5 vols. royal 8vo. *Leipzig,* 1848

† Gedächtnissrede auf Berzelius; auswärtiges Mitglied der Academie, von Herr Heinr. Rose, gehalten ... Akad. der Wissenschaften am 3 Juli, 1851. 4to. (*Abhandlung der Berlin Akad.* 1851.) *Berlin,* 1851

Berzelius und Hissinger. Afhandlingar i Fisik, Kemi och Mineralogie utgifne af W. Hissinger och J. Berzelius. 6 vols. *Stockholm,* 1806-1818

Vide also Hissinger and Berzelius.

† **Beseke,** Johann Melchior Gottlieb. *Born September 26, 1746, at Burg bei Magdeburg; died October 19, 1802, at Mitau.*

Ueber Elementärfeuer und Pflogiston als Uranfänge der Körperwelt, insbesondere über elektrische Materie. In einem Schreiben an Herr Achard. 8vo. 52 pp. *Leipzig,* 1786

† **Besio. G.** Nuova modificazione alla pila di Volta. Memoria. 8vo. 19 pp. 1 plate. *Lugano,* 1839

† **Bétancourt,** A. de. Nouvelle Télégraphie présentée par MM. Breguet et Bétancourt. (*Mém. de l'Instit.* tom. iii. *Bull. Soc. Philomath. A.* vi.)

In Gauss and Weber's Resultate, &c. for 1837, p. 14, is a footnote concerning a Notice communicated to Gauss by Humboldt, relative to a Wire (Drahtkelle) extended from Aranjuez to Madrid for telegraphic purposes, by Betancourt. Date, apud Moigno, p. 62, was 1787. **1787**

Vide also Breguet and Bétancourt.

Beuchot. Bibliographie de la France. 8vo.

* † **Bevan,** B. On register Rain Gauges. (*Phil. Mag. or Annals,* ii. 74.)
London, 1827

* † Remarks on Mr. J. Taylor's Rain Gauge. (*Phil. Mag. or Annals,* iii. 29.)
London, 1828

† **Bevis,** John. *Born October* 31, 1695, *at Old Sarum, Wilts; died November* 6, 1771, *in London.*

An Account of a Luminous Appearance in the sky, seen in London on March 13, 1734-5. (*Phil. Trans.* xli. 347, printed 1744.) *London,* 1739-40-41

On Coating Glass Plates with Tinfoil, &c. (Inventor of.) (*Vide* Watson, *Phil. Trans.* 1747.) **1747**

On Earthquakes. 8vo. **1769**

† **Bew,** Charles. Electrical Cure ; or, Opinions on the Causes and Effects of Ticdouloureux. 8vo. 94 pp. *London,* 1824

† **Beyer.** On Lightning-conductors, &c. 8vo. 2 editions. *Paris,* 1806-9

Bezaud (Bezault), L. Les Animaux et les Métaux deviennent-ils électriques par communication? (From *Poggendorff,* i. 146.) *Bordeaux,* 1749

† **Bezold,** W. von. Zur Theorie des Condensators. Inaugural Dissertation. 8vo. 60 pp. *Göttingen,* 1860

† Untersuchung über die elektrische Erregung der Nerven und Muskeln. 8vo. 330 pp. 2 plates. *Leipzig,* 1861

Ueber die physikale Bedeutung der Potential-Function in der Elektricitätslehre. 8vo. *München,* 1861

† **Bianchi,** G. Lettere . . . intorno l'Aurora Boreale veduta in Rimini addì 16 Dec. 1737. 12mo. 7 pp. (*Callogera, Raccolta,* xvii. 99.) *Venezia,* 1738

† Alcune Spiegazioni dell' Aurora o lume boreale. 12mo. 9 pp. (*Callogera, Raccolta,* xvii. 109.) *Venezia,* 1738

† Osservazioni intorno le Aurore Boreali vedutesi le sere de' 10 e 29 Mar. 1739. 12mo. 17 pp. 1 plate. (*Callogera, Raccolta,* xxi. 187.) *Venezia,* 1740

Sopra i piccoli moti apparenti osservati nei muri e nelle macchine della Reale Specola di Modena. Ricevuta 14 Dec. 1836. 4to. 55 pp. (*Ital. Soc. Mem.* xxi. 246.) *Modena,* 1837

Bianchi, Giovanni Batista. Lettre à Pivati. 4to. *Torino,* 1748

Bianchi, Iso. Elogio storico del Fromond. 4to. 64 pp. *Cremona,* 1781

† (*Note.—At p. 42 is Beccaria's letter to Fromond, dated Torino,* 27 *Gennajo,* 1754, *describing his experiment intended to prove that electric motions do not occur* in vacuo.)

† **Bianchini,** Dominicus. De Igne Electrico in naturam agente. 8vo. 15 pp.
Verona, 1787

† **Bianchini,** Giovanni Fortunato. *Born Dec.* 27, 1719, *at Chieti, Neapel; died Sept.* 2, 1779, *at Padua.*

Saggio di esperienze intorno la Medicina Elettrica atte in Venezia da alcuni amatori di Fisica. Al Sig. Ab. Nollet. 4to. 116 pp. (*Vide Anonym. Recueil* xx. *for French translation. An address by the "Amatori" to Nollet is signed by the initials N.N.N.N.*) *Venezia,* 1749

† Osservazioni intorno all' uso dell' Elettricità celeste, e sopra l'origine del Fiume Timavo. Riportate in due Lettere. 4to. 92 pp. 1 plate. *Venezia,* 1754

Bianchini—*continued.*

1. Lettera intorno un nuovo Fenomeno Elettrico (1753), all' Accad. Reale di Parigi, 39 pp. 1 plate.

2. Osservazioni intorno al Fiume Timavo scritte in una lettera al . . Guido Conte Cobenzi (1754), 41 pp.

† 3. Another letter follows, entitled "Epistola Reverendi Patris Petri Imperati Bononiam missa (preceded by two pages by Bianchini, and dated Castello Pucino, Sept. 1752.

(The first letter is dated Udine, 1753. Vassalli, p. 10, says, "printed in Mem. dell' Accad. 1764," and "La Lettera del Dottore sudetto fu scritta nel 1753," and contains Bianchini's account of the state of the conductors, &c., at the Castello Pucino in 1752. The third letter of 1602 contains an account (Latin) of the Ignis Fatuus called St. Helen's fire.)

Lettera intorno un nuovo Fenomeno Elettrico. 4to. (*Mem. Paris*, 1764.)
Paris, **1764**

On the Vertical Rod on the Chateau di Duino in the Friuli. 4to. (*Mémoires de l'Acad. pour* 1764, edit. orig. p. 44.)
Paris, **1764**

† **Bianchini**, Giuseppe. Parere sopra la cagione della morte della Signora Contessa Cornelia Zangari ne' Bandi Cesenate, esposto in una lettera al Sig. Co. Ottolino Ottolini. 8vo. 70 pp. Two editions. *Verona*, **1731-33**

† An Extract by Rolli, P., of an Italian Treatise, written by Bianchini, J., upon the Death of the Countess Cornelia Zangari ne' Bandi, of Cesena. To which are subjoined Accounts of the Death of Hitchell, J., who was burned to death by Lightning, and of Grace Pitt at Ipswich, whose body was consumed to a coal. 4to. 19 pp. (*Phil. Trans.* xliii. 447.) *London*, **1744-45**

† Parere sopra la cagione della Morte della Contessa Cornelia Zangari ne' Bandi Cesenate . . . esposto in una lettera al Conte Ottolino Ottolini. 8vo. 147 pp. 4th edition. *Roma*, **1758**
(This edition contains Cromwell Mortimer's Dissertation on Animal Heat (in French and Italian) from the Phil. Trans. Read 4th July, 1745.)

Bianconi, Giovanni Lodovico. *Born Sept.* 30, 1717, *at Bologna; died Jan.* 1. 1781, *at Perugia.*

† Due Lettere di Fisica al Maffei. 4to. 72 pp. 1 plate, and 110 pp.

1. Delle caraffe di vetro che scoppiano al cadervi dentro di alcune picciole materie e di altri vetri curiosi. Lettera prima.

2. Della Velocità del Suono. Altra lettera. *Venezia*, **1746**

† Lettre sur l'Electricité, écrite par M. Bianconi à M. le Comte Algarotti. Dated Augsbourg, 3 Nov. 1747. 8vo. 40 pp. *Amsterdam*, **1748**
(The printer or editor says that this forms a part of the 2nd vol. of the Journal des Savans d'Italie, and that the remarks are by the Journalists.)

† Sendschreiben über die Electricität an Hn. Grafen Algarotti. Ans dem Französichen übersetzt. 8vo. 44 pp. *Basle*, **1750**

† Portrait of. C. Bianconi, del. ; Benaglia, sculp.

† **Bianconi**, G. Cenni intorno all' origine ed ai progressi dell' arte Galvanoplastica. 8vo. *Bologna*, **1841**
(Nuovi Annali delle Scienze Naturali, Luglio, Bologna, 1841, mentions the goldsmith of Pavia, who knew the process of gilding very easily and cheaply by means of the pile in 1818.)

Bianconi, G. B. Intorno all' origine ed ai progressi dell' arte Galvanoplastica. (*Nuovi Annali delle Scienze Naturali*, tom. viii. p. 72.)

Bibliografia Italiana. (From *Athenæum*, July 15, 1871, p. 82.)

Bibliographie de la France, ou Journal Général de l'Imprimerie. Première Année, Nov. 1811, à Dec. 1812. A volume every year since. Pillet, Rue des Grands Augustins, No. 5, Editor.

Bibliotheca Americana. (*Eng. Cat.* 1864. p. 65.) *New York*, **1850-61**

Bibliothèque Médicale, par une Société de Médecins.
(*Sue. Hist.* iv. 206, *says that the authors "ont, à commencer au No. X., p. 94, et suir. produit une exposition abrégée des connaissances acquises jusqu'à ce jour sur le galvanisme, extraite des principaux ouvrages qui ont paru sur cette matière.*")

Bibliothèque, &c. (de Genève), consists of the following titles :
1. Bibliothèque Britannique . . . . . 1796 à 1815
2. „ Universelle . . . . . . 1816 „ 1835
3. „ „ {Nouvelle série . 1836–38} 1836 „ 1845
   {3e série . . 1839–45}
4. „ „ 4e série . . . . 1846 „ 1856 et seq.
5. Archives de l'Electricité (Supplément) . . . 1841 „ 1845
6. Archives des Sciences Physiques . . . . 1846 „ 1857 et seq.

† **Bibliothèque Italienne.** Ou Tableau des Progrès des Sciences et des Arts en Italie, par Giulio Gioberti, Vassalli Eandi, et Rossi. 5 vols. 8vo. (*Brugnatelli's copy.*)   *Turin*, 1803-4

Bibliothèque Médico-Physique du Nord. Vicat, editor. 8vo. Divided into three independent parts.   *Lausanne*, 1783

Biblioteca Germanica. Ridolfi, Bura, Santini, Configliachi, editori.   1822-23

Biblioteca Italiana. Ossia Giornale di Letteratura, Scienze, ed Arti, compilato da una Società di Letterati. 100 vols. 8vo.   *Milano*, 1816-40
(*Note.—The compilers were Monti, Breislak, Giordani, Acerbi Direttore, Gironi, Carlini, Fumagalli, G. Brugnatelli, Configliachi, Ferrario, Cotena, and Fantonelli.*)
Notice of "Navigazione col mezzo del Galvanismo," taken from the *Glasgow Mechanics' Magazine*, No. 23, October. 8vo. (*Biblioteca Italiana*, xli. 447.)
                              Milano, 1826

† Sul Monumento a Beccaria, &c. 8vo. 2 pp. (*Biblioteca Italiana* lxix. 397.)
                              Milano, 1833
† Sull' Herschell ed altri. Stelle cadenti. 8vo. 4 pp. (*Bibl. Ital.* lxxxii. 323.)
                              Milano, 1836
† Sul Dal Negro. 8vo. 2 pp. (*Bibl. Ital.* lxxxv. 121.)   *Milano*, 1837
† Dei Metallici ricoprimenti. 8vo. 2 pp. (*Bibl. Ital.* lxxxix. 126.)   *Milano*, 1838
† Reviews of Books and Papers relating to Electricity, &c.:
   Shorter Notices of Writings.   „
   Extracts from and abridged Accounts of Books.
   Reprints of Articles in other Journals.   „
   Accounts of Experiments and Observations.   „
   Accounts of Electrical Atmospheric Phenomena, as Auroræ, &c.
   Notices of various other matters referable to Electrical Science.
   Reflections, Remarks, Notes, &c.   „
   Necrologies of Electricians.
   8vo. 23 brochures. (*Bibl. Ital.* t. xxxiv. to xcix. inc.)   *Milano*, 1824-40
† De' trasporti Metallici operati dall' Elettrico. 8vo. 3 pp. (*Bibl. Ital.* xcviii. 269.)   *Milano*, 1840

Biblioteca Oltramontana, ad uso d'Italia ; colla Notizia dei Libri stampati in Piemonte.   *Torino*, 1786 ?

Bichat, Xav. Physiologische Untersuchung über Leben und Tod. Im Ausz. und mit Anmerkungen von Herold, aus dem Dänisch von Pfaff. 8vo.
                              Copenhagen, 1802-3
† Recherches Physiologiques sur la Vie et la Mort. 3rd edit. 8vo. 347 pp.
   (*Guitard says* 1801, "*the 1st edition is* 1800." *Vide Cerise's 2nd edition of* 1856.)   *Paris*, 1805
Anatomie générale appliquée à la Physiologie et à la Médecine. Nouvelle édition, précédée des Recherches Physiologiques sur la Vie et la Mort. 4 vols. 8vo.
                              Paris, 1812 or 1818

Bichat—*continued*.
Anatomie générale, précédée des Recherches Physiologiques sur la Vie et la Mort, avec des Notes de Maingault. 2 vols. 8vo. Portrait.   *Paris*, 1818

Anatomie générale. Nouvelle édition, avec des Additions par Béclard. 4 vols. 8vo.   *Paris*, 1821

Recherches Physiologiques sur la Vie et la Mort. 4e édition, augmentée, par F. Magendie. 8vo.   *Paris*, 1822

Recherches sur la Vie et la Mort, avec des Notes par Magendie. New edition. 8vo.   *Paris*, 1822

† Recherches Physiologiques sur la Vie et la Mort. Nouvelle édition, précédée d'une Notice sur la Vie et les Travaux de Bichat, et suivie de Notes, par Cérise. 12mo. 382 pp. 1 vignette.   *Paris*, 1856

† Bidone, Giorgio. *Born January* 19, 1781, *at Casal-Noceto, province Tortone, Piémont; died August* 25, 1839, *at Turin*. Description d'une nouvelle Boussole propre à observer les Mouvements de Rotation et de Translation de l'Aiguille aimantée, et Expériences . . . . Lecture le 28 Nov. 1807. 4to. 38 pp. 3 plates.   (*Mém. de Turin*, 1809-10.)   *Torino*, 1811

Bienvenu. On Mosaic Gold used in the Rubber of the Electric Machine about 1783 or 1784. 8vo. (*Lichtenberg's Mag.* ii. p. 211.)   *Gotha*, 1783 or 1784

Biester, J. P. De acu Magnet. &c. 4to. (From *Friedlander's Cat.* No. 163, 1867, p. 6.)   *London*, 1725

† Bigot de Morogues, Pierre Marie Sébastien Baron de. *Born April* 5, 1776, *at Orléans; died June* 15, 1840, *at Orléans*.
Anweisung zur Construction und Veranschlagung der Blitz-Ableiter . . nebst einem Nachtrage : über Taveniers Gewitters ableitende Säule . . 8vo. 102 pp.
                                                                         Glogau, 1834

† Mémoire Historique et Physique sur les Chutes des Pierres tombées sur la surface de la Terre à diverses époques. 8vo. 360 pp.   *Orléans*, 1812

† Chronological Catalogue of Stones and other large Masses, which are presumed to have fallen on the Earth. 8vo. 4 pp. From the *Journal des Mines*, xxxi. 430. (*Phil. Mag.* xliii. 448.)   *London*, 1814
(*This list begins in* 654 *before Christ, and ends in A.D.* 1812, *excepting one shower at Gideon in* 1451 *before Christ.*)

Bikker. Dissertatio de Natura Humana. (From *Van Troostwyk und Krayenhoff*, p. 162.)   *Lugd. Bat.* 1757

Bilharz, Th. Das Electrische Organ des Zitterwelses, anatomisch beschrieben. Fol.
                                                                         Leipzig, 1857

Billaudel. Prem. Mém. on the Meteor of 17th February, 1823. (*Bordeaux Acad.*)
                                                                         1823

Deux Mémoires sur la Météorologie and other matters. (*Bordeaux Acad.*)

On some Meteors observed at Bordeaux, and referred to Auroræ. (*Bordeaux Acad.* Séance 1829, p. 43.) *Vide* also Lacour.

† Billet, J. Des Condensations Electriques ; with Memoir on Light. 8vo. 7 pp. 1 plate. (*Ext. Acad. de Dijon.*)   *Dijon*, 1851

De quelques Appareils pour l'Electricité dynamique. 8vo.   *Paris*, 1854

Bina, Andrea. *Born Jan.* 1, 1724, *at Mailand*.
† Electricorum effectuum explicatio quam ex principiis Neutoniensis deduxit, novisque experimentis ornavit. 8vo. 157 pp.   *Batavii*, 1751

Elektrische Versuche, Gewitter und Regen betreffend. Briefe an Kostner. 8vo. (*Hamburg Mag.* xii.)   *Hamburg*, 1753

Lettera intorno all' Elettrizzazione dell' aria in occasione di tempo cattivo, &c.
                                                                         Perugia, 1753

Bina—*continued.*

† De Physicis experimentibus Chi. Wolfii, trans. 8vo. 2 vols. in 4 parts. 36 plates.
Venetiis, 1753-56
(*The 2nd part of vol.* ii. (1756) *contains Bina,* "*De Electricis effectibus,*
&c., ab interprete addita," 141 *pp.—viz., from p.* 215 *to* 356, 1 *plate.*
It is a 2nd edition of B. 10, *but not so called in the title.*)
Vide also Wolf and Bina.

Bineteau, M. A. Note sur l'Ozone atmosphérique. 8vo. 8 pp. *Lyon,* 1855

† **Biot,** Jean Baptiste. *Born April* 21, 1774, *at Paris; died Sept.* 8, 1862.
Some Articles on Electricity in the Bulletin de la Société Philomathique.
1801 to 1816
Rapport sur les Expériences de Volta. 4to. (*Mém. de l'Inst.* vol. v. 1802.)
Paris, 1802
(*The commission appointed in consequence of the First Consul's letter con-*
sisted of Laplace, Hallé, Coulomb, Hauy, and Biot. This commission
was appointed on the . . . ? The above report was made "le Messidor
suivant," 30th *June. The medal given to Volta is dated* 11 *Frimaire,*
an X. 2nd *December,* 1801.)

† Sur le Mouvement du Fluide galvanique. 4to. 11 pp. (*Soc. Philomath.* an 9,
No. 54, p. 45, par extrait.) *Paris,* 1802
Observation sur les Pierres météoriques. 4to. (*Soc. Philomath.* an 11, p. 129.)
Paris, 1803
Recherches sur cette question : Quelle est l'influence de l'oxidation sur les
effets de la Colonne électrique de Volta. 4to. (*Soc. Philomath.* an 11,
p. 120.) *Paris,* 1803

† Relation d'un Voyage fait dans le Département de l'Orne pour constater la
vérité d'un Météore observé à l'Aigle le 26 Floreal, an 11. 4to. 47 pp. 1 plate
(*i. e.,* carte des lieux.) (*Lu à l'Institut. le* 29 *Messidor, an* 11.) 1803

† Notice sur l'état actuel des Connaissances relatives au Galvanisme. 8vo. 3 pp.
(*Lu à la Séance Publique de l'Instit. National le* 1 *Messidor, an* 11.)
Paris, 1803

† Account of a Fire-ball which fell in the neighbourhood of Laigle : in a Letter to
the French Minister of the Interior. By C. Biot, Member of the National
Institute. Dated July 20, 1803. 8vo. 5 pp. (*Phil. Mag.* xvi. 224.)
London, 1803

† On the Formation of Water by compression ; with Reflections on the Nature of
the Electric Spark. Read before the National Institute. 8vo. 3 pp. (*Phil.*
Mag. xxi. 362.) *London,* 1805

† Traité de Physique. 8vo. *Paris,* 1816
Terrestrial Magnetism, by T. S. Evans. 8vo. *London,* 1817
Précis élémentaire de Physique expérimentale. 2 vols. 2nd edition. German
translation. *Vide* Fechner. *Paris,* 1818 and 1824
Translation of Fischer ; Physique mécanique, avec des Notes et un "Appendice
sur les Anneaux colorés . . . le Magnétisme," &c. 8vo. *Paris,* 1830
Mémoire sur la vraie Constitution de l'Atmosphère terrestre, déduite de l'expé-
rience, &c. 1841

† Mélanges scientifiques et littéraires. 3 vols. 8vo. *Paris,* 1858
Cenno Biografico dell' illustre G. B. Biot, redatto dal Professore P. Volpicelli,
4to. 4 pp. (*Atti dell' Acad. de' Nuovi Lincei Ann.* xv. Sess. iv. del 2 Marzo,
1862.) *Roma,* 1862
Vide also Gay-Lussac and Biot.
„ Humboldt and Biot, and Anon. Elect. 1821.

† **Biot and Becquerel.** Sur la Nature de la Radiation émanée de l'Etincelle
électrique qui excite la phosphorescence à distance. 4to. 5 pp. (*Comptes*
Rendus, 1839, 1 sem. viii. No. 7, p. 223.) *Paris,* 1839
Vide also Becquerel Biot, &c.

Biot et Cuvier.
　　Sur quelques Propriétés de l'Appareil galvanique. 4to. (Soc. Philomath. an 9.
　　p. 40. Sue, Hist. ii. 161, says : "Voyez aussi les Ann. de Chim. xxxix.
　　247. Elles contiennent une édition tout à fait différente de cet article.")
　　　　　　　　　　　　　　　　　　　　　　　　　　　　　　Paris, 1801

Biot, Edouard Constant. Born July 2, 1803, at Paris; died March 12, 1850, at
　　Paris.
†　Catalogue général des Etoiles filantes et des autres Météores observés en Chine
　　pendant vingt-quatre siècles. Depuis le viiᵉ siècle avant J.C., jusqu'au milieu
　　du xvii. de notre ère. Présenté à l'Acad. les 31 Mai et 26 Juillet, 1841. 4to.
　　224 pp. 1 plate (ou tableau circulaire.) (Acad. des Sciences, Savants étrangers,
　　tom. x.)   Paris, 1841 ?
†　Note supplémentaire au Catalogue des Etoiles filantes et des autres Météores
　　observés en Chine. 4to. 8 pp.   Paris, 1848

Birch.　Observations on Medical Electricity. 8vo. 2 editions.   1779-80
†　Letter to Adams on Medical Electricity. 8vo. Vide also Adams.   1792

† Birch, J.　Della Forza dell' Elettricità nella cura della soppressione . . . 4to.
　　　　　　　　　　　　　　　　　　　　　　　　　　　　　Napoli, 1778
　　(Forms part of Vivenzio's translation of Cavallo, &c.)
†　Considerations on the efficacy of Electricity in removing Obstructions with
　　ease. 8vo. 60 pp. (Volta's copy.)   London, 1779
†　Essay on the Medical Application of Electricity. 8vo. 57 pp.   London, 1803
　　(Formerly published in Adams's Electricity, now revised and reprinted by
　　itself.)

Birch, Thos.　On the luminousness of Electricity. 4to. (Phil. Trans. 1754.)
　　　　　　　　　　　　　　　　　　　　　　　　　　　　London, 1754

† Bird, G. (Captain).　Fall of a Meteoric Stone, 18th February, 1815, near the village
　　of Dooralla (in the territory belonging to the Pattialah Rajah,) India. 8vo.
　　2 pp. (Phil. Mag. lvi. 156.)   London, 1820

Bird, Golding.　Born 1815, in Norfolk.
†　Lectures on Electricity and Galvanism, in their physiological and therapeutical
　　relations, delivered at the Royal College of Physicians. Revised and extended.
　　12mo. 212 pp. Cuts.   London, 1849
　　On Medical Electricity. (Med. Gaz., June, 1847 ; and Guy's Hospital Reports,
　　1852, viii. pp. 139-46.)

† Birt, W. R.　Report on the Discussion of the Electric Observations at Kew. 8vo.
　　87 pp. 6 plates. (Report of the British Association for 1849.) London, 1850

† Birnbaum, H.　Das Reich der Wolken.   Leipzic, 1859

Bischof.　Vide Nöggerath and Bischof.

Bischoff.　Das Volta-Elektrometer (of water, &c.) "als Maass für die Thätigkeit
　　der Voltaischen Säule." (Physikisches Wörterbuch, 2 Abtheil. vol. iv.
　　p. 885.)

Bischoff, C. H. Ern.　Commentatio de usu Galvanismi in Arte Medica, speciatim
　　in morbis Nervorum Paraliticis. 8vo. 2 plates. (From Sue, Hist. iii. 67.)
　　　　　　　　　　　　　　　　　　　　　　　　　　　　　　Jena, 1801

Bischof, C. G. C.　Ueber die Magnetische Eigenschaft. einiger Gebirgsarten des
　　Fichtelgebirgs. 8vo. (Schweigger's Journal, xviii.)   Nürnberg, 1816
　　Untersuch über die Elektrometrische und Chemische Wirkungen der Voltaischen
　　Säule. 8vo. (Kästner's Archives, iv.)   Nürnberg, 1825
　　Elektrometrische Wirkungen der Voltaische Säule und Leitungsfähigkeit des
　　Glasses. 8vo. (Schweigger's Journal, xxxvi.)   Nürnberg, 1822
　　Beitrage zum Elektro-magnetism. 8vo. (Kästner's Archives, i.)
　　　　　　　　　　　　　　　　　　　　　　　　　　　　Nürnberg, 1824

Bischof—*continued.*

Ueber Volta's Fundamentalversuch. 8vo. (*Poggendorff's Annalen*, i.)
Leipzig, 1824

Ueber die Bitburger Meteoreisen. 8vo. (*Poggendorff's Annalen*, ii.)
Leipzig, 1824

Verfahren die Elektro-chemische Reihe der Metallen zu bestimmen. 8vo.
(*Schweigger's Journal*, lii.) *Nürnberg*, 1827

† **Bismark,** Graf Von. Telegraphen-Ordnung für die Correspondenz im Deutsch-
Oesterreichischen Telegraph-Verein, &c. Nebst den den innern Verkehr auf
den Preussischen Linien betreffenden zusatzlichen Bestimmungen. Reglement
für die Benutzung der Preussischen Eisenbahn-Telegraphen zur Beförderung
solcher Depeschen, welche *nicht* den Eisenbahndienst betreffen. Nebst den
übrigen Bestimmungen für den Expeditionsdienst im Telegraphischen Verkehr,
so weit solcher durch Preussische Staats und Preussische Eisenbahn-Telegraphen vermittelt wird. 8vo. 30 pp. *Berlin*, 1866

† Reglement über die Benutzung der innerhalb des Nord-deutschen Telegraphen-
Gebietes gelegenen Eisenbahn-Telegraphen zur Beförderung solcher Depeschen,
welche nicht den Eisenbahn-dienst betreffen. 8vo. 15 pp. *Berlin*, 1868

† Telegraphen-Ordnung für die Correspondenz auf den Linien des Telegraphen-
Vereins, &c. Nebst den den innern Verkehr auf den Linien des Nord-deutschen
Telegraphen-Gebietes und der innerhalb desselben gelegenen Eisenbahnen
betreffenden zusätzlichen Bestimmungen. Dezember, 1868. 8vo. 19 pp.
Berlin, 1868

Bizio, Bartolommeo. *Born at Costozza, near Venice; died September* 27, 1862, *at
Venice.*

Opuscoli Chimico-fisici. 8vo. tom. i. *Venezia*, 1827

† Intorno alle Molecole dei Corpi ed alle loro affinità dipendenti dalla Forza
repulsiva insita alle Medesime. 4to. 120 pp. (*Nel* i. *Mem. dell' Imp. Regio
Istit. Veneto di Scienze, Lettere ed Arti.*) *Venezia*, 1843

† Osservazioni intorno ad una condizionata particolarità della Grandine. 4to.
6 pp. (*Mem. dell' I. R. Istit. Veneto*, vi. part ii. p. 341.) *Venezia*, 1856

† **Bjerkander,** Clas. Auszug aus den zwanzigjährigen Witterungsbeobachtungen
zu Seara welche zeigen wie oft die Donnerwetter gewesen sind. 8vo. 3 pp.
(*Schwed. Akad. Abh.* xxxvii. 184.) *Leipzig*, 1775

† Fernere Anmerkungen über Donnerwetter. 8vo. 3 pp. (*K. Schwed. Akad. Abh.*
xli. 220.) *Leipzig*, 1779

Bjorn, Hans Outzen. *Born Dec.* 14, 1777, *at Astrup, Bröns Sogn, Torningslehn;
died May* 15, 1843, *at Frydenlund.*

De indole et origine Aërolithorum. 8vo. 88 pp. (*Heinsius*, vi. 85.) *Altona*, 1816
(*Contains many citations of books on the subject.*)

† **Black,** Jn. An attempt to establish a pure System of Mineralogy by the application
of the Electro-chemical Theory, by T. J. Berzelius. Translated from the
Swedish original. 8vo. 138 pp. (*Brugnatelli's copy.*) (*Vide* Berzelius.)
London, 1814

† **Blackwall,** J. On luminous Arches of the Aurora Borealis seen at Manchester
on the 1st and 26th December, 1828 ; and on that of the 29th of September, as
seen at Dublin. 8vo. 2 pp. (*Phil. Mag. or Annals*, v. 153.) *London*, 1829

† **Bladth,** P. J. Berattelse om tvänne aske-slag som traffat Svenska skeppet Stock-
holms-Slott, i Ost-Indien, an 1777. (*Vetensk Acad. Nya Handl.* 1780, p. 102.)
1780

Bericht von zwei Blitz-Schlägen, welche das Schwedische Schiff Stockholms-
Schloss in Ost-Indien, 1777, getroffen haben. 8vo. 14 pp. (*Neue Schwedische
Akademie Abhandlung*, i. 1780, p. 97.) Translation. *Leipzig*, 1780

Blainville. On Organs of the Torpedo. (From *Matteucci and Savi.*)

Blagden, Sir Ch. An Account of some Fiery Meteors. 4to. (*Phil. Trans.* lxxiv. part i. 1784.) *London,* 1784
(*Toaldo, in vol.* iv. *p.* 283, *of the* " *Completa Raccolta,*" *refers to this volume as* "*containing six articles on* '*Fiamma Volante,*' *Nos.* 8, 9, 10, 11, *and* 12, *containing various descriptions, &c., from different places concerning the globe* (*fire-ball*) *of the* 18*th of August,* 1783." *The last description is by Blagden, who thought fire-balls were electrical bodies.*)

Blanc. *Vide* Le Blanc.

Blanford. *Vide* Gastrell and Blanford.

† **Blaserna,** Pietro. Ueber den inducirten Strom der Nebenbatterie. (Vorgelegt 22 Juli, 1858.) 8vo. 46 pp. (*Aus den Sitzungsberichten* 1858 *der Mathem. Naturwiss. Classe der Kais. Akad. der Wissenschaften,* xxxiii. No. 24.)
Wien, 1858
Ueber den inducirten Strom der Nebenbatterie. 8vo. 10 pp. (*Aus den Sitzungsberichten* 1859 *der Mathem. Naturwiss. Classe der K. Akad. der Wissenschaften.*)
Wien, 1859
† Sullo sviluppo e la durata delle correnti d'induzione, e delle estra-correnti. Prima Memoria. 4to. 135 pp. 5 plates. (*Estratto dal Giornale delle Scienze Naturali ed Economiche,* vi. 1870.)
Palermo, 1870

† **Blaserna, Mach, und Peterin.** Ueber elektrische Entladung und Induction. (Vorgelegt 14 Juli, 1859.) 8vo. 50 pp. (*Aus den Sitzungsberichten* 1859, *der Kais. Akad. der Wissenschaften,* xxxvii. No. 20.) *Wien,* 1859

† **Blavier,** E. E. Cours théorique et pratique de Télégraphie Electrique, par E. E. Blavier, Inspecteur des Lignes Télégraphiques. 12mo. 467 pp. 6 plates.
Paris, 1857
Propagation de l'Electricité. Note sur la Réponse de M. Guillemin aux Observations de M. Gounelle. 8vo. 45 pp. *Nancy,* 1865
† Nouveau Traité de Télégraphie Electrique. Cours théorique et pratique à l'usage des Fonctionnaires, des Ingénieurs, &c. Tom. i. 8vo. 476 pp.
Paris, 1865
† Nouveau Traité de Télégraphie Electrique. Cours théorique et pratique à l'usage des Fonctionnaires, des Ingénieurs, &c. Tom. ii. 8vo. 479 pp.
Paris, 1867

Blesson. John Ludwig Urban. *Died Jan.* 19-20, 1861, *at Berlin.*
Magnetismus und Polarität der Thon-Eisensteine und deren Lagerstätte in Oberschlesien und den baltischen Ländern. 8vo. *Berlin,* 1816
† Ueber eine Verbesserung an Blitz-ableitern. 4to. 4 pp. 1 plate. *Berlin,* 1831

Bloch. Naturgeschichte der ausländischen Fische. Th. ii. 4to. *Berlin,* 1786

† **Block,** F. T. Dissertatio Physica altera de Electricitate. Thesis read 3rd February, 1753. 4to. 32 pp. *Duisburgi ad Rhenum,* 1753 ?

Blois. Bibliothèque Publique. Catalogue. Fol. MS. *Blois, n.d.*

Blodget. *Vide* Whipple, Blodget, &c.

Blondeau. Mémoires sur les Variations de l'Intensité magnétique ; sur les Rapports du Magnétisme avec l'Electricité, avec les différents Etats de l'Atmosphère, &c. (*Mém. de l'Acad. de Brest,* i. 421.) *Brest*
Mémoire sur l'Effet de deux Aiguilles aimantées placées l'une sur l'autre, lorsque librement suspendues elles se trouvent dans leur sphère d'activité réciproque, à peu près dans le même plan horizontal. (*Mém. de Brest,* i. 401.)
Brest

Blumenbach, Johann Friedrich. *Born May* 11, 1752, *at Gotha; died Jan.* 22, 1840, *at Göttingen.*
Institutiones Physiologiæ. 8vo. Ed. nova. *Göttingen,* 1798
Nachricht über Meteorsteine. 8vo. (*Voight's Mag.* vii. und *Gilbert's Ann.* xviii.) *Jena,* 1804

† **Blundell, W.** Telegraphic Companies considered as Investments, with Remarks on the superior Advantages of Submarine Cables. 8vo. 22 pp. *London*, 1869

† The Manual of Submarine Telegraphic Companies. 8vo. 64 pp. *London*, 1871

† **Bobierre, A.** Thèse de Physique. Des Phénomènes Electro-chimiques qui caractérisent l'Altération à la Mer, des Alliages employés pour doubler les Navires. 4to. 79 pp. 1 plate. *Nantes*, 1858

Boccone, P. Osservazioni Naturali. 12mo. (*Fosforo di Bologna*, and *Iron Magnetic Rings*, &c.) (*From Libri's Sale Cat.* 1861.) *Bologna*, 1684

Bock, F. S. Versuch einer kurzen Naturgeschichte des Preussischen Bernsteins, und einer neuen wahrscheinlichen Erklärung seines Ursprungs.
Königsberg, 1767

Böck, J. G. Beitrage zur Anwendung der Elektricität auf den menschlichen Körper. 8vo. *Erlangen*, 1791

Boeckel, Th. De l'Ozone comme Elément Météorologique. 8vo. 20 pp. (*Ext. de la Gazette Médicale de Strasbourg.*) *Strasbourg*, 1865

Bockmann. Ueber Blitzableiter. Eine Abhandlung auf höchsten Befehl bearbeitet. Neue Auflage von Wucherer. *Karlsruhe*, 1830

Ueber Blitzableiter. 3 Auflage von G. F. Wucherer. 8vo. *Carlsruhe*, 1839

Boeckmann, Johann Lorenz. *Born May* 8, 1741, *at Lubeck; died Dec.* 15, 1802, *at Carlsruhe.*

Description de l'Aurore Boréale. Extrait du 28 Juillet, 1780. 4to. (*Mém. Berlin*, 1780.) *Berlin*, 1780

Abhandlung über eine ganz neue Erscheinung an den sogenannten Glas-Bomben, nebst einer Anwendung auf die Entstehung gefrorner Fenster-Scheiben und einem Anhang von electrischen Sternen. 4to. (*Neue Abhandl. der Bairischen Akad. Philos.* iii. p. 1.) *München*, 1783

Abhandlung über die Anwendung der Electricität, bei Krankheiten. 8vo. *Durlach*, 1786

Beschreibung des Apparats für Luft-elektricität. 8vo. (*Gren's Journal*, i. 1790.) *Leipzig*, 1790

Beschreibung einiger neuen Werkzeuge zum Bestimmen der kleinsten Grade der Elektricität. 8vo. (*Gren's Journal*, i. 1790.) *Leipzig*, 1790

† Ueber die Blitzableiter. Eine Abhand. auf höchsten Befehl des Fürsten. 8vo. 80 pp. *Carlsruhe*, 1791

Versuch über Telegraphie und Telegraphen. 8vo. *Carlsruhe*, 1794

Bodde, Johann Bernhard. *Born Nov.* 24, 1760, *at Lette Herrschafs, Rheda; died July* 24, 1833, *at Münster.*

† Grundzüge zur Theorie der Blitzableiter. 8vo. 84 pp. *Münster*, 1809

Boddington. An accurate Statement of Facts relative to a Stroke of Lightning which happened on the 13th April, 1832. 8vo. *London*, 1832

Bodleian Library. Catalogue. MSS. *Oxford*
(*Collated partially by Sir Francis Ronalds in* 1847.)

Catalogus Librorum impress. Bibliot. Bodleianæ. 3 vols. fol. *Oxford*, 1843

Catalogue of Books added to the Bodleian Library in each of the years 1826 to 1836. Fol. *Oxford*

Boehm, J. Untersuchungen über das atmosphärische Ozon. 8vo. 34 pp. 1 plate. (*Aus den Sitzungsberichten* 1858, *der Kais. Akad. der Wissenschaften.*) *Wien*, 1858

Boerhaave. Essay on the Virtue of Magnetical Cures. From the Latin. 8vo. *London*, 1743

† **Boggio, Ig.** Indoratura ed inargentatura Elettrico-chimica. 8vo. 59 pp, *Novara*, 1848

66   BOH—BOI

† **Bohadsch,** Johann Baptist.   *Born* 1724, *at Prag; died Oct.* 16, 1768, *at Prag.*
    Dissertatio de utilitate Electrisationis in Arte Medica.   4to. 24 pp.   (*Volta's*
    copy.)   *Pragæ,* 1751
Bohadtch.   Treatise on Medical Electricity.   4to.   (*Phil. Trans.* xlvii. p. 351.)
                                                                    London
Böhm, Joseph George.   *Born March* 27, 1807, *at Rozdialowitz Böhmen, Kr.*
    Bunzlau.
Bohm and Kunes.   Magnetische und Meteorologische Beobachtungen zu Prag.
    Auf öffentliche Kosten herausgegeben von Jos. G. Böhm und Adalbert Kunes.
    Jahrgänge vom 1 Jan. 1850, bis 31 Dec. 1852. 3 vols. 4to.   *Prague,* 1853-55
Bohm.   (*Vide* Kriel and others.)
Bohnenberger, Gottlieb Christian.   *Born March* 4, 1732, *at Neuenberg, Wurtem-*
    berg; died May 29, 1807, *at Altburg.*
†   Beschreibung einer . . . Electrisir Maschine nebst einer neuen Erfindung, die
    Electrische Flaschen betreffen.   8vo. 76 pp. 6 plates.   (*Volta's copy.*)
                                                                    Stutgart, 1784
†   Fortgesetze Beschreibung, einer Electrisir Maschine von ganz neuer Erfindung·
    8vo. 110 pp. 6 plates.   (*Volta's copy.*)   *Stutgart,* 1786
†   Beschreibung einer Elektrisir Maschine und einiger neuen Versuche. 8vo.
    138 pp. 4 plates.   Zweite Fortsetzung.   (*Volta's copy.*)   *Stutgart,* 1786
†   Beschreibung einiger Elektrischer Maschinen und Elektrischer Versuche. 8vo.
    224 pp. 5 plates.   Dritte Fortsetzung.   (*Volta's copy.*)   *Stutgart,* 1788
†   Beschreibung einiger Elektrischer Maschinen und Elektrischer Versuche. 8vo.
    252 pp. 4 plates.   Vierte Fortsetzung.   *Stutgart,* 1789
†   Beschreibung einiger Elektrisir Maschinen und Elektrischer Versuche. 8vo.
    333 pp. 5 plates.   Fünfte Fortsetzung.   (*Volta's copy.*)   *Stutgart,* 1790
    Neue Gedanken über die Möglichkeit Elektrische Verstärkungs-flaschen weit
    stärker als bisher zu laden. 8vo.   (*Gren's Journ.* ii. 1790.)   1790
†   Beschreibung einiger Elektrischer Maschinen und Elektrischer Werkzeuge.
    8vo. 352 pp. 4 plates.   Sechste und lezte Fortsetzung.   (*Volta's copy.*)
                                                                    Stutgart, 1791
†   Beyträge zur theoretischen und praktischen Elektrizitätst-lehre. 8vo. 154 pp.
    Erste Stück.   *Stutgart,* 1793
†   Beyträge zur theoretischen und praktischen Elektricitäts-lehre. 8vo. 178 pp.
    1 plate. Zweites Stück.   *Stutgart,* 1793
†   Beyträge zur theoretischen und praktischen Elektricitäts-lehre. 8vo. 166 pp.
    2 plates. Drittes Stück.   *Stutgart,* 1794
†   Beyträge zur theoretischen und praktischen Elektricitäts-lehre.   Viertes Stück.
    8vo. 183 pp. 1 plate.   *Stutgant,* 1795
    Beyträge zur theoretischen und praktischen Elektricitäts-lehre.   Fünfte Stück.
                                                                    1795
†   Beschreibung unterschiedlicher Elektrizitätsverdoppler von einer neuen Ein-
    richtung nebst. . . . Versuchen. 8vo. 271 pp. 5 plates.   *Stutgart,* 1798
    Beschreibung und Gebrauch eines sehr empfindlichen Elektrometers, welches
    zugleich die Art der Elektricität anzeigt. 8vo. (*Gilb. Ann.* li.)   1815
Boily.   Iconographie de l'Institut Royal de France. . . . Portraits des Membres
    depuis 1814 jusq'en 1825. Lithog. par Boily. 4to. 190 portraits.   *Paris*
† **Bois,** Victor.   La Télégraphie Electrique. 1st edition. 8vo. 114 pp.   (*Forms part*
    of the Bibliothèque des Chemins de Fer.)   *Paris,* 1853
†   La Télégraphie Electrique. 2nd edition. 12mo. 128 pp.   *Paris,* 1855 *or* 1856
Boismont.   (*Vide* Brierre de Boismont.)
† **Boisbouvray,** Baron de.   Un mot sur l'Electricité. 8vo. 71 pp.   *Paris,* 1823
† **Boisgerard,** Jun.   On the Action of the Voltaic Pile upon the Magnetic Needle.
    Read in the Academy 9th November, 1820. 8vo. 4 pp. (*Phil. Mag.* lvii. 203.)
                                                                    London, 1821

Bois Reymond. (*Vide* Du Bois-Reymond.)

Boissac. (*Vide* Marqfoy and Boissac.)

† **Boissier,** H. Mémoire sur la Décomposition de l'Eau par les Substances métalliques. 4to. 7 pp. *Paris,* 1801

Bollenatus Burgundo-gallus. Theses physicæ de Meteoris ignitis. 4to. 8 pp.
1607

Bolley, Pompejus. *Born May* 7, 1812, *at Zindelberg.*
† Was ist ein elektrischer Telegraph. 8vo. Cuts. (*Aus dem Schweizerischen
Gewerbeblatt.*) *Aarau,* 1852

Bollinger, A. F. De effectu catalytico Rivi galvanici constantis. *Berol.* 1863

† **Bologna Academy.** Notizia del Nuovo Istituto dell' Accad. di Bologna, di
Corazza. 8vo. 207 pp. *Bologna,* 1780

Memoria storica sopra l'Università e l'Istituto delle Scienze di Bologna ; di
Mazzetti. 8vo. *Bologna,* 1840

1. Acad. Bonon. et Instit. Commentarii. 7 vols. (in 10 parts). 4to.
1731 *ad* 1791
2. Mem. dell' Istit. Nazionale Ital. 3 vols. (in 6 parts). 4to. 1806 *ad* 1813
3. Comment. Novi Acad. Scient. Bonon. 10 vols. 4to. 1834 *ad* 1850
4. Mem. dell' Accad. dell' Istit. di Bologna. 4to. 1850
(*Part of these are in the Library.*)

Rendiconto delle Sessioni dell' Accad. . . . dell' Istituto di Bologna. 8vo.
Bologna

† Programma del Premio Aldini. 8vo. 1 p. (*Giornale I. R. Ist. Lombardo,*
viii. 281.) *Milano,* 1843

† Programma dell' Accad. dell' Istituto di Bologna. 8vo. (*Giornale I. R. Ist.
Lombardo,* xiii. 408.) 1846
(*Not having conferred " il Premio Aldini sul Galvanismo" hitherto, the
Academy re-proposes the subject in new terms.*)

Programma dell' Accad. dell' Istituto di Bologna. Concorso al Premio Aldini
sul Galvanismo. 4to. (*Giornale I. R. Ist. Lombardo, nuova serie,* viii. 352.)
1856
(*Reference is made to the Academic premium given to Grimelli for a
memoir, " Sulla così detta Corrente della Rana," &c. in tom. x. of the
Nuovi Commentarii, and the Academy proposed, as a continuation for
1854, another Tema on the " Corrente Muscolare," and concerning Du
Bois-Reymond's theory.*)

Bologna Medical Society. Memorie della Società Medica di Bologna.
Bologna

Bologna Società Medico-Chirurgica (Memorie della). Seguito degli Opuscoli
da essa pubblicati. 4to. *Bologna*

Bologna Nuovi Annali delle Scienze Naturali. Dai Signori Alessandrini,
Bertolini, Gherardi, e Ranzani. 30 vols. 8vo. *Bologna,* 1834-54

Boltzmann, Ludwig. Ueber die Bewegung der Elektricität in krummen Flächen.
8vo. 8 pp. (*Zuchold's Cat.* 1866-67. *Aus den Sitzungsberichten der Kaiserl.
Akademie der Wissenschaften abgedruckt.*) *Wien,* 1865

† Ueber die elektrodynamische Wechselwirkung der Theile eines elektrischen
Stromes von verändlicher Gestalt. 8vo. 19 pp. 1 plate. (*Aus dem lx. Bde.
der Sitzungsberichten der Kaiserl. Akad. der Wissenschaften, Abth.* ii. *Juni
Heft,* 1869.) *Wien,* 1869

† **Bombay Magnetic Observatory.** Electrometer Observations (made by means
of F. Ronalds's Atmospheric Apparatus). Fol. 1 p. *Bombay,* 1850

Observations for 1847, under the superintendence of C. W. Montriou. Part ii.
4to. *Bombay,* 1851

Bombay Magnetic Observatory—*continued.*

Observations (Magnetical and Meteorological) made at the Hon. East India Company's Observatory at Bombay, in the year 1849. 4to. *Bombay,* 1852

Observations (Magnetical and Meteorological) made at the Hon. East India Company's Observatory, Bombay, under the superintendence of E. F. T. Ferguson, for the years 1854 and 1855. 4to. *Bombay,* 1855-56

Vide also Orlebar.

† **Bompass,** Charles C. An Essay on the nature of Heat, Light, and Electricity. 8vo. 266 pp. *London,* 1817

Bona e Corner. Relazione dell' andamento ed effetti del Fulmine che colpì il Campanile . . . di S. Francesco della Vigna (in Venezia), l'anno 1780, 24 Maggio. 4to. 8 pp. *Venezia,* 1780

(*This title is abridged from that of an official report made by Bona and Corner, officers of Artillery.*)

Bonaparte, C. L., Principe di Musignano. Iconografia della Fauna Italica. Fasc. 12. *Roma,* 1835

Bonaparte, Louis Charles Napoléon. *Born April 20, 1808, at Paris; died Jan. 9, 1873, at Chislehurst, Kent.*

Sur la Théorie de la Pile Voltaique. 4to. (*Comptes Rendus,* xvi. 1843.) *Paris,* 1843

† Décret qui institue un Prix en faveur de l'Auteur des découvertes qui rendront la Pile de Volta applicable avec économie, soit à l'Industrie, comme source de chaleur, soit à l'Eclairage, soit à la Chimie, soit à la Mécanique, soit à la Médecine pratique. 8vo. 2 pp. *Paris,* 1852

(*This article is contained in the "Bulletin des Lois de la République Française," No. 497, p. 554. The decree is dated 23d February, 1852.*)

Bonaparte, Premier Consul. Lettre au Ministre de l'Intérieur qui l'a transmise à . . . l'Institut. Ecrite le 26 Prairial, An. x. *Paris,* 1802

(*The annual galvanic prize medal of 3,000 francs value, and 60,000 francs as "encouragement" for an advance in electricity and galvanism comparable to that made by Franklin and Volta.*)

Bond. Apparatus for Astronomical Observations.

(*From Moigno. No particulars. "Présenté récemment à Ipswich. Employed at the Greenwich Observatory,* 1852.")

† **Bond,** B. R. The Hand-Book of the Telegraph ; a Comprehensive Guide to Telegraphy, Telegraph Clerks' Remembrancer, and Guide to Candidates for Employment in the Telegraph Service. 12mo. 68 pp. *London,* 1864

Bond, Henry. On Two Magnetic Poles. 1646 (?)

(*Extract from page 3 of his work entitled "The Longitude Found," and refers to his statement in the Seaman's Kalender about 30 years since, i. e. in about 1646. The above title is probably not exact.*)

The Variations of the Magnetic Needle predicted for many years following. 4to. (*Phil. Trans.* 1668.) *London,* 1668

The Undertakings of Mr. H. Bond, a famous teacher of the art of navigation, in London, concerning the Variation of the Magnetical Compass and the Inclination of the Inclinatory Needle. 4to. (*Phil. Trans.* 1672.) *London,* 1672

On the Variation of the Magnetical Compass and the Inclinatory Needle ; as the result and conclusion of 38 years' magnetical study. 4to. (*Phil. Trans.* an. 1673, p. 6065.) *London,* 1673

† On the Inclination of the Inclinatory Needle.

(*Churchman's Introduction,* p. 10, *says, "A small treatise in the time of Charles the Second, and dedicated to him."*)

Bond. (*Vide* Lovering and Bond.)

† **Bond,** Henry, Sen. The Longitude Found ; or a Treatise shewing an Easie and Speedy way, as well by Night as by Day, to find the Longitude, having but the Latitude of the Place, and the Inclination of the Magnetical Inclinatorie Needle. Printed by the King's Majesties Special Command. sm. 4to. Dedication to Charles the Second, 3 pp. Epistle to the Reader, &c. 4 pp. The King's Licence, &c. 3 pp. Text, 65 pp. 7 woodcuts, each on a separate leaf. (*Scarce.*) 1676

Bond, W. C. Description of the Electro-recording Apparatus in the Astronomical Observatory of Harvard College. Plate. (*History and Description of the Astronomical Observatory. Annals of ditto*, vol. i. part i. p. xlix.-lv.) 1856

Bondioli, Pietro Antonio. *Born* 1765 *at Corfu; died September* 16, 1808, *at Bologna.*

† Sopra l'Aurora Boreale. Memoria letta nell' Accad. di Padova 15 Dec. 1790. 8vo. 25 pp. (*Volta's copy.*) 1790(?)

† Sopra l'Aurora Boreale. Memoria. 8vo. 11 pp. (*Giornale Fisico-Medico di Brugnatelli*, i. 55. *Letta nell' Accad. di Padova*, 1790.) Pavia, 1792 (*This is a reprint of the above.*)

† Sopra le Aurore Boreali locali. Memoria. Presentata 19 Otto. 1801. 4to. 16 pp. (*Mem. Soc. Ital.* ix. 422.) Modena, 1802

† Elogio di Pietro Antonio Bondioli scritto da M. Pieri. Inserito nel tom. xv. della Soc. Ital. Seconda ed. 8vo. Treviso, 1812

† **Bonel,** A. Histoire de la Télégraphie. Description des principaux Appareils aériens et électriques. 12mo. 147 pp. plates. Paris, 1857

Bonelli, G.
Vide Du Moncel.

† Du Télégraphe des Locomotives de G. Bonelli. Système destiné à prévenir les collisions sur les Chemins de Fer. 8vo. 16 pp. Paris, 1856

† **Bonetti,** E. Di alcuni effetti dell' Elettrico sopra l'animale economia di G. Namias. 8vo. 5 pp. (*Giornale dell' I. R. Istit. Lombardo*, tom. iv. p. 214.) Milano, 1842

Bonifas. (*Vide* Ducala-Bonifas, M.)

Bonijol, L. Verfertigte, unter anderen Instrumenten auch eins, mittelst dessen, durch atmosphäriche Elektricität chemische Zersetzungen bewerkt werden. 8vo. (*Bibl. Univers.* xlv.) Geneva, 1830

† **Bonnefoy.** De l'application de l'Electricité à l'art de guérir. Dissertation inaugurale . . . Collége Royal de Chirurgie à Lyon. Read 30th November, 1782. 8vo. 163 pp. Lyon, 1782 (*There is also a German version.*)

De l'application de l'Electricité. Lyon, 1783

† **Bonnejoy.** Les Moyens pratiques de constater la mort par l'Electricité à l'aide de la Faradisation. 8vo. 32 pp. Paris, 1866

Bonnet, Charles. *Born March* 13, 1720, *at Genf. ; died May* 20, 1793, *at Genthod b. Genf.*

Contemplazione della Natura. Venezia, 1797

(*Elice Saggio, 2nd edition, p.* 29, *says that Bonnet "scriveva nell' anno* 1781, *che l'elettricità agisce sui nervi e sui muscoli . . . e che si riproducono in tal guisa le vibrazioni del cuore di una rana anche tre giorni dopo che é stato staccato dal di lei petto."* Tom. i. p. 169. *Di Bonnet.*)

† **Bonnycastle,** C. On the Distribution of the Magnetic Fluids in masses of Iron ; and on the Deviations which they produce in Compasses placed within their influence. 8vo. 11 pp. 1 plate. (*Phil. Mag.* lv. 446.) London, 1820

Bonnycastle, C., and another. On Mr. Bonnycastle's Dissertation on the Influence of masses of Iron on the Mariner's Compass, published in our 55th volume. 8vo. 5 pp. (*Phil. Mag.* lvi. 346.) *London,* 1820

Bonvicini, J. De Electrico Fluido in naturam agente. Thesis. 4to. *Venezia,* 1791

Bonzi. (*Vide* Beccari and Bonzi.)

† **Boquillon.** De l'Electrotypie, A. M. Quesneville, directeur. 8vo. 48 pp. (*Extrait de la Revue Scientifique et Industrielle du Dr.* Quesneville, tom. xi. 1842.) *Paris*

Borch, Michael Johann, Graf. *Born June* 30, 1753, *at* (*N. St.*) *Warkland; died January* 10, 1811, *at* (*N. St.*) *Warkland.*

† Memoria sopra il Fosforo marino. 4to. 8 pp. (*Atti dell' Accad. di Siena,* vi. 1781, p. 317.) *Siena,* 1781

Borchers. Anwendung eines kräftigen Magnets zur Ermittelung der Durchschlagsrichtung zweier Gegenorter. *Clausthal,* 1864

Borde, Jean Baptiste de la. (*Vide* La Borde.)

Bordé, F. Trattato elementare di Elettricità accomodato all' intelligenza comune. 8vo. 174 pp. 1 plate. *Modena,* 1843

Bordeaux Academy. Recueil des Dissertations qui ont remporté le prix de l'Académie Royale. 6 vols. 12mo. *Bordeaux,* 1715-39 (*This recueil contains three dissertations only on electricity, and they are printed each with a separate title-page, and are sold separately. Vide Lozeran-du-Fech,* 1726 ; *Sarrabat,* 1727 ; *and Quelmalz,* 1753.)

Recueil des Dissertations qui ont remporté le prix de l'Académie Royale. 3 vols. 4to. *Bordeaux,* 1741-79 (*This recueil contains three dissertations on electricity, and they are printed and sold separately. Vide Desagulier,* 1742 ; *Beraud,* 1748 ; *and Barberet,* 1750.)

Recueil des Ouvrages du Musée de Bordeaux. 1 vol. 8vo. 428 pp. *Bordeaux,* 1787

Séances de l'Académie Royale des Sciences, Belles Lettres, et Arts de Bordeaux. 8vo. Plates. *Bordeaux,* 1837

Recueil des Actes de l'Académie Royale de Bordeaux. Première Année, printed 1839, containing the "Actes" of 1838. *Bordeaux,* 1839-

Catalogue of the Library of the Imperial Academy. MS. fol. *Bordeaux, n. d.*

Bordeaux Chambre de Commerce. Catalogue des Livres composant la Bibliothèque de la Chambre de Commerce de Bordeaux, avec Supplément. 8vo. *Bordeaux,* 1852

Premier Supplément au Catalogue de la Bibliothèque de la Chambre de Commerce de Bordeaux, année 1855. Ouvrages reçus du 1 Jan. au 31 Déc. Fol. MSS. *Bordeaux,* 1855

Catalogue, containing additions. MSS. fol. *Bordeaux*

Catalogue of the Library of the Chamber of Commerce. No title. Unfinished. On cards. MSS. *Bordeaux*

Bordeaux Public Library. Catalogue. 8vo. *Bordeaux,* 1830
Supplément. 8vo. *Bordeaux,* 1847

† **Bordier,** Marcet. Notice d'un Fanal. 32 pp. 1 plate. *Paris,* 1823

† **Borelli,** G. Sull' Applicazione dell' Elettricità alla Navigazione. 8vo. 15 pp. 1 plate. *Torino,* 1855

Borenius, H. G. Om de framsteg som kunskapen om Jordmagnetismen gjort i synnerhet under de sednaste decennierna. 4to. 27 pp. (*Föredr. pa Finska Vetenskaps-Societetens årsdag den* 29 *April,* 1851. *Act. Soc. Fenn.* iii. 1852.) *Helsinfors,* 1852

Borgo, C. Vissuto dopo la metà del secolo, xviii. (*In his work, "Analisi ed Esame ragionato dell' arte della fortificatura delle piazze," is described his invention, called the Cifraparlante, which is exactly the telegraph (messo in voga) after him.*)

Born (Edlen von Born). (*Vide* Prag.)

Borough, William. A Discours on the Variation of the Compas or Magneticall Needle, wherein is mathematically shewed the maner of the observation, effectes, and application thereof, made by W. B. And is to be annexed to the new Attractive of R. N. 1581. *London,* 1581
   (*Note.—There is another edition, dated* 1585, *bound up with Norman,* 1585, *the preface bearing the same date as above.*)
   Vide Norman, Robert.

Borc, d'Antic Paul. *Born* 1726, *at Pierre Segude, Languedoc; died June,* 1784, *at Paris.*
   Mémoire sur la nature de la Matière électrique, et où l'on prouve que le verre n'est pas électrique par lui-même. 8vo. (*Mém. de l'Acad. de Dijon,* tom. ii. p. 29.) *Dijon,* 1774
   Œuvres contenant plusieurs Mémoires sur l'art de la Verrerie, sur la Faiencerie, la Poterie, l'art des Forges, la Minéralogie, l'Electricité, et la Médecine. 2 vols. 12mo. *Paris,* 1780

Boscha, J., jun. Het behoud van arbeids-vermogen in den galvanischen Stroom, Eene voorlezing voorgedragen in het Natuurkundig Gezelschapte Utrecht, 15 Januarij, 1858. Meet aanteckeningen. 8vo. 119 pp. *Leyden,* 1858

Boscovich, Ruggiero Giuseppe. *Born May* 18, 1711, *at Ragusa; died February* 15, 1787, *at Mailand. Poggendorf,* i. 246.)
   Dialoghi sull' Aurora Boreale. 4to. 47 pp. 1748(?)
   Sopra il Turbine. 8vo. *Roma,* 1749

Böse, C. H. Völlig entdecktes Geheimniss des Barth'schen Wetterparaskops. (*Reichs Anzeiger,* 1794.) 1794
   Vide also Albanus and Böse.

Böse, Georg Matthias. *Born September* 22, 1710, *at Leipzig; died September* 17, 1761, *at Magdeburg.*
   De attractione et electricitate, oratio inauguralis, quam in alma Matre Leucorea d. 17 Mart. 1738, Physicam ordinarie docturus habuit G. M. Böse. 4to. *Vitemburg,* 1738
   (*Note.—In his " Tentamina Electrica" this is called " Commentarius Primus Oratio inauguralis," &c. habita d. 17 March, 1738. It is not said to be the first Commentary in the title-page.*)
   Expériences sur l'Electricité. 4to. (*Mém. de Paris,* 1743, *Hist.* p. 45.) *Paris,* 1743
   De Electricitate. Commentarius ii. quo simul ad capessendos honores magisteriales, et lauream poeticam invitabatur, 3 Novembre. *Wittembergæ,* 1743
† Tentamina Electrica in Academiis Regalis Londinensi et Parisina primum habita, omni studio repetita quæ novis aliquot accessionibus locupletavit Geo. Mat. Böse. 4to. 96 pp. *Wittembergæ,* 1744
   (*This volume contains the Oratio inauguralis, or first Commentary, and the commencement of the second and third.*)
   De Electricitate inflammante et beatificante. Commentarius iii. Mai. 1744. *Wittembergæ,* 1744
   Die Elektricität nach ihrer Entdeckung und Fortgang mit poetischer Feder entworfen. 4to. *Wittenburg,* 1744
† Abstract of a Letter from Böse to De Maisan. Communicated by Baker from Ellis, and translated out of the Latin by Baker. Read May 23, 1745. 4to. 2 pp. (*Phil. Trans.* xliii. 419.) *London,* 1744-45
   Discours sur la lumière des Diamants, prononcé à Leipzig le 12 Mai, 1745. 4to. 12 pp. *Göttingen,* 1745
† Recherches sur la cause et sur la véritable théorie de l'Electricité. A Messieurs de la Société Royale de Londres. 4to. 56 pp. *Wittemberg,* 1745
   Recherches sur la cause et sur la véritable théorie de l'Electricité. 1746
   Tentamina Electrica hydraulicæ . . . utilia.Pars posterior. 4to. *Wittembcrg,* 1746

72   BOS—BOT

Böse, Georg Matthias—*continued.*

† Tentamina Electrica tandem aliquando hydraulicæ chymiæ et vegetabilibus
   utilia. Pars posterior. 4to. 47 pp.   *Wittembergæ*, 1747

   (*Note.—This volume contains his fourth and fifth Commentaries, the fourth
   written in 1746. Vide the 2nd page of his address.*)

† Extract of a Letter from G. M. Böse to W. Watson, on the Electricity of Glass
   that has been exposed to strong fires. Read April 6, 1749. 4to. (*Phil. Trans.*
   xlvi. 189.)   *London*, 1749-50

   Meteora heliaca, seu de maculis in sole deprehensis.   *Lipsiæ*, 1754

   L'Electricité, son origine et ses progrès. Poème traduit de l'Allemand par
   l'Abbé Joseph Ant. de C* * *. 12mo. 2 parts. 72 pp.   *Leipzig*, 1754

   (*This so-called translation is by Böse himself.*)

   Meteora heliaca seu de maculis in sole deprehensis.   *Lipsiæ* 1754

† * Sympathiam attractioni et gravitati substituit. . . . G. M. B. 4to. 20 pp.
                                                              Wittemburg

   De Terræ motus causis.   *Viteb.* 1756

   (*Vide* Anonymous, Electricity, 1744.)

† Bosellini, F. Spiegazione della Luce Boreale vedutasi il 16 Dec. 1737. 12mo. 27 pp.
   (*Callogera, Raccolta*, xvii. 141. *Before published.*)   *Venezia*, 1738

Bosscha, J. Dissertatio inaugurata de Galvanometro differentiali. *Ludg. Bat.* 1854
   Gleichzeitige Beförderung von drei oder vier Depeschen durch einen telegraphischen Leitungsdraht. (*Telegraphen-Vereins, Jahrg.* iii. s. 27 *und* 51,
   aus Verslagen en Mededeelingen der Koninck. Acad. Dl. iv.)

Bossi, Luigi. *Born February* 28, 1758, *at Milan ; died April* 10, 1835, *at Milan.*

† Bolide . . . nel Lario, 18 Mar. 1834. 8vo. 4 pp.(*Bibliot. Ital.* lxxiii. 355.)
                                                              Milano, 1834

Bostock, John. *Born* 1774 *at Liverpool ; died August* 6, 1846, *in London.*

† An Account of the History and present State of Galvanism. 8vo. 164 pp.
   2 plates.   *London*, 1818

Boston. (*Vide* American Academy.)

Böttger, R. Beyträge zur Physik und Chimie. Eine Sammlung eigener Erfahrungen,
   Versuche und Beobachtung. 1 Heft. 8vo.   *Frankfurt*, 1838

   Idem. 2 Heft. Neuere Beiträge.   1841

   Idem. 3 Heft. Materialien Vorlesungen.   1846

   Ueber die Einrichtung und Behandlung der Dobereinerschen Zündmaschine. 2
   Auflagen.   *Sondershausen*, 1838

   Mémoires où on trouve un mode d'argenture sur cuivre et laiton dans une solution ammoniacale d'azotate d'argent avec excès d'ammoniaque, à l'aide d'une
   pile distincte. (*L'Ami de l'Industrie de Frankfort, en Juillet,* 1840.)
                                                              Frankfort, 1840

Bottis, G. Breve relazione degli effetti di un Fulmine che cadde in Napoli il mese
   di Giugno del presente anno 1774 ; e alcune considerazioni sopra i medesimi.
   4to. 27 pp.   *Napoli*, 1774

Botto, Giuseppe Domenico. *Born April* 4, 1791, *at Moneglia b. Genua.*

† Esperienze sull' azione chimica delle correnti indotte dal Magnetismo terrestre e
   dai ferro-elettro-magneti con alcune Osservazioni sulla loro trasmissibilità
   nei conduttori liquidi, e sui fenomeni del disco di Arago. 8vo. 20 pp. 1 plate.
   (*Faraday's copy.*)   *Turin*, 1834

† Esperienze sull' azione chimica delle correnti indotte dal Magnetismo terrestre,
   e dai ferro-elettro-magneti con alcune Osservazioni. 8vo. 22 pp. 1 plate.
   (*Ext. Bibl. Univ.* lviii. 205.)   *Torino*, 1834

† Note on the application of Electro-magnetism as a mechanical power. 8vo.
   (*Taylor's Scientific Memoirs*, vol. i. part iv. p. 532. *Bibl. Univ.* vol. lvi.)
                                                              London, 1837

Botto, Giuseppe Domenico—*continued.*

Notizia sopra l'applicazione dell' Elettromagnetismo alla Meccanica. 8vo. 6 pp.
Torino, 1834

† Expériences sur les rapports entre l'induction électromagnétique et l'action électrochimique, suivies de considérations sur les machines électromagnétiques. Lu 15 Février, 1842. 4to. 18 pp. (*Mém. Torino,* 2nd series.) *Turin,* 1843

† Sur les lois de la chaleur degagée par le courant voltaïque et sur celles qui régissent le développement de l'Électricité dans la pile. Lu 9 Février, 1845. 4to. 18 pp. (*Acad. de Turin,* serie ii. tom. viii.) *Turin,* 1845

† Note sur un nouveau système de Télégraphie électrique. 4to. 4 pp. (*Poggendorff says Mém. Turin,* xi. 1851.) *Turin, n. d.*

† **Botto et Avogadro.** Mémoire sur les rapports entre le pouvoir conducteur des liquides pour les courants électriques et la décomposition chimique qu'ils en éprouvent. 4to. 39 pp. 1 plate. *Turin,* 1839

Boucher. (*Vide* Le Boucher.)

Boudet, P. Guide de l'Expéditeur des Dépêches Télégraphiques, avec les tarifs, pour la France et l'étranger. 12mo. *Paris,* 1869

† **Boudin.** Histoire physique et Médicale de la Foudre, et de ses effets sur l'homme, les animaux, les plantes, les édifices, les navires. 8vo. 31 pp. (*Ext. Annales d'Hygiène, &c.*) *Paris,* 1854

† De la Foudre considérée au point de vue de l'Histoire, de la Médecine légale, et de l'Hygiène publique. 8vo. 50 pp. *Paris,* 1855

† Histoire de la Foudre et des Paratonnerres. 8vo. 57 pp. cuts. (*Ext. Annales d'Hygiène.*) *Paris,* 1855

Boué, A. Chronologischer Katalog der Nordlichter bis zum J. 1856, sammt einer Bibliographie über diese Erscheinung. 8vo. 74 pp. (*Aus den Sitzungsberichten* 1856 *der Kaiser. Akad. der Wissenschaften.*) *Wien,* 1857

† Parallele der Erdbeben, der Nordlichter, und des Erdmagnetismus, sammt ihren Zusammenhang mit dem Erdplastik, &c. 8vo. 75 pp. (*Aus den Sitzungsberichten der Kaiserl. Akad. der Wissenschaften,* xxii. 8, 323.) *Wien,* 1857

† Etwas über Vulkanismus und Plutonismus in Verbindung mit Erdmagnetismus, sowie eines Aufzählungs-versuchs der submarinischen brennenden Vulkane. 8vo. 39 pp. (*Sitzungsb. der Mathem.-naturn.* Cl. lix. 1 Abth.)
Wien, 1869

† Ueber das gefärbte Seewasser und dessen Phosphorescenz im Allgemeinen. 8vo. 12 pp. (*Sitzungsb. der Mathem.-naturn.* Cl. lix. 2 Abth.) *Wien,* 1869

† Ueber die Geologie der Erdoberfläche, in Rücksicht auf die Vertheilung der Temperatur, der Aërolithen und der Oceane. 8vo. 47 pp. (*Sitzungsb. der K. Akad. Wien,* 1850, 1 Heft, p. 59.) *Wien,* 1850

† Chronologische Catalogue der Nordlichter bis zum Jahre 1856, sammt einer Bibliographie über diese Erscheinung. Vorgelegt im Januar, 1856. 8vo. 72 pp. (*Sitzungsberichte der K. Akad. der Wissenschaften,* vol. xxii. 1 Heft.) *Wien,* 1856

Bougainville, Louis Antoine, Comte de. *Born November* 11, 1729, *at Paris; died August* 31, 1811, *at Paris.*

† Some particulars of the Life of Count Bougainville. By M. Delambre. 8vo. 8 pp. (*Phil. Mag.* xliii. 371.) *London,* 1814

Journal de la Navigation autour du Globe de la frégate la Thétis et de la corvette l'Espérance, pendant les années 1824-5-6. Publié par ordre du Roi. Par M. le Baron de Bougainville, Cap. de Vaisseau, et Chef de l'Expédition. 2 vols. 4to. and atlas of 56 plates. vol. i. 742 pp. vol. ii. 351 pp. Vol. ii. has an additional title-page, and is edited by Ed. de la Touanne, lieutenant de la Thétis. Astronomical and Meteorological Observations in 2nd part of vol. ii. 165 pp., declinations included. 1837

Bouguer, Pierre. *Born February* 16, 1698, *at Croisic, Basse-Bretagne; died August* 15, 1758, *at Paris.*

Bouguer—*continued*.

† Remarques sur le Mémoire de M. Meynier touchant la meilleure Méthode d'observer sur mer la déclinaison de l'aiguille aimantée. 4to. 7 pp. (*Meynier's Mémoire is dated* 1732.)

† De la Méthode d'observer en mer la déclinaison de la Boussole. A remporté le Prix 1731. 4to. 67 pp. *Paris*, 1731
(*Bound with it is his undated work*, 7 *pages*, " *Remarques sur le Mémoire de M. Meynier touchant la meilleure Méthode d'observer sur mer la déclinaison de l'aiguille aimantée.*" *This* (*Bouguer*) *is in the Pièces de Prix de l'Acad. de Paris*, ii. *Mém.* 6.)

† Traité de Navigation, contenant la théorie et la pratique du pilotage. 8vo. 422 pp.
Paris, 1753

Vidit in America flammas e montibus exsilientes, dum certæ nubes (nimirum defectu electricæ) contra eosdem vento ferrentur. 4to. (*Mém de l'Acad. an* 1755, p. 281.) *Paris*, 1755

Traité de Navigation. (*From Van Swinden*, vol. ii. p. 101.)

Voyage au Pérou (p. 48.)
(*Note.*—*On the temperature of the atmosphere, at various heights, in reference to the formation of hail.*)

Bouilhet, H. Mémoire sur le Cyanure double de potassium et d'argent et sur son rôle dans l'argenture électro-chimique. 8vo. (*Ann. de Chim. et de Phys.* 3rd series, xxxiv. *Paris*

Bouillet, J. Description d'une Aurore Boréale singulière dans ses effets. 4to. (*Mém de Paris*, 1730.) *Paris*, 1730

De l'état des connaissances relatives à l'Eléctricité chez les anciens peuples de l'Italie. 8vo. 31 pp. *Saint Etienne*, 1862

Bouillon, Marquis de. His Mosaic Gold for Electric Rubbers. 4to. (*Observ. de Phys.* xxi.) *Paris*

Boulangé. (*Vide* Le Boulangé.)

Boullanger (or Anonymous). Mémoire sur l'Electricité. 8vo. 37 pp. (*Vide Nollet Essai*, 2nd edition, pp. 217, 240, 242.) *Paris*, 1746

Suite du Mémoire sur l'Electricité. 8vo. 30 pp. *Paris*, 1748

Lettre à l'Abbé Nollet sur l'Electricité. 8vo. 45 pp. *Londres*, 1749
(*For evidence of this work being by Boullanger vide page* 8 *of it.*)

† **Boullangère.** Traité de la cause et des phénomènes de l'Electricité. 2 vols. 8vo.
Paris, 1750

† **Boulu.** Traitement des Adénites cervicales chroniques au moyen de l'Electricité localisée. 8vo. 31 pp. (*Extrait du Rapport lu à l'Acad. Imp. de Méd. le* 22 *Ap.* 1856.) *Paris*, 1856

† **Bourdon, M. A.** Account of a Shower of Stones ; in a Letter from the Prefect of the Department of Vaucluse to the French Minister of the Interior, dated November 10, 1803. 8vo. 4 pp. (*Phil. Mag.* xvii. 271.) *London*, 1803

† **Bourdonnay,** Duclésio. Sur la distribution de l'Electricité à la surface des corps conducteurs. Thèse. 4to. 44 pp. *Paris*, 1840

Bourges, Secrétaire de l'Acad. de Bordeaux. Rapport sur les travaux de l'Académie, Séance 1835, 10 Sept. 8vo. (*Séances de l'Académie de Bordeaux pour* 1835, p. 48.) *Bordeaux*, 1835
(*Note.*—*Contains mention of Vallot's " Résultats ;" amongst others, " Le Rapport qui existe entre la pierre nommée par Pline Carchedoine* (sic) *et la Tourmaline.*")

Rapport sur les travaux de l'Académie, Séance 1837, 22 Sept. 8vo. (*Séances de l'Académie de Bordeaux pour* 1837, p. 83.) *Bordeaux*, 1837
(*Note.*—*Contains notices of two memoirs, as replies to a prize question on Lightning-conductors. The first is an anonymous one, which speaks of the forms of roofs and of metallic masses spread over the edifice, &c. The second is by Mermet, of Pau, who received a gold medal, but not the prize.*)

Bourges—*continued.*

Rapport sur les travaux de l' Acad. 1ʳ Trimestre 1838. 8vo. (*Actes de l'Acad. de Bord. pour* 1838, p. 84.) *Bordeaux,* 1839 (*Note.*—*Contains mention of Gachet's Notice sur deux échantillons de météorolites, one of which was from the meteorolite mentioned by Brard.*)

† **Bourgoin,** Edme. Electrochimie. Nouvelles recherches électrolytiques. 8vo. 36 pp. (*Ecole superieure de Pharmacie de Paris.*) *Paris,* 1868

Bourgouet. Rapport sur des expériences d'Electricité galvanique fortifiée. 8vo. (*Journal de Chim. de Van Mons,* No. iv. p. 73.) *Bruxelles* (*Vide* also Prevost, P.)

† **Boussac,** A. Précis de Télégraphie électrique et des connaissances mathématiques, physiques, et chimiques indispensables pour la télégraphie. 8vo. 503 pp. *Paris,* 1868

Boussingault. Expériences constatant l'efficacité des Lampes de Davy. Emploi de la lumière produite par la pile dans les lieux où l'atmosphère est sujette à devenir détonnante. 4to. (*Comptes Rendus,* xxi. 515.) *Paris,* 1845

Remarques à l'occasion d'une communication de M. Louyet sur l'emploi de la lumière produite par la pile pour l'éclairage des mines. 4to. (*Comptes Rendus,* xxii. 225.) *Paris,* 1846

Boussland. (*Vide* Humboldt and Boussland.)

Bouvier, de Jodoigne. Extrait d'une Lettre sur la substitution de la Pile de Volta au briquet physique. 8vo. (*Journal de Van Mons,* No. xl. p. 237.) *Bruxelles*

Expériences sur le remplacement des corps humides par une couche mince d'air dans la pile de Volta. 8vo. (*Journal de Van Mons,* No. xii. p. 300.) *Bruxelles*

† **Bowditch,** N. An Estimate of the height, direction, velocity, and magnitude of the Meteor that exploded over Weston, in Connecticut, Dec. 14, 1807. 4to. 24 pp. (*Mem. Amer. Acad.* old series, iii. part ii. p. 213.) *Cambridge, U.S.* 1815

† On the Variation of the Magnetic Needle. 4to. 7 pp. (*Mem. Amer. Acad.* old series, iii. part ii. p. 337.) *Cambridge, U.S.* 1815

† On the Method of computing the Dip of the Magnetic Needle in different latitudes, according to the theory of Biot. 4to. 5 pp. (*Mem. Amer. Acad.* old series, iv. part i. p. 57.) *Cambridge, U.S.* 1818

† On a Mistake which exists in the calculation of M. Poisson relative to the distribution of the electric matter upon the surfaces of two globes, in vol. xii. of the Mémoires de la Classe des Sciences Mathématiques et Physiques de l'Instit. Impérial de France. 4to. 2 pp. (*Mem. Amer. Acad.* old series, iv. part i. p. 307.) *Cambridge, U.S.* 1818

† On the Meteor which passed over Wilmington, in the state of Delaware Nov. 21, 1819. 4to. 11 pp. (*Mem. Amer. Acad.* old series, iv. part i. p. 295.) *Cambridge, U.S.* 1818

Vide also Dutrochet and Bowdich.

† **Bowdoin,** Js. Observations upon an hypothesis for solving the phenomena of Light ; with incidental Observations, tending to show the heterogeneousness of Light, and of the Electric Fluid, by their intermixture or union with each other. 4to. (*Mem. Amer. Acad.* i. 187.) *Boston,* 1785

Boyle, Hon. Robert. *Born January* 25, 1627, *at Lismore, County Cork, Ireland ; died December* 30, 1691, *in London.*

New Experiments concerning the relation between Light and Air in Shining Wood and Fish. 4to. (*Phil. Trans.*) *London,* 1667

Some Observations about Shining Fish. 4to. (*Phil. Trans.* 1672.) *London,* 1672

Experiments and Notes about the Mechanical Origin or Production of particular qualities, in several Discourses on a great variety of subjects, and, amongst the rest, of Electricity. 1676

The Aërial Noctiluca ; or, some New Phenomena, and a process of a fictitious self-shining substance. 8vo. *London,* 1680

Boyle—*continued.*

Experiments and Observations made upon the Icy Noctiluca ; to which is annexed a Chymical Paradox, grounded on new expeiiments, making it probable that chemical principles are transmutable. 12mo. (*R. S. Cat. of* 1839, p. 297.)
London, 1681-82

Experimenta et Observationes Physicæ, &c. 1691

† Philosophical Works, abridged by Shaw, ii. 506. 3 vols. small 4to. *London,* 1725

Bozolus (Professor, in Roma). On Discharge of the Leyden Jar as the active principle of the Electric Telegraph.
(*Vide Mariani, from a Memorandum given to me by Cavaliere Zantedeschi.*—F. R.) *Roma,* 1767

Brachet. *Vide* Becquerel and Brachet.

† **Brachet,** A. Solutions de l'éclairage Electrique produit par les courants de la pile. 8vó. 8 pp. *Paris,* 1858 (?)

† **Brae,** A. E. Electrical Communication in Railway Trains. The causes considered which have hitherto prevented its successful application. 4to. 36 pp. 4 plates *London,* 1865

† **Bragadin** (or Anonymous). Dubbii sull' efficacia de' Conduttori elett. 8vo. 122 pp. 1 plate. *Venezia,* 1795

† Risposta dell' autore dei Dubbii sull' efficacia dei Conduttori, alla giunta al Giornale Astrometeorologico del . . . Toaldo. (No title-page.) 8vo. 31 pp.

† **Brame,** E. Etude sur les Signaux des Chemins de Fer à double voie. Large 8vo. 262 pp. atlas of 18 plates folio. *Paris,* 1867

Brand. On Attraction of mixtures of Iron and Gold, Silver, Copper, Lead, &c. by the Magnet, and effect of Antimony. 8vo. (*Abhand. der Schwed. Akad.* xiii. 212.) *Hamburg, n. d.*

Brande, W. T. Experiments on the Effects of the Galvanic Current on Albumen. 4to. (*Phil. Trans.* 1809.) *London,* 1809

† On some Electro-chemical Phenomena. 8vo. 7 pp. 1 plate. (*Phil. Mag.* xliv. 124.) *London,* 1814

† Electro-chemical Decomposition of the Vegeto-alkaline Salts. 8vo. (*Phil. Mag. or Annals,* ix. 237.) *London,* 1831

The Dictionary of Science, Literature, and Art. Fourth edition, re-edited by the late W. T. Brande (the author) and George W. Cox, M.A. 3 vols. 8vo. *London,* 1872

Brande. (*Vide* Klingenstierna and Brande.)

† **Brandely,** A. Nouveau Traité des manipulations Electro-chimiques appliquées aux arts et à l'industrie. 8vo. 146 pp. 6 plates. *Paris,* 1848

† Die Operationen, Manipulationen und Geräthschaften der Electro-chemie in ihrer Anwendung auf Gold- Silber- und Bronce-arbeiten, Galvanoplastik und andere verwandte Gewerbe. Aus dem Französischen bearbeitet. 8vo. 179 pp. 10 plates.) *Weimar,* 1849

† **Brander,** Georg Friedrich. *Born November* 28, 1713, *at Regensburg; died April* 1, 1783, *at Augsburg.* Beschreibung eines magnetischen Declinatorii und Inclinatorii. 8vo. 72 pp. 2 plates. *Augsburg,* 1779

Brandes, Carl W. T. Sternschnuppen Beobachtungen im Aug. 1833 und deren Berechnung. (*Astronomische Nachrichten,* xvii. 1840.) 1840

Brandes, Heinrich Wilhelm. *Born July* 27, 1777, *at Groden bei Ritzebüttel ; died May* 17, 1834, *at Leipzig.* Ueber Sternschnuppen. 8vo. (*Voight's Mag.* vi. 1803.) *Jena o. Weimar,* 1803

Ueber die Geschwindigkeit, mit der ein vom Mond geworfn. Körper auf der Erde ankommen kann. 8vo. (*Voight's Mag.* v. 1803.) *Jena o. Weimar,* 1803

Ueber Feuerkugeln. 8vo. (*Voight's Mag.* v. 1803.) *Jena o. Weimar,* 1803

Ueber die Abweichung nach Westen bei steilrecht geschossenen Kugeln. 8vo (*Voight's Mag.* ix. 1805.) *Jena o. Weimar,* 1805

Ueber die Oscillationen der Drehwaage bei Coulomb's Versuchen. 8vo. (*Voight's Mag.* xii. 1806.) *Jena o. Weimar,* 1806

Brandes, Heinrich Wilhelm—*continued.*

Anleitung zu Sternschnuppen Beobachtungen. 8vo. (*Gilbert's Ann.* lxii. 1819.) *Leipzig,* 1819

Beiträge zur Witterungskunde, &c. *Leipzig,* 1820
Wie die anziehenden und abstossenden elektrischen Kräfte von der Entfernung abhängen. 8vo. (*Schweigger's Journal,* xxxv. 1822.) *Nürnburg,* 1822

Gesetzmässigkeit in der Bewegung der Sternschnuppen. 8vo. (*Poggendorff,* *Annal.* ii. 1824.) *Leipzig,* 1824

Beobachtungen über die Sternschnuppen, angestellt von mehreren Naturforschern und mit Untersuchungen über die Resultate derselben. 8vo. *Leipzig,* 1825
(*Note.—This is the first vol. of his Unterhaltungen für Freunde der Physik und Astronomie, of which work there are three vols.*)

Gesetzmässigkeit in der Bewegung der Sternschnuppen. 8vo. (*Poggendorff's* *Ann.* vi. 1826.) *Leipzig,* 1826

† Ueber Sternschnuppen. 8vo. (*Gilbert's Ann.* vi. viii. xiv. xviii. xxxvii. xlii. xlvii. lviii. lxii. lxxiii. lxxv.) *Leipzig*

† **Brandes,** R. Examination of a Gelatinous Substance found in a damp meadow ; as a contribution to the knowledge of Meteors called Shooting Stars. 8vo. 7 pp. (*Schweigger's Jahrbuch der Chem. N. R.* xix. 389. *Phil. Mag. or Annals,* iii. 271.) *London,* 1828

† **Branville,** P. de. Mémoire sur la pose des Cables sous-marins et sur les opérations préliminaires qui s'y rattachent. 8vo. 32 pp. (*Ext. des Mém. de la Soc. des Ingén. Civils.*) *Paris,* 1858
(*Vide* also Balestrini.)

Brard, Cyprian Prosper. *Born November* 21, 1788, *at l'Aigle, dép. Orne ; died November* 28, 1838.
Note sur une masse de Fer météorique. (*Bordeaux Acad. Sciences* 1829, p. 39.) 1829

† **Brasack,** F. Das Luftspectrum : eine prismatische Untersuchung des zwischen Platina-Elektroden überschlagenden elektrischen Funkens. 4to. 42 pp. 1 plate. (*Aus der Abhandlung der Natur. Gesellschaft zu Halle,* x. *besonders abgedruckt.*) *Halle,* 1866

Brasseur, A. J. Enseignement de la vraie doctrine du Magnétisme. Les principes expliqués d'après les effets. Mémoire adressé à l'Académie des Sciences. 12mo. 14 pp. *Paris,* 1860

Braun, C. J. H. E. Ueber das Nordlicht. *Nürnberg,* 1858

Braun, Josias Adam. *Born* 1712, *at Assche, südl. Brab. ; died October* 3, 1768, *at St. Petersburg.*
Observationes Meteorologicæ factæ Petropoli An. 1755 et consectaria. 4to. (*Novi Comment. Acad. Petrop.* vii. 388.) (*Vide Van Swinden, Receuil,* iii. 125, *on Atmospherico-Electrical Influence on the Needle.*) *Petropoli*

† **Braun,** J. S. Ueber die Anwendung des Lichts und der Elektricität in der Telegraphie, und der Construction electrischer Telegraphen. 4to. 38 pp. *Altenburg,* 1849

† **Bravais,** Auguste. *Born August* 28, 1811, *at Annonay, dép. Ardéche.*
Sur les Aurores Boréales vues à Bossekop et á Jupvig en 1838 et 1839 pendant le voyage de la Commission . . . du Nord. 8vo. 119 pp. 2 plates. (*Ext. des Voyages en Scandinavie, en Laponie, &c. Division, Aurores Boréales.*) (*Vide Gaimard.*) *Paris,* 1846

† Sur les variations de l'intensité magnétique horizontale observés à Bossekop en 1838 et 1839 pendant le voyage de la Commission Scientifique du Nord. 8vo. 60 pp. 1 plate. (*Ext. de Voyage en Scandinavie, &c. Division Magnétisme terrestre.*) *Paris,* 1847

† Historique des hypothéses faites sur la nature et la cause des Aurores Boréales. 8vo. 30 pp. (*Ext. des Voyages en Scandinavie, &c. Division Magnétisme terrestre,* tom. iii.) (*Vide Gaimard.*) *Paris,* 1856

78 BRA—BRE

Bravi. Sulla cagione dei venti irregolari ; con Appendice. *Bergamo, 1831*

Breda, Jacob van. Verhandeling over de Electriciteit van den Dampkring. (*Verhand. Maatsch. Harlem,* iv. 1788, *gekrönt von dieser Gesellschaft.*) *Haarlem,* 1788
Proewen genomen med twee elektrophores van verschillende soort, om de outekken met welke doorgans de meeste elektriciteit kon de verwekt worden. (*Allgemeene Vaderlandsche Letteroefeningen.*) *Haarlem,* 1799

Bredenberg, C. G. Bemerkungen vom rechten Gebrauche des Compasses, bey Aufsuchung der Eisenerze. 8vo. 2 pp. (*K. Schwed. Akad. Abhand.* xxii. 74.)
Hamb. v. Leipzig, 1760

† **Breguet** et **Betancourt.** Nouveau Télégraphe. (*Vide* Lagrange, Laplace, &c. Rapport.)

Breguet, Louis François Clément. *Born December* 22, 1804, *at Paris.*
Note sur un Appareil destiné à mesurer la vitesse d'un projectile dans différents points de sa trajectoire. 4to. (*Comptes Rendus,* xx. 157.) *Paris,* 1845

Remarques sur un passage qui se rapporte à la Communication (sur un appareil) dans une Note de M. Wheatstone sur le Chronoscope électromagnétique. 4to. (*Comptes Rendus,* xx. 1712.)
Note.—Moigno says *Séance de l'Acad. du* 9 Juin, 1845. *Paris,* 1845

† Sur la Télégraphie Electrique. 4to. (*Comptes Rendus,* xxi. 1119.)
Paris, 1845

Expériences faites au Télégraphe Electrique de Rouen. 4to. (*Comptes Rendus,* xxii. 743.) *Paris,* 1846

Remarques sur une modification dans la Construction des Télégraphes Electriques, recemment proposée par M. Dujardin. 4to. (*Comptes Rendus,* xxiii. 880.) *Paris,* 1846

Remarques à l'occasion d'une réclamation de priorité élevée par M. Morse concernant les Télégraphes Electriques. 4to. (*Comptes Rendus,* xxiii. 937.)
Paris, 1846

Appareil Electromagnétique pour le Télégraphe de Saint Germain. 4to. (*Comptes Rendus,* xxiii. 1082.) *Paris,* 1846

De l'induction par différents Métaux. 4to. (*Comptes Rendus,* xxiii. 1155.)
Paris, 1846

Moniteur Electrique pour les Chemins de Fer. 4to. (*Comptes Rendus,* xxiv. 428.)
Paris, 1847

Sur un accident arrivé au télégraphe de Saint Germain. 4to. (*Comptes Rendus,* xxiv. 980.) *Paris,* 1847

Appareils destinés à indiquer la vitesse, à chaque instant, du parcours d'un convoi sur Chemin de Fer ; et la durée du séjour dans les stations. 4to. (*Comptes Rendus,* xxix. 740.) *Paris,* 1849

Réclamation de priorité au sujet des Chronomètres à pointage de M. Abm. Louis Breguet. 4to. (*Comptes Rendus,* xxx. 206.) *Paris,* 1850

Manuel de Télégraphie Electrique. 1st edition. 12mo. *Paris*

Manuel de Télégraphie Electrique à l'usage des employés des Chemins de Fer. 12mo. *Paris,* 1851

Sur un Télégraphe Electrique mobile. 4to. (*Comptes Rendus,* xxxiv.) *Paris,* 1852

Sur les Télégraphes Electriques. 4to. (*Comptes Rendus,* xxxiv.) *Paris,* 1852

† Manuel de Télégraphie Electrique, 2nd edition. 12mo. 2 plates. *Paris,* 1853

Manuel de Telegrafia Electrica para el uso de los empleados en los caminos de hierro. Edicion Española, publicada por D. Balbino Cortés. 4to. 88 pp. 1 plate. *Madrid,* 1855

† Manuel de Télégraphie Electrique à l'usage des employés des Chemins de Fer. Revue et augmentée. 3rd edition. 12mo. 107 pp. 3 plates, and cuts. *Paris,* 1856

† Manuel de Télégraphie Électrique. Quatrième édition, revue, corrigée, et augmentée. 12mo. 252 pp. 4 plates. *Paris,* 1862

Breguet—*continued.*

Nouveau Transformateur. (*Vide* Du Moncel.)

† Télégraphie domestique. Instruction sur la pose . . . des Sonnettes électriques. 12mo. 61 pp. *Paris*, 1865

(*Vide* also Anon. Electric Telegraph, 1865.)

(*Vide* also Masson and Breguet fils.)

† **Breguet** et **Séré** (V. de) Télégraphe électrique, son avenir. Poste aux lettres électriques, Journaux électriques, suivi d'un aperçu theorique de la télégraphie. 8vo. 75 pp. *Paris*, 1849

Breitinger, David, Sen. *Born Nov.* 17, 1737, *at Schönholzerswyler, Thurgau; died Jan.* 30, 1817, *at Zurich.*

Reflexionen ob es wohl gerathen wäre, Strahlenableiter in unserer Stadt Zürich einzuführen. *Zurich*, 1776

Nachricht über das Einschlagen des Blitzes in einen Wetter-ableiter, nebst Berichtigung einiger Begriffe über die Wirkung der Ableiter. *Zurich*, 1786

† Ragguaglio d'un Fulmine caduto in un Conduttore. 4to. 3 pp. (*Opuscoli Scelti,* ix. 210.) *Milano*, 1786

Instruction für diejenigen, welche sich mit der Verfertigung und Visitation der Blitzableiter befassen. *Zurich*, 1825

Instruction über Blitzableiter im Canton Zürich. 4to. *Zurich*, 1830

† **Breintnall,** Jos. Notices of some Meteors observed at Philadelphia, extracted out of a Letter from him to P. Collinson, dated Philadelphia, May 9, 1738. 4to. 2 pp. (*Phil. Trans.* xli. 359.) *London,* 1739-41

Bremond, François de. *Born Sept.* 14, 1713, *at Paris; died March* 21, 1742, *at Paris.*

† Extract of a Letter from De Bremond to Mortimer, concerning a File rendered magnetical by Lightning. Translated from the French by T. S. Dated Paris June 4, 1740. 4to. 3 pp. (*Phil. Trans.* xli. 614.) *London,* 1739-41

Expériences Physico Mécaniques sur différens Sujets, et principalement sur la Lumière et l'Electricité, produites par le frottement des Corps. Traduites de l'Anglais de M. Hawksbee par feu M. de Bremond, de l'Acad. Royale des Sciences. Revues et mises au jour, avec un Discours préliminaire, des Remarques et des Notes, par M. Demarest. 2 vols. 8vo. *Paris,* 1754

† **Brenner.** Weitere Mittheilungen zur Elektrootiatrik. Separat-Abdruck aus der Petersb. med. Zeitschrift. 4to. 26 pp. *Petersburg,* 1863

† Zur Elektro-physiologie und Elektro-pathologie des Nervus acusticus. Separat-Abdruck aus der Petersb. medische Zeitschrift. 4to. 11 pp. *Petersburg,* 1863

† **Brenner,** R. Untersuchungen und Beobachtungen auf dem Gebiete der Elektro-therapie. 2 vols. 8vo. *Leipzig,* 1868-9

† **Brenta,** L. Fenomeni della visione . . . Descrizione di Pratici Esperimenti comprovanti la Forza attraente e respingente Elettro-magnetica entro l'occhio . . . 8vo. 39 pp. 3 plates. *Milano,* 1838

† Elettro-magneto-tipia . . . 8vo. 91 pp. 1 plate and portrait. *Milano,* 1840

Brera, V. L. Giornale di Medicina pratica, compilato da V. L. Brera. 12 vols. 8vo. (*Vide* also Brugnatelli and Brera.) *Padova,* 1812-17

Brescia Academy and Athenæum. (*Vide* Commentarii dell' Accad. . . . del Dipartimento, &c.)

Breslau Academy. Miscellanea curiosa, seu Ephemerides Medico-physicæ Academiæ Naturæ Curiosorum. 24 vols, 4to. *Norimb.* 1670-1706

Ephemerides Academiæ Cæs. Naturæ curiosorum, &c. 5 vols. 4to. *Norimb.* 1712-22

Breslau Academy—*continued.*

Acta Physico-medica Acad. Cæs. Leopoldino-Carolinæ Nat. Curios. 10 vols. 4to. *Norimb.* 1727-54

Nova Acta Physico-medica Acad. Cæs. Leopoldino-Carolinæ Nat. Curios. 8 vols. 4to. *Norimb.* 1754-91

† Verhandlungen der Leopoldino-Carolinæ Academiæ d. Naturforcher. 1 vol. 1818—*continued.*

Bressy, Joseph. Essai sur l'Electricité de l'Eau. 8vo. 178 pp. 2 plates. *Paris,* 1797

Breton. Nouvel Appareil Electro-magnétique. 4to. (*Comptes Rendus,* viii. 499, and ix. 242.) *Paris,* 1839

Breton Frères. Notice sur quelques perfectionnements apportés à un Appareil Electrique destiné à l'usage des Médecins. 4to. (*Comptes Rendus,* xviii. 527.) *Paris,* 1844

† **Breton et Beau de Rochas.** Théorie mécanique des Télégraphes sous-marins. Recherches sur les conditions de leur établissement. 8vo. 72 pp. 2 plates (*Ext. des Annales Télégraphiques,* Sept. et Oct. 1859.) *Paris,* 1859

† **Brett,** J. W. On the Origin and Progress of the Oceanic Telegraph, with a few brief Facts and Opinions of the Press. 8vo. 104 pp. *London,* 1858 (*Vide* also Patents.)

† **Brett and Little.** Compendium of the Improvements effected in Electric Telegraphs by Messrs. Brett and Little, with a description of their patent Electro-telegraphic Converser. . . 8vo. 64 pp. illustrated. *London,* 1847

Brettes. (*Vide* Martin de Brettes.)

Brewer and Delaroche. Essai sur le Galvanisme. Extrait du 8e vol. de la Bibliothèque Germanique Médico-chirurgicale. 8vo. *Paris,* 1802
(*This is their account of Grepengeiser's* "Recherches sur le Galvanisme," *translated from his* "Versuche.")

Brewster, Sir David. *Born Dec.* 11, 1781, *at Sedburgh, Roxburghshire, Scotland; died*

The Edinburgh Encyclopædia of Science, Literature, and Art. 18 vols. 4to. 533 plates. *Edinburgh,* 1810-30

† A Treatise on Magnetism, forming an Article in the 7th Edition of the Encyclopædia Britannica. 8vo. 363 pp. *Edinburgh,* 1838

A popular Treatise on Magnetism, from the 7th Edition of the Encyclopædia Britannica. 8vo. 107 illustrations. (*From Latimer Clark's Cat.*) *Edinburgh,* 1851

† On the Optical Phenomena and Crystallisation of Tourmaline, Titanium, and Quartz, within Mica, Amethyst, and Topaz. 4to. 7 pp. 1 plate. (*Faraday's copy.*) (*Trans. Royal Society, Edinburgh,* xx. Part iv.) *Edinburgh,* 1853

† Electricity. Article in the Encyclopædia Britannica, vol. viii. p. 523. 4to. 8th edition. 104 pp. 9 plates. *Edinburgh,* 1855

† On the Life Boat, the Lightning Conductor, and the Light-house. 8vo. (*Nor h British Review,* xxxii. 492. November, 1859.) 1859

† Galvanism. (*Edinburgh Encyclopædia,* vol. iv. p. 173.)

† Article on Electricity. 4to. 7th edition. 100 pp. 8 plates. (*Encyclopædia Britannica,* viii. 565.) *London.* (*This article is signed N. N. N.*)

† Article on Voltaic Electricity. 4to. 7th edition. 37 pp. (cuts.) (*Encyclopædia Britannica,* vol. xxi.) *London*

† Photographic Portrait of Sir David Brewster.

Vide Edinburgh Philosophical Journal.
„ Edinburgh Journal of Science. New Series.
„ London and Edinburgh Philosophical Magazine.
„ Ferguson (James) and Brewster.
„ Robison and Brewster.

Brézé, De. (Il Marchese.) Description de trois Machines Physico-chimiques.
About 1784
(The editors of the " Opuscoli" give a description of two of the machines, and refer to this book. One is an eudiometer, acting by the electric spark, exactly in Davy's manner.)

Briand. L'Electricité appliquée au Traitement curatif des Névralgies, des Rhumatismes, des Paralysies, &c. 12mo. 130 and 6 pp. *Paris et Rennes*, 1855

Brierre de Boismont. Diagn. Untersuchungen verschiedener Arten der allgemeinen Lähmung mittelst der localisirten Galvanisation. (*Annal Med. Phys.* 1850.) 1850

† **Briggs,** C. F., and **Meverick.** The Story of the Telegraph and History of the great Atlantic Cable. 8vo. 255 pp. Portrait and map. *New York*, 1858

† **Brigoli.** Lettera sopra la Macchina Elettrica ad un amico suo. Da Venezia li 25 Novembre, 1747. 4to. 8 pp. *Verona*, 1748

Brilhac. A Plate Electrical Machine. 2 plates. (Noben einander stehend.)

Brisbane, Sir Thos. Macdougal. *Born July* 23, 1773, *at Brisbane House, Ayrshire ; died Jan.* 27, 1860, *at Brisbane House, Ayrshire.*
Vide Makerstown Observations.

Brisbane, Sir Thomas Macdougall B. *Born* 1770.

Britannica, Encyclopædia. *Vide* Anons. Elect. 1823.

† **British Association Council.** Report on the existing Relations between the Kew Committee and the British Association. 8vo. 35 pp. *London*, 1870

British Association for the Advancement of Science. Report of the 1st and 2nd Meetings of the Association, at York in 1831, and at Oxford in 1832, including its Proceedings, Recommendations, and Transactions. 8vo.
London, 1833

Report of the	3rd Meeting at Cambridge in 1833.	8vo.	*London*, 1834
„	4th Meeting at Edinburgh in 1834.	8vo.	*London*, 1835
„	5th Meeting at Dublin in 1835.	8vo.	*London*, 1836
„	6th Meeting at Bristol in 1836.	8vo.	*London*, 1837
„	7th Meeting at Liverpool in 1837.	8vo.	*London*, 1838
„	8th Meeting at Newcastle in 1838.	8vo.	*London*
„	9th Meeting at Birmingham in 1839.	8vo.	*London*
„	10th Meeting at Glasgow in 1840.	8vo.	*London*
„	11th Meeting at Plymouth in 1841.	8vo.	*London*
„	12th Meeting at Manchester in 1842.	8vo.	*London*
„	13th Meeting at Cork in 1843.	8vo.	*London*
„	14th Meeting at York in 1844.	8vo.	*London*
„	15th Meeting at Cambridge in 1845.	8vo.	*London*
„	16th Meeting at Southampton in 1846.	8vo.	*London*
„	17th Meeting at Oxford in 1847.	8vo.	*London*
„	18th Meeting at Swansea in 1848.	8vo.	*London*
„	19th Meeting at Birmingham in 1849.	8vo.	*London*
„	20th Meeting at Edinburgh in 1850.	8vo.	*London*
„	21st Meeting at Ipswich in 1851.	8vo.	*London*
„	22nd Meeting at Belfast in 1852.	8vo.	*London*
„	23rd Meeting at Hull in 1853.	8vo.	*London*
„	24th Meeting at Liverpool in 1854.	8vo.	*London*
„	25th Meeting at Glasgow in 1855.	8vo.	*London*
„	26th Meeting at Cheltenham in 1856.	8vo.	*London*
„	27th Meeting at Dublin in 1857.	8vo.	*London*

82 **BRI—BRO**

British Association—continued.
Report of the 28th Meeting at Leeds in 1858. 8vo.   *London,* 1859
 „   29th Meeting at Aberdeen in 1859. 8vo.   *London,* 1860
 „   30th Meeting at Oxford in 1860. 8vo.   *London,* 1861
 „   31st Meeting at Manchester in 1861. 8vo.   *London,* 1862
† Minutes of Council. 8vo.   *London,* 1841-53
 Proceedings connected with the Magnetical and Meteorological Conference held
 at Cambridge in June, 1845. 8vo.   *London,* 1845

British Cyclopædia. Arts and Sciences. 2 vols. royal 8vo.   *London,* 1838

Brivazac. *Vide* Beaumont-Brivazac.

† **Brix, P. W.** Die Stadt-Telegraphen-Leitung in Berlin, zwichen der Centralstation
 im Königlichen Postgebäude und den von den funf Berliner Bahnhöfen abge-
 henden Telegraphen-Linien. 4to. 16 pp. 11 plates. (*Besonderer Abdruck aus
 der Zeitschrift des Deutsch-Oesterreichischen Telegraphen-Vereins.*)
     Berlin, 1857
 Zeitschrift des Deutsch-Oesterreichischen Telegraphen-Vereins. Redacteur, Von
 Dr. P. Wilde Brix. 16 Jahrg. 1869. 4to. 12 Hefte.   *Berlin,* 1869
 Annalen der Telegraphie. Herausgegeben, und redigert von Dr. P. W. Brix im
 Anschluss an die Zeitschrift des Deutsch-Oesterreichen Telegraphen-Vereins.
 4to.   *Berlin,* 1870

† **Brocklesby, T.** Elements of Meteorology for Schools, &c. 12mo. 2nd edition.
 240 pp. cuts. Electricity, p. 131.   *New York,* 1849

† **Brook, A.** Miscellaneous Experiments and Remarks on Electricity, the Air-pump,
 and the Barometer, with the description of an Electrometer of a new construction.   *Norwich,* 1789
 Vermischte Erfahrungen über die Electricität, Luftpumpe und Barometer, aus
 dem Englischen von Kuhn. 8vo.   *Leipzig,* 1790

† **Brooke, C.** An Account of the remarkable Magnetic Disturbance which continued from the 22d to the 25th October, 1847. (*Phil. Mag. for January,*
 1848.) (*Faraday's copy.*)   *London,* 1848
* † On the Automatic Registration of Magnetometers by Photography. No. iii.
 Received June 21, read June 21, 1849. 4to. 9 pp. (*Phil. Trans.* part i., *for*
 1850.)   *London,* 1850

Brougham, Lord. Tracts, Mathematical and Physical. Small vo.   1860

Brougham, Brewster, and Others. Natural Philosophy. 4 vols. 8vo. (*Library
 of Useful Knowledge.*)   *London,* 1835-40

† **Broun, J. A.** On the combined Motions of the Magnetic Needle, and on the
 Aurora Borealis. 8vo. 16 pp. 2 plates, (*Proceedings of the Royal Society
 of Edinburgh,* vol. ii. No. 39.) (*Faraday's copy.*)   *Edinburgh*
† Results of the Makerstown Observations. No. 1. On the relation of the Varia-
 tions of the horizontal Intensity of the Earth's Magnetism to the solar and
 lunar periods. 4to. 11 pp. 2 plates. (*Proceedings of the Royal Society of
 Edinburgh,* xvi. part ii.)   *Edinburgh,* 1846
† General Results of the Observations in Magnetism, &c. at Makerstown, in the
 Observatory of Sir T. Brisbane. 4to. 86 pp. 9 plates. (*Transactions of the
 Royal Society of Edinburgh,* ix, part ii.)   *Edinburgh,* 1850
† Report to Sir T. Brisbane on the completion of the publication of the Obser-
 vations at Makerstown. 4to. 16 pp.   *Edinburgh,* 1850
† On the Dust Figures observed on Plate Glass. (*Phil. Mag. for January,* 1851.)
     London, 1851
† On the combined Motions of the Magnetic Needle, and on the Aurora Borealis.
 8vo. (*Proceedings of the Royal Society of Edinburgh,* ii. No. 39.)
     Edinburgh

† **Browne, G. H.** Of the Trembling of the Nerves of a Frog after Death. Experiments by M. Du Verney in 1700. 8vo. (*Phil. Mag.* xviii. 285.)

London, 1804

(*Note.*—*He found this article in the Transactions of the Academy of Sciences at Paris for* 1700. *Reference is made to Gardiner in his Observations on the Animal Economy; to Luigi and Klagel; and to Gardini (this last on Lizardi), The above experiments by Verney are said to be the most remote instance of effects now called Galvanic.*)

† **Browne, Peter A.** An Essay on solid Meteors and Aërolites, or meteoric stones. 8vo. 38 pp. *Philadelphia,* 1844

† **Browning, J.** Part of a Letter from J. Browning to H. Baker, dated December 11, 1746, concerning the effect of Electricity on Vegetables. Read January 22, 1746-47. 4to. 3 pp. (*Phil. Trans.* xliv. 373.) *London,* 1746-47

Brown-Sequard, E. Die Lebensenergie herabsetzende Wirkung des electrischen Stromes. Versuche. (*Gaz. Méd. de Paris,* 1849, pp. 881-999) *Paris,* 1849

Sur les Résultats de la Section et de la Galvanisation du Nerf grand sympathique au Cou. 8vo. (*Ext. de la Gazette Médicale.*) *Paris,* 1854

† **Brück, R.** Electricité ou Magnétisme du Globe terrestre. 8vo. 298 pp. *Bruxelles,* 1851

† Electricité ou Magnétisme du Globe terrestre. Extrait des études sur les principes des Sciences physiques. Faits magnétiques proprement dits : variations de la déclinaison, de l'inclinaison, et des forces magnétiques dans le barreau aimanté librement suspendu, et causes de ces variations. Partie ii. 1 vol. 8vo. *Bruxelles,* 1855

† Electricité ou Magnétisme du Globe terrestre. Extrait d'études des Sciences physiques. Faits magnétiques proprement dits. Electricité atmosphèrique, Aurores, &c. Partie ii. vol. ii. 8vo. 542 pp. (This is the title in this copy, but in the preface, &c. it is called vol. iii.) *Bruxelles,* 1858

† L'Origine des Etoiles filantes. 8vo. 155 pp. 11 tables. *Bruxelles,* 1869

† Etude sur la Physique du Globe. Phénomènes atmosphériques. 8vo. 150 pp. 2 plates. *Bruxelles,* 1869

Brucke, E. Manifeste du Magnétisme du Globe et de l'humanité, ou résumé succinct du magnétisme terrestre et de son influence sur les destinées humaines. 8vo. (*Librairie Internationale.*) *Paris,* 1866

† Ueber den Einfluss der Strömesdauer auf die elektrische Erregung der Muskeln. (Vorgelegt 31 Oct. 1867.) 8vo. 9 pp. (*Sitzb. der Königl. Akad. der Wissenschaften,* lvi. Oct. Heft. 1867.) *Wien,* 1867

† Ueber das Verhalten entnervter Muskeln gegen discontinuirliche elektrische Ströme. 8vo. 4 pp. (*Sitzb. der Königl. Akad. der Wissenschaften,* lviii. 2 Abth. Juni Heft, 1868.) *Wien,* 1868

† Ueber die Reizung der Bewegungsnerven durch elektrische Ströme. 8vo. 16 pp. (*Sitzb. der Königl. Akad. der Wissenschaften,* 2 Abth. Oct. Heft, 1868.) *Wien,* 1868

Bruennow, F. De attractione moleculari. 4to. *Berolini,* 1843

Brugmans, Anton. *Born October 22, 1732, at Hantum, Friesland; died April 27 1789, at Gröningen.* (*Poggendorff,* i. 316.)

Tentamina Philosophica de materia magnetica ejusque actione in ferrum et magnetem. 4to. *Francqueræ,* 1765

Nieuwe manier om de magnetische Kragt der lichaamen te onderzoeken. (*Vaderlandsche Letteroefeninqen,* 1775 *et* 1776.) 1775-76

† Magnetismus, seu de affinitatibus magneticis. Observationes academicæ. 4to. 133 pp. 1 plate.

Note.—Van Swinden, Recueil i. 34, *quotes the title Magnetismus sive de Attractione magnetica Observationes.* 4to. *Groningæ,* 1777.

Lugd., Bat. 1778

84 **BRU**

Brugmans—*continued.*

Beobachtungen über die Verwandschaften des Magnets, aus dem Lateinischen übersetzt und mit einigen Anmerkungen vermehrt, von Eschenbach. 8vo. 167 pp.

Leipzig, 1781

† Philosophische Versuche über die magnetische Materie und deren Wirkung in Eisen und Magnet, aus dem Lateinischen des Brugmans, übersetzt und mit Anmerkungen und Zusätzen des Verfassers vermehrt. Herausgegeben von C. G. Eschenbach. 8vo. 309 pp. 6 plates. (*Volta's copy.*) *Leipzig,* 1784

Brugnatelli, Gaspare (Necrologia di.) *Born April 25, 1795, at Pavia ; died October 31, 1852, at Milan. Son of the celebrated Luigi Valentino Brugnatelli. Joined his father in editing the Giornale di Fisica, and joined Configliachi in the editorship after his death. Was at one time one of the directors of the Biblioteca Italiana—viz. from 1831 to 1840 inclusive,* vols. lxi. *to* xcix. (*Giornale I. R. Ist. Lomb.* new series, ix. 489. In 4to. 1856.)

† Dell' arte Galvano-plastica. 8vo. 11 pp. (*Biblio. Ital.* xcviii. 187.) *Milano,* 1840

† Metodo originale Italiano di Elettro-doratura esposta da G. Grimelli. Manuale dell' arte d'indorare e d'inargentare coi metodi elettro-chimici... di F. Selmi. 8vo. 8 pp. (*Giornale I. R. Ist. Lomb.* x. 266.) *Milano,* 1845

† Sui trasporti Elettrici, e le elettriche teorie di. 8vo. 2 pp. (*Giornale I. R. Ist. Lomb.* xii. 6.) *Milano,* 1845

Brugnatelli (G.), **Brunacci, e Configliachi.** Giornale di Fisica, Chimica, Storia-naturale, Medicina, ed Arti. Compilato da Gaspare Brugnatelli. Decade seconda. *Pavia,* 1818-27

Brugnatelli, Luigi Valentino. *Born February 14, 1761, at Pavia ; died October 24, 1818, at Pavia*

† Biblioteca fisica d'Europa, ossia Raccolta di Osservazioni sopra la fisica, matematica, chimica, storia naturale, medicina, ed arti. 22 vols. 8vo. (*The Electrical articles, &c. are in the Library.*) *Pavia,* 1788-91

† Annali di Chimica, ovvero, Raccolta di Memorie sulle Scienze, Arti, e Manifatture ad essa relative. 22 vols. 8vo. (*The Electrical articles are in the Library.*) *Pavia,* 1790-1805

† Giornale Fisico-medico, ossia Raccolta di Osservazioni sopra la fisica, matematica, chimica, storia naturale, medicina, chirurgia, arti, e agricoltura. Per servire di seguito alla Biblioteca fisica d'Europa. 20 vols. 8vo. (*The Electrical articles are in the Library.*) *Pavia,* 1792-96

Extrait d'une Lettre (de B.) sur la non-existence de la pile à charger. 8vo. (*Journal de Chim. de Van Mons,* xvi. 132.) *Bruxelles*

Expériences et Observations de Brugnatelli relatives à l'action de la pile sur diverses humeurs animales. 8vo. (*Journal de Chim. de Van Mons.* x. 114.) *Bruxelles*

† Memorie sull' elettricità animale inserite nel Giornale Fisico-medico del Sig. Brugnatelli. 8vo. 147 pp. (*All the Electrical Articles of this work are in the Library.*) *Pavia,* 1792

† Articolo di lettera al Valli sull' elettricità animale. 8vo. 2 pp. (*Annali di Chimica di Brugnatelli,* vii. 239.) *Pavia,* 1795

Osservazioni chimiche sopra l'ossi-elettrico. 8vo. 17 pp. 1 plate. (*Annali di Chimica di Brugnatelli,* xviii. 136.) *Pavia,* 1800

(*Note.—This contains his account of Bellissima cristallizzazione ottenuta sopra diversi metalli coll' ossi elettrico. He announces, in fact, the reduction, the transport, the precipitation, &c. of the metals dissolved on the negative pole of Volta's Electrometer. Cossa (p. 10, note) says that this Memoir was translated and printed in Van Mons' Journal,* vol. i. p. 1.)

On Electro-gilding. 8vo. (*Vide* his Elementi di Chimica, 4th edition, 1810.) (*Ann. de Chim.* xviii. 136 ; *and also in Van Mons' Journal de Chim.* i. 1.) *Paris,* 1806

Brugnatelli, Luigi—*continued.*

Extrait d'une Lettre de Brugnatelli sur des expériences qu'on prétend être contraires à l'identité des fluides électriques et galvaniques, datée de Paris du 3 Brumaire, an 10. (*Journal de Chim. de Van Mons,* ii. 216.)

Bruxelles, 1802

(*Note.—This is probably the proper title. It contains some experiments of Gautherot, made at the third meeting of the Commission of the Institute on Volta's "Identité" . . . held at " le cabinet de M. Charles.*")

Extrait d'une Lettre sur les Expériences prétenduement contraires à l'identité des Fluides Electrique et Galvanique. 8vo. (*Journal de Chim. de Van Mons,* i. 216.) *Bruxelles*

Extrait d'une Lettre. 1. Expériences Electro-galvaniques proposées par le premier Consul Bonaparte. 2. Construction d'une grande pile à Chaudrons. 8vo. (*Journal de Chim. de Van Mons,* i. 325.) *Bruxelles*

Extrait d'une Lettre contenant : 1. Vapeur vésiculaire. 2. Décomposition de l'Alcohol et de l'Oxiseptonique par la pile de Volta. 3. Analyse du Mercure fulminant. 8vo. (*Journal de Chim. de Van Mons* ii. 106.) *Bruxelles*

Extrait d'une Lettre contenant : 1. Propriété fulminante de l'acide nitrique. 2. Action de la pile sur diverses humeurs animales. 3. Conversion des rondelles de drap impregnées d'eau salée en savon de laine par l'action de la pile. 8vo. (*Journal de Van Mons,* iv. 143.) *Bruxelles*

Extrait d'une Lettre : 1. Sur une correction faite à la pile portative de Volta. 2. Sur une réforme de la nomenclature chimique. 8vo. (*Journal de Chim. de Van Mons,* i. 24.) *Bruxelles,* 1802

Extrait d'une Lettre contenant : 1. Nouvelles expériences galvaniques. 2. Découverte d'un sulfure d'antimoine oxidé rouge, sublimé. 8vo. (*Journal de Chim. de Van Mons,* i. 101.) *Bruxelles,* 1802

† Lettera al Sig. L. Targioni contenente un' Esposizione succinta delle principali scoperte o interessanti Osservazioni fatte nel corrente anno 1802, o poco prima. 8vo. 30 pp. (*Annali di Chim. di Brugnatelli,* xxi. 3.) *Pavia,* 1802

† Descrizione del Piliere del Volta, comunicata dal Brugnatelli al Van Mons. 8vo. 12 pp. 1 plate. (*Annali di Chim. di Brugnatelli,* xix. 77.) *Pavia,* 1802

† Notizie Letterarie : 4. Sul galvanismo. 5. Uso medico del piliere di Volta. 7. Piliere composto di un solo metallo, proposto da Davy. 9. Magnetismo del cobalto e del nikel, di Sage. 8vo. 8 pp. (*Annali di Chim. di Brugnatelli,* xix. 274, 277, 280, 281.) *Pavia,* 1802

† Fenomeni osservati colla Pila Voltiana. 8vo. 6 pp. (*Annali di Chim. di Brugnatelli,* xxi. 143.) *Pavia,* 1802
(*Note.—In this article, at p. 148, he says : "La maniera più spedita di vedere ripristinati alcuni termossidi metallici in soluzione, coll' azione della pila, si è per mezzo degli ammoniuro-metallici." This memoir was reproduced in Van Mons's Journal de Chim. lxxvi.*)

† Elementi di Chimica, &c. ii. *Pavia,* 1803
(The 4th edition, 1810, tom. i. p. 185. Electro-gilding, &c.)

Elementi di Chimica, &c. 2nd edition. 3 vols. 8vo. *Pavia,* 1804

Elementi di Chimica appoggiati alle più recenti Scoperte, Chimiche, e Farmaceutiche. 4 vols. 4to. (Figures.) *Pavia,* 1804

† Notizie delle principali Osservazioni e Scoperte fatte nella Chimica nell' anno 1804. 8vo. 35 pp. (*Annali di Chim. di Brugnatelli,* xxii. 1.) *Pavia,* 1805
(*Note.—He says, at p. 26 . . . "abbiamo fatto vedere come l'ammoniuro d'oro possa servire a dorare l'argento per mezzo della pila di Volta." At p. 16, Brugnatelli and Volta's Experiment upon Ritter's secondary pile.*)

† On Electro-gilding, headed Galvanism. 8vo. 1 p. *London,* 1805
(*Note.—In a letter to Van Mons, dated in 1803, he says : "I have lately gilt, in a complete manner, two large silver medals, by bringing them into communication, by means of a steel wire, with the negative pole of a Voltaic pile, and keeping them, one after the other, immersed in ammoniuret of gold, newly made and well saturated.*) (*Phil. Mag.* xxi. 187.)

Brugnatelli, Luigi—*continued.*

†   Nuove Scoperte Galvaniche del Ritter, comunicate dal Bernouilli al Van Mons. Esperienze Galvaniche del Medesimo. Osservazioni Chimico-galvaniche dell' Œrsted. 8vo. 17 pp. (*Annali di Chim. di Brugnatelli*, xxii. 77 to 92. Translated, with Notes by B . . ., from *Journal de Chim. de Van Mons*, No. 17.)   *Pavia*, 1805

†   Nota (di B.) sopra una Pila di sostanze vegetabili. 8vo. 4 pp. (*Annali di Chim. di Brugnatelli*, xxii. 301.)   *Pavia*, 1805

†   Osservazioni Chimico-galvaniche comunicate all' Istituto Nazionale d'Italia. 8vo. 44 pp. 1 plate. (*Annali di Chim. di Brugnatelli*, xxii. 257.)   *Pavia*, 1805

      (*Note.—This contains the experiments detailed in his* "Fenomeni Osservati colla Pila," *repeated on a larger scale.*)

†   Chemico-Galvanic Observations communicated to the National Institute of Italy. 8vo. 10 pp. (*Phil. Mag.* xxv. 57.)   *London*, 1806

Osservazioni sopra l'Identità di alcuni nuovi caratteri del Carbone con quelli dei metalli. 4to. (*Mem. dell' Istit. Naz. Italiano*, tom. i. part ii.p. 291.)   *Bologna*, 1806

Maniera di indorare le Medaglie ed i fini pezzi d'Argento col Galvanismo. (*Effemeridi Chim. Mediche di Milano*, 1807. Semestre i. 57.)   *Milano*, 1807

Discorso preliminare sulle Osservazioni e sulle Scoperte più importanti fatte nel 1807. 4to. (*Giorn. di Fis. Chim.* &c. i. 4-32.)   *Pavia*, 1808

Nota sopra un' Objezione fatta alla Teoria Termossigena, e sopra alcune Vegetazioni Metalliche ottenute col Galvanismo. 4to. (*Giorn. di Fis. Chim.*, &c. i. 28.)   *Pavia*, 1808

Singolare Degenerazione del cartone col Galvanismo. 4to. (*Giorn. di Fis. Chim.* &c. i. 146.)   *Pavia*, 1808

Sui Conduttori Elettrici applicati alla Pila Voltiana, detti Galvanici. Memoria dei Prof. Brugnatelli e Configliachi. 4to. (*Giorn. di Fis. Chim.* i. 147 e 338.)   *Pavia*, 1808

Nota sulla Decomposizione degli Alcali ottenuta dal Davy. 4to. (*Giorn. di Fis. Chim.* &c. i. 164.)   *Pavia*, 1808

Trattato elementare di Chimica generale, &c. 8vo. tom. iv.   *Pavia*, 1810

Di alcune leghe Metalliche ottenute col Galvanismo. 4to. (*Giorn. di Chim. Fis. e Storia Nat.* ix. 145.)   *Pavia*, 1816
      Note.—Cossa says: "*Questa memoria venne riprodotta per estratto nei seguenti giornali.*"
      Thomson's Annals of Philosophy, xii. 228. 1817.
      Journal de Pharmacie, iii. 425. Paris, Set. 1817.
      Giornale di Chimica, Fisica, e Storia Naturale, xi. 130. Pavia, 1818.

Osservazione di L. Brugnatelli sopra le leghe metalliche che si possono avere col Galvanismo per via umida. per servire di risposta all' Articolo del Sig. G. L., inserito negli *Annales de Chimie*, Fév. 1818. 4to. (*Giornale di Chimica, Fisica, e Storia Naturale*, xi. 130. 2nd decade.)   *Pavia*, 1818

Memoirs of the Life of Lewis Brugnatelli. 8vo. 6 pp. (*Phil. Mag.* liii. 321.)   *London*, 1819

Elogio del di Bizio. Letto nell' Ateneo di Venezia, 19 Luglio, 1827.   *Venezia*, 1832

†   Cenni sulla Vita e sugli Scritti di Luigi Valentino Brugnatelli, Prof. di Chimica nell' Università di Pavia, dal 1797 al 1818. Di A. Cossa. 8vo. 32 pp.   *Pavia*, 1857

†   Portrait of Brugnatelli, by Longhi.

Brugnatelli, L., e Brera, V. L. Commentarii Medici. (Opera Periodica.) 8vo.   1796 to 1799
      Note.—Follows the Giornale Fisico-Medico. "*Continuato dopo la parte 2 del tom. i. dal solo Brera.* (*Vide Vita di B. di Cossa, p.* 32.)

Brugnatelli, L., Brugnatelli, G. Brunacci, and Configliachi. Giornale di Fisica, Chimica, e Storia Naturale. (Decade prima.) 10 vols. 4to.
Pavia, 1808-17
(*Note.—Vols.* i. *to* vii. *are by L. Brugnatelli, Vol.* viii. *by L. and G. Brugnatelli, Vols.* ix. *and* x. *by G. Brugnatelli, Brunacci, and Configliachi. This work follows L. Brugnatelli's "Annali di Chimica,' &c. For the continuation of this journal, vide Brugnatelli, G., Brunacci, and Configliachi.* (*Seconda decade, &c.*)

Bruhns, C. Geschichte und Beschreibung der Leipziger Sternwarte, &c.
Leipzig, 1861
† Bestimmung der Längendifferenz zwischen Berlin und Wien auf Telegraphischem Wege ausgeführt. Von den Herren Forster und Weiss. Herausgegeben von Dr. C. Bruhns. 4to. 47 pp. (*Publ. der Konig Preuss. geodatis. Instituts.*)
Leipzig, 1871
Bestimmung der Längendifferenz zwischen Leipzig und Wien. iii. *Leipzig,* 1872

† **Bruhns e Förster.** Bestimmung der Längendifferenz zwischen den Sternwarten zu Berlin und Leipzig, auf Telegraphischem Wege ausgeführt im April, 1864. 4to. 73 pp. *Leipzig,* 1865

Brunacci. Vide Brugnatelli and others.

Brunel, Sir Marc Isambert. *Born April* 25, 1769, *at Hacqueville, Normandy; died December* 12, 1849, *at London.*

† **Brunel and Daniel.** Statement of Facts respecting Cooke and Wheatstone's relative Positions in connection with the Invention of the Electric Telegraph. Dated 27th April. 1 page, folded. *London,* 1841

Brunet, J. C. Manuel du Libraire. 3 vols. 8vo. *Paris,* 1810
Manuel du Libraire. 4 vols. 8vo. *Paris,* 1820
Nouvelles Recherches. Supplément. 3 vols. 8vo. *Paris,* 1834
Manuel du Libraire. 5 vols. 8vo. 4th edition. *Paris,* 1844
Manuel du Libraire et de l'Amateur des Livres. 6 vols. 12 parts, royal 8vo. 5th edition. *Paris,* 1860-4

Brunet, G. Annuaire des Sociétés savantes. Notice de. 8vo. 1846

† **Bruni, J. L.** Extract of a Letter from J. L. Bruni to H. Baker, concerning the Bologna Bottle. Read Jan. 31, 1744-5. 4to. 2 pp. (*Phil. Trans.* xliii. 272.)
London, 1744-5

Brunner, Carl, jun.? *Born June* 23, 1823, *at Berne.*
De ratione, quæ inter Fluidorum cohæsionem et Calorem aliasque vires Moleculares intercedit. 4to. Tabula adjecta. *Berolini,* 1846
Elektrische Lichterscheinungen ohne Donner. (*Fror. Not.* ix. x. 152.) 1848

† **Brunner, C.** (Sohn). Untersuchungen über die Cohäsion der Flüssigkeiten. 4to. 44 pp. 2 plates. 1849
(*Note.—He refers to Draper's (of New York) Theory of Electricity.*)

† **Brunnstrom, J. A.** Några thorier for Elektricitet. 8vo. 46 pp. *Lund,* 1864

† **Bruno, De.** Recherches sur la Direction du Fluide Magnétique. 8vo. 8 plates. 206 pp. *Paris,* 1785

† **Bruns, V. von.** Die Galvano-Chirurgie oder die Galvanokaustik und Elektrolysis bei chirurgischen Krankheiten. 8vo. 145 pp. *Tubingen,* 1870

Brussels Academy. Mémoire de l'Académie des Sciences et Belles Lettres de Bruxelles. 4to. *Bruxelles,* 1777-88
Nouveaux Mémoires. Idem. 4to. *Bruxelles,* 1820
Mémoire sur les Questions proposées. 4to. 1773 ?
Bulletin de l'Académie des Sciences et Belles Lettres de Bruxelles. 8vo.
Bruxelles

Brussels Academy—*continued.*
Observations Magnétiques et Météorologiques faites à des Epoques déterminées 4to. 5 parts ? (*Extrait de l'Acad. Roy. de Bruxelles*, xv. to xix. ?)
Bruxelles, 1844

† **Bryant, W.** Account of an electrical Eel ; or, the Torpedo of Surinam. 4to. 4 pp (*Trans. American Phil. Soc.* ii. 166. Old series.)   *Philadelphia*, 1786

Brydone, Patrick. *Born* 1741, *at Dumbarton, N.B.; died June* 19, 1818, *at Lennel House, Berwickshire.*
An Instance of the Electric Virtue in the Cure of a Palsy. 4to. (*Phil. Trans* part i. p. 392, 1757, and p. 695.)   *London*, 1757
Letter containing an Account of a Fiery Meteor seen on the 10th February, 1772, and also of some new Electrical Experiments. 4to. (*Phil. Trans.* 1773, p. 163.)
London, 1773

† Osservazioni sull' Elettricità. Lettera ix. 12mo. 16 pp. (*Scelta d' Opuscoli*, x. 71.)
Milano, 1775

Brydone. Voyage de Sicile et de Malte. (*Lightning of Volcanoes.*)

† **Buccio.** L'uso medico che si può fare dell' elettricità e del galvanismo. 8vo. 5 pp. (*Comment. Acad. Dipartimento del Mella per* 1811, p. 60.)   *Brescia*, 1812

† **Buch,** Leopold de. Extract of a Letter from M. Leopold de Buch, of Milan, to Professor Pictet, of Geneva, on the production of Muriatic Acid and Soda by the Galvanic Decomposition of Water. 8vo. 2 pp. (*Phil. Mag.* xxiv. 244.)
London, 1806

Buchan, A. Introductory Text-book of Meteorology. 8vo.   1871

Buchenau, F. Mittheilungen über einen interressanten Blitzschlag in mehreren Stielchen. 4to. 15 pp.   *Dresden*, 1867

† **Buchenroder, W. L.** von. An Account of the Earthquakes which occurred at the Cape of Good Hope during the month of December, 1809. Abridged by him. 8vo. 5 pp. (*Phil. Mag. or Annals*, ix. 71.)   *London*, 1831
(*Note.*—At p. 73 Reports of Meteors (explosive, &c.) seen at the time of Earthquake shocks, &c.)

Bucher. Einige gegen die Gewitterableiter gemachte Einwürfe beantwortet. 8vo.
Frankfort, 1790

Buchner, O. Die Feuermeteore, insbesondere die Meteoriten historisch und naturwissenschaftlich betrachtet. 8vo. 192 pp.   *Geissen*, 1859
Die Meteoriten in Sammlungen, ihre Geschichte, mineralogische und chemische Beschaffenheit. 8vo. 202 pp.   *Leipzig*, 1863
Zweites Quellenverzeichniss zur Literatur der Meteoriten. Ein Anhang zu Kesselmeyer, über den Ursprung der Meteorsteine. 4to. 19 pp. (*Abgedruckt aus den Abhandl. der Senckenbergischen naturforschenden Gesellschaft.*)
Frankfurt, 1863

† Die Construction und Anlegung der Blitzableiter, zum Schutze aller Arten von Gebäuden, Seeschiffen, und Telegrafen-stationen ; nebst Kosten Voranschlägen . . . . 8vo. 152 pp. mit einen Atlas von 6 Foliotafeln.   *Weimar*, 1867

Bucholz, C. F. Ueber die chemischen Wirksamkeit galvanischer Ketten, aus Metall-Lösungen, Wasser oder Säure, und Metallen. 8vo. (*Gehlen's Journ. für Chem. und Phys.* v. 1808.)   1808

Budge. On Irritation of the Medulla oblongata, and its Influence on the Brain, &c. (*Archiv von Roser und Wunderlich*, 1846.)   1846
On "ein Centralorgan für den Lendentheil des Sympathicus," and Electrisation thereof, &c. (*Virchow's Archiv*, 1859, p. 115.)   1859

† **Budge, J.** Ueber den Galvanischen Strom, welcher sich in der Haut des Frosches zu erkennen giebt. 8vo. 16 pp. (*Poggendorff's Annalen*, bd. cxi.)
Greifswald, 1860

Buerbaum, J. Die Elektro-magnetische Telegraphie, mit besonderer Berücksichtigung der ausgeführten Telegraphen-Systeme. 8vo.   *Berlin*, 1851

Buff, Heinrich. *Born May 23, 1805, at Rodelheim, Wetterau.*

Many Articles on Electricity, &c. in *Poggendorff, Ann.* 1829 to 1855.

† Zusammenhang der neuern Elektricitätslehre mit der Contact-theorie. 8vo. *Giessen,* 1842

† Ueber den Einfluss der Zwischenplatten in der Electrischen Säule. 8vo. 8 pp. (*Annal. de Pharm.* xxxii. 1 heft.) *Heidelberg,* 1840

† Ueber Volta-Electrische Quantität und Intensität. 8vo. 6 pp. (*Annalen der Pharm.* xxxii. 1 heft.) *Heidelberg,* 1840

† Ueber die trockene Electrische Säule. 8vo. 23 pp. (*Annal. der Chem. und Pharm.* xxxiv. 1 heft.) *Heidelberg,* 1840

† Ueber das Elektrolytische Gesetz. 8vo. 15 pp. (*Annal. der Chem. und Pharm.* lxxxv. 1 heft.) *Heidelberg,* 1842

† Zur Berührungselectricität. 8vo. 4 pp. (*Annal. der Chem. und Pharm.* lxxxiii. 2 heft.) *Heidelberg*

† Ueber die Electrische Beschaffenheit der Flamme. 8vo. 16 pp. (*Annal. der Chem. und Pharm.* lxxx. 1 heft.) (*Faraday's copy.*) *Leipzig*

Letters on the Physics of the Earth, by Dr. Hoffman. 12mo. *London,* 1851

Buffon, George Louis Leclerc. *Born Sept.* 7, 1707, *at Montbard, Bourgogne; died April* 16, 1788, *at Paris.* (*Poggendorff,* i. 338.)

Les Epoques de la Nature. 2 vols. (On Earthquakes, 4th Epoch.) *Paris,* 1780

† Traité de l'Aimant et de ses Usages. 4to. 208 pp. and tables. (This is the 5th volume of his *Hist. Nat. des Mineraux,* in 5 vols. 4to.) *Paris,* 1788

Bugge, Thomas. *Born Oct.* 12, 1740, *at Copenhagen; died Jan.* 15, 1815, *at Copenhagen.*

Articles in *Dansk. Vid. Selsk.* (*Poggendorff,* i. 339.)

Buissart. Mémoire sur les divers Avantages qu'on pourroit retirer de la Multiplicité des Conducteurs Electriques, ou Paratonnerres. Lu à l'Acad. d'Arras, 24 Avril, 1781. (*Saumlez, Phys. Supplem.* 1782. xxi. pp. 140-48.) 1782

Mémoire juridique sur les Conducteurs Electriques. (From *Van Swinden,* p. 137 ; *Van Troostwyk* and *Krayenhoff,* p. 241.)

Bulengerus. De Terræ motu et Fulminibus. (*Grævius, Thes. Antiq. Rom.* v. 519-42. lib. v.)

Bulletin International. Marié Davy et autres Auteurs. (Contains Observations of Storms, Tornadoes, &c.)

Bulletin des Sciences Mathématique, Astronomique, Physique, et Chimiques. 16 vols. 8vo. *Paris,* 1824-31

Bulletin des Sciences Technologiques. 18 vols. 8vo. *Paris,* 1824-31

Bullettino Meteorologico dell' Osservatorio del Collegio Romano. *Vide* Secchi.

† **Bulmerincq,** M. C. von. Beiträge zur ärztlichen Behandlung, mittelst des Mineralischen Magnetismus. Mit einer Vorrede von Heinrich Steffans. 8vo. 74 pp. *Berlin,* 1835

Bunsen, Jeremias. *Born Dec.* 8, 1688, *at Arolsen; died March* 11, 1752, *at Arolsen.*

Versuch wie die Meteore des Donners und Blitzes, item das Aufsteigen der Dünste, ingleichen des Nordscheins, aus elektrischen Wirkungen herzuleiten und zu erklären sind ; u.s.w. *Lemgo,* 1750

Erklärung der elektrischen und magnetischen Kräfte. 8vo. (From *Krunisz,* p. 28 ; *Nollet, Essai,* 4th edition, p. 222. *Poggendorff,* i. 340, says, "Frankfurt und Leipzig.) *Lemgo,* 1752

Bunsen, Robert Wilhelm. *Born March* 31, 1811, *at Göttingen.*

Anwendung der Kohle zu Voltaischen Battarien. (*Poggendorff, Ann.* liv. "in or before 1842.") *Leipzig*

N

Bunsen, R. W.—*continued.*
Bereitung einer Kohle als Ersatz des Platins in der Grove'schen Kette. 8vo.
(*Pogg. Ann.* lv. 1842.) *Leipzig,* 1842
(*This is Volta's discovery,* not *Bunsen's.* F. R.)
Ueber die Kohlenzinkkette. Inaugural Dissertation. *Marburg,* 1843
Verbesserte Kohlen-batterie. 8vo. (*Pogg. Ann.* lx. 1843.) *Leipzig,* 1843

Buntzen, Thomas. *Born Sept.* 5, 1776, *at Copenhagen; died* 1807, *at St. Petersburg.*
Beitrage zu einer künftigen Physiologie. *Kopenhagen and Leipzig,* 1803

† **Burci.** Passage of an Electrical Current . . . Medical. (*Vide* Matteucci.)
Dei Casi di Aneurisma nei quali può essere raccomandata l' ago-Elettro-
puntura, e dei Modi per regolarla . . . 8vo. 58 pp. *Pisa,* 1852

Burdach, K. F. Bericht von der anatomischen Anstalt zu Königsberg. 8vo.
Königsberg, 1822

Bureau des Longitudes. Annuaire publié par le Bureau des Longitudes.
23 vols. 18mo. *Paris,* 1822-52

Burgravius. Cura morborum . . . Magnet. 8vo. *Frankfort,* 1630

Burkhardt, Basle. Uber die Anwendung des Galvanismus bey Hysterie und Con-
vulsionen. Beobachtungen. 8vo. (*Allgemeine Medizinische Annalen, Junius,*
1802.) 1802

Burman, Erich J. De Mineræ Ferræ per Magnetem investigatione. *Upsaliæ,* 1728

Burnaby, A. Voyages dans l'Amérique Septentrionale. Traduit de l'Anglais.
(Conductors melted by Lightning.) *Lausanne,* 1778

† **Burnet.** On the Motion of Sap in Plants. Researches of Dutrochet on Endosmose
and Exosmose, who refers this influence to Electricity. At Royal Institution,
Evening Meeting, March 27, 1829. 8vo. 2 pp. (*Phil. Mag. or Annals,* v. 389.)
London, 1829

† **Burney, W.** Answers by Dr. Wm. Burney to the Queries proposed by John Farey,
Esq., sen. (*Phil. Mag. for June*), respecting Shooting Stars and Meteors.
8vo. 3 pp. (*Phil. Mag.* lviii. 22.) *London,* 1821

† Description of a Meteoric Phenomenon which appeared at Gosport, July 4, 1821.
8vo. (*Phil. Mag.* lviii. 78.) *London,* 1821

† On the Appearance of Meteors, Parhelia, and Paraselenæ, as Prognostics in
general of Wind and Rain. 8vo. 7 pp. (*Phil. Mag.* lviii. 127 and 198.)
London, 1821

† A large Meteor on the 11th December, 1821, observed at Gosport, 8vo. 1 p.
(*Phil. Mag.* lviii. 466.) *London,* 1821

Burnouf. (*Vide* Guillemin and Burnouf.)

† **Burq,** V. Métallo-thérapie. Traitement des Maladies nerveuses, &c., par les Applications Métalliques. Abrégé Historique, &c. 8vo. 48 pp. (Extrait de 22
Mémoires ou Notes aux deux Académies.) *Paris,* 1853

† **Burrell,** E. An Account of Two Northern Auroras, as they were observed at the
Vicarage of Sutton-at-Hone, in Kent. 4to. 1 p. (*Phil. Trans. for* 1717,
p. 584.) *London,* 1717

Bursen. On Electricity. 38 pp. *Leipzig,* 1752

Busse, Friedrich Gottlieb von. *Born April* 3, 1756, *at Gardelegen, Altmark; died*
Feb. 4, 1835, *at Freiberg.*

† Beruhigung über die neuen Wetterableiter. 8vo. 62 pp. *Leipzig,* 1791
Beschreibung einer wohlfeilen und sichern Blitzableitung, mit neuen Gründen
und Erfahrungen. 8vo. 1 plate. *Leipzig,* 1811

Bussolin, P. Memoria Chimico-docimastica esponente un Processo per separare il
rame dallo Zinco, applicabile pùre alla Miniera di rame (d'Agordo), e suoi
edotti, &c., di P. Bussolin. Capo . . . presso l'Imp. Reg. Zecca di Venezia.
8vo. (*Giornale da Rio,* lxvi. 156.) *Padova* 1828

Butet. Notice des principaux Résultats obtenus sur l'Electricité Galvanique, par les Expériences faites à l'Ecole de Médecine. (*Bulletin des Sciences de la Soc. Philomath.* No. 43, Vendémiaire, an 9.) *Paris,* 1801

Butschany, Matthias. *Born Feb.* 10, 1731, *at Altsohl, Ungarn ; died August* 2, 1796, *at Hamburg.* (From *Poggendorff,* i. 353.)
Dissertatio de Fulgure et Tonitru ex Phænomenis Electricis. 4to. pp. 1 et 2. (*Poggendorff,* i. 353.) *Göttingen,* 1757
Der Blitz entsteht nicht durch Entzündung einiger brennbaren Theilchen die in der Luft schweben, und ist auch kein Feuer. (*Beitrage zu Hannov. Magazin,* 1761.) *Hanover,* 1761
Eine Unvollkommenheit der Blitzableiter, nebst ihrer Verbesserung. 8vo. (From *Poggendorff,* i. 353.) *Hamburg,* 1787

Butterfield. On Magnetical Sand. 4to. (*Phil. Trans. for* 1698, p. 336.) *London,* 1698

Buvina, Michel. On the Influence exerted by the Magnet on Carpes, even at a distance. (*Buffon, Hist. Nat.* Edition of Sonnini, tom. xiii. *de l'Histoire des Poissons,* p. 103.)

† **Buys-Ballot.** Meteorologische Preisfrage ausgeschreiben von der Societät der Künste und Wissenschaften zu Utrecht, und Motivirung derselben. 8vo. 28 pp. (*Berl. Fortsch. der Phys.* iii. 565.) *Berlin,* 1847
(*At p.* 586, *Registering Apparatus for Atmospheric Electricity, &c., of Sir Francis Ronalds.*)

† **Buzengeiger.** Electrische Pendel Uhr. 8vo. (*Stutgard Morgen-Blätter,* Sept. 23, 1815.) 1815

† **Buzorini,** L. Luftelectricität, Erdmagnetismus und Krankheitsconstitution. 8vo. 227 pp. *Belle-Vue bei Constanz,* 1841

Buzzetti, C. Sullo stato Meteorico della Lombardia. 8vo. 2nd edition. 36 pp. (*Réfers to Kreil and Vedova's Magnetic Observations, &c.*) *Milano,* 1846

† **Buzzi,** F. Osservazione di un Amaurosi (gotta-serena) pituitosa curata coll' Elettricità. 4to. 2 pp. (*Opuscoli Scelti,* vi. 359.) *Milano,* 1783

† **Bywater,** T. An Essay on the History, Practice, and Theory of Electricity. 8vo. 127 pp. *London,* 1810
Essay on Electricity. (From *Bentley,* 1814 to 1839.)

C.

C., L. O. (*Vide* Anon. Electricity.)   **1813**

C * * *, J. (*Vide* Anon. Magnetism.)   **1695**

Cabanis, P. J. G. Rapports du physique et du moral de l'homme. 8vo. 2 vols. (*From Sue, Hist.* iv. 209.) *Paris*, 1805

Cabeus, Niccolas. *Born 1585, at Ferrara; died June 30, 1650, at Genua.* (*Poggendorff*, i. 355.)

Philosophia Magnetica, in qua magnetis natura penitus explicatur et omnium quæ hoc lapide cernunter causæ propriæ afferunter, &c., nova etiam Pyxis construitur, cum suo meridiano quæ propriam Poli elevationem cum suo meridiano ubique demonstrat; multa quoque dicuntur de electricis et aliis attractionibus, et eorum causis. Additis figúris variis tam cencis quam ligno incisis. Auctore Nicolo Cabeo, Ferrariensi, Societ. Jesu. *Prostant Coloniæ*, 1629

Meteorologicorum Aristotelis Commentaria et Quæstiones . . quibus universa fere phil. exper. exponitur. . . 4 vols. fol. (*Magnetic Mountains*, tom. i. p. 66. *Magn. and Tides*, ii. 68. *Thunder* (*Ancient Opinions on*), ii. 280, 284. *On Thunder elsewhere. On Electricity*, iv. 266 and 412. *Coloured Clouds*, iii. 216.) *Roma*, 1646

Cabot (Caboto, Gaboto), Sebn. Machte in 1497 . . eine Entdeckungs Reise in die westlichen Meere. . In 1498 bischiffte er auf einer zweiten Reise die Küsten der Vereinigten Staaten bis zur Spitze von Florida. Beobachtete auf diesen Reisen die Verschiedenheit der magnestichen Declination. 8vo. (*From Poggendorff*, i. 355.) *No indication as to publication.* Quaritch, Cat., June 25, 1871, p. 6, has : " *Cabot* (*S.*), *Memoir of, with a Review of the History of Maritime Discovery.*" **1831**

Cabry. (*Vide* Quetelet and others.)

Cadet, Jean Marc. *Born September 4, 1751, at Metz; died September, 1835, at Strasburg.*

On Lava attracted by the Magnet. (*Nova acta phys. med.* . . tom. iii.)

Cadet-de-Gassicourt, C. L. Dictionnaire de Chimie, &c. 8vo. 4 vols. (*From Sue*, iv. 204.) *Paris*, 1803

Cæsar. De bello Africano. Cap. 6. B.C. 44.

Caggiati. (*Vide* Linati and Caggiati.)

† **Cagnola.** Programmi per conferimento di premii dell' Istituto Lombardo, Accad. di Padova, Accad. di Torino, Istituto di Bologna. 4to. 6 pp. (*Giorn. dell' I. R. Ist. Lomb.* nuova serie, viii. 348. *Premio di Cagnola*, p. 349.) *Milano*, 1856

Cagnoli, Antonio. *Born September 29, 1743, at Zante; died August 14, 1816, at Verona.*

Osservazioni meteorologiche. 8vo. . . 7 vols. **1788-96**

Calamai, L. Osservazioni sull' anatomia della Torpedine. 8vo. (*From Atti della 7ma Adunanza degli Scienziati Ital.* Parte Seconda, p. iv.) **1845**

† **Calamai, Owen** and **de Martino.** (Sopra la sua.) "Memoria intitolata sull' anatomia della Torpedine, e sopra un gabinetto, &c. . . 4to. 5 pp. (*Atti 7 Adun. d. Scienziati Ital. pro 1745.*) *Napoli*, 1846

† **Calamini e Platteretti.** Delle pietre cadute nelle colli Parmigiane. Due lettere all' Editore. 4to. 4 pp. (*Nuova Scelta d'Opuscoli*, ii. 273.) *Milano*, 1807

† **Calcinardi Rizeri, P.** Tromba marina sul Lago di Garda. 8vo. (*Estratto di Lettere 8 Luglio e 1° Agosto.* (*Gior. dell' I. R. Istit. Lombur.* viii. 145.) *Milano*, 1843

Caldani, Floriano. *Born about* 1772, *at Bologna ; died April* 11, 1836, *at Padua.*

† Riflessioni sopra. . . Vasi assorbenti ed esperienze sulla Elettricità animale. 8vo. 182 pp. (*Lette nell' Acad.* . . *di Padova.*) *Padova,* 1792

† Osservazioni sulla membrana del Timpano e nuove ricerche sulla Elettricità animale. 8vo. 198 pp. 3 pls. (*Lette nell' Acad.* . . *di Padova, con un' Appendice.*) *Padova,* 1794

Lettera nella quale si esaminano alcune riflessioni circa le nuove ricerche sulla Elettricità animale publicate nel tomo xvii. degli Opuscoli Scelti. . . 8vo. 16 pp. *Padova,* 1795

† Lettera (prima) allo Spallanzani sull' Elettricità animale. Padova 10 Agosto, 1793. 8vo. 21 pp. (*Ann. di Chim. di Brugnatelli,* vii. 138.) *Pavia,* 1795

† Lettera seconda allo Spallanzani sull' Elettricità animale. 7 Ap. 1794. 8vo. 14 pp. (*Ann. di Chim. di Brugnatelli,* vii. 186.) *Pavia,* 1795

† Lettera terza all' Spallanzani sull' Elettricità animale. 8vo. 6 pp. (*Ann. di Chim. di Brugnatelli,* vii. 208.) *Pavia,* 1795

Lettera all' Olivi sull' Elettricità animale. 8vo. 27 pp. (*Ann. di Chim. di. Brugnatelli,* vii. 159.) *Pavia,* 1795

Traduzione degli Elementi de Chimica di Chaptall sulla 3a. edn. con . . alcune annotazioni del traduttore. 4to. 8 vols. *Venezia,* 1801

Caldani, Leopoldo Marc-Antonio. *Born November* 21, 1725, *at Bologna ; died December* 24, 1813, *at Padua.*

Sull' insensitivita ed irritabilità di alcune parti degli animali Lettera scritta al Sig. Haller. *Bologna,* 1757

Memorie lette nell' Acad. di Padova. 4to. *Padova,* 1804

† Memorie intorno alla vita ed alle opere, etc. 4to. 79 pp. portrait. (*Mem. della Soc. Ital.* xix. *Pt. fis.* p. 1, 1823.) *Modena,* 1823

In this Mem. works of his (*L. C.*) *and of* Haller *on Irritability, &c. are much referred to, particularly " sull' insensitività ed irritabilità di alcune parti degli animali : Lettera scritta al Sig.* Alb. Haller, *dated Bologna,* 1757.''

Caldecourt. (*Vide* Taylor and Caldecourt.)

† **Callan, N. J.** On the results of a series of experiments on the decomposition of water by the Galvanic Battery, with a view to obtain a constant and brilliant Lime Light. 8vo. 24 pp. (*Phil. Mag. February,* 1854.) *London,* 1854

† **Callaud, A.** Essai sur les Piles servant au développement de l'élect. Ouvr. couronné. 8vo. 54 pp. 1 plate. (*Ext. des Mémoires de la Société Impériale des Sciences, &c.* . . . *de Lille.*) *Paris,* 1860

Calogera, Angelo. Raccolta d'Opuscoli scientifici e filologici. 51 vols. of various dates. 12mo. Tom. i. 1728 ; tom. l. 1754 ; tom. li. 1757. *Venezia,* 1728-57

Note.—This collection was collated for Electricity, Magnetism, &c., 185—, *and the articles on those subjects were obtained by Sir Francis Ronalds.*

Nuova Raccolta d'Opuscoli scientifici e filologici. 42 vols. of various dates. 12mo. Tom. i. 1754 ; tom. xlii. 1787. *Venezia,* 1754-87

Cambridge Philosophical Society, Transactions of the. 4to. *Cambridge,* 1822

† **Cambridge U. S. University.** (*Vide* Farrar, 1839.)

Cambry, J. Traces du Magnétisme. 8vo. *La Haye,* 1784

Camerarius, R. J. De globo lucido meteoro. (*Ephem. Acad. Nat. Cur. Cent* 9 *et* 10.)

Camerer, J. W. Über das Einschlagen des Gewitters auf zwei mit Blitzableitern versehn Häusern. (*Tubing.-Blätter,* 1815, Bd. ii.) *Tubing.* 1815

† **Cameron,** Paul. Variation and deviation of the Compass rectified by azimuth and altitude tables. Likewise a treatise on Magnetism, and the deviations of the compass in iron ships, and the method of observing and correcting them by magnets. 3rd edition. 132 pp. *London,* 1868

† **Camin,** F. Da. Sull' Ago-puntura con alcuni Cenni sulla puntura elettrica. Lettere ed Osservazioni. 8vo. 45 pp. 1 plate. *Venezia,* 1834

† **Campos** Bautista, F. De la galvanocaustique chimique comme moyen de traitement des rétrècissements de l'urèthre. Thèse pour le Doctorat en Médecine soutenue 25 Fév., 1870. 4to. 162 pp. (*Faculté de Médecine de Paris.*)
Paris, 1870

Camus. (*Vide* Le Camus and Le Comus.)

Canali, Luigi. *Born* 1759, *at Perugia; died December* 8, 1841.

Sopra i paragrandini di Tholard. (*Giorn. Accad.* xix.)

Lettera . . sopra un' esperienza di Muschenbrock, . . e gli effetti dell' attrazioni e repulsioni elettriche. 4to. 18 pp. 1 plate. (*Opuscoli Scelti,* 4to, xviii. 55.) *Milano,* 1795

Questions sur la loi découverte par Volta relativement à l'élect. des vapeurs: (1) Cette loi a-t-elle du rapport aux autres lois propres de cette fluide . . . (2) Les anomalies . . en blessent-elles la vérité. 4to. (*Mem. de Turin,* tom. vi. pt. ii. p. 61.) *Torino.*

Canini, Giuseppe Maria Simone. *Born October* 31, 1720, *at Venice; died August* 1, 1796, *at Venice.*

† Dissertazione istorica che assegna e scopre il vero tempo in cui il Canini intraprese, e dichiarò il suo fisico studio sopra l'artifizial Magnetismo et sopra l'attivitò degli Effluvi di quello per gli occorsi utili e considerabili Effetti. 8vo. 64 pp. *Venezia,* 1775

Lettera apologetica sopra l'azione dell' effluvio magnetico sui nervi. (*From Poggendorff,* i. 370.) 1785

A pamphlet of 14 pages in 12mo, without other title than "*Agli Amatori dell' Esperienze fisiche,*" and without either date or place. The author describes his invention of an improved Magnet of "7 Corpi di finissimo Acciajo con vite, et placche d'Ottone connessi, in guisa, che formano un solo corpo, &c."

Canton, John. *Born July* 31, 1718, *at Stroud, Gloucestershire; died March* 22, 1772, *at London.*

On the Tourmaline. 8vo. (*The Gentleman's Magazine, September,* 1759, *and Phil. Trans., December,* 1759.) *London,* 1759

† **Cantoni,** Giovanni. Sunto di un corso di lezioni sui fenomeni elettrici e magnetici. 8vo. 114 pp. *Milano,* 1860

Sui fenomeni elettrici e magnetici ; compendio di elettrologia. 16mo. 142 pp. (*Zuchold's Cat.,* 1866, p. 178.) *Milano,* 1864

† Sulle correnti d'induzione. Nota seconda. 8vo. (*Rendiconti del R. Instit. Lombardo,* ii. 256, 1865.) *Milano,* 1865

† Cenni sul. Prof. Belli. (*Vide* Belli.)

Cantu, Cesare. Notizia di G. D. Romagnosi. 8vo. *Milano,* 1835

Caplin, J. T. F. Origin and use of the Electro-chemical Bath. 8vo.
London, 1856

† The Electro-chemical Bath for the extraction of . . . poisonous substances from the human body . . . and the relation of Electricity to the phenomena of Life, Health, &c. 8vo. 128 pp. *London,* 1856

† Selection of documents and autograph Letters in testimony of the Cures effected by the Electro-chemical Bath. 8vo. 140 pp. *London,* 1865

† Historical Records of the various affections cured by means of the Electro-Chemical Bath. 8vo. 284 pp. *London,* 1865

CAP—CAR 95

† **Capocci,** E. Seguito della Notizia alle scoperte di Melloni sul Calorico. 8vo. 17 pp. (*Bibliot. Ital.* lxxxix. 107.) *Milano,* 1838

† **Cappel,** A. J. L. Report on the Government Telegraph Department in Prussia. Fol. 54 pp. *London,* 1867

† **Capsoni.** Cenni storici sui Telegrafi elettrici. 8vo. 23 pp. *Milano,* 1850

Cara. Dissertation sur la Lumière et l'Electricité. 8vo. 86 pp. *London,* 1787

Caraffa, G. Sur la Tourmaline, lettre à M. Buffon. *Paris,* 1759

Caraffa. (*Vide* Noya-Caraffa.)

† **Caraman** (le Comte de). Observation sur un phénomène de l'atmosphère. Lu Nov. 1764. 4to. 3 pp. (*Toulouse Acad.,* 1st Serie, tom. i. *Hist.* p. 58). *Toulouse,* 1782

Carcassonne Bib. Publ. Catalogue. MSS.
This library contains few works on Electricity and Magnetism.

Cardanus, Giovanni. De fulgure. Liber unus. Fol. (*Cardani Geronimo Opera omnia,* 10 vols. folio, vol. ii.) *Lugd.* 1663

Cardell, F. P. Opusculum philosophicum de attractione magnetis. *Olomutiæ,* 1750

Carena, F. Expériences pour l'examen des deux principales théories de l'Electro-moteur de Volta. 8vo. (*Bibl. Ital.* iv. 63.) *Turin,* 1804

Carette, A. M. Géométrie du compas. (Translated from Mascheroni's Work.) *Paris,* 1798

† **Cari,** Gaetano. Nuovo conduttore spirale, con la teoria sua. 8vo. 42 pp. 1 plate. *Pistoja,* 1783

Carl, P. Repertorium für physikalische Technik für mathematische und astronomische Instrumentenkunde. 8vo. *Munchen,* 1865
† Die electrischen Naturkräfte des Magnetismus, die Electricität und der galvanische Strom, mit ihren hauptsächlichsten Anwendungen. 8vo. 314 pp. *Munchen,* 1871
† Über die bisher in Anwendung gebrachten galvanometrischen Einrichtungen. 8vo. 24 pp. 3 plates. (*Carl's Repertorium,* iii. 136.) *Oldenbourg*

† **Carli,** G. Dissertazione . . intorno alla Declinazione o Variazione della Calamita e Bussola nautica dal Polo. 4to. 32 pp. 1 plate. *Venezia,* 1747
† Lettera sopra l'elettricità animale, e l'apoplessia. 4to. (*Opuscoli Scelti,* 4to. xv. 302.) *Milano,* 1792

† **C. A. de C.** (*i.e.* de Carli), (Comte). Transunto di una lettera del C. A. de C. sull' elettricità animale rapporto alla generazione. 4to. 4 pp. (*Opuscoli Scelti,* 4to, xix. 66.) *Milano,* 1796

Carlini, Francesco. (*Born January* 8, 1783, *at Mailand; died August* 29, 1862, *at Badeort Crodo.*)
† Sulla legge della variazione oraria del Barometro. Memoria. 4to. 43 pp., 1 plate. (*Ital. Soc. Mem.* xx. 189.) *Modena,* 1828
† Sulla luce zodiacale. 8vo. 1 page. (*Giorn. dell' I. R. Istit. Lomb.,* 8vo. i. 24.) *Milano,* 1841
† Piano di osservazioni magnetiche contemporanee in varii punti del globo. 8vo. 1 page. (*Giorn. dell' I. R. Istit. Lomb.,* 8vo. xiii. 19.) *Milano,* 1846
† Rapporto intorno agli esperimenti dell' elettricità del Magrini. 8vo. 2 pp. (*Giorn. dell' I. R. Istit. Lomb.* 8vo. xiv. 19.) *Milano,* 1846

Carlisle, Sir Anthony. *Born February* 8, 1769, *at Stillington, Durham; died November* 2, 1840, *at London.*
Decomposition of Water. Exp. with Nicholson, April 30, 1800. 8vo. (*Nicholson's Journal,* iv.) *London,* 1800

† **Carmeli,** L. M. Ragionamento fatto sopra il Fenomeno apparso la notte del di 16 Dec. 1737. 8vo. 17 pp. (*Calogera, Raccolta . .* xviii. 463.) *Venezia,* 1738

Carmichael, J. S. Tentamen inaugurale medicum de paralysi. *Edinburgh* 1764
† On the Electric Fluids considered as different compounds of the Solar Rays.
8vo. 2 pp. (*Phil. Mag.* xli. 232.) *London*, 1813

Carminati, Bassiano. *Born about* 1750, *at Lodi; died at Milan.*
† Lettera al Galvani sull' Elettricità animale. 8vo. 7 pp. (*Giornale Fis. Med.*
di Brugnatelli, ii. 115.) *Paria*, 1792

Carmoy, —. Observations sur l'Electricité médicale. 8vo. (*Nouveaux Mém. de*
l'Acad. de Dijon.) *Dijon*, 1784
Réflexions sur les effets des commotions électriques relativement au corps
humain. 8vo. (*Nouv. Mém. de l'Acad. de Dijon*, 1785, 1er semestre, p. 112.)
Dijon, 1785
Lettre à M. Maret sur les effets des commotions électriques. 8vo. (*Nouv.*
Mém. de l'Acad. de Dijon, prem. semestre, p. 144.) *Dijon*, 1785
Réflexions sur l'électricité médicale. . . . (*Annales de la Soc. de Méd. de Mont-*
pellier, tom. vii. part i. p. 297.) *Montpellier.*

Carnevale-Arella, A. (*Vide* Arella.)

Caroccio, A. Sulla scoperta elettrica del Dott. A. Palagi ; discorso del Dott. A.
Caroccio. Articolo estratto dal Giornale di Bologna il Commercio, Anno
ii. No. 32, 33. . . .

† **Carpi, P.** Sull' influenza del magnetismo nelle chimiche combinazioni. Sperienze
del Carpi. Presentate 13th February, 1826. 4to. 8 pp. (*Mem. Soc.*
Ital. xx. 55, *Fisica.*) *Modena*, 1829

† **Carpue, J. C.** An introduction to Electricity and Galvanism ; with cases, showing their effects in the cure of diseases, to which is added a description of
Mr. Cuthbertson's plate electrical machine. Substance of lectures. 8vo.
112 pp. 3 plates. *London*, 1803

Carradori, Gioachino. *Born June* 6, 1758, *at Prato Toscana ; died Nov.* 24, 1818, *at*
Prato oder Pisa.
† Riflessioni sull' esperienze dei Pacts Van Troostwyk e Dieman, sulla decom-
posizione dell' acqua. . . . 8vo. 18 pp. (*Ann. di Chim. di Brugnatelli*, tom.
i. p. 1.) *Pavia*, 1790
† Lettere sopra l'Elettricità animale scritte a Felice Fontana. 8vo. 5 letters ;
the four first dated in 1792, 36 pp. ; the fifth dated in 1793, 23 pp. (*Brug-*
natelli's copy.) *Firenze*, 1793
† Articolo di lettera al Brugnatelli sull' elettricità animale e sopra alcuni nuovi
sali metallici. 8vo. 3 pp. (*Ann. di Chim. di Brugnatelli.* v. 27.)
Pavia, 1794
† Lettera sull' Elettricità animale . . . ad un amico. 8vo. 9 pp. (*Ann. di*
Chim. di Brugnatelli, viii. 140.) *Pavia*, 1795
† Articolo di Lettera sopra l'Elettricità animale . . . al Vitoni, B. 8vo. 5 pp.
(*Ann. di Chim. di Brugnatelli*, ix. 35.) *Pavia*, 1795
† Lettera sopra l'Elettricità animale scritta al Fontana (dated) da Prato, 24
Luglio, 1795. 8vo. 10 pp. (*Giorn. Fisico-Med.* An. viii. tom.) *Pavia*, 1795
† Esperienze ed Osservazioni sul fosforo delle Lucciole (Lampiris Italica). 8vo.
25 pp. (*Ann. di Chim. di Brugnatelli*, xiii. 41.) *Pavia*, 1797
† Lettera sopra la pretesa Elettricità animale scritta al Volta. 8vo. 7 pp. (*Ann.*
di Chim. di Brugnatelli, xv. 63.) *Pavia*, 1798
† Intorno alla causa della termossidazione dei metalli. . . Articolo di Lettera al
Fabbroni. 1 Agosto, 1801. 8vo. 11 pp. (*Ann. di Chim. di Brugnatelli*,
xxi. 31.) *Pavia*, 1802
† Annotazione all' opinione del Coulomb sul magnetismo universale. 8vo. 3 pp.
(*Ann. di Chim. di Brugnatelli*, xxi. 42.) *Pavia*, 1802

Carradori—*continued.*

† Osservazioni sopra alcune esperienze contrarie alla teoria di Lavoisier, fatte colla pila. 8vo. 9 pp. (*Ann. di Chim. di Brugnatelli,* xxi. 64.)
Pavia, 1802

Expériences et observations pour démontrer que la décomposition de l'eau par la pile . . n'est pas prouvée. 4to. (*Journ. de Phys.* An. xii. p. 20.)
Paris, 1804

† Esperienze per dimostrare che non è provata la decomposizione dell' acqua per mezzo della pila di Volta. 4to. 10 pp. (*Nuova Scelta d'Opuscoli,* 4to. i. 29.)
Milano, 1804

† Sperienze ed Osservazioni sopra l'Irritabilità della Lattuga . . (Presentata . . 19 Luglio, 1804.) 4to. 9 pp. (*Mem. Soc. Ital.* vii. 30. pt. i.)
Modena, 1805

† Istoria del Galvanismo in Italia o sia della contesa fra Volta e Galvani e Decisione ricavata dai fatti esposti dai due partiti. 8vo. 72 pp. *Firenze,* 1817

Memoria sull' azione del fluido elettrico, o galvanico come medicamento. (*Giorn. della Soc. Med. Chirurg. di Parma,* xiii. 33.)

† Notizie risguardanti la vita e studi del Dottore Giovacchino Carradori scritte dal . . . G. Raddi. Ricevute 3 Giugno, 1820. 4to. 8 pp. Portrait. (*Mem. della Società Ital.* tom. xix. p. 1.)
Modena, 1821

† **Carrara, M.** Sulla fosforescenza della Lucciola . . . 8vo. 14 pp. 1 plate. (*Bibliot. Ital.* lxxxii. 357.)
Milano, 1836

Carrié, Abbé, Hydroscopographie et metalloscopographie, ou art de découvrir les eaux souterraines et les gisements métallifères au moyen de l'électro-magnétisme. 8vo.
Paris, 1863

† **Carruthers.** (*Vide* Chambers & Carruthers.)

Cartheuser, Johann Friedrich. *Born September* 29, 1704, *at Hayne Grafsch, Stolberg; died June* 22, 1777, *at Frankfurt a. d. O.*

Dissertatio de incitamentis motuum naturalium externis. 4to. *Erfurt,* 1765

† **Cartwright** and **Sturgeon.** On electro-type from engraved copper plates. 8vo. 3 pp. 1 plate. (*Annals of Electricity,* v. 236.)
London, 1840

Casa. (*Vide* Della Casa L.)

Casali, G. De quorumdam vitrorum fracturis. Sermo alter, quo diluuntur objecta nonnulla quæ J. B. Scarella protulerat adversus sermonem primum. 4to. (*Comm. Bonon.* v. 1767.)
Bonon, 1767

† **Casari,** L. Sopra la grandine straordinaria caduta in Padova nel giorno 26 Agosto di questo anno. Memoria. 4to. 8 pp. (*Ann. del Reg. Lomb.-Veneto,* iv. 337.)
Padova, 1834

Sulle varie intensità delle correnti Voltiane . . . *Vienna,* 1835

† **Casati,** P. La Tromba parlante, le sue ragioni fisiche e matematiche. 4to.
Parma, 1673

Mechanicorum, libri viii. . . 4to. *Lugduni,* 1684

Caselli et **Bonelli.** Description des télégraphes de, par Du Moncel. (*Vide* Du Moncel.)
1863

Cassal, L. E. T. Essai sur les causes et Théorie du Magnétisme. 8vo. 120 pp. 3 plates.
Colmar, 1867

† **Cassano,** Prince, **Poleni,** Marquis, **Zanotti, De Revillas, Short** and **Fuller.** A collection of the observations of the remarkable Red Lights seen in the air on December 5, 1737, sent from different places to the Royal Society. 4to. (*Phil. Trans.* xli. p. 583 to 605.)
London, 1739-40-41

Casselmann, Wilhelm Theodor. *Born August 1, 1820, at Rinteln.*

Über den Einfluss des Gewitters auf die Drähte electromagnetischer Telegraphen.   8vo.   (*Poggendorff, Annalen,* lxxiii. 1848).   *Leipzig,* **1848**

Über die galvanische Kohlenzinkkette und einige mit derselben angestellten Beobachtungen.   (*Inaugural Dissertation.*)   8vo.   76 pp., 5 engraved tables. (*Heinsius von Schiller,* iii. 149.)   *Marburg,* **1843**

Cassini, Giovanni Domenico.   *Born June 8, 1625, at Perinaldo Grfsch. Nizza; died September 14, 1712, at Paris.*

Sur un nouveau phénomène ou sur une lumière céleste (Mar. 18, 1683).   (*The Zodiacal Light.*)   4to.   (*Anc. Mém. de Paris,* i. and viii. and x).   *Paris*

† Spina celeste meteora observata in Bologna, Marzo, 1668.   4to.   20 pp. 1 plate.   *Bologna,* **1668**

Cassini, Jacques.   *Born February 18, 1667, at Paris; died April 16, 1756, at Thury b. Clermont, in Beauvoisis.*

Observations sur la déclinaison de l'aimant.   4to.   (*Mém. de Paris,* vii. p. 508.)   *Pari.*

Observations de la déclinaison de l'aimant, faites à Londres en 1698.   4tos (*Mém. de Paris,* vii. p. 572.)   *Paris*

Observations de la déclinaison de l'aimant faites dans un voyage de France aux Indes Orientales, et dans le retour des Indes en France pendant les années 1703 et 1704.   4to.   (*Mém. de Paris,* 1705, *Mém.* p. 80.)   *Paris,* **1705**

Réflexions sur les observations de la variation de l'aimant faites dans le voyage du légat du Pape à la Chine l'an 1703.   (*Mém. de Paris,* 1705, p. 8.)
                                                                       Paris, **1705**

Réflexions sur la variation de l'aimant, observée par Houssaye, Capitaine commandant le vaisseau l'Aurore, pendant la campagne des Indes Orientales faite par l'escadre des vaisseaux, commandée par le Baron de Pallières en 1704 et 1705.   4to.   (*Mém. de Paris,* 1708, p. 173.)   *Paris,* **1708**

Réflexions sur les observations de la variation de l'aimant, faites sur le vaisseau le Maurepas, dans le voyage de la mer du Sud; avec quelques remarques de M. de la Verune, sur la navigation des côtes de l'Amérique et de la Terre de Feu.   4to.   (*Mém. de Paris,* 1708, p. 292.)   *Paris,* **1708**

De l'aurore boréale qui a paru le 16 Nov. de l'année 1729.   4to.   *Mém. de Paris,* An. 1729, *Hist.* p. 1, *Mém.* p. 321.)   *Paris,* **1729**

Observation sur la variation de l'aimant à la Guadeloupe.   4to.   (*Mém. de Paris,* vii. p. 455.)   *Paris*

Observation sur la variation de l'aimant à La Martinique.   4to.   (*Mém. de Paris,* vii. 456.)   *Paris*

† **Cassini,** Jacques Dominique.   *Born June 30, 1748, at Paris; died October 18, 1845, at Thury-sur-Clermont, Dép. Oise.*

De l'influence de l'équinoxe du printemps et du solstice de l'été sur la déclinaison et les variations de l'aiguille aimantée.   Lettre à l'Auteur du Journ. de Physique, Avril, 1784.   4to. 64 pp. 2 plates.   *Paris,* **1791**

Observation sur un météore en forme de globe de feu de la grandeur de la lune, vu en 1687 (d. 21 May).   4to.   (*Mém. de Paris,* ii. 74.)   *Paris*

Observations sur la déclinaison de l'aimant à Rome.   (*Mém. de Paris,* vii. 503.)   *Paris*

Déclinaison de l'aimant à Génes.   (*Mém. de Paris,* vii. 530).   *Paris*

† **Cassius Larcher, Daubincourt, et De Saintot.**   Précis succinct des principaux phénomènes du Galvanisme; suivi de la traduction d'un commentaire de J. Aldini sur un mémoire de Galvani ayant pour titre "Des forces de l'Electricité dans le mouvement musculaire" (ouvrage très rare et qui n'a point encore été traduit) : et de l'Extrait, d'un ouvrage de Vassalli Eandi ayant pour titre "Observations sur le fluide de l'Electromoteur de Volta."
8vo. 52 pp.   *Paris,* **1803**

† **Cassola, F.** Nuovi sperimenti sui raggi magnetici. 8vo. 3 pp. *(Bibliot. Ital.* lix. 129.) *Milano,* 1830

Castberg, P. A. Mémoire sur les effets du galvanisme appliqué aux sourds-muets, dans différents endroits de l'Europe. *(From Sue, Hist.* iv. 264).

Om Galvanismens mediciniska Nytte. *(Rafn's Nyt Bibl.* iv.)

Castelli, C. Dissertazione sull' origine delle straordinarie meteore dell' anno 1783 e sulla maniera d' impedire i fulmini e le grandini. 8vo. *(From MS. Catalogue, Padua Academy.)* *Milano*

† Congetture sull' origine dei nocevoli effetti della Brina, &c. 8vo. 24 pp. *(Volta's copy.)* *Milano,* 1793

† **Castro, M. F. de.** La Electricidad y los caminos de Hierro. . . Parte II. Los caminos de Hierro. 8vo. 504 pp. *Madrid,* 1858

Descripcion de un sistema de señales eléctricas para evitar Accidentes en los Caminos de Hierro, y exámen de los demás sistemas eléctricos propuestos. 4to. 216 pp. *Madrid,* 1858

† La Electricidad y los Caminos de Hierro. Descripcion y examen de los sistemas propuestos para évitar Accidentes en los Caminos de Hierro por medio de la electricidad precedidos da una reseña historico-elemental de esta ciencia y de sus principales aplicaciones. Impresa de Real Orden. Parte I. La Electricidad. 8vo. 573 pp. *(Ronalds' Teleg. alluded to at p.* 463.) *Madrid,* 1857

† L'Electricité et les Chemins de fer. Description et examen de tous les systèmes proposés pour éviter les accidents sur les Chemins de fer au moyen de l'électricité. Précédé d'un resumé historique élémentaire de cette science, et de ses principales applications. Publié par ordre du Gouvernement Espagnol. 8vo. 2 vols.: 1st. 4 and 610 pp.; 2nd. 542 pp. *Paris,* 1859

Cat. *(Vide* Le Cat.)

Catania Accademia Gioenia. Atti dell' Accademia Gioenia di Scienze naturali. 4to. 1833-44, comprising 15 vols. of the 1st series. *Catania,* 1833-44

Catherman, F. Vocabulaire télégraphique universel, ou moyen de converser entre personnes de différentes nations sans en connaître les langues . . . Suivi d'un système de signaux de nuit et de brume. 8vo. 32 pp. 1 plate. *Paris,* 1855

† **Cattaneo, F.** Riflessioni intorno alle azioni inducente e magnetizzante delle correnti elettriche . . . della scarica della boccia di Leida. 8vo. 13 pp. *(Giorn. dell' 1. R. Instit. Lomb.* iv. 43.) *Milano,* 1842

† **Cauderay, H.** Le Télégraphe entre l'ancien monde et le nouveau; suivi du Télégraphe sans Pile dans les bureaux intermédiaires. 8vo. 24 pp. 1 plate. *Lausanne et Paris,* 1861

Manuel pratique de télégraphie électrique. *Lausanne et Paris,* 1862

† **Caullet de Veaumorel.** Description de la machine électrique négative et positive de M. Nairne. Avec des détails de ses applications à la physique, et principalement à la médecine ; traduite de l'Anglois. 12mo. 179 pp. 5 plates. *Paris,* 1784

Caumond. De l'influence des méteores . . . sur la production des maladies. 4to. *(From Arago's Sale Cat.* p. 56.) *Montpellier,* 1802

Caumotte. Description d'un phénomène observé à Marseille le 15 Fév. 1730. 4to. 2 pp. *(Montpellier Acad. Hist. &c. par De Ratte,* vol. i. p. 332.) *Lyon,* 1766

Causland, Robert. Conjectures on some phenomena of the Barometer, &c. *Edinburgh,* 1788

† **Cavalleri, G. M.** Applicazione dell' elettricità alla cura della paralisi. Memoria. Letta 24 Agos. 1854. 4to. MS. Copia autentica della Mem. stampata in Milano. Tratta dagli Atti e Memorie del Accad. fisio-medico-statistica. 1854

Cavalleri, G. M.—*continued.*

† Osservazioni sulla cura della paralisi per mezzo dell' elettricità. Letta 19 Marzo, 1857. 8vo. 14 pp. (*Accad. fis.-med.-statistica di Milano.*)
Milano, 1857

† Della elettricità che sviluppano le varie stoffe degli abiti a contatto dell' aria, di loro stesse e del corpo umano. Memoria. Letta 8 Aprile, 1858, all' Accad. fisio-med.-statistica di Milano. 8vo. 16 pp. *Milano,* 1858

† Sulla luce problematica che manifestasi in tutto il cielo nel passaggio delle stelle cadenti in Agosto e Novembre, e di una proposta diretta a scoprirne l'origine. Nota. 8vo. 10 pp. (*Estratto dai Rendiconti del Reale Istituto Lombardo, Ch. di Scienz.* vol. iv. fasc. iii.) *Milano,* 1868

† Sulla Tromba che devastò le vicinanze di Monza il giorno 30 Giugno, 1865. Memoria. 8vo. 23 pp. (*Rendiconti del R. Instit. Lombardo,* ii. 231.)
Milano, 1865

Cavalli, Atagio. Lettere meteorologiche Romane. 8vo. (*Effect of Atmospheric Electricity on the Human Body.*) *Roma,* 1785

Cavallo, Tiberius. *Born March* 30, 1749, *at Naples; died December* 21, 1809, *at London.* (*From Poggendorff,* i. 405. *Phil. Mag.* xxxiv. 470 *says Dec.* 26.)

Extraordinary electricity of the atmosphere observed at Islington. 4to. (*Phil. Trans.* 1776.) *London,* 1776

† A complete Treatise on Electricity in theory and practice. 1st edition. 8vo. 412 pp. 3 plates. *London,* 1777

† Volständige Abhandlung der theoretischen und praktischen Lehre von der Electricitat; aus dem Englischen. (*V. I. S. Tr. Gehler.*) 3 editions. 8vo. 284 pp. 3 plates. *Leipzig,* 1779-83-85

† Trattato completo di Elettricità . . . tradotto con Addizioni e cangiamenti fatti dall' Autore. 8vo. 511 pp. 3 plates. *Firenze,* 1779

† Volledige Verhandeling over de Elektriciteit . . . wit het Engelsch vertaald, en verrykt met byvoegselen en verbeteringen, door den Schryver medegedeld; aan J. Th. Rossijn . . . 8vo. 339 pp. 3 plates. *Utrecht,* 1780

† Essay on the Theory and Practice of Medical Electricity. 1st edition. 8vo. 112 pp. 1 plate. *London,* 1780

† An Essay on the theory and practice of Medical Electricity. 2nd edition. 8vo. 124 pp. 1 plate. *London,* 1781

† Versuch über die Theorie und Anwendung der medicinischen Electricität. Aus dem Engl. übersetzt. 8vo. 34 pp. 1 plate. *Leipzig,* 1782

† Teoria e Pratica della Elettricità medica : e della forza dell' elett. nell cura . . . di G. Birch tradotta dall' Inglese . . . preceduta dall' Istoria elettrica . . . del Vivenzio. 4to. *Napoli,* 1784

† Volständige Abhandlung der theoretischen und praktischen Electricität nebst einigen Versuchen. Aus dem Engl. übersetzt. Dritte mit einigen Zusätzen des Übersetzers vermehrte Auflage. 8vo. 344 and 10 pp. 4 plates. *Leipzig,* 1785

† Traité de l'Electricité traduite de l'Anglois sur la 2me et dernière edition (par Silvestre). 8vo. 343 pp. 4 plates. *Paris,* 1785
Some of the notes are not designated as those of the translator, yet are such.

† Treatise on Magnetism, in theory and practice . . . 1st edition. 8vo. 343 pp. 2 plates. *London,* 1787

A complete Treatise, &c. on Electricity. 8vo. (*From Watt.*) *London,* 1787

Theoret. practische Abhandlung der Lehre von Magnet. aus dem Englischen. 8vo. *Leipzig,* 1788

Cavallo—*continued.*

† Supplement to the treatise on Magnetism. 8vo. 72 pp. 4 plates.
London, 1795

† A complete Treatise on Electricity in theory and practice . . . The 4th edition in 3 vols., containing the practice of Medical Electricity, &c. . . . The 3rd vol., new . . . contains Discoveries, &c. since the 3rd edition. 8vo.
London, 1795

† Volständige Abhandlung der . . . Elektricität . . . aus dem Englischen übersetzt und mit Anmerkungen und Zusatzen begleitet. 4to. vermehrte. . . Aufl. in 2 Bänden. (Von J. M. W. Baumann.) 2 vols. 8vo. *Leipzig,* 1797

Versuche über die Theorie und Anwendung der Medecin-Electricität aus dem Englischen. 8vo. *Frankfort u. Leipsig,* 1799

† Treatise on Magnetism in Theory and Practice. 3rd edition, with a Supplement. 8vo. 335 pp. 3 plates. *London,* 1800

A complete treatise on Electricity. . . New edition. 5th edition. 8vo.
(*From Watt.*) *London,* 1802

† Elements of Natural Philosophy. 4 vols. 8vo. *London,* 1803

Darstellung der Lehre der Elektricität des Galvanismus und Magnetismus aus dem Engl. 8vo. *Erfurt,* 1806

† **Cavallo** and **Cuthbertson.** The Articles : Electric, Electrical, Electricity, Electrometer, Electrophorous, Electrum, &c., in Rees's Cyclopædia, vol. xii. 4to. 51 pp. 55 plates. *London,* 1809

Cavendish, Hon. H. Life and Works of. By Dr. G. Wilson. 8vo.
London, 1851

An attempt to explain some of the principal phenomena of Electricity by means of an elastic fluid. 4to. (*Phil. Trans.* 1771.) *London,* 1771

† **Cazalis,** F. Lettres médicales sur la galvano-puncture. Traduites de l'Italien. 8vo. 15 pp. *Montpellier,* 1846

Cazélis. (*Vide* Masars-de-Cazélis.)

† **Cecchi.** Saggi di naturali esperienze fatte nell' Accademia del Cimento. 2nd edition. Folio, plates and portrait. *Firenze,* 1691

Appareil d'induction. Modification of Ruhmkorff's. (*La Ricreazione, journal des sciences physiques de Florence.*) *Florence.*

Armatures électro-aimants de M. Cecchi. (*In a note to La Ricreazione.*)

Cellio, Marco Antonio.

De terra magnete. (*From Watt.*) 4to. *Gryphisu,* 1692

Celsius, A. *Born November* 27, 1701, *at Upsala ; died April* 25, 1744, *at Upsala.* (*Poggendorff,* i. 410.)

† 316 Observationes de lumine boreali ab an. 1716 usque ad 1732 in Suecia habitæ. 4to. 48 pp. 1 plate. *Norimbergæ,* 1733

Meteorum ignitum die 3 Martii an. 1731 Upsaliæ observatum. (*Acta Litt. Sueciæ,* an. 1734, p. 78.) 1734

Observatio de lumine in barometro. (*Act. Litt. Suec.* 1734.) 1734

Observations of the Aurora Borealis made at London September 14 and October 4, 1735. 4to. (*Phil. Trans.* an. 1736, p. 241.) *London,* 1736

Observationes de lumine boreali ad circulum polarem habitæ, an. 1736. (*Acta Litterar. et Scient. Sueciæ,* an. 1737, p. 254.) 1737

Inclinatio acus magneticæ observata Torneæ ad latit. 65° 51 '. (*Acta Litterar. et Scient. Sueciæ,* an. 1738, p. 428.) 1738 ?

Observationes de lumine boreali anno 1739 factæ. 4to. (*Mem. de Mathemat. et de Phys.* iv. 137.) *Paris*

Celsius—*continued.*

† Anmerkungen über die stündlichen Veränderungen der Magnet-Nadel in ihrer Abweichung. 8vo. 4 pp. (*Schwedische Akad. Abhandl.* an. 1740, p. 45, vol. ii.)
Hamburg, 1740

† Von der Misweisung oder Abweichung der Magnetnädel von dem Nordstriche angemerkt in Upsal, von A. C. 8vo. 4 pp. (*K. Schwed. Akad. Abh.* ii. 161. *Swedish version.*)
Hamburg, 1740

† Neigung der Magnetnadel zu Upsal, von A. C. beobachtet. 8vo. 8 pp. *K. Schwed. Akad. Abh.* vi. 10. *Swedish version.*)
Hamburg, 1744

† Denkmal. Herrn Prof. Andreas Celsius. 8vo. 8 pp. (*K. Schwed. Akad. Abh.* viii. 143.)
Hamburg, 1746

Celsius and Hiortær. Observations of the needle at Upsala. Auroral influence. *Vide* Hiorta, pp. 36, 37, &c. (*Mém. de Suède,* tom. ix. p. 37.)   1746

Cèpède. (*Vide* La Cèpède.)

† **Ceppi.** Dissertatione serio-giocosa sull' Elettricità artificiale . . . 8vo. 345 pp. 13 plates.
Vercelli, 1784

Cerini, Giuseppe. Impossibilità fisico-chimica del paragrandine.   *Milano,* 1821

† **Cernuschi,** C. Descrizione del Telegrafo elettro-magnetico di Morse adottato anche a Milano . . 2nd edn. 8vo. 36 pp. and Table. 3 plates.
Milano, 1852

Cervelleri, Filippo. De galvanismi acupuncturæ magneticæ conjuncti nonnullis in nervorum morbis præstantia. Epistola. 8vo.   *Napoli,* 1839

De l'emploi de l'Electro-magnétisme dans les maladies des nerfs, etc. 8vo.
Naples, 1840

Cesaris, Giovanni Angelo. (*Born October* 30, 1749, *Casale Pusterlengo, Terri torio Lodigiano ; died April* 18, 1832, *at Mailand.*)

† Sul movimento oscillatorio e periodico delle fabbriche. 8vo. 12 pp. (*Effemeridi astronomichi di Milano per* 1813. *Appendice,* p. 105.)   *Milano,* 1812

Note.—*Placed here because such oscillations should affect* magnetic *Observations materially.*—F. R.

† Continuazione ⟨delle osservazioni sul movimento oscillatorio e periodico delle fabbriche. 8vo. 8 pp. (*Effemeridi astronomichi di Milano per* 1816. *Appendice.*)
Milano, 1815

Cesi, In. De Meteoris dissertatio. (*From Pogg.* i. 413.)   *Mantua,* 1700

† **Cézanne,** E. Le Câble transatlantique. 18mo. 72 pp. (*Conférences faites à la gare St.-Jean à Bordeaux.*)   1867

† **Challis,** J. Hydrostatat, &c. **Whewell.** Elektricität und Wärme. **Christie.** Erd Magnetismus. 8vo. 155 pp.   *Berlin,* 1838

Note.—*All from Report of third and fifth meetings of the British Association, translated by Kloden.*

† **Chalmers.** An account of an extraordinary Fire-ball bursting at sea, communi-cated by Chalmers. Read March 22nd, 1749. 4to. 2 pp. (*Phil. Trans.* xlvi. 366.)   *London,* 1749-50

† **Chalmers,** C. Thoughts on Electricity; with Notes of Experiments. Third edition. 8vo. 57 pp. 2 plates. (*First Edition published in* 1849.)
Edinburgh, 1851

† Electro-Chemistry with Positive Results; and Notes for inquiry on Geology, &c. . 8vo. 100 pp.   *London,* 1858

Chamberlayne, Jno. On the effect of Thunder and Lightning at Stampford, Courtney, in Devonshire. 4to. (*Phil. Trans.* 1712.)   *London,* 1712

Chambers, Ephraim. *Born . . . at Kendal, Westmoreland ; died May 15th, 1740, at Canonbury House, Islington.*

Cyclopædia, or an Universal Dictionary of Arts and Science. (*Poggendorff,* i. 117.) *London,* 1728

Dictionary of Arts, &c., Supplement by Rees. 4to. *London,* 1786

Chambers. An article founded upon Experiments, &c., exhibited to him by Mr. Wheatstone. (*Chambers' Edinburgh Journal, 25th July.*) 1840

† **Chambers,** W. and R. The Electric Telegraph. 8vo. 4 pp. (*Chambers's Journal,* 4th series, No. 348, August 27, 1870.) *London,* 1870

† **Chambers,** R. On Luminous Insects. Mention of a Paper read at the Linnæan Society, April 20, 1830. (*Phil. Mag. or Annals,* vii. 357.) 1830

Chambers and Carruthers. Cyclopædia of English Literature ; being a History, Critical and Biographical, of British Authors, from the earliest to the present times ; with specimens of their Writings. 2 vols. 8vo. (*From the Athenæum, July 29th,* 1871.) 1871

Chamolle. Les électro-moteurs. 16mo. 40 pp. *Besancon,* 1857

Changeux, P. N. Météorographie, ou l'art d'observer les phénomènes de l'atmosphère. *Paris,* 1781

Channing, W. F. Notes on the Medical Application of Electricity. 3rd Edition. 8vo. 199 pp. *Boston, U.S.,* 1852

† The Municipal Electric Telegraph . . . Application to Fire Alarms. 8vo. 28 pp. Cuts. (*American Journal of Science,* vol. xiii. 2nd series.)
New Haven, U.S., 1852

† **Chantrel,** J. Le Télégraphe électrique. 12mo. 71 pp. Cut. *Lille,* 1858

Le Télégraphe électrique. 3me édition. 18mo. 71 pp. 1 plate. *Lille,* 1865

† **Chapot,** L. Du Galvanisme en général . . (*Thesis.*) 4to. 57 pp.
Montpellier, 1842

† **Chapot-Duvert,** C. De quelques applications de l'électricité à la thérapeutique. Courants continus, Bains électrisés. 8vo. 69 pp. *Paris,* 1870

Chappe, Claude. *Born 1763, at Brûlon-le-Maine, Département de la Sarthe ; died January 23rd, 1805, Paris.*

Lettres (de Chappe et d'Aymer) sur le nouveau télégraphe (de Bréguet et Bétancourt.) 8vo. *Paris,* 1793

Chappe, Ignace Urbain Jean. *Born 1760, at Rouen ; died Jan. 26, 1829, at Paris.*

Histoire de la télégraphie. 1st edition. 2 vols. 8vo. *Paris,* 1824

† **Chappe,** l'aîné, Ign. U. J. Histoire de la télégraphie . . . précédée.

1. De l'Origine du télégraphe Chappe.

2. D'observations sur la possibilité de remplacer le télégraphe aérien par un télégraphe acoustique.

3. Des motifs que ont obligé les Chappe à demander leur retraite.

8vo. 268 pp. and atlas de 34 plates. *Le Mans,* 1840

(*On false magnetic telegraphs he refers (at p. 38) to Paracelsus, Maxwell, and Santinelli mentioned in the Dictionnaire des Sciences Médicales, verbo Magnétisme.*)

Chappe d'Auteroche, Jean. *Born 1722, March 2, at Mauriac-Auvergne ; died Aug. 1, 1769, at St. Lucar, California.*

Observations sur l'orage du 6 Août, 1767, et d'un coup de foudre qui s'est élevé de la terrasse de l'Observatoire. 4to. (*Mém. de Paris, 1767, Mém.* p. 344.)
Paris, 1767

Chappe d'Auteroche—*continued.*

† Voyage en Californie pour l'Observation de Vénus sur le disque du Soleil le 3 Juin, 1767. . . . Redigé et publié par M. Cassini fils. . . . 4to. 170 pp. 4 plates.

(*On Lightning* (*as ascending*), *p.* 31, *and reference to same subject in his Voyage en Sibérie. On Declination and Inclination, p.* 45.)   *Paris,* 1772
Voyage en Sibérie. (*Pogg.* i. 420, *says* 3 vols. 4to. *Paris,* 1763.)

Chappe, Robillard, and Sylvestre. Repetition of Galvani's and Valli's Experiments as Commissioners of the Soc. Philomatique. (*Bulletin des sciences de la Soc. Philomath. Mars,* 1793, No. 21.)   *Paris,* 1793

† Chappelsmith, J. An account of a Tornado near New Harmony, Indiana, April 30, 1852, with a map of the Track, &c. 4to. 12 pp. 1 map and 1 plate. (*Smithsonian Contributions,* vii.)   *Washington,* 1855

Chaptal, Jean Antoine Claude. *Born June* 4, 1756, *at Nogaret, Dép. de la Lozère; died July* 30, 1832, *at Paris.*

Observations sur l'influence de l'air et de la lumière dans la Végétation des Sels. 4to. (*Toulouse Acad.* 1re serie, tom. iii. p. 335.)   *Toulouse,* 1778

† Fall of Aerolites near Toulouse on the 10th April, 1812. 8vo. (*Phil. Mag.* xl. 315.)   *London,* 1812

Fragment de la Lettre . . . par Chaptal, de Montpellier le 11 Dec., 1786. (*Lyons Acad. Hist. par Dumas,* i. 166.)   1839

Charante, Nicolaus Henrik van. *Born* 1821, *at Rotterdam; died* 1847, *at Rotterdam.*

† Dissertatio physicus inauguralis continens Disquisitiones quasdam experimentales et theoreticas circa Magnetismum rotatione excitatum. 8vo. 80 pp. 1 plate. (*Acad. Lugd. Bat.,* 5 *Junii,* 1844.)   1844?

† Charault, R. Thèse. Recherches sur la déperdition de l'électricité statique par l'air et les supports. 4to. 36 pp.   *Paris,* 1860

Charleton, W. Disquisitiones duæ chymico-physicæ; prior, de fulmine. 12mo. (*Inquisitiones ii anatomico-physicæ; prior de fulmine.* . . .)   *London,* 1665

Charlton. A Ternary of Paradoxes, &c. . . . Magnetic Cure of Wounds, &c. 4to. (*From Watts.*)   *London,* 1650

Charpentier, F. P. *Died in* 1817. Erfand . . . eine neue Signal-Laterne. . . . &c.

† Chase, P. E. On the numerical relations of gravity and magnetism. Magnetic premium awarded Dec. 16, 1864. 4to. 20 pp. (*Trans. Amer. Phil. Soc.,* xiii. 117.) (*Faraday's copy.*)   *Philadelphia,* 1864

† Experiments upon the mechanical polarization of Magnetic Needles, under the influence of fluid currents or "Lines of Force." 8vo. 16 pp. (*Proceedings of the Amer. Phil. Soc. July* 21, *Oct.* 6, *and Nov.* 3, 1865.) *Philadelphia,* 1865.

Chaudron, Junot J. Notice sur la réduction, par voie électro-chimique, du silicium, &c., du tungstène, de l'aluminium et des autres métaux considérés jusqu'ici comme irréductibles, et sur leur application aux arts et à l'industrie. 8vo. 20 pp.   *Paris,* 1855

† Argyrolithe. . . . Réduction par voie électro-chimique . . . des métaux. 8vo. 216 pp.   *Paris,* 1855

Chaumont, (Secretary). Analyse des travaux de la classe des Sciences pendant l'année 1823. Notice de Marq. Victor "Plan d'Observations . . . pour connaître la marche diurne de l'Aiguille aimantée," p. 17. 8vo. (*Toulouse Acad.,* 2me série.)   *Toulouse,* 1830

† Analyse des travaux de la classe des Sciences pendant l'année 1824. Notice de Dralet Jan. 29, 1824, sur une Trombe du 25 Janv. (N.N.E. de Toulouse) avec Foudre et Eclair. 8vo. (*Toulouse Acad.* 2me série, tom. ii. p. 47.)
                    Toulouse, 1830

Chaumont—*continued.*

† Analyse des travaux de la Classe des Sciences pendant l'année 1835. Reference to Boisgeraud, "Quelques Observations sur la Grêle" (*with Thunder, &c., about 1834*). 8vo. pp. 99. (*Toulouse Acad.* 2me série, tom. iv. p. 96.)
Toulouse, 1837

† Analyse des travaux de la Classe des Sciences pendant l'année 1836. 8vo. (*Notice of Quatrefage's "Action de la Foudre sur des Etres organisés," Toulouse Acad.* 2me serie, tom. iv. p. 197.) *Toulouse*, 1837

Chauvancy. (*Vide* Reynold-Chauvancy.)

Chauveau (di Lyon). On contractions in the back of a frog produced from the actions called *réflexes*. . . . (*From Matteucci.*)

Chauvelot, S. Introduction à l'Electricité, avec des explications à nombre de phénomènes de chimie, de physique et économie animale.   *Madrid*, 1788

Chavarri y Rico y Sinobas. Memoria sobre el plan que podria adoptarse para verificar estudios meteorológicos en España, presentada al Director general de instruccion publica. 8vo. *Madrid*, 185- ?

Chemical Society. Instituted 1841.
1850 to 1862, Quarterly Journal. 15 vols. 8vo.
1863 to Monthly. *London.*

Chenot, Adn. Les chaudières à vapeur sont des machines électriques.
Paris, 1844

Cherbourg. Mémoires de la Société de Cherbourg. *Cherbourg*

Cherest, E. Notice sur la télégraphie électrique. 36 pp. 8vo.
Espinal, 1853

† **Chernak, L.** Dissertatio physica de theoria electricitatis Franklini . . 4to. 46 pp. (*Thesis read at Harlem, May 4, 1771.*) *Groningæ*, 1771

† **Cheron et Moreau-Wolf.** Des services que peuvent rendre les courants continus constants dans l'inflammation, l'engorgement et l'hypertrophie de la prostate. 8vo. 31 pp. (*Extrait de la Gazette des Hôpitaux.*) *Paris*, 1870

Chevalier. Patent Electric and self-recording Target. No mantlets required (*From the "Times," Oct. 15, 1861.*)

Chevallier, Jean Gabriel. (*Born September 13, 1778, at Mantes; died February 1848, at Paris.*)
Instruction sur les paratonnerres. *Paris*, 1823
(*Vide also* Salle.)

Chevreul, M. E. De la baguette divinatoire du pendule explorateur et des tables tournantes. 8vo. *Paris*, 1854

Chiaje. (*Vide* Delle Chiaje.)

† **Chigi, A.** Dell' Elettricità terrestre-atmosferica. . . 8vo. 170 and 1 pp.
Sienna, 1777

Lettera ad un amico sopra il Fulmine caduta, 18 Aprile, 1777, nella spranga . . torre del Palazzo . . di Sienna. *Sienna*, 1844

Children, John George. (*Born May 18, 1777, at Ferox Hall, Tunbridge; died January 1, 1852, at Halstead Place, Kent.*)
† An account of some experiments, performed with a view to ascertain the most advantageous method of constructing a Voltaic apparatus, for the purposes of chemical research. (From *Phil. Trans.* 1809, pt. i.) 8vo. 5 pp. (*Phil. Mag.* xxxiv. 26.) *London*, 1809
† Galvanic Battery, 20 pairs of copper and zinc plates, each 6 ft. and 2 ft. 8in., and experiments. 8vo. 2 pp. (*Phil. Mag.* xlii. 144.) *London*, 1813

Children—*continued.*

An account of some experiments with a large Voltaic Battery. 8vo. 7 pp.
(*Phil. Mag.* xlvi. 409.)   *London*, 1815

Chiminello, Vincenzo. (*Born June* 30, 1741, *at Marostica b. Vicenza ; died February* 16, 1815, *at Marostica.*) Nephew and Assistant to Toaldo at the Padua Observatory, and his successor there. He continued the Giornale Astrometeorol. after Toaldo's death in 1797 ; in 1818, or before, it was continued by the Abbé Busata. Chiminello published 68 memoirs.

*   Memoria sull' Igrometro che ottenne il premio dell' Acad. Teodoro-Palatina di
Manheim. (*Giorn. Enciclop. di Vicenza*, 1783.)   1783

†* Apologia dell Igrometro dell' Autore in risposta al Saussure. (*Giorn. Enciclop. di Vicenza.*)

†* Risultati di osservazioni barometriche per li quali si determina un doppio
flusso e reflusso cotidiano dell' atmosfera. 4to. 13 pp. 1 plate. 1 folding
table. (*Saggi di Padova*, i. 195.)   *Padova*, 1786

†* Ricerche sopra la causa piu efficace del doppio flusso e reflusso cotidiano dell'
atmosfera, ed altre cause concomitanti di questo fenomeno. 4to. 31 pp. 1
plate. (*Saggi di Padova*, i. 208.)   *Padova*, 1786

Relazione di una colonna di fuoco veduta a Ceneda per tre giorni consecutivi,
speditagli dall . . Zava G. di Ceneda. 4t ›. (*Padova Accad. Saggi*, ii. p. 9.)
                                                                  Padova, 1789

† Relazione di tre Auroræ Boreales singolari dell' Ottobre 1786. Letta all'
Accademia di Padova addi 7 Dec. dello Stesso anno. 4to. 4 pp. (*Opuscoli
Scelti*, in 4to, xv. 280.)   *Milano*, 1792

Relazione di un Arco luminoso osservato addi 5 Settembre, 1788, dal Chiminello. 4to. 4 pp. (*Mem. Soc. Ital.* vii. 153.)   *Verona*, 1794

Avvertenze per l'uso pratico dell' Ago magnetico. (*Giornale Astro-Meteorologico*, 1801.)   *Padova*, 1801

† Congetture sulla cagione delle diverse variazioni dell' ago magnetico dal Nord.
Ricevute, 3 Decem. 1803. 4to. 10 pp. (*Mem. Soc. Ital.* ix. 193.)
                                                                  Modena, 1804

Relazione di un' Aurora-australe osservata ai 17 Marzo 1803. (*Giornali Astrometeorologico*, 1804.)   *Padova*, 1804

Fenomeni del terremoto di Napoli dei 26 Luglio, 1805. (*Giornale Astrometeorologico*, 1806.)   *Padova*, 1806

† Risposta . . . al commento . . . nel Giornale Astrometeorologico, 1806, del Sig.
G. Scaguller. 12mo. 24 pp. (*On lightning conductors.*)   *Venezia*, 1806

Precauzione d'applicare il secondo conduttore ovvero l'Emissario per preservare
gli edifizii dai Fulmini. (*Giornale Astro-meteorologico*, 1806.)
                                                                  Padova, 1806

† Declinazione dell' ago magnetico in Padova negli anni 1806-7. 4to. 4 pp.)
(*Nuova Scelta d' Opuscoli*, in 4to, ii. 217.)   *Milano*, 1807

Biografia di. 8vo. (*Giorn. dell' Ital. Lettera De Rio*, serie 2, tom. xvii. p. 164.)
                                                                  Padova, 1818

† Della vita e degli studii dell' Abate Vincenzo Chiminello. Breve Memoria
estesa dal Fra. Bertirossi-Busata, inserita nel tomo xviii. degli Atti della
Soc. Ital. 4to. 16 pp.   *Modena*, 1819

ƒ Portrait by G. Prada dip., G. Bosa inc.

† **Chinale** e Compa. Paragrandine. Istruzione . . . 8vo. 25 pp. 1 plate. (*Estratti
del Propagatore.*)   *Torino*, 1828

Chiostri, L. Dell' influenza del suono sulla elettricità atmosferica. 8vo. 13 pp.
1 plate.   *Lucca*

Chladni Ernst Florens Friedrich. *Born November* 30, 1756, *at Wittenberg ; died
April* 3, 1827, *at Breslau.*

Üeber den Ursprung der von Pallas, gefundenen . . . Eisenmassen, und über
einige Naturerscheinungen. 4to. (*Volta's copy.*)   *Riga*, 1794

Chladni—*continued.*

† Über entgegengesetze Elektricität einer Katze. 8vo. (*Voight's Mag.* i. 1797.)
Jena, 1797

Über den Meteorsteinen von Bitburg. 8vo. (*Schweigger's Journ.* xliii. 1825.)
Nurnberg, 1825

Nouveau catalogue des Chutes de Pierres, &c. 16mo. 29 pp. (*Edited by Arago, in Annuaire pour 1826, par le Bureau de Longitudes.*) *Paris*, 1825

Merkwürdige meteorische Erschein-bei Saarbrück. 8vo. (*Poggendorff's Annalen,* vii. 1826.) *Leipzig*, 1826

Verzeichniss gefallener Stein-und Eisenmassen, Feuerkugeln, etc. (*In Gilbert's Annalen, many volumes. In Poggendorff's Ann.* vols. ii. vi. and vii. *and in Kerstner's Archiven,* vol. iv.)

Über die sprungweise gehende Bewegung mancher Feuerkugeln. 8vo.

Neues Verzeichniss der herabgefallenen Stein und Eisenmassen in chronologischer Ordnung. (*Vide also French version,* 1825.)

Chladni, E. F. F., Steininger and Næggerath. Further account of the great mass of native iron of Bitburg : with Dr. Chladni's observations. 8vo. 6 pp. ("*Schweigger's' Jl.* N.R. xvi. 385.") (*Phil. Mag. or Annals,* ii. 41.)
London, 1827

Chladni, E. F. F. und Schreibers. Über Feuer-Meteore und über die mit denselben herabgefallenen Massen, von E. F. F. Chladni, mit (p. 425) Anhang-Verzeichnisz der Sammlung von Meteor-Massen, welche sich im k.k. Hof-Mineralien-Cabinette in Wien befindet vom Director von Schreibers. 8vo. 434 pp. *Wien*, 1819

Nebst Zehn Steindrucktafeln und deren Erklärung von C. von Schreibers. Fol. viii. and 92 pp. 10 plates. *Wien*, 1820

Chompré, Nicolas Maurice. *Born September 23, 1750, at Paris; died July 24, 1825, at Ivry-sur-Seine.*

Account of an experiment made by the Galvanic Society of Paris, upon the formation of the Oxymuriatic Acid, and the separation of Soda from the Muriate of Soda, by means of the Pile of Volta. Communicated on the 15th of December, 1806, to the Galvanic Society of the French National Institute. 8vo. 7 pp. (*Phil. Mag.* xxviii. 59.) *London*, 1807

Chretien, L. Corso di osservazioni meterolog. nella Zona torrida . . . a bordo il Vesuvio nel 1843. 4to. *Napoli*, 1844

Christ, J. L. Von der merkwürdigen Witterung von der Entstehung des Nebels. (*Vide also p.* 482 "*Electric Mists.*") *Vienna*, 1783

† Christie, S. H. On the Magnetism of Iron arising from its rotation. Communicated April 20, 1825, by J. F. W. Herschell, Esq. Read May 12, 1825. 4to. 71 pp. 2 plates. (*Phil. Trans.* 1825.) *London*, 1825

Theory of the diurnal variation of the Magnetic Needle. 4to. 1827

On the mutual action of particles of Magnetic Bodies. 4to. 1827

† On the laws of the deviation of Magnetic Needles towards iron. Read June 5, 1828. 4to. 36 pp. (*Phil. Trans.* 1828, p. 325.) *London*, 1828

† On the Magnetic influence of the Solar rays. 4to. 18 pp. (*Phil. Trans.* 1828.) *London*, 1828

† Memoranda made during the appearance of the Aurora Borealis on the 18th of November, 1835. Read December 10, 1835. 4to. 4 pp. 2 plates lith. (*Phil. Trans. for* 1835.) *London*, 1835

† Experimental determination of the Laws of Magneto-electric Induction in different masses of the same metal, and of its intensity in different metals. 4to. 48 pp. 2 plates. (*Phil. Trans. for* 1833, p. 95.) *London*, 1838

108 CHR—CIL

Christie, S. H.—*continued.*

† On improvements in the instruments and methods employed in determining the direction and intensity of the Terrestrial Magnetic Force. 4to. 26 pp. (*Phil. Trans.* 1833.) *London*, 1833

On Magnetic attraction. 4to. (*From Lamont's Handbook.*) *Cambridge*

† Discussion of the Magnetical Observations made by Capt. Back, R.N., during his late Arctic Expedition. 4to. 39 pp. (*Phil. Trans.* part ii. 1836.) *London*, 1836

† Christie and Airy. Report upon a letter addressed by M. le Baron de Humboldt to H.R.H. the President of the Royal Society, and communicated by H.R.H. to the Council, June 9, 1836. 8vo. *London*, 1836 (*Vide also* Foster Parry *and* Christie. *Vide also* Challis.)

† Christophle, C. Histoire de la Dorure et de l'Argenture électro-chimiques. 8vo. 446 pp. *Paris*, 1851

† Christoffel, E. O. Dissertatio de motu permanenti electricitatis in corporibus homogeneis. 4to. 62 pp. 1 plate. *Berolini*, 1856

† Churchman, J. Three letters to Cassini on the relations of Magnetism and the Aurora. (*American Museum, a periodical publication.*) *Philadelphia*, 1788

† An explanation of the Magnetic Atlas or Variation Chart hereunto annexed ; projected on a plan entirely new. . . . 8vo. 4 pp. 2 tables. Chart (separate). *Philadelphia*, 1790

† The Magnetic Atlas or Variation Charts of the whole terraqueous globe; comprising a system of the variation and dip of the Needle, by which, the observations being truly made, the longitude may be ascertained. 4to. 80 pp. 3 charts, 2 coloured. *London*, 1794

† The Magnetic Atlas. . . . Fourth edition, with considerable additions. 4to. xviii. xxix. and 86 pp. 2 large charts and a small plate. *London*, 1804

Cicero. De Divin. I. xviii. A.U.C. 645. (*Lightning without audible Thunder.*)

Cigna, Giovanni Francesco. *Born July 2, 1734, at Mondovi ; died July 16, 1790, at Turin.*

† De analogia Magnetismi et Electricitatis. 4to. 25 pp. (*Miscellanea . . . Taurinensia,* tom. i. p. 43.) *Taurini*, 1759

† De motibus electricis experimentum. 4to. 3 pp. (*Miscellanea Taurinensia.*) *Taurini, n.d.*

† De novis quibusdam experimentis electricis. 4to. 41 pp. (*Miscellanea Taurinensia, Mélanges . . . de la Soc. R. de Turin pour les Années* 1762-5, tom. iii. p. 31.) *Taurini*, 1766

† De electricitate. 4to. 12 pp. (*Miscell. Taurin.* tom. v. p. 97.) *Taurini*, 1773

Lettres à M. Priestley sur la découverte de l'Electrophore. 1775

Lettre à M. La Grange sur la découverte de l'Electrophore. 1775

"Alcune note per provare che l'Elettroforo del Volta non era altro che il suo apparecchio moltiplicatore dell' elettricità, stampato nel tomo iii. della Società R. di Torino." 4to. *Torino*, 1775

† Memorie istoriche intorno alla vita ed agli studi di Gianfrancesco Cigna de Antonmaria Vassalli Eandi. Letta 14 Ap., 1822. 4to. 24 pp. (*In Mem. di Torino,* xxvi. p. xiii.) *Torino*, 1821

Cigogna, E. A. Saggio di Bibliografia Veneziana. 8vo. *Venezia*, 1847

Cilano de Maternus, G. C. De causis lucis borealis, &c. *Altona*, 1743

† **Cima, A.** Saggio storico-critico e sperimentale su le contrazioni galvaniche, e su le correnti elettro-fisiologiche. 8vo. 143 pp. *Cagliari*, 1846
Ricerche intorno alcuni punti di Elettro-fisiologia. (*Mem. Accad. . . . Istituto de Bologna*, viii.) *Bologna*, 1858
Cimento. (*Vide* Accademia del Cimento.)

† **Ciniselli, L.** Sulla elettro-puntura nella cura degli Aneurismi . . . 4to. 78 pp. *Cremona*, 1856
† Nuovo apparato elettro-motore ad un solo liquido ed a corrente costante. 4to. 11 pp. (*Estratto dalla Gazzetta di Cremona del* 15 *Maggio*, 1858, No. xx.) *Cremona*, 1858

† **Cioni e Petrini.** Lettera all' O. Targioni Tozzetti sul Galvanismo. 8vo. 18mo. (*Ann. di Chim. di Brugnatelli*, xxii. 322.) *Pavia*, 1805
† Memoria sull' azione chimica dell' elettricità nella decomposizione dell' acqua. 8vo. 55 pp. *Firenze*, 1805
† Letter of Messrs. Cioni and Petrini to Professor Pacchiani, of Pisa, on the supposed production of Muriatic Acid by Galvanism. 8vo. 4 pp. (*Abridged from Ann. de Chim.* lvi. 269; *Phil. Mag.* xxiv. 167.) *London*, 1806

Cipri, G. (de Palerme.) Découvertes physico-mécaniques. 8vo. (*From G. I. R. Istit. Lomb.* xiii. 108, *in* 8vo. 1846.) *Paris*, 1846

Circaud. Recherches sur la contraction de la fibrine du sang, par l'action galvanique. 4to. (*J. de Phys. Brumaire*, an. xi. pp. 402 and 468.) *Paris*, 1803

Ciresara, G. B. Lettera al . . Pigato sopra la convulsione elettrica. 4to. *Vicenza*, 1748

Cito, M. (Principe della Rocca). Esperimenti su la forza elettromotrice delle varie sostanze: pubblicati in occasione della settima riunione degli Scienziati Italiani. 4to. *Napoli*, 1845
Risposta alla lettera intorno alla forza elettro-motrice voltaica del . . . G. Grimelli. 8vo. (*From G. I. R. Istit. Lomb.* xiii. 108, 8vo.) *Napoli*, 1845

Civetti, G. L'Elettricismo. Poemetto . . . *Parma*, 1771

Civil Engineer and Architects' Journal. 4to. *London*

Civil Engineers, Institution of. Proceedings. (*Papers on Submarine Telegraphy, &c., are entered under Authors' names.*)

Claire-Deville. (*Vide* Ste. Claire-Deville.)

† **Clanny, W. R.,** M.D. Thunderstorm, account of, in letter to Tilloch, dated Hendon, July 7, 1808. 2 pp. (*Phil. Mag.* xxxi. 152.) *London*, 1808

† **Clark, Dr. D.** (*Vide* Austin and Clark.)

† **Clark, Sir John.** Part of a letter . . . to R. Gale. Dated November 6, 1731. (Odd impression made by Thunder on an Oak.) 4to. 2 pp. (*Phil. Trans.* xli. 325.) *London*, 1739-40-41

Clark, Latimer. *Born March* 10, 1822, *at Great Marlow, Bucks.*
† Experimental investigation of the laws which govern the propagation of the Electric current in long Submarine Cables. . . . Fol. 50 pp. 1 plate. (*Reprinted from Government Report on Submarine Cables*, 1861.) *London*, 1861
† On Electrical quantity and intensity. 8vo. 5 pp. (*Proc. Royal Inst. Great Brit.* 1861.) *London*, 1861
† On the Birmingham Wire Gauge. 8vo. 6 pp. and tables. (*Proc. Brit. Assn. for the Advancement of Science.*) *London*, 1867
† The Telegraphic Breviary, being a list of abbreviated Telegraphic addresses. 8vo. *London*, 1867
† Electrical tests of various recent submarine Telegraphic Cables (No. 2). Fol. 1 sheet, a table. *London*, 1868

Clark, Latimer—*continued.*

† An elementary treatise on Electrical Measurement for the use of Telegraphic Inspectors and Operators. 8vo. 175 pp. table. 1 plate, wood.
London, 1868

† On the Birmingham Wire Gauge. 8vo. 11 pp. (*Proc. Brit. Assn. for the Advancement of Science.*) (*Reprinted.*) *London,* 1869

† Traité élémentaire de la Mesure électrique à l'usage des Inspecteurs et Agents des Télégraphes, traduit de l'Anglais. (*Imprimerie Nationale.*) 8vo.
Paris, 1872

† On the Storms experienced by the Submarine Cable Expedition in the Persian Gulf, 1869. (*Journ. Met. Soc.*) *London,* 1873

† **Clark**, Latimer, and **Sabine**, Robert. Electrical Tables and Formulæ for the use of Telegraph Inspectors and Operators. 8vo. 285 pp. *London,* 1871

(*This work has been translated into Italian.*)

† **Clark**, W. New Compass. 8vo. 1 p. (*Phil. Mag.* lix. 467.) *London,* 1822

Clark, W. S. On Metallic Meteorites. An inaugural dissertation. 8vo. 79 pp. 3 plates litho. *Gottingen,* 1852

Clarke, D. K., Exhibited Machinery of 1862, a Cyclopædia of the Machinery represented at the International Exhibition. (*From Gilbert's Cat.* 1865.)

Clarke, E. M. Magneto-electric Machine. 8vo. (*London and Edinburgh Phil. Mag.* October, 1836, ix. 262.) *London,* 1836

Account of a series of experiments made with a large Magneto-electrical machine. 4to. (*Trans. and Proceed. London Elect. Soc.*) *London,* 1838

† **Clarke**, G. A treatise on the Magnetism of the Needle, &c. 8vo. 24 pp.
London, 1818

† Essay on the cause of the Magnetism of the Needle, with the reason of its being North and South. . . . 8vo. 20 pp. *Southwark,* 1825

Clarke, H., and **Maskelyne**. On a plan for properly observing Fire-ball Meteors. 8vo. (*Tilloch's Phil. Mag.* li. 130.) *London,* 1818

† **Clarus**, A. Geschichte des Galvanismus aus dem Französisch. von Sue, mit Anmerkungen und Zusätzen. Erste Theil. 8vo. 250 pp. *Leipsig,* 1802

† Geschichte des Galvanismus aus dem Französisch. von Sue, mit Anmerkungen und Zusätzen. Zweiter Theil. 8vo. 208 pp. (*Volta's copy.*) *Leipsig,* 1803

Clarus, J. C. A. Der Gewitter-Orkan, an. 8 Juli, 1819, und Einiges zur Naturgeschichte des Gewitters. 8vo. *Leipsig,* 1820

Claudian. Theod. Pulmanni craneburgii. Diligentia, et fide summa, è vetustis codicibus restitutus. 18mo. (*See poem, p.* 322, *Magnes.*) *Antverpiæ,* 1571

† **Clausius**, R. J. E. Die Lichterscheinungen der Atmosphēre. 8vo. 98 pp. 6 plates. (*Grunert's Beitz. zur meterol. Optik,* 4 *Hft.* 1850.) *Leipzig,* 1850

Anordnung der Elektricität auf sehrdünnen Platten und der beiden Belegungen der Franklin 'schen Tafel. 8vo. (*Poggend.* Ann. lxxxvi. 1852)
Leipzig, 1852

Mechan. Aequivalent elecktr. Entladung, &c. und dabei stattfindende Erwärm. des Leitdrahtes. (*Poggend. Ann.* lxxxvi. 1852.) *Leipzig,* 1852

Über die bei e. Leiter gethane Arbeit und erzeugte Wärme. 8vo. (*Poggend. Ann.* lxxxvii. 1852.) *Leipzig,* 1852

Uber die von Grove beobacht. Abhängigkeit des galvan Glühens von der Natur des umgebend Gasses. (*Poggend. Ann.* lxxxvii. 1852.) *Leipzig,* 1852

Anwend. der mechan. Wärmetheorie auf thermoelektr. Erscheinungen. 8vo. (*Poggend. Ann.* xc. 1853.) *Leipzig,* 1853

Über die Natur des Ozon. 8vo. (" *Vorgetragen in der naturforsch Gesellsch zu Zürich den* 8 *März,* 1858.") 1858

Clebsch, A.   Theorie der Elasticität fester Körper. 8vo. 424 pp. *Leipzig*, 1862

Clément-Mallet, J. J.   Documents pour servir à l'Histoire de la Télégraphie électro-magnétique. 8vo.   *Troyes*, 1850

Clerc.   Compte-rendu. 1er Sémestre de 1819. 8vo. (*Lyon's Acad. Comptes-rendus.*)
Lyon, 1819

(*Note.—Mention of a work received from le Comte de Lezai-Marnesia . . . sur les Paratonnerres et les Paragrêles—n. d.*)

† **Clerget.**   Rapport . . . Sonneries . . . de M. Mirand. (*Vide* Mirand.)

† **Close, D.**   Passage d'une lettre . . . sur les effets de la Foudre sur la chaîne du paratonnerre d'un vaisseau. 8vo. 2 pp. (*Toulouse Acad.* 4e série, tome iv. p. 483.)   *Toulouse*, 1854

† **Cochius, H.**   De luce electrica. Dissertatio inauguralis . . . 8vo. 66 pp.
Berol. 1861

† **Cochy-Moncan, A.**   Etude physiologique des Courants électriques. De l'influence des Courants électriques sur la circulation, et de quelques déductions thérapeutiques. 8vo. 112 pp.   *Paris*, 1870

† **Codazza, G.**   Sulle induzioni molecolari prodotte dalle ondulazioni longitudinali dell' etere. Mem. letta 25 Nov. 1852. 4to. 32 pp. (*Giorn. dell' I. R. Istit. Lombardo*, nuova serie, 4to. iv. 199.)   *Milano*, 1852

† Sulla polarizzazione rotatoria della luce sotto l'influenza delle azioni elettromagnetiche. Mem. Prima. 4to. 48 pp. (*Giornale dell' Istit. Lombardo*, nuova serie, iv. 491.)   *Milano*, 1853

† Sulla polarizzazione rotatoria della luce sotto l'influenza delle azioni elettromagnetiche. Mem. seconda. 4to. 44 pp. (*Giorn. dell' Istit. Lomb.* nuova serie, v. 299.)   *Milano*, 1854

† Considerazioni sulla possibilità dell' esistenza di un mezzo magnetico negli spazii vuoti. . . . Letta 6 Dec., 1855. 4to. 15 pp. (*Giornale dell' I. R. Istit. Lombardo*, nuova serie, 4to. viii. 247.)   *Milano*, 1855

† Sunto della prima parte. Sulle forze molecolari, &c. . . . Letta 5 Apr., 1860. 4to. (*Atti dell' I. R. Istit. Lombardo*, ii. 96.)   *Milano*, 1860

Coeman.   De acus magneticæ deviatione. 4to. (*From Watt.*) *Amsterdam*, 1668

Cohausen, Johann Heinrich.   *Born* 1665, *at Hildesheim; died July* 13, 1730, *at Münster.*

† Lumen novum phosphoris accensum, sive exercitatio physico-chimica de causa lucis in Phosphoris . . . exarata. 8vo. 306 pp. 5 plates and frontispiece.
Amsterdam, 1717

† **Cohn, F.**   Ein interessanter Blitzschlag beschrieben. 4to. 2 plates. (*Acad. Leop.* 1856, vol. xxvi. part i. p. 177.)   1856

Die Einwirkung des Blitzes auf Bäume. 4to. (*Acad. Leop.?*)

Colding, Ludwig August.   *Born July* 13, 1815, *at Arnakke b. Holbeck, Seeland.*

† Om magnetens Indvirkning paa det blödt Jern. 17 pp. 2 plates. (*Vidensk. Selsk. Skr.*)   *Kopenhaven*

Colepress, Samuel.   Account of some Magnetical experiments. 4to. (*Phil. Trans.* 1667, p. 502.)   *London*, 1667

Colla, Antonio.   *Born about* 1806; *died March* 8, 1857, *at Parma.*

† Aurora boreale . . 18 Ottob. 1836. 8vo. 2 pp. (*Bibliot. Ital.* lxxxiii. 465.)
Milano, 1836

† Fenomeni atmosferici. 8vo. 2 pp. (*Bibliot. Ital.* lxxxiii. 261.)   *Milano*, 1836

† Aurora-boreale . . 18 Feb. 1837. 8vo. 3 pp. (*Bibliot. Ital.* lxxxvi. 429.)
Milano, 1837

Notizia intorno ad agitazioni dell' ago magnetico osservate nel giorno 18 Aprile, 1842. 8vo.   *Parma*, 1842

Colla, Antonio—*continued.*

† Notizie meteorologiche relative agli anni 1841 e 1842. 12mo. 32 pp. (*Stelle Cadenti*, p. 29 ; *Ago Magnetico*, p. 23.)   *Parma*, 1842

Giornale Astronomico . . . per l'anno 1841, con Appendice di notizie astronomiche e meteorologiche di A. Colla. Meteorologo della D. Università di Parma.   *Parma*, 1841

Nota. Meteorologica, Aurora-boreale, Perturbazioni magnetiche, Bolide ed Areoliti. (*From Padua Acad. Nuovi Saggi.*)

Notizie meteorologiche relative agli anni 1841-2-3. 8vo.   *Parma*, 1844

Colladon, Jean Daniel. *Born December* 15, 1802, *at Genf.*

On his Multiplier, and atmospherical electricity. (*From Arella.*)

† **Collen, H.** On the application of the photographic Camera to Meteorological Registration. 8vo. 3 pp. 1 plate. (*Phil. Magazine*, xxviii. 73.)
                                          London, 1846

Note.—Collen claims a share in my inventions unjustly.—F.R.

† **Collette, J. M.** Le télégraphe imprimeur Hughes. Manuel principalement à l'usage des employés du télégraphe. 8vo. 82 pp. 6 plates.
                                          La Haye, 1871

Collezione di Opuscoli Scientifici e letterari, ed estratti d'opere interessanti. 8vo.
                                          Firenze, 1807

Collina, Abbondio. *Born at Bologna ; died December* 21, 1753, *at Bologna.*

† De acus nauticæ inventore. 4to. 11 pp. (*Comment. de Bonon. Scient.* . . *Instit.* tom. ii. part iii. p. 372.)   *Bononiæ*, 1747

† Considerazioni istoriche sopra l'origine della Bussola nautica nell' Europa e nell' Asia. 4to. 145 pp.   *Faenza*, 1748

Collinder. De fulguribus. 4to. (*From Watts.*)   *Upsal*, 1686

† **Colmet-d'Huart, De.** Nouvelle théorie mathématique de la chaleur et de l'électricité. 1re partie. Détermination de la relation qui existe entre la chaleur rayonnante, la chaleur de conductibilité, et l'électricité. 8vo. 142 pp. 1 plate.   *Luxembourg*, 1864

† Nouvelle théorie mathématique de la chaleur et de l'électricité. 2me partie. Théorie mathématique de la diathermansie, de la lumière et de la chaleur rayonnante naturelles et polarisées. Théorie mathématique de l'électricité statique en ayant égard aux découvertes des propriétés électriques des corps isolants faites par Faraday. 8vo. 148 pp. 3 plates (lith.)
                                          Luxembourg, 1865

Colquhoun, D. C. Inquiry into the origin of Animal Magnetism. 8vo. 2 vols.
                                          London, 1851

Columbus, Chris. "Entdecker der örtlichen Verschiedenheit d. magnet Declination, 1492, Sep. 13." (*From Pogg.* i. 467.)

† **Combes.** Rapport fait par M. Combes . . sur les appareils de télégraphie électrique de M. Regnault. 4to. 36 pp. 6 plates. (*Soc. d'Encouragement pour l'Industrie Nationale.*)   *Paris*, 1855

† **Comité des Fortifications,** &c. Instruction sur les Paratonnerres . . suivie des Rapports . . à l'Institut et à l'Académie des Sciences. Fol. 39 pp. 1 plate.
                                          Paris, 1808

† **Commentarii dell' Accademia di Scienze, Lettere, Agricultura, ed Arti del Departimento del Mella.** 8vo.
Vol. iv. for 1811 *was printed in* 1812.   *Brescia*, 1808

Commentarii dell' Ateneo di Brescia. 8vo.   *Brescia*, 1814 *to* 1851

Commentarii de rebus in scientia naturali et medicina gestis. 37 vols. 8vo.
Lipsiæ, 1752 *to* 1798
Decadis primæ, secundæ et tertiæ supplementum.   8vo.   15 parts. 6 vols.
Lipsiæ, 1763 *to* 1793
Commissaires chargés par le Roi de l'éxamen du Magnétisme Animal.
Rapport. 4to. 66 pp.   *Paris*, 1784
Sir Francis Ronalds says the Commissioners were Franklin, Majault, Le Roy,
Salin, Bailly, d'Arcet, Bailly, de Boey, Guillotin, Lavoisier. Borin died, and
was replaced by Majault.   Lavoisier was (probably) le Rapporteur, but
Franklin is also quoted in that capacity.

Comparoni. Memoria sulla Elettricità animale, 1802. 8vo.   (*Comment. dell'*
Accad. del Dipartimento del Mella, vol. i.)   *Brescia*, 1808

Compass Committee. *Vide* Liverpool Compass Committee.

Comus. (*Vide* Le Comus, Le Camus, and Ledru.)

Concato, L. Relazione sull' esito di varii casi di Colera trattati colla corrente
elettro-magnetica. 8vo.   *Padova*, 1855

† **Conceil, J. A.** Télégraphie nautique. 8vo. 2 plates.   *Dunkerque*, 1852
Documents officiels sur les expériences de la Télégraphie nautique. 8vo.
Dunkerque, n.d.

Concius, A. De succino. (*Pog.* i. 470.)   1660
Poggendorf refers in the same place to "Beschreib. über den ausser der Natur-
zeit. geschehenden Donnerschlag," without further indications.

Conférence maritime, tenue à Bruxelles pour l'adoption d'un système uniforme
d'observations météorologiques à la mer. Août et Sep. 1853.   1853

† **Conférence Télégraphique internationale de Vienne.** Documents de la
Conférence télégraphique internationale de Vienne. Fol. 459 pp.
Vienne, 1868

Configliachi, Pietro. *Born* 1779, *at Mailand ; died June* 27, 1844, *at Pavia.*
Sulla pretesa analogia fra alcuni fenomeni fisici. 4to. 63 pp. Memorie di P. C.
(*Giornale di Fisica* vi. 1813.)   *Pavia*, 1813
This is a first Memoir only, no second was printed.   It relates to Morichini's
discovery.
Esperienze dirette a verificare la proprietà del raggio violetto annunziata dal
Morichini di magnetizzare le punte di ferro. 4to. (*Originally a Memoir.*
Nel Giornale di Pavia, vi. 291.)   *Bologna*, 1813
† Traduzione, con note, di Häuy Pietre preziose. (*Vide* Häuy, iii. 261.) 8vo.
(*Volta's Copy.*)   *Milano*, 1819
Esperienze elettro-magnetiche. Mem. received on Jan. 4, 1821, by the Vene-
tian-Lombardy Inst., not printed in its Mems. (*Vide* vol. iv. p. 7.)
Ulteriori ricerche sull' elettro-magnetismo. Mem. received on May 24, 1821,
by the Venetian-Lombardy Inst., but not printed in its Mems. (*Vide* vol.
iv. p. 7.)
D'una Meteora luminosa. Mem. received on June 27, 1822, by the Venetian-
Lombardy Inst., but not printed in its Mems. *Vide* vol. iv. p. 7.
Sull' azione elettro-magnetica dei metalli. Mem. received on August 7, 1823,
by the Venetian Lombardy Inst., but not printed in its Mems. *Vide* vol. iv.
p. 7.
† Intorno alla costruzione dei Parafulmini. Memoria. (Ricevuta a di Feb. 3,
1830.) 4to. 12 pp. *Mem. Soc. Ital.* xx. 314, *Fisica.*)   *Modena*, 1829
† Elogio scientifico del Volta. 8vo. 56 pp. Portrait. (*Vide* Volta.)
Como, 1834
† Portrait. "Tratto da un disegno del Garavaglia Cornienti del Litog. Vas-
salli.   *Milano*
Short account of him. (No list of his works.) 8vo. 2 pp.   *Milano*, 1845
(*Vide* also Brugnatelli and others.)

Configliachi, Luigi (?). Osservazioni sulle Pile a secco. Memoir received on March 12, 1818, by the Venetian Lombardy Inst., not printed in its Memoirs (*Vide* vol. iv. p. 5.)

Congrés de France. Sessions.
1. . . . .
2. . . . .
3. . . . .
4. . . . .
5. Metz. . . . . 1837.
6. Clermond Ferrand . . . . 1839.
7. . . . .
8. . . . .
9. Lyon. 2 vols. . . . . 1842.
10. . . . .
11. . . . .
12. . . . .
13. . . . .
14. . . . .
15. . . . .
16. Rennes. 2 vols. 8vo. . . . . . . 1849.
17. Nancy. 2 vols. 8vo. . . . . . . 1850.
18. . . . .
19. . . . .
20. . . . .

† **Connel, A.** On the action of the Voltaic electricity on alcohol, ether, and aqueous solutions. 4to. 39 pp. 1 plate. (*Trans. Rl. Soc. of Edinburgh*, vol. xiii.) .
Edinburgh, 1835

† On the action of Voltaic electricity on pyroxilic spirit and solutions in water, alcohol, and ether. Read March 20, 1837. 4to. 27 pp. 1 plate. (*Trans. Rl. Soc. of Edinburgh*, vol. xiv.) (*Faraday's copy. Note by him at p. 25.*)
Edinburgh, 1837

† On the Voltaic decomposition of water. 8vo. 3 pp. (*Phil. Mag.* vii. 4th Series, p. 426, June, 1854.) *London*, 1854

Consoni, T. Varietà elettro-magnetica-animale negli oggetti che si muovono e relativa spiegazione. 8vo. *Firenze*, 1853

Constrand, S. A. Om magnetnalens misvisning i Stockholm. (*Vet. Acad. Handl.* 1817.) 1817

Conté, N. J. Erfand . . einen verbesserten Telegraphen. (*From Pogg.* i. 173.)

† **Contessi, Melandri.** Disquisizione sui paragrandini. 4to. 26 pp.
Treviso, 1826

† **Conti. A. S.** Riflessioni sull' Aurora boreale. 40 pp. 4to. *Venez.* 1739
(*This work is preceded by 3 pages " della Fata Morgana," by Angelucci.*)

Cooke, B. On the property of new Flannel sparkling in the dark. 4to. (*Phil. Trans.* 1747, p. 457.) *London*, 1747

† Part of two Letters from Cooke to Collinson concerning the sparkling of Flannel and the Hair of Animals in the dark. 4to. 5 pp. (*Phil. Trans.* xlv. 394.) *London*, 1748

† **Cooke. Thomas Fothergill.** Authorship of the practical Electric Telegraph of Great Britain ; or, the Brunel Award vindicated, in seven Letters, containing extracts from the Arbitration Evidence of 1841, edited in assertion of his brother's rights. 8vo. *London*, 1868

† Invention of the Electric Telegraph. The charge against Sir Charles Wheatstone, of tampering with the press, as evinced by a letter of the Editor of the *Quarterly Review* in 1855, reprinted from the *Scientific Review.* 8vo.
London, 1869

† **Cooke,** (afterwards Sir) William Fothergill. Telegraphic Railways ; or, the Single way recommended by safety, economy, and efficiency, under the safeguard and control of the Electric Telegraph. 8vo. 39 pp. 4 plates or cuts.
London, 1842

† The Electric Telegraph : was it invented by Professor Wheatstone? 8vo. 48 pp. *London,* 1854

† The Electrical Telegraph : was it invented by Professor Wheatstone? A Reply to Mr. Wheatstone's Answer. 8vo. 152 pp. *London,* 1856

† The Electric Telegraph : was it invented by Professor Wheatstone? Part ii. containing Arbitration Papers and Drawings. 8vo. 268 pp. 11 woodcuts on separate sheets. *London,* 1856

† The Electric Telegraph : was it invented by Professor Wheatstone? Part i. containing Pamphlets of 1854-6. 8vo. 282 pp. *London,* 1857

† The Electric Telegraph : was it invented by Professor Wheatstone? Third Thousand (or Edition). 8vo. 68 pp.
(*Vide* also Hamel and Cooke.) *London,* 1866

Notification as to the Earth being able to replace the Return Wire (conductor). Read in 1843. (*Journ. Society of Arts,* 1843.) *London,* 1843

Cookson, John. Of an extraordinary effect of Lightning in communicating Magneti 4to. (*Phil. Trans.* 1735.) *London,* 1735

Cooper. Experimental Magnetism. 8vo. (*From Young,* p. 436.) 1761

† **Cooper, C. C.** Identities of Light and Heat of Caloric and Electricity. 8vo. 96 pp. *Philadelphia,* 1848

Copenhagen Acad. Det Kongelige Norske Videnskabers selskabs Skrifter. 5 vols. 4to. *Kiobenhaven,* 1768-74

Nye Samling af det Kongelige Norske Vedenskabers selskabs Skrifter.
Kiobenhaven, 1784

Det Kongelige Danske Videnskabens selskabs Skrifter. 6 vols. 4to.
Kiobenhaven, 1801-18

Det Kongelige Danske Videnskabernes selskabs naturvidens kaberlige og mathematiske Afghandlinger. 2 vols. 4to. *Kiobenhaven,* 1824 & 1826

Note.—These Titles are selected from Muncke's Art Physik in Gehler's Worterbuch of 1833, *tom.* vii. 572. *The dates are not to be depended upon.*

Copenhagen Society. Acta Reg. Soc. Hafniensis. 4 vols. 8vo. *Hafn,* 1812

Nova Acta Reg. Soc. Hafniensis. 1 vol. 8vo. *Hafn,* 1819

† **Cornelius, C.** Erklärung der wichtigsten Erscheinungen der Elekt. unter Voraussetzung eines elektrischen Fluidums. 8vo. 79 pp. *Giessen,* 1846

† **Cornelius, C. B.** Zu Theorie der electromagnetischen Erscheinungen. 8vo. 17 pp. (*Zeitschrift für d. Gesammten Naturwissenchaften. Aug.* 1853, No. viii.) 1853

† Die Lehre von der Elektricität und dem Magnetismus. Versuch einer theoretischen Ableitung der gesammten magnetischen und elektrischen Erscheinungen. 8vo. 208 pp. *Leipzig,* 1855

Corner. (*Vide* Bona and Corner.)

Corniani, G. B. L'Aurora. Poemetto. 8vo. *Brescia,* 1779

Cornwall, Capt. A Table of Observations on the variations of the Compass in the Ethiopic Ocean, in the year 1721 and the beginning of the year 1722. 4to. (*Phil. Trans.* an. 1722, p. 53.) *London,* 1822

Cortambert. Sur le Galvanisme. 8vo. (*Mémoires de la Société Médicale d'émulation,* i. 232.) *Paris.*

Cortenovis, A. M. Dialoghi tre dell' elettricismo conosciuto dagli antichi ; etc. (*Giorn. letterar. di Venezia,* 1790.) *Venezia,* 1790

Cortes, Martin. Breve compendio de la esfera y de la arte de Navigar.
Cadix, 1546

Note.—Wherein the hitherto frequently denied inequality of Magnetic declination at various places was recognised.

Cosmos, Le. Revue encyclopédique des progrès des Sciences. Rédigé par Moigno. 8vo. *Paris,* 1852

Cosnier, Maloet, Darcet, &c. Rapport sur les avantages reconnus de la nouvelle méthode d'administrer l'électricité dans les maladies nerveuses, etc. . . . par M. Ledru, connu sous le nom de Le Comus. 18mo. 115 pp. *Paris,* 1783

† **Cossa, A.** Notizie relative alla storia dell' Elettro-Chimica. 8vo. 17 pp.
Pavia, 1858

Note.—Shows clearly Brugnatelli's priority in invention, and complete practical application of electro-plating, &c. p. 16.

Cossali, Pietro. *Born June 29th,* 1748, *at Verona; died December 20th,* 1815, *at Padua.*

Osservazione su di un fulmine accompagnato da strani fenomeni. (*Pogg.* i. 483.)

Soluzione generale del problema di determinare la capacità d'una botte circolare o concolare elettrica con i fondi uguali oppure disuguali. (*Pogg.* i. 483.)

† Sui barometri luminosi : con Appendice dimostrante nel Barometro una macchina elettrica singolare. Memoria. Ricevuta 22 Luglio, 1809. 4to. 28 pp. 1 plate. (*Mem. Soc. Ital.* xv. 76. part ii.) *Verona,* 1810

Sulla dipendenza dei movimenti del Barometro dall' elettricità si artificiale che naturale. (*Read on 20th July,* 1815, *in the Venetian-Lombardy I. R. Institution, but not printed in its Memoirs.*) 1815

Costa, M. A. Memorie sopra i mezzi di perfezionare la nostra conoscenza . . . dell' atmosfera, e descrizione di un istrumento . . . per osservazioni meteorologiche . . . 8vo. 50 pp. 1 plate. *Lucca,* 1839

† **Costa de Serda, E.** Essai d'un Réglement sur le service télégraphique en campagne. 8vo. 96 pp. *Paris,* 1866

† **Costantini, G. A.** (?) Difesa della . . Sentenzache i Fulmini discendono dalle nuvole . . . Riflessioni . . 4to. 184 and 12 pp. *Venezia,* 1749

† **Costanzia, G.** Nuova macchina elettrica. 4to. 2 pp. 1 plate. (*Opuscoli Scelti,* xii. 69.) *Milano,* 1789

† **Costard, G.** Part of a letter from Mr. G. Costard to Mr. T. Catlin, concerning a fiery Meteor seen in the air on July 14, 1745. 4to. 3 pp. (*Phil. Trans.* xliii. 522.) *London,* 1744-5

† **Cotes, R.** A description of the great Meteor which was seen on the 6th March, 171⅘ sent in a letter from . . . R. Cotes . . . to R. Danuye . . . 4to. 5 pp. 1 plate. (*Phil. Trans.* xxxi. for 1720-21.) *London,* 1723

Cotte, Louis. *Born October 20th,* 1740, *at Lyon; died October 4th,* 1815, *at Montmorency.*

† Traité de Météorologie. Reprinted. 4to. (*Acad. Paris.*) *Paris*

† Mémoires de Météorologie, suite. 4to. (*Acad. Paris. Reprinted.*) *Paris*

Observation sur des modifications de l'électricité des nuages, pendant un orage, et sur le trait de feu du tonnerre, qui partoit en même temps de la terre et du nuage. 4to. (*Mém. de Paris,* an. 1769, *Hist.* p. 19.)
Paris, 1769

Observation que la neige, tombant d'une nuée orageuse, est, comme la pluie, un véritable conducteur de l'électricité. 4to. (*Mém. de Paris,* an. 1772, p. i. *Hist.* p. 16.) *Paris,* 1772

Table of 134 Auroræ observed at Montmorency in twelve years, 1768-79.
Paris, 1783

Cotton, E. Of a considerable Loadstone digged out of the ground in Devonshire. 4to. (*Phil. Trans.* an. 1667, p. 423.) *London*, 1667

Cotugno, Domenico. *Born January 29, 1736, at Riva, Naples; died October 6, 1822, at Naples.* (*Poggendorff*, i. 417.)

† Said to have received a violent shock on dissecting a Mouse. (*In Giornale Enciclop. di Bologna, foglio*, viii. 1786.) *Bologna*, 1786

Note.—This is stated in a letter of one page from Cotugno to Vivenzio, dated Napoli, 2 Ottobre, 1784, appended to Vivenzio's Teoria e Pratica dell' Elett. Med. translated from Cavallo, p. 157, 1784. Volta did not believe the account. Sue, Hist. du Galv. i. 1, referring to the Journal Encyclopédique de Bologna, gives a somewhat different account from that of Vassalli, and from that of Cotugno's letter of October 2, 1784, to Vivenzio. Sue refers also to the Journal de Physique, xli. 57, on the subject.

† **Couche.** Sur le télégraphe des trains de M. Bonelli, &c. . . . 8vo. 24 pp. (*Ext. Annales des Mines*, vol. vii.) *Paris*, 1856

† **Coudret, J. F.** Recherches médicales physiologiques sur l'Electricité animale . . . 8vo. 496 pp. and table. 3 plates. *Paris*, 1837

† **Coudriniere.** Osservazione sulla luce dell' acqua del mare. 12mo. 4 pp. (*Scelta d'Opuscoli*, 12mo. xiv. 107.) *Milano*, 1776

Coulomb, Charles Augustin. *Born June 14, 1736, at Angoulême; died August 23, 1806, at Paris.*

Résultats des différentes méthodes employées à donner aux barres d'acier, le plus haut degré de magnetisme. 4to. (*Mém. de l'Institut.* vi. 1806.) *Paris*, 1806

Recherches sur la meilleure manière de fabriquer les aiguilles aimantées. . . . 4to. (*Mém. de Matémat. et de Phys.* ix. 165.) *Paris*, 1777

† Recherches théoriques et expérimentales sur la force de torsion et sur l'élasticité des fils de métal. Application de cette théorie à l'emploi des métaux dans les arts et dans différentes expériences de physique. Construction de différentes balances de torsion pour mesurer les plus petits degrés de force. Observations sur les lois de l'élasticité et de la cohérence. 4to. 41 pp. 2 plates. (*Mém. de Paris*, an. 1784, *Mém.* p. 229.) *Paris*, 1784

Description d'une boussole, dont l'aiguille est suspendue par un fil de soie. 4to. (*In Mém. de Paris*, an. 1785, *Mém.* p. 560.) *Paris*, 1785

† Mémoires sur l'Electricité et le Magnétisme. Extraits des Mémoires de l'Acad. R. des Sciences de Paris, publiés dans les années 1785 à 1789, avec planches et tableaux. 4to. *Paris*, 1785-9

† Détermination théorique et expérimentale des forces qui ramènent différentes aiguilles aimantées à saturation, à leurs méridiens magnétiques. Lu le 26 prairial, an. vii. 4to. 1 plate. (*Mém. de l'Institut.* 1re. cl. tom. iii. p. 176.) *Paris*, 1799

† Nouvelle méthode de déterminer l'inclinaison d'une aiguille aimantée. 4to. 20 pp. (*Mém. de l'Instit. Nat.* tom. iv. *Sc. Math. et Phys. Mém.* p. 565.) *Paris*, 1799

Mém. sur l'Electricité et le Galvanisme. 4to. 1810

Experiments on the causes which determine a body to take one kind of Electricity rather than another. (*Biot's Physiq.* tom. ii.)

Mémoire sur le Magnétisme. 4to. (*Soc. Philomath.* an. 10, pp. 101, 114.) *Paris*, 1802

Nouveau moyen pour mesurer l'inclinaison de l'aiguille aimantée. 4to. (*Société Philomath.* an. 3, p. 53.) *Paris*, 1795

(*Note.—Arella says that Biot had possession of Coulomb's MSS.*)

Coulvier-Gravier. Recherches sur les Météores. 8vo. 1850

† Catalogue des Globes filants observés de 1841 à 1853. 4to. 1 plate.
Paris, 1854

† Note sur le retour des Etoiles filantes aux 9, 10, 11 Août. 4to. 4 pp. 1 plate. (*Ext. Ann. Chin.* 3 série, vol. xliii.) *Paris*, 1855

† Recherches sur les Météores et sur les lois qui les régissent. 8vo. 372 pp. 16 plates (lith.). *Paris*, 1859

† Suite du Catalogue des Globes filants (Bolides) observés à l'observatoire météorologique du Luxembourg du 3 Septembre, 1853, au 10 Novembre, 1859. 4to. 15 pp. (*Ext. des Ann. de Chim.* 3me série, 1860.) *Paris*, 1860

Précis des recherches sur les Météores et sur les lois qui les régissent. 12mo. 182 pp. 12 plates. *Paris*, 1863

† Les Etoiles filantes. 8vo. 14 pp. (*Extr. de la Revue Maritime et Coloniale*, Fév. 1864.) *Paris*, 1864

Précis des recherches sur les Météores et sur les lois qui les régissent. 12mo. 182 pp. 2 plates. *Paris*, 1866

Recherches sur les Météores et sur les lois qui les régissent. 8vo. 272 pp. 1 plate. (*From Zuchold's Cat.* 1866, p. 68.) *Paris*, 1866

† Lettres sur les Etoiles filantes. 12mo. 147 pp. *Paris*, 1866

† **Coulvier-Gravier et Saigey.** Recherches sur les Etoiles filantes. Introduction Historique. 4to. 192 pp. *Paris*, 1847

Courcelles, de. Dissertatio de electricitate, summo viscosam disobstruente materiam remedio. 4to. (*From Guitard and Holbeck*, p. 52.) 1756

Courtnay. (*Vide* Bache and Courtnay.)

† **Couvier, George.** Rapport sur le galvanisme fait à l'Institut national. 4to. 4 pp. (*Mém. des Sociétés Savantes et Littér.* i. 132, *and from Sue, Hist.* ii. 151, *and in Journ. de Phys.* vii. 318.) *Paris*, 1801

† **Cowper,** (Poet.) Letter to Tilloch, from J. S. S., containing an Extract from the 3rd vol. p. 178, in a letter to the Rev. John Newton, of Cowper's Correspondence as published by Hayley. This extract contains an account of two Fireballs which burst "*on the steeple, or close to it*," at Olney. 8vo. (*Phil. Mag.* xix. 296.) *London*, 1804

Coxe, Dr. J. R. Use of Galvanism as a Telegraph : in an extract of a Letter from Dr. J. Redman Coxe, Professor of Chemistry, Philadelphia. 8vo. (*From Vail*, p. 128, *in Thomson's Annals*, 1st series, vii. 162.) *London*, 1816

Cozzi, Andrea *Born about 1795, at Florence ; died November* 27, 1856, *at Florence.*

† Sulle applicazioni della forza elettro-chimica della pila all' analisi dei sali metallici disciolti in liquidi organici vegeto-animali. Memoria. 8vo. 10 pp. (*From Pogg.*) *Firenze*, 1835

Storia dei più grandi progressi della Scienza elettrica. Discorso . . nel suo priv. Instituto. 8vo. *Firenze*, 1837

Sulle applicazioni della forza elettro-chimica . . all' analisi dei Sali. 8vo. 7 pp. *Firenze*, 1842

† **Cradock, Z.** The appearance of a fiery Meteor as seen by Cradock, communicated by M. H. Baker. (Read June 7, 1844.) 4to. 1 p. (*Phil. Trans.* xliii. 78.) *London*, 1744-5

† **Crahay, J. G.** Mémoire sur les Oscillations diurnes du Baromètre. 15 pp. 2 plates. 4to. *Bruxelles*, 1843

Sur l'emploi de le fonte de fer dans la confection d'aimans artificiels. (*Bulletin de Bruxelles Clas. des Sciences*, 1853, p. 406.) *Bruxelles*, 1853

† **Craig, Wm.** On the influence of the variations of Electric Tension as the remote cause of Epidemic and other Diseases. 8vo. 436 pp. *London*, 1859

Cramer, Johann Andreas. *Born December* 14, *1710, at Quedlinburg ; died December* 6, *1777, at Berggiesshübel b. Dresden.*

Cramer, Joh. And. Elementa docimasticæ.

Anfangs gründe d. Probirkunst Neue Auflösung, von J. F. A. Gottling.
Leipzig, **1794**

Cramer, Joseph Anton. *Born February* 12, *1737, at Paderborn ; died December* 21, *1794, at Hildesheim.*

Bericht von einer am 5 Nov. 1784, zu Hildersheim beobachteten feurigen Lufterscheinung. *Bremen,* **1785**

† Über die Entstehung des Nordlichts. 8vo. 109 pp. *Bremen,* **1785**

Cramer, Gabriel. *Born July* 31, *1704, at Geneva ; died January* 4, *1752, at Bagnolo b. Nismes.*

† An account of an Aurora Borealis attended with unusual appearances, in a letter from the learned Cramer to Jurin J., dated Geneva, February 20, 1730, N.S. 4to. 4 pp. (*Phil. Trans.* xxxvi. 279.) *London,* **1729-30**

† **Craufurd,** R. Stone from the clouds on the 5th April, 1804, at Possil, about three miles from Glasgow. 8vo. 4 pp. (*Phil. Mag.* xviii. 371.) *London,* **1804**

Crause (Krause), R. W. De fulmine tactis. *Jenæ,* **1694**

Crell, L. F. F. Chemisches Journal. 6 Bde. *Helmstadt,* **1778-81**

Chem. Archiv. 2 Bde. *Helmstadt,* **1783**

Neues Chem. Archiv. 8 Bde. *Helmstadt,* **1783-94**

Neustes Chem. Archiv. 1 Bd. 8vo. *Helmstadt,* **1798**

Chemische Annalen. 40 Bde. 8vo. *Helmstadt,* **1784—1803**

Beiträge zu d. Chem. Ann. 6 Bde. 8vo. *Helmstadt,* **1785-99**

Crescimbene, G. Pensieri sulla vitale elettromozione. 8vo. *Bologna,* **1838**

Creve, Johann Caspar Ignaz Anton. *Born October* 28, *1769, at Coblenz ; died July* 7, *1853, at Eltville, Nassau.*

Beiträge zu Galvanis Versuchen üb. d. Kräfte d. thierischen Elekt. auf d. Bewegung d. Muskeln. 8vo. 104 pp. *Frankfurt and Leipzig,* **1793**

Phénomènes du Galvanisme. (*Sue,* i. 227, *says " Voyez les Mémoires de la Soc. med. d'émulation,* i. 236.") *Paris*

Von der thierischen Electricität. (*Schreiben der Berliner Gesells. Naturf. Freunde,* B. xi. p. 113.) *Berlin,* **1794**

† Vom Metallreize, ein neu entdeckt untrügl Prüfungsmittel des wahren Todes. 8vo. 226 pp. 1 plate. *Leipzig,* **1796**

Vorläufige Bekanntmachung der Natur des Metallreizes entdeckt von Prof. Creve. (*Salzburg med. chirurg. Zeitung,* i. 49, *July,* 1796.) *Salzburg,* **1796**

De Galvanismi in praxi medica usu.
Sue gives an " *Extrait d'un rapport fait sur cet ouvrage à l'Ecole de Médecine de Paris,*" par Hallé et Sue, le 24 prairial, an. 11, . . and refers to the Jl. de la Soc. de Méd. in 8vo. xviii. 216.

† **Crimotel de Tolloy.** Electricité, Galvanisme, et Magnétisme appliqués aux maladies nerveuses, &c. 42 pp. 8vo. *Paris,* **1853**

Crœse, G. De Fulmine. 4to. (F *Watts.*) *Amsterdam,* **1659**

Croissant, à Laval. Machines électriques en papier. (*Vide* Du Moncel, " Exposé," 2nd edit. p. 399.)|

† **Croker.** An account of a Meteor seen in the air in the Daytime, on Dec. 8, 1733. 4to. 2 pp. (*Phil. Trans.* xli. 186.) *London,* **1739-40-41**

† **Croker, T. H.** Experimental Magnetism, or the truth of Mr. Mason's discoveries in that branch of Natural Philosophy. 8vo. 2 plates.   *London*, 1761

*† **Crookes, Wm.** A Handbook of the Waxed Paper process in Photography. 8vo. 55 pp. (*Vide also* Samuelson and Crookes.)   *London*, 1857

Cross. (*Vide* Singer and Cross.)

Crosse, Andrew. Born June 17, 1784, at Broomfield; died July 6, 1855, at Broomfield.

† Experiments in Voltaic Electricity. 8vo. 6 pp. (*Phil. Mag.* xlvi. 421.)   *London*, 1815

† Memorials scientific and literary of And. Crosse the Electrician; edited by Mrs. Crosse. 8vo. 360 pp.   *London*, 1857

Crova, A. Mémoire sur les lois de la force électro-motrice de polarisation. Propositions de chimie données par la Faculté des sciences de Montpellier. 4to. 45 pp. Pl.   *Metz*, 1862

Cruger, P. Disputatio de motu magnetis. (*From Pogg.* i. 501.)   *Leipzig*, 1615

† **Cruickshanks, W.** On Galvanic Electricity. 8vo. (*Nicholson's Journ.* iv.)   *London*, 18—

† **Crusell, G.** Über den Galvanismus, als chemisches Heilmittel gegen örtliche Krankheiten. Mit 3 Zusätzen. 4 Hefte. 8vo. 158 pp.   *St. Petersburg*, 1841-3

† **Cruwys, Samuel.** A description of an Aurora Borealis seen on the 6th February, 1720-1, at Cruwys-Morehard, in Devonshire. 4to. 5 pp. (*Phil. Trans.* xxxi. for 1720-21, p. 186.)   *London*, 1721

Csapo. (*Vide* Jedleck and Csapo.)

Culley, R. S. A Handbook of practical Telegraphy. 8vo.   *London*, 1863

A second Edition was published in 1867.
A third   „   „   „   „   1868.
A fourth   „   „   „   „   1870.
A fifth   „   „   „   „   1871.

On Printing Telegraphs. With Fac-similes and Explanation of Wheatstone's Automatic, Caselli's Pantelegraph, Bonelli's, and other systems. (*Popular Science Review*, No. xi., *April*.)   *London*

Cumming, James. Born October 24, 1777, at London; died 1861.

† On the connexion of Galvanism and Magnetism. Read before the Cambridge Philosophical Society, April 2, 1821. 4to. 1 plate.   *Cambridge*, 1822

† On the application of Magnetism as a measure of Electricity. 5 pp. 8vo. (*Phil. Mag.* lx. 253.)   *London*, 1822

On the development of Electro-Magnetism by Heat. 4to. 2 plates.   *Cambridge*, 1824

† On the relative Polarities of the Metals as developed by Heat. A paper read by him on April 28th, 1823, at the Cambridge Phil. Soc., and in its Proceedings for 1823. 8vo. 2 pp.   *Edinburgh*, 1824

† Table of Thermo-Electrics. 8vo. 1 p. (*Edinb. Phil. Journal*, x. 185.)   *Edinburgh*, 1824

† Manual of Electro-dynamics, chiefly from Demonferand, with notes and additions. 8vo. 291 pp. 7 plates.   *Cambridge and London*, 1827

Cuneus, N. A reputed discoverer of the Leyden Phial.

Note.—Beyond the assertion of D'Alibard (pp. 30-32) that Musschenbrock and Allaman, citizens of Leyden, communicated the experiment made by Cuneus to the Académie Royale des Sciences de Paris at the beginning of 1746, I can find no such communication in Reuss or other authorities.—F. R.

*† Cunningham. On the motions of the Earth, &c. &c.   281 pp. 8vo.
London, 1834

† Cuper. La Réforme télégraphique et la loi du 27 Juin, 1861.   8vo. 73 pp.
Paris, 1866

† Curioni, G.   Ricerche analitiche sul Bolide caduto a Trenzano il 12 Nov. 1856.
4to. 8 pp. (*Atti dell' I. R. Istit. Lomb.* i. 398. *Letta* 24 *Luglio,* 1859.)
Milano, 1858

Curtet. Lettre à Van Mons, sur quelques nouveaux phénomènes galvaniques. 8vo.
(*Journ. du Van Mons, No.* vi. p. 272; *also in Journ. de Phys.* an. xi. 54.)
Bruxelles, 1803

Curtis, P. Luigi Maria.   De Saggio sull' Elettricità naturale diretto ad ispiegare
i movimenti &c. de Vulcani.   8vo. 88 pp.   Napoli, 1780

Curtius. (*Vide* Donkin Curtius.)

Custine, Comte de.   Discours sur les propriétés de la Lumière, du Feu, et du
Fluide électrique.   8vo. (*Mém. de la Soc. de Nancy,* ii. 60.)   Nancy

Cuthbertson, John.   *Born probably in England; living in* 1816.

Description et usage d'un condensateur d'électricité, et explication du mode
d'agir du fluide électrique dans l'appareil de Volta. 8vo. (*Jour. de Chim.
de Van Mons,* No. xii. p. 313.)   Bruxelles.

Eigenschappen van de Elektriciteit.   8vo. 3 vols.
Amsterdam, 1769, 1782, 1793

Vollständige Abhandlung der theorischen und praktischen Lehre . . . aus dem
Englischen.   Dritte mit Zusätzen des übersetzers (Gehlers des Jüngern),
vermehrte Auflage.   3rd edition. 8vo.   Leipzig, 1785

Abhandlung von der Elektricität . . . aus dem Holländischen. 8vo.
Leipzig, 1786

† Beschreibung einer Elektrisirmaschine und einigen damit von T. R. Deiman
und A. Pacts von Troostwyk angestelten Versuchen. Aus dem Holländischen.   8vo. 110 pp. 2 plates.   Leipzig, 1790

Abhandlung von der Electricität, &c., aus dem Holländischen. Fortsetzung.
8vo.   Leipzig, 1796

Vollständige Abhandlung der Theorie und praktischen Lehre der Elektricität
nebst einigen Versuchen; aus dem Englischen mit Aumerkungen und
Zuschuss. 8vo. 2 vols.   Leipzig, 1797

† Copy of a letter from Mr. Cuthbertson to Dr. Pearson, communicating an
important and curious distinguishing property between the Galvanic and
Electric fluids. 8vo. 1 p. (*Phil. Mag.* xviii. 358.)   London, 1804

† On a Distinguishing Property between the Galvanic and Electric Fluids. 8vo.
3 pp. (*Phil. Mag.* xix. 83.)   London, 1804

† On the Decomposition of Water by Galvanism. 8vo. 2 pp. (*Phil. Mag.*
xxiv. 170.)   London, 1806

† Practical Electricity and Galvanism. 1st edition. 8vo. 271 pp. 9 plates.
London, 1807

† An account of a new method of increasing the charging capacity of coated
Electrical Jars. Discovered by John Wingfield, Esq. 8vo. 4 pp. (*Phil.
Mag.* xxxvi. 259.)   London, 1810

† Practical Electricity and Galvanism. 2nd edition, with corrections and additions. 8vo. 9 plates.   London, 1821

(*Vide* also Cavallo and Cuthbertson.)

Cuvier, G. L. C. (*Vide* Biot and Cuvier.)

P

Cuypers, C. Exposé d'une méthode par laquelle on rend les disques de verre . . . des machines électriques capables d'exciter l'électricité dans une atmosphère humide. . . . (p. 10). Manière de faire bons caissons . . . Description d'un électrophore perpétuel, &c. . . . 8vo. 38 pp.

La Haye, 1778

† **Cyclopædia,** The. (By Rees.) Notice concerning. 8vo. 7 pp. (*Phil. Mag.* lvi. 218.)

London, 1820

† **Czermak, J. N.** Der electrische Doppelhebel. Eine Universal-Contact-Vorrichtung, zur exacten Markirung des Moments, in welchem eine beliebige Bewegung biginnt, oder ihre Richtung ändert. 8vo. 26 pp. 1 plate.

Leipzig, 1871

123

D.

† **Dachaur.** G. Ozon. Eine gedrängte Zusammenstellung bisher gewonnener Resultate. 8vo. 204 pp. *München*, 1864

* **Dagron.** Traité de photographie microscopique. 18mo. 36 pp. *Paris*, 1863.

Daguin. Traité de physique. *Paris*, 1858

D'Agut. MSS. Mémoire de la périodicité des Etoiles filantes. (*Actes Bordeaux Acad.*) *Bordeaux*, 1839

Dahl, F. Handbok i Galvansplastic för Kunstnärer, Industrie-Idkare och Wänner of Numismatiken. Efter egen Erfarenhet samt de nyaste och bästa utländska Köllar utarbetad. 4to. 180 pp. 3 plates. *Carlshamm*, 1852
Galvanoplastik. (From Zuchold, 1856, p. 209.)

† **Dakin,** G. Proposed improvements in the construction of the Cylinder Electrical Machine, and accompanying apparatus. 8vo. 6 pp. (*Phil. Mag. or Annals*, viii. 251.) *London*, 1830

Dakbijla ka Prakaran. An Explanation of the Electric Telegraph. 8vo. 96 pp. *Agra*, 1855

Dalance, (or D * * *). Traité de l'Aimant : divisé en deux parties ; la première contenant les espériences, et la seconde les raisons que l'on peut rendre. Par M. D * * *. 12mo. 140 pp., and 33 other plates. (*Rare.*)
Amsterdam, 1687
This work is criticised in the Acta Eruditorum, Aug. 1687, p. 424.

† Traité de l'Aimant. 4to. *Liege*, 1691

Dalberg, F. von. Meteorcultus der Alten. 8vo. (*Vide* Martin, p. 116.)
Heidelberg, 1811
Über den Meteorcultus d. Alten vorzügl. in Bezug auf Steine, die vom Hemmel gefallen ; e. Beytr. zu Alterthumskunde. 8vo. *Heidelberg*, 1811

† **Dalibard.** Expériences et Observations sur l'Electricité faites . . . par Franklin. Trad. de l'Angloise. 1re edition. 12mo. 222 pp. 1 plate. *Paris*, 1752

† Expériences et Observations sur l'Electricité faites . . . par Franklin. Trad. de l'Anglois. Seconde edition, revue . . augmentée . . . avec notes . . . par Dalibard. 12mo. 2 vols. 1 plate. *Paris*, 1756

Mémoire, lu à l'Acad. Royale des Sciences, le 13 Mai, 1752. Expériences et Observations sur le Tonnerre relatives à celles de Philadelphie. (*From D'Alibard, Exp. and Obser. par Franklin*, ii. 67.)

This memoir does not appear to have been printed in the Memoirs of the Academy. Bouguer rendered an account of his and Delor's experiments to the Academy on the 19th (?). Dalibard gives his memoir at length in the above (vol. ii. p. 67 *et seq.*), *but says nothing about Delor's.*

† **D'Almeida, J.** C. Thèse, présentée à la Faculté des Sciences à Paris . . . Sur la décomposition, par la pile, des sels dissous dans l'eau, Soutenue le 12 Août, 1856. . . . 4to. *Paris*, 1856

Dal Negro Salvatore. *Born Nov. 12, 1768, at Venice ; died March 12, 1839, at Padua.*

✝ Nuovo metodo di costruire macchine elettriche. . . . 4to. 112 pp. 1 plate. *Venezia*, 1799
Note.—Poggendorff (ii. 265) says 1801 and Padova which is probably a mere reprint.

† Dell' elettricismo idro-metallico, &c. *Opuscolo* i. 4to. 116 pp. 1 plate. *Padova*, 1802

124   DAL

Dal Negro Salvatore—*continued.*

† Dell' elettricisimo idro-metallico. *Opuscolo* ii. 4to. 84 pp.   *Padova*, 1803

† Descrizione di un nuovo Elettrometro ed alcune esperienze relative alla carica
   della colonna voltiana. Presentata 14 Maggio, 1804. 4to. 12 pp. 1 plate.
   (*In Mem. Soc. Ital.* xi. 623.)   *Modena*, 1804

† Nuovi esperimenti elettro-magnetici. Memoria (i.) estratta dal fascicolo 6
   Nov. e Dec., 1831, degli *Annali* delle Scienz. del Regno Lombardo-Veneto.
   4to. 16 pp. 1 plate. (*Not finished.*)   *Padova*, 1831

† Nuove esperienze ed osservazioni elettro-magnetiche. Memoria (i.) Letta 21
   Giug. e 10 Lug., 1831. 4to. 19 pp. 1 plate. (*Accad. di Padova Nuovi Sagg.*
   vol. iii. p. 353.)   *Padova*, 1831

† (Lettera.) Al Sig. Dott. Amb. Fusinieri Direttore. . . . Padova il di 20
   Aprile, 1832. Nuovi esperimenti relativi all' azione del magnetismo sulle
   spirali elettro dinamiche, e descrizione di una nuova batteria elettromotrice.
   4to. 3 pp. (*Ann. del Reg. Lomb.-Veneto,* ii. 109.)   *Padova*, 1832

† Nuove esperienze ed osservazioni elettro-magnetiche. Mem. ii. 4to. 16 pp.
   1 plate. (*Inscritta nel bimestre* v. vol. ii. *degli Annali delle Scien. del Regno
   Lombardo-Veneto.*)   *Padova*, 1832

† Esperimenti diretti a confermare le nuove proprietà degli elettromotori del Volta
   scoperte del . . Dal Negro. 4to. 11 pp. (*Annali delle Scienze del Regno
   Lombardo-Veneto, Bimestre* ii. vol. iii. 1833.)   *Padova*, 1833

† Nuove proprietà degli elettromotori elementari scoperte dal Prof. Dal Negro.
   Presentate all' Accademia di Scienze . . . di Padova il 26 Marzo, 1833.
   4to. 2 pp. (*Ann. del Reg. Lomb.-Veneto,* iii. 105.)   *Padova*, 1833

† Seconda serie di esperimenti diretti a confermare maggiormente l'efficacia dei
   perimetri delle lamine metalliche costituenti gli elementi elettrici del Volta,
   scoperta dal . . . Dal Negro. 4to. 4 pp. (*Ann. delle Scienze del Reg. Lomb.-
   Veneto,* iii. 228.)   *Padova*, 1833

† Nuova macchina elettro-magnetica, immaginata dall' Ab. Salv. Dal Negro.
   4to. 14 pp. 1 plate. (*Annali delle Scienze del Regno Lombardo-Veneto,* 10
   Marzo, 1834, tom. iv. 67 pp.)   *Padova*, 1834

† Terza serie di esperimenti, che servono di fondamento ad un principio che rende
   ragione delle nuove proprietà degli Flettromotori del Volta. . . . 4to. 13 pp.
   (*Annali delle Sci. Reg. Lombardo-Veneto,* tom. iv. p. 324.) *Padova*, 1834

† Nuovi esperimenti sul magnetismo temporario. Memoria iii. . . . letta il di
   25 Aprile, 1835, all' Accademia di Padova. 4to. 11 pp. (*Ann. delle Scienze
   del Reg. Lomb.-Veneto,* v. 165.)   *Padova*, 1835

† Nuovi esperimenti che confermano l'influenza della reciproca distanza dei
   perimetri dei due metalli sull' efficacia degli elettro-motori. 4to. 8 pp.
                                                                 Padova, 1835

  Dinamo-magnetometro, immaginato dal Prof. Dal Negro. Memoria ricevuta
  9 Marzo, 1837. 4to. 12 pp. 1 plate. (*Mem. Soc. Ital.* xxi. 323.)
                                                                 Modena, 1837

† Sopra le proprietà dei perimetri dei due metalli costituenti gli elementi
   Voltiani. Memoria letta all' Accademia di Padova li 26 Giugno, 1838. 4to.
   13 pp. 1 plate. (*Nuovi Saggi* . . . *Accad. di Padova,* tom. iv. p. 389.)
                                                                 Padova, 1838

† Descrizione degli Arieti elettro-magnetici tanto semplici, quanto composti
   immaginati dal . . S. Dal Negro. 4to. 16 pp. 1 plate. (*Mem. estratta
   dagli Ann.* . . *del Reg. Lomb.-Veneto.* viii. *Bim.* i. *Gennajo. e Feb.* 1838.)
                                                                 Padova, 1838

†   Portrait by F. de Martiis incise.

* Dalrymple, D. Meteorological, &c. observations on the Climate of Egypt. 8vo.
                                                                 London, 1861

Dalton, John. *Born September 5, 1766, at Eaglesfield, near Cockermouth, Cumberland; died July 27, 1844, at Manchester.*

† On the height of the Aurora Borealis above the surface of the earth, particularly one seen on the 29th of March, 1826. 8vo. 11 pp. (*Phil. Trans.* 1828, part ii.; *Phil. Mag.* or *Annals,* iv. 418.) *London,* 1828

*† Meteorological Observations and Essays, with Appendix. 2nd edition. 8vo. *Manchester,* 1834

Chemical Philosophy. 2 vols. 8vo. *London,* 1842

† Memoirs of the Life and scientific researches of, by W. C. Henry. 8vo. 249 pp. 3 plates and portrait. (*Cavendish Society.*) *London,* 1854

D'Amécourt. (*Vide* Ponton D'Amécourt.)

Da Murello, A. G. Dello svolgimento dell' elettricità nell' economia animale vegetale. Discorso accademico. 8vo. (*From Da Rio, Giornale.*)
Torino, 1805

Dana, James Freeman. *Born September 23, 1793, at Amherst, New Hampshire, U.S.; died April, 1827, at New York.*

† New electrical Battery. 8vo. 1 p. (*Phil. Mag.* lv. 468, and *Silliman's Journal,* i. 1819.) *London,* 1820

† **D ** Angelstown.** Nouvel éxamen des causes des Aurores boréales. 8vo. 30 pp. 1 plate. *Paris,* 1836

Daniell, John Frederic. *Born March 12, 1790, at London; died March 13, 1845, at London.*

Meteorological Essays and Observations. 1st edit. 8vo. *London,* 1823

On a new Register-Pyrometer for measuring the expansions of solids, and determining the higher degrees of temperature upon the common Thermometric Scale. 8vo. 9, 12, and 8 pp. 1 plate. (*In Phil. Mag.* or *Annals,* x. 191, 268, 350.) *London,* 1831

† On Voltaic Combinations. In a letter addressed to Ml. Faraday. . Read February 11, 1836. 4to. 18 pp. 2 plates. (*Phil. Trans. vol. for* 1836, p. 107.) *London,* 1836

† Further observations on Voltaic Combinations. Read April 6, 1837. 4to. 21 pp. 1 plate. (*Phil. Trans. for* 1837, pt. i.) (*Faraday's copy.*) *London,* 1837

† Fourth Letter on Voltaic Combinations, with reference to the mutual relations of the generating and conducting surfaces. Read January 25, 1838. 4to. 16 pp. 1 plate. (*Phil. Trans.* 1838, pt. i.) (*Faraday's copy.*) *London,* 1838

An Introduction to the study of Chemical Philosophy. 1st ed. 8vo.
London, 1839 ?

† Fifth Letter on Voltaic Combinations, with some account of the effects of a large constant Battery; and on the Electrolysis of secondary compounds. Read May 30, 1839. 4to. 24 pp. (*Phil. Trans. for* 1839, pt. i. p. 89.) (*Faraday's copy.*) *London,* 1839

† On the Electrolysis of Secondary Compounds. In a letter addressed to Michael Faraday. . . Read June 13. 4to. 16 pp. (*Phil. Trans. for* 1839, p. 97.) *London,* 1839

† Second Letter on the Electrolysis of Secondary Compounds, addressed to Michael Faraday. . Read May 21. 4to. 16 pp. (*Phil. Trans. for* 1840.) *London,* 1840

† Sixth Letter on Voltaic Combinations. Read April 28, 1842. 4to. 19 pp. (*Phil. Trans. for* 1842, pt. ii.) (*Faraday's copy.*) *London,* 1842

† Applications of Electro-Magnetic Force. 8vo. 7 pp. (*Ext. from Daniell's Chem. Philosophy,* 1843.) (*Roget's copy.*) *London,* 1843

† An Introduction to the study of Chemical Philosophy. 2nd ed. 8vo.
London, 1843

* Meteorological Essays and Observations. 3rd ed. 2 vols 8vo. *London,* 1845

† **Daniell and Miller.** Additional researches on the Electrolysis of Secondary Compounds. 4to. 19 pp. (*Phil. Trans. for* 1844, pt. i. (*Faraday's copy.*)
London, 1844
(*Vide* also Brunel and Daniell.)

* **Danti,** Ignatius. Anemographia. M. Enatii Dantis. 4to. 26 pp.
Bononiæ, 1578
(*A vertical rod (with vane) carries (below) a horizontal toothed wheel which acts upon a vertical wheel on a spindle carrying an Index.*)

Danzig Society. Versuche und Abhandlungen der naturforsch Gesellschaft in Danzig. 4to. Plates. *Danzig u. Leipzig,* 1747 *to* 1756
Neue Sammlung von Versuchen u. s. w. 4to. *Danzig,* 1778
Neuere Schriften der naturforschenden Gesellschaft in Danzig. 4to. 2 vols.
Halle, 1821-28

Daquin, Joseph. *Born* 1757, *at Chambéry; died July* 12, 1815, *at Chambéry.*
Essai météorologique . . par Toaldo. Nouv. Edn. augmentée. . . On y a joint tradn. françs. des Pronostiques d'Aratus. 4to. (*Trans. from 2nd Ed. of Toaldo Saggio.* . . *Almc. electy.* p. 265.) *Chambery,* 1784

† **D'Arbas.** Description d'un Météore singulière, qui parut aux lieux de Marliac, &c., le 13 Juin, 1787. 4to. (*Mém. de l'Acad. de Toulouse,* 1 Serie, tom. iv.)
Toulouse, 1790

*† **Darcet.** Neue Erfindung metallene Gussabdrücke mit Gyps-Form zu machen. 8vo. 44 pp. 2 plates. *Tubingen,* 1806

Darcet, Jean (d'Arcet). Rapport sur l'Electricité dans les malades nerveuses.
Paris, 1783

Mémoire sur l'art de dorer le bronze au moyen de l'amalgame d'or et mercure.
Paris, 1830
Die Kunst d. Broncevergoldung. 8vo. 126 pp. 6 plates. *Frankfurt,* 1833

Darcet. (*Vide* Cosnier, &c.)

Da-Rio, Nicolò. *Born August* 15, 1755, *at Padua; died February* 13, 1845, *at Padua.*
Giornale della Italiana Letteratura. 66 vols. 8vo. (*Indices* in vols. xxxi. and lxi.) (*Published anonymously.*) *Padova,* 1802 *to* 1828

Darguier. (*Vide* Marcorelle and Darguier.)

Darling, J. Cyclopædia bibliographica. 8vo. *London,* 1854

Darling, Dr. (*Vide* Dods.)

Darondeau, B. Cours de régulation des Compas professé a l'Ecole du Génie Maritime.

† Notice sur les résultats des expériences relatives aux perturbations du compas à bord des navires en fer, faites par G. B. Airy . . à la demande du bureau de l'amirauté; traduite de l'anglais par M. B. Darondeau, ingén. hyd. de marine. 8vo. 27 pp. (*Annales Maritimes et Coloniales.*) "Rapport à S. E. le Ministre de la Marine sur une Mission accomplie en Angleterre pour étudier les questions relatives à la régulation des compas." *Paris*
Sur l'emploi du compas étalon et la courbe des déviations à bord des navires en fer et autres. 8vo. 40 pp. 2 plates. *Paris*

† **Darondeau et Chevalier.** Observations météorologiques et magnétiques faites pendant le voyage autour du monde sur la corvette la Bonite, commandé p. M. Vaillant. 4 tomes. 8vo. Pls. *Paris,* 1840-46

† In Voyage autour du monde . . sur la Bonite en 1836 et 1837 . . commandée par Vaillant. Partie physique. Observations Météorologiques. 2 vols. 8vo. 1 plate (in vol. i.) *Paris,* 1840-1

In Voyage autour du monde . . sur la Bonite en 1836 et 1837 . . commandée par Vaillant. Partie physique. Observations Magnétiques. 2 vols. 8vo. 4 plates. *Paris,* 1842-46

Daubencourt et De Santiot. Précis succinct des principaux phénomènes du galvanisme, suivi de la traduction d'un commentaire de J. Aldini, sur un mémoire de Galvani, ayant pour titre "Des forces de l'électricité dans le mouvement musculaire," et l'extrait d'un ouvrage de Vassalli Eandi ayant pour titre "Expériences et observations sur le fluide de l'électromoteur de Volta." 8vo. *Paris,* 1803

(*Vide also* Larcher, Daubencourt, Cassius *and others.*)

† **D'Arrest.** Om den magnetiske Declinations seculaire variation i Kjobenhaven. 8vo. 11 pp. *Kjobenhaven,* 1859

Daubeny, C. G. B. Introduction to the Atomic theory. 2nd Edition. 8vo. 1850

† **D'Aubuisson.** Official Report of a fall of Aërolites near Grenade. 8vo. 5 pp. (*Phil. Mag.* xliv. 100.) *London,* 1814

† **Dauriac,** P. La Télégraphie électrique, son histoire. . et ses applications en France, en Angleterre aux Etats-Unis, &c. . suivis d'un Guide de l'Expéditeur . . 8vo. 120 pp. *Paris,* 1864

Daven, Fulgenzio, or **Fulgenzio,** Daven (di Savoja.) "Una latina dissert. della teoria e uso della elettricità." *Vienna,* 1762

(*Vide* Bauer, Fulgentius.)

† **D'Avezac.** Anciens témoignages historiques relatifs à la Boussole. 8vo. 11 pp. *Paris,* 1858

† Aperçus historiques sur la Boussole et ses applications à l'étude des phénomènes du Magnétisme terrestre. 8vo. 16 pp. (*Lue à la Soc. Géograph.* 20 *Avril,* 1860. *Ext. Bulln. Soc. de Géograph. Av. et Mai,* 1860.) *Paris,* 1860

David. Dissertation de l'identité du fluide nerveux et du fluide électrique. *Paris,* 1830

† Thèse . . . sur la théorie electro-chimique. 48 pp. 8vo. *Douai,* 1836

† **Davidson,** Robert. (*Vide* Forbes.) On Motive Power by Electro-Magnetism. 1840

* **Davies, A. M.** Meteoric Theory of Saturn's Rings. (*From the Athenæum, Dec.*) 1871

† **Davies, E.** An account of what happened from Thunder in Carmarthenshire; partly from the woman's mouth that suffered by it, partly from what was observed by others; communicated to the Royal Society by Eames J., as he received it in a letter from Davies E., dated Pencarreg, Saturday, Dec. 6, 1729-30. 4to. 5 pp. (*Phil. Trans.* xxxvi. 444.) *London,* 1729-30

† **Davies, T. S.** Geometrical investigations concerning the phenomena of Terrestrial Magnetism. Read Feb. 5th and 12th, 1835. 4to. 27 pp. (*Phil. Trans. for* 1835.) *London,* 1835

† Geometrical investigations concerning the phenomena of Terrestrial Magnetism. On the number of points at which a Magnetic Needle can take a position vertical to the Earth's surface. Read Feb. 4, 1836. 2nd series. 4to. 32 pp. 7 plates. (*Phil. Trans.* 1836.) *London,* 1836

† **Daviet de Foncenex,** F. Récit d'une foudre ascendante eclatée sur la tour du fanal de Villefranche. (*Biblioteca oltramontana,* 1789.) 1789

† **Davis, Dl.** Book of the Telegraph. 8vo. 44 pp. Cuts. (*Preface dated* 1847 ; *the first edition was in* 1842.) *Boston, U.S.,* 1851

† A Manual of Magnetism. . . . 4th edition. 8vo. 322 pp. 2 plates. *Boston, U.S.* 1852

Davy, Edmund. Mentioned shortly Gilding and Silvering by Galvanism in 1830. From Martin, vol. ii. p. 15. (*Phil. Trans.* 1831.) 1831

Davy, Dr. John. *Born May* 24, 1791, *at Penzance.*

Some observations on Atmospheric Electricity. 4to. (*Trans. Royal Society of Edinburgh,* 1836.) *Edinburgh,* 1836

Davy, John. Experiments on the Torpedo. (*Phil. Trans.* 1832, 1834, *and* 1841.)

Davy, Sir Humphry, Bart. *Born Dec.* 17, 1778, *at Penzance, Cornwall ; died May* 29, 1829, *at Geneva.*

An account of some Galvanic combinations, formed by the arrangement of single metallic plates and fluids, analogous to the new Galvanic apparatus of Mr. Volta. 4to. (*Phil. Trans.* 1801, p. 397.) *London,* 1801

† Outlines of a view of Galvanism. 8vo. 16 pp. (*Chiefly extracted from a course of lectures on the Galvanic phenomena, read at the . . . Royal Institute.*) (*Roget's copy.*) *London,* 1801

Sur l'Electricité développée par le contact de diverses substances. (*Société Philomath.* an. 10, p. 111.) *Paris,* 1802

Outlines of Lectures on Chemical Philosophy. 8vo. *London,* 1804

† The Bakerian Lecture on some chemical agencies of Electricity. 8vo. 16, 16 and 14 pp. (*Phil. Mag.* xxviii. pp. 3, 104, and 220.) *London,* 1807

† Nuove osservazioni sulla scomposizione degli Alcali. 4to. 5 pp. (*Nuova Scelta d'Opusc.* ii. p. 282. *Account by Amoretti.*) *Milano,* 1807

† Scoperte Galvaniche del Sig. Davy. Transunto del rapporto della Commissione del Galvanismo, letto all' Istit. 4 Gen. 1808. (*Read* 1808.) 4to. 11 pp. (*Nuova Scelta d'Opusc.* ii. 190.) *Milano,* 1807

† The Bakerian Lecture. On some new phenomena of chemical changes produced by Electricity, particularly the decomposition of the fixed Alkalis, and the exhibition of the new substances which constitute their bases ; and on the general nature of Alkaline bodies. 8vo. 18, 12, and 9 pp. (*From Phil. Trans.* 1808. *Phil. Mag.* xxxii. pp. 1, 101, and 146.) *London,* 1808

† Description of a new Eudiometer, invented by H. Davy, for the combustion of oxygen and hydrogen gases, by R. Knight. 8vo. 1 page, 1 plate. (*Phil. Mag.* xxxi. 3.) *London,* 1808

† Experiments on the Alkalies, repeated on the 14th and 28th January, 1808, before the Askesian and Minerological Society. (*Phil. Mag.* xxix. 372.) 1808

† Electro-chemical researches on the decomposition of the Earths ; with observations on the metals obtained from the Alkaline Earths, and on the amalgam procured from Ammonia. (*From Phil. Trans. for* 1808, part ii. *Phil. Mag.* xxxii. 193.) 8vo. 31 pp. *London,* 1808

† The Bakerian Lecture. An account of some new analytical researches for the nature of certain bodies, particularly the Alkalies, Phosphorus, Sulphur, Carbonaceous matter, and the Acids hitherto undecomposed ; with some general observations on Chemical Theory. 8vo. 10 pp. (*Phil. Mag.* xxxiii. 479.) *London,* 1809

† Researches on the Oxymuriatic Acid, its nature and combinations ; and on the elements of the Muriatic Acid. With some experiments on Sulphur and Phosphorus made in the Laboratory of the Royal Institution. 8vo. (*In Phil. Mag.* xxxvi. 352 and 404. *Phil. Trans. for* 1809, part ii.) *London,* 1810

† The Bakerian Lecture for 1809. On some new Electro-Chemical researches on various objects, particularly the metallic bodies from the Alkalies and Earths, and on some combinations of Hydrogen. 8vo. 13 pp. (*Phil. Mag.* xxxv. 401, xxxvi. 17 and 85. *Phil. Trans. for* 1810, part i.) *London,* 1810

† Elements of Chemical Philosophy. Part i. vol. i. 8vo. xvi. and 511 pp. 12 plates. (*Phil. Mag.* xl. 145.) (*Roget's copy.*) *London,* 1812

Elements of Agricultural Chemistry. 4to. (*Experiments on action of Electricity in Vegetation.*) *London,* 1813

† Elementi di filosofia chimica, . . tradotti dall' Inglese in Francese dal Van Mons e in Italiano dal Dr. G. . . . con Note dei Sig. Prof. L. V. Brugnatelli e P. Configliachi. Volume i. 8vo. 275 pp. (*Volta's copy.*) *Milano,* 1814

Davy, Sir Humphry, Bart.—*continued.*

† Elementi di filosofia chimica, tradotti sulla versione francese fatta dal Van Mons, e comentati da G. Moretti e G. Primo. 2 volumi. 8vo. 1 vol. 410 pp. 4 plates ; 2nd vol. 391 pp. 8 plates. (*Volta's copy.*) *Milano,* 1814

Elemente d. chemischen Theils d. Naturwissenschaft, aus d. Engl. von Fried. Wolf. 1r bd. 1ste Abtheil. 8vo. *Berlin,* 1814

† Elementi di filosofia chimica tradotti dall' Inglese in Francese dal Van Mons e in Italiano, dal Dr. G. ; . . . con Note dei Sig. Prof. L. V. Brugnatelli e P. Configliachi. Volume ii. 8vo. *Pavia,* 1816

(This volume contains (at the end) Davy's Introduzione alla filosofia chimica contenente la storio della chimica aggiunta all' edizione Pavese in due volumi). (*Volta's copy.*) *Pavia,* 1816

Beiträge zur Erweiterung des Chemischen Theils der Naturlehre. Aus dem Englischen übersetzt. 8vo. *Berlin,* 1820

† On the Magnetic phenomena produced by Electricity. . . . 8vo. 8 pp. (*Phil. Trans. for* 1821, part i. *Phil. Mag.* lviii. 43.) *London,* 1821

† Further researches on the Magnetic phenomena produced by Electricity, with some new experiments on the properties of electrified bodies in their relations to conducting powers and temperature. 8vo. 10 pp. (*Phil. Mag.* lviii. 406.) *London,* 1821

† On the Electrical phenomena exhibited *in vacuo.* 8vo. 8 pp. (*Phil. Trans. for* 1822, pt i. *Phil. Mag.* lx. 179.) *London,* 1822

† Preservation from Lightning. (*Portable conductor.*) 8vo. 1 p. (*Phil. Mag.* lix. 468.) *London,* 1822

† The Bakerian Lecture. On the relations of Electrical and Chemical changes. 1827. 8, 11, and 10 pp. (*Phil. Mag. or Annals,* i. pp. 31, 94, and 190.) *London,* 1827

† Six Discourses delivered before the Royal Society at their anniversary meetings on the award of the Royal and Copley Medals, preceded by an Address to the Society on the progress and prospects of Science. 4to. *London,* 1827

† Account of some Experiments on the Torpedo. 8vo. 4 pp. (*Phil. Trans. for* 1829, part i. *Phil. Mag. or Annals.* vi. 81.) *London,* 1829

† Life of Sir Humphry Davy . . . by John Ayrton Paris. (*Phil. Mag. or Annals,* x. 214, 379, 426.) *London,* 1831 *Note.—This is a review of Paris's book.*

Memoirs of, by John Davy. 8vo. 2 vols. *London,* 1836

† Works edited by John Davy. 8vo. 9 vols. *London,* 1839-40

Fragmentary Remains, Literary and Scientific, with a Sketch of his Life, edited by his Brother. 8vo. 1858

Davy. (*Vide* Marié-Davy.)

Debbe, C. W. Die Influenz-Elektrisirmaschine von Holz. (Programm d. Realsch. zu Bremen.) 4to. 7 pp. *Bremen,* 1870

De Boismont. (*Vide* Brierre de Boismont.)

De Bougainville. (*Vide* Bougainville.)

De Boillon. (*Vide* Boillon.)

Debout. Bulletin général de thérapeutique. (*From Guitard.*) *Paris,* 1852

De Branville. (*Vide* Branville.)

De Bremond. (*Vide* Bremond.)

† **Debrun,** L. P. Mémoire sur les Trombes qui ont paru au-dessus de Paris, le 16 Mai, 1806, adressé à l'Institut. 8vo. 28 pp. *Paris,* 1807

De Bruno. (*Vide* Bruno.)

De Caraman. (*Vide* Caraman.)

De Castro. (*Vide* Castro.)

Dechales, C. F. M. Mundus mathematicus. 4 vols. fol. 2nd edition. (*From Van Swinden*, i. 38, *and from Poggendorf.*) *Lyons*, 1690

† Decharme, M. C. Météorologie. Note sur les éclairs phosphoréscents observés à Angers, le 25 Juillet, 1868. 8vo. 6 pp.) *Angers*, 1870

† Decharmes, C. Visites à l'Exposition Universelle. Electricité. 8vo. 15 pp. (*Extrait de la Picardie ; Revue littéraire et scientifique.*) *Amiens*, 1856

† Météorologie. De l'Ozone. Lu à l'Académie d'Amiens, 15 Décembre, 1855. 8vo. 12 pp. (*Extrait de la Picardie ; Revue littéraire et scientifique.*) *Amiens*, 1856

† Application de la lumière électrique à la Microscopie. 8vo. 7 pp. *Amiens.*

† Météorologie. Application du Télégraphe électrique à la prédiction du temps. 8vo. 6 pp. *Amiens*, 1856

† Nouvelle pile électrique. Nouvelles hélices, etc. . . . 8vo. 9 pp. *Amiens*, 1856

De Colnet. (*Vide* Colnet.)

De Curtis. (*Vide* Curtis.)

De Custine. (*Vide* Custine.)

De Dree. (*Vide* Dree.)

De. *Note.—All names having the prefix* DE *will be found under the first letter of the second portion of the name, as above.*

De Foncenex. (*Vide* Darret de Foncenex.)

De Gassicourt. (*Vide* Cadet de Gassicourt.)

Degaulle (or Degault), Jean Baptiste. *Born July 5, 1732, at Attigny, France ; died April 13, 1810, at Honfleur, France.*

Description et usage d'un nouveau compas azimutal. *Havre*, 1779

Instruction sur la manière de vérifier les boussoles. (*From Poggendorff*, i. 535.) *Honfleur*, 1803

Deimann, J. R. Geneeskundige proeven en beschouwingen over de gunstige werking der electriciteit in verschellende ongestoladeden. (*From Poggendorff*, i. 1555.) *Amsterdam*, 1779

Von den guten Wirkungen der Elektricität auf verschiedene Krankheiten mit Zusätzen &c. von Kuhn. 2 vols. 8vo. (*From Dove*, p. 243.) *Kopenhagen*, 1793

Note.—The 2nd part or vol. consists entirely of Kuhn's additions to the Dutch original.

Versuche, welche die Zerlegung des Wassers durch die elektrischen-Funken näher bestätigen. 8vo. 1796

Deiman mit Paets van Troostwijk. Beschryving van eine electrisir-machine en van proefnemingen mit dezelve. 4to. *Amsterdam*, 1790

(*Vide* also Troostwijk.)

De la Borde. (*Vide* La Borde.)

De la Fond. (*Vide* Sigaud de la Fond.)

De la Garde. (*Vide* La Garde.)

De la Hire. (*Vide* La Hire.)

† Delalande, J. On the Stones said to have fallen from the Atmosphere. 8vo. 3 pp. (*Phil. Mag.* xvii. 228.) *London*, 1803

† Delalot-Sevin. Systèmes Delalot-Sevin pour l'éclairage et le chauffage par l'électricité à l'aide du photogène et du thermogène. 8vo. 39 pp. 4 plates. *Besançon*, 1861

DEL 131

† **Delamarche, A.** Eléments de télégraphie sous-marine. 8vo. 83 pp.
Paris, 1858

† Elemente der unterseerischen Telegraphie nach dem französsichen frei bearbitet . . . mit Anmerkungen, &c. . . . Von C. Veichelmann. 8vo. 108 pp. 1 plate. *Berlin*, 1859

De la Métherie. (*Vide* La Métherie.)

De la Motte. (*Vide* La Motte.)

Delaroche. (*Vide* Brewer and Delaroche.)

De la Rive. (*Vide* La Rive.)

De la Rue. (*Vide* La Rue.)

De la Salle. (*Vide* La Salle.)

Delaunay, Louis. *Born* 1740, *in the Netherlands; died* 1805 (*about*), *at Vienna.* (*Poggendorff*, i. 540.)

Lettre sur la Tourmaline. 4to. 1782

Minéralogie des Anciens. 2 vols. 8vo. (*From Poggendorff*, ii. 540.)
Bruxelles, 1803

† **Delaunay, Veau.** Letter of M. Veau Delaunay, M.D., to M. Delamétherie, upon the production of the Muriatic Acid by means of Galvanism. 8vo. 2 pp. (*Phil. Mag.* xxvii. 260.) *London*, 1807

Manuel de l'électricité. 8vo. 270 pp. 13 plates. *Paris*, 1809

Delaval, E. H. Electrical Experiments and Observations. 4to. (*Phil. Trans.* 1739.)
London, 1739

An account of the effects of Lightning, &c. 4to. (*Phil. Trans.* 1764.)
London, 1764

† **Delesse, A.** Sur le pouvoir magnétique des minéraux et des roches. 8vo. 59 pp. (*Annales des Mines*, xiv. 81.) *Paris*, 1849

† Sur le pouvoir magnétique des Roches (suite). 8vo. 25 pp. (*Ext. du* tom. xv. 4 série, *des Annales des Mines*, 1849.) *Paris*, 1849

† **Delezenne, M.** Notes sur la polarisation. 8vo. 300 pp. 5 plates. *Lille*, 1835

*† Notes sur la Polarisation. 8vo. *Lille*, 1835

† Notions élémentaires sur les phénomènes d'induction. 8vo. 132 pp. 2 plates. (*Ext. de la Soc. R. de Lille. Mém. Séance du* 20 *Déc.* 1844.) *Lille*, 1845

† Additions aux notions élémentaires sur les phénomènes d'induction. 8vo. 153 pp. 2 plates. *Lille*, 1848

† Expériences sur les Piles sèches. 8vo. 24 pp. (*Ext. des Mém. de la Soc. R. des Sciences . . . de Lille.*) *Lille*

Sur la constitution et la suspension des nuages. (*From Gauthier-Villars Cat. Juin*, 1855.)

Delille. Nouvel examen de la Phosphorescence de l'Agarie et de l'Olivier. (*Bulletin de la Soc. d'Agricult. de l'Hérault, Fév.* 1837.) *Montpellier*, 1837

Phosphorescence des Plantes. 8vo. *Montpellier*

Delius, Heinrich Friedrich von. *Born July* 8, 1720, *at Wernigerode; died October* 22, 1791, *at Erlangen.*

Gedanken von der anziehenden elektrischen Kraft. *Wernigerode*, 1744

Della Bella, Giovannantonio. *Born August* 30, 1730, *at Padua; died November* 24, 1823, *at Padua.* (*Poggendorff*, i. 139.)

Trattato sopra l'utilità dei conduttori elettrici. (*Pogg.* i. 139.)

Noticias historicas, e praticas à circa do modo de defender os edificios dos ontrages dos raios. 8vo. *Lisboa*, 1773

Memoria i. ii. sobre a força magnetica. 4to. (*Mem. da Acad. Real das Sc. de Lisboa*, t., i. 35 and 116 pp.) *Lisboa*, 1780

Note.—Lamont, Handbook, p. 427, *says Della Bella discovered the law of magnetic attraction and repulsion before Coulomb.*

Della Casa, L. Considerazioni sulla elettricità atmosferica, ciel sereno ; e sopra alcuni fenomeni che ne dipendono. Memoria . . . estratta dal vol. v. delle Mem. dell' Acad. delle Scienze dell' Istituto di Bologna. 4to.
Bologna, 1854

Sulla causa delle Correnti indotte nei circuiti metallici. (*Mem. dell' Accad. delle Scienze dell' Istituto di Bologna*, vol. vii. 1857-58.) *Bologna*, 1857-58

Della Pausa elettrica. (*Accad. delle Scienze dell' Istit. di Bologna*, p. 71, an. 1858.) *Bologna*, 1858

Neue Methode, die meteorologischen Instrumente graphisch. zu machen. (*Mem. Accad. dell' Scienze dell' Istit. di Bologna*, vol. ix. 1858.)
Bologna, 1858

† **Dell' Acqua, C.** Sul modo di ben servirsi dell' apparato elettro-magnetico di Wolff perfezionato dal fisico-macchinista C. Dell' Acqua. 8vo. 12 pp.
Milano, 1857

† Norme pratiche per ben costruire ed applicare i Parafulmini. 4to. 37 pp. 1 plate (lith.) *Milano*, 1859

Della Torre. (*Vide* Torre, G. M. Della.)

Delle Chiaje. On the organs of the Torpedo. (*From Matteucci et Savi.*)

Dellmann, Friedrich. *Born* 1805.

† Abhandlung : Uber ein neues Elektrometer. 4to. 24 pp. *Coblenz*, 1842

† Über die Gesetzmässigkeit und die Theorie des Elektricitäts-Verlustes. 4to. 28 pp. *Kreuznach*, 1864

† Über Luft Elektricität, Nebel, und Höhenrauch &c. 8vo. 20 pp.
Kreuznach, 1870

† **Del Muscio, G.** Dissertazione con cui si risponde a varj dubbj promossi contro la teoria dell' elettricismo del Franklin dal . . . Poli nelle sue " Riflessioni intorno agli effetti di alcuni fulmini. 8vo. 66 pp. *Napoli*, 1774

Delor. (*Vide* Dalibard.)

Demarco, S. Il fluido elettrico applicato a spiegare i fenomeni della natura.
Ancona, 1771

De Martino. (*Vide* Calamai &c.)

Demonferrand, Jean Baptiste Firmin. *Born* 1795 ; *died Jan.* 22, 1844, *at Paris.*

† Manuel de l'électricité dynamique, ou traité sur l'action mutuelle des conducteurs électriques et des aimants. 8vo. 210 pp. 5 plates. *Paris*, 1823

† Trattato d'Elettricità dinamica, ossia dell' azione delle correnti elettriche tra loro e sopra le calamite, &c. . . . opera . . . tradotta, per la prima volta, dal Silv. Gherardi, con Note ed Aggiunte. 8vo. 244 pp. 6 plates.
Bologna, 1824

† Handbuch der dynamischen Elektricität. . . Bearbeitet von M. G. T. Fechner. 8vo. 220 pp. 5 plates. *Leigzig*, 1824
(*Vide* also Ampère.)

† **Demonville.** Mémoire explicatif des phénomènes de l'aiguille aimantée, pour faire suite à la question de longitude, etc. 8vo. 48 pp. 2 plates. *Paris*, 1833

Dempp, K. W. Vollstandiger Unterricht in der Tecnik Blitzableitersetzung, nach 66 Modellen. 8vo. *Munchen*, 1842

Dempster, G. An account of the Magnetic Mountains of Cannay, an island of ten or twelve miles. (*Trans. Soc. Antiquaries of Scotland*, i.)

Denys, G. L'art de naviguer perfectionnée par la connaissance de la variation de l'imant , ou Traité de la variation de l'aiguille aimantée. (*From Poggendorff*, i. 550.) *Dieppe*, 1666

† **Denza, Franc.** Le aurore polari del 1869, ed i fenomeni cosmici che le accompagnarono. Memoria. 8vo. 42 pp. *Torino*, 1869

Denza, Franc.—*continued.*

† Aurore boreale et autres phénomènes météorologiques observés en Piémont, le 3 janvier, 1870. 8vo. 7 pp. *Torino,* 1870

† Norme per osservazioni delle meteore luminose ; con in fine il modulo pel registro delle stelle cadenti. 8vo. 29 pp. *Torino,* 1870

Deratte. (*Vide* Roucher Deratte (or De Ratte).

Deregis, G. Sopra gli istrumenti di osservazione ad indicazione continua. *Novara,* 1852

† **Derham, W.** An account of some Magnetical experiments and observations. 4to. (*Phil. Trans.* xxiv. for 1704-5, pp. 2136 and 2138.) *London,* 1706

† Extract of several letters from different parts of Europe, relating to the Aurora Borealis seen Oct. 19, N.S. 1726. 4to. 4 pp. (*Phil. Trans. for* 1728, p. 453.) *London,* 1728

† Of the Meteor called the Ignis Fatuus, from observations made in England by Derham and others in Italy, communicated by Sir Thomas Derham, Bart., 4to. 11 pp. (*Phil. Trans.* xxxvi. 204.) *London,* 1729-30

† A letter to Sir Hans Sloane containing a description of some uncommon appearances observed in an Aurora Borealis by Derham. 4to. 2 pp. (*Phil. Trans.* xxxvi. 137.) *London,* 1729-30

† **Dering, G. E.** Improvements in the Electric Telegraph. 8vo. 11 pp. 1 plate. (*Specification of Patent, &c.*) *London,* 1850-1

De Rochas. (*Vide* Breton and De Rochas.)

Desaguliers, Jean Théophile. *Born March* 12, 1683, *at La Rochelle ; died February* 29, 1744, *at London.*

Course of Experimental Philosophy. 4to. 1 vol. (*From Kruniz,* p. 34. *Watt,* p. 299.) 1725

There are many editions. The edition of 1719 *contains no electricity. The first edition of his own seems to be that of* 1725 ; 2*nd,* 1727 ; 3*rd,* 1734 ; 4*th,* 1745 ; 5*th,* 1763. (*Vide Watt.*) *There is a French version of* 1753 *in* 4*to.*

An attempt to solve the phenomena of the rise of Vapours, formation of Clouds, and descent of Rain. 4to. (*Phil. Trans.* an. 1729, p. 6.) *London,* 1729

An account of some Magnetical experiments. 4to. (*From Reuss, Repert.* iv. 375. In *Phil. Trans.* 1738, p. 385.) *London,* 1738

† Some thoughts and experiments concerning Electricity. 4to. 8 pp. (*Phil. Trans.* xli. 186.) *London,* 1739-40-41

† Experiments made before the Royal Society, Feb. 2, 1737-8, by J. T. D., and Feb. 9 and Feb. 16. 4to. 7 pp. (*Phil. Trans.* xli. 193.) *London,* 1739-40-41

† An account of some Electrical experiments made before the Royal Society on the 16th of Feb. 1737-8. 4to. 9 pp. 1 plate. (*Phil. Trans.* xli. 200.) *London,* 1739-40.41

† An account of some Electrical experiments made at His Royal Highness the Prince of Wales's House at Cliefden on the 15th April, 1738, where the Electricity was conveyed 420 feet in a direct line. 4to. 2 pp. (*Phil. Trans.* xli. 209.) *London,* 1739-40-41

† Some things concerning Electricity. 4to. 4 pp. (*Phil. Trans.* xli. 634.) *London,* 1739-40-41

Desaguliers, Jean Theophile—*continued.*

† An account of some Electrical experiments made before the Royal Society the 22nd of Jan., 1740-41. 4to. 3 pp. (*Phil. Trans.* xli. 637.)
London, 1739-40-41

† Electrical experiments made before the Royal Society, March 15, 1740-1. 4to. 2 pp. (*Phil. Trans.* xli. 639.) *London,* 1739-40-41

† Several Electrical experiments, made at various times, before the Royal Society by J. T. D. 4to. 6 pp. (*Phil. Trans.* xli. 661.) *London,* 1739-40-41

Account of some Electrical experiments. 4to. (*Phil. Trans.* 1739, No. 266, p. 200; 1741, Nos. 268, 269, 270, pp. 637, 639, and 661; 1742, p. 14.)
London, 1739-41-42

Some further observations, &c. 4to. (*Phil. Trans.* abd. viii. 435.)
London, 1741-2

Some conjectures concerning Electricity, and the rise of vapours. 4to. (*Phil. Trans.* 1742, p. 140.) *London,* 1742

He supposed particles of " Pure Air " to be electric bodies, and always in a vitreous state.—Qy. if by " Pure Air " he meant the oxygen of the atmosphere (as Priestley did), has not this old theory some relation with the modern Ozone theory (electrical oxygen)?—F. R.

† Dissertation sur l'Electricité des corps: qui a remporté le prix . . . Acad. Rle. de Bordeaux. 4to. 28 pp. *Bordeaux,* 1742

† Dissertation concerning Electricity. 8vo. 50 pp. *London,* 1742

French version of his Course of Experimental Philosophy. 4to. 1753

(*Vide* also Eames, John.)

Desains, Edouard François. *Born July 26, 1812, at St. Quentin, Dép. de l'Aisne.*

† Thèse de Phys. Sur la mesure des Courants électriques. 4to. 22 pp. 1 plate.
Paris, 1837

De Saintot. (*Vide* Cassius.)

Desbordeaux. Sur le moyen d'obtenir un courant constant avec la pile de Wollaston. 4to. (*Comptes Rendus,* xix. 273.) *Paris.*

Descartes (Cartesius) René Du Perron. *Born March 31, 1596, at La Haye, Touraine; died February 11, 1650, at Stockholm.*

Principia philosophiæ, iv. § 33. 4to. *Amstelodami,* 1656

Hat eine philosophische Theorie des Magnetismus gegeben, wonach schraubenformige Ströme vom Nordpol zum Südpol gehen sollen. Principia philosophiæ, iv. 133. (*From Lamont's Handbook,* p. 430.)

Principia philosophiæ. Pt. iv. *Amstelodami,* 1664

Note.—Poggendorff (i. 555) says : " Discours de la méthode pour bien conduire sa raison et chercher la vérité dans les sciences ; plus la Dioptrique, les Météores et la Géométrie. 1 vol. 4to. Leyd. 1637. Latinisch, und vom Verf vermehrt, Amstelod. 1644."

De Saint Amans. (*Vide* Saint Amans.)

Deshais. (*Vide* Anon. Medical Electricity, 1763.)

Deschanel. (*Vide* Privat-Deschanel.)

Deschanel. Natural Philosophy: an elementary treatise. Translated and edited, with extensive additions, by Prof. Everett, D.C.L., of Queen's College, Belfast. In four parts. 719 engravings. Part iii. Electricity and Magnetism. *London,* 1872

Desgranges, (Secretary). Compte rendu des travaux de l'Acad. 2me an. 4me Simestre. 8vo. (*Bordeaux Acad. Actes,* 2me an. pour 1839.) *Paris,* 1840

Deshais. (*Vide* Sauvages, 1749.)

Desmarets, Nicolas. *Born September 16, 1725, at Soulaines, Dép. de l'Aube; died September 28, 1815, at Paris.*

† Expériences physico-méchaniques . . traduites de M. Hawksbee par feu M. de Brémond . . revues et mises au jour, avec un Discours préliminaire des Remarques et des Notes par M. Desmarets. 2 vols. 12mo. 7 planches.
Paris, 1754
(*Vide* Hawksbee.)

Desmortiers, Lebouvier. Observations sur le danger du Galvanisme dans le traitement des maladies. 4to. (*Journ. de Phys.* 1801, p. 467.) *Paris,* 1801

† Examen des principaux systèmes sur la nature du fluide électrique; et sur son action dans les corps organisés et vivants. 8vo. 360 pp. 2 plates.
Paris, 1813

Désormes, C. B. Expériences et observations sur les phénomènes physiques et chimiques que présente l'appareil de Volta. 8vo. (*Ann. de Chim.* xxxvii. 1801, p. 284.) *Paris,* 1801

Désormes et Hachette. Mémoire pour servir à l'histoire de cette partie de l'électricité qu'on nomme Galvanisme. 8vo. (*Ann. de Chim.* xliv. 1802.)
Paris, 1802

Doubleurs de l'Electricité. 8vo. (*Ann. de Chim.* xlix. 1804.) *Paris,* 1804

† **Desparquets.** L'électricité appliquée au traitement des malades. Manuel pratique. 8vo. 107 pp. *Paris,* 1862

† **Despine.** Sur les Grêles en 1844. 4to. 5 pp. (*Ext. de la Gazette de l'Association Agricole,* No. 23, 3me Année.) *Turin,* 1845

† Recherches sur les Grêles auxquelles sont exposés les Etats de terre-ferme de S. M. le Roi de Sardaigne. 4to. 22 pp. (*Gazette de l'Association Agricole,* 1844, Nos. 29 and 30.) *Turin,* 1844 ?

† Observations sur les Grêles tombes en 1840 dans les Etats de Terre Ferme de S. M. le Roi Sardaigne, d'après les renseignements recueillis par la Commission supérieure de statistique. 4to. 39 pp. (*Turin Acad. Mém.* 2me serie, t. vii.) *Turin*

† **Desplats,** V. Physique Médicale. Lois générales de la production et de la propagation du courant Electrique. Thèse. 8vo. 81 pp. *Paris,* 1863

Despretz, César Mansuète. *Born May 4, 1791, at Lessines, Belgium; died March 15, 1863, at Paris.*

Traité de physique. *Bruxelles,* 1837

Sur l'action galvanique des métaux. 4to. 31 pp. *Paris*

† Note sur la déviation de l'aiguille aimantée, par l'action des corps chauds et froids. 4to. 6 pp. (*Ext. des Comptes Rendus,* xxix. Séances des 27 Août et 10 Sep. 1849.) *Paris,* 1849

† Note relative à l'Electricité développée dans la contraction musculaire, &c. 4to. 6 pp. (*In Comptes Rendus,* 1849, 1re série, xxviii. No. 22, p. 653.)
Paris, 1849

Sur la correction qu'il faudrait apporter à la formule exprimant la valeur de l'intensité électrique avec la boussole des tangentes. 4to. (*Comptes Rendus,* xxxv. 449. *Vide* Wiedemann, ii. 196.) *Paris,* 1852

Des applications de l'électricité aux arts et à l'industrie. 8vo. 320 pp. 8 plates. (*In Brux. Acad.* 1859.) *Bruxelles,* 1859

Desrousseaux, E. Sources de l'électricité, force attractive et répulsive, gravitation des astres suivant une nouvelle théorie. Découverte d'une nouvelle force. . . . 8vo. 20 pp. *Paris, 1864*

† Sources de l'électricité, force attractive et répulsive, gravitation des Astres suivant une nouvelle théorie. 2me edition. 8vo. 36 pp. *Paris, 1864*

† L'électricité devoilée, et son rôle dans la matière. Nouveau système du monde basé sur le calorique et l'électricité agissant comme forces motrices. 2me édition. 8vo. 107 pp. *Paris, 1868*

† **Dessaignes.** Extract from a memoir on Phosphorescence, presented to the Institute by Dessaignes (Journ. des Mines, xxvii. 213), on April 5, 1809. 8vo. 11 pp. (*Phil. Mag.* xxxviii. 3.) *London, 1811*

On the Phosphorescence of the compressed Gases. 8vo. 2 pp. (*Phil. Mag.* xliv. 313.) *London, 1814*

† **Detienne.** A peculiar construction of the conductor of an Electric Machine for increasing the action thereof. 4to. (*Jour. de Phys. Juillet,* 1775.) *Paris, 1775*

Metodo per avere . . . un perfetto isolamento. 12mo. 6 pp. (*Scelta d'Opuscoli in* 12mo. xxiv. 96.) *Milano, 1776*

De Thury. (*Vide* Héricart de Thury.)

De Tolley. (*Vide* Crimotel de Tolley.)

Deux-Ponts-Bérigny, E. A. Observations faites à Versailles avec le papier dit ozonométrique de M. Schönbein, pendant le mois d'Août, 1855. 8vo. 12 pp. *Paris, 1856*

Deverdun, De Borda, e Pingré. Voyage (par ordre du Roi) . . . la frégate Flora en 1771-2 . . . 4to. 2 vols. *Paris, 1778*

Deville. (*Vide* Ste. Claire-Deville.)

† **Devillers.** Le Colosse aux pieds d'Argile. 8vo. s.l. 174 pp. (Mesmerism.) (*Volta's copy.*) 1784

† **Devincenzi, G.** Procédé de gravure électrochimique . . . 4to. 2 pp. (*Ext. des Comptes Rendus de l'Acad. séance* 5 *Nov.* 1855.) *Paris, 1855?*

† **Diamilla-Muller, D.** Physique du globe. Recherches sur le magnétisme terrestre. 32 pp. 9 plates. *Torino e Firenze, 1870*

*† **Dickinson, R.** Prospectus of a new system of Beaconing. (Not Electrical.) 8vo. 7 pp. 1 plate. (*Phil. Mag.* l. 443.) *London, 1817*

Dieman. (*Vide* Carradori, 1790.)

Dienger, Jos. Über die Gleichgewichtslage einer Magnetnädel, die unter dem Einflusse eines Magneten steht, und über magnetische Curven. (*Grunert's Arch.* xii. 307.)

Über din Verlust an Elektricität durch die Luft. (*Grunert's Archiv,* xi. 1848.) 1848

Dietrich, A. Die Elektricitätsverhältnisse der Atmosphäre und der Erdoberfläche unter dem Einfluss der Eisenbahnen und der elektrischen Telegraphie. 8vo. 38 pp. *Dresden, 1858*

Dietrich, P. F. von. Sur les conducteurs des édifices anciens. 4to. (*Schrift. Gesellsch. nurf Fr. in Berlin,* xxv. 1784.) *Berlin, 1784*

Digby, Sir Kenelm. *Born July* 11, 1603, *Gothurst, Buckinghamshire; died June* 11, 1665, *at London.*

Demonstrationes immortalitatis animæ . . . 8vo. *Frankfurt, 1664*

There are several English editions, the first in 1644 : " *De attractionibus electricis earumque causis, in seiner Demonstratio immortalitatis animæ Francof,* 1664, *einer Uberset. eines* 1644 *zu Paris, in engl. Sprache erscheinenet Werks.*" *Geoloth, Bibliot.* 2 *Stuck, p.* 269, *refers to pp.* 219-22 *for the above.*

Digges, Leonard. *Born at Digges Court, Kent ; died about 1573.*

Prognostication everlasting of right good effect ; or Choice Rules to judge the Weather by the sun, moon, and stars, etc.   1555

† **Digges, Thomas.** *Died August 24, 1595, at London.*

A Prognostication everlasting of right good effecte, &c. . . . Published by Leonard Digges, lately corrected and augmented (by T. D.) his Son. 4to. (*From Watt. Watt gives the dates 1576-78-92.*)   *London*, 1592 ?

Digney. Trasmettitore automatico di Digney (Telegrafia.) (*From Matteucci.*)

Dijon Academy. Mémoires de l'Académie de Dijon. 8vo.   *Dijon*, 1739-84

Dilthey. De electro et Eridano. 4to.   *Darmstadt*, 1824

Dingler, J. F. &c. Polytechnisches Journal u. s. w. herausgegeben erst von ihm allein von 1820 to 1831, Bd. i.—xxxix. dann bis 1840, Bd. lxxviii., mit sienem Sohn Emil Max, darauf bis zur Gegenwart von Letzterem. 8vo.
Stuttgart, 1856

Dingley, Robert. *Born   ; died July 9, 1781.*

Observations on Thunder. 12mo.   *London*, 1658

D'Innare, C. F. Gedanken über Vulkane, Erdbeben und gegenwärtige Witterung. 8vo.   *Frankfurt u. Leipzig*, 1783

† Von der Elektricität (erster Theil). 8vo. 481 pp. 16 plates. *Frankfurt*, 1784

Dionysius, Areopagita. Rites et observationes antiquæ.   A.D. 50

Dirks, H. Perpetuum mobile. Search for self-moving power during the 17th, 18th, and 19th centuries. . . . 8vo.   *London*, 1861

† **Dircks, H.** Contributions towards a history of Electro-Metallurgy, establishing the origin of the art. 8vo. 102 pp.   *London*, 1863

Diruff, C. J. Ideen z. Naturerklär. d. Meteor od. Luftsteine. 8vo.
Göttingen, 1804

Disney, R. W. Dissert. de electricitate. 8vo. (*From Guitard.*)
Leyde, (*London ?*) 1790

Dispan. On Electrum.   *Toulouse*, 1827

Dittmar, S. G. Der Polarschein oder das Nordlicht nach einer neuen naturgemässen Theorie erklart.   *Berlin*, 1831

Ditton. Longitude and latitude found by the inclinatory and dipping needle. 8vo.   *London*, 1719

Longitude and Latitude found by the inclinatory and dipping needle. With the laws of Magnetism discovered by R. Norman. 8vo.   *London*, 1721

Divisch, Procopius. *Born August 1, 1696, at Senftenberg Mähr ; died December 21, 1765, at Prendiz.*

"Erfand einen Wetter ableiter. Beschrieben in Pelzel's Abbild, böhm u. mähr. Gelehrt. Bd. iii." (*From Poggendorff*, i. 580.)

Längst. verlangte Theorie von der meteorologischen Elektricität. 8vo.
Tübingen, 1768

Dixon, Pat. Account of a cure of a Dumb Person. (*London Magazine.*)
London, 1752

Doat. A Pile. Described by Du Moncel. Exposé. 2nd edition, p. 462.

† **Dobelli, F.** La Bussola e le Aurore boreali. 12mo. 32 pp.   *Milano*, 1867

† **Dobbie, W.** An attempt to explain the phenomenon known by the name of the Aurora Borealis. 8vo. 9 pp. (*Phil. Mag.* lvi. 175.)   *London*, 1820

† On the cause of the Magnetic power of the Poles of the Earth. 8vo. 4 pp. (*Phil. Mag.* lxi. 252.)   *London*, 1823

Dobereiner, J. W. Über Electricität, Wärme, und Licht. 8vo. (*Schweigger's Journ.* xxiii. 1818.)   *Nürnburg,* 1818

Einfache Voltasche Kette von langer Thätigkeit. 8vo. (*Gilbert's Ann.* lxviii. 1821.)   *Leipzig,* 1821

Phyto-elektr.-chem. Versuche u. s. u. 8vo. (*Gilbert's Ann.* lxxii. 1822.)   *Nürnburg,* 1822

Über Eschwegit u. merkwürd, Verändr. d. Holzes durch Blitz. 8vo. (*Gilbert's Ann.* lxxiii. 1823.)   *Leipzig,* 1823

Lichterscheinungen bei Kristallisationen. 8vo. (*Schweigger's Jour.* xli. 1824.)   *Nurnburg,* 1824-8

Die neuesten u. wichtigsten physikalisch-chem. Entdekungen. 4to. *Jena,* 1824

On combustion of Hydrogen by means of spongy Platinum, and his Electrical theory concerning.

> *Poggendorf,* i. 582, *has " Über neu entdeckte höchst merkwürdige Eigenschaften des Platins und die pneumatisch-capillare Thätigkeit gesprungener Gläser."* (*Schweigger's Journ.* xxxviii. and xxxix. *und Gilbert's Ann.* lxxiv. and lxxvi.)

> *Poggendorff* (i. 584) *after many quotations of his (D.'s) works in Schweigger's Journ. and Gilbert's Annalen, adds : " Fernere Aufsütze im Reichs-Anzeiger* (1803), *Algem. Anzeiger d. Deutschen* (1812), *in Tromsdorff's Journ., Buchner's Repert., Kastner's Archiv, Hänle's Mag., Brande's Arch., u. s. w."*

Dods, J. B. Electro-Psychology. Treatise by Dr. Darling. 12mo. *London,* 1851

Electro-Psychology. Lectures. 12mo.   *New York,* 1852

† The philosophy of Electrical Psychology . . 12 Lectures. Stereotype edition. 8vo. 252 pp.   *New York*

Dodson, James. (*Vide* Mountain and Dodson.)

† **Dodwell R.** An illustrated Handbook to the Electric Telegraph. 2nd edition. 8vo. 80 pp.   *London, n.d.* (1861 ?)

† **Dollond, P. and J.** Directions for using the Electric machine made by P. and J Dollond. 8vo. 24 pp. 1 plate.   *London,* 1761

† **Dollond, J.** The Life of John Dollond, F.R.S., inventor of the Achromatic Telescope. 8vo. 6 pp. portrait. (*Phil. Mag.* xviii. 47.)   *London,* 1804
Imputed inventor of the plate electrical machine.

D'Hombre-Firmas, Louis Augustin, Baron. *Born about* 1785, *at Alais; died March* 5, 1857, *at Alais.*

On Alorolithes. 8vo. (*Montpellier Acad. Encontre.*)   *Montpellier,* 1809

Recueil de Mémoires et d'Observations de Physique, de Météorologie, d'Agriculture, et d'Histoire Naturelle. 8vo. 6 vols.   *Nimes,* 1839-47

† Recueil de Mémoires, etc.   1823-42-44

† Effets de la Foudre. (*Résumé des travaux de l'Académie de Toulouse pendant* 1839-40-41.)   *Toulouse,* 1843

Disjouval. (*Vide* Quatremere-Disjouval).

Domin, J. F. von. Sono campanarum fulmina promoveri potius quam prohiberi.   *Quinque Ecclesiis,* 1786

Commentatio de electricitate medica in Regio Musæo Quinque Ecclesiensi instituta.   *Quinque Ecclesiis,* 1790

Commentatio altera de electricitate medica, &c. (*From Poggendorff,* i. 589.)   *Pestini,* 1793

Ars electricitatis. (*From Poggendorff,* i. 589.)   *Pestini,* 1794

Ars electricitatem ægris tuto adhibiendi. (*From Poggendorff,* i. 589.)   *Pestini,* 1796

Lampadis electricæ optimæ notæ descriptio, &c. (*From Poggendorff,* i. 589.)   *Pestini,* 1799

Donaggio, O. Regolatore elettro-magnetico. 8vo. *Milano, 1857*

† Donker, Curtius. Commentatio de convenientia atque differentia effectuum tensionis electricæ et fluxus electrici. 4to. 1 plate. 53 pp.
Lugd. Bat. 1823

Donndorff, Johann August. *Born March 23, 1754, at Quedlinberg ; died November 22, 1837, at Quedlinberg.*

Schreib. an den Grafen von Borke über Elektricität. 8vo. *Quedlinberg,* 1781

Abhandlung über Elektricität, Magnetismus, Feuer, &c. 8vo.
Quedlinberg, 1783

† Die Lehre von der Elektricität, theoretisch und pracktisch aus einander gesetz. 8vo. 2 vols. 7 large plates. *Erfurt,* 1784

† Donné, Al. Recherches sur les influences . . . météorologiques sur les piles sèches. 8vo. 16 pp. 1829

† Recherches sur quelques unes des propriétés chimiques des sécrétions, et sur les courants électriques qui existent dans les corps organisés. 8vo. 19 pp.
Paris, 1834

† Donovan, M. Observations and experiments concerning Mr. Davy's hypothesis of Electro-Chemical Affinity. 8vo. (*Phil. Mag.* xxxvii. pp. 227 and 245.)
London, 1811

† Observations on the inadequacy of the hypothesis at present received to explain the phenomena of Electricity. 8vo. 3 pp. (*Phil. Mag.* xxxix. 396.)
London, 1812

† Reflections on the inadequacy of the principal hypothesis to account for the phenomena of Electricity. 8vo. (*Phil. Mag.* xliv. 334 and 401.)
London, 1814

† On the inadequacy of Galvanic hypothesis. . . . Paper read at the Kirwenian Society of Dublin on April 19. 8vo. 2 pp. (*Phil. Mag.* xlv. 381.)
London, 1815

† A new theory of Galvanism. Paper read at the Kirwenian Society, Dec. 28, and January 11. 8vo. 3 pp. (*Phil. Mag.* xlv. 154.) *London,* 1815

† Observations on a Paper by G. A. de Luc, containing some remarks on Mr. Donovan's reflections concerning the inadequacy of Electrical hypothesis. By M. Donovan. 8vo. 4 pp. (*Phil. Mag.* xlv. 200.) *London,* 1815

† On the Origin, Progress, and Present state of Galvanism. An Essay read at the Kirwenian Society of Dublin on February 22, March 8, and March 22. 8vo. pp. (*Phil. Mag.* xlv. 222 and 308.) *London,* 1815

† Second reply to Mr. De Luc's observations on a Paper entitled "Reflections on the inadequacy of the principal hypothesis to explain the phenomena of Electricity." 8vo. 2 pp. (*Phil. Mag.* xlvi. 13.) *London,* 1815

† Answers to remarks on Mr. Donovan's "Reflections on the inadequacy of Electrical hypothesis." 8vo. 2 pp. (*Phil. Mag.* xlvii. 167.) *London,* 1816

† On a new method of obtaining pure Silver; with observations on the defects of former processes. 8vo. 4 pp. (*Phil. Mag.* xlvii. 204.) *London,* 1816

† Essay on the Origin, Progress, &c. . . of Galvanism. 8vo. 390 pp. 1 plate.
Dublin, 1816

† On certain improvements in the construction of Galvanometers: on Galvanometers in general, and on a new instrument for measuring the relative force of Magnetism in compound needles intended to be nearly astatic. (*Read* May 22, 1848). 4to. 18 pp. (*Trans. Royal Irish Academy,* xxii. part iii. p. 233.) (*Faraday's copy.*) *Dublin,* 1851

Donovan, M.—*continued.*

On Galvanometric Deflections produced by attrition and contact of metals under certain circumstances. Read Jan. 8, 22, and Feb 12, 1849 4to. 34 pp. (*Trans. Royal Irish Academy*, xxiii.) *Dublin*, 1854

Doppelmayer, Johann Gabriel. *Born* 1671, *at Nürnberg ; died December* 1, 1750, *at Nürnberg.* (*Poggendorff*, i. 593.)

† Neuentdeckte Phenomina der electrischen Kraft. u. des dabei . . . erscheinenden Liecht. 4to. 88 pp. 5 plates. *Nurnberg*, 1744

Über das elektrische Licht. 1749

On Electricity. (*From Watt.*)

† **Doppler.** Über eine Reihe markscheiderischer Declinations Beobachtungen aus der Zeitung 1735-36. 8vo. 4 pp. (*Sitzb. d. k Akad.* 1849, v. und vi. Heft. p. 1.) *Wien*, 1849

† Bemerkungen und Anträge, die Einsendungen magnetischer Beobachtungsdaten aus Joachimsthal, Freiberg, Pribram, Leoben, Ischl, und Salzburg betreffend. 8vo. 32 pp. (*Sitzungib. d. k. Akad.* 1850, iv. Heft. p. 337.) *Wien*, 1850

† Über die neuester Zeit in Freiberg in Sachsen aufgefundenen Declinationsbeobachtungen aus alterer Zeit. 8vo. 3 pp. (*Sitzungsb. d. k. Akad. d. Wiss. Wien*, vii. Bd. i. Heft. p. 160.) *Wien*, 1851

Über eine bisher unbenutze Quelle magnetischer Declinations Beobachtungen. *Wien*, 1860

Doppler, Christian. *Born November* 29, 1803, *at Salzburg ; died March* 17, 1853, *at Venice.*

Uber d. wahrscheinliche Ursache d. Electricitäts-Erregung durch Berühr. (*Wien polytechnich Jahrb.* 1834. xviii.) *Wien*, 1834

Über e. merkwürd Eigenthümlichk. d. elect. Spannung. 8vo. (*Baumgartner und J. v. Holger's Zeitschr.* v. 1837.) *Wien*, 1837

Beitr. z. Elektricitäts-Erreg. durch Reibung. (*Hessler's encyclop. Zeitsch.* 1842.) 1842

† Versuch einer auf rein mechanischen Principien sich stützenden Erklärung der galvano-elektrischen und magnetischen Polaritäts-Erscheinungen. Fol. 20 pp. 2 plates. *Wien*, 1849

D'Orbissan (Président). Trombe de terre et Aurore boréale (le 8 May, 1780), dans la lettre qu'il écrivit le 13 Mai 1780 à l'Acad. 4to. (*Toulouse Acad.* 1re série, vol. ii. p. 24, *Hist.*) *Toulouse*, 1784

Dorotea, D. L. (*Vide* Accademia degli Aspiranti, &c.)

Dorville, E. Monographie de la pile électrique ; sa forme, ses applications, ses perfectionnements. 8vo. 24 pp. *Paris*, 1857

† Resumé des travaux qui ont précédé l'union de l'Amérique et de l'Europe par l'électricité au moyen d'un câble sous-marin transatlantique. 8vo. 16 pp. *Paris*, 1857

Double, F. J. De l'influence galvanique sur le sang. (*Recueil périodique de la Soc. de Méd.* xvi. pluviose, an. xl. p. 65.) *Paris*, 1803

Dove, Heinrich Wilhelm. *Born October* 6, 1803, *at Siegnitz.*

Über Hygrometeore ; üb. d. Gewitter üb. mittl. Luftströme ; ub. Barometr. Minima. 8vo. (*Poggend. Ann.* xiii. 1828.) *Leipzig*, 1828

Correspondirende Beobachtungen über die regelmässigen stündlichen Veränderungen und über die Perturbationen der magnetischen Abweichung im mittleren und östlichen Europa . . . mit einem Vorworte von Alexander von Humboldt. 8vo. 35 pp. 2 plates. *n.d.*

Uber gleichzeit. Störung in d. tagl. Variat. und Declinat. d. magnet. Kraft. 8vo. (*Poggendorff, Ann.* xx. 1830.) *Leipzig*, 1830

Dove, Heinrich Wilhelm.—*continued.*

Über elektro-magnet. Anzich. und Abstoss. 8vo. (*Poggendorff, Ann.* xxviii. 1833.)
Leipzig, 1833

Magneto-elektr. Electromagnete. 8vo. (*Poggendorff, Ann.* xxix. 1833.)
Leipzig, 1833

Discontinuat. d. Blizes ; opt. Eigensch. d. Diopsids. 8vo. (*Poggendorff, Ann.* xxxv. 1835.)
Leipzig, 1835

Apparat fur geradlin, circular, und elliptie Polarization. 8vo. (*Poggendorff, Ann.* xxxv. 1835.)
Leipzig, 1835

Versuche ub. Circularpolarization. 8vo. (*Poggendorff, Ann.* xxxv. 1835.)
Leipzig, 1835

Magneto-elektr. Apparat usw. 8vo. (*Poggendorff, Ann.* xliii. 1838.)
Leipzig, 1838

Thermo-säule für constante Ströme. 8vo. (*Poggendorff, Ann.* xliv. 1838.)
Leipzig, 1838

Magnetisme. Bericht der Berl. Akad. 1839

Inducirte Ströme, die galvano-metr. gleich wirken, physiolog. aber ungleich. 8vo. (*Poggendorff, Ann.* xlix. 1840.) *Leipzig,* 1840

Gesetz d. Stürme. 8vo. (*Poggendorff, Ann.* lii. 1841.) *Leipzig,* 1841

Über d. durch Magnetisiren d. Eisens mittelst Reibungs elektricität inducirt. Ströme. (*Poggendorff, Ann.* liv. 1841.) *Leipzig,* 1841

Magnetism. d. sogenannt unmagnet. Metalle. 8vo. (*Poggendorff, Ann.* liv. 1841.) *Leipzig,* 1841

† Über Induction durch electromagnetisirt Eisen. 4to. 91 pp. 1 plate. *Berlin,* 1842

Über d. Gegenstrom zu Anfang und zu Ende eines primären. 8vo. (*Poggendorff, Ann.* lvi. 1842.) *Leipzig,* 1842

Ob der Funke bei Unterbrech. e. elektr. Stroms im Moment der Unterbrech oder später erscheine? 8vo. (*Poggendorff, Ann.* lvi. 1842.) *Leipzig,* 1842

Über d. inducirten Ströme in massiv. Eisen und eisern Drahtbündeln bei Annäherung e. Stahlmagnets. 8vo. (*Poggendorff, Ann.* lvi. 1842.) *Leipzig,* 1842

† Untersuchungen im Gebeite der Inductions Elektricität. Eine in der Akad. der Wissensch. zu Berlin gelesene Abhandlung. 4to. 96 pp. 1 plate. (*Faraday's copy.*) *Berlin,* 1842

† Literatur des Magnetismus und der Elektricität. 8vo. 37 pp. (*In his Repertorium de Physik,* tom. v. p. 152.) *Berlin,* 1844

Ladungs strom d. elektr. Batterie. 8vo. (*Poggendorff, Ann.* lxiv. 1845.) *Leipzig,* 1845

† Über Wirkungen aus der Ferne. Eine am 1 März im Vereine fur wissenschaftliche Vortrage gehaltene Vorlerung. 8vo. 44 pp. *Berlin,* 1845

† Das Gesetz der Stürme in e. Bezichung zu der algemein Bewegung der Atmosphere 3te schr. vermehrte Auflage. 8vo. 346 pp. 2 maps. *Berlin,* 1866

Über Ströme von Flaschen saulen. 8vo. (*Poggendorff, Ann.* lxxii. 1847.) *Leipzig,* 1847

† Über Elektricität. Eine am 26 Feb., im Vereine für Weissenschaftl. Vorträge gehaltene. 8vo. 38 pp. *Berlin,* 1848

Methode gespannt Saiten u. elast. Federn, mittels e. Elektromagnets, in tonende Schwingg. usw. zu versetzen. 8vo. (*Poggendorff, Ann.* lxxxvii. 1852.) *Leipzig,* 1852

Über das elektr. Licht. 8vo. (*Poggendorff, Ann.* lxxxvii. 1852.) *Leipzig,* 1852

Dove, Heinrich Wilhelm.—*continued.*

† Über das Gesetz der Stürme. 8vo. 1st edition, 115 pp. 1 map. (*Besonderer Abdruck aus des Verfassers "Klimatologische Beiträge."*) *Berlin,* 1857

The Law of Storms : considered in connection with the ordinary Motions of the Atmosphere. With diagrams and charts of storms. 8vo. Translated, with the Author's sanction and co-operation, by Robert H. Scott, M.A. Trin. College, Dublin. *London,* 1862

The Law of Storms : considered in connection with the ordinary Movements of the Atmosphere. With diagrams and charts of storms. 2nd edition, 8vo. 330 pp., entirely revised and considerably enlarged. Translated by Robert H. Scott. *London,* 1862

Die Stürme der gemässigten Zone : mit besonderer Berücksichtigung der Stürme des Winters 1862-63. 8vo. 120 pp. *Berlin,* 1863

† Das Gesetz der Stürme : in seiner Beziehung zu den algemeinen Bewegungen der Atmosphäre. *Dritte* sehr vermehrte Auflage. 8vo. 346 pp. 2 plates.
Berlin, 1866

Many Papers, &c. in the " Monatsbericht d. Berlin Akademie," the greater part of which are in *Poggendorff's Annalen.*

† Über den Sturm vom 17 November, 1866. 4to. 50 pp. 2 plates. (*Abhandl. d. k. Akad. d. Wissen. zu Berlin,* 1867.) *Berlin,* 1867

† Gedächtnissrede auf Alex. von Humboldt gehalten in der öffentl. Sitzung d. Konigl. preuss. Akad. d. Wissnch. zu Berlin am 1 Juli dem Leibnitztage des Jahres 1869. 8vo. 31 pp. *Berlin,* 1869

Dove and Moser. Repertorium der Physik. 8vo. 7 vols.
Treidlander, Cat. 1871, p. 12, *says : "Unter Mitwirkung von Lejeune-Dirichlet, Jacobi, Neumann, Riess, Seebeck, etc.* 8 Bde. Berlin, 1837-49. 8vo. *Vollständiges Exemplar,* 7.20." *Berlin,* 1837-49

† **Drago,** Raffael. Sulla relazione dei fenomeni meteorologici colle variazioni del magnetismo terrestre. 2nd edition, 8vo. 88 pp. *Genova,* 1870

Dralet, On a Trombe, with thunder, &c. *Toulouse,* 1830

Draper. J. W. *Born May* 5, 1811, *near Liverpool.* (*Articles in Journal of Franklin Institute, and Silliman's Journal.*)

Drebbel, C. De natura elementorum ; quomodo venti, pluviæ, fulgura, tonitrua, ex iis provocentur et quibus serviant usibus. *Hamburg,* 1621

† **Dree, De.** On the Stones said to have fallen at Ensisheim, and at other places. 8vo. 8 pp. From the Memoir of M. De Dree. (*Phil. Mag.* xvi. 289.)
London, 1803

† Account and Description of a Stone which fell from the clouds in the Commune of Sales, near Ville-Franche, in the Department of the Rhone. From a Memoir of M. De Dree, read in the National Institute, April 11, 1803. 8vo. 8 pp. (*Phil. Mag.* xvi. 217.) *London,* 1803

Drescher, L. Die Electromagnetische Telegraphie: oder leichtfassliche u. specielle Beschreibung der vorzüglichsten electromagnetischen Telegraphen-Apparate u. d. Anwendung derselben in d. Praxis. 4to. 38 pp. *Cassel,* 1848

† Die Electromagnetische Telegraphie . . . und die Anwendung derselben in der Praxis. 2 Abdruck. 4to. 38 pp. 4 plates. *Cassel,* 1849

† **Dropsy,** J. de Cracovie. Electro-thérapie, ou application médicale pratique de l'Electricité ; basée sur de nouveaux procédés. Ouvrage présenté au concours décrété par S.M. l'Empereur des Français. 8vo. 170 pp. *Paris,* 1857

† **Drummond,** T. On Meteoric Stones. 8vo. (*Phil. Mag.* xlviii. 28.) *London,* 1816

Dub, Christophe Julius. *Born August* 16, 1817, *at Berlin.*

De ancoris electromagnetum, Dissertatio. 8vo. 42 pp. *Berolini,* 1848

† Über die anziehende Wirkung der Elektromagnete. 8vo. (*Poggendorff, Ann.* lxxx. u. lxxxi. 1850, u. lxxxv. 1852.) *Leipzig,* 1850, 1852

Dub, Christophe Julius.—*continued.*

Gesetze der Anziehung hufeisenförmiger Elektromagnete. (*Poggendorff, Ann.* lxxxvi. 1852.) *Leipzig,* 1852

Gesetze der Anziehung stabförmiger Elektromagnete. 8vo. (*Poggendorff, Ann.* xc. 1853.) *Leipzig,* 1853

Über Elektromagnetische Spiral-Anziehung. 8vo. (*Poggendorff, Ann.* xiv. 1855.) *Leipzig,* 1855

† **Dub, J.** Die Gesetze des Electromagnetismus im weichen Eisen. 4to. 42 pp. *Berlin,* 1853

† Der Electromagnetismus. 8vo. 516 pp. *Berlin,* 1861

Über die Gesetze d. Vertheilung des Magnetismus in Elektromagneten. *Berlin,* 1862

† Ueber den Einfluss der Dimensionen des Eisenkernes auf die Intensität der Elektromagnete. Eine Experimental-Untersuchung. 8vo. 48 pp. *Berlin,* 1862

† Die Anwendung des Elektromagnetismus mit besonderer Berücksichtigung der Telegraphie. 8vo. 645 pp. *Berlin,* 1863

Dublin Philosophical Journal and Scientific Review. 8vo. *Dublin,* 1825

Dublin Royal Irish Academy. Transactions of the Royal Irish Academy. 4to. *Dublin and London,* 1788

Index to the Transactions of the Royal Irish Academy from 1786 to the present time. By N. Carlisle. 4to. (*From R. S. Cat.* p. 592) *London,* 1813

Proceedings of the Royal Irish Academy. 8vo. (*From Cat. R. S.* 1839, p. 592.) *Dublin,* 1836

† Scientific Papers read before the Royal Irish Academy, and published in its proceedings. Vol. i. 8vo. *Dublin,* 1866

Dubois. Lettre, etc. (in Tableau des Sciences). (*Pogg.* i. 607.) *Paris,* 1776

† **Dubois, E.** De la déviation des Compas à bord des navires, et du moyen de l'obtenir à l'aide du Compas de déviations. 8vo. 56 pp. *Paris*

Dubois-Reymond, Emil H. *Born November* 7, 1818, *at Berlin.*

Über den sogenannt. Froschstrom und den Elektromot. Fische. 8vo. (*Poggendorff, Ann.* lviii. 1843.) *Leipzig,* 1843

Quæ apud veteres de piscibus extant argumenta. (*Poggendorff,* i. 228.) *Berolini,* 1843

† Untersuchungen über thierische Electricität. Vol. i. 8vo. *Berlin,* 1848

Sur l'Electricité animale. 8vo. *Berlin,* 1850

† Note sur la loi du courant musculaire, etc. 8vo. 9 pp. *Paris,* 1850

† Note sur la loi qui préside à l'irritation électrique des nerfs, etc. 8vo. 11 pp. *Paris,* 1850

† Über thierische Bewegung. Rede gehalten im Verein fur wissenschaftliche Vorträge am 22 Feb. 1851. 8vo. *Berlin,* 1851

Untersuch. über Thier Elektricität. (*Berlin Acad.* 1852-3.) *Berlin,* 1852-53

† Animal Electricity. Edited by H. B. Jones. 8vo. 214 pp. *London,* 1852

† On C. Matteucci's letter to Bence Jones, editor of an abstract of Du Bois-Reymond's Researches in Animal Electricity. 8vo. 41 pp. *London,* 1853

Über Zitterwelse und thierische Elektricität. 8vo. *Berlin,* 1853

† Theilte die dritte Fortsetzung seiner Untersuchungen über thierische Elektricität mit. 8vo. 46 pp. (*Monatsb. d. Akademie der Wissenschaften, Berlin,* 6 *und* 20 *Januar,* 1853, *Gesammt sitzungen der Akademie.*) *Berlin,* 1853

† Uber Ströme die durch Andrücken feuchter Leiter an Metall-Leiter entstehen 8vo. 14 pp. (*Faraday's copy.*) 8vo. 14 pp. *Berlin,* 1854

Dubois-Reymond, Emil H.—*continued.*
Verfahren, um feine galvanometr. Versuche einer grossen Versammlung zu
zeigen. 8vo. (*From Poggendorff*, i. 228.)   *Leipzig,* 1855
† Theilte die zweite Fortsetzung seiner Untersuchungen über thierische Electri-
cität mit. 8vo. 30 pp.   *Berlin*
† Untersuchungen über thierische Electricität. 8vo.   *Berlin,* 1860
† Beschreibung einiger Vorrichtungen und Versuchsweisen zu Elektrophysiolo-
gischen Zwecken. 4to. 89 pp.   *Berlin,* 1863
† Nachtrag, 23 Juni, 1870. Gesammtsitzung der Akademie.   *Berlin,* 1870

Poggendorff (i. 228) *says, at the end of his list of Dubois' works : "Antheil
an den Jahresberichten : Die Fortschritte der Physik,* 1847-48-50.

Dubois-Reymond and Beets. Zur Theorie der Nobilischen Farbenringe. 8vo.
(*Poggendorff, Ann.* lxxvi.)   *Leipzig,* 1846

Duboscq, J. Note sur un régulateur électrique. 4to. (*Comptes Rendus,* xxxi.
807.)   *Paris,* 1850
† Description d'un appareil de lumière électrique. 8vo. 4 pp.   *Paris,* 1853

Du Bourg. (*Vide* Barbeu du Bourg.)

Ducala-Bonifas, M. Histoire naturelle du monde.   *Genève,* 1782

Duchemin, M. Les bouées électriques. 8vo. 8 pp.   *Paris,* 1866

Duchenne, G. B. (de Boulogne). Modèle et description d'un appareil électromagnétique à double courant. 4to. (*From Comptes Rendus,* p. 174.) *Paris,* 1849
† Recherches sur l'état de contractibilité et de sensibilité électro-musculaires
dans les paralysies du membre supérieur. 8vo.   *Paris,* 1850
† Recherches sur les propriétés physiologiques et thérapeutiques de l'électricité
de frottement, de l'électricité de contact, et de l'électricité d'induction. 8vo.
26 pp.   *Paris,* 1851
Appareil volta-faradique à double courant. 8vo. 20 pp. (*Comptes Rendus,*
xxviii. 268, 1849.)   *Paris,* 1851
† De la valeur de l'électrisation localisée comme traitement de l'atrophie musculaire progressive. 8vo. 24 pp. (*Extrait du Bulletin de Thérapeutique,* 1853,
pp. 295, 407, et 438.)   1853
† De l'influence de l'électrisation localisée sur l'hémiplégie, rhumatisme de la
face, etc. 8vo. 27 pp.   *Paris,* 1854
† De l'électrisation localisée et de son application à la physiologie, à la patho-
logie, et à la thérapeutique. 8vo. 926 pp. cuts.   *Paris,* 1855
Note sur quelques nouvelles propriétés différentielles des courants d'induction
du' premier et du second ordres. 8vo. 11 pp. (*Communiqué à l'Académie de
Médecine le* 18 *Mars,* 1856.)   *Paris,* 1856
Recherches électrophysiologiques et pathologiques sur les muscles qui meuvent
le pied.   *Paris,* 1856
† De l'électrisation localisée, et de son application à la pathologie et à la thérapeutique. 2me édition, entièrement refondue. 8vo. 1046 pp. 1 plate.
                                                                 Paris, 1861
† Physiologie des mouvements démontrée à l'aide de l'expérimentation électrique
et de l'observation clinique et applicable à l'étude des paralysies et des
déformations. 8vo. 872 pp.   *Paris,* 1867
Examen critique des principales méthodes d'Electrisation. 3me édition. 8vo.
111 pp.   *Paris,* 1870
De l'Electricité localisée et son application à la Pathologie et à la Thérapeu-
tique, etc. 3me édition, refondue. 2 parts. 8vo.   *Paris,* 1872
On localised Electrisation and its application to Pathology and Therapeutics.
Translated from the French, with additional notes and observations, by John
N. Radcliffe, M.R.C.S.
(*Vide* also Erdmann and Duchenne.)

Ducom. De la puissance et des effets des Ouragans, Typhons, et Tornados, des régions tropicales. Manœuvres à faire à bord des navires pour recevoir ces tempêtes . . . 8vo. 23 pp. 3 plates. *Bordeaux*, 1851

† **Ducos,** J. B. Résultats cliniques ou pratiques de la méthode électro-thérapeutique. 18mo. 64 pp. *Paris*, 1855

† Résultats cliniques ou pratiques de la méthode électro-thérapeutique du Dr. J. B. Ducos. 5me édition. 16mo. 64 pp. *Paris*, 1865

Due. (*Vide* Hansteen and Due.)

Du Fay, C. F. *Born September* 14, 1698, *at Paris; died July* 16, 1739.

Concerning Electricity. (*Phil. Trans. Abr.* vol. viii. p. 393.) 1733.4

Hist. de l'Electricité, &c. (*Mém. de l'Acad. Paris, Hist.* p. 4 ; *Mém.* pp. 23, 73, 233, 457) *Paris*, 1733

Histoire de l'Electricité, &c. (*Mém. de l'Acad. Paris, Hist.* p. 1 ; *Mém.* pp. 341, 503.) *Paris*, 1734

† A letter . . . to the Duke of Richmond and Lenox concerning Electricity (Dec. 27, 1733.) 4to. 9 pp. (*Phil. Trans.* xxxviii. for 1733-34.) *London*, 1735

Histoire de l'Electricité. (*Mém. de l'Acad. Paris, Hist.* p. 1 ; *Mém.* pp. 86, 307.) 1737

On Grey's Experiments (pendulum.) (*Danzig Mem.* vol. i. p. 226.) 1737

On Grey's Experiments (pendulum.) (*Mém. Acad. Paris ; Mém.* p. 436.) 1737

Versuche und Abhandlungen von der Electricität derer Körper aus dem Franz. ubers. 8vo. 311 and 17 pp. 1 plate. (*His Life is included.*) *Erfurt*, 1745

Anmerkungen über verschiedene mit dem Magnet angestellte Versuche übers. 8vo. *Erfurt*, 1748

An fluidum nervorum sit electricum. *Montpel.* 1749

Thèses de Physiologie. Le Fluide nerveux, est-il un fluide électrique? 12mo. 44 pp. (*Recueil sur l'Elect. Méd.* ii. 405.) *Paris*, 1763

(*Vide* also Anon. Medical Elect. 1763.)

Du Fech. (*Vide* Lozeran du Fech.)

Dufour, L. Recherches microscopiques sur l'étincelle électrique. 8vo. (*Bibl. Univ.* 1855 ; *Pogg. Ann.* xcix.) *Genève*, 1855

De la tenacité des fils métalliques qui ont été parcourus par des courants galvaniques. 8vo. (*Bibl. Univ.* 1855 ; *Poggend. Ann.* xcix.) *Genève*, 1855

De l'influence de la température sur la force des aimants. 8vo. (*Bibl. Univ.* 1856.) *Genève*, 1856

† Recherches sur les rapports entre l'intensité magnétique des barreaux d'acier et leur température. 8vo. 52 pp. *Lausanne*, 1858

† **Du Fresnel.** Deuxième Mém. sur l'Electricité Galvanique appliquée aux affections chroniques de l'estomac, des intestins et de la vessie. 4to. 11 pp. *Paris*, 1847

(*Vide* also Fresnel, A.)

Duhalde, Jean Baptiste. *Born Feb.* 1, 1674, *at Paris; died August* 18, 1743, *at Paris.*

Description de l'Empire de Chine. (*Ancient Compasses used on land.*) (*From Becquerel and Becquerel E.* 1859.) 1738

Duhamel du Monceau, Henry Louis. *Born* 1700, *at Paris; died July* 22, 1781, *at Paris.*

Façon singulière d'aimanter un barreau d'acier au moyen auquel on lui a communiqué une force magnétique, quelques fois triple de celle qu'il auroit, si on l'eut aimanté à l'ordinaire. 4to. (*From Reuss, Repert.* iv. 376. *In Mém. de Paris,* an. 1745, *Hist.* p. 1, *Mém.* 181.) *Paris*, 1745

Duhamel du Monceau, Henry Louis.—*continued.*

Observation d'une mine de fer attirable par l'aimant. 4to. (*From Reuss, Repert. iv.* 371. *In Mém. de Paris,* an. 1745, *Hist.* p. 47.) *Paris,* 1745

Différents moyens pour perfectionner la boussole. 4to. (*From Reuss, Repert. iv.* 378. *In Mém. de Paris,* an. 1750, *Hist.* p. 1, *Mém.* 154.) *Paris,* 1750

Observation d'une grande variation dans le déclinaison de l'aiguille aimantée le 6 Sept. 1771. 4to. (*Mém. de Paris,* an. 1771, *Hist.* p. 32.) *Paris,* 1771

Description de pleusieurs boussoles qui sont établies dans le parc de Denainvilliers pour observer les variations de l'aiguille aimantée tant en déclinaison qu'en inclinaison. 4to. (*Mém. de Paris,* an. 1772, part ii. *Mém.* p. 44.) *Paris,* 1772

Duhamel, Jean Baptiste. *Born* 1624, *at Vire, Normandy ; died August* 6, 1706, *at Paris.* (*Poggendorff,* i. 616.)

De meteoris et possilibus. 4to. *Paris,* 1660

Philosophia vetus et nova in regia Burgundi pertracta. 4 vols. 12mo. And also 6 vols. 12mo. (*Lib.* v. *De vi Electrica.*) *Paris,* 1678, 1681, and 1700

Philosophia veteri et nova. 4to. *Nurnberg,* 1682

De corp. affect. lib. i. cap. iii. §1.

> *Gralath. Elect. Bibliot.* 2 *Stuck, p.* 273, *says: "De viribus electricis. In Operibus Philosophicis,* tom. ii. *de corporum affectionibus,* lib. ii. cap. vii. p. 200— 205, *Edit. Noriberg,* 1681, 4to."

† **Dujardin.** Télégraphie électrique du Dr. Dujardin. 8vo. 11 pp. 2 plates. *Lille,* 1851 ?

Dujardin, A. Many articles, &c., in the Comptes Rendus on Electric Telegraphs, vols. v. to xxx. 1837 to 1850

Dujardin, A. de Lille. Sur une nouvelle machine électro-magnétique. 4to. (*Comptes Rendus,* xviii. 837, xxi. 528 and 892, xxii. 554, xxiii. 261.) *Paris,* 1844-45-46

Note sur une batterie électro-magnétique, au moyen de laquelle on peut obtenir des courants d'induction très-puissants. 4to. (*Comptes Rendus,* xxi. 1181.) *Paris,* 1845

† **Dujardin,** C. Traité élémentaire de l'Electricité extrait en partie du grand ouvrage de M. Becquerel, et renfermant un Article pratique sur les Paratonnerres par M. Gay-Lussac. s.a. 12mo. *Paris*

Dujardin, P. A. J. Télégraphie électro-acoustique, nouvelle nomenclature des signaux de Chappe par Dr. Dujardin. *Lille*

Dulk, Friedrich Philipp. *Born November* 22, 1788, *at Schirwindt, East Prussia ; died December* 14, 1851, *at Königsberg.*

Über Elektromagnetismus. 54 pp. *Königsberg,* 1824

Dumas. (*Vide* Prevost, J. L., and Edwards, W. F.)

Dumas, C. L. De magnetismo animali, &c. . judicium medicum ? 4to. 3 pp. *Montpelii,* 1790

Dumas, Jean Baptiste. *Born July* 15, 1800, *at Alais, Dep. Gard.*

† Rapport sur les nouveaux procédés introduits dans l'art du doreur par MM. Elkington et De Roultz. Commissioners Thenard, D'Arcet, Pelouze, Pelletier, Dumas. 4to. 23 pp. (*Extract des Comptes Rendus . . de l'Acad.* 29 Nov. 1841, xiii. 998.) (Dumas, rapporteur.) *Paris,* 1841

Dumas, Jean Baptiste—*continued.*

Phénomènes qui accompagnement la contraction de la fibre musculaire (mit
J. L. Prevost). *Paris,* 1823

† Rapport à l'Empereur du Ministre de l'instruction publique, adjugeant le prix
de 50,000 frs. institué en 1852, en faveur de l'auteur des applications les
plus utiles de la Pile de Volta ; et à l'effet d'ouvrir un nouveau concours
pour une troisième période de 5 ans. 8vo. 37 pp. *Paris,* 1864

Réflexions et Observations (sur le Galvanisme).

Du Menil, A. P. J. Geschichtl.-wissenschaftl. Darstell. d. Stoichometrie u. Elek-
trochemie. 8vo. *Hannov.* 1824

Du Moncel, Th... Vcte. *Born March 6, 1821, at Paris.*

† Des Observations Météorologiques. 8vo. 63 pp. 1 plate. *Cherbourg,* 1851

† Mémoire sur le télégraphe imprimeur. 8vo. (*Mém. de l'Acad. de Cherbourg,*
an. 1851.) *Cherbourg,* 1851

Note sur l'électro-magnétisme et ses applications. 8vo. 12 pp.
Valognes, 1852

† Mémoire sur le Magnétisme statique et le Magnétisme dynamique. 8vo. 72 pp.
Paris et Cherbourg, 1852

† Anémographe électrique. "L'Electricité appliquée aux Sciences." 8vo. 43
pp. 1 plate. *Paris,* 1852

Mémoire sur les électro-moteurs. 8vo. 40 pp. *Paris,* 1852

† Des électro-moteurs. 2me édition, revue et augmentée. No. 3, "L'électricité
appliquée aux sciences." 8vo. 60 pp. *Paris,* 1852

Sur un moteur électro-magnétique. 4to. (*Comptes Rendus,* xxxiv. 1852.)
Paris, 1852

Anémographe électrique à onze fils de M. Th. du Moncel. (*Journal l'Union,*
26 *Janvier,* 1853.) *Paris,* 1853
*He gives the article in full, and all his inventions are fully described in his
"Exposée," and various works printed separately.*

† Considérations nouvelles sur l'électro-magnétisme. 8vo. 54 pp. 1 plate.
Paris, 1853

Sur les moniteurs électriques des chemins de fer. (*Journal de l'Arrondissement
de Valognes, Mai,* 1853.) 1853

Réactions des aimants sur les corps magnétiques non-aimantés, ces réactions
considérées comme des effets statiques. 4to. (*Comptes Rendus,* xxxvi. 385.)
Paris, 1853 ?

Nouveau système d'inflammation à distance des substances inflammables par
le courant d'un pile de Daniel et des conducteurs très fins. 4to. (*Comptes
Rendus,* xxxvii. 953.) *Paris,* 1853

† Exposé sommaire des principes . . . de l'électricité. 8vo. 77 pp.
Paris, Cherbourg, 1853

† Exposé des applications de l'électricité. 1er volume, Applications physiques
et mécaniques. 1re édition, 8vo. 280 pp. *Paris,* 1853

† Exposé des applications de l'électricité. 2e vol. 1re édition. 8vo. 378 pp.
Paris, 1854

† Théorie des éclairs. 8vo. 46 pp. *Cherbourg,* 1854

Télégraphe imprimeur. (*Vide his Exposé,* tom. ii. 2e édition, p. 130.) 1854

Note sur l'explosion des mines par l'électricité. 4to. (*Comptes Rendus,* xxxix.
649-51.) *Paris,* 1854

† Coup d'œil sur l'état des applications mécaniques et physiques de l'électricité.
8vo. 37 pp. (*La Revue Contemporaire.*) *Paris,* 1855 ?

Du Moncel—*continued.*

† Notice sur l'appareil d'induction électrique de Ruhmkorff et les expériences que l'on peut faire avec cet instrument.   1re édition.   8vo. 152 pp. cuts.
Paris, 1855

† Moniteur automatique des chemins de fer, imaginé en Mai 1853.   8vo. 16 pp.
Caen, 1855

† Exposé des applications de l'électricité.   Tom. i. et. ii. 2e édition, 8vo.
Paris, 1856

† Exposé des applications de l'électricité.   Applications mécaniques, physiques, et physiologiques.   Tom. iii. 2e édition, 8vo. 454 pp. 4 plates.   *Paris*, 1857

† Notice sur l'appareil d'induction électrique de Ruhmkorff, et les expériences que l'on peut faire avec cet instrument.   2nd or 3rd edition, 8vo. 222 pp.
Paris, 1857

† Ruhmkorff's Induction-Apparat u. die damit anzustell.   Versuche.   Nach d. Franz. v. C. Bromeis u. J. F. Bockelmann.   8vo. 176 pp.   *Frankfürt*, 1857

† Notice historique et théorique sur le tonnerre et les éclairs.   8vo. 54 pp.
Paris, 1857

† Etudes du magnétisme et de l'électro-magnétisme au point de vue de la construction des électro-aimants.   8vo. 268 pp. 1 plate.   *Paris*, 1858

† Revue des applications de l'électricité en 1857 et 1858.   4th vol. of *Exposé*. 8vo. 616 pp. 3 plates.   *Paris*, 1859
This is called "Suite à l' Exposé des applications de l'électricité," 2e série.

† Notice sur l'appareil d'induction-électrique de Ruhmkorff, suivie d'un mémoire sur les courants induits.   4e édition (rare), 8vo. 400 pp.   *Paris*, 1859

† Mémoire sur les courants induits des machines magnéto-électriques.   8vo. 33 pp.   *Paris*, 1859

† Etude des lois des courants électriques au point de vue des applications électriques.   8vo. 201 pp.   *Paris*, 1860
(This is called, on the cover of the pamphlet only, " Suite à l' Exposé des applications de l'électricité," 3e série.)

† Recherches sur la non-homogénéité de l'étincelle d'induction.   8vo. 115 pp.
Paris, 1860

† Recherches sur les transmissions électriques à travers le sol dans les circuits télégraphiques.   8vo. 31 pp.   (*Annales Télégraphiques*, tom. iii. p. 465.)
Paris, 1861

† Recherches sur les constantes des piles voltaïques.   2e édition, augmentée. 8vo. 32 pp.   *Cherbourg*, 1861

† Recherches sur l'électricité.   8vo. 62 pp.   (*Extrait des Mémoires de l' Académie de Caen.*)   *Caen*, 1861

† Exposé des applications de l'électricité.   Tome v. Revue des découvertes faites de 1859 à 1862.   8vo. 244 pp.   *Paris*, 1862

Exposé des applications de l'électricité, tom. v. 2d fascicule, Revue des découvertes faites en 1859, 1860, 1861 et 1862, *fin.*   2nd edition. 8vo. 245-560 pp.   *Paris*, 1862

† Traité théorique et pratique de télégraphie électrique, à l'usage des employés télégraphistes, des ingénieurs, des constructeurs, et des inventeurs.   8vo. 613 pp. 3 plates.   *Paris*, 1864

† Notice sur l'appareil d'induction électrique de Ruhmkorff, 5e édition, entièrement refondue.   8vo. 400 pp.   *Paris*, 1867

Notice sur le câble transatlantique.   8vo.   *Paris*, 1869
In the *Comptes Rendus* (No. 436) M. Moncel describes his experiments on the Voltaic battery.

† Recherches sur les meilleures conditions de construction des électro-aimants. 8vo. 124 pp.   *Paris et Caen* 1871
(*Vide " La Science,"* journal edited by Du Moncel.)

Du Moncel—*continued.*

† Description des télégraphes électro-chimiques de MM. Caselli et Bonelli. 8vo. 36 pp. 1 plate. (*Annales Télégraphiques*, Mai et Juin, 1863.)
Paris, 1863

Dumont, Arist. Mémoire sur l'application de la télégraphie électrique au relations sommaires des habitants des grandes villes. 4to. (*Comptes Rendus*, xxxi. 449.)
Paris, 1850

Dumont d'Urville, Jules Sébastien César. *Born May 23, 1790, at Condé sur Noireau, Département du Calvados ; died May 8, 1842, Paris, Versailles.*

Voyage de découvertes de *l'Astrolobe*, exécuté pendant les années 1826-29. Observations nautiques, météorologiques, hydrographiques, et physiques.
Paris, 1834

Voyage au Pole Sud et dans l'Océanie, sur les corvettes *l'Astrolabe* et la *Zélée*, exécuté par ordre du Roi pendant les années 1837 à 1840, sous le commandement de J. Dumont d'Urville. 8vo. 34 vols.
Paris

Physique, par MM. Vincendon Dumoulin et Coupvent-Desbois. 8vo. 4 vols. plates.
Paris, 1842

Magnetical declinations on board the " Astrolobe," observed by Kosmann, chef de Timonerie, are given in the tables of " Routes des Corvettes," in vol. i.

† **Dunbar, W.** Description of a singular phenomena at Bâton Rouge, dated June 30, 1800. (Read January 16, 1801.) 4to. 1 p. (*Trans. American Phil. Society*, vol. vi. p. 25, part i.)
Philadelphia, 1804

Duncan, A. Medical cases, with remarks.
Edinburgh, 1778

† **Dunn, J.** Description of the whirlwind at Scarborough on the 24th June, 1823. 8vo. 8 pp. (*Edinburgh Phil. Journal*, x. 11.)
Edinburgh, 1824

Dunn, Samuel. *Born at Crediton, Dorsetshire ; died 1792.*

Magnetic Atlas. (*Fine Work.*) (*From Young*, p. 442.)
London, 1776

New Atlas of the Mundane system.
1788

Duperrey, Louis Isidore. *Born October 21, 1786.*

Voyage de la *Coquille* pendant les années 1822-25. Physique. 6 vols.
Paris, 1826

† Observations du pendule invariable, de l'inclination et la déclinaison de l'aiguille aimantée faites dans la campagne de la corvette la *Coquille* pendant les années 1822-25. 8vo. 32 pp.
Paris, 1827

Du magnétisme terrestre. 8vo.
1834

† Notice sur la position des poles magnétiques de la terre. 4to. 8 pp.
Paris, 1841

† Réduction des observations d'intensité du magnétisme terrestre faites par M. de Freycinet durant le cours du voyage de la corvette l' *Uranie*. 4to. 12 pp. (*Extrait des Comptes Rendus de l' Académie*, tom. xix. Séance du 2 Septembre, 1844.)
Paris, 1844

Du Petit Thouars, Abel. Voyage autour du monde sur la frégate *La Vénus* pendant les années 1836-39, publié par ordre du Roi par Abel du Petit Thouars, capitaine de vaisseau. 8vo. 11 (?) vols. atlas folio.
Paris, 1840-4—

Voyage autour du monde sur la frégate *Vénus* pendant les années 1836 à 1839. Partie physique, par De Tressan. 5 vols. 8vo. atlas in folio.
Paris, 1842 *to* 1844

Duport, Saint-Clair. Production des métaux précieux au Mexique.
Paris, 1843

† **Dupotet.** An introduction to the study of Animal Magnetism. 8vo. (*Roget's copy.*) *London*, 1838

† **Du Pré.** Articolo di lettera al Brugnatelli. 8vo. 5 pp. (*Ann. di Chim. di Brugnatelli*, ix. 156.) *Pavia*, 1795

Sur l'explication des phénomènes de la chaleur, etc. 8vo. 30 pp. *Paris*, 1828

Duprez, François Joseph Ferdinand. *Born October 21, 1807, at Ghent.*

† Mémoire en réponse à la question suivante : On demande un examen approfondi de l'état de nos connaissances sur l'électricité de l'air, et des moyens employés jusqu'à ce jour pour apprécier les phénomènes électriques qui se passent dans l'atmosphère. 4to. 134 pp. *Bruxelles*, 1843

† Statistique des coups de foudre qui ont frappé des paratonnerres, ou des édifices et des navires armés de ces appareils. Mémoire présenté . . . le 5 Décembre, 1857. 4to. 63 pp. (*In Bruxelles Académie*, tom. xxxi. 1859.) *Bruxelles*, 1859

Duprez d'Aulnay, L. Dissertation sur la cause physique de l'électricité. *Paris*, 1746

Dupuis, C. F. Erfand den Telegrafen den Chappe verbessert. 1788

Dupuytren, C. Faits particuliers, et notes sur le galvanisme. *About* 1801

Durand. Introduction à l'électricité. Se trouve en France chez Durand, neveu. 12mo. 144 pp. *Madrid*, 1788

Durand, C. Inquisitiones physico-medicæ circa electricitatem positivam. 4to. 8 pp. *Monspelii*, 1786

† **Durand,** F. A. Théorie électrique du froid, de la chaleur et de la lumière. 8vo. 36 pp. *Paris*, 1863

Durand, l'Abbé. Le Franklinisme refuté. 12mo. 69 pp. *Paris*, 1788

† **Dureau,** A. et **Moreau,** E. L. Des poissons électriques. Exposé anatomique et physiologique, par A. Dureau. Des causes et éléments de production de l'électricité de la Torpille, par E. Lemoine-Moreau. 8vo. 34 pp. *Paris*, 1868

Etude sur les poissons électriques, la Torpille, le Gymnote, la Raie, le silure trembleur. 8vo. 20 pp. *Paris*

Du Roi, J. P. Üeber die Wiener Stahlmagnete. (*Braunschweig, Anzeig* 1775 St. p. 13.) *Braunschweig*, 1775

Neue elektr. Versuche über d. Ausbrüten d. Hühner und Schmetterlings-Eier, über d. Aufkeimen d. Samen u. d. Wuchs d. Pflanzen. *Braunschweig*, 1779

† **Du Sable,** La télégraphie météorologique en Angleterre. 8vo. 38 pp. *Paris*, 1863

Dutens, Louis. *Born Jan. 15, 1730, at Tours ; died May 23, 1812, at London.*

*† Abhandlung von den Edelstein. Aus dem Franzos. übers. 8vo. (*Baer's Catalogue*, No. vi. *Naturw.* p. 35, 1867.) *Nurnberg*, 1779

Abrégé chronologique pour servir à l'histoire de la physique.

Recherches sur l'origine des découvertes, &c. 2 vols. 8vo. *Paris*, 1766

† Origine des découvertes, &c. 3rd edition. 4to. *London*, 1796

† **Dutertre.** Des aurores et de quelques autres météores. 8vo. 8 pp. *Le Mans*, 1822

Du Tour, Etienne François. *Born in 1711, at Riom, Basse Auvergne ; died 1784, at Riom.*

On Flame, &c. (*Letter to Nollet.*) 1745

Du Tour, Etienne Francois—*continued.*

Explication de deux phénomènes de l'aimant, sur les différences qu'apportent les secousses données à un carton sur lequel on étend de la limaille de fer à l'arrangement de cette limaille présentée à la pierre d'aimant. 4to. (*Mém. de Mathém. et de Phys.* i. 375.) *Paris,* 1750

Mémoire sur la manière dont la flamme agit sur les corps électriques. 4to. (*Mémoire de Mathématique et de Physique,* ii. 246.) *Paris,* 1755

† Recherches sur les différents mouvements de la matière électrique. 12mo. 318 pp. 4 plates. *Paris,* 1760

Observation sur le tourbillon magnétique. 4to. (*Mémoire de Mathématique et de Physique,* iii. 233.) *Paris,* 1760

Sur l'électricité en moins. 4to. (*Mémoire de Mathématique et de Physique,* iii. 244.) *Paris,* 1760

Discours sur l'aimant. (*Pièces de Prix de l'Académie de Paris,* tom. v. *Mém.* ii. p. 49.)

Mémoires sur les effets électriques du tonnerre tombé près de Riom en Auvergne 4to. (*Mém. de Paris,* an. 1766, *Hist.* p. 37.) *Paris,* 1766

De la nécessité d'isoler les corps que l'on électrise par communication, et des avantages qu'on corps, convenablement isolé, retire du voisinage des corps non électriques. 4to. (*Mémoire de Mathématique et de Physique,* ii. 516.) *Paris,* 1775

(*Vide* also Euler and others.)

Dutrochet, R. J. H. Nouvelles recherches sur l'endosmose et l'exosmose. (*From Poggendorff,* i. 633.) *Paris,* 1828

(*Vide* also Burnet.)

† **Dutrochet** and **Bowdich.** On the heights of two Meteors. 8vo. 1 p. (*Phil. Mag.* lvi. 397.) *London,* 1820

Duvert, *vide* Chapot-Duvert.

† **Dwight,** S. E. Notice of a fire-ball on March 21, 1813. 8vo. 2 pp. (*Phil. Mag.* or *Annals,* iii. 74.) *London,* 1828

Dyckhoff. Expériences sur l'activité d'une pile de Volta dans laquelle les corps humides sont remplacés par des couches minces d'air. 8vo. (*Journal de Chimie de Van Mons,* No. xi. p. 190.) *Bruxelles*

E.

E, B. (*Vide* Anon., Electricity, 1804.)

Eames, John. Extract from the Journal-books of the Royal Society concerning Magnets having more poles than two; with some observations by Désaguliers on the same subject. 4to. (*Philosophical Transactions* for 1738, p. 383.)
London, 1738

Eandi, Giuśeppe Antonio Francesco Geronimo. *Born October* 12, 1735, *at Salluzo; died October* 1, 1799, *at Turin.*

† Memorie istoriche intorno gli studi del Beccaria. 8vo. (*Vide* Beccaria.)
Torino, 1783

† Sur l'électricité dans le vide. Lu le 10 Mai, 1790. 4to. 11 pp. (*Mémoires de Turin,* an. 1790-91, printed 1793, p. 7, tom. v.)
Turin, 1793

Résolution des questions suivantes de l'électricité. 1. L'air, est-il électrique par frottement? 2. La lumière excitée par le frottement dans les corps, est-elle électrique? 3. Les corps résineux, décélent-ils de l'électricité par la chaleur et la fusion? 4to. (*Mémoires de Turin,* vi. pt. i. p. 1, 1802.)
Turin, 1804

† Notice sur la vie et les ouvrages d'Eandi, par A. M. Vassalli-Eandi. Lue le 3 pluviose, an. xi. 4to. 75 pp. (*Mémoires de Turin,* an. x. et xi. 1802-3, printed xii. 1804, p. 1.)
Turin, 1804

Eandi et Vassalli-Eandi. Physicæ experimentalis lineamenta : ad subalpinos. 8vo.
The 2nd vol. at p. 149, *Volta's " Atmospheric Apparatus." The work has copious notes, a learned historical preface, and a Physical Bibliography.*
Torino, 1793-94

Eandi, Vassalli Antonio Maria. (*Vide* Vassalli-Eandi.)
Antonio Maria Vassalli was nephew of Giuseppe Antonio Eandi, and added the name of his uncle to his own, thus calling himself Antonio Maria Vassalli-Eandi.

Ebel, Johann Gottfried. *Born October* 6, 1764, *at Züllichau, Schles.; died October* 7 (8), 1830, *at Zürich.*
Manuel du voyageur en Suisse.

Ebensperger, J. L. Die Einrichtung des Telegraphen. Eine gemeinfassliche Belehrung. 8vo. 84 pp.
Zwickau, 1854

Eberhard, Christoph. *Born* 1675*; died* 1730, *at Halle.*
Versuch einer magnetischen Theorie. 4to.
Leipzig, 1720
Specimen theoriæ magneticæ, quo ex certis principiis magneticis ostenditur, vera universalis methodus inveniendi longitudinem et latitudinem, confectum a Ch. Eberhard, Londini, Oct. 31, 1718, nunc vero juris publici factum a S. B. W. Lips.
Lips. 1720
Note.—From Poggendorff, i. 640,*who adds: " This Anonym. gab davon auch e. deutsche Uberset. Lipzig,* 1820."

Eberhard, Johann Peter. *Born December* 2, 1727, *at Altona; died December* 17, 1779, *at Halle.*
Gedanken von Feuer und den damit verwandten Körpern, dem Licht und der elektrischen Materie, &c. 8vo. *From Krunisz,* p. 39, *and from Poggendorff,* i. 640, *and Hensius,* i. 722.)
Halle, 1750
Gedanken von d. Ursachen d.Gewitter u. ihrer Aehnlichkeit mit d. Electricität. (*Wochenth. Halle 'sche Anzeig,* 1754.)
Halle, 1754
Von den Ursachen d. Nordlichts und dessen Aehnlichkeit mit d. Elektricitat. (*Wochenth. Halle 'sche Anzeig,* 1758.)
Halle, 1758
Vorschläge zur bequemern und sicheren Anlegung der Pulvermagazine. 8vo. (*From Kuhn* (1866), p. 273.)
Halle, 1771

Ebert, F. A. Algemeines bibliographisches Lexikon. 2 vols. 4to.
Leipzig, 1821-30
A translation (English) in 4 vols. 8vo. *Oxford*, 1837

† **Ebner, Freiherr. von.** Ueber die Anwendung der Reibungs-Electricität zum Zünden von Sprengladungen. 8vo. 27 pp. 5 plates. (*Aus den Sitzungsbn. der k. Akad. der Wissensch.* xxi. Bd. 1 Hft.) *Wien*, 1806

† **Eckeberg, C. G.** Bericht von einem Nordscheine welcher auf der Ausreise auf dem Schiffe der ostindischen Gesellschaft Sophia Albertina d. 30 Jan. 1755, bemerket worden. 8vo. 4 pp. (*K. Schwed. Akad. Abh.* xix. 58.)
Hamburg and Leipzig, 1757

† **Ecker, A.** Eine Beobachtung uber die Entwicklung d. Nerven d. elektr. Organs von Torpedo Galvanii. 10 pp. 1 plate. (*Zeitschrift. fur wissensch. Zoologie,* i. Bd. 1 Heft, Leipzig, 1848.) *Leipzig*, 1848

Eckhardt, C. Beiträge zur Anatomie und Physiologie. (*From Meyer.*)

Ecole de Médecine de Paris. (*Vide* Hallé.)

Economista. (*Vide* L'Economista.)

Edinburgh Encyclopædia. By Sir D. Brewster. 18 vols. 4to.
Edinburgh, 1810-30

Edinburgh Journal of Science. New series, by D. Brewster. 8vo.
Edinburgh, 1831-32

Edinburgh Philosophical Journal, exhibiting . . . progress of discovery in Natural Philosophy . . . by D. Brewster and Jameson.
Edinburgh, 1819-24
Idem. by Jameson. The whole in 14 vols. 8vo. *Edinburgh*, 1824-26

Edinburgh New Philosophical Journal. Conducted by R. Jameson (and L Jameson). 8vo. . . . to 1854. *Edinburgh*, . . . 1854
Idem. New series. 8vo. 1855 to . . . *Edinburgh*, 1855 . . .
Exhibiting a view of the progressive discoveries . . . in the Sciences and Arts. New series. Editors—T. Anderson, Sir W. Jardine, J. H. Balfour ; for America, H. D. Rogers. *London and Edinburgh*

Edinburgh Royal Physical Society. Instituted 1771. Proceedings of. 8vo.

Essays and observations, Physical and Literary. 8vo. 3 vols.
Edinburgh, 1754-71
Transactions of the Royal Society of Edinburgh. 4to. *Edinburgh*, 1788
Proceedings of the. 8vo. *Edinburgh*, 1832

Edlin von Born. (*Vide* Prag.)

Edlund, Erik. *Born March* 14, 1819, *at Nerike.* (*Works on Electricity, &c., in Danish and other Journals. Vide Pogg.* i. 643.)

Edmé. (*Vide* St. Edmé.)

Edwards, William Frederic. *Born April* 14, 1777, *at Jamaica ; died* 1842, *at Paris.* (*Poggendorff,* i. 644.)

† De l'influence des agens physiques sur la vie. 8vo. 654 pp. *Paris*, 1824

Note.—An appendix contains matter on Electricity compiled by Prevost and Dumas.

† **Eeles, H.** Letter concerning the cause of Thunder. 4to. (*Phil. Trans.* an. 1752, p. 524.) *London*, 1752

Philosophical Essays, in several letters to the Royal Society. 8vo. 189 pp.
London, 1771

Note.—Latimer Clark's Cat. says : " An edition was published at the same time in Dublin from same type, except the title-page."

Egeling. Diss. de electricitate. 4to. *Ultrai*, 1759

Egell, A. Observationes in historiam phosphororum naturalium. *Wirceb*. 1773

R

154 EGG—ELE

† **Egger.** Phys. u. statist. Beschreibung von Island-Nordlichts. 1 Theils. 1
Abtheit. p. 245. 8vo. *Kopenhagen, 1786*

† **Ehrenberg, C. G.** Das Leuchten d. Meeres. Neue Beobachtungen nebst Uber-
sicht der Hauptmomente der geschichtlichen Entwicklung dieses merkwur-
digen Phenomens. Ein in d. Königl. Akad. d. Wissensch. im April, 1834,
gelhaltener Vortrag, mit einigen Zusätzen gedruckt im October, 1835.
Berlin, 1835

Über d. Meteorstaubfall i. Schles. am 31 Jan. 1848. 8vo. *Berlin*

† Über d. sehr merkwürdigen Passatstaub-Fall . . . am 17 Feb. 1850, auf den
höchsten Gotthard-Alpen der Schweiz. 8vo. 22 pp. *Berlin*

† Über d. rothen Meteorstaubfall im Aufang d. Jahres, 1862, in den Gasteiner
d. Rauriser Alpen u. dei Lyon. 8vo. 26 pp. *Berlin, 1862*

Ehrmann, Friedrich Ludwig. *Born 1741, at Strasbourg; died February 17, 1800, at
Strasbourg.*

† Description et usage de quelques Lampes à Air inflammable. 8vo. 35 pp.
1 plate. *Strasbourg, 1780*

Gebrauch einer Elekt-Lampe. Aus dem Frans. von ihm selbst übersetst, mit
Anmerk. und einem Anhang. 12mo. *Strasbourg, 1780*

Eimmant. (*Vide* Volckamer.)

Eisenlohr, Otto. *Born September 3, 1806, at Carlsruhe; died July 25, 1853, at Bad
Antogast.*

Untersuchungen über den Einfluss des Windes auf den Barometerstand . . .
und die vershiedenen Meteore. 112 pp. *Leipzig, 1837*

Eisenlohr, W. Anleitung z. Ausführung u. Visitation d. Blitzableiter.
Carlsruhe, 1848

Eisentraut. Dissert. de corporum electricorum vi attractiva . . .
Herbip. 1748

† **Ekeberg, C. G.** Beobachtungen der Neigung der Magnetnadel auf einer Reise
nach und von Canton 1766 u. 1767. 8vo. 4 pp. (*K. Schwed. Akad. Abh.*
xxx. 238.) *Leipzig, 1768*

† Beobachtungen mit dem Neigungscompasse auf einer Seereise von Götheborg
nach Canton in China: und von da zuruck 1770 u. 1771. 8vo. 15 pp. (*K.
Schwed. Akad. Abh.* xxxiv. 254.) *Leipzig, 1772*

† Beobachtungen mit dem Neigungscompass auf einer Seereise nach Canton in
China und wieder rückwärts. 8vo. 6 pp. (*K. Schwed. Akad. Abh.* xxxvii.
306.) *Leipzig, 1775*

Ekling. Improvements on Bain's Apparatus, which at the beginning of 1846 was
introduced on the line between Edinburgh and Glasgow, &c. (*Described in the
Eisenbahnzeitung,* 1847, pp. 99; *Jahr.* 1848, p. 21.)

† **Eklöf, J. H.** Försok att för atskilliga orter i Europa bestämma Norrskenets
arliga periodicitet. Foredr för Vet. Soc. d. 5 Dec. 1842. 4to. 17 pp. (*Acta
fennica.* ii. 1843.) *Helsingfours* (?) 1842

† **Electric Telegraph Company.** Notice of instantaneous communication with
all the principal towns of the kingdom. 4to. 2 pp. (*Cooke, editor.*)
London, 1850
(*Ten shillings for twenty words: maximum charge.*)

† **Electric Telegraph and Railway Review.**
(*Vide* Lundy, editor.)

† **Electric Telegraph Department in India.** First Report of the Operations,
from Feb. 1, 1855, to Jan. 31, 1856. 1856
(*Vide* O'Shaughnessy.)

Electrical Magazine. Vol. i. 8vo. *London, 1843*
(*Vide* C. V. Walker.)

Electrical Society. *Vide* London Electrical Society, *and also* Society of Telegraph Engineers.

Electrician, The. A weekly Journal of Telegraphy, Electricity, &c. 4to.
London, 1862

Elias, P. *Born April* 18, 1804, *at Amsterdam.*

Beschrijving eener Maschine ter anmending van het Elektro-magnetism as beweegkracht. *Haarlem,* 1842

† Verslag van eenige op Natuurlijke Magneten genomen proeven. 8vo. 7 pp.
Haarlem, 1851

(*Overgedrkt uit het Tijdschrift voor de Wisen Natuurkundige Wetenschappen, uitgegeven doorde Erste Klasse van het Kon-Ned. Instituut, Deel.* v. *Bladzijde,* 134.)

Over het vermogen der magneto-elektrische machine. (*Verslagen etc. d. k. Acad. v. Wetenschappen,* xi. 1861.) 1861

Elice, Ferdinando. *Born April* 19, 1786, *at Loano, Genua.*

Lettera sulla niuna influenza del suono delle campane per attrarre il fulmine.
Genova, 1817

† Saggio sull' Elettricità. 1st Ed. 8vo. 71 pp. *Genova,* 1817

† Account of his Essay on Electricity. Printed 1817, at Genoa. 8vo. 2 pp. (*Phil. Mag.* i. 453.) 1817

† Influenza delle campane sul fulmine. 8vo. 3 pp. (*Bibl. Ital.* xxxii. 423.)
Milano, 1823

† Saggio sull' Elettricità. Seconda Edizione. 100 pp. 8vo. (*Brugnatelli's copy.*) *Genova,* 1824

(*His plate electric machine for the two Electricities described* (*very shortly*) *at* p. 55. *One side thickly coated with* " *ceralacca* " (*sealing-wax*).) 1824

† Osservazioni sull' Istruzione dei Parafulmini approvata dalla R. Acad. delle Scienze di Parigi il di 23 Aprile 1823 e pubblicata nel 1824. 8 pp. 8vo.
Genova, 1826

† Lettera del . . F. Elice . . al sig. C. F. 8vo. 3 pp. (*Bibl. Ital.* xlv. 136.)
Milano, 1827

Note.—On the effects of lightning on the tower of the lighthouse at Genoa, Jan. 4th, 1827.

Istruzione sui Parafulmini. Lettera . . al P. C. Dentone. 8vo. 24 pp.
Genova, 1839

Istruzioni sui Parafulmini ; lettera . . indirizzata al Pittore C. Dentone. Seconda Edizione con aggiunte dell' autore. 8vo. 32 pp. *Genova,* 1841

Lettere al Sig. C. Dentone. 1. Scoperta Scintilla elettrica ottenuta dal caffè e da altri semi 9 Marzo 1842. 2. Elettricità ottenuta da parecchie sostanze 16 Marzo 1842. 8vo. 2 pp. *Genova,* 1842

Nuovo metodo per eccitare l'elettricità collo schioppo, e proposta di un fulmine artificiale. 8vo. 3 pp. *Genova,* 1843

Sull' elettricismo eccitato collo Schioppo. Lettera seconda 18 Marzo 1843. (*To Dentone.*) 8vo. 7 pp. *Genova,* 1843

Osservazioni sui Parafulmini. Luglio 1843 e Nota sulla conducibilità di parecchi metalli per l'elettrico, e proposta di fare le punte dei parafulmini di palladio. 1 Agosto 1843. 8vo. 11 pp. *Genova,* 1843

Notizie elettriche, All' Illust. Sig. Cav. L. Foppiani. 8vo. 12 pp.
Genova, 1844

† Osservazioni ed esperienze sull' Elettricità All' . . Foppiani. 8vo. 16 pp. (*Brugnatelli's copy.*) *Genova* 1844

† **Elkington et de Ruolz.** Sur les nouveaux procédés introduits dans l'art du Doreur par MM. Elkington et de Ruolz. Rapport fait à l'Acad. des Sciences, par M. Dumas. (*Vide* Dumas.) 4to. 1841

Ellery, R. L. T. Monthly Record of results of observations in Meteorology, Terrestrial Magnetism, &c., taken at the Melbourne Observatory during February, 1872. (*From Athenæum*, July 29, 1872.)

† **Ellicott, Andrew.** Account of an extraordinary Flight of Meteors (commonly called Shooting-stars), . . as extracted from his Journal, in a Voyage from New Orleans to Philadelphia. 4to. 1 p. (*Trans. Am. Phil. Soc.*, Old series, vol. vi. part i. p. 28.) *Philadelphia*, 1804

Ellicott, John. *Died in* 1772, *at London.*

Letter on weighing the strength of Electrical Effluvia. 4to. (*Phil. Trans.* 1746, p. 96.) *London*, 1746

† Several Essays towards discovering the laws of Electricity, communicated to the Royal Society by J. Ellicott, and read on the 25th February, 1747, and at two meetings soon after. Read February 25, 1747-48. 4to. 30 pp. (*Phil. Trans.* xlv. 195.) *London*, 1748

(*Latimer Clark's Catalogue says*: "*To which is prefixed part of a letter from Nollet to Martin Folkes, President of the Academy at Paris.* 8vo. 38 pp.)

Ellinger, A. Beiträge zur Erläuterung d. Vorstellung von Wetterwolken und Blitzen. (*Munchen Denkschr.* 1803-6.) *Munchen*, 1803-6

Ellinger, R. Beiträge ub. d. Einfluss d. Himmels-Körper auf. uns. Atmosph. 2 and 3 Heft. 8vo. *Munchen*, 1814-15-16

Elliott, Captain. Meteorological Observations made at the Observatory at Singapore in 1841. Roy. 4to. 45 pp. *Madras*, 1850

† **Elliot, C. M.** Magnetic survey of the Eastern Archipelago. Read January 16, 1851. 4to. 45 pp. 8 plates. 2 large charts. (*Phil. Trans.* part i. for 1851.) *London*, 1851

† **Elliot, R. J.** On the Magnetic combinations and the action of Selenic Acid on Methyl-Alkohol. Inaugural Dissertation. 8vo. 44 pp. *Gottingen*, 1862

† **Ellis, F.** On Governor Ellis's discovery of the action of cold on Magnetic Needles. 8vo. 2 pp. (*Phil. Mag.* lx. 340.) *London*, 1822

Ellis, J. De Dionæa muscip. ; planta irritabili nuper ¨detecta, ad Linné epistol. a. d. Engl. ubers mit Anmerk, vor J. C. Dr. Schreber. 4to. *Erl*, 1771

† **Elsner, F. C. L.** Die galvanische Vergoldung und Versilberung, sowohl matt als glänzend, so wie die Verkupferung, Verzinnung, Verbleiung, Verzinkung, Bronzirung, Verplatinirung und Vernickelung metallener Gegenstande auf demselben Wege. Nach einigen Erfahrungen bearbeitet. 8vo. 280 pp. 1 plate. 2 figs. *Berlin*, 1843

Neue Erfahrungen bei d. galvan. Vergoldung. *Berol*. 1845

Elvius, Peter (Pehr). *Born* 1710, *at Upsala; died September* 27, 1749, *at* (*A. St.*) *Ekholmsound.*

† Geschichte der Wissenschaften von den Aenderungen bey Abweichung der Magnet-Nadel. 8vo. 12 pp. (*Schwedische Akad. Abhandl.* an. 1746, p. 89, vol. ix.) *Hamburg*, 1746

† Historisk berättelse om electriciteten. 8vo. (*Vetenskaps Acad. Handlinger*, 1747, *and Reuss*, iv. 346.) *Stockholm*, 1746

Geschichte der Wissenschaften. Von der Elektricität. 8vo. 7 pp. (*K. Schwed. Akad. Abh.* ix. 179.) *Hamburg*, 1746

Elvius, Peter (Pehr).—*continued.*
Historisk berättelse om andringar i magnetnalens missvisning. 8vo. (*Vetenskaps Acad. Handlinger,* 1747.) *Stockholm,* 1747

† Lage von Gothenberg durch astronom. Beobachtungen bestimmt von P. E. 8vo. 16 pp. (*K. Schwed. Akad. Abh.* x. 300, *Abweichung des Magnets,* p.311.) *Hamburg and Leipzig,* 1748

† **Emmett,** J. B. On an Electrical Phenomenon. 8vo. 3 pp. (*Phil. Mag. or Ann.* v. 170.) *London,* 1829

Emory, E. H. Observations, astronomical, magnetic, and meteorological, made at Chagres and Gorgona, Isthmus of Darien, and at the city of Panama, New Grenada. 4to. *Cambridge (U. S.),* 1850

† **Emory,** W. H. Astronomical, Magnetical, and Meteorological Observations made at Panama. Communicated by Bond, August 8, 1849. (*Mems. American Acad.* New series, v. part i. p. 1.) 4to. 24 pp. *Boston,* 1853

Enciclopedia Circolante Ital. e Straniera. . . . Progressi, scoperte, ec. Foglio periodico settimanale.

Encke. Über d. Bestimm d. Längen-Unterschiedes zwiche d. Sternwart. v. Brüsel und Berlin abgeleitet auf telegraph. Wege. 4to. 67 pp. (*Berlin Acad.* 1858.) *Berlin,* 1858

† **Encke,** J. F. Über die magnetische Declination in Berlin. 4to. 21 pp. (*Abhandl. Akad. Berlin,* 1857.) *Berlin,* 1857

Encontre, Daniel. *Born* 1762, *at Nismes ; died Sep.* 16, 1818, *at Montpellier.*
Communication, &c., of the "Relation de la chute de deux Aérolithes par MM. Pages et D'Hombre Firmas." 8vo. (*Montpellier Acad. Rec. des Bulletins,* iii. 52.) *Montpellier,* 1809

Encyclopædiæ and Cyclopædiæ.
Encyclopædia of British Arts. (*Vide* Barlow, B.)
British Cyclopædia.
Encyclopædia Britannica.
Encyclopædia Metropolitana.
The Cyclopædia by Rees
British Encyclopædia, by Nicholson.
English Enclyclopædia.
Iconographic Encyclopædia, I.
London Encyclopædia, I.
Nichols' Cyclopædia of the Physical Sciences.
Penny Cyclopædia.
Chambers's Cyclopædia.
Enclyclopædia of Edinburgh.
Brewster's Encyclopædia.

Encyclopædia Britannica.
7th edition, edited byNapier. 4to. *Edinburgh and London,* 18—
Articles. Aurora-borealis, by , vol. iv.
Electricity, by Sir D. Brewster, vol. viii.
Voltaic Electricity, by Sir D. Brewster, vol. xxi.
Magnetism, by Sir D. Brewster.
8th edition, edited by Traill. 4to. *Edinburgh and London,* 1853-61
Articles. Electricity, by Sir D. Brewster, vol. viii.
Aurora Borealis, by R. Jameson, vol. iv.
Beccaria, by Thomas Young, vol. iv.
Chemistry, by William Gregory, vol. vi
Franklin, B., by A. Nicholson, vol. x.
Magnetism, by Sir D. Brewster, vol. xiv.
Meteorology, by Sir John Herschell, vol. xiv

Encyclopædia Britannica—*continued.*
Articles. Sir Isaac Newton, by Sir D. Brewster, vol. xvi.
Polar Regions, by Sir T. Richardson, vol. xviii.
Telegraph, by W. Thomson, vol. xxi.
Voltaic Electricity, by Sir D. Brewster, vol. xxi.
Terrestrial Magnetism, by Sir John Herschell.

Encyclopædia of Edinburgh. Conducted by D. Brewster. Part i. (*Phil. Mag.*
xxvi. 146.)   *Edinburgh*, 1808

Encyclopædia Londinensis.

Encyclopédie Méthodique. Article Paratonnerre.   1782

Encyclopædia Metropolitana. Edited by Smedly and Rose. 59 parts, each
21s., 1817-45 ; bound in 30 vols. 4to.   *London*, 1849

The Divisions, containing Pure and Mixed Sciences, with 125 plates, Treatises,
by Airy, Barlow, Lardner, Phillips, Senior, Herschell, Daubeny, Maurice,
and others. 6 vols. 4to.   1829-36

Pure and Mixed Sciences, containing Grammar, Logic, Rhetoric, Geometry,
Arithmetic, Algebra, Theory of Numbers, Trigonometry, &c., &c., Mechanics,
Hydrodynamics, Pneumatics, Optics, Astronomy, Magnetism, Electro-
Magnetism, Electricity, Galvanism, Heat, Light, Chemistry, Sound, &c.,
by the following authors, Barlow, Capt Kater, Herschell, Linn, Roget. 2
vols. 4to. plates. (*Spon's Catalogue*, 1865.)   1829

Experimental Philosophy. 4to. Vol. ix.   *London*, 1848

Encyclopædie der Physik. (Karsten ?)

Encyklopadie, Allgemeine, der Physik. Bearbeitet von C. W. Brix, G. Decher,
F. C. O. v. Feilitzsch, F. Grashof, F. Harms, H. Helmholtz, G. Karsten,
H. Karsten, C. Kuhn, J. Lamont, J. Pfeiffer, E. E. Schmid, F. Schulz, L.
Seidel, G. Weyer, Herausgegeben von Gustav Karsten. 8vo.
1. Band. Allgemeine Physik von G. Karsten, F. Harms und G. Weyer. Pp.
449—560 c. figg. xyl.
19. Band. Fernewirkungen des galvanischen Stroms, von F. v. Feilitzsch. Pp.
641—752 c. figg. xyl.
20. Band. Angewandte Electricitätslehre von C. Kuhn. Pp. 1009—1104 c.
figg. xyl.

Ende, Ferdinand Adolphe. *Born* 1760, *at Celle ; died about* 1817, *at Mannheim.*
Über Massen und Steine die aus dem Monde auf d. Erde gefallen sind. 4to.
(*Gilbert's Ann..xviii.*)   *Braunschweig*, 1804

Endter, J. A. Theatrum sympatheticum auctum ; exhibens varios auctores de
Pulvere sympathetico quidem Digbæum, Straussum, Papinicum et Mohyum :
de unguento vero amatorio : Goclemum, Robertum, Helmontium, &c. . . .
4to. 722 pp.   *Norimbergæ,* 1662

Engel. Bibliotheca Selectissima sive Catalogus Librorum in omni genere Scientiarum Rarissimorum. 8vo.   *Bernæ,* 1743

† **Englefield,** Sir H. C. Some particulars respecting the Thunderstorm at London
and in its vicinity on the 31st of August, 1810. 8vo. 4 pp. (*Phil. Mag.*
xxxvi. 349.)   *London,* 1810

English Cyclopædia.
Edited by Charles Knight. 8vo. 22 vols.   *London,* 1856-62
Arts and Sciences, 8 vols.
Biography, 6 vols.
Geography, 3 vols.
Natural History, 4 vols.
Synoptical Index, 1 vol.

ENN—ERD 159

† **Ennemoser, J.** Der Magnetismus, &c., nach der allseitigen Beziehung seines Wesens, seiner Erscheinungen, Anwendung, und Enträthselung, in einer geschichtlichen Entwickelung von allen Zeiten und bei allen Völkern wissenschaftlich dargestellt. 781 pp. 8vo. *Leipzig*, 1819

† Der Magnetismus im Verhaltnisse zu Natur und Religion. 8vo. 546 pp.
Stuttgart und Tubingen, 1842

Enner, Von. Über die Anwendung der Reibungs-Electricität zum Zünden von Sprengladungen. 4to. 3 plates. (*Sitzungsber der . . . d. k. Akad. d. Wissensch. zu Wien*, xxi. 85-111.) *Wien*, 1856

† **Enschede.** Dissertatio de calore qui excitatur de electricitate. 4to. 79 pp.
Lugd. Bat. 1834

Entrecasteaux, J. A. B. Voyage à la recherche de La Pérouse rédigé par de Rossch. The Atlas rédigé par Beautemps Beaupré. 4to. 2 vols. Atlas of 40 plates. *Paris*, 1808

Epoandro Napili ossia **Leonardo Papini.** Dissertazione (elettricità.)
Faenza, 1752

Epp, Franz Xavier. *Born December 25*, 1789.

Problemata electrica. 8vo. *Monachii*, 1773

† Abhandlung von dem Magnetismus der naturlichen Electricität. 8vo. 128 pp. 2 plates. *Munchen*, 1777

† **Epple, B.** Die electrische Telegraphie im Allgemeinen und ihre Anwendung hauptsächlich im deutsch-österreichischen Telegraphen-Verein, in Belgien, Dänemark, Frankreich, dem italienischen Staaten der Schweiz, &c. 12mo. 134 pp. 7 plates. *Kempten*, 1855

† **Erckmann, Jules.** Etablissement des lignes électriques sous-marines sans câbles sous-marins. 8vo. 14 pp. *Paris*, 1861

Considérations sur l'origine de l'Electricité. Mémoire présenté à l'Académie des sciences de Paris. 4to. 11 pp. *Paris*, 1866

† **Erckmann et Agmini.** La lithomalakie électrique, nouvelle methode pour ramollir et rendre friable les pierres, calculs et gravelle contenus dans la vessie. 8vo. 11 pp. *Paris*, 1863

† **Erdman mit Richter.** Dritter Bericht über Elektrotherapie von A. Erdman mit Zusätzen von H. E. Richter. 4to. 29 pp. (*Med. Jahrb. Sep. Abd. Dresden.*)

(*Vide* also Richter and Erdman.)

Erdmann. Die Anwendung der Electricität in der Medecin. 3rd Aufl.

† Nachträgliche Bemerkungen zur Abhandlung des Hrn. Prof. v. Kobell über Vervietfältigung von Zeichnungen durch Galvanismus Seil 151. 8vo. 2 pp. 1 plate.

† **Erdmann, B. A. mit Duchenne.** Die örtliche Anwendung der Elektricität in der Physiologie, Pathologie, und Therapie mit Zugrundlegung von Duchenne de Boulogne de l'electrisation localisée, &c. . . . Zweite vielf. umgerb. Auflage. 2nd edition. 8vo. 266 pp. Cuts. *Leipzig*, 1858

Erdmann, J. F. von. Utrum aqua per electricitatem columnæ a cel. Volta inventæ in elementa sua dissolvatur? *Viteb*, 1802

Description d'un nouvel appareil galvánique très-actif. 8vo. (*Journ. de Chim. de Van Mons*, No. xii. p. 288.) *Bruxelles*, 1802

Uber d. Wasserzersetzung durch d. Volta'schen Säule. 8vo. (*Gilbert's Ann.* xi. 1802.) *Leipzig*, 1802

Galvan-elektrische und medicin Versuche. 8vo. (*Gilbert's Ann.* xi. 1802.)
Leipzig, 1802

Galvanische Versuche im Wiener Irrenhause angestellt. (*Horn's Arch. f. medecin Erfahr*, vi. 1804. 1804

160   ERF—ERM

Erfurt, Academia Moguntina Scientiarum. Acta, an. 1776-1795. 12 vols.
4to.   *Erfurti,* 1777-1796
Nova acta, ab an. 1797. 4 vols. 8vo.   *Erfurti,* 1799-1809 ?

Erman, Georg Adolph. *Born May* 12, 1806, *at Berlin.*

† Essai sur la direction et intensité de la force magnétique à St. Petersbourg.
Lu à l'Acad. le 11 Juin, 1828. 4to. 12 pp. (*Mém. des Sav. étrang.* t. i.)
                                                                 1828

Magnet Beobb in Russland. 8vo. (*Poggend. Ann.* xvi. and xvii.)
                                                      Leipzig, 1829

† Bericht an. d. Kön. Acad. . . . über die Fortsetzung seiner magnetischen
Beobachtungen in Russischen Assien, durch den grossen und atlantischen
Ocean. 8vo. 38 pp. 1 map.   *Berlin,* 1830

Gestalt d. erdmagnet. Linien. 8vo. (*Poggend. Ann.* xxi.)   *Leipzig,* 1831

Magnet. Declinat. Inclinat. und Intensität von Berlin. 8vo. (*Poggend. Ann.*
xxi.)   *Leipzig,* 1831

Reise um die Erde, durch Nord-Asien und d. beiden Oceane, in d. J. 128-29-
30 ausgefürht. In e. hist. und e. physikal. Abtheil. dargest.
   Abthl. 2. Physikalische Beobachtungen, 2 Bde :—
   Bd. I. Ortsbestimmungen und Declinations Beobachtungen auf d. festen
   Lande. 8vo.   *Berlin,* 1835
   Bd. II. Inclinationen und Intensitäten Declinations Beobachtungen auf
   d. See. Periodische Declinations veranderungen. 8vo.   *Berlin,* 1841

Declination zu Irkutzk und Einfluss e. Erdhebens darauf. 8vo. (*Poggend.
Ann.* xxxix.)   *Leipzig,* 1836

De inclinat. virium magnet. mensura. 4to.   *Berol.* 1839

Über d. Aufstellung eines Inclinat. auf einem Schiffe. (*Astr. Nachr.* xvi.)
                                                                 1839
Über Sternschnuppenbeobachtungen. (*Astr. Nachr.* xvii.)   1840

Archiv für wissenschaftliche Kunde von Russland. Herg. v. A. Erman. 8vo.
                                                      Berlin, 1841

Vergleich seiner Magnet Beob. mit d. Theorie d. Erdmagnetismus von Gauss.
(*Astr. Nachr.* xix.)   1842

On the continuance of magnetical and meteorological Observations. 8vo.
(*Report Brit. Assoc. for* 1845.)   *London,* 1845

Magnet. Declinat. Inclinat, u. Intensität, v. Berlin. 8vo. (*Poggend. Ann.*
lxviii.)   *Leipzig,* 1846

Magnet. Beobachtungen und geogr. Ortsbestimmungen daselbst (*i.e.*) following
Bemerkk üb. d. Herschel'sche Actinometer usw. (*Astr. Nachr.* xxxv.)
                                                                 1853

Erman, Paul. *Born February* 29, 1764, *at Berlin ; died October* 11, 1851, *at Berlin.*
Uber d. elektroskop. Phœnomene d. Volta'schen Säule. 8vo. (*Gilbert's Ann.*
viii.)   *Leipzig,* 1801
Sur les phénomènes électrometr. de la colonne de Volta. 8vo.   *Paris,* 1801

Über d. elektroskop. Phœnomene d. Volta'schen Saule. 8vo. (*Gilbert's Ann.* x.)
                                                      Leipzig, 1802

Versuche e physischen Theorie d. Volta'schen Saule. 8vo. (*Gilbert's, Ann.* xi.)
                                                     !*Leipzig,* 1802

Skeptische Beiträge z. atmosphäri Elektrometrie. (*Gilbert's Ann.* xv.)
                                                      Leipzig, 1803

Über d. Entlad d. Volta'schen Säule durch Vermittlung e. beträchtl. Strecke
eines Stroms. 8vo. (*Gilbert's Ann.* xiv. 385.)   *Leipzig,* 1803

Über d. fünffache Verscheidenheit der Körper in Rucksicht auf galvanisches
Leitungsvermögen. Gokrönnt vom National Institut. . . . .) 8vo.
(*Gilbert's Ann.* xxii.)   *Leipzig,* 1806

Erman, Paul.—*continued.*

Über elektrish. geographische Polarität, permanente elektrische Ladung, und magnetisch-chemische Wirkungen. 8vo. (*Gilbert's Ann.* xxvi.)
Leipzig, 1807

† Extract of a memoir upon two new classes of Galvanic Conductors. From Ann. de Chim. lxi. 713. 8vo. 8 pp. (*Phil. Mag.* xxviii. 297.)   *London*, 1807

Über muscular-Contraction. 8vo. (*Gilbert's Ann.* xl.)   *Leipzig*, 1812

† Versuch einer Zurückführ d. mannigfalt. Erscheinungen d. elektr. Reizung auf einen einfachen chemisch-physischen Grundsatz. (Vorgelesen 20 Mai, 1813.) 4to. 16 pp. (*Abhandl. Berl. Acad.* 1812-13.)   *Berlin*, 1812

† Über d. wechselseit. Einfluss von Elektricität u. Wärmethätigkeit. 4to. 11 pp (*Abhandl. Berl. Akad.* 1814-15.)   *Berlin*, 1814

† Bemerk. üb. d. Verhältniss d. unmagnet. Eisens zur tellurische. Polaritat. Vorgelesen, 1 Dec. 1814. 4to. 26 pp. (*Abhandl. Berl. Akad.* 1814-15.)
Berlin, 1815

† Vorläuf. Bemerkk. über d. durch blosse geomtr. Ungleichheit der Berührungs-flächen erregte elektr. Spannung. Vorgelesen den 3 August, 1817. 4to. 12 pp. (*Abhandl. Berl. Akad.* 1816-17.)   *Berlin*, 1817

† Über e. eigenthümlich reciproke Wirkung d. zwei entgegengestetzen elektr. Thätigkeiten. Vorgelesen 11 Feb. 1819. 4to. 26 pp. (*Abhandl. Berlin Akad.* 1818-19.)   *Berlin*, 1819

† Üb. e. eigenthümlich reziproke Wirkung der 2 entgegengesetzt. electr. Thatigkeiten. Vorgelesen am 11 Feb. 1819. 4to. 26 pp. 1 plate. (*Abhandl. Berlin Acad.* 1818-19, p. 351.)   *Berlin*, 1818-19

† Umrisse zu den phys. Verhältnissen des von Ærsted entdekten elektro-chemischen Magnetismus skizzirt. 8vo. 112 pp. 1 plate.   *Berlin*, 1821

Merkwürd. magnet. Beobachtung. 8vo. (*Poggend. Ann.* ix.)   *Leipzig*, 1827

† Über die magnetischen Verhaltnisse der Gegend. von Berlin. (" Zusammenge-tragen aus mehreren den Jahreen, 1826-28 angehörigen Berichten.") 4to. 59 pp. 2 plates.   *Berlin*, 1828

† Beiträge zur Monographie des Marekanit Turmalin und brasilianischen Topas in Bezug auf Elektricität. 4to. 22 pp. (*Abhandl. Berlin Akad.* 1829.)
Berlin, 1829

Der 3 August u. d. Granit-Schale-zur öffentl. Sitzung d. Akad. 4to.
Berlin, 1831

† Erzeug. von Elektromagnetismus durch blosse Modification der Vertheilung d. Polarität in einem unbewegt Magnet. Gelesen 25 Oct. 1832. 4to. 16 pp. (*Abhandl. Berlin Akad.* 1832.)   *Berlin*, 1832

On the influence of Friction upon Thermo-Electricity. 8vo. (*Report Brit. Assoc.* 1845.)   *London*, 1845

† Gedächtnissrede auf P. Erman (von) Du Bois-Reymond. 4to. 27 pp. (*Gehalten in d. K. Akad. zu Berlin,* 7 *Juli,* 1853.)   *Berlin*, 1853

Erman, P. and Marechaux. Über d. trockne Volta'sche Säule. 8vo. (*Gilbert's Ann.* xxii.)   *Leipzig*, 1806

† **Ermerins, Jacob.** De lege repulsionis electricæ. (Thesis.) 4to. 54 pp. 2 tables.
Lugd. Bat. 1827

† **Ersch, J. S.** Handbuch der deutschen Literatur, seit der Mitte des 18 Jahrhund. bis auf die neueste Zeit. . . . des 2 Bands 1 des ganz Werke fünfte Abtheilung 1st edition. 8vo.   *Amsterdam*, 1813

Handbuch . . . neue Ausg. . . . 4 Th. in 8 Bde. 8vo.   *Leipzig*, 1822-40

† Litteratur der Mathematik, Natur-und-Gewerbskunde, &c. . . Neue fortge-setzte Ausgabe von F. W. Schweigger-Seidel. 8vo. 1739 pp.
Leipzig, 1828

Eschenbach, Christian Gotthold.   *Born November* 24, 1753, *at Leipzig ; died November* 10, 1831, *at Leipzig.*

H.   Ant.   Brugman's . . .   Beobachtungen über die Verwandschaften des Magnets. a. d. latin. über. und mit einigen Anmerkungen vermehrt.   8vo.
167 pp.   *Leipsig,* 1781

Eschenmayer, Carl Adolph von.   *Born July* 4, 1770, *at Neurenberg, Wurtemberg ; died November* 17, 1852, *at Kirchheim unter Teck.*

Versuch die Gesetze magnetischer Erscheinungen aus Sätzen der Naturmetaphysik mithin a priori zu erklären entwickeln.   8vo.   *Tubingen,* 1798

† Versuch d. scheinb. Magie des Thier Magnetismus aus phisiologischen und psychischen Gesetzen zu erklären.   8vo.   180 pp.
                                                                Stuttgart and Tubingen, 1816

Eschke, E. A.   Galvanische versuche.   *Berlin,* 1803

Erfahrungen uber den Galvanismus.   8vo.   *Berlin,* 1803

L'Esprit des Journaux.   Contains articles on Thunder, &c., as *e.g.* pour October, 1721, pp. 132 and 333, &c.

Espy, James Pollard.   *Born May* 9, 1786, *at Washington, U.S. ; died January* 24, 1860, *at Cincinnati, U.S.*

On Waterspouts.   (*Franklin Institut.* xvii. 1836.)

Deductions from observations made and facts collected on the path of the Brunswick Spout of June 19, 1835.   4to.   (*Transactions of the American Philosophical Society,* New series, vol. v.)   *Philadelphia,* 1837

The Philosophy of Storms.   8vo.   552 pp.   *Boston,* 1841

Esser, Ferd.   Abhandl. üb. Blitzableiter.   8vo.   *Munster,* 1785

† **Etenaud,** A.   Guide des directeurs de station et des stationnaires chargés de bureaux de l'Administration des Lignes Télégraphiques.   8vo.   252 pp.
                                                                          Le Puy, 1860

† **Eton,** R.   Description of a double exciting cylinder Electrical machine.   Letter to the Editor.   8vo.   2 pp.   1 plate.   (*Annals of Electricity,* vii.   81.)   (*Roget's copy.*)   *London,* 1841

Ettingshausen, A. von.   Über d. Einricht. u. d. Gebrauch seiner magneto-elektr. Maschine.   (*Bericht über d. Naturforscher-Versammlung zu Prag,* 1837.)   *Prag,* 1837

Über einen Satz von Gren das elektr. Potential betreffend.   (*Sitzungsber. d. Wiener Akad.* vol. i. 1848.)   *Wien,* 1848

Anfangsgründe der Physik.   8vo.   with 150 woodcuts.   *Wien,* 1860

Ettingshausen mit Baumgartner.   Zeitschrift fur Physik und Mathematik. x. Bde.   8vo.   *Wien,* 1826-32

Euler, Johann Albrecht.   *Born November* 27 (A. St.), 1734, *at St. Petersburg ; died September* 6 (A. St.), 1800, *at St. Petersburg.*

Théorie de l'inclinaison de l'aiguille magnétique confirmée par des expériences. 4to.   (*Mémoires de Berlin,* 1755, p. 117.)   *Berlin,* 1755

Des cerfs volants.   4to.   (*Mémoires de Berlin,* 1756.)   *Berlin,* 1756

† Disquisitio de causa physica electricitatis ab Acad. Scient. Imp. Petrop· præmio coronata . . . die Sept. 6, 1755 : una cum aliis duabus dissertationibus de eodem argumento.   4to.   28 pp.   *Petropoli,* 1757

Recherches sur la cause physique de l'électricité.   4to.   *Mémoires de Berlin,* 1757, p. 125.)   *Berlin,* 1757

Nachr. v. einer magnet. Sonnenuhr.   4to.   (*Schr. Kurlayrisch Acad.* v. 1768.)
                                                                          Munchen, 1768

† **Euler, Frisi et Beraud.** Dissertationes selectæ. Jo. Albi. Euleri, Pauli Frisii, et Laurentii Beraud, quæ ad Imperialem Scientiarum Petropolitanam Academiam an. 1755 missæ sunt, cum Electricitatis causa, et theoria, premio proposito, quæreretur. 8vo. 204 pp. 1 plate.   *Petropoli et Lucæ*, 1757

Euler, Leonhard. Born *April* 15, 1707, *at Basle; died September* 7, 1783, *at St. Petersburg.*

Théorie nouvelle de l'aimant. 4to.   (*Pièces de Prix de l'Académie de Paris.*)
    Paris, 1744

Recherches physiques sur la cause de la queue des comètes, de la lumière boréale et de la lumière zodiacale. 4to. (*Mémoires de Berlin*, an. 1746, p. 117.)   *Berlin*, 1746

Opuscula varii argumenti. Nova Theoria Magnetis. 3 vols. 4to.
    Berlin, 1746 and 1751

Opuscula varii argumenti. Tom. iii. 4to.   *Berlin*, 1746-50-51

Dissertatio de magnete. 4to. (*Pièces de Prix de l'Académie des Sciences de Paris*, v. *Mém.* ii.)   *Paris*

De observatione inclinationis magneticæ dissertatio. (*Pièces de Prix de l'Académie de Paris*, v. *Mém.* ix. p. 63.)
    Poggendorff (i. 702) *says*, " *Sur l'inclinaison de l'aimant*, 1743, *Gekront von d. Paris Académie über in deren Pièces de Prix gedruckt.*"

Recherches sur la déclinaison de l'aiguille aimantée. 4to. (*Mémoires de Berlin*, an. 1757, p. 175.)   *Berlin*, 1757

Opera posthuma mathematica et physica anno 1849 detecta quæ ed. P. H. et Nic. Fuss. 2 vols. 4to.   *Petropoli*, 1862

Corrections nécessaires pour la théorie de la déclinaison magnétique proposée dans le vol. xiii. (1757) des Mémoires de l'Académie. 4to. (*Mémoires de Berlin*, an. 1766, p. 213.)   *Berlin*, 1766

Lettres à une Princesse d'Allemagne sur quelques sujets de physique et de philosophie. 3 vols. 8vo.   *St. Petersburg*, 1768-72

Breife. 8vo.   *Leipzig*, 1769

Breife. 8vo.   *Leipzig*, 1774

Conjectura circa naturam aeris pro explicandis phænomenis in atmosphæra observatis. 4to. (*Nov. Act. Petrop.* iii. part i. 1779.)   *Petropoli*, 1779

Letters translated by Hunter, ii. 34. 8vo. 2 vols.   *London*, 1795

† Portrait. Lithographie d'après Mdlle. Morice par E. Sandrier.

Euler, D., and J. Bernouilli and Dutour. Pièces de Prix de l'Académie de Paris. Boussoles. 4to.   *Paris*, 1748

† **Eustis, H. L.** The tornado of August 22, 1851, in Waltham, West Cambridge, and Medford, Middlesex County, Massachusetts. 4to. 8 pp. with map. (*Mems. American Academy*, New series, v. pt. i. p. 169.)   *Boston*, 1853

Evans. Catalogue of engraved British Portraits. 8vo.   *London*

Evans, T. S. On the laws of Terrestrial Magnetism in different latitudes. 8vo. 5 pp. (*Phil. Magazine*, xlix. 3, p. 95.)   *London*, 1817

† **Evans, F. J.** Chart of the lines of equal Magnetic declination in the South Polar regions and in the Pacific Ocean. Epoch 1859. No. ii. Fol. 1 sheet map, coloured.
    Taken from the Admiralty charts and completed . . . according to the theory of Gauss.   *London*

† Chart of the lines of equal Magnetic declination in the North Polar regions and in the Atlantic and Indian Oceans. Epoch 1859. No. i. Fol. 1 sheet map, coloured.
    Taken from the Admiralty charts and completed by Mr. F. J. Evans according to the theory of Gauss.   *London*, 1859

Evans, F. J.—*continued.*

† Chart of the Curves of equal Magnetic variation, 1858.   Sheet 3 ft. 4½ in. by
2 ft. 2 in.   *London*, 1859
*Reduced to that epoch from numerous observations, made by officers of H.M.
Navy, between 1850 and 1858 ; as also from various magnetic surveys, undertaken of late years by the British and foreign Governments. (The observations at sea being corrected for the effects of the ship's iron.)*

† An elementary Manual for the deviations of the Compass in iron ships.   8vo.
143 pp. 5 plates.   *London*, 1870

† **Evans and Smith.** Admiralty Manual for ascertaining and applying the deviations of the Compass caused by the iron in a ship. Published by order of
the Admiralty. 2nd edition. 8vo. 108 pp. 6 separate engravings.
                                                                London, 1862

† Admiralty Manual for ascertaining and applying the deviations of the Compass
caused by the iron in a ship. Published by order of the Admiralty. 8vo.
166 pp.   *London*, 1863

Über die Deviationen des Compasses, welche durch das Eisen eines Schiffes
verursacht werden.   Nach dem Englischen deutsch bearbeitet von F.
Schaub. 8vo. 152 pp. 6 plates.   *Wien*, 1864

Exhibition, London, 1851 and 1862. Official, descriptive, and illustrated Catalogue
of the Great Exhibition of 1851, together with the reports of the Jurors.   5
vols. Royal 8vo. Illustrated with hundreds of fine woodcuts. *London*, 1851

Official, descriptive, and illustrated Catalogue. 3 vols. Royal 8vo. Illustrated
with hundreds of fine woodcuts.   *London*, 1851

Official Catalogue, illustrated. 4 vols. 8vo.   *London*, 1852

Ditto, with reports. 6 vols. 4to.   *London*, 1852

Reports of the Juries on the subjects of the thirty classes into which the
Exhibition was divided. 8vo.   *London*, 1852

Official illustrated British Catalogue. 2 vols. 8vo.   *London*, 1862

Official illustrated Foreign Catalogue. 2 vols. 8vo.   *London*, 1862

Official illustrated Catalogue of the Exhibition of 1862, comprising both the
English and Foreign divisions. 4 vols. Royal 8vo.   *London*, 1862

The illustrated Catalogue of the British Industrial department of the International Exhibition of 1862. 2 vols. Royal 8vo.   *London*, 1862

Illustrated Catalogue. 3 vols. Royal 8vo.   *London*, 1862

A Cyclopædia of Machinery represented at the International Exhibition of
1862. Edited by D. H. Clarke. Royal 8vo.   *London*, 1864

Reports of the Juries of the International Exhibition of 1862. Royal 8vo.

† **Exposition universelle de 1867.** Rapports publiés sous la direction de M.
Mich. Chevallier. 8vo. tom. ii. and viii.   *Paris*, 1868

Exleben. Au Prof. Lichtenberg, die seltsame Wirkung eines Wetterstrahls auf
ihn betreffend. 8vo. (*In Cotting. Mag. J. i. 1780, St. i. pp. 104-8.*)
                                                              Göttingen, 1780

† Anfangsgründe der Naturlehre mit Verbesser, &c. v. Lichtenberg.   6th
edition. 12mo.   *Göttingen*, 1794

Exley, T. Principles of Natural Philosophy, or a new theory founded on gravitation and applied in explaining the properties of matter and the phænomena
of chemistry, electricity, galvanism, magnetism, and electro-magnetism.
8vo. 478 pp.   *London and Bristol*, 1829

Eyck, S. S. (Van ?).   Over de magnetische proeven van Oersted. (*Allgem.
Konst-en Letterbode,* 1821, and in *Bibl. Universelle,* 1821.)   1821

Mémoire sur l'électro-magnétisme.   1822

Over het electromagnetismus.   *Harlem*, 1823

Eyck. (*Vide* Tan, or Ten Eyck.)

Eydam, Immanuel. *Born February 20, 1802, at Jena; died February 9, 1847, at Berka.*

† Die Erscheinungen der Electricität und des Magnetismus in ihrer Verbindung mit einander. 8vo. 3 plates. *Weimar,* 1843

Eynard. Mémoire sur l'électrophore. *Lyon,* 1804

Mémoire sur les théories de l'électricité. *Lyon,* 1805

Eytelwein, J. A. (*Vide* Gilly and Eytelwein.)

F.

F, E. (*Vide* Anon. Meteorolog. Phen.)   1821

Fabbroni, A. Vitæ Italorum doctrina excellentium. 20 vols. 8vo.
                                              Pisis et Lucæ, **1778 to 1805**

Fabbroni, Giovanni Valentino M. *Born Feb.* 13, 1752, *at Florence; died Dec.* 17, 1822, *at Florence.*

† Articolo di lettera . . . sopra i prodigi di Pennet. . . . . 8vo. 3 pp. (*Ann. di Chim. di Brugnatelli*, ii. 316.)   *Pavia*, **1791**
Dell' azione chimica dei Metalli nuovamente avvertita. (*Atti della Reg. Soc. Economica di Firenze, ossia dei Georgofili*, vol. iv. p. 349, 1801.)
                                                        Firenze, **1801**
Memoria intorno alla causa della termossidazione dei diversi metalli, mediante il loro contatto; letta alla Soc. dei Georgofili di Firenze.   **1793**
Sur l'action chimique de différents métaux entr'eux à la température de l'atmosphère, et sur l'explication de quelques phénomènes galvaniques. (*From Ital. Soc. Mem.* xx. p. 26; *Lombardi's Elogio di Fabbroni.*) *Paris*, **1799**

† Riflessioni sulle pietre cadute del Cielo. . . Lettera à Brugnatelli. 8vo. 4 pp.
(*Ann. di Chim. di Brugnatelli*, xxi. 277.)   *Pavia*, **1802**

† Lettera al Conte Niccolò Da Rio di Padova (del 18 Maggio, 1805). 8vo. 1 p.
(*Giorn. dell' Ital. Lettera del Dal Rio*, ix. 97.)   *Padova*, **1805**

Sulla tromba galvanica. 4to. (*Giorn. di fisica*, iii. 1810.)   *Pavia*, **1810**

† Elogio del Cav. Giovanni Fabbroni scritto da A. Lombardi. 4to. 30 pp. (*Mem. Soc. Ital.* xx. p. 1.)   *Modena*, **1828**

Fabre et Kunneman. Machines élect. en caoutchouc. (*From* Du Moncel.)

Fabré-Palaprat, et La Beaume. Du Galvanisme appliqué à la médecine, &c.
. . . Avec des notes sur quelques remèdes auxiliaires, par La Beaume. . . .
† trad. de l'anglais et précédé de remarques, &c. . . . par R. B. Fabré Palaprat. 8vo. 438 pp.   *Paris*, **1828**

Fabricius, F. Dansk Bogfortegnelse for Aarene 1841-1858. 4to. 252 pp.
                                              Kiobenhavn, **1861**

Fabri, Honoré (Honoratus). *Born* 1606(?), *at Le Bugey, Diöcese Belley; died Mar.* 9, 1688, *at Rome.*

Physica seu scientia rerum corporearum. . . . 4to.   *Lugd.* **1669-1671**

Fabri, Ruggero. Osservazioni microscopiche della scintilla elettrica. 4to. (*Atti d. Pont. Accad. dé Nuovi Lincei*, p. 16, an. 1858.)   *Roma*, **1858**

Sulla induzione elettrostatica. (*Atti d. Pont. Accad. dé nuovi Lincei*, p. 405, an. 1858.)   *Roma*, **1858**

Fabris, Nicc. *Born* 1739; *died* 1801.

"Erfand eine Uhr ohne Rader und Gewichte, durch einen Magnet bewegt."
(*Poggendorff*, i. 714.)

Fabrizio d'Acquapendente. De visione, voce, et auditu. (*Poggendorff*, i. 714.)
                                                        Venet. **1600**

Fabroni, A. d'Arezzo. On shocks received from a (live) Cat. (*In the Anthologia*, 1830, p. 173.)   **1830**

Fabroni. Sur l'irritation métallique.
        Sue (*Hist. p.* 229) *gives an* "*Extrait de l'ouvrage de Fabroni sur l'irritation métallique,*" *but no date or place.*

Fahlberg, Samuel. *Born about 1755, in Norway ; died after 1834, at Barthélemy ?*
Anmarkningar rorande orcanen 1792 pa. On St. Barthelemi, saint baro-
meterns och electricitetens förhallande vid detta tillfalle. (*Vetensk Acad.
Nya Handl.* an. 1794, s. 275.) *Stockholm,* 1794
Beskrifning ofver elektriska alen Gymnotus electricus. (*Vet. Acad. Nyr.
Handl.* 1801.) *Stockholm,* 1801

Fait. (*Vide* McFait.)

Falconer, W. Observations on the knowledge of the Ancients respecting Electricity. . . . (*Mem. of the Society of Manchester,* iii. 278.) *Manchester*

† Osservazioni sulle Notizie degli Antichi intorno alla elettricità del . . . Falconer comunicata dal Dr. Percival. (*Translation.*) 4to. 9 pp. (*Opus. Scelt.*
xiv. 274.) *Milan,* 1791

Falconieri. (*Vide* Landi and Falconieri.)

Falconet, C. Dissertation historique et critique sur ce que les anciens out cru de
l'Aimant. 4to. (*Mém. de littérature tirés des Registres de l'Acad. Rle. des Inscrip.* . . . *depuis* 1711 *and* 1747, tom. iv. *Paris,* 1746.) *Paris,* 1746

Sur les bétyles. 4to. (*Acad. des Inscrip.* xii. *Mém.* p. 515, *et seq.*) *Paris*

Fantasti, G. C. Opinione sopra la cagione della morte della Sig. Contessa
Cornelia Zangari ne' Bandi (Cesenate). Esposta in una lettera al Sig.
Marchese Don Garzia di Toledo. 8vo. *Verona,* 1731

Faraday, Michael. *Born Sept.* 22, 1791, *at Newington, London ; died Aug.* 25, 1867,
at Hampton Court, Middlesex.

† Description of an Electro-Magnetical Apparatus for the exhibition of Rotatory
Motion. 8vo. 9 pp. 1 plate. (*Arago's copy.*) *London,* 1822

† Chemical Manipulation, being Instructions to Students in Chemistry. 8vo.
1830

† On Light and Phosphorescence. Experiments of Pearsall. Restoration of
phosphorescent power by Electrical discharges to bodies which had been
deprived of it by Calcination. 8vo. 1 p. (*Phil. Mag.* ix. 318.)
London, 1831

On the Magneto Electric Spark and Shock, and on a peculiar condition of
Electric and Magneto-Electric Induction. 8vo. (*Phil. Mag.* 1834, v. 349.)
London, 1834

On the influence of Induction. . . . *Vide* Jenkin. (*Phil. Mag.* v. 349, 444.)
London, 1834

† Experimental Researches in Electricity. Reprinted from the Phil. Trans.
8vo. 574 pp. 8 plates. *London,* 1839

Latimer Clark's Cat. says : " *There is a second edition of this work dated* 1849,
but it appears to be only a reprint, since the preface is still dated 1839."

Chemical Manipulation. 8vo. *London,* 1843

† Experimental Researches in Electricity. Reprinted from the Phil. Trans.
Vol. ii. 8vo. 302 pp. 5 plates. *London,* 1844

† On Electric Induction. Associated cases of Current and Static Effects. 8vo.
11 pp. *London,* 1854

On Subterraneous Electro-Telegraphic Wires. 8vo. (*Phil. Mag.* ser. iv. vol.
vii.) *London,* 1854

Une solennelle et rassurante adhésion, 1 Mars, 1854. 8vo. 2 pp. (*Faraday's
copy.*) *Paris,* 1854

Experimental Researches in Electricity. Reprinted from the Phil. Trans.
Vol. iii. 8vo. 588 pp. 4 plates. *London,* 1855

Experimental Researches in Chemistry and Physics. (*From Quaritch's Cat.,*
No. 202, 1870.) 1859

Faraday, Michael.—*continued.*

Lectures on the various Forces of Matter, and their relations to each other.
12mo. 59 woodcuts.   1860

Zuchold (1860, *p.* 55) *says : " Edited by William Crookes, London."*

Six Lectures on various Forces of Matter. 3rd edition. 8vo. *London,* 1862

† Faraday as a Discoverer. By John Tyndall. 8vo. 171 pp. 2 portraits.
London, 1868

Eloge historique de Michel Faraday. Lu . . . 10 Mai, 1868, Institut
Impérial de France (par Dumas). 4to. 58 pp.   *Paris,* 1868

6 diverse Abhandlungen über Electricität, Magnetismus, &c. aus verschiedenen
eng. Zeitschriften. (*From Weber, Cat. Berlin,* 1870.)

† **Faraday and Schonbein.** On a peculiar Voltaic condition of Iron, &c. . . .
(Several articles.) 8vo. 14 pp. and 2 pp. (*London and Edinburgh Magazine
for July and August,* 1836.)   *London,* 1836

Faraday, M. and Sturgeon, W. Supplementary note to Faraday's eleventh series
of Experimental Researches. 8vo. 5 pp. (*Annals of Electricity,* iv. 229
and 231.)   *London,* 1839

Faraday and Riess. On the action of Non-conducting Bodies in Electric
Induction. 8vo.   1856

† **Fardely, W.** Die Galvanoplastik. 8vo. 47 pp. 1 plate.   *Mannheim,* 1842

Sur les télégraphes electriques. 4to. (*Comptes Rendus,* xviii. 792.) *Paris,* 1844

Der electr. Telegraph. 8vo.   *Mannheim,* 1844

Der Zeiger-Telegraph für den Eisenbahndienst dargestellt, &c. . . . nebst
practischen Angaben über die Behandlung der Apparate, &c. 8vo. 47 pp.
8 plates.   *Mannheim,* 1856

† **Farey, J.** A series of queries addressed to Dr. Burney, of Gosport, regarding
Shooting Stars and Meteors, with some suggestions on the same subject to
the Astronomical Society of London for making these phenomena available
in settling the longitudes of places, and towards extending our knowledge of
the very numerous planetary and satellitic bodies composing the Solar system.
8vo. 6 pp. (*Phil. Mag.* lvii. 346.)   *London,* 1821

† **Farey, J.** sen. On Shooting Stars and Meteors. 8vo. 4 pp. (*Phil. Mag.* lviii. 183.)
London, 1821

Faria, De e Arago. Breve compendio o tratado sobre a Electricidade. 4to.
2 plates.   *Lisboa,* 1800

† **Fario e Zantedeschi.** Esperienze intorno alle correnti elettro-fisiologiche negli
animali a sangue caldo. Memoria i. 8vo. 40 pp. (*Articolo estratto dal vol.*
iii. *del Memoriale della Medicina Contemporanea.*) *Venezia,* 1840

† **Farquharson, J.** On a definite arrangement, and order of the appearance and
progress, of the Aurora Borealis, and on its height above the earth. (Read
February 29, 1829.) 4to. 18 pp. (*Phil. Trans. for* 1829, p. 103.)
London, 1829

† Report of a geometrical measurement of the height of the Aurora Borealis
above the earth. 4to. 14 pp. (*Phil. Trans.* 1839, p. 267.) *London,* 1839

† **Farrar, J.** An account of the violent and destructive Storm of September 23,
1815. 4to. 6 pp. (*Mem. Amer. Acad.* Old series, iv. part i. p. 92.)
Cambridge, U.S. 1818

† An account of a singular Electrical Phenomenon observed during a snow-
storm accompanied with thunder. 4to. 5 pp. (*Mem. Amer. Acad.* Old series,
iv. part i. p. 98.)   *Cambridge, U.S.* 1818

Elements of Electricity and Magnetism, &c. By Lovering.
Cambridge, U.S. 1842

Farrar, J.—*continued.*

Elements of Electricity, Magnetism, and Electro-Magnetism. (Second part of a course of Natural Philosophy, &c.) *Cambridge, U.S.* 1826

† Elements of Electricity, Magnetism, and Electro-Dynamics. . . . 2nd part of a course of Natural Philosophy, at Cambridge, New England. 8vo. 376 pp. 6 plates. *Boston*, 1839

Fatio (Facio, Faccio) de Duillier, N. Lettre à Cassini sur une lumière extraordinaire qui parait dans le ciel depuis quelques années. *Amsterdam*, 1686

† **Fau, J.** Manipulations électrotypiques . . . Traduit de l'Anglais sur la 10me éd. de Walker et augmenté de Notes, &c. 12mo. 116 pp. *Paris*, 1843

Manipulations électrotypiques trad. de Walker. 6me édition Française, 18me édition Anglaise. 6me édition. 16mo. 156 pp. *Paris*, 1861

Faulwetter, Carl Alexander. *Born June 29, 1745, at Nürnberg; died May 15, 1801, at Nürnberg.*

† Kurze Grundsätze der Elektricitäts-Lehre. 3 vols. 8vo. 27 plates. *Nürnberg*, 1793

† Kurze Grundsätze der Elektricitäts-Lehre. 1794. The 5th part (or 4th vol.) 8vo. 440 pp. *Nürnberg*, 1794

† **Faure, G.** Conghietture fisiche intorno alla cagione dei fenom. nella macchina elettrica. 4to. 140 pp. *Roma*, 1747

Favre, Pierre Antoine. *Born February 20, 1813, at Lyon.*

† Thèse de physique. Recherches thermiques sur les courants hydro-électriques. 4to. 28 pp. *Paris*, 1853

Favre. (*Vide* Mallet-Favre.)

Fay. (*Vide* Du Fay.)

† **Faye, H. A. E. A.** Sur un moyen de soustraire les pendules astronomiques à l'influence des variations de la température et de la pression atmosphérique. 4to. (*Comptes Rendus*, 1847, xxv. p. 375.) *Paris*, 1847

Fayet. (*Vide* Nougarede de Fayet.)

Fea, C. Relazione dell' Aurora-boreale veduta in Roma, &c. . . . li 3, 4, 5, 6, e seguenti d'Agosto, con Osservazioni critiche. 8vo. 17 pp. *Roma*, 1831

Appendice alla Relazione sudetta, in risposta a Tre oppositori. 8vo. 13 pp. *Roma*, 1831

Fearnley, C. (*Vide* Hansteen and Fearnley.)

† **Feburier.** Mémoire sur quelques propriétés du fluide électrique considérées dans leur rapport avec la Végétation. 8vo. 94 pp. (*Printed by order of the Soc. d'Agriculture.*) *Paris*

Fech. (*Vide* Lozeran-du-Fech.)

Fechner. Lois expérimentales de l'intensité des courants dans un circuit fermé. 4to. *Leipzig*, 1831

† Maassbestimmungen über die galvanische Kette. 4to. 260 pp. 1 plate. *Leipzig*, 1831

† De nova methodo magnetismum explorandi, &c. 4to. 25 pp. *Leipzig*, 1832

De magnetismo variabili qui chalybi actione galvanica inducitur commentatio, &c. 4to. 24 pp. *Lipsiæ*, 1835

Fechner, Gustav Theodor. *Born April 19, 1801, at Gross-Särchen, near Muskau Lausitz.*

† Handbuch der dynamischen Elekt. (*Trans. of Demonferand's Manual.*) *Leipzig*, 1824

Lehrbuch der Experimental-physik oder Erfahrungs-naturlehre; zweite Auflage der deutschen Bearbeitung, mit Hinzufügung der neuen und einheimischen Entdeckungen. 2nd edition. 5 vols. (*Pog.* i. 727.) 1828-30

s

Fechner, Gustav Theodor—*continued.*

Lehrbuch des Galvan. u. Electro chemie. 8vo.   *Leipzig,* 1829

† Elementarlehrbuch des Elektromagnetismus nebst Beschreibung der haupsachlichsten electro-magn. Apparate. 8vo. 157 pp. 4 plates.   *Leipzig,* 1830

Repertorium d. Experimental-physik. 8vo. 3 vols.   *Leipzig,* 1832

Many papers in Schweigger's Journal, vols. xlix. 1829 to lxix. 1833.
Many others in Poggendorff's Annalen, vols. xli. 1837 to lxiv. 1845.

† De variis intensitatem vis Galvanicæ metiendi methodis. Disputatio . . . 4to.
     Lipsiæ, 1835

† Über die physikalische und philosophische Atomenlehre. 8vo. 210 pp.
     Leipzig, 1855

Uber die physikalische und philosophische Atomlehre. 2 vermehrte Auflage.
8vo. 360 pp.   *Leipzig,* 1864

Feddersen, Bernhard Wilhelm.   *Born March* 26, 1832, *at Schleswig (Stadt).*

† Beiträge zur Kenntniss des elektrischen Funkens. Inaugural dissertation.
8vo. 35 pp. (*Poggend. Ann.* ciii. 1858.)   *Kiel,* 1857

Uber die eigenthuml. Stromtheilung bei Entladung d. Leidener Batterie. 8vo.
(*Poggendorff, Ann.* cxv. 1862.)   1862

† **Feddersen,** W. Über elektrische Wellenbewegung vorgelegt von Hankel. 8vo.
4 pp.   *Stockholm,* 1859

† Die oscillatorische elektrische Entladung und ihre Grenze. Vorgelegt von
Hankel. 8vo. 7 pp.   1861

† Uber die elektrische Flaschenentladung. 8vo. 31 pp. 1 plate. (*Poggendorff's*
Ann. cxiii. 437.)   *Leipzig,* 1862

† Über die elektrische Flachenentladung. 8vo. 40 pp. 1 plate. (*Poggendorff's*
Ann. cxvi. 132.)   *Leipzig,* 1862

† **Feilitzsch,** F. C. O.   Many papers in Poggendorff's Annalen, 1844-45, &c.

† **Feilitzsch,** F. von. On the physical distinction of Magnetic and Diamagnetic
bodies. Communicated by Mr. Faraday. 8vo. 7 pp. (*Phil. Mag. January,*
1851.)   *London,* 1851

† Erklärung der diamagnetischen Wirkungsweise durch die Ampère'sche Theorie.
Erste Abhandl. 1852. Zweite Abhandl. 1853-58. (*Poggend. Ann.* lxxxvii.)
     Leipzig, 1852-53

† Die Lehre von den Fernwirkungen des galvanischen Stromes. Elektromagnetismus, Elektrodynamik, Induction und Diamagnetismus. 8vo. 834 pp.
3 plates.   *Leipzig,* 1865

Felbiger, Johann Ignatz von.   *Born January* 6, 1724, *at Gross-Glogau ; died May*
17, 1788, *at Pressburg.*

† Die Kunst Thuerme, oder andere Gebäude, vor den schadlichen Wirkungen des
Blitzes, durch Ableitung zu bewahren (augebracht an einem Thurm). 8vo.
110 pp. 1 plate.   *Breslau,* 1771

Vorschlaege wie Nordlichter zu beobachten. (*From Poggendorff,* i. 730.)
     Sagan, 1771

Wie Nordlichter zu beobachten. 4to. (*From Young,* p. 482.)   *Sorau,* 1772
(*Heinsius says* "*Vorschlage wie* . . . *and* 1771.'')

Über die Witterung. 4to. 1 plate.   *Sagan,* 1773
(*Poggendorff* (i. 730) *says :* " *Anleitungen jede Art von Witterung in Karten zu*
bezeichnen, usu. 1773.)

Felbiger, Johann Ignatz von—*continued.*

Wie weit gewähren wohl Gewitterableiter, Sicherheit fur umstehende Gebaude?
Pressburg, **1787**

Feldt, L. De transitu nonnullarum stellarum . . . tabula. Adjecta adhuc sunt
de electricis in atmosphæra phænomenis &c. obss. *Brunsbergæ,* **1844**

Felici. On the laws of Induction. (*From Matteucci.*)

Felkel, A. Wahre Beschaffenheit des Donners. *Wien,* **1780**

† **Fellens,** G. B. Manuale di Meteorologia. . . . 12mo. 155 pp. 1 plate.
Milano, **1832**
(*There is an edition, Paris,* 1833, *in* 18mo.)

Feller, C. G. Dissertatio de therapia per electrum. *Lipsiæ,* **1785**

Fenwick. Reflections on calcareous manure and electric fluids, &c. (*From Watt.*)
1798

Fenwick, T. A treatise on subterraneous Surveying and the variation of the
Magnetic Needle. 2nd edition. 8vo. *London,* **1823**

† **Fenwick,** T., and **Baker.** Elementary and practical treatise on subterraneous
Surveying and the Magnetic Variation of the Needle. 3rd edition. 8vo.
160 pp. (*With Baker's method of conducting.*) *London,* **1861**

Ferguson, James. *Born* 1710, *at Keith, Banffshire, Scotland; died November* 16,
1776, *at Edinburgh.*

† An introduction to Electricity. 1st edition. 8vo. 140 pp. 3 plates.
London, **1770**

An introduction to Electricity. 2nd edition. 8vo. (*Reprint.*) *London,* **1775**

An introduction to Electricity. 3rd edition. 8vo. 140 pp. 3 plates. (*Reprint.*)
London, **1778**

† Introduzione alla elettricità traduzione dall' Inglese. 8vo. 144 pp. 3 plates.
Firenze, **1778**

† An introduction or lectures on Electricity. New edition, corrected with an
appendix, &c., by C. F. Partington. 8vo. 102 pp. 1 plate. *London,* **1825**

† **Ferguson and Brewster.** Essays and treatises on astronomy, electricity, &c.,
with an appendix by Sir D. Brewster. New edition. 8vo. 382 pp. 14 plates.
Edinburgh, **1823**

† **Ferguson,** R. M. Electricity. 8vo. 273 pp. (*Chambers' Educational Course.*)
London and Edinburgh, **1866**

Ferguson, E. F. T. (*Vide* Bombay Magnetical Observatory.)

Fernaux, Coupans. Bibliothèque Asiatique et Africaine; ou catalogue des
ouvrages relatifs à l'Asie, &c. En 2 parties. 8vo. *Paris,* **1841**

Ferrara, Francesco. Sopra l'ambra siciliana. 8vo. *Palermo,* **1805**

† **Ferrari** (*i.e.* D. F.) Alcune nozioni di elettro-metallurgia. 4 pp. (*In l'Economista,* Anno iv. vol. i. pp. 31 and 142.) *Milano,* **1846**

Ferrari, Girolamo. *Born* 1794 (*about*), *at Vigevano.*

† Metodo elettro-chimico per stagnare. Articolo di lettera di G. Ferrari farmacista a Vigevano. 8vo. 2 pp. (*Bibl. Ital.* xcviii. 268.) *Milano,* **1840**

† Sulla galvanizzazione, osservazioni ed esperienze del farmacista Gerolamo
Ferrari, comunicate alla regia Accademia delle scienze di Torino. 8vo.
2 pp. (*Bibl. Ital.* tom. c. p. 275.) *Milano,* **1840**

† Teoria della grandine. Folio. 3 pp. (*L'Economista,* Anno iv. vol. ii. p. 172.)
Milano, **1846**

Ferrari. (*Vide* Resti-Ferrari.)

Ferrario, O. Corso di chimica generale. *Milano,* **1837-40**

† Processi dei Sig. Elkington per vestire metalli. 8vo. 4 pp. (*In Giorn. I. R. Ist-
Lomb.* i. 273.) *Milano,* **1841**

Ferrario, O.—*continued.*

†   Apparato semplice per dorare l'argento, il rame, l'acciajo, e per riprodurre rilievi galvano-plastici, di R. Boettger. 8vo. 10 pp. (*In Giorn. I. R. Ist. Lomb.* ii. 275.)   *Milano,* 1841

† **Ferri, P.**   Riflessioni sopra gli Argomenti addotti dal . . . Maffei . . . intorno la Formazione del Fulmine. 4to. 52 pp.   *Vicenza,* 1748

Fershey, C. G.   Observations upon the Meteors of August. 4to. (*Trans. Ann. Phil. Soc.*, New Series, vol. vii.)   *Philadelphia,* 1841

Ferussac, André Etienne (Baron). *Born December* 30, 1786, *at Chartron, Dép. Tarn and Garonne ; died January* 21, 1836, *at Paris.*

† **Ferussac and others.** Bulletin des sciences mathématiques,astronomiques,physiques et chimiques. *Begins in* 1824. 16 vols. 8vo. plates. (*Muncke in Gehler*, vii. 564, *says* 1823–31.)   *Paris,* 1824-31
(*The other writers are Ampère, Cauchy, Dupin, and Damoiseau.*)

†   Bulletin des sciences téchnologiques. 19 vols. 8vo. plates.   *Paris,* 1824-31

Feuillée, Louis. *Born* 1660, *at Mane, near Sisteron, Provence ; died April* 18, 1732, *at Marseilles.*

Observations pour la variation de l'aimant. 4to. (*Mémoires de Paris,* 1704.)   *Paris,* 1704

Observations de la variation et de l'inclination de l'aimant à Cartagena, à Portobella et à Porto-Cabello. 4to. (*Mémoires de Paris,* 1708; *vide Whiston Longit.* p. xxv.)   *Paris,* 1708

Observations de la variation et de l'inclination de l'aimant à Conception et à Coquimbo. 4to. (*Mémoires de Paris,* 1711.)   *Paris,* 1711

Journal d'Observations physiques et mathématiques. Indes occidentales. 4to. 2 vols. (*MSS. Catalogue, Bordeaux Académie.*)   *Paris,* 1714

Observation sur une pluie de sable dans la mer atlantique précédée d'une aurore boréale. 4to. (*Mémoires de Paris,* 1719.)   *Paris,* 1719

† **Fick.**   Untersuchungen über electrische Nervenreizung. 4to. 51 pp.
         Braunschweig, 1864

† **Fickel.**   Der mineralische Magnetismus als grosses Heilmittel. 8vo. 32 pp.
         Leipzig, 1836

† **Fieber, F.**   Compendium der Elektrotherapie. 8vo. 146 pp.   *Wien,* 1869

Fiedler, K. G.   Merkwürdige Blitzschlaege. 8vo. (*Gilb. Ann.* lxviii.)   1846

Fiedler, K. W.   Über Blitzrochren in Deutschland und Ungarn.

† **Field, H. M.**   History of the Atlantic Telegraph. 8vo. 364 pp.
         New York, 1866

† **Fierstemann.**   Conducting power of different fluids for Voltaic Electricity. 8vo. 1 p. (*Phil. Mag. or Annals* iv. 383.)   *London,* 1828

Figuier, Louis Guillaume. *Born February* 15, 1819, *at Montpellier.*

†   Exposition et histoire des principales découvertes scientifiques modernes. 4th edition. 3 vols. 12mo.   *Paris,* 1855

†   Exposition et histoire des principales découvertes scientifiques modernes. Tome iv. contenant la machine électrique. 12mo. 532 pp.   *Paris,* 1857

†   Les applications nouvelles de la science à l'industrie et aux Arts en 1855. 12mo. 428 pp. cuts.   *Paris,* 1857

†   L'année scientifique et industrielle. Troisième Année. 2 vols. 12mo.
         Paris, 1859

Filiasi, Jacopo. *Born* 1750 (*about*), *at Venice ; died February* 18, 1829, *at Venice.*

†   Memoria sulle procelle che annualmente sogliono regnare nelle maremme Veneziane. 8vo. 43 pp. (*Giorn. Fis. Med.* iv. 152.)   *Pavia,* 1792

† **Finlaison,** John, and **Bain.** An account of some remarkable applications of the Electric fluid to the useful arts, by Mr. Alex. Bain, with a vindication of his claim to be the first inventor of the Electro-Magnetic Printing Telegraph, and also of the Electro-Magnetic Clock. By John Finlaison. 8vo. 127 pp. 5 plates. *London,* 1843

Firmin. Natural History of Surinam.

† **Firminger,** T. Account of the meteor seen on the evening of Sunday, November 13, 1803; with some observations on the best means of ascertaining the altitude, bearing, magnitude, distance, and velocity of such phenomena. 8vo. (*Phil. Mag.* xvii. 279.) *London,* 1803

Fischer, Daniel. Relatio de fulgure tonitru et fulmine. De insolito quodam phænomeno Kesmarkini die 10 Aug. 1717. *Breslau*

Fischer, D. G. von. Notice sur la Sibérite, ou la Tourmaline rouge de Sibérie 4to. *Mosk,* 1813

Notice sur la Sibérite, ou la Tourmaline rouge de Sibérie. 4to. *Mosk,* 1818

Fischer, Ernst Gottfried. *Born July 17, 1754, at Hoheneiche, near Saalfeld ; died January 27, 1831, at Berlin.*

Über Telegraphie. (*G. N. Fischer's Teutsche Monatsschrift,* 1795.) 1795

Lehrbuch d. méchanischen Naturlehre. 8vo. *Berlin,* 1805

† Physique méchanique trad. par Biot. 8vo. 8 plates. *Paris,* 1806

Beschreibung d. Volta'schen Eudiometers. 8vo. (*Neue Schriften d. Gesellsch. naturf. Freund. in Berlin,* i. 1807.) *Berlin,* 1807

Lehrbuch d. Mechanischen Naturlehre. Zweite gnenzlich umgearbeitete und verm. Aufl. 2 vols. *Berlin,* 1819

Physique Traduite de l'allemand, avec Notes par Biot. 8vo 1819

Über den Ursprung der Meteorsteine. 8vo. (*Poggendorf,* i.752, *says in Abhandl. d. Berliner Acad. f.* 1820 *u.* 1821.) *Berlin,* 1820

Lehrbuch d. mechanischen. Naturlehre 3e Aufl. 2 vols. 8vo. 7 plates. *Berlin,* 1827

Mechan. Naturlehre im Auszuge f. d. höh. Schulunterricht entworfen v. E. F. August. 8vo. *Berlin,* 1829

Physique méchanique trad. par Biot revue et augm. 8vo. 12 plates. *Paris,* 1830

Lehrbuch d. mechan. Naturlehre neu bearb. v. E. F. August. 2 vols. 4to. verm. u. verb. Aufl. *Berlin,* 1837

† Praktische Anleitung zur Verfertigung &c. Künstlicher Magnete .. so wie die neueste Entdeckung denselben die höchste Anziehungs kraft zu erhalten . usu. 8vo. 53 pp. 2 plates. *Heilbroun,* 1833

Practical Treatise on Medical Electricity.

Fischer, Johann Andreas. *Born November 28, 1667, at Erfurt; died February 13, 1729, at Erfurt.*

De magnetismo macro-et microcosmi. *Erfurth,* 1687

Fischer, J. B. De fœno sub combustione per fulminis ignem in massam seu scoriam calcariam redacto. 4to. (*Nov. Act. Acad. Nat. Cur.* iii. 1733.) *Bresiau,* 1733

Fischer, J. C. Geschichte der Physik seit der Wiederherstellung der Kunste und Wissenschaften bis auf die neuesten Zeiten. 8 vols. 8vo. *Göttingen,* 1801-8

Fischer, J. N. Beweis das das Glockenlauten bey Gewitter schadlich sey. 8vo. *München,* 1784

174   FIS—FLE

† Fischer, K. C. F.   Über ·d. gegenseitige Einwirk. v. Elektromagneten, Stahlmagneten u. deren Anker.   Ein Beitrag zur Lehre, v. Magnetismus von
   Director Dr. Fischer.   4to. 29 pp.   *Nordhausen*, 1842

Fischer, Nicolaus Wolfgang. *Born January* 15, 1782, *at Gross Meseritz, Mähren ;
   died August* 19, 1850, *at Breslau.*

   Krit. Erscheinungen, die als. Wirk. d. galvan. Action erklärt. worden sind.
   usw. 4to. (*Abhandl. d. Berlin Akad. fur* 1814-15.)   *Berlin,* 1814-15

   Unterschied zwicken chem. u. galv. Erscheinungen. 8vo. (*Gilbert's Ann.*
   lxxii.)   *Leipzig,* 1822

   Uber d. merkw. Verhalt. d. salzsauren Zinns zum Wasser usw. 8vo. (*Schweigger's Jl.* xxxix.)   *Nürnburg,* 1823

† Das Verhaltniss der chemischen Verwandschaft zur galvanischen Electricität
   in versuchen dargestelt. 8vo. 238 pp.   *Berlin,* 1830

   Über d. Nutzanwendung d. Galvanismus in med.-gerichtl. Hinsicht. (*Hufeland's Journal der Heilkunde,* vol. lxx.)   1830

   Über Ozon. 8vo. (*Poggendorf's Ann.* vols. lvi. u. lxxvi.)
                                                          Leipzig, 1845 u. 1849

† Fisher, G.   On the Errors in Longitude as determined by chronometers at sea,
   arising from the action of the iron in the ships upon the chronometers. 8vo.
   9 pp. (*Phil. Mag.* lvii. 249.)   *London,* 1821

† Magnetical experiments made principally in the south part of Europe and in
   Asia Minor, during the years 1827 to 1832. 4to. 16 pp. (*Phil. Trans.*
   1833.)   *London,* 1833

† Fitzinger. Über den Proteus angumus der Autoren. 8vo. 13 pp. (*Sitzungsb.
   d. k. Akad. d. Wiss. Wien,* 1850, ii. Bd. iii. Heft, p. 291.)   *Wien,* 1850

† Fitzroy, Admiral.   Barometer Manual. 4th ed. Board of Trade. 8vo. 18 pp.
                                                            London, 1861

† Barometer and Weather Guide. Board of Trade. 8vo. 31 pp. *London,* 1861

   The Weather Book ; a manual of practical meteorology. 8vo. 470 pp. 16
   illustrations.   *London,* 1862

† Fizeau, Armand Hippolite Louis. *Born September* 23, 1819, *at Paris.*

   Contre-épreuve en cuivre d'une image photographique, obtenue au moyen des
   procédés galvanoplastiques, sans altération de l'image originale. 4to.
   (*Comptes Rendus,* xii. 401.)   *Paris,* 1841

   Examen des expériences, faites en 1848 et 1849, aux Etats-Unis, par MM.
   S. C. Walker et O. M. Mitchel, pour déterminer la vitesse de propagation
   de l'électricité. 4to.   *Paris,* 1851

   (*Moigno says Mém. présenté à l'Acad. des Sciences le* 13 *Janvier,* 1851. *Not in
   Poggendorf or in the Comptes Rendus.*)

   Sur les machines électriques inductives et sur un n › ι   facile d'accroître leurs
   effets. 4to. (*Comp. Rend.* xxxvi.)   *Paris,* 1853

Fizeau, A. H. L. et Foucault. Sur l'intensité de la lumière émise par le charbon
   dans l'expérience de Davy. 8vo. (*Ann. de Chim. et Phys.* Ser. 3, xi. 1844.)
                                                            Paris, 1844

Fizeau, A. H. L. et Gounelle. Sur la vitesse de la propagation de l'électricité.
   4to. (*Comp. Rend.* xxx. 1850.)   *Paris,* 1850

Flagg. H. C. Observations on the Numb Fish, or Torporific Eel. 4to. 4 pp.
   (*Trans. Amer. Phil. Soc.,* Old Series, vol. ii. p. 170.) *Philadelphia,* 1786

Flaugergues, P. P. Traité des machines électro-dynamiques.   *Paris,* 1840

Fleischhauer, J. H. Die Naturkräfte im Dienste des Menschen. Gemeinfassliche naturwissenschaftliche Vorlesungen. 3. Vorlesung. Die elektro-
   magnetische Telegraphie. Nach den besten Quellen bearbeitet. 8vo. 56 pp.
                                                            Thüringen, 1851

Fleischhauer, J. H.—*continued.*

Die elektromagnetische Telegraphie. Nach den besten Quellen bearbeitet. 8vo. (*In "Die Naturkräfte im Dienste des Menchen." Gemeinfassliche naturwis senschaftliche Vorlesungen.*) *Langensalza,* 1855

Die Meteore. 2 Abtheilung. Die Elektro- und Aëro-Meteore und die Atmosphäre als deren Herd, &c. Mit einer Zugabe "Das Barometer als Wetterglas." 8vo. vi. and 89 pp. 1 plate lith. (*In "Die Naturkräfte im Dienste des Menschen. Gemeinfassliche naturwissenschaft Vorlesungen. 2nd verbesserte Auflage.*") *Langensalza,* 1855

Fleuriau de Bellevue. *Born about* 1761 ; *died February* 9, 1852, *at La Rochelle.*

Mémoire sur les Pierres météoriques, et notamment sur celles tombées près de Jonzac, au mois de Juin 1819. 24 pp. 2 plates. *Paris,* 1821

† On Meteoric Stones. Paper read in 1820 at the Academy, particularly on those which fell near Jonzac, in the Department of Charente. 8vo. 2 pp. (*Phil. Mag.* lviii. 456.) *London,* 1821

Flinders, Matthew. *Born* 1760 *at Donnington, Lincolnshire.*

Concerning the differences in the Magnetic Needle on board the "*Investigator,*' arising from an alteration in the direction of the ship's head. 4to. (*Phil. Trans.* 1805.) *London,* 1805

A voyage to Terra Australis . . . in the years 1801-1803, &c. in the ship "*Investigator.*" 4to. 2 vols. 9 plates, and an atlas in folio containing 18 maps and 10 plates. *London,* 1814

Floderus, M. M. Om dubbel och frydubbel Telegrafering. Akademisk Afhandling. 8vo. 20 pp. 1 plate. *Upsala,* 1857

Florence. Acad. del Cimento.

Vide Saggi. 1666
    Do. 1667
    Do. 1684
    Do. 1684
    Do. 2nd edition 1691
    Do. 4th edition 1711
    Do. Tentamina 1731
    Do. 1761
    Do. In Tozzetti 1780
    Do. Gazzeri 1841
    Do. Edi. Classiche . . . . —

Florence Antologia. (*Vide* Antologia di Firenze.)

Florence Museum. Annali del Museo Imperiale di Fisica e Storia Naturale di Firenze. 2 vols. 4to. plates *Firenze,* 1808-9

Archivis meteorologico centrale Italiano nell' I. R. Museo di Fisica e Storia naturale. 1a Pubblicazione. *Firenze,* 1858

† **Florian, Joseph.** Beobachtungen über Magnetabweichungen. 8vo. 10 pp. (*Sitzungsb. d. k. Akad. Wissn. Wien* 1850. iv. Heft, p. 370.) *Wien,* 1850

† **Foissac, P.** Rapports et discussions de l'Académie Royale de Médecine sur le magnétisme animal . . . avec des notes par P. F. 8vo. 554 pp. *Paris,* 1833

† **Folkes, Martin.** An account of the Aurora Borealis seen at London on the 30th March last, as it was anxiously observed by Martin Folkes, Esq. 4to. 3 pp. (*Phil. Trans. for* 1717.) *London,* 1717

Follini, Giorgio. *Born about* 1751 ; *died August* 2, 1831, *at Turin.*

† Teoria elettrica brevemente esposta . . . 8vo. 164 and 4 pp. 2 plates. *Ivrea,* 1791

† Ragguaglio della guarigione di un ostinato tumore operata per via dell' Elettricità. 8vo. 8 pp. *Ivrea,* 1791

Follini, Giorgio—*continued.*

† Sul passaggio del Fulmine . . . delli 6 Agosto, 1795, nel Tempio di S. Andrea in Vercelli. 8vo. 49 pp. 1 plate. *Vercelli,* 1795

† Memoria fisica sull' uso del fuoco elettrico in medicina. 8vo. 76 pp.
Casale, 1798 (?)

Physicæ elementalis elementa. *Taurini,* 1823

Fond. (*Vide* Sigaud-de-la-Fond.)

Fonda. Sopra la maniera di preservare gli edifizii dal Fulmine. *Roma,* 1770

Fontana. Ricerche filosofiche sopra la fisica animale. *Firenze,* 1775

Opuscoli scientifici. 8vo. *Firenze,* 1783

Fontana, Felice. *Born April* 15, 1730, *at Pomarolo, near Roveredo, Tyrol; died January* 11, 1805, *at Florence.*

Descrizioni ed usi di alcuni stromenti per misurare la salubrità dell' Aria.
Firenze, 1774

† Articolo di lettera all' Ab. G. Mangelli. Firenze, 10 Nov. 1792. 8vo. 3 pp.
(*Giorn. Fis. Med.* iv. 116.) *Pavia,* 1792

Lettere sopra l'elettricità animale. 1793

Fontana, Gregor. Disquisitiones physico-mathematicæ nunc primum editæ. 4to.
384 pp. 3 plates. *Papiæ,* 1780

*† Disquisitiones physico-mathematicæ (nunc primum editæ). 4to. 384 pp.
Papiæ, 1780

Fontenelle. *Vide* Julia-Fontenelle.

† **Fonvielle, W. de.** Eclairs et Tonnerre. 12mo. *Paris,* 1867

† **Forbes, Eli.** An account of the effects of Lightning on a large rock in Gloucester, in a letter . . . to . . . M. Cutler (dated July 3, 1783). 4to. 4 pp. (*Mems. of the American Acad.,* Old Series, i. 253, part ii.) *Boston,* 1785

Forbes, James David. *Born April* 20, 1809.

History of Natural Philosophy. 4to. *Edinburgh*

† Account of some experiments in which an Electric Spark was elicited from a Natural Magnet. 4to. 9 pp. 1 plate. (*Edinb. Trans.* tom. xii.) (*Arago's copy.*) *Edinburgh,* 1832

† An account of some experiments on the Electricity of Tourmalin, &c., when exposed to heat. 4to. 14 pp. (*Edinb. Trans.* tom. xiii.) (*Arago's copy.*)
Edinburgh, 1834

† Account of some experiments made in different parts of Europe on Terrestrial Magnetic Intensity, particularly with reference to the effect of Height. 4to. 29 pp. 1 map. (*Edinb. Trans.* tom. xiv.) (*Arago's copy.*)
Edinburgh, 1839

† Account of some additional experiments on Terrestrial Magnetism made in different parts of Europe in 1837. 4to. 9 pp. (*Trans. R. S. Edinb.* vol. xv. part i.) *Edinburgh,* 1840

Review of the Progress of Mathematical and Physical Science. 4to.
London, 1858

* General Index to the *British and Foreign Medical Review.* 8vo. *London,* 1849

† **Forbes, P.** On the application of Electro-Magnetism as a motive power; in a letter to Faraday. 8vo. 2 pp. (*Annals of Electricity,* v. 239.) *London,* 1840

Forbin, Comte de. Les Mémoires du Comte de Forbin en 1646. Edition de 1740.
Amsterdam, 1740

Force, P. Record of Auroral Phenomena observed in the higher northern latitudes, compiled by Peter Force. 4to. 118 pp. *Washington,* 1856

† **Forsach, J. A.** Katechismus der elektrischen Telegraphie. 8vo. 58 pp.
Leipzig, 1852

Handleiding tot de Kennis der electrische Telegraphie. Naar het Hoogduitsch door M. J. van Oven. 12mo. *Utrecht,* 1853

† Handbuch der electrischen, galvanischen, magnetischen und electromagnetischen Telegraphie. 8vo. 161 pp. 45 plates. *Wien,* 1854

Forskal, P. Beobachtungen über d. Silurus electricus, über das Leuchten des Meeres, &c., auf semer Reise in Arabien 1761-63. (*From Poggendorff,* i. 776, " Deren Früchte gesammelt sind von K. Niebuhr in Flora ægyptiaco-arabica Havn, 1775 : Descriptiones annualum.")

Forsten Vershour. (*Vide* Vershour.)

† **Forster, B.** Description of a method of fitting up in a portable form the Electric Column, lately invented by J. A. De Luc. Also an account of several experiments made with it. 8vo. 6 pp. 1 plate. (*Phil. Mag.* xxxv. 205.)
London, 1810

† On De Luc's Column. 8vo. (*Phil. Mag.* xxxv. 317, 399, and 468 ; do. xxxvi. 74, 317, and 472.) *London,* 1810

† On De Luc's Electric Column. 8vo. 4 pp. (*Phil. Mag.* xxxvii. 197.)
London, 1811

† On the Aurora Borealis. 8vo. 3 pp. (*Phil. Mag.* xli. 263.) *London,* 1813

† On constructing Electric Columns. 8vo. 2 pp. (*Phil. Mag.* xlvii. 265.)
London, 1816

† Description of an electrical instrument called " The Thunder-storm Alarum." 8vo. 2 pp. 1 plate. (*Phil. Mag.* xlvii. 344.) *London,* 1816

† Luminous Belt of September 29th. 8vo. 2 pp. (*Phil. Mag.* iv. 463.)
London, 1828

† **Forster, Dr.** Aurora Borealis lately observed at Boreham, in Essex. 8vo. 1 p. (*Phil. Mag.* iv. 317.) *London,* 1828

Forster, Johan. Georg. Adam. *Born November* 27, 1754, *at Nassenhuben ; died January* 11, 1794. (*Poggendorff,* 1566.)

Forster and Lichtenberg, G. C. Göttingisches Magazin für Wissenschaft u. Litteratur 3 Jahrgang. 8vo. *Gottingen,* 1780-82

Vide also Lichtenberg.

Forster Johann Reinhold. *Born October* 22, 1729, *at Dirschau ; died December* 16, 1798, *at Halle.*

Observations made during a voyage round the world, on Physical Geography, &c. 4to. 649 pp. of text. (*Opuscoli Scelti in* 4to. ii. 385, *on Auroræ.*)
London, 1778

Observations on a Voyage round the world. 4to. *London,* 1778

† Osservazioni . . . Vermi lucenti . . . 4to. 3 pp. (*From Rozei,* 1783, *Luglio.*) (*Opus. Scelt.* vi. 419.) *Milano,* 1783

" Nachricht von einer an verschieden Körpern, auf einem mit Wachstuche überzogenen Tische, nach ofterer Entzundung entzündlicher Luft, und wenigem Elektrisiren, am folgenden Morgen bemerkten Anhaufung des Harzstaubes und Bärlappstaubes. 8vo. (*Gott. Mag.* iii. 1783, St. 4, pp. 573-5.) *Göttingen,* 1783

Über d. Natur d. Feuers u. d. Elektricität. (*Crell. Neue Entdeck.* xii. 1784.)
1784

† **Forster, L. Von.** Allgemeine Bauzeitung heraus gegeben v. L. V. Foster.
Wien, 1836-47

Forster, Richard. An account of a Meteor seen at Shefford, in Berkshire, October 20, 1759, with some observations on the weather of the preceding winter. 4to. (*Phil. Trans.* 1759, p. 299.) *London,* 1759

178   FOR—FOS

Forster, Thomas Ignatius Maria. *Born November* 9, 1789, *at London; died about* 1850.

† Notice of a Memoir on Meteors of various sorts. By T. I. M. Forster, F.R.A.S., &c. Bruges, 1846. 8vo. (*Phil. Mag. 3rd Series,* xxxi. p. 219.)
London, 1847

†. **Forster,** T. On De Luc's Electric Column. 8vo. 2 pp. (*Phil. Mag.* xxxvii. 424.)
London, 1811

† On the influence of the Atmosphere in certain diseases. 8vo. 3 pp. (*Phil. Mag.* xxxviii. 68.)
London, 1811

Researches about Atmospheric Phenomena and . . . Meteorological Journals, &c. 8vo.
London, 1813

† Observations on a Fiery Meteor. 8vo. 1 p. (*Phil. Mag.* xliii. 26.)
London, 1814

† Observations on certain luminous Meteors called Falling Stars. 8vo. 3 pp. (*Phil. Mag.* lvii. 418.)
London, 1821

† On simultaneous Thunder-storms. 8vo. 2 pp. (*Phil. Mag.* lx. 195.)
London, 1822

†. Researches about Atmospheric Phenomena and Calendar, &c. 3rd edition. 8vo. 6 plates.
London, 1823

† On the Aurora Borealis of 26th September. 8vo. 2 pp. (*Phil. Mag. or Annals,* iii. 75.)
London, 1828

† On the Zodiacal Light of the 29th of September, as it appeared from Chelmsford. 8vo. 2 pp. (*Phil. Mag. or Annals,* iv. 389.)
London, 1828

Forster. (*Vide* Bruhns and Forster.)

Fortin of Brest. Renowned for improvements in the Declinatorium and Inclinatorium. (*Poggendorff,* i. 778.)

Fortis, Giovanni Battista genannt Alberto. *Born November* 11, 1741, *at Padua; died October* 21, 1803, *at Bologna.*

† Lettera del . . . Fortis allo Spallanzani su gli sperimenti di Pennet. 4to. 15 pp. (*Opus. Scelt.* xiv. 259.)
Milano, 1791

† Elogio di Alberto Fortis di C. Amoretti. Avviso dell' Editore. 4to. 16 pp. e 1 p. (*Nuova Scelt. d'Opusc.* ii. pp. 425 e 440.)
Milano, 1807

Fortschritte (die) des Physik, dargestellt von der physikalischen Gesellschaft zu Berlin. Redigirt von G. Karsten, A. K. Kronig, &c. 8vo.
Berlin, 1855-58

Foster, Parry, and Christie.

† 1. Observations on . . . the Needle at the Whale Fish Islands, by Foster. 2 pp.

2. Magnetical Observations at Port Bowen, in 1824-5 . . . by Parry and Foster. 44 pp.

3. Abstract of the daily Variation of the Needle, No. 2, by Foster. And Abstract of Results . . . 7 pp. (*i.e.* Parry and Foster's Observations on . . intensity.)

4. Dip of the magnetic Needle observed at Woolwich and at different stations within the Arctic Circle, by (?) 3 pp.

5. Observations on the diurnal changes in the position of the horizontal Needle, under a reduced directive power, at Port Bowen, 1825, by Foster. Communicated January 12, 1826. 4 plates. 48 pp.

6. A comparison of the diurnal changes of intensity in the dipping of horizontal Needles at Port Bowen, by Foster. Communicated February 25, 1826. 11 pp.

† 7. Account of the repetition of Christie's experiments on the magnetic properties imparted to an iron Plate by rotation, at Port Bowen, in May and June, 1825, by Foster; together with Christie's remarks thereon. 2 plates. 18 pp. (*From Phil. Trans.*) (*Arago's copy.*)
London, 1826

Foster, H. A comparison of the changes of magnetic intensity . . . in the dipping and horizontal Needles at Trewrenburgh Bay. 4to. 9 pp. (*Phil. Trans.* p. 303, 1828.) *London,* 1828

Foster, G. C. (*Vide* Lardner, 1866.)

Fothergill, John. *Born October 12, 1712, at Carr End, Yorkshire; died December 26, 1780, at London.*

† An extract of John Fothergill his Essay upon the origin of Amber. Read March 1, 1743-4. 4to. 5 pp. (*Phil. Trans.* xliii. 21.) *London,* 1744

Account of the Magnetical machine contrived by the late Dr. Gowin Knight. 4to. (*Phil. Trans.* 1776, p. 591.) *London,* 1776

† **Fotheringhame, T.** Directions for ascertaining and counteracting the Local Attraction of the Mariner's Compass by day and by night on board all vessels (steamers included) at sea, or in a roadstead, and on land, with cases . . . to which is added a few practical hints on the Variation of the Compass and the utility of the Lead and Charts, with illustrations. 4to. 60 pp. *London,* 1837

Foucault, Jean Bernard Léon. *Born September 19, 1819, at Paris; died February 11, 1868, at Paris.*

Appareil destiné a rendre constante la lumière émanant d'un charbon placé entre les deux poles d'une pile. 4to. (*Comptes Rendus,* xxviii. 68. *Paris,* 1849

Rapport sur son Mémoire sur l'Appareil à lumière constante . . . par Dumas. 4to. (*Comptes Rendus,* xxviii. 120. *Paris,* 1846

Appareil Photo-électrique. 4to. (*Comptes Rendus,* xxviii. 698.) *Paris,* 1849

† De la chaleur produite par l'influence de l'aimant sur les corps en mouvement. 4to. 2 pp. *Paris,* 1855

Etude sur le pendule à oscillations électro-continues. 8vo. 20 pp. *Amiens,* 1856

Rheotome. (*Vide Du Moncel.*)

(*Vide* also Fizeau and Foucault.)

Fourcroy, Antoine François de. *Born June 15, 1755, at Paris; died December 16, 1809, at Paris.*

Discours à la séance publique de l'Ecole de Médecine, Paris. 4to. *Paris,* 1801

† Memoir on the Stones which have fallen from the Atmosphere, and particularly near Laigle, in the department of L'Orne, on the 26th of April last. Read by C. Fourcroy in the Institute, June 19th, 1803. 8vo. 7 pp. (*Phil. Mag.* xvi. 299.) *London,* 1803

† Mémoire sur la nature de la fibre musculaire, et sur le siège de l'irritabilité. *Paris.*

(*Vide* also Vauquelin).

† **Fourcy, E. de.** Rapport sur Appareil de Prudhomme. (*Vide* Prudhomme.)

Fournet. Observations sur la distribution des Orages dans le département du Rhône. 8vo. *Lyon,* 1841

† Faits pour servir à la théorie de la Grêle. 4to. 10 pp. (*An. . . . Soc. Agricult.* vi. 131.) *Lyon,* 1843

Aperçus sur le magnétisme des minéraus et des roches. (*Ann. de la Soc. d'Agriculture, &c. de Lyon,* 1848.) *Lyon,* 1848

† **Fournet et Benoit.** Grêles du département du Rhône, Dégats, périodicité, directions des orages à grêles. 8vo. 95 pp. 2 plates. *Lyon,* 1869

Fowler, Thomas. *Born January 22, 1736, at York; died June 22, 1801, at York.*

A remarkable case of the morbid effects of Lightning successfully treated. (*Medical and Philosophical Commentary by a Society of Physicians in Edinburgh,* vol. vi.) *Edinburgh*

A case of an obstinate Quartan Ague cured by Electricity. (*Mem. Med. Soc. of London,* iii.)

† **Fowler,** Richard. Experiments and Observations relative to the influence lately discovered by M. Galvani . . . 8vo. 176 pp.   *Edinburgh,* 1793

Fowler. (*Vide* Monro and Fowler.)

Fownes, G. Manual of Chemistry, by Jones and Hoffman. 8th edition. 12mo.
London, 1861

Manual of Elementary Chemistry, theoretical and practical. 9th edition. Thick 12mo.   1863

Fox, R. W. Description and use of a Dipping-needle Deflector, invented by Robt. Were Fox, Esq. By T. B. Jordan, instrument maker. 8vo. 10 pp. 1 plate. (*Annals of Electricity.*)   *London*

† On the variable intensity of Terrestrial Magnetism, and the influence of the Aurora Borealis upon it. 4to. 9 pp. (*Phil. Trans. for* 1831, p. 199.)
London, 1831

† Notice of some experiments on Subterranean Electricity made in Penzance Mine, near Falmouth. 8vo. 6 pp. (*Trans. Royal Cornwall Polytechnic Society.*)   *Falmouth*

† **Fozembas.** Mémoire sur l'électro-moteur. Dédié à MM. les Médecins. 8vo. 16 pp.   *Paris,* 1833

† **Fragneau.** Appareil, accidents, chemins de fer. (*Vide* Abria.)   *Bordeaux,* 1853

Franceschi, G. La Elettricità animale nuovo elemento filosofico della medicina. 8vo. 96 pp.   *Ancona,* 1841

Franchot. " Pendule a mouvement continu de M. Franchot."
Note.—This was applied to improve Foucault's illustration of the Earth's rotation. Moigno describes the principle, &c., but gives no notice of where and when it was first described.

† **Francis. G.** Electrical experiments, &c. 8vo.   *London,* 1844
Electrical experiments. 5th edition. 8vo. 91 pp. Engravings and cuts.
London, 1850

Franklin, Benjamin. *Born January* 17, 1706, *at Governor's Island, near Boston ; died April* 17, 1790, *at Philadelphia.*

There are numerous editions and reprints of Franklin's Electrical works, &c. The date of his first letter is the 28th July, 1747. (*Vide Priestley, Hist.* 5th edition, p. 142.)

Letter concerning the Effects of Lightning. (*Phil. Trans.* 1751, p. 289.)
London, 1751

New experiments and observations on Electricity, &c., in letters to and edited by Collinson. Part i. 4to.   *London,* 1751

† Expériences et Observations sur l'Electricité. Traduit de l'anglais. 1st edit. 8vo. 222 pp. 1 plate.   *Paris,* 1752

Letter concerning an Electrical Kite. (*Phil. Trans.* 1752, p. 505.)
London, 1752

† New Experiments and Observations in letters to Collinson. Enlarged by Part ii.
1752

† New Experiments and Observations communicated to Collinson (and read at the Royal Society). To which are added a Paper, by Canton, and another (in Defence of Franklin against Nollet) by Colden. Part iii. 4to. 44 pp.
London, 1754

† New Experiments and Observations in letters to Collinson. Enlarged by Part iii. 4to. 154 pp. 1 plate.   *London,* 1754

Note.— This volume contains the second edition of Parts i. *and* ii., *and an edition not named of Part* iii. (*all dated* 1754). *The second letter of this second edition (to Collinson) should contain (at* p. 12) *a footnote which appeared in the first edition, wherein Franklin very kindly acknowledges a first suggestion by Hopkinson relative to the electrical property of Points in dissipating a charge. Kuhn (N. Entdeck. f.* 19) *refers to the Experiment of Hopkinson, and gives particulars not mentioned by Franklin, but no indication of where first described.*

Franklin, Benjamin—*continued.*

† New Experiments and Observations . . . in several letters to Collinson.
Part II. 2nd edition. 4to. 23 pp. *London* (?) 1754

Supplemental Experiments. Part iii. 4to. *London*(?) 1754

Electrical Experiments, made in pursuance of those by John Canton, with
explanation. 4to. (*Phil. Trans.* 1755, p. 300. *London,* 1755

† Expériences et Observations sur l'Electricité . . . Traduit de l'angloise.
Revue, corrigée, et augmentée d'un supplément du même auteur, avec des
notes et des expériences nouvelles, par M. D'Alibard. 2e édition. 2 vols.
12mo. 1 plate. *Paris,* 1756

† Briefe v. d. Elektricität a. d. Engl. 8vo. *Stockholm,* 1758

Briefe v. d. Elektricität uberst. nebst Anmerkungen von Wilcke. 8vo. xxiv.
and 354 pp. 1 plate. *Leipzig,* 1758

An account of the effects of Electricity in paralytic cases. 4to. (*Phil. Trans.*
1758, p. 695.) *London,* 1758

† New Experiments and Observations on Electricity made at Philadelphia . . .
and communicated in several letters to P. Collinson. Part i. 3rd edition.
4to. 86 pp. 1 plate. 1760

Physical and Meteorological Observations, Conjectures, and Suppositions.
4to. (*Phil. Trans.* 1765, p. 182.) *London,* 1765

New Experiments and Observations in letters to Collinson, enlarged by letters
and papers on other subjects. 4to. 500 pp. 1766

† Experiments and Observations on Electricity . . . to which are added letters,
&c., on Philosophical subjects; the whole corrected, &c. . . and now *first*
collected into one volume. 4th edition. 4to. 496 pp. *London,* 1769

† Letter from Dr. B. Franklin to D. Hume, Esq., on the method of securing
houses from the effects of Lightning. 8vo. 15 pp. *Edinburgh,* 1771

† Œuvres de Franklin, traduites de l'anglais, sur la 4e édition, par Barbeu
Dubourg, avec des additions nouvelles. . . 2 vols. 4to. *Paris,* 1773

† Experiments and Observations on Electricity . . . to which are added letters
and papers on Philosophical subjects. The whole corrected, &c. . . . and
now collected into one volume. 5th edition. 4to. 514 pp. *London,* 1774

Political, Miscellaneous, and Philosophical pieces (" not contained in the preceding collection.") 4to. (*From advertisement to his Complete Works,* 1866.)
1779

Sämmtliche Werke, aus dem Engl. und Franz. übersetz. nebst, des franz.
übersetzers B. Dubourg, Zusätzen, und mit einigen Anmerk. versehen
von G. T. Wenzel." *Dresden,* 1780

Lettere di Beniamino Franklin a Giambattista Beccaria volgarizzate dal Conte
Prospero Balbo. 8vo. 8 pp. *Torino,* 1783

*Note.—There are six letters printed at the end of Eandi's " Memorie istoriche
intorno gli studi del Padre Giam. Beccaria."*

† Opere filosofiche di B. Franklin nuovamente raccolte, e dall' originale Inglese
recate in lingua Italiana. 8vo. 125 pp. 3 plates. *Padova,* 1783

Philosophical and Miscellaneous Papers. 8vo. 1787

Erweitertes Lehrgebäude der natürlichen Elektricität. 8vo. *Wien,* 1790

† Conjectures concerning the formation of the Earth, &c., in a letter to the
Abbé Soulavie, Sept. 22, 1782. 4to. 5 pp. (*Trans. Amer. Phil. Soc.* iii. 1.)
Philadelphia, 1793

† Queries and Conjectures relative to Magnetism, and the theory of the Earth ;
in a letter to Mr. Bodain, read Jan. 15, 1790. 4to 4 pp. (*Trans. Amer.
Phil. Soc.* Old Series, vol. iii. p. 10.) *Philadelphia,* 1793

Franklin, Benjamin—*continued.*

Meteorological Imaginations and Conjectures.   (*Mem. of the Soc. Manchester.*)
                                                                      Manchester

Works.   2 vols.   12mo.   (*From Watt.*)   *London,* 1793

† The complete Works of B. Franklin.   3 vols.   8vo.   1806

Works, Political, Philosophical, and Miscellaneous, with his Memoirs of his
   Life, and Private Correspondence.   6 vols.   8vo.   1818

Franklin's Memoirs, Posthumous and other Writings, by W. T. Franklin.
   3 vols.   4to.   (*From Phil. Mag.* p. 61.)   1819

Autobiography and Life, by H. H. Weld.   Illustrated.   8vo.
                                                             London, 1849

Works of, with Notes and Life, by Jared Sparks.   New edition.   10 vols.   8vo.
                                                          Boston, U.S., 1850

Autobiography of, by Jared Sparks.   *Boston, U.S.,* 1856

Autobiography.   Edited from his manuscript, by J. Bigelow.   8vo.
                                                        Philadelphia, 1868

Works.   2 vols.   8vo.   (*From Young,* ii. 112.)   *London*

Works, consisting of his Life, &c.   2 vols.   12mo.   (*From Watt.*)

Franklin's Memoirs, Posthumous and other Writings, by W. T. Franklin.
   New edition.   8vo.   (*From Watt.*)

† Portrait by lithograph.   Lit. di G. Ricordi, P. Fontana.

† **Franklin,** B. and others.   Rapport des Commissaires chargés par le Roi de
   l'examen du Magnétisme animal.   The signatures are Franklin, Majault,
   Le Roy, Sallin, Bailly, D'Arcet, De Bory, Guillotin, and Lavoisier.   4to.
   66 pp.   *Paris,* 1784

† Report of Dr. B. Franklin and other Commissioners, charged by the King of
   France with the examination of the Animal Magnetism . . . Translated
   from the French, with an historical introduction.   8vo.   108 pp.
                                                             London, 1785

Franklin, C.   Erweitertes Lehrgebäude der Elektricität.   8vo.   *Wien,* 1790

Franklin-Institut.   Journal of the Franklin Institut. of the State of Pennsylvania
   for the promotion of the Mechanical Arts.   1st series, 1826-27, 4 vols. ;
   2nd, 1828-1840, 26 vols. ; 3rd, 1841-1856, 32 vols.   8vo.
                                                        Philadelphia, 1826-56

Franklin, Sir John.   *Born April 16, 1786, at Spilsby ; died June 11, 1847, at Point
   Victory, on the north-west coast of King William IV. Island.*

Franklin, Georg.   Declaratio phænomenorum juxta methodum scholasticam.
                                                          Oenipont, 1747

De electricitate ejusque phenomenis.   *Oenipont,* 174—

Franz, Jh.   Dissertatio de natura Electri.   *Vindob.* 1751

Frauenhofer, Joseph.   *Born March 6, 1787, at Straubing ; died June 7, 1826, at
   Munich.*

On Lines, &c., in the Electric Spectra.   (*Abhandlungen der K. Bayerischen
   Akad. der Wissenschaften,* 1814 e 1815.)   *München,* 1814 *or* 1815

Détermination du pouvoir réfringent et dispersif, &c.   (*From Dove,* 227.)

Frecksel.   Bemerkungen über Blitzschläge.   1819

† **Fredholm,** K. A.   Om meteorstenfallet vid Hessle den 1 Januari 1869.   8vo.
   43 pp.   *Upsala,* 1869

Freitel.   Télégraphe imprimeur.   (*From Du Moncel.*)   1855

† **Freke, John.** An Essay to show the Cause of Electricity, &c. . . . In a letter to Mr. W. Watson. 8vo. 64 pp. *London,* 1746

† Treatise on the nature . . of Fire, in Three Essays. I. Showing the Cause of Vitality. . . . II. On Electricity. III. Showing the Mechanical Cause of Magnetism. . . 196 pp. 8vo. *London,* 1752 (*Vide* also Anon. Elect. 1748.)

† **Fremery, N. C.** Dissertatio philos. inauguralis de Fulmine. 4to. 96 pp. 1 plate. *Lugdini, Batav.* 1790

Frémy, Edmond. *Born Feb.* 28, 1814, *at Versailles.*

† Sur l'Oxygène et l'Ozone. Deuxième Conférence du 10 Avril, 1866, sous le patronage de l'Impératrice. Au bénéfice de la Société de Secours des Amis des Sciences, fondée par le Bn. Thénard. 8vo. 32 and 50 pp. *Paris,* 1866

† **Fremy et Ed. Becquerel.** Recherches électro-chimiques sur les propriétés des corps electrisés. 8vo. 44 pp. 1 plate. *Paris,* 1852

French Meteorological Society. Annuaire de la Société méteorologique de France. *Paris,* 1857

Frenzel, J. S. T. De Torpedine veterum genere raia. 4to. *Wittenberg,* 1777

Fresnel, Aug. Œuvres complètes publ. par MM. Henri de Sénarmont, Edm. Verdet, et L. Fresnel. 3 vols. 4to. *Paris,* 1868-70

Fresnel, A. J. Sur les essais ayant pour but de décomposer l'Eau avec un aimant. 8vo. (*Annales de Chim. et Phys.* xv. 1820.) *Paris,* 1820

Fresnel. (*Vide* Du Fresnel.)

Freycinet, Claude Louis Desaulses de. *Born Aug.* 7, 1779, *at Montélimart; died Aug.* 18, 1842, *at Landgut Freycinet, near Loriol, dép. Drôme.*

Voyage de découvertes aux Terres Australes sur les corvettes le Géographe, &c. . . pendant . . 1800-1-2-3-4 sous le command du Capt. de vaisseau N. Baudin.

Navigation et Géographie. . redigé par M. Louis Freycinet. 4to. *Paris,* 1815 (*Vide* also Peron and Freycinet.)

† Account of the French voyage of Discovery and Circumnavigation performed in 1818, 1819 and 1820. (*Phil. Mag.* lvii. 20.) 8vo. 9 pp. *London,* 1821

† Voyage autour du monde sur les corvettes l'Uranie et la Physicienne pendant . . 1817, 1818, 1819 and 1820. Magnétisme terrestre. 4to. 342 pp. 1 map. *Paris,* 1842

Freygang, W. Von. Gedanken über die Luftsteine a.d Franz. *Göttingen,* 1805

Friberg, O. Dissertatio gradualis de pyxide nautica: præside A. Celsio. 4to. *Upsaliæ,* 1743

Friedlander. Expérience avec l'appareil galvanique de Volta. *Paris,* 1801 (*Vide* also Pfaff, C. H.)

† **Friesach, K.** Geographische und magnetische Beobachtungen in Nord- und Süd-Amerika, angestellt in den Jahren 1856 und 1857 . 44 pp. (*Sitzungsberechte der Wien Acad.,* vol. xxix. No. 9.) *Wien,* 1856-7

† Astronom u. magnet. Beobachtungen in Amerika angestellt 1857, 1858, und 1859. 8vo. 42 pp. (*Wiener Sitzungsberichte,* 1852.) *Wien,* 1859

Friese. Theoria Galvanismi. 8vo (*From Dove,* p. 205.) *Bonn,* 1842

Frieten. Theoria Galvanismi. (*From Lempertz Cat.* No. lxxxiv. 1867.) *Bonnæ,* 1842

Frisi, Paolo. *Born April 13, 1728, at Milan; died Nov. 22, 1784, at Milan.*

Nova electricitatis theoria.   *Milano,* 1755

Una sua lettera al Dott Lami.   *Firenze,* 1755

† De existentia et motu Ætheris seu de theoria electricitatis, ignis, et lucis. Dissertatio. 4to. 66 pp. 1 pl.   *Petropoli,* 1755

† De causa electricitatis. Dissertatio. 8vo.   *Petrop. et Lucæ,* 1757

Dissertationum variorum . . in quo habentur (No. 3.)   De natura et motu Ætheris . . . quærum specimen anno 1755. Imps. Petropolitanæ Scient. Acad. jussu et suffragio editum est (p. 178, part i.), et De Electricitate (p. 212, part ii.)   2 vols. 4to.   *Lucæ,* 1759 *and* 1761

† Opuscoli filosofici. 8vo. 118 pp. (*Consists of five writings; the second only electric, viz.,* " *Dei Conduttori Elettrici,*" 22 pp.)   *Milano,* 1781

Opera. 3 vols. 4to.   1782

† Operette scelte di Paolo Frisi Milanese, con le Memorie Storiche intorno al medesimo scritte da Pietro Verri. 12mo. 419 pp. Portrait. (*Trans.* vol. clxiii. *of the* " *Biblioteca scelta di Opere Italiane antiche e moderne,* p. 26, " *Dei Conduttori Eletrici.*")   *Milano,* 1825

(*Vide* also Euler, Frisi and Beraud.)

† **Frisiani, P.** Memoria di Gauss tradotta, &c.

† Ricerche sul Magnetismo terrestre. Memoria (*Prima Letta* 10 *maggio,* 14 *Giugno e* 23 *Agosto* 1860.) Printed in 1860. 4to. 35 pp. (*Memorie del R. Istit. Lomb.* viii.)   *Milano,* 1860 ?

† Sulle sue "Ricerche sul Magnetismo terrestre." 4to. 1 p. (*Atti dell' I. R. Istit. Lomb.* ii. 98 & 101.)   *Milano,* 1860

† Ricerche sul Magnetismo terrestre. (*Memoria Seconda. Letta* 22 *Agosto,* 21 *Novembre, e* 5 *Decembre* 1861, 9 *e* 23 *Gennazo* 1862.) 36 pp. (*Memorie del R. Istit. Lomb.* ix.)   *Milano,* 1862 (?)

Fritch. (*Vide* Kreil.)

† **Fritsch, K. Von.** Ueber die Störungen des täglichen Ganges einiger der wichtigsten meteorologischen Elemente an Gewittertagen. 8vo. 75 pp. 1 plate. (*Aus. den Sitzungsberichten* 1859 *der k. Akad. der Wissenschaften,* xxxviii. *Bde.* 633.)   *Wien,* 1860

† Üb. d. Mitwirkung elektr. Strome bei d. Bildung einiger Mineralien. Inaugural Dissertation. 8vo. 51 pp.   *Gottingen,* 1862

4 Abhandl. über Gewitter u. Orkane. 8vo. (*Wien. Akad.*)   *Wien*

Frobesius, Johann Nicolaus. (*Born January* 7, 1701, *at Goslar; died September* 11, 1756, *at Helmstädt.*)

† Nova et antiqua luminis atque Aurora-borealis spectacula. 4to. 160 pp. *Helmstadii,* 1739

Froment. Sur un instrument électrique à lame vibrante. 4to. (*Comptes Rendus,* xxiv. 428.)   *Paris,* 1847

M. Pouillet donne de vive-voix quelques détails sur la construction d'un télégraphe électrique, soumis par M. Froment au jugement de l'Académie. 4to. (*Comptes Rendus,* xxx. 562.)   *Paris,* 1850

" Compteurs électro-chronométriques de M. Froment." (*From Du Moncel and Moigno.*)

" Im J. 1852 im Auftrage der Sardinischen Regierung einen dem Morse'schen Telegraphen ähnlichen Schreib apparat construirt." 8vo. (*Polytech. Jour.* clix. 173.)   *Stuttgart,* 1853

Electro-moteurs. (*From Du Moncel.*)

† **Frommhold, C.** Der constante galvanische Strom modificirbar in seinem Inten-sitäts- und Quantitätswerth. Nachtrag zur Electrotherapie von C. Fromm-hold. 8vo. 66 pp.   *Pesth,* 1867-67

Fromondi, L. Meteorologicorum libri sex. 1st edition. 4to. *Antverpiæ,* 1627
(*Libri Sale Cat.* p. 328, *says he employed pulsations of the heart to calculate the distance of Thunder.*)

Froriep, Ludwig Friedrich von. *Born January 15, 1779, at Erfurt; died July 28, 1847, at Weimar.*

Dissertatio de methodo neonatis asphycticis succurrendi. *Jena,* 1801

Notizen aus d. Gebiet d. Natur-und-Heilkunde. 50 vols. 4to.
Weimar, 1822-36

† **Froriep,** L. F. and **Froriep,** R. Neue Notizen, id. 40 vols. 4to.
Weimar, 1836-45

Froriep, R. Beobachtungen über die Heilwirkung der Electricität bei Anwendung des magneto-electrischen Apparates. Erstes Heft. Die rheumatische Schwiele. 8vo. 292 pp. *Weimar,* 1843

† On the therapeutic application of Electro-magnetism to the treatment of rheumatic and paralytic affections, translated from the German by R. M. Lawrance. 8vo. 205 pp. *London,* 1850

Electro-magnetism in Rheumatic affections. 2nd edition. 8vo. *London,* 1852

Frytsch, M. De Meteoris, cum emendat. J. Hagii. 8vo. *Witeb.* 1583

Fuchs, J. C. Von einem merkwürdigen Wetterschlage in Potsdam. (*Allerneueste Mannigfaltigkeiten* 1782.) 1782

Zusätze und Ergänzungen der Nachricht von einem merkwürdigen Wetterschlage in Potsdam. . . . (*Allerneueste Mannichfaltigkeiten,* J. ii. Th. iii.)

Fulgenzio. (*Vide* Daven Fulgenzio.)

Fulke (or Fulco), Dr. W. A goodly gallery with a most pleasaunt Prospect into the garden of naturall Contemplation, to beholde the naturall causes of all kind of Meteors, as well fyery and ayrey, as watry & earthly, &c. &c. 8vo.
London, 1571

A goodly gallery with a most pleasaunt Prospect into the garden of naturall Contemplation, to beholde the naturall causes of all kind of Meteors, as well fyery & ayrey, as watry & earthly, &c. &c. 8vo. *London,* 1634

Meteors, or a plain Description of all Kinds of Meteors, as well Fiery as Airey, as Watry as Earthly. 12mo. 1670

† **Fuller,** Jn. jun. A letter from J. F. to Sir Hans Sloane, Bart., concerning the Fire Ball seen and described by Lord Beauchamp, in Sussex. 4to. 2 pp. (*Phil. Trans.* xli. 871.) *London,* 1739-40-41

Funk, Christlieb Benedict. *Born July 5, 1736, at Hartenstein, Grafschaft Schönberg; died April 10, 1786, at Leipzig.*

Die nördliche und südliche Erdoberfläche auf die Ebene des Aequators projicirt. *Leipzig,* 1781

Furstenau, J. H. De electricitate. Programma. 4to. *Rintel,* 1744

Fusinieri, Ambrogio. (*Born July 9, 1773, at Vicenza; died January 14, 1853, at Vicenza.*)

Nuovi fenomeni magnetici. 4to. (*Giorn. di Fisica, &c.* Dec. ii. vol. v. 1822.)
Pavia, 1822

Sopra il trasporto di materia ponderabile nelle scariche elettriche. 4to. (*Giornale di Pavia, Bimestre* vi. 1825.) *Pavia,* 1825

Singolari effetti prodotti da una scarica di elettricità atmosferica. 4to. (*Giorn. di fisica, etc.,* Dec. ii. vol. ii.) *Pavia,* 1825

† Sopra il trasporto di materia ponderabile nella folgore. 4to. 19 pp. (*Giornale, di Pavia, Bimestre* v. 1827.) *Pavia,* 1827

186 **FUS**

Fusinieri, Ambrogio—*continued.*

† Aggiunta di nuove osservazioni, istituite, anche sulle piante, alla Memoria sopra il trasporto di materia ponderabile nelle folgori. 4to. 8 pp. (*Giornale di Pavia, Bimestre* vi. 1827.) *Pavia*, 1827

Estratto di alcune Memorie sopra il trasporto di materia ponderabile nelle scariche elettriche delle macchine ordinarie e nelle folgori, e conseguenze che ne derivano contro la teoria degli imponderabili, del Dott A. . F. . 8vo. 38 pp. (*Giorn. delle Ital. Letter. del Da Rio*, vol. lxv. p. 155.)
Padova, 1828

† Memoria iii. Sopra il trasporto di materia ponderabile nelle Folgori. 4to. 14 pp. (*Ann. del Rea. Lomb.-Veneto*, vol. i. 291. *Continuazione della Memoria* iii. . .) *Padova*, 1831

† Annali delle scienze del Regno Lombardo-Veneto. Opera periodica di alcuni collaboratori. 4to. (*Nearly all Articles on Electricity in Library*.)
Padua, Vicenza & Venezia, 1831-45

(*Note.*—*This Journal is called " in un certo modo, una continuazione. . . del Giornale di Pavia.* (*dei*) *Configliachi e Brugnatelli.*")

† Estratto di alcune Memorie pubblicate negli anni 1826-27 e -30 dal Nobili sopra i Colori che acquistano i metalli sotto l'azione della pila di Volta, &c. . . 4to. 5 pp. (*Ann. Sci. Reg. Lomb.-Veneto*, i. 165.) *Padova*, 1831

† Breve notizia di nuovi Esperimenti elettro-magnetici del Prof. Dal |Negro. 4to. 3 pp. (*Ann. del Reg. Lomb.-Veneto*, i. 278.) *Padova*, 1831

Remarks on Matteucci, Marianini, and Namis (Namias?) " sopra la scossa elettrica delle rane." 4to. (*Fusinieri's Annali*, i. 342, 2nd part.)
Padova, 1831

† Nota. (Sopra Esperienze del Dal Negro nei giorni 30 e 31 Marzo 1832.) 4to. 3 pp. (*Ann. del Reg. Lomb.-Veneto*, ii. 106.) *Padova*, 1832

† Sopra la nuova Batteria magneto-elettrica del Sig. Prof. Dal-Negro. Notizia del Direttore di questi Annali. 4to. 2 pp. (*Ann. del Reg. Lomb.-Veneto*, ii. 139.) *Padova*, 1832

† Sopra il magnetismo di movimento; e risultati di alcuni esperimenti. . . 4to. 18 pp. (*Ann. del Reg. Lomb.-Veneto*, ii. 189.) *Padova*, 1832

† Risposta alla Lettera del Zamboni. 4to. 3 pp. *Padova*, 1834

† (Notizia sulle) Nuove osservazioni sopra le apparenze elettro-chimiche. Del
† Prof. L. Nobili. Firenze, 24 Dicem. 1833.

(Sopra) Note sul magnetismo. Del Prof. L. Nobili. Firenze, 10 e 14 Gennaio, 1834. 4to. 7 pp. (*Ann. del Reg. Lomb.-Veneto*, iv. 164 and 167.)
Padova, 1834

† Sopra la grandine straordinaria caduta in Padova nel giorno 26 Agosto di quest' anno. Nota. (*In Ann. del Reg. Lomb.-Veneto*, iv. 236.) *Padova*, 1834

† Risposta ad un Opuscolo stampato (Settem. 1834) a Firenze dal F. Rossellini sopra un articolo del Fusinieri nel 2do. Bimestre, 1834, degli Ann. della Scienza del Reg. Lomb.-Veneto. 4to. (*Ann. del Reg. Lomb.-Veneto*, iv. Fasc. v.) *Padova*, 1834

† Relazione di alcune esperienze del Magrini. 4to. 4 pp. (*Ann. del Reg. Lomb.-Veneto*, v. 220.) *Padova*, 1835

† Della influenza reciproca di più calamite, riguardo all' intensità dei loro magnetismi, &c. (*Ann. della Scienza del Reg. Lomb.-Ven.* tom. v.) *Padova*, 1835

† Memoria sul Magnetismo temporario delle barre di ferro dolce, per influenza del magnetismo terrestre; sulle loro posizioni di neutralità; e sopra qualche altra esperienza magnetica. 4to. 16 pp. 1 plate. (*Ann. del Reg. Lomb.-Veneto*, v. 306.) *Padova*, 1835

† Risposta alla lettera 2da. del Zamboni sulla teoria elettro-chimica delle pile voltiane. 4to 11 pp. (*.Ann. del Reg. Lomb.-Veneto*, vi. 31.) *Padova*, 1836

Fusinieri, Ambrogio—*continued.*

† La pila del Zamboni d'accordo colla teoria elettro-chimica. Memoria in risposta anche ad un opuscolo del Zamboni " *sull' argomento delle pile secche contro la teoria elettro-chimica.* Verona, 1836." Parte i. 28 pp. e ii. 22 pp. (*Ann. del Reg. Lomb.-Veneto*, vi. 293.)   *Padova,* 1836-37

† (Sopra lo scritto) " *Sulle proprietà elettriche ed elettro-fisiologiche della torpedine ; del padre Santi Linari.*" (Indicatore Sanese No. 50, 13 Dec. 1836.) 4to. 3 pp.   (*Ann. del Reg. Lomb.-Veneto*, vii. 123.)   *Padova,* 1837

† Scintilla elettrica tratta dalla pila termo-elettrica dall' Antinori e dal Santi Linari. 4to. 1 page.   (*Ann. del Reg. Lomb.-Veneto*, vii. 131.)   *Padova,* 1837

† (Analisi della Memoria del Marianini.)   " *Sulla teoria degli elettromotori.* Memoria iv.   Esame di alcune esperienze addotte dal Faraday per provare che la elettricità voltaica nasce dall' azione dei liquidi sui metalli, ec.* Modena, 1837. 4to. 11 pp.   Continuazione e fine dell' analisi di una Memoria del Marianini. 4to. 32 pp.   (*Ann. del Reg. Lomb.-Veneto* vii. 192 and 262.   *Padova,* 1837

† (Notizia di.)   " Sulla dispersione delle due elettricità " del Belli.   (*Bibl. Ital. tomi* 85 *e* 86.)   4to. 2 pp.   (*Ann. del Reg. Lomb.-Veneto*, vii. 203.)   *Padova,* 1837

† Risposta . . ad un articolo inserito nella Biblioteca ̇Universale, Jan. 1836, 196, circa le esperienze del Belli sulla dispersione delle due elettricità. (*Ann. del Reg. Lomb.-Veneto*, viii. 134.)   *Padova,* 1838

† Circa la elettricità della Torpedine e di tutti gli animali in generale secondo il Matteucci. 4to. 3 pp.   (*Ann. del Reg. Lomb.-Veneto*, viii. 239.)
                                                        Padova, 1838

† Sopra la costituzione della scintilla elettrica ; riflessioni in confronto di alcuni articoli del " Corso elementare di Fisica " del G. Belli. 4to. 6 pp. (*Ann. del Reg. Lomb.-Veneto*, viii. 284.)   *Padova,* 1838

† Notizia (dell'opera).   Sulle proprietà delle correnti magneto-elettriche (di) De la Rive (nella) Bibliothèque Univers.   1838, Mars, p. 134, et Avril, p. 366. 4to. 4 pp.   (*Ann. del Reg. Lomb.-Veneto*, viii. 290.)   *Padova,* 1838

Delle correnti elettriche secondarie costituite elementi ponderabili, e dipendenti dalla forza di espansione della materia attenuata. 4to.   (*Fusinieri's Annali*, ix. 1.)   *Venezia,* 1839

Reflessioni sopra il Magnetismo transversale prodotto dalle correnti elettriche. 4to. 7 pp.   (*Fusinieri's Annali*, ix. 19.)   *Venezia,* 1839

Sopra un nuovo apparecchio del Sig. Gauss per le osservazioni della intensità del magnetismo terrestre Riflessioni. 4to. 4 pp. (*Fusinieri's Annali*, ix. 75.)
                                                        Venezia, 1839

Sulle variazioni delle forze delle calamite per le loro influenze reciproche, e sulle calamite composte. 4to. 4 pp. (*Fusinieri's Annali*, ix. 79.)   *Venezia,* 1839

† Circa i principii generali del magnetismo tranversale prodotto dalle correnti elettriche. Discussione del Fusinieri; con Risposta a due Articoli del Zantedeschi nella Gazzetta di Venezia 13 e 27 Mag. 4to. 28 pp.
                                                        Venezia, 1839

† Esame dei fondamenti della teoria matematica dell' elettro-magnetismo. Matematico lettere ad suo. 4to. 10 pp.   *Padova,* 1840

† Discussione esperimentale sulle leggi delle attrazioni e repulsioni elettro-magnetiche, e della loro differenza dalle deviazioni scoperte da Œrsted. (Appendice A, p. 29.) 4to. 29 pp. (*Ann. del Reg. Lomb.-Veneto*, tom. x.)
                                                        Padova, 1840

† Incidente avvenuto a causa dell'esame dei fondamenti della teoria matematica dell' elettro-magnetismo, intrapreso da A. Fusinieri nel Bim. ii. 1840 degli Annali. 4to. xii. pp. (*Ann. del Reg. Lomb.-Veneto*, x.)   *Vicenza,* 1840

Risposta . . ad una lettera del G. Bellavitis nel Poligrafo di Nov. e Dec. 1840. 4to. 18 pp.   (*Fusinieri's Annali*, vol. x.)   *Venezia,* 1840

188   FUS–FUT

Fusinieri, Ambrogio—*continued.*

† Sopra alcuni fenomeni meteorologici che hanno rapporto collo sviluppo di elettricità e del calorico nativo dei corpi. Memoria. Ricevutta di 12 Decembre, 1837. 4to. 28 pp. (*Mem. Soc. Ital.* xxii. 48.)   *Modena,* 1841

† Sopra il trasporto di materia ponderabile nelle scariche elettriche. Memoria Estratta dal Giornale di fisica ec. di Pavia, Bim. vi. Seconda edizione in Appendice al Bim. vi. 1842, degli Annali delle Scienze del Regno Lombardo-Veneto. 4to. 11 pp.   *Padova,* 1843

† Sul termo-elettricismo del mercurio. Del moto vorticoso o a spirale scoperto dal Prof. Zantedeschi nell'arco luminoso fra due punte di carbone collocate ai poli della pila. 4to. 1 page. (*Ann. del Reg. Lomb.-Veneto* xiii. 217.)
                                                              Vicenza, 1844

† Memorie sperimentali di meccanica molecolare. 4to. 310 pp. 6 plates.
                                                              Padova, 1844

† Sopra alcuni effetti meccanici delle correnti galvaniche. 4to. 15 pp. (*Ann. del Regno Lomb.-Veneto,* xiv. 183.)   *Vicenza,* 1845

† Sulle ossidazioni interne delle coppie saldate di zinco e di ram che entrano a componere la pila di Volta. Memoria. 4to. 15 pp. (*Ann. del Reg. Lomb.-Veneto,* xiv. p. 243.)   *Vicenza,* 1845

† Memorie sopra la Luce, il Calorico, la Elettricità, il Magnetismo, l'Elettro-Magnetismo ed altri oggetti. 4to. 354 pp.   *Padova,* 1846

† Memorie di Meteorologia. 4to. 251 pp. Tables.   *Padova,* 1847

† Biografici cenni sopra Ambrogio Fusinieri, di Capparozzo. 8vo. 36 pp.
                                                              Vicenza, 1854

† Elogio di Ambrogio Fusinieri, letto all'Accad. Olimpica di Vicenza, 4 Feb. 1855 dal Dr. F. S. Beggiato, Presidente. 4to. 29 pp.   *Vicenza,* 1855

(*Vide* also Zamboni and Fusinieri.)

Fuss, Nicolaus von. *Born January* 30, 1755, *at Basle; died January* 4 (*N. St.*) 1826, *at St. Petersburg.*

† Observ. et Expér. sur les Aimants artificiels, principalement sur la meilleure manière de les faire. 4to. 2 plates. 38 pp. (*Acta Petrop.* ii. 35.)
                                                        Petersburg, 1778

Futerand, Pouzin. De l'influence d'un Orage sur trois Epilepsies vermineuses. 8vo. (*Montpellier, Acad. Recueil des Bulletins,* vol. vi. p. 345.)
                                                        Montpellier, 1815?

G.

G., W. M. (*Vide* Anon. Electricity.)   1823

G., S. (*Vide* Anon. Electricity.)   1838

Gabler, Matthias. *Born February 24, 1736, at Spalt, Franken; died March 30, 1805, Wembdingen.*

Theoria Magnetis. 8vo. (*From Young*, p. 437.)   *Ingolst.* 1781

Gachet. Notice sur deux Météorolites-Echantillons. (*From Bordeaux Acad.*)   *Bordeaux,* 1738

Gadd, Peter Adrian. *Born April 12, 1727, at Birkala; died August 11, 1797, at Abo.*

De originaria corporum electricitate. 4to.   *Aboæ,* 1769
(*Note.—Poggendorff*, i. 826, *says*: "*Obss. chimico-physicæ de originaria corporum mineralium electricitate.*")

† **Gaeta R.** Lettera ad un suo amico, sulle scariche elettriche senza arco conduttore. 4to. 12 pp. (*Opusc. Scelt.* vi. 247.)   *Milano,* 1783

Gaiffe. Constructed a little apparatus by which a small pump can be set in motion by an electro-magnetic motor. (*From Carl*, p. 287.)

Gaillard. Expériences galvaniques. (*Mém. de la Société médicale d'Emulation*, i. 235.)   *Paris*

† **Gaimard, P.** Voyages en Scandinavie, &c., pendant les années 1838-39-40 sur la Recherche . . . *Magnétisme terrestre*, par Lottin, Bravais, Lilliehook, Siljestrom, Meyer, Delaroche, Poncié, le Capitaine Tabvere, &c. 3 vols. 4to. 8 plates.   *Paris*

† Voyages en Scandinavie, &c., pendant les années 1838-39-40 sur la Recherche. *Aurores boréales*, par Lottin, Bravais, Lilliehook, et Siljestrom. 1 vol. 4to. 12 plates.

† Voyages en Islande et Groenland, pendant les années 1835 et 1836 sur la Recherche. *Physique*, par M. Victor Lottin. 1 vol. 4to.   *Paris,* 1838

Galilei, Galileo. *Born February 18, 1564, at Pisa; died January 8, 1642, at Arcetri, Tuscany.*

Opere. 3 vols. 4to. (*From Arago's Sale Cat.* p. 63.)   *Firenze,* 1718

Opere di. (Tom. iii. p. 355. A letter dated November 16, 1607, on piece of Loadstone.) (*Zantedeschi in Bibliot. Ital.* lxxxvi. 135.)   *Padua,* 1744

Galizi, D. Dissertazione dell' Aurora-boreale. 12mo. (*Calogera Nuova Raccolta* xxxix. 64.)   *Venezia*

† **Galle, J. G.** Über die am 11 Decbr., 1852, in Schlenin beobachtete Feuerkugel. 4to. 7 pp. (*Acad. Breslau*, 1853, p. 187.)   *Breslau,* 1853

† Über die Bahn des am 30 Januar, 1868, beobachteten und bei Pultusk im Königreiche Polen als Steinregen niedergefallenen Meteors durch die Atmosphäre. 8vo. 41 pp.   *Breslau,* 1868

† Katechismus der elektrischen Telegraphie. 8vo. 191 pp.   *Leipzig,* 1855

Katechismus der elektrischen Telegraphie. 2. Vermehrte und verbesserte Auflage. 8vo. 212 pp.   *Leipzig,* 1859

Katechismus der elektrischen Telegraphie. 3. Vermehrte und verbesserte Auflage. 8vo. 232 pp.   *Leipzig,* 1864

† **Galle, Ludw.** Katechismus der elektrischen Telegraphie. Von Ludw. Galle. Vierte wesentlich verm u. verb. Aufl. bearb. v. Prof. Dr. Karl Ed. Zetzsche. 4to. 372 pp.   *Leipzig,* 1870
(*This forms part, No.* 21, *of Weber's Illustrirte Katechismen.*)

† **Galle, M.** Beyträge zur Erweiterung und Vervollkommnung der Elektricitätslehre. 8vo. 280 pp. 3 plates. *Salzburg,* 1813

    Beiträge zur Erweiterung . . . der Elektricitätslehre. 2 vols. 8vo. (*From Dove,* p. 215.) *Salzburg,* 1816

† **Galletti, B. et Jounin, A.** De l'électricité en général et de ses applications en particulier. *Première Partie.* 8vo. 31 pp. *Paris,* 1844

Gallini, Stefano. *Born March 22, 1756, at Venice; died May 26, 1828, at Padua.*

†   Se e quanto il fluido elettrico o galvanico influisca nella produzione de' fenomeni della vita, sopra tutto ne' corpi animali. Memoria. Ricevuta li 12 Giugno, 1819. 4to. 22 pp. (*Mem. della Soc. Ital.* xviii. 232.) *Modena,* 1820

Gallitzin, Dmitri Alexewitsch Fürst. *Born December 21, 1738, in Russia; died March 21, 1803, at Brunswick.*

    Method of distinguishing the Pos. and Neg. states of Electricity by means of " poudre de *résine* répandue sur un carreau de verre." 4to. (*Mem. de l'Acad. Imp. et Roy. de Bruxelles,* tom. iii. p. 14. *Bruxelles*
    (*Note.—Is this the experiment described by him in his " Lettre sur quelques objets de l'Electricité, 1778 " ?*)

†   Lettre sur quelques objets d'Electricité addressée à l'Académie Imple. des Sciences de St. Pétersbourg. 4to. 16 pp. (*Accad. Petrop.*) *Petersburg,* 1778

†   Lettre sur quelques objets de l'Electricité. 4to. 10 pp. 1 plate. (*Volta's copy.*) *La Haye,* 1778

†   Nuovo fenomeno dell' Elettricità. 4to. 1 p. (*Opusc. Scelti,* ii. 305.) *Milano,* 1779

    Sendschreiben a. d. Akad. zu St. Petersburg, über einige Gegenstande d. Elektr. 8vo. 3 plates. *Munster,* 1780

    Sur l'Electricité. 4to. (*Journ. de Phys.* xxi. 1782.) *Paris,* 1782

    Sur la forme des conducteurs électriques. 4to. (*Journ. de Phys.* xxii. 1783.) *Paris,* 1783

    Sur quelques expériences de M. Achard. 4to. (*Journ. de Phys.* xxii. 1783.) *Paris,* 1783

†   Observations sur les conducteurs. Addressée à l'Acad. de Bruxelles. Date July 6, 1778, La Haye. 4to. 10 pp. 1 plate. (*Volta's copy.*) *La Haye,* 1778 ?

†   Observations sur l'Electricité naturelle par le moyen d'un Cerf Volant. 4to. 6 pp. 1 plate. (*Addressée à l'Acad. de Petersb. Date September 25, 1778.*) (*Volta's copy.*)

Gallucci, G. P. Modus fabricandi horaria mobilia permanentia cum acu magnetica. *Venet.* 1596

† **Galton, F.** The Telotype, a printing electric Telegraph. 8vo. 32 pp. 4 plates. *London,* 1850

Galvani, Luigi (Aloisio). *Born September 9, 1737, at Bologna; died December 4, 1789, at Bologna.* (*Pogg.* i. 839.)

†   Propositiones de Electricitate, &c. 4to. 23 pp. 1 plate. *Bononiæ,* 1786

†   De viribus electricitatis in motu musculari. Commentarius : cum *Aldini* dissertatione et notis. 4to. 80 pp. 3 plates. *Mutinæ,* 1792

†   Lettera al Carminati sull' elettricità animale. 8vo. 15 pp. (*Giorn. Fis. Med.* ii. 131.) *Pavia,* 1792

†   Transunto della dissertazione del . . . Galvani . . . sulle forze dell' Elettricità nei moti muscolari. 4to. 29 pp. 1 plate. (*Opusc. Scelt.* xv. 113.) *Milano,* 1792

    (*Note.—This translation is by Soave, who adds footnotes, the last of which contains : "Articolo di lettera, degli 8 Marzo (1792), dell' Aldini.*")

    Abh. ü. d. Kräfte d. their Elekt. auf d. Bewegung d. Muskeln u.d. Schriften v. Valli, Carminati u. Volta üb. diesen Gegenstand. überst. v. J. Mayer. 8vo. *Prague,* 1793

Galvani, L.—*continued.*

Dell' uso e dell' attività dell' arco conduttore nelle contrazioni dei muscoli.
8vo. 168 pp. *Bologna,* 1794

*Note.—This was erroneously supposed to be Aldini's work. The Supplement
is not dated. It must have been printed in 1794, or before the 12th January,
1795, for F. Caldani, in his "Lettera . . circa le nuove ricerche" . .
dated 12th January, 1795, refers to it.* (*Vide Aldini.*)

† De viribus electricitatis in motu musculari. Commentarius, 4to. 56 pp. 4
plates. (*Comment. Bonon. Scient.* vii. 363.) *Bononiæ,* 1798

† Memorie sulla elettricità animale . . al L. Spallanzani . . Aggiunte alcune
elettriche esperienze di G. Aldini. 4to. 105 and 3 pp. 2 plates. (*Brugnatelli's
copy.*) *Bologna,* 1797

Memorie . . alli Spallanzani per le Stampe del Sassi Bologna, anno 1797.
Modena, 1841

† Opere edite ed inedite del Prof. Luigi Galvani raccolte e publicate per cura
dell' Accad. delle Scienze dell' Istit. di Bologna. (*Edited by Gherardi, who
added an Appendix, or 2nd volume.*) 4to. 120 and 505 pp. Portrait and
Autographs. *Bologna,* 1841

Aggiunta alla collezione delle opere di Galvani. (*Edited by Gherardi.*) 4to.
Bologn.a, 1842

A German translation, with some additional observations, by Voight (of
Galvani's works).

† Elogio di Luigi Galvani detto da Mle. Medici nel Teatro anatomico dell' antico
Archiginnasio di Bologna. . . 4to. 29 pp. *Bologna,* 1845

Predieri, E. Di alcuni Autografi del Galvani ultimamente rinvenuti. 4to.
Bologna, 1862

Eloges historiques de, par Alibert, &c. (*Vide Alibert.*) *Paris*

† Portrait Incisione. Marchi fecit.

† Galvanic Society (Paris). Communications to, by Winckler, Abrial (a Mem.
of Pfingsten), Mojou, Nauche, Gautherot, Izarn, and the Commission of
Medical Application (of Galvanism). 8vo. 2 pp. (*Phil. Mag.* xv. 281.)
London, 1803

† On 28th May made, for the second time, some experiments on a large scale at
the Veterinary School of Alfort. (*Phil. Mag.* xvi. 90.) 1803

† Notice of Experiments made by the Galvanic Society of Paris on the discovery
announced by M. Pacchiani, of the composition of the Muriatic acid. 8vo.
5 pp. (*Ann. d. Chim.* lvi. ; *Phil. Mag.* xxiv. 172.) *London,* 1806

† Account of its Proceedings during the years 1804 and 1805, read at its sitting
6th February, 1806. 5 pp. (*Phil. Mag.* xxiv. 183.) *London,* 1806

† Notices of Experiments made by the Galvanic Society of Paris, drawn up by
M. Riffault, one of the members. 8vo. 3 pp. (*Ann. d. Chim.* lvi. 61.)
(*Phil. Mag.* xxv. 260.) *London,* 1806

Gamaches, Etienne Simon. *Born about 1672, at Meulan ; died February* 7, 1756,
at Paris.

Dissertation sur le Méchanisme de l'Electricité. (*Mém. de Treves,* Janv. 1753.
vol. ii. art. 9 ; and in his *Dissertations Littéraires et Philosophiques,* à Paris.
1755. 12mo.) 1755

† Gamble, J. Observations on Telegraphic experiments, or the different modes
which have been or may be adopted for the purpose of distant communication. 4to. 20 pp. 1 plate. *London, n. d.*

Gandolfi, B. Lettera al Sig. D. Morichini sull' ottima ed economica costruzione
delle macchine elettriche. (*Antologia Romana,* 1797.) *Roma* 1797

Gandolfi, B.—*continued.*
Dissertazione sopra le condizioni necessarie perché una macchina elettrica sia
capace del massimo effetto. (*Giorn. Lett. di Napoli*, 1802.)   *Napoli*, 1802
(*Vide* also Medici and Gandolfi.)

Gann. (*Vide* Mackrell, Gann and Pollock.)

† **Gannett, C.** An account of a curious and singular appearance of the Aurora-
borealis on the 27th March, 1781. 4to. 4 pp. (*Mem. Amer. Acad.* Old
Series, vol. ii. part i. p. 136.)   *Boston*, 1793

† An historical Register of the Aurora-borealis from August 8, 1781, to
August 19, 1783. 4to. 7 pp. (*Mem. Amer. Acad.* Old Series, i. part ii.
p. 327.)   *Boston*, 1785

† **Ganot, A.** Trattato elementare di fisica sperimentale ed applicata, e di meteorologia . . 7a edizione Italiana, sulla 10a edizione originale del 1862, aumentata
di 44 nuove figure e dei più recenti studii . . per cura de C. Hajech e. V.
Masserotti. 8vo. 944 pp.   *Milano*, 1862

Traité élémentaire de Physique expérimentale et appliquée et de Météorologie,
suivi d'un recueil nombreux de problèmes. 18mo. 857 pp.   *Paris*, 1862

Traité élémentaire de Physique expérimentale et appliquée et de Météorologie.
1864

Physics experimental and applied. Edited by E. Atkinson. For the use of
Colleges and Schools.

Garcés de Marsilla, D. A. Manual de telegrafía electrica para uso de empleados en los telégrafos eléctricos militares de Calaluña. 8vo. 40 pp. 1 plate.
Madrid, 1853

Gardane. Conjectures sur l'Electricité médicale, avec des recherches sur la
colique métallique. 12mo. (*From Manduit*, p. 249.)   1768

Conjectures sur l'Electricité médicale, avec des recherches sur la colique
métallique. (*From Guitard.*)   *Paris*, 1778
(*Note.—Van Troostwyk and Krayenhoff say at p. 159, 1768, and at p. 227,*
1778.)

Garde. (*Vide* La Garde, De.)

Gardini, Anton. Beschreibung eines sehr empfindlichen Electrometers. (*Mayer's
Samml. Physik. Aufsätze der Gesellsch. Böhmischer. Naturf.*, tom. iii. p. 393.)

Gardini, Giuseppe Francesco. *Born January 22, 1740, at Vascagliana, Province of
Asti, in Piedmont ; died May 15, 1816, at Asti.*

† L'applicazione delle nuove scoperte del fluido elettrico agli usi della ragione-
vole Medicina. . . Gandini. 8vo. 248 pp.   *Genova*, 1774

† De effectibus electricitatis in Homine Dissertatio. 8vo. 141 pp. *Genuæ*, 1780

† De influxu electricitatis atmosphericæ in Vegetantia. Dissertatio ab Acad.
Lugdunensi præmio donata an. 1782. 8vo. 157 pp.
Augustæ Taurinorum, 1784

† De electrici ignis natura. Dissertatio. 4to. 236 pp. 1 plate. *Mantuæ*, 1792

† De electrici ignis natura ed. et præf. est J. Mayer. 8vo. 197 pp. 1 plate.
Dresden, 1793

Abh. v. d. Natur. d. elek. Feuers. a.d. Lat. von J. G. Geisler. 8vo.
Dresden, 1793

Descrizione d'un istrumento proprio per conoscere l'elettricità tanto giornaliera
quanto spontanea degli uomini, &c. (*Giorn. scientifico, &c. Torino*, 1789.)
Torino, 1789

Esperimenti fatti nel mese . . 1789 sopra l'elettricità spontanea degli uomini,
&c. (*Giorn. scientifico, &c. Torino*, 1789.)   *Torino*, 1789

Sulla differenza dell' azione dell' elettricità della macchina e della pila. 1806

De ratione qua distribuitur ignis electricus in diversorum corporum superficie.
Turin, 1812

† **Gardini,** Guido. Esercitatione fisica tra li Sig. Studenti . . di Susa sulla elettricità, &c. . . 12mo. 45 pp. *Susa ?* 1798

† **Gardner, D.** Galvanic Experiments, by Mr. D. Gardner, Lecturer on Chemistry
at the City Dispensary (to Mr. Tilloch). 8vo. 2 pp. (*Phil. Mag.* xxv. 364.)
London, 1806

Garipuy. Mémoire sur un coup de Tonnerre arrivé près la ville de Castres. . .
Réflexions sur les conducteurs électriques. (Lu 11 Avril, 1782.) 4to.
(*Toulouse Acad.* 1re série, tom. ii. p. 188. *Toulouse,* 1784

Garn, J. A. De Torpedine recentiorum genere anguilla. 4to. *Witteb.* 1778

Garnault. (*Vide* Harris.)

Garnier, P. Modification apportée à l'indicateur dynamomètre de Watt. Nouveau
mécanisme pour les communications télégraphiques au moyen de l'électricité.
4to. (*Comptes Rendus,* xxi. 526.) *Paris,* 1845

Note sur un nouveau système d'Horloges électrochrones. 4to. (*Comptes
Rendus,* xxv 271, and xxix. 189.) *Paris,* 1847 *and* 1849

Télégraphe à Cadran. (*Table Comptes Rendus,* p. 235.)

† Cylindre automatique pour la composition et la transmission des signaux
télégraphiques de Morse. M. P. Garnier contre M. G. Marqfoy, Arrêt de
la Cour Impériale de Paris. 8vo. 15 pp. 1 plate. (*Extrait du Journal
d'Invention.*) *Paris,* 1858

Garo (Professeur à Turin). Une lettre . . électricité. (*From Nollet.*) *Turin,* 1753

Garrat, A. C. Electro-physiology and Electro-therapeutics. 8vo.
Boston, U.S., 1860

† Medical Electricity, embracing Electro-physiology and Electricity as a Therapeutic, &c. . . 3rd edition, revised, &c. 8vo. 1103 pp. *Philadelphia,* 1866

Garzoni, Leond. Veneziano. A work (much praised) on the Nature of Magnetism.
Before that of Gilbert

(*Note.—Faure says (in reference to electrical matter, or having relation thereto):
" Il Trattato del P. Leonardo Garzoni lodato dal Cabeo girò solo per le mani
dei dotti inedito e non compiuto." "Morte interveniente," says Cabœus.*)

† **Gasc,** J. P. Mémoire sur l'influence de l'Electricité dans la fécondation des
plantes et des animaux, &c. . . 4to. 63 pp. *Paris,* 1823

Gasserus. (*Vide* Peregrinus, Peter.)

Gassendi, Pierre. *Born January 22, 1592, at Champtercier, near Digne ; died Oct.
24, 1655, at Paris.*

Opera omnia ed. hab. Lugduni Bat. Folio. *Lugduni,* 1658

Gassicourt. (*Vide* Cadet de Gassicourt.)

† **Gassiot,** J. P. An account of Experiments made with the view of ascertaining
the possibility of obtaining a Spark before the circuit of the Voltaic battery
is completed. (Read December 19, 1839.) 4to. 10 pp. (*Phil. Trans. for
1840,* part i. p. 183.) (*Faraday's copy.*) *London,* 1840

† On certain phenomena connected with the Spark from a secondary coil. 8vo.
2 pp. 1 plate. (*Proc. London Elec. Soc.*) *London,* 1841

On the Polarity of the Voltaic battery. (Read November 15, 1842.) 8vo.
7 pp. (*Proc. London Elec. Soc.*) *London,* 1842 ?

† A description of an extensive series of the Water battery, with an account of
some experiments made in order to test the relation of the electrical and
chemical actions which take place before and after completion of the Voltaic
circuit. 4to. 14 pp. 1 plate (*Phil. Trans. for* 1844, part i. p. 39.) (*Faraday's
copy.*) *London,* 1846

† On the decomposition of Water under pressure by the Galvanic battery. 8vo.
2 pp. (*Report of the British Association for* 1853.) *London,* 1854

Gassiot, J. P.—*continued.*

†   The Bakerian Lecture. On the Stratifications and dark Band in electrical discharges as observed in the Torricellian vacua. 4to. 16 pp. 1 plate. (*Phil. Trans.* 1858.)   *London,* 1858

†   Stratified discharges in Vacuum tubes. 8vo. 4 pp. 1 plate coloured. (*Intellectual Observer,* March 1866, p. 81.)   *London,* 1866

Gastaldi, L.   Saggio di telegrafia mobile a bandiere atta a far servizio si in terra che in mare non men di giorno che di notte. 8vo.   *Torino,* 1855

† **Gastrell and Blanford.**   Report of the Calcutta Cyclone of the 5th October, 1864. 8vo. 150 pp. 8 plates and charts.   *Calcutta,* 1866

† **Gatteschi, G.**   Saggio sul magnetismo esposto nelle sue lezioni. 8vo. 112 pp. (*Volta's copy.*)   *Pisa,* 1818

Gattoni, Giulio Cesare.   *Born . . . ; died,* 1809.

†   Lettera all' editore sui Fulmini di Ritorno. (Data Como, 15 Luglio, 1808.) 4to. 14 pp. 1 plate (lith). (*Nuova Scelt. d' Opusc.* ii. 289 & 310.)   *Milano,* 1807

Gattoni, G. S.   Studies on Electricity, &c. (*From Lombardi,* vol. ii. p. 74.) 1768

Gaubil, Ant.   Observations mathématiques, astronomiques, géographiques, chronologiques, et physiques, tirées des anciens livres, ou faites nouvellement aux Indes et à la Chine par les Pères de la Comp. de Jésus. Vol. ii.   *Paris,* 1829

Gaubius, J. D.   (*Vide* Allamand.)

† **Gaugain, J. M.**   Mémoire sur une nouvelle boussole des tangentes. (Présenté à l'Académie des Sciences, 24 Janvier, 1853.) 8vo. 6 pp. (*Ann. de Chim.* 3me série, xli. 66.)   *Paris,* 1854

†   Traduction, Préface, et Notes. 8vo. (*Vide Ohm, Théorie des Courants.*)   *Paris,* 1860

Gauguier.   (*Vide* Gonon and Gauguier.)

Gauss, Karl Friedrich.   *Born April* 30, 1777, *at Brunswick; died February* 23, 1855, *at Göttingen.*

†   Intensitas vis magneticæ terrestris ad mensuram absolutam revocata. 4to. 44 pp.   *Göttingen,* 1833

Nachrichten über d. magnet. Observatorium in Göttingen. (*Götting. gelehrt. Anzeig.* 1834.)   *Göttingen,* 1834

    Note.—Gauss and Weber (*printed* 1838), *p.* 15, *say :* " *Die erste öffentliche Erwähnung dieser Versuche* (*i.e. application of their single connexion, &c. to telegraphic purposes*) *findet man in den Gött. gel. Anz.* 1834. *S.* 1273, *and Vergl. Schumacher's Jahrbuch fur* 1836."

Intensitas vis magneticæ terrestris ad mensuram absolutam revocata. 4to. 44 pp.   *Gottingen,* 1834

    Note.—Poggendorff, i. 854, *says, In Comment. recent. Soc. Gott.* viii. 1832-37, (*and*) *deutsch in Poggend. Ann.* xxviii.

Erdmagnetismus und Erdmagnetometer. (*In Schumacher's Jahrbuch,* 1836.)   1836

" Manche Ergänzungungen zu dem bereits beschriebenen Systeme (Telegraphie)." (*Sitzung d. Gesellschaft d. Wissen.* 19 *September* 1837.)   *Göttingen,* 1837

Vide also Gauss and Weber, Resultate for 1837, p. 15, &c. for an "Aufsatz" containing the principal contents of the " öffentlicher Vorlesungen" in the Sitzung, September 19, 1837.

†   Misura assoluta dell' intensità della forza magnetica terrestre, Memoria tradotta e commentata da Paolo Frisiani. 4to. 197 pp. 1 plate. (*Forms the Primo Supplemento alle Effemeridi astronomiche di Milano.*)   *Milano,* 1838

Ein neues Hülfsmittel für die magnetischen Beobachtungen. 4to.   *Göttingen,* 1838

Gauss, Karl Friedrich.—*continued.*
Theorie d. Erdmagnetismus vergliechen mit A. Erman's Beobachtungen.
(*Astr. Nachr.* xix. 1842.) 1842

† Zum Gedächtniss. Von W. S. v. Waltershausen. 4to. 108 pp. *Leipzig*, 1856
Seine sämmtl. Werke, her ausgegeben von d. Königl. Gesellsch. d. Wiss. in
Göttingen werden. 7 Bde. 4to. umfassen Erschienen. Bd. v. Mathemat.
Physik. . . . 4to. 1863
(*Vide* also Fusinieri.)

† **Gauss und Weber.** Resultate aus den Beobachtungen des Magnetischen Vereins
im Jahre, 1836. Idem im Jahre, 1837. Idem im Jahre, 1838.
Leipzig, 1837-8-9

† Resultate aus den Beobachtungen des magnetischen Vereins in den Jahren,
1836-7-8-9-40-41 Jährlich. 6 vols. 8vo. and 3 vols. Atls. in 4to.
Leipzig, 1837-43

† "Nach den Elementen der Theorie entworfen. Supplement zu den Resultaten
aus den Beobachtungen des magnetischen Vereins unter Mitwerkung
von C. W. B. Goldschmidt. 36 pp. 18 maps and 4 tables.
Leipzig, 1840
(*Vide* also Abria.)

Gauteron. On Tourbillon. 4to. (*Montpellier Acad.*) *Montpellier*, 1778

Gautherot, N. Recherches sur le Galvanisme. (*Méms. des Soc. Savantes et
Littér.* i. p. 471.) 1802 (?)
"Rapport fait à l'Institut, le 21 fructidor d. 1802 par Fourcroy et Vanquelin."
(*On 5 Mem. of Gautherot.*) *Pau*, 1802

Mémoire sur le fluide galvanique. (*Méms. des Soc. Savantes et Littér.* i. pp.
164 and 168.)

† Adhesion of the ends of two wires in contact with the upper and lower ends of
the Pile in 1801. Notice translated from the Ann. de Chim. xxxix. 209.
8vo. 1 p. (*Phil. Mag. or Annals,* iv. 458.) *London*, 1828

† **Gautherot and Nauche.** Two Experiments in Galvanism (by Gautherot). Facts
in regard to the application of Galvanism in cases of Hemiplegia (by Nauche).
8vo. 3 pp. (*Phil. Mag.* xv. 368.) *London*, 1803

Gauthier (or **Gautier**), J. L. Dissertatio inauguralis de irritabilitatis notione,
naturâ, et morbis. 8vo. 190 pp. *Hales*, 1793

Gautier, J. Mémoire sur l'aimant. 8vo. (*Poggend.* i. 958, *says in Mém. Soc.
Nancy,* ii.) *Nancy.*

Gavard, Capt. Notice sur le Diagraphe. 8vo. 4 folding plates. *Paris*, 1839

† **Gavarret, T.** Lois générales de l'électricité dynamique. 4to. 151 pp.
Paris, 1843

† Traité de l'électricité. Tom. i. 12mo. 595 pp. *Paris*, 1857
† Traité de l'électricité. Tom. ii. 12mo. 604 pp. *Paris*, 1858
† Lehrbuch der Elektricität. Deutsch bearbeitet von Dr. R. Arendt, Erster
Th. 8vo. 463 pp. *Leipzig*, 1859
† Lehrbuch der Elektricität. Deutsch bearbeitet von Dr. R. Arendt. Zweiter
Th. 8vo. 524 pp. *Leipzig*, 1860
† Télégraphie électrique. 12mo. 428 pp. *Paris*, 1861
† Telegrafia elettrica di J. Gavarret con aggiunte tratte da E. E. Blavier,
traduzione Italiana sull' originale Francese del 1861 con preliminari e note
di Ales. Magrini. 8vo. 395 pp. *Milano*, 1862
(*Vide* also Abria.)

Gay. System of relation (electric) between Stations and Trains in motion, &c.
(*From Du Moncel, Exposé,* p. 175.)

196   GAY—GAZ

Gay-Lussac, Louis Joseph. *Born December 6, 1778, at St. Léonard, Limousin ;
died May 9, 1850, at Paris.*

† Account of his last aerostatic ascent, given to the . . National Institute. 8vo.
3 pp. (*Phil. Mag.* xx. 83.)   *London,* 1804

† Account of an Aerostatic Voyage performed by M. Gay-Lussac, on the 29th of
Fructidor, year 12, and read in the National Institute, Vendémiaire 9th,
year 13. 8vo. 8 pp. (*Phil. Mag.* xxi. 220.)   *London,* 1805

† Instructions sur les Paratonnerres . . . suivies des Rapports . . . à l'Institut
. . et à l'Acad. des Sciences. Fol. 39 pp. 1 plate. (*Vide full account of this
work at Comité des Fortifications.*)   *Paris,* 1808

† Instructions sur les Paratonnerres. (Signé) Poisson, Lefevre, Gineau,
Dulong, Fresnel, et Gay-Lussac rapporteur. 4to. 31 pp. 2 plates.
                                                               Paris, 1824
† Instruction sur les paratonnerres. 8vo. 51 pp. 2 plates.   *Paris,* 1824

† Discours de M. Becquerel (sur) lui . . . 4to. 5 pp. (*Institut. National Acad.
des Sciences.*)   *Paris*

† **Gay-Lussac and Biot.** Account of an Aerostatic Voyage performed by Messrs.
Gay-Lussac and Biot. Read in the French National Institute, August 27,
1804. 8vo. 9 pp. (*Phil. Mag.* xix. 371.)   *London,* 1804

† **Gay-Lussac et Humboldt.** Mémoire sur la Torpille. 8vo.   *Paris*

  Observations sur l'intensité et l'inclinaison des forces magnétiques, faites en
  France, Suisse, Italie et Allemagne. (*Mém. d'Arcueil,* vol. i. 1807.
                                                               Paris, 1807
† **Gay-Lussac et Pouillet.** Instruction sur la Paratonnerres adoptée par l'Acad.
des Sciences. 12mo. 130 pp. Cuts.   *Paris,* 1855

† **Gay-Lussac et Thenard.** Decomposition of the Alkalies. Repetition of Davy's
experiments, &c. 8vo. 2 pp. (*Phil. Mag.* xxxii. 88.)   *London,* 1809

† Recherches physico-chymiques faites sur la Pile. 2 vols. 8vo. *Paris,* 1811
† Recherches physico-chimiques.   *Paris,* 1811

† **Gay-Lussac, &c.** Anleitung zur verfertigung und Benutzung der Blitzableiter
. . . aus dem Französischen übersezt. 43 pp. 2 plates.   *Strasburg,* 1824

  (*Vide* also Poisson and others.)

Gazette Salutaire. Contains many writings on Medical Electricity, from 1776 to
1782, and perhaps more.

† **Gazzaniga,** C. L. Di un miglioramento da farsi alle macchine elettriche. . . .
Lettera al Direttore. (*Con Nota del Fusinieri.*) 4to. 4 pp. (*Ann. del. Reg.
Lomb.-Veneto,* ii. 359.)   *Padova,* 1832

† Saggio di esperienze elettriche fatte coll'uso della macchina elettrica a collettori
conjugati. 4to. 36 pp. 1 plate. (*Ann. del. Reg. Lomb.-Veneto,* iii. p. 311.)
                                                               Padova, 1833

† Sunto di sperienze . . . elettriche magnetiche. 8vo. 5 pp. (*Bibl. Ital.* lxxvi.
159. *This forms part of an article at p. 156, headed "Fata-morgana . . .
Sunto di sperienze ottiche, elettriche magnetiche."*)   *Milan,* 1834

† Appendice al Saggio d² esperienze elettriche fatte coll'uso della macchina
elettrica a collettori conjugati, inserito nel bimestre v. e vi. 1833, degli
Annali delle Scienze del Regno Lombardo-Veneto. 4to. 14 pp. 1 plate.
(*Ann. . . del. Reg. Lomb.-Veneto,* vi. 79.)   *Padova,* 1836

† Sul magnetismo temporario, esperimenti del . . Cesare Gazzaniga. 4to. 7 pp.
1 plate. (*Ann. del Reg. Lomb.-Veneto,* vi. 287.)   *Padova,* 1836

† Nuove esperienze elettriche. 4to. 25 pp. (*Ann. del Reg. Lomb.-Veneto,* vii. 215.)
                                                               Padova, 1837

† Delle magneti penetrative permanenti. (*To be continued.*) 4to. 22 pp. 1
plate. (*Ann. del Reg. Lomb.-Veneto,* viii. 263.)   *Padova,* 1838

Gazzaniga, C. L.—*continued.*

Delle magneti penetrative permanenti. Della misura della forza della magnete dedotta dai movimenti di translazione. Continuazione. Vedi la prima parte, p. 263, vol. viii. Annali. 4to. 12 pp. (*Fusinieri's Annali,* ix. 26.)
Venezia, 1839

Continuazione della Memoria sulle Magneti penetrative permanenti. 4to. 10 pp. 1 plate. (*Fusinieri's Annali,* ix. 66.) *Venezia,* 1839

† Sopra il magnetismo terrestre. Nota. Continuazione e fine alla Nota. 4to. 21 pp. 3 tables. *Ann. del Reg. Lomb.-Veneto,* xii. 207 and 245.)
Vicenza, 1842

† Elettro-statica. 4to. 15 pp. 1 plate. (*Article of the Encyclopedia Italiana, Fascicolo* xxvi.)

† Della Corrente. 4to. 20 pp. 2 plates. (*Articolo estratto dall' Enciclopedia Italiana, Fascicolo* xviii. et xx.)

Gazzeri, G. Aggiunte ai Saggi di naturali Esperienze fatte nell' Accad. del Cimento, p. 4. *Firenze,* 1841

Gazzoni, L. *Died Venice,* 1592. Poggendorff, i. 864, says : "Aus einer unedirten Schrift desselben, soll nach Cabeo's Angabe, G. B. Porta seine Kentnisse vom Magnet entnommen haben."

Cabeus in his Philosophia Magnetica (fol. 1629), *Preface, calls him Garzonius, and mentions Porta in the manner indicated, and others also.*

Gehlen. A. F. Neues allgem. Journ. d. Chemie. 8vo. 9 vols. (*Journ. für Chemie Physik u. Mineralogie.*) *Berlin,* 1803-6

Uber d. Misch d. Meteorsteine. 8vo. (*Schweigger's Journ.* vi. 1812.) 1812

Uber d. Elektrochem. System u. s. w. 8vo. ·(*Schweigger's Journ.* xii. 1814.)
Nürnberg, 1814

Über d. Reduct. d. Metalle durch einander u. die Lichterscheinungen dabei. 8vo. (*Schweigg. Journ.* xx. 1817.) *Nürnberg,* 1817

Gehler, Johann Samuel Traugott. *Born Nov.* 1, 1751, *at Gorlitz ; died Oct.* 16, 1795, *at Leipzig.*

Phys. Wörterbuch . . 6 vols. 8vo. *Leipzig,* 1787-95

Phys. Wörterbuch od Vers. e. Erklär. des Vornehmst. a. d. Naturl. 6 vols. 8vo.
Leipzig, 1799-1801

Physikal. Wörterbuch. neu bearb. v. Brandes, Gmelin, Horner, Littrow, Muncke, Pfaff. 11 vols. 8vo. *Leipzig,* 1825-45

Vol. i.	1825	From Heinsius, vii. 267.
Vol. ii.	1826	
Vol. iii.	1827	
Vol. iv. Part 1	1827	
„ „ 2	1828	
Vol. v. Part 1	1829	
„ „ 2	1830	
Vol. vi. Part 1	1831	From Schultz, i. 262.
„ „ 2		
Vol. vii. Part 1	1833	
„ „ 2	1834	
Vol. viii.	1836 1837	
Vol. ix. Part 1	1838	
„ „ 2	1839	
„ „ 3	1840	From Schultz, ii. 304.
Vol. x. Part 1	1841	
„ „ 2	1842	
„ „ 3	1844	From Schultz, iii. 278.
Vol. xi.	1845	

198   GEH—GEN

Gehler, Johann Samuel Traugott—*continued.*
Dass xi. Bd. Sach u. Namen-register, mit ergänzten Zusätzen v. Muncke, &c.
(*From Poggendorff,* ii. 238.)
Neues Wörterbuch.
Geiger, C. F.   Abhandlung über den Galvanismus und dessen Anwendung. 8vo.
Leipzig, 1803
Geiger, Philipp Lorenz.   *Born Aug.* 30, 1785, *at Freinsheim ; died Jan.* 19, 1836, *at Heidelberg.*
Dissertation sur le Galvanisme et son application à la Médecine. 8vo. 32 pp.
1802 *or* 1803
Geisler, J. G.   (*Vide* Gardini, 1793.)
Geissler of Bonn.   Tubes for Luminous effects of Ruhmkorf.   (*From Carl.*)
Gellert, C. E.   De densitate mixtorum ex metallis et semi-metallis factorum. 4to.
Petropol, 1751
(*From Van Swinden, " Analogia,"* p. 26, vol. i., *who refers to* tom. xiii.
p. 382 *of the Commentar. Acad. Petrop. for Gellert's Experiments,* 15 *and*
16. *Reuss says,* p. 392, *" Gellert, Experiments on the action of a small
Magnet being greater than that of a larger on a mixture of Iron and other
metals : and property of Antimony in weakening the attraction when mixed
with Iron.*)

† Gellibrand, H.   A Discourse mathematical on the Variation of the Magnetic
Needle, together with the admirable Diminution lately discovered.   Sm. 4to.
22 pp.   *London,* 1635
(*Discovery of secular variation of declination.*)   (*Very scarce.*)

Gemignani, L.   Della vitale elettromozione.   8vo.   *Lucca*

† Gemminger, M.   Electrisches Organ von Mormyrus und Schwanzskelet. von Eryx.
(*Dissertatio Inauguralis.*)   1 plate, 15 pp. 8vo.   *Munchen,* 1847

Geneva Archives des Sciences physiques et naturelles.   Nouvelle Période.
9 vols. 8vo.   *Geneva,* 1858-60

†   Archives de l'Electricité.   Supplément à la Bibliothèque Universelle de
Genève.   5 vols. 8vo.   *Geneve,* 1841-45

Archives des Sciences physiques.   33 vols. 8vo.   *Geneve,* 1846-56

Geneva Bibliotheque.   Bibliothèque Britannique en 2 séries (*i.e.* divisions)
intitulées (1) Littérature, (2) Sciences et Arts.   1e Série. 144 vols. 8vo.
60 vols. de Littérature, 60 de Sciences et Arts, 20 d'Agriculture, 4 de
Tables = 144.   *Geneva,* 1796-1815

Bibliothèque Universelle des Sciences, Belles-Lettres, et Arts faisant suite à
la Bibliothèque Britannique redigée à Genève par les auteurs de ce dernier
recueil.   2de Série. 136 vols. 8vo.   60 vols. de Litt., 60 de Sciences
et Arts, 14 d'Agriculture, 2 de Tables = 136.   (*From Cat. des Livres de la
Société de Lecture de Genève* p. 530.)   *Geneve,* 1816-35

Bibliothèque Universelle de Genève.   Nouvelle Série.   Third series. 39 vols.
8vo.   *Geneva,* 1836-45

Bibliothèque Universelle.   Littérature et Sciences.   Quatrième Série.   Fourth
series.   33 vols. 8vo. (*continued after* 1856).   *Geneve,* 1846-56

Catalogue de la Bibliothèque Publique de Genève rédigée par Louis Vaucher.
8vo. 948 pp.   In two parts.   *Geneve,* 1834

MSS. Catalogue de la Bibliothèque Publique de Genève   Fol.   *Geneva.*

Geneva Institution.   Bulletin de l'Institut National Genévois. 8vo.
Geneva, 1853
Mémoires de l'Institut National Genévois. 4to. (*From Cat. des Livres de la
Société de Lecture de Geneve.   Mémoires continued after* 1853)   *Geneve,* 1853

Geneva Société de Lecture. Catalogue des livres de la Société de Lecture de Genève. (Première partie) with Supplement. 8vo. 560 pp. and 90 pp. Tables. *Geneve, 1839*

Catalogue des livres de la Société de Lecture de Genève. Seconde partie. 8vo. 432 pp. and 55 pp. Tables. *Geneve, 1857*

Geneva Société de la Physique. Mémoires de la Société de la Physique et d'Histoire naturelle de Genève. 4to. Plates. (*The thirteenth vol. printed in* 1854.) *Geneve, 1821*

† **Genevois, J. P.** Théorie de la Grêle et moyens de la prévenir. 4to. 1 plate. 92 pp. et Table. *Turin, 1838*

Teoria della Grandine e mezzi sperimentali dei prevenirla. 8vo. 108 pp. *Torino, 1840*

Genhart, R. Ætiologia de Magnetismo naturali, &c. 4to. 1 plate. 30 pp. *Turici, 1803*

Genoa Acad. Memorie dell' Accad. delle Scienze e Belle Arte. 4to. *Genova, 1809*

Gentil. Voyage dans les Mers de l'Inde, &c. (Tom. ii. p. 659, *Ascending Lightning*. *Vide* Bertholon, p. 160.)

Experiments, &c. in Vienna in 1853 on Contrary Currents in the same wire, exhibited to the Vienna Academy. Application to the line from Vienna to Linz whilst the question was under discussion in Paris in 1853-4, at French Institute. He was Director of the Austrian Telegraphs. (*From Magrini's Notizie Istoriche, &c.*)

Gentleman's Magazine. *London.*

Geoffroy Saint-Hilaire, E. Sur l'Anatomie comparée des organes électriques de la raie torpille, du gymnote étourdissant, et du silure trembleur. (*Annal. du Muséum*, vol. i. An. xi.) ; *Paris, 1803* (?)

† Memoir on the comparative anatomy of the Electric Organs of the Torpedo, the Gymnotus electricus, and the Silurus electricus. (*Translation from Annals du Muséum National*, No. 5.) 8vo. 11 pp. 1 plate. (*Phil. Mag.* xv. 126.) *London, 1803*

† **Georg, F.** Den Elektromagnetiske Telegraph Popular fremstilling med. sœrligt Hensyn til Damer. 24 pp. 8vo. *Kjöbenhavn, 1857*

Geographical Society of London. (*Vide* Royal Geographical Society.)

Gerard, F. Méthode de déterminer la différente capacité des métaux d'accumuler le fluide électrique. 8vo. (*Journal du Chim. de Van Mons*, No. 4, p. 30.) *Bruxelles*

Gerbi, R. Corso di Fisica. 5 vols. 8vo. 1,800 pp. 18 plates. *Pisa, 1823-25*

† **Gerboin, A. C.** Recherches expérimentales sur un nouveau mode de l'Action élect. 8vo. 1 plate. 358 pp. *Strasbourg, 1808*

Quelques expériences galvaniques. (*Mém. des Soc. Savantes et Littér.* ii. 199.) *Paris*

Gerdes, F. Beskrifning pä et skydrag, observeradt uti Mälaren och Mariæbergs sundet den Aug. 10, 1790. (*Vetensk Acad. Nya Handl.* A. 1790, S. 208) *Stockholm, 1790*

Beschreibung eines Wolken-Zuges, der in Mäler und Mariabergs, Sunde ist beobachtet worden den 10 Aug. 1790. (*In Neue Schwedische Akad. Abhandl.* J. 1790, S. 196.) *Stockholm, 1790*

Gerhard, C. A. De l'action de l'électricité sur le corps humain et de son usage dans les paralysies. 4to. (*Mém. de Berlin*, 1772, p. 141.) *Berlin, 1772*

Sur les principes de la Tourmaline. 4to. (*Mém. de Berlin*, 1777.) *Berlin, 1777*

Observations sur la physique. (*From Van Troostwyk and Krayenhoff.*) 1779 (*Note.—Poggendorff,* i. 881, has " *Triga dissertationum physico-medicarum.* 8vo. *Berol,* 1763.")

Gerke, Francis Clemens.   Pract. Telegraphist od. d. electro-magnet.   Telegraph nach d. Morsiche Systeme.   8vo. 144 pp.   *Hamburg, 1851*

Der Electro-Magnetismus als Maschinen-Triebkraft.   Versuch einer Lösung des Problems.   8vo. 16 pp.   1 plate.   *Hamburg, 1857*

† **Gerling,** Becker, und **Andera** (?)   Graphische Darstellung der magnetischen Declination zu Marburg auf u. Jahre 1843.   Fol. 7 plates.   *Marburg*

Gerling, C. L.   Beschreibung eines neuen Hütchens zur Aufhängung der Magnetnadel in Compässen.   (*From Lamont, Handbook,* p. 433, *in Schr. d. Gesellsch. zu Bejörderung d. gesammt. Naturwiss. in Marburg,* vol. i.)   *Marburg, 1823*

† Corresponding changes of the diurnal magnetic declination observed at Marburg (zu Hesse.)   Latitude 50° 48′ 47″ N.; Longitude 8° 46′ 25″ E.   4to. 28 pp.   *Washington, 1854*

Gersdorff, A. T. Von.   Ueber einige elektrische Versuche.   8vo. (*Neue Schr. der Gesellsch. Naturf. Freunde zu Berlin,* ii. 247.)   *Berlin, 1799*

† Über meine Beobachtungen der atmosphärischen Elektricität zu Messersdorf in der Oberlausitz; nebst einigen daraus gezogenen Resultaten.   4to. 108 pp. 15 plates.   *Görlitz, 1802*

Gersdorf und **Knebel.**   Anzeiger d. nothwendigsten Verhaltungs regeln bey nahen Gewittern, &c. . .   1st edition. 8vo.   *Görlitz, 1798*

Anzeige d. nothwendigsten Verhaltungs regeln bey nahen Gewittern, &c. . .   2nd edition. 8vo.   1800

† **Gerspach,** E.   Histoire administrative de la 'Télégraphie aérienne en France.   8vo. 116 pp.   (*Extrait des Anns. Télégraphiques.*")   *Paris, 1861*

Gesellschaft Deutscher Naturforcher und Aerzte.   Amtliche Berichte über d. Versammlungen de Gesellschaft. . . 4to.

Gesellschaft Naturforschender Freunde.   (*Vide* Berlin Society.)

Gesellschaft Med. in Mainz.   Galvanische u. elektrische Versuche au Menschen u. Thier Körpern, &c.   4to.   *Frankfort-on-Maine, 1804*   (*The society was Kircher, Molitor, Ruf, Skack, jun., Wenzel and Wittmann.*)

Gesner, Conrad.   De raris et admirandis herbis, quæ sive quod noctu lucent, sive alias ob causas Lunariæ nominantur, Commentariolus.   Et obiter de aliis etiam rebus quæ in tenebris lucent.   4to.   *Tiguri (Zurich),* 1555
De lunariis herbis et rebus noctu lucentibus.   8vo.   1669

Gessner, J. Mathias.   De electro veterum.   (*Comment. Soc. Götting.* iii. 67, 1753.)   *Göttingen, 1753*

† **Gete,** C. J.   Anmerkungen von einem seltsamen Verhalten des Seecompasses an einer Stelle in den nylandischen scheeren.   8vo. 5 pp. 1 map. (*K. Schwed. Akad.* Abh. xii. 298.)   *Hamb. u. Leipzig,* 1751

† **Geutebruck,** J. G.   Erörterungen und Wünche in Hinsicht auf Blitzableiter.   Zwei Vorlesungen in der Naturforschenden Gesellschaft der Osterlands.   8vo. 48 pp. 1 plate.   *Leipzig, 1828*

Gherardi, G.   Ragionamento accademico sopra varie questioni fisico-elettriche.   4to. 24 pp.   *Montalboddo, 1810*

† **Gherardi,** S.   Tradne. di Demonferand, Manuel, con note ed *aggiunte* del traduttore.   8vo. (*Much additional matter.*)   *Bologna,* 1824

Alcune esperienze sopra le nuove correnti e le scintille magneto-elettriche.   Lettera al Nobili.   8vo. 17 pp.   *Firenze,* 1832

Dissertatio de aliquot experimentis recens physicum inventum Faradayi spectantibus, ac de præcipuis nonnullis voltianorum profluviorum (*i.e.* delle correnti del Volta) proprietatibus.   2 parts. 4to. 38 pp. and 1 errata. (*Ext. Nov. Act. Acad. Inst. Bonon.* Tom. ii.)   *Bonon.* 1840

(*In a foot-note at* p. 1 *is said* " *Priorem hanc Dissertationis partem Academiæ iegijam inde ab ejus conventu diei* 5 *Aprilis* 1832.)

Gherardi, S.—*continued.*

Commentarium de novo quoddam apparatu magneto-electrico. Lect. 17 Jan. 1833. 4to. 26 pp. 1 plate. (*Ext. Nov. Act. Acad. Inst. Bonon.* tom. ii.)
1840 ?

Disquisitio de magnete electrico recentiori ad usus chimicos et de profluviis magnetico-electricis ad hoc aptioribus. 4to. 26 pp. 1 plate. (*Ext. Act. Acad. Inst. Bonon.* tom. iii.)
1840 ?

De quadam appendice ad galvanometrum multiplicans, et de ejus usu in profluviis variis, ac præcipue Faradyeis expendendis: habita . . 5 Mart. 1835. 4to. 42 pp. (*Ext. Act. Acad. Inst. Bon.* tom. iii.)
1840?

Rapporto sui manoscritti del . . Galvani, legati all' Accad. . . di Bologna . . per Aldini . . con Appendice. . . su di alcuni passi risguardanti l'elett. animale. . . Dalla Collezione dell' opere del. . Galvani pubb. dall' Accad. . . dell' Ist. di Bologna. 4to. 106 pp.
Bologna, 1840

Osservazioni sopra alcune esperienze elettro-magnetiche di Nobili.

De electricitate et magnetismo animadversiones variæ. (*From Lamont.*)
Bononiæ, 1841

Ghirlanda. Intorno agli effetti del galvanismo nell' emiplegia con perdita di memoria. (Read) 7 Maggio 1835. (*Treviso Athenæum,* vol. v. p. v. *in Elenco delle letture. . . fatte rol.* . 1833-34, *a tutto* 1843-44.)

† **Ghisi, L. A.** Telegrafia elettrica, ossia Descrizione dei Telegrafi elettro-magnetici loro modo di agire, &c. . . Seconda Edn. ampliata e corretta. 4to. 76 pp. 1 plate (lith.)
Milano, 1850

† Descrizione di due nuovi telegrafi elettrici, aggiunti nella seconda ediz. alla edizione prima della lezione sulla telegrafia elettrica. 8vo. 16 pp. 1 plate.
Milano, 1850

† Nuove teorie sulla formazione della Grandine, sull' origine dell' elettricità nella pila, esposte in due lezioni. 8vo. 31 pp. 1 plate.
Lodi, 1869

Giessman, Fr. Über d. Steinregen. (*From Ersch* 1828, p. 260.)
Wien, 1803

Giquet. Sur l'electrum d'Homère. (*Revue archéol.* 15 *Juillet* 1859, 16e *année,* p. 235-41.)
Paris, 1859

† **Gilbert, Davies.** On the luminous Belt of September the 29th. 8vo. 2 pp. (*Phil. Mag. or Annals,* iv. 453.)
London, 1828

† **Gilberti, Guil.** De magnete magnetisque corporibus, et de magno magnete tellure ; Physiologia nova, plurimis et argumentis et experimentis demonstrata. Ed. prima. Folio (sm.) 254 pp.
Londini, 1600
(*Kuhn,* p. 7, *refers to an edition printed at Amsterdam,* fol. 1600.)

† Tractatus ; sive Physiologia nova de Magnete. . . Omnia nunc . . recognita . . opera et studio Wolfgangi Lochmans. 2nd edition. 4to. 232 pp. 12 plates.
Sedini, 1628

† Tractatus ; sive Physiologia nova, de Magnete . . opera et studio D. Wolfgangi Lochmani. 3rd edition. 4to. 232 pp. 12 plates.
Sedini, 1633
(*This seems to be a mere reprint of the edition of* 1628.)

De mundo nostro sub lunario Philosophia nova. Opus posthumum, ab auctoris fratre collectum. . . 4to. 316 pp. 1 plate. (*The 3rd and 4th books are on Meteorology, on Magnetism in reference to the Motion of the Moon, &c.*)
Amsterdam, 1651

† **Gilbert, Isaiah.** A method of affording relief to persons injured by Lightning. In a letter from Mr. Isaiah Gilbert to the Rev. Mr. Steele. 8vo. 2 pp. (*Phil. Mag.* xvii. 306.)
London, 1803

Gilbert, L. W. Annalen der Physik. 30 vols. 8vo.
Leipzig, 1797-1808
Titel als Neue Folge. 76 vols. and 1 vol. Register.
1809-24
(*Fortgesetzt in Poggendorf's Annalen der Phys. und Chemie. Herausgegeben zu Berlin.*)

U

Gilbert, L. W.—continued.

† Neue Versuche mit trockner elekt. Säule (enthaltend): (1) Zamboni. Brief
  un Prof. Pictet (p. 182). (2) Jager Säule (p. 187), Buzengeiger, Uhr (188),
  u. Ramis (p. 189). (3) Behrens und Bohnenberger's Electrometer (p. 90).
  (4) Auszug aus einem Schreiben d. von Jager an Prof. Gilbert (p. 95).
  (Gilbert's Ann. et Phys. li. 182.)   Leipzig, 1815

Untersuch. über d. Einwirk d. geschloss. elektrischen Kreises auf d. Magnetnadel. 8vo. (Gilbert's Ann. lxvi. 1820.)   Leipzig, 1820

† Gilii, F. L. Memoria fisica sopra il Fulmine Caduto in Roma sulla casa dei P. P.
  Filippini di S. Maria in Vallicella, detta comunemente la chiesa nuova nel
  di 26 Novembre 1781. 12mo. 28 pp.   Roma, 1782

† Breve ragionamento sopra il conduttore . . innalzato sulla Basilica di Sta.
  Maria degli Angioli. . . 8vo. 22 pp.   Roma, 1793

† Gillet-de-Laument. Account of an Aerolite which fell in Moravia, and a mass
  of Native Iron which fell in Siberia. 8vo. 2 pp. (Phil. Mag. xlvii. 174.)
                                                               London, 1816

Gillet et Saintard. Sur un nouveau système de Télégraphie électrique. 4to.
  (Comptes Rendus, xx. 1573.)   Paris, 1845

† Gillies, &c. Account of some Electrical Phenomena, recently communicated to the
  Royal Society of Edinburgh. 8vo. 3 pp. (Phil. Mag. xlv. 69.) London, 1851

Gilliss. J. M. Magnetical and Meteorological Observations made at Washington,
  &c. 8vo.   Washington, 1845

† The U. S. Naval Astronomical Expedition to the Southern hemisphere during
  the years 1849, 1850, 1851, 1852. Vol. vi. Magnetical and Meteorological
  Observations under the direction of J. M. Gilliss, Superintendent. 4to.
  420 pp.   Washington, 1856

Gilly und Eytelwein. Kurze Anleitung, auf welche Art Blitzableiter an d.
  Gebäuden anzubringen sind. 8vo. 3 plates.   Berlin, 1798

Kurze Anleit. &c. (Blitzableiter) 8vo. 3 plates.   Berlin, 1802

Kurze Anleit auf welche Art. Blitzableiter an d. Gebäuden anzubringen sind.
  3e Aufl. 8vo.   Berlin, 1819

Gilpin, G. Observations on the variation and on the dip of the Magnetic needle,
  made at the apartments of the Royal Society, between 1786 and 1805,
  inclusive. 4to. (Phil. Trans. 1806.)   London, 1806

Ginge, A. Om nordly sets indflydelse paa magnetnaalens declination. (Skrifter
  det Kiöbenhavenske Selsk. Nye Saml. iii. 531.)   Kiobenhavn

Gintl, W. Der transportable Telegraph für Eisenbahnzüge. 8vo. 8 pp. 3 plates
  (Aus den Sitzungsberichten der math.-naturw. Cl. der k. Akad. der Wiss.
  Wien vi. Bd. iv. Hest. p. 460.)   Wien, 1851

Der Elektro-Chemische Schreib-Telegraph auf die gleichzeitige Gegen-Corres-
  pondenz an einer Drahtleitung angewendet. 8vo. 18 pp. 6 plates. (Aus
  den Sitzungsberichten 1854 der k. Akad. der Wissenschaften.) Wien, 1855

† Giobert, G. A. Articolo di lettera al Brugnatelli sopra diversi argomenti Fisico-
  Chimici. 8vo. 4 pp. (Gior. Fis. Med. i. 188.)   Pavia, 1792

Invention d'une pile de charge, ou batterie galvanique, et isolement des fonctions de la pile. 8vo. (Bibliot. Ital. iv. 121. Extrait du Jl. de Chim. de
  Van Mons, xiv. 200.)   Turin, 1804

† Recherches sur l'action que le fluide galvanique exerce sur différents fluides
  aériformes. (Lues le 6 prairial, an. 11.) 4to. 13 pp. (Mémoires de Turin,
  Anns. 12 et 13.)   Turin, 1805
  (Vide also Julio and others.)

Gioberti. (Vide Michelotti and Gioberti.)

Gioja, Flavio. Pilot of Amalfi; falsely called Inventor of the Compass in 1302 or
  1303. (Vide Klaproth. Letter to Humboldt, Paris, 1834.)

Giordano, G. Batoreomotro. 4to. *Napoli, 1863*

Giorgi, Eus. Über Blizableiter. (*Vide Majocchi Annali,* viii. 178.)

Giorgini, G. Elettro doratura Italiana. (*Indicatore economico* 17 *Maggio.*)
Modena, 1844

Giornale Agrario della Toscana ed Atti dei Georgofili. *Firenze, 1827*

Giornale Enciclopedico di Vicenza. (The title from 1783 to 1784 is *Nuovo Gle. Enciclop.* Contains works of Toaldo, &c.) *Vicenza, 1779 to 1784*

Giornale di Fisica. . . (*Vide* Brugnatelli, L. and others, 1808-17, and Brugnatelli, G. and others, 1818-27.)

Giornale di Pavia. In continuation by Brugnatelli, G. and Configliachi.

Giornale dell' Italiana Letteratura. 66 vols. 8vo. (*Vide* Da Rio for some Electrical Articles.) *Padova, 1802-28*

Giornale Letterario Scientifico Modenese. (Often referred to by Grimelli.)

Giornale Letterario ossia Progressi dello Spirito umano nelle Scienze, &c. . . Contains some Articles on Electricity. 4to. (Anon., Toaldo, Garducci printed in 1783.) 1783

Giornale dei Letterati d'Italia. Zeno, Editor. 12mo. *Firenze, 1710*

(*The Editor of the " Storia Giornale " says at* p. 261 : " In tom. iv. part iv. art. 6, in tom. v. part i. art. 2, and part ii. art. 6 in 1748, *I Giornalisti danci l'Istoria della Elettricità.*")

Giornale Toscano di Scienze, Med. Fis. e Nat. *Pisa, 1840*

† **Giovene, G. M.** Osservazioni elettrico-atmosferiche e barometriche, insieme paragonate Presentata. . . (17 Pratile, Ann. vii.) 5 Guigno, 1798. 4to. 34 pp. (*Mem. Soc. Ital.* viii. part i. 85 pp.) *Modena, 1799*

† A ppendice alle osservazioni elettrico-atmosferiche e barometriche comparate. "Un mero accidente impedì che essa Appendice fosse colle Osservazioni Medesime dell' Autore nel tomo viii. Nota del Segretario." (*Ricuperata il di* 21 *Ottob.* 1801.) 4to. 7 pp. (*Mem. Soc. Ital.* ix. 438.) *Modena, 1802*

An Experiment on the natural production of Salts by means of natural Electromotors. (*Bibliot. Ital. in* vol. xli. p. 238.) 1826

Di alcuni fenomeni meteorol. della Puglia Peucezia. Memoria postuma ricevuta 22 Maggio, 1837. 4to. (*Ital. Soc. Mem.* xxii. p. 1, Pte. Fisica.)
Modena, 1841

† Elogio storico del Canonico Arciprete Giuseppe Maria Giovene scritto dall' A. Tripaldi. 4to. 33 pp. (*Mem. della Soc. Ital.* tom. xxii. p. 1.) *Modena, 1841*

† **Giquel, E.** Traité de déviation des Compas à bord des navires, renfermant de grandes quantités de fer. 8vo. 140 pp. 5 plates. *Paris, 1868*

Girarbon. Machine Electrophorique. (*From Du Moncel, who gives no title, date, &c.*)

† **Girardi, M.** Saggio di osservazioni anatomiche intorno agli organi elettrici della Torpedine. Al Gio. Gott. Walter, Berlino. 4to. 18 pp. 2 plates. (*Mem. Soc. Ital.* iii. 553.) *Verona, 1786*

Girardin. On an Electric Machine. "Zur pos. u. neg. Electricität." (*Nouv. de la Répub. des Lettres et des Arts,* 1779.) 1779

De l'application de l'Electricité aux . . . maladies. (*From Guitard.*)
Paris, 1823

Girou de Buzareingues, L. F. C. Influence des déboisements sur le système météorologique du Dép. de l'Aveyron. (*Feuille villageoise de l'Aveyron,* 1835.) *Aveyron, 1835*

† **Gisborne, Fr.** Telegraphic communication with India. 8vo. 16 pp.
London, 1868 ?

† **Gisler, Nils.** Beschreibung eines Glanzes in der Luft mit Donner, der in Westnorrlande den 13 Jan. 1763, wahrgenommen worden. 8vo. 3 pp. (*K. Schwed. Akad. Abh.* xxv. 65.) *Hamb. and Leipzig, 1763*

Giuli, G. Sull' influenza che sembrano avere le correnti elettriche per ristabilire la salute in alcune malattie. 8vo. *Bologna*, 1840

† Della influenza che sembrano avere le correnti elettriche per ristabilire la salute in alcune malattie, dietro l'uso dei bagni d'acqua salina, ed in ispecie di quelle di Monte Catini in Toscana. Mem. letta alla prima riunione degli Scienziati Italiani, in Pisa, Ottobre, 1839. 4to. 12 pp. (*Ann. del Reg. Lomb.-Veneto*, x. 30.) *Padova*, 1840

Nuovo metodo per scuoprire il ferro nelle acque minerali, anche in quantità minima. 8vo. (*Ital. Soc. Mem.* xxiii. (11). *Inserita nel* tom. vi. *dei Nuovi Annali delle Scienze naturali Bologna.*) *Bologna*, 1842

† **Giuli, G.**, e **Santi-Linari.** Lettera del Giuli al Fusinieri e Lettere del Santi-Linari al Giuli. 4to. 2 pp. (*Ann. del Reg. Lomb.-Veneto*, ix. 200.) *Venezia*, 1839
(*This last letter concerns many of his experiments in electro-magnetism, &c.*)

† **Giulio.** Sur les effets du fluide galvanique appliqué à différentes plantes .. Extrait d'un mémoire lu à la classe des sciences exactes de l'Acad. le 29 Pluviose, an. xi. 8vo. (*Biliot. Ital. (and Julio, &c.*), i. 28.) *Turin*, 1803

Rapport presénté à l'Acad. de Turin, sur le tête et le tronc de trois hommes. (*Volta's copy.*) 1802

† Rapport lu à l'Acad. de Turin, sur la puissance stimulante de l'électricité ordinaire et du galvanisme. 8vo. 6 pp. (*Bibliot. Ital.* iv. 25.) *Turin*, 1804

† **Giulio e Rossi.** Estratto di alcune esperienze fatte dai Sig. Giulio e Rossi. Movimento del Cuore di animali. (Per mezzi galvanici.) 8vo. 6 pp. (*Gior. Fis. Med. di Brugnatelli*, tom. i. dell an. 1793, p. 82.) *Pavia*, 1793

De excitabilitate contractionum in partibus musculosis involuntariis ope animalis electricitatis. 4to. (*Mem. d. Turin*, xi. 34, pro 1792-99.)
(*Vide* also Julio, &c. and Vassalli-Eandi.) *Turin*, 1800

Glanvill, Josh. Sceptis scientifica; or Confest Ignorance the Way to Science. (*From T. F. Cooke's Authorship of the Practical Elect. Tel.* p. 75.) 1665
Note.—This work contains a very evident derivation from Faminius Strada's mere poetical fantasy, relative to communication between two friends, at a distance from each other, by means of magnetic needles, &c., having no conducting connexion.

Gloesener. Mémoire sur une Horloge magnéto-électrique, sur les Télégraphes électriques, et sur un appareil magnéto-électriques. 4to. (*Comptes Rendus*, xxvi. 336, and xxvii. 23.) *Paris*, 1848

Elektro-magnetischer Chronograph. 1849

† Recherches sur la Télégraphie électrique. 4to. 124 pp. 12 plates. *Liége*, 1853

Télégraphe chimique. 8vo. *Liége*, 1853

Télégraphie à aiguille perfectionnée. 8vo. 1 plate. 1857

† Traité général des applications de l'Electricité. Tom. i. 8vo. 18 plates. *Paris et Liége*, 1861

Traité général des applications de l'Electricité. 2 vols. 8vo. Plates. *Paris*,1863

"**Glosterian.**" (*Vide* Anon. Meteorolog. Phen. various.) 1823

Glover, J. Over d. oosprong d. vreemde delfstoffen en h. regenen v. steenen, en proeve eener geschid. d. menschh. naar de rede. 8vo. 94 pp. *Hard.* 1805

† Electrical Shocks from a Cat. 8vo. 2 pp. (*Phil. Mag.* lx. 467.) *London*, 1822

† **Gmelin,** Christian Gottlob. *Born October* 12, 1792, *Tubingen; died May* 13, 1860, *at Tubingen.*

Experimenta electricitatem quæ contactu evolvitur spectantia. 8vo. *Tubingen*, 1820

Analys. d. Turmalius von Karingsbricka. 8vo. (*Schweigger's Journ.* xxxi. 1821.) *Nürnburg*, 1821

Gmelin, Christian Gottlob.—*continued.*

Über d. Coagulat. d. Eiweisses durch galvan. Electricität. 8vo. (*Schweigger's Journ.* xxxvi. 1822.) *Tübingen*, 1822

† Analyses of Tourmalines. 8vo. 2 pp. (*Phil. Mag. or Annals*, iii. 460.)
London, 1828

Handbuch der Chemie. Fortsetzung v. List, Lehmann u. Rochleder. Parts 45 to 58, vol. iii. 8vo. *Heidelberg*, 1858-61

Gmellin, J. F. Einleitung in die Chemie. 2te Abth. *Nürnburg*, 1780

Geschichte der Chemie seit dem Wiederaufleben der Wissenschafter bis an das Ende der 18 Jahrhund. 8vo. *Göttingen*, 1797-99

† Prælectio de columnæ metallicæ a Volta inventæ effectibus chimicis. Habita, 13 Nov. 1803. 4to. 55 pp. (*Commentat. Soc. Göttingensis*, xv. *Phys.* p. 38.)
Göttingen, 1800-3

† **Gmellin**, Ferd. Gottlob. Diss. sistens observat. physicas chimicas de Elect. et Galv. 8vo. 62 pp. *Tübingen*, 1802

Gmelin, Leopd. Handbuch d. theoret. Chemie. 2 vols. 8vo.
Frankfurt a. M., 1817-19

Handbuch d. theoret. Chemie. 2nd edition. 2 vols. *Frankfurt*, 1821-22

Handbuch d. theoret. Chemie. 2 vols. 8vo. 3rd edition. *Frankfurt*, 1827-29

Über e angebl. meteorische Masse. 8vo. (*Gilbert's Ann.* lxxiii. 1823.)
Leipzig, 1823

† Art "Phosphor" in Gehler's Wörterbuch. 1ste Abtheil. viii. Bd. 8vo. 8 pp.
Leipzig, 1833

Versuch einer elektro-chemisch. Theorie. 8vo. (*Poggendorff, Ann.* xliv. 1838.)
Leipzig, 1838

Handbook of Chemistry by H. Watts. 15 vols. 8vo. *London*, 1848-61

Handbook of Chemistry, translated and edited by Henry Watts. 1871

Gmelin and Schaub. Effets chimiques de la colonne métallique, par M. Gmelin expériences galvaniques de M. Schaub. (*Magazin Encyclopédique*, tom. vi. p. 201.)

† **Göbel**, Severin. De Succino. Libri duo. (*Each alternate page only is numbered.*) 12mo. 60 pp. *Turin*, 1558

At the end of this are a few pages with the title " De Bitumine et cognatis ei, Naphtha."

Goclenius, R., jun. Tract. de magnetica curatione vulnerum, citra ullam super-stitionem et dolorem et remedii applicationem, &c. Marpurgi, 1608. 8vo.
(*From Watt.*) *Marp.* 1609

† Mirabilium naturæ . . . Adjecta est in fine brevis et nova Defensio magneticæ curationis vulnerum ex solidis principiis. 12mo. 303 pp. *Francofurti*, 1625

Godigno, N. De Abissinorum rebus. (p. 67.) (*From Zantedeschi, Trattato*, i. 209.)
1615

Godin. Table alphabet. . . . de l'Acad. Rle. 4 vols. 4to. 1666—1730

† **Goebel**, A. Untersuchung eines am 29 April, 11 Mai, 1855, auf Oesel niederge-fallenen Meteorsteines. 4to. (*Archiv. f. d. Naturk. Liv.-Esth. u. Curlands.* Bd. i. Dorp. 1854.) *Dorpat*, 1855

† Über Aerolithenfälle in Russland aus früheren Jahrhunderten. Lu le 20 Décembre, 1866. 4to. 15 pp. (*Bulletin de l'Acad. Imp. de St. Pétersb.* xi. 527, 1867.) *Pétersbourg*, 1867

† **Goldschmidt**, B. Results of the daily observations of Magnetic Declination during six years at Göttingen. 8vo. 12 pp. 1839

† Üb. d. magnet. Declination in Göttingen. 8vo. 31 pp. *Göttingen*, 1845

† **Golubew, A.** Über die Erscheinungen, welche elektrische Schläge au den sogenannten farblosen Formbestandtheilen des Blutes hervorbringen. (Vorgelegt in der Sitzung am 16 April, 1863.) 8vo. 18 pp. 1 plate. (*Sitzb. d. k. Akad. d. Wissensch. April-Heft Jahrg.* 1868, lvii.) *Wien,* 1868

† **Gonella, G. B.** Osservazioni sugli apparati elettro-telegrafici. 8vo. 24 pp. *Torino,* 1850

† **Gonon, E.** Mémoire sur le système télégraphique nouveau . . . Lu à l'Acad. 12 Fev. 1844. 4to. 102 pp. 1 plate. *Paris,* 1844

† Des Télégraphes aériens et électriques. Questions. 4to. 107 pp. 1 plate. *Paris,* 1845

† **Gonon et Gauguier.** Lettres sur l'importante question de la Télégraphie. 4to. 40 pp. *Paris,* 1850?

† **Goransson, B.** Observationer öfver Magnetiska Inclinationen i Lund. 8vo. 30 pp. *Lund,* 1865

Gordon, Andreas. *Born July 12, 1712, at Gofforach, co. Angus, Scotland; died August 22, 1751, at Erfurt.*

Phænomena electricitatis exposita. 8vo. 88 pp. 2 plates. *Erford,* 1744

Tentamen explicationis electricitatis. 88 pp. *Erford,* 1745

Philosophia. 3 parts. 8vo. *Augsburg,* 1745

Versuche einer Erklärung der Eletricität. 8vo. *Erford,* 1745

† Versuche einer Erklärung der Eletricität. Tweyte Auflage mit neuen Versuchen und Instrumenten vermehret. 8vo. 112 pp. 3 plates. *Erfurt,* 1746

† Phænomena electricitatis exposita. 3rd edition. 8vo. 88 pp. 3 plates. (*Appendix ad Physicam.*) *Erford,* 1746

Epistola ad amicum Eisentrantium. (*From Krunisz.*) *Erford,* 1748

Phisicæ experimentalis elementa (in usum Acad. Erfordiæ. 2 vols. 8vo.) (*From Poggendorff,* i. 928.) *Erfordiæ,* 1751-53

Gordon, J. E. New Anemometer for force and direction at any distance from the Vane. (*Phil. Mag. for January,* 1872.)

† **Gore, G.** Theory and practice of Electro-deposition. . . . 8vo. 104 pp. (*Orr's Circle of the Sciences,* p. 891.) *London,* 1856?

† On the electrical relations of Metals, &c. in fused substances. 8vo. 6 pp. (*Phil. Mag. for June,* 1864.) *London,* 1864

† On the properties of electro-deposited Antimony. 8vo. 10 pp. (*Journal of the Chemical Society.*) *London,* 1843

(*Vide* also Anon. Electro-Metallurgy.)

Gorno, P. (Notice of) Memoria sui Temporali. 8vo. (*Comment. Ateneo Brescia, vol. pr. in* 1824, p. 34.) *Brescia,* 1824

(Notice of or) Memoria sulla Elettricità delle Nubi. 8vo. (*Comment. Ateneo Brescia, vol. printed in* 1825 *for* 1824, p. 72.) *Brescia,* 1825

† Sui corpi idio elettrici, Memoria. 8vo. 10 pp. (*Comment. dell' Ateneo di Brescia per l'ann. accad.* 1827. *Printed in* 1828, p. 55.) *Brescia,* 1828

† Sul freddo dei Temporali. Notizia della sua Memoria. 8vo. 19 pp. (*Bibliot. Ital.* xcix. 84, *and in Comment. dell' Ateneo di Brescia, vol. for* 1830, *printed in* 1831, p. 124.) *Brescia,* 1831

† Nuove informazioni sull' origine del freddo dei Temporali. Memoria. 8vo. 4 pp. (*Bibliot. Ital.* xcix. 84, *and in Comment. dell' Ateneo di Brescia, vol. for* 1832, *printed in* 1833, p. 41.) *Brescia,* 1833

† **Gostling, W.** Letter to P. Collison concerning the Fire Ball seen by Lord Beauchamp in Kent. Dated Dec. 13, 1741, *Canterbury.* 4to. 2 pp. (*Phil. Trans.* xli. 872.) *London,* 1739-40-41
(*Vide* Beauchamp.)

Goth. G. Ub. d. Hagelstürme in Steiermark im Juli 1846. Nebst Haidingers Nachtrag zu dieser Abhandlung.

Gottingen, Societas Scient. Commentarii Soc. Reg. Scient. Götteng. 4to. 28 vols. *Göttingen*, 1752-1808

Teutsche Schriften von der Kön. Societät de Wissensch. zu Göttingen. 8vo. *Göttingen*, 1771

Gottingische Gelehrte Anzeigen. 8vo. (*From Libri Sale Cat.* p. 380.) *Göttingen*, 1814-25

Gottingisches Magazin. (*Vide* Lichtenberg u. Forster, 1780-1785.)

† **Gottschalk.** Tarif für telegraphische Depeschen von Bremen nach allen Stationen d. deutsches terreischen Telegraphen Vereins und nach den ubrigen europaeischen und aussereuropaeischen Staaten. 16mo. 50 pp. *Bremen*, 1868

† **Gouge, A.** Beiträge zur Lehre vom Magnetismus Bericht einer aus Mitgliedern der K.K. Gesellschaft der Ärzte zu Wien bestehenden Commission erstattet. 8vo. 198 pp. *Wien*, 1846

† **Gough, J.** Remarks on Hygrometry, and the Hygrometer of J. Berzelius. 8vo. 6 pp. 1 plate. (*Phil. Mag.* xxxiii. 177.) *London*, 1809

Gould, Benjamin Apthorp. *Born September* 27, 1824, *at Boston, U.S.*

† On the velocity of the Galvanic Current in telegraphic wires. In a Report to Bache. 8vo. 28 pp. 1 plate. (*American Journal of Science*, 2nd series, xi. Jan. 1851.) *New Haven, U.S.* 1851

Goulier, C. M. Etudes géometriques sur les étoiles filantes. 8vo. 154 pp. 2 plates. *Metz*, 1868

† **Gounelle, E.** Observations sur les Expériences de M. Guillemin. 8vo. 28 pp. 1 plate. *Paris*, 1863 (*Vide* also Fizeau and Gounelle.)

† **Gourdon.** Notice sur l'Elect. employée comme moyen thérapeutique. 8vo. 12 pp. *Paris*, 1838

† **Govi,** Gilb. Romagnosi e l'Elettro-magnetismo. Ricerche storiche. 8vo. 16 pp. (*Estratto dagli Atti della Reale Accad. delle Scienze di Torino*, vol. iv. 7 *April*, 1869.) *Torino*, 1869

(*Note.—Govi shows clearly at page* 8, &c. *that Romagnosi's experiment had no connection with Oersted's, and was not magnetical. The needle was repelled by mere electric action (of a pile). It was in fact a modification or repetition of Milner's experiment in* 1783, *or of Peltier's, which was a repetition of Milner's.—F. R.*)

† L'Aurora boreale. 8vo. 4 pp. (*Estratto dalla Gazzetta Ufficiale del* 28 *Ottobre*, 1870.) *Firenze*, 1870

Gradis, B. Dorure et Argenture électro-chimiques. 8vo. 24 pp. *Paris*, 1851

† **Graham, D.** (Major). Observations of the Magnetic Dip made at several positions, chiefly on the South-western and North-eastern frontiers of the United States ; and the Declination at two positions on the River Sabine, in 1840. Read August 16, 1844. 4to. 52 pp. (*Trans. Amn. Society*, New Series, vol. ix.) *Philadelphia*, 1846

An account of observations made of the variation of the Horizontal Needle at London, in the latter part of the year 1722 and the beginning of the year 1723. 4to. (*Phil. Trans.* 1724, p. 96.) *London*, 1724

Observations of the Dipping Needle made at London in the beginning of the year 1723. 4to. (*Phil. Trans.* 1725, p. 332.) *London*, 1725

† Some observations, made during the last three years, of the quantity of the variation of the Magnetic Horizontal Needle to the Westward. Read April 21, 1748. 4to. 2 pp. (*Phil. Trans.* xlv. 279.) *London*, 1748

† **Graham, T.** On the finite extent of the Atmosphere. 8vo. 3 pp. (*Phil. Mag.* or *Annals*, i. 107.) *London*, 1827

Elements of Chemistry, by Watts. 2 vols. 8vo. *London*, 1850-59

208   GRA

† **Grailich, Jos., and Lang, V.** Untersuchungen über die physicalischen Verhaltnisse Krystallisirter Körper. 1. Orientirung der optis. Elasticitätsaxen in den
Krystallen des rhombischen Systems. Vorgel. in d. Sitz. vom 12 Juni, 1857.
8vo. 75 pp. 7 plates. (*Sitzb. Wien Acad.* vol. xxvii. 1 Heft.) *Wien,* 1857

† Untersuchugen üb. d. Physicalischen Verhaltnisse Krystallisirter Korper. 2.
Orientirung der magnetischen. Verhaltnisse in Krystallen des rhombischen
Systems Vorgetragen in d. Sitzung vom 8 July 1858. 8vo. 25 pp. (*Sitzungsb.
Wien Acad.* vol. xxxii. No. 21.)   *Wien,* 1858 ?

† Untersuchungen üb. d. pbysikalischen Verhältnisse Krystallizirter Körper.
4. ∪ber die Beziehung zwischen Krystallform, Substanz und physakilischen
Verhalten. 8vo. 82 pp. Vorgetragen in der Sitzung am 7 Oct. 1858.
(*Sitzungsb. d. Wien Acad.* xxiii. No. 27.)   *Wien,* 1858 ?

Gralath, Daniel. *Born June* 8, 1739, *at Danzig; died August* 10, 1809, *at Danzig.*

† Geschichte der Elektricität. 1. Abschnitt. 4to. 130 pp. 1 plate. (*Versuche u.
Abhandl. der Naturforsch. Gesellschaft in Danzig,* Th. i. S. 175.)
                                                              Danzig, 1747

† Nachricht von einigen elektrischen Versuchen. (Balance or scales used in 1746.)
4to. 29 pp. 1 plate. (*Versuche u. Abhandl. der naturforschenden Gesellschaft
in Danzig,* Th. i. No. xiii. p. 506.)   *Danzig,* 1747

Geschichte der Elektricität ii. Abscnitt. 4to. 106 pp. 1 plate. (*Versuche u
Abhandl. der Naturforsch. Gesellschaft in Danzig,* Th. ii. S. 355.)
                                                              Danzig, 1754

† On Leyden Phial. (*Danzig Mem.* ii. 433.)

† Gralath's account of Experiments merely from Priestley. (*Danzig Memoirs,*
vol. ii. p. 438.)

† Electrische Biliothek. St. i. 4to. 28 pp. (*Versuche u. Abhandl. Naturforsch.
Gesellsch. in Danzig,* Th. ii. S. 525.)   *Danzig,* 1754

† Electrische Bibliothek. St. ii. 4to. 64 pp. (*Versuche u. Abhandl. der Naturforsch. Gesellsch. in Danzig,* Th. iii. S. 265.)   *Danzig,* 1756

† Geschichte der Electricität. iii. Abschnitt. 4to. 69 pp. (*Versuche u. Abhandl.
Naturforsch. Gesellschaft in Danzig,* Th. iii. S. 492.)   *Danzig,* 1756

Grandami (Grandamicus), Jacques. *Born* 1588, *at Nantes; died February* 12.
1672, *at Paris.*

† Nova demonstratio immobilitatis terræ patita, ex virtute magnetica. 4to
170 pp. Plates.   *Flexiæ,* 1645

Grantham. (*Vide* Liverpool Compass Committee.)

Grapengieser, C. J. C. Versuche den Galvanismus, zur Heilung einiger Krankheiten anzuwenden angestellt und beschrieben von G. 1st ed. 8vo. Plates.
                                                              Berlin, 1801

† Versuche den Galvanismus zur heilung einiger Krankheiten anzuwenden. 2
Auflage. 8vo. 230 pp. 2 Plates.   *Berlin,* 1802

Graperon. Sur un nouveau galvanomètre, avec une notice sur quelques faits
galvaniques. 8vo. (*Journ. du Galv.*)   1803

Réflexions sur une expérience galvanique. 8vo. (*Journ. du Galv.* No. iii.
p. 101.)   *Paris*

S'Gravesande, Wilhelm Jacob. *Born September* 27, 1688, *at Herzogenbusch; died
February* 28, 1742, *at Leyden.*

† Œuvres philosophiques et mathématiques. Publieés par S. Allamand.
2 vols. 4to.   *Amsterdam,* 1774

† Eléments de Physique, par Joncourt. 2 vols. 4to. plates.   *Leide,* 1846

Gravier. (*Vide* Coulvier-Gravier.)

Gray, Stephen. *Born end of seventeenth century, in England; died February 15, 1736, at London.*

New Electrical experiments. 4to. (*Phil. Trans. Abd.* vi. part ii. p. 7.) 1720

† An account of some new Electrical experiments. 4to. 4 pp. (*Phil. Trans.* xxxi. *for* 1720-21.) *London*, 1723

Farther experiments concerning Electricity. (*Phil. Trans. Abd.* vi. part iv. p. 96 *in the vol. marked* vii.) *London,* 1731

Farther experiments concerning Electricity. (*Phil. Trans. Abd.* vi. part ii. p. 24.) *London*, 1731

Concerning the Electricity of water. (*Phil. Trans. Abd.* vi. part ii. p. 22.) *London*, 1731

More experiments concerning Electricity. (*Phil. Trans. Abd.* vi. part ii. p. 9.) *London*, 1731

† A letter to Mortimer, containing several experiments concerning Electricity, by Gray, S., dated Charter House, February 8, 1730-31. 4to. 27 pp. (*Phil. Trans.* xxxvii. 18.) *London*, 1731

† A letter concerning the Electricity of Water, from Gray to Mortimer, C., dated Charter House, January 20, 1731-32. 4to. 4 pp. (*Phil. Trans.* xxxvii. 227.) *London*, 1731-32

† A letter from Gray to Mortimer containing a farther account of his experiments concerning Electricity. Dated Charter House, June 7, 1732. 4to. 7 pp. (*Phil. Trans.* xxxvii. 285.) *London*, 1731-32

† Two letters from Gray to Mortimer, containing farther accounts of his Experiments concerning Electricity. First letter dated Charter House, October 15, 1732. 4to. 11 pp. (*Phil. Trans.* xxxvii. 397.) *London*, 1731-32

Experiments and Observations upon the Light, &c. (*Phil. Trans. Abd.* viii. 395.) 1734-35

Some experiments relating to Electricity. (*Phil. Trans. Abd.* viii. 401.) 1735

Experiments and observations upon the Light that is produced by communicating Electrical Attraction to animal or inanimate bodies, together with some of its most surprising effects. 4to. (*Phil. Trans.* 1735, pp. 16 and 166.) *London*, 1735

Concerning the Revolutions of small Pendulous Bodies, &c. (*Phil. Trans. Abd.* viii. 403.) 1735-36

Electrical Experiments taken from his mouth by Dr. Mortimer. (*Phil. Trans. Abd.* viii. 404.) 1735-36

Lettre de Gray à Mortimer. 4to. (*Translation from Phil. Trans. by Bremond.*) *Paris*, 1738

† A letter from S. Gray to Mortimer . . . containing some experiments relating to Electricity. 4to. 5 pp. (*Phil. Trans.* xxxix. *for* 1735-36, p. 166.) *London*, 1738

† Mr. Stephen Gray . . . his last letter to Granville Wheeler, Esq. . . . concerning the revolutions which small pendulous bodies will, by Electricity, make round larger ones from west to east, as the planets do round the sun. Dated February 6, 1735-36. 4to. 1 p. (*Phil. Trans.* xxxix. for 1735-36, p. 220.) *London*, 1738

† An account of some Electrical experiments intended to be communicated to the Royal Society by Mr. S. Gray, taken from his mouth by C. Mortimer, M.D., R.S. Secr. on February 14, 1735-36, being the day before he died. 4to. 4 pp. (*Phil. Trans.* xxxix. for 1735-36, p. 400.) *London*, 1738

Greatrakes, V. A brief account of Mr. V. Greatrakes, and divers of the strange cures by him lately performed. 4to. *London*, 1666

Green, George. *Born July* 14, 1793, *at Nottingham; died March* 31, 1841, *at Sneinton.*

An Essay on the application of Mathematical Analysis to the theories of Electricity and Magnetism. 4to. 72 pp. *Nottingham*, 1828
(*Latimer Clark says in his Catalogue,* "*one of the most important works ever written on Electricity.*"—F.R.)

210   GRE—GRI

Green, J. Electro-Magnetism. 12mo. 210 pp. Plates.   *Philadelphia,* 1827

Green, W. P. Selection of papers on the subject of fixed Lightning Conductors
to the masts of H.M.'s Navy, constructed so as to pass from the truck to
the keelson . . . illustrated by engravings, &c. 8vo.   *London,* 1824

On Lightning Conductors for Ships.   1828

† **Greenwood, J.** An account of an Aurora Borealis seen in New England on the
22nd October, 1730, by Greenwood, J. Communicated in a letter to the
late Dr. Rutty. Dated Harvard College, October 24, 1730. 4to. 15 pp.
1 plate. (*Phil. Trans.* xxxvii. 55.)   *London,* 1831-32

Greg, R. P. An Essay of Meteorites. 8vo. 40 pp. (*Phil. Mag. November-
December,* 1854.)   *London,* 1854

† An Essay on Meteorites. 8vo. 40 pp. (*Probably a reprint.*) *Manchester?* 1855

† Catalogue of the collection of Meteorites belonging to R. P. Greg, Esq. 8vo.
8 pp.   *Manchester,* 1865

Gregorio, D. Lettera intorno all' elettricità.

Greimble, Dissertatio physica de genu progressu et effectibus fulminis. 12mo.
Agust. Vind. 1759

† **Greiss, C. B.** Über d. Magnetismus der Eisenerze. 8vo. 18 pp. (*Poggendorff's
Ann.* xcviii. 1856.)   *Wiesbaden,* 1856

† **Greiss, C.** Zur Geschichte d. Magnetismus. 4to.   *Wiesbaden,* 1861

Grellois, E. Notice sur les observations ozonométriques faites pendant neuf mois,
en 1855 et 1856, à la pointe du Sérail, à Constantinople. 4to. 13 pp.
(*Ext. de l'Annuaire de la Soc. Météorol. de France,* tom. v. p. 58, séance 14
Av. 1857.)   *Versailles,* 1857

† Ozonomètre. Résultats de quelques expériences faites à Thionville. 4to.
7 pp. (*Ext. de l'Annuaire de la Soc. Météorol. de France,* tom. v. p. 183.
Séance du 14 Juillet, 1857.)   *Versailles,* 1857

Gren, Friedrich Albert Carl. *Born May 1, 1760, at Bernburg; died November 26,*
1798, *at Halle.*

Apparat, durch d. elektr. Funken brennbare u. lebens luft aus d. Wasser zu
erhalten. 8vo. (*Gilbert's Ann. d. Physik,* ii. 1790.)   *Leipzig,* 1790

Journal der Physik. 8 vols. 8vo.   *Leipzig,* 1790-93

Neues Journal der Physik. 4 vols. 8vo.   *Leipzig,* 1795-97

Grundriss der Naturlehre, § 1408. 8vo.   *Halle,* 1797

† Grundriss der Naturlehre (6 Aufl.) hersg. von K. W. G. Kastner. 8vo.
Halle, 1820

(*Vide* also Prevost, 1794.)

Grewingk, Schmidt. Uber die Meteoritenfälle von Pillistfer, Buschhof und
Igast in Liv - und Kurland. 8vo. 140 pp. 2 plates and map.
Dorpat, 1864

Gribel, Eduard Wilhelm. *Born May 6, 1816, at Stettin; died July 18, 1856, at
Reichenbach.*

† De relatione actionum caloris et electricitatis. (Dissert. inaugur.) 4to. 28 pp.
Berol, 1837

† **Grieb, C. F.** Die Wunder der elektrischen Telegraphie. 16mo. 3 plates.
Stuttgart, 1850

Griffith, J. W. (*Vide* Plücker, 1847.)

† **Grigolato, G.** Alcune ricerche teoriche sulla origine e natura dell' Ozono. Memoria
4to. 12 pp. (*Mem. . . dell' Ateneo di Treviso,* vol. v. p. 71.) *Padova,* 1847

† **Grillenzoni, C.** Di alcuni nuovi esperimenti del Dr. A. Palagi di Bologna, sulle variazioni elettriche a cui vanno soggetti i corpi scostandosi dal suolo, o da altri corpi, ovvero, accostandosi ad essi ; ricordo del Dott. C. Grillenzoni. 8vo. 8 pp. (*Estratto dalla Gazzetta Medica Italiana-federativa-Toscana*, tom. iii. ser. ii.)

Grimaldi, F. M. Physico-Mathesis de lumine coloribus et iride aliisque adnexis libri ii. (*Opus. posth.*) 4to. *Bononiæ*, 1665

(Gatteschi refers to his discovery of the magnetism produced by holding a bar of iron perpendicularly (in 1550) in this work. Grimaldi died in 1663.)

† **Grimaldi, G.** Dissertazione (viii.) di . . . sopra al primo Inventore della Bussola. 4to. 25 pp. (*Saggi di Dissertazioni accademiche . . . nell' Accad. Etrusca . . . di Cortona*, tom. iii. p. 195.) *Roma*, 1741

† **Grimelli, G.** Osservazioni ed esperienze elettro-fisiologiche dirette ad instituire la elettricità medica. 8vo. 336 pp. 1 plate. (*Called also " Elettricità Fisio-logico-Medica."*) *Modena*, 1839

Lettera all' . . Amici intorno alle contrazioni che produconsi nell' atto di chiudere, non che in quello di aprire, il circuito prettamente nerveo-muscolare della rana. *Modena*, 1842

Lettera al . . . Bufalini intorno alle contrazioni che ottengonsi nell'atto, di chiudere e di aprire il circuito composto, per l' una parte, di tessuti nerveo muscolari, per l'altro, di un sistema elettromotore idro-metallico. (*Memoria sul Galvanismo.*) *Modena*, 1842

Seguito di osserv. e di esper. Elettro-fisiologiche. *Modena*, 1843

Metodo italiano di elettro-doratura . . . esposto . . . 1843-44. *Modena*, 1844

† Metodo originale Italiano di Elettro-doratura. 8vo. 18 pp. *Modena*, 1844

Memoria concernente varie notizie sperimentali elettrometallurgiche e elettromediche, letta . . . 19 Dicembre, 1843, alla Reale Accad. . . di Modena. 8vo. (*Foglio di Modena 8 Feb.* 1844.) *Modena*, 1844

Elettro metallurgia dichiarata e svolta ad uno dei Direttori dei Nuovi Ann. delle Scienze . . che si pubblicano in Bologna. Tom. x. 1843. (*Foglio di Modena 9 Maggio*, 1844. *Modena*, 1844

† Storia scientifica ed artistica dell' Elettro-metallurgia originale Italiana con un Saggio teorico-pratico di elettro-metallurgia piana e solida; e un' Appendice lessicologica. 8vo. 192 pp. *Modena*, 1844

† Memoria sul Galvanismo premiata dall' Accad. . . . dell' Istit. di Bologna. 4to. 193 pp. *Bologna*, 1849

Prospetto delle Mem. elettriche e magnet. pubblicate dal Prof. Cav. Stefano Marianini. *Modena*

Grimm, Mélange d'observations galvaniques. 8vo. (*Journal de Chim. d. Van Mons*, No. x. p. 82.) *Bruxelles*

† **Grimm, C.** De maris nocturna lucis emissione. Dissert. inauguralis. 4to. 19 pp. *Hanover*, 1840

Grimm, J. Über die Namen des Donners. Eine academische Abhandlung vorgelesen am 12 Mai, 1863. 4to. 28 pp. *Berlin*, 1855

Grindat. (*Vide* Strype.)

† **Grinwis, C. H. C.** De verdeeling der electriciteit over het oppervlak eens geleiders. 8vo. 72 pp. *Utrecht*, 1858

212   GRI–GRO

Grinwis, C. H. C.—*continued.*

† Wiskundige theorie der wrijvings electriciteit.   (In 2 Stücken.)   8vo.   266 pp.
Utrecht, 1869

In Guthe's Bibliotheca, 2 *Heft, July bis Dec.* 1869, p. 149, *appears the entry :*
" *Grinwis C. H. C. : Wiskundige theorie der wrijvings electricitat.* 2e *stük.
Royal* 8vo. (viii. *bl. en bl.* 161—266, *mit* 2 *tusschen den tekst gepl. houts-
neefig. Utrecht.* (*J. L. Beijers.*)''

Grischow, Augustin Nathanael.   *Born September* 29, 1726, *at Berlin ; died June* 4,
1760, *at St. Petersburg.*

† Observatio insoliti luminis australis Petropoli habita.   4to.   (*Novi Comment.
Acad. Petrop.* iv. *pro* 1752-53, *printed* 1753, p. 68.)   *Petropoli,* 1752-53

Observatio insoliti luminis australis Petropoli habita.   4to.   (*Novi Commentar.
Acad. Petropolit.* iv. 474.)   *Petropoli,* 1758

Grischow.   (*Vide* Lomonosow.)

† **Griselini, F.**   Lettera al Padre D.Angelo Calogiera, intorno l'elettricità, e alcune
particolari esperienze della medesima.   12mo. 1 plate.   *Venezia,* 1747

Observations sur la Scolopendre marine luisante, e la Baillouviana.   8vo.
32 pp.   *Venice,* 1750

Gronau, Karl Ludwig.   *Born June* 7, 1742, *at Berlin ; died December* 8, 1826, *at
Berlin.*

Bemerkungen über Nebel u. Nordschein.   (*Schr. d. Gesellsch. naturf.* Fr. vi.
1725.)   1785

Bemerkungen über Gewitter. (*Schr. d. Gesellschaft naturf.* Fr. ix. 1789.)   1789

Beobachtung der Witterung d. Mark Brandenburg.   1 Theil.   8vo.
Berlin, 1794

Über d. vom Himmel gefallenen Steine.   8vo.   *Berlin,* 1804

Grones, G.   Memoria sulla Procella avvenuta in Venezia.   24 Giugno. 1822.
Letta nell' Ateneo di Treviso 6 Agosto.   8vo. 59 pp.   *Venezia,* 1823

Gronov.   On the Gymnotus electricus.   (*Phil. Trans.* lxv. pt. i. pp. 94 and 102 ;
pt. ii. p. 395.)   1760

† **Gros, Baron T. B. L.**   Lettre sur la Télégraphie électrique.   8vo. 31 pp. 18 plates.
Paris, 1856

Anschauliche Darstellung der electrischen Telegraphie zur Verständlichung
des grossen Publicums.   In einem Briefe an eine Dame.   Nach dem Französischen.   8vo. 38 pp. 18 plates.   *Weimar,* 1857

† Anschauliche Darstellung der elektrischen Telegraphie zur Verständlichung
der grossen Publicums.   In einem Briefe an eine Dame.   Nach dem
Französischen.   Zweite, vermehrte Auflage.   8vo. 54 pp. 18 plates.
Weimar, 1862

Grosier.   Mémoires d'une société considérée comme corps littéraire . . . ou
Mémoires des Jésuites sur les Sciences.   3 vols.   4to.   *Paris,* 1792

Gross, Johann Friedrich.   *Born May* 5, 1732, *at Nagold, Würtemberg ; died February*
5, 1795, *at Stuttgart.*

† Elektrische Pausen.   8vo. 136 pp. 1 plate.   (*The plate is in the title-page.*)
Leipzig, 1776

Précis des Pauses électriques.   4to.   (*Journ. de Phys.* x. 1777.)   *Paris,* 1777

† Grundsäze der Blizableitungs kunst geprüft, und durch einen merkerwürdigen
Fall erläutert.   Nach dem Tode des Verfassers heraus gegeben von J. F. W.
Widenmann.   8vo. 228 pp. 1 plate.   *Leipzig,* 1796

Aufsatze in d. "Tübingen Berichten von gelehrten Sachen." (*From Pogg.* i. 959.)

Grosser, J. H.   Fascicul. tentaminum phys. med. electricorum, c. not.   8vo.
(*From Heinsius,* ii. 188.)   *Würzbg.* 1786

Grosser, J. M. von. Phosphorescentia adamantum novis experimentis illustrata ; üb. d. Leuchten d. Diamenten in Finstern nach neuen Versuchen. 8vo.
Wien, 1777

† **Grossi, L.** Ritratti daguerotipi. 8vo. 1 p. (*Giornale I. R. Istit. Lomb.* iv. 278.)
Milano, 1842

Grothuss, Theodor. *Born January* 20, 1785, *at Leipzig ; died March* 14, 1822, *at St. Geddutz.* (*Poggendorff also calls him Christian Johann Dietrich.*)

† Mémoire sur la décomposition de l'Eau et des corps qu'elle tient en dissolution à l'aide de l'Electricité galvanique. 12mo. 22 pp. (*Volta's copy.*)
Rome, 1805

† Memoir upon the decomposition of Water, and of the bodies which it holds in solution, by means of Galvanic Electricity. 8vo. 10 pp. 1 plate. (*Phil. Mag.* xxv. 330.)
London, 1806

De l'influence de l'électricité galvanique sur les végétations métalliques. 8vo. (*Ann. Chim.* lxiii.)
Paris, 1807

Chem-galvan. Beobacht. 8vo. (*Gehlen's Jorn.* 1808.)
Berlin, 1808

Über d. elektricität die sich bei andern Zustanden d. Wassers entwickelt. 8vo.
Berlin, 1810

Uber d. Grenze d. Verbrennlichkeit entzündl. Gasgemenge bei abnehmenden Dichtigkeit, und über d. Farben d. elektr. Funkens in verschiedenen Mitteln. 8vo. (*Schweigger's Jour.* iii. *und* iv. 18₁1-12.)
Nürnberg, 1811-12

Versuche u. Ideen über Brennen ; über elektr. Leitungsfähigkeit verschiedener Gase ; über Acidität; über Morichini's Versuche ; u.s.w. 8vo. (*Schweigger's Journ.* ix. 1813.)
Nürnberg, 1813

Über d. chem. Wirksamkeit d. Lichts u. d. Electricität, u.s.w. ; merkwürdige galvan. Wasserzersetzung, u.s.w. 4to. 175 pp.
1818

Verstärkungsmittel d. elektr. Wirk. bei gewöhnl. Elektrisirmaschinen. 8vo. (*Scherer's Allgem. Nord. Ann. di Chemie,* iv. 1820.)
Berlin, 1820

Merkwürd Zersetz. d. Wassers. durch Wasser im Kreise d. Volta'schen Säule. 8vo. (*Schweigger's Journ.* xxviii. 1820.)
Nürnberg, 1820

Phys.-chemis. Forschungen. 1r. Bd. 4to.
Nürnberg, 1820

Zwei newe den Elektro-magnetismus betreff. Thatsachen. 8vo. (*Schweigger's Journ.* xxxi. 1821.)
Nürnberg, 1821

Untersuchung eines, am 30 Juni 1820, im Dünaburger Kreise herabgefallenen Meteorsteins. 8vo. (*Gilbert's Ann.* lxvii. 1821.)
Leipzig, 1821

† On the Phosphorescent organ of the Lampyris Italica. 8vo. 1 p. (*Phil. Mag.* lix. 67.)
London, 1822

Also many Papers in German journals which may contain available Electric matter. (*Vide Poggendorff,* i. 959-60.) (*Vide also Grove, W. R.* 1845.)

Grove, William Robert. *Born July* 11, 1811, *at Swansea.*

† On the inaction of Amalgamated Zinc in Acidulated Water. 8vo. 4 pp. (*Phil. Mag.* xv. Aug. 1839, p. 81.)
London, 1839

† On a small Voltaic Battery of great energy ; some observations on Voltaic combinations, &c., and forms of arrangement; and on the inactivity of a Copper positive Electrode in Nitro-sulphuric Acid. 8vo. 6 pp. (*Phil. Mag.* xv. *for October,* 1839, p. 287.)
London, 1839

† On some Phenomena of the Voltaic disruptive discharge. 8vo. 5 pp. (*Phil. Mag.* xvi. *for June,* 1840, p. 478.)
London, 1840

† Voltaic reaction, or the Phenomenon usually termed Polarization. 8vo. 2 pp. (*Ann. of Elect.* iv. 502.)
London, 1840

† On some Electro-Nitrogurets. 8vo. 8 pp. (*Phil. Mag.* xix. *for August* 1841, p. 97.)
London, 1841

Grove, William Robert.—*continued.*

† On a Gaseous Voltaic Battery. 8vo. 4 pp. (*Phil. Mag.* xxi. *for Dec.* 1842.)
London, 1842

† On the Gas Voltaic Battery. Experiments made with a view of ascertaining the Rationale of its action and its application to Eudiometry. Read May 11, 1843. 4to. 20 pp. 1 plate. (*Phil. Trans. for* 1843, pt. ii. p. 91.) (*Faraday's copy.*) *London,* 1843

† On Grothus's theory of Molecular decomposition and recomposition. 8vo. 2 pp. (*Phil. Mag. for Nov.* 1845.) *London,* 1845

† On the Gas Voltaic Battery. Voltaic action of Phosphorus, Sulphur, and Hydrocarbons. Read June 19, 1845. 4to. 8 pp. 1 plate. (*Phil. Trans. for* 1845, pt. ii. p. 351.) (*Faraday's copy.*) *London,* 1845

† On the correlation of the Physical Forces : being the substance of a course of Lectures delivered in the London Institution in the year 1843. 8vo. 52 pp. (2nd edition, 1850, *vide Rowsell Cat.* N. 166, 1870. There is a third edition, 1855, *vide Zuchold*, p. 167 ; and a fourth of 1862, *vide Engl. Cat.* p. 318.) (*Faraday's copy.*) *London,* 1846

† Sur certains phénomènes d'ignition Voltaique, etc. . . traduit de l'Anglais par Louyet. 8vo. 27 pp. 1 plate. (*Ext. Bulletin du Musée de l'Industrie,* 4e livre, 1847.) *Paris,* 1847

† On certain phænomena of Voltaic ignition and the decomposition of Water into its constituent gases by Heat. Read Nov. 19, 1846. 4to. 21 pp. 1 plate. (*Phil. Trans. for* 1847.) (*Faraday's copy.*) *London,* 1847

† On the Effect of surrounding Media on Voltaic ignition. Read Dec. 14, 1848. 4to. 11 pp. (*Phil. Trans. for* 1849, pt. i. p. 49.) (*Faraday's copy.*) *London,* 1849

† On the Electro-chemical Polarity of Gases, Read April 1, 1852. 4to. 15 pp. 1 plate. (*Phil. Trans. for* 1852, p. 87.) (*Faraday's copy.*) *London,* 1852

† On the Electro-chemical Polarity of Gases. 8vo. 18 pp. 1 plate. (*Phil. Mag.* *Suppl.* vol. iv. p. 498.) *London,* 1852

† On some anomalous cases of Electrical decomposition. 8vo. 7 pp. (*Phil. Mag. for March,* 1853.) *London,* 1853

† On the Electricity of the Blow-pipe flame. 8vo. 4 pp. (*Phil. Mag. for January,* 1854.) *London,* 1854

† Experiments on Voltaic reaction. 8vo. 4 pp. (*Phil. Mag.*) *London*

(*Vide also* Matteucci and Grove.)

Grummuth. On Electric Light in vacuo. (*In Danzig Memoirs,* i. 417.)

Grundig, Christoph Gottlob. *Born September 5, 1707, at Grosdorfhayn, near Freiberg ; died August 9, 1780, at Freiberg.*

Neue Versuche nützlicher sammlungen. . . 48 Thle. od 4 Bd. 8vo. *Schneiberg,* 1748-64

Archiv der Mathematik und Physik mit besonderer Rücksicht auf das Bedürfniss der Lehrer herausg. Vol. i. to xxv. 8vo. (*From Dulau, Germ. Cat.* 1864, p. 1. Poggendorff (i. 966 says) : " Bis jetzt 29 Thle. 8vo. Greifswald 1841-57.") *Greifswalde,* 1841-55

Grynæus, Simon Jun. Commentarii duo : De ignitis meteoris unus ; alter, De cometarum, causis, &c. . . *Basil,* 1580

† **Guadagni,** G. Dissertazioni due sopra le Aurore boreali. 12mo. 104 pp. 1 plate. (*Calogera Raccolta,* vol. xxx. p. 1.) *Venezia,* 1744

Gualandris, Angelo. *Born* 1750, *at Padua ; died December* 13, 1788, *at Mantua.*

† Narrazione epistolare . . del Turbine avvenuto nel Mantovano . . il 9 d'Agosto 1785. (Data) Mantova, 29 Agosto 1785. 4to. 19 pp. *Mantova,* 1787

† **Guarini,** G. (Conte). Dell' elettricità naturale. Risponderà il Sig. Conte Gior. Guarini. 4to. 23 pp. *Modena,* 1779

† **Guazzi, A.** Transunto del Ragguaglio d'un Fulmine caduto presso Casalmaggiore con danno di tre persone. 4to. 3 pp. (*Opusc. Scelt.* xiv. 301.) *Milano,* 1791

Gube, F. Die Ergebnisse der Verdunstung und des Niederschlages nach Messungen an neuen, zum Theil registrirenden Instrumenten auf der königl. meteorologischen Station Zechen bei Guhraw. Mit einem Vorworte von (H.) Dove. 8vo. 51 pp. 4 plates. *Berlin,* 1864

Guden, Philipp Peter. *Born* 1722, *at Bockenem, near Hildesheim; died March* 7, 1794, *at Münden.*

† Von der Sicherheit wider die Donner stralen. 8vo. 200 pp. *Göttingen,* 1774

Guebhard. A. De la Lumière électrique. 32mo. 51 pp. *Saint-Germain,* 1867

Guéneau de Montbéliard, P. *Born April* 2, 1720, *at Semur Auxois; died Nov.* 28, 1785, *at Semur.*

Sur les Paragrêles. 8vo. (*Mém. de l'Acad. de Dijon,* viii. 1776, 14 Novembre.) *Dijon,* 1776

Guereo. Compte Rendu, 2e Série, 1819. 8vo. *Lyon,* 1819

Guericke, Otto von. *Born Nov.* 20, 1602, *at Magdeburg; died May* 11, 1686, *at Hamburg.*

Prodromo . . Invenzioni nuove . . del P. Franc. Lana. Fol. *Brescia,* 1670

† Experimenta nova (ut vocantur) Magdeburgica de vacuo spatio, primum a R. P. Gasp. Schotto . . nunc vero ab ipso auctore perfectius editur variisque aliis experimentis aucta . . . Folio. 244 pp. numbered. Portrait, and 20 other plates. (*Electricity,* pp. 147 to 150.) *Amsterdam,* 1672

† Dies F. Otto v. Guericke u. sein Verdienst. 8vo. 54 pp. *Magdeburg,* 1862

Guerin, R. T. Phénomènes Electro-dynamiques. Action mutuelle des fils conducteurs de courants électriques. Thèse soutenue . . Nov. 1828. 8vo. 44 pp. 1 plate. *Paris,* 1828

Guettard, Jean Etienne. *Born Sept.* 22, 1715, *at Etampes; died Jan.* 8, 1786, *at Paris.*

Sur une manière " d'imprégner le fer de sel, de façon que le fer reste attirable par l'Aimant."

† **Gueyton, A.** L'art de la Galvanoplastie à l'usage de MM. les Orfèvres, Bijoutiers et Bronziers. 8vo. 34 pp. *Paris,* 1855

Guidotti. (*Vide* Sgagnoni and Guidotti.)

Guidotti, G. B. Memoria fisico-chimica sulle pietre cadute. 8vo. *Parma,* 1808 ?

† **Guillemin, C. M.** Thèse. Propagation des courants dans les fils télégraphiques. 4to. 66 pp. 1 plate. *Paris,* 1860

Recherches expérimentales sur l'induction Volta électrique. 4to. 69 pp. *Montpellier,* 1861

† Recherches expérimentales sur la transmission des Signaux télégraphiques. 47 pp. 1 plate. *Paris,* 1863

† **Guillemin et Burnouf.** Mémoire sur la propagation des Courants élects. Résumé des expériences faites à Nancy du 23 Sep. au 10 Oct. 1859. 8vo. 16 pp. 1 plate. (*Extrait des Mémoires de l'Acad. de Stanislas.*) *Nancy,* 1860 ?

† **Guisan, F. S.** De gymnoto electrico. Commentatio. 4to. 1 plate, 34 pp. *Tubingen,* 1819

† **Guitard, M. T.** Histoire de l'Eléctricité médicale. 8vo. 396 pp. 6 plates. *Paris and Toulouse,* 1854

† **Guitard, T.** Précis d'electrothérapie médico-chirurgicale. 12mo. 324 pp. *Paris*

Gull, W. On the value of Electricity as a remedial agent. (*Guy's Hospital Reports,* ii. Ser. viii. 80, part i.) *London,* 1852

† **Gundolf.** Der elektro-magnetische Telegraph. Aus den physikalischen Grundlehren allgemein fasslich dargestellt. 8vo. 33 pp. 1 plate. *Paderborn,* 1851

216
GUN—GUY

Günther. Etwas vom Elektrophor. 8vo. (*Abhandl. d. Hallischen Naturforsch Gesellsch.* Bd. i. p. 63.) *Dessau u. Leipzig,* 1783

† **Günther, J. J.** Die Atmosphäre und ihre vorzüglichen Erscheinungen, nach den Grundsäzen der neuern Meteorologie, nach eigenen und Anderer Beobachtungen bearbeitet. 8vo. 253 pp. 1 plate. *Frankfurt,* 1835

Gurney, Goldsworthy. A Series of Lectures upon the Elements of Chemical Science, comprising the basis of the new theory of Crystallization, and diagrams to illustrate the elementary combinations of Atoms, particular theories of Electrical influence and of Flame . . description of his Blow-pipe, &c., &c. 8vo. (*Phil. Mag.* lxii. 301.) *London,* 1823

† **Gustavson, Col.,** Ex-King of Sweden. Reflexions upon the Phænomenon of the Aurora-Borealis, and its relation with the Diurnal Movement. (*Phil. Mag.* lviii. 312.) *Frankfort,* 1821

† **Gutle, J. C.** Beschreibung verschiedener Elektrisir-maschinen zum Gebrauch fur Schulen. 8vo. 312 pp. 12 plates. *Leipzig u. Nürnberg,* 1790

Beschreibung eines math. physikal Maschinen und Instrumenten Kabinet. . 1st Stuck. 8vo. *Leipzig u. Nürnberg,* 1794

Ditto, 2nd stuck. *Leipzig u. Nürnberg,* 1790

Beschreibung des grossen elektrisch. Zauberspiegels. *Nürnberg,* 1792

Beschreibung verschiedener Elektrisir maschinen. 8vo. *Nürnberg,* 1794

*† Zaubermechanik od. Beschreibung mechanischer Zauberbelustigungen. 8vo., 40 plates. *Nürnberg,* 1794

Kleine Electricitätslehre. *Nürnberg,* 1798

† Lehrbuch der praktischen Blitzableitungskunst. . ols Fortsetzung der " Theoretischen Blitzableitungslehre." 8vo. 446 pp. 16 plates. *Nürnberg,* 1804

Algemeine Sicherheitsregeln für Jederman bey Gewitter. 8vo. *Nürnberg,* 1805

† Fasslicher Unterricht wie man sich bei Gewittern vor den . Wirkungen des Blitzes ohne Blitzableiter sicher . . verwahren Kann. 8vo. 140 pp. *Nürnberg,* 1805

Beschreibung verschiedener Elektrisirmaschinen. 3 Thl. (*Der 1te und 2te hat also den Titel-Beschreibung eines mathematischer physikalischenInstrumenten-Kabinets, &c. d. 3te Beschreibung electrischer Instrumente.*) 8vo. *Nürnberg,* 1806

† Lehrbegriffe fur den gemeinen Mann über Elektricität und Blitzableitung. 8vo. 67 pp. *Nürnberg,* 1811

Beschreib. u. Abbild. e. neu eingericht. sehr wirksamen elektr. einfachen Glasscheibenmaschine, z. Hervorbningung, beider Elektricitäten. 8vo. 1 plate. *Nürnberg,* 1811

† Neue Erfahrungen über die beste Art Blitzableiter anzulegen. 8vo. *Nürnberg,* 1812

Neue wissenschaftliche Erfahrungen, Enteckungen und Verbasserungen, &c. . 8vo. 272 pp. 4 plates. *München,* 1826

† **Gutle und Luz.** Unterricht vom Blitz und den Blitz-und-Wetter Ableitern, zur Belehrung und Bernhigung sonderlich der ungelehrten, und des gemeinen Mannes von F. Luz neu bearbeitet von J. K. Guthe. Erster theil. 8vo. 222 pp. 1 plate. *Nürnberg,* 1804

Gutzmann, F. Über den Steinregen. 4to. *Wien,* 1803

Guyot, A. Tables meteorological and physical. *Washington,* 1858 (*Vide* also Smithsonian Institution.)

† **Guyot, J.** De la Télégraphie de jour et de nuit. 8vo. 214 pp. 3 plates. *Paris,* 1840

Guyton. Effets présumés du galvanisme dans le règne minéral. 1862

H.

H., J. (*Vide* Anon. Meteorolog. Phen. various.)   **1818**

Haarlem Hollandsche Maatschappij, der Weetenschappen.   Naturkundige
Verhandelingen van het. . .

   (*Note.*—*Muncke says* 1775 *to* 1817, 8 vols. *Royal Society Cat. says* 2 vols.
   1775, *and* 41 vols. 1759-93, *also* vols. xiii. *and* xxiii. 1824-36.)

Haaxman. (*Vide* Vliet.)

† **Hachette,** Jean Nicolas Pierre. *Born May* 6*th,* 1769, *at Mézières ; died January*
   16*th,* 1834, *at Paris.*

   Sur le Galvanisme.   (*Journ. de l'Ecole polytéchn.* iv. 1802.)   *Paris,* **1802**

Hachette, J. N. P. Précis des lecons sur le calorique et l'Electricité.   8vo.
   80 pp.   *Paris,* **1805**

   Sur les prétendus poles électriques de la terre et sur le prétendu magnétisme
   des Piles.   8vo.   (*Ann. de Chim.* lxv. 1808.)   *Paris,* **1808**

† Programmes de Physique, ou Précis de Leçons sur les principaux Phénomènes
   de la Nature. . .   8vo. 248 pp. 6 plates.   *Paris,* **1809**

   Sur la formation des tubes fulminaires.   8vo.   (*Ann. de Chim.* xxxvii. 1828.)
                                                                        Paris, **1828**

   De l'action chimique produite par l'induction électrique : décomposition de
   l'eau.   8vo. (*Ann. de Chim.* li. 1834.)   *Paris,* **1834**

   (*Vide* also Desormes et Hachette.)

Hachette et Thenard. De l'inflammation des métaux par la pile galvanique.
   4to.   (*From Sue,* ii. 345.)   *Paris,* **1801**

† **Hachette et Ampère,** Sur les expériences électro-magnétiques de MM. Oersted
   et Ampère.   4to. 8 pp. (*Journal de Phys., Septembre* 1820.)   *Paris,* **1820**

Hacker, P. W. Zur Theorie des Magnetismus.   *Nürnberg,* **1856**

   Über das Gesetz des Magnetismus, wie er sich bei der Tragkraft hufeisenförmiger Magnete und bei der Schwingungs dauer geradlieniger Magnete zu
   erkennen gibt (*Abhandl. d. naturw. Gesellsch. zu Nürnberg,* i. 51-80 *und*
   135-142.)   *Nürnberg.*

† **Hackwitz.** "Unter den seit dem Jahre 1840 errichteten Werkstatten fur Galvanoplastik dürfte die in Berlin unter der Leitung des Herrn v. Hackwitz besti-
   hende zuerst genannt werden." (*From Martin,* 1856, p. 13.)

† **Hädenkamp.** Über die Wirkung des durch eine Drathspirale gehenden elektrischen Stroms auf eine in der Spirale befindliche weiche Eisenmasse
   1852-54.   (*Crelles. Journ.* xliv. 1852.)   **1852**

Haen, A. de. Ratio medendi in nosocomio pract. (i. 234, 199).   15 vols. 8vo.
   (*From Krunitz,* p. 177.)   *Vienna,* **1757**

† **Haedenkamp.** Über d. Gezetze d. Erscheinungen d. Lichts in Krystall-
   schen Körpern.   4to.   *Hann,* **1846**

† **Haeser.** De radis lucis violacei vi magnetica commentatio.   4to.   26 pp.
                                                                        Jenæ, **1832**

† **Hagemann, H.** Eudiometriæ histor., qua describuntur et comparantur præcipui
   apparatus et modi examinandi aeris atmosphærici titulum Oxygenicum.
   4to. 4 plates.   *Groen.* **1829**

† **Hager, G.** Memoria sulla Bussola orientale Letta all' Università di Pavia.
   31 pp. Plates.   *Pavia,* **1809**

† **Haggren.** Memoria sui fiori lampeggianti, tratta dallo Svezzese (da Soave).
   4to. 2 pp. (*Opusc. Scelti,* xii. 141.)   *Milano,* **1789**

Hahn, Friedrich von. *Born in* 1741, *at Landgut, Neuhaus, Holstein; died October* 9, 1805, *at Remplin, near Malchin.* (*Poggendorff,* i. 994.)

Bemerkungen über die Neigungs-Nadel. 4to. (*Schriften der Berliner Gesellsch. Naturf. Freunde,* x. (*Beobacht.* iv.) p. 355.) *Berlin,* 1792

Bemerkungen über die Entstehung der Feuer Kugeln. 4to. (*Neue Schriften der Gesellsch. Naturf. Freunde,* ii. 222.) *Berlin,* 1799

Preface to Schilling's "De Lepra commentationes." (*Vide Schilling, G. W. and Hahn.*)

† **Hahnreider,** E. A. Bestimmung d. absolut. Intensität d. magnetischen Erdkraft. 4to. 12 pp. 1 plate. *Meseritz,* 1844

† **Haidinger,** W. Ritter Von. Niedrigste Höhen von Gewitter wolken (Zwei Fälle in Erinnerung gebracht). 8vo. 10 pp. (*Aus den Sitzungsberichten* 1852, *der k. Akad. der Wissenschaften abgedruckt.* Vol. ix. ii. Heft.) *Wien,* 1853

Serpentin mit magnetischer Polarität. (*Jahrb. der geologischen Reichsanstalt,* 1857, s. 806.) *Wien,* 1857

Das Meteoreisen von Sarepta. 8vo. 12 pp. 2 plates. (*Aus den Sitzungsberichten der kais. Akad. der Wissenschaften.*) *Wien,* 1862

† Herrn Director Julius Schmidt's Feuermeteor vom 18 October 1863. 8vo. 2 pp. (*Sitzungsb. d. k. Akad. d. Wiss. Wien,* 1863, xlviii. p. 559.) *Wien,* 1863

† Ein Meteor des 10 August, 1863 (vorgelegt 8 Oct. 1863). 8vo. 2 pp. (*Sitsungsb. d. k. Akad. d. Wiss. Wien,* 1863, xlviii. iii. Heft, p. 309.) *Wien,* 1863

† Das Carleton-Tucson-Meteoreisen in k. k. Hof-Mineralien-Cabinete (vorgelegt 8 Oct. 1863). 8vo. 8 pp. 1 plate. (*Sitsungsb. d. k. Akad. d. Wiss. Wien,* 1863, xlviii. iii. Heft, p. 301.) *Wien,* 1863

† Parnallee. Dritter Bericht. (Der Meteorsteinfall von Parnallee bei Madura in Hindostan). 8vo. 9 pp. (*Sitzungsb. d. k. Akad. d. Wiss. Wien,* 1863, xlvii. 420.) *Wien,* 1863

† Der Meteorstein von Tourinnes-la-Grosse, bei Tirlemont im k. k. Hof-Mineralien-Cabinete. 8vo. 5 pp. (*Sitzungsb. d. k. Akad. d. Wiss. Wien,* 1864, xlix. i. Heft. p. 123.) *Wien,* 1864

† Sternschnuppen, Feuerkugeln und Meteoritenschwärme ein Zusammenhange betrachtet. 8vo. 11 pp. (*Sitzungsb. d. k. Akad. d. Wiss.* 1864, *Wien,* xlix. i. Heft, p. 6.) *Wien,* 1864.

† Mittheilungen der Herren Baron Paul Des Granges, seiner Photograpien, von Sautorin, und Sternwarte-Directors Julius Schmidt, über Feuermeteore, Meteorsteinfalle, und über die Rillen auf dem Monde, aus Athen. 8vo. 6 pp. (*Sitzb. Wien Akad.*) *Wien*

† Die Meteoriten des k. k. Hof-Mineralien-Cabinets am 1 Juli 1867, und der Fortschritt seit 7 Jänner 1859. 8vo. 10 pp. (*Sitzb. mathem.-naturw. Cl.* lvi. ii. Abth. *Akad. Wien,* p. 175.) *Wien*

† Der Meteorit von Simonod. 8vo. 4 pp. (*Sitzb. d. mathem. & phys. Cl. Wien Akad.* liv. 127.) *Wien*

† Die Localstunden von 178 Meteoritenfällen. 8vo. 8 pp. *Wien*

† Ein Meteorfall bei Trapezunt am 10 December, 1863. 8vo. 5 pp. (*Abdr aus den Sitzungsberichten der k. Akad. der Wissenschaften.*) *Wien,* 1864

† Ein vorhomerischer Fall von zwei Meteoreisenmassen bei Troja. 8vo. 8 pp. (*Abdruck aus den Sitzungsberichten der kaiserl. Akad. der Wissenschaften.*) *Wien,* 1865

† Der Meteorit von Turakina, Wellington, Neuseeland. 8vo. 3 pp. (*Abdruck aus den Sitzungsberichten der kaiserl. Akad. der Wissenschaften.*) *Wien,* 1865

† **Haidinger, W.** Ritter Von—*continued*.

Der Meteorsteinfäll am 9 Juni, 1866, bei Knyahinya. (Zweiter Bericht) Vorgelegt. . . am 11 October, 1866. 8vo. 48 pp. 3 plates. (*Sitzb. d. k. Akad. d. Wissensch.* ii. Abth. Octob. Heft, 1866, Bde. liv.) *Wien*, 1866

† Der Meteorsteinfäll am 9 Juni 1866, bei Knyahinya nächst Nagy Berenzna im Ungher Comitate. 8vo. (*Aus dem* liv. Bde. *d. Sitzb. d. k. Akad. d. Wissens.* ii. Abth. Juli Heft, Jahrg. 1866.) *Wien*, 1866

† Die Tageszeiten der Meteoritenfälle verglichen. Zweite Reihe. Vorgel. . . 31 Jänner, 1867. 8vo. 10 pp. 1 table. (*Sitzb. d. k. Akad. d. Wissensch.* ii. Abth. Jan. Heft, Jahrg. 1867, Bde. lv.) *Wien*, 1867

† Die Tageszeiten der Meteoritenfälle verglichen. Vorgelegt. . . 17 Jänner, 1867. 8vo. 14 pp. 1 table. (*Sitzungsb. d. k. Akad. d. Wissensch.* ii. Abth. Jahrg. 1867, Bde. lv.) *Wien*, 1867

Der Meteorsteinfäll von Slavetice in Croatien am 22 Mai 1868. Vorläufiger Bericht. 8vo. 7 pp. *Wien.*

† Elektrische Meteore am 20 October, 1868, in Wien beobachtet. 8vo. 9 pp. 1 chromolith. (*Sitzungsb. d. k. Akad. d. Wiss.* lviii. ii. Abth. 1868.) *Wien*, 1868

Der Meteorsteinfäll vom 30 Jan. 1868, unweit Warschau. Nebst einem Anhang in Bezug auf den angeblichen Meteorsteinfall in Baden-Baden. 8vo. 8 pp. (*Aus d. Sitzungsber. d. k. Akad. d. Wissensch.*) *Wien*, 1868

† Der Meteorsteinfäll am 22 Mai 1868, bei Slavetice. *Zweiter* Bericht. Vorgelegt in d. Sitzung. am 3 Dec. 1868. 8vo. 12 pp. 1 plate. (*Sitzb. d. k. Akad. d. Wissensch.* ii. Abth. Heft Jahr. 1868, lviii.) *Wien*, 1868

† Die südwestlichen Blitzkugeln am 20 Octbr. 1868. Nachträge zu der Mittheilg. am 5 Novbr. 8vo. 2 pp. (*Sitzb d. k. Akad. d. Wiss.* Dec. Heft 1868 lviii. Bde.) *Wien*, 1868

† Ein kugelförmiger Blitz am 30 Aug. 1865, gesehen zu Feistritz bei Peggau in Stiermark. 8vo. 4 pp. (*Sitzb. d. k. Akad. d. Wiss.* Dec. Heft, 1868, lviii. Bde.) *Wien*, 1868

† Licht, Wärme und Schall bei Meteoritenfällen. Bemerkungen. 8vo. 50 pp. (*Sitzungsb. d. k. Akad. d. Wiss.* lviii. ii. Abth.) *Wien*, 1869

Der Meteorit. von Goalpara in Assam, nebst Bemerkungen über die Rotation der Meteoriten in ihrem Zuge. 8vo. 2 plates. *Wien*, 1869

† Hessle, Rutlam, Assam. 3 neue Meteoriten. Vorläufiger Bericht. 8vo. 7 pp. (*In Sitzungsb. d. k. Akad. d. Wiss.*) *Wien*, 1869

† Bemerkungen zu Herrn Dr. Stanislaus Meunier's Note über den Victorit oder Enstatit von Deesa. Preise für aufzusuchende Meteorsteine aus altbekannten Fällen, von welchen unsere Museen noch nichts besitzen. 8vo. 6 pp. (*Sitzungsb. d. k. Akad. d. Wiss.* Jan. Heft, Bde. lxi. Jan. Heft.) *Wien*, 1870

† Die zwei homerischen Meteoreisenmassen von Troja Nachtrag zu den Mittheilungen über dieselben vom 6 October, 1864. 8vo. 8 pp. (*Sitzb. d. k. Akad. d. Wissensch.* ii. Abth. Jan. Heft 1870, lxi. Bde.) *Wien*, 1870

† Der Ainsa-Tucson-Meteoreisenring in Washington und die Rotation der Meteoriten in ihrem Zuge. 8vo. 16 pp. 1 plate. (*Sitzb. d. k. Akad.* lxi.) *Wien*, 1870

† Herrn Director Julius Schmidt's Beobachtung der Meteore in der Nacht des 13-14 November, 1866. 8vo. 4 pp. (*Sitzb. d. mathem.-naturwiss,* Cl. liv. p. 771.) *Wien*

† **Hajech.** Sul telegrafo stampante ideato dal C. Mezzanotte. 4to. 1 p. (*Atti dell' I. R. Istit. Lomb.* ii. 102.) *Milano*, 1860

Haldat du Lys, Charles Nicolas A. de. *Born December* 24, 1770, *at Bourmont, Lorraine; died November* 26, 1852, *at Nancy.*

Recherches sur l'incoercibilité du fluide magnétique. (*Mém. de l'Acad. de Nancy*, 1830.) *Nancy*, 1830

Haldat du Lys, Charles Nicolas A. de.—*continued.*

Extraits du Précis des travaux de la Société Royale des Sciences . . . de Nancy pendant les an. 1829 à 1832; publiés en 1833. Articles de M. le Dr. de Haldat, Sec. de cette Académie. 8vo. 18 pp. *Nancy,* 1833

Notice sur la vitesse avec laquelle s'exerce l'influence magnétique. (*Mém. de l'Acad. de Nancy,* 1838.) *Nancy,* 1838

1. Note sur la condens. de la force magnét. vers les surfaces des Aimants.
2. Expér. nouvelles sur le Magnét. par Rotation. Lettre à M. Arago.
3. Observ. d'une Couronne autour de la Lune.
8vo. (*These three works compose a brochure of* 16 pp.) *Nancy,* 1838

† Recherches sur quelques phénomènes du magnétisme, le fantôme magnétique, et sur la diffraction complexe. 8vo. 45 pp. (*Ext. des Mém. de la Soc. Royale* . . . *de Nancy pour* 1839.) *Nancy,* 1840

Recherches sur la généralité du magnétisme, ou complément des expériences de Coulomb sur le même sujet. (*Mém. de Nancy,* 1841.) *Nancy,* 1841

† Recherches sur la puissance motrice et l'intensité des courants de l'Elect. dynamique. 8vo. 54 pp. *Lyon,* 1842

Sur la force coercitive de la polarité des aimants sans cohésion. (*Mém. de l'Acad. de Nancy,* 1845.) *Nancy,* 1845

† Hist. du Magnétisme dont les phénomènes sont rendus sensibles par le mouvement. 8vo. 49 pp. 1 plate. *Nancy,* 1845

† Deux mém. sur le Magnétisme. 8vo. 40 pp. 1 plate. (*These are* (1) *Recherches sur l' Universalité de la Force magnétique,* p. 3; (2) *Recherches sur l'application de la Force magnétique,* p. 33.) *Nancy,* 1846

† Essai historique sur le magnétisme et l'universalité de son influence dans la nature. 1st edition. 8vo. 21 pp. *Nancy,* 1849

† Essai historique sur le Magnétisme et l'universalité de son influence dans la nature. 2de ed. revisée et corrigée. 8vo. 29 pp. *Nancy,* 1850

† Note sur le fantôme magnétique. 8vo. 12 pp. (*Ext. des Mém. de la Société* . . . *de Nancy.*) *Nancy,* 1850

† Exposition de la doctrine magnétique, ou Traité philosophique, historique, et critique du magnétisme. 8vo. 320 pp. 4 plates. *Nancy & Paris,* 1852

† Nouvelles recherches sur l'attraction magnétique . . . à l'appui d'un mém. sur l'universalité du Magnétisme. 8vo. 19 pp. (*Ext. Soc. Royale de Nancy.*) *Nancy*

† Recherches sur la cause du Magnétisme par Rotation, and (at p. 12) Recherches sur la généralité du Magnétisme. 8vo. 53 pp. 1 plate. (*Lamont, Handb. says* 1841.) *Nancy,* 1841

† **Hales,** S. Extract of a letter from Hales, S. to Hall, W. concerning some Electrical Experiments. Read June 30, 1748. 4to. 3 pp. (*Phil. Trans.* xlv. 409.) *London,* 1748

Some considerations on the causes of Earthquakes. 4to. (*Phil. Trans.* 1750.) *London,* 1850

Hall. Ephemerides. An. 1777. Appendix auroræ theoria. 8vo. *Vindoboni,* 1777

Dissertatio de Electricitate. (*From Manduit,* p. 111.)

† **Hall,** J. Suggestion for establishing a Telegraphic intercourse between London and Dublin. 8vo. 2 pp. (*Phil. Mag.* xxxiv. 124.) *London,* 1809

† A Meteor of uncommon magnitude and brilliancy seen in the vicinity of Middlebury, United States, on the evening of 17th June. Scientific account of. 8vo. 1 p. (*Phil. Mag.* lii. p. 236.) *London,* 1818

Hall. (*Vide* Palmer and Hall.)

Hall, Marshall. On the condition of Muscular Irritability in the paralytic muscles. (*Med. Chir. Trans.* Ser. ii. t. iv.)

Hallaschka, F. I. C. Dissertatio de phenomenis electro-magneticis, &c.
Pragæ, 1822

Hallé, J. N. Rapport au nom de la commission nommée pour répéter les expériences sur le galvanisme. 4to. (*Soc. Philomath.* an. 1798, p. 131.)
Paris, 1798

Exposition abrégée des principales expériences répétées par M. Volta. (*Bulletin des Sciences de la Soc. Philomathique,* an. 10, No. 58.)

Compte rendu à l'Institut National sur le Galvanisme. *Paris*

("*Rapport fait au nom d'une commission nommée par l'Institut, pour examiner et vérifier les phénomènes galvaniques.*" *This commission was composed of Coulomb, Sabathier, Pelletan, Charles, Fourcroy, Vanquelin, Guyton, and Hallé. Venturi of Modena joined it . . . and Humboldt also, "pour répéter celles (experiments) qui sont contenues dans l'article* vi. *et qui ont été faites au mois de prairial de l'an.* 6.")

Détail des expériences faites à l'Ecole de Médecine de Paris, sur le Galvanisme. (*Jour. de Médecine de Corvisart, &c.* . . i. Nivose. an. 9, p. 351.) *Paris,* 1801

Les résultats des expériences faites par le Cit. Hallé sur les malades qu'il a traité dans les Cabinets de l'Ecole de Médecine de Paris. (*Jour. de la Soc. Philomatique,* Messidor, an. ix. (1801.) *Paris,* 1801

† **Hallencreutz, D.** Beobachtung an Gewitterwolken welche Blitze gegen einander geben zu Pello innerhalb des Polarkreises. 8vo. 3 pp. 1 plate. (*K. Schwed. Akad. Abh.* xxxv. 85.) *Leipzig,* 1773

Haller, Albrecht von. *Born October* 16, 1708, *at Bern ; died December* 12, 1777, *at Bern.*

Observation sur la sensibilité des nerfs et des tendons. 4to. (*Mém. de Paris,* ann. 1753, *Hist.* p. 136.) *Paris,* 1753

Elementa Phisiologiæ corp. humani. 8 tomi. 4to. (*From Volta.*)
Lausan, 1757-78

Disputationes ad morborum historiam et curationem facientes. (Vol. i. p. 19 *et seq.*) 7 vols. 4to. *Lausan,* 1757-60

Halley, Edmund. *Born October* 29, 1656 (*A. St.*), *at Haggerston, near London ; died January* 14, 1724 (*A. St.*), *at Greenwich.*

† Theory of the Variation of the Magnetical Compass. 4to. 14 pp. (*Phil. Trans.* 1683, p. 208.) *London,* 1683

Theoria variationis pixidis magneticæ. (*Opusc. Act. Erudit. Lips.* tom. i. p. 244.)

† Theory of the Variation of the Magnetical Compass. 4to. 12 pp. (*Phil. Trans.* vol. xiii. 1683, p. 208.) *London,* 1683

† An account of the cause of the change of the Variation of the Magnetical Needle ; with an hypothesis of the Structure of the internal parts of the Earth. 4to. 16 pp. 1 plate. (*Phil. Trans.* 1693, p. 563.) *London,* 1694

Tabula nautica. Fol. (*From Watt.*) 1700

A General Chart shewing, at one view, the Variation of the Compass, &c. 1701

(*Poggendorff,* i. 1095, *says :* "*Erste Declinations karte, Frucht seiner Reisen von 1698-1700.*")

Sur la déclinaison de l'aimant. 4to. (*Mém. de Paris,* 1701, *Hist.* p. 9.)
Paris, 1701

Sur la déclinaison de l'aimant. 4to. (*Mém. de Paris,* 1706, *Hist.* p. 3.)
Paris, 1706

† Some remarks on the Variations of the Magnetical Compass, published in the Memoirs of the Royal Academy . . . with regard to the general Chart of those Variations made by E. Halley ; also concerning the true longitude of the Magellan Straits. 4to. 4 pp. (*Phil. Trans.* 1714, p. 165.)
London, 1714

222 HAL–HAM

Halley, Edmund—*continued.*

† An account of several extraordinary Meteors or Lights in the Sky. 4to. 6 pp. (*Phil. Trans.* 1714-15-16, p. 159.) *London*, 1714

A description of the Phenomena of March 6, 1716, as it was seen on the ocean near the coast of Spain ; with an account of the return of the same sort of appearance on March 31 and April 1 and 2 following. 4to. (*Phil. Trans.* 1716, p. 430.) *London*, 1716

† An account of the late surprizing appearance of the Lights seen in the air on the 6th March, 1716 ; with an attempt to explain the principal phenomena thereof. 4to. 23 pp. 2 plates. (*Phil. Trans.* 1716, p. 406, T. p. xxix. for 1714-15-16.) *London*, 1717

† An account of the extraordinary Meteor seen all over England on the 19th March 1719, with a demonstration of the uncommon height thereof. 4to. 13 pp. (*Phil. Trans.* 1719, p. 978.) *London*, 1719

An account of the phenomena of a very extraordinary Aurora Borealis, seen at London on November 10, 1719, both morning and evening. (*Phil. Trans.* 1719, p. 1099.) *London*, 1719

Observation sur les coups de Tonnerre multipliés et extraordinaires. 4to. (*Mém. de Paris*, 1731, *Hist.* p. 19.) *Paris*, 1731

Tabulæ nauticæ variationes magneticas denotantes. With an account of the Improvements made therein by W. Mountain. Fol. (*From Watt.*) *London*, 1758

Hallstrom, Gustav Gabriel. *Born November 25, 1775, at Ilmola-Socken, Wasa-Län; died June 2, 1844, at Helsingfors.*

Dissertatio de variationibus declinationis magneticæ diurnis ; et animadversiones circa hypotheses ad explicandas variationes diurnas excogitatas. *Abo*, 1803

† De apparitionibus Auroræ borealis, in septentrionalibus Europæ partibus. 4to. 14 pp. (*Pogg. says In Act. Soc. Scient. Fenn.* ii. 1847.)

Hamberger, Georg Erhard. *Born December 21, 1697, at Jena; died July 22, 1755, at Jena.*

De partialitate acus magneticæ. 4to. (*From Lamont, Handb.* p. 435.) *Jena*, 1727

Elementa physices methodo mathem. . . . conscritta . . . cap. x. 3 Ausg. 576 in scholiam. 8vo. (*From Krunitz*, p. 154, *and from Waitz.*) *Jena*, 1761

Gralath's account of Hamberger's experiment on the Attraction of Bodies by the Barometer tube and mercury. (*Danzig Mém.* ii. 426.)

Hamburgisches Magazin · · · **oder gesammelte Schriften zum Unterricht und Vergnugen.** 28 vols. 8vo. *Hamburg*, 1746-63

Neues Hamb. Magazin oder Fortsetzung gesammelter Schriften. 20 vols. *Hamburg*, 1767-84

Hamel. (*Vide* Duhamel, J. B.)

Hamel, T. The Life, or some account, of Schilling von Canstadt, in which Schilling's Electric Telegraph and mining apparatus for exploding are mentioned. 4to. (*Bull. Acad. Petersb.* ii. 1860.) *Petersburg*, 1860

Die Entstehung der galvanischen und elektromagnetischen Telegraphie. 8vo. (*Mélanges physiques et chim. tirés du Bulletin de l'Acad. de St. Pétersbourg*, iv. *Livraison* 2 *ou* 3, pp. 227—284.) *St. Pétersbourg*, 1860

† **Hamel and Cooke.** Historical account of the introduction of the Galvanic and Electro-magnetic Telegraph into England by Dr. Hamel ; read before and reprinted by the Society of Arts, with comments thereon by Wm. F. Cooke. 12mo. 79 and xvi. pp. 1 cut. *London*, 1859

Hamilton, Hugh. *Born March 26, 1729, at Grafsch, Dublin; died Dec.* 1, *1805, at Ossory.*

Phil. essays . . observations, &c. on the Aurora, &c. 12mo. *London,* 1767

† Osservazioni e congetture sopra la natura dell' Aurora boreale e le code delle Comete. 12mo. 59 pp. (*Scelta d'Opuscoli,* xxxi. 3.) *Milano,* 1776

† **Hamilton, W.** Particulars respecting Hail-storms in the West Indies. 8vo. 3 pp. (*Phil. Mag.* xliv. 191.) *London,* 1814

Hammer, H. G. Die Electricität als fortlaufend bildende und erhaltende Kraft, von ihrem atomistischen Ursprung bis zur vollständigen Ausbildung der Organe. 8vo. 144 pp. *Dresden,* 1855

Hanaw. Nachricht aus St. Petersburg.

(*Note.—Poggendorff* (ii.634) *has under the head "Richman," H.* (*Hanov.*)*"Nachricht aus St. Petersburg vom* 3 *Aug. von dem* . . *Todesfall d. H. Prof. Richmann mit phys. Anmerk. begleitet."* *Frankfurt gelehrt Zeitung,* 1753.)

Erläuterte Merkwürdigkeiten. (*From Van Swinden,* i. 416. *Refers to magnet* p. 354.)

Hanke, J. Horologium nocturnum magneticum, &c. (*From Pogg.* i. 1011.) *Olomut.* 1683

† **Hankel,** Die Gesetze der Krystallelektricität. 4to. 28 pp. *Halle,* 1840

Hankel, H. Über die Construction eines Electrometers. Messungen d. Anziehung zwischen einem Eisenkern und eine ihn umgebende elektrische Spirale.

Leipzig, 1851

Hankel, Wilhelm Gottlieb. *Born May 17,* 1814, *at Ermsleben Reg. Bez. Merseburg.*

† De thermo electricitate crystallorum. 8vo. 32 pp. 1 plate. *Halle,* 1839

Many articles on Electricity, Magnetism, &c. in Poggendorff's Annalen, vols. xix. (1840) to ciii. (1858). (*Pogg.* i. 1011.)

† Über die Magnetisirung von Stahlnadeln durch den elektrischen Funken und den Nebenstrom desselben. 4to. 28 pp. *Halle,* 1845

† Grundriss der Physik. 8vo. 327 pp. *Stuttgart,* 1848

† Mittheilung einiger Versuche üb. d. Elektricität der Flamme u. die hierdurch erzeugten elektrischen Ströme. 4to. 15 pp. (*Poggendorff,* Ann. lxxxi. (1850). *Leipzig,* 1850

† Messungen d. Abstossung d. krystall Wismuths durch d. Pole eines Magnets mittelst. d. Drehwage. 8vo. 20 pp. (*König. Sässische Gesellsch. d. Wissen.*) *Leipzig,* 1851

† Elektrische Untersuchungen. Erste Abhandlung über die Messung der atmosphärischen Elektricität, nach absoluten Maassen. 4to. 219 pp. 2 plates. (*König. Sächsische Gesellschaft der Wissensch.* vol. v.) *Leipzig,* 1856

† Elektrische Untersuchungen Zweite Abhand. Über die thermoelektrischen Eigenschaften des Boracites. 4to. 101 pp. (*König. Sächische Gesells. der Wissensch.* vol. vi.) *Leipzig,* 1857

† Elektrische Untersuchungen Dritte Abhandl. Über Electricitäts erregung zwichen Metallen und erhitzten Salzen. 4to. (*König. Sächische Gesells. der Wissen.* vol. vi.) *Leipzig,* 1858

† Electrische Untersuchungen Vierte Abhandl. Über das Verhalten der Weingeist-flamme in elektrischen Beziehung. 4to. 79 pp. (*König. Sächische Gesells. d. Wissen.* vol. vi.) *Leipzig,* 1859

† Electrische Untersuchungen Fünfte Abhandl. Maassbestimmungen der elektromotorischen Kraft. Erster Theil. 4to. 52 pp. (*König. Sächische Gesells. d. Wissen.* vol. ix.) *Leipzig,* 1861

† Elektrische Untersuchungen. Abhandl. vi. Maassbestimmungen der elektromotorischen Kräfte. Part ii. 4to. 106 pp. (*König. Sachischen Gesells. d Wissen.* vol. xi.) *Leipzig,* 1865

Hankel, Wilhelm Gottlieb—*continued*

Neue Theorie der elektr. Erscheinungen (2 Abhandl.) 8vo. 1865

† Elektrische Untersuchungen, Abhandl. vii. Über die thermoelektrischen Eigen schaften des Bergkrystalles. 4to. 70 pp. 2 plates. (*Abhandl. d. K. S. Gesells. d. Wissen.* xiii. *Leipzig.*) *Leipzig*, 1866

† Elektrische Untersuchungen, Abhandlung viii. Uber die thermoelektrischen Eigenschaften des Topases. 4to. 96 pp. 4 plates (coloured). (*Abhandl. d. mathem. phys. Classe d. K. Sächsische Gesells. d. Wissen.* ix. *Bandes.* (No. iv.) *Leipzig*, 1870

† Elektrische Untersuchungen Neunte Abhandlung. Über die thermoelektrischen Eigenschaften des Schwerspathes. 69 pp. 8vo. (*Abhandl. d. Mathem. phys. Classe d. K. Sächsischen Gesells. d. Wissenschaft,* vol. x. No. iv.) *Leipzig*, 1872 (*Note.—There are two title-pages; one of them without the designation of the Royal Saxon Society, by which it was printed.*)

† Elektrische Untersuchungen Zehnte Abhandlung. Über die Thermoelektrischen Eigenschaften des Aragonites. 70 pp. 3 plates. (*Abhandlung. d. K. Sächsischen Gesells. d. Wissenschaft,* No. 5, vol. x.) *Leipzig*, 1872

† **Hänle, C. F.** Die Ursache der innern Erdwärme der Entstehung des Erdplaneten, der Feuerkugeln, Sternschnupfen und Meteorsteine. 8vo. 78 pp. *Lahr*, 1851

† Galvano-Epikalymmatik oder hydroelektrische Metallüberziehung Vergoldung, Versilberung, &c. auf galvanischem Wege. Nebt dem Neuesten . . über Galvanoplastik, &c. . . 2te sehr vermehrte Auflage. *Lahr*, 1857

Hannemann, J. L. De fulminis effectu miro. (*Miscell. Acad. Nat. Cur.*) 1685

De pisce torpedine ejusque proprietatibus admirandis. (*From Pogg.* i. 1012.)

Hanover. Hannoversche gelehrte Anzeigen-nützliche Sammlungen, Beiträge und Hannoversches Magazin seit 1750 nach einander in 4to. herausgegeben. (*From Reimarus Vom Blitz,* 1778, p. xliii.)

Hanover Magazin. (Contains many articles on Thunder, Lightning, &c. in 1781, 1782, &c.)

† **Hanow, M. C.** Verschiedene neue Versuche mit den gläsernen Spring-Kölbchen 4to. 10 pp. 1 plate. (*Versuche d. Naturf. Gesellsch. in Danzig,* i. 534.) *Danzig*, 1747

† Erläuterte Ursachen der Versuche mit den Spring-Kölbchen. 4to. 69 pp. (*Versuche d. Naturf. Gesellsch. in Danzig,* iii. 328.) *Danzig*, 1756

Hansteen, Christopher. *Born Sept.* 26, 1784, *at Christiania.*

† Untersuchungen über den Magnetismus der Erde, übersetzt von P. F. Hanson. Erster Theil. Die mechanischen Erscheinungen des Magneten mit Anhang enthalt. Beobachtungen der Abweichung und Neigung der Magnetnadel. 4to. 502 and 148 pp. 5 plates, atlas 7 plates. *Christiania*, 1819

† Ueber die tägliche Veränderung der Intensität des Erd-Magnetismus, und den Magnetismus vertikalstehender Körper. Vorgel. im März, 1821. *Christiania*, 1821

† Observations on Magnetism. 8vo. 4 pp. (*Phil. Mag.* lix. 248.) *London*, 1822

† About 11 articles in the "Mag. for Naturvidenskaberne" edited by him in conjunction with Lundh and Maschmann (1823 to 1853); also about 6 in "Schweigger's Journal," 1813 to 1827, and about 10 in Poggendorff's "Annals" (1825 to 1855.)

† On the Polar lights, or Auroræ Boreales and Australes. 8vo. 11 pp. 1 plate. (*Phil. Mag. or Annals,* ii. 334.) *London*, 1827

† Periodisk forandring i Jordens magnetiske Intensitet, som er afhængig af Maanebanens Beliggenhed. 8vo. 34 pp. (*Nyt Magasin for Naturvidenskaberne,* ii.) *Christiania*, 1839

Hansteen, Christopher—*continued.*

† Magnetiske Jagttageleser, anstillende paar et Togt i Middelhavet med den Norske Corvet Ornen i Sommeren 1840, af Capne. Konow og Valeur ; middeelte af Chr. Hansteen. 8vo. 26 pp. (*Nyt Mag. for Naturvidenskaberne*, vol. iii.) *Christiania*, 1841

† Magnetiske Jagttageleser paar en Reise igjennem Danmark og en Deel af det nordlige Tydskland i Sommeren 1839. 8vo. 43 pp. 1 plate (curves). (*Nyt Mag. for Naturvidenskabene*, vol. iii. p. 227.) *Christiania*, 1842

† Bidrag til Bestimmelsen af forskjellige Constanter for Christiania. 8vo. 84 pp. 2 plates. (*Nyt Mag. for Naturvidenskabene*, vol. iii.) *Christiania*, 1842

† Disquisitiones de mutationibus quas patitur momentum acus magneticæ. 4to. 44 pp. 1 plate. *Christiania*, 1842

† Bidrag til Bestemmelsen af forskjellige Constanter. (*Fortsat fra 3 die Bds. forste Hefte* 584.) 8vo. 21 pp. (*Nyt Mag. for Naturvidenskabene*, vol. iii.) *Christiania*, 1842

† Magnetiska Terminsiagttageleser i Christianias magnetiska Observatorium, middeelte af C. H. 8vo. 11 pp. (*Nyt Mag. for Naturvidenskabene*.) *Christiania*, 1842

† Magnetiska Jagttageleser anstiblede paa et Ovelsestogt med den Norske Corvet Ornen til Lissabon og de Azoreske oer i Sommeren 1841, af Expeditionens Officierer, middeelte af C. H. 8vo. 5 pp. (*Nyt Mag. for Naturvidenskabene*, vol. iii.) *Christiania*, 1842

† Minimum af Magnetnaalens Inclination i Christiania. (*Tillæg til Undersogelesen* S. 273-282.) 8vo. 3 pp.(*Nyt Mag. for Naturvidenskabene*, vol. iii.) 8vo. 3 pp. *Christiania*, 1842

Interpolations formula for Magnetnaalens Misvisw. og Helding, &c. 4to. *Christiania*, 1844 ?

† Magnetiske Jagttageleser anstillede paa forskjellige Söereiser i Atlanterhavet og Middelhavet af den Norske Marines Officener, samt paa en Reise til Stockholm ; meddelta af C. H. 8vo. 15 pp. (*Nyt Mag. Naturvidenskabene*, vol. iv.) *Christiania*, 1845

† Meteorologiske Constanter for Christiania. 4to. 51 pp. and 5 pp. (*Nyt Mag. fur Naturvid.* v. 374 and vi. 51 pp.) *Christiania*, 1845-51

† Sur la diminution de l'inclinaison magnétique en Europe. (Lettre adressée le 22 Septembre, 1853, à M. Quételet par M. Hansteen.) 8vo. 18 pp. (*Acad. Rle. de Belgique : Ext. du* v. xx. No. 10, *des Bulletins*.) *Bruxelles*, 1853

† Sur les Aurores Boréales et sur l'inclinaison magnétique à Bruxelles. Lettres de M. Hansteen communiquées par M. A. Quételet, Sec. de l'Acad.) 8vo. 24 pp. (*Extr. du* tom. xxi. No. 5 *des Bulletins*.) *Bruxelles*, 1855

† Den magnetiske Inclinations Forandring i den nordlige tempererte Zone 4to. 71 pp. 1 chart. *Kjobenhavn*, 1855

† Den magnetiske Inelinations Forandringer i den nordlige og sydlige Halvkugle 4to. 46 pp. (*Fortsaettelse af Afhandlingen i Kgl. Danske Videnskabernes Selskabs Skriæftez fente Rkke*, 4de *Bands*, 1ste *Hefte*, 1856, s. 99—167.) *Kjobenhavn*, 1857

† Den magnetiske Inclinations periodiske Forandringer. Till K. vet. Akad. inlemnad d. 3 Mars, 1857. 4to. 22 pp. 1 plate. (*K. Vet. Akad. Handl.* ii. No. 2.) *Kjobenhavn*, 1857

† Physikalske Meddelelser ved. A. Arndtsen, Efter Foranstaltning af det akademiske Collegium, udgivne af Chr. Hansteen. Universitäts-Program for 2det Semester, 1858. 4to. 75 pp. 2 pl. *Christiania*, 1858

† Das magnetische System der Erde. 8vo. 4 plates. 36 pp. (*Zeitschr. f. pop. Mitth.* i. 33.) *Altona*, 1859 ?

† Sur les Aurores Boréales et sur l'inclinaison Magnétique à Bruxelles ; Lettres de M. Hansteen communiquées par M. A. Quetelet. 8vo. 24 pp. (*Acad. Royale de Belgique. Extr. du* tom. xxi. No. 5 *des Bulletins*.) *Bruxelles*

† Jordmagnetiske iattagagelser. 4to. 8 pp. *Upsala*, 1866

† **Hansteen und Due.** Resultate magnetischer astronomischer und meteorologischer Beobachtungen, auf einer Reise nach dem östlichen Sibirien in den Jahren 1828-30. Anhang . . Beobacht . . 4to. 189 pp. *Christiania,* 1863

† **Hansteen, C.** und **Fearnley, C.** Die Universitäts-Sternwarte in Christiania. 4to. 3 plates. *Christiania,* 1849

Happach, Lr. H. F. Gf. Beobacht. u. Erklär. merkwürd Naturerscheinn. 8vo. *Quedlinb.,* 1812

† **Harding.** Observations on the Variations of the Needle. 4to. 11 pp. (*Dublin Acad.* (?) vol. iv.) *Dublin,* 1791

† **Hardy, C.** Description of an improved Telegraph. 8vo. 6 pp. 1 plate. (*Phil. Mag.* xxxiii. 343.) *London,* 1809

Hardy, E. Aperçu sur la Théorie du magnétisme terrestre de M. Pariset. 8vo. 12 pp. *Paris,* 1862

Hare, Robert. *Born January 17, 1781, at Philadelphia; died May 15, 1858, at Philadelphia.*

A new Theory of Galvanism . . Calorimotor, &c. 8vo. 17 pp. (*From Dove,* p. 185.) *Philadelphia,* 1819

† A new Theory of Galvanism, supported by some Experiments and Observations made by means of the Calorimotor, a new galvanic instrument; also a new mode of decomposing Potash extemporaneously. 8vo. 10 pp. 1 plate. (*Phil. Mag.* liv. 206.) *London,* 1819

† A Memoir on some new modifications of Galvanic apparatus, with Observations in support of his theory of Galvanism. 8vo. 11 pp. 1 plate. (*Phil. Mag.* lvii. 284.) *London,* 1821

† Description of an electrical Plate machine; the plate mounted horizontally, and so as to show both negative and positive Electricity. 8vo. 2 pp. (*Phil. Mag.* lxii. 8.) *London,* 1823

† An Essay on the question, Whether there be two Electrical fluids, according to Du Faye, or one according to Franklin? 8vo. 5 pp. (*Phil. Mag.* lxii. 3.) *London,* 1823

† An Essay on the question whether there be two Electrical fluids or one? Also a description of an Electrical Plate Machine . . &c. 8vo. 10 pp. 1 plate. (*Originally in the Phil. Mag.* lxii. pp. 3 and 8, &c.) 1823

† On the construction and applications of the improved Sliding-rod Eudiometer, and of the Volumescope 8vo. 9 pp. (*Phil. Mag. or Annals.* vi. pp. 114 and 171.) *London,* 1829

† An examination of the question . . (relative to) Mechanical Electricity and the Galvanic fluid. 8vo. 11 pp. *Philadelphia,* 1836

† Experimental observations, &c., with theoretical suggestions respecting the causes of Tornadoes, Falling Stars, and the Aurora Borealis. 4to. 60 pp. 1 plate. *Philadelphia,* 1836

Description of an Electrical machine, with a plate 4 ft. in diameter, so constructed as to be above the operator; also of a Battery discharger employed therewith, and some observations on the causes of the diversity in the length of the Sparks erroneously distinguished by the terms Positive and Negative. 4to. (*Amer. Phil. Soc. Trans.* new series, vol. v.) *Philadelphia,* 1837

Sundry improvements in apparatus or manipulation. 4to. (*Amer. Phil. Soc. Trans.* new series, vol. v.) *Philadelphia,* 1837

On the causes of the Tornado, or Water-spout. 4to. (*Amer. Phil. Soc. Trans.* new series, vol. v.) *Philadelphia,* 1837

† Engraving and description of a Rotatory-multiplier, or one in which one or more needles are made to revolve by a galvanic current. 4to. (*Trans. Amer. Phil. Soc.* new series, vol. vi. p. 343.) *Philadelphia,* 1839

† Improved process for obtaining Potassium. Read Dec. 7, 1838. 4to. 2 pp. (*Trans. Amer. Phil. Soc.* vol. vi. p. 341.) *Philadelphia,* 1839

Hare, Robert—*continued.*

† Engraving and description of an apparatus for the decomposition and recomposition of Water. . . Read Dec. 7, 1838. 4to. 2 pp. (*Trans. Amer. Phil. Soc.* new series, vol. vi. p. 339.) *Philadelphia,* 1839

† Communication faite à la Société Philosophique Américaine . . 1839, au sujet des Trombes, et relativement à un mémoire de M. Peltier, sur la cause de ces Météores, par R. Hare. . . Accompagnée de la traduction d'un mémoire sur les causes des Trombes, publié par le même auteur dans les Trans. de la susdite Société, vol. v. 8vo. 12 pp. *Philadelphia,* 1840

Of the conclusions arrived at by the Committee of the Academy (Paris) agreeably to which Tornadoes are caused by Heat. 8vo. 89 pp.
Philadelphia, 1852

Of the conclusion arrived at by a Committee of the Academy of Sciences of France, agreeably to which Tornadoes are caused by Heat. 2nd edition. 8vo.
Philadelphia, 1852

† A Memoir on some new modifications of Galvanic apparatus, with observations in support of his new theory of Galvanism. 8vo. 17 pp. 1 plate. (*n. d.*)

† A letter to Prof. Faraday on certain Theoretical opinions. 8vo. (*Amer. Journal of Science,* No. 1, vol. xxxviii.) *New York?*

Description of an apparatus for deflagrating Carburets, Phosphurets, and Cyanides in vacuo, or in an atmosphere of Hydrogen; with an account of some results obtained by these and other means, especially the isolation of Calcium. 4to. (*Trans. Amer. Phil. Soc.* new series, vol. vii.)
Philadelphia, 1841

† **Hare and Allen.** Account of a Tornado which, towards the end of August, 1838, passed over the suburbs of the City of Providence, in the State of Rhode Island, and afterwards over a part of the village of Somerset; also an extract of a letter on the same subject from Z. Allen, Esq., of the City of Providence. 4to. 5 pp. (*Trans. Amer. Phil. Soc* new series, vol. vi. p. 297.)
Philadelphia, 1839

Harriot (or **Hariot**), Thomas. *Born* 1560, *at Oxford; died July* 2, 1621, *at London.*

On Magnetic Variations. (*From Gilbert, and vide Pogg.* i. 1019.)
(*Note.—One of the " Viri docti qui in longinquis navigationibus variationis magneticæ differentias observaverunt."*)

† **Harles,** C. F. Andeutungen einer pathogisch. Elektrologie und insbesondere einiger vorzugsweise electrischen Krankheiten. 4to. 80 pp. (*Abhandl. der Physik. Medic. Soc. zu Erlangen,* B. i. s. 52.) *Frankfürt,* 1810
Der Republikanismus in der Naturwissenschaft und Medicin auf der Basis, und unter der Aegide, des Electricismus. 8vo. (*From Heinsius,* vi. 335.) *Bonn,* 1819

Harpez de la Force. Proposal of a Prize Essay. (*Danzig Mem.* i. 261.)

Harrington, R. (or Anonymous). A new system of Fire, &c., &c.; also an elucidation, &c. . . Electricity. 8vo. *London,* 1796

Harrington. Experiments and observations on Volta's Pile, &c. 8vo.
Carlisle, 1801
The death-warrant of the French system of Chemistry, &c. 8vo. *London,* 1804

Harris, John. *Born about* 1667; *died September* 17, 1719.

Navigantium . . Bibliotheca, or a complete collection of Voyages, &c. . . Introduction . . Loadstone. 8vo. (*From Watt.*) *London,* 1702

Navigantium . . Bibliotheca, or a complete collection of Voyages, &c. . . Introduction . . Loadstone. 8vo. (*From Watt.*) *London,* 1705

† **Harris,** Joseph. An account of some Magnetical observations made in the months of May, June, and July, 1732, in the Atlantick, or Western Ocean; as also the description of a Water-spout. Communicated by Mr. George Graham, F.R.S. 4to. 5 pp. 1 plate. (*Phil. Trans.* xxxviii. for 1733-34, p. 75.)
London, 1835

* † **Harris,** Salem. Of the influence of Solar and Lunar attraction on Clouds and Vapours. 8vo. 6 pp. *(Phil. Mag.* xxxvi. 58.) *London,* 1810

Harris (afterwards Sir), William Snow. Electrical Conductors for Ships. Experiment in Plymouth Harbour. 8vo. 7 pp. *(Phil. Mag.* lx. 231.) *London,* 1822

† Observations on the effects of Lightning on floating bodies, &c. with an account of a new method of applying fixed and continuous Conductors of Electricity to the Masts of Ships. Letter to Sir T. B. Martin. 4to. 89 pp. 5 plates. *London,* 1823

(*Note.—The illustration accompanies plate* i. *The lines on the paper originally consisted of gold leaf. . . A discharge has been passed over the gold leaf to show by its deflagration the course of the electric matter.*)

On the Laws of Electric accumulation. 8vo. (*Trans. of the Plymouth Instit.* Nov. 1825.) *Plymouth,* 1825

† On the relative powers of various metallic substances as Conductors of Electricity. Read Dec. 14, 1826. 4to. 7 pp. 1 plate. (*Phil. Trans.*) *London,* 1827

† Experimental inquiries on Electrical accumulation. 8vo. 53 pp. 4 plates. (*From the Trans. of the Plymouth Institution.*) *Plymouth,* 1828

Inquiries concerning the Laws of Magnetic forces, containing a description and drawing of a new Magnetical instrument for measuring these forces. 4to. (*Edinb. Phil. Trans.* April 1829.) *Edinburgh,* 1829

† On the utility of fixing Lightning Conductors on Ships. 8vo. 23 pp. (*Roget's copy.*) *Plymouth,* 1830

† On the transient Magnetic state of which various bodies are susceptible. 4to. (*Phil. Trans. for* 1831, part i.) *London,* 1831

On the influence of Screens in arresting the progress of Magnetic action, and on the power of masses of iron to control the attractive force of a Magnet. 4to. (*Phil. Trans. for* 1831.) *London,* 1831

On a new Electrometer, and on the heat excited in Metallic bodies by Voltaic Electricity. 4to. (*Edin. Phil. Trans.* Dec. 1831.) *Edinburgh,* 1831

On the transient Magnetic state of which various substances are capable. Communicated by Davies Gilbert, Esq. 4to. 24 pp. 2 plates. *London,* 1831

On the employment of vibrating Magnetic bars in experimental inquiries relative to the Earth's magnetic intensity. 8vo. (*British Association Report for* 1832, vol. ii.) *London,* 1832

On the investigation of Magnetic intensity by the oscillation of the horizontal needle. 4to. (*Edinb. Phil. Trans.* Jan. 1834.) *Edinburgh,* 1834

† On some elementary laws of Electricity (first series), containing an account of several new instruments and discoveries in Electricity. 4to. 33 pp. 3 plates. (*Phil. Trans. for* 1834.) (*Faraday's copy.*) *London,* 1834

Three short papers on Electricity. 8vo. (*British Association Reports for* 1835.) *London,* 1835

† A series of Papers on the defence of Ships and Buildings from Lightning. 8vo. 46 pp. (*Nautical Magazine,* xxv.) *London,* 1835

On the nature of Electrical repulsion. 8vo. (*British Association Reports for* 1836.) *London,* 1836

† Inquiries concerning the Elementary Laws of Electricity. Second Series. 4to. 36 pp. 2 plates. (*Phil. Trans. for* 1836, part ii.) *London,* 1836

A series of three Papers, termed Illustrations of cases of damage by Lightning in the British Navy. (*Nautical Magazine for* 1838.) *London,* 1838

† State of the question relating to the protection of the British Navy from Lightning, by the method of fixed Conductors of Electricity, as proposed by Mr. Snow Harris. With appendix. 8vo. *Plymouth,* 1838

Harris (afterwards Sir), William Snow—*continued.*

The Bakerian Lecture. Inquiries in Electricity (third series), containing a further investigation of the Phenomena of Electrical induction and attraction. 4to. (*Phil. Trans. for* 1839, part ii.) *London,* 1839

† On Lightning Conductors, and on certain principles in Electrical science; being an investigation of Mr. Sturgeon's experimental and theoretical researches in Electricity, published by him in the *Annals of Electricity,* &c. 8vo. 12 pp. 1 plate. (*Phil. Mag for* Dec. 1839, p. 463.) (*Faraday's copy.*) *London,* 1839

† Commission appointed to inquire into the plan of W. S. Harris. Copy of the report and evidence relating to the protection of Ships from the effects of Lightning. Ordered by the House of Commons to be printed 11th Feb. 1840. Folio. 96 pp. 12 plates. *London,* 1840

On the course of the Electrical discharge, and on the effects of Lightning on certain ships of the British Navy. 8vo. (*From his Record of Phil. Papers,* p. 4.) (*Edinb. and Lond. Phil. Mag.* Feb. and March, 1840.) *London,* 1840

† On Lightning Conductors, and the effects of Lightning on H.M.'s ship *"Rodney"* and certain other ships of the British Navy; being a further examination of Mr. Sturgeon's Memoir on Marine Lightning Conductors. 8vo. 12 pp. 1 plate. (*Annals of Electricity,* iv. 484.) *London,* 1840

† Extract of a letter from W. S. Harris to W. Sturgeon. 8vo. 13 pp. (*Anns. of Electricity,* v. 208.) (*Roget's copy.*) *London,* 1840

On the supposed Electro-magnetical effects of Marine Lightning Conductors. (*Nautical Magazine,* Enlarged Series, No. 2, vol. for 1841.) 1841

On the specific conductive capacity of various Electrical substances. 4to. (*Phil. Trans. for* 1842.) *London,* 1842

Observations on the action of Lightning Conductors. (*Proc. London Elec. Soc. for* 1842.) *London,* 1842

† On the effects of Lightning on the British ship "*Underwood.*" 8vo. 8 pp. (*Nautical Mag. for* June 1842.) (*Roget's copy.*) *London,* 1842

† On the nature of Thunderstorms, and the means of protecting Buildings and Shipping against . Lightning. 8vo. *London,* 1843

A theoretical and practical view of Thunderstorms, and the protection of Buildings and Ships from Lightning. (*From his Record of Phil. Papers,* p. 6.) *London,* 1843

† On Damage by Lightning in the British Navy. 8vo. 66 pp. (*Extract from the Nautical Magazine,* 1843.) *London,* 1843

† Meteorology of Thunderstorms at Sea, with analytical deductions, and a history of the effects of Lightning on 210 ships of the Royal Navy. 8vo. *London,* 1844

Note.—The first part was printed in the "Nautical Magazine" after the second part containing the history of cases had been completed. The first part has 18 pp.; the second part is entitled "Damage by Lightning in the British Navy," and has 66 pp.

† Remarkable instances of the protection of certain Ships of H.M.'s Navy from the destructive effects of Lightning; collected from various authorities. 8vo. 18 pp. *Plymouth,* 1844

Remarkable instances of defence of certain Ships of the Royal Navy from the destructive agency of Lightning, with practical and theoretical deductions. (*From his Record of Phil. Papers,* p. 6.) *London,* 1846

Letter to the Secretary of the Incorporated Society for building Churches, &c. on the Preservation of Public Buildings from Lightning. 8vo. (*From his Record of Phil. Papers,* p. 6.) *Plymouth,* 1847

A Public Official Letter to the India Board, dated June 21, 1847, relative to a Board Order requiring all Transports to be fitted with his Conductors. (*From his Record of Phil. Papers,* p. 7.) 1847

Harris (afterwards Sir), **William Snow**—*continued.*

† Remarkable instances of the protection of certain Ships . . from the destructive effects of Lightning. . . 8vo. 61 pp. 2 plates. *London*, 1847

Instructions for the application of permanently fixed Conductors in H.M.'s ships, drawn up for the use of H.M.'s dockyards. Printed by order of the Lords Commissioners of the Admiralty. (*From his Record of Phil. Papers.*) *London*, 1848

Rudimentary Electricity. 1st edition. 12mo. 160 pp. *London*, 1848

† Letter to the Earl of Wilton on returns . . . relative to . . . fixed Metallic Conductors employed in H.M.'s Navy. 8vo. 35 pp. *Plymouth*, 1849

† Letter on the Preservation of Public Buildings from . . . Lightning (revised) addressed to the . . . Society for building Churches, &c. dated December, 1847. 8vo. 12 pp. *London*, 1850

† On the relative Cost and Efficiency of permanent and temporary forms of Lightning Conductors as applicable to the defence of the Royal Navy. 8vo. 27 pp. *Plymouth*, 1850

† Rudimentary Magnetism. . . Parts i. and ii. 12mo. 159 pp. *London*, 1850

† Remarkable instances of the Preservation of certain Ships of the Royal Navy from Lightning. Abridged from Official and other authenticated Reports. 8vo. 19 pp. *Plymouth*, 1850

Rudiments of Electricity. New edition. 12mo. *London*, 1851

† Rudimentary Magnetism. . . Part iii. 12mo. 186 pp. *London*, 1852

† Destruction of Merchant Ships. Shipwreck by Lightning. 8vo. 6 pp. (*Nautical Magazine for November,* 1852.) *London*, 1852

† Papers relating to Harris's Lightning Conductors and the destructive effects of Lightning on Ships. Fol. 11 pp. 2 plates. *London*, 1852

† A record of (his Philosophical Papers, &c. 8vo. 7 pp. *Plymouth*, 1852? (*Note.—Latimer Clark's Catalogue says probably* 1852.)

† Papers relative to Harris's Lightning Conductors. Appendix, with Addendum of 1 sheet. Fol. 37 pp. *London*, 1852

† Review of the History and Progress of the general system of Lightning Conductors . . in the Royal Navy. 8vo. 10 pp. (*Reprinted from the Nautical Magazine for March,* 1853.) 1853

† Shipwrecks by Lightning. Copies of papers relative to Shipwrecks by Lightning as prepared by Sir Snow Harris, and presented by him to the Admiralty. Fol. 82 pp. 5 plates. *London*, 1854

† On a general law of Electrical Discharge. 8vo. 21 pp. 1 plate. (*Phil. Mag. for May,* 1856.) *London*, 1856

† Rudimentary treatise on Galvanism and the general principles of Animal and Voltaic Electricity. . . 12mo. 215 pp. *London*, 1856

† On certain phenomena of Electrical Discharge. 8vo. 6 pp. (*Phil. Mag. for August,* 1856.) *London*, 1856

† Leçons élémentaires d'Electricité . . . traduites et annotées par Garnault. 12mo. 264 pp. cuts. *Paris*, 1857

† Researches in Statical Electricity. (No. 1.) 8vo. 27 pp. (*Phil. Mag.* 1857.) *London*, 1857

† On some special laws of Electrical Force. 8vo. 4 pp. (*Phil. Mag. for August,* 1857.) *London*, 1857

† Rudimentary Electricity ; being a concise exposition of the general principles of Electrical Science. . . Fifth edition. 12mo. 195 pp. 1 cut separate. *London*, 1859

Instructions for the application of permanently-fixed Conductors in H.M. ships, drawn up for the use of H.M. Dockyards. Printed by order of the Lords Commissioners of the Admiralty.

Harris (afterwards Sir), William Snow—*continued*.
† New Steering Compass, by Mr. Snow Harris. . . . Constructed by W. C. Cox, optician, Devonport. 8vo. 3 pp. (*Faraday's copy*.) *Plymouth*

On a new and improved form of the Mariners' Compass employed in several ships of the Royal Navy and in H.M. yacht, and in much repute in the merchant service. *From his Record of Papers*, p. 5. (*Reports of the Royal Polytechnic Society of Cornwall.*)

† A treatise on Frictional Electricity, in Theory and Practice. Edited, with a Memoir of the author, by Charles Tomlinson, F.R.S. 8vo. 291 pp.
London, 1867

† Rudimentary Electricity: showing the general principles of Electrical Science and the purposes to which it has been applied (by the late Sir W. S. Harris). Sixth edition (with additions). 12mo. 200 pp. *London*, 1868

Rudimentary Magnetism: a concise Exposition of the general principles of Magnetical Science, and the purposes to which it has been applied. Second edition, revised and enlarged, by H. M. Noad, Ph.D., F.R.S. &c.
London, 1872

† **Harrsch,** F. L. Pyrotechnia sublimis sæculi primævi vel liber meteorarum. 4to. 131 pp. 3 plates. *Viennæ*, 1778

Harsu, Jacques de. *Born* 1730, *at Geneva ; died* 1784, *at Geneva.*

Recueil des effets salutaires de l'Aimant, dans les maladies. 12mo. 276 pp.
Genève, 1783

Hartmann, George. *Born February* 9, 1489, *at Eckoltsheim, bei Bamberg ; died April* 9, 1564, *at Nürnberg.* (*Pogg.* i. 1023)

Briefwechsel mit dem Herzog Albrecht von Preussen. 8vo. (*Dove's Repertorium*, ii. 129.)

"Entdeckte (obwohl unvollkommen) diamagn. Inclination, so wie anderere Eigenschaften des Magnets (S. seinen Briefwechsel mit dem Herzog Albrecht von Preussen im. J. 1544. *Berlin*, 1544

(*Lamont, Astronomie u. Erdmagnetismus*, p. 250, *says, " Um eben diese Epoche entdeckte G. Hartmann . . . dass eine an ihrem Schwerpunkte aufgehängte Nadel nicht wagerecht sich stelle, sondern wie er sich ausdrückt ' unter sich zeige.'" . . . Carl. Elekt. Naturkraft, p. 8, mentions this letter. Dove in his Catalogue, p. 274, says he made the discovery in 1543.*)

† **Hartmann,** Johann Friedrich. *Born* . . . ; *died May* 30, 1800.

Abhandlung von der Verwandschaft und Ähnlichkeit der elect. Kraft mit den erschrecklichen Luft-Erscheinungen. 8vo. 253 pp. 1 plate. *Hannover*, 1759

Beschreibung eines elektr. Glockenspiels. 8vo. (*Hamb. Mag.* xxiv. 1759.)
Hamburg, 1759

Merkwrd. elektr. Versuche mit d. Verstärkungsflasche. 8vo. (*Hamburg Mag.* xxiv. 1759.) *Hamburg*, 1759

Verbesserter Versuch seines künstl. elektr. Blitzes. 8vo. (*Hamb. Mag.* xxiv. (1759.) *Hamburg*, 1759

De electricitate plumæ Psittaci notata quadam. 4to. (*Nov. Act. Nat. Cur.* t. iv. s. 76-82.) *Nüremberg*

Versuche über d. Erderschütterung. 8vo. (*Hamb. Mag.* xxv. (1761.)
Hamburg, 1761

Anmerkungen über die nöthige Achtsamkeit bey Erforschung der Gewitter-Elektricität, nebst Beschreibung eines Electricität-Zeigers. Vorgelesen den 7 April, 1764. 4to. 57 pp. 2 plates. (*Soc. des Wissen. Göttingen.*)
Hannover, 1764

Versuche im Lufbleeren Raume. 8vo. 3 plates. (*Saxtorf*, tom. ii. p. 508, *says,* " *Electrische Experimente im lufbleeren Raume Hanover.*") *Hannover*, 1766

Hartmann, Johann Friedrich—*continued.*
Die angewandte Elect. bei Krankheiten des menslichen Körpers. 8vo. 1770

Newen Erklarung der Entstehungsart der Donnerwetter. (*Göttingischen gemein. Abhandl. von J.* 1775.) *Göttingen,* 1775

Natürl. Luftelektricität d. Atmosphäre, tabell. entworfen. 8vo. *Hannover,* 1779

Encyclopädie der elekt. Wissenschaften ; als ein Vorbereitung zu nehrer Kent. der Elect.; tabellarich entworfen. 4to. 256 pp. (*In Bibl. Genevæ.*)
Bremen, 1784

Brachium paralyticum mediante electricitate persanatum. 4to. (*Nov. Acta Acad. Nat. Curios.* iv. 126.) *Norimbergæ*

Hartmann, Philipp Jacob. *Born March* 26, 1648, *at Stralsund; died March* 28, 1707, *at Königsberg.*

An account of Amber. (*Phil. Trans. Abd.* ii. 473.)

Succini Prussici physica et civilis historia. 8vo. (*From Kunitz,* p. 62.) *Erfurt,* 1677

† Succini Prussici physica et civilis historia cum demonstratione... 8vo.
291 pp. 3 plates. *Francofurti,* 1677

Succincta, succini Prussici Hist. et Demonstratio. 4to. (*From Watt.*)
London, 1679

Exercitatio de generatione mineralium, vegetabilium, et animalium in aere, occasione annonæ et telæ cælitus delapsarum 1686 in Curonia. 4to. (*Miscell. Acad. Nat. Curios. Dec.* 2, *An.* 7, 1688, *Append.* p. 1.)
Norimberg, 1688

† Succinta succini Prussici Historia et Demonstratio. 4to. 48 pp. (*Phil. Trans.* xxi. 5.) *Berol.* 1699

† An account of several curiosities relating to Amber, lately sent to the Royal Society from Philippus Jacobus Hartmannus (author of the Account of it published last Transaction), and which are now in their repository at Gresham College. 4to. 2 pp. (*Phil. Trans.* xxi. *for* 1699, p. 49.)
London, 1700

De sudore unius lateris. 4to. *Hal,* 1751

Hartsoeker, Nicolaus. *Born March* 26, 1656, *at Gouda ; died December* 10, 1725, *at Utrecht.*

Principes de Physique. *Paris,* 1696

Conjectures physiques. 4to. 14 and 371 pp. *Amsterdam,* 1706

† Suite des Conject. physiques. 4to. 147 pp. 1 plate. *Amsterdam,* 1708

† Eclaircissements sur les Conject. 4to. 189 pp. *Amsterdam,* 1710

Suite des Eclaircissements sur les Conject. 4to. 104 pp. *Amsterdam,* 1712

2de Partie de la Suite des Conj. 4to. 156 pp. 11 plates. *Amsterdam,* 1712

Cours de Physique. 4to. (*From Cat. Bibl. Lyon.*) *La Haye,* 1730

† **Harvey,** G. Remarks on the influence of Magnetism on the Rates of Chronometers. 8vo. 11 pp. (*Edinb. Phil. Journal,* x. p. 1.) *Edinburgh,* 1824

Harward, S. Discourse of several kinds and causes of Lightning. 4to. (*From Watt.*) *London,* 1604

† **Harzer,** F. Die Magnet-Electricität als motorische Kraft. Practische Anwendung des Electro-Magnetismus auf Telegraphie, so wie auf den Betrieb der Uhren und anderer Maschinen. 2 vermehrte Auflage. 8vo. 236 pp. 18 plates. (*Neuer Schauplatz der Künste und Handwerke,* 175 Bd.)
Weimar, 1854

† **Hasper,** W. Galvanoplastik. Gründliche Anleit. für Buchdrucker, Schriftgiesser, Kupferstecher, &c... 8vo. 56 pp. *Carlsr.* 1858

Hassencamp, J. M. Wie ein Ort durch Wetterableiter zu sichern.
Rinteln, 1782

Von den grossen Plätzen d. Strahlableiter, u. ihrer vortheilhaftesten Einrichtung zur Beschützung gauzer Städte. *Rinteln,* 1684

Hassenstein. Chemisch Electrische Heilmethode. *Leipzig*, 1853

Hassenstein, C. H. Het magnetismus en de electriciteit, anbevolen als een krachtig, wetenechappelijk en rationeel geneesmiddel. Naar de 3 Hoogduitsche uitgave bewerkt, door een M.D. 8vo. 18 plates.
Amsterdam, 1852

† Das elektrische Licht. Erläuternde und kritische Besprechung seiner Benutzung zur Beleuchtung von Strassen, Nebst Beschreibung der neuesten Regulatoren zur Erhaltung eines ganz gleichmässigen Lichtes. 8vo. 194 pp. 15 plates. *Weimar,* 1859

Hassius, C. G. Carmen elegiacum de electricitatis corporum stupenda utilitate. (*"Auctore Christ Günth. Hassio."*) (*From Krunitz,* p. 63.) *Leipsig,* 1744

Hatcher, W. H. Electric Telegraph, account of. 12mo. (*From English Cat.* 1864, p. 342). *London,* 1847

† **Hatchett,** C. An Analysis of the Magnetical Pyrites; with remarks on some of the other Sulphurets of Iron. 8vo. 15 and 9 pp. (*Phil. Mag.* xxi. 133, 213.) *London,* 1805

† **Hatchett.** On the Electro-Magnetic Experiments of MM. Œrsted and Ampère. 8vo. 10 pp. (*Phil. Mag.* lvii. 40.) *London,* 1821

Hauch, Adam Wilhelm von. *Born September 26, 1755, at Kopenhagen; died February 26, 1838, at Kopenhagen.*

A lu devant *l'Acad.* . . . de *Copenhagen* un Mémoire. (On decomposition of water by the Galvanic Pile.) (*Vide Sue,* ii. 256, *who refers to the Journ. de Chimie de Van Mons,* i. 109, *about* 1800.)

On Galv. parallelismus cum elect. diversitas. (*Nordischen Archives,* ii. 2 St. p. 1.)

Forsog til et forbedret Udlade-Electrometer. 4to. (*Vidensk. Selsk. Skrift. Ny Samml.* iv. 1793.) *Kopenhagen,* 1793

Beskrivelse af en ny Luftpröver eller Eudiometer. 4to. (*Vidensk. Selsk. Skrift. Ny Samml.* 1793.) *Kopenhagen,* 1793

Om Luft-Electricitet. (*Skandinav. Museum,* ii. 1798.) *Kopenhagen ?* 1798

Von der Luftelekt. besonders mit Anwendung auf Gewitterableiter. 8vo. *Kopenhagen,* 1800

Oversigt af Ligheden og Uligheden imellem Galvanismen og Electriciteten, etc. (*Bibl. for Physik, Medicin og Ockon,* xix. 1800.) *Kopenhagen,* 1800

Zwo. physikal Abhand. (1) von Gewitter in Winter in Norweg. (2) Resultat v. Versuch. 8vo. *Kopenhagen,* 1801

Hauff, Johann Karl Friedrich. *Born April 21, 1766, at Stuttgart; died October 24, 1846, at Brussels.*

De nova methodo naturam et leges phæn. elect. Galvanismi cognoscendas. 4to. *Marburg,* 1803

(*Poggendorff* (i. 1033) *says,* "*Quæ a Galvano cognomen vortita sunt investigandi. Comment.*")

Neuer galvan. Apparat. 8vo. (*Gilbert's Ann.* xv. 1803 u. xviii. 1804.) *Leipzig,* 1803-4

Hauksbee, Francis. *Died about 1713.*

† Experiments on the Production and Propagation of Light from the Phosphorus *in vacuo,* made before the Royal Soc ety. 4to. 2 pp. (*Phil. Trans.* xxiv. for 1704-5, p. 1865.) i *London,* 1706

† Several Experiments on the Mercurial Phosphorus made before the Royal Society, at Gresham College. 4to. 7 pp. 1 plate. (*Phil. Trans.* xxiv. for 1704.5, 2129.) *London,* 1706

Hauksbee, Francis—*continued.*

† Several Experiments on the Attrition of Bodies *in vacuo.* Made before the Royal Society at Gresham College. 4to. 11 pp. 1 plate. (*Phil. Trans.* xxiv. *for* 1704-5, p. 2165.) London, 1706

† An account of an Experiment touching the production of a considerable Light upon a slight attrition of the hands on a glass globe exhausted of its air; with other remarkable occurrences. 4to. (*Phil. Trans.* 1706, p. 2277.)
London, 1706

An account of an Experiment made before the Royal Society of Gresham College touching the extraordinary Electricity of Glass, producible on a smart attrition of it; with a continuation of experiments on the same subject and other phenomena. 4to. (*Phil. Trans.* 1706, p. 2327.)
London, 1706

Several Experiments showing the strange effects of the Effluvia of Glass producible on the motion and attrition of it. 4to. (*Phil. Trans.* 1707, p. 2372.) London, 1707

An account of the repetition of an Experiment touching motion given to bodies included in a glass, by the approach of a finger near its outside; with other Experiments on the Effluvia of Glass. 4to. (*Phil. Trans.* 1708, p. 82.) London, 1708

An account of an Experiment touching the production of Light within a glass globe, whose inward surface is lined with sealing-wax, upon an attrition of its outside. 4to. (*Phil. Trans.* 1708.) London, 1708

An account of some Experiments touching the Electricity and Light producible on the attrition of several bodies. 4to. (*Phil. Trans.* 1708, p. 87.)
London, 1708

An account of an Experiment showing that an object may become visible through such an opaque body as pitch, in the dark, while it is under the circumstances of attrition and a vacuum. 4to. (*Phil. Trans.* 1709, p. 391.)
London, 1709

† Physico-mechanical Experiments on various subjects. 1st edition. 4to. 7 plates. 194 pp. London, 1709

An account of an Experiment touching an attempt to produce Light on the inside of a globe glass lined with melted flowers of sulphur; as in the experiments of sealing-wax and pitch. 4to. (*Phil. Trans.* 1709, p. 439.)
London, 1709

An account of an Experiment concerning an endeavour to produce a Light thro' a metallick body, under the circumstances of a vacuum and attrition. 4to. (*Phil. Trans.* 1711, p. 328.) London, 1711

An account of Experiments concerning the proportion of the power of the Loadstone at different distances. 4to. (*Phil. Trans.* 1712, p. 506.)
London, 1712

† Esperienze fisico-mechaniche sopra varj soggetti contenenti un racconto di diversi stupendi fenomeni intorno la Luce e l'Elettricità . . . Opera . . . tradotta dall' idioma Inglese. 4to. 162 pp. 7 plates. Firenze, 1716

† Physico-mechanical Experiments on various subjects . . . to which is added a supplement containing several new experiments not in the former edition. 2nd edition. 8vo. 336 pp. 8 plates. London, 1719

† Expériences mécaniques, trad. par de Bremond et mises au jour, avec un discours preliminaire, des remarques, et notes, par Demarest. 2 vols. 12mo. 7 plates. Paris, 1754
(*Vide* also Taylor, B.)

Hausenius, Christian August. *Born June* 19, 1693, *at Dresden; died May* 2, 1743, *Leipzig.*

Hausenius, Christian August—*continued.*

† Novi profectus in hist. electricitatis, post obitum auctoris, ex MSS. ejus editi.
Præmissa est commentatiuncula de vita et scriptis viri. 4to. 49 pp.
Lipsiæ, 1743

Novi profectus in hist. electricitatis post obitum auctoris, ex MSS. ejus editi.
Præmissa est commentatiuncula de vita et scriptis viri. 8vo. (*From Heinsius,* ii. 291.) *Lipsiæ et Quedb.* 1745

† **Hausenius,** C. A., et **De Sanden.** Novi profectus in hist. electricitatis, post obitum auctoris, ex MSS. ejus editi. Accessit V. C. Hen. De Sanden. Dissertatio de Succino electricorum principe, quam ed. et de vita B. Hausenii præfatus est Jo. Ch. Gottsched. 8vo. 128 pp. 1 plate. *Lipsiæ,* 1746

Hausen. On the revival of Hawkesby's Globe. (*Danzig Mem.* i. 278.)

Hausmann, J. F. L. Über die Polarität der Harzer Granitfelsen. 8vo. (*Crell's Chem. Ann.* 1803, ii. 207.) *Helmstadt,* 1803

Über die durch Molecularbewegungen in starren leblosen Körpern bewirkten Formveränderungen. (*Abhandl. d. Gesellsch. d. Wiss. zu Gött.* vi. 1853-55 ; vii. 1856-57.) *Gottingen,* 1853-55-56-57

Hautefeuille, Jean de. *Born March* 20, 1647, *at Orléans; died October* 18, 1724, *at Orléans.*

Nouveau moyen de trouver la déclinaison de l'aiguille aimantée avec une grande précision. 4to. *Paris,* 1683

Balance magnétique, avec des réflexions sur une balance inventée par Perrault, &c. 4to. *Paris,* 1702

Nova ratio accuratissime investigandi acus magneticæ declinationem. (*Opusc. Act. Erudit. Lips.* tom. i. p. 293.)

Hauy, René Just. *Born February* 28, 1743, *at St.-Just, Dép. Oise; died June* 3, 1822, *at Paris.*

Mémoire sur les propriétés électriques de plusieurs minéraux. 4to. (*Mém. de Paris Acad.* 1785, *Mém.* p. 206.) *Paris,* 1785

† Exposition raisonnée de la Théorie de l'Electricité et du Magnétisme . . . d'après les principes de M. Œpinus. 8vo. 238 pp. 4 plates. *Paris,* 1787

Observations sur la structure des crystaux appeles zéolithes, et sur les propriétés électriques de quelques uns. 4to. (*Mém. de l'Inst. Nat.* an. 4, tom. i. *Sc. Mathém. et Phys.* p. 49.) *Paris,* 1795

Observations sur les Aimants naturels. 4to. (*Soc. Philomat.* an. 5, p. 34.) *Paris,* 1796

† Darstellung der Theo. d. Elek. und d. Magn. üb. v. Murhard, mit Anmerkungen. 8vo. 310 pp. 7 plates. *Altenburg,* 1801

Traité de physique. 1st edition (?). 8vo. *Paris,* 1803

† Observations sur l'électricité des substances métalliques. 4to. (*Soc. Philomat.* an. 12, p. 191.) *Paris,* 1804

Observations on the Electricity of Metallic Substances. 8vo. 4 pp. (*Phil. Mag.* xx. 120.) *London,* 1805

Traité de physique. 2nd edition. 2 vols. 8vo. 1806

† Elementary Treatise. (*Translated by Gregory.*) 2 vols. 8vo. plates. *London,* 1807

† On the Electricity of Minerals. 8vo. 5 pp. 1 plate. (*Phil. Mag.* xxxviii. 81.) *London,* 1811

Über d. Elektricität d. Mineralkörper. Uebersetzt von K. C. Leonhard. 8vo. *Frankfürt,* 1812

Traité des caractères physiques des pierres précieuses. 8vo. 3 plates. *Paris,* 1817

236 HAU—HEI

Hauy, René Just—continued.
Sur la veitu magnétique considerée comme moyen de reconnaître la présence
du fer dans les minéraux. (Mém. Mus. Hist. Nat. iii. 1847, p. 169.)
Paris, 1817

† Trattato dei caratteri fisici delle pietre preziose. Traduzione con Note del
Configliachi. 8vo. 227 pp. 3 plates. Milano, 1819

Neue Beobachtungen u. Studien üb. d. Elektric. d. Mineralkörper; a. d.
Franz. m. Anmerk. v. J. J. Nogerath. 8vo. Bonn, 1819

Traité élément. de physique. 3rd edition. 2 vols. 8vo. Paris, 1821

Traité de Minéralogie. Paris, 1822

† Sur l'électricité, produite dans les minéraux, à l'aide de la Pression. 4to. 6 pp.
(Mém. du Muséum, tom. iii.) Paris

Hauy, V. Mémoire historique sur les télégraphes en général, et sur les diverses
tentatives faites pour en introduire l'usage en Russie. 8vo. 56 pp.
Petersburg, 1810

Hawksley. Variations, Magnetic. (From Churchman, Introd. p. xv.). 1745

Hayez, E. De meteorum intervallis a superf. terræ mensur. 4to. 182—?

† Head, F. (Sir). Stokers and Pokers; or, the North-Western Electric Railway,
Telegraph, &c. New edition. 8vo. London, 1861

† Hearder, J. N. On the application of Cast Iron as a substitute for Steel, in the
construction of very powerful permanent Magnets, with a specimen of a
Cast Iron Magnet of great power, and a detail of some peculiar phenomena
connected with its magnetic properties. 8vo. 13 pp.

Hearn. On the cause of the discrepancies observed by Mr. Baily with the
Cavendish apparatus for determining the mean density of the Earth. 4to.
(Phil. Trans. 1847, p. 217.) London, 1847

† Heathcote. Extract of a Letter to Mr. Flamsteed, from Cabo Cars Castle, on
the Coast of Guinea, the 14th Dec. 1683, concerning the Tide on that
Coast, Variation of the Needle, &c. 4to. 1 page. (Phil. Trans. xiii. 1683,
p. 578.) London, 1683

† Hédouin, Paul. L'Electricité appliquée au sondage des Mers. Les Câbles
électriques sous-marins. I. Sondage des Mers. II. Appareils déjà connus
et appliqués. III. Electro-Barathromètre. 8vo. 81 pp. 6 plates.
Paris, 1870

Heer. (Vide Vorsselman de Heer.)

Heeren, Friedrich. Born August 11, 1803, at Hamburgh.
† Der elektrische Telegraph. 8vo. 32 pp. (Besonderer Abdruck aus dem technischen
Worterbuch von K. Karmarsch und F. Heeren.) Prag, 1854

† Hehl. Anleitung zur Errechtung und Untersuchung der Blitzableiter fur Bauver-
standige, Bau- und Feuerbeschauer und Gebaude-Inhaber. 8vo. 54 pp.
Stuttgart, 1827

† Hehl, Joh. Über electro-dynamische Vertheilung. 4to. 12 pp. 1 plate. (Programm
der häreren Gewerbeschule in Cassel.) Cassel, 1841

Heidenreich. Phys.-chem. Untersuchungen des Bluts durch die electrische
Säule. (Neue medicin. Zeitung, 1847, No. 31.) 1847

† Heidmann, J. A. Vollständige . . . Theorie der Elektricität. 2 vols. 8vo. 5
plates. Wien, 1799

Vollständ. Theorie der Elektricität. 2 vols. 8vo. Wien, 1803

Zuverl. Prüfungsart z. Bestimm. d. wahr. von d. Scheintode; nebst einer
physiologischen Erfahr. aus d. Anwend. d. verstärk, galvanischen Electricität
d. theorischen Organismus. 8vo. Wien, 1804

† Vollständige auf Versuche, &c. gegründete. . . Theorie der Galvanischen
Elektricität. . . i. Band. 8vo. 258 pp. 2 plates. Wien,1806

Heidmann, J. A.—*continued.*

†   Theory of Galvanic Electricity, founded on experience. Abridged by M. Guyton from Ann. de Chim. lxi. p. 70. 8vo. 8 pp. (*Phil. Mag.* xxviii. p. 97.)   *London*, 1807

Heidmann, J. H. Observations physico électriques. (*Journal de Chimie*, tom. vi. p. 190.)

† **Hein, Theod.** Analyse eines Meteoriten aus Dacca in Bengalen. . . 8vo. 4 pp. (*Sitzb. d. k. Akad. d. Wissensch.* ii. Abth. Oct.-Heft, Jahrg. 1866, Bd. lvi.)   *Wien*, 1866

Heine. Beobachtungen über Lähmungszustände der unteren Extremitäten und deren Behandlung.   *Stuttgart*, 1840

Heineken, C. Meteorological Register kept at Funchal in Madeira in the year 1826 ; with some prefatory observations on the climate of that island, &c. 8vo. 13 pp. (*Phil. Mag. or Annals*, ii. 362 and 411.)   *London*, 1827

Heinrich, Placidus. *Born October* 19, 1758, *Schierling, Regenkreis, Bayern; died January* 18, 1825, *Regensberg.*

Ueber die Wirkung des Geschützes auf Gewitterwolken. 4to. (*Neue Abhandl. der Baierischen Akad. Philos.* v. p. i.)   *München*, 1789

Uber d. Schäffer'schen Pendelversuche. 8vo. (*Gilb. Ann.* xxvii. 1807.)   *Leipzig*, 1807

† Die Phosphorescenz der Körper. . . 1ste Abhandl. Von der durch Licht bewerkten Phosphorescenz. "Dritter Abschnit Phosphorescenz durch Bestrahlung mittelst. der electrischen Lichts." 4to. 132 pp. (*Volta's copy.*)   *Nüremberg*, 1811

Versuche mit Phosphor. 8vo. (*Schweigg Journ.* iv. 1812.)   *Nürnberg*, 1812

Die Phosphorescenz der Körper. . . 2te Abh. 4to. 312 pp. (*From Journ. de Phys.* lxxiv. pp. 158 and 307.   *Nüremberg*, 1812

Uber Phosphorescenz im Pflanzen. u. Thierreich. 8vo. (*Schweigg. Journ.* xiii. 1814.)   *Nürnberg*, 1814

Die Phosphorescenz der Koerper. 2 vols. 4to. (*From Journ. de Phys.* lxxxv. 158 and 307.)   *Nüremberg*, 1814

Über Zamboni's Säule. 8vo. (*Schweigg. Journ.* xv. 1815.)   *Nürnberg*, 1815

Die Phosphorescenz der Körper. . . 3tte. Abhand. 4to. 424 pp.   *Nüremberg*, 1815

Die Phosphorescenz der Körper. . . 4te. Abhand. 4to. 596 pp.

Phosphorescenz durch Erwarmung u. durch Bestrahlung. 8vo. (*Schweigg. Journ.* xxix. 1820.)   *Nüremberg*, 1820

† **Heinrichsen, H.** Ideen üb. d. wechselseitige Elektricitäts-verhältniss zwischen dem thier. Organismus u. d. aussern Natur, mit Entfaltung zweier, bisher übergangener, alle Processe d. Lebens bedingender Naturkrafte. 8vo. 330 pp.   *Leipzig*, 1839

Das Wesen d. Wechselfiebers d. Fallsucht u. d. Blennorrhoe, dargest. von Seiten ihrer elektr. Natur, &c. . . 8vo.   *Leipzig*, 1839

Heinsius, W. Allgemeines Bücher-Lexikon, od. vollständ Alphabet. Verzeichniss aller v. 1700 b. zu. Ende 1827 erschien. Bücher, welche in Deutchland u. in den durch Sprache u. Literatur. damit verwandten Ländern gedruckt worden sind xx. 7 vols. Welcher die v. 1822 b. Ende 1827 erschien Bücher u. d. Berichtigungen früh. Erscheinungen enthält. Hrsg. v. C. G. Kayser. M. 2 Anh. 4to.   *Leipzig*, 1828-29

Algemeines Bücher-Lexikon od volständ. alphabet. Verzeichnisz aller von 1700 bis Ende 1867 erschienen Bucher, welche in Deutchland u. in den durch Sprache u. Literatur damit verwandten Landern gedruckt worden sind. Nebst Augabe der Druckörter, der Verleger, der Preise ze. 14, Bd. welche die von 1862 bis Ende 1867 erscheinen Bücher u. die Berichtiggn früherer Ercheinungen enthalt. Hrsg. v. *Karl Rob. Newmann*, 8 u. 9 *Lfg.* 4to.   *Leipzig*, 1869

Heintz, W. H.   Eigenthümlich elektroskopische Zuständ d. Glases.   8vo.
(*Poggend. Ann.* lix. 1843.)   *Leipzig,* 1843

Heinze, Johann, Georg.   *Born April 23, 1719, at Suhla Henneberg ; died December*
28, 1801.

Neue elekt. Versuche mit der von Marum erfundenen Electrisir Machine, und
dem vom Schoffer bekannt gemachten elektricitats Kager in einem Schreiben
an Hrn. Gondela, bekannt gemacht.   4to.   *Oldenberg,* 1777

† Heis, Ed.   Die period. Sternschnuppen und d. Resultate der Erscheinungen,
obgelcitet aus den'während der letzten 10 Jahre zu Aachen angestellten
Beobachtungen.   4to.   40 pp.   1 plate.   *Coln,* 1849

† Die Feuerkugel, welche am Abende des 3 December 1861 in Deutschland
gesehen worden ist.   8vo.   18 pp.   1 plate.   (*Abdruck aus der "Wochenschrift
für Astronomie.*)   *Halle,* 1862

† Die grosse Feuerkugel, welche am Abende des 4 März 1863, in Holland,
Deutschland, Belgien und England gesehen worden ist Nebst einer Karte.
8vo.   56 pp.   *Halle,* 1863

† Grosse Feuerkugel in der Nacht des 10-11 März in Westfalen und Hanover.
8vo.   12 pp.   *Halle,* 1866

Helfenzrieder, Johann Evangelist.   *Born December 9, 1724, at Landsberg on Lech ;
died March,* 1803.

Erläuterung der Weise, wie die Sonnenund Monds-Strahlen zu Gestaltung
des Nord-Lichts. . . heranfgebracht werden. . . (*Acta Acad. Moguntinæ*
An. 1778-79, p. 281.)   1778

† Vorschlag. . . die Blitzableiter zei vorbessern.   8vo.   15 pp.   *Salzburg,* 1785

A new invention in Lightning Conductors.   8vo.   (*Abhand. eine Privat-Gesell-
schaft,* vol. i. No. 12.)   *München,* 1792

Handgriffe bey Errichtung eines Blitzableiters von verbesseter Art.   8vo.
(*Abhand. einer Privat-Ges. in Ober-Deutschland,* Th. i. p. 193.)   *München,* 1792

Hell (commonly Höll), Maximilian.   *Born May 15, 1720, at Schemnitz, Ung. ; died
April 14, 1792, at Vienna.*

Ephemerides ad Meridian.   Vindobon.   8vo.   *Vienna,* 1757-91

Anleitung zum nützlichen Gebrauch der kunstlichen Stahlmagnete.   8vo.
Wien, 1762

An introduction towards the useful employment of artificial Magnets.   (*From
Lamont, Handb.* p. 436.)

Auroral phænomena (in daylight) at Wardhuis, 3rd July, 1769.   (*Acta Lipsiensia pro* 1770, p. 82.)   *Lipsiæ,* 1770

Auroræ boreales, theoria nova a Maxim. Hell . . . edita anno 1776.   8vo.
Vindobonæ, 1776

Ephemerides.   An. 1777.   Appendix Auroræ theoria.   8vo.   *Vindeborni,* 1777

Hellant, Anders.   *Born . . . ; died November 23, 1789.*

Magnet-nalens misvisning, etc. i Upsala.   8vo.   (*Vet. Acad. Handb.* 1740.)
Stockholm, 1740

† Abweichungen der Magnet-Nadel in den nördlichsten Theilen Schwedens. (von
A. H. beobachtet).   8vo.   2 pp.   (*Schwedische Akad. Abdhandl.* An.   1756,
vol. xviii. p. 68.)   *Hamb. & Leipz.* 1756

† Abweichung der Magnet-Nadel an mehreren Stellen innerhalb des Nördlichen
Polar-Kreises, beobachtet.   8vo.   4 pp.   (*Schwedische Akad. Abhand.* An.
1777, vol. xxxix. p. 285.)   *Leipzig,* 1777

Observations of the Needle at Franeker (Auroral Influence).   (*Mem. de Swede*
tom. xviii. p. 68.)

HEL 239

Heller, Theodor Aegidius. *Born November,* 1759, *at Reulbach*; *died October* 19, 1810, *at Fulda.*

Über d. Verhalt. trockner u. feuchter Luft bei elektr. Erscheinungen. 8vo. (*Gren's New Journ.* ii. 1795.) *Leipzig,* 1795

Beobach. d. atmospär. Elektricität. 8vo. (*Gren's New Journ.* iv. 1797.) *Leipzig,* 1797

Uber d. magnet Mittelpunct d. wirchen Eisens u. dessen Verändr. 8vo. (*Gilbert's Ann.* iv. 1800.) *Leipzig,* 1800

Uber d. Leitungs vermogen d. Wassers; u. üb. d. Licht. elektr. Funkens. 8vo. (*Gilbert's Ann.* vi. 1800.) *Leipzig,* 1800

Entdeckte Veränderungen des .. Magnetismus in ihrem Zusammenhange mit den Ständen der Sonne und des Mondes (Bericht der Münch. Akad.) 1809

Hellwag, Christoph. Friedrich. *Born March* 6, 1754, *Calw, Würtemburg ; died October* 16, 1835, *Eutin.*

† **Hellwag,** C. F. und **Jacobi,** M. Erfahrungen über die Heilkräfte des Galvanismus, und Betrachtungen über desselben chemische und physiologische Wirkungen, mitgetheilt von Helwag ..; und Beobachtungen bey der medicinischen Anwendung der Voltaischen Säule, von Jacobi. 8vo. 184 pp. 1 plate. (*Volta's copy.*) *Hamburg,* 1802

Helmert. Die geschichtliche Entwickelung der heutigen Telegraphie. 4to. *Dresden.*

Helmholtz, H. L. T. Über d. Methode Kleinste Zeittheile zu messen, u. ihre Anwendung auf physiolog. Zwecke. (*Müller's Archiv,* 1852.) 1852

Über d. Dauer u. d. Verlauf der durch Strömesschwankungen nducirten elektr. Ströme. 8vo. (*Pogg. Ann.* lxxxiii. 1851.) *Leipzig,* 1851

Gesetze d. Vertheilung elektr. Ströme in Körperlich. Leitern, mit Anwendung auf thierische elektrische Versuche. 8vo. (*Pogg. Ann.* lxxxix. 1853.) *Leipzig,* 1853

Helmont, Johann Baptist van. *Born* 1577, *at Brussels; died December* 30, 1644, *at Vilvorde.*

De magnetica .. vulnerum naturali et legitima curatione. 8vo. *Paris,* 1621

† Ortus medicinæ id est Initia physicæ inauditæ progressus medicinæ novus in morborum ultionem ad vitam longam. Auctore J. B. Van Helmont. .. edente auctoris filio, F. M. Van Helmont cum ejus Prefatione, ex Belgica translata. 4to. 800 pp. portrait. *Amsterodami,* 1648

At p. 89, *On Thunder, &c.* P. 90, *Visiones mirabiles in montibus altis.* p. 91 *Historia Tonitrui, &c. At* p. 746, *De Magnetica vulnerum curatione...* 34 pp. *Poggendorff* (i. 1060) *says " u. viele spätere Ausgaben." (At* p. 6. *his differential Thermometer :* vide *Leslie.*)

Ternary of Paradoxes, Magnetic Cure of Wounds, Nativity of Tartarian Wine, &c. Translated by Dr. W. Charlton, &c. 4to. (*From Spon's Cat.* 1866: *vide also Charlton.*) *London,* 1650

Helmuth, J. H. Beobach. eines im Süden leuchtenden Bogens. *Braunschweig.* *Anzeig,* 1777, S. 95.) *Braunschweig*

Von d. wohlthätig. Erfindung d. Blitzableiters. (*Braunschw. Anzeig,* 1777, S. 55.) 1777

Über d. Entstehung des Nordlichts. *Braunschweig,* 1777

† **Helvetius,** J. F. Disputatio philosophica de Magnete. (Thesis.) 4to. 14 pp. *Lugd. Bat.* 1677

† **Helvig,** C. G. Bemerkungen über Blitz und Donner, nebst Vermuthungen über das Entstehen der Luft-Erscheinungen. 8vo. 32 pp. 1 plate. (*Gilbert's Ann. d. Physik,* li. S. 2, S. 10.) *Leipsig,* 1815

† Helvig (General). Direct method of ascertaining the velocity of Cannon-balls. 8vo. 1 p. (*Phil. Mag. or Annals,* iv. 65.)   *London,* 1828

Helvig, Major Swedois. Bourguet, Hermann, and Grapengeiser. Composition of Electric Pile, &c.   *Paris,* 1801 ?

Hemmer, Johann Jacob. *Born* 1733, *at Horbach; died May* 3, 1790, *at Mannheim.*

Sur l'Electricité des Métaux. 4to. (*Observ. sur la Phys. Jul.* 1780, p. 50.)   *Paris,* 1780

Beschreibung einiger merkwürdiger Wetterschläge. 4to. (*Commentat. Acad. Theodoro-Palatinæ* iv. *Phys.* p. 87.)   *Mannheim,* 1780

Zergliederung des beständigen Elektrizitäts-Trägers. 4to. (*Commentat. Acad. Theodoro-Palatinæ* iv. *Phys.* p. 94.)   *Mannheim,* 1780

On Experiments with an Electrophorus. (*Mém. de l' Acad. de Mannheim,* vol. iv. p. 112.)

Glückliche Wirkung des elektrischen Feuers bey einer vieljährigen Lähmung. 4to. (*Commentat. Acad. Theodoro-Palatinæ* iv. *Phys.* 116.)   *Mannheim,* 1780

Descriptio instrumentorum Societatis Meteorologicæ Palatinæ, etc. 4to.   *Mannheim,* 1782

Versuche über d. Thau. (*Pfalzbayr. Beiträge, Hft.* xi. 1782.)   1782

Kurzer Begriff u. Nutzen d. Wetterableiter, u.s.w. 8vo.   *Düsseldorf,* 1782

Kurze und deutliche Anweisung wie man, durch einen in jedem Orte wohnenden Schmied, oder andere in Metall arbeitende Handwerker, eine sichere Wetterableitung mit sehr geringen Kosten in allerhand Gebäuden anlegen lassen kann. 8vo.   *Friedrichstadt,* 1783

Von Wetterstrahlen, welche Thiere treffen aber nicht tödten. 4to. (*Commentat. Acad. Theodoro-Palatinæ* v. *Phys.* p. 150.)   *Mannheim,* 1784

Elektrische Versuche mit belegten Thieren. 4to. (*Commentat. Acad. Theodoro-Palatinæ.*)   *Mannheim,* 1784

De fulminis ictibus in campanas, quæ pulsantur, ubi electricitas nubium ac fulminis theoria, nova et uberiore luce perfunduntur. 4to. (*Commentat. Acad. Theodoro-Palatinæ* v. *Phys.* p. 237.)   *Mannheim,* 1784

Über d. Glockenläuten bey Gewittern. 4to. (*Commentat. Acad. Theodoro-Palatinæ* v. 1784.)   *Mannheim,* 1784

Anleitung Wetterableiter . . anzulegen. 8vo.   *Offenbach,* 1786

† Anleitung Wetterleiter . . anzulegen. 2nd edition. 8vo. 232 pp.   *Mannheim,* 1788

Verhaltungsregeln u.s.w.   1789

† De electricitate flammæ. 4to. (*Commentat. Acad. Theodoro-Palatinæ* vi. *Phys.* p. 23.)   *Mannheim,* 1790

† Gutta serena electricitate feliciter sublata. 4to. (*Commentat. Acad. Theodoro-Palatinæ* vi. *Phys.* p. 47.)   *Mannheim,* 1790

De variatione acus magneticæ tempore Auroræ borealis. 4to. (*Commentat. Acad. Theodoro-Palatinæ* vi. *Phys.* 317.)   *Mannheim,* 1790

Conductorum fulmineorum vim egregiam tribus recentioribus exemplis docet. 4to. (*Commentat. Acad. Theodoro-Palatinæ* vi. *Phys.* 516.)   *Mannheim,* 1790

Beobachtung einer Wettersäule nebst Erläuterung. 4to. (*Commentat. Acad. Theodoro-Palatinæ* vi. *Phys.* p. 533.)   *Mannheim,* 1790

Merkwürd. Erschein. an einer vorüberziehend. Wolke. 4to. (*Commentat. Acad. Theodoro-Palatinæ* vi. 1790.)   *Mannheim,* 1790

Unterr. z. sicherst Anleg. d. Wetterableiter. 8vo.   *Mannheim,* 1808

Rathgeber wie man sich vor Gewittern in unbewaffneten Gebäuden verwahren soll. 8vo. 1 plate.   *Mannheim,* 1809

Hemmer, Johann Jacob—*continued.*

On the analogy of Electricity and Magnetism. (*Rheinische Beitrag zu Gelehrsamkeit for* 1781, 5es Cahir.)

† Variæ curationes electricæ. 4to. (*Commentat. Acad. Theodoro-Palatinæ* v. *Phys.* 321.) *Mannheim*

Nachricht von den in Kurpfalz angelegtern Wetterleiten. 4to. (*Commentat. Acad. Theodoro-Palatinæ* iv. *Phys.* p. 21.) *Mannheim*

Enarrationes conductorum fulminis superiore quinquennio variis in locis a se positorum. 4to. (*Commentat. Acad. Theodoro-Palatinæ* v. *Phys.* p. 295.) *Mannheim*

De electricitate animali, ubi in spontaneam præcipue inquiritur. 4to. (*Commentat. Acad. Theodoro-Palatinæ* vi. *Phys.* p. 119.) *Mannheim*

Nachricht von einigen merkwürdigen Wetterschlägen. 4to. (*Commentat. Acad. Theodoro-Palatinæ* vi. *Phys.* p. 324.) *Mannheim*

† **Hemmern,** Giacomo. Breve istruzione sul conduttor elettrico, e suoi vantaggi. . . 4to. 10 pp. (*Opusc. Scelti,* vi. 388.) *Milano,* 1783

Henckel, Johann Friedrich. *Born August* 11, 1679, *at Merseburg ; died Jan.* 26 1744, *at Freiberg.*

Pyritologia oder Kieshistorie Neue Ausgabe . . 2s Aufl. 8vo. 13 plates.
Leipzig, 1754

Pyritologie ou histoire naturelle de la Pyrite ; on y a joint le Flora Saturnisans, ou l'auteur démontre l'alliance qui se trouve entre les Végétaux et les Minéraux, traduits de l'Allemand par d'Holbach. 2 vols 4to. Folding plates. 1760

† **Hengstlin,** H. Bietticano. Seleno-tropium maris, seu dissertatio physico-mathematica de æstus marini proprietatibus magneticis in Luna repertis, præside. Dn. Joh. Lud. Mælingo. 8vo. 12 pp. 1 plate. *Tubingæ,* 1682

† **Henle,** F. G. J. Über Narcine, eine neue Gattung electrischer Rochen, nebst e Synopsis d. electrischen Rochen. 4to. 44 pp. 4 plates. *Berlin,* 1834

† **Henle und Meissner.** Bericht über Fortschritte der Anatomie und Physiologie. 8vo. 1866

Henley, William. *Died about* 1779.

† An account of the death of a person destroyed by Lightning in the Chapel in Tottenham-Court-road, and its effects on the building ; as observed by Mr. Wm. Henley, Mr. Edward Nairne, and Mr. Wm. Jones. 4to. 8 pp. 1 plate. (*Phil. Trans.* lxii. 133.) *London,* 1773

Experiments concerning the different efficacy of pointed and blunted rods in securing buildings against the stroke of Lightning. 4to. (*Phil. Trans.* 1774, p. 133.) *London,* 1774

Electricity, a stimulant for the recovery of persons apparently drowned, or otherwise suffocated. (*Trans. of the Humane Soc.* vol. i. p. 63.) *London*

(*Vide also* Ronayne and Henley.)

Henley, W. T. Télégraphe Electrique, dans lequel les piles sont remplacées par des électro-aimants. 4to. (*Comptes Rendus,* xxx. 412.) *Paris,* 1850

List, &c. of articles (Magnets, &c.) exhibited by him at the Paris Exhibition in 1853. 4to. 5 pp. *Paris and London,* 1853

Henn. De Amperi principiis in phænomenor. electromagnet. doctrina propositis. 8vo. 1850

Henrici, Friedrich Christoph. *Born August* 26, 1795, *at Osnabrück.*

† Über die Elekt der Galv. Kette. 8vo. *Göttingen,* 1840

Many articles, &c., on Galvanic Electricity, &c. from 1841 to 1850, in *Poggendorff's Annalen,* lii. to lxxx.

Henry Jos. On some modifications of the Electro-magnetic apparatus. 8vo.
    (*Trans. of the Albany Institute,* i. 22.)   *Albany,* 1831

† Contributions to Electricity and Magnetism. No. i. Description of a Galvanic
    Battery for producing Electricity of different intensities. 4to. 6 pp. 1 plate.
    (*Trans. Amer. Phil. Soc.* v.)   *Philadelphia,* 1835

† On the influence of a Spiral Conductor in increasing the intensity of Electricity
    from a galvanic arrangement of a single pair. 8vo. 8 pp. (*Taylor's Scient.*
    Mem. i. 540.)   *London,* 1837

† Contributions to Electricity and Magnetism. No. ii. On the influence of a
    Spiral Conductor in increasing the intensity of Electricity from a galvanic
    arrangement of a single pair. Read Feb. 6, 1835. 4to. 9 pp. (*Trans.*
    Amer. Phil. Soc. v.)   *Philadelphia,* 1837

† Contributions to Electricity and Magnetism. No. iii. On Electro-dynamic
    induction. Read Nov. 2. 1838. 4to. 35 pp. (*Trans. Amer. Phil. Soc.* vol.
    vi. p. 17, Article ix.)   *Philadelphia,* 1839

† On a new method of determining the Velocity of Projectiles. 8vo. (*Proc.*
    Amer. Phil. Soc. iii. 165.)   *Philadelphia,* 1843

Contributions to Electricity and Magnetism. 4to. (*Trans. Amer. Phil. Soc.*
    new series, vol. viii.)   *Philadelphia,* 1843

Method of protecting from Lightning buildings covered with metallic roofs
    8vo. (*Proc. of Amer. Phil. Soc.* iv. 179.)   1845

† Report concerning a letter of S. D. Ingram to R. Patterson relative to the
    effect of Thunder on Telegraphic wires. 8vo. 9 pp.   1846

† Extracts from the Proceedings of the Board of Regents of the Smithsonian
    Institution, in relation to the Electro-magnetic Telegraph. 8vo. 39 pp.
                                  1857

† On the application of the principle of the Galvanic Multiplier to Electro-
    magnetic apparatus, and also to the development of great magnetic power
    in soft Iron, with a small galvanic element. 8vo. 12 pp. (*American Journal*
    of Science and Arts, xix. No. 2.)   *Newhaven, U.S.*

† **Henry** and **Ten Eyck.** Powerful Electro-magnet. 8vo. 2 pp. (*Phil. Mag. or*
    Annals, x. 314.) (*This is a short account of the Magnet, taken from Silliman's*
    Journal.)   *London,* 1831

Henry, William. *Born Dec.* 12, 1775, *at Manchester; died Sept.* 2, 1836, *at Pendle-*
bury, near Manchester.

Elements of Experimental Chemistry. 1st edition. 2 vols. 8vo. *London,* 1799

† On the formation of Muriatic Acid by Galvanism. To the Editor of the *Phil.*
    Mag. dated Manchester, July 23, 1805. 8vo. 3 pp. (*Phil. Mag.* xxii. 183.)
                              London, 1805

† Description of an apparatus for the Analysis of the compound inflammable Gases
    by slow combustion; with experiments on the Gas from coal, explaining its
    application. Communicated by H. Davy. 8vo. 18 pp. 1 plate. (*Phil-*
    Trans. 1808, part ii. ; *Phil. Mag.* xxxii. 277.)   *London,* 1808.

† Additional experiments on the Muriatic and Oxymuriatic Acids. 8vo. 7 pp.
    Phil. Mag. xl. 337.)   *London,* 1812

† On Sir Humphrey Davy and Dr. Wollaston. (*From* 11*th edition of Henry's*
    Chemistry, 1830.) 8vo. 2 pp. (*Phil. Mag. or Annals,* vii. 228.) *London,* 1830

Henshaw, T. Some experiments and observations upon May-dew. 4to. (*Phil.*
    Trans. 1665.)   *London,* 1665
    (*Note.—Quere as to Electricity.*)

Hensing, J. T. De germinatione metallica artificiali ejusdemque novo invento
    4to.   *Giss.* 1718

† **Henwood,** W. J. On the Electric currents observed in some metalliferous veins.
    8vo. 20 pp. (*Extracted from the "Mining Review." Reprinted from the*
    "*Annals of Electricity," with additions, and an Appendix by the Author.*)
                                 London, 1837

Herberger, Johann Eduard. *Born . . . at Speyer ; died March 12, 1855, at Würtzburg.*

Systematisch-tabellarische Übersicht der chemisch. Gebilde organ. Ursprungs mit genauer Angube ihrer Eigenschaften zc. 1 Liefer. die elektro-positiven, 2 Liefer. die elektro-negativen organchem. Gebilde. Folio. 2 vols.
Nürnberg, 1831-36

Herbert, Joseph Edler von. *Born September 2, 1725, at Klagenfurt ; died March 28, 1794, at Vienna.*

Theoriæ phænomenorum electricorum . . 1st edition. 8vo. *Wien,* 1772

Von der vortheilhaftesten Reibung, die Elektricität in heftigerem Grade zu erregen. (*Beitrage zu verschieden. Wissenschaften,* 1775.) 1775

† Theoriæ phænomenorum electricorum. . . Editio altera aucta. 8vo. 246 pp. 6 plates. *Vindobonæ,* 1778

Dissertatio de vi electrica aquæ. *Labacii,* 1778

Herder, Siegmund August Wolfgang. *Born Aug. 18, 1776, at Bückeburg ; died Jan. 29, 1838, at Freiburg.*

Diagnostische prakt. Beiträge zur Zuveitezing der Geburts hulfe. (*From Guitard.*) *Leipsig,* 1803

Herepath, J. Mathematical Physics. 2 vols. 8vo. *London,* 1847

Herger. Die Systeme der magnetischen Curven, isogonen und isodynamen nebst anderweitigen empirischen Forschungen über die magnetisch polaren Kräfte ausgeführt in 37 grossen graphischen Darstellungen auf 31 Tafeln und erläutert unter den Auspicien des Herrn Hofrath, Dr. Schottin von J. Ernst Herger, nebst einem Vorworte vom Herrn Dr. G. Erman. Fol. 33 plates = 37 figs. 59 pp. *Leipsig,* 1844

Héricart de Thury, Louis Etienne François. *Born June 3, 1776, at Paris ; died Jan. 15, 1854, at Rome.*

† Meteoric iron in France. "Le Globe." 8vo. 1 page. (*Phil. Mag. or Annals* iv. 457.) *London,* 1828

De l'influence des arbres sur la Foudre ; et ses effets. 8vo. 27 pp. *Paris,* 1838

Herlicius, D. (Herlich, &c.) Tractatus de fulmine et aliis impressionibus, prodigiis et miraculis. Vom Blitz, Donner und allerbei Feurzeichen, u.s.w. 4to. *Starg,* 1604

Herman, C. G. Mechanischer verbesseter Wind-Regen- u. Trockenheits beobachter. 8vo. (*2nd edition is dated* 1793.) *Freiburg,* 1789

Moniteur électrique.
Note.—Du Moncel says it is an application of other instruments and invented by Breguet.

Hermann, Leonhard David. *Born June 27, 1670, at Massel ; died May 1, 1736, at Massel.*

Maslographia ; oder Beschreibung der schlesisch Maszel. 4to. *Brieg,* 1711

Heroart, J. F. Admiranda ethnicæ theologiæ mysteria propalata, &c. Ubi lapidem magnetem antiquissimis nationibus pro Deo cultum commonstratur. 4to. *Ingolstadii,* 1623
Note.—Attempt to prove that the properties of the loadstone were well known to the Ancients.

Herodotus. On Electron, &c. in his General History. (*From Poggendorff,* i. 1086.)

Herrmann, R. In naturali Magnetismo in chalybem inducendo quanto momento sit tempus. *Breslau,* 1863

† **Hermelin, S. G.** Über das Verhalten des Magnets in Gruben. 8vo. 4 pp. (*K. Schwed. Akad. Abh.* xxix. 329.) *Leipsig,* 1767

† **Herschel, A. S.** Detonating Meteors of February and November. 8vo. 3 pp. (*Intellectual Observer.* No. 50, p. 99.) *London,* 1866

Herschel, Sir John Frederick William, Bt. *Born March 7, 1792, at Slough, near Windsor.*

Repetition of Arago's experiments on the Magnetism, &c. during the act of Rotation. 4to. *(Phil. Trans.* 1825.) *London,* 1825

Light and Sound. Also Heat, Electro-magnetism, Electricity, Galvanism, Chemistry. 4to. 34 plates. *(From Encyclop Metropolitana)* *London,* 1827-30

Preliminary discourse on the study of Natural Philosophy. 12mo. 1831

Letter to Arago from the Cape on Falling Stars. 8vo. *(Annales de Chimie,* 1836.) *Paris,* 1836

† A Manual of Scientific Enquiry, prepared for the use of Her Majesty's Navy, and adapted for Travellers in general. 8vo. 488 pp. 2 plates. *London,* 1849

Physical Geography from the Encyclopædia Britannica. Post 8vo. 441 pp. *London,* 1861

Terrestrial Magnetism. Article in Encyclopædia Britannica, 8th edition.

*† Art. Meteorology from Encyclopædia Britannica. Fcap. 8vo. *London,* 1861

* Meteorology. From the (8th edition of the) Encyclopædia Britannica. 2nd edition. 12mo. 288 pp. 3 plates. *London,* 1862

Outlines of Astronomy. Essays from the Edinburgh and Quarterly Reviews; with Addresses and other pieces. 6th edition.

(Vide also Royal Society.)

Herschell and Main. Manual of Scientific Enquiry. 3rd edition. 8vo. *London,* 1864

† Hervieu. Essai sur l'électricité atmosphérique. . . 8vo. 266 pp. *Paris,* 1835

Hesse, Jul. Erfahrungen und Beobachtungen über die Anwendung des magneto-elektrischen Rotations-Apparats in verschiedenen Krankheiten. *New Brandenburg,* 1843

† Hessel, J. F. C. Versuche über Magnet-Ketten, u. ub. d. Eigenschaften d. Glieder derselben. 8vo. 301 pp. 3 plates. *Marburg,* 1844

Hessler, F. Über d. Einfluss d. materiell. Beschaffenheit d. Prisma auf d. chem. Wirk. des durch dasselbe zerlegt. Sonnenlicht. *(Baumgartner's Zeitschrift f. Physik,* iii. 1835.) *Wien,* 1835

Über einen elektromagnet. Inductions-Apparat. 8vo. *(Abhandl. d. Böhm. Gesellsch. d. Wiss.* 5 *Folge,* i. 1841.) *Prag,* 1841

Uber d. galvan. Kohlenzinksäule d. Prof. R. Bunsen. 8vo. *(Abhandl. d. Böhm. Gesellsch. d. Wiss.* 5 *Folge* ii. 1843.) *Prag,* 1843

† Hessler and Pisko. Lehrbuch der Technichen Physik. 3te Auflage. 2 vols. 8vo. 1373 pp. *Wien,* 1866

1 Bd. Mechanik, Akustik u. Elekt. Von J. F. Hessler.
2 Bd. Schluss d. Elek. Optik u. Wärme. Von Hessler und F. J. Pisko.

Heuffers. Manner of constructing the Pile. 8vo. *(From Sue,* iv. 233.) *Hulm.*

Heurlin, S. De actione electricitatis in corpora organica. *Lund,* 1776

Heusinger, Johann Michael. *Born Aug.* 24, 1690, *at Sundhausen, Gotha; died Feb.* 24, 1751, *et Eisenach.*

Diss. de noctiluca mercuriali. " Seu de luce quam argentum vivum in tenebris fundit." 4to. *Gissæ,* 1716

Hevelius, Johann. *Born Jan.* 28, 1611, *at Danzig; died Jan.* 28, 1687, *at Danzig.*

Letter containing an observation of the Variation of the Magnetic Needle at Danzig in the year 1670, &c. . . 4to. *(Phil. Trans.* 1670, p. 2059.) *London,* 1670

De variatione acus magneticæ. *(Opusc. Act. Erudit. Lips.* tom. i. p. 103.)

† **Hiffelsheim.** Des applications médicales de la pile de Volta. 8vo. 152 pp.
Paris, 1861

† **Higgins, W.** Description and analysis of a Meteoric Stone which fell in the county of Tipperary, in Ireland, in the month August, 1810. 8vo. 6 pp. (*Phil. Mag.* xxxviii. 262.) *London,* 1811

† Experiments and observations on the Atomic Theory and Electrical Phenomena. 8vo. 180 pp. *London,* 1814

† **Higgins, W. M.** Alphabet of Electricity. 16mo. 113 pp. *London,* 1834

† **Highton, E.** The Electric Telegraph ; its history and progress. 12mo. 179 pp.
London, 1852

Reports on Telegraphic Wires. Fol. *London,* 1856-8

Consideration of the probability of the success of the Atlantic Telegraph Cable. 8vo. *London,* 1857

† **Highton, Henry and Edward.** Specification of the Patent for improvements in Electric Telegraphs. Sealed Jan. 25, 1848. 8vo. 37 pp. 3 plates.
London, 1849

Hildebrand (or " A. H."). Schreiben eines Geistlichen zu Wien an einen seiner Freunde zu Presburg, von dem immerwährenden Elektrophor ; aus dem Französischen mit Anmerk. überst. *Wien,* 1776

(*Note.*—*Socin says that Jacquet is the author of the original French work and Hildebrand the* (" A. H.") *translator as above.*)

Hildebrandt, Georg Friedrich. *Born June* 5, 1764, *at Hanover ; died March* 25, 1816, *at Erlangen.*

Galvan Wasserzersetz-Apparat. 8vo. (*Gilbert, Ann.* xxi. 1805.) *Leipzig,* 1805

Volta'sche Säule aus drei Metallen. 8vo. (*Gilbert's Ann.* xxx. 1808.)
Leipzig, 1808

Über d. Unabhängigkeit d. Erregung d. Galvanismus v. d. Unterschiede d. Oxidabilität d. Erreger. 8vo. (*Gehlen's Nec. Journ.* vi. 1808.) 1808

Uber d. Unterschied d. Lichts beider Electricitäten in verdünnter Luft. 8vo. (*Schweigger's Journ.* i. 1811.) 8vo. *Nürnberg,* 1811

Über d. verschieden. Wirksamk. verschiedener Metalle in Erzeng d. elektr. Spitzenlichts. 8vo. (*Schweigger's Journ.* xi. 1814.) *Nürnberg,* 1814

† **Hill, John.** Letter to Tilloch, on Cuthbertson's and on Brooke's modes of increasing the charge of Electrical Jars and Batteries. 8vo. 2 pp. (*Phil. Mag.* xxxvii. 79.)
London, 1811

† **Hill, Rowland.** Improvement in the method of forming Electrical Planispheres. 8vo. 2 pp. (*Phil. Mag.* lii. 293.) *London,* 1818

Hilliard, Jno. Fire from Heaven : concerning a man burnt to ashes by Lightning. 4to. (*From Watts.*) *London,* 1613

Hinrichs, G. Die elektromagnetische Telegraphie, sammt den nöthigen Kenntnissen aus der Physik leichtfasslich wissenschaftlich dargestellt. 8vo. 2 plates. Map. *Hamburg,* 1856

Hintler, A. Meteora, seu corpora sublimia physicæ considerationi subjecta. 4to.
Salisburgum, 1758

Hiorter. (*Vide* Celsius and Hiorter.)

Hipp. Notice sur les horloges électriques et particulière sur celles établies à Genève en 1861. *Neuchatel,* 1864

Télégraphe autographique. (*From Du Moncel.*)

"Der Hipp'sche Schreibapparat ohne Relais." (*From Kuhn.*)

Copirtelegraphen. (*Kuhn, p.* 1007, *gives no account of the source of quotation, but an account of the thing itself.*)

Invented or described several apparatus, as alluded to by Du Moncel, Kuhn, &c. but I can find no references to their sources of information. He obtained a Medal at the Exposition of 1853.

(*Vide* also Hartmann, Description du télégraphe de M. Hipp.)

Hire. (*Vide* La Hire, De.)

Hirt. Der Tempel Salomonis. 4to. (*From Martin*, p. 307.)   *Berlin*, 1803

Hisinger, W. *Born Dec.* 22, 1766, *at Elfstorps, Bruk; died June* 28, 1852, *at Skinskattiberg.* (*Poggendorff*, i. 1112.)

Forsok med elektriska stapelns verkan på. Djur-och Växtämmen (Dl. i.) (*Afhandl. i Fisik. Kemi och Mineralogi*, Dl. i.)   (*Quere : Is this a* 1st *part, or vol. ?*)   *Stockholm*

Untersokn af benzo ësyrade salters forhallende till jordarter och metalloxider. (*Afhandl. i Fisik, Kemi och Mineralogi*, Dl. iii.)   *Stockholm*, 180—

Hisinger and Berzilius. Forsok med. elektr. stapelns verkan på salter, &c. 8vo. (*Afhandl. i Fisik, Kemi och Mineralogi*, Dl. i.)   *Stockholm*, 1806 ?
Poggendorff says that Hisinger published many articles in this periodical between 1806 *to* 1818.

Hisinger and Gahn. Forsok att med. elektriska gnistor fran en vanlig Elektricitets-machin sonderdela vatten. (*Afhandl. i Fisik, Kemi och Mineralogi*,Dl. i.)   *Stockholm*, 1806
(*Vide* also Berzelius and Hisinger.)

† **Hittorf, W.** De joutum migrationibus electrolyticis. Pars prima. Comment. Chem. Phis. 4to. 20 pp. 1 plate (lith.)   *Monasterii Guestphalorum*, 1853

† **Hjortberg, G. F.** Sechsjährige Versuche mit der Elektricität an unterschiedlichen Kranken. 8vo. 11 pp. (*K. Schwed. Akad.* Abh. xxvii. 200.)   *Leipzig*, 1765

† Beschreibung und Abzeichnung der Werkzeuge die bey Anwendung der Elektricität auf Kranke mit Nutzen sind gebraucht worden von G. F. H. 8vo. 7 pp. 1 plate. (*K. Schwed. Akad.* Abh. xxvii. 280.)   *Leipzig*, 1765

† Auszug aus des . . G. F. Hiortbergs Tagebuche über die von ihm 1766 angestellten Versuche, die Elektricität gegen allerley Krankheiten zu brauchen. 8vo. 4 pp. (*K. Schwed. Akad.* Abh. xxx. 99.)   *Leipzig*, 1768

Hjorter, Olof Peter. *Born* 1696, *at Jämtland; died April* 25, 1750, *at Upsala.*

Om magnetnälens atskilliger ändringer, som af framl. Prof. Andr. Celsius blifvit jaktagne och sedan vidare observerade. (*Vetensk. Akad. Handl.* 1747.)   1747

† Von der Magnet-Nadel mannichfaltigen Veränderungen, welche durch den verstorbenen A. Celsius sind in Acht genommen und nachgehends weiter beobachtet worden. 8vo. 15 pp. (*Schwedische Akad. Abhandl. A.* 1747, p. 30, vol. ix.)   *Hamburg*, 1747
Note.—Van Swinden (iii. 239), *refers to Hiorter's suspicion relative to Bars magnetised during an Aurora, &c., and quotes Hiorter in other places on other subjects.*

† **Hoadly, Benjamin.** *Born February* 10, 1705, *at London; died August* 10, 1757, *at Chelsea.*

Hoadley and Wilson. Observations on a Series of Electrical Experiments by Dr. Hoadley and Mr. Wilson. 4to. 76 pp.   *London*,1756

Betrachtungen über eine Reihe elektrischer Versuche.   *Leipzig*, 1763

Betrachtungen über eine Reihe von Versuchen; nach der 2n von Wilson verbesserten . . Ausgabe; a. d. Engl. 8vo. (*From Saxtorf*, ii. 509.)   *Leipzig*, 1763

† **Hodgson, J.** A Method of correcting the variation of the Mariner's Compass. 8vo. 4 pp. 1 plate. (*Phil. Mag.* xxxix. 370.)   *London*, 1812

Hofmann, B. Einladungs schrift de moralitate circa electricitatis experimenta, præsertim fulmina. Folio.   *Mersberg*, 1754

Hoffmann, Johann Christian. *Born* 1768, *at Schlettau, near Lauchstadt.*

Praktische. . . Anleitung. . . Elektrisermachinen zu bauen. 8vo.   *Leipzig*,1795

Anweisung gute Elektrisirmaschinen zu bauen, &c.   *Leipzig*, 1798

† **Hoffmann,** Johann Christian—*continued*
Anviisning til ved Hjælp af den galvaniske Kobberudskilning at maug-foldiggiöre en med Pen eller Ridfefjeder udfort Skrift eller Tegning. 8vo. 22 pp. 1 plate. *Kjobenhavn,* 1842

† Anweisung zum Vervielfältigen mit der Feder oder Reissfeder, durch Hülfe der galvanischen Kupferausscheidung. 8vo. 22 pp. 1 plate.
Kopenhagen, 1842

Hoffmann, J. J. Igns. Beschreibung eines sehr einfachen u. sehr wirksamen. Papier-Elektrophors. *Aschaffenb.* 1847

Hoffman. (*Vide* Buff, H. 1851.)

† **Hogstrom,** P. Anmerkung über die Thiere die aus den Wolken kommen. 8vo.
Hamburg, 1749

Holger. (*Vide* Baumgartner.)

Höll. (*Vide* Hel.

Holland, Sir Henry, Bart. Essays on Scientific and other Subjects contributed to the *Edinburgh and Quarterly Reviews.* *London,* 1862

Hollandsche Maatschappij voor Wetenschappen. A Society similar to the Smithsonian Institute at Washington, for the purpose of exchanging the publications and periodicals of the Learned Societies.

† **Hollick,** F. Neuropathy... An explanation of the action of Galvanism, Electricity, and Magnetism in the cure of Disease. 12mo. 193 pp.
Philadelphia, 1847

Hollis. On Electricity and Magnetism. 8vo. (*From Roorbach, Bibl. Amer of* 1852.)

Hollmann, Samuel Christian. *Born December* 3, 1696, *at Stettin ; died September* 4, 1787, *at Göttingen.*

† Epistola (ad Mortimerum Sec. R. S.) de subitanea congelatione, de igne electrico, &c. Read Jan. 10, 1744-5. 4to. 10 pp. (*Phil. Trans.* xliii. 239.)
London, 1744-5
Of Electrical Fire. (*Phil. Trans.* x. 271.) 1744-5

† De Barometrorum cum Aëris et Tempestatum mutationibus consensu. Read April 23, 1749. 4to. 10 pp. (*Phil. Trans.* xlvi. p. 101.) *London,* 1749-50

Holm, P. Stuurmans rechtwyzende compas. *Amsterdam*

† **Holmes.** Magneto-electric Light as applicable to Lighthouses. 8vo. 34 pp. 1 plate. *London,* 1862

† **Holmgren,** K. A. Recherches relatives à l'influence de la température sur le Magnétisme. 4to. 20 pp. (*Ups. Acad.* 1855.) *Upsala,* 1855

Recherches sur l'influence de la température sur le Magnétisme. 4to. 2 plates. (*Ups. Acad.* 1859.) *Upsala,* 1859

† Rön angaende Magnetismus inverken pa Warmeledningen hos fasta kroppar. 8vo. 27 pp. (*Akademisk Afhandling.*) *Stockholm,* 1861

Holsbeek, H. van. Appareil électro-galvanique de M. O'Connell. 8vo. 1 plate.
Bruxelles, 1859

† Compendium d'électricité médicale. 2e édition, . . augmentée, &c. 8vo. 680 pp. *Bruxelles et Paris,* 186

† (*Vide* also Annales de l'Electricité Médicale.) 8vo.

Holtz, —. An Electrical Machine formed on the principle of Influence. 1864

† **Holzmann,** C. H. A. Mechan. Arbeit, erforderlich z. Erhalt. e. elektr. Stroms. 8vo. (*Poggendorff's Ann.* xci. 1854.) *Leipzig,* 1854
Über d. Polarisat. e. elektr. Stroms. 8vo. (*Poggendorff, Ann.* xcii. 1854.)
Leipzig, 1854

Holzmann, M. Praktisch-elektrische Versuche und Aufschlüsse aus geprüften Erfahrungen . . 8vo. 72 pp. 1 plate. *Ingolstadt,* 1837

Holzmuller G. Über die Anwendung der Jacobi-Hamilton'schen Methode auf den Fall der Anziehung nach dem elektrodynamischen Gesetz von Weber. Hallische Inaugural-Dissert. 8vo. 23 pp. *Halle,* 1870

Hombre-Firmas. (*Vide* D'Hombre-Firmas, p. 138.)

† **Home,** Everard. Hints on the subject of Animal Secretions. From Phil. Trans. for 1809, part ii. Communicated by the Society for the improvement of Animal Chemistry. 8vo. 6 pp. (*Phil. Mag.* xxxv. 108.) *London,* 1810

Homer, Odyss. xx. 139.
Note.—Lightning without audible thunder. (*From Young,* p. 483.)

Hompeck, Anton. *Died after* 1770, *at Vienna.*
Abhandl. v. d. elekirisn. Abstossung. 8vo. (*From Krunitz,* p. 69.) *Wien,* 1765

Hooke, Dr. Robert. *Born July* 18, 1635, *at Freshwater, Isle of Wight ; died March* 3, 1703, *at London.*
Variation (Magnetic). (*From Churchman.*) 1674
His Optical Telegraph. (*Vide Poggendorff,* i. 1138, *and* 420 *C. Chappe. Vide also Böckmann, J. L., Über Telegraphie, &c., Carlsruhe,* 1794. *Hooke's Telegraph was made known (by him) long before Claude Chappe's.*) 1684
Posthumous Works of, containing his Cutlerian lectures, and other philosophical discourses (concerning) an hypothetical explication of . . Magnetism, &c. His Life prefixed. Folio. *London,* 1705

Hopkins. On the connexion of Geology with terrestrial Magnetism. 8vo. 24 plates. *London,* 1844

† **Hopkins,** Evan. Notice sur la dépolarisation des Navires de fer, par M. Evan Hopkins. . . Traduit de l'anglais par F. A. B. Craufurd. 8vo. 17 pp. 1 plate. *Paris,* 1867

Hopkinson. (*Vide* Rittenhouse and Hopkinson.)

Hopkinson, Thos. *Born April,* 1709, *at London ; died* 1751, *at Philadelphia.*
On the effects of Points in Electricity. (*Vide* Franklin, 1754.)

† **Hoppe,** Martin. Über das Gewitter. 4to. 18 pp. (*Program des Fürstlich. Hedwigschen Gymnasium, in Neustettin* . . 10 *und* 11 *April* 1865). *Neustettin,* 1865

Horace. Od. i. 34.
Note.—Lightning without audible thunder. (*From Young,* p. 483.)

Horbye. Sur les phénomènes d'Erosion en Norvège. (*Atti. I. R. Inst. Lomb.* i. 125.) *Christiania,* 1857

Horing. Beobachtungen über Heilwirkungen der Electricität bei Anwendung des magnet-electr. Rotations-Apparates. (*Würtembergisches Correspondenzblatt,* 1846, No. 28, 29.) 1846

† **Horion.** De phænomenis electro-chimicis. 4to. 58 pp. 1 plate. 1829

Horn, H. Das Wirken der Elektricität in den Organismen. 12 Heft. 8vo. 44 pp. *München,* 1867

Hornsman. (*Vide* Œrsted and others.)

Horner. Erste Hälfte des Art. Magnetismus des neuen Gehln. Wörterb. 1825 *to* 1845?
Note.—He also wrote the Articles "Ablenkung d. Magnetnadel," "Abweichung d. Magnetnadel."

Horner, Johann Kaspar. *Born March* 21, 1774, *at Zurich ; died Nov.* 3, 1834, *at Zurich.*
Bemerkung über Blitzableiter u.s.w. *Zürich,* 1816
Über Wasserhosen u. Erdtromben. 8vo. (*Gilbert's Ann.* lxxiii. 1823.) *Leipzig,* 1823
Verbesserung d. Schmalkalderschen Bussole. 8vo. (*Gilbert's Ann.* lxxv. 1823.) *Leipzig,* 1823

Hornes, M. Ueber den Meteorsteinfall bei Ohaba im Blasendorfer Bezirke in Siebenbürgen, in der Nacht zwischen dem 10 u. 11 October, 1857. 8vo. 8 pp. (*Aus den Sitzungsberichten* 1858 *der mathem. naturw. Classe der kais. Akad. der Wissenschaften.*) *Wien,* 1858

Horsburgh, James. *Born September* 23, 1762, *at Elie, Fifeshire, Scotland; died May* 14, 1836, *at Herne Hill, near London.*

* Atmospheric Register for indicating Storms at sea. Fol oblong.
  London, 1816

† Compendium . . . of Winds . . . Luminous Appearance . . . of the Sea. Magnetism, &c. . . 8vo. 52 pp. *London,* 1817

† On the Insulated or Safety Compass lately invented by Mr. Jennings. 8vo. 3 pp. (*Phil. Mag.* liii. 365.) *London,* 1819

† **Hortentz, A. B.** On production of Muriates by the Galvanic Decomposition of Water. 8vo. 1 page. (*Phil. Mag.* xxiv. 91.) *London,* 1806

Hough, G. W. The Galvanic Battery. (*Monot. Anzeig.* 1870, p. 100.) *Albany*

† **Houink, J.** Diss. Phys. de theoria elementi apparatus voltaici. 8vo. 189 pp. *Groningæ,* 1835

† **House, R. E.** History of the Invention of the Electric Telegraph. Abridged from the works of L. Turnbull and E. Highton, with remarks on R. E. House's American Printing Telegraph and the claims of S. F. B. Morse as an inventor. 12mo. 130 pp. *New York,* 1853

† **Houzeau, J. C.** Sur les étoiles filantes périodiques du mois d'aout, et, en particulier, sur leur apparition de 1842. Présenté 3 Aout, 1844. 4to. 54 pp. (*Mem. sav. étrang. Acad. Brux.* xviii. 1845.) *Bruxelles,* 1845

Howard, E. Copernicans convicted, &c. 8vo. (*From Watt.*) *London,* 1705

Howard, Luke. *Born November* 28, 1772, *at London.*

*† On the Modifications of Clouds, and on the principles of their Production, Suspension, and Destruction, being the substance of an Essay read before the Askesian Society in the Session 1802-3. 11, 14, and 7 pp. (*Phil. Mag.* vol. xvi. pp. 97 and 344; and vol. xvii. p. 5), 3 plates, 7 figs. *London,* 1803

*† On the best means for conducting Meteorological Observations in different Places and Climates, so as to produce some uniformity in the modes of obtaining and summing up the results. 8vo. 4 pp. (*Phil. Mag.* lvii. 81.) *London,* 1821

* Essay on the Modifications of Clouds. 8vo. *London,* 1832

*† Seven Lectures on Meteorology. 2nd edition revised. 12mo. 218 pp. *London,* 1843

† **Howldy, T.** Influence of Atmospheric Moisture on an Electric Column composed of discs of zinc and silver. 8vo. 5 pp. (*Phil. Mag.* xliii. 241.) *London,* 1814

† Influence of Atmospheric Moisture on an Electric Column composed of discs of zinc and silver. 8vo. 2 pp. (*Phil. Mag.* xliii. 363.) *London,* 1814

† On the Franklinian Theory of the Leyden Jar, &c.; with remarks on Mr. Donovan's Experiments. 8vo. 8 pp. (*Phil. Mag.* xlvi. 401.) *London,* 1815

† On certain Electrical Phenomena. 8vo. 8 pp. (*Phil. Mag.* xlvii. 285.) *London,* 1816

† Remarks on Mr. Sturgeon's Paper "On the Inflammation of Gunpowder by Electricity." 8vo. 3 pp. (*Phil. Mag. or Annals,* i. 343.) *London,* 1827

† **Hoxton, Walter.** An account of an unusual agitation in the Magnetical Needle, observed to last for some time, in a voyage from Maryland, by Hoxton, W., communicated in a letter to Papillon, D. Dated September 2, 1724. 4to. 2 pp. (*Phil. Trans.* xxxvii. 53.) *London,* 1731-32

† The Variation of the Magnetic Needle as observed in three voyages from London to Maryland. 4to. 5 pp. (*Phil. Trans.* xli. 171.) *London,* 1739-40-41

z

Hoy, James. On the height of the Aurora Borealis from the Earth. 8vo. (*Tilloch's Phil. Mag.* li. 422.)   *London,* 1818

Huart. (*Vide* Colmet-d'Huart.)

Hube, Mich. Über die Ausdünstung und ihre Wirkungen in der Atmosphäre. 8vo.   *Leipzig,* 1790

† **Huber,** Joh. Jac. De aëre atque electro-œconomiæ animali famulantibus et imperantibus cogitationes tumultuariæ, quas demonstrationibus anatomicis præmittit. 4to. 52 pp.   *Casselis,* 1747

Hubner, J. G. Gedanken v. d. Magnetnadel. 8vo.   *Halle,* 1772

† **Hubner,** L. Abhandlung über die Analogie der elektrischen und magnetischen Kraft. 4to. 34 pp. (*Neue Abh. der Baier, Akad. Philos.* ii. 351.)   1780

Hues, Robert. *Born about* 1553, *at Harford ; died May* 24, 1632.

On Magnetic Variations. (*From Gilbert,* p. 7.)

Hufeland, C. W. Diss. sistens usum vis electricæ in asphyxia experimentis illustratum. 4to.   *Gottingue,* 1783

Journal de Médecine pratique. 8vo.   *Berlin*

Journal d. pract. A. K.

(*Note.—Contains Quesnel, Galvanische Versuche,* vol. xiii. 4 s. St. s. 130.)

† **Hughes,** David Edward. Expériences sur la forme et la nature des électro-aimants. 8vo. 11 pp. 3 plates. (*Annales Télégraph.* 1864.)   *Paris,* 1864

† Le télégraphe imprimeur.   1871

† **Hugueny,** C. A. Nouvelles considérations sur les agens généraux moteurs de l'action universelle. 8vo. 1 plate. 73 pp.   *Strasbourg,* 1834

Considérations générales, &c. 3me étude. 8vo. 51 pp.   *Strasbourg,* 1842

† **Hulme,** N. Experiments and observations on the Light which is spontaneously emitted, with some degree of permanency, from various bodies. 4to. (*Phil. Trans.* 1800, part i. p. 161.)   *London,* 1800

(*Note.—This is one of many articles in Reuss's Repertorium,* vol. iv. p. 252 *et seq. on Light which should be considered or examined in reference to Electricity.*)

A continuation of the experiments and observations of the Light which is spontaneously emitted from various bodies ; with some experiments and observations on Solar Light when imbibed by Canton's phosphorus. 4to. (*Phil. Trans.* 1801, p. 403.)   *London,* 1801

Humboldt. Friedrich Heinrich Alexander (Freiherr von). *Born Sept.* 14, 1769, *at Berlin.*

Aphorismi ex doctrina physiologiæ chimicæ plantarum : Anhang zu Floræ Fribergensis specimen, &c.   *Berol,* 1793

(*Note.—Aphorismi " Deutsch von Fischer mit Zusätzen von Hedwig und Ludwig. Leipzig,* 1795.)

† Versuche über die gereizte Muskel und Nervenfaser : nebst, Vermuthungen über den chemischen Process des Lebens in der Thier u. Pflanzen welt. 8vo. 2 vols.   *Posen u. Berlin,* 1797

† Expériences sur le Galvanisme et, en général, sur l'irritatiou des fibres muscu-laires et nerveuses. Trad. de l'Allem. avec additions par Jadelot. 8vo. 530 pp. 8 plates.   *Paris,* 1799

Inclinaisons et déclinaisons de l'aiguille aimantée observées depuis Vendémiaire jusqu'en Germinal An. 7 à Paris, Nimes, Montpellier, Perpignan, Gironne, Barcelone, Cambrils, Madrid, Valence, Medina del Campo, Guaderana, Ferol, Océan Atlantique entre l'Europe, l'Amérique et l'A*f*rique. (*Société Philomath. An.* 7. p. 27.)   *Paris,* 1797

Humboldt, Friedrich Heinrich Alexander—*continued.*

Lettre de M. Humboldt à M. Loder sur l'application du Galvanisme à la médecine pratique. (*La Bibliothèque Germanique,* iv. Messidor, An. viii. p. 301.)   1800

Observations de Zoologie, &c. 4to.   *Paris*

Versuche, Muskel und Nervenfaser. 2 vols. 8vo. *Posen and Berlin,* 1797 & 1799

Über d. merkwürd. magnet. Polarität einer Gebirgskuppe von Serpentinstein. 8vo. (*Grens. Neu Journ.* iv. 1797.)   *Leipzig,* 1797

Versuche ü. d. chem. Zerleg. d. Luftkreises. 8vo.   *Brescher,* 1799

Über d. unterirdischen Gasarten, &c. 8vo. 2 plates.   *Brescher,* 1799

† Lettera su una pietra serpentina verde dotata d'una fortissima polarità magnetica. 4to. 1 page. (*Opusc. Scelti,* xxi. 126, *from Ann. de Chimie,* xxii.)
                                                                Milano, 1801

Versuche üb. d. electrischen Fische. 8vo.   *Jena,* 1806

† Observations sur l'Anguille électrique (Gymnotus electricus, *Linn.*) du nouveau continent. 4to. 44 pp. 1 plate. (*Many quotations.*)   *Paris,* 1806

Ansichten d. Natur. 2 Bde. 12mo.   *Tübingen,* 1808

Sur l'électricité animale. 4to. (*Soc. Philomath.* tom. i. p. 92.)   *Paris*

Physique générale et relation historique. 4to.   *Paris,* 1811-25

Sur les poissons électriques. 8vo. (*Ann. Chim. Phys.* xi. 1819.)   *Paris,* 1819

Ansichten d. Natur, mit wissenschaftl. Erläuterungen. 2 Bde. 2te. Ausg. 16mo.   *Stuttgart,* 1826

Beob. d. magnet. Intensität u. Inclinat. auf d. Reise nach u. in Amerika. 8vo. (*Pogg. Ann.* xv. 1829.)   *Leipsig,* 1829

Mittel um die Egründ. einig. Phänomene d. tellurisch. Magnetism zu erleichteren. 8vo. (*Pogg. Ann.* xv.)   *Leipsig,* 1829

† Sur le Magnétisme polaire d'une montagne de Chlorite schisteuse et de Serpentine. Lettre à Becquerel. 8vo. 7 pp.   *Paris,* 182-?

Fragmen. de géologie et de climatologie Asiatiques. 2 vols. 8vo. *Paris,* 1831

Über einige elektromagnet Erscheinungen u. d. verminderten Luftdruck unter den Tropen. 8vo. (*Pogg. Ann.* xxxvii. 1836.)   *Leipzig,* 1836

Asie centrale, Recherches sur les chaines de Montagnes et la Climatologie comparée. 3 vols. 8vo.   *Paris,* 1843

Central-Asien. Untersuchungen über die Gebirgsketten und die vergleichende Klimatologie. 2 vols. and map.   *Berlin,* 1844

Cosmos, Entwurf einer physischen Weltbeschreibung. 4 vols. 8vo.
                                                Stuttgart u. Tubingen, 1845

Aspects of Nature, translated by Mrs. Sabine. 2 vols. square. *London,* 1848

Ansichten der Natur, mit wissenschaftl. Erläuterungen. 2 vols. 3rd edition. Verb und verm ausg. 8vo.   *Stuttgart,* 1849

Cosmos, translated by Mrs. Sabine. 4 vols. 8vo.   *London,* 1849-58

Alexander von Humboldt ein biographisches Denkmal von Prof. Dr. H. Klencke. 8vo. 252 pp. 1 plate.   *Leipsig,* 1851

Cosmos. Essai d'une description physique du monde, traduit par Ch. Galuski. 4 vols. 8vo.   *Paris,* 1848-59

Kleinere Schriften. Vol i. Geognostische u. physikalische Erinnerungen. 8vo.   *Stuttgart,* 1853

Atlas to the same. Umrisse von Vulkanen. 12 plates 4to. *Stuttgart,* 1853

† Atlas to Humboldt's Kosmos. 42 maps of physical geography, &c., by T Bromme. Fol.   *Stuttgart,* 1854

Humboldt, Friedrich Heinrich Alexander—*continued.*

Briefe über Humboldt's Kosmos. Ein Commentar für gebildete Laien. 4 vols. 8vo. *Leipzic,* 1855-60

Reise in die Aequinoctialgegenden des neuen Continents. 4 vols. 8vo.
Stuttgart, 1859-60

Alexander von Humboldt, sein wissenschaftliches Leben und Wirken . . von Wittwer. 8vo. 440 pp. Portrait and autograph. *Leipsig,* 1861

Personal Narrative, by Williams. 8vo. *London*
(*Note.—In* vol. iii. *he gives an account of atmospheric phenomena witnessed at Cumana* . . *by Bonpland on the 12th Nov.* 1799, *viz. thousands of bolides* (*fire-balls*) *and falling stars succeeded each other during four hours*) . . (*Vide Phil. Mag.* lii. 312.)

Kosmos oder physische Erdbeschreibung. 5 vols. 8vo.

Commentar über Humboldt's Kosmos von Cotta und Schaller. 4to. 8vo.

Umrisse von Vulkanen aus den Cordilleren von Quito und Mexiko. 4to. pls.

† Portrait. Lithograph by Boorbach (of Berlin) from a photograph.

Gedächtnissrede auf. (*From Dove.*) 1869

† **Humboldt und Biot.** On the variations of the terrestrial magnetism in different latitudes. By Messrs. Humboldt and Biot. Read by M. Biot in the French National Institute, 26th Frimaire, An. 13 (17th December, 1804). 8vo. 10 pp.
(*Phil. Mag.* xxii. 248, 249.) *London,* 1815

Humboldt und Bonpland. Untersuchungen über die Geographie des neuen Continents, bearbeitet von Oltmanns. 2 vols. 8vo. *Paris,* 1810

Voyage dans l'intérieur de l'Amérique méridionale fait en 1799-1804. In d. grossen Aufl. 17 vols. fol. u. 11 vols. 4to. Part i. Physique générale et relation historique. 3 vols. 4to. Paris, 1811-25, oder 13 vols. 8vo., 1816-31. Part ii. Zoologie, &c.

† **Humboldt and Gay-Lussac.** Experiments on the Torpedo, by Messrs. Humboldt and Gay-Lussac. Extracted from a letter from M. Humboldt to M. Berthollet, dated Rome, 15 Fructid. Year 13. 8vo. 5 pp. (*From Ann. de Chim.* No. 166.) (*Phil. Mag.* xxiii. 356.) *London,* 1806
(*Vide* also Gay-Lussac.)

Observations sur l'intensité et l'inclinaison des forces magnétiques faites en France, en Suisse, en Italie, et en Allemagne ; avec Gay-Lussac. 8vo.
(*Mémoires d'Arcueil,* i. 1807.) *Paris,* 1807
(*Vide* also Bétancourt.)

Hume, A. The learned Societies and Printing Clubs of the United Kingdom, with a Supplement by A. J. Evans. 12mo. *London,* 1853

Hume, G. L. Essay on Chemical Attraction. 8vo. *London,* 1836

Humphreys, J. D. The Electro-Physiology of Man, with practical illustrations, &c. 12mo. *London,* 1843

Hunt, R. Atlantic Telegraphy. 8vo. (*Popular Science Review,* No. 17, 1866.)
London, 1866
(*Note.—This number contains also " The Applications of Photography.*")

Hunt. (*Vide* Walker and Hunt.)

Hunter, John. *Born Feb.* 13, 1728, *at Kilbride, Lanarkshire ; died Oct.* 16, 1792, *at London.*

On the numerous Nerves, &c., of the Electric Organs of the Torpedo. 4to.
(*Phil. Trans.*) *London,* 1773-1775

Anatomical observations on the Torpedo. 4to. (*Phil. Trans.* 1773.)
London, 1773

An account of the Gymnotus Electricus. 4to. (*Phil. Trans.* 1775.)
London, 1775

First perceived the Galvanic sensation of Light in the experiment on the Eyes.
(*Opusculi Scelti,* xxii. 364.)
(*Note.—Becquerel makes the mistake of saying that Sulzer saw this light.*)

Hupsch. Untersuchung des Nordlichts. 8vo. (*Opusc. Scelti* (*in* 4to.) tom. ii. p. 14.) *Cologne,* 1778

Hunter. (*Vide* Euler, L.)

Hutchins, Thomas. *Born* 1730 (*about*), *at Monmouth County, New Jersey ; died April* 20, 1788.
Experiments on the Dipping Needle, made by desire of the Royal Society. 4to (*Phil. Trans.* 1775, p. 129.) *London,* 1775

Huth, J. G. Über die chemische und elektrische Wirkungs-Weite einer Volta-ischen Säule. (*Neue Schr. die Berliner Gesellch. Natur Freunde,* iv. 161, 1803) *Berlin,* 1803
Expériences et observations sur l'électricité galvanique ; distance à laquelle la pile exerce son action. 8vo. (*Journ. d. Chim. de Van Mons, No.* vi. p. 289.) *Bruxelles*

† **Hutton, C.** On the Origin of Stones that have fallen from the Atmosphere. 8vo. 9 pp. (*Phil. Mag.* xxii. 71.) *London,* 1805

† **Huxham, John.** A Letter from J. Huxham to C. Mortimer containing observations on the Northern Lights seen Feb. 16 and 16, 1749-50. Read June 21, 1750. 4to. 2 pp. 1 part of a plate. (*Phil. Trans.* xlvi. 472.) *London,* 1749-50

I.

Iconographic Encyclopædia of Science. 4 vols. 8vo., and 2 vols. 4to. of plates.
New York, 1852
(*Note.—Trübner says, "systematically arranged by G. Heck, 500 sq. plates by Germans. Text translated by Baird."*)

† **Ideler, J. L.** Über d. Ursprung d. Feuerkugeln u. d. Nordlichts. 98 pp.
Berlin, 1832

† Meteorologia veterum Græcorum et Romanorum. 8vo. 31 pp. *Berolini*, 1832
Untersuchungen über den Hagel, und die elektrischen Erscheinungen in unserer Atmosphäre. 8vo. (*Poggendorff's Ann.* xxvi. 1832.) *Leipzig*, 1832

† Untersuchungen über den Hagel und die elektrischen Erscheinungen in unserer Atmosphäre. Nebst einem Anhange . . 8vo. 1 plate, 148 pp.
(*Poggendorff's Ann.* xxvii. 1833.) *Leipzig*, 1833
Aristotelis meteorologicorum, libri iv. 8vo. 2 vols. *Leipzig*, 1834-36

Imhof, Maximus. Born July 26, 1758, at Reissbach; died April 11, 1817, at *Munich*.
Theoria electricitatis recentioribus experimentis stabilita . . 8vo. 123 pp.
Haydhusii, 1790
Was hat die heutige Arzneykunde seit einem halben Jahre in Rücksicht einer Zueckmässigen Anwendung d. Elektricität auf Kranke gewonnen?
4to. *München*, 1796
Analyse zweier baierisch. Meteorsteine. 8vo. (*Gilb. Ann.* xviii. 1804.)
Leipzig, 1804

† Über das Schiessen gegen heranziehende Donner- und Hagel-Gewitter. (*Read 28th March*, 1811.) 4to. 24 pp. *München*, 1811

Theoretisch prakt. Anweisung zur Anlegung der Blitzableitern. 8vo.
München, 1816

Imperial Dictionary, English, Technological, and Scientific. Adapted to the Present State of Literature, Science, and Art. Upwards of 2,500 engravings on wood. 2 vols. imperial 8vo. (*From "Athenæum," March* 4, 1872.)

Indian (East) Telegraphs. (*Vide* O'Shaughnessy.)

Ingalls, T. On the Luminous appearance of the Ocean. 8vo. (*Trans. of the Albany Institute*, i. 8.) *Albany, U.S.*, 1831

Ingenhousz, Jan. Born Dec. 8, 1730, at Breda, Holland; died Sept. 7, 1799, at *Bowood, near London* (?).

Experiments on the Torpedo. 4to. (*Phil. Trans.* 1775.) *London*, 1775
Versuche mit Pflanzen. (Vol. i.) 2 vols. 8vo. *Vienna*, 1778
A ready way of lighting a candle by a very moderate Electrical Spark. 4to. (*Phil. Trans.* An. 1778, p. 1022.) *London*, 1778
Electrical experiments to explain how far the phenomena of the Electrophorous may be accounted for by Dr. Franklin's theory of Positive and Negative Electricity. 4to. (*Phil. Trans.* An. 1778, p. 1049.) *London*, 1778
Improvements in Electricity. 4to. (*Phil. Trans.* 1779.) *London*, 1779
On some new methods of suspending Magnetic Needles. 4to. (*Phil. Trans.* An. 1779, p. 537.) *London*, 1779
Experiments upon Vegetables . . 8vo. (*From Cat. Roy. Soc.* p. 313.)
London, 1779
Expériences sur les végétaux, trad. de l'Anglais (by the Author himself). 8vo.
Paris, 1780
Sur divers mouvemens du fluide électrique. (*Journ. Phys.* xvi. 1780.)
Paris, 1780
Anfangsgründe der Electricität aus dem Englischen mit Anmerkungen von Molitor. 8vo. 134 pp. *Wien*, 1781

Ingenhousz, Jan.—*continued.*

Vermischte Schriften physisch. med. Inhalts; übersezt u. herausg. v. N. K. Molitor. 2 vols. 8vo. *Wien,* 1782-4

† Nouvelles expériences et observations sur divers objets de physique. 8vo. 489 pp. 5 plates. *Paris,* 1785

De l'influence de l'électricité atmosphérique sur les végétaux. 4to. (*Journ. Phys.* xxxii. 1788.) *Paris,* 1788

Effet de l'électricité sur les plantes. 4to. (*Journ. Phys.* xxxv. 1789.) *Paris,* 1789

Versuch mit Pflanzen. (*From Young,* p. 424.) 1790

Elements of Electricity. (*From Young,* p. 432.)

Physikalische Schriften. (*From Exleben,* p. 530.)

Inglefield. A new theory of the Physical Causes of Terrestrial Magnetism. 8vo. *London,* 1851

† **Inglis,** Gavin. On the Theory of Water-spouts. 8vo. 7 pp. (*Phil. Mag.* lii. 216.) *London,* 1818

Innare. (*Vide* D'Innare.)

Innes, R. Miscellaneous Letters, &c. Aurora. 4to. *London,* 1732

† **Innocenti,** G. Preesistenza dell' acido muriatico e della soda nell' acqua galvanizzata. 4to. 10 pp. (*Nuova Scelta d'Opusc.* ii. 96.) *Milano,* 1807

L'Institut. Journal des Académies et Sociétés scientifiques. (*Vide* Arnault.)

Institut National. Rapport fait . . sur les exps. de Volta. 4to. *Paris*

Inventions, Patents for. (*Vide* Woodcroft.)

Inventions, Patents for. The Society of Telegraph Engineers have received from Her Majesty's Commissioners of Patents a complete collection of Specifications relating to Electricity, Magnetism, &c.

Irish Academy. (*Vide* Dublin.)

Isnard. Electrical Apparatus for blowing up Rocks under water. (*From Elice, Saggio,* 2nd ed. p. 61.)

Istituto Lombardo. Giornale dell' I. R. Istituto e Bibl. Italiana, 1841-47. 16 vols. 8vo.

Memorie dell' I. R. Istituto e Bibl. Italiana, 1843-... 4to.

Giornale dell' I. R. Istituto e Bibl. Italiana, 1847-56. 9 vols. 4to.

Atti dell' I. R. Istituto e Bibl. Italiana, 1858-... 4to.

(For each of these Collections *vide* "Lombardy I. R. Institution.")

Istituto Lombardo-Veneto. Memorie dell'I. R. Ist. Lombardo-Veneto.

Italian Society. Memorie di Matematica e Fisica della Società Italiana delle Scienze, 1782 to —. 4to. Tom. i. to vii. 1782—1794, printed at Verona; tom. viii. to xiii. 1799—1807, at Modena; tom. xiv. to xvii. 1809—1815-16, at Verona; tom. xviii. to xxv. 1820—1854, at Modena.

Italy. (*Vide* Accademia Italiana.)

Itier. Electro-plastic. (*From Congres . . de France,* 9*me Sess. à Lyon,* i. 421.) 1841

*† **Ivory,** J. On the Hygrometer by Evaporation. 8vo. 8 pp. (*Phil. Mag.* lx. 81.) *London,* 1822

Izarn, Guiseppe. *Born Jan.* 10, 1766, *at Cahors, Dép.Lot; still living,* 1834 (St.F.)

† Des pierres tombées du ciel. Lithologie atmosphérique, présentant la marche et l'état actuel de la Science sur le phénomène des pierres de foudre, pluies de pierres, pierres tombées du ciel, &c., plusieurs observations inédites, communiquées par MM. Pietat, Sage, Darcet, et Vanquelin ; avec un Essai de théorie sur la formation de ces pierres. 8vo. 421 pp, *Paris,* 1803

Lettre a M. de la Métherie sur l'effet galvanique des disques métalliques oxidés. 4to. (*Journ. de Phys.* An. xi. p. 157). *Paris,* 1803

Manuel du Galvanisme . . 1805. 8vo. 304 pp. 6 plates. *Paris,* 1805

† Manuale del Galvanismo adattato alla fisica, alla chimica, e alla medicina, tradotto dal Francese. 8vo. 206 pp. 6 plates. *Firenze,* 1805

J.

J., P. N. (*Vide* Anon. Meteorolog. Phen. Aerolites.) **1814**

† **Jackson.** Description of a Galvanic Battery on an improved construction, invented by Mr. Jackson ; communicated by James Miller, M.D. 8vo. 5 pp.
(*Phil. Mag.* xli. 308.) *London,* 1813

Jackson, Charles Thomas. *Born June* 21, 1805, *at Plymouth, Massachusetts.*
Réclamation de priorité relative à l'invention d'un télégraphe électro magnétique attribué a M. Morse. 4to. *Paris*

Electro-magnetic Telegraph. (*Boston Post, Jan.* 1849.) *Boston, U.S.* **1849**
(*Note.— He claims the invention of it.*)

† **Jacobæus,** H. W. Du mouvement imprimé à l'aiguille aimantée par l'influence subite de la lumière du soleil avec une théorie nouvelle fondée sur des recherches faites par H. W. Jacobæus. 8vo. *Copenhague,* 1856

Jacobi. Expériences faites à Dorpat en 1836 ou 1837 pour connaître la limite de la vitesse avec laquelle l'électricité se développe dans les conducteurs.
(*Bullétin Scientifique de l'Acad. de St. Pétersbourg,* iii. 333.) *Pétersbourg,* 1838

(*Vide* also Anon., Electro-Metallurgy, 1843.)

Remarques relatives à un Mémoire de M. Pouillet sur des appareils destinés à mesurer la vitesse des projectiles. 4to. *Paris,* 1845

Erfahrungen über die Heilkraft des Galvanismus. (*From Martens.*)

† **Jacobi,** C. G. J. Sur quelques points de la galvanométrie. 4to. 6 pp. (*Compt. Rendus Séance, Sept.* 8, 1851, p. 277, tom. xxxiii. No. 10.) *Paris,* 1851

Jacobi, H. Note sur quelques expériences avec une cible électro-magnétique. 8vo.
(*Mélanges Phys. et Chim. du Bull. de l'Acad. de St. Pétersb.* tom. v. liv. 5.)
St. Pétersbourg, Riga, and Leipzig, 1863

Jacobi, Moritz Hermann Von. *Born Sept.* 21, 1801 (*N. St.*) *at Potsdam.*
Die erste Nachricht von seinem Boot. (*In die Augsburger Allgem. Zeitung,* 30 Okt. 1839.) **1839**

Einige Notizen über galvanische Leitungen. (*Bull. Phys. Math. de St. Petersburg,* i. 129.) *Petersburg,* 1842

Rapport fait à l'Acad. des Sciences de St. Pétersbourg sur la dorure galvanique.

† Mem. sur l'application de l'électro-magnétisme au mouvement des machines.
8vo. 54 pp. 1 plate. *Potsdam,* 1835
"Das erste genügende Resultat von Jacobi's Arbeiten (in Galvano-plastic) wurde am 17 Oct. 1837 der Akad d. Wissensch. in Petersburg vorgelegt."
(*From Martin and Pogg.* i. 1177.)

Expériences electro-magnétiques formant suite au Mém. sur l'application de l'électro-magnétisme au mouvement des machines de 1835. (*Bull. Scient. Acad. Pétersb.* ii. 1837.) *Petersburg,* 1837

† On the application of Electro-Magnetism to the movement of machines. 29 pp. 1 plate. *London,* 1837

Über Becquerel's einfache Sauerstoffekette. 8vo. (*Pogg. Ann.* xl. 1837.)
Leipzig, 1837

† Die Galvanoplastik od. d. Verfahren coharentes Kupfer in Platten od nach sonst gegeb Formen unmittelbar aus Kupfer auflösungen auf galvan Wege zu produciren. Nach d. auf Befehl d. Gouvernements in Russ Sprache bekannt gemachten Originale. 8vo. *Petersburg and Berlin,* 1846

† La Galvanoplastica ossia processo per ottener immediatamente in via Galvanica, lastre o altre date forme solide di Rame dalle soluzioni di questo metallo . . Versione dal Tedesco del G. Guissani. 8vo. 69 pp. 1 plate.
Milano, 1841

Jacobi, Moritz Hermann Von—*continued.*

Sur la pile à effet constant du Prince Bagration. (*Bull. Phys. Math. Acad. Petersb.* ii. 1844.) *Petersburg,* 1844

Account of Prince Bagration's Battery. (*L'Institut.* . *No.* 530 22 Fev. 1844.)

Description d'un Télégraphe éléctrique naval, établi sur la frégate à vapeur le "Polkan." 4to. *St. Petersburg,* 1856

About fifteen articles on Galvanism, Electro-Magnetism, and Galvanoplastics in the Bull. Scient. Acad. Petersburg, 1837—1842 ; and about seventeen articles on the same subjects in the Bull. Phys. Math. Acad. Petersb. 1843 —1857.

† Galvano-plastic. Exposition universelle de 1867 à Paris. Rapports du Jury International, tom. viii. p. 123. 8vo. 33 pp. *Paris,* 1868

(*Vide* also Hellwag and Jacobi.)

Jacopo. (*Vide* Riccati-Jacopo.)

Jacquet de Malzet, Louis Sébastien. *Born* 1715, *at Switzerland or Nancy ; died August* 17, 1800.

† Précis de l'Electricité, ou extrait expérimental et théorique des Phénomènes électriques. 8vo. 235 pp. 7 plates. *Vienna,* 1775

Lettre d'un Abbé de Vienne à un de ses amis de Presbourg, sur l'Electrophore perpétuel. (Published anonymously.) (*From Volta,* "*Collezione,*" tom. i. pt. i. p. 159.) (*Vide Hildebrand for the German translation.*) *Vienna,* 1775

Schreiben eines Geistlichen zu Wien an einen seiner Freunde zu Presburg von dem immerwährenden Elektrophor. aus dem Franzocn. mit Anmerkn. übersetzt. (Published anonymously.) (*Vide* Hildebrand.) *Wien,* 1776

† **Jacquin.** In Letter to M. Fisher on Davy's experiments with battery on the Alkalies. 8vo. 2 pp. (*Phil. Mag.* xxxvi. 73.) *London,* 1810

† **Jadelot,** J. F. N. Expériences sur le Galvanisme . . de Humboldt. Traduction avec des additions. 8vo. 530 pp. 8 plates. *Paris,* 1799

† **Jager.** Versuch einer Vergleichung des Turmalins mit den trockenen electrischen Säulen. Geschrieben im Nov u. Dec. 1816. 8vo. 48 pp. *Leipzig,* 1817

† **Jahn,** A. Vorläufige Beurtheilung d. Contact u. chemischen Hypothese d. Galvanismus ein Programm . . 8vo. 50 pp. *Dresden,* 1842

Jahrbücher der K. K. Central Anstalt f. Meteorologie und Erdmagnetismus. (*Vide* Kreil, Jelinek, Fritsch.)

Jalaguier. Des effets de l'Electricité sur les corps vivants. (Thèse.) *Montpellier,* 1836

Jallabert, Louis. *Born July* 26, 1712, *at Geneva ; died March* 11, 1768, *at Beguin, near Geneva.*

Sur une trombe vue sur le Lac de Genève. 4to. (*Mémoires de Paris,* 1742.) *Paris,* 1741-42

La guérison d'un paralytique par le moyen de l'Electricité. 4to. (*Mémoires de Paris,* 1748.) *Paris,* 1748

† Expériences sur l'Electricité, avec quelques conjectures sur la cause de ses effets. 8vo. 304 pp. 3 plates. *Genève,* 1748

† Expériences sur l'Electricité, avec conjectures. 2e édition. Sm. 8vo. 379 pp. 3 plates. *Paris,* 1749

(*This edition contains (at* p. 363) "*Lettre de Sauvages*" *of June* 25, 1749.)

† Experimenta electrica usibus medicis applicata. Oder Versuche über die Électricität, aus denen der herrliche Nutzen derselben in der Arzneywissen-schaft . . zu ersehen . . Denen zu Ende beygefügt . . Sauvages . . Schreibner an . . Bruhier . . Aus dem Franz. übersetzt. 8vo. 312 pp. 3 plates. *Basle,* 1750

(*Vide* also Anon. Medical Electricity, 1763.)

Jallabert u. **Ammersin.** Versuche über die Electricität, mit einem Zusatze von der eigentlichen Electricität des Holzes, von Ammersin.   8vo.   *Basel*, 1771

† **James,** Sir H.   Instructions for taking Meteorological Observations, with Tables for their correction, and Notes on Meteorological phenomena.   Roy. 8vo. 52 and 34 pp.   22 plates.   *London*, 1861

Two Maps of the World, with lines of magnetic declination ; and four Charts of the Stars, on the geometrical projection of two-thirds of the sphere.

Jameson, Robert.   *Born July* 11, 1774, *at Leith ; died April* 17, 1854, *at Edinburgh.*

(*Vide* Edinburgh Phil. Journal, and Edinburgh New Phil. Journal.)

† **Jameson,** W. W.   Storm at Rotterdam in August, 1823.   (Reference to anonymous account in *Phil. Mag.* p. 232 *ante.*)   8vo. 1 p.   (*Phil. Mag.* lxii. 315.)
                                                                    London, 1823

Jamin, J. C.   Sur les mouvements imprimés par un aimant aux liquides traversés par les courants.   8vo.   (*Ann. Chim. Phys.* xliii. 1855.   *Paris*, 1855

Janin de Combe Blanche, Jean.   *Born January* 11, 1730, *at Carcassonne.*

† Mémoires et Observations anatomiques, physiologiques et physiques, sur l'œil, &c.   *Lyon*, 1772

Sur les causes de la mort subite et violente . . les victimes . . rappelées à la vie.   8vo.   *Paris*, 1772

(*Note.—A man restored after having been hung.*)

On the use of Electricity in black Cataract (schwartzen Staare).   17 cases.
                                                                    Paris, 1773

Jaques de Vitry.   (*Vide* Vitry, Jas.)

Jedlick et **Csapo.**   Pile monstre ; perfectionnement dans les piles de Bunsen. (*From Du Moncel, Expo é,* 2nd edition, pp. 82 and 466.)   1855

Jeffreys, J.   An inquiry into the laws governing the two great powers, Attraction and Repulsion.   8vo.   (*Extracted from the Journal of the Asiatic Society for September,* 1833.)   *London*, 1833

† **Jelgersma,** W. B.   Specimen physicum continens experimenta quædam circa phænomena, Lagenæ Leidensis; præcipue circa effectus caloris in eadem. 4to. 44 pp.   *Franequeræ*, 1775

Specimen experimentorum Lagenam Leidensem spectantium.   8vo. 205 pp.
                                                             Franequeræ, 1776

† Specimen physicum inaugurale de caloris influxu in electricitatem . . 4to. 54 pp.   *Franequeræ*, 1776

Jelinek.   (*Vide* Kreil.)   1841

Jelinek, C.   Beitrage zur Construction selbstregistrirender meteorologischer Apparate.   Roy. 8vo. 42 pp. 9 plates.   (*Sitzungsbericht. Wien Acad.* v. 1850, ii. *Abtheil.*)   *Wien*, 1850

(*N.B.—The instrument of Wheatstone, reported (at p.* 24) *to be at the Kew Observatory, was never there.—F. R.*)

† Uber die jährliche Vertheilung der Gewitter tage, nach den Beobachtungen an den meteorischen Stationen in Oesterreich und Ungarn.   8vo. 9 pp. (*Sitzb. d. k. Akad. d. Wiss.* lxi.)   *Wien*, 1870

† Uber die Leistungen e an der KK. centralanstalt fur Meteorologie u. E. befindlichen registrirenden Thermometers v. Hipp.   8vo. 13 pp.   *Wien*, 1870

Uber die Sturme des November und December, 1866.   8vo. 32 pp. 4 plates.
                                                                    Wien, 1867

† **Jencken,** J. F.   Treatises on light, colour, electricity, and magnetism.   Translated by H. D. Jencken.   8vo. 232 pp.   *London*, 1869

† **Jenkin,** Fleeming.   Précis of a Lecture on the construction of Telegraphic Lines, delivered by Fleeming Jenkin at the Royal Engineer Establishment, Chatham, Jan. 2nd, 1863, written from the Lecturer's notes by C. S. Beauchamp, and corrected by the Lecturer.   8vo. 18 pp. 16 figures.   *London*, 1863

Jenkin (and **Masson** ?)   On the influence, by induction, of an Electric current on itself. 8vo.   *London*, 1834

   (*From Faraday, who says he published "a brief account of these results," &c., in the Phil. Mag. for* 1834, v. 349, 444.)

Jenkins. On the Magneto-electric shock. 8vo. (*Phil. Mag.* 1834, vol. v. p. 349.)
                                                             London, 1834
   (*This is Faraday's account of Jenkins' experiments and apparatus.*)

† **Jest, C.** Notizia di un nuovo apparato per esplorare l'azione dell' elettromagnetismo sui corpi. 8vo. 2 pp.   *Torino*, 1847

† **Jest, E. F.** Cenni sulla nuova macchina idro-elettrica d'Armstrong e sulla nuova pila di Bunsen. 8vo. 11 pp. 1 plate.   *Torino*, 1844

†   Macchina idro-elettrica di Armstrong e nuova Pila di Bunser. Folio. 5 pp. 1 plate. (*L' Economista Ann.* iii. 1845, vol. i. p. 23.)   *Milano*, 1845

Jesuits at Pekin. Magnetic Observations. *Vide* Entrecasteaux, Voyage à la recherche de "La Pérouse." (*Mém. Acad. Pétersbourg*, viii. 276.)

   The Jesuits at Pekin made in the year 1755, an experiment with plates of glass, &c. (on charge, or induction, &c.), an account of which was sent to Petersburg, and published in the Memoirs of the Academy there, vol. viii. p. 276. (*Seventh vol. of the new Commentaries of the Academy of St. Petersburg.*)

   (*Vide* also Grosier.)

Jobert (de **Lamballe.**) Etudes sur le système nerveaux.   *Paris*, 1838
   On Medical Electricity. (*Bulletin général de Thérapeutique*, tom. xxiii.)

†   Des appareils électriques des poissons électriques. La. 8vo. 104 pp. 11 plates.
                                                       Paris, 1858

Johnson, E. J. On the influence which Magnetic Needles exercise over each other. 4to. (*Phil. Trans.* 1834.)   *London*, 1834

†   Report of Magnetic Experiments tried on board an iron steam vessel, by order of the Right Honourable the Lord Commissioners of the Admiralty. . . Accompanied by plans of the vessel, and tables showing the horizontal deflection of the Magnetic Needle at different positions on board, together with the dip and magnetic intensity observed at those positions, and compared with observations made on shore, with the same instruments. Addressed to Charles Wood, Esq., M.P. &c., and communicated by Captain Beaufort, R.N., F.R.S., Hydrographer to the Admiralty, by command of the Right Honourable the Lords Commissioners of the Admiralty. 4to. 22 pp. 1 plate. (*Phil. Trans.* 1836.)   *London*, 1836

†   Practical illustrations of the necessity for ascertaining the Deviations of the Compass, with explanatory diagrams, and some account of the Compass system now adopted in the Royal Navy ; Notes on Magnetism, &c., published under the sanction of the . . Lords Commissioners of the Admiralty. 4to. 85 pp. 2 plates.   *London*, 1847

†   Practical illustrations of the necessity for ascertaining (the) Deviation of the Compass. . . 2nd edition. Published under sanction of the . . Admiralty. 8vo. 174 pp. Chart and 2 plates.   *London*, 1852

Johnston. Physical Atlas, illustrating the geographical distribution of Natural Phenomena. 2 vols.   1848

† **Joly, A.** Nouveau système de Signal fixe avec répétiteur mu par électricité à l'usage des Compagnies de Chemins de fer. 4to. 3 pp. 1 plate.
                                                 Grenoble, 1858

† **Jones, A.** Historical sketch of the Electric Telegraph. 8vo. 192 pp.
                                                 New York, 1852

Jones, Henry Bence. *Born December* 31, 1813, *at Thorington Hall, Suffolk; died* 1873.

Animal Electricity : being an abstract of the discoveries of Dr. Dubois Raymond. Edited by H. B. Jones. 8vo.   London, 1853 ?
(*Latimer Clark's Catalogue says* 1852.)

Jones, G. (*Vide* Perry.)

† **Jones, G.** Observations on the Zodiacal light, from April 2, 1852, to April 22, 1855, made chiefly on board the United States steam frigate Mississippi, during her late cruise in Eastern Seas, and her voyage homeward ; with conclusions from the data thus obtained.   4to. 705 pp.   (*Forms the 3rd vol. of the United States Japan Expedition.*)   Washington, 1856

Jones, Thomas. On his Reflecting Compass (in principle identical with Schmalkalder's), and his Metallic Spectacles. (*Gilb. Ann.* liv. pp. 197, 308.)

† Description of a new Reflecting Compass (*i.e.* Kater's.) 8vo. 3 pp. 1 plate.
(*Phil. Mag.* xlvi. 7.)   London, 1815

† **Jones, W.** Discourses on the Natural History of the Elements. 4to. 10 plates.
London, 1781

† Essay on Electricity . . by the late G. Adams. The 5th edition, with corrections and additions by William Jones. 8vo. 594 pp. 6 plates.
London, 1799

† Six Letters on Electricity. 8vo. 68 pp.   London, 1800
Jones. (*Vide* Rittenhouse and Jones.)

Jordan. C. J. Engraving by Galvanism. Letter. 8vo. (*Mechanic's Magazine for June,* 1839.)   London, 1839

(*Note.—Dicks calls this the "earlie   published account of the manipulation requisite for obtaining casts by galvanic action ;" totally ignoring, or perhaps being ignorant of, Brugnatelli's complete processes in the year* 1802.)

Jordan T. B. Remarks on Electro-Metallurgy. (*8th Annual Report of the Royal Cornwall Polytechnic Institution.*)

Jorg. Electricität als allein Ursache aller Naturerscheinungen. 12 pp.
Neuburg, 1864

Jori, B. Saggio di fenomeni elettrochimici nella repristinazione dell' argento da diverse sue soluzioni saline col mezzo di ferro solo, ed accoppiato ad altri metalli. 8vo. 27 pp.   Milano, 1844

Joule, James Prescott. (*Born December* 24, 1818, *near Manchester.*)

† Description of an Electro-Magnetic Engine. Letter to the Editor dated August 30, 1839. 8vo. 3 pp. (*Ann. of Elect.* iv. 203.)   London, 1839

† On Electro-magnetic forces. 8vo. 8 pp. Plate. (*Ann. of Elect.* iv. 474.)
London, 1840

† On Electro-magnetic forces. 8vo. 12 pp. 1 plate. (*Ann. of Elect.* v. 187.)
London, 1840

† Description of a new Electro-Magnet. Letter to the Editor. 8vo. 2 pp. 1 plate. (*Ann. of Elect.* v. 431.)   London, 1841

† On Voltaic apparatus. 8vo. 6 pp. (*Proceedings Lond. Elect. Society.*)
London, 1842

† On the Heat evolved during the Electrolysis of Water. 8vo. 20 pp. and table. (*Literary and Phil. Society, Manchester,* vol. vii. pt. ii. 2nd series.) (*Arago's copy.*)   Manchester, 1843

† On the calorific effects of Magneto-Electricity, and on the mechanical value of Heat. (British Association, at Cork, August 21, 1843.) 8vo. 30 pp. (*Phil. Mag.* Series iii. xxiii. for 1843.)   London, 1843

(*Vide* also Scoresby, W., and Joule, 1846.)

Joule, James Prescott—*continued.*

†   Account of Experiments demonstrating a limit to the Magnetizability of Iron. 8vo. 20 pp. (*Phil. Mag. for October and December*, 1851.)   *London*, 1851

†   Account of Experiments with a powerful Electro-Magnet. 8vo. 5 pp. (*Phil. Mag. for January*, 1852.)   *London*, 1852

†   On the Fusion of Metals by Voltaic Electricity. Read March 4, 1856. 8vo. 4 pp. (*Mem. of the Literary and Phil. Society of Manchester, Session* 1856-7. Vol. xiv.)   *Manchester*, 1857

Jounin. (*Vide* Gallette and Jounin.)

† **Jourdain, A.** Considérations théoriques sur les condensateurs électriques. (Jourdain *Répétiteur*.) 8vo. 16 pp.   *Poitiers*, 1858

Journal der Chemie von Scherer. 10 vols. 8vo.   *Berlin*, 1798-1802

  Neues allgemeines Journal d. Chemie von Gehlen. 6 vols. 8vo. *Fortset. desselben sind.*   1803-5

  1. Journal für die Chemie u. Phys. von Gehlen. 9 vols. 8vo.   1806-10

  2. Neues Jour. für Chemie u. Phys. von Schweigger. 60 vols. 8vo.   *Nürnb.* 1811-30

  3 Neues Jahrbuch der Chem. u. Phys. von Schweigger-Seidel. 8vo.   *Halle*, 1831

Journal de Chimie. Par J. B. Van Mons. 8vo.   *Bruxelles*, 1802

Journal de Mathématiques. Vols. i. to xviii. 4to.   *Paris*, 1836-53

Journal für die reine und angewandte Mathematik. Herausgegeben von A. L. Crelle. Vols. xix. to lii. 4to.   *Berlin*, 1839-56

† **Journal of Natural Philosophy, Chemistry, and the Arts.** By William Nicholson. 5 vols. 4to.   *London* 1787-1802

  36 vols. 8vo.   *London*, 1802-13

Journal of Pure and Applied Mathematics 8vo.   *London*, 1855-62

Journal of the Franklin Institute of the State of Pennsylvania. Edited by T. P. Jones. 8vo.   *Philadelphia*

Journal de Physique. Tableau du travail annuel de toutes les académies de l'Europe ; ou Observations sur la physique, &c. . . par Rozier, J. Vol. i. 4to.   *Paris*, 1772

  Observations sur la physique, &c. . . . par Rozier. Vols. ii.—xii. 4to. *Paris*

  Observations sur la physique, &c. . . . par Rozier et Monge. Vols. xiii.— xxvi. 4to.   *Paris*

  Observations sur la Physique, &c. . . . par Rozier, Monge, et De la Métherie. Vols. xxvii.—xliii. 4to.   *Paris*

  Journal de Physique, &c. . . . par De la Métherie. Vols. xliv.—xlvii. (or Vol. i.—iv.) 4to.   *Paris*

  Journal de Physique, &c. . . . par De la Métherie (excepting vols. lxxvi. and lxxvii. in which he was aided by Ducrotay de Blainville). Vols. xlviii.— lxxxiii. 4to.   *Paris*

  Journal de Physique, &c. . . . par De la Métherie et Ducrotay. Vol. lxxxiv. 4to.   *Paris*

  Journal de Physique, &c. . . . par Ducrotay de Blainville. Vols. lxxxv.— xcvi. 4to. (Being together 96 vols.)   *Paris*

  (*Note.—Many of these are in the Library.*)

Journal des Savans. 77 vols. 4to.   *Paris*, 1665-1748

  Continuation. 34 vols. 4to.   *Paris*, 1749-92

  Nouvelle Continuation. 28 vols. 4to.   *Paris*, 1816-45

Jugel, J. G. Physica mystica et phys. sacra sacratiss. Offenbarung der magnet.
Anziehungskraft. 8vo.   *Berlin,* 1782

Julia-Fontenelle, J. S. E. Sur les combustions humaines spontanées. (Revue
médicale.) 8vo.   1828

Julien, A. S. Sur quelques phénomènes volcaniques mentionnés par les auteurs
chinois. (*Compt. Rend.* 1840.) 4to.   *Paris,* 1840

Julien, J. Markvärdigt norrsken och magnetnälens rorelser d. 4 Apr. 1791,
observerad in Uleåborg. (*Vetensk. Acad. Handl.* 1793.)   1793

Julio (or Giulio), Gioberti, Vassalli-Eandi et Rossi. Bibliothèque Italienne, ou
Tableau des progrès des sciences et des arts en Italie. 5 vols. 4to.
                                                                  Turin, 1803-4

Jundzill. Du Télégraphe des locomotives de G. Bonelli, destiné à prévenir les
collisions sur les chemins de fer. 8vo.   1856

Jungnitz, L. A. Aphorismen über d. Lehre von d. Elektricität. 8vo.
                                                                   Breslau, 1794

Aphorismen von der Lehre über die Elektricität. 8vo.   *Breslau,* 1796

Über d. Erfolg. d. Blitzfeuer auf d. Schneekoppe. 8vo.   *Breslau,* 1805

Darstell. d. Erfolgs d. auf d. Schneekoppe v. H. v. Lindener 1805 angestellten
u. an mehreren Orten beobachteten Blitzfeuer. 8vo.   *Breslau,* 1806

Junoblowiskiana Society. Proposal for a Prize, to be given in 1795, on Matters
related to the Experiments of Galvani, Valli, Volta, and others. Prize
announced in 1793. 8vo. (*Comment. de rebus in Scient. Nat. et in Med.*
gestis, xxxvi. 54.)   *Leipzig,* 1793?

Jurgensen. Hen bevaegelsen af electr. Stromme. 8vo. 3 pp. *Kjobenhvn.* 1856

K

† **Kamtz** (or **Kaemtz**), Ludwig Friedrich. *Born January* 11, 1801, *at Treptow a. d. Riga, Pommern.*

Dissertatio mathematico-physica de legibus repulsionum electricarum mathematicis. 8vo. 29 pp. 1 plate. *Halle,* 1823

Über d. Gesetz nach welchem d. elektro-magnet. Kraft d. Schliessdrahts d. Voltaischen Säule durch Schweigger's Multiplicator verstärkt wird. 8vo. (*Schweigg. Journ.* xxxviii. 1823.) *Nürnberg,* 1823

† On the law according to which the Electro-magnetic power of the connecting wire of the Voltaic pile is augmented by Schweigger's Multiplier. (Schweigger and Meineke's Neues Journal, viii. 100.) 8vo. 9 pp. (*Phil. Mag.* lxii, 441.) *London,* 1823

Über Lichtmeteore. 8vo. (*Schweigg. Journ.* xlv. 1825.) *Nürnberg,* 1825

Über Nordlichter. 8vo. (*Schweigg. Journ.* lii. and lxi. 1828 and 1831.) *Nürnberg,* 1828 *and* 1831

Über d. Elektricität beim Contact animal. u. vegetabil. Substanzen unter sich und mit Salzen. 8vo. (*Schweigg. Journ.* lvi. 1829.) *Nürnberg,* 1829

Lehrbuch der Méteorologie. 3 vols. 8vo. *Halle,* 1831-2-6

Beschreibung eines auf d. Faulhorn beob. Gewitters. 8vo. (*Schweigg. Journ.* lxix. 1833.) *Nürnberg,* 1833

† Vorlesungen über Meteorologie. 8vo. 591 pp. 6 plates. *Halle,* 1840

Cours complet de météorologie, traduit et annoté par Martins. 12mo. *Paris,* 1843

Complete Course of Meteorolog 12mo. (*From English Cat.* 1864. p. 419.) *London,* 1844

† Resultate Magnetischer Beobachtungen in Finnland. Gelesen 6 Oct. 1848. 4to. 86 pp. 1 plate. (*Mém. des Sav. Étrang.* tom. vi. *and in Bull. Phys. Math. Acad. S. Petersb.* vii. 1849.) *Paris,* 1849

† **Kahm** (or **Kalm**), P. Einige im nördlichen America beobachtete Nordscheine. 8vo. (*Journ.* 1730, 1731, 1737, 1739, 1741, 1746, 1748, 1749, 1750.) (*Schwedische Akad.* Abhandl. an. 1752, p. 153.) *Stokholm,* 1752

Kaiser, C. G. Über elektro-chem. Vergoldung, Versilberung u. Verplatinirung. (*Kunst u. Gewerbeblatt d. polytech. Vereins f. d. Königr.-Bayern,* 1842.) 1842

† **Kalm,** P. (*Vide* Kahm.)

† **Kane,** E. K. Magnetical observations in the Arctic Seas, made in 1853-4-5. Reduced and discussed by Schott. 4to. or Fol. 66 pp. 2 plates. (*Washngt. Smiths. Inst.* 1858. *Contributions.*) *Washington,* 1858

Karlinski. (*Vide* Kreil and others.)

Karlinski. (*Vide* Bohm and Karlinski.)

Karsten, D. L. G. Die magnetische Sprechmaschine. (*Berlin Monatschrift,* 1789.) *Berlin,* 1789

Karsten, Gustav. *Born November* 24, 1820, *at Berlin.*

Uber elektrische Abbildungen. 8vo. (*Pogg. Ann.* lvii. 1842 and lviii. lix. 1843.) *Leipzig,* 1842-43

† Imponderabilium præsertim electricitatis. Theoria dinamica cum appendice de imaginibus quæ luce, calore, electricitate procreantur. 4to. 47 pp. 2 plates. *Berolini,* 1843

Die Fortschritte der Physik. Dargestellt von der physikalischen Gesellschaft zu Berlin. Redigirt von G. Karsten. 8vo. *Berlin,* 1847-53

Karsten, Karl Johann Bernhard. *Born November* 26, 1782, *at Bützow, Mecklenburg ; died August* 22, 1853, *at Berlin.*

Uber d. elektrische Polarisirung des Flüssigen als das Wesen aller galvan. Thätigkeit d. Ketten u. s. w. 4to. (*Abhandl. Berlin Acad.* 1833.)
                      Berlin,

Über d. Wirkungsart d. einfach. galvan. Kette. (*Monatsb. d. Acad. Berlin,* 1836.)   *Berlin,* 1836

Über Contact-Elek. Schreiben an A. Humboldt. 8vo. 150 pp. 1 plate.
                      Berlin, 1836

Über Feuer-Meteor und über ein merkwürdiger Meteormassenfall. bei Thorn. 4to. (*Abhandl. Berlin Akad.* 1853.)   *Berlin,* 1853

Karsten, Wenceslaus Johann Gustav. *Born December* 15, 1732, *at New Brandenburg, Mecklenburg Strelitz ; died April* 17, 1787, *at Halle.*

Anleitung z. gemmeinnützl. Kenntn. d. Natur. . . . 8vo.   *Halle,* 1783

Kast, J. J. De magnete. 4to.   *Argentorati* (*Strasburg*), 1683

† **Kastner,** Adalb. Telegraphen-Kalender f. d. Jahre 1869, herausg. von Adalb. Kästner 4 Jahrg. 8vo. 90 pp.   *Wien,* 1869

Kastner, Karl Wilhelm Gottlob. *Born October* 31, 1783, *at Greifenberg, Pomm.; died July* 13, 1857, *at Erlangen.*

Observationes de Electromagnetismo.   *Erlangen,* 1821

† Observationes de Electromagnetismo. 4to. 10 pp.   *Erlangen,* 1822

Archives für die gesammte Naturlehre. 18 vols.   *Nürnburg,* 1824-29

Archives für Chemie und Meteorologie.   *Nürnburg,* 1830

(*Poggendorff says* 1830 *to* 1835, 9 *vols. Schmidt, Cat.* cccxxx. 1865, *says :* "*Archives f. d. gesammte Naturlehre u. Meteorologie* 27 *Bde.* (*alles was erscheinen est*) 1824-35." *And* "*Archives f. Chemie u. Meteorologie,*" 1-9 *Bd.*)

Viele Aufsätze u. Notizen in Archiv für d. gesammte Naturlehre, Archiv für Chemie u. Meteorologie, Trommsdorff's Journ., Voight's Magazin, Schweigger's Journ. u.a.m.

(*Some of these may relate to Electricity, &c.*)

* † **Kater,** H. Description of an improved Hygrometer. By Lieutenant Henry Kater, of His Majesty's 12th Regiment. 8vo. 4 pp. 1 plate. (*Phil. Mag.* xxvii. 322.)   *London,* 1807

† On the best kind of Steel and Form for a Compass Needle. Phil. Trans. for 1821, part i. 8vo. 15 pp. (*Phil. Mag.* lix. 359.)   *London,* 1822

† On the Luminous Zone observed in the Heavens on the 29th of September last. 8vo. 2 pp. (*Phil. Mag. or Annals.* iv. 337.)   *London,* 1828

Kayser, C. G. Index locupletissimus librorum, qui inde ab anno 1750 usque ad annum 1832 in Germania et in terris confinibus prodierunt. Vollständ. Bücher-Lexicon, &c. enthalt. alle v. 1750 bis z. Ende des Jahres 1832 in Deutschland u. in d. angrenzenden Landern gedruckten Büchern xx. 6 vols. 4to.   *Leipzig,* 1834-5-6

Sachregister 2 Hälften. 4to.   *Leipzig,* 1837-8

Novus index locupletissimus librorum qui inde ab an. 1833 usque ad an. 1840 in Germania et in terris confinibus prodierunt. Neues Bücher-Lexikon, enth. alle von 1833 bis Ende 1840 gedruckten Bücher zc., nebst Nachträgen u. Berichtigungen früherer Erscheinungen. 2 vols. 4to. *Leipzig,* 1841-2

Bücher Lexikon, 1750-2. 12 vols. 4to.   *Leipzig,* 1834-54

Index locupletiss. librorum qui inde ab an. 1841 usque ad an. 1846 in Germania et in terris confinibus prodierunt. Neues Bücher-Lexikon, 9th u. 10th Thl. oder neues Bücher-Lexikon 3 u. 4 Thl. 1841 bis Ende 1846. 4to.
                      Leipzig, 1848

† **Kayser, E.** Beobachtungen d. magnetisch Declination in Danzig und Bemerkungen daz. 8vo. 27 pp. *Danzig,* 1864

Kazwini (a Persian). On Aerolites, in a book entitled Adschaibel Machlukat
(Miracle of the Creation). Notice in the Bibl. Brit. No. 304. (*From Nuova
Scelta d Opuscoli* (in 4to.) ii. 333.)
(*He died in* 1275.)

† **Keferstein, W.** Beitrag zur Geschichte der Physik der elektrischen Fische. Sm.
8vo. 18 pp. (*Nachrichten von der G. A. Universität und der Konigl. Gesellschaft
der Wissen. zu Göttingen Januar* 31, No 3, 1859.) *Göttingen,* 1859

Keferstein, W. M. mit **D. Kupffer.** Über d. elektr. Organe von Gymnotus
electricus und Mormyrus oxyrhynchus. (*Henle u. Pfeuffer's Zeitschr. f. rat.
Med. Newe Folge,* iii. 1858.) 1858

Keil. Modification of Saxton's Apparatus. (*From Meyer.*)

† Der mineralische Magnetismus in physikal. physiolog. u. theraput.
Beziehung, &c. 8vo. 56 pp. 1 plate. *Erfurt,* 1846

Keir, J. On the Crystallisation observed in Glass. 4to. (*Phil. Trans.* 1776.)
London, 1776

† **Keller, F.** Sopra alcune proprietà della propagazione della corrente elettrica nei
fili telegrafici dedotte dalla teoria di Ohm. Nota. 4to. 33 pp. (*Annali di
Matematica pura ed applicata* ii. No. 5, 1859.) *Roma,* 1859

Keller, F. A. E. Des ouragans, tornados, typhons et tempêtes. Typhons de
1848 ; typhon de 1849. 8vo. 95 pp. 2 plates. *Paris,* 1862

Kelsch, M. De fasciis quibusdam magnæ claritatis in cælo visis. (*Commerc.
Litt. Norimb.* An. 1734.) *Norimb.* 1734

Observ. de aurora boreali. (*Commerc. Litt. Norimb.* an. 1734.) *Norimb.* 1734

Kemp. On amalgamated Zinc plates in Galvanic arrangements. 8vo. *Jameson's
Phil. Jl., December,* 1826.) *Edinburgh,* 1826
(*Note.—This is not the exact title.*)

Kemp, M. Description of a new Galvanic pile and trough. 8vo. (*Jameson's
Journ.* vi. 1828.) 1828

Kendall, A. Magnetic Variations. (*From Gilbert, De Magneti,* 1600, p. 7.)

Kenngott, Adf. Ueber Meteoriten oder die meteorischen Stein- und Eisenmassen.
Ein öffentlicher Vortrag gehalten am 19 Februar 1873 in Zürich.
Leipzig, 1863

† Ein Dünnschliff einer Meteorsteineprobe von Knyahinya. 8vo. 8 pp. 1
plate. (*Sitzungsber K. Akad. d. Wiss.* ii. Abth. Mai-Heft, 1869.) *Wien,* 1869

Kent (Inglese). Zinco amalgamato quale elemento elettro positivo.
(*Note.—From Grimelli, Storia,* p. 102.)

† **Kerner, T.** Galvanism and Magnetism as Restoratives. 3rd edit. translated
from the German. 8vo. 39 pp. *Cannstatt,* 1858

† **Kerz, P.** De electro-magnetismi vi et usu. *Bonn,* 1846

Kesselmeyer, P. A. Ueber den Ursprung der Meteorsteine. Versuch eines
Quellenverzeichnisses zur Literatur über Meteoriten Von Otto Buchner.
(Abgedruckt aus den Abhandlungen der Senckenbergischen naturforschen-
den Gesellschaft.) 4to. 18 pp. 3 plates. *Frankfurt a. M.* 1860

Kessler, K. G. De motu materiæ electricæ ut causa efficiente motuum et sensuum in corpore animato. 8vo. (*From Kuhn, Hist.* ii. 50. Poggend. (i. 1250)
says 1748, *and Deutsch* 1749, *and calls him Kesler.*) *Uratisb.* 1747

Die Bewegung der elek. Materien als die wirkende Ursache der Bewegungen
und Empfindungen in lebendigen Körpern erklaret. 8vo. *Landshut,* 1749

Vires medicamentorum ab igne electrico pendere adserit. 8vo.
Landeshutæ 1750

† **Kestler, J. S.** Physiologia Kircheriana experimentalis. . . . Fol. *Amsted.* 1680

† **Kew Committee.** Report of the Kew Committee of the British Association for the Advancement of Science, for 1859-1860. 8vo. 15 pp.   *London.*

† **Kew and Lisbon Observatories.** Photo-lithographic impressions of traces produced simultaneously by the Self-recording Magnetographs at Kew and Lisbon. . . Fol. 18 sheets.   *London and Portobello, n. d.*

Kiechl, F. Kravogl's zweiter Elektromotor. 4to. 13 pp. 1 plate. *Insbruck,* 1871
Versuche zur Bestimmung des calorischen Aequivalentes der Elektricität. 8vo. 19 pp. (*Sitzungsb. d. K. Akad. d. Wiss.*)   *Wien.*

Kielman, K. A. Systematische Darstellung aller Erfahrungen über die einzelnen Metalle. 2 vols. 4to.   *Arau,* 1807

Kielmaver, K. F. Versuche über d. sogenannte animal. Elektricität. 8vo. (*Gren's Journ.* viii. 1794.)   *Leipzig,* 1794
Dissertatio sistens observationes physicas et chimicas de electricitate et galvanismo. 8vo.   *Tubingen,* 1802
Dissertatio sistens experimenta quædam influxum electricitatis in sanguinem et respirationem spectantia. 8vo.   *Tubingen,* 1810
Examen experimentorum quorumdam effectus magnetis chimicos spectantium. 8vo.   *Tubingen,* 1813

Kienmayer, Franz von. (*Died May 30, 1802, at Vienna.*)
Sur une nouvelle manière de préparer l'amalgame électrique, et sur les effets de cet amalgame. 4to. (*Jour. Phys.* xxxiii. 1788. p. 97.)   *Paris,* 1788

† Lettera . . sopra una nuova maniera di preparar l'amalgama elettrico, e i suoi effetti. 4to. 8 pp. (*Opusc. Scelti,* xii. 3.)   *Milano,* 1789

Kierski, M. Dissertatio de electricitatis in praxi medica usu. 8vo. 30 pp.
Berolini, 1854

Kilian. Versuche über Restitution der Nerven-Erregbarkeit nach dem Tode (*From Meyer.*)   *Giessen,* 1857

† **King, Ed.** Remarks concerning Stones said to have fallen from the Clouds both in these days and in ancient times. 4to. 34 pp. 1 pl.   *London,* 1796

Kingsley. (*Vide* Silliman and Kingsley.)

Kinkelin. Über die Bewegung eines Magnetischen Pendels. (*Grunert's Archiv.* xxviii. 456.)

Kinnersley, Eb. New Experiments in Electricity. 4to. (*Phil. Trans.* 1763.)
London, 1763
On some Electrical experiments made with Charcoal. 4to. (*Phil. Trans.* 1773.)
London, 1773

Kinzel. Über Diamagnetismus. 4to.   *Ratibor.* 1855

Kiobenhaven Selskab. Skrifter som udi det Kiobenhaven Selskab ere Fremlagde
Kiobenhaven, 1745

Kirby. Analysis of Electricity and Fire. 8vo.   1777
(*Latimer Clark's Catalogue says "An Analysis of the electrical fire. . . together with an account of an uncommon effect of lightning. . . 1777.*)

Kircher, Athanasius. (*Born May 2, 1601, at Geysa near Fulda ; died October 30, 1680, at Rome.*)
Ars magnesiæ, hoc est disquisitio . . . de natura, viribus et prodigiosis effectibus magnetis. . . . 4to.   *Herbipoli,* 1631
(*Note.—He says in his Elenchus at end of his Magnes. folio ed.* 1654, *" Magnesia in Herbipoli,* 1630.)

† Magnes; sive de arte magnetica. Opus tripartitum. 4to. 916 pp. 2 plates.
Roma, 1641

Kircher, Athanasius—*continued.*

† Magnes ; sive de arte magnetica. Ed. secunda post Romanam multo correctior.
4to. 797 pp. Many plates. *Coeln*, 1643
(*He says in his Elenchus at end of his Magnes.* (*folio ed.* 1654) 1641.)

Prælusiones magneticæ. (*From Pogg.* i. 1259.) *Romæ*, 1645

† Magnes ; sive de arte magnetica. . . 3rd ed. Fol. 618 pp. (Lib. iii. 451: "*De magnetismo electri, seu electricis attractionibus, earumque causis.*) *Roma*, 1654

† Magneticum naturæ regnum, sive Disceptatio physiologica de triplice in natura rerum magnete. . . 4to. 136 pp. *Roma*, 1667

† Magneticum naturæ regnum. 12mo. 201 pp. *Roma*, 1667

Mundus subterraneus. In xii. libri digestus. 2 vols. Fol. Maps and plates.
(*From Cat. Bibl. Tours " Ed. la plus complete.")* *Amsted.* 1678

Physiologia Kircheriana experimentalis. Quam ex vastis operibus Kircherianis extraxit et red. J. S. Kestler. *Amstedolami*, 1680

Musæum Kircherianum. . . à S. Philippo Bonanini. Fol. *Romæ*, 1709

Musæum Kircherianum. . . notis illustratum. . . Fol. 2 vols. Atlas.
Romæ, 1763

Kirchhoff, G. R. Über Durchgang eines Stroms durch eine Ebene, besonders eine kreisformige. 8vo. (*Poggend.* Ann. lxiv. 1845 u. lxvii. 1846.
Leipzig, 1845-46

About six articles in Poggend. Ann. from vol. lxvii. 1846 to vol. lxxxi. 1850.

Über d. inducirt Magnetismus eines unbegrenzten Cylinders von weichem Eisen. 8vo. (*Crelle's. Jour.* xlviii. 1854.) 1854

Démonstration des lois d'Ohm, fondée sur les principes ordinaires de l'électricité statique. 8vo. (*Ann. de Chim.*) *Paris*, 1854

Uber d. Bewegung d. Elektr. in Drähten. 8vo. (*Crelle's Journ.* c. 1857.)
1857

Uber d. Bewegung d. Elektr. in Leitern. 8vo. (*Crelle's Journ.* cii. 1857.)
1857

Kirchhoff, Nicolaus Anton Johann. (*Born September* 23, 1725, *at Itzehoe ; died September* 10, 1800, *at Hamburg.*)

Zurüstung, die Wirkung der Gewitterwolken darzustellen. 8vo. (*Gött. Mag.* J. i. 1780, St. ii. pp. 322-26.) *Göttingen*, 1780

Beschreibung einer Zurüstung, welche die auziehende Kraft der Erde gegen die Gewitterwolke &c. . . beweiset . . nebst e. Beschraib. verschn. . .
Maschin. 8vo. 56 pp. 1 plate. *Berlin, Nicolai, and Hamburg*, 1781

Kirchmaier, G. C. De fulmine et tonitru. *Viteberg*, 1659

De luce, igne ac perennibus lucernis. 4to. *Viteberg*, 1676

Noctiluca constans et per vices fulgurans diutissima quæsita nunc reperta. 4to.
Viteberg, 1676

De luce igne ac perennibus lucernis. (*Miscell. Acad. Nat. Cur.* 1677.) 1677

De phosphoro et natura lucis, nec non, de igne, commentatio epistolica. 4to.
Viteberg, 1680

De lampade volante. (*Miscell. Acad. Nat. Cur.* 1685.) 1685

De ignium miraculis, locisque semper ardentibus. *Viteberg*, 1693
(*Note.—These are all from Pogg.* i. 1261.)

Kirchmaier, Sebn. De filis meteoricis, vulgo filamentis Mariæ. 4to.
Viteberg, 1666

Kirchmaier, Theod. De virgula divinatrice. 4to. (*From Pogg.* i. 1262.)
Viteberg, 1678

Kirchvogel, A. B. De natura electricitatis aereæ. 1767
(*Vide* also Bauer and others.)

Kirkwood, Daniel. Meteoric Astronomy : a Treatise on Shooting-Stars, Fire-Balls, and Aerolites.   *London and Philadelphia,* 1867 ?

Kirtz (?) Krunitz. Dissertatio sistens electricitatis in medecina usum et abusum. *(From Guitard.)*   1787

Kirwan, R. A description of a new Anemometer. 8vo. 5 pp. 1 plate. *(Phil. Mag.* xxxiv. 247.)   *London,* 1809

Klaproth, Julius. *Born October* 11, 1783, *at Berlin ; died August* 27, 1835, *at Paris.*

† Lettre à Humboldt sur l'invention de la Boussole. 8vo. 138 pp. 3 plates.
    Paris, 1834

Klaproth, Martin Heinrich. *Born December* 1, 1743, *at Wernigerode ; died January* 1, 1817, *at Berlin.*

Über meteorische Stein u. Metallmassen. 8vo. *(Gehlen, Journ. f. Chem.* i. 1803.)   1803

Beiträge zur chem. Kenntniss de Mineral-korper. 6 vols. 8vo.
    Berlin, 1795-1815

† Des masses pierreuses et métalliques tombées de l'atmosphère. Lu à l'Acad. 27 Janvier et 10 Mars 1803. 4to. 28 pp. *(Mém. de l'Acad. R. de Berlin, for* 1803.)   *Berlin,* 1803

† Dissertation on Stones and Iron Masses falling from the heavens, &c. 8vo. 2 pp. Read 18th January, 1803, in R. Acad. of Sciences at Berlin. *(Phil. Mag.* xv. 182.)   *London,* 1803

Untersuch d Meteorsteins von Lissa. *(Gehlen, Journ. f. Chem. u. Phys.* viii 1809.   *Berlin,* 1809

Klein, G. Dissert. de metallorum irritamento veram ad explorandam mortem. 4to.   *Moguntiæ (Mainz),* 1794

Gaubius published an observ. of Klein (on a cure by Electricity) in the Haarlem Soc. Mem. *(Mém. de la Soc. de Haarlem,* tom. i.)   *Haarlem*

Kleist, Ewald Georg Von *Born . . . . ; died December* 11, 1748, *at Cöslin.*

Invented the Leyden Phial October 11, 1745 *(Kleist'sche oder Leyden Flasche). (Vide Poggendorff, Wörterb.* i. 1271, *for some names having reference thereto ; and a history of the invention, &c.)*

Krünitz, Priestley's Hist. p. 53, *says, in reference to Kleist's invention, Priestley errs when he states that his account was communicated to an Academy at Berlin and copied from its Register by Gralath ; but Kleist had on the 4th November,* 1745, *given to Dr. Lieberkuhn in Berlin an account of his new experiments, and afterwards he, under date the 18th of this month, received the answer from him that the experiments were perfectly new and remarkable , he then addressed it, with Dr. Lieberkuhn's remarks thereon, on the 28th November, to Paul Swietlickt, Fellow of the Danziger-Naturforschenden Gesellschaft, who communicated the same to the Society, and it was from the Register of this Society that the experiments were taken by Gralath in the order and in the words which the Prelate himself had employed.*

Account of the Leyden Phial sent to Dr. Lieberkuhn. *(In Danzig Mem.* i. 407.)
    1745

Klindworth, J. A. Kurze Beschreibung eines der groszten Elektrophore von Herr Hofmechanicus. 8vo. *(Lichtenberg's Mag.* i. 35—45.) *Gotha,* 1781-5

Klingenstierna, Samuel. *Born Aug.* 18, 1698, *at Tollefors near Linkjöping ; died October* 26, 1765, *at Stockholm.*

Dissertatio de Electricitate. Part i. 4to.   *Upsal,* 1740
Dissertatio de Electricitate. Part ii. 4to.   *Upsal,* 1742
Tal om de naysta rön vid Electriciteten.   *Stockholm,* 1755

Klingenstierna et **Brande.** Dissertatio de magnetismo artificiale.
    Stockholm (Holmiæ), 1752

Klingert, K. H. Construirte . . . eine trockne Säule mit Uhrwerk. 8vo. (*Gilbert Annalen,* liii. 1816.) *Leipsig,* 1816

† **Klinkhardt, C. H.** Der Magnet. Eine Erklärung der merkwürdigsten Erscheinungen des mineralischen Magnetismus. 8vo. 66 pp. *Leipzig,* 1835

Klinkosch, Joseph Thaddäus. *Born October* 24, 1734, *at Prague; died April* 16, 1778, *at Prague.*

Schreiben, den thierischen Magnetismus und die sich selbst nieder stehende elektrische Kraft betreffend. 8vo. (*Abhandl. einer Privat-Gesell. in Böhmen,* ii. p. 171 : cf. B. iii. s. 223.) *Prag,* 1776

Beschreibung d. Volta'schen Elektrophors. *Ubersetzung,* 1777

Beschreibung eines Electricitäts-Trägers ohne Harz und Glas. (*Abhandl. einer Privat-Gesellsch. in Böhmen,* iii. 391.) *Prag,* 1777

On the Electrophorus. (*Mém. de l'Acad. de Prague,* iii. 218. *Böhm, Privat-Gesell.* Abhandl. iii. 213—226.) *Prague*

Kloden. (*Vide* Challis, &c.)

Klœrich, F. W. Versuche über d. Wirkungen d. Magnets in Vertreibung d. Zahnschmerzen. (*Götting. Anzeigen,* 1765.) *Göttingen,* 1765

Von dem medicin Gebrauch d. Magnets im funften Jahrhundert. (*Götting. Anzeigen,* 1766.) *Göttingen,* 1766

Klugel, Georg Simon. *Born August* 19, 1739, *at Hamburg; died August* 4, 1812, *at Halle.*

Beschreib. d. Wirkung. ein. heftig. Gewitters d. 12 Juli 1789 zu Halle, nebst Erklärung d. Entstehung d. Gewitters.

Dsslb. Angebd : 2 div. Schriften üb. Wetterleiter. *Halle,* 1789

(*Note.*—*These two entries are from Schmidt's Cat. No.* ccxxxx. *des Antiquarischen Bücherlagers,* 1865.)

† Beschreibung der Wirkungen eines heftigen Gewitters . . . 8vo. 64 pp. *Halle,* 1789

Knebel. (*Vide* Gersdorf von, and Knebel.)

Knight, Charles. English Cyclopædia Arts and Sciences, 8 vols.

Knight, Gowin. *Born* . . . ; *died Ju ie* 9, 1772, *at London.*

† An account of some Magnetical F cperiments, shewed before the Royal Society by Knight, G. on Thursday th 3 15th November, 1744. Read November 15, 1744. 4to. 6 pp. (*Phil. Tra is.* xliii. p. 161, *pro* 1744-5, printed 1746.) *London,* 1744-45

† A letter from Knight, G. to the President; concerning the Poles of Magnets being variously placed. Read April 4, 1745. 4to. 3 pp. (*Phil. Trans.* xliii. 361.) *London,* 1744-45

† A collection of the Magnetical Experiments communicated to the Royal Society by Knight, G. in the years 1746-47. Read Feb. 19, 1746-47, July 2, 1747, and Dec. 17, 1747. 4to. 17 pp. (*Phil. Trans.* xliv. 656.)

(*Note.*—*These three papers are entitled as follows :—*
I. *An account of some Magnetic Experiments exhibited before the Royal Society the* 19th *February,* 1746 (*and read then*). *This is a report of the President, &c., at* p. 656.
II. *An account of some new Experiments lately made with artificial Magnets, date June* 4, 1747. *Read July* 2, 1747, p. 662.
III. *Some further Experiments relating to the general phenomena of Magnetism. Read December* 17, 1747, p. 665. *Erroneously marked* iv.) *London,* 1746-47

An attempt to demonstrate that all the phænomena in nature may be explained by Attraction and Repulsion, wherein the attractions of Cohesion, Gravity and Magnetism are shown to be one and the same. 4to. 91 pp. *London,* 1748

Knight, Gowin—*continued.*

Remarks on Waddell's Letter concerning Lightning destroying the Polarity of a Compass. (*Phil. Trans.* an. 1749, p. 111.) *London,* 1749

† A description of a Mariner's Compass contrived by Knight, G. Read July 5, 1750. 4to. 8 pp. (*Phil. Trans.* xlvi. 505.) *London,* 1849-50

An attempt . . . Cohesion, Gravity, and Magnetism . . . . the same, and the phenom. of the latter explained. 8vo. (*From Watt.*) *London,* 1754

† A collection of some papers formerly published in the Phil. Trans. relating to the use of Dr. Knight's Magnetic Bars, with some Notes and Additions. 8vo. 23 pp. *London,* 1758

Magnetical Machine. (*Vide* Fothergill.) 1776

(*Vide* also Waddel and Knight.)

† **Knoblauch, H.** Über das Verhalten krystallisirter Körper zwischen elektrischen Polen. 8vo. 13 pp. (*Königl. Akad. der Wissenschaft. zu Berlin mitgetheilt am* 1 *Mai,* 1851.) *Berlin,* 1851

Knoblauch, K. H., und Tyndall. Über d. Verhalten Kristallisirter Körper zwischen den Polen einer Magneten. 2 Abhandlungen. 8vo. *Leipzig,* 1850

† **Knobloch, M.** Der Galvanismus in seiner techn. Anwendung seit d. Jahre 1840. 8vo. 116 pp. *Erl,* 1842

† **Knobloch, W.** Über Meteorerscheinungen. Populärer Votrag, gehalten zu Warschau, am 11 Marz, 1868. 8vo. 29 pp. *Berlin,* 1868

Knochenhauer, Karl Wilhelm. *Born April* 10, 1805, *at Potsdam.*

† Über die Veränderungen, welche der Entladungsstrom einer elektrischen Batterie erleidet, wenn mit dem Schliessungs drathe eine zweite Batterie in Verbindung gesetztwird. 8vo. 77 pp. 1 plate. (*Sitzungsb. Wien Akad.* i. 1852. *Dated Meiningen,* 1848.) *Wien,* 1852

† Notiz über den Widerstand des Eisendrathes in electrischen Strome. 8vo. 4 pp. (*Sitzungsbericht d. Wien Acad.* xv. 1855.) *Wien,* 1855

† Über die inducirte Ladung der Nebenbatterie im ihrem Maximum Zweite Abhandlung. Vorgelegt in der Sitzung, vom 20 April 1854. 8vo. 29 pp. (*Sitzungsb. Wien Akad.* xv. No 2.) *Wien,* 1855

† Versuche mit einer getheilten Batterie. Vorgelegt, 8 Oct. 1857. 8vo. 48 pp. (*Sitzungsberichten d. Wien Acad.* xxvii. No. 2.) *Wien,* 1857

† Versuche zur Theorie des Condensators. 8vo. 19 pp. (*Sitzungsb. Wien Akad.*) *Wien*

Uber die Gesetze des Magnetismus nach Ampere's Theorie. 8vo. (*Poggend. Ann.* xxxiv. 481.) *Leipzig*

About 30 Articles in Poggend. Ann. vol. xxxiv. 1835, to vol. civ. 1858.

About 6 Articles in Sitzungsberichte d. Wiener Acad. vol. i. 1848 to vol. xxii. 1857.

† Über die inducirte Ladung der Nebenbatterie in ihrem Maximum. 8vo. 58 pp. (*Aus den Sitzungsberichten* 1853 *der K. Akad. Wissenschaften,* x. 219.) *Wien,* 1853

Beiträge zur Elektricitätslehre. 8vo. 128 pp. 1 plate. *Jena,* 1854

† Ueber die gemeinsame Wirkung zweier elektrischer Ströme. 8vo. 36 pp. 1 plate. (*Aus den Sitzungsberichten* 1855, *der mathematisch naturwissenschaftlichen Classe der K. Akad. der Wissenschaften,* xviii. No. 1, 1855.) *Wien,* 1856

† Beobachtungen über zwei sich gleichzeitig entladende Batterien. 8vo. 18 pp. (*Aus den Akad. der Wissensch.* Bd. xxv. s. 71.) *Wien,* 1857

† Ueber den elektrischen Zustand der Nebenbatterie während ihres Stromes. 8vo. 42 pp. (*Akad. der Wissenschaften,* xxxiii. No. 25.) *Wien,* 1858

Knochenhauer, Karl Wilhelm—*continued.*

† Ueber den Strom der Nebenbatterie. 8vo. 14 pp.. (*Aus den Sitzungsberichten* 1859, *der mathem. naturwiss. Classe der Kais. Akad. der Wissenschaften,* xxxiv. No. 2.) *Wien*, 1859

† Ueber die Theilung des elektrischen Stroms. 8vo. 29 pp. (*Aus den Sitzungs- berichten,* 1859, *der mathem. naturwiss. Classe der K. Akad. der Wissenchaften,* xxxvi. No. 16.) *Wien,* 1859

Ueber das elektrische Luftthermometer. 8vo. 62 pp. (*Aus den Sitzungsberichten* 1860, *der K. Akad. der Wissenschaften.*) *Wien*, 1860

Ueber den Zusammenhang des Magnetismus mit den Oscillationen des Batterie- stromes. 8vo. 19 pp. (*Abdruck aus den Sitzungsberichten der K. Akad. der Wissenschaften.*) *Wien,* 1864

Sammlung v. 11 Abhdl. üb. electr. Batterien. 8vo. 400 pp. (*Wien Acad.*) *Wien*

4 Abhandlungen über das electrische Luftthermometer. 8vo. 150 pp. (*Wien Acad.*) *Wien*

† **Knor, M.** Erste Bericht der gymnastisch-orthopädischen und electrischen Heilan- stalt in München. 4to. 32 pp. 4 plates. *München*, 1860

Knorr, Ernst. Über elektr. Abbildungen u. Thermographien. 8vo. (*Pogg. Ann.* lxi.—lxiii. 1844.) *Leipzig*, 1844

Der Tastengyrotrope u. seine Anwendung in d. Physik u. Telegraphie. 8vo. (*Pogg. Ann.* xc. 1853.) *Leipzig*, 1853

Kobell, Franz Von. *Born July* 19, 1803, *at Munich.*

Analyse eines Magneteisenerzes von Arendal u. d. Franklinits. 8vo. (*Schweigg. Journ.* lxiv. 1832.) 1832

(On) Krystallelectricität and his invented Gems bartelectroscopes.

Über d. Fortschritte d. Galvanographie u.s.w. (*Gelehrt. Anzeig d. Münch. Acad.* 1834.) *München*, 1834

† Ueber eine neue Anwendung der galvanischen Kupferpräcipitation zur Ver- vielfältigung von Gemälden und Zeichnungen in Tuschmanier durch den Druck. 8vo. 7 pp. (*Jour. f. prakt. Chemie,* tom. xx.) 1840

Note.—This is followed (at p. 155) *by a " Nachschrift " signed by Erdmann ; and Erdmann wrote " Nachtragliche Bemerkungen " to it in a separate article.*

† Die Galvanographie, eine Methode gemalte Tuschbilder durch galvanische Kupferplatten im Drucke zu vervielfältigen. (1st edit.) 4to. 18 pp. 7 pls. (*Akad. München.*) *München*, 1842

† Die Galvanographie (von ihm erfunden). (1st ed.) 8vo. *Nürnb.* 1842

Über d. Anlaufen d. Kupferkieses unter Einfl. d. galvan Stroms. (*Gelehrt, Anzeig d. Münch. Acad.* 1843.) *München*, 1843

Die Galvanographie eine Methode Tuschbilder u. Zeichnungen durch gal- vanische Platten im Drucke zu verfielfältigen. 2s. verm Aufl. mit Abhild. des galvan. Apparats u. galvanograph. Proben. 8vo 21 pp. 4 plates. *München*, 1846

Über d. galvan Leitungsfähigkeit d. Mineralien. (*Gelehrt, Anzeig d. Münch. Acad.* 1850.) *München*, 1850

† Über die Bildung galvanischer Kupferplatten vorzüglich zum Zweck der Gal- vanographie mittelst des Trommel-Apparates. 4to. 33 pp. (*Abhandl. d. K. Akad. d. Wissen.* ii. Cl. vi. Bd. ii. Abth. p. 347.) *München*, 1851

Koelle, Aug. Uber d. Wesen die Erscheinung d. Galvanismus. Oder Theorien d. Galvanis. u. d. geistigen Gährung, mit Andeutungen üb. d. materiellen Zusammenhang d. Naturreiche. 8vo. *Stuttgart*, 1825

Koenberg, B. H. Vernünftige Gedanken von den Ursachen der Electricität. 4to. *Wismar*, 1746

Koestlin, C. H. Examen experimentor. quorumd. effectus magnetis chimicos
spectant. 4to.   *Tubingen,* 1813

† **Koetteritz, S.** Lehrbuch der Electrostatik. 8vo. 335 pp.   *Leipzig,* 1872

† **Kohl, F.** Die optisch-mechanische und elektro-magnetische Telegraphie 2te
Aufl. 8vo. 47 pp. 3 plates.   *Leipzig,* 1850

Köhler, J. G. Die Salze aus dem elektro-chemischen Standpunkte betrachtet.
8vo.   *Prague,* 1839

Kohlrausch, R. H. Sur les phénomènes électroscopiques d'une pile voltaique
dont le circuit est fermé. 8vo.   *Paris,* 1854?

Sur la proportionnalité de la force électromot. et de la tension électrique d'un
élément voltaique. 8vo.   *Paris,* 1854?

† Theory of the electric residue in the Leyden Jar. 8vo. 45 pp.   (*Phil. Mag.*)
                                                           London, 1854

Üb. d. elektrisch Vorgänge bei d. Elektrolyse.   *Marburg,* 1855

About twelve articles on Electricity in Poggend. Ann. lxxii. 1847 to xcviii.
1856.

† **Kohlrausch und Weber.** Elektrodynamische Maasbestimmungen, insbesondere
Zurückführung der Stromintensitätsmessungen auf mechanisches Maas.
(Abh. iv.) 4to. 72 pp. (*Aus den Abhandl. der Mat. Phys. Classe der Konigl.
Sächsischen Gesellschaft der Wissenschaften.* Band iii.)   *Leipzig,* 1856

Kohlreif, G. A. Über d. Verbesserung d. Electrisir-Maschinen ; vorzüglich des
Reibzeugs an denselben aus einem Schreiben . . . an den Herausgeber.
8vo. (*Lichtenberg's Magaz.* i. Stück 3, p. 104. 1st edit. of the Mag.)
                                                           Gotha, 1782

Über Verbesserung d. Electrisir-machinen, vorzüglich der Reibungs an denselben. 8vo. (*Lichtenb. Mag.* i. Stück 3, p. 101.)   *Gotha,* 1785

Sollte die Elektricität wirklich die Wärme verursachen, und sollte diese Wärme
eine Wirkung der Zersetzung des Elementarfeuers und Phlogistons seyn?
8vo.   *Weimar,* 1787

Empfehlung d. Gewitterstangen zum Nutzen d. Ackerbaus. 8vo. (*Hannov.
Magaz.* 1879.)   *Hannover,* 1789

Apparat zur Luftelektricität. 8vo. (*Gren's Journ.* i. 1791, p. 219, 385.)
                                                           Leipzig, 1791

Kolke, H. Von. Über eine neue Methode, die Intensität des Magnetismus zu
bestimmen, nebst einigen mit Hülfe derselben gefundenen Resultaten.

† De nova magnetismi intensitatem metiendi methodo, &c. 8vo. 37 pp. 1 plate.
                                                           Bonnæ, 1848

† **Kolk (Van der), H. G. S.** Specimen physicum de methodis quibus resistentia galvanica determinatur, metallorum imprimis . . . (pro gradu Doctoratus).
La. 8vo. 119 pp.   *Trajecto ad Rhenum,* 1860

(*Vide* also Schröder.)

Koller, Sternschnuppen-Beobachtungen im J. 1839. 4to. (*Ann. Wien Sternwarte* xx.)   *Wien,* 1839

Bericht über d. meteorolog. u. Magnet Bestimmung zu Kremsmünster im J.
1842. (*Jahrb. d. ober-österr Museal, Vereins Francisco-Carolinum zu Linz.*)
                                                           Wien, 1843

Über die Berechnung periodischer Natur-Erscheinungen. 4to. (*Denkschr.
Wien Acad.* vol. i.)   *Wien,* 1849

Astronom. meteorolog. u. magnet Beobbachtungen in Schumacher's Astr.
Nachr. Bd. viii.—xx. (*Lamont's Ann. f. Meteorologie u. Erdmagnetism u.
Gauss u. Weber's Result aus d. Beobb. d. magnet Vereins.*)

Kolliker, A. Zur Lehre von der Contractilitat der menschlichen Blut und Lymphgefässe. (*Prager Vierteljahrschrift*, 1849, B. vi. Heft i.) *Prag,* 1849

Uber d. Endigungen der Nerven in der Muskeln de Frosches. 4 plates.
Leipzig, 1862

Ueber die Contraction der Lederhaut des Menchen und Thiere. (*Zeitschrift fur wissensch. Zoologie,* Band ii.)

Konig, Jul. Beiträge zur Theorie der elektrischen Nervenreisung. 8vo. 10 pp.
Wien, 1870

Konstantinoff. On the Chronoscope. (*Vide* Moigno and Breguet.)
(*Note.—Konstantinoff seems to be the original inventor.—*F. R.)

† **Kopp, E.** Considerazioni sulla differenza . . fra la forza . . del Vapore e l'elettro-magnetismo. . . . Fol. 4 pp. (*L'Economista,* anno iii. vol. ii. p. 118.)
Milano, 1845

Kopp, J. H. Dissertatio sistens tentamen de causis combustionis spontaneæ in corpore humano pactæ. 8vo.
Jenæ, 1800

Ausführl. Darstell. d. Selbstverbrenn d. menschl. Körpers. 8vo.
Frankf.-a.-M. 1811

(*Vide* also Liebig and Kopp.)

† **Kostlin, K. H.** Dissertatio physica experimentalis de effectibus electricitatis in quædam corpora organica. (Thesis.) 4to. 36 pp
Tubingæ, 1775

Koten, J. H. Van. De galvanische stroom, toegepast op electro-magnetische telegrafen en uurwerken. 8vo. viii. and 147 pp. 4 plates lith. (Hassels. 2 fl. 50 ct.) (*From Zuchold's Cat.* 1856, p. 68.)
Amsterdam, 1856

De galvanische stroom, toegepast op electro-magnetische telegrafen en uurwerken. 8vo. 150 pp. 4 plates.
Amsterdam, 1859

† De elektro-magnetische telegrafie, in de voornaamste tijdperken van hare ontwikkelingen en haar teegenwoor digstandpunt; benevens de elektromagnetische seinklokken en uurwerken. 8vo. 283 pp. 8 plates.
Amsterdam, 1862

† **Krafft,** Georg Wolfgang. *Born July 15, 1701, at Tuttlingen, Wirtemb.; died June 12, 1754, at Tübingen.*

De viribus attractionis magneticæ experimenta 4to. 12 pp. plate p. 276. (*Comment. Acad. Petropol.* xii. 1850.)
Petropol, 1740

Prælectiones in physicam theoreticam. (*From Van Swinden.*) *Tubingen,* 1750

(*Poggendorff,* i. 1309, says, " *Prælectiones academicæ publicæ in Physicam theoreticam,* iii. plates, *Tubing.* 1753-4.")

Observationes meteorologicæ an. 1750 Tubingæ factæ. 4to. (*On the Aurora in day-time,* p. 403.) (*Novi Comment. Acad. Petrop.* v. 400.)
Petropoli

(*Vide* also Nyerup and Kraft.)

Krafft, Wolfgang Ludwig. *Born August 25, 1743, at St. Petersburg; died November 20, 1814, at St. Petersburg.*

Declinatio acus magneticæ Kiovii observata. 4to. (*Novi Comment. Acad. Petropol.* xv. mp. 586.)
Petropoli, 1771

Expositio declinationis acus magneticæ in variis Imperii Russici regionibus observatæ. 4to. (*Novi Comment. Acad. Petropolitanæ,* xvii. Hist. p. 52, *Mem.* p. 695.)
Petropoli, 1773

Experimenta acu magnetica Petropoli instituta. 4to. (*Novi Comment. Acad. Petropolit.* xix. *Hist.* p. 68, *Mem.* p. 610.)
Petropoli, 1775

Tentamen theoriæ electrophori. 4to. (*Acta Acad. Petrop.* 1778, part i. *Hist.* p. 70, *Mem.* 154.)
Petropoli, 1778

Expér. sur le Phosphore de M. Canton. 4to. (*Acta Acad. Petr.* i. 1778.)
Petropoli, 1778

Krafft, Wolfgang Ludwig—*continued.*
Annotationes circa constructionem et usum acus inclinatoriæ et determinatio inclinationis magneticæ Petropoli ad finem anni 1778. 4to. (*Acta Acad. Petropolitanæ*, an. 1778, part ii. *Mem.* p. 170.) *Petropoli,* 1778
Observation d'une aurore australe vue à St. Pétersbourg le 6-17 Fevr. 1778. (*Acta Acad. Petropolitanæ*, an. 1778, part i. *Hist.* p. 45.) *Pétersbourg,* 1778
Über ein hypothet Gesetz d. Neigung d. Magnetnadel an vershiedn Orten d. Erde. 4to. (*Mem. Acad. St. Petersb.*) *Petersb.* 1809

Kramer, Anton. Johan de. *Born July 21, 1806, at Milan ; died September 25, 1853, at Tremezzo um Comer See.* (*Giorn. I. R. Ist. Lomb.* Nuova S. ix. 165, in 4to. 1856.)
Über einen neuen durch Einfluss d. Erdmagnetismus wirksam elektromagnet Apparat. 8vo. (*Poggend. Ann.* xliii. 1838.) *Leipzig,* 1838
† Nuovo apparato rotatorio elettro-magnetico messo in moto dal magnetismo terrestre. Nota comunicata alla Bibliot. Ital. da Ant. D. Kramer di Milano. 8vo. 6 pp. 1 plate. (*Bibl. Ital.* lxxxix. 163.) *Milano,* 1838
† Sopra le copie in rame del Tito Polito. 8vo. 1 page. (*Giorn. I. R. Ist. Lomb.* i. 30.) *Milano,* 1841

Kramer, G. E. Über Telegraphen-Schreib-apparate. 8vo. (*Dingler's Polytech. Journ.* cxix. 1851.) *Stutgart,* 1851
Über d. Project. d. galvan. Uhren in Berlin. 8vo. (*Dingler's Polytech. Journ.* cxxi. 1851.) *Stutgart,* 1851
Telegraph. (*Vide* Moigno, p. 441, Du Moncel, &c.)

† **Kramer e Belli.** Sulla produzione dell' Ozono per via chimica, di C. F. Schonbein. Traduzione. Parte prima. 8vo. 66 pp. (*Giorn. I. R. Ist. Lomb.* ix. 397.) *Milano,* 1844
† Sulla produzione dell' Ozono per via chimica di C. F. Schonbein. Traduzione. Continuazione e fine. 8vo. 54 pp. (*Giorn. I. R. Ist. Lomb.* x. 201.) *Milano,* 1845
† L'Ozono non è acido . . . nitroso-nitrico. (Nota spedita dall' Autore Schonbein ai Tradutori posteriormente al Congresso.) 8vo. 9 pp. (*Giorn. I. R. Istit. Lomb.* x. 258.) *Milano,* 1845

Kratter, H. Versuch einer Entwickelung der Gründbegriffe, der Meteorsteine. . . 8vo. *Wien,* 1825

Kraus. (*Vide* Reitlinger and Kraus.)

Kravogle. Electromagnetische Motor. 8vo. 16 pp. 2 plates. *Wien,* 1868

Krazenstein, Christian Gottlieb. *Born January 30, 1723, at Wernigerode ; died July 6, 1795, at Copenhagen.*
Abhandlung üb. den nuzen der elektricität in der artzneywissenschaft. 2nd edition. 8vo. *Halle,* 1745
(*Poggendorff,* i. 1364, adds : " *Unter d. Titel-Physikal Breife vom Nutzen d. Elektr. u.s.w. Ib.* 1746 u. 1772.")
† Theoria electricit. more geometrico explicata. 4to. 62 pp. 1 plate. *Halle,* 1746
Progr. quo historiam restitutæ loquelæ per electrificationem recenset. 4to. *Havniæ,* 1753
Vorless. üb. d. Experim. Physik. 1st edition. *Kopenhagen,* 1758
De transmut. aquæ in terram. Diacrisis hypotheseos Franclinianæ de electricit. 4to. *Havniæ,* 1778
† Vorlesungen üb. d. Experim. Physik. 4th edition. *Kopenhagen,* 1781
Vorlesungen über exp. Physik. 5 Aufl. vermehrte. 12mo. 232 pp. (*Volta's copy.*) *Kopenhaven,* 1782
† Vorlesungen über die Experimental Physik (vermerhte Auflage.) 6th edition. 8vo. 232 pp. *Kopenhaven,* 1787
(*In this 6th edition he says that the book is an extract-abbreviation from his Latin "System der Naturlehre."*)

Kratzenstein, C. A. Tentamen resolvendi problema geographico-magneticum ab Acad. Petrop. propositum. 4to. *Petropoli,* 1798

Krayenhoff, Cornelius Rudolph Theodor von. *Born June* 2, 1758, *at Nymwegen; died November* 24, 1840, *at Nymwegen.*
De l'application de l'électricité. . . (*Vide* Troostwyk, Van.) 1788

Krecke, F. W. C. Description de l'observatoire météorologique et magnétique à Utrecht. *Utrecht,* 1848

Kreil, Karl. *Born November* 4, 1798, *at Ried.*
† Descrizione degli Apparati magnetici e dei metodi con cui si eseguiscono le osservazioni. 4to. 65 pp. 1 plate. *Milano,* 1838

Note.—This is a description by Kreil of the Apparatus used at Milan, and is appended to Frisiani's translation of Gauss, "Misura assoluta," p. 133, *in the Primo Supplemento alle Effemeridi di Milano,* 1838.

Magnet. und meteorol. Beob. zu Prag Järhlich (seit Aug.) 1839-48

Astro-meteorolog. Jahrbuch fur Prag. 4 jahrg. 1842-5

† Magnetische und geographische. Ortbestimmungen in Böhmen. Ausgeführt in den Jahren 1843-5 von K. Kreil. 4to. 95 pp. 2 plates. *Prag,* 1846

† Lettera a P. Frisiani sulle osservazioni magnetiche fatte in Lombardia 4to. 3 pp. (*Giorn. dell' I. R. Istit. Lomb.* Nuova Serie, i. 111.) *Milano,* 1847

Jahrbücher für Meteorologie und Erdmagnetismus von Kreil. 8 vols. 4to. *Wien,* 1848-61

Beschreibung meteorolog. Autographen-Instrumente. 4to. (*Sitzungsber. d. Wien Acad.* iii. 1849.) *Wien,* 1849

† Ueber den Einfluss der Alpen auf die Äusserungen der magnetischen Erdkraft. Folio. 46 pp. 4 plates. (*Extr. f. Sitzb. d. Wien Acad.* ii. 1849.) *Wien,* 1849

† Apparate für magnetische Beobachtungen. 8vo. 3 pp. (*Sitzungb. d. Akad. d. Wissensch.* 1850, i. Heft. p. 129.) *Wien,* 1850

† Mittheilung über das auf der Prager Sternwarte aufgestellte Inductions Inclinatorium, und über ein authographes Thermometer aus Zinkstangen. 8vo. 5 pp. (*Aus d. Juniheft des Jahrg.* 1850 *d.* . . . *K. Akad. d. Wissen. besonders abgedruckt.*) *Wien,* 1850

Entwürf eines Meteor-Beobachtungs-system für die Österreich Monarchie (*From Atti I. R. Ist. Lomb.* ii. 114.) *Wien,* 1850

Uber magnet Variations-Instrumente. 4to. (*Sitzungsber. d. Wien Acad.* iv. 1850.) *Wien,* 1850

† Beschreibung der, an der Kaiserl. Königl. Sternwarte zu Prag, aufgestellten selbstverzeichnenden meteorologischen Instrumente : Windfahne, Winddrucke, Regen- und Schneemesser. 4to. 8 pp. 2 plates. *Prag,* 1851

† Berichte über d. Central-Anstalt für Meteorologie u. Erdmagnetismus. 4to. (*Sitzungsb. der Wien Acad.* viii. and ix. 1852.) *Wien,* 1852

Einfluss des Mondes auf die magnetische Declination, und auf die Intensitat der horizontalen Componente der magnetischen Erdkraft. 2 vols. 8vo. *Wien,* 1852-3

† Über d. Einfluss d. Mondes auf d. magnet Declination. La. 4to. 47 pp. (*Denkscrift d. Wien Acad.* iii. 1852.) *Wien,* 1852

Vortrag üb. d. Central-Anstalt d. Meteor u. Erdmagnetismus. 1852
Der Magnetismus auch ein Räthsel.

† Magnetische und geographische Ortsbestimmungen im südöstlichen Europa und einigen Küstenpunkten Asiens. 8vo. 22 pp. (*Sitzungsb. d. Wien Acad.* xxxvi. No. 16.) *Wien*

† Einfluss des Mondes auf die horizontale Componente der magnetischen Erdkraft. Gelesen . . . am . . . 11 März, 1852. La. 4to. 56 pp. (*Denkschriften Wien Acad.* v. 1853.) *Wien,* 1853

Kreil, Karl—*continued.*

Jahrbücher der K. K. central Anstalt für Meteorologie und Erdmagnetismus.
Von Karl Kreil . . . herausgegeben durch die Kaiserlich Akad. der Wissech.

I. Bd. Jahrgang 1848-9, printed in 1854, 420 pp. 2 plates (wood).
II. Bd. Jahrgang 1850, printed in 1854, 257 pp.
III. Bd. Jahrgang 1851, printed in 1855, 252 pp. Anhang von
Fritsch. 44 pp.
IV. Bd. Jahrgang 1852, printed in 1856, 352 pp. Anhang von Fritsch.
Table, &c. 54 pp. At p. 213 (of this vol. iv.) Perspective View
and Plan of the House, &c. At p. 215 Perspective View of the
interior of the old Observatory.
V. Bd. Jahrgang 1853, printed in 1858
VI. Bd. Jahrgang 1854, printed in 1859.
VII. Bd. Jahrgang 1855, printed in 1860.
VIII. Bd. Jahrgang 1856, pp. 131 and 145, printed in 1861.
IX. Bd. Jahrgang.
X. Bd. Jahrgang.
XI. Bd. Jahrgang.
XII. Bd. Jahrgang 1867, printed in 1869.
4to.

This is the 4th vol. of the "Neue Folge." The authors are Jelenck and Fritch.

Mehre Aufsätze in Poggend. Ann.

† Kurz Abriss der Entstehungs- und Entwicklungs geschichte des magnetischen
Vereins, und nähre Beleuchtung des Standpunktes welchen Prag darin
einnimt. 4to. 31 pp.

† Bestimmungen einiger Längenunterschiede mittelst des elektromagnetischen
Telegraphen. 8vo. 15 pp. (*Aus d. fünften Heft d. Sitzungsberichte d. K.
Akad. d. Wissensch. besonders abgedruckt.*) *Wien*

† Resultate aus den magnetischen Beobachtungen zu Prag. 4to. 3 plates.
44 pp. (*Aus dem viii. Bde. der Denkschriften der . . . K. Akad. d. Wissensch.
besonders abgedruckt.*) *Wien*, 1855

† Magnetische und geograp. Ortbestimmungen an den Küsten des Adriatischen
Golfes im J. 1854. 4to. 46 pp. 1 plate. (*Aus dem x. Bande der Denk-
schriften . . . de K. Akad. d. Wissensch. in Wien besonders abgedruckt.*)
Wien, 1855

Erste Ergebnisse der magnetisch. Beobachtungen in Wien. 4to. *Wien*, 1856

† Anleitung zu den magnetischen Beobachtungen. 2 vermehrte Auflage. La.
8vo. 216 pp. (*Als Anhang zum xxxii. Bande der Sitzungsberichte der K. Akad.
der Wissenschaften.*) *Wien*, 1858

† Magnetische u. geograph. Ortbestimmungen im südöstlichen Europa u. einigen Küstenpunkten Asiens. 4to. 8 plates. 94 pp. (*Vorgelegt 24 June,
1859. K. Akad d. Wissensch.*) *Wien*, 1862

† **Kreil u. Fritsch.** Magnetische und geograph. Ortsbestimmungen im österreich.
Kaiserstaate. 1—5 Jahrg. pro 1846-47-48-50-51. 4to. 2 vols. (*Vide
Heinsius von Schiller, vol. ix. for further particulars, p. 535.*)
Prag, 1848-9-50-1-2

Magnetische und geographische Ortsbestimmungen in Oesterreich. 4to. 5 vols. ?
Prag, 1848-52

† **Kreil and others.** Magnetische und Meteorologische Beobachtungen zu Prag, in
Verbindung mit mehreren Mitarbeitern ausgefürt und auf öffentliche Kosten herausgegeben.

1841-47, Jahrgänge i.-vii. vom 1839-46 von Kreil, K.
1848-51, Jahrgänge viii.-x. vom 1847-49 von Kreil und Jelinck.
1853-55, Jahrgänge xi.-xiii. 1850-52 von Bohm und Künes.
1856-62, Jahrgänge xiv.-xxii. 1853-61 von Bohm und Karlinski.
1863-66, Jahrgänge xxiii. 1862. . . Von Bohm und Moritz Allé.
4to. Vol. i. 1 plate ; ii. 1 pl.; iii. 2 pls.; iv. 1 pl.; x. 2 pls. *Prag*

(*Note.—Each Jahrgang is a volume.*)

† **Kreil e Vedova.** Osservazioni sull' intensità e sulla direzione della forza magnetica istituite negli anni 1836, 1837, 1838 all' I. R. Osservatorio di Milano da O. Kreil e P. della Vedova. 4to. 341 pp. (*Forms the " Secondo Supplemento alle Effemeridi di Milano.*") *Milano*, 1839

Kress, G. L. Die Galvanoplastik für industrielle und Kunstlerische Zwecke. Resultate 26 jähriger Erfahrungen. 8vo. 112 pp. *Frankfurt-a.-M.* 1867

Kries, Friedrich Christian. (*Born October* 18, 1768, *at Thorn, West Prussia; died June* 28, 1849, *at Gotha.*)

Over de oorzaken der aardbevingen. 8vo. *Utrecht*, 1820

Von den Ursachen der Erdbeben ; eine Preisschrift zur Beantwortung der Frage ; Welches sind die nächsten Ursachen der Erdbeben ? Müste man die elektrische oder die galvanische Kraft mit unter diese Ursachen zählen, oder sind die Erscheinungen, welche man nicht selten bey Erdbeben wahrnimmt, für Mitwirkungen der nëhmlichen Ursache zu halten. 8vo. *Utrecht*, 1820

Über Muncke's Erklär. d. elektromagnet. Erscheinungen. 8vo. (*Gilb. Ann.* lxxi. 1822.) *Leipzig*, 1822

Von d. magnet. Erscheinungen. 8vo. (*From Poggendorff*, i. 1320.) *Leipzig*, 1827

† Von den Ursachen der Erdbeben und von Magnetischen Erscheinungen, &c. (Electy.) Zwey Preisschriften. 8vo. 151 pp. 1 plate. *Leipzig*, 1827

Uber e. Wetterschlag auf d. Leuchtthurm von Genua. 8vo. (*Pogg. Ann.* xii. 1828.) *Leipzig*, 1828

De nexu inter terræ-motus, vel montium ignivomorum eruptiones, et statum atmospheræ. (Gekronte Preisschrift.) 4to. (*Nova Acta Soc. Jablonowsk* iv. 1832.) 1832

† **Kromhout, J. H.** Projet d'un Diastimètre électrique pour les Batteries de Côte. 8vo. 24 pp. 4 plates. *La Haye*, 1867

Kruger. Meditations on Electricity. (*Vide* Priestley.)

Change of colour in bodies by Electricity. (*Danzig Mem.* i. 417.)

Kruger, Georg. Prodromus auroræ boreæ seu historiæ meteorologicæ teutonico-curlandicæ &c. d. i. Vortrab. teutscher und Kurlandischer Gewitter-historie, durch die wahre naturliche Astrologie bewehret u.s.w. 4to. *Riga*, 1700

Kruger, Johann Gottlob. (*Born June* 15, 1715, *at Halle; died October* 6, 1759, *at Brunswick.*)

Zuschrift an seine Zuhörer, worinn er ihnen seine Gedanken von der Elektricität mittheilet, und ihnen zugleich seine kunftigen Lectionen bekannt macht. 8vo. *Halle*, 1744

Gedanken von d. Elektricität u.s.w. 2nd edit. 8vo. *Halle*, 1745

Geschichte d. Erde in d. allerältesten Zeit. 8vo. *Halle*, 1746

(*Poggendorff*, i. 1322, *says : " Angehängt ist Abhandl. von d. Elektricität und darin wird die erste öffentl. Nachricht von d. Kleist'schen Verstärkungsflasche gegeben.*")

Abhandlung von der Elect. ; als ein Anhang bey dessen Geschicht der Erde . . nebst einigen Breifen des Von Kleist. . 8vo. *Halle*, 1746

Gedanken von d. Ursache d. Erdbeben u.s.w. 8vo. (*From Poggendorff*, i. 1323. *Halle*, 1756

Diss. de electricitatis Musschenbroc kianæ in sanandis morbis efficacia. 4to. (*From Poggendorff*, i. 1323.) *Helmstadt*, 1756

Krumme, W. De conditione magnetica compositionum quarumdam cupri. Dissertatio. 8vo. 27 pp. *Bonnæ*, 1858

Krunitz, Johann Georg.   (*Born March* 28, 1728, *at Berlin; died December* 20, 1796, *at Berlin.*)

† Verzeichnis der vornehmsten Schriften der Electricität und den electrischen Curen. 8vo. 200 pp.   *Leipzig,* 1769

† Priestley's Geschichte der Elektricität (Ubersetzung.). 4to.   *Berlin,* 1772 (*Vide* also Kirtz.)

Krziwaneck, J.   De electricitate acupunctura Perkinismo et frictione. 8vo.
                                                                      Prag, 1839

Kuhlberg, A.   Analyse und Beschreibung der Meteorite von Neft, Honolulu, Lixua und eines im Gouvernement Jekatherinoslaw gefallenen Meteoriten. 8vo. 34 pp. 2 plates.   (*Aus dem Archiv fur die Naturkunde Liv. Esth. und Kurlands abegrdkt.*)   *Dorpat,* 1865

† **Kuhn,** C.   Handbuch der angewandete Electricitätslehre. 8vo. 1396 pp. (*Karsten's Encyclopædie, i.e. Encycl. Allgemeine.*)   *Leipzig,* 1866

Kuhn, Karl Gottlob.   (*Born July* 13, 1754, *at Spergau, near Merseburg ; died June* 19, 1840, *at Leipzig.*)

Traité de l'électricité. (*From* Kuhn, 1873, p. 89.)   1771

† Geschichte der medicinischen und physikalischen Elektricität und den neuesten Versuche. . . Erster Theil. 8vo. 278 pp. 4 plates.   *Leipzig,* 1783

† Geschichte der medicinischen und physikalischen Elektricität und den neuesten Versuche. . . Zweiter Theil. 8vo. 392 pp. 2 plates.   *Leipzig,* 1785

Anwendung und Wirksamkeit der Elektricität zur Erhaltung &c. . . der Gesundheit des menschlichen Korpers ; aus dem Franz des Berthollon übersetzt ; und mit newen Erfahrungen bereichert und bestatiget. 8vo. 2 Th. Plates.   1788-9

Etwas über die Kuren des Hrn. Grafen von Thun aus physikalischen und medicinischen Gesichtspunkten betrachtet.   *Leipzig,* 1794

† Die neuesten Entdeckungen in der physikalischen und medicinischen Elektricität. Als eine Folge der Geschichte. . . Erste Theil. 8vo. 273 pp.
                                                                      Leipzig, 1796

† Die neuest Entdeckungen in der physikalischen und medicinischen Elektricität. . Als eine Folge der Geschichte. . . Zweiter Theil. 8vo. 240 pp. 1 plate.   *Leipzig,* 1797

Über Blitzableiter ; Über d. Principien d. Telegraphie im Allgem. u. jene d. elektr. Telegraphie insbesondere. "In d. Gelehrt Anzeigen d. Königl. bayer Acad. von 1851 u. 1852 ;" *or,* in "Astronom Kalender 1850-52."   1851-2

(*Note.—It seems uncertain which this entry belongs to, or whether to both. Kühn says : " The Royal Medical Society of Paris has published in the 2nd part (Theil) of its Writings, the account, which I had given to it, of the results of my treatment of 82 patients, and in the Historical Part (Theil) of the 3rd vol. the continuation of my cures.*")

(*Vide* also Reitlinger and Kühn.)

† **Kuhnert,** E. H.   Über die Quelle der galvanischen Elect. Inaugural Dissertation. 8vo. 49 pp. 1 plate.   *Marburg,* 1842

Kulenkamp, N.   Die Elektricität in Bremen.   1743

† **Kulp,** E.   Lehrbuch der Experimental-Physik. 3 Bd.; Die Lehre von der Elektricität und dem Magnetismus. 8vo. 542 pp.   *Darmstadt,* 1862

Kumpel, J. A. F.   De magnetismo. 4to.   *Jenæ,* 1788

Kunemann, Machines (elect.) en caoutchouc. (*From Du Moncel, who gives no other particulars.*)

† **Kunneman.** (*Vide* Fabre and Kunneman.)

Kunes. (*Vide* Kreil and others, and also Bohm and Künes.)

† **Kunze,** Karl Sebastian Heinrich.   *Born February* 2, 1774, *at Kiel; died May* 30, 1820, *at Flensberg.*

† Besch. e. elek. Apparatus für Schulen. zur Erklärung der Gewittermoterie.
8vo. 56 pp. 1 plate.   *Hamburg,* 1796

Schauplatz d. gemeinnützigst Machinen, nach J. Leupold u. Anderen bearbeitet. 3 Bde. 8vo.   *Hamburg,* 1796-1802

Neue elektrische Versuche. 4to. (*From Young,* p. 416.)

Kunze, F. H.   Einige Bermerk üb. d. Galvan. in phys. chem. und med. Hinst. 8vo.   *Hamburg,* 1804

Kunzek, A.   Atmosphärische Elektricität. 8vo. (*Leicht fassliche Darstellung d. Meteorol.* p. 174.)   *Wien,* 1847

Kupffer, Adolphe Theodor.   *Born Jan.* 6, 1799, *at Mitau.*

Theorie des phénomènes électro-dynamiques, uniquement déduits de l'expérience. 4to.   *Paris,* 1826

Variation d. magnet Intensitat in Kassan u. Einfluss d. Nordlichts auf d. Magnetnadel. 8vo. (*Poggend. Ann.* x. 1827.)   *Leipzig,* 1827

Über d. Vertheil d. Magnetism in Magnetstäben. (*Poggend. Ann.* xii. 1828.)   *Leipzig,* 1828

Wärme-Einflusses auf d. Magnetnadel. 8vo. (*Poggend. Ann.* xvii. 1829.)   *Leipzig,* 1829

† Instructions pour faire des Observations météorologiques et magnétiques. 8vo. 83 pp. 2 plates.   *Petersburg,* 1836

† Recueil d'Observations magnétiques faites à St. Pétersbourg, et sur d'autres points de l'empire de Russie ; par A. T. Kupfer et ses collaborateurs, et publié par ordre de l'Acad. en Déc. 1857. 4to. 717 pp. 3 plates.
   Petersburg, 1837

Annuaire magnétique et météorologique du Corps des Ingénieurs des Mines de Russie, 1837-46, 10 vols   8vo.   *St. Petersburg,* 1839-49

Observations météorologiques et magnétiques faites dans l'étendue de l'empire de Russie. 20 vols. 4to.   *Pétersbourg,* 1837-39-48 ?

† Sur les Observations magnétiques fondées par ordre des Gouvernements d'Angleterre et de Russie sur plusieurs points . . 8vo. 11 pp. (*Rapport addressé à l'Acad. des Sciences de Pétersbourg.*)   1840

De l'influence de la température sur la force magnétique des barreaux. (*Bull. Phys. Math. Acad. Petersb.* i. 1843.)   *Petersburg,* 1843

† Rapport sur les Observations faites à l'Observatoire magnétique . . de Helsingfors. 8vo. 13 pp.   *Petersburg?* 1848

Annales de l'Observatoire physique central de Russie, an. 1847-54. 8 vols. 4to.   *Pétersbourg,* 1850-56

(*Vide* also Keferstein and Kupffer.)

Annales de l'Observatoire physique central de l'empire de Russie publié par ordre de l'Empereur, ans. 1851-55. 9 vols. 4to.   1853-59

Kuyper. (*Vide* Cuypers.)

Kyper (Kieper), A.   Disp. de fulmine quod a. 1636 . . turrim nitrariam aulicam Regiomonti percussit.   1637

L.

Labarte. L'émaillerie dans l'antiquité et au moyen age. 8vo. *(From Martin,* p. 105.) *Paris,* 1857

La Borde, Jean Baptiste de. *Died* 1777, *at Colancelle.*
Le Clavissin électrique; avec une nouvelle théorie du méchanisme et des phénomènes de l'électricité. 8vo. 164 pp. *Paris,* 1761

Laborde. Note sur un nouveau Télégraphe électrique, dont les indications sont données au moyen du son. 4to. *(Comptes Rendus,* xxi. 526.) *Paris,* 1845

† **Labrosse, F.** Dés aimatation des Navires en fer d'après la méthode de E. Hopkins. Trad. de l'Anglais. Réduction des Déviations des Compas et résultats des expériences à bord du Northumberland et de la Charente. 8vo. 80 pp. 2 plates. 1868

Lacam. Thoughts on Magnetism. 8vo. *(From Young,* p. 437.)

† **Lacassague et Thiers.** Nouveau système d'éclairage électrique, ses avantages, ses instruments, &c. . . avec Pièces justificatives; Rapport de la Société d'Encourag. Programme de la Commission du Ministre de la Marine, &c. 8vo. 142 pp. *Paris et Lyon,* 1857

† **La Cépède.** Essai sur l'électricité naturelle et artificielle. 2 vols. 8vo. vol. i. 375 pp.; vol. ii. 389 pp. *Paris,* 1781

† Théorie des Comètes, pour servir au système de l'Electricité universelle, suivie d'une lettre critique sur l'attraction. 8vo. 71 pp. *London et Paris,* 1784

Lacoste. *(Vide* St. Martin and Lacoste.)

Lacour, Sec. Acad. Bordeaux. Rapport sur les travaux de l'Acad. pendant 1822. 8vo. *(Séances de l'Acad. de Bord. pro* 1822.) *Bordeaux,* 1822?
(Contains references to two Mem. on Meteorology and Electricity by Billandel.)
Rapport sur les travaux de 1 Acad. pendant 1823. 8vo. *(Séances de l'Acad. de Bord. pro* 1824.) *Bordeaux,* 1823
(Contains references to a 1st Mem. on the Meteor of 17th Feb. 1823 and to a 2nd Mem. on Meteorology.)

† **Lacroix, F.** Essai sur l'Induction électrique. 8vo. 45 pp. 1 plate. *Grenoble,* 1864

† **Lacy, H. de.** Du Galvanisme médical. Conseils aux malades sur l'emploi du Galvanisme. . . 8vo. 32 pp. *Paris,* 1849

† **Lafollye.** Appendice à un Mémoire sur un nouvel appareil électrique. 4to. 14 pp. Lithographed MSS. *Bordeaux,* 1857

La Garde, De. Elettricità celest. 8vo. *Modena,* 1753

Lagrange, Laplace, &c. Rapport sur un nouveau télégraphe, des Cit. Bréguet et Bétancourt. Dated "21 germinal an. 6." 4to. 11 pp. *(Mém. de l'Institut* iii. 22. *Cl. Sci. Mathém.)* *Paris,* 1798

Lagrave. Expériences galvaniques sur les aveugles. 4to. *(Journ. de Phys.* an. xi. p. 159.) *Paris,* 1830

Expériences tendantes à prouver que les lois du galvanisme semblent différer de celles de l'électricité. 4to. *(Journ. de Phys.* Ventose, an. xi. p. 233. *Extrait.)* *Paris,* 1830

Expériences galvaniques tendantes à prouver que nous avons deux fluides dans l'économie animale, l'un positif, et l'autre négatif, qui paraissent produire dans leur ensemble l'agent de la vitalité. 4to. *Paris,* 1803

Expériences sur le galvanisme de la pile de Volta, plongée dans l'eau. 4to. *(Journal de Phys.* xi. p. 472 and xi. p. 140.) *Paris,* 1803

Expériences galvaniques tendantes à prouver que l'oxide qui se forme sur la surface des disques métalliques de la pile, ne rend pas absolument nul l'effet de son action . . qu'il a même la propriété de remplacer les rondelles de drap mouillé, et même de conserver son action 15 à 20 jours. 4to. *Paris,* 1803

La Hire, De. Letter concerning a new sort of magnetical Compass, with several curious magnetical experiments. 4to. (*Phil. Trans.* 1687, p. 344.)
London, 1687

Lair. Essai sur les combustions humaines produits par un long abus des liqueurs spiritueuses, &c. *Paris,* 1800

Lalande, Joseph Jérome le François de. *Born July 11, 1732, at Bourg-en-Bresse; died April 4, 1807, at Paris.*
Observations sur les nouvelles méthodes d'aimanter, et sur la déclinaison de l'aimant. 4to. (*Mém. de Paris.*) *Paris,* 1761
Lettre sur un Météore extraordinaire adressée à Messrs. les Auteurs du *Journal des Savans.* *Paris,* 1771
Une notice sur la découverte du Galvanisme. (*Journ. des Savans,* Nov. 1792.)
Paris, 1792
Abrégé de l'Astronomie.
Note.—On Aurora, p. 101 *et seq.* says that "*les Aurores boréales électrisent les pointes isolées placées dans de grande tubes de verre.*"

† **Lallemand.** Thèse . . études sur les lois de l'Induction à l'aide de la Balance électro-dynamique. 4to. 42 pp. 1 plate. *Paris,* 1851

Lambert, Alexandre. Historique de la télégraphie, ses phases, ses systèmes divers. 8vo. 44 pp. *Paris,* 1862

† **Lambotin.** Observations sur les Pierres méteoriques. 8vo. 7 pp. *Paris,* 1814

Lambron, E. Etudes expérimentales sur le dégagement d'électricité dans les eaux sulfureuses de Bagnères de Luchon. 8vo. 47 pp. (*Extrait des Annales de la Société d'hydrologie médicale de Paris,* tom. ii.) *Paris,* 1865

Lameillière. (*Vide* Lavialle de Lameillière.)

Lamé, Gabriel. *Born July 22, 1795, at Tours.*
Cours de physique. 8vo. 2 vols. *Paris,* 1837

La Métherie, Jean Claude de. *Born September 4, 1743, at Clayette, Mâconnais; died July 1, 1817, at Paris.*
Essai analytique de l'air pur, et des différentes espèces d'air. 8vo. *Paris,* 1785
(*Note.—There is a 2nd edit. of 1788, 2 vols. Kahn N. Entdeck.* i. 15, *refers to the German version of this 2nd edit.* vol. i. pp. 297-310, *on the subject of the electric spark being a real combustion.*)
Réflexions sur l'Elect. animale. 4to. (*Journ. de Phys.* xlii. p. 252.) *Paris*
Discours préliminaire. 4to. (*Journ. de Phys.* liii.) *Paris*
(*Note.—Contains his Résumé sur l'Hist. du Galvanisme.*)
Discours préliminaire. 4to. (*Journ. de Phys.* liv.) *Paris*
(*Note.—Contains his continuation of Résumé sur l'Hist. du Galvanisme.*)

† Estratto di una Lettera . . . a Brugnatelli sopra diversi argomenti. 8vo.
(*Ann. di Chim. di Brugnatelli,* xix. 156.) *Pavia,* 1802
Suite des expériences galvaniques sur l'irritabilité de la fibrine et la décoloration du sang. 4to. (*Journ. de Phys.* Pluviose, an. xi. p. 161.) *Paris,* 1803

Lami. (*Vide* Lamy.)

† **Laming,** Richard. On the primary forces of Electricity. 8vo. 24 pp. (*Phil. Mag.* June, 1838.) *London,* 1838

† De l'application des axiomes de la mécanique, &c. . . . au phénom. de l'élect. 8vo. 28 pp. 1 plate. *Paris,* 1839

Lamont, Johann. *Born December 13, 1805, Braemar, Scotland.*
Ueber das Verhältniss der magnetischen Intensitäts und Inclinations-Störungen. 8vo. 12 pp. (*Sitzungsb. d. K. Bayer Akad. d. Wiss.* vom. 14 Juni, 1862, p. 76.) *München,* 1862

† Über das Verhalten des Nadelmagnetismus bei Temperaturänderungen. (*Gelehrte Anz. herausg. von Mitgl. der Bayer Akad. d. Wissensch.* xiii. 1005.)
München, 1841

B B

Lamont, Johann.—*continued.*

Discorso d'apertura sull' Osservatorio magnetico eretto nella Specola Reale di Monaco. 4to. (*Mem. Soc. Ital.* tom. xxiii.) *Monaco,* 1841

Bestimmung d. Horizontal Intensität des Erdmagnetismus. 4to. 50 pp.
München, 1843
Über galvanisch registrirense Uhren. (*Lamont, Astronom. Kalender,* 1852, pp. 150-158.) 1850

Astronom. Kalender für das Königreich Bayern mit Beiträgen von Prof. Kuhn, Meister, Pollak, Jahrgang, 1850, 1851, 1852-53.

Jahrisbericht der Münchener Sternwarte.

† Über däs magnetische Observatorium der K. Sternwarte bey München. Eine offentliche Vorlesung gehalten in der förmlichen Sitzung der K. Akad. . . am 25 Aug. 1840. 4to. 56 pp. *München,* 1841

† Bestimmung der Horizontal Intensität des Erdmagnetismus aus absoluten Maas. 4to. 53 pp. 1 plate. (*Abhandl. d. Mathem. Phys. Cl. d. K. Bayeris. Akad. d. Wissen.* iii. 619, pro 1842 and 1843.) *München,* 1842

† Annalen für Meteorologie, Erdmagnetismus und verwandte Gegenstände redigert von Grunert, Koller, Kreil, Lamont, Plienenger, Stieffel; herausgegeben von Dr. Lamont. Jahrgang, 1842, 1 Heft. 8vo. 1 plate.
München, 1842

† Resultate der magnetischen Beobachtungen in München währ. der dreijährigen Periode 1840-1-2, 4to. 16 pp. 1 plate. (*Abhandl. d. Math. Phys. Cl. d. K. Bayer. Akad.* iii. 671, pro 1842 and 1843.) *München,* 1843

Magnetismus der Erde. 8vo. (*Repertorium der Physik.*) *Berlin,* 1846

† Resultate des magnet. Observatoriums (*sic*) in München während der dreijährigen Periode 1843-45. 4to. 117 pp. 1 plate. *München,* 1847

† Handbuch d. Erdmagnetismus. 8vo. 264 pp. 6 plates. *Berlin,* 1849

† Beschreibung der an der Münchener Sternwarte zu den Beobachtungen verwendeten neuen Instrumente u. Apparate. 4to. 102 pp. 8 plates.
München, 1851

† Beobachtungen des meteorologischen Observatoriums auf dem Hohenfreissenberg von 1792-1850, herausg. von Dr. J. Lamont. 8vo. 787 pp. (1*ste Supplementband zu den Annalen der Münchener Sternwarte.*) *München,* 1851

Astronomie und Erdmagnetismus. 8vo. 289 pp. 3 plates. (*Aus der Neuen Encyclopedie fur Wissenschaften und Kunst besonders abgedruckt.*)
Stutgart, 1851

Bestimmung der Horizontal-Intensität des Erdmagnetismus nach absolutem Maase. 4to. (*Abhandl. Acad. Münch.* iii. 1851.) *München,* 1851

† Magnetische Ortsbestimmungen ausgeführt an verschiedenen Puncten des Königreichs Bayern, und an einigen auswärtigen Stationen. 1. Theil enthaltend die allgemeinen Grundlagen zur Bestimmung des Laufs der Magnetischen Curven in Bayern. 8vo. 199 and cccciv. pp. 18 plates.
München, 1854

† Magnetische Karten von Deutschland und Bayern, nach den neuen Bayerischen und Oesterreichischen Messungen. Fol. 16 pp. 6 maps.
München, 1854

Theorie der Magnetisirung des weichen Eisens durch den galvanischen Strom. 8vo. 1854
† Magnetische Ortbestimmungen ausgeführt an verchiedenen Puncten des Königreich Bayern, &c. 8vo. 191 and ccxcii. pp. 26 lith. Tapelen.
München, 1856

Het magnetismus der aarde, populair beschreven. Uit het Hoogduitsch vertaald door W. F. Kaiser, &c. Met eene voorrede en een bijvoegsel, over het magnetismus der zon en maan, door F. Kaiser. 8vo. 112 pp. 1 plate.
Zwolle de Erven, 1856

Lamont, Johann.—*continued.*

Resultate aus den an d. K. Sternwarte veranstaltenen meteorolog. Untersuchungen. 4to. *München*, 1857

† Untersuchungen über die Richtung und Stärke des Erdmagnetismus an verschiendenen Puncten des südwestlichen Europa... 4to. 198 and cxv. pp. 13 plates. *München*, 1858

† Untersuchungen über die Richtung und Stärke des Erdmagnetismus in Nord-Deutchland, Belgien, Holland, Dänemark im Sommer des Jahres 1858, ausgeführt und auf öffentliche Kosten herausgegeben. 4to. 91 and xlv. pp. 3 plates. 6 maps. *München*, 1859

Der Erdstrom und der Zusammenhang desselben mit dem Magnetismus der Erde. 4to. 74 pp. 1 plate. *Leipzig*, 1862

† Beitrag zu ein mathem. Theorie d. Magnetismus. 8vo. 18 pp. (*Sitzungsb. d. K. Bayer Akad. d. Wissens.* vom. 11 Juli, 1862, p. 103, ii. Heft ii.) *München*, 1862

† Ueber die zehnjährige Periode in der täglichen Bewegung der Magnetnadel, und die Beziehung des Erdmagnetismus zu den Sonnenflecken. 8vo. 11 pp. (*Sitzungsb. d. K. Bayer Akad. d. Wissen.* vom 14 Juni, 1862, p. 66, ii. Heft ii.) *München*, 1862

† Der Erdstrom und die Telegraphenströme. 8vo. 32 pp. (*Zeitschrift de Östterreichischen Gesellschaft fur Meteorologie*, ii. 1. *Prolenum* No. 1, 1st January, 1867.) *Wien*, 1867

† Handbuch des Magnetismus. 8vo. 468 pp. (*Allgemeine Encyclopædie der Physik*, xv. Band, (*i.e.*) *completed in* vol. xv.) *Leipzig*, 1867

(*Note.—Pp.* 1 *to* 203, *Bd.* ii. *printed in* 1863; *pp.* 209 *to* 256, *Bd.* vii. *printed in* 1864; *pp.* 257 *to* 416, *Bd.* vii. *printed in* 1864; *pp.* 417 *to* 469, *Bd.* xv. *printed in* 1867.)

Many observations, &c. in the Academies of Munich, Brussels, &c., and in *Poggend. Ann.*

Lamothe, Sec. de l'Acad. de Bordeaux. Table méthodique des publications de l'Acad. de Bordeaux depuis son origine. 8vo. (*Actes de l'Acad. de B.* 9me *Ann.* p. 751.) *Bordeaux*, 1847

† **La Motte, H. J.** De. Erfahrung von einem durch die Electricität gehobenen krampfichten Mutterbeschwer, angestellet. 4to. 8 pp. (*Versuche d. Naturf. Gesellsch. in Danzig*, ii. 552.) *Danzig*, 1754

Lamoureux, J. V. F. Account of a Fall of Stones at Agen on the 5th Sept., 1814. (Transmitted to the Institut. with specimens.) 8vo. 1 page. (*Phil. Mag.* xliv. 316.) *London*, 1814

Lampadius, Wilhelm August. *Born August* 8, 1772, *at Hehlen Herzogth, Brunswick; died April* 13, 1842, *at Freiberg.*

† Versuche und Beobachtungen über die Elektricität und Wärme der Atmosphere, nebst einer Theorie der Luftelektricität, &c... 8vo. 200 pp. 3 leaves of Tables. *Leipzig*, 1793

(*Note.—The title-page dated* 1805 *is false.—F. R.*)

Versuche und Beobachtungen über Elektricität und Warme der Atmosphere, nebst einer Theorie der Luftelektricität nach De Luc, &c... 8vo. *Leipzig*, 1804

† System. Grundriss d. Atmosphärologie nebst Litteratur derselben. 8vo. 392 pp. *Freybourg*, 1806

Beyträge zur Atmosphärologie. Ein Nachtrag zum Grundrisz d. Atmosphärologie. 8vo. *Freybourg*, 1817

† Grundriss d. Electro chemie. 8vo. or 12mo. *Freybourg*, 1817

Many articles in the Journals of Crell, Tromsdorff, Gehlen, Schweigger, Gren, Gilbert, Kastner, Erdmann, and others. Also some translations.

Lamy, François. *Born 1636, at Schloss Montereau, Perche; died April 11, 1711, at Abtei, St. Denis. (Poggendorff, i. 1363.)*

Conjectures physiques sur deux colomnes (*sic*) de Nuë qui ont parus depuis quelques années; et sur les plus extraordinaires effets du Tonnerre: avec une explication de ce qui s'est dit jusqu'icy des Trombes de mer: et une nouvelle addition, ou l'on verra de quelle manière le Tonnerre tombé nouvellement sur une Eglise de Lagni a imprimé sur une nappe d'autel une partie considérable du Canon de la Messe. 12mo. 241 pp. 3 plates.

Paris, 1689

A German account of the extraordinary effects of Lightning at the Church of Lagni, printed in the *Hamburg Mag.* iii. 226, and taken from Lamy's French work "Conjectures Phys.".. dated 1696, 12mo. is given in Bauer, Abhandl. 1770, p. 161.

(*Note.—This German account contains a copy of "la partie considérable du Canon de la Messe," which was found printed by lightning upon the altar-cloth, and also of certain parts in red ink not thus reprinted.*)

Lamy, M. Note sur les courants électriques engendrés par le Magnétisme terrestre. 8vo. 16 pp. (*Extrait des Mémoires de la Société Impériale des Sciences, de l'Agriculture et des Arts de Lille.*) *Lille,* 1858

Lana (Lana Terzi) Francesco de. *Born December 13, 1631, at Brescia; died February 26, 1687, at Rome.*

Observationes mutationis declinationum magneticarum in eodem locum (*sic*), simul cum inventione qua ipsæ declinationes exactius in posterum observari possunt. (*Acta Nova Acad. Philexoticorum,* No. x.) 16—

Prodromo di alcune inventioni nuove. Folio. *Brescia,* 1670
(*Note.—Method for writing in cypher for telegraphs, and many other inventions.*)

"Eine Mittheilung von ihm in Betreff der Aufhängung von Magnetnadeln an Seidenfäden." (*Acta Erud.* 1686, p. 560.) 1686

Nova methodus construendæ pyxidis magneticæ, &c. (*Acta Nova Acad. Philexoticorum,* No. xi.) 168—

Magisterii naturæ et artis. Third part, 1692, 22nd book, pp. 287-312 (Parmæ) de motu quem vocant attractionis electricæ. 3 vols. Folio.
Parma and Brixia, 1684-92

Lande. (*Vide* Lalande.)

† **Landgrebe, Georg.** Quædam de electro-chimismo. Dissertatio inauguralis. 8vo. 44 pp. *Marburgi,* 1825

† Naturgeschichte der Vulcane und die damit in verbindung stehenden Erscheinungen. 2 vols. 8vo. *Gotha,* 1855

† **Landi, T. and Falconieri, C.** New process for laying the Submarine Telegraphic Cable, proposed by Thos. Landi and Chas. Falconieri. 8vo. 8 pp. 1 plate.
Paris

Lando. (*Vide* Mongiardini and Lando.)

† **Landois, H.** Exposé des causes de la coloration des corps et des lois qui régissent la reproduction des couleurs, et Traité de l'électricité, du calorique, de la lumière, suivi de quelques mots sur le magnétisme animal. 8vo. 35 pp. 1 plate. *Paris,* 1857

† **Landriani, J. B.** Nova electricitatis theoria, quam .. in Universitate D. Alexandri .. J. B. Landriani propugnabat .. 8vo. 91 pp. 1 plate.
Mediolani, 1755

Landriani, Marsiglio. Ricerche .. salubrità dell' aria. 8vo. 92 pp. 2 plates.
Milano, 1775

† Osservazioni sulla poca o nessuna affinità che ha l'umido aereo colle materie resinose, e specialmente collo zolfo. 12mo. 11 pp. (*Scelta d'Opuscoli in* 12mo. xv. 102.) *Milano,* 1776

LAN 285

Landriani, Marsiglio.—*continued.*

† Lettera al . Volta. (On an improved Electrophorus.) 12mo. 14 pp. 1 plate.
(*Scelta d' Opuscoli in* 12mo. xix. 73.) *Milano,* 1776

† Gli effetti del Fulmine caduto la sera del 25 Agosto, 1780, nel campanile e
Monastero di S. Vincenzo al Castello in Milano. 4to. 6 pp. (*Opus.
Scelti,* iii. 328.) *Milano,* 1780

† Dell' utilità dei Conduttori elettrici. 8vo. 304 pp. 1 plate. *Milano,* 1784

Abhandlung vom Nutzen der Blitzableiter. Auf Befehl der Gubernims
herausgegeben. Aus dem Italianischen von G. Muller. 8vo. *Wien,* 1786

† Altra ricaduta del propagatore . . ossia ultima risposta contro la difesa dei
Paragrandini. Letta all' Ateneo di Venezia. 8vo. 60 pp. *Milano,* 1826

Von einigen Entdeckungen in der thierischen Elektricität. (*Mayer's Samml.
Phys. Aufs. der Gesellsch. Böhmischer Naturf.* B. iii. p. 384.) *Prag*

Ueber die magnetische Eigenschaft des Kobolt-Königs. (*Mayer's Samml.
Physik. Aufsäze der Gesellsch. Böhmischer Naturf.* iii. 388.) *Prag*

Lane, Timothy. *Born June,* 1734; *died July* 5, 1807.

† Description of an Electrometer, invented by Mr. Lane, with an account of
some experiments made by him with it, in a letter to B. Franklin . . 4to.
12 pp. 1 plate. (*Phil. Trans.* lvii. p. 451.) *London,* 1768

"Stellte eine Reihe chemisch magnetischer Versuche an." 8vo. (*Monthly
Magazine, Dec.* 1805.) *London,* 1805

† On the Magnetic attraction of Oxides of Iron. From the *Phil. Trans.* for
1805. 8vo. 3 pp. (*Phil. Mag.* xxiii. 253.) *London,* 1806

Lang, Victor Von. Untersuchungen über die physicalischen Verhältnisse krystal-
liserter Körper. 1. Orientirung der optis. Elasticitätsaxen in den
Krystallen des Rhombischen Systems. Zweite Reihe. 8vo. 45 pp. 5 plates.
(*Sitzungb. Wien Acad.* vol. xxxi. No. 18.) *Wien,* 1858

† Über der Einstatit im Meteoreisen von Breitenbach. 8vo. 9 pp. 1 plate.
(*Sitzungsber. K. Akad.* lix. Bde. ii. Abth. Apl. Heft, 1869. *Vorgelegt*
29 *April,* 1869.) *Wien,* 1869
(*Vide* also Grailich and Lang.)

Langberg, Lorenz Christian. *Born March* 18, 1810, *at Christiansund; died March*
21, 1857, *at Christiania.*

† Jagttagelser over den magnetiske Intensitet paa forskjellige Steder af Europa.
4to. (small). 25 pp. (*Nyt. Magaz. f. Naturvidenskab.* vol. v.)
Kjöbenhavn, 1848

Magnetiske Jagttagelser paa en Reise i Christiansands Stift i Sommeren 1848.
4to. (*Nyt. Magaz. f. Naturvidenskab.* vi. 1851.) *Kjöbenhaven,* 1851

Langenbucher, Jacob. *Born at Augsburg; died* 1791, *at Augsburg.*

Beschreib. e. Elektrisirmachine. (*From Young,* p. 431.) 1778

† Beschreibung einer beträchtlich verbesserten Elektrisirmachine, sammt vielen
Versuchen . . 8vo. 268 pp. 8 plates. *Augsburg,* 1780

Richtige Begriffe vom Blitz und von Blitzableitern. 8vo. *Augsburg,* 1783

Praktische Elektricitäts-lehre. 8vo. *Augsburg,* 1788

Langguth. Ueber die Bewegung der Electricität in Körpern, welche eine constante oder mit der Richtung veränderliche Leitungsfähigkeit besitzen. 4to.
24 pp. *Greifswald,* 1865

Langguth, G. A. *Born December* 26, 1754, *at Wittenberg; died February* 9, 1814
at Wittenberg. (*Poggendorff,* i. 1370.)

De torpedine veterum genere raja. 4to. (*From Dove,* p. 207.) *Wittemberg,* 1784

Opuscula historiam naturalem spectantia. (*From Dove,* p. 205.)
Wittemberg, 1784

Langguth, G. A.—*continued.*

De torpedine recentiorum genere anguilla.　4to.　(*From Dove*, p. 205.)
Wittemberg, 1788

Magnetische u. meteorologische Beobachtungen.　8vo.　(*Gilbert, Ann.* xiii.
1803.)　　　　　　　　　　　　　　　　　　　　　　　　　　　　　*Leipzig*, 1803

Langhansen, C.　De aurora boreali d. 17 Martii observata.　(*From Poggendorff*, i.
1371.)

Langsdorf, W.　Silber als Einheit fur d. Messung d. elektrisch. Leitungswiderstandes.　　　　　　　　　　　　　　　　　　　　　　　　　　*Heidelberg*, 1851

Langworthy.　View of the Perkinian Elect.　8vo.　　　　　　　　　　　1798

† **Lanjuinais, T. D.**　Sur le mémoire sur la Boussole des Orientaux, par T. Hager.
8vo.　8 pp.　(*Extrait du Moniteur*, No. 231, 1809.)　　　　　*Paris*, 1809

Lanteires, Jean.　Essai sur le Tonnerre considéré dans ses effets moraux sur les
hommes, et sur un coup de foudre remarquable ; suivis des notes communi.
quées à l'auteur par M. le Professeur Saussure, à Genève.　8vo.
Lausanne, 1789

La Parola.　Foglio di Scienze, Arti, e Belle-lettere, &c.　4to.　*Bologna*, 184-?

La Peyrouse, Philippe.　(Picot de la Peyrouse.)　*Born October* 20, 1744, *at
Toulouse ; died October* 18, 1818.

† Description d'un Météore singulier du 24 Juillet, 1790.　4to.　2 pp.　(*Toulouse
Acad.* 1re série, iv. 189.)　　　　　　　　　　　　　　　　　*Toulouse*, 1790

La Place, Cyrille Pierre Théodore.　*Born November* 7, 1793, *on the sea.*

Voyage autour du Monde par les mers de l'Inde et de la Chine ; exécuté sur
la corvette de l'Etat la "*Favorite*," pendant les années 1830, 1831, et 1833.
5 vols. et atlas.　8vo.　　　　　　　　　　　　　　　　　　*Paris*, 1833-39

Campagne de circumnavigation de la frégate "*l'Artémése*" pendant les années
1837-38-39-40, sous le commandement de M. Laplace, capitaine de vaisseau,
publié par ordre du Roi . .　8vo.　　　　　　　　　　　　　　*Paris*, 1841

La Place, P. S. et Lavoisier, A. L.　Sur l'électricité qu'absorbent les corps qui
se réduisent en vapeurs.　4to.　(*Mém. Paris*, 1781.)　　　　*Paris*, 1781

† **Lapostolle.**　Traité des Parafoudres et des Paragrêles en cordes de paille . .　8vo.
320 pp. &c., also 3 supplements.　　　　　　　　　　　　　*Amiens*, 1820

† Trattato sul modo di preservare le abitazioni dal Fulmine e le campagne dalla
Grandine.　Opera volgarizzata da Bodei.　8vo. 189 pp. 1 table.
Milano, 1821

† **Laprade.**　Mémoire . . sur la question . . quel sont les effets que produisent
les Orages sur les animaux . .　8vo. 140 pp.　(*Read in the Soc. de Med.,
Bruxelles*, in 1809.)　　　　　　　　　　　　　　　　　　　*Bruxelles*

Larcher, Daubencourt, et Zanetti, aîné.　Observations chimiques faites (par
L. &c.) sur différents liquides animaux soumis à l'action galvanique; lues
à l'Institut.　8vo.　(*Ann. de Chim.* xlv. p. 195.)　　　　　　　*Paris*

(*Vide* also Cassius and others.)

Lardner, Dionysius.　*Born April* 3, 1793, *at Dublin ; died April* 29, 1859, *at Paris.*
(*Poggendorff*, i. 1379 and 1580.)

(*Note.*—*It is difficult to find the dates, &c. of the several editions of his various
works.—F. R.*)

† Manual of Electricity, Magnetism, and Meteorology.　Vol. i. 12mo. 439 pp.
(*Vide* Lardner and Walker for vol. ii.)　　　　　　　　　*London*, 1841

Handbook of Philosophy and Astronomy.　3 vols. 1851-53.　Vol. ii. Heat,
Electricity, and Magnetism.　　　　　　　　　　　　　　*London*, 1851-3

Lardner, Dionysius.—*continued.*

Museum of Science and Art. 12 vols. Post 8vo.   *London*, 1854-6
Vol. I. contains : Meteoric Stones and Shooting Stars.
    III.   ,,   Electric Telegraph.
    IV.   ,,   Ditto (concluded).
    X.   ,,   Electro Motive-power—Thunder, &c., and the Aurora.

Populäre Lehre von den electrischen Telegraphen. . . Für angehende Telegraphisten, Eisenbahnbeamte, Techniker, etc. deutsch bearbeitet von Carl
Hartmann. 8vo. 137 pp. 5 plates.   *Wiemar*, 1856

Handbook of Natural Philosophy (4 vols.) Vol. iv. Electricity, Magnetism,
and Acoustics. 12mo.   *London*, 1856

† The Electric Telegraph popularised. 8vo.   *London*, 1859

† Handbook of Electricity, Magnetism, and Acoustics. Seventh thousand.
Edited by G. C. Foster. 8vo. 442 pp.   *London*, 1866

† Lardner and Walker. Manual of Electricity, Magnetism, and Meteorology.
Vol. ii. 12mo. 550 pp.   *London*, 1844

La Rive, Auguste Arthur De. *Born October 9, 1801, at Geneva*

† De l' action qu'exerce le globe terrestre sur une portion mobile du circuit voltaïque. (Lu 4 Septembre, 1822, Ampère assistait.) (*Bibliot. Universelle*,
Sept. 1822.) (*From Arago's copy.*)   *Genève*, 1822

† Mémoire sur quelques-uns des phénomènes que présente l' électricité voltaïque
dans son passage à travers les conducteurs liquides. 8vo. 32 pp. (*Ann. de
Chim. et de Phys.* 1825.) (*Extrait.*)   *Paris*, 1825

† Recherches sur le mode de distribution de l'électricité dynamique dans les
corps qui lui servent de conducteurs. (Read December, 1823.) 4to. 21 pp.
(*Mém. de la Soc. de Gen.* tom. iii. pt. i. p. 109.)   *Genève*, 1825

Analyse des circonstances qui déterminent le sens et l'intensité du courant
électrique dans un élément voltaïque. 8vo. (*Ann. de Chim.* xxxvii. 225.)
                                                 Paris, 1828

† On the power of Water and Bromine in conducting Electricity. 8vo. 2 pp.
(*Phil. Mag. or Annals*, iii. 151, and *Quarterly Journal*, 1828, xxxv. 161.)
                                                 London, 1828

† Recherches sur une propriété particulière des conducteurs métalliques de l'é-
lectricité. (Mém. lu 22 Juin, 1826.) (*Geneva Soc. de Phys.* iii. part ii.)
                                                 Genève, 1826

Sur les effets calorifiques de la pile. 8vo. (*Bibl. Univ.* xl. 1829.) *Genève*, 1829

† Action of Sulphuric Acid upon Zinc, and causes producing Electricity. (Letter
from De la Rive to the French Academy, read by Arago.) 8vo. 4 pp. (*Phil.
Mag. or Annals*, viii. 298.)   *London*, 1830

† Esquisse historique des principales découvertes faites dans l'électricité depuis
quelques années. 8vo. 239 pp. (*Bibliothèque Univ. de Genève*, 1833.)
                                                 Genève, 1833

(*Note.—This " coup d'œil "* . . *is in his Archives de l' Electricité*, tom. i. p. 5,
1841.) (*Faraday's copy.*)

De l'électricité développée par le frottement des métaux. 8vo. (*Bibl. Univ.*
lix. 1835.)   *Genève*, 1835

† Recherches sur la cause de l'Electricité voltaïque. 4to. 174 pp. 1 plate (in
2nd part). In 3 parts. (*Extrait des Mém. de la Société de Physique de Genive.*)
1st part read Nov. 20, 1828, printed in tom. iv. p. 285, 1828.
2nd part read   , printed in tom. vi. p. 149, 1833.
3rd part read April 16, 1835, printed in tom. vii. p. 457, 1836.
The 3rd part read also in *Acad. de Paris*, July 22, 1834.
The volume (3 parts) seems to be a reprint (with date 1836) and with *variations* in 3 parts.   *Genève*, 1836

La Rive, Auguste Arthur De.—*continued.*

De l'Electricité développée par la désoxydation de certaines substances minérales. 8vo. (*Bibl. Univ.* Nouv. Sér. i. 1836.) *Genève,* 1836

† Researches into the cause of Voltaic Electricity. 8vo. 26 pp. *London,* 1837

† De l'influence qu'exerce la chaleur sur la facilité que le courant électrique possède à passer d'un liquide dans un métal. 8vo. 8 pp. (*Tiré de la Bibliothèque Universelle de Genève,* Février, 1837.) *Geneva,* 1837

† Des mouvements vibratoires que déterminent dans les corps, et essentiellement dans le fer, la transmission des courants électriques et leur action extérieure. 8vo. 35 pp. (*Extrait des Archives de l'Electricité,* No. 17. *Supplément à la Bibliothèque Universelle de Genève.*) *Genève*

† Recherches sur les propriétés des courants magnéto-électriques. 4to. 56 pp. (*Mém. lu à la Société de Physique de Genève, le* 16 *Avril,* 1837.) *Genève,* 1839

† Notice sur un procédé électro-chimique ayant pour objet de dorer l'argent et le laiton. 8vo. 19 pp. (*Bibl. Univ.* Fév. 1840.) (*Arago's copy.*) *Genève,* 1840 (*Note.—No allusion is made to Brugnatelli, the real inventor.*—F.R.)

† Mémoire sur quelques phénomènes chimiques qui se manifestent sous l'action des courants électriques développés par Induction. Lu en deux fois—le 4 Septembre, 1838, et le 17 Mai, 1840. 4to. 74 pp. (*Geneva Soc. de Phys.* ix. pt. i. p. 161.) *Genève,* 1841

† Archives de l'Electricité. Supplément à la Bibliothèque Universelle de Genève. 5 vols. 8vo. Plates. *Genève,* 1841—45

† Coup d'œil sur l'état actuel de nos connaissances en électricité. 8vo. 30 pp. (*Extrait des Archives d'Electricité,* tom. i.) *Genève,* 1841

† Nouvelles recherches sur les propriétés des courants électriques, discontinus et dirigés alternativement en sens contraires. 8vo. 80 pp. (*Extrait des Archives de l'Electricité,* 1841.) *Genève,* 1841

(*Vient de paraître également dans la première partie du* tom. ix. *des Mémoires de la Société de Physique de Genève, sous le titre de "Quelques phénomènes qui se manifestent sous l'action des courants électriques développés par induction." J'y ai fait seulement quelques additions et quelques changements de forme, &c.*)

† Des effets calorifiques de l'Electricité. 8vo. 8 pp. (*Archives de l'Electricité,* tom. ii. p. 501.) (*Arago's copy.*) *Genève,* 1842

† Observations sur une Note de M. Poggendorff relative à l'hypothèse d'un contre-courant dans la Pile . . 8vo. 10 pp. (*Archives de l'Electricité,* No. 6, tom. ii. p. 481.) (*Arago's copy.*) *Genève,* 1843

Sur l'éclairage des Mines au moyen de la lampe électrique. 4to. (*Comptes Rendus,* xxi. 634.) *Paris,* 1845

De l'action combinée des courants d'induction et des courants hydro-électriques. (Mémoire lu 21 Mars, 1844.) 4to. 27 pp. (*Geneva Soc. de Phys.* xi. pt. i. p. 225.) *Genève,* 1846

† Quelques recherches sur l'Arc voltaïque, et sur l'influence qu'exerce le magnétisme soit sur cet arc, soit sur les corps qui transmettent les courants électriques discontinus. Communiqué à la Société Royale de Londres, 19 Nov., 1846. 8vo. 30 pp. (*Bibl. Univ. Soc. de Phys.* iv. 345.) *Genève,* 1846

† Recherches sur les phénomènes moléculaires qui accompagnent la production de l'Arc voltaïque entre deux pointes conductrices. 4to. 5 pp. (*Compte Rendus,* xxii. No. 17.) *Paris,* 1846

† Researches on the Voltaic Arc, and on the influence which Magnetism exerts both on this Arc and on bodies transmitting interrupted electric currents. (Read Jan. 7, 1847.) 4to. 13 pp. (*Phil. Trans. for* 1847. pt. i.) (*Faraday's copy.*) *London,* 1847

† A Treatise on Electricity in theory and practice. Vol. i. 8vo. 564 pp. (*Translation from the French MSS., by C. V. Walker.*) *London,* 1853

La Rive, Auguste Arthur De.—*continued.*

Mémoire sur la cause des Aurores Boréales. (Communiqué à la Société en Déc. 1848 et Nov. 1853.) 4to. (*Geneva Soc. de Phys.* xiii. 373.) *Genève,* 1854

† Traité d'électricité théorique et appliquée. Tome i. 8vo. 620 pp. (*Contains matter written subsequently to Walker's translation of the MSS. in* 1853.) *Paris,* 1854

Des expériences de M. P. Volpicelli sur la polarité Electro-statique. Note de M. De la Rive. 4to. (*Extrait de la Bibliothèque Universelle de Genève, Archives des Sciences Physiques,* tom. xxviii. 4e Série, No. 112, Avril 1855, p. 265.) *Genève,* 1855

† A Treatise on Electricity in theory and practice. Vol. ii. 8vo. 892 pp. (*Translation by C. V. Walker.*) *London,* 1856

† Traité d'Electricité théorique et appliquée. Tome deuxième. 8vo. 856 pp. *Paris,* 1856

† Traité de l'Electricité théorique et appliquée. Tome troisième. 8vo. 788 pp. *Paris,* 1858

† A Treatise on Electricity in theory and practice. Vol. iii. 8vo. 818 pp. (*Translation by C. V. Walker.*) *London,* 1858

† Nouvelles recherches sur les Aurores Boréales et Australes, et description d'un appareil qui les reproduit . . 4to. 29 pp. 1 plate. (*Ext. des Mém. de la Soc. de Phys. et d'Hist. Nat. de Genève* xvi. 2e partie.) *Genève,* 1862

† Bibliothèque Universelle de Genève. 4e Série. Archives de l'Electricité. (*Supplément aux Archives des Sciences Phys. et Naturelles.*) (*Vide* Genève.)

† Recherches sur les phénomènes qui caractérisent la propagation de l'électricité dans les fluides élastiques très-raréfiés. 4to. 43 pp. *Genève,* 1863

Sur la Polarisation rotatoire magnétique des liquides. 8vo. *Genève,* 1870

† **La Rive et Marcet.** Quelques observations de physique terrestre faites à l'occasion de la perforation d'un puits artésien. (Lu 18 Avril, 1834.) 4to. 26 pp. 1 plate. (*Geneva Soc. de Phys.* vi. 503. *Magnetic Experiments at* p. 526.) *Genève,* 1833

(*Vide* also Macaire and De la Rive.)

† **Laroque.** Note sur des Eclairs de forme inusitée, observés à Toulouse pendant l'orage du 16 Juillet, 1850. 8vo. 3 pp. (*Toulouse Acad.* 3e Série, vi. 349.) *Toulouse,* 1850

† Recherches sur le Magnétisme. Première partie. (Lu 10 Août, 1854.) 8vo. 9 pp. (*Toulouse Acad.* 4e Série, iv. 423.) *Toulouse,* 1854

† Proposal of Prize, by the Acad. for "Recherches sur l'Electricité atmosphérique" for 1848. 8vo. (*Toulouse Acad.* 4e Série.) *Toulouse,* 1855

Larrey. Galvanic experiments on the Human Body. (*Bulletin de la Société Philomatique,* Mai et Juin, 1793, Nos. 23 et 24.) *Paris,* 1793

† **Larrey (le Baron.)** Notice sur l'efficacité du Moxa, et sur les inconvénients du Galvanisme dans certaines névroses. . . 4to. 22 pp. (*Acad. des Sciences,* xviii. 417.) 1840

† **Lartigue.** Observations sur les Orages dans les montagnes des Pyrénées. 8vo. 7 pp. (*Ext. Comptes Rendus,* 3 Déc. 1855.) *Paris,* 1855

† **La Rue, W. De.** On the structure of Electro-precipitated Metals. 8vo. 6 pp. 1 plate. (*Journal Chemical Soc.* Article cxxx. p. 300 of vol.) *London*

La Salle, A. De. Relation du voyage exécuté pendant les années 1836 et 1837 sur la corvette la "*Bonite.*" 2 vols. 8vo. *Paris,* 1845-52

La Salle, De, Darondeau et Chevallier. Voyage autour du monde exécuté pendant les années 1836 et 1837 sur la corvette la "*Bonite,*" commandée par M. Vaillant. . . Publié par ordre du Roi. Relation du voyage. 8vo. Atlas in folio. *Paris*
(*Vide* also Darondeau et Chevallier.)

† **Lassone. De.** Dissertation sur les effets de l'électricité, avec un détail des expériences faites à ce sujet sur des Paralytiques. 12mo. 10 pp. (*Recueil sur l'Electricité Médicale*, p. 245, tom. i.) *Paris*, 1763

† **Lasteyrie, F.** L'électrum des anciens, était-il de l'émail ? Dissertation sous forme de réponse à . . Labarte. 8vo. 78 pp. *Paris*, 1857

† **Laterrade, A.** De la Grêle, et des moyens d'en combattre les effets. 8vo. 42 pp. *Condom*, 1847

† De la Grêle, et des moyens d'en combattre les effets. 2e édition, corrigée, etc. 8vo. 64 pp. *Paris*, 1848

† **Lathrop, Dr. John.** Fatal effects of Lightning. In a letter to . . Joseph Willard . . 4to. 7 pp. (*Mem. Amer. Acad.* Old Series, ii. pt. ii. p. 85.) *Charlestown*, 1804

† An account of the effects of Lightning on the house of Jn. Mason, in Boston, in a letter to Joseph Willard. . . 4to. 4 pp. (*Mem. Amer. Acad.* Old Series, ii. pt. ii. p. 91.) *Charlestown*, 1804

† Effects of Lightning on several persons in the house of Samuel Cary, of Chelsea, August 2, 1799, in a letter to John Davis. 4to. 4 pp. (*Mem. Amer. Acad.* Old Series, iii. pt. i. p. 82.) *Cambridge, U.S.* 1809

† Effects of Lightning on the house of Capt. D. Merry, and several other houses in the vicinity, on the evening of the 11th May, 1805, in a letter to John Davis. 4to. 6 pp. (*Mem. Amer. Acad.* Old Series, iii. pt. i. p. 86.) *Cambridge, U.S.* 1809

Latini, Brunetto. Le Trésor. De l'origine et de la nature de toutes choses. (*From Pogg.* i. 1383) *Paris*, 1260

Il tesoro. (*Translation in Italian.*) *Treviso*, 1474

(*Note.—Contains one of the oldest documents on the knowledge in Europe of the Compass.—F.R.*)

Laugier, André. *Born August 1, 1770, at Lisieux ; died April 19, 1832, at Paris.*

† Extract of a Memoir, by A. Laugier, on a new principle in Meteoric Stones. Read in the French National Institute, March 10, 1806. 8vo. 3 pp. (*Phil. Mag.* xxvi. 11.) (*Ann. de Chim.* lviii. 261.) *London*, 1807

† Expériences propres à confirmer l'opinion émise par des naturalistes sur l'identité d'origine entre le fer de Sibérie et les pierres météoriques, ou aerolithes. 4to. 12 pp. (*Geneva Acad.* iii. 341.) *Geneva*, 1818 ?

† Analysis of an Aerolite which fell at Jonsac 13th June, 1819. 8vo. 2 pp. (*Ann. de Chimie et de Phys.* xiii. 441, *and Phil. Mag.* lvi. 157.) *London*, 1820

† **Laugier, and others.** Meteorite which fell near Ferrara in 1824 (Analysis of). 8vo. 1 p. (*Phil. Mag. or Annals* ii. 70.) *London*, 1827

Laument. (*Vide* Gillet-de-Laument.)

Lausanne Society. Mémoires de la Société Physique de Lausanne. 4to. *Lausanne*, 1784 *to* 1790 ?

Laval, A. F. Voyage de la "*Louisiane*," fait par ordre du Roi en 1720. 4to. *Paris*, 1728

Laval. (*Vide* Croissant à Laval.)

Lavant. Sur un tremblement de terre, et sur des effets singuliers, etc., de la foudre. 4to. (*Toulouse Acad.* 1re Série, ii. 15 Hist.) *Toulouse*, 1784

† **Lavergne.** Rapport sur l'union générale . . contre la Grêle. 8vo. 31 pp. (*Société d'Agriculture de la Gironde.*) *Bordeaux*, 1846

† **Laverine.** Su alcune cure fatte in Como col voltaismo medico. 4to. 4 pp. (*Opus. Scelti*, xxii. 132. *Translation by Amoretti.*) *Milano*, 1803

Lavialle de Lameillière. Documents législatifs sur la télégraphie électrique en France, comprenant les lois, exposé des motifs, rapports et résumés des discussions aux Chambres, etc. ; précédés d'une introduction historique, 1841-54. 8vo. 396 pp. *Paris*, 1865

† Documents législatifs sur la télégraphie en France (for 1855-1864). 8vo. *Paris*

† **Lavini.** Analisi chimica esplorativa e proporzionale di un Meteorite caduto nel mese di Luglio 1840 a Cereseto nelle vicinanze di Casale e Moncalyo. (Letta il 20 Decem. 1840.) 4to. 9 pp. (*In* Serie ii. tom. iii.)

Lavinij, A. Sp. Meteorologia sopra una caduta di aeroliti. (*From Schmidt's Cat.* ccxxxx. 1865.)

Lavoisier, A. L. Opuscules physiques et chimiques. 8vo. (2nd edition, 1801.)
Paris, 1774

† Rapport des Commissaires chargées par le Roi de l'examen du magnétisme animal. 4to. 66 pp. *Paris*, 1784

Traité élémentaire de chimie présenté dans un ordre nouveau, et d'après les découvertes modernes. 2 vols. 8vo. (3rd edition, 1801.) *Paris*, 1789

Mémoires de Chimie. 2 vols. 8vo. *Paris*, 1803
(*Vide* also Laplace and Lavoisier.)

† **Lawrance, R. M.** On the application and effect of Electricity and Galvanism in the treatment of cancerous, nervous, rheumatic, and other affections. 8vo. 101 pp. *London*, 1853

Electro-Magnetism in the treatment of Rheumatic and Paralytic affections. 8vo. (*From the "Times,"* Nov. 28, 1853.) *London*

† Gout and Rheumatism : and the curative effects of Galvanism. 2nd edition. 12mo. 196 pp. *London*, 1855

Paralysis : its treatment by Galvanic Electricity. (*From the "Times,"* Feb.17, 1857.) *London*, 1857 ?

Galvanism : its medical applications and uses. (*From the "Times,"* Feb. 17, 1857.) *London*

† On localised Galvanism applied to the treatment of Paralysis and Muscular Contractions. 8vo. 164 pp. *London*, 1858

† **Lawson, H.** Atmospherico-Electrical Observations from 24th to 26th December, 1821. 8vo. 1 p. (*Phil. Mag.* lix. 71.) *London*, 1822

Lawson, Jos. *Meteorological Register,* 1831-1842 inclusive. 8vo. *Washington*, 1851

La Science. Journal edited by Du Moncel.

† **La Beaume.** Remarks on the History, &c. . . Medical efficacy of Electricity . . Observations on Galvanism. Second edition, enlarged. 8vo. 368 pp. 2 plates. *London*, 1820

† On Galvanism, with observations on its chemical properties and medical efficacy in chronic diseases. 12mo. 271 pp. 3 plates. (*For French translation vide Fabré-Palaprat.*) *London*, 1826

† On Galvanism, and its efficacy in the cure of Indigestion, &c. . . 12mo. 36 pp. *London*, 1848

(*Vide* also Fabré-Palaprat and La Beaume.)

Lebaillif. On repulsion of Bismuth and Antimony by the Magnet. (*From Matteucci.*)

Le Blanc. Rapport sur l'Electricité Médicale. 1819

† **Le Boucher, A.** Guide télégraphique à l'usage des fonctionnaires, employés, et agents de cette administration, contenant un dictionnaire raisonné de toutes les questions et matières se rattachant au service. 8vo. 232 pp. *Paris*, 1863

† **Le Boulangé, P.** Mémoire sur le Chronographe électro-balistique. 8vo. 78 pp. 1 plate. *Paris*, 1864

Lebouvier-Desmortiers. (*Vide* Desmortiers.)

Le Camus. Ergo a fluido electrico vita, motus, et sensatio. 59 pp.   (*From Bonne-foy*, p. 59.)   *Paris*, 1761

Dissertatio an a fluido electrico vita, motus, et sensatio. (*From Krunitz*, p. 159.)
Paris, 1761
(*Vide* also Le Comus and Ledru.)

Le Cat, Claude Nicolas. *Born September* 6, 1700, *at Blérancourt, Picardie; died August* 20, 1768, *at Rouen.* (*Poggendorff*, i. 1400.)

Mémoire sur l'Electricité.   Lu dans une séance de l'Académie de Rouen le 12 Juillet, 1746.   1746

† A Memoir on the Lacrymæ Batavicæ, or Glass-drops, the tempering of Steel, and effervescence accounted for by the same principle.   Translated from the French by T. S.   4to.   (*Phil. Trans.* xlvi. 175.)   *London*, 1749-50

Physical Essay on the Senses.   8vo.   1750

Premier Mémoire sur l'Electricité.   Lu à l'Assemblée de l'Académie Royale des Sciences de Rouen en 1745.

(*Gralath makes a very slight reference* (*at p.* 312) *to a gold-leaf experiment by Le Cat, and to his opinion relative to the motions of the planets, contained in this Memoir.   Mangin, in his* (*anonymous*) *Histoire de l'Electricité* (vol. i. 84), *refers to a Memoir of Le Cat, read* 12*th July,* 1746, *at Rouen, containing the celebrated gold-leaf experiment, and gives a full account of it, &c.*)   *Londres*, 1751

Dissertation sur le principe de l'action des muscles.   4to.   *Berlin*, 1754

A posthumous Memoir of his on Spontaneous Animal Combustion, said to have been presented to the Société Médicale d'Emulation by Sig. Lair, Secretary of the Agricultural Society of Caen, who published in 1800 an Essay. (*Brera's Gle. de Med. Prat.* iv. 316.)   *Verona*, 1813

† Traité de l'existence, de la nature, et des propriétés du Fluide des Nerfs, et principalement de son action dans le mouvement musculaire, ouvrage cou-ronné en 1753 par l'Académie de Berlin; suivi des Dissertations sur la sensibilité des Méninges, des Tendons, etc., l'Insensibilité du Cerveau, la Structure des Nerfs, l'Irritabilité Hallerienne, etc.   8vo.   331 pp.   6 plates.
Berlin, 1765

Leche, Jn. Auszug der zwölfjährigen Witterungs Beobachtungen zu Abo, sechstes und letztes Stück.   8vo.   (*K. Schwed. Akad.* Abh. xxv. 275.)
Hamburg and Leipzig, 1763

Lechman. On the Tourmaline.   (*Berlin Acad. prob.* 1756.)   *Berlin*, 175-?

Leclanché. Notizen über elektrische Säulen und deren Anwendung beim Tele-graphenwesen mit besonderer Berücksichtigung der Säule "Leclanché," mit Mangan hyperoxyd und einer Flüssigkeit.   8vo.   15 pp.   *Wien*

Le Comus (i.e. **Le Dru**).   Permeability of Glass.   (*Roux. Journal de Med.* Sept. et Octobre, 1774.)   1774

† Nuove esperienze elettriche fatte 5 Feb. 1775. . . Altre esperienze elettriche fatte 27 Feb. 1775.   12mo.   13 pp.   (*Scelta d'Opuscoli* (in 12mo) xiii. 101 and 107.)   *Milano*, 1776

† Nuove esperienze elettriche fatte 4 Aprile e Giug. 1775.   12mo.   28 pp.   (*Scelta d'Opuscoli* (in 12mo.) xiv. 3, 11, 15, 21.)   *Milano*, 1776

† Osservazioni esperienze . . su l'elettricità medica.   Continuazione dell' analisi elettrica sulle sostanze animali.   12mo.   16 pp.   (*Scelta d'Opuscoli*, xxi. 70.)
Milano, 1776

(*Vide* also Ledru and Le Camus.)

Lecoq, Henri.   *Born April* 14, 1802, *at Avesnes* (*Nord*).   (*Poggendorff*, i. 1402.)
(*Note.—Another Henri Lecocq born in* 1810.)

Observations at Clermont, &c. on the Puy-de-Dome in July 28, 1835, &c., of Hailstones.   Concurrence of Electricity in formation of Hail, &c.

L'Economista. Giornale di Agricoltura, &c. . . Cattaneo Direttore. Folio.
Milano, 1843-47

† **Lecount, P.** Description of the changeable Magnetic properties possessed by all iron bodies . . . effects on Ships' Compasses. 8vo. 55 pp.
London, 1820

Ledru (called Comus), Nicolas Philippe. *Born* 1731, *at Paris ; died October* 6, 1807, *at Paris.*

Verfertigte das Inclinatorium zur Nordpol-Expedition des Capt. Phips.

Sur le traitement des épileptiques. 1783

(*Vide* also Le Comus, Le Camus, and Cosnier.)

† **Lee, A.** An account of the effect of Lightning on Two Houses in the city of Philadelphia, in a letter from A. Lee to James Bowdoin (dated July 29, 1781). 4to. 6 pp. (*Mem. American Acad.* Old Series, i. 247, part ii.)
Boston, U.S. 1785

† **Leeson.** Experiments for ascertaining the amount of pressure which would stop the electro-chemical decomposition of water by Voltaic Electricity of a certain tension. (Notice by Editor.) 8vo. 1 p. (*Ann. of Elect.* iv. 238.)
London, 1839

† **Lefroy, J. H.** Preliminary Report on the Observations of the Aurora Borealis made by the non-commissioned officers of the Royal Artillery, at the various Guard-rooms in Canada. 8vo. 10 pp. (*Phil. Mag.* 3rd Series, xxxvi. p. 457.)
London, 1850

† On the application of Photography to the self-registration of Magnetical and Meteorological instruments. 8vo. 16 pp. 1850

† On the irregular fluctuations of the Magnetical elements at the stations of Magnetical Observation in North America. 8vo. 13 pp. 1 plate. (*Proceedings of the American Association for the Advancement of Science,* 1851.)
Albany, U.S. 1852

† Second Report on Observations of the Aurora Borealis, 1850. 8vo.
Toronto, 1852

† **Lefroy and Richardson.** Magnetical and Meteorological Observations at Lake Athabasca and Fort Simpson, by Captain J. H. Lefroy ; and at Fort Confidence in Great Bear Lake by Sir John Richardson. Large 8vo. 391 pp. 2 plates. *London,* 1855

Legallois. Résumé succinct sur le galvanisme. (*Journ. du Galvanisme,* No. ii. p. 49, et No. iii. p. 97.) *Paris*

† **Legnazzi, L.** Relazione sulla applicazione dell' Elettricità alla macchina Jacquard, Osservazioni. 8vo. 16 pp. *Milano,* 1855

Legros. (*Vide* Onimus and Legros.)

Lehaitre. Une instruction théor. et prat. sur les Paragrêles. *Bourg,* 1825

Lehman, C. F. Wider d. Zeichensprache d. Hrn. Bergstrasser, &c. 8vo. 7 plates. *Danzig,* 1795

Lehmann, Johann Gottlob. *Died January* 22, 1767, *at St. Petersburg.*

† Abhandlung von Phosphoris. Sm. 4to. 36 pp. *Dresden u. Leipzig,* 1730

† De cupro et orichalco magnetico. 4to. 22 pp. (*Novi Comment. Acad. Petrop.* xii. *pro* 1766 et 1767. Imp. 1768, pp. 38 et 368.) *Petropoli,* 1766-67

Von magnet Theilen im Sande. (*Mém. de la Soc. de Haarlem,* tom. xi. part i.)
Haarlem, 1769

† **Lehmann, W. O. W.** Nogle oplysninger om Telegrafanlæget imellem Helsinger og Hamborg. 8vo. 41 pp. *Odense,* 1856

† **Lehot, C. J.** Observations sur le Galvanisme et le Magnétisme. 8vo. 8 pp. (*In Ann. de Chim.?*)

† Mémoire sur le Galvanisme. Lu à l'Institut Nat. le 26 Frimaire, an. 9. 4to. 15 pp. (*Journ. de Phys.* Pluviose, an. 9, lii. 135.) *Paris*, 1801

† **L'Hermite.** Phénomènes chimiques produits par les courants électriques. Thèse présentée à l'école de Pharmacie de Paris. 4to. 64 pp. *Paris*, 1847

Leibnitz. Opera Philosophica, quæ extant Latina, Gallica, Germanica, omnia instruxit, J. E. Eardmann. 8vo. *Berolini*, 1840

† **Leidenfrost, L. C.** Specimen inaugurale physicum sistens nonnulla miscellanea experimenta circa electricitatem. 4to. 36 pp. *Duisburgi ad Rhenum*, 1781

† **Leipzig.** Acta eruditorum Lipsiensia. 4to. 50 vols. *Lipsiæ*, 1682-1731

Nova acta eruditorum. 43 vols. *Lipsiæ*, 1732-1776

Actorum erud. supplementa. 10 vols. *Lipsiæ*, 1692-1734

Ad Acta nova erud. Supplementa. 8 vols. *Lipsiæ*, 1735-1757

Indices generales auctorum et rerum. 4to. 6 vols. *Lipsiæ*, 1692-1745
The whole consists of 117 *vols.*

Societatis naturæ scrutatorum Acta. 4to. *Lipsiæ*, 1822

Allegemeines Magazin der Natur-Kunst, und Wissenschaft. *Leipzig*

Leipzig Society. Abhandlungen der mathematisch-physischen Classe der König-lich Sächsischen Gesellschaft der Wissenschaften. 4to. *Leipzig*

Berichte über die Verhandlungen der Königl. Sächsischen Gesellschaft der Wissenschaften zu Leipzig. Mathematisch-physiche Classe. 8vo. *Leipzig*

† **Leithead.** Electricity: its nature, operation, &c., in the Phænomena of the Universe. 12mo. 399 pp. *London*, 1837

Lemaire. Magazino magnetico. (*Where first described?*) (*Vide Gatteschi.*)

† **Leman, S.** Considérations sur les Pierres, les masses de fer, et les poussières, dites météoriques. 8vo. 46 pp. (*Ext. du Nouveau Dict. d'Hist. Nat.* tom. xxvi.) *Paris*, 1818

Lémery, Louis. *Born January* 25, 1667, *at Paris; died June* 9, 1743, *at Paris.*

Observation sur une pierre de l'Isle de Ceylon qui attire et repousse différents corps, mais d'une manière différente de l'aimant. 4to. (*Mém. de Paris*, an. 1717; *Hist.* p. 7.) 4to. *Paris*, 1717

Lemoine-Moreau. (*Vide* Dureau.)

Lemolt et Archereau. Nouvelle pile de Bunsen. (*Vide* Moigno.)

Le Monnier, Louis Guillaume. *Born June* 27, 1717, *at Paris; died September* 7, 1799, *at Montreuil.*

A letter from Paris, containing some new Experiments in Electricity. 4to. (*From Krunitz.*) *London*, 1746

Recherches sur la communication de l'électricité. 4to. (*Mém. de Paris*, an. 1746, *Méms.* p. 447; *and in Phil. Trans.* an. 1746, p. 290.) *Paris*, 1746

Observations sur l'électricité de l'air. 4to. (*Mém. de Paris*, an. 1752, *Hist.* p. 8, *Mém.* p. 233.) *Paris*, 1752

Le Monnier, P. Observations et expériences sur les propriétés de l'aimant. 4to. (*Mém. de Paris*, an. 1733, *Hist.* p. 13.) *Paris*, 1733

Variations de l'aimant à Paris. 4to. (*Mém. de Paris*, 1770, 1771, *Mém.* pp. 459 (1771), 93, 95.) *Paris*, 1770-71

Recherches sur les variations horizontales de l'aimant. 4to. (*Mém. de Paris*, an. 1772, pt. i. *Mém.* p. 157.) *Paris*, 1772

Suite des recherches sur les variations de l'aimant aux chaines des montagnes en Normandie et d'abord dans l'Apennin. 4to. (*Mém. de Paris*, an. 1772, pt. ii. *Mém.* p. 457.) *Paris*, 1772

Le Monnier, P.—*continued.*

Articles "Aimant," et "Aiguille aimantée" in the French Encyclop. (*From Lamont, Handb.* p. 439.)

Remarques sur la carte suédoise de l'inclinaison de l'aimant, publiée à Stockholm dans le trimestre de Juillet des Actes de l'Acad. an. 1768. 4to. (*Mém. de Paris,* an. 1772, pt. ii. *Mém.* p. 461.) *Paris,* 1772

Mém. sur la variation de l'aimant en 1772 et 1773. 4to. (*Mém. de Paris,* an. 1773, *Mém.* p. 440.) *Paris,* 1773

Mém. sur la variation de l'aimant en 1773 et 1774 au jardin du Temple et à l'Observatoire Royal. 4to. (*Mém. de Paris,* an. 1774, *Mém.* p. 237.) *Paris,* 1774

† Loix du Magnétisme pour indiquer les courbes magnétiques comparées aux Observations. Parts i. and ii. 8vo. 168 and 40 pp. 3 maps. *Paris,* 1776 and 1778

Extraits des Mémoires de l'Acad. de Suède, au trimestre des trois derniers mois de l'année 1775 sur les observations magnétiques. 4to. (*Mém. de Paris,* an. 1777, *Mém.* p. 88.) *Paris,* 1777

Construction de la boussole dont on a commencé à se servir en Août 1777. 4to. (*Mém. de Paris,* an. 1778, *Hist.* p. 36, *Mém.* p. 66.) *Paris,* 1778

Réflexions sur les observations de la déclinaison et variation de l'aimant dans l'Océan Atlantique faites à la mer. 4to. (*Mém. de Paris,* an. 1779, *Mém.* p. 378.) *Paris,* 1779

Le Normand, Louis Sébastien. *Born May 25, 1757, at Montpellier.*

† Sull' utilità dei Parafulmini e Paragrandini per l'agricoltura. Seconda edizione. 8vo. 19 pp. *Milano,* 1823

† **Lenz, E.** Über die Leitungsfähigkeit des Goldes, Bleis, und Zinnes, für die Elektricität bei verschiedenen Temperaturen ; als Zusatz zu der in desen Memorien . . T. ii. p. 631 enthaltenen Abhandlung über die Leitungsfähigkeit 5 andetrer Metalle. (Read Apr. 5th, 1836.) 4to. 17 pp. (Mem. vi. Ser. *Sc. Math.* tom. iii. 1st part, p. 439, *Acad. de St. Pétersbourg.*) *St. Petersburg,* 1836 ?

† Über die Kraft eines Magneten in Beziehung zur Kraft der einzelnen Magnete, aus welchen er zusammengesetzt ist. 4to. 4 pp. (*Bulletin scientifique,* Mem. vi. Ser. *Sc. Math.* tom. iii.) 1833

† Über die Gesetze der Wärme-entwickelung durch den galvanischen Strom. 4to. 22 pp. 1 plate. (*Aus d. Bulletin de la Classe Phys. Math. de l'Acad. Imp. de Sc. de St. Pétersb.* tom. i. Nos. 14, 15, 16, *besonders abgedruckt.*) (*Faraday's copy.*) *St. Petersburg,* 1843

† Bericht über die magnetische. Expedition in der Umgegend der Insel Jussary. 8vo. 4 pp. (*Aus dem Bulletin,* tom. ii. pp. 440-443. *Quere of the Société Impériale des Naturalistes de Moscow.*) *Moscow,* 1860

(*Vide* also Lütke.)

Lenz, Heinrich Friedrich Emil. *Born February 12, 1804, at Dorpat.*

Four papers on Electricity and Magnetism in Mem. Acad. Pétersb. Series vi. vol. i. 1831, ii. 1833, iii. 1838.

Eight Articles in the Bullet. Scient. de l'Acad. de Pétersb. vol. i. 1836, ii. 1837, iii. 1838, iv. and v. 1838-39, v. 1839, vi. 1840, ix. 1842.

Further Articles on Galvanism, Electro-Magnetism, &c., in the Bull. Physico-Math. de l'Acad. de Pétersb. vol. ii. 1844, iii. 1845, v. 1847, vii. 1849, x. 1852, xii. 1854, and xvi. 1858.

† Über d. Bestimm. d. Richtung d. durch electrodynamiche Vertheilung erregten galvan. Ströme. 8vo. 10 pp. 1 plate. (*Poggend. Ann.* xxxi. 1834.) *Leipzig,* 1834

† **Lenz, Heinrich Friedrich Emil.**—*continued.*

Über d. Gesetze nach welchen der Magnet auf eine Spirale einwirkt wenn er ihr plötzlich genähert oder von ihr entfernt wird, u.s.w.   8vo.   (*Poggend. Ann.* xxxiv. 1835.)   *Leipzig*, 1835

Über die praktische Anwendung des Galvanismus.   8vo.   (*Elect. Magn. Telegraph* (68 pp.) *of Schilling v. Canstadt described in this work.*)   *Petersburg*, 1839

Lenz, R.   Untersuchung einer unregelmässigen Vertheilung des Erdmagnetismus im nördlichen Theile des finnischen Meerbusens.   4to.   38 pp.   3 maps.   (*Mémoires de l'Académie Impériale des Sciences de St. Pétersbourg*, vii. Série, tom. v. No. 3.)   *St. Petersburg, Leipzig*, 1862

Magnetische Beobachtungen an einigen Punkten der finnländischen und esthländischen Küsten.   8vo.   (*Mélanges physiques et chimiques tirés du Bulletin de l'Académie Impériale des Sciences de St. Pétersbourg*, tom. vi. 5 *livraison, St. Pétersbourg*, 1866, pp. 620-629.)   *Leipzig*, 1866

Leonardi, Camilli.   Speculum lapidum.   *Venetiis*, 1502
(*Note.*—*Mentions the polarity of the Magnet. There is an English translation, also an edition of* 1516, *Venice, in* 4to.)

Leonelli, Zecchini.   *Born* 1776, *at Cremona ; died October* 12, 1847, *at Corfu.*

† Démonstration des causes des phénomènes électriques, ou théorie de l'électricité prouvée par l'expérience.   8vo.   95 pp.   *Strasbourg*, 1813

Leonhard, Karl Cäsar Von.   *Born September* 12, 1779, *at Rumpenheim ; died January* 23, 1862, *at Heidelberg.*

Handbuch der Oryktognosie (p. 83).   (*From Dove*, p. 357.)   1825

Leotandus, Vincent.   *Born* 1595, *at La Val Louise, near Embrun ; died June* 13, 1672, *at Embrun.*

Magnetologia in quâ exponitur nova de magnetis philosophia.   4to.   420, 10, and 6 pp.   *Lugduni sumpt. L'Anisson*, 1668

† **Leprince.**   Nouvelle théorie de l'Aurore Boréale.   8vo.   15 pp.   1 plate.   *Versailles, (before)* 1817

Lerebours, T.   *Born* 1794, *at Paris.*

† **L....., J.** (*i.e.*) Lerebours, J.   Traité de Galvanoplastie.   1st edition.   8vo.   122 pp,   *Paris*, 1843

† Traité de Galvanoplastie.   2de édition, revue et augmentée.   8vo.   142 pp.   *Paris*, 1845

Le Noble.   Aimants artificiels d'une très grande force.   4to.   (*Mém. de Paris*, an. 1772, pt. i. Hist. p. 17.)   *Paris*, 1772

† Rapport sur les Aimants presenté par M. l'Abbé Le Noble.   4to.   15 pp.   1783

(*Vide* also Andry and Thouret.)

† **Le Noble d'Autun.**   Découverte sur le Galvanisme, comme cause des sons.   4to.   36 pp.   *Milan and Paris*, 1803

Lermier.   Observations sur le Déboisement . . . en France suivies de quelques souvenirs des hautes montagnes.   8vo.   (*Actes de l'Acad. de Bord.* 6me An. p. 459.)   *Bordeaux*, 1845

† **Leroux, F. P.**   Etudes sur les machines Electro-Magnétiques et Magneto-Elect.   8vo.   23 pp.   (*Ext. des Annales du Conservatoire Imp. des Arts et Métiers.*)   *Paris*, 1856?

† Les machines magneto-électriques françaises et l'application de l'électricité à l'éclairage des phares.   Deux leçons faites à la Société d'Encouragement.   4to.   76 pp.   2 plates.   (*Extrait du Bulletin de la Soc. d'Encouragement pour l'Indust. Nat.* an. 1867.)   *Paris*, 1868

Le Roy, Julien.   Usage d'un nouv. cadran universel à boussole et propre à tracer les méridiennes.   *Paris*, 1734

Le Roy, Jean Baptiste. *Born . . . at Paris ; died January 20, 1800, at Paris.*
Many Memoirs in the Mém. d. Sav. Etrang. *(Mém. d. Paris,* 1753-73.)

† Estratto d'una lettera al Rozier sulla scintilla osservata dal Walsh nella Anguilla tremante. 4to. 2 pp. *(Scelta d'Opus.* Nuov. ed. iii. 88. *Trans. by Soave, printed in the 12th ed. of the Scelta in* 1777, vol. **xxvi.** *with Remarks.)*
Milano, 1784

† Lettera al Rozier su i Parafulmini. 4to. 2 pp. *(Scelta d'Opuscoli,* Nuova ed. ii. 222. *Translated by Fromond. It was printed in the 12th in ed.* 1776, vol. **xviii.** *The original French version in the Journ. de Phys.* vol. ii.)
Milano, 1782

† **Le Roy d'Etiolle,** J. Sur l'emploi du Galvanisme dans les hernies étranglées et les étranglements internes. 8vo. 8 pp. *(Ext. des Archives générales de Médecine. Mém. lu à l'Acad. Rle. de Médecine.)* *Paris*

Le Sage, George Lewis. *Born January* 9, 1676, *at Colombière, near Couches ; died February* 5, 1759, *at Geneva.*

Des corps terrestres et des Metéores. 8vo. *(From Pogg.* i. 1433.) *Genève,* 1730

Le Sage, George Lewis, jun. *Born June* 13, 1724, *at Geneva ; died November* 9, 1803, *at Geneva.*

Essai de chimie mécanique. 4to. 1758

Lettre à M. Prevost de Genève, Berlin, 22 Juin, 1788. *(From Moigno.)*
(Note.—On the subject of " une correspondance prompte, etc., entre deux endroits éloignés, au moyen de l'électricité.")

Traité de physique rédigé d'après les notes de M. Le Sage par M. Prevost.
Geneva, 1818

Leschau, B. G. Grundzüge der reinen Electricitätslehre. 8vo. *Wien,* 1826

Leschevin. Memoir upon a process employed in the ci-devant Maçonnais of France, to avert showers of Hail and to dissipate Storms. By M. Leschevin, Chief Commissary for Gunpowder and Saltpetre at Dijon. *(From Millin's Magazin Encyclopédique for* 1806, tom. ii. p. 5.) 8vo. 7 pp. *(In Phil. Mag.* **xxvi.** 212.) *London,* 1807

† **Leslie,** Sir John. *Born April* 16, 1766, *at Largo, Fifeshire, Scotland ; died November* 3, 1832, *at Coates, near Largo.*

* On Davy's reference to Van Helmont's Thermometer. 8vo. 3 pp. 3 figures. *(From notice in Phil. Mag.* xl. 236.) *London,* 1812

* Anonymous or A. B. On the Differential Thermometer. 8vo. 2 pp. 3 figs. p. 329. *(Reply to Leslie.) (Phil. Mag.* xl.) *London,* 1813

* Description of an Atmometer, and an account of some Photometric, Hygrometric, and Hygroscopical Experiments. 8vo. 9 pp. *(Phil. Mag.* xlii. 44.) *London,* 1813

† Observations on Electrical Theories. 8vo. 39 pp. 1 plate. *Edinburgh,* 1824 *(Note.—Read at the R. Soc. Edinb.* 1792, *but not printed in its Transactions. Written in* 1791.)

On the Inefficacy of Lightning Conductors. *(From Arella, who says that this statement appears in Fernsac, Scienze fisiche matemat.* 1829, p. 130.) 1829

* Treatises on Natural and Chemical Philosophy. 8vo. *Edinburgh,* 1838

Geometrical Analysis. *(From Dove, p. 256.)*

† **Lesure,** A. Thèse (soutenue le 17 Nov. 1857). Expériences relatives à l'action des courants électriques sur les nerfs. 4to. 34 pp. *Paris,* 1857

† **Letheby,** H. An account of the dissection of a Gymnotus Electricus, &c., together with reasons for believing that it derives its electricity from the Brain and Spinal Cord; and that the nervous and electrical forces are identical. La. 8vo. 21 pp. 3 plates. *(Proceedings of the London Electrical Society. Read August* 16, 1842.) *London,* 1842

Letheby, H.—*continued.*

† An account of the dissection of a second Gymnotus Electricus; together with a description of the Electrical phænomena and anatomy of the Torpedo. 8vo. 18 pp. 5 plates. (*Proceed. Lond. Elect. Soc. Read June 17, 1843.*)

London, 1843

Luchtenburg, Duc N. de. Sur la composition du pyrite magnétique de Bodenmais. 8vo. *St. Pétersbourg, Riga, and Leipzic,* 1864

Leupold. Traité de physique. 1823
(*Note.—The first volume only was printed, containing matter on Electricity and Magnetism, &c.*)

Leutmann, Johann Georg. *Born November 30, 1667, at Wittemberg; died 1736, at Frühjahr, St. Petersburg.*
Instrumenta meteorognosiæ inservientia. 8vo. (*From Kruniz, p. 156.*)
Wittemberg, 1725

† **Leuwenhoeck,** A. Van. Part of a Letter dated April 5, 1697, giving an account of several Magnetical experiments; and of one who pretended to cure or cause diseases at a distance, by applying a sympathetic powder to the urine. 4to. 10 pp. 1 plate. (*Phil. Trans.* xix. *for* 1695-6-7, p. 512.) *London,* 1698

Le Verrier. Note sur un système régulier d'observations météorologiques établi en France, par les soins de l'administration des lignes télégraphiques et de l'Observatoire Impérial de Paris. 4to. *Paris,* 1856

Levi, G. Di alcuni fenomeni osservati sulla cute di un paraplegico, &c. . . . (*From Zantedeschi, Trattato,* ii. 508.) *Firenze,* 1844

† **Levis,** A. de. Fenomeno magnetico descritto dal P. De Levis. 4to. 3 pp. (*Opusc. Scelti,* xvi. 69.) *Milano,* 1793

Levy, Marc. Observations sur la magnétologie. (*From Pogg.* i. 1442.) *Paris,* 1844

† **Lewis,** R. An account of an Aurora Borealis seen in New Zealand on the 22nd October, 1730, by R. Lewis. Communicated in a letter to P. Collinson. Dated Annapolis in Maryland, December 10, 1730. 4to. 2 pp. (*Phil. Trans.* xxxvii. 69.) *London,* 1731-2

† **Lewthwaite,** John. New form of an experiment in Electro-Magnetism. 8vo. 1 p. (*Phil. Mag. or Annals,* ii. 459.) *London,* 1827

† **Leymarie.** Une nouvelle manière d'envisager le système cristallin de la Tourmaline. 8vo. (*Toulouse Acad.* 3me série.) *Toulouse,* 1850

Lezay-Marnézia, Claude François Adrien, Marquis de. *Born August 24, 1735 at Metz; died November 9, 1800, at Paris.*
Sur les paratonnerres et les paragrêles. (*Vide* Clerc.)

Liais, E. Mémoire sur un bolide observé dans le département de la Manche le 18 Novembre, 1851. *Cherbourg,* 1852

† Mémoire sur la substitution des Electro-moteurs aux machines à vapeur . . et horloge. 8vo. 24 pp. *Paris,* 1852

Sur la détermination du centre de gravité d'un barreau aimanté. (*Mém. de la Soc. de Cherbourg,* iv. 220.) *Cherbourg*

Pendule électromagnétique. (*Mém. de la Soc. de Cherbourg,* ii. 294 and iv. 205.) *Cherbourg*

Système d'éclairage électrique. (*From Du Moncel.*)

Libes, Antoine. *Born July 2, 1752, at Béziers; died October 25, 1832, at Paris.*
Théorie de l'électricité appuyée sur des faits, confirmée par le calcul. 4to.
Paris, 1800
Traité . . de physique. (Electricity by pressure.) 8vo. (*From Dove,* p. 229.)

† **Libri,** G. B. I. T. Histoire des Sciences mathématiques en Italie. 4 vols. 8vo.
Paris, 1838-48

Saggio d'esperienze elettro-metrici del Marianini. 8vo. *Milano ?*
His Library was sold by auction by Leigh and Sotheby on the 23rd April, 1861. (*Catalogue collated for Electricity by F. R.*)

Licetus, Fortunatus. *Born October 3, 1577, at Rapallo in Genuesischen; died May 17, 1657, at Padua.*

† Pyronarcha sive de Fulminum natura deque Februm origine. Sm. 4to. 126 pp. 1 fig. *Patavii,* 1634

† Litheosphorus; sive de lapide Bononiensi lucem in se conceptam et ambienti claro mox in tenebris mire conservante. Liber Fort. Licetus (Genuensis). . 4to. 280 pp. *Utini,* 1640

Lichtenberg, George Christoph. *Born July 1, 1744, at Ober-Ramstädt, near Darmstadt; died February 24, 1799, at Göttingen.*

De nova methodo naturam ac motum fluidi electric investigandi. Commentatio Prior. 4to. 15 pp. 4 plates. *Göttingen,* 1778

De nova methodo naturam ac motum fluidi electrici investigandi. Comment. ii. 8vo. *Göttingen,* 1779

An Dr. Exleben; die seltsame Wirkung eines Wetterstrahls auf ihn (E.) betreffend. 8vo. *(Götting. Mag. J. i. S. ii. pp. 216-20.) Göttingen,* 1780

Über Pelletier's Anwendung d. Elektricität zur Erkennung mineral Körper. 12mo. *(Crelle's Ann.* 1786.) 1786

Anmerk. zu Exlebens Anfangsgrund der Naturlehre. (On Volta's Condenser.) 12mo. *Göttingen,* 1794

† Über Gewitterfurcht und Blitzableitung. 8vo. *Göttingen,* 1802

† Neueste Geschichte der Blitzableiter. 8vo. *Göttingen,* 1803

† Vorschlag den Donner auf Noten zu setzen. 8vo. *(Lichtenberg's Mathem. und Phys. Schriften, &c.* i. 478.)

† Versuche zur Bestimmung der zweckmässigsten Form der Gewitterstangen. 8vo. *Göttingen,* 1803

Beobachtung der Magnet-Nadel am Harze. *(Bergbaukunde,* tom. ii. p. 127.)

† Vermischte Schriften (9th vol, &c.) 9 vols. 12mo. *Gotha,* 1800-1805

† **Lichtenberg,** G. C. und **Forster.** Göttingisches Magazin der Wissenschaften und Litteratur. 8vo. 6 vols. *Göttingen,* 1780-5

† **Lichtenberg,** Ludwig Christian. *Born 1738, at Ober-Ramstädt, near Darmstadt; died March 29, 1812, at Gotha.*

Verhaltungsmaass regeln bey nahen Donnerwetter nebst d. Mitteln sich gegen d. schädl Wirkungen d. Blitzes in Sicherheit zu setzen. 1 Aufl. 8vo. 1 pl. *Gotha,* 1774

Verhaltungs regel bey nehen Donnerwettern; nebst d. Mitteln, sich gegen d. schädl. Wirkungen d. Blitzes in Sicherheit zu setzen. 2e Aufl. 1775. 8vo. 78 pp. 1 plate. *Gotha,* 1775

† Magazin für das Neueste aus der Physik und Naturgeschichte. 8vo. 3 vols. *Gotha,* 1781-86

Lichtenberg und **Voight.** Magazin für das Neueste aus der Phys. 8vo. 9 vols. *Gotha,* 1786-1799

(*Vide* also Michaelis and Lichtenberg.)

Lichtenstein. Galv. differentia ab electricitate. 8vo. *(Loder's Journ. fur die chirurgie,* iii. 508.) *(From Sue, Hist.* iv. 262.)

Lieberkuhn. Fixed spirit of wine. *(Sauvages œuvres diverses,* ii. 49.)

Lieberkuyn. On Electricity. *(From Nollet.)*

Liebig, Justin Von. Zur Beurtheil d. Selbstverbrenen d. menschl. Körpers. 8vo. *Heidelberg,* 1850

Liebig and **Kopp.** Annual Reports on the Progress of Chemistry, Physics, Mineralogy, and Geology, from 1847 to 1850. 4 vols. 8vo. 1849-53

Jahresbericht über die Fortschritte der Chemie, Physik, &c.

Liebig, Poggendorf, &c. Handwörterbuch der reinen und angewandten Chemie von Liebig, Poggendorf, Wöhler, &c. *Braunschweig,* 1857-63

Liebknecht, Johann Georg. *Born April 23, 1679, at Wasungen Hessen; died September 17, 1749, at Giessen.*

Dissertatio de noctiluca merc. seu de luce quam argentum vivum in tenebris fundit. 4to. (*From Kruniz*, p. 156.)   *Giessen*, 1716

Pharus seu de prodigiis ignis cœlestibus ut vulgo vocantur ex omni ævo collectis diss. hist. math. &c. 4to.   *Giessen*, 1721

Lieutaudi. Magnetologia. 4to. (*From Dove.*)   *Lugdini, Bat.* 1668

† **Lilienthal.** Zur Galvanoplastik. 4to. 6 pp.   *Konigsberg*, 1849

† **Lille,** C. W. De Lunæ in declinationem acus magneticæ vi (Dissert. Acad.) . . . ad Imp. Alexand. in Fenniæ Univers. 4to. 15 pp.   *Helsingforsiæ*

Limmer, C. P. De Tonitru.

De Magnete ejusque effectibus, &c.
(*These two "Disputations" are amongst more than sixty which were held at Zerbst between 1686 and 1709.*) (*From Pogg.* i. 1463.)

Linari-Santi. *Born November* 1, 1777; *died July,* 1858, *at Naples.*

Sur les propriétés électriques et électrophysiologiques de la torpille. 8vo. (*Bibl. Univ.* Ser. ii. tom. viii. 1837 et xviii. 1838.)   *Geneve*, 1837-8

† Lettera del Santi Linari al Giulij. (Dated October 19, 1838.) 4to. 1 p. (*Fusinieri Ann. Scient. Reg. Lomb.-Veneto,* ix. 200, 1839.)   *Venezia*, 1839

† Cenni di nuove indagini sulle proprietà elettriche della Torpedine. 8vo. (*Bibliot. Ital.* xcii. 258.)   *Milano*, 1839

Sull' elettricitá animale. (*Rendiconto dell' Accad. di Napoli,* ii. 1843.)
                      Napoli, 1843

(*Vide* also Giuli and Linari-Santi, and Palmeri and Linari-Santi.)

† **Linati,** P. Intorno agli effetti della corrente elettrica continua sulle funzioni del gran-simpatico. Memoria. 8vo. 25 pp.   *Parma*, 1857

† Degli studi elettro-fisiologici presso l'alta antichità. Memoria. 20 pp.
                      Parma, 1858

† **Linati e Caggiati.** Intorno gli effetti della corrente elettrica continua sulle funzioni del Gran-simpatico. Nuove Esperienze. 4to. 34 pp. (*Estratto dal Tempo Giorn. di Medicina, &c.* fasc. viii. ix. x. Ag. Sett. e Otto.)
                      Firenze, 1858

Linck, Johann Wilhelm. *Born December* 25, 1760, *at Leipzig; died December* 25, 1805, *at Leipzig.*

De Raja Torpedine. 8vo.   *Lips.* 1788

† **Lindhult,** I. Kurzer Auszug aus tägl. Verzeichnisse. . . . Krankheiten die durch Elekt. sind gelindert, &c. 8vo. 4 pp. (*K. Schwed. Akad.* Abh. xiv. 312.)   *Hamburg and Leipzig,* 1752

† Fortsetzung der Nach. von Krankheiten die durch Elekt. geheilt werden. 8vo. 13 pp. (*K. Schwed. Akad.* Abh. xv. 141.)   *Hamburg and Leipzig,* 1753

Lindig, F. Quomodo mutantur vires electricæ cum temperatura. (Dissert. inauguralis.) 8vo. 35 pp.   *Berolini,* 1864

Linemann, A. De iride. De meteoris ignitis. De igne elementari. Memoria sæcularis seu Collectio observat. astronom.   1644

Linguet, S. N. H. Mém. manuscrit pour le départment de la Marine sur les moyens d'établir des signaux par la lumière.   1782

Linné (Linnæus), Carl von. *Born May 24, 1707, at Råshult, Småland; died January 10, 1778, at Upsala.*

Wilkens Anmerkung zu vorhergehenden Aufsäzen. 2 pp. (*K. Schwed. Akad. Abh.* xxiv. 291.)

† Beschreibung des Ungewittervogels. 8vo. 4 pp. 1 plate. (*K. Schwed. Akad. Abh.* vi. 93.)   *Hamburg*, 1745

† Leuchtende Würmer aus China vom H. Geheimrath Raben eingegeben und von C. Linnæus beschreiben. 8vo. 6 pp.

De Geer, C. Auslegung und Anmerkungen über die 5 und 6 Zeichnung der 1 Tafel, welche die chinesischen leuchtenden Insekten vorstellet. 8vo. 2 pp.
† 1 plate. (*K. Schwed. Akad.* Abb. viii. 61 pp.)   *Hamburg*, 1746

Flora Zeylanica. (On the Tourmaline.) 8vo. 240 pp. 4 plates. *Stockholm*, 1747

Diss. sistens consectania electrico-medica. 4to.   *Upsal*, 1754
    (*Note.—A German translation entitled " Einige Nachrichten von dem Nuzen der Elect. in der Arzneykunst" in Hamb. Magaz.* xix. Bd. 3, St. 1757-8, S. 325-335.*)

On Light on the Flowers of the Cupricium (Tropæolum). (*Acta Holmiensiæ*, vol. xxiv. p. 292.)   1762

† **Linnæa**, Elizabeth Christina. Vom Blitzen der indianischen Kresse. 8vo. 3 pp.
    (Signed by C. Linnæus.)   *Hamburg and Leipzig*, 1762

Linnstrom, H. Schwedisches Bücher-Lexikon, 1830-65. 8vo. 96 pp.
                                                         Leipzig, 1869?

† **Lion**, Moise. Electricité statique. Histoire et recherches nouvelles. 8vo. 190 pp.
    7 plates.   *Paris*, 1868

Lippens, P. " L'appareil Lippens destiné spécialement au service télégraphique des chemins de fer a été décritsen 1852 dans le Annales des Travaux Publics de Belgique, tom. xi."

† Télégraphie des chemins de fer, des mines, &c. . . . Appareils, sonneries, et accessoires de P. Lippens. 8vo. 32 pp. 2 plates.   *Bruxelles*, 1856

Lipschitz, Rudolph Otto Sigismund. *Born May 14, 1832, at Königsberg.*

† Determinatio status magnetici viribus inducentibus commoti in ellisoide. 4to.
    25 pp.   *Berolini*, 1853

Über d. Anwendung eines Abbildungs princips auf d. Theorie d. Vertheilung d. Electric. 4to.   *Berlin*, 1862

L'Iride. Giornale di Scienze, Lettere ed Arti per la Sicilia. (*Biblioteca Italiana*, xxvi. 111.)   *Palermo*, 1822

Lisbon Acad. Memorias da Acad. Real das Sciencias de Lisboa 12 vols. ? 4to.
                                                              1797 ?

Hist. Mem. da Acad. 2de Serie.   1843

Literary and Philosophical Society of New York. (*Vide* New York.)

Litta, Alfonso Agostino. *Born at Milan; died October 3, 1781.*

† Riflessioni sulla capacità de' Conduttori elettrici esposte in una lettera al Volta.
    4to. 3 pp. (*Opus. Scelti*, i. 340.)   *Milano*, 1778

Little. (*Vide* Brett and Little.)

Littré. (*Vide* Ampère.)

† **Littrow**, K. Von. Sternschnuppen u. Kometen. Geschichte der Entdeckung des Zusammenhanges zwischen diesen beiden Gattungen von Himmels-Körpern. 8vo. 41 pp.   *Wien*, 1867

Liverpool Compass Committee. First Report of the Committee on the Deviations of the Needle in Iron and other Vessels, occasioned by Induction or Polar Magnetism, by Messrs. Yates and Grantham, in which the results of that Committee's operations are described.   1855

Second Report of the Liverpool Compass Committee to the Board of Trade.
                                                                          1857

Third Report of the Liverpool Compass Committee to the Board of Trade.
                                                                          1861

Liverpool Literary and Philosophical Society. Founded 1812. Proceedings in 8vo. (*From English Cat.* 1864, p. 864.)   1844 *to* —

Livy. Hist. Romana. Lib. i. cap. 20, lib. i. cap. 22, A.D. 16. (*Phil. Trans.* xlviii. part i. p. 211.)

Lixis, E. Mémoire sur un Bolide, &c. 8vo.   *Cherbourg,* 1852

Ljubimoff, Nik. Das Grundgesetz der Electro-Dynamik. 8vo. 170 pp.
                                                                    Moscow, 1856

Lloyd, Humphrey. *Born April* 16, 1800, *at Dublin.*

† An attempt to facilitate Observations of Terrestrial Magnetism. Read October 28, 1833. 4to. 15 pp.   *Dublin,* 1833

† Further development of a Method of observing the Dip and the Magnetic Intensity at the same time. 4to. 13 pp. (*Trans. of the Royal Irish Acad.* vol. xvii.)   *Dublin,* 1836

† On the mutual action of Permanent Magnets, considered chiefly in reference to their best relative position in an Observatory. 4to. 20 pp. 1 plate. (*Trans. Royal Irish Acad.* xix. part i.)   *Dublin,* 1840

† Supplement to a paper on the Mutual Action of Permanent Magnets. 4to. 10 pp. 1 plate. (*Trans. Royal Irish Acad.* xix. part ii.)   *Dublin,* 1841

† Account of the Magnetical Observatory of Dublin, and of the Instruments and methods of observation employed there. 4to. 54 pp. 5 plates. (*Faraday's copy.*)   *Dublin,* 1842

† On a new instrument for the Measurement of the Inclination and its changes. 8vo. 16 pp. (*Arago's copy.*)   *Dublin,* 1842

† Account of the Induction Inclinometer and of its Adjustments. 8vo.
                                                                    London, 1842

† On the determination of the Intensity of the Earth's Magnetic Force in absolute measure. 4to. 16 pp. (*Trans. Royal Irish Acad.* xxi. part i.)
                                                                    Dublin, 1843

† An account of a method of determining the total Intensity of the Earth's Magnetic Force in absolute measure. 8vo. 8 pp. (*Proceedings Royal Irish Acad.* Jan. 24, 1858.)   *Dublin,* 1848

† Remarks on the Theory of the Compound Magnetic Needle. 8vo. 3 pp. (*Proceedings Royal Irish Acad.* May 22, 1848.)   *Dublin,* 1848

† On certain questions connected with the Reduction of Magnetical and Meteorological Observations. 8vo. 4 pp. (*Proceedings Royal Irish Acad.* June 12, 1848.)   *Dublin,* 1848

† Circular for the information of the British Colonial Magnetical Observatories. 8vo. 7 pp.   *London,* 1848

† Results of observations made at the Magnetical Observatory of Dublin during the years 1840-3. First series : Magnetic Declination. Read May 11 and 25, 1846. 4to. 25 pp. 3 plates. (*Trans. Royal Irish Acad.* xxii. Part i.)
                                                                    Dublin, 1849

† On the induction of soft Iron, as applied to the determination of the changes of the Earth's Magnetic Force. 8vo. 11 pp. (*Proceedings Royal Irish Acad.* Nov. 1850.)   *Dublin,* 1850

Lloyd, Humphrey.—*continued.*

† On the position of the Isogonal Lines in Ireland, deduced from the observations of Captain Sir James Clark Ross, R.N. 8vo. 5 pp. (*Proceedings Royal Irish Acad.* Nov. 30, 1850.) *Dublin,* 1850

† On the influence of the Moon upon the position of the freely-suspended Horizontal Magnet. 8vo. 11 pp. (*Proceedings Royal Irish Acad.* Feb. 1853.) *Dublin,* 1853

† On the Magnetic influence of the Moon. 8vo. 3 pp. (*Proceedings Royal Irish Acad.* Dec. 1853.) *Dublin,* 1853

† Notes on the Meteorology of Ireland, deduced from the observations made in the year 1851. Read June 27 and December 12, 1853. 4to. 88 pp. 4 plates. (*Trans. Royal Irish Acad.* xxii.—*Science.*) *Dublin,* 1854

† On Earth Currents, and their connection with the phenomena of Terrestrial Magnetism. 8vo. 6 pp. 1 plate. (*Proceedings Royal Irish Acad.* Nov. 11, 1861, p. 318.) *Dublin,* 1861

† On Earth Currents in connection with Magnetic Disturbances. 8vo. 5 pp. (*Proceedings Royal Irish Acad.* April 28, 1862, p. 392.) *Dublin,* 1862

† On the probable causes of the Earth Currents. 8vo. (*Proceedings of the Royal Irish Acad.* June 23, 1862, p. 442.) *Dublin,* 1862

† On Earth Currents, and their connection with the diurnal changes of the Horizontal-Magnetic Needle. Read Nov. 11 and 30, 1861. 4to. 27 pp. 2 plates. (*Transactions of the Royal Irish Acad.* vol xxiv.—*Science.*) *Dublin,* 1862

† Observations made at the Magnetical and Meteorological Observatory at Trinity College, Dublin, under the direction of the Rev. Humphrey Lloyd. Vol. i. (for) 1840-1843. 4to. 512 pp. 8 plates. *Dublin,* 1865

† On the variations of the Magnetic declination at Dublin. 8vo. 11 pp. (*Proceedings Royal Irish Academy.*) *Dublin*

† On the corrections required in the measurement of the Magnetic declination. 8vo. 11 pp. (*Proceedings Royal Irish Academy.*) *Dublin*

† Description of the Theodolite-Magnetometer. 8vo. 7 pp. (*Proceedings Royal Irish Academy.*) *Dublin*

† Observations made at the Magnetical and Meteorological Observatory, Trinity College, Dublin, under the direction of Humphrey Lloyd. Vol. ii. (for) 1844-1850. 4to. 541 pp. 2 plates. *Dublin,* 1869

† **Lloyd, Sabine, and Ross.** Observations on the directions and intensity of the Terrestrial Magnetic force in Ireland. 8vo. 45 pp. 1 plate. (*Report of the British Association for* 1835.) *London,* 1836

Lobb, Harry. A Popular Treatise on Curative Electricity, to sufferers from paralysis, rheumatism, neuralgia, and loss of nervous and physical power. 12mo. 75 pp. *London,* 1867

† On the curative treatment of Paralysis and Neuralgia, and other affections of the nervous system with the aid of Galvanism. 2nd edition. *London*

Lobe, W. De vi corporum electrica. 4to. *Lugd. Bat.* 1743

† **Lochmann, Wolfgang.** (*Vide* Gilberti.)

Locke, John. *Born February* 29, 1792, *at Lempster, New Hampshire; died July* 10, 1856, *at Cincinnati.*

† On the Magnetic Dip at several places in the State of Ohio, and on the relative horizontal magnetic intensities of Cincinnati and London . . in a letter to John Vaughan, Esq. Read June 15, 1838. 4to. 7 pp. (*Trans. Amer. Phil. Soc.* new series, vol. vi.) *Philadelphia,* 1839

Observations to determine the Horizontal Magnetic Intensity and Dip at Louisville, Kentucky, and at Cincinnati, Ohio. 4to. (*Trans. Amer. Phil. Soc.* new series, vol. vii.) *Philadelphia,* 1841

Locke, John—*continued.*

† Observations made in the years 1838-39-40-41-42 and -43, to determine the Magnetic Dip and the Intensity of Magnetic force in several parts of the United States. Read April 19, 1844. 4to. 46 pp. 3 plates. (*Trans. Amer. Phil. Soc.* new series, vol. ix.)   *Philadelphia*, 1846

Report on the invention and construction of his Electro-Chronograph. 8vo.
     Cincinnati, 1850

† Observations on Terrestrial Magnetism. Accepted for publication July, 1851. 4to. 30 pp. (*The Smithsonian Contributions*, vol. iii.)   *Washington*, 1852

Lofft, Capel. On supposed Lunar Projectiles; on Aërolithes; and on Notices of Comets. 8vo. (*Tilloch's Phil. Mag.* li. 203.)   *London*, 1818

     (*Vide* also Acton, J.)

Lohmeier, P. De fulmine. (*From Pogg.* i. 1491.)   *Rint.* 1676

† **Lohmeier u. Pohl.** Leitfaden zum Selbstunterricht in den Aufangsgründen d. Telegraphen-Wesens, für Telegraphen-Candidaten, &c. . . 8vo. 158 pp.
     Berlin, 1870

Loiseau. Galvanomètre à orientation fixe. (*From Du Moncel, no date, &c.*)

Lo-Luz, Robert. Recherches sur les influences solaires et lunaires pour prouver le Magnétisme universel, &c. . 8vo.   *Paris*, 1788
     (*Note.—Poggendorff*, i. 1491, *says* : *Lo-Looz, Robert de.* 2 vols. 8vo. *London and Paris*, 1788.)

Lombardi, Antonio. *Born September* 22, 1768, *at Modena.*

Storia della letteratura Italiana nel secolo xviii. 6 vols. 12mo. *Venezia*, 1832
     (*Note.—There is a quarto edition, Milano, and also an octavo, uniform with Tiraboschi.*)

† **Lombardy I. R. Institution.** Giornale dell' I. R. Istituto Lombardo di Scienze, Lettere ed Arti, e Biblioteca Italiana, compilata da varii dotti nazionali e stranieri. 16 vols. 8vo. (*Electrical Articles in Library.*) *Milano*, 1841-47
     (*Note. —The " Giornale dell' I. R. Istituto Lombardo e Biblioteca Italiana" (in 8vo.) is the suite of the journal the "Biblioteca Italiana," which appeared from 1816 to 1840 inclusive. In the following year this became an organ of the Lombardy Institution, and modified its title in the above manner. It was thus published from 1841 to 1847 inclusive (still in 8vo.) 16 vols. In a part of this period, i.e. in 1843 and 1845, were published 2 vols. of the "Memorie dell' Istituto Lombardo" in 4to. In 1847 the Giornale began to appear alone, under the title " Giornale dell' I. R. Istituto Lombardo e Biblioteca Italiana," in 4to. This was discontinued in 1848 and 1849, but its publication was renewed from 1850 until 1856 inclusive. At about this last time it changed its title, or rather ceased to exist, and in 1858 began a new journal, entitled "Atti dell' I. R. Istituto Lombardo,' in 4to. In 1852 was published the 3rd volume of the " Memorie," and in 1861 the 8th volume. It continued after 1861.*)

Memorie dell' I. R. Istituto Lombardo di Scienze Lettere ed Arti. Vol. i. 1843 ; ii. 1845 ; iii. 1852 ; iv. 1854 ; v. 1856 ; vi. 1856 ; vii. 1858 and 1859 ; viii. 1859-60-61. 4to. (*This volume is also called* vol. ii *della Serie* ii.)
     Milano, 1843

Giornale dell' I. R. Istituto Lombardo di Scienze, Lettere ed Arti, e *Biblioteca Italiana*, nuova serie (in 4to.) 9 vols. (in 46 fascicoli) in 4to. *Milano*, 1847-56

Atti dell' I. R. Istituto Lombardo di Scienze, Lettere ed Arti. 4to.
     Milano, 1858

Lombroso, C. Azione degli Astri e delle Meteore sulla Mente Umana.   *Milan*

Lomond, On Transmission of Electric signs. (*From Young.*)

Lomonosow, Michael Wassiljewitsch. *Born* 1711, *at Denissowskaja ; died April* 4, 1765, *at St. Petersburg.*

Lomonosow, M.—*continued.*

Or ationis de Meteoris electricis explicationes. 4to. 9 pp. 3 plates. *Petropoli*

Lomonosow and Grischow. Oratio de Meteoris vi electrica ortis. Academicorum nomine ad sermonem M. Lomonosow respondet, et de meteoris insolitis sibi observatis disserit A. M. Grischow. 4to. 68 pp. (in all). 4 plates. (*Acad. Imp. Scient. Petropoli.*) *Petropoli*, 1755

London Journal of Arts and Sciences, edited by Newton. 8vo.
London, 1820 to —

London and Edinburgh Phil. Magazine and Journal of Science, by D. Brewster, R. Taylor, and R. Phillips. Vols. i.-xi. 8vo. *London*, 1832-50

London, Edinburgh and Dublin Philosophical Magazine and Journal of Science, by Brewster, Kane, and Francis. 8vo. (Fourth series of the *Phil. Mag.*) *London*, 1851-8

† **London Electrical Society.** The Transactions and the Proceedings of, from 1837 to 1840. 1 vol. 4to. 206 and 20 pp. 5 plates. (*Faraday's copy.*)
London, 1841

† Proceedings of the London Electrical Society. Edited by C. V. Walker, Hon. Sec. 8vo. 565 pp. 16 plates. *London*, 1843

London Encyclopædia. 22 vols. 8vo. *London*, 1839

London Royal Society. (*Vide* Royal Society.)

London Royal Institution. (*Vide* Royal Institution.)

† **Longet, F. A.** Recherches expérimentales sur les conditions nécessaires à l'entretien et à la manifestation de l'irritabilité musculaire, avec application à la Pathologie. La. 8vo. 36 pp. *Paris*, 1841

† Anatomie und Physiologie des Nervensystems : aus dem Französichen übersetzt und mit d. Ergebnissen . . bis auf die Gegenwart ergänzt und vervollständigt ; von Dr. J. A. Hein. 1st Band. Ein von dem Franz. Inst. gekrönte Preischrift. 8vo. 730 pp. 4 plates. *Leipzig*, 1847

† Anatomie und Physiologie des Nervens Systems : aus dem Franz. . . 2e Band. 8vo. 579 pp. 4 plates. *Leipzig*, 1849

† **Longet et Matteucci.** Sur la relation qui existe entre le sens du courant électrique et les contractions musculaires dues à ce courant. (Prem. Mem.) 8vo. 15 pp. *Paris*, 1844

Longobardi, Niccolo. Jesuit ; one of the Missionaries to China. *Born about* 1565 ; *died December* 11, 1655, *at Pekin.* (*Poggendorff*, i. 1494.)

† **Longridge, J. A.** On submerging Telegraphic Cables. 8vo. (*Proceedings of the Institution of Civil Engineers,* xvii.) *London*, 1858

Lonicerus, Janus. Compendium de meteoris ex Aristotelo, Plinio, et Pontano, &c. Libri iv. 8vo. *Francof.* 1548

Loomis, Elias. Observations to determine the Magnetic Dip at various places in Ohio and Michigan . . in a letter to S. C. Walker, Esq. 4to. (*Trans. Amer. Phil. Soc.* new series, vol. vii.) *Philadelphia*, 1841

† On the Storm which was experienced throughout the United States about the 20th December, 1836. Read March 20, 1840. 4to. 39 pp. 3 plates. (*Trans. Amer. Phil. Soc.* new series, vol. vii.) *Philadelphia*, 1841

Additional Observations of the Magnetic Dip in the United States. 4to. (*Trans. Amer. Phil. Soc.* new series, vol. vii.) *Philadelphia*, 1841

Observations to determine the Magnetic Intensity at several places in the United States, with some additional observations of the Magnetic Dip. 4to. (*Trans. Amer. Phil. Soc.* new series, vol. viii.) *Philadelphia*, 1843

Observations of the Magnetic Dip in the United States. 4th series. 4to. (*Trans. Amer. Phil. Soc.* new series, vol. viii.) *Philadelphia*, 1843

Loomis, Elias.—*continued.*

Supplementary Observations on the Storm which was experienced throughout the United States about December 20, 1836. 4to. (*Trans. Amer. Phil. Soc. new series*, vol. viii.) *Philadelphia*, 1843

† On two Storms which were experienced throughout the United States in the month of February, 1842. Read May 26, 1843, 4to. 24 pp. 13 maps. (*Trans. Amer. Phil. Soc.* new series, vol. ix.) *Philadelphia*, 1846

† Experiments on the Electricity of a plate of Zinc buried in the earth. 8vo. 11 pp. (*American Journal of Science, &c.* 2nd series, vol. ix.) *New Haven*, 1850

Determination of difference of Longitudes. Project for observations relative to researches on the laws of Storms (Ouragans, Hurricans) in North America. Letter to Sabine, August 2, 1847. 8vo. (*Phil. Mag.* Nov. 1847.) *London*, 1847

† On certain Storms in Europe and America, December, 1836. 4to. 26 pp. 13 maps (coloured). (*Smithsonian Contributions to Knowledge.*) *Washington*, 1860

Loon, Van. Lightning Column; or a Mirror for the North Sea. Folio. (*From Watts.*) *Amsterdam*, 1649

† **Lor, De.** (*Vide* Beccaria, Lettre à Nollet. Traduit par De Lor.)

Where were his Experiments with Lightning Conductors first published, and when?

Note.—*Nollet* (p. 12) *quotes* "*la · Gazette du 27 Mai,*" *in which it is said that le Sieur Bouguer rendered to the Académie Royale an account of this Experiment the 19th et tout de suite. Nollet adds: " La Compagnie reçut la date et la narration du fait, et les fit déposer dans ses Registres pour en faire honneur . . à Messrs. Dalibard et Delor." Mangin, Hist.* i. 187, *refers to the account in the "Gazette de France " du 27 Mai,* 1752, "*que nous relaterons ci-après.*"

Lorberg. Zur Theorie der Bewegung der Elek. in Leitern. *Ruhrort* (?)

Lorenz, Otto. " The Catalogue of all French publications from 1840 to 1865, compiled by the German bookseller Lorenz, settled in Paris, is at last completed, having been interrupted by the involuntary flight of the editor from Paris, about a year ago. In the absence of any comprehensive catalogue since Quérard, which reaches only to 1839, this is a great boon to librarians, booksellers, and persons who desire to refer to the publications of French authors. The arrangement is alphabetical, under the name of the author; in anonymous works, under the first substantive of the title. Each author's list is preceded by a short biographical notice." (*From Athenæum*, Sept. 9, 1871.) 1871

Catalogue général de la Librairie Française pendant vingt-cinq ans (1840-65). 4 vols. 8vo. *Paris*, 1867-71

† **Lorenzini, S.** Osservazioni intorno alle Torpedini. 4to. 136 pp. 5 plates. *Firenze*, 1678

† **Lorenzo de Medici.** Portrait. Caronni dis. et incis.

Lorgna, Antonio Maria. *Born* 1736, *at Verona; died June* 28, 1796, *at Verona.*

† Sopra una Fulminazione di terra. Lettera al Volta. (Verona, 15 Mag. 1781.) 4to. 7 pp. (*Opus. Scelti*, iv. 235.) *Milano*, 1781

Lettera (al Toaldo) sui Parafulmini. Verona, 14 Mag. 1778. (On an insulated Conductor for safety and for observations.) 2 pp. Risposta (*to the above*, *with the notice* "19 Mag. rec." 2 pp.) (Toaldo, G.) *Padua*, 1840

(*The above two articles are bound together with an Address to some friends (or Dedication), and signed Gaetano dott. Sorgata e Jacopo Prof. Cecconi (of* 1 *page). The whole forms a brochure, without any proper title-page. It is said in the Dedication that the two letters were found (in MS.) in the Biblioteca del Seminario.*)

Lorimer, John. *Born* 1732 ; *died July* 13, 1795, *at London.*

Description of a new Dipping-needle. 4to. (*Phil. Trans.* an. 1775, p. 79.)
London, 1775

† A concise Essay on Magnetism ; with an account of the declination and inclination of the Magnetic Needle. . . 4to. 34 pp. Portrait, and 6 other plates. *London*, 1795

Löscher, M. G. De phænomeno septentrionali luminoso. *Viteb.* 1721

De novo phosphoro æthereo. (*From Poggendorff*, i. 1486.)

† **Loschmidt, J.** Ableitung des Potentiales bewegter elektrischer Massen aus dem Potentiale für den Ruhezustand. 8vo. 8 pp. (*Sitzb. d. K. Akad. Wissensch.* ii. Abth. Juli Heft, Jahg. 1868. lviii.) *Wien*, 1868

Die Elektricitäts bewegung der galvanischen Ströme. 8vo. 5 pp. (*Sitzungsb. d. K. Akad.*) *Wien*, 1869

Loss, P. De fulmine in genere cum auctario. (*Pogg.* i. 1500.) *Gedani*, 1636

Lottin, Victor Charles. *Born October* 26, 1795, *at Paris ; died February* 18, 1858, *at Paris.*

Sur les aurores boréales. (*Annal. Maritim.* lix. 1839.) 1839

† Physique dans le voyage en Islande, &c. sur la "Recherche." (*Vide* Gaimard.)

Lottner, K. L. E. Versuch einer mathemat. Theorie d. elektr. Residuum in d. Leidner Flasche. (*Prgm. d. Realschule*, 1855.) *Lippstadt* (?), 1855

Über d. Zweckmässigste Combination galvan. Ketten. (*Schlomich u. Witzschel, Journ. f. Math. u. Phys.* ii. 1857.) 1857

Louis, Antoine. *Born February* 13, 1723, *at Metz ; died May* 20, 1792, *at Paris.*

Observations sur l'Electricité, ou l'on tâche d'expliquer son Mécanisme et ses effets sur l'économie animale. 12mo. 175 pp. *Paris*, 1747

Lous, Christian Karl. *Born* 1724, *at Copenhagen ; died* 1804, *at Copenhagen.*

Forsög til et nagt Misviisnings-Instrument. 4to. *Havn*, 1767

† Tentamina experimentorum ad compassum perficiendum, &c. 4to. 130 pp. 8 plates. *Hafniæ*, 1773
(*Vide* also two following entries.)

Lous, C. E. Beskrifning over et nyt opfunden söe-inklinations-compass, tillige med nogle anmarkninger over dette slagsinstrumenter. 4to. (*Skrifter det Kiöbenhavnske Selsk.* xii. 93.) *Kiöbenhavn*, 1779

On misviisningens her i Kiöbenhavn befundne forandring i de sidste forlöbne 50 Aar, samt middelmissviisningen aar for aar. 4to. (*Skrifter det Kiöbenhavnske Selsk. Nye Saml.* iii. 161.) *Kiöbenhavn*, 1788

Louyet, Paulin Laurent Charles Evalery. *Born January* 28, 1818, *at Mons ; died May* 3, 1850, *at Ixelles, near Brussels.*

"M. Louyet écrit qu'il a proposé, dès l'année 1838, l'emploi de la lumière produite par la Pile pour l'éclairage des Mines." 4to. (*Comptes Rendus,* xxii. 225.) *Paris*, 1846

† Recherches expérimentales sur le zincage voltaïque du fer. 8vo. 7 pp. (*Acad. Rle. de Belgique,* tom. xiii. No. 3 *des Bulletins.*) (*Arago's copy.*)

Love, G. (or Anon.) Renseignements sur l'application de l'électro-magnétisme aux machines locomotives. 8vo. 32 pp (*Extrait des Annales de Chemins de Fer*, No. 24, *et de la Revue Scientifique et Industrielle.*) *Paris*, 1851

† **Love, G. H.** Essai sur l'identité des Agents qui produisent le Son, la Chaleur, la Lumière, l'Electricité, &c. 8vo. 296 pp. *Paris*, 1861

Lovens. On Electricity as a living force. (*From Young*, p. 435.)

Lovering, Joseph. *Born December 25, 1813, at Charlestown, near Boston, U.S.*

Farrar's Electricity and Magnetism. *Cambridge, U.S.* 1842

† An account of the Magnetic Observations made at the Observatory of Harvard University, Cambridge (U.S.) Communicated by Joseph Lovering. 4to. 76 pp. (*American Acad. Mems.* new series, vol. ii. p. 85.) *Cambridge, U.S.* 1846

† **Lovering and Bond.** An account of the Magnetic Observations made at the Observatory of Harvard University, Cambridge (U.S.) Communicated by Joseph Lovering. 4to. 84 pp. 6 plates. (*American Acad. Mems.* new series, vol. ii. p. 1.) *Cambridge, U.S.* 1846

† **Lovett, R.** The subtil medium proved. . . 8vo. 141 pp. *London,* 1756

† Sir J. Newton's Æther realised, or the 2nd part of the subtil medium proved. 8vo. 77 pp. *London,* 1759

† The Reviewers reviewed, and Appendix. 8vo. 41 pp. *Worcester,* 1760

† Philosophical Essays . . in 3 parts. . . To which is subjoined, by way of Appendix, a clear account of the Variation of the Needle. 8vo. 525 pp. 4 plates. *Worcester,* 1766
(*The Appendix is entitled " A brief theory of the North Magnetic Pole and the Mariners' Compass-Needle."*)

† The Electrical Philosopher. . . To which is subjoined a Postscript, containing strictures upon the uncandid animadversions of the Monthly Reviewers. 8vo. 290 pp. 2 plates. *Worcester,* 1774
(*Note.—Latimer Clark says, " to which is subjoined a Postscript containing strictures of the Monthly Reviewers on those essays."*)

A Letter to the Monthly Review, &c. 8vo.

Elements of Natural Philosophy. 8vo. (*From Bent. of* 1811.)

(*Vide* also Anon. Magnetism, 1766.)

Low, Sampson. Classified Catalogue of School, College, Classical, Technical, and General Educational Works in use in Great Britain, arranged according to subjects. 8vo. *London,* 1871

† The English Catalogue of Books published from January, 1835, to January, 1863, comprising the contents of the "London" and the "British" Catalogues, and the Principal Works published in the United States of America and Continental Europe, with the dates of Publication, in addition to the size, price, and publisher's name. 8vo. 91C pp. *London,* 1864

† British Catalogue from 1837 to 1852. 1853
(*Note.—Collated for Electricity, &c.*)

Lowe, E. J. Treatise on Atmospheric Phenomena. 8vo. *London,* 1847

Lowe. (*Vide* Scoffern & Lowe.)

Löwenörn, P. von. Uber den Magnet. "Ein Beytrag z. Erklärung sowohl der Abweichung als Neigung d. Magnetnadel zc, aus d. Dänischen von J. A. Markussen." 8vo. *Kopenhagen,* 1802

Lower. Electricity rendered useful. (*Medical.*) 8vo. (*From Dove,* p. 242.) *London,* 1760

Lowitz. (*Vide* Volta, 1788.)

† **Lowndes, F.** Observations on Medical Electricity, containing a Synopsis of all the Diseases in which Electricity has been recommended or applied with success. 8vo. 51 pp. *London,* 1787

Observations on Medical Electricity. 8vo. 1789

Utility of Medical Electricity. 1791

* **Lowndes, W. T.** Bibliographer's Manual of English Literature. 4 vols. 8vo.
London, 1834

Lowthorp, J. (*Vide* Royal Society.)

Lozeran du Fech, Louis Antoine. *Died* 1755.

† Dissertation sur la cause et la nature du Tonnerre et des Eclairs. Et Lettre de l'Auteur à M. Sarrau, Séc. de l'Acad. viii. pp. 12mo. 100 pp. (*Accad. de Bordeaux.*) *Bordeaux,* 1726

Observation d'un phénomène céleste. (*Mem. de Trévoux,* 1730.) 1730

Dissertation sur la lumière septentrionale. (*Mem. de Trévaux,* 1732.) 1732

Lubbock, Sir John William, Bart. *Born March* 26, 1803, *in London ?*

On Shooting Stars. 8vo. 12 pp. 2 plates. (*Phil. Trans.* vol. xxxii. 1848, and xxxv. 1849.) *London,* 1848

Luc, Jean André, De. *Born February* 8, 1727, *at Geneva ; died November* 7, 1817, *at Windsor.*

† Modifications . . . 4to. *Genève,* 1772

† Nouv. Idées sur la Météorologie. 8vo. 2 vols. *Londres,* 1786

Observations sur un ouvrage intitulé Lithologie atmosphérique suivies de nouvelles réflexions sur les Pierres tombées du ciel ou de l'atmosphère, et de quelques remarques sur la lettre de M. Biot . . publiée dans le No. 182 de la Bibliothèque Britannique. 8vo. *Genève,* 1803

† Introduction à la physique terrestre par les fluides expansibles, &c. 8vo. 2 vols. *Paris,* 1803

† Traité élémentaire sur le fluide éléctrico-galvanique. 8vo. 2 vols. 15 plates. *Paris & Milan,* 1804

† On the variable action of the Electric Column. 8vo. 6 pp. (*Phil. Mag.* xliv. 248.) *London,* 1814

† Observations on Mr. Donovan's Reflections on the inadequacy of the principal hypotheses to account for the phenomena of Electricity. 8vo. 12 pp. (*Phil. Mag.* xlv. 97.) *London,* 1815

† A Reply to Mr. Donovan's Observations, &c. on Mr. De Luc's paper, published in our number for February. 8vo. 5 pp. (*Phil. Mag.* xlv. 329.) *London,* 1815

Bemerkungen über elektrische Bewegungen und deren Wirkung auf Spitzen ; desgleichen über Blitz, Donner und die sogenannten Wetter-Ableiter. (*Neue Schriften der Gesellsch. Naturf. Freunde,* ii. 137.) 8vo. *Berlin*

† Death of, announced, with short account of his Life, and list of his principal works. 8vo. 3 pp. (*Phil. Mag.* l. 392.) *London,* 1817

Lucan. On conducting Lightning. (*From Becquerel, Hist.*)

Lucca Accad. Atti della Rl. Accad. Lucchese. 8vo. (*The 13th volume is dated* 1845.) *Lucca*

⸸ **Lucca, De.** Recherches sur l'iode dans l'air, dans l'eau, &c. 8vo. 16 pp. (*Ext. Journ. de Pharmacie.*) *Paris,* 1854

Lucretius, Titus Carus. *Born* B.C. 99 ; *died about* B.C. 56.

De rerum natura, lib. vi. v. 1040. (*From Ingenhouz, Nouv. Expt.* p. 328. *Magnets.*)

Lüdecke. (*Vide* Lüdicke.)

† **Lüder, G.** De methodis demonstrandi declinat. magnetis variam et inconstantem. (Duæ dissertationes.) Theses. 4to. 56 pp. 1 plate. *Vitemberga,* 1718

⸸ **Lüders, J.** Das Nord-oder-Polarlicht, wie es ist, und was es ist. Eine Zusammenstellung von Thatsachen über dasselbe, und diesem verwandte Erscheinungen der Atmosphäre. Nach Beobachtungen im Westen der Vereinigten Staaten von Nord-America. 8vo. 45 pp. *Hamburg,* 1870

† **Ludersdorff.** Effect of the Voltaic Pile upon Alcohol. 8vo. 1 p. *(Phil. Mag.* lx. 155.) *London,* 1822

† **Ludewig, J.** Der Bau von Telegraphenlinien mit besonderer Berücksichtigung der Vorschiften der Telegraphen-Verwaltung des Norddeutschen Bundes, so wie der neuen Maasordnung. Für Telegraphen-Beamte und Techniker Zweite vermehrte und umgearbeitete Auflage. 8vo. 327 pp. *Leipzig,* 1870

Lüdicke, August Friedrich. *Born October 6, 1748, at Oschatz; died December 12, 1822, at Wilsdruf.*

Comment. de attractionis magnetum naturalium quantitate. 4to. *Viteb.* 1779

Beschreib. e. wenig Rostbaren galvan. Batterie. 8vo. *(Gilbert, Ann.* ix. 1801.) *Leipzig,* 1801

Versuche mit e. magnetisch. Batterie. 8vo. *(Gilbert, Ann.* ix. u. xi. 1802.) *Leipzig,* 1802

Versuche mit d. trockn. elektr. Säule. 8vo. *(Gilbert, Ann.* l. 1815.) *Leipzig,* 1815

Einfluss d. Magnetism auf. d. Krystallisation d. Salze. 8vo. *(Gilbert, Ann.* lxviii. 1821.) *Leipzig,* 1821

Ludolff, Christian Friedrich. *Born March 5, 1707, at Berlin; died October 22, 1763, at Berlin.*

Kindled (by electric sparks) spirit of Trobenius. *(Mém. de l'Acad. Royale à Berlin,* 1744.) 1744

(Priestley says he was the first to fire inflammable substances, but does not say if he published his discovery himself.)

Ludolf, C. F. (le jeune.) Mémoire sur l'electricité des Baromètres. 4to. *(Mém. de Berlin,* an. 1745, p. 12.) *Berlin,* 1745

(" Mémoire sur l'Electricité des Baromètres, par M. Ludolff le Jeune, et traduit du Latin, published at Berlin in 1751."—Latimer Clark's Catalogue.)

Ludolph, H. Vollst. Einleit. in die Chymie. 8vo. *Frankfort,* 1752

† **Lüdgte, Rob.** Über den Einfluss mechanischer Veränderungen auf die magnetische Drehungsfähigkeit einiger Substanzen. Thesis. 8vo. 33 pp. *Berlin,* 1869

Ludwig, Christian. Boschäftigte sich mit Electricität. *(Gehler's Wörterb.)*

Lulolfs, Johann. *Born August 5, 1711, at Zütphen; died November 4, 1768, at Leyden.*

† De Aurora Boreali. (Dissertatio .. inauguralis.) 4to. 83 pp. *Trajecti ad Rhenum,* 1731

† **Lullin, A.** Dissertatio physica de Electricitate quam .. præside H. B. de Saussure .. publice tueri conabitur Ams. Lullin, Sept. 26. 8vo. 55 pp. *Geneva,* 1776

Lunardi, Otto. Tre anonime Dissertationi latine comparvero 1755 in Roma su l'electricità. Due di quelle sono del P. Otto Lunardi che le fece defendere da due Nobili Conviti dero Sem. Rom. L'autore della terza, che fu defesa in Casa del Conte Soderini mi è ignoto. This 3rd is by Ct. Soderini himself.

Lundborg, J. M. De electricitate atmosphæræ. 4to. *Lund,* 1791

† **Lundy, T. E.** The Electric Telegraph and Railway Review, edited by T. E. Lundy. Fol. *London,* 1870

(Note.—The first five numbers are called Elect. Telegraph Review only. The last number issued seems to be No. 51.)

Lunn. Elect. Article in the Encycl. Metropolitana. 4to. 124 pp. *London*, 1830

Lussac. (*Vide* Gay-Lussac.)

Lütke, Friedrich Benjamin von. *Born September* 17, 1797 (*A. St.*), *at St. Petersburg.*

† **Lütke,** F. B., and **Lenz,** E. Beobachtungen des Inclination und Intensität der Magnetnadel angestellt auf einer Reise um die Welt auf den Sloop Seniawin in den Jahren 1826-7-8-9 vom Capt. F. B. Lütke berechnet und bearbeitet von E. Lenz. (Gelesen den 12 October, 1834.) 4to. 36 pp. (*Bullet. Scient. Acad. Petersb.* i. 1836.)   *Petersburg*, 1836

Luz, Johann Friedrich. *Born August* 2, 1744, *at Obernbreit, Frank.* ; *died July* 20, 1827.)

† Unterricht vom Blitze u. v. Blitz u. Wetter Ableitern. 8vo. 1 plate. (*From Kuhn of* 1866, p. 279.)   *Nuremberg*, 1783

Lehrbuch d. theor. prakt. Blitzableitungs lehre neu bearbeitet von Gutle, 1te Theil. 8vo.   1804

Unterricht vom Blitz und den Blitz- und Wetter Ableitern . . . neu bearbeitet von Gutle. 1st Theil. 8vo. 220 pp. 1 plate.   *Nurnberg*, 1804

Lyon, John. *Born* . . . . ; *died June* 30, 1817, *at Dover.*

† Experiments and Observations made with a view to point out the errors of the present received theory of Electricity. 4to. 280 pp. 2 plates. *London*, 1780

† Further proofs that Glass is permeable by the Electric effluvia, and that the Electric particles are possessed of a polar virtue. 4to. 80 pp. 2 plates.
                                        London, 1781

† Remarks on the leading proofs offered in favour of the Franklinean System of Electricity. 8vo. Plate.   *London*, 1791

Account of several new and interesting phænomena discovered in examining the bodies of a man and four horses killed by Lightning near Dover. 8vo.
                                        London, 1796

Lyons Acad. Delandine, President. Compte Rendu des travaux de l'Acad. pendant le 2e Semestre, 1804. 8vo. (*Comptes Rendus pour* 1804.) *Lyon*, 1806

Comptes Rendus des travaux de l'Acad. pendant les années 1804 à 1841. 8vo. (*The Compte Rendu for* 1806 *printed in* 1803, *that for* 1841 *printed in* 1843.)   *Lyon*, 1809, 1843

De la Prade, Président. Compte Rendu des travaux de l'Acad. pendant le 1er Semestre, 1823. 8vo. (*Comptes Rendus pour* 1823.)   *Lyon*, 1825
(*Note.—Describes further Experiments of Mollet on Decompositions by the Pile. Mentions a Mem. of Tabareau on Light, Magnet. and Elect. &c.*)

Histoire de l'Acad. Royale des Sciences, Belles-lettres et Arts de Lyon par T. B. Dumas, Séc. perpétuel. 8vo. 2 vols. (*From its estab. in* 1700 *to* 1838, *inclusive.*)   *Lyon*, 1839

Mémoires de l'Acad. Royale des Sciences, Belles-lettres et Arts de Lyon. 1st vol. 1845, 416 pp. (Section des Sciences) ; 2nd vol. 1847, 429 pp. 2 vols. La. 8vo.   *Lyon*, 1845-7

Comptes Rendus et Extraits des Procès verbaux des Séances de l'Acad. Rle. des Sciences, Belles-lettres et Arts de Lyon. 8vo. 132 pp.   *Lyon*, 1847

Mémoires de l'Acad. Nationale des Sciences, Belles-lettres et Arts de Lyon. Nouv. série. Classe des Sciences, 1st vol. 1851. La. 8vo.   *Lyon*, 1851

Lyon. Bibliothèques du Palais des Arts. Catalogue général par ordre alphabétique des Bibliothèques du Palais des Arts, de l'Acad., et des Sociétés d'Agriculture, de Médecine, de Pharmacie, et Linnéenne. (*Par Monfalcon.*) Fol. 242 pp.   *Lyon*, 1844

Lyon. Bibliothèque Lyonaise de M. Coste. Catalogue de la Bibliothèque Lyonaise de M. Coste, redigé et mis en ordre par Vingtrimer, son bibliothécaire. La. 8vo. 840 pp. *Lyon*, 1853

Lyon. Bibliothèque de la Ville. Catalogue de la Bibliothèque de la Ville de Lyon MSS. Fol. *Lyon*

Lyon Congrès. Congrès scientifique de Lyon, 9me Session, 1841. Ecrits d'Auteurs Lyonais sur divers sujets. 8vo. *Lyon*, 1841

Congrès scientifique de France, 9me Session, tenu à Lyon en Septembre, 1841. Tome i. Procès verb. des Sect. Tom. ii. Mémoires. 2 vols. 8vo.
Lyon et Paris, 1842

Lyon. Société Nationale d'Agriculture. Annuaire des Sciences physiques, etc., publié par la Société Nationale d Agriculture. 11 vols. La. 8vo. Plates. *Lyon*, 1838 *to* 1848

M.

M . . . (*Vide* Anon. Medical Elect. 1763.)

M . . . (*Vide* Anon. Thunderstorms, 1783.)

Maas, Anton Jacob. *Born June 13, 1795, at Maestricht.* (*Poggendorff*, ii. 2.)
(*Note.—Papers in Bull. Acad. Bruxelles.*)

Macaire, Isaac François. *Born July 21, 1796, at Geneva.*
Sur la phosphoréscence des Lampyres. 8vo. (*Ann. Chim. et Phys.* xvii. 1821.)
Paris, 1821

† **Macaire et De La Rive.** Expériences pour servir à l'histoire de l'acide muria-
tique (hydrochlorique). 4to. 9 pp. (*Mém. de la Soc. de Genève*, tom. ii. pt.
ii. p. 61. *Lu 19 Juin* 1823.) *Genève,* 1824

† **McCallum,** D. The Globotype Telegraph : a recording instrument, by which
small coloured balls are released, one by one, and made to pass over a series
of inclined planes, by the force of their own gravity. 8vo. 32 pp. 1 Cut.
London, 1856

† **Marcartney,** J. Observations upon Luminous Animals. 8vo. 12 pp and 14 pp.
2 plates. (*Phil. Mag.* xxxvii. pp. 24 and 93. *Originally in Phil. Trans. for*
1810, pt. ii.) *London,* 1811

McCullagh. (*Vide* Anon. Magnetism, 1778.)

MacCulloch. Traités et Essais, Essai sur l'origine de la boussole. (*Trad. dans le*
Moniteur, No. du 27 Octobre, 1853, p. 1194.) *Paris,* 1853

† **Macdonald.** On the North-west Magnetic Pole. 8vo. 6 pp. (*Phil. Mag.* lvii. 88).
London, 1821

On the discovery of a North-west Magnetic Pole. (In the Gentleman's Mag. for
July, 1821.) 8vo. 5 pp. (*Phil. Mag.* lviii. 99.) *London,* 1821

Macdonald, John. Observation of the diurnal variation of the Magnetic needle
at Fort Marlborough in the Island of Sumatra. 4to. (*Phil. Trans.* 1796, p.
340.) *London,* 1796

Observations of the diurnal variation of the Magnetic needle at St. Helena.
4to. (*Phil. Trans.* 1798, p. 397.) *London,* 1798

† **Macfait,** E. Observations on Thunder and Electricity. 8vo. (*Essays and Obser-*
vations, Phys. and Liter. i. 189.) *Edinburgh,* 1754

† **McGregor,** W. Questions on Magnetism, Electricity, and practical Telegraphy,
for the use of students. 12mo. 90 pp. *London,* 1868

† **Macintosh,** C. Electricity produced by separation of parts. 8vo. 1 page
(*Edinb. Phil. Journ.* x. 185.) *Edinburgh,* 1824

Macintosh, John. Facts and figures relative to Submarine Telegraphy as a branch
of commercial enterprise. 8vo. 36 pp. 1 cut. *London,* 1866

Mackay, The theory and practice of finding the Longitude at sea or on land ;
to which are added various methods of determining the Latitude of a place,
and variation of the Compass, with new tables. 3rd edit. 2 vols. Imp. 8vo.
Plates. *London,* 1810

Mackenzie. Electro-physiological Experiments. 8vo. (*Lancet,* 6 March, 1858.)
London, 1858

† **Mackrell,** Gann, and **Pollock.** On the action upon the Galvanometer by arrangement of coloured liquids in a U tube. 8vo. 50 pp. *London,* 1850

† **Maclear,** T. On a Luminous Arch seen at Biggleswade, on the 23rd of March.
8vo. 3 pp. (*Phil. Mag. or Annals,* v. 373.) *London,* 1829

McRea. (*Vide* Turnbull and McRea.)

Macusson (or D. F. A. M. R. D. C.) Dissertation sur le Feu boréal. 12mo.
111 pp.   1733

† **Madison,** James. Experiments on Magnetism communicated in a letter to
Thomas Jefferson. (Read May 4, 1798.) 4to. 6 pp. (*Trans. Amer. Phil. Soc.*,
old series, vol. iv. p. 323.)   *Philadelphia,* 1799

Madras Observatory. (*Vide* Taylor, Worster and Jacob.)
(*Vide* Goldingham and Taylor.)

Madrid Acad. Memorias de la Real Acad. de Ciencias. 4to. 9 plates.
                                                                Madrid, 1850-51

Madrid. (*Vide* also Academia de Madrid.)

Maffei, Scipione de. *Born June* 1, 1675, *at Verona; died February* 11, 1755, *at
Verona.*

† Della formazione de' Fulmini Trattato del Sig. Mar. S. Maffei raccolta di
varie sue lettere, in alcune delle quali si tratta anche degli Insetti rigeneran-
tisi, e de' Pesci di mare sui monti, e più a lungo dell' Elettricità. (Edited by
Tumermanni, 15 Letters.) 4to. 189 pp.   *Verona,* 1747
(*Note.—This is said by Grimelli (Storia) to contain all the anterior letters on
Lightning published by him, i.e. all before* 1746, *the date of the censorship.*)

Tre Lettere. 4to.   *Verona,* 1748

Gedanken von dem Blitz, aus dem Ital. übers. 8vo.   *Frankf. u. Lipz.* 1758

A letter without title-page, date, or place.   "Al suo carissimo . . amico il P.
D. Ippolito Bevilaqua." 4to. 8 pp. (*It concerns the case of La Signora Contessa Cornelia Baudi Cesena, &c.*)

Opuscoli letterari, con alcune sue lettere edite ed inedite, con Ritratto. 8vo.
                                                                Venezia, 1829
Elogio di Pindemonte.   *Verona*

Portrait. P. Anderloni dis. ed. inc.

† **Maffioli** di Udine. On Lightning Conductors. (*Giornale d'Italia,* 25 Agosto,
1770.)   1770

Magolotti, Lorenzo. *Born December* 13, 1637, *at Rome; died March* 1, 1712, *at
Florence.*

† Saggi di naturali Esperienze fatte nell' Accad. del Cimento. 1st edit. Fol.
(*Vide Accad. del Cimento.*)   *Firenze,* 1666
(*Note.—Libri Sale Cat.* p. 480, *says this edition was printed for private dis-
tribution and is rare, it having been replaced by a new one dated* 1667.)

† Saggi . . Accad. del Cimento. 2nd edit. Fol.   *Firenze,* 1691
Lettere scientifiche, ed erudite. Portrait. 4to.   *Firenze,* 1721

† Saggi di naturali esperienze fatte nell' Accad. del Cimento descritte dal Conte
Lorenzo Magalotti. In questa Edizione si aggiunge la sua vita scritta dal
Sig. D. M. Manni. 8vo. 192 pp. 28 plates.   *Venezia,* 1761

Saggi . . Accad. del Cimento. 2 vols. 8vo.   *Milano, n. d.*
(*Note.—This belongs to an Edizione delle opere classiche Italiane, Divisia Delle
opere di Lorenzo Magalotti.*)

(For Life *vide* Accademia del Cimento.)

Magarotto, A. Assertiones tres ex physica publice propugnandæ. 12mo. 94 pp.
                                                                Patavii, 1791
(*Note.—The* 1st *is* "*Ductores electrici ad ædificia a fulminibus tutanda plurimum
conferunt, præsertim si acuminati sint.*" *The* 2nd *is* "*Vis electricitatis est
potissima terræmotus causa. The* 3rd *has no reference to Electricity.*)

† Franklini Theoria de electricitatis principio. 8vo. 158 pp. 2 plates.
                                                                Patavii, 1805

† Discorso sull' origine della Gragnuola con alcune riflessioni sulli Paragrandini.
8vo. 35 pp.   *Vicenza,* 1825

MAG 315

Magazin der neuesten Erfindungen. Von Poppe, Kuhn, und Baumgartner. Neu Folge.

Magazin für Naturvidens Kaberne. Udgivet of Professors Lunth, Hansteen, og Maschmann. *Christiania, 1823*
(*Vide* Muncke in Gehler's Wörterbuch, vii. 566.)
Magazin für d. Neueste aus. d. Physik u. Naturgeschichte, &c. 8vo. 12 vols.
Gotha, 1781-99
(*Vide* Lichtenberg and Voight.)

† **Magellan.** Est. d'una lettera sui conduttori. (*From Rozier*, Juin, 1782.) 4to. 3 pp. (*Opusc. Scelti*, v. 304.) *Milano, 1782*
† Articolo di lettera al Landriani intorno alla elettricità, &c. 4to. 3 pp. 1 plate. (*Translation by Arnocelli.*) (*Opusc. Scelti*, vi. 296.) *Milano, 1783*
(*Note.—Perforation of a glass tube by Electricity, and other experiments.*)
In his edition of Cronstedt's Mineralogy, On a Tourmalin in M. Steigliz's Cabinet at Leipzig.

† **Maggi, P.** Saggio d'una teoria matematica delle induzioni elettrodinamiche in seguito alla teoria fisica degli stessi fenomeni pubblicata nel giornale di Firenze lo scorso ottobre dal . . Nobili. Letto dal . Maggi il 1 Mag. alla conversazione . presso il . G. G. Orti. 1833. 8vo. 32 pp. and 26 pp.
Verona, 1833
Sopra una nuova proprietà del ferro dolce magnetizzato. Memoria letta . . 17 Feb. 1850 all' Accad. di Agricolt. . di Verona. 8vo. 15 pp. *Verona, n.d.*
Sopra un probabile uso geognostico del filo Voltaico. Nota del Dott. P. Maggi letta . . 18 Aprile, 1850, all' Acad. di Agricolt. . di Verona. 8vo. 9 pp.
Verona, n.d.

† **Maggiotto, F.** Lettera di . . al Prof. G. Toaldo sopra una nuova costruzione di Macchina-Elettrica. (Data) Venezia 15 Febbrajo, 1781.) 8vo. 12 pp. 1 pl.
Venezia,1781
† Saggio sopra l'attività della Macchina elettrica costrutta da lui ed alcuni riflessi intorno l'elett. fluido. 8vo. 28 pp. 1 plate. *Venezia, 1781*
† Considerazioni elettriche. 12mo. 20 pp.
† (1) Transunto di una Lettera al Toaldo sopra la costruzione, e gli effetti di una nuova Macchina Elettrica. 3 pp. *Venezia, 1731*
(2) Esperienze che provano l'attività della nuova Macchina elettrica. 3 pp.
(3) Considerazioni sopra il fluido elettrico. 7 pp. 4to. 13 pp. *Milano, 1783*
† Considerazioni elettriche. 4to. 9 pp. (*Opusc. Scelti*. v. 409.) *Milano, 1782*

Magliozzi. Notizie storiche intorno l'invenzione e l'uso della bussola, presso tutti i popoli antichi e moderni. 4to. *Napoli, 1849*

Magnat. De l'arrêt, à distance, des convois, en cas de rencontre prochaine.

Magnus. Fortführung v. Flüssigkeiten durch Electricität.

Magnetofilo, Il. Giornale della Società magnetica di Torino, pubblicato sotto la direzione di E. Allis. Anno i. vol. i. 8vo. 408 pp. *Torino, 1855*

Magnier. Nouveau manuel complet de télégraphie . par Ch. Walker . trad. de l'Anglais par D. Magnier. Suivi d'une Appendice. 18mo. 194 pp. 2 plates.
Paris, 1851

Magnin, Arthur. Le feu du ciel, histoire de l'électricité et de ses principales applications. Idées des anciens, premières observations, machine éléctrique, bouteille de Leide, paratonnerre, &c. 3e édition. 8vo. 240 pp. *Tours, 1866*

Magnus, Heinrich Gustav. *Born May 2*, 1802, *at Berlin.*
† De vi ancoræ in electro-magnetes et chalybo-magnetes. 8vo. 28 pp.
Berolini, 1836
† Über thermo-electrische Ströme. Gelesen 20 März, 1851. 4to. 1 plate. (*Berlin Acad. "Phys. Kl.* 1851.") *Berlin, 1851*
† Über directe und undirecte Zersetzung durch den galvanischen Strom. 8vo. 28 pp. (*Pogg. Ann.* civ. 1858.) *Leipzig, 1858*
Seven or more Articles in Pogg. Ann. vol. xxviii. 1833 to vol. civ. 1858.

† **Magrini, Ales.** Apparato di rotazione continua fondata sulle azioni attrattive e ripulsive delle correnti di un solenoide su di una barra magnetica, la quale possa muoversi soltanto parallelamente a se stessa. (Letta 9 Agosto, 1860.) 4to. 6 pp. 1 plate. *Milano*, 1860

Telegrafia elettrica di Gavarret, &c. (*Vide* Gavarret.)

† **Magrini, L.** Sopra l'elettro-magnetismo e le recenti scoperte del prof. Ab. Salv. Dal Negro. 8vo. 44 pp. 2 plates. *Padova*, 1834

† Indagini circa l'azione chimica delle calamite temporarie e risultamenti esperimentali che manifestano la proprietà dei ferri prismatici di magnetizzarsi più che i cilindrici in parità di massa. 4to. 8 pp. (*Mem. estratta dagli Ann. del Reg. Lomb.-Veneto* v. *Bimestre di Marzo e Aprile, Maggio e Guigno*, 1835.) *Padova*, 1835

† Nuovo motore elettro-magnetico. Invenzione di Luigi Magrini. . Memoria letta all' Imp. R. Accad. di Scienze . di Padova . . 15 Dec. 1835. 4to. 8 pp. 1 plate. *Padova*, 1836

† Nuovo metodo per determinare la legge delle attrazioni e ripulsioni magnetiche secondo le distanze. Singolari fenomeni osservati negli aghi di acciajo magnetici e non magnetici in corso di un temporale. Sperienze che si oppongono alla ipotesi di Oersted circa la causa della deviazione dell' ago magnetico per mezzo della corrente elettrica. Memoria letta, e relativi esperimenti eseguiti . . 24 Maggio, 1835, all' Imp. R. Accad. di Scienze . . di Padova da L. Magrini. 4to. 14 pp. 1 plate. *Padova*, 1836

† Nuovo metodo per rendere costante l'efficacia d'un elemento galvanico. Estratto da una Memoria letta . 8 Maggio, 1837, all' Ateneo di Venezia di L. Magrini. 4to. 3 pp. 1 plate. *Padova*, 1837

† Telegrafo elettro-magnetico pratica e a grande distanze immaginato ed eseguito da L. Magrini. 8vo. 86 pp. 4 plates. *Venezia*, 1838

† Intorno la influenza della lunghezza dei circuiti sulla intensità delle correnti Voltiani. Estratto della Memoria letta . 18 Decem. 1837 nel veneto Ateneo relativa al nuovo Telegrafo elettro-magnetico di Luigi Magrini. 4to. 11 pp. *Padova*, 1838

† Caso recentissimo di una stitichezza straordinaria vinta coll'azione del Galvanismo per opera di L. Magrini. 4to. 5 pp. *Venezia*, 1839

Relazione di alcune sue esperienze elettro-magnetiche eseguite colla limatura di ferro, 31 Luglio, 1857 (dal Namias). 4to. *Venezia*, 1839

† Sulla trasmissione di correnti voltaiche disuguali e contrarie da uno stesso conduttore. Estratto. 8vo. 2 pp. (*Giorn. I. R. Ist. Lomb.* ix. 28). *Milano*, 1844

† Three Notices of Declarations, &c. concerning (1) "correnti elettrichi," (2) Nuovi lavori sperimentali sull' elettricità tellurica Parte prima, and (3) Idem Parte seconda. (*Giorn. I. R. Ist. Lomb.* x. 1845, pp. 34, 35, 37.) *Milano*, 1844

† Sulla Elettromozione tellurica. Sunto di nuovi lavori esperimentali eseguiti dal Prof. L. Magrini in occasione del sesto congresso scientifico, mediante il grande apparato fatto costruire dalla città di Milano. 8vo. 69 pp. (*Estr. dal* vol. vii. *del Politecnico di Milano*.) *Milano*, 1845

† Expériences sur la force électro-motrice tellurique exécutées . avec l'appareil que la Ville de Milan fit construire. 4to. 8 pp. *Paris*, 1845

† Esperienze sulla forza elettro-motrice e sulla conduttività tellurica . (*Cenni estratti da un lavoro lungo di cui si sta preparando la pubblicazione.*) 8vo. 31 pp. *Milano*, 1845

† Osservazioni intorno all' esperienze sul Telegrafo elettrico eseguite sulla linea da Parigi a Rouen. 8vo. 2 pp. *Milano*, 1846

† Esperienze dell' elettricità-tellurica di Magrini. 1 p.

MAG

317

Magrini, L.—*continued.*

Intorno le sperienze telluro-elettrichi di Magrini. 2 pp. 8vo. (*Giorn. I. R. Istit. Lomb.* xiv. 5, 6.) *Milano*, 1846

† Sulla tromba che devastò l'I. R. Parco di Monza il 13 Marzo 1846. 8vo. 26 pp. 5 plates. (*Giorn. I. R. Istit. Lomb.* xv. 206.) *Milano*, 1846

† Risposta alle considerazioni di G. Belli sulle Trombe di terra e di mare. 8vo. 9 pp. (*Giorn. I. R. Istit. Lombardo*, xvi. 63.) *Milano*, 1847

† Replica alle considerazioni sulle Trombe di terra e di mare e ad una sperienza per imitarle di G. Belli. 8vo. 8 pp. (*Giorn. I. R. Istit. Lombardo*, xvi. 316.) *Milano*, 1847

† Sulla Tromba che devastò il territorio di Orzinovi il 26 Luglio, 1847. 4to. 8 pp. (*Giorn. dell I. R. Istit. Lomb.* Nuova Serie,i. 97.) *Milano*, 1847

† Communications of several kinds were made by him to the I. R. Institute, and shortly mentioned by the Secretary in the year 1847. 1847

† Osservazione, &c. sopra un fenomeno avvertito dal Sig. Dubois Reymond. 4to. 6 pp. (*Giorn. dell I. R. Ist. Lomb.* Nuova Serie, ii. 41.) *Milano*, 1850

† Sopra un modo particolare di polarizzazione dei coibenti, svelato dai fenomeni dell' elettricità vindici. (Letta 28 Gennajo, 1851.) 4to. 9 pp. (*Giornale del I. R. Istit. Lombardo*, tom. ii. Nuova Serie, p. 291.) *Milano*, 1851

† Ricerche sulla natura del principio elettrico ; ossia sperienze tendenti a provare che il principio delle vibrazioni può essere applicato anche ai fenomeni elettrici. 4to. 37 pp. 4 plates. (*Giorn. dell' I. R. Istit. Lombardo*, Nuova Serie, tom. iii. p. 318.) *Milano*, 1852

† Nota sul Riassunto generale di telegrafia (Milano, 1854) comunicata a questo I. R. Istituto, 9 Marzo, 1854. 4to. 5 pp. (*Giorn. dell' I. R. Istit. Lombardo*, Nuova Serie, tom. v. p. 503.) *Milano*, 1844

† Sulla particolarità del legno di trasmittere una corrente elettrica più facilmente nel senso delle fibre che trasversalmente. Nota. 4to. 4 pp. (*Giorn. del I. R. Istit. Lomb.* Nuova Serie, vi. 187.) *Milano*, 1844

† Intorno i guasti cagionati dal Fulmine al Telegrafo elettrico nella notte dal 18 a 19 Maggio pp. sulla linea da Milano a Vercelli ; e, nel giorno successivo, sulla linea da Milano a Brescia ; e intorno al modo di premunirsi contro questa azione della elettricità atmosferica. 4to. 13 pp. 1 plate. (*Estratta dal Giorn. dell' I. R. Istit. Lomb.* tom. vi. fas. 31 and 32.) *Milano*, 1854

† Sugli effetti dell' arco Voltiano nell' olio di Trementina. (Letta 9 Nov. 1854.) 4to. 7 pp. (*Estratta dal Giorn. dell' I. R. Istit. Lomb.* tom. vi. fas. 54 and 55.) *Milano*, 1855

† Notizie storiche, considerazioni ed esperimenti sul quesito : possono in uno stesso filo coesistere due correnti contrarie, e trasmettersi simultaneamente due dispacci elettrici in senso opposto? (Letta 8 Feb. 1855.) 4to. 16 pp. (*Estratta dal Giorn. dell' I. R. Istit. Lomb.* tom. vii. fas. 37 and 38.) *Milano*, 1855

† Monografia delle principali Trombe terrestri osservate in Lombardia dopo il 1845, e considerazioni sui loro effetti e sulla loro origine. 8vo. 96 pp. 1 plate. (*Estr. dalla Rivista Ginnasiale*, an. 5, 1858.) *Milano*, 1858

† Sopra alcuni singolari effetti della scarica elettrica ; e specialmente sul trasporto delle molecole del condensatore e dell' eccitatore. 4to. 1 p. (*Atti dell' I. R. Istit. Lomb.* i. 67.) *Milano*, 1858,

† Rivendicazione della scoperta di un fenomeno di polarità elettro-magnetica. 4to. 3 pp. (*Atti dell' I. R. Istit. Lomb.* i. 206.) *Milano*, 1858

† Sopra una proposta del Perani di scaricare le nubi dell' elettricità. 4to. 2 pp. (*Atti del I. R. Istit. Lomb.* i. 221.) *Milano*, 1868

† Sulla Procella che colpì la città e i contorni di Milano il 30 Luglio 1858.

† Sull' applicazione di palloni metallici da scaricare le nubi dall' elettricità.

† Sulla formazione della Grandine. 4to. 6 pp. (*Atti dell' Istit. Lomb.* i. 224-7.) *Milano*, 1858

Magrini, L.—*continued.*

Sulla scoperta di correnti elettriche continue, a circuito aperto, confermata dalle recenti sperienze di Wheatstone. 4to. 4 pp. (*Atti dell' I. R. Ist. Lomb.* i. 250.) *Milano*, 1858

Sulla scintilla elettrica ottenuta per l'influsso del magnetismo temporario mediante l'apparato di Ruhmkorff. 4to. 1 p. (*Atti dell' I. R. Istit. Lomb.* i. 269.) *Milano*, 1858

† Cenni sopra una nuova forma di pila Voltiana. 4to. 2 pp. (*Atti dell' I. R. Istit. Lomb.* i. 390.) *Milano*, 1858
(*Note.*—*The date of the volume is not the date of the article.*)

† Osservazioni sulla Grandine caduta a Milano il 25 Luglio 1859. 4to. 1 p. (*Atti dell' I. R. Istit. Lomb.* i. 399.) *Milano*, 1858
(*Note.*—*The date of the volume differs from the date of the Fascicolo containing the article.*)

† Sopra una macchina elettrica di speciale costruzione e sopra alcuni fenomeni per essa ottenuti. 4to. 16 pp. 3 plates. (*Letta nelle adunanze dell' Istituto Lombardo del 20 Nov.* 1856, *e del 5 Genn.* 1858.) (*Estratto Mem. dell' Istituto,* vol. vii. fasc. 4.) *Milano*, 1858

Metodo facile e economico di costruire Macchine elettriche molto potenti. Letta nell' Adunanza del 20 Nov. 1856. 4to. 7 pp. 3 plates. (*Mem. I. R. Istit. Lomb.* vii. 249.) *Milano*, 1859

Sopra alcuni effetti della scarica che danno indize di trasporto delle molecole dei conduttori. Nota letta dell' Adunanza nel 5 Genn. 4to. 6 pp. (*Mem. I. R. Ist. Lomb.* vii. 257.) *Milano*, 1859
(*Note.*—*Although the volume is dated* 1859, *the Fas.* iv. *in which this and the preceding paper appears is dated* 1858. *These form a latter part of the Brochure presented Marz.* 23, 1861, *entitled " Sopra una Macchina elettrica speciale costruzione,"* &c. 16 pp. 3 plates.)

† Intorno ad alcuni fenomeni d'induzione elettro-magnetica ottenuti coll' apparato di Ruhmkorff. Nota letta 10 Maggio 1860. 4to. 4 pp. *Milano*, 1860

† Intorno un compito Osservatorio meteorologico proposto da Ales. Volta sino dal 1791. 4to. 6 pp. 1 plate. (*Atti dell' I. R. Istit. Lomb.* ii. 239.) *Milano*, 1860

† Sopra un metodo di togliere alle nubi maggiore copia di elettricità che coll, ordinario parafulmine. Nota letta 25 Agost. 1859. 4to. 3 pp. *Milano*, 1860

† Sulla priorità dell' osservazione che la presenza del ferro dolce in una spirale aumenta l'effetto della scintillazione, &c. 4to. 1 p. *Milano*, 1860

† Sui manoscritti, &c. di Volta. 4to. 1 p. (*Atti dell' I. R. Istit. Lomb.* ii. 102.) *Milano*, 1860

† Sopra un caso di polarizzazione del vetro. 4to. 1 p. (*Atti dell' I. R. Istit. Lomb.* ii. 102.) *Milano*, 1860

† Notizie biografiche e scientifiche su Alessandro Volta esposte dal Dr. L. Magrini. Lette in varie tornate del 1861 del R. Istituto Lombardo di Scienze, Lett. ed Arti. 4to. 53 pp. 1 plate. *Milano*, 1861
(*Note.*—*This Collection comprise the following Articles.* (1) *Sui manoscritti inediti di Volta communicazione fatta* . 28 *Giugno,* 1860, vol. ii. p. 234. (*of*) *Atti del R. Istit. Lomb.* (2) *Intorno un compito osservatorio meteorol. proposto da Volta. Comunic. fatta* 7 *Feb.* 1861, vol. ii. p. 239 (*of Id.*) (3) *Sugli strumenti meteorol.&c. Discorso inedito di Volta,* vol. ii. p. 246 (*of Id.*) *The two last articles may be considered as one.* (4) *Notizie biografiche e scientifiche su Volta desunte dai suoi Autografi recentemente rinvenuti, communicate* . 21 *Feb. e* 7 *Marzo,* 1861, vol. ii. p. 254 (*of Id.*)

Ragguaglio dei Lavori. Tornata 7 Feb. 1861.
(*Note.*—*He is said to have obtained interesting notices concerning a meteorological Observatory proposed by Volta in a " Carteggio " of Aug.* 1791, *given to the R. D. Magistrato, and now existing in the Archivi governativi di Milano.*)

Magrini, L.—*continued.*

Comunica alcune osservazioni sopra un fenomeno elettrico non ancora avvertito. 4to. (*Sparks of a new kind from an Electrical Machine.*) 1861

† Sulla meteora che nella sera del 4 Marzo, 1861, colpiva la cattedrale di Milano ; e sulla riforma de' suoi Parafulmini. Memoria. 4to. 11 p.
Milano, 1861

† Risultati di nuove sperienze elettro-magnetiche e diamagnetiche. Con avvertimento del Codazza. 8vo. 8 pp. (*Rendiconti del R. Istit. Lomb.* ii. 129.)
Milano, 1865

† **Mahmoud,** Effendi (Director of the Cairo Observatory). Observations et recherches sur l'intensité magnét. et sur les variations . . de 1829 à 1854. 8vo. 23 pp. (*Ext. du* tom. xxi. No 9, *des Bull. de l'Acad.*) *Bruxelles,* 1854

† Mémoire sur l'état actuel des Lignes Isocliniques et Isodynamiques dans la Grande-Bretagne, la Hollande, la Belgique et la France. Presenté à . . l'Acad. Rl. de Belgique le 8 Nov. 1856. 4to. 47 pp. 1 plate. (*Ext. du* tom. xxix. *des Mém. Couronnés et Méms. des Savants Étrangers, Acad. Rl. de Belgique.*) *Bruxelles,* 1856

Mahndel. Sur les pierres de foudre. 4to. (*Acad. des Inscrips.* xii. *Hist.* 163.)
Paris

† **Mahne, E. H.** De magnetismo animali. 8vo. 68 pp. *Gandavi,* 1829

Mahon, Charles Viscount. *Born August* 3, 1753, *at Geneva ; died December* 15, 1816, *at Chevening, Kent.*

† Principles of Electricity, containing divers new Theorems and Experiments, together with an Analysis of the superior advantages of high and pointed Conductors. . 4to. 263 pp. 6 plates. *London,* 1779

† Principes de l'Electricité . . trad. de l'Angl. par l'Abbé N. . . . . 8vo. 250 pp. 6 plates. *Londres,* 1781

Grundsäte d. Elekt. a. d. Engbn. mit Anmerk. v. Saeger. 8vo. 6 plates.
Leipzig, 1798

† **Maille, P. H.** Nouvelle théorie des Hydrométéores suivie d'un mémoire sur l'Electricité atmosphérique et d'un autre sur la Pluviométrie. 8vo. 372 pp. 2 plates. *Paris,* 1853

Maimbray. At Edinburgh electrified two Myrtle trees, during the whole month of October, 1746. (*From Priestley, Hist.* 4th edition, p. 137.)

Main. (*Vide* Herschell & Main.)

† **Main, J.** On the Phænomena of Waterspouts. 8vo. 3 pp. (*Phil. Mag.* or *Annals,* iii. 114.) *London,* 1828

Mainz Medicinischen Privatgesellschaft. Galvanis. und. elekt. Versuche. 4to. 50 pp. *Frankfort,* 1829

Mairan, Jean Jacques d'Ortous de. *Born November* 26, 1678, *at Béziers ; died February* 20, 1771, *at Paris.*

† Dissertation sur la cause de la lumière des Phosphores et des Noctiluques : (qui a remporté le prix de l'Acad. Rle. . . de Bordeaux pour l'an. 1717). 12mo. 54 pp. *Bordeaux,* 1717

Sur les effets de la chute du Tonnerre sur un arbre. 4to. (*Mém. Par.* 1724.)
Paris, 1724

Description de l'Aurore boréale du 26 Sept. 1726, observée au chateau de Breuillepont, village entre Pacy et Ivry, diocèse d'Evreux. 4to. (*Mém. de Paris,* an. 1726, *Hist.* p. 3, *Mém.* p. 198.) *Paris,* 1726

Traité physique et historique de l'Aurore boréale. 4to. (*Mém. de Paris,* an. 1731, *Mém.* p. 1, an. 1732, *Hist.* p. 1.) *Paris* 1731
(*Poggendorff* ii. 17, *says* 1731, 1732, 1747, *and* 1751.)

Mairan, Jean Jacques d'Ortous de.—*continued.*

Observations de quelques aurores boréales, qui ont paru cet automme 1731 à Breuillepont en Normandie, diocèse d'Evreux. 4to. (*Mém. de Paris,* an. 1731, p. 379.) *Paris,* 1731

Journal d'observations des aurores boréales qui ont été vues à Paris ou aux environs, dans le cours des années 1732 et 1733. Avec plusieures observations de la lumière zodiacale dans les mêmes années. 4to. (*Mém. de Paris,* an. 1733, *Hist.* p. 23, *Mém.* 477.) (*From Young,* p. 489.) *Paris,* 1733

Traité de l'Aurore boréale. 4to. *Paris,* 1733

Journal d'observations des aurores boréales, qui ont été vues à Paris ou aux environs, à Utrecht, et à Pétersbourg, dans le cours de l'année 1734; avec quelques observations de la lumière zodiacale. 4to. (*Mém. de Paris,* an. 1734, *Mém.* p. 567.) *Paris,* 1734

Traité . . . de l'Aurore boréale suite aux Mém. de l'Acad. Royale des Sciences, an. 1731. 12mo. *Amsterdam,* 1735

Système de L. Euler sur la cause de la queue des comètes, de l'aurore boréale et de la lumière zodiacale. 4to. (*Mém. Paris,* 1747.) *Paris,* 1747

Eclaircissemens sur le traité physique et historique de l'Aurore boréale. 4to. (*Mém. de Paris,* an. 1747, *Hist.* p. 32, *Mém.* p. 363.) *Paris,* 1747

Précis de la suite des Eclaircissmens du traité de l'Aurore boréale. 4to. (*Mém. de Paris,* an. 1751, *Hist.* p. 40.) *Paris,* 1751

† Traité de l'Aurore boréale revue et augmenté. 2e. Ed. 4to. (*First printed in Acad. de Paris,* 1731.) *Paris,* 1754

† **Maisiat,** M. Mém. sur quelques changements faits à la Boussole et au rapporteur, suivi d'un nouv. instrument, nommé grammomètre. 8vo. 178 pp. 8 plates. *Paris,* 1818

Maistre, X. (Comte de). Sur la cause qui fait surnager une aiguille d'acier sur la surface de l'eau. 4to. (*Mém. d. Turin,* série ii. 1841.) *Torino,* 1841

Majocchi, Giovanni Alessandro. *Born at Cremona; died October* 27, 1854, *at Turin.* (*Poggendorff,* i. 20.)

Dei Paragrandini. Ragionamento. 8vo. 14 pp. Pirotta. (*Inserito nella Gazetta di Milano,* 12 e 13 Luglio, 1823.) *Milano,* 1823

Sull' incertezza della meteorologia sulla necessità e sul modo di stabilire i fatti per isciogliere il problema dell' utilità dei Paragrandini Ragionamento. 8vo. 40 pp. *Milano,* 1824

Istruzione teorica e pratica sui Parafulmini. 8vo. 114 pp. 1 plate. *Milano,* 1826

Elementi di fisica. 8vo. 352 pp. *Milano,* 1826

† Estratto d'una lettera del. 8vo. 5 pp. (*Bibl. Ital.* xliv. 300.) *Milano,* 1826

† Parafulmini (Articolo comunicato.) 8vo. 5 pp. (*Bibl. Ital.* xlix. 294.) *Milano,* 1828

† Sulle correnti magneto-elettriche e sulla calamita elettrica. Lettera scritta al . . Carlini. 8vo. 19 pp. 1 plate. (*Bibl. Ital.* lxvii. 184.) *Milano,* 1832

† Galvanometro universale a forza variabile. 4to. 7 pp. 1 plate. (*Ann. del Reg. Lomb.-Veneto,* viii. 61.) *Padova,* 1838

† Discorso detto . . il giorno 14 Luglio 1838 nel teatro di fisica dell' I. R. Liceo di Milano al terminare le lezioni dell' anno scolastico. 4to. 10 pp. (*Ann. del Reg. Lomb.-Veneto,* viii. 205.) *Padova,* 1838

† Annali di Fisica, Chimica, &c. The first 20 vols. have the title "Annali di Fisica, Chimica, e Matematiche; col Bullettino di Industria, &c. . . The vols. 21 to 28 are entitled "Annali di Fisica, Chimica, e Scienze accessorie coi Bullettini di Farmacia, &c." 28 vols. 8vo. With plates. *Milano,* 1841-47

Majocchi, Giovanni Alessandro—*continued.*

† Appendice agli "Elementi di Fisica," del D. Scinà. 8vo. 98 pp. 1 plate.
Milano, 1842

† Delle condizioni necessarie alla produzione della corrente voltaica. Memoria
Seconda. 8vo. 29 pp. (*In his Annali di Fisica* xx. 1845.) *Milano,* 1846

† Nuove sperienze e considerazioni sull' origine della corrente elettrica nella
pila. Mem. terza. 8vo. 54 pp. (*In his Annali di Fisica*, tom. xxiv. 1846.)
Milano, 1846

Majocchi e Selmi. Annali di Fisica Chimica e Scienze affini coi Bullettini di
Farmacia e di Tecnologia. 8vo. *Torino,* 1850 *to* —

Majus (i.e. May), Heinrich. Disp. de tonitru. (*Pogg.* ii. 21.) *Marp.* 1673

Disp. de fulmine. 4to. (*Pogg.* ii. 21.) *Marp.* 1673

Makerstoun Observations. Observations in Magnetism and Meteorology made
at Makerstoun, in Scotland, in the Observatory of General Brisbane, 1841—
1846. 4to. 4 vols. (*Edinb. Phil. Trans.* xvii.—xix.) *Edinburgh,* 1845-49

Mako (von Kerek Gede), Paul. *Born July* 9, 1723, *at Jäss-Apath ; died August* 19,
1793, *at Pesth.*

† Dissertatio physica de natura, et remediis Fulminum. 8vo. 100 pp.
Goritiæ, 1773

† Physikal Abhandlung vom Nordlicht. 8vo. (*Auch in d. Beitr. zu verschiedn.
wiss.* 8vo. *Wien* 1795.) *Wien,* 1773

Dissertationes physicæ. 8vo. *Wien,* 1781

Mako, P. und Retzer. Physikalische Abhandlung von den Eigenschaften des
Donners, und den Mitteln wider das Einschlagen. Verfaszt von P. Mako
und von J. E. von Retzer in das Deutsche übersetzt. 1st ed. 8vo. 125 pp.
1 plate (*i.e.* frontispiece). *Wien,* 1772

Physikalische Abhandlung von den Eigenschaften des Donners, u. d. Mitteln
wider das Einschlagen. Verfaszt von P. Mako, und von J. E. Retzer, in das
Deutsch. übersetzt. 2te. Aufl. (*Von Retzer aus d. latin Orig. das erst.* 1773
erschien.) *Wien,* 1775

Malacarne, Michele Vincenzo Maria. *Born September* 28, 1744, *at Saluzzo ; died
September* 4, 1816, *at Padua.*

Osservazioni meteorolog. fatte e scritte da G. V. Malacarne e comunicate
alla Bibl. fis. d'Europa. 12mo. *Pavia,* 1789

Elogio di Ruggeri. 8vo. 1817

† An engraved Portrait of G. Asioli, inc. (*Also a small engraved Portrait from
the above.*)

Malfanti, G. Le méteore. *Genoa,* 1586

Mallemans (Mallement) de Messanges, C. *Born* 1653, *at Beaune, Bourgogne ;
died April* 17, 1723.

Nouveau système de l'Aiman, à M. l'Abbé Dangeau. 4to. 16 pp. 2 plates.
(*From Lamont, Handb.* p. 439.) *Paris,* 1680

Mallet-Favre, Jacques André. *Born September* . . . 1740, *at Geneva ; died January*
30, 1790, *at d'Avully, near Geneva.* (*Poggendorff,* ii. 28.)

Observationes variæ in Laponia ad Ponoi institutæ, a. 1769. 4to. (*Nov.
Comm. Acad. Petr.* xiv. 1770. *Comprising De acus magneticæ declinatione
Ponoi, an.* 1769, *observata.*) *Petropoli,* 1770

De acus magneticæ declinatione Ponoi an. 1769 observata. 4to. (*Novi Commentar. Acad. Petropol.* xiv. pt. ii. p. 33.) *Petropoli*

Mallet. (*Vide* Clement-Mallet.)

Malzet. (*Vide* Jacquet de Malzet.)

Manchester Society. Memoirs of the Literary and Philosophical Society of Manchester. 1 vol. 8vo. *London*, 1785
Memoirs of the Literary and Philosophical Society of Manchester. 5 vols. 8vo. (*Continued.*) *Warrington*, 1789-96

† Manent. Découverte . . . circulation de l'électro-magnétisme, le principe des sciences, etc. 8vo. 125 pp. *St. Gaudens*, 1857

Mangin. *Born at Langres ; died about 1772.*
Question nouvelle . . sur l'électricité, proposée . . par . . l'Acad. de Dijon et troitée par . . de Mangin. *Paris*, 1749
† Histoire générale et particulière de l'Electricité. (Anon.) 3 vols. 12mo. 1 plate. (*Comprises, in* vol. ii. p. 13, *his Question Nouvelle. See above.*) 1752

Manheim Acad. Theodoro-Palatina. Historia et Commentationes. 11 vols.? 4to. *Manhemii*, 1766

Manheim Societ. Meteorolog. Palat. Ephemerides Societatis meteorologicæ Palatinæ. 10 vols. 4to. *Manheim*, 1781-92

Mann, Theod. Augustin. *Born June* 22, 1735 (*N. St.*), *in Yorkshire; died February* 23, 1809, *at Prague.*
* Sur les Marées aériennes, &c. 8vo. *Bruxelles*, 1792

Manni. Life of Magalotti.
(*Vide* Accademia del Cimento.)

Manteufel, Count. Used " vases et gobelets de porcelaine de Saxe et du Japon " mounted as usual for elect. machines, which were very effective. Some time before 1752. (*From Mangin, Hist.* i. 31).

Mantua Acad. Memorie della Reale Accademia di Scienze, Belle-lettere ed Arti di Mantova. 4to. *Mantova*, 1795

Manuzio. Degli Elementi e di molti loro effetti. 4to. *Venezia*, 1557

Manzini, Carlo Antonio. Della sicura incertezza nella declinazione dell'ago magnetico dal meridiano. 4to. *Bonon*, 1650
Sulla declinatione dell' ago magnetico dal meridiano, etc. 4to. *Bologna*, 1650

Maraschini, G. B. G. *Born* 1774, *at Scio ; died September* 26, 1825.
Sulle combustioni spontanee del Corpo umano. 8vo. 37 pp. *Vicenza*, 1822

Marat, Jean Paul. *Born May* 24, 1743, *at Boudry, Cant. Neufchatel ; died July* 13, 1793, *at Paris.* (*Assassinated by Charlotte Corday.*)
† Découvertes sur le feu l'électricité et la lumière. 2nd ed. 8vo. 38 pp. *Paris*, 1779
Recherches physiques sur le feu. 8vo. *Paris*, 1780
† Recherches physiques sur l'électricité. 8vo. 461 pp. 5 plates. *Paris*, 1782
† Entdeckungen über d. Licht. durch eine Reihe neuer Versuche bestätigt aus el Franzos von Chr. E. Weigel mit Anmerk. 8vo. xiv. and 166 pp. *Leipzig*, 1783
† Mémoire de l'électricité médicale. 8vo. 111 pp. *Paris*, 1784
Phys. Forschungen über das Feuer,Übersetzt vonWeigel. (*Fm. Wiegel*, p.444.)
† Physische Untersuchungen über Electricität, a. d. Franz. 1782, mit Anmerkun-gen übersetzt von C. F. Weigel. 5 Kupf. 8vo. 660 pp. *Leipzig*, 1784
Observ. de M. l'Amateur Avec à M. l'Abbé Sans. Elect. Med. 8vo. 33 pp. *Paris*, 1785
Œuvres de Marat. 8vo. 1788
(*Note.—A title-page seems to have been printed, for affixing to a collection of his works previously published.*)

Maravigna. Du Galvanisme et de l'Electricité développée, etc. *Catania*, 1825

Marcel. On fragments of, and powdered Magnets. (*From Van Swinden.*)

† **Marcel,** A. An abstract of a letter, written in Dutch, to the illustrious Royal Society of London by Marcel, A. Communicated by Desaguliers, J. T. 4to. 5 pp. (*Phil. Trans.* xxxvii. 294.) *London*, 1731-32
(*Note.—Communication of Magnetism to Iron and Steel without the help of any Loadstone.*)

Marcet. (*Vide* La Rive De, and Marcet.)

Marchetti, Giov. De phosphoris quibusdam, ac præsertim de bononiensi. 4to. (*Comment. Bonon.* vii. 1794.) *Bonon*, 1791

Marchisio, A. Sopra un nuovo liquido indoratore. Lettera .. al Sig. Prof. S. Marianini. (*Indicatore Economico*, 21 *Giugno* 1844, *Modena*.) *Modena*, 1844

Marcilla, A. G. de. Tratado de telegrafia eléctrica. 4to. Plates. *Madrid*, 1851

† **Marco,** Felice. Principii della teoria meccanica dell' elettricità e del magnetismo cioè la costituzione della materia ponderabile svelata. 8vo. 230 pp. *Torino*, 1867

Marcorelle. Observation de la déclinaison de l'aiguille aimantée faite à Toulouse, le 27 Septembre, 1750. 4to. (*Mém. de Mathém. et de Phys.* ii. 612.) *Paris*

Description d'une trombe de terre .. à Escole le 15 Juin 1785. 4to. (*Toulouse Acad.* 1re Série, iii. 114.) *Toulouse*, 1788

Déclinaison de l'aiguille aimantée observée à Toulouse depuis le commencement de 1747 jusqu'à la fin de 1756. 4to. (*Mém de Math. et de Phys.* iv. p. 117.) *Paris*

Marcorelle and Darguier. Observations sur 3 Aurores boréales : 2 by Marcorelle, at Narbonne, 29th February, 1780, 1 by Marcorelle and Darguier, at Narbonne and Toulouse, on the 28th-29th July, 1780. 4to. (*Toulouse Acad.* 1er Série, ii. 20, *Hist.*) *Toulouse*, 1782

† **Marcus,** S. Ueber eine neue Thermosäule. 8vo. 6 pp. 1 plate. (*Abd. aus den Sitzungsberichten der Kais. Akad. der Wissenchaflen.*) *Wien*, 1865

Maréchaux, Peter Ludwig. *Born December* 28, 1764, *at Prenzlau; died* (?). (*Poggendorff*, ii. 46.) (*Papers in Gilbert's Ann.* x. to xxvii.) (*Vide* also Erman and Maréchaux.)

Marelli del Verde (Conte). Spiegazione della carica e scarica del Quadro Frankliniano. 4to. (*Opuscoli Scelti*, viii. 11.) *Vercelli*, 1785
† Spiegazione della carica, e scarica del quadro Frankliniano esposta. 4to. 12 pp. (*Opus. Scelti.* viii. 157.) *Milano*, 1785

Mareska, Joseph Daniel Benoît. *Born September* 9, 1803, *at Ghent; died March* 31, 1858, *at Ghent.*

† De legibus mathematicis electricitatis dynamicæ. 4to. 19 pp. 1 plate. *Gand.* 1826

Maret, Hugues. *Born October* 6, 1726, *at Dijon; died June* 11, 1785, *at Dijon.*

Description d'un météore observé à la chartreuse de Dijon le 20 Juillet 1779. 8vo. (*Nouveaux Mém. de l'Acad. de Dijon.*) *Dijon*, 1783

Margollé et Zurcher, Les météores. Ouvrage illustré de 23 vignettes sur bois par Lebreton. 18mo. 333 pp. *Paris*, 1864
† Les Météores. 8vo. 332 pp. *Paris*, 1865

Les Météores. 2nd ed. *Paris*, 1867

Les Météores. 3rd ed. revue et augmentée. 18mo. 304 pp. (*Bibliothèque des Merveilles.*) *Paris*, 1870

Marherr, Philipp Ambrosius. *Born* 1738, *at Vienna; died March* 28, 1771, *at Prague.*

† Von der Wirkung der Luft-electricität in dem menschlichen Körper aus dem Lat. 8vo. 32 pp. *Chur. u. Linden*, 1780

Prgrm. de electricitatis æreæ in corpus humanum actione. 4to. *Vienna*, 1766

324 MAR

Marherr und **Kirchvogel.** Von der Wirkung. d. Luft Elek.
Chur. u. Linden, 1767
(*Vide* also Bauer and others.)

Mariani Parthenii, Josephi. Electricorum, libri vi. (*From a memorandum (or notice*) *given to Ronalds by the Caval. F. Zantedeschi et Padua in* 1860, *who adds that, at* pp. 33, 34, " *si descrive in quest' opera il telegrafo elettrico scintillante, nel quale era impiegata l'elettricità di attrito e della bottiglia di Leida. Le sillabe e le parole venivano formate con gruppi di scintille di convenzione.*")
Romæ (excudebat generosus Salamoni), 1757

Marianini, Pietro. *Born June* 30, 1787, *at Zeme, near Novara; died March* 20, 1855, *at Mortara.*

† Proposta dell' Elettromotore voltaico siccome Pato-scopio. Presentata alla terza Riunione degli Scienziati Ital. in Firenze, Settem. 1841. 8vo. 11 pp. (*Giorn. Med.* tom. xv. p. 473.) *Venezia,* 1841

† Sopra il fenomeno che si osserva nelle calamite temporarie di non cessare totalmente, nè quasi totalmente, l'attrazione fra la calamita e l'ancora, quando, al cessar della corrente nel filo conduttore avvolto alla calamita, si conserva l'ancora ad essa applicata. Memoria. 4to. 24 pp. 1 plate. (*Mem. Soc. Ital. delle Scienze,* tom. xxv. p. i.) *Modena,* 1851

† Sopra l'aumento di forza assorbente che si osserva in un' elica elettro-dinamica quando è circondata da un tubo di ferro. Nota. 8vo. 8 pp. *Modena,* 1852

Sopra alcune fogge di calamite artificiali armate e sopra alcuni metodi per Magnetizzare. Memoria. (*Padua Accad. Rivista.*) *Torino,* 1856

† Sperienze relative alla dipendenza che l'attrazione tra calamita e ferro trae dal magnetismo che in questo si produce per influenza. Nota. Seguita da altra nota (del medesimo) su Alcune Sperienze puramente magnetiche relative al fatto che l'intensità del magnetismo di una calamita artificiale è maggiore quando l'ancora vi è applicata. 4to. 15 pp. *Modena,* 1869

Marianini, Stefano Giovanni. *Born January* 5, 1790, *at Zeme, near Novara, Piedmont.*

† Saggio di esperienze elettrometriche. 8vo. 206 pp. *Venezia,* 1825

Sul rapporto che esiste fra l'energia degli elettromotori ed i loro effetti. (*Mem. letta all' Ateneo-Veneto,* 20 Marzo, 1823.) *Pavia,* 1825

Memoria sulla perdita di tensione che soffrono gli apparati voltiani quando si tiene chiuso il circolo ; e sul riacquistare ch'essi fanno la tensione primitiva quando si sospende la comunicazione fra i poli. 4to. 20 pp. (*Esercit. Ateneo-Veneto,* i. 293.) *Venezia,* 1827

Nuovo Galvanometro moltiplicatore proposto e descritto da S. M. 4to. 3 pp. 1 plate. (*Esercitazioni . . dell' Ateneo di Venezia,* tom. i. p. 318.)
Venezia, 1827

† Memoria sopra la scossa che provano gli animali nel momento che cessano di fare arco di comunicazione fra i poli di un elettromotore, e sopra qualche altro fenomeno fisiologico dell' Elettricità. 8vo. 32 pp. *Venezia,* 1828

Sulla Teoria degli Elettromotori Voltiani.
His First Memoir on this subject is in tom. xx. Mem. Soc. Ital. 1832, 4to. Read 1830.

Second Memoir printed at Venice in 1830. 8vo.

Third Memoir in *Ann. Reg. Lomb.-Veneto,* 1836. 4to.

Fourth Memoir in tom xxi. *Mem. Soc. Ital.* 1837. 4to.

Fifth Memoir in his *Mem. di Fis. Speriment.* an. 2, 1838. 8vo.

† Sixth Memoir in his *Mem. di Fis. Speriment.* an. 2, 1838. 8vo.

Marianini, Stefano Giovanni—*continued.*

Sei memorie di fisica. (*Elenco dei libri mandati nei*, 1841-4.)   *Modena*

† Memoria sopra la teoria chimica degli elettromotori Voltiani semplici e composti. 8vo. 63 pp.   *Venezia*, 1830

Sur une analogie qui existe entre la propagation de la lumière et de l'électricité. 4to. (*Bibl. Universel*, xliii.)   *Turin*, 1830

† Memoria sopra le scintille eccitate per entro i liquidi dagli elettromotori. 4to. 9 pp.   *Padova*, 1831

Nota sopra il Salto della scintilla a traverso della fiamma. 2 pp. (*Ann. del Reg. Lomb.-Veneto*, i. 128 and 137.)   *Padova*, 1831

† Memoria sopra il fenomeno che presenta un arco metallico, di non egual superficie ne' suoi estremi, quando serve a tradurre l'elettricità da un fluido ad un altro della stessa natura. (Letta all'Ateneo di Treviso.) 4to. 11 pp. (*Ann. del Reg. Lomb.-Veneto*, tom. i. p. 281.)   *Padova*, 1831

† Sopra la teoria della pila. Memoria. Ricevuta 3 Feb. 1830. 4to. 13 pp. (*Mem. Soc. Ital.* xx. 347.)   *Modena*, 1832

† Lettera al Direttore di questi Annali sopra un principio di azione chimica prodotta alla superficie dei metalli dalle correnti faradiane. 4to. 2 pp. (*Ann. del Reg. Lomb.-Veneto*, ii. p. 144.)   *Padova*, 1832

† Nota sopra la facoltà elettromotrice del mercurio. 4to. 4 pp. (*Ann. del Reg. Lomb.-Veneto*, iii. 217.)   *Padova*, 1833

† Memoria di alcune paralisi curate coll' elettricità mossa dagli apparati voltaici ; con un Appendice sopra un nuovo fenomeno elettro-fisiologico. Parte prima, p. 17. Continuazione e fine dell Memoria. Parte seconda e Appendice, p. 68. 4to. 24 pp. (*Ann. del Reg. Lomb.-Veneto*, iii. 17 and 68.)   *Padova*, 1833

Sopra le contrazioni muscolari ed alcune sensazioni prodotte dalle correnti elettriche. Memoria letta all' Ateneo-Veneto li 4 Marzo, 1833. Parte prima, 9 pp. Continuazione e fine della Memoria. Parti seconda e terza, 10 pp. 4to. 19 pp. (*Ann. Reg. Lomb.-Veneto*, iv. pp. 32-57.)   *Padova*, 1834

† Memoria sopra il fenomeno elettro-fisiologico delle alternative voltiane. Letta all' Ateneo-Veneto, 26 Maggio, 1834. 4to.

Continuazione e fine della Memoria. 4to. (*Ann. del Reg. Lomb.-Veneto*, iv. 2 and 241.)   *Padova*, 1834

† Sopra la causa alla quale il Peltier attribuisce le contrazioni che provano gli animali quando s'interrompe il circolo voltiano di cui fanno parte. Lettera all'Accad. Reale di Parigi. 4to. 5 pp. (*Ann. del Reg. Lomb.-Veneto*, v. 301.)   *Padova*, 1835

† Sopra la teoria degli elettromotori. Memoria iii. Risposta alle osservazioni del Parrot. 4to. 12 pp. (*Ann. del Reg. Lomb.-Veneto*, vi. 13.)   *Padova*, 1836

† Sulla teoria degli Elettromotori. Memoria iv. Esame di alcune sperienze addotte dal Sig. Faraday, &c. Ricevuta li 22 Luglio, 1836. 4to. 41 pp. (*Mem. Soc. Ital.* xxi. 205.)   *Modena*, 1837

† Mem. di fisica speriment. scritta dopo il 1836, An. 1, 1837. 8vo. 144 pp.   *Modena*, 1838

† Mem. di fisica speriment. scritta dopo il 1836, An. 2, 1838. 8vo. 160 pp. 1 pl.   *Modena*, 1838

† 5a Memoria sugli Elettromotori. 8vo.   *Modena*, 1838

(*Note.—This is the Anno Secondo of his Memorie di fisica . . scritte dopo 1836, under the title "Sulla teoria degli elettromotori Voltaici Mem. v. Esame di alcune osservazioni e di alcuni ragionamenti recati a difesa della teoria chimica della pila."*)

Marianini, Stefano Giovanni—*continued.*

† 6a Memoria sugli Elettromotori. (*Belli in Bibliot. Ital. C.* 188.) **1838**
(*Note —Sulla teoria degli elettromotori Voltaici Mem.* vi. *Esame e difesa
della teoria del contatto, con una digressione sopra una Bottiglia di Leiden
la quale, mediante il contatto semplice si carica da se stessa.*)
" Un cenno del congegno che formò il soggetto della digressione sopra una
bottiglia di Leiden, p. 88, della Teoria degli elettromotori . Mem. v. fu
dato in una lettera al Pietro Marianini del 10 Maggio 1838, e venne poi pubblicato nella Gazzetta piemontese, No. 232 (1838.)" **1838**

† Mem. di fisica speriment. scritta dopo il 1836, An. 3, 1839. 8vo. 160 pp.
Modena, 1839

† Metodo per ottenere i bassi-rilievi in rame senza apposito elettromotore
voltaico. 8vo. 17 pp. *Novara,* 1840

† Mem. di fisica speriment. scritta dopo il 1836, An. 4, 1840. 8vo. 144 pp.
Modena, 1841

† Cenno di esperimenti elettro-grafici. Lettera diretta al . Pietro Marianini di
Mortara da Stefano Marianini. (Dated) Modena, 29 Agosto, 1841.
(*Estratto dal fascicolo* ii. *del Giornale Letterario Scientifico Modenese, Agosto*
1840.) Elettrografia. Problema, Dato uno scritto o disegno in metallo,
stamparlo sulla carta col mezzo dell' elettricità voltaica, e senza far uso
d'inchiostro. Modena, 19 Aprile, 1841. (*Estratto dalla Gazzetta Piemontese,*
No. 96, Anno 1841.) 8vo. 3 pp. *Vigevano,* 1841

† Di due Paralisi curate coll' elettricità voltaica. Memoria. Letta alla . .
terza Riunione degli Scienziati Ital. in Firenze il 21 Sett. 1841. 8vo. 6 pp.
(*Giorn. Med.* tom. xv. p. 484.) *Venezia,* 1841

† Metodo di sperimentare la conducibilità per l'elettrico dei metalli. Memoria
letta il 3 Luglio all' Accad. . di Modena. 8vo. 2 pp. (*Giorn. I. R. Istit.
Lomb.* ii. 287.) *Milano,* 1841

Memorie di fisica sperimentale. **1842**

Problemi di Magnetismo nell' Albo offerto dalla Rle. Accad. . . di Modena
agli Augti. Sposi Francesco e Aldegonda Principi Ereditarii.
Modena, 1842

† Memoria sull' indebolimento che avviene nel magnetismo d' un ferro quando
si fa scorrere su d'una calamita debole in modo da magnetizzarlo, se non lo
fosse, nel medesimo senso in cui già si trova magnetizzato. 4to. 30 pp.
(*Mem. Soc. Ital.* xxiii. Parte Fisica.) *Modena,* 1843

† Memoria di alcune analogie e di alcune discrepanze osservate tra le azioni
magnetizzanti della boccia di Leida, della coppia voltaica, e della calamita.
4to. 19 pp. (" *Inserita nella parte fisica del Tomo* xxiii. *delle Mem. della Soc.
Ital.*") *Modena,* 1843

† Memoria sopra la corrente che nasce in un filo metallico chiuso, quando si
sospende la corrente Voltaica che passa vicina e parallela ad esso. Letta
alla R. Accad. . . di Modena . . 5 Agosto, 1841. 4to. 11 pp. (" *Un
sunto era letto* 27 *Settem.* 1841, *alla terza Riun. degli Scienziati.*")
Modena, 1844

Memoria sul magnetismo dissimulato, e sopra alcuni fenomeni da esso derivati.
4to. 20 pp. (*Mem. Soc. Ital.* xxiii. (*Parte Matematica.*) *Modena,* 1844

Lettera al . . Grimelli intorno alla elettro-metallurgia originale ital. &c. . .
(*From Zantedeschi, Trattato* ii. 485.) *Modena,* 1845

† Storia di una sensazione particolare che provava una paralitica quando veniva
elettrizzata . . Ricevuta 4 Maggio, 1845. 4to. 10 pp. (*Mem. della Soc.
Ital.* xxiii. Pte. Matem.) *Modena,* 1846

† Sopra l'azione magnetizzante delle correnti elettriche momentanee. Memoria
vii. Influenza del conduttore liquido fatto attraversare dalla scarica della
boccia di Leida nella magnetizzazione da essa prodotta nel metallo attorno
al quale si fa circolare la scarica stessa. 4to. 27 pp. (*Mem. Soc. Ital.*
tom. xxiv. pt. ii.) (*Faraday's Copy.*) *Modena,* 1846

MAR

327

† **Marianini,** Stefano Giovanni—*continued.*

Sull' azione magnetizzante delle correnti elettriche momentanee. Memoria viii. Dell' influenza del ferro, attorno a cui circola una scarica elettrica, nella magnetizzazione di altro ferro, attorno al quale circola pure la scarica medesima. 4to. 24 pp. (*Mem. Soc. Ital.* tom. xxiv. pt. ii. (*Faraday's copy.*) *Modena,* 1847

† Sopra l'azione magnetizzante delle correnti elettriche momentanee. Memoria ix. Sull' influenza che nella magnetizzazione del ferro operata dalla scarica elettrica esercitano i metalli, attorno ai quali si fa circolare la scarica medesima. 4to. 19 pp. (*Mem. Soc. Ital.* tom. xxiv. pte. ii.) *Modena,* 1847

Prospetto delle Memorie elettriche, &c. (*From Grimelli, Storia,* p. 133.)

Sulla proprietà posseduta in particolar modo dai corpi umidi di assorbire l'elettricità dagl'isolanti solidi elettrizzati quando si trovano a contatto con essi. Memoria. 4to. (*Inserita nella parte* ii. *del tomo* xxv. *delle Memorie della Società Ital. delle Scienze residente in Modena.*) *Modena,* 1854

Dell' induzione leido-magnetico-elettrica. 8vo. (*Il Nuovo Cimento,* iv. 1856.) 1856

† Sulla probabile esistenza di una nuova analogia fra l'Elettricità e la Luce ; ossia se il fluido del Franklin abbia a riguardarsi costituito da più fluidi, i quali non posseggano. 4to. 13 pp. (*Mem. della Soc. Ital.* ii. ser. i. 303.) *Modena,* 1862

† **Marianini,** S. e **Marianini,** P. Teorema di elettro-magnetismo e sua dimostrazione. Con una Nota di P. Marianini. 8vo. 2 pp. and 3 pp. 1841

(*Note.—This brochure is headed "Dalla Gazzetta Piemontese,"* No. 190, del 1841. *Fisica sperimentale.*)

† Elettro-plastica. Modi di fare coll' elettricità voltaica impronte o stampi della maggior esattezza, atti a produrre in rame i bassi rilievi metallici. Estratto da una lettera del Marianini Stefano scritta al fratello suo Pietro addi 14 di Febbrajo, 1841. Nota di Pietro, con Aggiunta. (*Estratto dalla Gazzetta Piem.* No. 119, *del* 1841.) 8vo. 6 pp. *Vigevano,* 1841

† **Marié, A.** De l'application de l'Electricité à la Thérapeutique. 8vo. 77 pp. *Montpellier,* 1854

† **Marié-Davy.** Premier Mémoire sur l'Electricité. 4to. 24 pp. (*Montpellier Acad.* i. 13, *Sect. Sciences.*) *Montpellier,* 1847

† Deuxième Mémoire sur l'Electricité. 4to. 7 pp. (*Montpellier Acad.* i. 37.) *Montpellier,* 1847

† Troisième Mémoire sur l'Electricité. 4to. 15 pp. (*Montpellier Acad.* i. 45.) *Montpellier,* 1847

Instructions sur l'installation et l'usage des instruments de Météorologie.

† Note sur les Instruments de mesure de l'Electricité statique. 4to. 15 pp. (*Montpellier Acad.* i. 315.) *Montpellier,* 1850

† Cinquième Mémoire sur l'Electricité. 2e partie. 4to. 19 pp. (*Montpellier Acad.* ii. 147.) *Montpellier,* 1852-53

† Première note sur l'action de l'Electricité sur le système nerveux. 4to. 16 pp. (*Montpellier Acad.* ii. 265.) *Montpellier,* 1852-3

† Première note sur la nouvelle machine Electro-magnétique. 4to. 13 pp. 2 pl. (*Montpellier Acad.* ii. 442.) *Montpellier,* 1854

† Recherches théoriques expérimentales sur l'Electricité, considérée au point de vue mécanique. 8vo. *Paris,* 1861-2

† Météorologie : les mouvements de l'Atmosphère et des Mers considéré au point de vue de la prévision du temps. 8vo. 498 pp. 24 maps. *Paris,* 1866

(*Vide* also Bulletin International.)

† **Marini, P.** Relazione Memoria sul Fulmine caduto in Brescia nella torre della Paletta 1803, e Sul Parafulmine costrutto sopra al palazzo della Loggia da lui 1805. 8vo. (*Comment. dell' Accad.. del Dipartimento del Mella*, tom. i. *Elenco delle Memorie*.)   *Brescia*, 1808

Marini, T. De electricitate cælesti, sive, ut alii vocant, naturali. 4to. 12 pp. (*Comment. de Bonon. Scient. Instit.* tom iii. p. 205.)   *Bononiæ*, 1755

Mariotti, Prospero. *Born* 1703, *at Perugia; died October* 1767, *at Perugia.*
† Lettera scritta ad una dama sopra la cagione de' fenomeni della macchina elettrica. 4to. 26 pp.   *Perugia*, 1748

Marius (Mayr), Simon. Prognosticon astrologicum das ist ausführliche Beschreibung d. Gewitter, samt andern natürlichen Zufällen, aufs J, 1607.
        Onolsbach, 1606

Maritime Conférence. (*Vide* Conférence.)

Marklin, G. Catalogus disputationum in Acadd. Scandinaviæ et Finlandiæ a 1778-1820, acced. supplem. ad catal. Lideniarum. 4 vols. 8vo. *Upsaliæ*, 1820

Marmier, S. Storia della letteratura in Danimarca ed in Svezia, versione del . F. de Bardi. 2 vols. 8vo.   *Firenze*, 1841

† **Marmora, A.** (della). L'Istmo di Suez e la Stazione telegrafico-elettrica di Cagliari. Ragionamento, per far seguito alle sue Questioni marittime spettanti alla Sardegna. 4to. 23 pp. 1 map.   *Torino*, 1856

Marni. Sulla formazione dell' Acido muriatico per mezzo della Pila . .   *Firenze?*

Marqfoy, G. De l'abaissement des taxes télégraphiques en France.   *Paris*, 1860
Nouveau système d'appareils électriques destiné à assurer la sécurité des chemins de fer. 8vo. 64 pp. 6 plates.   *Bordeaux*, 1857
Mémoire sur les essais des ponts en tôle par l'Electricité. 8vo. 16 pp. (*Extrait des Annales Télégraphiques*, Juillet et Août, 1858.)   *Paris*, 1858
Notice élémentaire sur la Télégraphie Electrique. 8vo. 29 pp. (*Extrait des Instructions de la Compagnie des Chemins de Fer du Midi*.)   *Paris*, 1858
† Discours sur la Télégraphie électrique prononcé à la Société Philomatique de Bordeaux, 22 Mars, 1861. 8vo. 13 pp.   *Bordeaux*, 1861
† Des réformes nécessaires en Télégraphie.   8vo. 166 pp.   *Paris*, 1866

† **Marqfoy et de Boissac.** Application de l'Electricité aux Annonces d'incendies. Projet présenté à Mons. le Maire (de Bordeaux), &c. 8vo. 24 pp. 1 plate.
        Bordeaux, 1860

Marqué, Victor. On Magnetic Declination, Thunder, &c. (*Acad. de Toulouse*.)
        Toulouse, 1827

Marrigues à Montfort l'Amaury. Suite de la guérison de la Paralysie d'après la méthode de Mons. l'Abbé Sans. 63 pp.   1773

Marsault, J. P. L. Electro-calorique, ou les Paratonnerres réformés . . où l'on trouve un nouveau plan de paratonnerre, etc. 18mo. 76 pp.   *Niort*, 1852

Marseille, Acad. Mémoires de l'Académie des Sciences, Belles-lettres, et Arts de Marseille. 8vo.   *Marseille*, 1846 *to* —
Mémoires publiés par l'Académie de Marseille. 8vo.   *Marseille*

Marseille Bibl. Publ. Catalogue de la Bibliothèque publique de Marseille. MS. Folio. (*Collated for Electricity, &c.* 1857.)   *Marseille, n.d.*

† **Marsh, J.** On a particular construction of M. Ampère's rotating Cylinder. 8vo. 2 pp. 1 plate. (*Phil. Mag.* lix. 433.)   *London*, 1822

Marshall, C. (or C. M.) On "an Expeditious method of conveying Intelligence." Letter from Renfrew, 1753. (*Scots' Magazine*, xv. p. 73, 1753.)   1753
(*Note.—In the Athenæum,* No. 1932, *Nov.* 5, 1864, *is said :* "*In the Scots' Magazine* (vol. xv. p. 73) *of February. The reference should be* ' *March*,' *and that we know nothing more of his identity than the initials C. M.*" *Du Moncel,* tom. ii. p. 2, *referring to the Cosmos du* 17 Feb. 1854, *says :* "*Voici ce document tel que nous le trouvons et qui avait été publié en Février,* 1753, *dans le* xv. *volume,* p. 88, *du Scots' Magazine.*" *He then gives the French translation and the original date, viz.* 1st Feb. 1753. (*Vide also Cornhill Mag.* vol. ii. p. 65.)

Marsigli, L. F. Conte. Del fosforo minerale. Italico: latine. 4to.   1702

Martens, Franz Heinrich. *Born November 4, 1778, at Wismar; died May 11, 1805, at Jena.*

† Volstandige Anweisung zur therapeutischen Anwendung des Galvanismus; nebst einer Geschichte, &c. . 8vo. 336 pp. (*Volta's copy.*)
Leipzig and Weisenfelt, 1803

Uber den Galvanismus als Heilmittel u. in Krankheit. anwendbar. 8vo.
Leipzig, 1803

Abbildung u. Beschreibung einer sehr bequemen in der Rocktasche tragbaren Voltaischen Säule. 4to.   *Leipzig,* 1803

Description d'une Pile de Volta. 4to. 2 plates.   *Leipzig,* 1803 ?

Beschreibung einer Volta'schen Säule. (*From Pogg.* ii. 61.)   *Leipzig,* 1803

Martens, Martin. *Born December 8, 1797, at Maestricht.*

† Réflexions sur la théorie électro-chimique de l'affinité et la composition moléculaire des corps. 4to. 15 pp. (*Nouv. Mém. Acad. Brux.* tom. x. 1837.)
Bruxelles, 1837

† Mémoire sur la Pile galvanique et sur la manière dont elle opère les décompositions des corps. 4to. 47 pp. (*Nouv. Mém. Acad. Brux.* tom. xii. 1839.)
Bruxelles, 1839

† Recherches sur la passivité des Métaux, et sur la théorie de la Pile voltaïque. 8vo. (*Ext. Acad. Roy. de Bruxelles,* tom. viii. No. 10 *des Bulletins* to tom. xi.)
Bruxelles, 1840-44

Diverses notices sur la passivité du Fer et des Métaux, et sur l'action de la Pile galvanique. 8vo. (*Bull. Acad. Brux.* vii. viii. ix. x. xi. 1840-44.)
Bruxelles, 1840-44

† Notice sur la théorie de la Pile voltaïque. 8vo. 12 pp. (*Ext. Bulletin Acad. Brux.* ix. No. 3.)   *Bruxelles,* 1842

† Notice sur l'action chimique des courants galvaniques. 8vo. 19 pp. (*Acad. Roy. de Bruxelles,* tom. ix. No. 7 *des Bulletins.*)   *Bruxelles,* 1842

Mémoire sur la Pile galvanique et ses décompositions. 4to.   *Bruxelles,* 1845

Sur les variations de la force électromotrice du Fer. 8vo. (*Bull. Acad. Brux.* xii. 1845.)   *Bruxelles,* 1845

† Recherches sur les variations de la force électromotrice du Fer. 4to. 46 pp. (*Nouv. Mem. Acad. Brux.* tom. xix. 1845.)   *Bruxelles,* 1845

De la théorie électro-chimique dans ses rapports avec la loi des substitutions. 8vo. (*Bull. Acad. Brux.* xvii. 1850.)   *Bruxelles,* 1850

Sur les Piles à acides et alcalis séparés par des corps poreux. 8vo. *Bulletin Acad. Brux.* xviii. 1851.)   *Bruxelles,* 1851

Essai sur la nature et l'origine des différentes espèces de brouillards secs. 8vo.
1851

Sur les décompositions Electro-chimiques. 8vo. (*Bull. Acad. Brux.* xix. 1852.)
Bruxelles, 1852

Martin. (*Vide* Roger-Martin.)

Martin. (*Vide* Saint-Martin.)

† **Martin.** Mémoire sur un coup de Tonnerre qui a éclaté dans l'Eglise de St. Nicolas de Toulouse . . 17 March, 1787. Lu 29 Mai, 1788. 4to. 9 pp. (*Toulouse Acad.* 1re série, iv. 100.)   *Toulouse,* 1790

Martin, Adam Georg. *Born March 8, 1812, at Vienna.*

† Repertorium der Galvanoplastik und Galvanostegie. 2 Bde. 8vo. 298 and 201 pp.   *Wien,* 1856

(*This work contains very many references to writings, patents, &c., but very few indications of the sources from which they are derived, or even of the dates.*)

E E

Martin, A. R.   Meteorologiska observationer, giorde på en resa til Spitzbergen. 8vo. (*Vetensk. Acad. Handl.* 1758, *p.* 307.)   *Stockholm,* 1758

† Natürlicher Phosphorus; oder Versuche mit Fischen und Fleische, so im Finstern leuchten. 8vo. 5 pp. (*K. Schwed. Akad.* Abh. xxiii, 224.)
Hamburg and Leipzig, 1761

Martin, Benjamin.   *Born* 1704, *at Warplesdon, Surrey; died February* 9, 1782, *at London.*

† An Essay on Electricity. 8vo. 40 pp.   *Bath and London,* 1746

† A Supplement containing Remarks on a Rhapsody. . . 8vo. 38 pp. *Bath,* 1746

Essay on Electricity. 8vo. (*From Priestley's Catalogue, and Young,* p. 415.)
Bath, 1748

† Supplement to the Philosophia Britannica. Appendix i. containing new Experiments in Electricity, and the Method of making Artificial Magnets. 8vo. 32 pp.   *London,* 1759

Martin.   (*Vide* Anon. Electricity, 1748.)

Martin de Brettes.   Mémoire sur un projet de chronographe électro-magnétique, et son emploi dans les expériences d'artillerie. 4to. (*Mém. de l'Acad.* xxv. 751.)   *Paris,* 1847

† Mémoire sur un projet de chronographe électro-magnétique, et son emploi dans les expériences de l'artillerie. 8vo. 75 pp. 2 plates.   *Paris,* 1849

† Des artifices éclairants en usage à la guerre; et de la lumière électrique. 8vo. 168 pp. 3 plates.   *Paris,* 1852

† Etudes sur les appareils électro-magnétiques destinés aux expériences d'artillerie en Angleterre, en Russie, en France, en Prusse, en Belgique. 8vo. 388 pp. 6 plates.   *Paris,* 1854

Physique appliquée. Projet de câble télégrapho-magnétique. 8vo. *Paris,* 1856

† Appareils chrono-électriques à induction. Application aux expériences balistiques. 8vo. 205 pp. 4 plates.   *Paris,* 1858

Instruction pratique pour l'usage du pendule balistique à induction. 8vo. 179 pp. 3 plates.   *Paris,* 1863

Martin, H.   Du succin, de ses noms divers et de ses variétés suivant les anciens. 4to. (*Mém. présenté à l'Acad. des Inscript.* tom. vi. 1re série, 1re partie.)
Paris, 1860

De l'Aimant de ses noms divers et de ses variétés suivant les anciens. 4to.
Paris, 1861

Les attractions electriques: observations et théories des anciens. *Rome,*1864-5

Les attractions et les répulsions magnétiques d'après les observations des anciens. 4to.   *Rome,* 1864-5

Observations et croyances des anciens sur les phénomènes lumineux de l'électricité atmosphérique. (*La Revue Archéologique,*1865-1866.) *Paris,*1865-6

(*Note.—This was reprinted with a second part in his* " La Foudre, &c.," *the title of the whole* (2 *parts*) *being* "*La Foudre et le Feu Saint-Elme dans l'Antiquité.*"

† La foudre, l'électricité et le magnétisme chez les anciens. 12mo. 418 pp.
Paris, 1866

Observations et théories des anciens sur les attractions et les répulsions magnétiques et électriques. 4to.   *Rome,* 1865

Martin, Jean Emile.   Nouv. école électro-chimique ou chimie des corps pondérables et impondérables. 2 vols. 8vo.   *Paris,* 1858-9

Martini, K. C.   Antheil d. Erdmagnetismus an d. Beschaffenheit d. Metal-Lagerstätten. 8vo. (*Gilbert's Ann.* lxxii. 1822.)   1822

Martino, Giamb. San.   (*Vide* San-Martino.)

† **Martins-Matzdorff, J.** Katechismus der Galvanoplastik. 8vo. 104 pp.
Leipzig, 1868

† **Martyn, John.** A Letter from John Martyn to Machin concerning an Aurora Australis, seen March 18, 1738-9, at Chelsea, near London. 4to. 3 pp. (*Phil. Trans.* xli. 840.) *London,* 1739-40-41

† An Account of an Aurora Australis seen Jan. 23, 1749-50, at Chelsea, by John Martyn. Communicated by M. Folkes. Read Jan. 25, 1749-50. 4to. 3 pp. (*Phil. Trans.* xlvi. 319.) *London,* 1749-50

† A Letter from John Martyn to the President concerning an Aurora Australis seen Feb. 16, 1749-50. Read Feb. 22, 1749-50. 4to. 1 p. (*Phil. Trans.* xlvi. 345.) *London,* 1749-50

Marum, Martin Van. *Born March 20, 1750, at Gröningen; died December 26, 1837, at Haarlem.*

Verhandeling over het Electrizeeren door M. Van Marum. in uolke de Beschryving en Afbeelding van ene nieuw uitgevondene Electrizec Machine benevens enige meuwe proeven uitgedagt en in 't werk gestold deer den Auteur, en Gerhard Kuyper. 8vo. 96 pp. 1 plate. (*See German Translation.*)
Gröningen and Amsterdam, 1776

† Abhandlung über das Elektrisiren enthaltend die Beschriebung und Abbildung einer neuerfundenen Elektrisirmaschine nebst einigen neuen Versuchen welche von dem Verfasser mit Hülfe des H. Gerhard Kuyper . ausgedacht und ins Werk gestellt worden sind. Aus dem Holländischen übersetzt von J. W. Möller. 12mo. 102 pp. 2 plates. *Gotha,* 1777

† Antwoord op de Vrag. Door præven te toonen, welke Luchtverhevelingen van de werking der Natuurlijke Electriciteit afhangen, &c. . ? 4to. 74 pp. 2 plates. (*Verhandl. Genootsch. Rotterdam,* vi. 1781.) *Rotterdam,* 1781

† Description d'une très-grande machine électrique placée dans le Muséum de Teyler . et des experiments. 4to. 205 pp. 6 plates. (*In French and Dutch.*) (*Volta's copy.*) *Haarlem,* 1785
(*Note.—This work appeared first in the Teyler's Tweede Genootschap.*)

† Transunto della descrizione della grandissima Macchina elettrica posta nel Museo di Teyler a Harlem, e degli sperimenti con essa fatti. 4to. 16 pp. 1 plate. (*Opusc. Scelti,* ix. 41.) *Milano,* 1786

Beschr. e ungemein grossen Elektrissir Maschine u. d. damit im Tylerschen Museum . angest. Versuche; a. d. Holl. 4to. *Leipzig,* 1786 *and* 1798

Antwoord op de vraag. Welk is de aart van de verschillende schadelyke en verstikkende witdampingen van Moerassen, Moderpoden, Secreten Riolen, &c. mit Van Troostwyk. (*Verhandl. Genootsch Rotterdam,* viii. 1787.)
Rotterdam, 1787

Sur les paratonnerres. 4to. (*Journ. Phys.* 1787, xxxi.) *Paris,* 1787

Expériences concernant quelques metéores électriques. 4to. (*Journ. Phys.* xxxi. 1787.) *Paris,* 1787

† Première continuation des expériences faites par le moyen de la Machine électrique Teylerienne. 4to. 231 pp. 10 plates. *Haarlem,* 1787

Description des frottoirs élect. (*Journ. des Phys.* xxxiv. 1789 and xxxviii. 1791.) *Haarlem,* 1789-91

Second Letter to Landriani on his Rubbers. 4to. *Paris,* 1791

Description, &c. Continuations. 1re et 2nde. 1795

Seconde Continuation des expériences faites par le moyen de la Machine électrique Teylerienne. 4to. 391 pp. 10 plates. *Haarlem,* 1795
(*Note.—This work first appeared in the Teyler's Tweede Genootschap,* St.9,Bl.1. *Vide Reuss, Repert.* iv. 351.)

Description de quelques appareils chimiques nouveaux ou perfectionnés de la fondation Teylerienne, et des expériences faites avec ces appareils. 4to.
Hearlem, 1798

Marum, Martin Van.—*continued.*

† Lettre à Volta contenant des expériences sur la colonne électrique, faites par lui et le Prof. Pfaff dans la laboratoire de Teyler à Haarlem en Nov. 1801. 8vo. 33 pp. 1 plate. *Haarlem*, 1802

(1) Letter to Ingenhouz.

    (*Note.—Kuhn, N. Endeck.* i. 104, *refers to his letter to Ingenhouz, containing remarks about the use of baked wood with the glass insulators of conductors.*)

(2) Letter to Landriani.

(3) Description of Rubber.

Letter to De la Mètherie on Death occasioned by Lightning. 4 pp. (*Opusc. Scelti,* xiv. 210.)

    (*Poggendorff has* "*Sur la cause de la mort des hommes, &c. tués par l'élect. ou par la foudre. Journ. Phys.* xxxviii. 1791.")

Lettre à Nauche sur la décomposition de l'eau avec l'appareil élect. ordinaire 8vo. (*Journ. du Galv.* xi. Cahier, p. 187.) *Paris*

Lettre sur une Machine élect. 4to. (*From Young.*)

    (*Poggendorff has* " *sur une nouv. Machine électrique. Journ. Phys.* xxxviii. 1791, *et* xl. 1792.)

Marum en **Troostwyk.** Antwood op de Vraag ; op te geeven den besten toestel van den electrophore, de byzondere verschynzelen van dit electrish werkting proefkundig te verklaaren, en aan te wüzen, welk nieuw licht hetzelve aan de leere der electriciteit toegebracht heeft, 1783.

                             Rotterdam, 1783

Sur la cause de l'électricité des substances fondues et refroidies. 4to. (*Journ. Phys.* xxxiii. 1788.) *Paris,* 1783

(*Vide also* Troostwyk and Marum.)

Marum, and **Œrsted.** Expériences faites avec les nouveaux appareils galvaniques de Ritter.

Il y a joint l'imitation de quelques unes de ces expériences avec l'appareil électrique ordinaire. (*Journ. de Chim. de Van Mons,* No. xiv. p. 212.)

† **Marx,** G. Disputatio Physica. De electricitate corporum, quam præside J. L. Schurer, &c. 4to. 28 pp. *Argentorati,* 1755

Marzari, G. On Conductors for Lightning. Account of his Admonitions, &c. (*Treviso Athenæum,* vol. ii. p. 73.)

On formation of Hail. (*Id.* vol. iii. p. 53.)

His theory, &c. on Hail. (*Id.* vol. iii. p. 84.)

On Paragrandini. (*Id.* vol. iii. p. 146.)

Discourse on his death. (*Id.* vol. iv. p. i.) 4to. 22 pp. *Padova,* 1834

† **Marzari,** G. e **Toaldo,** G. Memoria Descrizione d'una Tempesta di Fulmini a Castel. Franco . 25 Aprile, 1786. Letta 8 Feb. 1787.

† **Marzari,** G. (or Anon.) Maniera pratica di fare li conduttori ai campanili, alle chiese, ed alle case, descritta per uso dei fabbri, falegnami, e muratori, &c. . stampata per ordine del Magistrato Excell. Alla Sanità. 4to. 37 pp. *Venezia,* 1787

Masars (de Cazéles.) Mémoire sur l'électricité médicale. 8vo. 122 pp. *Toulouse and Paris,* 1780

† Second Mémoire sur l'électricité médicale et Histoire du traitment de quarante deux malades. 8vo. 311 pp. *Toulouse et Paris,* 1782

Trois recueils d'observations sur les effets qu'il a obtenus de l' électricité appliquée aux maladies, tantot seule, tantot combinée avec les rem èdes qu'employe la médecine, selon qu'il y a été déterminé par les circonstances. 4to. (*Mém. de Toulouse,* tom. ii. *Histor.* p. 59.) *Toulouse,* 1784

Extrait d'un Mémoire sur l'électrisation par bain, par souffle, et par aigrettes. Lu le 26 Juil. 1785. (*Mém. de Toulouse,* iii. p. 365. *Toulouse,* 1788

Recueil de Mémoires et Observations tiré des MSS. de M. Mazaré. 8vo. 182 pp. *Béziers,* 1792

MAS 333

Mascagni, Paolo. *Born February 5, 1755, at Castelletto, near Siena; died October, 1815, at Florence.*

Lettera ad uno de' Quaranta della Società Italiana sul Galvanismo. 8vo. 14 pp. (*Ann. de Chim. di Brugnatelli*, xxii. 308.) *Pavia,* 1805
(*This is nearly identical with the following "Lettera," excepting that Brugnatelli has added notes and altered the nomenclature.*)
Lettera ad uno de' Quaranta della Società Italiana delle Scienze. 8vo. 16 pp. 1805
Portrait. R. Focosi del. L. Rados inc.

Mascheroni, Lorenzo. *Born May 14, 1750, at Castagnetto, near Bergamo; died July 30, 1800, at Paris.*

Maniera di misurare l'inclinazione dell' ago calamitato. 8vo. 23 pp. 1 plate.
Bergamo, 1782
La geometria del Compasso. 8vo. *Pavia,* 1797
Géomètrie du Compas. trad. de l'Italien par A. M. Carette. 8vo. *Paris,* 1798

Mascuelli, G. De medicina magnetica. (*Vide* Maxwell, for German Translation.)
Francfort, 1613

Masenello, A. Regola di pronostico sul Temporale. 8vo. 19 pp. *Padova,*1833

Maskelyne, Nevil. *Born October 5, 1732, at London; died February 9, 1811, at Greenwich.*

Published something (manifesto) on the subject of Falling Stars, inviting observations, in 1783, which Toaldo reprinted in the Giornale Enciclopedico di Vicenza for the Toaldo Raccolta, vol. iv. p. 253.
(*Vide* also Clarke and Maskelyne.)

Massé, J. De l'Electricité en Thérapeutique. 8vo. 48 pp. *Paris,* 1850
Quelques mots sur l'appareil galvano-élect. portatif du Prof. Récamier. 8vo. 32 pp. *Paris,* 1851

Masselotte, fils E. La Dorure au mat par le procédé pyro-électrique realisée sans danger pour la santé des Ouvriers. 8vo. *Paris,* 1867

Masson, Antoine Philibert. *Born August 22, 1806, at Auxonne, Département de la Côte d'Or.*

Recherches sur l induction exercée par un courant sur lui-même. 8vo. (*Ann. de Phys. et de Chim.* 2me série, lxvi. 5 (and 28), 1837.) *Paris,* 1837
(*Du Moncel says* 1836.)
Effets physiologiques des courants induits. 8vo. (*Ann. de Chim. et de Phys.* lxvi. 26.) *Paris,* 1837
Théorie physique et mathématique des phénomènes electro-dynamiques et du magnétisme. Thèse soutenue . . . 1838. 8vo. 90 pp. 2 plates.
Paris, 1838
Note sur un essai de télégraphe électrique, fait au collége de Caen. 4to. (*Comptes Rendus*, vii. 88.) *Paris,* 1838
Sur le fluide électrique à l'état de diffusion dans le vide. 4to. (*Comptes Rendus*, vii. 671.) *Paris,* 1838
Thèse. Des phénomènes électro-dynamiques. 8vo. *Paris,* 1838
Etudes de Photométrie électrique. (1er et 2d Mémoires.) 8vo. 68 pp. 2 plates.
(*Ext. des Anns. de Chim.* 3me série, xiv. *Communiqués à l'Acad.* 19 Fév. e 5 Août, 1844.) *Paris,* 1845
Notice sur la photométrie électrique. 8vo. (*Annales de Phys. et de Chim.* 1847. *Paris,* 1847
Etudes de Photométrie électrique. (3me Mémoire.) 8vo. 51 pp. 1 plate. (*Ann. de Chim.* . . . 3me série, tom. xxx.) *Paris,* 1850
Etudes de Photométrie électrique. Sur la lumière produite par les courants voltaïques dans l'air et dans les liquides. (Cinquième Mémoire.) 4to. 4 pp. (*Comptes Rendus*, 1851, ier semestre, xxxii. No. 5.) *Paris,* 1851

Masson, Antoine Philibert.—*continued.*
† Note sur la lumière électrique. 4to. 3 pp. (*Ext. des Comptes Rendus . . . de l'Acad.* xxxvi. Séance du 7 Fév. 1853. (*Faraday's copy.*)   *Paris,* 1853
† Mémoire sur l'étincelle électrique. (Couronne par la Soc. Hollandaise des Sciences à Haarlem.) 4to. 94 pp. 2 plates.   *Haarlem,* 1854
† Notice sur les travaux scientifiques de M. A. Masson. 4to.   *Paris,* 1858 ?
Masson et Breguet fils. Description d'un nouveau télégraphe électrique. 4to. (*Comptes rendus,* vii. 710.)   *Paris,* 1838
† Mémoire sur l'induction. 8vo. 24 pp. 1 plate. (*Ann. de Chim. . .* 3me série, tome iv. et *Comptes Rendus,* xiii. 426, 1841.) (*Arago's copy.*)
    Paris, 1842
Massuet, Pierre. *Born November* 10, 1698, *at Mouzon on Meuse; died October* 6, 1776, *at Lankeren.*

Essai de Physique par M. P. Van Muschenbrock, avec une description de nouvelles sortes de machines pneumatiques, et un recueil d'expériences par I. V. M. Trad. du Holland. par M. Pierre Massuet. 2 vols. 4to. Portrait and plates.
    Leide, 1751
Eléments de la philosophie moderne. 2 vols. 12mo. Figs. *Amsterdam,* 1752
Maternus. (*Vide* Cilano de Maternus.)
† **Mathey,** A. O. Traité pratique de dorure et argenture galvanique appliquées à l'horlogerie. 8vo. 22 pp. 1 plate. (*Ext. du Technologiste,* tom. xvi. 1854-5.)   *Paris,* 1855
Mathieu. Un Mémoire sur le galvanisme. 4to. (*Recueil de la Soc. d'Agriculture et Sciences d'Autun,* an. x. p. 21.)   *Autun,* 1802
Mathiot, G. On the Electrotyping operations of the United States Coast Survey. 8vo. 15 pp. 1 plate. (*Extract from the American Journal of Science and Arts,* xv. 2nd series, 1853.) (*Roget's copy.*)   *New Haven, U.S.* 1853
Matteucci, Carlo. *Born June* 20, 1811, *at Forli, Papal State.*
    (*His papers in periodicals, &c., are innumerable. Vide Pogg.* ii. 79, 80, and 81.)
† Discorso sul periodo dei Temporali. 8vo. 9 pp. (*Arago's copy.*)
    Firenze, 1827
† Del Temporale. Discorso. 8vo. 67 pp.   *Bologna,* 1828
Discorso sulle pietre meteoriche. 8vo. 12 pp.   *Bologna,* 1828
Influenza dell' elettricità terrestre sui Temporali.   *Bologna,* 1829
Sulla decomposizione dei Sali metallici per la pila. 8vo.   *Forli,* 1830
Influenza del calore sul Magnetismo. 8vo. (*Fusinieri's Annali,* i. 425.)
    Forli, 1831
† Sopra alcuni punti di meteorologia elettrica. 4to. 3 pp. (*Ann. . . del Reg. Lomb.-Veneto,* ii. p. 334.)   *Padova,* 1832
† Sulle secrezioni animali. Discorso. 4to. 3 pp. (*Ann. . . del Reg. Lomb.-Veneto,* ii.)   *Padova,* 1832
† Sulle correnti elettro-magnetiche di Faraday. Osservazioni. 4to. 3 pp. (*Ann. . . del Reg. Lomb.-Veneto,* iii. 185.)   *Padova,* 1833
† Sulla digestione. Cenni. 4to. 3 pp. (*Ann. . . del Reg. Lomb.-Veneto,* iii. 220.)   *Padova,* 1833
† Sulle correnti elettromagnetiche di Faraday. Osservazione. 8vo. 6 pp.
    Forli, 1833
† Memoria sopra la formazione della Grandine. Memoria. 8vo. 14 pp.
    Pesaro, 1834
Sur l'électricité animale. Mém. présenté à l'Acad. de Bruxelles. 8vo. 4 pp.
    Florence, 1834
† Analisi d'una nuova specie di calcolo intestinale. 4to. 3 pp *Ann. . . del Reg. Lomb.-Veneto,* iv. 117.)   *Padova,* 1834

Matteucci, Carlo.—*continued.*

Sul passaggio della corrente elettrica attraverso i liquidi. Memoria. 8vo.
Firenze, 1835

† Sur la propagation du courant électrique dans les liquides. 8vo. 89 pp.
Paris, 1837

† Recherches physiques, chimiques, et physiologiques sur la Torpille. 8vo.
39 pp. (*Tiré de la Bibliothèque Univ. de Genève,* Novembre, 183 7.)
Genève, 1837

Application of the Electric Current in a case of Tetanus. 8vo. (*Bibliot.
Univers.* Mai, 1838.) *Genève,* 1838

† De la force chimique du courant Voltaïque. 3me Mémoire. (Dated Ravennæ,
14 Mai, 1839.) 8vo. 12 pp. (*Arago's copy.*) *Paris,* 1839 ?

† De la diffusion du courant électrique dans les liquides. (Dated Ravennæ,
1 Juin, 1839.) 8vo. 4 pp. (*Arago's copy.*) *Paris,* 1839

Elettricità fisiologica-medica. (*From Zantedeschi, Trattato,* ii. 354.) *Modena,* 1839

† Essai sur les phénomènes électriques des animaux. 8vo. 88 pp. 1 plate.
Paris, 1840

De l'induction du courant électrique de la Bouteille de Leyde. 8vo. 20 pp.
(" *Tiré de la Bibliothèque Universelle de Genève.*") (*Arago's copy.*)
Genève, 1840 ?

† Deuxième Mémoire sur le courant électrique propre à la Grenouille et sur
celui des Animaux à sang chaud. 8vo. 43 pp. *Paris,* 1841

† "Il M. mostra alla Sezione (fisica) . . . e descrive un istrumento da lui
chiamato Induzionmetro differenziale." 4to. 1 page. (*Scienziati Ital.
Congress.* 1841. *Sezio. fis.* p. 234.) *Firenze,* 1841

Lettera all'Amici (Prof.) Rana. *Modena,* 1842

† Sur la Phosphoréscence excitée par la lumière solaire, par l'Etincelle électrique
et par les flammes du phosphore, du potassium et du sodium. 8vo. 16 pp.
(*Bibliot. Universelle,* Juillet, 1842.) (*Arago's copy.*) *Genève,* 1842

Lezioni di fisica date nell' I. R. Università di Pisa. 1st edition. 3 vols. 8vo.
Pisa, 1841-2
(*The second volume appropriated to Electricity, and has* 716 *pages and* 7 *plates.*)

† Ricerche fisico-chimiche della Fosforescenza delle Lucciole. 8vo. 16 pp.
(*Estr. dalle Miscellanee di Chimica, Fisica, &c. Giornali che si publicano in Pisa,*
anno i. 1843.) *Pisa,* 1843
(*Not electrical.*)

† Traité des phénomènes électro-physiologiques des Animaux ; suivi d'études
anatomiques sur le système nerveux et sur l'organe électrique de la Torpille
par P. Savi. 8vo. 348 pp. 6 plates. *Paris,* 1844

† Fenomeni fisico-chimici dei corpi viventi. Lezioni. 1st edition. 8vo. 186 pp.
Pisa, 1844

† Lezioni di Fisica . . date nell' I. e R. Università di Pisa. Seconda edizione
corretta ed ampliata. 8vo. 473 pp. 10 plates. (*Faraday's copy.*) *Pisa,* 1844

† Mesure de la Force Nerveuse développée par le Courant électrique. 8vo. 15 pp.
1 plate. (*Ext. des Ann. de Chim. et de Phys.* 3me série, tom. xi.)
Paris, 1844

Note sur la conductibilité de la terre pour le courant électrique. 4to.
(*Comptes Rendus,* xviii. 1032 and xxii. 86.) *Paris,* 1844-46

Sur l'emploi de la Terre comme conducteur pour le télégraphe électrique. 4to.
(*Comptes Rendus,* xx. 1431.) *Paris,* 1845

† Account of a new experiment in Electro-Statical Induction . . . in a letter to
Dr. Faraday. 8vo. 2 pp. (*Phil. Mag.* for April, 1845.) *London,* 1845

Sur l'état électrique des corps cohibens. 4to. (*Comptes Rendus.*) *Paris* 1846

Matteucci, Carlo.—*continued.*

Leçons sur les phénom. phys. et chim. des corps vivants. 18mo. *Paris,* 1846

† Lezioni sui fenomeni fisico-chimici dei corpi viventi. 2nd edition. 8vo. 328 pp.
Pisa, 1846

† Ricerche Elettro-Fisiologiche. Corrente muscolare : Prima Memoria, p. 268, 19 pp. Della Corrente propria della Rana : Seconda Memoria, p. 286, 8 pp. Sulla Contrazione indotta : Terza Memoria, p. 356, 23 pp. 4to. (*Mem. della Soc. Ital.* tom. xxiii. *Matem.*) *Modena,* 1846

† "Leçons sur les phénomènes physiques des corps vivants par C. Matteucci. Ed. française publiée, avec des additions considérables sur la 2me édition italienne." 12mo. 406 pp. Cuts. *Paris,* 1847

† Lectures on the Physical Phenomena of Human Beings, translated by Pereira; with corrections and additions by Matteucci and notes by Pereira. 8vo. 435 pp. *London,* 1847

† Sopra la Fosforescenza del Mare, e dei Pesci. Dated Viareggio, 20 Agosto, 1847. 8vo. 10 pp. (*Arago's copy.*) *Pisa,* 1847

Sur l'influence du magnétisme sur le pouvoir rotatoire de quelques corps. 8vo. (*Ann. de Chim. et de Phys.* xxiv. 1848.) *Paris,* 1848

Sur la perte de l'électricité dans l'air plus ou moins humide. 4to. (*Comptes Rendus,* xxix. 305.) *Paris,* 1849

Extrait d'un mémoire sur la conductibilité de la terre. 4to. (*Comptes Rendus,* xxx. 774.) *Paris,* 1850

Lezioni di fisica. Terza edizione, Napolitana sulla terza di Pisa intieramente rifusa ed ampliata di nuove lezioni da R. C. con un supplemento diviso in due lezioni. . per Giuseppe M. Paladini. With atlas. 8vo. *Napoli,* 1850

† Manuale di Telegrafia elett. 12mo. 205 pp. 1 plate. *Pisa,* 1850

† Réponse aux deux dernières lettres de M. de Bois-Reymond inserées dans les Nos. 17 et 18 des Comptes-Rendus de l'Acad. &c. . . . Présentée à l'Acad. des Sciences, Pisa, 4 Mai, 1850. 8vo. 8 pp. *Florence,* 1850

Lezioni di fisica date nell' I. R. Università di Pisa. 4th edit. 3 vols. 4to.
Pisa, 1851

† Manuale di Telegrafia elettrica. Seconda editione con aggiunte. 8vo. 224 pp.
Pisa, 1851

† Recherches expérimentales sur la propagation du courant électrique dans la terre. 4to. (*Comptes Rendus,* xxxii. 511 ; *L'Instit.* 1851, p. 115.) *Paris,* 1851

† Lezioni di Elettricità applicate alle arti industriali all' economia domestica e alla terapeutica. 8vo. 199 pp. *Torino,* 1852

† Dei recenti progressi della Telegrafia elettrica. Discorso letto nell' Università di Pisa il 18 Nov. 1852. 8vo. 20 pp. (*Estr. dal Crepuscolo,* an. iv. No. xix.)
Milano, 1853

† Lettre de C. Matteucci à Mr. H. Bence Jones, éditeur d'une brochure On Animal Electricity ou Extrait des Découvertes de M. Du Bois-Reymond. 8vo. 16 pp. *Florence,* 1853

† Cours spécial sur l'Induction, le Magnétisme de rotation, le Diamagnétisme, et sur les relations entre la force magnétique et les actions moléculaires. 8vo. 278 pp. 2 plates. *Paris,* 1854

† Sui fenomeni fisici e chimici della contrazione muscolare. Ricerche. Prolusione al corso dei fenomeni fisico-chimici dei corpi viventi. Letta il 10 Marzo, 1856. 8vo. 33 pp. (*Estr. del Nuovo Cimento,* an. ii. vol. iii.) *Torino,* 1856

† Lezioni di Elettro-fisiologia. Corso dato nell' Università di Pisa nell' anno 1856. 8vo. 88 pp. *Torino,* 1856

† Electro-Physiological Researches. Physical and Chemical Phænomena of Muscular Contraction. 10th series, part i. Communicated by Michael Faraday. Received June 12 ; read June 19, 1856. 4to. 25 pp. (*Phil. Trans.* 1857.) *London,* 1857

Matteucci, Carlo.—*continued.*

† Cours d'électro-physiologie professé à l'Université de Pise en 1856. 8vo.
177 pp. 2 plates.   *Paris,* 1858

Sui fenomeni elettro-magnetici sviluppati dalla torsione. Ricerche esperimentale. 8vo. (*Nuovo Cimento,* vii. 66, 1858.)   *Torino,* 1858

† Sulla forza elettro-motrice secondaria de' Nervi e di altri tessuti organici.
Mem. 8vo. 14 pp. (*Estr. dal Nuovo Cimento,* vol. xi. Fascicolo 1.) *Pisa,* 1860

† Manuale di Telegrafia elettrica. 8vo. 394 pp. 3 tables. 6 plates.
                                                                Torino, 1861

† Corso di elettro-fisiologia in sei lezioni date in Torino. Raccolte steno-
graficamente e rivedute dall' autore. 8vo. 142 pp.   *Torino,* 1861

† Sur les phénomènes physiologiques des animaux.   *Paris*

† Mém. sur la propagation de l'Elect. dans les corps isolants, solides, et gaz-
eux. 8vo. 45 pp. (*Ext. Ann. de Chim.* 3me série, xxviii.)   *Paris*

† Mém. sur la propagation de l'Electricité dans les corps solids isolants. 2me
partie. 8vo. 39 pp. 1 plate. (*Ext. Ann. de Chimie,* 3me série, xxvii.) *Paris*

† Mém. sur le dévelopement de l'élect. dans les combinaisons chimiques, et sur
la théorie des piles formées avec un seul métal et deux liquides différents.
8vo. 20 pp. (*Ext. des Annales de Chimie,* tom. xxxiv.)   *Paris*

† De la méthode expérimentale dans l'étude des phénomènes de la vie. 4to.
20 pp.   *Florence*

† Recherches sur le courant propre de la Grenouille et des animaux à sang chaud.
Mémoire qui fait suite aux Recherches sur les phénomènes électriques de la
Torpille et sur le courant propre de la Grenouille. 8vo. 36 pp. (*Ext. des
Archives de l' Electricité . . . de Genève.*)

La pila di Volta. Lettura. 16mo. 48 pp.   *Firenze,* 1867?

Note sur la résistance de la terre. 8vo (*Ann. des Chim. et Phys.* (3), xli. 173.)
                                                                     Paris
Discorso sull' Influenza della Elettricità nella formazione delle Meteore acquee.

† **Matteucci, Piria,** &c. Il nuovo Cimento. Giornale di Fisica, di Chimica e delle
loro applicazioni alla medicina, &c. 8vo.   *Torino e Pisa,* 185—?

Matteucci and Grove. On the Electricity of Flame. 4to. (*Phil. Trans.*)
                                                          London, 1854
(*Vide* also Longet and Matteucci.)

(*Vide* also Zannotti and Matteucci.)

† **Matteucci, P.** De Aurora boreali anni MDCCXXXVIII. 4to. 1 p. (*Comment. de
Bonon. Scien. Instit.* tom. ii. part iii. p. 496.)   *Bononiæ,* 1747

† **Matthæus de Sadeler, A.** Disputatio . . inauguralis de Meteoris, subjecit,
3 Feb. 4to. 22 pp.   *Lugd. Bat.* 1684

Matthiessen, L. Ueber die Anordnung der Electricität auf isolirten Leitern
von gegebener Form und die Methoden der Messung des Bindungs-
coeffizienten mit dem elektrischen Verstärkungsapparate. Eine experimentelle Untersuchung. 4to. 24 pp. 1 plate.   *Jever,* 1861

† **Mattia, G. A. de.** Cenni sullo sviluppo elettrico dal Vapore. Memoria. 8vo.
58 pp.   *Padova,* 1845

Mattioli. (*Vide* Maffioli.)

Matzdorff. (*Vide* Martins-Matzdorff.)

Matzenauer, E. Erdmagnetismus und Nordlicht. Ein Versuch ihren Zusammenhang mit Zugrundelegung der P. T. Meissner'schen Wärmelehre zu
erklären. 2 vermehrte Auflage. 8vo. 31 pp.   *Innsbruck,* 1861

Mauchart, Johan David. Fulminis admirandus agendi et incedendi modus; ubi
simul incidenter de lachrymis vitreis. (*Ephem. Acad. Nat. Cur.* cent. iii. et
iv. 1715.)   1715

Mauduit. Mém. 1. Sur l'électricité considérée relativement à l'économie animale et à l'utilité dont elle peut être en médecine. (*Hist. et Mém. de la Soc. Roy. de Médec.* 1776, *Mém.* p. 461.)   *Paris*, 1776

Mém. 2. Tentative pour découvrir dans quel rapport les différentes substances sont conductrices du fluide électrique. (*Hist. et Mém. de la Soc. Roy. de Médec.* 1776, *Mém.* p. 514.)   *Paris*, 1776

Mémoire sur le traitement électrique administré à quatre-vingt deux malades. (*Hist. et Mém. de la Soc. Roy. de Médec.* 1777 et 1778, *Mém.* p. 199, 1779, *Hist.* p. 187.)   *Paris*, 1777

Mémoire sur les effets généraux, la nature, et l'usage du fluide électrique, considéré comme médicament. 8vo. (*Hist. et Mém. de la Soc. Roy. de Médec.* 1777 et 1778, *Mém.* p. 432.)   *Paris*, 1778

Mémoire sur les différentes manières d'administrer l'électricité et observations sur les effets que ces divers moyens ont produits. (*Hist. et Mém. de la Soc. Royale de Méd.* 1780 et 1781, *Mém.* p. 264.)   *Paris*, 1780-1

Mémoire sur les effets de l'électricité employée dans la cure des tremblemens causés par les vapeurs du mercure, de la paralysie qui succède à la colique des peintres, des rhumatismes graves et invétérés, de la sciatique, des mouvemens spasmodiques et des engelures. (*Hist. et Mém. de la Soc. Roy. de Médec.* 1782 et 1783, *Mém.* p. 160.)   *Paris*, 1783

† Mém. sur les différentes manières d'administrer l'élect. et observations sur les effets que ces divers moyens ont produits. 1st edition. 4to. 152 pp. 2 plates. Lu 2 Déc. 1783. (*Ext. des Mém. de la Soc. Roy. de Méd.*)   *Paris*, 1784

† Mémoire sur les différentes manières d'administrer l'électricité, et observations sur les effets qu'elles ont produits. Extrait des Mém. de la Soc. Royale de Médecine. Imprimé par ordre du Roi. Called 2nd edition. 8vo. 301 pp. 2 plates.   *Paris*, 1784

† Précis des Journaux tenus pour les malades qui ont été électrisés pendant l'année 1785 et des mémoires sur le même objet adressés à la Soc. Royale de Méd. pendant la même année. Travail servant de suite au Mém. sur les différentes manières d'administrer l'élect. 8vo. 46 pp.   *Paris*, 1786

Maunior. (*Vide* Schaw.)

† **Maurice.** De l'électricité médicale. 12mo. 252 pp.   *Paris*, 1810

† **Mauritius.** Notiz über eine einfache Vorrichtung zur Bestimmung der magnetischen Declination. 8vo. (*Pogg. Ann.*)   *Leipzig*

† Versuche über den Magnetismus bei verschiedenen Temperaturen. 8vo. (*Pogg. Ann.* cxx.)   *Leipzig*

Uber Vertheilung des Magnetismus im Innern v. Magneten.   *Coburg*, 1864

† **Maurocordatos, N. A. G.** Thèse de Chimie. Sur les décompositions qu'on opère par la Pile. 4to. 15 pp.   *Paris*, 1839

† Thèse sur les causes de la production de l'élect. dans la Pile. 4to. 20 pp.   *Paris*, 1839

Maurolycus, F. Problematica mechanica, cum appendice et ad magnetem, et ad pixidem nauticam pertinentia. 4to.   *Messanæ*, 1613

Maury, Matthew Fontaine. *Born January* 14, 1806, *at County Spottsylvania, Virginia.*

† On the probable relation between Magnetism and the Circulation of the Atmosphere. 4to. 17 pp. (*From Appendix to the Washington Astronomical Observations for* 1846.)   *Washington*, 1851

Physical Geography of the Sea. 6th edition, with important Addenda, including Jansen's Ozone Experiments, which cast unexpected light upon Atmospheric Circulation. 8vo. 360 pp.   *London*, 1857

The Physical Geography of the Sea, and its Meteorology. 11th edition revised, being the 2nd edition of the author's reconstruction of the work. 490 pp.   *London*

Mauss. Système contrôleur (élect.) 4to. (*Comptes Rendus*, 11 Août, 1845.)
Paris, 1845

† **Mawgridge, R. A** . . relation of the . . . effects of an unusual clap of Thunder and Lightning. 4to. 2 pp. (*Phil. Trans.* xix. for 1695-6-7, p. 782.)
London, 1690

† **Maxwell, H.** Observations on Trees as Conductors of Lightning. Communication dated June 21, 1787. 4to. 2 pp. (*Mem. Amer. Acad.* old series, ii. part i. p. 143.)
Boston, U.S. 1793

† **Maxwell, W.** (**Maxvellus**). Medicina magnetica. Libri iii. Edente Georgio Franco. 12mo. 200 pp.
Francofurti, 1679

† Drey Bücher der magnetischen Artzney-kunst . . . von Guillelmo Maxvello. Durch Georgium Francken . . . übersetzet. 12mo. 14 + 8 + 282 + 8 pp.
Frankfurt, 1687

De medicina magnetica. 12mo. (*From Watt.*)   1687

† Drei Bücher der magnetischen Heilkunde . . . herausgegeben von G. Frank, aus dem Lateinischen. 12mo. 248 pp. (*Forms part of "Kleiner Wunder-Schauplatz . . . von Scheible,* 3e Theil.)   *Stuttgart*, 1855

May, W. Verhall der uitwerkinge van eenen enkelden blixemslag, voorgevallen op een van's lands oorlogscheepen, in het jaar 1749. (*Verhandel. van het Maatsch. te Haarlem,* xii. 391.)
Haarlem

On the effect of Lightning on Ship's Compasses. (*Verhandel. der Holland Maatsch. der Weetench. te Haarlem,* xii. bl. 391.)

May, Heinrich. (*Vide* Mapis.)

† **Maycock, J. D.** On the excitement of Voltaic Plates, in reply to Mr. De Luc's objections to the doctrines maintained by the Author. 8vo. (*Phil. Mag.* xlviii. 165 and 255.)
London, 1816

† **Mayer.** Notice sur la Télégraphie électrique en général et . . . sur la Télégraphie électrique souterraine. 8vo. 43 pp.
Paris, 1851

Mayer, And. Diss. sistens acus inclinatoriæ phenomena. 4to. *Gryphiswald*, 1777

† **Mayer, A. F. J. C.** Spicilegium observationum anatom. de Organo electrico in Raiis anelectricis et de Hæmatozois. 4to. 17 pp. 3 plates.   *Bonnæ*, 1843

Die Pacinischen Körperchen. Eine physiolog. Abhandlung. 4to.
Bonn, 1844

Mayer, F. A. Ein paar Worte über ein paar Druidenbäume im Konigr. Baiern. 8vo. 2 plates lith.
Eichstadt, 1826

† **Mayer, G. F.** Il pendolo elettromagnetico del prof. Dal Negro descritto. 8vo. 7 pp. 1 plate. (*Poligrafo di Verona,* ii. 97.)
Verona, 1836

Mayer (Meyer), **F. C.** De luce boreali. 4to. (*Comment. Acad. Petrop.* i. 1726.)
Petropoli, 1726

Mayer, Johann. *Born February 6, 1754, at Prague; died June 5, 1807, at Prague.*

Über d. Elektricität d. Vögel. 8vo. (*Abh. Privatgesellsch. Bohm.* v. 1782.)
Prag, 1782

Sämml. physik Aufsätze besond d. Köhm. Naturgesch. betr. v.e. Gesellsch. Böhm. Gelehrten. 4 vols. 8vo.
Dresden, 1791-4

Abhandlungen, &c. v. Galvani, Valli, Carminati u. Volta über die Kräfte d. thier. Elek. 8vo. 4 plates.
Prag, 1793

Beyträge z. Geschichte d. meteorischen Steine in Böhmen. 8vo. (*From Ersch. Young,* p. 427, *says* 2 *vols.*)
Dresden, 1805

Mayer, Johan Tobias (sen.) *Born February* 17, 1723, *at Marbach; died February* 20, 1762, *at Göttingen.*

Opera inedita. Vol. i. 4to.
Gottingen, 1775
(*Herausgegeben von Lichtenberg enthaltend.* (6) *Theoria magnetica.* (7) *Computus declinationum et inclinationum magneticarum ex theoria nuper exhibita deductus.*)

Mayer, Johann Tobias, Sohn. *Born May 5, 1752, at Göttingen; died November 30, 1830, at Göttingen.*

Ob. es nöthig sei eine zurückstossende Kraft anzunehmen. 8vo. (*Gren's Journ.*) vii. 1793.) *Leipzig*, 1793

Anfangsgründe d. Naturlehre. 8vo. *Göttingen*, 1801 (*The 6th edition is dated 1827.*)

† Commentatio de usu accuratiori acus inclinatoriæ magneticæ, recitata . . . 11 April, 1814. 4to. 38 pp. 1 plate. (*Comment. Soc. Gött. recent.* iii. 1814-15.) *Göttingen*, 1814 or 1815

† Commentatio . . . super legem actionis vis electricæ repulsivæ in distantiam; experimenta et disquisitiones. Prælecta . . . die 13 Julii, 1822. 4to. 30 pp. 1 plate. (*Comm. Soc. Gött. recent.*) *Göttingen*, 1819

Mayer, Joseph. *Born June 5, 1752, at Prague; died October 24, 1814, at Vienna.* Beobb. über d. Leuchten d. Adriat Meers. 8vo. (*Abh. Bohm. Gesellsch. d. Wiss.* 1st Folge, Bd. i. 1785.) *Prag*, 1785

Ueber die magnetische Kraft des Krystallisirten Eisen-Sumpf-Erzes. (*Bohm. Gesellsch. der Wissensch.* 1788, p. 238.) *Prag*, 1788

Mayer. (*Vide* Zantedeschi and Mayer.)

Mayne. Description of public and private Observatories. *London*

† **Mayne**, Z. Letter, 1694, concerning a Spout of Water that happered at Topsham on the river between the sea and Exeter. 4to. 4 pp. 1 plate. (*Phil. Trans.* xviii. for 1694.) *London*, 1694

Mayr, G. Abhandlung üb. Elektricität und sichernde Blitz-Ableiter für jedes Gebäude für Reise- u. Frachtwagen, Schiffe, Bäume u. Denkmaler. Nebst e. Anh. über Hagel-Ableiter. Geprüft (2 Aufl.) neu. u. verbess. 12mo. *Munchen*, 1839

Mazari. (*Vide* Volta and Mazari)

Mazéas, Guillaume. *Born 1712, at Landerneau, Bretagne; died 1776, at Vannes.*

Letters concerning the success of the experiments in France of verifying the conjectures of B. Franklin, upon the analogy of Thunder and Electricity. Translated from the French by J. Parsons. 4to. (*Phil. Trans.* 1752, p. 534.) *London*, 1752

(*This letter contains De Lor's experiment with a bar of iron 99 feet high, on May 18, 1752.*)

Observations upon the Electricity of the Air, made at the Château de Maintenon during the months of June, July, and October, 1753. Translated from the French by James Parsons. 4to. (*Phil. Trans.* 1753, p. 377.) *London*, 1753

† **Mazzacane.** Lettere sopra il Elettricismo. 4to. 159 pp. *Napoli*, 1780

Mazzera. De electricitate animali et galvanismo. (And four other dissertations) 8vo. *Taurini*, 1802

Mazzolari, G. A Poem in Latin verses on Electricity. In six books. *Romæ*, 1767

Mazzucato, A. Adjutore al **Magorotto**. Disputatio de Frankliniano electricitatis principio. . . . (Thesis.) 138 pp. *Patavii*, 1804

Meade, William. *Born in Ireland; died August 29, 1833, at Newburgh.*

Outline of the origin and progress of Galvanism. 8vo. (From Watt.) *Dublin*, 1805

Mechanics' Magazine, Museum Journal, and Gazette, containing an account of all the inventions and improvements in steam engines, fire engines, machinery of all kinds, railroads, canal and river engineering, working of collieries, mining, shipbuilding, propelling, agricultural machinery, waterworks, boring, &c. &c.; also Chemistry, Metallurgy, Magnetism, Photography, Electricity, Mathematics, &c. &c. from 1823-58. 8vo. 1823-7- ?

Medical Society of London. Instituted 1773. Transactions, 1797—1805. 6 vols.; 1810, 1 vol.; 1817, 1 vol.; 1846, new series, vol. i. 8 parts, 1847.

Medical and Chirurgical Royal Society of London. Instituted 1805. Transactions. Vols. i. to xlv. Proceedings.

† **Medici, Cosimo de.** Portrait. incisione. Longhi dis.; Parma inc.

† **Medici e Gandolfi.** Esperienze (galvaniche) fatte in Bologna sul sangue. 4to. 11 pp. (*Opus. Scelti*, xxii. 331.) *Milano*, 1803

† **Medici, Lorenzo de.** Portrait. Caronni inc.

Medina, Pedro de. Arte de navegar. *Valladolid*, 1545
    (*Note.—He denied the variation of the Compass.*)

† **Mehler, Dr. F. G.** Über e. m. den Kugel- u. Cylinder-functionen verwandte Function u. ihre Anwendung in der Theorie der Electricitäts vertheilung. 4to. 30 pp. (*Aus Jahrsbericht d. Gymnasium, besonders abgedruckt*, Ostern, 1870.) *Elbing*, 1870

Méhu. (*Vide* Sestier and Méhu.)

† **Meibauer, R. O.** Der Novemberschwarm der Sternschnuppen. 8vo. 57 pp. *Berlin*, 1868
    (*Forms part of his " Ueber die physische Beschaffenheit unseres Sonnensystems,"* 2 Theil.)

Meidinger, J. Fred. Born 1726; died 1777.
    (*Said to have written on the action of Electric Fire on Metals and Minerals. Pogg.* ii. 102.)

Meidinger, H. Ueber voltametrische Messungen. Inauguraldissertation zur Erlangung der venia legendi. 8vo. 27 pp. *Giessen*, 1854

Meier. Merkw. Geschichte der magnetisch hellsehenden Auguste Müllerin in Carlsruhe : herausg. mit einer Vorrede von C. C. v. Klein. 8vo. *Stuttgart*, 1808

Meissner, Ferd. De Electricitate. 8vo. *Vratisl.* 1767

† **Meissner, G.** Neue Untersuchungen über den elektrisirten Sauerstoff. 4to. 110 pp. 2 plates. (*Abhandlungen d. K. Gesells. d. Wissen. zu Göttingen*, vol. xiv.) *Göttingen*, 1869

† Untersuchungen über die elektrische Ozonerzeugung, und über die Influenz-Elektricität auf Nicht-Leitern. 4to. 109 pp. 1 plate. (*Abhandl. d. K. Gesells. d. Wissen. zu Göttingen*, tom. xvi.) *Göttingen*, 1871
    (*Vide* also Henle and Meissner.)

† **Meissner und Meyerstein.** Über ein neues Galvanometer, Electrogalvanometer genannt. 8vo. 18 pp. 1 plate. (*Henle und Pfeufer Zeitschrift*, 3e R. Bd. xi.) *Leipzig*, 1859

Melander apud Stromer. Brief von Simon Melander. 8vo. 1 p. (*K. Schwed. Akad. Abh.* xvi. 158.) *Hamb. u. Leips.* 1754

Melandri, Contessi Girolamo. *Born March* 29, 1784, *at Bagnacavallo, near Ravenna ; died February* 25, 1833, *at Padua.*

Trattato elementare di chimica generale e particolare. 8vo. *Padova*, 1826 to —

† Disquisizione sui Paragrandini. Letta nell' Ateneo di Treviso il 15 Dec. 1825. 4to. 26 pp. (*Inserita nel* vol. x. *del Gle. sulle Scienze . . delle provincie Venete.*) *Treviso*, 1826

† Considerazioni critiche sopra l'efficacia del Paragrandine metallico. . . . 8vo. 37 pp. *Firenze*, 1827

† Metodo di preservare le fodere di rame delle navi dalla corrosione, proposto dal Sig. G. Melandri Contessi, e preceduto da esami ed analisi chimiche instituite sopra pezzi di fodera di rame rapidamente corrosa dalle acque del mare. 4to. 18 pp. (*Ann. del Reg. Lomb.-Veneto*, ii. 78.) *Padova*, 1832

Melbourne and other Observatories. (*Vide* Neumayer.)

342 MEL—MER

Melchior, Jn. Alb. Diss. de noctilucis. 4to. *Franequeræ*, 1742

Diss. de Electricitate. 4to. *Duisburgi ad Rhenum*, 1751

Mellarde (of Turin). On Medical Electricity. Letter to the Acad. di Bologna. (*Pivati's Exp. &c.*) 1749

Melloni, Macedonio. *Born April 11, 1798, at Parma; died August 11, 1854, at Portici.*

† Analisi delle tre Memorie pubblicate dal Faraday intorno alle azioni delle calamite e delle correnti elettriche sulla luce polarizzata e sulla massima parte dei corpi ponderabili. Parti 1a e 2da. 8vo. 51 pp. *Napoli*, 1846
(*La prima parte di questa Analisi è estratta dal fascicolo 33, e la parte seconda dal 34, del Museo di Scienze e Letteratura; giornale napolitano; Anno iii; 25 Giugno e 29 Luglio, 1846.*)

† La Thermochrose, ou la coloration calorifique. Première partie. 8vo. 357 pp.
Naples, 1850

Ricerche intorno al Magnetismo delle Rocce. Memoria i. Sulla polarità magnetica delle lave. 4to. (*Mem. dell' Acad. di Napoli*, i. 121.) *Napoli*, 1853

† Ricerche intorno al Magnetismo delle Rocce. Memoria ii. Sopra la calamitazione delle lave in virtu del calore e gli effetti dovuti alla forza coercitiva di qualunque roccia magnetica. 4to. 29 pp. (*Mem. dell' Accad. di Napoli*, i. 141. (*From Lamont, Handb.* p. 440.) (*Faraday's copy, with marks by him.*)
Napoli, 1853

Sur l'aimantation des roches volcaniques. 4to. (*Comptes Rendus*, xxxvii. 229, 1853.) *Paris*, 1853

Sur le magnétisme des roches. 4to. (*Comptes Rendus*, xxxvii. 966, 1853.)
Paris, 1853

Discorso necrologico per l'illustre Macedonio Melloni compilato dal Prof. P. Volpicelli. 4to. 4 pp. (*Estratt dall. Sess.* vi. del 20 Agosto, 1854, p. 298.)
Roma, 1854

† Sopra alcuni fenomeni di Elettricismo statico e dinamico recentemente osservati da Faraday nei conduttori de' Telegrafi sotterranei e sottomarini. 8vo 9 pp.

Elettroscopio di Melloni. 4to. (*Accad. di Napoli*, 1854.) *Napoli*, 1854

Delle due memorie sul Magnetismo delle rocce Estr. del Prof. Volpicelli. 4to.
(*Atti Accad. Nuovi Lincei*, vii. 145, 1853.) *Roma*, 1854

† Elogio storico di M. Melloni recitato . . Accad. di Napoli . . 1 Dec. 1854, dal A. Nobili. Estratto dal Rendiconto della Società Rle. Borbonica Accad. delle Scienze semestre 1854. 4to. *Napoli*, 1855

† Ricerche intorno al magnetismo delle Rocce. Memorie presentate all' Accad. . . nel 1853. 4to. 43 pp. (*Memorie per le Scienze naturali presentate dai Socii nell' An.* 1853.) *Napoli*, 1857

† **Menard de la Groye.** Account of a work entitled "Storia de' Fenomeni del Vesuvio avvenuti negli anni 1821, 1822, e parte del 1823," &c. : "History of the Phœnomena of Vesuvius during the years 1821, 1822, and part of 1823 ; accompanied with observations and experiments by Monticelli and Covelli. 8vo. 5 pp. (*Phil. Mag.* lxiii. 46.) *London*, 1824

Menil. (*Vide* Du Menil.)

Menon (of Angers). Influence de l'Electricité sur la végétation.

In his letters to Réaumur mentions experiments on Onions and Ranunculus, made in the winter of 1748. He made many other experiments. Was Correspondant de l'Acad.

Mentzel, Menzel Chn. De lapide bononiensi in obscuro lucenti. 8vo.
Bilefeldiæ, 1673

Mercer. Lettre à Franklin du New Brunswick, le 11 Novembre, 1752. 1752

Meredith. Considerations on the utility of Conductors . . 8vo. 1 plate.
London, 1789

† **Merget.** (Deux) Thèses de Physique et de Chimie. (1) Sur quelques cas particulièrs de la décomposition de l'eau. (2) Sur la formation d'empreintes métalliques des corps conducteurs, par l'emploi des courants électriques. 4to. 42 pp. (*Présentées à l'Académie de Paris,* 30 Août, 1849.) *Paris,* 1849

† Etude sur les travaux de Romas . . précédé du rapport de M. Abria. Couronné par l'Acad. 8vo. 80 pp. (*Actes de l'Acad. de Bordeaux,* 15e année.)
Bordeaux, 1853

Mermet, A. C. Memoir on Lightning Conductors. Prize question, Bordeaux Academy. 8vo. *Bordeaux,* 1837

Mermet. (*Vide* Bourges.)

† **Mesmer, A.** Mémoire sur la découverte du Magnétisme animal. 8vo. 85 pp.
Genève et Paris, 1779

Précis historique des faits relatifs au Magnétisme animal jusque en Avril, 1781. 8vo. *Londres,* 1781

† **Mesnard, J. B.** Lettre aux Parisiens sur les Télégraphes, et notamment sur le Télégraphe E. Gouon. 8vo. 63 pp. *Paris,* 1846

Messanges. (*Vide* Mallemans de Messanges.)

† **Metcalf, S. L.** A new theory of Terrestrial Magnetism. 8vo. 158 pp.
New York, 1833

Meteorological Committee. Report to Parliament for the year 1867.

Meteorological Society. Meteorological Society's Transactions. Vol. i. 8vo.
1839

Métherie, De La. (*Vide* La Métherie.)

Metterkamp, D. C. Über Blitzableitungen, gegen Busse's Theorie. 8vo.
Leipsig, 1812

Metzger, Johann Jacob. *Born March 28, 1783, in Canton Schaffhausen; died June 12, 1853, at Wagenhausen.*

Verbesserung d. elektr. Schreibemaschine u. Doppelflasche. (*Verhandl. d. Schweiz Naturf. Gesellsch. Aarau,* 1823.) *Aarau,* 1823

His electrical plate machine. (*From Elice, Saggio,* 2nd edition, p. 55.)

† **Meunier, S.** Etude sur les alliages météoriques de fer et de nickel. 8vo. 8 pp. (*Cosmos des 22 et 29 Août,* 1868.) *Paris,* 1868

† Note über den Krystallisirten Enstatit aus dem Meteoreisen von Deesa. 8vo. 3 pp. (*Sitzb. d. K. Akad. d. Wissensch.* ii. Abth. Janne Heft, 1870, lxi. Bde.)
Wien, 1870

Meunier, Victor. De l'orfévrerie électro-chimique. Histoire et description. 8vo. 279 pp. *Paris,* 1861

Meurer, H. Abhandl. v. d. Blitze u. d. Verwahrungsmitteln dag. 4to.
Trier, 1791

Abhandlung von d. feurigen Lufterscheinungen, &c. 8vo. *Trier,* 1793

Meverick. (*Vide* Briggs and Meverick.)

Meyer. Chymische Versuche, &c. French version by Dreux. *Paris,* 1766

Memorie sull' elettricità animale. . . Translation of Galvani by Aldini, 1792, with commentary by Aldini ; some treatises on the same subject, and some observations by Moscati and Vesci added to the preface. (*Vide Sue,* i. 27.)
1792

Meyer, Com. L'arte di rendere i fiumi navigabili in varj modi, etc. *Roma ?* 1696 (Part ii. " *Diversi segreti per conoscere la bontà dei metalli, e la virtù della calamita, &c.*")

344 MEY—MIC

Meyer, Johann Friedrich. *Born October* 24, 1705, *at Osnabrück ; died November* 2, 1765, *at Osnabrück.*
Chymische Versuche zur mehreren Erkenntinsz des ungelöschten Kalchs der elastischen und elektrischen materie, &c. . . 8vo. *Hanover,* 1764

Meyer, Johann Karl Friedrich. *Born* 1733, *at Stettin ; died February* 20, 1811, *at Stettin.*
Versuche mit der von Pallas in Siberien gefundenen Eisenstufe, nebst einigen allgem. Erfahr vom Eisen. 8vo. *(Beschaft. d. Gesellsch. Naturf. Freunde,* ii. 1776.) *Berlin,* 1776
Versuche mit der von dem H. Prof. Pallas in Siberien gefundenen Eisenstufe nebst einigen allgem. Erfahr vom Eisen. 8vo. 30 pp. *(Beschäftigung d. Berlin Gesellsch. Naturförd Freunde,* iii. 1777, p. 385.) *Berlin,* 1777
Versuche mit der von Pallas in Siberien gefundenen Eisenstufe, &c. nebst einigen allgem. Erfahr vom Eisen. 8vo. *(Schriften Gesellsch. Naturf. Freunde,* i. 1780.) *Berlin,* 1780

† **Meyer,** H. von (zu Frankft.) Über die Turmaline von der Insel Elba. 8vo. 17 pp. *(Archiv. f. d. Ges. Natural,* xiv. 3, p. 342.)

Meyer, Moritz. Ueber die Behandlung der Neuralgien durch Electricität. *(Deutsche Klinik,* 1857, No. 9.) 1857

† Die Electricität in ihrer Anwendung auf die practische Medicin. 2nd edition. 8vo. 373 pp. *Berlin,* 1861

† Die Electricität in ihrer Anwendung auf practische Medecin. Dritte . . vermehrte Auflage. 8vo. 423 pp. *Berlin,* 1868

† Electricity, in its Relations to Practical Medicine. Translated from the third German edition, with Notes and Additions. By Wm. A. Hammond, M.D. 8vo. 497 pp. *New York,* 1869

† **Meyer,** W. H. Theodor. Bestimmungen über die Intensitat des frein Magnetismus in Künstl. Magneten nebst Untersuchungen über Coercitirkraft. 8vo. 31 pp. 2 plates. *Marburg,* 1857

Beobachtungen über das geschichtete elektrische Licht sowie über den merkwürdigen Einfluss des Magneten auf dasselbe nebst Anleitung zur experimentellen Darstellung der fraglichen Erscheinungen. 4to. 29 pp. 4 plates. *Berlin,* 1858

Meyerstein. (*Vide* Meissner and Meyerstein.)

† **Meynier.** Mémoire sur le sujet du Prix proposé par l'Académie Royale en l'an 1729, touchant la meilleure méthode d'observer sur mer la déclinaison de l'Aiguille aimantée, ou la variation de la Boussole. 4to. 96 pp. 4 plates. *Paris,* 1732

† **Michaelis.** Briefwechsel zwischen Michaelis und Lichtenberg über die Absicht oder Folgen der Spitzen auf Salomons Tempel. 8vo. *(Gœtingischer Mag.* 3e année.) *Gœttingen,* 1783
(*Note.—Martin says that this is also in G. C. Lichtenberg's Physikalische und Mathem. Schriften,* tom. iii. p. 251-301. *Gœttingen,* 1803, *in* 12mo.)
(*Vermischte Schriften,* 1800-1805. 9 vols. 12mo. *Gotha.*)

† **Michaelis,** G. A. Über d. Leuchten d. Ostsee nach eigenen Beobachtungen, nebst einigen Bemerkungen über diese Erscheinung in andern Meeren. Mit e Vorworte v. Pfaff. 8vo. 52 pp. 2 plates in 4to. *Hamburg,* 1830

† **Michaelis** und Lichtenburg. Briefwechsel über die Absicht oder Folgen der Spitzen auf Salomon's Tempel, 1783. 8vo. (*Lichtenberg's Math. und Phys. Schriften,* iii. 251, *and in Gott.Mag.* i. iii. (1873), St. v. p. 735-68.) *Göttingen,* 1804

† **Michaud.** Observations sur les trombes de mer vue de Nice, en 1789, le 6 Janvier et le 19 Mars, par M. Michaud. 4to. 20 pp. 1 plate. (*Mém. de l'Acad. Royale de Turin,* an. 1788-9, p. 3.) *Torino,* 1790

Histoire complète des Télégraphes depuis leur origine. . . 12mo. 204 pp. 3 plates. *Genève,* 1853

MIC—MIE 345

Michell, Jno. Erfand die Drehwaage (Vor. Coulomb). . . (*Vide Cavendish, Phil. Trans.* 1798.)

† Mitchell, J. Treatise on Artificial Magnets. 1st edition. 8vo. 81 pp. 1 plate.
Cambridge, 1750

† Treatise on Artificial Magnets. 2nd edition, corrected, &c. 8vo. 78 pp. 1 plate.
Cambridge, 1751

Michelotti, Vittorio. *Born April* 19, 1774, *at Turin; died April* 4, 1842, *at Turin.*

† Saggio intorno ad alcuni fenomeni elettro-magnetici e chimici. Approvato 3 Giugno, 1821. 4to. 16 pp. (*In Mem. di Torino,* tom. xxvi. p. 365.)
Torino, 1821

† Descrizione di una particolare batteria Voltiana. Letta 9 Decemb. 1821. 4to. 7 pp. 1 plate. (*Mem. di Torino,* tom. xxvi. p. 433.)
Torino, 1821

† Continuazione del Saggio intorno ad alcuni fenomeni elettro-magnetici, ecc. Letta 17 Marzo, e 14 Aprile, 1822. 4to. 30 pp. 1 plate. (*Mem. di Torino,* tom. xxvii. p. 1.)
Torino, 1823

† Michelotti e Gioberti. Osservazioni sopra qualche fenomeno elettrico che si manifesta pendente l'infuocamento del platino spugnoso prodotto dal gaz idrogeno. Lette 28 Mar. 1824, 4to. (*Mem. di Torino,* tom. xxx. p. 189.)
Torino, 1826

(*Vide* also Vassalli, Rossi and Michelotti and Rossi and Michelotti.)

† Middendorff, A. von. Sibirische Reise : Meteorolog. geotherm. u. magnet Beobachtungen. 4to. 194 pp. 4 plates. *St. Petersburg*, 1848

Middeldorpf, A. T. Die Galvano-caustik, ein Beitrag zur operativen Medicin.
Breslau, 1854

Middleton, C. A new and exact Table, collected from several observations, taken in four voyages to Hudson's Bay in North America from London ; showing the Variation of the Magnetic needle, or sea-compass, in the pathway to the said Bay, according to the several latitudes and longitudes, from the years 1721 to 1723. 4to. (*Phil. Trans.* 1726, p. 73.)
London, 1726

† A new and exact Table, collected from several observations taken from the year 1721 to 1729, in nine voyages to Hudson's Bay in North America by C. Middleton; showing the Variation of the Compass according to the latitudes and longitudes under-mentioned, accounting the longitude from the meridian of London. Communicated by B. Robins. 4to. 5 pp. (*Phil. Trans.* xxxvii. 71.)
London, 1731

† Observations on the Weather, in a voyage to Hudson's Bay in North America in the year 1730, by C. Middleton. 4to. 3 pp. (*Phil. Trans.* xxxvii. 76.)
London, 1731-2

† Observations of the Variation of the Needle and Weather, made in a voyage to Hudson's Bay in the year 1731. Communicated to the Royal Society by B. Robins. 4to. 7 pp. (*Phil. Trans.* xxxviii. *for* 1733-34, p. 127.)
London, 1733-4

The use of a new Azimuth Compass for finding the Variation of the Compass or Magnetic needle at sea, with greater ease and exactness than any ever yet contrived for that purpose. 4to. (*Phil. Trans.* 1738, p. 395.) *London*, 1738

An observation of the Magnetick needle being so affected by great cold that it would not traverse. 4to. (*Phil. Trans.* 1738, p. 310.) *London*, 1738

The effects of Cold ; together with observations of the longitude, latitude, and declination of the Magnetic needle at Prince of Wales's Fort upon Churchill River, in Hudson's Bay, North America. 4to. (*Phil. Trans.* 1742, p. 157.)
London, 1742

† Miége, B Vade-mecum pratique de télégraphie électrique à l'usage des employés. Deuxième édition, revue, corrigée, etc. 12mo. 119 pp. *Paris*, 1862

† Guide pratique de télégraphie électrique, ou Vade-mecum pratique à l'usage des employés des lignes télégraphiques. . . 12mo. 148 pp. *Paris*

F F

† **Miége et Ungerer.** Vade-mecum pratique de télégraphie électrique à l'usage des employés du télégraphe. 12mo. 68 pp. 2 plates. *Paris,* 1855

Miettinger, Ludwig. Brevis meteorum explicatio. *Græcii,* 1698

† **Mikelli, A.** Le stelle cadenti. Lezione tenuta all' Ateneo di Venezia . . 32mo. 64 pp. ("*Forma il* vol. xxxv. *della Scienza del Popolo.*") *Milano,* 1888

Mikozski. (*Vide* Zaliwski.)

Milan. Società Patriotica. Memorie della Società Patriotica di Milano. 4to. (*Opuscoli Scelti,* xii. 1.) *Milano*

Milani. Coup d'œil géné sur l'histoire de l'Electricité. 8vo. *Valence,* 1853

† **Milani, G.** Corso elementare di fisica e meteorologia. Vol. vi. L'elettricità dinamica e l'elettro magnetismo. 8vo. 303 pp. *Milano,* 1869

Milano, Mich. Cenni geolog. sul tenimento di Massa Lubrense. Darin . . Teorica del Tifone e tromba di origine sotto-marina. 4to. *Napoli,* 1820

Miles, Henry. *Died February* 12, 1763.

† A Letter from Miles, H. to Baker, H. of firing Phosphorus by Electricity. Read March 7, 1744-5. 4to. 4 pp. 1 plate. (*Phil. Trans.* xliii. 290.) *London,* 1744-5

† A Letter from Miles, H. to the President; containing observations of Luminous Emanations from human bodies and from brutes, with some remarks on Electricity. 4to. 6 pp. (*Phil. Trans.* xliii. 441.) *London,* 1744-5

Electricity of a cane (*sic*) of black Sealing-wax and Brimstone. 4to. (*Phil. Trans.* 1746, p. 27.) *London,* 1746

Electrical Observations. 4to. (*Phil. Trans.* 1746.) *London,* 1746

Several Electrical Experiments. 4to. (*Phil. Trans.* 1746, pp. 53 and 158.) *London,* 1746

Concerning Electrical Fire. 4to. (*Phil. Trans.* 1846, p. 78.) *London,* 1746

Concerning the Electricity of Water. 4to. (*Phil. Trans.* 1746, p. 91.) *London,* 1746

† A letter from Miles, H. to the President concerning the Storm of Thunder which happened June 12, 1748. Read June 23, 1748. 4to. 5 pp. (*Phil. Trans.* xlv. 383.) *London,* 1748

† A letter from Miles, H. to Baker, H. concerning an Aurora Borealis seen January 23, 1750-1. Read February 22, 1749. 4to. 3 pp. (*Phil. Trans.* xlvi. 346.) *London,* 1749-50

† **Militzer, H.** Über die Verwendung einer gemeinschaftlichen Batterie für vielfache Schliessungkreise. 8vo. 41 pp. *Wien,* 1866

Über die Bestimmung der Constanten eines galvanischen Elementes. 8vo. 9 pp. (*Sitzb. d. Mathem.-naturw. Cl.*) *Wien,* 1869

Miller, William Allen. Elements of Chemistry, Theoretical and Practical. Revised edition, complete in 3 vols. 8vo.

Part I.—Chemical Physics.
Part II.—Inorganic Chemistry.
Part III.—Organic Chemistry.

† **Miller, G.** Observations on the Theory of Electrical Attraction and Repulsion. 4to. 14 pp. (*Dublin Acad.* vii. 139.) *Dublin,* 1799

Miller, Gerh. And. Schreiben an einen guten Freund von d. Ursache u. d. Nutzen d. Elektricität u.s.w. 4to. *Weimar,* 1746

Miller. (*Vide* Daniell and Miller.)

Milliet, Dechales C. F. Cursus, seu mundus mathematicus. Edit. aucta, &c. opera, A. Varcin. 4 vols. Folio. *Lugduni,* 1690
(*Contains Tract. de Magnete, Tract. de Meteoris, &c.*)

Mills, I. An Essay on the Weather, and directions for preserving lives and buildings from . . . Lightning. 8vo. *London,* 1770

Milly, Compt. Mémoire sur la réduction des chaux métalliques, par le feu électrique. Lu à l'Acad. de Paris le 20 Mai, 1774. 4to. (*Rozier, Obser.* iv. 146-50.) *Paris,* 1774
(*Note.—Many controversial, &c. articles by Comus, Fontana, Sigaud de la Fond, Brisson and Cadet, Rouelle and d'Arcet, appeared subsequently in Rosier's, &c. on this subject.*)

Milner, Thomas. *Born* 1719 ; *died September* 13, 1797, *at Maidstone*.

† Experiments and Observations in Electricity. 8vo. 111 pp. 2 plates. *London,* 1783
(*Note.—Magnetic Needle, &c. forming an Electrometer,* Fig. 13, p. 35.)

† **Minotto,** G. Di alcuni scientifici lavori fattisi recentemente in Venezia. 4to. 1 p. *Venezia,* 1840

† Sull' economia della Pila, e su alcune applicazioni della Galvano-plastica. Memoria letta all' I. R. Istituto delle provincie Venete il 29 Maggio, 1841. 8vo. 8 pp. (*Giorn. I. R. Ist. Lomb.* ii. 121.) *Milano,* 1841

† **Mirand.** Rapport fait, par M. Clerget, au nom du Comité des Arts économiques sur des Sonneries Electro-télégraphiques de M. Mirand. 4to. 18 pp. 1 plate. (*Bull. Société d'Encouragement pour l'Industrie.* . . 2me série, No. 6.) *Paris*

Appareil d'induction. (*From Du Moncel.*)

† **Miranda,** P. Redactor. Informe de la Rle. Acad. de Ciencias sobre Telegrafia elettrica presentado a la misma por una comision especial . . . y approbado en sesion gen. de 29 Diciem. de 1854. 4to. 184 + 2 pp. 5 plates. (*Mem. Acad. Madrid,* iii. part i.) *Madrid,* 1856

Mirus, C. E. Vom Nordlicht u. dessen Ursachen u.s.w., nach Mairan. *Oberlausitz, Nachlese,* 1770 ?

Von d. Art u. d. Gesetzen, nach welchen d. magnet. Kraft wirkt. *Oberlausitz, Nachlese,* 1770 ?

Missionaires de Pékin. Méms. concernant l'Histoire des sciences, &c. . . . par les Missionaires de Pékin. More than 12 vols. 4to. *Paris,* 1776-8 ?

† **Mississippi Railroad Explorations.** Appendices A to N, astronomical, magnetical, climatological, barometrical, and other Observations. 4to. 288 pp. *Washington,* 1856

Mitchel. (*Vide* Walker and Mitchel.)

Mitsching. Anleitung der Witterung, &c. 4to. 2 plates. (*From Ersch,* p. 116.) *Zittau,* 1802

Mochetti, Francesco. *Died March* 15, 1839, *at Como.*

† Lettera al P. Configliachi. (On the destroyed Lightning Conductor raised by Gattoni in Como ; and other matters.) 8vo. 13 pp. *Como,* 1814

Modena Academy. Reale Accademia di Scienze, Lettere, ed Arti di Modena.

Modena Literary Journal. Giornale letterario scientifico Modenese.
(*Vols.* v. vi. vii. *contains articles by Grimelli on Electro-Gild ng, &c. Vide* p. 35 *of his Storia.*)

Foglio periodico di Modena.
(*Vide Grimelli's Storia, as above. This may or may not be the same periodical as the above. It is often quoted by him, but it is impossible to get at his titles of articles, &c. correctly. The work itself must be consulted.*)

Moigno, François Napoléon Marie. *Born April 20, 1804, at Guémené, Dep. Morbihan.*

On Wheatstone's Submarine Electric Conductor, &c. (*L'Epoque*, Oct. 1846.)
1846

† Traité de Télégraphie électrique. 1st edition. 8vo. 420 pp. 16 plates.
Paris, 1849

† Note sur l'apparition constante de la lumière au pole négatif de la pile. 4to. 2 pp. (*Compt. Rendus,* xxx. 1850.) *Paris,* 1850

† Traité de Télégraphie électrique. 2de édition. Refondue. 8vo. 632 pp. Atlas of 22 plates. *Paris,* 1852

Cosmos : Revue encyclopédique hebdomadaire des progrès des sciences. 8vo. *Paris,* 1852

† Les accidents de chemin de fer et l'électricité. Système de Signaux élect. de M. Tyer. 12mo. 14 pp. *Paris,* 1856

Manuel de la Science. Annuaire du Cosmos. Vols. i. and ii. 18mo. *Paris,* 1859

† Les éclairages modernes. (Conférence.) 12mo. 118 pp. *Paris,* 1867

La France Scientifique (Cosmos). *Paris,* 1871

† **Moilin,** T. Traité élémentaire . . . de Magnétisme. 12mo. *Paris,* 1869

Mojon, Benedetto. *Born 1784, at Genua ; died June, 1849, at Paris.*

† Histoire académique du magnétisme animal, par Burdin et Dubois. 8vo. 14 pp. (*Giorn. I. R. Istit. Lomb.* i. 425.) *Milano,* 1841

Mojon, B. jun. Sur l'application de l'électricité dans la chlorose. 1845

Mojon, Benoit. Réflexions sur la théorie des sécrétions développées au moyen de l'électricité animale. 8vo. (*Journal du Galvanisme,* xi. cahier, p. 168.) *Paris*

Molard, C. P. Description des machines et des procédés spécifiés dans les Brévets d'invention. 4to. *Paris,* 1812

† **Molenier,** J. Essai sur le méchanisme de l'Electricité. 8vo. 108 pp. *Bordeaux,* 1768

† **Molin.** Falsità di un esperimento di Matteucci. 8vo. 10 pp. (*Sitzb. d. K. Akad. d. Wissen.* 1851, vi. Bd. iii. Heft, p. 313.) *Wien,* 1851

† **Molinier,** V. Notice sur l'usage de la Boussole au xiii. siècle, &c. . . . Lu 16 Mai, 1850. 8vo. 17 pp. (*Toulouse Acad.* 3me série vi. p. 193.) *Toulouse,* 1850

Molitor, Nicolaus Karl. *Born 1754, at Meisenbach ; died September 27, 1826, at Mainz.*

Joh. Ingenhousz. Anfangsgrunde der Elect. aus dem Englischen . . . Vorbericht des Übersetzers. 8vo. 134 pp. *Wien,* 1781

Joh. Ingenhousz. Vermischte Schriften. 1784

"Machte, mit andern Mainzer Aerzten, galvan. Versuche au Hingerichteten Galvanisch-elektr. Versuche an Menschen- u. Thierkörpern."
Frankfort-on-Maine, 1804

Moll, Gerrit (Gerhard). Electro-magnetische Proeven. 8vo. *Amsterdam,* 1830

Electro-magnetic Experiments. 8vo. (*Brewster's Journal of Science,* series ii. vol. iii. 1830.) *London and Edinburgh,* 1830

Sur la formation d'aimants artificiels au moyen du galvanisme. (*L'Institut,* No. 13, 1833, and *Pogg.'s Annalen,* xxiv.) *Paris,* 1833

Moll, Gerrit, and **Van Beck.** Expériences électro-magnétiques. 3 Aufsätze. 4to. (*Journ. Phys.* xcii. 1821.) *Paris,* 1821

Moll, Gerrit. **Van Rees,** and **Van den Bos.** Über d. Magnetisiren d. Stahls durch Maschinen-Elektricität. 8vo. (*Gilb. Ann.* lxxii. 1822.) *Leipzig,* 1822

Möller, J. W. Translation of Van Marum's Verhandeling over het Electrizeren.

Mollet, Joseph. *Born November 5, 1758, at Aix, Provence; died January 30, 1829, at Lyon.*

On the deviation of the Needle by action of the Pile. 8vo. *Lyon*, 1821

† Deux Mémoires : 1er, Décomposition de l'eau par la Pile Voltaïque ; 2d, Action du Courant Voltaïque sur l'Aiguille aimantée. 8vo. 9 pp. (*Acad. de Lyon.*) *Aix*, 1821

A third Memoir on the Decomposition of Water by the Pile. 8vo. *Lyon*, 1821

Cours élémentaire de physique expérimentale. 2 vols. 8vo. Plates. *Lyon et Paris*, 1822

† Mémoire sur la composition, et sur l'action de la Pile Voltaïque. 8vo. 16 pp. (*L'Acad. de Lyon*, Mai, 1823.) *Lyon*, 1823

Further experiments on the Pile, and Decompositions by it. 8vo. *Lyon*, 1823

† **Molossi,** P. Sui mezzi praticati dal. . Tholard . . . onde preservare le campagne dalla Tempesta. Osservazioni. 8vo. 24 pp. 1 plate. *Milano*, 1823

Molyneaux, Samuel. Account of the strange effects of Thunder and Lightning. 4to. (*Phil. Trans.* 1708.) *London*, 1708

† **Molyneux,** William. A demonstration of an error committed by common surveyors in comparing of surveys taken at long intervals of time, arising from the Variation of the Magnetic Needle. 4to. 5 pp. 1 plate. (*Phil. Trans.* xix. for 1695-6-7, p. 625.) *London*, 1695-6-7

Moncan. (*Vide* Cochy-Moncan.)

Moncel. (*Vide* Du Moncel.)

Monge. Gasp. Sur l'effet des étincelles excitées dans l'air fixe. 4to. (*Mém. Paris*, 1786.) *Paris*, 1786

Précis des leçons sur le calorique et l'électricité. 8vo. *Paris*, 1805

† **Mongiardini.** Dell' applicazione del galvanismo alla medicina. Memoria prima. Letta alla Soc. Med. di Emulazione, 16 Dec. 1802. 8vo. 25 pp. *Genova*, 1803

† **Mongiardini** e **Lando.** Sul Galvanismo. Memoria Seconda. (Letta alla Soc. Med. di Emulazione il 24 Giugno, 1803.) 8vo. 21 pp. *Genova*, 1803

Moniteur, Le. Contains articles on Electricity, &c. in 1808. &c. (*Vide* Seyffer, pp. 509-10.) 1808, &c.

Monier. On the Leyden Phial.

Monnier. (*Vide* Le Monnier.)

† **Monro,** Alexander. *Born March 20, 1733, at Edinburgh; died October 2, 1817, at Edinburgh.*

An Attempt to determine by Experiments how far some of the most powerful Medicines, viz., opium, ardent spirits, and essential oils, affect animals by acting on those nerves to which they are primarily applied, &c. Read 1761. 8vo. 74 pp. (*Essays and Observations*, iii. 292.) *Edinburgh*, 1771

† Experiments on the Nervous System . . Animal Electricity. 4to. 43 pp. *Edinburgh*, 1793

Experiments relating to Animal Electricity. 4to. (*Trans. of the Royal Society of Edinburgh*, iii.) *Edinburgh*, 1794

Monro u. **Fowler.** Abhandlung ueber thierische Electricität und deren Einfluss auf das Nervensystem, aus dem Englischen. 8vo. *Leipzig*, 1795

Mons, Jean Baptiste Van. *Born November* 11, 1765, *at Brussels ; died September* 6, 1842, *at Löwen.*

Uber thier. Elekt. und Galvanisme. 18 pp. *Brüssel,* 1798

† Natuurkundig vertoog over de verschynzels van het Galvanismus of dierlyke Electriciteit Voorgelezen. 8vo. 18 pp. (*Brugnatelli's copy.*)
Antwerpen, 1800

Censura commentarii a Wieglebo nuper editi, cui titulus, *de vaporis aquei in aerem conversione.* 4to. *Bruxelles,* 1801

Sur l'électrité animale. 12mo. 1802

† Estratto di una lettera del Van Mons a Brugnatelli sopra esperienze galvaniche. 8vo. 4 pp. (*Ann. di Chim. di Brugnatelli,* xix. 153.)
Pavia, 1802

† Estratto di una lettera del . Van Mons a Brugnatelli. 8vo. 6 pp. (*Ann. di Chim. di Brugnatelli,* xxi. 239.) *Pavia,* 1802

Journal de Chimie. 8vo. *Bruxelles,* 1802

† Principes d'électricité en confirmation de la théorie . de Franklin ; adressés dans une lettre à Brugnatelli. 8vo. 316 pp. *Bruxelles,* 1803

Lettre à Bucholz sur la formation des Métaux en général, et une particulier de ceux de Davy, &c. *Bruxelles,* 1810

† Grundsätze der Elektricität in einem Briefe an . Brugnatelli a. d. Französischen von Wurger. 8vo. 236 pp. *Marburg,* 1812

Vorrede und Zusätze zu Davy's Eléments de Philosophie chimique. 8vo. 2 vols. *Bruxelles,* 1813-16

† On Meteoric Stones, &c. (In his correspondence with the editor.) 8vo. 4 pp. (*Phil. Mag.* xlvi. 313.) *London,* 1815

Principes élémentaires de chimie philosophique ; avec applications générales de la doctrine des proportions determinées. 8vo. *Bruxelles,* 1818

† Quelques particularités concernant les Brouillards de différente nature Présenté à l'Acad. en Avril, 1827. 4to. 39 pp. *Bruxelles*

† Sur la réduction des alcalis en métal. 4to. 4 pp. *Bruxelles,* 1823

Sur les charges électriques. (*Bull. Acad. Brux.* i. *et* ii. 1835.) 1835

Sur une particularité dans la manière dont se font les combinaisons par le Pyrophore. Lu . . 4 Juillet, 1835. 4to. 8 pp. (*Acad. de Bruxelles,* xi.) *Bruxelles,* 1838

† Efficacité des métaux compacts et polis dans la construction des Pyrophores. Lu Juillet 4, 1835. 4to. 10 pp. (*Acad. de Bruxelles,* xi.) *Bruxelles,* 1838

Sur les charges électriques. 4to. (*Bull. Acad. Brux.* i. et ii. 1835.) *Bruxelles,* 1835

Sur l'électricité du Sucre. 4to. (*Bull. Acad. Brux.* vi. 1839.) *Bruxelles,* 1839

Mém. sur la question proposée, &c. à Bruxelles. (Tonnerre.)

Extrait d'une lettre de Van Mons à Brugnatelli, sur la décomposition de l'eau par la pile.

Sur l'identité des deux fluides. 8vo. *Bruxelles*

Montalbani, Ovidio. De illuminabili lapide bononiensi, epistola. 4to. *Bononiæ,* 1634

Brontologia, cioè, dicorso del tuono. 4to. *Bononiæ,* 1644

Montanari, Geminiano. *Born June* 1, 1633, *at Modena ; died October* 13, 1687, *at Padua.*

† La fiamma volante gran meteora veduta sopra l'Italia la sera del 31 Marzo, 1676. Speculazioni fisiche e astronomiche, in una lettera all' Trad. Gonzaga. 4to. 95 pp. 1 plate. *Bologna,* 1676

Montanari, Geminiano.—*continued.*

† Le forze d'Eolo. Dialogo . sopra gli effetti del Vortice o sia Turbine detto . La Bisciabuova che il 29 Luglio 1686, ha scorso e flagellato molte Ville . dé, territorii di Mantova, Padova, Verona, &c. Opera postuma. 12mo. 288 pp. 1 plate. *Parma,* 1694

Des moyens á l'aide desquels on peut distinguer les névroses des lésions dites organiques. Thèse. *Paris,* 1838

Montbéliard. (*Vide* Guéneau de Montbéliard.)

Monteclara, C. M. A Latin Thesis on Galvanism. *Taurini,* 1802?

† **Montéty.** Quelques mots sur l'électricité. Thèse. 4to. 29 pp. *Montpellier,* 1842

† **Montferrand.** A Memoir on Electro-dynamic Phenomena. Read on Feb. 3, 1823, at the Paris Acad. 8vo. 2 pp. *London,* 1823

Montgomery, J., and others. Lives of Eminent Literary and Scientific Men of Italy, Spain, and Portugal. 3 vols. 8vo. *London,* 1835-37

Montigny, Charles. *Born January* 8, 1819, *at Namur.*

Sur un phénomène électrique. (*Bull. Accad. Brux.* vii. 1841.) *Bruxelles,* 1841

Note sur la vitesse du bruit du tonnerre. 8vo. 13 pp. *Bruxelles*

† **Montpellier Acad. des Sciences.** Histoire de la Société Royale des Sciences établie à Montpellier, avec les Mémoires de Mathématiques et Physique tirés des registres (par De Ratte). *Lyon,* 1766; *Montpellier,* 1778

Recueil des Bulletins publié par la Société libre des Sciences et Belles Lettres de Montpellier. 1803-13

Acad. des Sciences et Lettres de Montpellier. Mémoires de la Section des Sciences. 4to. Plates. 1847-55, &c.

† Extraits des Procès Verbaux des Séances de la Section des Sciences. 8vo. *Paris,* 1847, *&c.*

Montpellier Bibl. Pub. " Catalogues méthodiques . des livres scientifiques légués à la Ville de Montpellier, par Aug. de St. Hilaire." Folio. MS. *Montpellier,* 1855-6

Catalogue général des livres de la Bibliothèque publique de la Ville de Montpellier. MS. *Montpellier*

Catalogue of books in the Nouvelle Bibliothèque de Montpellier. Fol. *Montpellier*

Montpellier Ecole de Médecine. Catalogue of books in this establishment.

Montriou, C. W. Observations made at the Magnetical and Meteorological Observatory at Bombay for the year 1847, under the superintendence of C. W. Montriou. Part ii. 4to. 504 pp. 14 plates. *Bombay,* 1851

† **Moody,** S. Effects of Lightning on the House of Silas Moody, of Arundel, in Maine, August 17, 1807, in a letter to Dr. Eliot. 4to. 1 p. *Cambridge, U.S.* 1809

Moos, M. Dissertaz. inaug. e Tesi. Applicazioni dell' Elettricità. 8vo. 15 pp. *Padova,* 1847

Morales. (*Vide* Vasquez-y-Morales.)

† **Moran,** F. Dissertat. inaug. quædam de physiologia ex effectibus electricitatis orientia complectens. 8vo. 25 pp. *Edinburgh,* 1820

† **Moratelli,** G. Memorie fisico chimiche. 8vo. 433 pp. 3 plates. *Venezia,* 1805

Moreau. (*Vide* Duncan et Moreau.)

Moreau, Wolf. (*Vide* Cheron.)

Moreau, J. L. Expériences galvaniques. (*La Décade Philosophique,* Ann. xi. No. xxxiv.)

† **Moretti** e Primo. Trad. di Davy. (*Vide* Davy's " Elementi," 1814.)

Morgan, George Cadogan. *Born* 1754, *at Bridge End, Glamorganshire; died November* 17, 1798, *at Southgate.*

Morgar, George Cadogan.—*continued.*

† Lectures on Electricity. 2 vols. 12mo.   *Norwich,* 1794

† Vorles. üb. d. Elektricität a. d. Engl. m. Anmerk. 8vo.   *Leipzig,* 1798

† Vorlesungen ub. d. Elektricität aus dem Engl. mit einigen Anmerk. 8vo.
345 pp. 2 plates.   *Leipzig,* 1790

Morgan and Barber. Account of the Aurora-Borealis seen near Cambridge
October 24, 1847, September 21, 1846, and March 19, 1847. 8vo. 12 plates
coloured.   *Cambridge*

Morgan, J. Dissertatio de usu electricitatis in re medica. (*From Guitard.*)
    Edinburgh, 1815

Morichini, Domenico Pini. *Born September 23, 1773, at Civitantino; died
November 19, 1836, at Rome.*

Mem. sopra la forza magnetizzante del lembo estremo del raggio violetto.
(*From Zantedeschi,* ii. 214.)   *Roma,* 1812

Seconda Memoria sopra la forza magnetizzante del lembo estremo del Raggio
violetto. 8vo. 32 pp. 1 plate. (*Volta's copy.*)   *Roma,* 1813

Biografia di (di Jandelli).   *Roma,* 1837

† Elogio storico del Cavaliere Domenico Morichini scritto dal socio e segretario
A. Lombardi. 4to. 10 pp. Portrait. (*Mem. della Soc. Ital.* xxvi. p. 3.)
    Modena, 1844

† Raccolta degli scritti editi ed inediti del Cav. D. Morichini. 2 vols. 8vo.
    Roma, 1852

Morin, Jean. *Born 1705, at Meung-sur-Loire, Orléanais; died March 28, 1764, at
Chartres.*

Nouvelle Dissertation sur l'électricité des corps. 200 pp. 12mo. *Chartres,* 1748

Réplique à Mons. l'Abbé Nollet sur l'électricité. 12mo. 48 pp.
    Chartres and Paris, 1749

Morin, R. Inesattezza d'un esperimento di Matteucci. Considerazioni ed esperimenti del Dr. R. Morin. 8vo.   *Venezia,* 1851

Morin, S. Du magnétisme et des sciences occultes. 8vo. 532 pp. *Paris,* 1860

Morlet. Recherches sur les lois du Magnétisme terrestre. 55 pp. *Paris,* 1837

† **Morlet, C. A.** Sur la détermination de l'Equateur magnétique, et sur les changements qui sont survenus dans le cours de cette courbe depuis 1776 jusqu'à
nos jours. 4to. 52 pp. 1 map.   *Paris,* 1832?

Moro, Anton Lazzaro. *Born 1687, at San Vito Frioul; died 1764.*

Lettera (ossia Dissert.) sopra la caduta dei fulmini dalle nuvole, al Marchese
Maffei. 12mo. 131 pp.   *Venezia,* 1750

† **Morozzo, Carlo Lodovico.** *Born August 5, 1744, at Turin; died July 2, 1804, at
Collegno, near Turin.*

Sur une aurore-boréale extraordinaire observée à Turin le 29 Février, 1780.
Lu le 3 Mars. 4to. 11 pp. 2 plates.   *Turin,* 1786

    (*He adds two letters to Beccaria: the first concerning an Aurora observed by him
on the 15th June,* 1780, *dated 18th June,* 1780; *the second dated the* 11th
August, 1780. *Refers to Beccaria's intended work on Aurora.* (*Ext. Mem.
Acad. de Turin,* 1784-5, tom. i. pt. ii. pp. 328-338.)

Expériences sur la fiole de Bologne. 4to. (*Mém. de l'Acad. de Turin,* iii.
pp. 449 *and* 464, *i.e.* viii. 449.)   *Turin,* 1788

Sopra alcuni fenomeni dei Fosfori Bolognesi nei differenti fluidi aeriformi. 4to.
19 pp. *Mem. della Soc. Ital.* v. 440 (1790).   *Torino,* 1790

De la lumière phosphorique que plusieurs pierres donnent en les frottant avec
une plume ou avec un épingle; et particulièrement sur la phosphoréscence
de la Tremolite et de la Cyanite; suivi de quelques observations sur l'Electricité positive ou négative de différentes pierres. 4to. (*Mém. de l'Acad.
Royale de Turin,* vi. 140-149.)   *Turin,* 1801?

Observations sur des épingles d'acier soumises à l'action de la pile de Volta.
4to. (*Mom. de Turin,* tom. xiv. p. 89, *Hist.*)   *Torino,* 1805?

Vita di Carlo Ludovico Morozzo; del Balbo, Conte P. 4to.   *Torino,* 1811

Morren, A. Sur les phénomènes lumineux que présentent quelques milieux raréfiés pendant et après le passage de l'étincelle électrique. 8vo. 23 pp.
Marseille, 1862

† **Morrison, R. J.** Report on his Portable Magnetic Electrometer. Read on 21st April before Elect. Soc. 8vo. 8 pp.
London, 1838

Morse, Samuel Finlay Breese. *Born April* 27, 1791, *at Charlestown, Massachusetts ; died April* 2, 1872, *at New York.*

Telegraphic Signals and Dictionary, exhibited 2nd September, 1837, and 4th September. 8vo. (*Journal of Franklin Institute,* Nov. 1837 ; *also in Silliman's Journal,* Jan. 1838, xxxiii. 185-187.) *Philadelphia,* 1837
Signalisirungsmethode.
(*Kuhn gives* (p. 853) '' *eine getreue Copie der am* 4 *Sept.* 1837, *zum ersten Male, zu Stande gebrachten Morse'schen Telegraphirungsmethode vor.'' Quere title.*)

Lettre sur son Télégraphe Electro-magnétique. (*Comptes Rendus,* vii. 593 ; xxii. 745, 1004.) *Paris,* 1838 *and* 1846

Experiments made with 100 pairs of Grove's battery, passing through 160 miles of insulated wire. In letter dated New York, Sept. 4, 1843. 8vo. (*Silliman's Journal,* 1843.) *Newhaven, U.S.* 1843

Lettres concernant l'étendue des lignes de télégraphes électriques établies aux Etats-Unis, et du grand usage qu'on fait, dès aujourd'hui, de ce mode de communication. (*Comptes Rendus,* xxiii. 545, 716.) *Paris,* 1846

Sur un appareil destiné à remédier aux effets de l'affaiblissement du courant dans les longues lignes de télégraphes électriques ; réclamation de priorité pour cette invention à l'égard de M. Breguet, fils. 4to. (*Comptes Rendus,* xxiii. 986.) *Paris,* 1846

(*Vide* Vail, 1847, '' *The American Electro-Magnetic Telegraph,'' for full accounts of Morse's Telegraph up to* 1847, *and reports, letters, &c. of Morse. Du Moncel gives further particulars.*) 1847

Télégraphe. *Forster's, Bauzeitung,* 1848, p. 229.) 1848

Die Relais. (*Silliman's Journal,* 1848, v. 58.) 1848

The Electro-Magnetic Telegraph. Reply to Jackson's article of January. (*Boston Morning Post,* May, 1849.) *Boston, U.S.* 1849

Beschreibung des Morse'chen Schreibapparates . . wie dieselbe in den Telegraphenanstalten Preussens benutzt worden ist . . 4to. (*Brix. Zeitschrift.* i. 193. *Vide also* p. 268.) *Berlin,* 1854

Paris Universal Exposition, 1867. Reports of the U.S. Commissioners. Examination of the Telegraphic Apparatus and the Processes in Telegraphy By S. F. B. Morse, United States' Commissioner. 8vo. 166 pp.
Washington, 1869

On his Telegraph, &c. 8vo. (*Phil. Mag.* xlvi. 264.)

† Morse's Patent. Full exposure of Dr. Chas. Jackson's pretensions . . 8vo. 66 pp. *New York,* 185—

Mortenson. Dissertatio de electricitate ; sub præsidio Klingensternæ. Pars prior. 4to. *Upsal,* 1740

Dissertatio de electricitate ; sub præsidio Klingensternæ. Pars posterior. 4to. *Upsal,* 1742

† **Mortimer, C.** An observation of the lights seen in the air, an Aurora-Australis, on March 18, 1738-9, at London, by Mortimer, C. 4to. 2 pp. (*Phil. Trans.* xli. 839.) *London,* 1739-40-41

† A letter to Folkes, M., from Mortimer, C., concerning the Natural Heat of Animals. Read July 4, 1745. 4to. 8 pp. (*Phil. Trans.* xliii. 473.)
London, 1744-45

Morveau, Louis Bernard Guyton de. *Born January* 4, 1737, *at Dijon ; died Jan.* 2, 1816, *at Paris.*

† Transunto di una lettera del De Morveau al De Montbeillard, sull' influenza del fluido elettrico nella formazione della Grandine. 12mo. 10 pp. (*Scelta d'Opuscoli,* xxxiii. 60.) *Milano,* 1777

354 MOS– MOU

Moscati, Pietro. *Born January,* 1739, *at Castiglione delle Stiviere ; died January* 19, 1824, *at Milan.*

† Lettera al Landriani sopra alcune nuove elettriche vegetazioni. 4to. 8 pp. 1 plate. (*Opus. Scelti,* iv. 410.) *Milano,* 1781

† Descrizione dell' Osservatorio meteorologico eretto (in Milano) al fine dell' anno 1780. 4to. 26 pp. 1 plate. (*Mem. Soc. Ital.* v. 356.) *Verona,* 1790 (*Vide* also Meyer.)

* † **Moseley, W. M.** Some account of the Solar spots which appeared during the year 1816. 8vo. 5 pp. (*Phil. Mag.* xlix. 182.) *London,* 1817

* † On the Solar spots. In reference to the "Correspondent" in *Phil. Mag.* Jan. last. 8vo. 2 pp. (*Phil. Mag.* lvii. 149.) *London,* 1821

† **Moser, L.** Über d. neueren magnetischen Entdeckungen. 8vo. 30 pp. *Königsberg,* 1834

† Über d. Erscheinungen des Magnetismus d. Erde. 8vo. 35 pp. *Königsberg,* 1834

Moser. (*Vide* Dove and Moser, and Riess and Moser.)

Mosmann, G. Unterhaltung über die electro-magnetische Telegraphie in der Schweiz. 8vo. *Schaffhausen,* 1852

Unterhaltungen über die elektro-magnetische Telegrafie in der Schweitz. 8vo. 10 pp. 1 plate. *Schaffhausen,* 1852

Mossotti, Ottaviano Fabrizio. *Born April* 18, 1791, *at Novara.*

Sur les forces qui régissent la constitution intérieure des corps. 4to. (*At* p. 9, *on Distribution of Electricity.*) *Turin,* 1836

Su di una proposta fatta dal Prof. Belli alla 2da riunione degli Scienziati Ital. e publicata nel tom. 100 della Biblioteca Ital. 8vo. (*Mem. Estratta dal Cimento, Giorn. de Pisa,* 1844.) *Milano,* 1844

Sulle forze che regolano la costituzione interna de' Corpi. Mem. (dedicato a Plana.) (*From Arella, Storia,* vol. ii. p. 131.)

† Discussione analitica sull' influenza che l'azione di un mezzo dielettrico ha sulla distribuzione della elettricità alla superficie di più corpi elettrici disseminati in esso. 4to. 26 pp. (*Mem. Soc. Ital.* xxiv. pt. ii.) (*Faraday's copy.*) *Modena,* 1846

† Comunicazione all' viii. adunanza italiana dei cultori delle scienze naturali. 8vo. 9 pp. (*Estratta dal Cimento,* an. v. fasc. Marzo-Aprile.) *Pisa,* 1847

Most, G. F. Über die grosen Heilwirkungen des in unsern Tagen mit Unrecht vernachlässigten Galvanismus. (*From Meyer.*) *Luneberg,* 1823

Motte, De La. (*Vide* La Motte.)

† **Motte, J. B.** Responsio ad questionem . . quæ est origo verisimilior electricitatis in columna electrica, et qua ratione compositio et decompositio corporum hujus ope obtinentur ? (Præmio ornata. 1821.) 4to.

Motz, A. J. G. W. von. Bijdrage tot militaire telegraphie. 8vo. 36 pp. *Vlissengen,* 1868

Mouilleron. Télégraphe électrique. (*Vide* Du Moncel.)

Mouilleron et Antoine. Elect. Uhrer. (*Vide Kuhn.*)

Moulon, A. De. De positivæ electricitatis, vel per excessum, a negativa, vel per defectum, necessitate instituendi discriminis, in therapeutico usu ejusdem fluidi. Disceptatio hab. in C. R. Archygymnasio Patavino . . 8vo. 31 pp. *Patavii,* 1826

Mountaine, William. *Died May* 2, 1779.

An account of some extraordinary effects of Lightning, July 16, 1759, with some remarks by Gowin Knight. 4to. (*Phil. Trans.* an. 1759, p. 294.) *London,* 1759

Mountaine, William.—*continued.*

Some observations on the Variation of the Magnetic needle, made on board the *Montagu* man-of-war in the years 1760-62, by David Ross. 4to. (*Phil. Trans.* an. 1766, p. 216.) *London,* 1766

Letter on Robt. Douglass's 1719 Observations. 4to. (*Phil. Trans.* an. 1776, p. 18.) *London,* 1776

Mountain, W. and **Dodson, J.** An attempt to point out in a concise manner the advantages which will accrue from a periodical review of the Variation of the Magnetic needle throughout the known world. 4to. (*Phil. Trans.* an. 1754, p. 875.) *London,* 1754 Chart.

(*Note.*—*Van Swinden says :* "*Les auteurs* pour 1744 *ont publié un nouveau pour* 1756.")

Letter concerning the variation of the Magnetic needle, with a set of tables annexed, which exhibit the result of upwards of 50,000 observations, in six periodical reviews, from the year 1700 to the year 1756, both inclusive ; and are adapted to every five degrees of latitude and longitude in the more frequented oceans. 4to. (*Phil. Trans.* an. 1757, p. 329.) *London,* 1757 ?

† An account of the methods used to describe Lines on Dr. Halley's Chart. 4to. 16 pp. *London,* 1784

(*There was also an edition of* 1758.)

Mousson, Joseph Rudolph Albert. *Born March* 17, 1805, *at Solothurn.*

Über e. subjective Lichterscheinung. 8vo. (*Pogg. Ann.* xxxix. 1836.) *Leipzig,* 1836

Über d. richtende Kraft d. Magnete. 4to. *Zurich,* 1846

† Coup d'œil historique sur le développement de l'électricité par la vapeur d'eau. 8vo. 15 pp. (*Soc. d'Hist. Nat. de Zurich,* November 30, 1846, tom. iv.) *Zurich,* 1846

† Über d. Elektricität bei Dampfbildung. (*Mittheilungen d. naturf. Gesellsch. zu Zürich,* i. 1847.) *Zurich,* 1847

Über die Veränderungen des galvanischen Leitungswiderstandes der Metalldrähte. Sm. 4to. 91 pp. 1 plate. (*N. Denkschr. d. allgem. Schweiz. Gesellsch.* xiv. 1854.) *Zurich* (?), 1854

† **Mousson, A.** Note sur les courants électriques développés dans l'action chimique des liquides les uns sur les autres. 8vo. 6 pp. (*Bibl. Univ.*) (*Arago's copy.*)

Mouzin. De l'emploi de l'Aimant dans le traitement des maladies. 8vo. *Paris,* 1843

† **Mullaly, John.** The Laying of the Cable, or the Ocean Telegraph ; being a complete and authentic narrative of the attempt to lay the Cable across the entrance to the Gulf of St. Lawrence in 1855, and of the three Atlantic Telegraph Expeditions of 1857 and 1858. 8vo. 329 pp. *New York,* 1858

† **Muller.** Einiges über d. Leitungswiderstand d. Metalle. 4to. 24 pp. 1 plate. *Wesel,* 1857

Muller, Von. (*Vide* Sonntag, A.)

† **Muller, A.** Entwickelung der Gesetze des Elektro-Magnetismus. 8vo. 44 pp. *Braunschweig,* 1850

† **Muller, C. H.** Übersetzung von Singer. (*Vide* Singer.) Galvanische Versuche. 8vo. (*Gilb. Ann.* vii. 1801.) *Leipzig,* 1801

Über Lapostolle's Blitzableiter. 8vo. (*Gilb. Ann.* lxviii. 1821.) *Leipzig,* 1821

Muller, D. G. Over de werking der dampkrings-electriciteit en de beveiliging der gebonen daartegen. 8vo. 25 pp. *Amsterdam,* 1857

Muller, Ferdinand. Über die Relation zwischen magnetischer Inclination und Horizontalintensitat. 8vo. mit 1 karte. (*Mélanges Phys. et Chim. tirés du Bull. de l'Acad. . . . de Petersb.* vii. pp. 15—71.) *Petersburg and Leipsig,* 1867

Muller, Gerhard Andreas. *Born February 23, 1718, at Ulm ; died February 26, 1762, at Gie sen.*

Schreiben an einen guten Freund von der Ursache u. dem Nutzen d. Electricität. 4to.　　　　　*Wien,* 1746

(*Note.—Musschenbrock, Institutiones, says,* p. 239, *"Schreiben von der Ursache der Elekt.*

† **Muller,** H. Ueber einige Verhältnisse der Netzhaut bei Menchen und Thieren. (With Kolliker, Notiz üb. d. elekt. Nerven des Malapterurus.) 8vo. 10 pp.
Paris

Muller, Johannes H. J. De barometri anomaliis quibusdam in prognosticis tempestatum. (*From Poggendorff,* ii. 222.)　　　　　1730

† Kurze Darstellung d. Galvanismus. Nach Turner, mit Benutz d. Original-Abhandlungen Faraday's bearbeitet. 8vo. 101 pp.　　*Darmstadt,* 1836

Principles of Physiology and Meteorology. 8vo. 2 plates.　　*London,* 1847

Bericht über die neuesten Fortschritte der Physik. In ihrem Zusammenhang dargestellt. 1—10 Lfg. (1 Bd.) 8vo.　　*Braunschweig,* 1849-51

† Entwicklung d. Gesetze des Electromagnetismus. 8vo. 44 plates.
Braunschweig, 1850

Magnetisirung von Eisenstäben durch d. galvan. Strom. 8vo. (*Poggend. Ann.* lxxix. 1850.)　　　　　*Leipzig,* 1850

Grundrisz d. Physik und Meteorologie. 2nd edition. 8vo.
Braunschweig, 1850

Uber d. Sättigungspunct d. Elektromagnete. 8vo. (*Poggend. Ann.* lxxxii. 1851.)　　　　　*Leipzig,* 1851

Zur Theorie d. diamagnet. Erscheinungen. 8vo. (*Poggend. Ann.* lxxxiii. 1851.)　　　　　*Leipzig,* 1851

Grundrisz d. Experimental-Physik. . . . 3rd edition. 8vo.
Braunschweig, 1852

Über d. Magnetisirung von Stahl u. Eisen durch d. galvan. Strom. 8vo. (*Poggend. Ann* lxxxv. 1852.)　　　　　*Leipzig,* 1852

Zur Theorie d. Elektromagnetischen. Maschinen. 8vo. (*Poggend. Ann.* lxxxvi. u. lxxxvii. 1852.)　　　　　*Leipzig,* 1852

Lehrbuch d. Kosmichen Physik. 8vo. With Atlas.　*Braunschweig,* 1856

Untersuchungen über Elektromagnetismus. 8vo. (*Poggend. Ann.* cv. 1858.)
Leipzig, 1858

Lehrbuch der kosmischen Physik. (Müller-Pouillets Lehrbuch der Physik und Meteorologie. 3. Band.) 2, durch einen Anhang bereicherte Ausgabe der 2. Auflage. Mit 316 in den Text eingedruckten Holzstichen und 1 Atlas von 33 Stahlstich-Tafeln, zum Theil in Farbendruck.
Braunschweig, 1865?

Lehrbuch der Kosmischen Physik 3 umgearb. u. verm Auflage. 4to.
Braunschweig, 1872

Physiology, by Baily (vol. i. p. 64, &c.), on the Torpedo.

Muller, Jos. Schreiben an Born über die in Tyrol gefundene Tourmaline. 4to.
Wein, 1773

Muller. (*Vide* Diamilla-Müller, D.)

Muller, K. A. Verhaltung d. Karnstoffs im Galvan. Strom. (*Erdmann's Journ.* lvii. 1852.)　　　　　1852

Muller, M. Beschreibung d. Sturmfluthen an den Ufern d. Nordsee u. der sich darin ergieszenden Ströme u. Flüsse am 3ten u. 4n. Febr. 1825. 8vo.
Hanover, 1825

Description abrégée du procédé d'amalgamation à Halsbrucke près de Freyberg en Saxe. 8vo.
Freyberg, 1831

† **Muller-Pouillet.** Lehrbuch d'Physik und Meteorologie. 7 Aufl. 2 vols. 8vo.
Braunschweig, 1868-9

Muller von Reichenstein, Franz Joseph. *Born* 1740, *at Vienna; died October* 12, 1825, *at Vienna.*

Nachr. von den in Tyrol entdeckten Tourmalinen an Born. 4to. 2 plates.
Wien, 1778

Mumenthaler, Joh. Jacob. Mechanic and optician. *Born in* 1729, *at Langenthal Canton Berne; died in* 1813.

Made, about the year 1778, very effective Electric machines and Electrophori of a kind of paper specially adapted to the purpose.

(*He was originally a bookbinder.*)

Munchow, K. D. von. Über Volta's Fundamentalversuch. 8vo. (*Poggend. Ann.* vol. i. 1824.)
Leipzig, 1824

Munck af Rosenschold, P. S. Forsök att. reducera frictions electriciteten till bestamda lagar.
Lundæ, 1837

† Artis mensoriæ electricæ fundamenta mathematica. (Thesis.) 8vo. 40 pp. 1 plate.
Lundæ, 1838

† Artis mensoriæ electricæ fundamenta mathematica. 8vo. 40 pp. 1 plate.
Lundæ, 1839

(*This is an amended version of his work (Thesis) of* 1838.)

Muncke, Georg Wilhelm. *Born November* 28, 1772, *at Hillingsfeld, near Hameln; died October* 17, 1847, *at Grotskmehlen.*

Merkwürd. elektr. Versuche. 8vo. (*Gilb. Ann.* vol. xli. 1812.) *Leipzig*, 1812

† Physikalische Abhandl. einiger Versuche zu Erweiterung der Naturkunde. 450 pp.
Giessen (?), 1816

Über Lichtpolarisirung u. d. Oersted'schen Versuch. 8vo. (*Gilb. Ann.* lxvi. 1820.)
Leipzig, 1820

Über Elektricitäts wund Wärmelehre. 8vo. (*Schweigg. Journ.* xxx. 1820.)
Nürnberg, 1820

Über Meteorsteine. 8vo. (*Schweigg. Journ.* xxx. 1820.) *Nürnberg*, 1820

Versuche über d. Elektromagnetismus. 8vo. (*Gilb. Ann.* lxx. u. lxxi. 1822.)
Leipzig, 1822

Über d. bei Ovelgönne gefundene angebl. Meteormasse. 8vo. (*Gilb. Ann.* lxxiii. 1823.)
Leipzig, 1823

Neue magnet. Beobacht. am Messing. 8vo. (*Pogg. Ann.* vi. 1826.)
Leipzig, 1826

Über Nordlichter. 8vo. (*Schweigg. Journ.* lii. 1828.) *Nurnberg*, 1828

Handbuch der Naturlehre (1st and 2nd parts). 8vo. *Heidelberg*, 1829

Thermoelektrische Beobachtungen, mitgetheilt in der Versammlung der Aerzte und Naturforcher zu Hamburg. (*From Lamont*, p. 441.) 1830

Vollendung des von Horner begonnen Artikel. Magnetismus, in der neuen Bearbeitung von Gehler's Phys. Wörterbuch. 8vo. *Leipzig*

† (Art.) Nordlicht in Gehler's Wörterbuch, vii. 1st Abth. 8vo. 156 pp.
Leipzig, 1833

† (Art.) Magnetismus in Gehler's Wörterbuch. Bd. vi. 2es Abth. (*Vide* Horner.) 8vo. 558 pp.
Leipzig, 1836

Über thermische Säulen. 8vo. (*Pogg. Ann.* xlvii. 1839.) *Leipzig*, 1839

Muncke, Georg Wilhelm.— *continued.*

Über Wiederherstell. d. Kraft geschwächter Magnete. 8vo.   (*Pogg. Ann.* l. 1840.)   *Leipzig,* 1840

Eine thermo-elektr. Beobacht. 8vo.   (*Pogg. Ann.* lii. 1841.)   *Leipzig,* 1841

Mittel die Wirk Volta'scher Säulen zu verstarken. 8vo.   (*Pogg. Ann.* liii. 1841.)   *Leipzig,* 1841

Elektrische Telegraphen. (*Physik Wörterbüch,* xi. 587—595.)   1845

† Art. Physik in Gehler's Wörterbüch, vol. vii. p. 493.

Electromagnetismus. Artikel des neuen Gehlerschen Wörterbuchs.

Mundt. Electrical Machine of strips of Silk   (Streifen von Seide.)   (*Grens' Journ. d. Phys.* vii. 319.)

Munich. Bayer Kunst und Gewerbblätter.   *München*

Anzeiger fur Kunst und Gewerbefleiss in Bayern.   *München*

Munich Academia. (*Vide* Bavarian Acad.)

Munich Privat Gesellschaft. Abhandlungen einer Privat Gesellschaft. 8vo.
                                                            München, 1792

Munter, F. Über d. v. Himmel gefall. Steine d. Alten, in Vergleich m. d. in neuern Zeiten herabgefall. Steinen a. d. Dänischen, v. T. A. Markussen. 8vo.
                                                            Kopenhagen, 1805

Über die vom Himmel gefallenen Steine, Bæthylien genannt, par Marckhavren (du Danois). 8vo.   *Kiobenhaven und Leipzig,* 1805

Explication de l'inscription d'un autel étrusque de Cortone.   (*Diss. Philosoph. et Histoire de la Société Royale de Danemark,* i. 3.)   *Kiopenhafen,* 1823

Murello. (*Vide* Da Murello.)

Murhard, F. W. A.   Versuch einer historisch-chronolog. Bibliographie d. Magnetismus.   *Cassel,* 1797

Murphy, Patrick. *Died December* 1, 1847, *at London.*

† Rudiments of Gravity, Magnetism, and Electricity, in their agency on the Heavenly Bodies. 8vo.   *London,* 1830

Murphy, Robert. *Born in co. Cork, Ireland; died March* 12, 1843.

† Elementary Principles of Electricity. 8vo. 145 pp. *London and Cambridge,*1833

† **Murray,** Sir James. Electricity as a cause of Cholera or other Epidemics, and the relation of Galvanism to the action of Remedies. 12mo. 160 pp. *Dublin,* 1849

Murray, John. *Died July* 22, 1820, *at Edinburgh.*

Elements of Chemistry. 8vo. 2 vols. 1040 pp. 3 plates. *Edinburgh,* 1810

On the Discoloration of Silver by a hard-boiled Egg. 8vo. 2 pp. (*Phil. Mag.* xliii. 62.)   *London,* 1814

On Coating objects with Plumbago for receiving Galvanoplastic depositions.
(*Note.—I recollect his mention to me of it.—F.R.*)

† **Murray,** J. Observations on Electrical and Chemical terms. 8vo. 3 pp. (*Phil· Mag.* xliii. 175.)   *London,* 1814

† On Electrical phenomena ; and on the new substance called Iode. 8vo. 3 pp. (*Phil. Mag.* xliii. 270.)   *London,* 1814

† On the phenomena of Electricity. 8vo. 4 pp. (*Phil. Mag.* xlv. 38.)
                                                            London, 1815

† On Galvanic Troughs. Experiment. 8vo. 1 p. (*Phil. Mag.* l. 145.)
(*Note.— Exposure of the plates to the atmosphere restored their power of heating a Wire, after it had ceased.*)   *London,* 1817

Murray, J.—*continued.*

† Voltaic action (effect of high temperature on), and Meteoric stone acted upon by the Oxy-hydrogen Blow-pipe (and other matter). 8vo. 3 pp. (*Phil. Mag.* 1. 312.) *London*, 1817

Elements of Chemical Science as applied to the Arts and Manufactures and to Natural Phenomena. 294 pp. (*From Phil. Mag.* lii. p. 60.) *London*

† On Aphlogistic phenomena and the Magnetism of Violet Light. 8vo. 4 pp. (*Phil. Mag.* liii. 268.) *London*, 1819

† Translation of description of a Storm Compass in the Correspondance Astronomique. 8vo. 1 p. (*Phil. Mag.* liii. 468.) *London*, 1819

† On Aërolites. 8vo. 5 pp. (*Phil. Mag.* liv. 39.) *London*, 1819

† On the decomposition of Metallic Salts by the Magnet. 8vo. 3 pp. (*Phil. Mag.* lviii. 380.) *London*, 1821

† On a Magnet effecting a decomposition of Nitrate of Silver. 8vo. 1 pp. (*Phil. Mag.* lviii. 387.) *London*, 1821

† On the Heat produced by Chlorine: and on a singular effect produced by Lightning. 8vo. 2 pp. (*Phil. Mag.* lx. 61.) *London*, 1822

† Notice of some new Galvanic experiments and phenomena. 8vo. 3 pp. *London*, 1822

† Some experiments connected with the relations of Caloric to Magnetism. 8vo. 1 p. (*Phil. Mag.* lxi. 207.) *London*, 1823

† On a phenomenon developed in Chemical Action. 8vo. 2 pp. (*Phil. Mag.* xi. 394.) *London*, 1823

† Note on the influence of Heat on Magnetism, &c. 8vo. 1 p. (*Phil.Mag.*lxii.74.) *London*, 1823

† Experiments on the Light and Luminous Matter of the Lampyris noctiluca. Read at the Linnæan Society, December 16, 1823. 8vo. 1 p. (*Phil. Mag.* lxii. 456.) *London*, 1823

† Experiments on the deviation of the Magnetic Needle as effected by Caloric, &c. 8vo. 2 pp. (*Phil. Mag.* lxiii. 130.) *London*, 1824

† Lightning Conductors. 8vo. 2 pp. (*Annals of Electricity*, vii. 82.) *London*, 1841

† **Murray, John** (of Glasgow). Experimental Researches on the Light of the Glow-worm. 12mo. 177 pp. *Glasgow*, 1826

† Treatise on Atmospheric Electricity. 12mo. 149 pp. 1 plate. *London and Edinburgh*, 1830

† Description of a new Lightning Conductor. 8vo. 1 plate. 63 pp. *London*, 1833

Om platina's magnétismus. 8vo. *Stockholm*, 1775

Musschenbrock, Petrus Van. *Born March* 14, 1692, *at Leyden; died September* 19, 1761, *at Leyden.*

† De viribus magneticis. 4to. (*Phil. Trans.* an. 1725, p. 370.) *London*, 1725

Epitome elementorum physico-mathematicorum. 8vo. 1726

Physica experimentalis et geometrica de Magnete, Tuborum capill. &c. Dissertationes 4to. 685 pp. 10 plates. *Leyden*, 1729

† Tentamina experimentorum naturalium captorum in Academia del Cimento, quibus commentarios, nova experimenta, et orationem, de methodo instituendi experimenta physica addidit. 4to. *Lugd. Bat.* 1731

† Ephemerides Meteorologicæ, Barometricæ, Thermometricæ, Magneticæ, Ultra-jectinæ, conscripta à Muschenbrock P. Ultrajectinæ anno 1729. 4to. 28 pp. 2 plates. (*Phil. Trans.* xxxvii. 357.) *London*, 1731–2

Musschenbrock, Petrus .Van.— *continued.*

†   An Abstract of a Letter to Desaguliers concerning Experiments made on the Magnetic Sand. 4to. 6 pp. (*Phil. Trans.* an. 1734, p. 297.)   *London,* 1734

Elementa physices. 8vo.   *Leyden,* 1734

Natural Philosophy by Colson. 2 vols. 8vo.   *London,* 1744

Grundlehre d. Naturwissenschaften und einige neue Zusätze des Verfassers. übersetze von Gottsched. 8vo.   *Stockholm,* 1747

Institutiones physicæ, conscriptæ in usus academicos. 8vo. 29 plates. 743 pp.   *Lugduni Bat.* 1748

Essai de Physique . traduit du Hollandois par Mr. Pierre Massuet. 4to. 2 vols.   *Leyden,* 1751

†   Dissertatio physica experimentalis de Magnete Lugduni Batavor. anno 1729 edita nunc vero auditoribus oblata. 4to. 283 pp. 10 plates.   *Viennæ,* 1754

†   Physicæ experimentalis et geometricæ de Mag. Tuborum capill. Dissertationes. 4to.   *Wien,* 1756

Tentamina experimentorum . nova experimenta add.   *Vienna,* 1756

Observation sur un météore lumineux de forme ovale et à queue. 4to. (*Mém. de Paris,* an. 1756.)   *Paris,* 1756

†   Introd. ad philosophiam naturalium. 2 vols. 4to.   *Lugd. Bat.* 1762

†   Compendium physicæ experimentalis ad usus academicos. 8vo.   *Leyden,* 1762

On the Leyden Phial. (*Danzig Mem.* ii. 433.)

Muzriano, M. Modo di scoprire vari moti vcri ed apparenti delle macchie del Sole. 12mo.   *Venezia,* 1758

Mylius, C. Nachrichten und Gedanken von der Elektricität des Donners. 8vo.   *Berlin,* 1752

Mynster, O. H. Grundträkken af Elektricitätslären og Magnetismen.   *Kiobenhaven,* 1806

N.

N. . . . (L'Abbé.) (*Vide* Anon. Elect. 1781.)

N. R. (*Vide* Anon. Miscel. 1821.)

Nahmmacher, G. C. Mechanismus der Kunstl. Elektricität verglichen mit electrischen Naturbegebenheiten und der Elekt. d. Turmalins u. d. Zitteraals. 8vo. (*From Heinsius*, iii. 15.) *Berlin*, **1791**

Nairne, Edward. *Died* 1806, *in London.*

Experiments on two Dipping Needles, which needles were made agreeable to a plan of Mr. Mitchell, and executed for the Board of Longitude. 4to. (*Phil. Trans.* 1772, p. 496.) *London*, **1772**

Electrical Experiments made with a machine of his own workmanship, a description of which is prefixed. 4to. (*Phil. Trans.* 1774, p. 79.) *London*, **1774**

† Experiments on Electricity, being an attempt to show the advantage of elevated pointed Conductors. Read at the Royal Society, June 18th and 25th, 1778. 4to. 40 pp. 4 plates. (*Phil. Trans.* 1778, p. 823.) (*Volta's copy.*) *London*, **1779**

An account of the effect of Electricity in shortening Wires. 4to. (*Phil. Trans.* 1780, p. 334.) *London*, **1780**

Letter—containing an account of Wire being shortened by Lightning. 4to. (*Phil. Trans.* 1783, p. 223.) *London*, **1783**

† Description and use of Nairne's Patent Electrical Machine, &c. 8vo. 68 pp. 5 plates. *London*, **1783**

† Description and use of Nairne's Patent Electrical Machine. . . . (Printed for Nairne and Blunt.) 8vo. 68 pp. 5 plates. *London*, **1787**

Description, &c., of an Electrical Machine. 8vo. (*From Watt.*) *London*, **1796**

Namias, Giacinto. *Born April* 10, 1810, *at Venice.*

† Sulle ragioni fisiologiche di alcuni fenomeni che presentano gli animali sottoposti all' azione dell' elettricità. Lettera al S. Marianini. 8vo. 15 pp. *Padova*, **1831**

† Di alcuni effetti dell' elettrico sopra l'animale economia, &c. . . . 8vo. 31 pp. (*Giornale per servire.* . . . *Patologia*, tom. xv. p. 442.) *Venezia*, **1841**

† Della elettricità applicata alla Medicina. Memoria IIa. 8vo. 30 pp. (*Estratta dal Giornale Veneto di Scienze Mediche.*) *Venezia*, **1851**

† Sui principii elettrofisiologici che devono indirizzare gli usi medici della elettricità e sui metodi più acconci a giovarsene nelle singole malattie. Studii del dott. N. 8vo. 148 pp. 1 plate. (*Giornale Veneto di Scienze Mediche*, serie seconda, tom. xv. p. 3.) *Venezia*, **1860**

Nancy. Société des Sciences, Lettres et Arts. Précis des Travaux. 8vo. (*From Cat. Roy. Soc.* p. 598.) *Nancy*

Napier, James. On Electrical Endosmose. (*Chem. Soc. Mem. and Proceed.* iii. 28.)

† A Manual of Electro-Metallurgy. . . . 8vo. 142 pp. (*Encycl. Metropolitana.*) *London*, **1851** (*Note.—Second and third editions were published in* 1852 *and* 1857 *respectively.*)

† **Napione, Carlo Antonio Galeani.** *Born at Turin; died in* 1814 *at Rio Janeiro.* Memoria sul Lincurio. 4to. 14 pp. *Roma*, **1795**

Naples Accad. Atti della Reale Accad. delle Scienze e Belle-Lettere di Napoli, dalla fondazione sino all' anno 1787. 4to. 370 pp. 20 plates.
Napoli, **1788**

Atti della Reale Accad. delle Scienze. Sezione della Società Reale Borbonica di Napoli. 4to. *Napoli*, **1819**

Rendiconto delle adunanze, e de' lavori dell' Accad. delle Scienze. Sezione della Società Reale Borbonica di Napoli. Fascicolo 1o. 4to. *Napoli*, **1842**

Memorie della Reale Accad. delle Scienze dal 1852 in avanti. 4to. *Napoli*

Naples Accad. degli Aspiranti. Annali dell' Accad. degli Aspiranti naturalisti di Napoli.

Bullettini dell' Accad. . . . 8vo. *Napoli*, **1842**

Esercitazioni accad. dell' Accad. . . . *Napoli*

Naples. Progresso delle Scienze Lettere ed Arti. *Napoli*

Nardi. De rore disquisitio phys. 4to. *Florent.* **1642**

Nasman, G. Det omgifvande mediets inflytande på galvaniska strömmar i isolerade ledningstrådar. Akademisk Afhandling. 8vo. 21 pp.
Upsala, **1860**

† **National Academy of Sciences.** Report of, for the year 1863 (by Bache, A.D.) 1864. 8vo. 118 pp. 7 plates. *Washington*, **1864**
(*Note.—This report contains, at p. 23, the " Report of the Chairman (Bache), of the Compass Committee of the National Academy of Sciences, Jan.* 1864," 73 *pages.*)

Naturf. Gesellschaft in Danzig. (*Vide* Danzig Society.)

Nauche, Jacques. *Born May* 18, 1776, *at Vigeois, Dép. Corrèze.*

De l'application du galv. à la rétention d'urine, suite de la paralysie de la vessie. 8vo. (*Journ. du Galv.* No. ii. p. 66, et No. iii. p. 122.) *Paris*, **1803**

Mémoires sur le galvanisme. 2 vols. 8vo. (*From Poggendorf*, ii. 256.)
Paris, **1804-5**

Des effets chimiques de la pile, sur l'air atmosphérique ; de la non transmission des effets de la pile dans le vide. 8vo. (*Journ. du Galv.* No. ix. pp. 49 and 51.)
Paris

Journal du Galvanisme, de Vaccine, &c. par une Société de Physiciens, &c. Redigé par J. Nauche. . . Présid. de la Société Galvanique. Tom. i. 1803, 344 pp. 1 plate ; tom. ii. 1803, 256 pp. 2 plates. **1803** *to —*
(*Vide* also Gautherot and Nauche.)

Naumann, Karl Friedrich. *Born May* 30, 1797, *at Dresden.*

† Kritische Untersuchung der allgemeinen Polaritätsgesetze. 8vo. 221 pp.
Leipzig, **1822**

Grundriss d. Krystallographie. 8vo. *Leipzig*, **1825**

Magnetische Beob. in Norwegen. 8vo. (*Poggend. Ann.* iii. u. iv. 1825.)
Leipzig, **1825**

Lehrbuch der reinen und angewandten Krystallographie. 2 vols. 8vo. 17 plates. *Leipzig*, **1830**

Über einen merkwürd. Blitzschlag. 8vo. (*Poggend. Ann.* xxxv. 1835.)
Leipzig, **1835**

Anfangsgründe d. Krystallographie. 25 plates. *Dresden u. Leipzig*, **1841**

Anfangsgründe d. Krystallographie. 8vo. 25 plates lith. *Leipzig*, **1850**

Grundriss d. Krystallographie. 8vo. (*From Pogg.* ii. 257.) *Leipzig*, **1852**

Anfangsgründe d. Krystallographie. 2 Aufl. 8vo. (*From Pogg.* ii. 257.)
Dresden, **1854**

Elemente d. theoretischen Krystallographie. 8vo. (*From Pogg.* ii. 257.)
Leipzig, **1856**

Nauman, M. E. A. Kritische Untersuchungen d. allgem. Polaritätsgesetze. 8vo.
Leipzig, 1822

Nautical Magazine or Journal. *London, commencing* 1832
(*Note.—Contains articles by Sir W. S. Harris on Electricity in* 1834-38-41-43. *Continued.*)

Nautoniez, Gvillavme De. Mecometrie de L'eymant : c'est-à-dire la manière de mesurer les longitudes par le moyen de l'eymant . . . Davantage y est montrée la Déclinaison guideymant pour tous lieux . . . De l'invention de Guillavme de Nautoniez Sieur de Castelfranc en Languedoc . . . Fol. plates. (*Very rare and curious book.*) *Venes.* 1603
(*Note.—The copy from which this* abridged *Title was taken is in the possession of Dr. Desbarreaux Bernard, Bibliothécaire de l'Académie des Sciences de Toulouse. Other copies have other dedications printed expressly for various sovereigns. Dr. Bernard was about to publish a full account of this work in* 1857.)

Nauwerck, C. L. Versuche neuer Erklärung u. Folge der jetzigen Witterung auf Oekonomie anwendbar, mit meteorolog. Bemerkk, die Gewitter ableiter betreffend. 8vo. *Dresden u. Leipzig*, 1787

Navarro, y Abel de Veas. Physica electrica o compendio, &c. &c. 8vo.
Madrid, 1752

† **Navez.** Application de l'Electricité à la mesure de la vitesse des Projectiles. 8vo. 2 plates. *Paris*, 1853

† Rapport sur les Expériences faites à Liége en 1851-2, au moyen d'un appareil électro-balistique, système Navez. 8vo. 63 pp. 3 plates. *Paris*, 1855

† Instruction sur l'appareil électro-balistique du Capitaine Navez. 8vo. 190 pp. 7 plates. *Paris*, 1859

Neale. Directions, &c. 8vo. (*Electrical Machine.*) (*From Watt.*) *London,* 1747

Nebel, W. B. Dissertatio physica de Mercurio lucente in vacuo, quam . . . publico erud. examini submittit A.D. 24 Mart. 1719. W. B. Nebel, Auctor. (*Sub pres. J. Bernouilli.*) 4to. 73 pp. *Basileæ*, 1719

Nebel, Daniel Wilhelm. *Born January* 1, 1735, *at Heidelberg ; died July* 3, 1805, *at Heidelberg.*

† Dissertatio inauguralis philosophica de magnetæ-artificiali. 4to. 67 pp.(*Trajecti ad Rhenum.*) 1756
De electricitatis usu medico. 4to. *Heidelbourg*, 1758

Necker, Louis. *Born* 1730, *at Geneva ; died* 1804, *at Geneva.*

Theses physicæ de Electricitate. Sub pres. Jolaberti Sept. 27, 1847. 4to. 16 pp.) *Geneva*, 1747

Le règne minéral ramené aux méthodes de l'histoire naturelle. (*From Wartmann, p.* 201.)

Necker de Saussure, Louis Albert. *Born April* 10, 1786, *at Geneva.*

Sur quelques rapports entre la direction générale de la stratification et celle des lignes d'égale intensité magnétique dans l'hémisphère boréal. 8vo. (*Bibl. Univ Fév.* 1830, *from Pogg.* ii. 262.) *Genève*, 1830

Résumé des observations d'Aurores boréales faites en Ecosse. 4to. (*Comptes Rendus*, xii. 1841.) *Paris*, 1841

Needham, John Tuberville. *Born September* 10, 1713, *in London ; died December* 30, 1781, *at Brussels.*

Some new Electrical experiments lately made at Paris. 4to. (*Phil. Trans.* 1746, p. 247.) *London*, 1746

Recherches sur la question si le son des cloches pendant les orages fait éclater la foudre en la faisant descendre sur le clocher, etc. (*Mém. de Brux.* iv. 1783.) *Bruxelles*, 1783

Needham, John Tuberville.—*continued.*

Recherches sur les moyens les plus efficaces d'empêcher le dérangement produit souvent dans la direction naturelle des aiguilles aimantées, par l'électricité de l'atmosphère. (*Mém. de Bruxelles*, iv. 73.)　　*Bruxelles*, 1783

Neeff, Christian Ernst. *Born August 23, 1782, at Frankfurt a. M. ; died July 15, 1849, at Frankfurt a. M.*

Nachschrift zu Buch's Brief über Elektromagnetismus. 8vo. (*Schweigg. Journ.* xxxi. 1821.)　　*Nürnberg*, 1821

Beschreib. u. Anwend. d. Blitzrades. 8vo. (*Poggend. Ann.* xxxvi. 1835.)　　*Leipzig*, 1835

Beob. von Sternschnuppen. 8vo. (*Poggend. Ann.* xxxix. 1836.)　　*Leipzig*, 1836

Description d'un appareil électro-magnétique. (*Comptes Rendus*, viii. 406.)　　*Paris*, 1839

Über d. Verhaltniss der electr. in Wärme Polarit. (*Schweigg. Journ.* vol. lxix.)

Über einen neuen Magnetelektromotor. 8vo. (*Poggend. Ann.* xlvi. 1839.)　　*Leipzig*, 1839

Über den Moderator (zu seinem Magnetelektromotor). 8vo. (*Poggend. Ann.* l. 1840.)　　*Leipzig*, 1840

† Über das Verhältniss der elektrischen Polarität. zu Licht u. Wärme. 8vo. 16 pp. (*Breslau Antiq. Cat.* No. 52, 1858, page 30, and in *Schweigger's Journ.* lxvi.)　　1845

Negro. (*Vide* Dal Negro.)

† **Nelis, De.** Account of some further electrical experiments by M. De Nelis, of Mechlin. 8vo. 4 pp. (*Phil. Mag.* xlviii. 127.)　　*London*, 1816

Nervander, Joh. Jacob. In doctrinam electromagnetismi momenta.　　*Helsingfours*, 1829

Sur un galvanomètre à chassis cylindrique, par lequel on obtient immédiatement et sans calcul la mesure de l'intensité du courant électrique, etc. 8vo. (*Ann. Chim. Phys.* lv. 1833.)　　*Paris*, 1833

Untersuchungen über die tägliche Veranderungen d. magnet. Declination. (*Bulletin scientifique Acad. St. Pétersb.* vi. 1840.)　　*St. Petersb. & Leipzig*, 1840-42

† Minnes-tal öfver Johan Jacob Nervander (Hallet på Finska Vetenskaps-Societetens Arshögtid den 29 April 1848) af Herr Gust. Borenius. (*Acte fennica.*)　　*Helsingfours*, 1848

† **Netoliczka, E.** Über unterseeische Telegraphenkabel. 4to. 19 pp. 1 plate. (*Note.*—*List of Authors on submarine cables, at end.*)　　*Graz*, 1867

Netto, F. A. W. Lehrbuch der Geostereoplastik. 8vo. (*Buer's Cat.* No. vi. 1867, p. 94.)　　*Berlin*, 1826

Anweisung zur Galvanoplastik . . . nach Spencer, Jacobi, und v. Kobel. 8vo. 64 pp. 2 plates.　　*Quedlinburg u. Leipzig*, 1840

Neufeld, O. Ozon et quæ concludi possint e notis adhuc viribus ejus physicalibus, chimicis et physiologicis in ulteriores effectus in animantium naturas. Dissertatio. 8vo. 46 pp. (*Regiomonti Pr.*)　　1852

Neuffer, C. Anleitung zur Aufbauung der galvanischen Säule, und zur Anwendung derselben auf verschiedene Krankheiten. 8vo.　　*Ulm*, 1802

Neumann. Eine Zusammenstellung der bis jetzt über d. Apparat, bekannt gewordenen Arbeiten u. eigene Experimental-Untersuchungen von C. H. Pfaff. 8vo. 2 plates.　　*Nurnberg*, 1819

† **Neumann, Carl.** Explicare tentatur quomodo fiat ut lucis planum polarisationis per vires electricas vel magneticas declinetur. Dissertatio inauguralis. 4to. 13 pp.　　*Halle*, 1858

Neumann, Carl.—*continued.*

Die Principien der Electrodynamik. Mathemat Untersuch. 4to.

Tubingen, 1868

Theorie der Elektricitäts- und Wärme-Vertheilung in einem Ringe. Halle, Buchhandlung des Waisenhauses. 8vo. 51 pp.

Über das Gleichgewicht der Wärme u. die Elektricität in e. Körper.

Halle, 1861 or 1863

† Die magnetische Drehung der Polarisationsebene des Lichts. Versuch einer mathematischen Theorie. 8vo. 82 pp. *Halle,* 1863

† Theorie der Elektricitäts- und Wärme-Vertheilung in einem Ringe. 8vo. 51 pp. *Halle,* 1864

Neumann, Casp. De ambra grysea. 4to. (*From Heinsius,* i. 42.) *Dresden,* 1736

Neumann, E. Die Elektricität als Mittel z. Untersuchung d. Geschmacks sinnes s. a. (*Aus. d. Konigsberger med. Jahrb.* Bd. iv.) *Königsberg*

Neumann, Franz Ernst. *Born September* 11, 1798, *at Ukermark.*

Ueber eine neue Eigenschaft der Laplace'schen $Y(_n^m)$ und ihre Anwendung zur analytischen Darstellung derjenigen Phänomene, welche Functionen der geographischen Länge und Breite sind. 4to. (*Schumacher's Astron. Nachrichten,* xv. 313.) *Altona*

Entwickl. der in ellipt. Coordinaten ausgedrückten reciprok. Entfern. zweier Puncte in Reihen, welche nach d. Laplace'schen $Y(^n)$ fortschreiten, u. Anwend. dieser Reihen z. Bestimm. d. magnet. Zustandes eines Rotations-Ellipsoids u.s.w. 8vo. (*Crelle's Journ.* xxvi. 1843.) 1843

† Die mathematischen Gesetze der inducirten elektrischen Ströme. (*Read* October 27, 1845.) 4to. 87 pp. (*Aus der Berlin Acad. von* 1845 *besonders abgedruckt.*) *Berlin,* 1846

† Uber ein allegemeines Princip der mathematischen Theorie inducirter elektrischer Ströme. Vorgelegt . . . am 9 Aug. 1847. 4to. 71 pp. 1 plate. (*Berlin Acad. Phys. Abhdl.* 1847.) *Berlin,* 1848

† Recherches sur la Théorie mathématique d'induction. Lu à l'Académie de Berlin, le 27 Octobre, 1845. Traduit par A. Bravais. 4to. 66 pp. (*Journal de Mathématiques Pures et appliquées,* tom. xiii. Avril 1848.)

Paris, 1848

Neumann, J. G. Krystallin, Structur des Meteoreisens von Braunaw. 4to. 1 plate. *Wien,* 1819

Neumann, K. A. Über hölzerne Kochgeräthschaft, über eine Volta'sche Walzenbatterie u.s.w. 8vo. (*Scherrar's Journ.* viii. ix. 1802.) *Berlin,* 1802

Der verwünschte Burggraf, ein Meteorolith in Ellbogen. 8vo. (*Gilb. Ann.* xlii. 1812.) *Leipzig,* 1812

Über Meteorolithen u.s.w. 8vo. (*Gilb. Ann.* xliv. 1813.) *Leipzig,* 1813

† **Neumayer,** G. Results of the Magnetical, Nautical, and Meteorological Observations made and collected at the Flagstaff Observatory, Melbourne,. and at various stations in the Colony of Victoria, March, 1858, to February, 1859. Fol. 262 pp. 3 plates. *Melbourne,* 1860

Results of the Magnetic Survey of the Colony of Victoria executed during the years 1858-64. 4to. *Mannheim,* 1869

† Bericht über das Niederfallen eines Meteorsteines bei Krähenberg, Kanton Homberg, Pfalz. 8vo. 113 pp. (*Sitzb. d. K. Akad. d. Wissensch* ii. Abth July, 1869, lx. Bd.) *Wien,* 1869

† **Neve,** T. Part of a Letter from Neve, T. to Mortimer, C. concerning an Aurora Australis seen March 18, 1738-39, at Chelsea, near London. 4to. 1 p. (*Phil. Trans.* xli. 843.) *London,* 1739-40-1

† **Newall, R. S.** Observations on the present condition of Telegraphs in the Levant, &c., with especial reference to the concession of the line between the Dardanelles and Alexandria, and to the Convention between Austria and England with regard to the Line between Ragusa and Egypt. Fol. 63 pp. 1 map. *London,* 1860

Newton, Sir Isaac. *Born December* 25, 1642, *at Whoolsthorpe; died March* 20, 1726, *in London.*

Observation that excited Glass attracted Light Bodies on the unrubbed side. 4to. (*Birch, History of Royal Society,* vol. iii. p. 260.) *London,* 1675

† Optics. pp. 314 and 327. Two theoretical Queries on Electricity. 8vo. (*From Priestley,* p. 14.) *London*

New York, Literary and Philosophical Society of. Transactions of the Literary and Philosophical Society of New York. 4to. *New York,* 1814

† **Niccolas.** Relazione di due cure mediche fatte col mezzo della elettricità. 4to. 2 pp. (*Opusc. Scelti,* vi. 198.) *Milano,* 1783

Nichols, Prof. Cyclopædia of the Physical Sciences, comprising Acoustics, Astronomy, Dynamics, Electricity, Heat, Magnetism, Meteorology, &c. Maps and illustrations. Large 8vo. *Roxburghe* (*Note.—There is a second and third edition.*)

Nicholson, William. *Born* 1753, *at London; died May* 21, 1815, *at London.*

† Journal of Natural Philosophy. 5 vols. 4to. 36 vols. 8vo. *London,* 1787-1813

British Encyclopædia. 6 vols. 8vo. (*From Bent. of* 1811.) *London*

Natural Philosophy. 2 vols. 8vo. (*From Bent. of* 1811.) *London*

† Death of, announced by the Editor. 1 p. (*Phil. Mag.* xlv. 396.) *London,* 1815

Nicholson and Carlisle. Experiments in Galvanic Electricity. (*Nicholson's Journal,* iv. 179.)

Nicklès, François Joseph Jérôme. *Born October* 30, 1820, *at Erstein* (*Bas Rhin*).

Note on his Electro-magnetic operations on the Wheels of Carriages traversing Rail-roads. 4to. (*Comptes Rendus,* ann. 1851.) *Paris,* 1851

† Sur le zinc amalgamé des piles. 8vo. 12 pp. (*Extrait du Journ. de Pharmacie,* Avril, 1852, xxii.) *Paris,* 1852

† Sur un nouveau Système d'Electro-Aimants. 8vo. 7 pp. 1 plate. (*Extrait des Annales de Chim. et Phys.* 3e Série.) *Paris,* 1853

† Les électro-aimants circulaires. Thèse. 4to. 27 pp. 1 plate. *Paris,* 1853

† Recherches sur l'aimantation. 8vo. 7 pp. 1 plate. (*Extrait des Mém. de l'Acad. de Stanislas.*) *Nancy,* 1855?

Nouveau procédé d'aimantation des roues de locomotives. (*Revue des Sociétés savantes,* Mai, 1859.) 1859

† Les électro-aimants et l'adhérence magnétique. 8vo. 302 pp. 5 plates. *Paris,* 1860

Note on Falling Stars. (*Revue Archéologique,* Avril, 1866, pp. 295-6.) *Paris,* 1866

Nicolai, J. C. W. Uber Blitzableiter Programm. 1796

Das Nordlicht Programm. (*Pogg.* ii. 283.) 1807

Nicolas, Pierre François. *Born December* 26, 1743, *at St. Mihiel en Barrois; died April* 18, 1816, *at Caen.*

Avis sur l'électricité comme remède dans certaines maladies. *Nancy,* 1782

Nicollet, J. N. Observations of the Magnetic Dip, made in the United States in 1841. 4to. (*Trans. American Society,* New Series, vol. viii.) *Philadelphia,* 1843

† **Nippoldt, W. A.** Untersuchungen über den galvanischen Widerstand der Schwefelsäure bei verschiedenen Concentrations-graden. Inaugural-Dissertation . . 8vo. 40 pp. *Frankfurt a. Maine*, 1869

† **Nivelet.** Electricité médicale. De l'électrisation géné alisée ou d'une méthode . d'appliquer l'électricité au traitement des maladies internes. 8vo. 107 pp. (*In Reinwald's Cat. for* 1861.) *Nancy et Paris*, 1860

† Mémoire sur la différence d'action physiologique et chimique des pôles positif et négatif dans les courants voltaïques ou continus et dans les courants intermittents d'induction. Suivi de deux Lettres . . 8vo. 31 pp.
Paris et Commercy, 1861

† Guide pratique du médecin électricien ou théorie des appareils Volta-magnétiques . . 8vo. 54 pp. 3 plates. *Paris et Commercy*, 1862

Application de l'électricité d'induction au traitement de la goutte sciatique, et des névralgies en général. 8vo. 1863

† **Nixon, J.** Meteor observed September, 1828, at Horton, in Ribblesdale. 8vo. 1 p. (*Phil. Mag. or Annals*, iv. 316.) *London*, 1828

† **Noad, H. M.** Course of Eight Lectures on Electricity, Galvanism, Magnetism, and Electro-Magnetism. (First edition.) 12mo. 382 pp. 1 cut. *London*, 1839

† Lectures on Electricity; comprising Galvanism, Magnetism, Electro-magnetism, and Thermo-electricity. Second edition enlarged. 8vo. 457 pp.
London, 1844

† A Manual of Electricity : including Galvanism, Magnetism, Diamagnetism, Electro-dynamics, Magneto-electricity, and the Electric Telegraph. 4th edition entirely re-written. Part i.—Electricity and Galvanism. 8vo. 522 pp. 1 plate. *London*, 1855

† A Manual of Electricity: including Galvanism, Magnetism, Diamagnetism, Electro-dynamics, Magneto-electricity, and the Electric Telegraph. 4th edition entirely re-written. Part ii.—Magnetism and the Electric Telegraph. 8vo. 910 pp. 1 cut. *London*, 1857

The Inductorium, or Induction Coil: being a Popular Explanation of the Electrical Principles on which it is constructed. Third edition. 8vo. Engravings. *London*, 1858

† The Induction Coil ; being an Explanation of the Principles of its Construction, with a Description of Experiments illustrative of the Phenomena of the Induced Current. Second edition. 12mo. 109 pp. 1 cut. *London*, 1866

† Student's Text-book of Electricity. 519 pp. 8vo. *London*, 1866

Nobili, Ant. Memoria sulle stelle cadenti . . Letta 14 Agos. 1838. 4to. 28 pp. (*Accad. di Napoli*, tom. v. part i. p. 119.) *Napoli*, 1845 ?

Nobili, Leopoldo. *Born* 1784, *at Trassilico, near Reggio (Modena) ; died August* 5, 1835, *at Florence.*

L'identità dell' attrazione molecolare coll' attrazione astronomica. 4to. 4 pl. *Modena*, 1819

† Introduzione alla meccanica della materia. 8vo. 194 pp. 7 plates. *Milano*, 1829

† Nuovi trattati sopra il Calorico l'Elettricità e il Magnetismo. 8vo. 401 pp 8 plates. *Modena*, 1822

† Sul confronto dei circuiti elettrici coi circuiti magnetici. 8vo. 48 pp.
Modena, 1822

Sur la lumière des Aurores Boréales, imitée par une expérience électromagnétique. 8vo. (*Bibl. Univ.* xxv. 1824.) *Genève*, 1824

† Questioni sul Magnetismo. 8vo. 181 pp. 4 plates. *Modena*, 1824
(*Copies with the spurious title-page dated* 1838 *were issued after his death in* 1835.)

Nobili, Leopoldo.—*continued.*

Sur le courant électrique de la grenouille. 8vo. (*Bibl. Univers.*)
Geneva, 1827

Sopra un Galvanometro con nuove aggiunte. Mem. presentata alli 31 Luglio, 1826. 4to. 13 pp. 1 plate. (*Ital. Soc. Mem.* xx. p. 173.) *Modena,* 1828

Analisi . . . degli effetti elettrofisiologici della rana. *Reggio,* 1829

Termo moltiplicatore, ossia termoscopio elettrico. (*From Zantedeschi, Trattato,* i. p. 334.) *Reggio,* 1829

† On the Colours produced upon Steel-plates. Exhibited and explained at a Friday evening meeting at the Royal Institution, January 30, 1829. 8vo. 1 page. *London,* 1829

Estratto di alcune Memorie pub. dal Nobili. (On colours of metals under the action of the Pile, &c.) 4to. *Padova,* 1831

† Teoria fisica dell' Induzione elettro-dinamica. 8vo. 35 pp. 1 plate.
Firenze, 1832

† Nuove osservazioni sopra le apparenze elettro-chimiche, &c. 8vo. 12 pp. 1 plate. (*Arago's copy.*) *Firenze,* 1833

† Neue Beobachtungen über d.elektrochemischen Figuren, die electro-dynamischen Gesetze, u.d. innern Mechanismus d. voltaischen Säule. 8vo. 17 pp. 1 plate. (*Poggendorff's Annalen d. Phys.* xxxiii.) *Leipzig,* 1834

† Descrizione di due nuove pile termo-elettriche, &c. 8vo. 20 pp.
Firenze, 1834

† Sopra l'Elettricità animale. 8vo. 4 pp. *Firenze,* 1834

† Note sul Magnetismo—
1. Sopra la distribuzione del magnetismo nell' interno delle Calamite. (Dated 10 Gennajo, 1834.)
2. Sopra la forza coercitiva. (Dated 14 Gennajo, 1834.) 8vo. 7 pp. for both. (*Arago's copy.*) 1834

† Memorie ed Istrumenti edite ed inedite. 2 vols. 8vo. 14 plates. *Firenze,* 1834 (*In* vol. ii. *is an enumeration of the Memorie ed Istrumenti, which were probably printed originally at Reggio in 1826 to 1830.*)

† Memoria su l'andamento e gli effetti delle correnti elettriche dentro le masse conduttrici. 8vo. 51 pp. *Firenze,* 1835

Nota sopra alcuni perfezionamenti. (*From Zantedeschi, Trattato,* ii. 182.)

Elogio storico del Cav. Prof. Leopoldo Nobili letto dal V. Antinori alla Soc. Columbaria . . . 24 Gennajo, 1836. *Firenze,* 1836

† Cenni storici sulla vita e sulle principali scoperte del Cav. Leopoldo Nobili scritti dal E. Giorgi, Ricevuti 1 Agosto, 1837. 4to. 16 pp. Portrait. (*Mem. Soc. Ital.* tom. xxii. p. 100.) *Modena,* 1839

Elogio funebre del Prof. Leopoldo Nobili scritto dall' Ab. G. Castelli. (*From Grimelli, Storia,* p. 85.) *Firenze,* 1841

† Elogio storico del Cav. Professore Leopoldo Nobili scritto dal F. Bordè. 8vo. 62 pp. *Modena,* 1847

† **Nobili ed Antinori.** Sopra la forza elettromotrice del Magnetismo. 8vo. 15 pp. (*Nell' Antologia,* No. 131, Nov. 1831.) (*Arago's copy.*) *Firenze,* 1831

Descrizione delle nuove calamite elettriche ed osservazioni sulle medesime, Mag. 20, 1832. 8vo. *Firenze,* 1832

Nuovo condensatore elettro-dinamico del Nobili, Mag. 8, 1832. 8vo.
Firenze, 1832

Sulla sensibilità del termo-moltiplicatore del Nobili, Mag. 11, 1832. 8vo. (Tutti Estratti dall' Antologia, No. 136. 25 pp. per tutti 3o.) *Firenze,* 1832

† Sopra le forza elettromotrice del Magnetismo. Estratta dall' Antologia, No. 131, November, 1831. Dated Dal Museo, 1832. 4to. 10 pp. *Ann. del Reg. Lomb.-Veneto,* ii. 96.) *Padova,* 1832

On the electromotive force of Magnetism, with Notes by Faraday. 8vo. (*Antologia di Firenze,* No. 131, *and Phil. Mag.* June, 1832.) *London,* 1832

Noble. (*Vide* Le Noble.)

Noble. (*Vide* Le Noble d'Artion.)

† **Nocetus, C.** De Iride et Aurora-boreali. Carmina . . . cum Notis Boscovich.
4to. 127 pp. 2 plates.   *Romæ*, 1747

Noel, F. Observationes mathematicæ et physicæ in India et China factæ ab anno
1684 ad annum 1708, una cum mappa stellarum australium, &c. 4to.
(*From Whiston's Longitude*, p. xxiii.)   *Prague*, 1710

Nöggerath, Jacob. *Born October* 10, 1788, *at Bonn.*

Neue Beobachtungen. 8vo. (*Vide* Haüy.)   *Bonne*, 1819

† Über die am 7ten Mai, 1822, zu Bonn niedergefallenen Hagelmassen, &c.
4to. 16 pp. 1 plate.

† On Meteoric Iron from Mexico. In a Letter to Dr. Chladni. 8vo. 2 pp.
(*Phil. Mag. or Annals*, ii. 46.)   *London*, 1827

Über d. magnet. Polarität zweier Basaltfelsen in d. Eifel, u.s.w. 8vo.
(*Schweigg. Journ.* lii. 1828, p. 224.)   *Nürnberg*, 1828

Die Bruchhauser Steine am Issenberg. (*Karsten's Arch. f. Min.* iii. 1831.)
                                                                                                                                      1831

Nöggerath mit **Bischof.** Meteoreisen von Bitburg. 8vo. (*Schweigg. Journ.*
xliii. 825.)   *Nürnberg*, 1825

† **Nöggerath and Reuss.** On the Magnetic Polarity of two Rocks of Basalt near
Nürnburg in the Eifel, with some observations on the extension of Basalt in
that district; drawn up from the observations of Bergmeister Schulze, of
Düren. On the Magnetic Polarity of two Basaltic Rocks in the lordship of
Schröckenstein. By M. Reuss, of Berlin. 8vo. 5 pp. (*Schweigg. Journ. and
Phil. Mag.* viii. 174.)   *London*, 1830
(*Vide* also Chladni and others.)

Nollet, Jean Antoine. *Born November* 19, 1700, *at Pimpre, Diocése Noyon; died
April* 24, 1770, *at Paris.*

Programme, ou Idée générale d'un Cours de Physique expérimentale. 12mo.
190 pp.   *Paris*, 1738
(*There was probably a* 4to. *edition of* 1735.)

Expériences physiques faites à Bordeaux en 1741 divisées en 16 Leçons. 4to.
MS. (*From the Cat. of the Bordeaux Pub. Lib.*)   *Bordeaux, n. d.*

† Leçons de Physique expérimentale. 1er édit. 1er et 2d vol. 12mo. Paris,
1743; 3me vol. 12mo. Paris, 1745; 4me vol. 12mo. Paris, 1748; 5me vol.
12mo. Paris, 1755; 6me vol. 12mo. Paris, 1764. (*Les Propriétés de l'Aimant*,
tom. vi. p. 160; *Electricité*, tom. vi. p. 234.)
(*His conjectures as to the identity of Lightning and Electricity*, tom. iv. p. 314,
in 1748. *There are various other sets, composed of vols. of different editions,
dates, &c.*)

Conjectures sur les causes de l'Electricité des corps. 4to. (*Mém. de Paris*, 1745,
Hist. p. 4, *Mém.* p. 107.)   *Paris*, 1845

Conjectures tirées de l'expérience.   1745-46
(*Forms the third part of his Essai*, p. 138, *and is extracted from two Memoirs
which he read at the Acad. de Paris in April*, 1745, *and " après Pâques "
in* 1846. *The Memoir read April* 5, 1745, *is entitled " Conjectures sur les
Causes de l'Electricité." The Memoir read " après Pâques " in* 1846 *is perhaps
that entitled " Observations sur quelque snouveaux phénomènes d'électricité."*)

On a Blue-glass Machine. (*Acad. de Paris.*)   *Paris*, 1745

Observations sur quelques nouveaux phénomènes d'électricité. 4to. (*Mém. de
Paris*, 1746, *Hist.* p. 1, *Mém.* p. 1.)   *Paris*, 1746

The contents of his first Memoir, read in the Accad. Reale, 1746. (*Mém. de
Trevoux, Art.* 69 of June, 1746.)   1746

On the Leyden Phial. Letter to S. Wolfe, dated Versailles, May 8. (*Dantzig
Mem.* tom. i. p. 409. Forms part of *Graloth's Geschichte d. Elekt.*)   1746

Nollet, Jean Antoine.—*continued.*

† Essai sur l'Electricité des corps. 1er édit. 12mo. 227 pp. 4 plates. *Paris,* 1746

† Essai sur l'Electricité. 4to. 131 pp. 1 plate. *Madrid,* 1747

† Saggio intorno all' Elettricità de' Corp. trad. dal Francese. Aggiuntevi alcune esperienze ed osservazioni . . . del Sig. Gugl. Watson. 12mo. 254 pp. 5 plates. *Venezia,* 1746

Abhdl. v. d. Elektricität d. Körper a. d. Frz. 8vo. *Erfurt,* 1748

† Part of a Letter from Abbé Nollet to Folkes, M. concerning Electricity. Translated from the French by Stack, T. Read February 11, 1747-8. 4to. 8 pp. (*Phil. Trans.* xlv. 187.) *London,* 1748

Eclaircissements sur plusieurs faits concernant l'Electricité. (4 Méms.) 4to. (*Mém. de Paris,* 1747, *Hist.* p. 1, *Mém.* p. 102, 149, 207 (1848), 164.) *Paris,* 1847-8

Vorles. über die Elementarphysik a. d. Frz. 9 vols. 8vo. *Erfurt,* 1748-72

† Recherches sur les causes particulières des phénomènes électriques. 1re édition. 12mo. 444 pp. 8 plates. *Paris,* 1849

† Extract of a letter from Nollet to Charles, Duke of Richmond, accompanying an Examination of certain Phænomena in Electricity, published in Italy, by the same, and translated from the French by W. Watson. Read March 29, 1750. 4to. 30 pp. (*Phil. Trans.* xlvi. 368.) *London,* 1749

† Ricerche sopra le cause particolari de' fenomeni elettrici, &c. 8vo. 334 pp. 8 plates. *Venezia,* 1750

Examination of certain Phænomena in Electricity. 4to. (*Phil. Trans.* 1750, p. 368.) *London,* 1850

Essai sur l'Electricité des Corps. 2e edition. 12mo. 272 pp. 4 plates. *Paris,* 1750

On extracting Electricity from the Clouds. 4to. (*Phil. Trans.* 1752, p. 553.) *London,* 1750

Comparaison raisonnée des plus célèbres phénomènes d'électricité tendant à faire voir que ceux qui nous sont connus jusqu'à présent, peuvent se rapporter à un petit nombre de faits qui sont comme les sources de tous les autres. 4to. (*Mém. de Paris,* an. 1753, *Hist.* p. 6, *Mém.* 429 and 475.) *Paris,* 1753

Sur la prétendue distinction des électricités en plus et en moins. Partie i. 4to. (*Mém. de Paris,* an. 1753, *Hist.* p. 6, *Mém.* p. 475 ?) *Paris,* 1753

† Lettres sur l'Electricité. Première partie (ou vol.), 1re édition. 12mo. 264 pp. 4 plates. *Paris,* 1753

Suite du Mém., dans lequel j'ai entrepris d'examiner si l'on est bien fondé à distinguer des électricités en plus et en moins. 4to. (*Mém. de Paris,* an. 1755, *Mém.* p. 293.) *Paris,* 1755

Observation sur un météore en manière de colonne de feu qui embrasa une écurie et tua quatre chevaux. 4to. (*Mém. de Paris,* an. 1759, *Hist.* p. 35.) *Paris,* 1759

Observation sur une Aurore Boréale complète; vue à Upsal. 4to. *Mém. de Paris,* an. 1759, *Hist.* p. 37.) *Paris,* 1759

† Lettres sur l'Electricité. Deuxième Partie (ou vol.), 1re édition. 12mo. 284 pp. 4 plates. *Paris,* 1760

Lettere intorno all' Elettricità . . . del Sig. Abate Nollet . . tradotte dal Francese. (2 parts only.) 2 vols. 8vo. 1st vol. 228 pp. 4 plates. 2nd vol. 256 pp. Plates. *Napoli,* 1761

Nouvelles expériences d'Electricité faites à l'occasion d'un ouvrage publié depuis peu en Angleterre par Robert Symmer. 4to. (*Mém. de Paris,* an. 1761, *Hist.* p. 10, *Mém.* p. 244.) *Paris,* 1761

Réflexions sur quelques phénomènes cités en faveur des électricités en plus et en moins. Part i. 4to. (*Mém. de Paris,* an. 1762, *Hist.* p. 10, *Mém.* p. 137 and 270.) *Paris,* 1762

Nollet, Jean Antoine.—*continued.*
Mémoire sur les effets du Tonnerre comparés à ceux de l'Electricité avec quelques considérations sur les moyens de se garantir des premiers. 4to. (*Mém. de Paris*, an. 1764, *Hist.* p. 1, *Mém.* p. 408.) *Paris,* 1764

Sendschreiben v. d. Electricität a. d. Fiz. 8vo. *Erfurt,* 1765

Application curieuse de quelques phénomènes d'Electricité. 4to. (*Mém. de Paris*, an. 1766, *Hist.* p. 1, *Mém.* p. 323.) *Paris,* 1766

† Lettres sur l'Electricité. Troisième Partie (ou vol.) 1re édition. 12mo. 295 pp. 4 plates. *Paris,* 1767

Vergleichung d. Wirkung des Donners mit der Electricität. 8vo. *Prag. Hohenberg,* 1769

Lettres sur l'Electricité. Troisième Partie (ou vol.) 2me édition. 12mo. 295 pp. 4 plates. *Paris,* 1770

L'abbé Nollet de Pimprez, par Lecot. 8vo. 76 pp. *Noyon,* 1856

Mitteis, H., Abbé Nollet in seiner Stellung gegen B. Franklin. 4to. 12 pp. (*Programm, &c., des Kais. Konigl. Obergymnasiums zu Laibach, fur* 1856.) *Laibach,* 1856

N. . . . (l'Abbé). (*Vide* Anon. Elect. 1781. *Vide* also Anon. Elect. 1745.)

Nolloth, M. S. On the submergence of the Atlantic Telegraph Cable. 8vo. 15 pp. Diagram. *London*
(*Note.*—*This pamphlet is contained with others in a volume lettered " Papers and Pamphlets on Submarine Telegraphs,* 1857-61.*"*)

Nordisches Archiv. für d. Natur. u. Arzeneiwissenschaft. (*Vide* Pfaff and others.)

Norman, Robert.
(*Note.*—*No authentic account discoverable. His first dipping needle was made in* 1576. *Vide his " New Attractive."*)

The Newe Attractive, containyng a short discourse of the Magnes or Lodestone, and amongst other his vertues, of a new discovered secret and subtill propertie concernyng the Declinyng of the Needle, touched therewith, under the plaine of the Horizon. Now first found out by Robert Norman, Hydrographer. 4to. (*Black Letter, scarce.*) *London,* 1581
(*Note.*—*Mr. Latimer Clark has a copy of this work in his collection.*—EDITOR.)

The Newe Attractive . . . newly corrected and amended by M. W. B., 1585. (*From Watt,* p. 709.) *London,* 1585

The New Attractive . . . newly corrected and amended by M. W. B. *London,* 1596
(*Note.*—*A copy of this edition is in the Royal Society.*—ED.)

The New Attractive, with the application thereof for finding the true variation of the Compass, by W. Burrowes. (*Vide* Borough, W.) 4to. (*From Libri's Sale Cat.* 573.) *London,* 1614

† The New Attractive : shewing the nature, propertie, and manifold vertues of the Loddstone; with the declination of the Needle, touched therewith, under the plaine of the Horizon. Found out and discovered by Robert Norman. Reprinted 1720. 8vo. (*Separately paged and published with Whiston's Longitude.*) *London,* 1720

Normand. (*Vide* Le Normand.)

The North Atlantic Telegraph. (*Vide* Royal Geographical Society.)

† **Norton, W. A.** On Terrestrial Magnetism. 8vo. 35 pp. (*American Journal,* 2nd series, vol. iv.)

Nöschel, A. Über einen interessanten Hagelfall im Kaukasus beobachtet. 8vo. (*Mélanges phys. et chim. tirés du Bull. de l'Acad. Imp. de St. Pétersbourg,* tom. v. *Livraison* 6 *et dernière,* pp. 658 and 660. 1 plate lith.) *St. Petersburg, Riga, and Leipzig,* 1861-4

Noton, B. Register of the Pluviometer at Bombay, in the year 1828. 8vo. 1 page. (*Phil. Mag. or Annals* vii. 14.) *London,* 1830

† **Nougarede de Fayet.** De l'Electricité dans ses rapports avec la Lumière, la Chaleur, &c. 8vo. 112 pp. *Paris,* 1839

Novelle di Firenze. Novelle letterarie pubblicate in Firenze. (Lami, Dr. G., Editor.) 37 vols. 4to. 1740-76.

Novellucci. His Electric Plate Machine described by the Marq. Ridolfi. (*Vide Antologia di Firenze,* Agosto, 1824, p. 159.)

Noya-Caraffa, Giovanni Duc de. *Born* 1715, *at Naples ; died July* 8, 1768, *at Naples.*

Lettre sur la Tourmaline à Buffon. 4to. 35 pp. 1 plate. (*Vide Priestley, Hist.* 5th edition, p. 271, *who says the Letter was presented to the Royal Society, and refers to Phil. Trans.* li. pt. i. p. 396.) *Paris,* 1759

† **Nuneberg.** Osservazioni su gli effetti dell' elettricità nella vegetazione. 12mo. 3 pp. (*Scelta d' Opuscoli,* xvii. 113.) *Milano,* 1776

† **Nuova Scelta d'Opuscoli.** Interessante sulle scienze. 2 vols. 4to. (*All the Elect. Articles are in the Library.*)

Nuovo Cimento, Giornale di Fisica, Chimica, &c. Direttori (dei vol. i. a x.) C. Matteucci e R. Piria (dei vol. xi. a —) C. Matteucci, R. Piria, et G. Meneghini. 8vo. *Torino and Pisa,* 1855 *to* —

(*Note.*—58 *numbers of various dates between* 1855 *and* 1864 *are in the Library.*) (*Faraday's copy, with notes by him.*)

Nyerup and **Kraft.** Almindeligt Litteratur-lexicon for Danmark, Norge, og Island, &c. 4to. *Kiobenhaven,* 1820

Nyevelt. (*Vide* Zuylen van Nyevelt.)

Nysten, Pierre Hubert. *Born October* 30, 1771, *at Lüttich ; died March* 3, 1818, *at Paris.*

Article, Electricité et Galvanisme, in the Dictionnaire des Sciences Médicales, published by him and Hallé.

† Nouvelles expériences galvaniques sur les organes musculaires de l'homme et des animaux à sang rouge. 8vo. 144 pp. *Paris and Strasbourg,* 1803

† Neue an den muskulösen Organen des Menschen und rothblutiger Thiere angestellte galvanische Versuche. Aus dem Französischen von C. F. Dorner. 8vo. 104 pp. Table. *Tubingen,* 1804

† Recherches de Physiologie pour faire suite à celles de Bichat sur la vie et la mort. 8vo. 427 pp. *Paris,* 1811

Nystrom, C. A. Rechen-Aufgaben aus der Elektricitätslehre, besonders fur Telegraphen-Beamte. 8vo. *Berlin,* 1862

373

O.

† **Oberbeck, A.** Über die sogenannte Magnetisirungs-constante. Inaugural Dissertation. 8vo. 40 pp. 2 plates. *Berlin*, 1868

† **Obermayer, A.** von. Experimentelle Bestimmung des Leitungswiderstandes in Platin-Blechen. 8vo. 16 pp. 1 plate. (*Sitzb. d. K. Akad. d. Wissensch.* ii. Abth. Juli, 1869, lx. Bde.) *Wien*, 1869

Über die Anwendung eines Elektromotors zur stroboskopichen Bestimmung der Tonhöhe. 8vo. *Wien*, 1871

† **O'Connell, A.** On the action of Voltaic Electricity in pyroxylic spirit, and the solutions in water, alcohol, and æther. 4to. 28 pp. 1 plate. (*Trans. R. Soc. Edinb.* xiv. pt. 1.) *Edinburgh*, 1837

Oersted, Hans Christian. *Born August 14, 1777, at Rudkjöbing, of Langeland; died March 9, 1851, at Copenhagen.*

Beschreibung einer Galvanischen Batterie ohne Platten. 8vo. 6 pp. (*Voight's Mag.* iii. 412.)

(*Note.—Sue*, iv. 263, says *Oerst*ıd in *Voight's Magazine*, iii. B. p. 412 : "*Novus Apparatus per amalgamata et liquores solventes.*" *Seyfer*, p. 126, refers to *Voight's Mag.* and gives a description of Oersted's Exper. but not the title of Oersted's Paper on Voight.) *Weimar*, 1801

Ansicht der chemischen Naturgesetze. *Berlin*, 1802
(*From Dove*, p. 218. *Pfaff* says 1812 ; *Heinsius* says 1812 (*Raelschulb*, v. 12). *Poggendorf* adds : "*Durch d' neweren Entdeckungen gewonnen,*" 1812 ; and "*Franzosisch von Marcel de Serres, Paris,* 1813." This work has a very different title. The date 1802 is one of Dove's numerous blunders.)

Suite des expériences de M. Ritter sur les phénomènes galvaniques, communiquée par M. Oersted. 4to. (*Bull. de la Soc. Philomatique*, No. lxvii. an. xi. p. 128. *Also in Journ. de Phys. and in Journ. du Galv.*) *Paris*, 1803

Materialien für einer Chemie des Neunzehnten Jahrhunderts. 1 Stück. *Regensburg*, 1803

Description d' une battérie galvanique sans plaques. 8vo. (*Journ. de Chim. de Van Mons*, No. iv. p. 68.) *Bruxelles*

Suite des expériences de J. G. Ritter à Jena sur les phénomènes galvaniques. (*Philomath. An.* 11, p. 128.) *Paris*, 1803

Galvanochemische Bemerkk. 8vo. (*Gehlen, N. Journ.* iii. 1804.) 1804

Om Overeestemmelsen imellem de electriske Figurer og de organiske Former. (*Skandinav. Literatur-Selskabs Skrifter*, vol. i. 1805.) 1805

† Chemico-Galvanic Observations. 8vo. 2 pp. (*Phil. Mag.* xxiii. 129.) *London*, 1806

Forsög i Anledning af nogle Steder i Winterl's Skrifter. 8vo. (*Nyt Biblioth. f. Physik, etc.* ix. *auch Gehlen's Journ. f. Chem.* i. 1806.) *Kiobenhaven*, 1806 ?

Über Ritter's Ladungssäule u. ein neues Metallthermometer. 8vo. (*Gehlen, N. Journ.* vi. 1806.) 1806

*† Gesammelte Schriften, deutsch von Kannegiesser. 6 vols. (*Portrait*.) *Leipzig*

Über Simon's neues Gesetz für elektr. Atmosphärenwirkung. 8vo. (*Gehlen's N. Journ.* viii. 1808.) 1808

† Ansicht der chemischen Naturgesetze durch d. neueren Entdeckungen gewonnen. 8vo. 1 plate. 298 pp. (*Vide following for translation by Marcel de Serres.*) (He refers in the Introduction to several other previous works of his, in Gehlen's Journal, the Journal de Phys. &c.) *Berlin*, 1812

† Recherches sur l'identité des forces chimiques et électriques. Traduit de l'Allemand par Marcel de Serres. 8vo. 258 pp. 1 plate. *Paris*, 1813

Oersted, Hans Christian.—*continued.*

Loven for de elektriske Virkningers Sväkkelse ved Afstanden. 4to. (*Oversigt over det Kongl. danske Videnskabernes Selskabs Forhandlinger*, 1814-15.)
Kiobenhaven, 1814-15

Über Contact-Elektrisität. (*Schweigg. Journ.* xx. 1817.) *Nürnberg*, 1817

† Experimenta circa effectum conflictus electrici in acum magneticam. 4to. 4 pp.
Hafniæ, 1820

Neuere elektromagnet. Versuche. 8vo. (*Schweigg. Journ.* xxix. 1820.)
Nürnberg, 1820

† Notice, headed " Galvanic Magnetism," of his Apparatus (Voltaic) exhibiting Magnetic Polarity. 8vo. 1 p. (*Phil. Mag.* lvi. 394.) *London*, 1820

Betracht. über den Elektromagnetismus. (*Schweigg. Journ.* xxxii. *u.* xxxiii. 1821.) *Nürnberg*, 1821

Versuche über Zamboni's zweigliedrige galvan. Kette. (*Schweigger, Journ.* xxxii. *u.* xxxiii. 1821.) *Nürnberg*, 1821

Galvanomagnetiske Undersogelser. 1821-22. (*Schweigger, Journ.* xxviii.?)
1821-22

† On Zamboni's discovery concerning the Electricity produced by the contact of a single solid Conductor with a single fluid Conductor. (In a letter.) 8vo. 1 p. (*Phil. Mag.* lix. 462.) *London*, 1822

Et nyt galvanomagnetisk Forsög. (*Tidskrift for Naturvidenskaberne*, i.)
Kiobenhaven, 1822

Efterretning om nogle nye, af Fourier og Oersted anstillede thermoelectriske Forsög. 4to. (*Oversigt Over det Kongl. danske Videnskabernes Selskabs Forhandlinger*, 1822-23.) *Kiobenhaven*, 1822-23

Sur quelques nouv. expériences Thermo-électriques par MM. Fourier et Oersted. 8vo. (*Ann. Chim. Phys.* xxii. 1823.) *Paris*, 1823

En ny thermo-electrisk Kjäde. 4to. (*Oversigt Over det Kongl. danske Videnskabernes Selskabs Forhandlinger*, 1823-24.) *Kiobenhaven*, 1823-4

Om Glodningen ved en galvanisk Strom med Hensyn til de Fraunhoferske Linier. 4to. (*Oversigt Over det Kongl. danske Videnskabernes Selskabs Forhandlinger*, 1823-24.) *Kiobenhaven*, 1823-24

Bemarkninger om Nordlysets Theorie. 4to. (*Oversigt Over det Kongl. danske Videnskabernes Selskabs Forhandlinger*, 1823-24.) *Kiobenhaven*, 1823-24

Om en Forbedring ved Nobili's electro-magnetiske Multiplicator. 4to. (*Oversigt Over det Kongl. danske Videnskabernes Selskabs Forhandlinger*, 1825-26.)
Kiobenhaven, 1825-26

Om Brugen af den elektromagnetiske Multiplicator til Sölvpröve. 4to. (*Oversigt Over det Kongl. danske Videnskabernes Selskabs Forhandlinger*, 1826-27.)
Kiobenhaven, 1826-27

Ein Kapitel aus der elektromagnet. Probirkunst. 8vo. (*Schweigg. Journ.* lii. 1828.) *Nürnberg*, 1828

Elektromagnetiske Forsög för at udfinde om galvaniske Redskaber kunne bruges til at frembringe meget stärke Magnete. 4to. (*Oversigt Over det Kongl. danske Videnskabernes Selskabs Forhandlinger*, 1828-29.)
Kiobenhaven, 1828-29

Et electromagnetiske Forsög som strider imod Ampère's Theorie. 4to. (*Oversigt Over det Kongl. danske Videnskabernes Selskabs Forhandlinger*, 1829-30.)
Kiobenhaven, 1829-30

Betragtninger over Forholdet imellem Lyden Lyset, Varmen og Electriciteten. 4to. (*Oversigt Over det Kongl. danske Videnskabernes Selskabs Forhandlinger*, 1829-30.) *Kiobenhaven*, 1829-30

Detection of Alloy in Silver by the Magnetic Needle. 8vo. 1 p. (*Phil. Mag.* or *Annals*, viii. 230.) *London*, 1830

† Forklaring over Morgen-og Aftenröden. 4to. (*Oversigt Over det Kongl. danske Videnskabernes Selskabs Forhandlinger*, 1830-31.) *Kiobenhaven*, 1830-31

Oersted, Hans Christian.—*continued.*

Forklaring over Faraday's magnetisk-electriske Opdagelse. 4to. (*Oversigt Over det Kongl. danske Videnskabernes Selskabs Forhandlinger*, 1831-32.)
Kiobenhaven, 1831-32

Utfaldet af magnetiske Jagtagelser i Kjobenhaven efter Guss's Jagtagelsesmaade. 4to. (*Oversigt Over det Kongl. danske Videnskabernes Selskabs Forhandlinger*, 1834-35.)
Kiobenhaven, 1834-35

Om Skypompen. 4to. (*Oversigt Over det Kongl. danske Videnskabernes Selskabs-Forhandlinger*, 1836-37.)
Kiobenhaven, 1836-37

Über d. Wettersäule. (*Schumacher's Astron. Jahrbuch*, 1838.) 1838

Om et nyt Elektrometer. (*Forhandl. Skandinavsk. Naturforskare*, &c. 1840, auch Pogg. Ann. liii. 1841.) 1840

Om Haarrörsvirkningerne. 4to. (*Oversigt Over det Kongl. danske Videnskabernes Selskabs Forhandlinger*, 1840.)
Kiobenhaven, 1840

Om Vaegtstangselectrometret. 4to. (*Oversigt Over det Kongl. danske Videnskabernes Selskabs Forhandlinger*, 1840.)
Kiobenhaven, 1840

Om Grove's Apparat. 4to. (*Oversigt Over det Kongl. danske Videnskabernes Selskabs Forhandlinger*, 1841.)
Kiobenhaven, 1841

Beretning om magnetiske Observationer. 4to. (*Oversigt Over det Kongl. danske Videnskabernes Selskabs Forhandlinger*, 1841.)
Kiobenhaven, 1841

Om Galvanoplastiken. 4to. (*Oversigt Over det Kongl. danske Videnskabernes Selskabs Forhandlinger*, 1842.)
Kiobenhaven, 1842

Om Faraday's diamagnetisk Forsög. 4to. (*Oversigt Over det Kongl. danske Videnskabernes Selskabs Forhandlinger*, 1847.)
Kiobenhaven, 1847

Forsög om Baerekraften af den polytechniske Laeranstalts store Electromagnet. 4to. (*Oversigt Over det Kongl. danske Videnskabernes Selskabs Forhandlinger*. 1847.)
Kiobenhaven, 1847

† Précis d'une série d'expériences sur le diamagnétisme. 8vo. 8 pp. (*Naturlehre d. Shönen.* Hamb. 1848 ?) (*Faraday's Copy.*)
Copenhagen, 1848

Beretning om hans Undersogelse over Diamagnetismen. 4to. 8 pp. (*Oversigt Over det Kongl. Danske Videnskabernes Selskabs Forhandlinger*, 1848.)
Kiobenhaven, 1848

† Melloniske Apparat. 8vo. 8 pp. (*Oversigt Over det Kongl. Danske Videnskabernes Selskabs Forhandlinger . . . af H. C. Oersted*, 1 Aut. 1848.)
Kiobenhaven, 1848

† Resultater af Undersögelser over Diamagnetismen. 4to. 8 pp. (*Oversigt Over det Kongl. Danske Videnskabernes Selskabs Forhandlinger*, 1849.)
Kiobenhaven, 1849

† Diamagnetismen, &c. 8vo. 3 pp. (*Oversigt Over det Kongl. Danske Videnskabernes Selskabs Forhandlinger . . . af H. C. Oersted*, 1 Aut. 1849.)
Kiobenhaven, 1849

† Der Geist in der Natur. Deutsch von K. L. Kannegieszer. Nebst einer biograph. Skizze von P. L. Moller. 8vo. 200 pp. Portrait. Leipzig, 1850

Soul in Nature, and other works, translated by Horner. 8vo. (*From Spon's Cat.*)
London, 1852

Forsög. og Bemaerkninger over den galvaniske Elektricitæt. (*Nyt Biblioth. f. Physik, etc.* Bd. i. auch *Scherer's Nord. Arch.* ii.)
Kiobenhaven

Beretning om nogle Forsög hvorved godtgjöres at man kan pröve Solvets keenhed ved den electro-magnetiske Multiplicator. (*Ursin's Magaz. f. Kunstnere, etc.* vol. i.)

Et Kapitel af den electro-magnetiske Probeerkunst med en stannographeret Tegning. (*Ursin's Magaz. f. Kunstnere, etc.* vol. ii.)

Many more articles in various Danish journals, and in other places, some of which are doubtless Electrical, &c.
(*Vide* also Anon. Elect. 1821.)

Oersted, Horneman, and Reinhardt. Tidgscrift for Naturvidenskoberne. 8vo.
Kiobenhaven, 1822 ?

† **Oettingen, A.** Von. Über das Laden d. Leidener Batterie durch Induction, und über die Entladung der Batterie durch das Inductorium. 4to. 26 pp. 1 plate. *Dorpat*, 1862

Der Rückstand der Leidener Batterie als Prüfungs-mittel für die Art der Entladung. 8vo. *Dorpat*, 1862

† Über das Laden der Leidener Batterie durch Induction und über die Entladung der Batterie durch das Inductorium. 4to. 27 pp. 1 plate lith. *Dorpat*, 1862

Ohm, Georg Simon. *Born March* 16, 1787, *at Erlangen; died July* 7, 1854, *at Munich.*

Vorläufige Anzeige d. Gesetzes nach welchem Metalle die Contact-Elektricitat leiten. 8vo. (*Poggend. Ann.* iv. 1825.) *Leipzig*, 1825

Theorie der Elektroscopischen Erscheinungen d. Säule. (*Poggend. Ann.* vi. u. vii. 1826.) *Leipzig*, 1826

Bestimm. des Gesetzes, nach welchem die Metalle die Contact-Elektricität leiten, nebst einem Entwurfe zu einer Theorie des Voltaischen Apparats. 8vo. (*Schweigg. Journ.* xlvi. 1826.) *Nürnberg*, 1826
(*Note.—This article contains his important law.*)

† Die galvanische Kette mathematisch bearbeitet. 8vo. 245 pp. 1 plate. *Berlin*, 1827

Einige elektr. Versuche. 8vo. (*Schweigg*. *Journ.* xlix. 1827.) *Nürnberg*, 1827

Nachträge zu seiner mathemat. Bearbeitung d. galvan. Kette. 8vo. (*Kastner's Archiv.* xiv. 128.) *Nürnberg*, 1828

Theoret. Herleitung d. Gesetzes, nach welchem sich d. Erglühen von Metalldrähten, durch d. Galvan. Kette, richtet, &c., u. nähere Bestimm. d. Modification, die d. elektr. Strom durch d. Einfluss von Spitzen erleidet. 8vo. (*Kastner's Archiv.* xvi. 1829.) *Nürnberg*, 1829

Experiment. Beitrage zu einer Vollstandigen Keuntniss d. elektromagnet. Multiplicators. 8vo. (*Schweigg. Journ.* lv. 1829.) *Nürnberg*, 1829

Nachweisung e. Übergangs von d. Gesetze d. Elektricitäts verbreitung zu dem der Spannung. 8vo. (*Kastner's Archiv.* xvii. 1829.) *Nürnberg*, 1829

Gehorcht d. hydro-elektr. Kette den von d. Theorie ihr vorgeschrieb. Gesetzen oder nicht? 8vo. (*Schweigg. Journ.* lviii. 1830.) *Nürnberg*, 1830

Versuche zu e. näheren Bestimm. der Natur der unipolar Leiter. 8vo. (*Schweigg. Journ.* lix. u. lx. 1830.) *Nürnberg*, 1830

Versuche über der Zustand der geschloss. einfachen galvan. Kette u.s.w. 8vo. (*Schweigg. Journ.* lxiii. 1831.) *Nürnberg*, 1831

An Thatsachen fortgeführte Nachweise d. Zusammenhangse d. mannigfalt. Eigenthumlichkeiten galvan. Ketten. 8vo. (*Schweigg. Journ.* lxiv. 1832.) *Nürnberg*, 1831-2

Über eine verkannte Eigenschaft d. gebundenen galvan. Elektricität. 8vo. (*Schweigg. Journ.* lxv. 1832.) *Nürnberg*, 1832

Zur Theorie d. galvan. Kette. 8vo. (*Schweigg. Journ.* lxvii. 1833.) *Nürnberg*, 1833

† Beiträge zur Molecular-Physik, Bd. i. 4to. *Nürnberg*, 1849

Grundzüge der Physik als Compendium zu seinen Vorlesungen. 1 Abtheilung: Allgemeine Physik. 8vo. 193 pp. *Nürnberg*, 1853

Grundzüge der Physik als Compendium zu seinen Vorlesungen. 2 Abtheilung. Besondere Physik. 8vo. 195 to 564 pp. *Nürnberg*, 1854

† Denkrede auf die Akademiker Dr. T. Siber und Dr. Georg Simon Ohm im Auszuge vergetragen . . . von Dr. Lamont. 4to. 36 pp. *München*, 1855

† Denkrede auf die Akademiker Dr. T. Siber und G. S. Ohm . . . von Dr. Lamont. 4to. *München*, 1855

Ohm, Georg Simon.—*continued.*

Die Dreieinigkeit der Kraft, Gemeinverständliche Vorlesungen über Himmels-u. Erdkunde. 2nd ed. 8vo. 345 pp. *Nürnberg,* 1860

Théorie mathématique des Courants électriques. Traduction. Préface et Notes de J. M. Gaugain. 8vo. 202 pp. *Paris,* 1860

† **Ollers** — (of Bremen). Letter from . . . to Baron Von Zach on the Stones which have fallen from the Heavens. (*From Algemeene Konst en Letter Bode,* "No. 17, 1803.") 8vo. 5 pp. (*Phil. Mag.* xv. 289.) *London,* 1803

Oldenburg, H. Remarks on Loadstones and Sea-Compasses. 4to. (*Phil. Trans.* year 1667, 413.) *London,* 1667

The Variation of the Magnetick Needle predicted for many years following. 4to. (*Phil. Trans.* year 1668, p. 789.) *London,* 1668

† **Olfers,** J. F. M. Die Gattung Torpedo in ihren naturhist. u. antiquar. Beziehung erläutert. 4to. 35 pp. 3 plates. *Berlin,* 1831

† **Oliver,** A. (Salem). A Theory of Lightning and Thunder Storms. 4to. 28 pp. (*Trans. Amer. Phil. Soc.* Old Series, ii. p. 74.) *Philadelphia,* 1786

Theory of Water-spouts. 4to. 17 pp. (*Trans. Amer. Phil. Soc.* Old Series, ii. p. 101.) *Philadelphia,* 1786

Oliviero, A. Der Instrumenten-Apparat zur Galvanokaustik, wie ihn Middeldorph in letzter Zeit anwendete. (Inaug Diss.) 8vo. 40 pp. 5 plates. *Breslau,* 1868

Olmsted, Denison. *Born June* 18, 1791, *at East Hartford, Connecticut; died May* 13, 1859, *at New Haven.*

† On the recent Secular Period of the Aurora Borealis. Folio. 52 pp. (*Published by the Smithsonian Institution, May,* 1856.) *Washington,* 1856

On Hail . . . (*From Bellani in Bibliot. Ital.* xcix. 83.)

Onimus et **Legros.** Traité d'Electricité médicale. Recherches physiologiques et cliniques. 8vo. *Paris,* 1871

† Traité de l'Electricité médicale. 8vo. 802 pp. 141 engravings. *Paris,* 1872

Onofrio, M. A. Fortges. Bericht, v. Sc. Brieslock u. Ant. Winspeare ; nebst e. meteorol. Abh. v. Hagel, &c. e. Anweis. Hagelableiter zu verfert. u. e. Untersuch. d. Frostableiters d. Hn. v. Bienenberg, nach d. Grunds. d. Electr. v. M. A. d'Onofrio ; a. d. Ital. 4to. *Dresden,* 1795

† **Oppermanno.** Dissert. physico medica, &c. 4to. (*From Krunisz,* p. 92. *Not in Pogg. or Heinsius.*) *Ratisberg,* 1746

Opuscoli Matematici e Fisici di diversi autori. 4to. *Milano,* 1832

Opuscoli Scelti sulle scienze e sulle arti. 22 vols 4to. (*Nearly all the Electrical articles are in the Library.*)

Nuova collezione d'opuscoli scientifici compilata per cura dei signori G. Bruni Fr. Cardinali, Fr. Orioli, Fr. e Raff. Tognetti. 5 vols. 4to. *Bologna,* 1817-24

(*Note.—The 1st vol.* (1817) *contains Poggioli on the influence of the Magneti rays on Vegetation, and Orioli in reply.*)

Orbissan. (*Vide* D'Orbissan.)

Orchoski. Veränderungen der Achse d. Erde. 1812

Orioli, F. Sopra un nuovo metodo di conoscere le più piccole correnti elettriche degli elettromotori. *Bologna,* 1824

† Regole pratiche da seguitarsi per armare le campagne contro alla Grandine in supplemento alla Dissertazione . . . Lette il 13 Maggio, 1824, alla Soc. Agraria di Bologna. 8vo. 16 pp. *Bologna,* 1824

Del Parsgrandine. Dissertazione. Letta il 15 Gennajo, 1824, alla Soc. Agraria di Bologna. 8vo. 32 pp. *Bologna,* 1824

Orioli, F.—*continued.*

Dissertation. Traduite par Chavannes. Feuille du Canton de Vaud, tom. xii. p. 4. *Bologna*, 1824

Nuove osservazioni intorno gli effetti dei Paragrandini metallici. 8vo. 12 pp. 1 map. *Bologna*, 1825

† De' Paragrandini metallici. Nuovo discorso. Letto il 10 Marzo, 1825, alla Soc. Agraria di Bologna. 8vo. 51 pp. 1 table. 1 plate. *Bologna*, 1825

† De' Paragrandini metallici. Discorso Quarto. Letto alla Soc. Agraria di Bologna, 16 Marzo, 1826. 8vo. 118 pp. *Bologna*, 1826

† Brevi considerazioni sulla risposta dell' . . Acad. Rle. delle Scienze di Parigi a sua Eccell. il Ministro dell' Interno in Francia intorno i Paragrandini; con un' Appendice. 8vo. 24 pp. *Bologna*, 1826

Orléans, le Duc de, et **Robert.** On Hail (?), in Account of their Balloon Ascent. (*Gazzetta di Milano,* 19 Agosto, 1834.) 1834

Orléans Société des Sciences, Phys. &c. Bulletin des Sciences Phys. et Méd. et d'Agricult. . . . Tom. i. 8vo. *Orleans*, 1810
(*Note.—The 7th vol. was printed in 1813. The Society instituted April 18, 1809.*)

Orléans Société Royale des Sciences, &c. Annales de la Soc. des Sciences, Belles lettres et Arts d'Orléans. 8vo. (Tom. i. for 1818 printed in 1819; the tom. xiv. for 1835 printed in 1836.) *Orleans*, 1819

Orleans Public Library. Catalogue (MS.) (*Collated.*)

† **Orlebar, A. B.** Observations made at the Magnetical and Meteorological Observatory at Bombay, April to December, 1845, and printed by order of the . . . East India Company, under the superintendence of A. B. Orlebar. Fol. 276 + 73 pp. 13 plates. *Bombay*, 1846
Observations made at the Magnetical and Meteorological Observatory at Bombay, in the year 1846. 4to. *Bombay*, 1849

Orliaguet, M. Essai sur le paratonnerre et le paragrêle. 8vo. 19 pp. *Limoges and Paris,* 1865

Orosius, P. *Born in the 5th century.*

Historiarum, lib. vii. (adversus Paganos.) (*Vide* Barrington's version.) (*From Watt.*)

† **Orsolato, G.** Sopra alcuni getti di luce provenienti dai calici dei fiori del Tropeolum majus. Lettera al Fusinieri. 4to. 5 pp. (*Ann.* . . *del Reg. Lomb. Veneto,* xiv. 35.) *Vicenza*, 1845

† **Osann, Gottfried Wilhelm.** *Born October 26, 1797, at Weimar.*

Dissertatio de natura affinitatis chimicæ. 4to. (*From Heinsius.*) *Jena*, 1822
Ten Articles in Poggendorff's Annalen. Vol. viii. 1826, to vol. cvi. 1859, on Platinum, Phosphorescence, Meteoric Iron, Electricity, Ozone, &c.

† Versuche über Phosphorescenz durch Isolation, und Beschreibung eines neuen Photometers. 8vo. 22 pp. 1 plate. (*Pogg. Ann.* xxxiii. 405, 1834.) *Leipzig,* 1834

† Die Anwendung des hydroelektrischen Stromes als Aetzmittel. 8vo. 23 pp. 1 plate. *Würzburg,* 1842

† Grundzüge der Lehre von dem Magnetismus und der Elektricität. 8vo. 183 pp. *Würzburg,* 1847

† Erfahrungen in dem Gebiet des Galvanismus gesammelt. Mit Abdrücken von galvanisch geätzten Zinnplatten. 8vo. 76 pp. *Erlangen,* 1853

† Die Kohlenbatterie in verbesserter Form. 8vo. 24 pp. *Erlangen,* 1857

† **O'Shaughnessy, W. B.** Account of the Electric Telegraph. Full and precise practical instructions for the Construction of Lines. 8vo. *London,* 1853

† First report on the operations of the Electric Telegraph Department in India, from 1st February, 1855, to 31st January, 1856. Fol. 79 pp. 24 plates. 1856

Ostertag, Johann Philipp. *Born May* 29, 1734, *at Idstein, Nassau-Weilburg ; died November* 24, 1801, *at Regensburg.*

Commentatio philologico-physica de Jove Elicio, ad solemnia Gynnasii Weilburgensis die 25 Julii, 1775, celebranda omnes litterarum fautores invitat J. P. Ostertag. 4to. 15 pp.   *Wezlariæ,* 1775

Prgrm. von d. Blitzableitern. 4to. (*From Pogg.* ii. 337.)   *Regensburg,* 1781

De auspiciis et acuminibus. (*From Martin.*)   *Ratisbonne,* 1779

Archäologische Abhandlung über die Blitz-Ableiter ; und die Kenntnisse der Alten von der Electricität. 4to. (*Neue Abhandl. der Baierischen Akad.* iv. 113.)   *München,* 1785

Kleine Schriften.   *Salzbach,* 1810

Antiquarische Abhandl. über Gewitterelektricitat.   *Salzbach,* 1810

† **Oudry.** Applications en grand de la Galvanoplastie et de l'Electro-métallurgie. 8vo. 18 pp. (*Exposition Univ. de* 1867, *Rapports du Jury international,* tom. viii. p. 154.)   *Paris,* 1868

Oven, J. Van. Het electro-magnetismus en de electro-magnetische telegraaf. Nieuwe (titel) uitgave. 8vo. 109 pp.   *Amsterdam,* 1857

Ovid. Metam. lib. 3, fast. v. Numa and Jupiter Elicius. (*Vide* Dutens, p. 150.)
                                                                 A.D. 17

Owen. (*Vide* Calamai and others.)

† **Owsjannikow, Ph.** Ein Beitrag zur Kenntniss der Leuchtorgane von Lampyris-Noctiluca. 4to. 11 pp. 1 plate. (*Mems. de l'Acad. Imp. des Sciences de St. Petersb.* vii. série. tom. xi. No. 17.)   *Leipzig and St. Petersburg,* 1868

Ozanam, Jacques. *Born* 1640, *at Bouligneux, near Villars in Bresse ; died April* 3, 1717, *at Paris.*

Recreations, &c. 4 vols. 8vo. (Several editions.)   *Paris,* 1724
(*Note.—Poggendorff,* ii. 341, *says* 2 vols. 8vo. *Paris,* 1694 ; *Nouv. ed.* 4 vols. 8vo. *Id.* 1724. *Von Montuela,* 4 vols. 8vo. 1778. *Von Hutton,* 4 vols. 8vo. 1803.)

P.

Pacchiani, Francesco Giuseppe. *Born October 4, 1771, at Prato; died March 31, 1835, at Florence.*

Nuovi esperimenti sul proposito della decomposizione dell' acqua relativamente al polo positivo e negativo della pila del Volta, etc. (*Letta alla Soc. dei Georgofoli di Firenze,* Settembre, 1804.) **1804**

Osservazioni tendenti a provare non esser vera la proprietà comunemente creduta inerente al polo positivo della colonna elettrica del Volta cioé di svinuppare dall' acqua l'ossigeno, ne tampoco la contraria inerente al polo negativo di svilluppare l'idrogeno. . . (*Letta alla Soc. dei Georgofili di Firenze,* Aug. 1804.) *Firenze,* 1804 (?)

† Sull' acido muriatico ottenuto per l'azione della pila Voltiana. Estratto di varie lettere su questo argomento pubblicate. 4to. 9 pp. (*Nuova Scelta d'Opuscoli,* i. 277.) *Milano,* 1804

† Tre Lettere (sopra l'Acido muriatico)—
  (1) al Pignotti (data) Pisa, 9 Maggio, 1805, 12 pp.
  (2) al Fabbroni ,, Pisa, 10 Giugno, 1805, 14 pp.
  (3) al ,, ,, Firenze, 9 Luglio, 1805, 14 pp. 8vo. 40 pp. **1805**

† Lettera sopra i principi costituenti dell' ossimuriatico (al Pignotti).
Lettera 2da. del medesimo.
Lettera 3da. del medesimo. 8vo. 32 pp. (*Ann. Chim. di Brugnatelli,* xxii. pp. 125,134, 144.) *Pavia,* 1805

† " Letter . . to Laurence Pignotti," dated Pisa, May 9, 1805. 8vo. 5 pp. (*Phil. Mag.* xxii. 179.) *London,* 1805
  (*Note.—This is a translation of the first of the three letters, and is copied from No. 3 of the Edinb. Med. and Surg. Journal of the 1st July,* 1805. *It concerns the supposed "formation of Muriatic Acid by Galvanism," and is accompanied by short remarks by the Editors of the Edinb. Med. and Surg. Journal.*)

† Extract of a New Letter of Dr. Francis Pacchiani, Professor of Natural Philosophy in the University of Pisa, to M. Fabbroni, upon the Composition of the Muriatic Acid. 8vo. 5 pp. (*Phil. Mag.* xxiv. 176.) *London,* 1806
  (*Note.—This is a translation of the second of the three letters (not quite complete), first into French and then into English.*)

Pacini (di Pistoja). Nuovi organi scoperti nel corpo umano. (*From Beckensteiner's Trans. of Henlé and Kalliker's Corpuscules de Pacini,* p. 27.) **1840**

† **Pacini,** F. Sopra l'Organo elettrico del Siluro elettrico del Nilo . . 8vo. 23 pp. 1 plate. (*Annali delle Scienze Naturali Luglio,* 1846. *Letta all' Acad. . . di Bologna,* March 26, 1846.) (*Arago's copy.*) *Bologna,* 1846

† Sulla struttura intima dell' Organo elettrico del Gimnoto e di altri pesci elettrici . . 8vo. 35 pp. (*Letta alle R. Accad. dei Georgofili* . . 19 Settemb. 1852.) *Firenze,* 1852

† **Pacinotti,** L. Aurora-boreale (18 Feb. 1837). 8vo. 4 pp. (*Bibliot. Ital.* lxxxv. 462.) *Milano,* 1837

Padua Accad. Saggi scientifici e letterarj dell' Accad. di Padova. 3 vols. 4to. tom. i. 1786, ii. 1789, iii. 1794. *Padova,* 1786-94

Memorie dell' Accademia di Scienze, Lettere, ed Arti di Padova. 4to. (1 vol. only). *Padova,* 1809
  (*This has been called also un Saggio. It contains no Mem. on Electrical matter.*)

Nuovi Saggi della Cesareo-Regia Accademia di Scienze, Lettere ed Arti di Padova. 6 vols. 4to. (Tom. i. 1817, ii. 1825, iii. 1831, iv. 1838, v. 1840, vi. 1847.) *Padova,* 1817-47

Padua Accad.—*continued.*

Rivista Periodica dei Lavori della I. R. Accademia di Scienze ed Arti di Padova. Redattore G. F. Sponga. 8vo. (About 8 vols. printed up to 1860.)
Padova, 1852 *to* —

Catalogo dei Libri esistenti presso l'I. R. Accademia di Scienze, Lettere ed Arti di Padova. MSS. *Padova*, 1853

Pagani, O. M. Aggiunta alla storia del Sonnambulo publicata dal Sig. Dott. G. Pigati, col racconto della di lui guarigione, per mezzo della virtù elettrica. 8vo. 14 pp. (*From Zaccaria, Storia*, iii. 267.) *Vicenza*, 1751

On Medical Electricity. (*From Kruniz*, p. 185.) 1751

Page, Charles G. *Born June* 28, 1812, *at Salem, Mass.*

About twenty-eight Articles in Silliman's Journal, vols. xxvi. 1834, to xlix. 1845. On New Electrical Instruments, Electro-Magnetism as a moving power, new Magnetic Machine, Magnetic Multiplier, Improved form of Saxton's Machine, &c. &c., and about eighteen articles in Silliman's Journal, New Series, vol. i. 1846, to vol. xiii. 1852. On Electric Inductions, Secondary spark, Galvanic current, Trevelyan's bars, Electro-magnetism as a moving power, Magnetic Telegraph, &c. &c. 8vo. *New Haven, U.S.* 1834-52

Electro-magnetic Apparatus and Experiments. 8vo. (*Silliman's Journal*, xxxiii. 1838.) *New Haven, U.S.* 1838

New Magnetic Electrical Machine of great power. 8vo. (*Silliman's Journal*, xxxiv. 1838.) *New Haven, U.S.* 1838

On a Mass of Iron temporarily magnetised, producing shocks and sounds (first observed). (*Silliman's Journal*, xxxii. and xxxiii.) 1838?

An improved form of Saxton's Magneto-electric Machine. 8vo. (*Silliman's Journal*, xlviii. 1845.) *New Haven, U.S.* 1845

A Phenomenon discovered by Page, and described in the Bibl. Univ. New series, tom. x. p. 398, relative (quere) to the Torpedo.

† **Page and Rittenhouse.** A Letter to D. Rittenhouse from J. Page, and from the former to the latter, concerning a remarkable Meteor seen in Virginia and Pennsylvania on the 31st October, 1779. 4to. 4 pp. (*Trans. Amer. Phil. Soc.* Old Series, vol. ii. pp. 173 and 175.) *Philadelphia*, 1786

Pages. On Aerolithes. 8vo. (*Vide* Encontre.) *Montpellier*, 1809

Pagni, G. Del Pantelegrafo Caselliano. *Firenze*, 1858

† **Paisley**, Lord. An account of a Treatise entituled Calculations and Tables relating to the attractive virtue of Loadstones, &c. Printed anno 1729. 4to. 6 pp. (*Phil. Trans.* xxxvi. 245.) *London*, 1729-30

† **Palagi**, A. Nuove sperienze sull' elettricità telluro-atmosferica. 8vo. 3 pp. (*Rendiconto . . R. Accad. dei Georgofili*, p. 419, 16 Gennajo, 1853.)
Firenze, 1853

† De quelques expériences nouvelles sur les variations électriques que subissent les corps lorsqu'ils s'éloignent du sol ou d'autres corps, et lorsqu'ils s'en rapprochent. Gazzetta Medica Italiana, tom. iii. sect. ii. 8vo. 6 pp. (*Ext. Bibl. Univ.* Genève, Juillet, 1853, xxiii. 4e série.) *Bologna*, 1853

Sulle variazioni elettriche a cui vanno soggetti i corpi allor che si allontano si avvicinano fra di loro . . (Estratto dai Nuovi Annali delle Scienze naturali di Bologna. Fascicolo di Settem. e Ottob. 1853.) 8vo. *Bologna*, 1853

Dell' azione elettrica della terra facente parte d'un circuito elettrico. (*Rendiconto dell' Accad. delle Scienze dell' Ist. di Bologna*, p. 72, 1858.) *Bologna*, 1858

(*Vide* also Bertelli, &c.)

Palaprat. (*Vide* Fabré-Palaprat.)

† **Pallas**, E. M. M. De l'influence de l'Electricité atmosphérique et terrestre sur l'organisme . . 8vo. 353 pp. *Paris*, 1847

Pallas, Peter Simon.　*Born September* 22, 1741, *at Berlin ; died September* 8, 1811, *at Berlin.*

Reise durch verschiedene Provinzen d. russ. Reichs.　3 vols.　8vo.　(*From Pogg.*
　ii. 348.)　　　　　　　　　　　　　　　　　　　　　*Petersburg,* 1771-76

Reise durch verschiedene Provinzen des Russischen Reichs.　3 vols. 4to.
　　　　　　　　　　　　　　　　　　　　St. Petersburg, 1773-1801

D'une masse de fer natif trouvée en Sibérie.　4to.　(*Act. Acad. Petrop.* i. 1778,
　and Phil. Trans. 1776.)　　　　　　　　　　　*Petropoli,* 1778
　　　　　　　　　　　　　　　　　　　　　　　London, 1776

Voyages . . traduit par Gauthier de Peyronie.　5 vols.　4to.　*Paris,* 1788

Bemerk. auf e Reise in d. südl. Statthalterschaften d. Russ. Reichs in. d.
　J. 1793 u. 1794.　2 Bde.　8vo.　　　　　　　　　*Leipzig,* 1799-1801

Voyages du Professeur Pallas dans plusieurs provinces de l'empire de Russie,
　etc. . . Traduit de l'allemand par Gauthier de la Peyronie. . Revue et Notes
　par Lamarcq, etc.　New edition.　8 vols.　8vo.　Plates.　　　*Paris,* 1803

Voyages dans les années 1793 et 1794 . . Traduit par Delaboulaye.　2 vols.
　4to.　Atlas de 55 plates.　(28 vignettes.)　　　　　　　*Paris,* 1805

† **Palm,** G. A.　Der Magnet im Alterthum.　4to. 34 pp.　(*Programm des König.*
　Würtembergischen . . *Semmais Maulbronn.*)　　　　　*Stuttgart,* 1867

Palma. (Siciliano)　Ricerche medico-elettriche.　(*From Bertholon.*)　　　1749

Palmer, W.　Illustrations of Electrotype.　4to.　　　　　　　*London,* 1841

Palmer and Hall.　Magneto-electric Machine.　(*Vide* Mayer.)

Palmieri, Luigi.　*Born April* 21, 1807, *at Faicchio. Terra di Lavoro.*

Elementi di fisica esperimentale di Pouillet . . . 3a. Edne. voltata in Ital.
　con Note e Giunte.　4 vols. 8vo.　　　　　　　　*Napoli,* 1839-41

Elettricità atmosferica ; continuazione degli " studii meteorologici," &c. (*Polio-*
　rama pittoresco, Anno xv. No. 23-25.)　　　　　　　　　　1807 ?

Alcune sperienze sull' induzione del magnetismo terrestre. 8vo.　*Napoli,* 1841

Télégraphe Electro-magnétique.　(*Vide* Moigno.)

Nuovo apparecchio d'induzione tellurica.　4to. 16 pp. 1 plate.　(*Rendiconto dell'*
　Accad. di Napoli, iv. 1845.)　(*Faraday's copy.*)　　　　　　*Napoli,* 1845

Esperienze ed Osservazioni di meteorologia elettrica.　Sunto di una Memoria
　letta alla Rle. Accad. . di Napoli nel Nov. del 1850 . . . 8vo.　(*Giorn. I. R.*
　Istit. Lomb. Nuov. Ser. 4, ii. 346.)　　　　　　　　　　*Napoli,* 1850

Elettricità atmosferica, continuazione degli studii meteorologici fatti sul Real
　Osservatorio Vesuviano.　4to.　　　　　　　　　　　*Napoli ?* 1854

Annali del Reale Osservatorio meteorologico Vesuviano compilati da Luigi
　Palmieri.　Anno i. 1859.　4to.　(*Zuchold's Cat.* 1860, p. 53.)　*Napoli,* 1859

Palmieri, L. e Santi Linari, P.　Nuove esperienze sull' induzione del magne-
　tismo terrestre.　4to.　(*Estratto dal Rendiconto della Rle. Accad. . . . di Na-*
　poli, Anno 1842, No. 5)　　　　　　　　　　　　*Napoli,* 1842

Descrizione della batteria magneto-elettro-tellurica.　(*Rendiconto dell' Accad. di*
　Napoli, iii. 1844.)　　　　　　　　　　　　　　　*Napoli,* 1844

† Telluro-Elettricismo.　La scossa e la decomposizione dell' acqua per mezzo
　delle correnti indotte dal magnetismo terrestre.　Nota.　Elettro-Magnetismo.
　Intorno ai fenomeni d'induzione delle calamite temporarie : sperienze.　4to.
　3 pp. + 1 p.　(*Ann. delle Scien. del Reg. Lomb.-Veneto,* xiii. 114 & 116.)
　Extracted from the *Rendiconto delle Adunanze* . . *della R. Accad. delle Scien.*
　No. 9, Mag. e Giug. 1843, p. 173, *di Napoli ?*)　　　　　*Vicenza,* 1844

† La scintilla delle correnti indotte dal magnetismo terrestre.　8vo. 1 page)
　(*Giorn. I. R. Istit. Lomb.* ix. 43 ; *Rendiconto dell' Accad. di Napoli,* iii. 1844.
　　　　　　　　　　　　　　　　　　　　　　　Milano, 1844

PAL—PAR   383

Palmstedt, Carl. Några underrattelser om det levrande exemplar af Gymnotus electricus, som forvarades uti Adelaide Gallerie i London . . . (*Skandena Naturforsk. motets Forhand*, 1842.)   1842

† **Palmstiern, N.** Nachrichten von einigen merkwürdigen Donnerschlägen. 8vo. 14 pp. (*K. Schwed. Akad. Abh.* xi. 118.)   *Hamburg & Leipsig*, 1749

Paltrinieri, J. (di Modena.) Expériences sur le fluide électro-magnétique utilisé par l'action et la réaction simultanément, dans son application comme force motrice au mouvement des machines. 4to.   *Paris*, 1845

Expériences sur le fluide électro-magnétique utilisé par l'action et la réaction simultanément, dans son application comme force motrice au mouvement des machines. 8vo.   *Paris*, 1845

† Esperienze sul fluido elettro-magnetico . . . sua applicazione qual forza motrice . . . Fol. 6 pp. 1 plate. (*L'Economista*, An. iii. vol. ii. p. 198.)
                                                        Milano, 1845

Panizzi. Dissertazioni. Three : the 2nd on Hail and Conductors ; the 3rd on Thunder. (*From Opuscoli Scelti*, xi. 4.)   *Mantova*, 1788

Pansner, Jn. Hen. Der Pyrotelegraph. (*From Pogg.* ii. 352.)   *Jenæ*, 1801

Papin, N. Raisonnements touchant la Salure, etc. . . . de la mer, auquel est ajouté un traité de la lumière de la mer. 12mo. 156 pp.   *Blois*, 1647

† **Paquerée, A.** Des accidents sur les chemins de fer, et les moyens de les prévenir. Première partie. 4to. 12 pp. 1 plate.   *Bordeaux*, 1856

† Des accidents occasionnés par l'application de quelques procédés industriels, et des moyens de les prévenir. Deuxième partie. Description d'un manomètre régulateur. 4to. 13 pp. 1 plate.   *Bordeaux*, 1856

Paracelsus. *Born* 1493 ; *died* 1541.
(For titles of various editions of his works *vide Poggendorff*, ii. 357.)

† Sein Leben und Denken Drei Bucher von Lessing M.B. 8vo. 250 pp. 1 portrait.
                                                        Berlin, 1839

† **Paris, J. A.** (*Vide* Davy, Sir H.)

† **Paris, Académie des Sciences.** Histoire et Mémoires de l'Académie des Sciences de Paris. 4to. 167 vols.   *Paris*, 1666-1790
(*Note.—A portion of the collection is in the Library.*)

Paris, Académie Royale des Sciences de l'Institut. Mémoires de l'Académie Royale des Sciences de l'Institut de France. 4to.   *Paris*, 1816

Mémoires présentés par divers Savants à l'Académie Royale des Sciences de l'Institut de France. Sciences Mathématiques et Physiques. 4to.
                                                        Paris, 1827

Comptes Rendus hebdomadaires des Séances de l'Académie des Sciences. 4to.
                                                        Paris, 1835

† Table générale des Comptes Rendus des Séances de l'Académie des Sciences, publiée par les Secrétaires perpétuels (pour les Tomes i. à xxxi. 3 Août 1835 à 30 Décembre 1850). 4to. 1018 pp.   *Paris*, 1853

Comptes Rendus hebdomadaires . . . Supplément. 4to.   *Paris*, 1856

Paris, Bibliothèque Royale. Catalogue des livres imprimés et des Manuscrits de la Bibliothèque du Roi. 10 vols. fol.   *Paris*, 1739-53

Paris, Institut de France. Mémoires de la classe des sciences mathématiques et physiques. 4to.   *Paris*, 1796 to —

Mémoires de l'Institut National des Sciences et des Arts. 4to. 72 vols.
                                                        Paris, 1798 to —
Compte Rendu à la Classe des Sciences mathématiques et physiques de l'Institut National des premières expériences faites en Florial et Prairial de l'an (1797) par la Commission nommée pour examiner et vérifier les phénomènes du Galvanisme. 4to. 107 pp.   *Paris*.

Rapport . . . sur les expériences du cit. Volta, Commissaires Laplace Coulomb, Halle . . Biot, etc.   *Paris*, 1801

Paris Imperial Library. Catalogi librorum impressorum Bibliothecæ Regiæ. Tomus ix. R. Philosophi . . . Fol. MSS. *Paris, n. d.*
(*Collated for Electricity, Magnetism, &c.* Single Authors between December 20, 1853, and March 23, 1854.)

Catalogi librorum impressorum . . . Supplementum novum. Tom. i. R. Philosophi in 4to. Fol. MSS. *Paris*
(*Collated for Electricity, Magnetism, &c.* Single Authors between December 20, 1853, and March 23, 1854.)

Catalogi librorum impressorum . . . Supplementum novum. Tom. ii. R. Philosophi in 8vo. Fol. MSS. *Paris*
(*Collated for Electricity, Magnetism, &c.* Single Authors between December 20, 1853, and March 23, 1854.)

Paris, Société d'Encouragement pour l'Industrie Nationale. Bulletin de la. 4to. *Paris,* 1802

Table of contents of the Bulletin. 4to. *Paris,* 1802-18

Paris, Société Philomathique. Bulletin des Sciences de la Société Philomathique. 4to. 3 vols. *Paris,* 1791-1804

Nouveau Bulletin Ed. 4to. *Paris,* 1832 *to* —

(*Vide* also Société Philomathique.)

(*Vide* also Société Galvanique de Paris.)

† **Pariset, M. G. H.** Recherches sur le Magnétisme terrestre. 8vo. 152 pp. 1 plate. *Paris,* 1862

Park, Mungo. *Born September 10, 1771, at Fowlshiels, near Selkirk ; died end of the year 1805, on the Niger.*

† Travels in Africa. Rennel's Chart of Variations. 4to. 2nd edition. *London,* 1799

† **Parker, T. A.** Polar Magnetism. The cause of Polar Magnetism, &c. . . . 8vo. 34 pp. 1 plate. (*A paper read before the American Institute.*) *New York, U.S.* 1868

† Second Lecture on Polar Magnetism. Its astronomical origin ; its period of revolution, &c. 8vo. 33 pp. *New York, U.S.* 1869

† **Parkinson, J. C.** The Ocean Telegraph to India. A Narrative and a Diary. 8vo. 328 pp. 9 plates. *Edinburgh and London,* 1870

Parliamentary Papers.
General Index to the Reports from Committees of the House of Commons 1715-1801. Fol. *London,* 1803

Catalogue of Papers printed by order of the House of Commons 1731-1800. Fol. *London,* 1807

General Indices to do., 1801-26, 1801-32, 1832-44. Fol. *London,* 1829-33-45
(*Note.—Since 1845 a separate Index has been published to the Papers of each Session.*)

Catalogue of Parliamentary Reports, and a Breviate of their contents 1696-1834. (Indexes to the subject-matters of the Reports 1801-34.) Fol. *London,* 1834

† Lightning Conductors. Fol. (*Vide* Harris.) *London,* 1841-7

† Telegraphs and Telegraphic Companies generally. *London,* 1854-63

† Atlantic Telegraph. 1857-63

† Mediterranean, Red Sea, and East India Telegraphs. *London,* 1858-66

† Report of the Joint Committee of Privy Council for Trade and the Atlantic Telegraph Committee. *London,* 1862

† Submarine Telegraphic Company Correspondence, &c. Fol. *London,* 1859

† Packet and Telegraphic Contracts. Report, 1859. First Report, 1860. Second Report, 1860. Appendix to Second Report. Third Report, 1860. Index to three Reports. Fol. *London,* 1859-60

Parliamentary Papers.—*continued.*

† Dardanelles and Alexandria Telegraph. Fol. *London,* 1863

Electric Light. Fol. *London,* 1863-66

† List of Parliamentary Papers for the Sessions from 1836 to 1863 inclusive. 8vo. 411 and 96 pp. *London,* 1864

† Lists of Parliamentary Papers for the Sessions of 1864, 1865, and 1866. 8vo. *London*

† Canada and Pacific Telegraph. Fol. *London,* 1864

† Malta and Alexandria Telegraph. Fol. *London,* 1861-4-5

† Deviation of Compasses. Fol. *London,* 1866

† Special Report from the Select Committee on the Electric Telegraphs Bill; together with the Minutes of evidence taken before them. Ordered by the House of Commons to be printed 16th July, 1868. Fol. 253 pp. *London,* 1868

† Electric Telegraphs. Return to an order of the Hon. the House of Commons, dated 3rd April, 1868, for Copy of Reports to the Postmaster-General, by Mr. Scudamore, upon the Proposal for Transferring to the Post-office the Control and Management of the Electric Telegraphs throughout the United Kingdom. (Mr. Chancellor of the Exchequer.) Ordered by the House of Commons to be printed 3rd April, 1868. 8vo. 166 pp. 11 maps. *London,* 1868

Parola. (*Vide* La Parola.)

Parrot, Georg Friedrich. *Born July* 15, 1767, *at Mömpelgarde; died July* 8 (*A. St.*), 1852, *at Helsingfors.*

Über Galvanismus u. Verbesser, d. Volta'schen Säule. (*Voight's Mag.* iv. 1802.) 1802

Skizze e. Theorie d. galvan. Elektricität u. der durch sie bewirkten Wasserzersetz. 8vo. (*Gilb. Ann.* xii. 1802 ; *and from Seypfer,* p. 200.) *Leipzig,* 1802

Beitr. z. galvan. Electkricität. 8vo. (*Gilb. Ann.* xxi. 1805.) *Leipzig,* 1805

Übersicht d. Systems d. theoret. Physik. 2 vols. 8vo. *Dorpat,* 1809-11

This work had also the title " Grundriss de Physik d. Erde u. Geologie." 3 vols. 8vo. *Riga and Leipzig,* 1813

Über d. Oxydation d. Metalle im Wasser. (*Grindels Russ. Jahrb. f. Chem. u. Pharm.* ii. 1810.) 1810

Uber d. Zambonische Säule. 8vo. (*Gilb. Ann.* lv. 1817.) *Leipzig,* 1817

Über d. Gesetze d. elektr. Wirk. in d. Ferne. 8vo. (*Gilb. Ann.* lx. 1819.) *Leipzig,* 1819

Über d. Sprache d. Elektricitätsmesser. 8vo. *Gilb. Ann.* lxi. 1819.) *Leipzig,* 1819

Entretiens sur la Physique. 6 vols. 1819-24 ? *Dorpat,* 1822

Über Aberrationen d. Magnetnadel auf Schiffen u.s.w. (*Naturwiss. Abhandl. aus Dorpat,* i. 1823.) *Dorpat and Berlin,* 1823

Sur les Phénomènes de la Pile Voltaïque. 8vo. (*Ann. Chim. et Phys.* xlii· 1829.) *Paris,* 1829

Observations relatives au Mém. de Marianini sur la théorie chimique des Electromoteurs Voltaïques. 8vo. (*Ann. Chim. et Phys.* xlvi. 1831.) *Paris,* 1831

Le Télégraphe basé en tous points sur les principes de la Physique. 4to. (*Mém. Acad. Pétersb.* Ser. vi. tom. i. 1838.) *Petersburg,* 1838

† Notice sur les Aurores Boréales. Lu 23 Déc. 1836. 4to. 18 pp. (*Mém. 6me Série Sciences Mathém.* tom. iii. 1re Partie.) *Petersburg,* 1838

Parrot, Georg Friedrich—*continued.*
† Nouvelles expériences en faveur de la théorie chimique de l'électricité. Lu 16 Juin, 1837. 4to. 30 pp. (*Mém. 6me Série Sciences Mathém.* tom. iii. 1re partie.) *Petersburg,* 1838
† Essai sur le procès de Végétation Métallique et de la Cristallisation. Lu 10 Octobre, 1840. 4to. 82 pp. 5 plates. (*Mém. 6me Série Sciences Mathém.* tom. v. 1re partie. *The plates are marked* tom. iv.) *Petersburg,* 1841
Über d. Einfluss verschiedn. Lichtflammen auf. d. Spann. d. Zambonischen Säule. (*Pander's Beitr. z. Naturk.* i.)
Handbuch der Phys. ii. 554. (*From Dove,* p. 172.) *Dorpat,* 1822?

† **Parry,** (afterwards Sir) W. E. Journal of a Voyage . . . with Appendix containing the Scientific and other Observations. 4to. 489 pp. *London,* 1821
(*Note.—This is the first Voyage, in* 1819 *and* 1820. *The second was described in a separate work, printed in* 1824, *performed in* 1821 *to* 1823; *and an Appendix was printed in* 1825. *The third, printed in* 1826, *performed in* 1824 *and* 1825. *The fourth, or Narrative, printed in* 1828, *performed in* 1827. *The Four Voyages in* 6 *vols.* 8vo. *printed in* 1833.)
Magnetical Observations at Port Bowen. 4to. (*Phil. Trans.* 1826.) *London,* 1826
Observations for determining the Dip of the Magnetic Needle at Woolwich. 4to. (*Phil. Trans.* 1826.) *London,* 1826

Parson, G. Elementary Magnetism and the local attraction of Ships' Compasses, adapted for the use of Navigators, with a Table of the true Azimuth of the mean place of the North Pole Star at intervals of one hour of sidereal. time. 8vo. *London and Sunderland,* 1864

Partington. (*Vide* Ferguson, James.)

Partsch, Paul Maria. *Born June* 11, 1791, *at Vienna ; died October* 3, 1856, *at Vienna.*
† Die Meteoriten, oder vom Himmel gefallene Steine u. Eisenmassen im k. k. Hof-Mineralien-Cabinette zu Wien. 8vo. 162 pp. 1 plate and table. *Wien,* 1843
Katalog. der Bibliotheck des k. k. Hof-Mineralien-Cabinetes zu Wien. 4to. *Wien,* 1851
Über d. Meteoreisen von Rosgatà u.s.w. mit F. Wöhler. 8vo. (*Sitzungsberichte d. Wien Acad.* viii. 1852.) *Wien,* 1852
Über d. Meteorstein unweit Mezö-Madaras in Siebenbürgen. 8vo. (*Sitzungsberichte d. Wien Acad.* xi. 1853.) *Wien,* 1853

Partsch & Wöhler. Analyse des Meteoreis, von Rasgala in Neugranade, von Wöhler in Gottingen. 8vo. (*Sitzungsb. d. K. Akad.* 1852.) *Wien,* 1852

† **Pascalis,** P. A. Mémoire sur l'Electricité Médicale. 8vo. 43 pp. *Faris,* 1819

† **Pasley,** C. W. Project for the establishment of Telegraphs on a new construction. 8vo. (*Phil. Mag.* xxix. 205 and 292.) *London,* 1808
† Description of the French Telegraphs used on the coast of Flanders, &c.; with Observations on the same, and a plan of a Polygrammatic Telegraph on a new construction. 8vo. 3 pp. 1 plate. (*Phil. Mag.* xxxv. 339.) *London,* 1810

Pasquali, V. Dissertatio de electricitatis potestate in animalem œconomiam. 8vo. 20 pp. *Padova,* 1823

Pasumot, Fra. Observations sur les effets de la foudre dans une maison de Paris. 1784

Pater, Paul. Duo phænomena rarissima, alterum crux in Luna, alterum meteorem ignitum. (*From Pogg.* ii. 374.) *Jenæ,* 1681

Paterson, William. On a new Electrical Fish, Tetrodon Electricus. 4to. (*Phil. Trans.* 1786.) *London,* 1786

† **Patterson, R.** An Easy and Accurate Method of finding a True Meridian Line, and thence the Variation of the Compass. Read April 7, 1786. 4to. 9 pp. (*Trans. Amer. Phil. Soc.* Old Series, vol. ii. p. 251.) *Philadelphia*, 1786

† An improvement on Metallic Conductors, or Lightning Rods : in a letter to D. Rittenhouse, honoured with the Magallanic Premium . . . December, 1792. Read November 5, 1790. 4to. 4 pp. (*Trans. Amer. Phil. Soc.* Old Series, vol. iii. 321.) *Philadelphia*, 1793

Patterson, R. M. On Electricity from Steam. 8vo. (*Silliman's Journ.* xi. 1841.) *Philadelphia*, 1841

Paucker, Magnus Geo. Von. Dissertatio inauguralis de nova explicatione phænomeni electricitatis corporum rigidorum. 4to. (*From Pogg.* ii. 377.) *Dorpat*, 1813

Paul. Mém. sur les signaux de jour, &c. 4to.

Pauli, Adrian. De succini natura. (*From Pogg.* ii, 378.) *Danzic*, 1614

Paulian, Aimé Henry. *Born July 23, 1722, at Nismes ; died about 1802, at Manduel, near Nismes.*

Nouvelles conjectures sur les causes des phénomènes électriques. 4to. *Nimes*, 1762
(*Note.—This quotation seems to be erroneous. Paulian does not refer to a book with this title in his "Electricité soumise à un nouvel examen. . Avignon*, 12mo. 1868, *at* p. 14 *of which his* 2nd *Lettre is entitled "Conjectures nouvelles sur les causes physiques des phénomènes électriques. Réponses à quelques objections de M. Nollet contre ces conjectures." He seems here to refer to the article Electricity in his Dictionnaire de Physique*, 4to. *Avignon*, 1761, 8th ed.)

† L'électricité soumise à un nouvel examen . 12mo. 286 pp. 2 plates. *Avignon*, 1768

Paulsohn, P. On Electric Cures. (*From Troili.*) 1754

† **Paura, R.** Correnti elettro-chimiche misurate in diversi liquidi e solidi organici tolti dagli animali viventi. La. 4to. 84 pp. *Napoli*, 1849

Payssè. Lettre à " Elle contient une expérience relative au galvanisme qui a été faite en l'an vi. au mois de Brumaire." (*Journal de la Société des Pharmaciens*, 1re an. p. 100.) *Paris*

† **Pazienti, A.** Dell' azione chimica della luce, del calorico, dell' elettrico, e del magnetismo sopra i corpi inorganici. Dissertazione. La. 8vo. 158 pp. 1 pl. *Padova*, 1846

† Programma di un lavoro sugli studii elettrici in Italia. 8vo. 28 pp. *Venezia*, 1850

† Ricerche sulla conducibilità elettrica. 8vo. 12 pp. *Venezia*, 1851

† Sulla scintilla elettrica prodotta dalla pila Voltiana. Nota. 8vo. 11 pp. *Venezia*, 1852

† Sopra l'induzione elettrica ; osservazioni. 8vo. 12 pp. *Venezia*, 1852

† Nozioni elementari della elettricità esposte. 8vo. 87 pp. 4 plates. *Vicenza*, 1855

† Intorno ad un Regolatore pegli Apparecchi d'induzione volta-elettrica (nota) comunicata all' Accad. Olimpica di Vicenza . . 11 Marzo 1855. 8vo. 4 pp. (*Giornale Veneto di Scienza Medicale*, vol. v. serie ii.) *Venezia*, 1855

† Nozioni elementari di elettrometria e di Reometria elettrica. 8vo. 37 pp. *Vicenza*, 1859

† Nozioni elementari di Fisica. Parte Quarta. Elettricità. La. 8vo. 127 pp. 2 plates. *Vicenza*, 1859

Pearson, George. *Born 1751, at Rotterham, York ; died November 9, 1828, in London.*

† Sperimenti . per accertare la natura del Gas prodotto dalle scariche elettriche attraverso l'acqua. 4to. 8 pp. 1 plate. *Milano*, 1801

† Some Account of George Pearson, M.D., F.R.S., &c. &c. With a Portrait from an original painting. 8vo. 4 pp. (*Phil. Mag.* xv. 274.) *London*, 1803

Peart, Edward. *Born about* 1756 ; *died November* 26, 1824, *at Butterwick, near Gainsborough.*

Elementary Principles of Nature. (*Containing references to Magnetism and Electricity.*) 8vo. *Gainsboro'*, 1789

Versuche über die Urstoffe der Natur und ihre Gesetze. Nebst einer Vorrede und einem Anhange über die doppelte elektrische Materie. 8vo. *Leipsig*, 1791

† On Electricity, with occasional observations on Magnetism . . and Electric Atmospheres. 8vo. *Gainsboro'*, 1791

† On Electric Atmospheres . and Letter to Read . 8vo. 81 pp. *Gainsbor.'*, 1793

Physiology, or an Attempt, &c. 8vo. *London*, 1798

Péclet, Jean Claude Eugène. *Born February* 10, 1793, *at Besançon ; died December* 7, 1857, *at Paris.*

† Mémoire sur la détermination des coefficients de conductibilité des métaux par la chaleur. 8vo. 9 pp. *Paris*, 1841

† Mémoire sur le développement de l'électricité statique pendant le contact des corps. 8vo. 29 pp. (*Extr. des Annales de Chim. et Phys.* an. 1841, 3me série.) *Paris*, 1841

† Condensateur électrique. 8vo. 4 pp. Plate. (*Extr. des Annales de Chim. et Phys.* an. 1841, 3me série). *Paris*, 1841

† Mémoire sur un nouveau Galvanomètre. 8vo. 4 pp. Plate. *Paris*, 1841

Essai historique sur l'électricité. (*From Zantedeschi,* i. 189.)

Peddington. Essai sur la loi des Tempêtes, &c. . Trad. de l'angl. par Fornier Duplan. 8vo. *Paris*, 1845

† **Pedersen, P.** Almeenfattelig Veirlære. 8vo *Kjobenhavn*, 1854

† **Peel, W.** Production of Muriate of Soda by the Galvanic decomposition of Water. Letter to Tulloch. 8vo. 2 pp. (*Phil. Mag.* xxi. 279 and xxvii. 82.) *London*, 1805

† On the production of Muriates by the Galvanic decomposition of Water : with a second letter on the subject from Mr. W. Peel, of Cambridge (dated February 22). 8vo. 4 pp. *London*, 1805

† Third Communication from Mr. W. Peel, of Cambridge. On the production of Muriates by the Galvanic and Electric decomposition of Water. 8vo. 3 pp. (*Phil. Mag.* xxiii. 257.) *London*, 1809

P.,G.(i.e.Peignot, Gabriel.) Essai chronologique sur les Hivers les plus rigoureux depuis 396 an. avant J. C. jusqu'en 1820 (incl.) suivi de quelques Recherches sur les effets les plus singuliers de la Foudre, depuis 1676 jusqu'en 1821. . 8vo. 240 pp. *Paris and Dijon*, 1821

Peissel, J. O. Theses recent. philos. controv. de meteore lucido singulari. 4to. *Vitemberg*, 1731

Pelchrzim, T. Über electrische Uhren, Maschinen-Bewegung durch Electromagnetismus, Wasserzersetzung durch galvanische Einwirkung, Galvanoplastik, electrisches Licht und Minensprengung durch den galvanischen Strom. Ein Vortrag, gehalten im wissenschaftlichen Verein von Officieren zu Potsdam, als selbstständige Fortsetzung eines Vortrags über electromagnetische Telegraphie. 12mo. 36 pp. 1 plate. *Potsdam*, 1854

Télégraphe à Cadran. (*Vide* Moigno, 2nd edition, p. 426.)

Über elektrische Telegraphie. *Potsdam*, 1853

Pelison. Blizfanger. (*Abhandl. d. naturforsch. Gesells. zu Berlin* v. Jahr 1792, x.)

† **Pell, John.** A Letter of remarks on Gellibrand's Mathematical Discourse on the Variation of the Magnetic Needle. 1635

Pellegrini, A. De fluido electrico ejusque doctrinæ progressu. Notiones historicæ. Dissertatio inauguralis. 8vo. 39 pp. 1858

† **Pellis, Ph.** Exposé d'une simplification du télégraphe électrique écrivant. 8vo. 16 pp. *Bordeaux*, 1857

Peltier, Jean Charles Athanase. *Born February 22, 1785, at Ham, Dep. de la Somme; died October 27, 1845, at Paris.*

† Nouvelles expériences sur la caloricité des courants électriques. 8vo. 16 pp. 1 plate. (*Ann. Chim. Phys.* lvi.) *Paris*, 1834

† Observations sur quelques causes d'erreur dans les mesures des tensions électriques, et description d'un nouvel électromètre. 8vo. 11 pp. 1 plate. (*Extr. des Ann. de Chim. et Phys.* Août, 1836.) *Paris*, 1836

† Observation sur les multiplicateurs et sur les piles thermo-électriques.

† An article in the "Institut" concerning his Experiments confirming Belli's results, on Dissipation of Electricity, 20th July, p. 237.

† Observations sur la structure des Muscles et Expériences sur la Contraction. 8vo. 8 pp. (*Ext. des Ann. des Sciences Naturelles*, Février, 1838.)*Paris*, 1838

† Recherches expérimentales sur les quantités d'actions dynamiques et statiques que produit l'oxidation d'un milligramme de Zinc, &c. 8vo. 26 pp. (*Ann. de Chim.* lxvii. 422.) (*Arago's copy.*) *Paris*, 1838

† Mémoire sur la formation des tables des rapports qu'il y a entre la force des courants électriques et la déviation des aiguilles des Multiplicateurs, &c. 8vo. 89 pp. (*Ext. Ann. de Chim.* vol. lxxi. p. 225.) *Paris*, 1839

† Notice des faits principaux et des instruments nouveaux, ajoutés à la science de l'électricité. 4to. 8 pp. *Paris*, 1839

† Météorologie ; Observations et recherches expérimentales sur les causes qui concurrent à la formation des Trombes. 8vo. 442 pp. 3 plates. *Paris*, 1840 ; *Bruxelles*, 1841

† Recherches sur la cause des phénomènes électriques de l'atmosphère, &c. 8vo. 49 pp. 1 plate. (*Ext. Ann. de Chim.* 3me série, tom. iv.) *Paris*, 1842

† Mémoire sur les diverses espèces de Brouillards. Lu le 2 Juillet, 1842 4to. 25 pp. (*Acad. Rle. de Bruxelles*, tom. xv. 2me partie, *Des Mems. Couronnés.*) *Bruxelles*, 1842

† Essai sur la coordination des causes qui précedent, produisent et accompagnent les phénomènes électriques. 4to. 71 pp. 1 plate. *Paris*, 1844

† Météorologie électrique. Première partie. 8vo. 52 pp. (*Ext. des Archives d'Electricité*, vol. iv. p. 173.) *Genève*, 1844

† Notice sur la vie et les travaux scientifiques de J. C. A. Peltier. 8vo. 472 pp. Portrait. *Paris*, 1847

† Lettre à M. Quetelet sur l'électricité atmosphérique. Communiquée à l'Acad. de Bruxelles, 5 Janv. 1850. 8vo. 11 pp. 1 plate. (*Ext. du Bulletin de l'Acad.* vol. xvii. No. Janv. 1850.) *Paris*, 1850

† **Peluso,** A. Saggio sugli effetti dell' elettrico nell' umano organismo. 8vo. 123 pp. *Milano*, 1839

Penada, Jacopo. *Born December 11, 1748, at Padua ; died February 23, 1828, at Padua.*

Tavole meteorologiche. 2 vols. 4to. *Padova*, 1807-8
(*An Appendix to the vol.* 1808 *contains notices of the number of days of* " Tuono " *and of* "Fulmini," *and how many Iridi and Aurorœ boreales appeared in the course of the ten years from* 1796 *to* 1805.)

† **Pennes,** J. A. Notice sur le Bain Electro-Chimique. 16mo. 36 pp. *Paris*

Penny Cyclopædia of the Society for the Diffusion of Useful Knowledge, complete with both Supplements. 30 vols. in 17. Imp. 8vo. cloth. *London*, 1733-58

Penny Cyclopædia. 29 vols. Sm. fol. *London*, 1860

Penrose, F.　Treatise on Electricity.　8vo.　40 pp.　　*London and Oxford*, 1752

† 　An Essay on Magnetism.　8vo.　40 pp.　　　　　*Oxford*, 1753

Pepys, W. H.　A new Eudiometer, accompanied by experiments.　4to.　(*Phil. Trans.* 1807.)　　　　　　　　　　　　　　　　　*London*, 1807

† 　Description of the apparatus invented for the decomposition of the Alkalis under Naphtha, by Galvanism.　8vo.　1 page.　1 plate.　(*Phil. Mag.* xxxi. 241.)　　　　　　　　　　　　　　　　　　　*London*, 1808

† 　Description of a Mercurial Voltaic Conductor.　8vo.　2 pp.　1 plate.　(*Phil. Mag.* xli. 15.)　　　　　　　　　　　　　　　*London*, 1813

Description of a new construction of the Voltaic Apparatus.　8vo.　(*Quart. Journ. of Science*, i. 1816.)　　　　　　　　　　　*London*, 1816

An account of an Apparatus for performing Electro-Magnetic Experiments.　4to.　(*Phil. Trans.* 1823.)　　　　　　　　　　*London*, 1823

† **Pepys, W. H.** (jun.)　Description of Mr. Pepys's large Galvanic Apparatus.　8vo.　3 pp.　1 plate.　(*Phil. Mag.* xv. 94.)　　　　　　*London*, 1803

Description of a newly-invented Galvanometer.　8vo.　(*Phil. Mag.* June 1801, vol. x. 1801.)　　　　　　　　　　　　　　　*London*, 1801

† **Perard, Louis.**　Etude sur les procédés suivis pour déterminer les Eléments du Magnétisme terrestre.　4to.　194 pp.　2 plates.　(*Mém. ou réponse à la Question . . par l'Acad. Rle. de Belgique pour* 1870.)　　*Bruxelles*, 1871

† **Percival, P.**　An account of a Luminous Appearance in the air at Dublin, on January 12th, 1819-20.　4to.　2 pp.　1 plate.　(*Phil. Trans.* xxxi. for 1720-21, p. 21.)　　　　　　　　　　　　　　　　　　*London*, 1723

Percival, Thomas.　*Born September* 29, 1740, *at Warrington ; died August* 30, 1804, *at Manchester.*

Medical and Experimental Essays.

† **Pereda y Martinez, S. de.**　La naturaleza al alcance de los niños.　Bosquejo de algunos fenómenos físicos y cuerpos naturales.　8vo.　　*Madrid*, 1864

Perego, Antonio.　*Born* 1787, *at Oldaniga in the Milanese.*

Relazione sul Fulmine caduto in Iseo, il 17 Mag. 1833.　(*Comment. Ateneo Brescia*, vol. printed in 1834 pro 1833.)　　　　　*Brescia*, 1834

Sul la elettricità che per filtrazione si sviluppa nel Mercurio.　Memoria.　　　　　　　　　　　　　　　　　　　　　　*Milano*, 1841

† Sull' elettricità que si sviluppa nel mercurio coll' immersione, con una nota sul potere deferente del vetro, &c.　8vo.　20 pp.　(*Ann. di Fisica di Majocchi*, fasc. 16, 1842.)　(*Arago's copy.*)　　　　　　　　*Milano*, 1842

† Dell' elettrizzamento dei metalli ed altri corpi.　8vo.　10 pp.　(*Comment. Ateneo di Brescia*, per 1842, p. 68.)　　　　　　　　　　　*Brescia*, 1842

† Dell' elettricità di alcune sostanze animali e delle pietre, e quindi di un nuovo elettroscopio.　8vo.　10 pp.　(*Comment. Ateneo di Brescia*, per 1842, page 77.)　　　　　　　　　　　　　　　　　　　*Brescia*, 1842

Intorno ai processi meccanici atti a sviluppare, nei corpi solidi, l'elettricità statica ; e di alcune applicazioni che ne derivano.　8vo.　32 pp.　　　　　　　　　　　　　　　　　　　　　　　　*Brescia*, 1843

† Nuove esperienze elettriche.　Memoria.　8vo.　7 pp.　(*Comment. Ateneo di Brescia.*)　　　　　　　　　　　　　　　　　*Brescia*, 1845

Note intorno a qualche fenomeno elettrico.　8vo.　12 pp.　(*Atti delle Adunanze dell' I. R. Istituto Veneto*, del 10 Agosto 1846.)　　　*Venezia*, 1846

Descrizione dei danni cagionati dalla caduta di un fulmine in Mompiano provincia di Brescia.

PER 391

† **Peregrinus, Petrus.** Petri Peregrini Maricurtensis. De magnete, seu Rota perpetui motus, libellus. Sm. 4to. 43 pp. Figured title page. Preceded by " A. P. Gasseri Præfatio in Epistolam Petri Peregrini de Magnete."
Augsburgi, 1558
(*Very scarce and curious.* This work has long been attributed to the fallacious *Adsigerius.*)

(*Vide* Wenckebach, Bertelli, and Cavallo, T.)

† **Peretti, Pietro.** Born February 2, 1781, at Castagnole, Piedmont.

Sopra un lavoro chimico presentato alla Sezione tecnica dell' I. R. Accad. di Bavaria da Schonbein col titolo. Rapporto dell' Ossigeno ozonizato con la chimica pratica. *Roma,* 1857

† **Perewoschtschikow, Demetrius.** Born 1790, at Scheschkejew, Gouv. Pensa.

Pérévoschtchikoff (see above). On Formation of Hail. (*Bulletino della Soc. Imp. dei Naturalisti di Mosca,* 1829.) 1829

Perkins. (*Vide* Anonymous, " Sur les Tracteurs," and Anon. " Recherches sur le Perkinisme.")

(*Vide* Sue, Histoire du Perkinisme.) 8vo. (*Sue, Hist.* iv. 286.) *Paris,* 1805

† **Perkins, B. D.** The influence of Metallic Tractors on the human body. 8vo.
London, 1798

The influence of Metallic Tractors on the human body . . lately discovered by Dr. Perkins of North America . . By B. D. Perkins, son of the discoverer. 2nd Ed. 8vo. 99 pp. *London,* 1799

† **Perkins, Jno.** Conjectures concerning Wind and Waterspouts, &c. 4to. 13 pp. *Trans. Amer. Phil. Soc.* ii. 335.) *Philadelphia,* 1786

Péron, François. Born August 22, 1775, at Cerilly, Dép. Allier ; died December 14, 1810, at Paris.

Voyages de découvertes aux Terres Australes, sur les corvettes la Géographie, &c. pend. 1800-1-2-3-4, rédigé par M. F. Péron, naturaliste. Tome Premier. 4to. Plates. *Paris,* 1807

† **Peron et Freycinet.** Voyages de découvertes aux Terres Australes, sur les corvettes la Géographie, &c. pend. 1800-1-2-3-4. Historique, tome second, rédigé en partie, par feu F. Péron e continué par Ls. Freycinet, cap. de Frégate. 4to. Plates. *Paris,* 1816

Perrière de Reiffé, Jacques Charles François. Born at Pays d'Aunis ; died December 13, 1776.

† Mécanismes de l'électricité et de l'univers. Sm. 8vo. 2 vols. 1st 365 pp., 2nd. 340 pp. 3 plates and table. *Paris,* 1856

† **Perrot.** Lettre (à l'Acad.) sur les procédés électro-chimiques de dorure, d'argenture, de zincage, &c. 8vo. 15 pp. *Paris,* 1845

† Lettre de M. Perrot sur les procédés électro-chimiques de dorure, d'argenture, de zincage, &c. (A M. le Président de l'Acad.) 8vo. 15 pp. *Paris,* 1845

Perrot, S. sous chef des Lignes télégraphiques. Almanach des Lignes télégraphiques ou Manuel . . de télégraphie électrique. *Paris,* 1854

† Almanach des Lignes télégraphiques Année 1855. 12mo. 108 pp.
Paris, 1855

Perry, M. C. Narrative of the Expedition of an American Squadron to the China Seas and Japan in 1852-53, under Commodore M. C. Perry. Compiled by Hawks. 4to. Plates coloured, &c. and maps. (*Observ. on Zodiacal light by G. Jones.*) *Washington,* 1856

Perry, Stephen Joseph. Papers in the Phil. Trans. on Terrestrial Magnetism, also of a Magnetic Survey of France in the Summers of 1868-69.

Person, Charles Cléophas. *Born May* 3, 1801, *at Mussy-sur-Seine ; died in Dép. Aube.*

Sur l'hypothèse des courants électriques dans les nerfs. 8vo. (*Journ. de Physiol. Expér.* 1830, x. 216.) *Paris,* 1830

† Théorie du Galvanisme. Thèse présentée à la faculté de médecine de Paris, 7 Jan. 1831. 4to. 36 pp. 1 plate. *Paris,* 1831

Résumé d'un Mémoire sur les résultats du Galvanisme appliqué dans les maladies des yeux, &c. *Bordeaux,* 1843

Traitement galvano-punctural de l'amaurose. (*Journ. des Connaissances médico-chirurgicales,* Nov. 1853.) 1853

Peschard, Albert. Application de l'Electricité aux Grandes Orgues. 8vo. 28 pp. *Caen,* 1865

Peschel, C. F. Elements of Physics, by E. West. 3 vols. Fcp. 8vo. *London,* 1845

Peslin. (*Vide* Raulin and Peslin.)

† **Peters,** C. A. F. Über die Bestimmung des Lagenunterschiedes zwichen Altona und Schwerin ausgeführt in 1858 durch galvanische Signale. 4to. 267 pp. 2 plates. *Altona,* 1861

Petersburg Imp. Acad. Bulletin Scientifique, publié par l'Académie . . . et rédigé par son Secrétaire perpetuel. 4to. *St. Petersburg and Leipzig*

Transactions. 9 vols. 8vo. (*There are many other titles.*) *St. Petersburg,* 1815-21

Recueil des Actes des Séances publié par l'Acad. des Sciences. 4to.

Academia Scientiarum Petropolitana Commentarii (pro) 1726-1746. 17 vols. 4to. *Petropoli,* 1728-51

Novi Comment. 1747-1775. 21 vols. 4to. *Petropoli,* 1750-76

Acta . . . 1777-1782. 12 vols. 4to. *Petropoli,* 1778-86

Nova Acta (pro) 1783-1803. 15 vols. *Petropoli,* 1787-1806

Mémoires ; avec l'Histoire de l'Acad. (each vol. in 2 parts) (pro) 1803–1826. 11 vols. *Petropoli,* 1809-30

Petetin, Jacques Henry Désiré. *Born* 1744, *at Lons-le-Saulnier ; died February* 27, 1808, *at Lyon.*

Mém. sur la découverte des phénomènes que présentent la Catalepsie et le Somnambulisme, &c. Partie première. 8vo. 62 pp. 1787

Mém. sur la découverte des phénomènes de l'affection Hystérique essentielle. Seconde partie. 8vo. 126 pp. 1787

† Nouveau mécanisme de l'Electricité. 8vo. 300 pp. 10 plates. *Lyon,* 1802

† Théorie du Galvanisme ; ses rapports avec le nouveau mécanisme de l'Electricité, publié en l'an 10. 8vo. 44 pp. *Paris et Lyon,* 1803

† Electricité animale prouvée par la découverte des phénomènes physiques et moraux de la Catalepsie hystérique, et par les bons effets de l'électricité artificielle. 8vo. 156 pp. *Lyon,* 1805

† Electricité animale prouvée par la découverte des phénomènes physiques et moraux de la Catalepsie Hystérique et de ses variétés. 8vo. 382 pp. Portrait. (*A "Notice sur la vie de Petetin" by an Anon. with separate paging, is annexed,* 121 pp.) *Paris,* 1808

Observations sur les effets de l'Electricité dans le Traitement de la Catalepsie, du Tetanos, et de l'Astme Convulsif, Symptomes d'Affection Hystérique essentielle. (*Actes de la Soc. de Santé de Lyon, depuis l'an* 1-5, p. 230.) *Lyon*

Petit. Compte Rendu des travaux pendant 1805. 8vo. (*Lyons Acad. Comptes Rendus.*) *Lyon,* 1806

Petit, Frédéric. *Born July 16, 1810, at Muret (Haute Garonne).*

† Mémoire sur le Bolide du 27 Octobre et sur une conséquence remarquable qui paraît résulter de son apparition. 8vo. 16 pp. (*Toulouse Acad.* 3me Série, i. 303.) *Toulouse,* 1844

† Note sur les rapports qui existent entre les époques des apparitions d'Etoiles Filantes et les Températures terrestres observées à Toulouse. 8vo. 6 pp. (*Toulouse Acad.* 3me série, ii. 47.) *Toulouse,* 1846

† Notice sur un Bolide, 19 Août, 1847, à Paris et à la Chapelle. Lu 17 Fév. 1847. 8vo. 7 pp. (*Toulouse Acad.* 3me série, iv. p. 105.) *Toulouse,* 1848

† Méthode pour déterminer la parallaxe et le mouvement des Bolides. 8vo. 55 pp. (*Toulouse Acad.* 3me série, v. 53.) *Toulouse,* 1849

† Mémoire sur le Bolides du 5 Janvier, 1837. 8vo. 13 pp. *Toulouse,* 1851 (*Note.—He refers to his publication in October, 1844, in the Comptes Rendus relative to this " Bolide ".*)

† Observations sur une Chute de Pluie, par un Ciel serein. 8vo. (*Toulouse Acad.* 4me série, i. 234.) *Toulouse,* 1851

† Reply (alluded to) of Petit to Leverrier on the subject of Bolides. 8vo. (*Toulouse Acad.* 4me série, i. 319.) *Toulouse,* 1851

† Lecture d'une Lettre . . . Chute de Pierres à Escalquens. 8vo. (*Bulletin Toulouse Acad.* 4me série, i. 320.) *Toulouse,* 1851

† On Asteroides and High Temperature. 8vo. (*Bulletin Toulouse Acad.*) *Toulouse,* 1851

† Mémoire sur la parallaxe et le mouvement du Bolide du 6 Juillet, 1850. 8vo. 5 pp. (*Toulouse Acad.* 4me série, ii. 80.) *Toulouse,* 1852

† Mémoire sur le Bolide de 5 Juin, 1850. 8vo. 6 pp. (*Toulouse Acad.* 4me série, iii. 327.) *Toulouse,* 1853

† Verbal communication on the Zone of Asteroides which he says the earth now traverses. 8vo. (*Toulouse Acad.*) *Toulouse,* 1853

† Mémoire sur le Bolide du 21 Mars, 1846. 8vo. 7 pp. (*Toulouse Acad.* 4me série, iv. 439.) *Toulouse,* 1854

† Note sur l'Inclinaison et la Déclinaison Magnétiques à Toulouse. Read 31st August, 1854. 8vo. 1 page. (*Toulouse Acad.* 4me série, iv. 459.) *Toulouse,* 1854

† Note sur l'Inclinaison et Déclinaison Magnétiques à l'Observatoire de Toulouse. 8vo. (*Toulouse Acad.*) *Toulouse,* 1855

† Note sur les Bolides du 24 Décembre 1850 et du 2 Avril 1852. 8vo. 7 pp. (*Toulouse Acad.* 4me série, v. 130.) *Toulouse,* 1855 (*Vide* also Abbadie.)

Petit, P. A letter about the Loadstone, where chiefly the suggestion of Gilbert, touching the circumvolution of a globous magnet, called Terrella; and the variation of the variation is examined. 4to. (*Phil. Trans.* 1667, p. 502.) *London,* 1667 (*Note.—Brewster (Treatise on Magnetism, p. 8) refers to Petit's Observation on Variation in 1630 and 1660 in Paris.*)

† **Petiton.** Notice sur les Aurores Boréales. 8vo. 29 pp. *Paris,* 1840

* **Peto,** J. Essay on Spontaneous Combustion. 18mo. *London,* 1843

Petrequin. On Cure of " Sacchi Aneurismatici " by his method. (Electrified needles inserted. *Vide* Matteucci.)

Petrie. Régulateur à lumière intermittente. (*From Du Moncel.*)

Petrie. (*Vide* Staite and Petrie.)

Petrina, F. A. Entdeckungen im Galvano-Voltaismus. (*Baumgartner's Zeitschrift f. Phys.* v. 1837.) *Wien,* 1837

Versuche über Galvano-Voltaismus. (*Holger's Zeitschrift,* vol. i. 1840, ii. 1842.) 1840

I I

Petrina, F. A.—*continued.*

Five Articles in Poggendorff's Annalen, from vol. xlix. 1840, to vol. lxiv. 1845.

Theorie grossplattiger galvanischer Elemente. (*Bericht über d.Naturf.-Versamml. zu Gratz.*) *Gratz*, 1844

Einige merkwürd. Erschein. bei Grove'schen Elementen. (*Bericht über d. Naturf.-Versamml. zu Gratz.*) *Gratz*, 1844

Magneto-electr. Maschine. vortheilhaftester Einrichtung für Aerzte, 1844. *Linz*, 1844

Neue Therie d. Elektrophors g. neues Harzkuchen-Elektroskop. 4to. 1 plate. *Prag*, 1846

(*This is also in the Abhh. d. Böhm. Gesellsch. d. Wiss.* Folge v. Bd. iv. 1847.)

Seven articles in the Abhand. d. Böhm. Gesellsch. d. Wiss. from 1847 to 1853, on Electricity, Magnetism, Electro-Magnetism, &c.

Elektromagnet. Telegraph auf d. österr. Eisenbahnen. 4to. *Prag*, 1848

Beiträge zur Physik. 8vo. 16 pp. 1 plate lith. (*Aus den Sitzungsberichten 1854 der Kaiserl. Akad. der Wissenschaften.*) *Wien*, 1854

Über d. Coexistenz zweier einen Leiter in entgegengesetzen Richtungen durchlaufenden Ströme. 8vo. (*Pogg. Ann.* xcviii. 1856.) *Leipzig*, 1856

Four Articles in the Sitzungs-berichte d. Wien Acad. vol. iii. 1849 to xiii. 1854.

Petrini, Pietro. *Born December* 8, 1785, *at Pistoja; died December* 8, 1822, *at Pisa.* (*Vide* Cione and Petrini.)

Petroff, Bazilius. Notice sur les expériences galvano-voltaïques (russisch). 1803

Recueil de nouv. expériences relatives à l'electricité (russisch). 1804

Petsholdt, G. P. A. De calamitis et lithanthracibus. *Dresden v. Leipzig*, 1841

† **Petzholdt, A.** Die galvanische Vergoldung. Versilberung, Verkupferung für den Techniker, &c. . . Zweite umgearbeitete Auflage. 8vo. 88 pp. 1 plate. *Leipzig*, 1843

Petzholdt, J. Addressbuch deutcher Bibliotheken. Dritte Ausgabe. 12mo. *Dresden*, 1848

Peyrouse. (*Vide* La Peyrouse.)

† **Peytel.** Explication du phénomène des Trombes. 4to. 4 pp. (*Montpel. Acad. Sect. Sciences*, i. 105.) *Montpellier*, 1848

† **Peytavin, J. B.** Essai sur la constitution physique des fluides électriques et magnétiques. 8vo. 227 pp. 1 plate. *Nantes*, 1830

Pfaff. (*Vide* Michaelis, G. A.)

Pfaff. (*Vide* Schweigger and Pfaff.)

Pfaff, Christian Heinrich. *Born March* 2, 1773, *at Stuttgart; died April* 23, 1852, *at Kiel.*

De electricitate animali. 8vo. *Stutgart*, 1790 ?

Dissertatio inauguralis medica de electricitate, sic dicta animali. 8vo. *Stutgard*, 1793

De electricitate animali. 8 plates. (*From Exleben*, p. 533.) *Stutgard*, 1793

Extrait. 8vo. (*Ann. d. Chemie*, xxxiv. 307.) *Paris* (*Note.—This is probably translated from some one of his numerous articles in Gilbert, &c.*)

Abhandlung über die sogenannte thierische Electricität. Beytrage, &c. (*Gren's Journ.* tom. viii.) 1794

Pfaff, Christian Heinrich.—*continued.*

† Über thierische Elektricität und Reizbarkeit. Ein Beytrag zu den neuest Entdeckungen. 8vo. 398 pp. *Leipzig,* 1795

Nouvelles expériences galvaniques. 4to. (*Soc. Philomath.* an. 4, tom. ii. p. 181.) *Paris,* 1796

Nordisches Archive für d. Natur u. Arzeneiwissenschaft. 4 Bde. 8vo. *Kopenhagen,* 1799-1804

Mémoire sur les expériences de Humboldt. (*Archives du Nord pour la physique et la Médecine rédigées par . . Pfaff à Kiel et Scheel à Copenhajne : dans le 1r Cahier vers l'automne de* 1799.) *Copenhagen and Kiel,* 1799

About fifteen Articles on the Pile, the dry Pile, &c. (*Gilbert's Annalen,* vol. vii. 1801 to vol. lxxiv. 1821.)

Versuche über d. Anwend. d. Elektricität d. Volta'schen Säule bei Taubstummen. 8vo. *Kopenhagen,* 1802

Die neust. Endeckungen Französ. Gelehrten, u.s.w. 12 Hfte. Mit Friedlander. 8vo. *Leipzig,* 1803

Französ. Annal. für d. allgem. Naturgeschichte Physik, Chemie, u.s.w. 4 Hefte. Mit Friedlander. 4to. *Hamburg,* 1804

Vermischte chem. Bemerk.; zwei Briefe. 8vo. (*Gehlen, N. Journ. f. Chem.* v. 1805.) 1805

Über Pacchiani's Darstell. von Salzsäure aus Wasser durch d. Volta'sche Säule 8vo. (*Gehlen, Journ. f. Chem. u. Phys.* ii. 1806.) 1806

† Upon the formation of the Muriatic Acid. By M. Pacchiani. Extract of a Letter from Professor Pfaff, of Kiel, to M. Berthollet. 8vo. 2 pp. (*Ann. de Chim.* lx. 314.) (*Phil. Mag.* xxvii. 338.) *London,* 1807

Neues Nordisch. Archive, u.s.w. Mit Scheel und Rodolphi. 4 Heft. 8vo. *Frankfort,* 1807-8

† Some further remarks upon the supposed formation of Muriatic Acid in Water by the influence of the Galvanic Pile. 8vo. 2 pp. (*From Ann. Chim.* lxii. 23 ; *Phil. Mag.* xxix. 19.) *London,* 1808

Über d. galvan. Verhalt d. feuchten Leiter mit d. trocknen u.s.w. in Säulen mit besondr. Hinsicht auf Ritter's elekt. System. 8vo. (*Gehlen, Journ. f. Chem. u. Phys.* ii. 1808.) 1808

About twenty Articles on Voltaic Electricity, &c. in Schweigger's Journal, vol. i. 1811 to vol. lxiv. 1832.

† Über und gegen den thierischen-Magnetismus, &c. 8vo. 184 pp. *Hamburg,* 1817

Über d. eigenthüml. Elektricität d. menschl. Körpers. (*Meckel's Arch. f. Physiol.* iii. 1817.) 1817

† Der Electromagnetismus. Eine historisch-kritisch. Darstellung der bisherigen Entdeckungen auf dem Gebiete desselben ; nebst eigenthümlichen Versuchen. 8vo. 288 pp. 8 plates. *Hambourg,* 1824

† Der Elektromagnetismus, ein histor.-krit. Darstellung d. bisher. Entdeckk. auf d. Gebiete dess.; nebst eigenthüml. Versuchen. 8vo. 8 plates. *Hamburg,* 1824

Über Blitz und Blitzableiter. 8vo. (*Gehler's Phys. Wörterbuch neu bearb.*) *Leipzig,* 1825

Rapport sur les expériences qui confirment l'identité du fluide électrique avec le fluide qui fait naître les phénomènes galvaniques répétées par M. Pfaff, au cabinet de physique de l'École centrale du département de la Dyle. 8vo. (*Journ. de Chim. de Van Mons,* No. 2, p. 191.) *Bruxelles*

(*Repetition, by Pfaff, of Volta's Experiments on the identity of Electricity and Magnetism.*)

Er war Mitarbeiter am Neu Gehlerschen Wörterbuch.

Pfaff, Christian Heinrich.—*continued.*

† Revision der Lehre vom. Galvano-Voltaismus. . . . " Mit besond. Rücksicht auf Faraday's, De la Rive's, Becquerel's, Karsten's u. a. neueste Arbeiten üb. dies. Gegenstand." 8vo. 227 pp. 1 plate. (*From Dove*, p. 172, and T. p.)
Altona, 1837

About Eleven Articles on Voltaic Electricity, Electromagnetism, &c. in Poggendorf's "Annalen," vol. xl. 1837, to vol. liii. 1841.

† Parallele der chemischen Theorie u. der Volta'schen Contacttheorie der galvanischen Kette, mit besond. Rücksicht auf die neuesten Einwürfe Faraday's, Gmellin's, u. Schönbein's gegenaletztere, nebst allgem. Betrachtungen über das Wesen einer phys. Kraft u. ihrer Thätigkeit. Nebst e. Anhang, Beschreibung u. Abbildung eines sehr compendiosen u. zu theoretischen Versuchen sowohl, als heilkundiger Anwendung zweckmässig eingerichteten electro-magnetischen Inductions-Apparates von . . A. W. Cramer.
Kiel, 1845

Pfaff mit Andern. (*Vide* Nordisches Archiv. d. Natur- d. Arzeneiwissenschaft u. Neues nordisch Archiv, u.s.w. *From Poggendorf*, ii. 425.)

Pfaff, C. H. and Säule. The Articles Galvanism, and the Articles on Electricity in the new edition of Gehler's "Physikalisches Wörterbuch." (*From Wartmann, in Archives de l'Electricité*, i. 59.)

Pfaff (Johann?) Wilhelm (Andreas?). *Born Dec.* 5, 1774, *at Stutgart; died June* 26, 1835, *at Erlangen.*

Darstellung des Voltaismus nebst einem Anhang v. Versuchen. 8vo.
Stutgart, 1802?

† Der Voltaismus dargestellt nebst einem Anhang von Versuchen. 8vo. 116 pp.
Stutgart, 1803

† Übersicht über den Voltaismus und die wichtigsten Sätze zur Begründung einer Theorie desselben. 8vo. 127 pp. *Stutgart*, 1804

Die Umkehrung d. Volt. Pole durch Hrn. Pohl, u.s.w. 8vo. (*From Poggendorf*, ii. 429.) *Nürnberg*, 1827

† Die gesammt Naturlehre für das Volk und seine Lehrer. 8vo. 576 pp. 6 plates. *Leipzig, &c.* 1834

Pfeiffer, C. Der elektrische Telegraph. Eine gemeinfassliche Belehrung über das Wesen, die Einrichtung und die verschiedenen Arten der elektrischen Telegraphen, sowie über die Erregung, Fortleitung und Geschwindigkeit des elektrischen Stromes. 8vo. 142 pp. 4 plates. *Leipzig*, 1862

† Handbuch der elektro-magnetischen Telegraphie nach Morseischen System. Ein theoretisch-praktischer Leitfaden für angehende-Telegraphisten, &c. 8vo. Atlas von 16 Tafeln. *Weimar*, 1865

† Procentische Zusammensetzung des Meteorsteins von Paralle bei Madura in Ostindien. Vorgelegt am 15 Mai, 1863. 8vo. 4 pp. (*Sitzungsb d. k. Akad.* xlvii. 460.) *Wien*, 1863

† Pfluger, E. Ueber die durch constante Ströme erzeugte Veränderung des Motorischen Nerven. (*Medicinische Central Zeitung vom 15 Marz u. 16 Juli*, 1856.
1856

On the "Nn. Splanchnici :" their influence on the Motions of the Guts, &c. (*Vide* Meyer.) 1856

Verändr. d. Erregbarkeit e. Nerven durch e. constant elektr. Ström. (*Monatsberichte d. Berlin Akad.* 1858.) *Berlin*, 1858

Untersuchungen über die Physiologie des Elektrotonus. 8vo. 5 plates.
Berlin, 1859

Pflueger, E. F. W. Disquisitiones de sensu Electrico Commentatio. (Thesis) read 21 Mar. 1860. 4to. 16 pp. *Bonnæ*, 1860

Philadelphia. Journal of the Franklin Institute of the State of Pennsylvania, for the Promotion of the Mechanical Arts. 8vo. 62 vols. *Philadelphia*, 1826-56

Philadelphia, U. S. (*Vide* American Phil. Soc.)

Philadelphia, American Musæum. A periodical. Contains Churchman, 3 Letters of his to Cassini on the relations of Magnetism and the Aurora. Letters printed in October, 1788. 1788

† Philipeaux. Etudes sur l'électricité appliquée . . . au traitement des paralysies. 8vo. 106 pp. *Paris and Lyon*, 1857

† Philipp, D. Alphabetisches Sachregister der wichtigsten technischen Journale für den Zeitraum vom 1 Januar bis 31 Juni, 1866. 8vo. 55 pp. *Berlin*, 1866

† Alphabetisches Sachsregister der wichtigsten technischen Journale, für den Zeitraum vom 1 Januar bis 30 Juni, 1869. 8vo. 48 pp. and 55 pp.
Berlin, 1869-70

Phillipps, H. C. Code of Naval Signals. 8vo. *London*, 1835

Phillips, J. A. Manual of Metallurgy. Post 8vo. (*Encyclopædia Metropolitana*.)
London, 1852

Phillips, John. *Born December 25, 1800, at Marden, Wiltshire.*

On a Modification of the Electrophorus. 8vo. (*Phil. Mag.* Series iii. vol. ii.)
London, 1833

† Magnetic Phenomena in Yorkshire. 8vo. 21 pp. (*Extract from the Proceedings of the Yorkshire Philosophical Society.*) *York*, 1854

(*Vide* Rivot and Phillips.)

Phillips, Sir Richard. *Born 1778, in London ; died May 11, 1851, in London.*

† Electricity and Galvanism explained on the mechanical theory of Matter and Motion. 8vo. 6 pp. (*Phil. Mag.* lvi. 195.) *London*, 1820

Reference to his Annals of Philosophy, in which he mentions an experiment upon a young Poplar, whereby it would seem that copper was imbibed in the branches, &c. from a solution placed at its roots, and that it was precipitated on a knife used to cut off a branch. (*Bibliog. Ital.* xxvii. 107.) 1822

† Phillips, R. Electrical Formation of Crystallized Sulphuret. 8vo. 3 pp. (*Phil. Mag. or Annals*, vii. 226.) *London*, 1830

Philosophical Magazine Phil. Mag. Comprehending the Various Branches of Science. . . . By A. Tilloch. Vols. i.—lx. 1798-1822

Idem, . . . by Tilloch and Taylor. Vols. lx.—lxv. 1822-25

Idem, . . . by Taylor. Vols. lxvi.—lxviii. 1826

Phil. Mag. ; or, Annals of Chemistry, Mathematics, Astronomy, . . . and General Science ; New and United Series of the Phil. Mag. and Annals of Philosophy, by Taylor and Phillips. 1827-32

London and Edinburgh Phil. Mag. and Journ. . . . by Taylor, Phillips, and Brewster. Third Series. 1833-50

London, Edinburgh, and Dublin Phil. Mag., by Brewster, Kane, and Francis. All in London. 8vo. 1851 *to present date*

Philosophical Transactions. (*Vide* Royal Society.)

Phipson. Ueber die Phosphorescenz bei den Mineralien, Pflanzen und Thieren. Aus dem Französischen bearbeitet und mit den neuesten Erfahrungen bereichert von Johannes Müller. 8vo. 46 pp. *Berlin*, 1858

Phipson, T. Wonders of the Wire. A Manual of the various Electro-Magnetic Telegraphs at present in use. 8vo. 28 pp. *London*, 1847?

Phipson, T. L. De la Phosphoréscence en général et des Insectes Phosphoriques en particulier. 8vo. *Bruxelles*, 1858

Phosphorescence ; or, The Emission of Light by Minerals. 12mo. *London*, 1862

Phosphorescence ; or, The Emission of Light by Minerals, Plants, and Animals. . . . 8vo. 225 pp. Coloured Frontispiece. (*See above.*) *London*

† Meteors, Aerolites, and Falling Stars. 8vo. Plates. 1867

† **Phœnix, J.** Letter to Tilloch on Electricity. 1 p. (*Phil. Mag.* xxx. 154.)
London, 1808

Pianciani, Giambattista. *Born October 27, 1784, at Spoleto.*

Theoriæ electro-dynamicæ synopsis. *Romæ*, 1825

Specimina Meteorologica. (*From Bellani in Bibliot. Ital.* xcix. 97.)
Romæ, 1828

Istituzioni Fisico-Chimiche. 4 vols. 8vo. *Roma*, 1834-4-5
(*Note.—In six Books the fourth Elect. &c., fifth and sixth Meteorol. &c.*)

Memoria intorno alla Grandine. (*Bibliot. Ital.* xcix. 97.) *Roma*, 1835

Saggio sui fenomeni d'induzione magneto-elettrica. Letto all'Accad. dei Lincei,
8 Agosto, 1836. 8vo. 30 pp. 1 plate. (*Estratto dal Giornale Accademico,*
tom. lxix.) *Roma*

Elementi di fisico-chimica. *Napoli*, 1840-1

† Saggio d'applicazione del principio dell' induzione elettro-dinamica a' fenomeni
elettro-fisiologici e, in particolare, a quelli delle Torpedini. Ricevuto 7
Maggio, 1838, and Appendice 7 Giúgno, 1838. 4to. 41 pp. (*Mem. Soc.*
Ital. xxii. 7.) *Modena*, 1841

† Esperienze e congetture sulla forza magnetica. Ricevute 6 Marzo, 1840. 4to.
17 pp. (*Mem. Soc. Ital.* xxii. 210.) *Modena*, 1841

† Necrologico cenno intorno al R. P. Gio. Batt. Pianciani compilato dal Prof.
Paolo Volpicelli. 4to. 12 pp. (*Estratto dagli Atti dell' Accad. pontif.*
dei Nuovi Lincei, an. xv. . . 1862.) *Roma*, 1862

Elenco degli scritti del p. Gio. B. Pianciani . . . per cura del sig. principe D.
B. Boncompagni. 4to. 12 pp. *Roma*, 1862

Picard, Jean. *Born July 21, 1620, at La Flèche, Anjou ; died July 12, 1682, at*
Paris.

Observations sur la lumière du baromètre. (*Anc. Mém. Paris,* tom. ii.)

Expérience faite à l'observation sur le Baromètre touchant au nouveau phénomène. (*Anc. Mém. Paris,* tom. x.)

On an Observation of the Needle at Paris, when its direction was exactly
North, *i.e.* in 1666.

Article Grêle in Encyclopédie Méthodique. (*From Bibliot. Ital.* xcix. 42.)

† **Picchioni, L.** Della sostanza odorifera che dall' aria atmosferica e dall' acqua
svincolano le correnti elettriche. 8vo. 5 pp. (*Bibliot. Ital.* xcvi. 404.)
Milano, 1839

† **Piccinelli, G.** Transunto dell' elettricità medica cavato dalle opere inglesi di
Adams e di Nairne. 4to. 27 pp. 1 plate. (*Opus. Scelti,* viii. 310.)
Milano, 1785

Pick, H. Elektrometer und Galvanometer. Ein Beitrag zur Erweiterung der
Kenntniss dieser Instrumente. 4to. 22 pp. (*Jahresbericht über das k. k.*
Akademische Gymnasium in Wien während des Schuljahres 1854-55.)
Wien, 1855

† **Picke.** A Walk round the Exhibition. 8vo. (*Contains description of Ronalds'*
Apparatus.) *Paris*, 1855

Pickel, Georg. *Born November 20, 1751, at Sommerach a. Main ; died July 20,1838,*
at Würzburg.

† Experimenta physico-medica de electricitate et calore animali . . . Thesis . . .
defendit . . . Georgius Pickel . . . die 1 Septembris, 1778. 12mo. 134 pp.
2 plates. (*Volta's copy.*) *Wiceburgi*, 1788
(*Note.—This little work was much approved by Volta, as Compendium, &c. ;*
and this copy has many ticks (√) of his.)

Der Rettungs-Apparat z. Wiederbelebung v. Scheintodten, u.s.w. (*From Pog-*
gendorff, ii. 444.) 1812

Abhandl. über Blitzableiter, u.s.w. (*From Poggendorff,* ii. 444.) 1821

Pickel, Georg.— *continued.*

Er verbesserte d. elektr. Feuerzeug u. d. elektr. Pistole. (*Vide Altes Gehler-sches Wörterbuch*, ii. 109 u. iii. 511.)

† Art. Pistole elektrische. 8vo. (*Gehler's Wörterb.* vii. 573.)

Pickel, Ignaz Balth. Authent. Nachr. v. e unweit Eichstadt vom Himmel gefal-lenen Meteorstein. (*Moll's Ann. d. Berg u. Huttenk.* iii. 1805.) 1805

† **Pickering**, R. A scheme of a Diary of the Weather; together with draughts and descriptions of Machines subservient thereunto; inscribed to the President and Fellows of the Royal Society by Pickering, R. Read May 3, 1741. 4to. 17 pp. (*Phil. Trans.* xliii. 1.) *London*, 1744-5

Pictet, Marc Auguste. *Born July 23, 1752, at Geneva; died April 19, 1825, at Geneva.*

† "Has furnished the Imperial Institute of France with an account of two Meteorolites, one of which fell on board of a ship; a novel circumstance." (*Phil. Mag.* xli. 71.) *London*, 1813

On Atmospheric Electricity in relation to Magnetic Disturbances during Auroræ at Umba in 1769, &c. (*Nov. Comment. Petropol.* xiv. p. ii. p. 88.) 1769?

Piddington, H. Sailor's Horn Book for the Law of Storms, with Transparent Storm Cards, &c. 4to. 1848

*† The Sailor's Horn Book of the Law of Storms. 3rd edition. 8vo. (*From English Cat.* p. 603.) *London*, 1860

Piderit, Johann Rudolph Anton. *Born August 18, 1720, at Pyrmont; died August 2, 1791, at Cassel?*

Dissertatio inauguralis philosophica de Electricitate. 4to. (*Gralath, Bibliot.* 2 Stück, p. 323.) *Marburg*, 1745

Dissertatio ii. De Electricitate. 4to. (*Gralath, Bibliot.* 2 Stück, p. 323.) *Marburg*, 1746

Pidoux. (*Vide* Trousseau and Pidoux.)

† **Piehl.** Handbog for det Telegraphenyttende Publicum. 8vo. 244 pp. *Helsingfors*, 1856

Pierce, B. On the Perturbations of Meteors approaching near the Earth . . . in a letter to S. C. Walker, Esq. 4to. (*Trans. Amer. Phil. Soc.* New Series, vol. viii.) *Philadelphia*, 1843

† **Pierre.** Einige Bemerkungen über magnetische und diamagnetische Erscheinun-gen. 8vo. 23 pp. (*Sitzungsb. d. k. Akad. d. Wissen.* 2 Heft. p. 37.) *Wien*, 1850

† **Pierre**, V. Kravogl's Elektromagnetischer Motor. 8vo. 16 pp. 2 plates. (*Sitzb. d. k. Akad.* Martz Heft, 1868, lvii.) *Wien*, 1868

Pignacca, A. Della corea osservata in Pavia nell' Ospitale e nella Clinica medica dall' anno 1848 al 1854. 8vo. *Pavia*, 1856

† **Pigram**, W. Successful application of the Magnet, employed to extract a frag-ment of iron out of the human eye, which had been lodged there about five months. 8vo. 3 pp. (*Phil. Mag.* xxxii. 154.) *London*, 1808

Pilatre de Rozier. Sur la cause de la Foudre. (*Journ. Phys.* xvi. 1780.) *Paris*, 1780

Sur des Expériences Electriques. 4to. (*Journ. Phys.* xvii. 1781.) *Paris*, 1781·

Pilger, F. Versuche, durch den Galvanis. die Wirkung verscheid. Gifte (q.) und Arzneimittel . . . der Nerven zu prufen. 8vo. *Giessen u. Darmstadt*, 1801

Pilkington, James (Bishop of Durham). The Burning of St. Paul's Church in London in 1561, on the 4th June, by Lightning. 8vo. *London*, 1561?

† Works of; edited for the Parker Society by James Scholefield. 8vo. 703 pp. *Cambridge*, 1842

(*Note.—At* p. 481, *in a note* (2), *is said that* "*Strype in his Life of Grindal,* pp. 53-55, fol. *London*, 1710, *gives a full account of the circumstances of the burning of St. Paul's and of the Bishop's Sermon.*" *See above.*)

*† Works edited by James Scholefield. 8vo. *London*, 1842

† **Pinaud, A.** Mémoire sur quelques appareils propres à simplifier la démonstration des phénomènes électro-dynamiques. 8vo. 14 pp. 1 plate. (*Toulouse Acad.* 2e série, vi. 129.) *Toulouse*, 1843

† Mesure de courants électriques-galvano-plastiques. Images Daguerriennes. 8vo. (*Résumé des travaux de l'Académie (Toulouse) pendant* 1839-40-41.) *Toulouse*, 1843

† Mémoire sur la coloration, par l'Electricité, des papiers impressionables à la lumière, et sur une nouvelle classe d'empreintes électrographiques. Lu 13 Juillet, 1843. 8vo. 14 pp. (*Toulouse Acad.* 3e série, i. 146.) *Toulouse*, 1844

† Note sur une disposition nouvelle de la Pile à courant constant, et sur ses effets chimiques. 8vo. 10 pp. (*Toulouse Acad.* 3e série. ii. 169.) *Toulouse*, 1846

† Programme d'un cours élémentaire de Physique . . 7me édition, revue, corrigée et augmentée. 8vo. 427 pp. 9 plates. *Toulouse et Paris*, 1853

† Abrégé de Télégraphie électrique. 8vo. 30 pp. 2 plates. *Toulouse*, 1853

Pinazzo, A. Dissertazione sopra alcuni buoni fisici effetti che nascono dai temporali. *Mantova*, 1788

† **Pine, T.** On the connection between Electricity and Vegetation (resumed from page 253). 8vo. 11 pp. (*Annals of Electricity*, iv. 421.) (*Roget's copy.*) *London*, 1840

† **Pinson, P. C.** Essai sur quelques applications de l'Electricité à la Médecine. Thèse. 4to. 44 pp. *Paris*, 1857

Piria. (*Vide* Matteucci, Piria, &c.)

Pisko, F. J. Lehrbuch der Physik für Untergymnasien. 8vo. 240 pp. *Wien*, 1865

(*Vide* also Hesler and Pisko.)

Pistoj, Candido. *Born about* 1736, *at Siena; died about* 1780, *at Siena.*

† Articolo di Lettera scritta all' Abate Rozier. Siena, 25 April, 1777. 4to. 4 pp. (*Scelta d'Opusc.* nuova edit. vol. iii. p. 255. *Printed in* 12mo. *edition in* 1777. *Sopra un Fulmine caduto in Siena, ai* 18 Aprile, 1777.) *Milano*, 1784

† **Pistollet,** Account of a Meteoric Stone which fell in the environs of Langres, in Moravia. 8vo. 5 pp. (*Phil. Mag.* xlvii. 349.) *London*, 1816

Pitaro. (*Vide* Andria, A. N.)

Pivati, Giovanni Francesco. *Born* 1689, *at Padua; died* 1764, *at Venice.*

† Della elettricità medica. Lettera del . . Pivati al . . Zanotti. Sm. 4to. 53 pp. (*See French translation*, 1750, *in* 12mo. *in Calogera's Raccolta, Venice.*) *Lucca*, 1747

† Riflessioni fisiche sopra la medecale elettricità. 4to. 166 pp. *Venezia*, 1749

Lettre sur l'Electricité Médicale à Zanotti. 12mo. 40 pp. *Paris*, 1750

(*Vide* also Anon. Medical Electricity, 1763.)

Nuovo Dizionario scientifico e curioso sacro-profano. 10 vols.? Folio. The last, 10th vol. is dated 1751. *Venezia*

Pixii, Hyp. Fils. Machine magnéto-électrique. 4to. (*Acad. des Sciences*, Sept. 3, 1832?) (*Ann. de Chimie for* July, 1832.) *Paris*, 1832

Nouveaux appareils d'électro-magnétisme. 8vo. *Paris*

† Instruction pour remonter l'appareil Magnéto-élect. et Note . . 4to. 4 pp. 1 plate. *Paris*, 1832

Plana, Giovanni Antonio Amedeo. *Born November* 8, 1781, *at Voghera.*

Mémoire sur la découverte de la loi du choc direct des corps durs. . . publiée en 1667 par A. Borelli, et sur les formules générales du choc excentrique des corps durs ou élastiques, etc. ; suivi d'une Appendice où l'on expose la théorie des oscillations et l'équilibre des barreaux aimantés. 4to. (*Mém. de Turin*, série ii. vol. iii. 1844.) *Torino*, 1844

Plana, Giovanni Antonio Amedeo.—*continued.*

† Mémoire sur la distribution de l'Electricité à la surface de deux sphères conductrices complètement isolées. 4to. 331 pp. (*Ext. des Mém. de l' Acad.* 1845.)
Turin, 1845

† Mémoire sur la Théorie du Magnétisme. (*Astr. Nachr.* xxxix. 1854.) 1854
Cenno biografico di Giovanni Plana compilato dal Prof. P. Volpicelli. 4to.
6 pp. (*Estratto dagli Atti dell' Accad. pontif. de' Nuovi Lincei* . . Feb. 14, 1864,
tom. xvii. p. 169.) Roma, 1864

Mémoire sur la distribution de l'Electricité à la surface intérieure et sphérique d'une sphère creuse de métal, et à la surface d'une autre sphère conductrice électrisée que l'ont tien isolée dans sa cavité. Lu . . 28 Avril, 1854. 4to. 39 pp. (*In Mém. de l'Acad. de Turin,* série ii. tom. xvi. p. 57.)
Turin, 1857 ?

Mémoire sur l'application de l'équilibre magnétique à la détermination du mouvement qu'une plaque horizontale de cuivre tournant uniformément sur elle-même, imprime par réaction . . à une aiguille aimantée. 4to. 98 pp. (*Acad. de Turin,* 1858, vol. xvii. série ii.) Turin, 1858

Planta, Martin. *Born* 1727, *at Süss, Unter-Engadin ; died March,* 1722, *at Haldenstein.*

In Allg. deutsche Biblioth. B. xxiv. Anh. 4, Abth. p. 549. 1760
(*Poggendorff,* ii. 465, says : "*Ist als Erfinder d. Glosscheiben-Elektrisir-Maschine zu betrachten ; da er sich deren, schon* 1755, *bediente, viel früher also als Ingenhousz* (1764), *und Ramsden* (1766). *Siehe Allgem. Deutsche Bibliothek,* Anhang zu Bd. xiii.-xxiv. Abth. iv. s. 549; *Exlebens Naturlehre* 6 Aufl. s. 461; *Amsteins Biogr. von Planta im Neuen Sammler fur Bünden Jahrg.* 1808.")

Plantamour, Emile. *Born May* 24, 1815, *at Geneva.*

Résultats des Observations magnétiques faites à Genève dans les années 1842-43. 4to. 52 pp. 3 plates. (*Mém. lu à la Soc. de Phys. de Genève,* tom. x. 1843 or 1844.) Genève, 1844
Détermination télégraphique de la différence de longitude entre les observatoires de Genève et de Neuchatel. 4to. Genève, 1864

Planté, G. Recherches sur la Polarisation voltaïque. 8vo. 32 pp. 1 plate.
Paris, 1859

Plata, F. M. (di Trapani). Dissertatio de electricitate . . publice disputationi exposita. Sm. 4to. 74 pp. Panormi (*Palermo*), 1749

Plateau, J. A. F. De l'action qu' exerce sur une Aiguille aimantée un Barreau aimanté tournant dans un plan et parallèlement au-dessous de l'Aiguille. 8vo. (*Quetelet Correspond. Mathém.* vi. 1830.) Bruxelles, 1830

† Sur un Problème curieux .de Magnétisme. 4to. 37 pp. (*Mém. Acad. Rle. de Belgique,* tom. xxxiv.) Bruxelles, 1864

Plater, Felix. De meteoris in genere et speciatim de ignitis. (*From Poggendorff,* ii. 467.) 1640

Planmann, A. Diss. Hypotheses quasdam de causis electricitatis perstringens. Pt. ii. Abo, 1772

Platteretti. (*Vide* Calamini und Platteretti.)

Playfair, John. *Died July* 20, 1819, *at Edinburgh.* (*Phil. Mag.* liv. 78.)

† Outlines of Natural Philosophy. 2 vols. 8vo. (*From Poggendorff,* ii. 470.)
1812-16

† Article on Magnetism in Encyclop. Britannica. (*From Lamont. Handb.* p. 443.)

† Magnetising power of Violet Rays. 8vo. 1 p. (*Phil. Mag.* liii. 155.)
London, 1817

Pless u. Pierre. Beiträge zur Kenntniss des Ozons, &c., und des Ozongehaltes der atmosphärischen Luft. 8vo. 27 pp. (*Aus den Sitzungsberichten* 1856, *der K. Akad. der Wissenschaften.*) Wien, 1857

402 PLI—PLU

Plieninger. Über die Blitzableiter, ihre Vereinfachung und die Verminderung ihrer Kosten. Nebst einem Anhang über das Verhalten der Menchen bei Gewittern. Eine gemeinnfassliche Belehrung fur die Verfertiger der Blitzableiter, sowie fur die Hausbesitzer. Im Auftrage der k. Centralstelle des landwirthschaftlichen Vereins in Würtemberg verfasst. 8vo. 114 pp. 3 plates. 1835

Pliny, Caius Secundus. *Born* A.D. 23, *at Comum (Como), or Verona; died August* 25, 79, *at Vesuvius.*

Hist. Nat. lib. i. cap. 20; lib. xxviii. cap. 4. Tourmalin, lib. xxxvii. cap. 7. (*Vide Watson in Phil. Trans.* xlviii. part ii. p. 211.) A.D. 79

Hist. Mundi, lib. ii. cap. 4 (Death of Tullius Hostilius). (*From Becquerel's Hist.* p. 2.)

Plot, R. A Catalogue of Electrical bodies. 4to. (*Phil. Trans.* 1698, vol. xx. p. 384.) *London,* 1698

Plobsheim. (*Vide* Zorn von Plobsheim.)

Plücker, Julius. *Born July* 16, 1801, *at Elberfeld.*

Über d. Ohm'sche physikal. Gesetz. (*Crelle's Journ.* xxxv. 1847.) 1847

About twenty-five articles on Magnetism, Diamagnetism, &c., in Poggendorff's Annalen, lxxii. 1847 to vol. cx. 1860 inclusive.

† Ueber die Abstossung der optischen Axen der Krystalle durch die Pole der Magnete. 8vo. 29 pp. (*Poggend. Ann.* lxxii. 1847, p. 315.) *Leipzig,* 1847

† On the Repulsion of the Optic Axis of Crystals by the Poles of a Magnet (Translation). 8vo. 23 pp. 1 plate. (*Scientific Mems.* vol. v. pt. xix. p. 353. *Translated from Poggend. Ann.* vol. lxxii. No. 10, Oct. 1847.) *London,* 1847

† Ueber das Verhältniss zwischen Magnetismus und Diamagnetismus. 8vo. 8 pp. (*Poggend. Ann.* lxxii. 1847, p. 343.) *Leipzig,* 1847

† On the Relation of Magnetism to Diamagnetism. Translated by Dr. T. W. Griffith (*from Poggendorff's Annalen,* October, 1847. *With notes of the Translator*). 8vo. 7 pp. (*Taylor's Scientific Mems.* vol. v. part xix. p. 376.) *London,* 1847?

† Experimental-Untersuchungen über die Wirkung der Magnete auf gasförmige und tropfbare Flüssigkeiten. 8vo. 33 pp. 1 plate. (*Poggend. Ann.* lxxiii. p. 549.) *Leipzig,* 1848

† Ueber die verschiedene Zunahme der magnetischen Anziehung und Diamagnetischen Abstossung bei zunehmender Kraft des Elektromagneten. 8vo. 7 pp. (*Poggend. Ann.* lxxv. 1848, p. 413.) *Leipzig,* 1848

† Ueber ein einfaches Mittel, den Diamagnetismus schwingender Körper zu verstärken. Diamagnetische Polarität. 8vo. 6 pp. (*Poggend. Ann.* lxxiii. p. 613.) *Leipzig,* 1848

† Ueber Intensitäts Bestimmung der Magnetischen und Diamagnetischen Kräfte. 8vo. 57 pp. (*Poggend. Ann.* lxxiv. p. 321.) *Leipzig,* 1848

† Ueber das Verhalten des abgekühlten Glases zwischen den Magnetpolen. 8vo. 3 pp. (*Poggend. Ann.* lxxv. 1848, p. 108.) *Leipzig,* 1848

† Ueber das Gesetz, nach welchem der Magnetismus und Diamagnetismus von der Temperatur abhängig ist. 8vo. 13 pp. (*Poggend. Ann.* lxxv. 1848, p. 177.) *Leipzig,* 1848

† On the Magnetic Relations of the Positive and Negative Optic Axis of Crystals. In a letter to, and communicated by, Dr. Faraday. 8vo. 2 pp. (*Phil. Mag. for June,* 1849.) *London,* 1849

† Ueber die neue Wirkung des Magnets auf einige Krystalle, die eine vorherrschende Spaltungs-Fläche besitzen. Einfluss des Magnetismus auf Krystall-Bildung. 8vo. 11 pp. (*Poggend. Ann.* lxxvi. 1849, p. 576.) *Leipzig,* 1849

† Enumeratio novorum phænomenorum a se in doctrina de magnetismo inventor. 4to. 28 pp. 1 plate. (*Faraday's copy.*) *Bonnæ,* 1849

Plücker, Julius.—*continued.*

† Ueber das magnetische Verhalten der Gase. 8vo. 22 pp. (*Poggend. Ann.* lxxxiii. 1851, p. 87.) *Leipzig,* 1851

Numerische Vergleichung des Magnetismus des Sauerstoffgases und des Magnetismus des Eisens. 8vo. 7 pp. (*Poggend. Ann.* lxxxiii. 1851, p. 108.) *Leipzig,* 1851

Ueber das magnetische Verhalten der Gase weitze Mittheilung. 8vo. 20 pp. (*Poggend. Ann.* lxxxiv. p. 161.) *Leipzig,* 1851

† Ueber die Theorie des Diamagnetismus, die Erklärung des Ueberganges magnetischen Verhaltens in diamagnetisches und mathematische Begründung der bei Krystallen beobachteten Erscheinungen. 8vo. 34 pp. 1 plate. (*Pogg. Ann.* Bd. lxxxvi.) *Leipzig,* 1851

† Commentatio de crystallorum et gazorum conditione magnetica, qualis hodie intelligitur. 4to. 34 pp. *Bonnæ,* 1854

† On the Magnetic Induction of Crystals. Read April 23, 1857. 4to. 45 pp. (*Phil. Trans.* 1858.) *London,* 1858

Détermination du pouvoir magnétique et diamagnétique des corps. (*De la Rive, "Traité"* . . . tom. i. p. 579.)

On the Magnetic Induction of Crystals. 4to. 46 pp. (*Phil. Trans.* 1858.) *London,* 1859

† **Plucker, J. u. Beer.** Ueber die magnetischen Axen der Krystalle und ihre Beziehung zur Krystallform und zu den optischen Axen. 8vo. 48 pp. (*Poggend. Ann.* lxxx. 1850, p. 115.) *Leipzig,* 1850

Ueber die magnetischen Axen der Krystalle und ihre Beziehung zur Krystallform und den optischen Axen. 8vo. 33 pp. (*Pogg. Ann.* lxxxii. 1851, 42.) (*A Krystalle, deren optische Axen in der symmetrischen Ebene liegen. Fortsetzung, von* Bd. lxxxi. S. 162.) *Leipzig,* 1851

Plutarch. Vitæ. Numa's (Religious Ceremonies). (*Vide Watson, Phil. Trans.* xlviii. part i.) *Lysander,* A.D. 120

Plymouth Institution. Transactions of the. 8vo. *Plymouth,* 1830

Poëy, A. Sur les tempêtes électriques et la quantité de victimes que la foudre fait annuellement aux Etats-Unis d'Amérique et à l'île de Cuba. 8vo. *Versailles,* 1855

† Météorologie. Des caractères physiques des Eclairs en Boules et de leur affinité avec l'état sphéroidal de la matière. 8vo. 7 pp. (*Ext. du Journal La Science,* No. du 7 Juin.) *Paris,* 1855

† Analyse des hypothèses anciennes et modernes que ont eté émises sur les Eclairs sans tonnerre . . . accompagnée d'une description des Eclairs sans tonnerre observés sous diverses latitudes. La. 8vo. 64 pp. *Versailles,* 1856

† Couleurs des Etoiles et des Globes filants observés en Chine pendant 24 siècles depuis le vii. siècle avant Jésus-Christ jusqu'au milieu du xvii. siècle de notre ére. 4to. 3 pp. (*Ext. des Comptes Rendus de l'Acad.* tom. xliii. p. . , Séance du 15 Déc. 1856.) *Paris,* 1856

† Couleur des Etoiles et des Globes filants observés en Angleterre de 1841 à 1855. 4to. 3 pp. (*Ext. des Comptes Rendus de l'Acad.* tom. xliii. p. . . , Séance 29 Dec. 1856.) *Paris,* 1856

† Couleurs des Globes filants observés à Paris de 1841 à 1853, avec l'indication des trainées, des fragments, etc. diversement colorés observés tant en Chine qu'en Angleterre. 4to. 4 pp. (*Ext. des Comptes Rendus de l'Acad.* tom. xliv. p. . . , Séance du 12 Jan. 1857.) *Paris,* 1857

† Analyse des hypothèses . . . qui ont été émises sur les Tonnerres sans Eclairs par un ciel parfaitement serein ou dans le sein des nuages ; et Relation des Tonnerres sans Eclairs observés sous diverses latitudes en particulier à la Havane, &c. 4to. 29 pp. (*Ext. de l'Annuaire de la Soc. Météorologique de France,* tom. iv. p. 113, Séance du 11 Nov. 1856.) *Versailles,* 1857

Poey, A.—*continued.*

† Répartition géographique de l'universalité des Météores en zones terrestres, atmosphériques, solaires ou lunaires, et de leurs rapports entre elles. 8vo. 24 pp. *Paris*, 1858

Relacion de los Trabajos físicos y météorologicos hechos por A. Poëy tanto en la Habana como en Europa. 8vo. 40 pp. *Paris*, 1858

Description de deux magnifiques Aurores-boréales observées á la Havane. La. 8vo. 4 pp. (*Ext. de l'Annuaire de la Soc. Météores de France*, tom. vii. p. 189, Séance 8 Nov. 1859.) *Versailles*, 1859

† Sur les Eclairs sans tonnerre observés à la Havane pendant l'année 1859 dans le sein des cumulo stratus de l'horizon. La. 8vo. or 4to. 3 pp. (*Ext. de l'Annuaire de la Soc. Météorol. de France*, tom. viii. p. 75, Séance 12 Juin, 1860.) *Versailles*, 1860

† Travaux sur le Météorologie, la Physique du globe en général, et sur la Climatologie de l'île de Cuba et des Antilles ; publiés par A. Poëy. 8vo. 24 pp. *Versailles*, 1861

† Relation historique et théorie des images photo-électriques de la Foudre observées depuis l'an 360 de notre éra jusqu'en 1860. Deuxième édition, revue, &c. 12mo. 110 pp. *Paris*, 1861

† Sur la neutralité de la force électro-magnétique de la terre et de l'atmosphère observée à la Havane durant les Aurores-boréales de 1859. La. 8vo or 4to. 16 pp. (*Ext. de l'Annuaire de la Soc. Météorol. de France*, tom. ix. p. 42, Séance 19 Fév. 1861.) *Versailles*, 1861

† Bibliographie cyclonique. Catalogue comprenant 1,008 ouvrages, brochures, et écrits qui ont paru, jusqu'à ce jour, sur les ouragans, et les tempêtes cycloniques. 2e édition, corrigée et considérablement augmentée. 8vo. 96 pp. *Paris*, 1866

Etoiles filantes, observées à la Havane du 24 Juillet au 12 Août (1863), et remarques sur le retour périodique du mois d'Août. 4to.

Poggendorff, Johann Christian. *Born December* 29, 1796, *at Hamburg.*

Physisch-chemische Untersuchungen z. näheren Kenntniss d. Magnetismus d. Volta'schen Säule. (*Oken's Isis*, 1821. *Darin sein Elektro-magnet Multiplicators, &c.*) 1821

Annalen der Physik und Chemie, herausgeben von Poggendorff. 8vo. *Leipzig*, (*commenced*) 1824

Über einige Magnetisirungs Erscheinungen. 8vo. 55 pp. 1 plate. (*From Weber Cat. Berlin* (1870), p. 42, and T. p. in *Pogg. Annalen*, 1838, Bd. xlv. p. 353.) *Leipzig*, 1838

Gedächtnissrede auf T. J. Seebeck. 4to. (*Abh. d. Berlin Acad. f.* 1839.) *Berlin*, 1839

Twelve Articles (about) on Magnetism and Electricity in his "Annalen" from 1824, vol. i. to 1845, vol. lxvi. together with many remarks, extracts, small notices, &c. in the same period, some of which may relate to the same subjects.

Various Articles on Electricity, &c. 8vo. (*Handwörterbuch d. reinen. u. angewandten Chemie.*) 1846

† Über d. Verhalten d. Quecksilbers bei seiner electromagnetischen Rotation. 8vo. 10 pp. (*Monatsberichte d. Berlin Acad.* 1848.) *Berlin*, 1848

† Lebenslinien zur Geschichte der exacten Wissenschaften seit Wiederherstellung derselben. 4to. 14 pp. 3 plates. *Berlin*, 1853

Beiträge zur Kenntniss der Inductions-Apparate und ihrer Wirkungen. 8vo. (*Pogg. Ann.* 1854.) *Leipzig*, 1854

Beobachtungen über Inductions-Electricität. 8vo. (*Pogg. Ann.* 1855.) *Leipzig*, 1855

Poggendorff, Johann Christian.— *continued.*

Neue Verstarkungsweisen d. Inductionsstroms. 8vo. (*Pogg. Ann.* 1855.)
Leipzig, 1855

† Über die galvanischen Ketten aus zwei Flüssigkeiten und zwei einander nicht berührenden Metallen. 8vo. 42 pp. (*Pogg. Ann.* xlix.) *Leipzig*

Thirty-seven Articles (about) on Magnetism and Electricity in the "Monatsberichte d. Berlin Acad." from 1839 to 1856.

† Biographisch-Literarisches Handwörterbuch .zur Geschichte der exacten Wissenschaften, enthaltend Nachweisungen über Lebensverhältnisse und Leistungen von Mathematikern, Astronomen, Physikern, Chemikern, Mineralogen, Geologen, u.s.w. aller Völker und Zeitern. 2 vols. 4to.
Leipzig, 1863

Von dem Gebrauch der Galvanometer als Messwerkzeuge. 8vo.

Poggi, F. L'elettricità ne' suoi rapporti fisici, chimici, meccanici alla portata di tutti, esposizione de' principali fenomeni e delle più recenti applicazioni. 8vo. 100 pp. 2 plates. *Modena,* 1856

† **Poggioli, M. P.** Nouvelle application de l'Electricité par frottement sans commotion. 8vo. 15 pp. (*Mémoire lu à l'Institut le* 31 Oct. 1853.)
Paris, 1853

Pohl, Georg Friedrich. *Born February* 24, 1788, *at Stettin ; died June* 10, 1849, *at Breslau.*

Über d. Zusammenhang d. Magnetismus mit d. Elektricität u.d. Chemismus. 8vo. (*Gilb. Ann.* lxix. 1821, und lxxi.) *Leipzig,* 1821

Beitr. z. näheren Kenntn. d. Elektromagnetismus. (*Oken's Isis,* 1822.) 1822

Versuche über d. Einwirk d. Erdmagnetismus auf bewegliche Elektromagnete, u.s.w. 8vo. (*Gilb. Ann.* lxxiv. u. lxxv. 1823.) *Leipzig,* 1823

Über d. polare Thätigkeit d. flüssig Leiters in d. galvan. Kette. 8vo. (*Kastner's Archiv.* ii. u. iii. 1824.) *Nürnburg,* 1824

Über d. Phänomena d. elektr. Ladung. 8vo. (*Kastner's Archiv.* vi. 1825.)
Nürnberg, 1825

† Der Process der Galvanischen Kette. 8vo. 430 pp. *Leipzig,* 1826

Zur Lehre vom Elektromagnetismus. 8vo. (*Kastner's Archiv.* ix. 1826 u. xi. 1827.) *Nürnberg,* 1826-7

Eight Articles on Galvanic Electricity, Electro-magnetism, &c. in Poggendorff's "Annalen," vol. viii. 1826 to vol. liv. 1841.

Der Gyrotrop u. d. Siderophor als zur Anstell. galvan. Versuche besonders geeignete Apparate. 8vo. (*Kastner's Archiv.* xiii. xiv. 1828.)
Nürnberg, 1828

† Ansichten und Ergebnisz über Magnetismus, Elektricität und Chemie. Ein Bericht an das grössere naturwissenschaftliche Publicum. 8vo. 88 pp. (*Auszugsweise vorgetragen in der deutschen Naturforscher und Aerzte zu Berlin.*)
Berlin, 1829

† Der Elektromagnetismus theoretisch-praktisch dargestellt. 1 Theil. 8vo. 292 pp. 3 plates. *Berlin,* 1830

Übers. der Verhandlungen der schlesischen Gesellschaft für vaterländische Kultur, p. 85. 1841
(*This appears under the head "Wirkung der Erde auf den Schliessungsdraht."*)

Grundlegung der drei Kepler'schen Gesetze. *Breslau,* 1843

† Der Elektromagnetismus u. die Bewegung d. Himmelskörper in ihrer gegenseit Beziehung dargelegt. 8vo. 95 pp. *Breslau,* 1846

Über das Wesen der Elektricität und Schwere. Offener Brief an Dove. 8vo.
Breslau, 1848

Pohl, Joseph. Tentamen physico experimentale in principiis peripateticis, fundatum super phænomenis electricitatis. 8vo. *Pragæ*, 1747

Tentamen . . . phænom. electricitatis. Editio alter. 8vo. *Pragæ*, 1750

Pohl. (*Vide* Lohmeier and Pohl.)

Poisson, Siméon Denis. *Born June* 21, 1781, *at Pithiviers, Dép. Loiret ; died April* 25, 1840, *at Paris.*

Observations sur les substances minérales que l'on suppose tombées du ciel sur la terre. 4to. (*Société Philomathique*, 11, p. 180, 1803.) *Paris*, 1803

Sur la distribution de l'electricité à la surface des corps conducteurs. Premier et second Méms. 4to. (*Mém. de l'Institut.* 1811.) *Paris*, 1811

Mémoire sur la distribution de l'Electricité dans une sphère creuse électrisée par influence. 4to. (*Bulletin de la Soc. Philomat.* p. 49.) *Paris*, 1824

† Mém. sur la théorie du Magnétisme. 4to. 31 pp. (*Mém. Acad. Rle. des Sciences, Paris*, tom. v. p. 247, lu 2 Fév. 1824.) *Paris*, 1824

† Second Mémoire sur la théorie du Magnétisme. 4to. 45 pp. (*Mém. Acad. Rle. des Sciences, Paris*, tom. v. p. 488, lu le 27 Dec. 1824.) *Paris*, 1825

Solution d'un problème relatif au Magnétisme terrestre ; avec un Préambule. 4to. (*Bulletin de la Soc. Philomat.* 1825, p. 82, and 1826, p. 19.) *Paris*, 1825-6

† Mém. sur la théorie du Magnétisme en Mouvement. Lu le 10 Juillet, 1826. 4to. 130 pp. (*Mém. de l'Acad. des Sciences*, vi. 441.) *Paris*, 1826

Mém. sur les déviations de la boussole produites par le fer des vaisseaux. 4to. (*Paris Acad.* 1840.) *Paris*, 1840

† Mém. sur les déviations de la Boussole produite par le fer des vaisseaux. Lu 4 Juin, 1838. 8vo. 66 pp. (*Mém. de l'Acad.* tom. xvi. *Ext. des Connaiss. des Temps.*) *Paris*, 1838 ?

† Cat. de ses ouvrages et mém. scientifiques. 4to. 28 pp. *Paris*, 1815

† **Poisson, Gay-Lussac,** &c. Instruction sur les Paratonnerres, adoptée par l'Acad. 23 April, 1823. Publié par ordre du Ministre de l'Intérieur. 8vo. 51 pp. 2 plates. *Paris*, 1824

† Anleitung zur Verfertigung und Benutzung der Blitzableiter. 8vo. 43 pp. 2 plates. (*Translation, without name of Translator.*) *Strasbourg*, 1824

Poissonnier Pierre Isaac. *Born July* 5, 1720, *at Dijon ; died September* 15, 1798, *at Paris.*

† **Poissonnier** et Cæteri. Rapport des Commissaires de la Société Royale de Médecine nommés par le Roi pour faire l'examen du Magnétisme animal. 4to. 39 pp.
(*Note.—Poissonnier, Caille, Mauduit, and Andry. They refer to Paracelsus, Van Helmont, Goclenus, Maxwell, &c.,* p. 3.) *Paris*, 1784

Polansky, Thaddäus. *Born March* 13, 1713, *at Hradisch Mähren ; died October* 12, 1770, *at Olmütz.*

† Dissertatio physico-experimentalis . . de tonitruo, fulgore, seu corruscatione contra sensa et opiniones antiperipateticorum. 12mo. 156 pp. 1 plate. *Olonuucii*, 1747

† **Polcastro,** G. B. Notizia sopra la conversione dal . . Pacchiani, dell' acqua in acido muriatico ossigenato, per mezzo del fluido galvanico o idrometallico . . alli Redattori . . 8vo. 8 pp. (*Giorn. Ital. Letter. del Dal Rio*, x. 182.) *Padova*, 1805

Poleni, G. (il Marchese). *Born August* 23, 1683, *at Venice ; died December* 22, 1761, *at Padua.*

Dissertatio de barometris, &c. (die erste in dessen *Miscellaneis*.) 4to. *Venetia*, 1709

† **Poleni, J.** Observatio Auroræ borealis visæ nocte insequente diem 16 Decembris, anno 1737, habita Patavii ab T. P. . . 4to. 8 pp. *Patavii*, 1738

† Observatio Auroræ borealis, visæ . . 16 Decembris, anno 1737, habita Patavii. 12mo. 13 pp. (*Calogera, Raccolta d' Opuscoli*, vol. xvii. p. 1.) (*See above.*)
Venezia, 1738
Observatio Auroræ borealis visæ nocte insequente diem 27 Martii, 1739, habita Patavii. 4to. (*Commentar. Acad. Petropolitanæ*, viii. 440.) *Petropol*

† **Polettini, L.** Le meraviglie della Telegrafia elettrica, ossia Telegrafo elettromagnetico Americano . . Opera di Alf. Vail . . tradotta dall' Inglese al Francese da Ipp. Valtemare e voltata in Italiano de L. Polettini : con Note, &c. La. 8vo. or 4to. 169 pp. 5 plates (in 2 parts.) *Verona*, 1850

Poli, Giuseppe Saverio. *Born October* 24, 1746, *at Molfetta ; died April* 7, 1825, *at Naples.*

† La formazione del Tuono, della Folgore, &c. . . 8vo. 152 pp. *Napoli*, 1772

† Riflessioni intorno agli effetti di alcuni fulmini. 8vo. 148 pp. *Napoli*, 1773
(*Note.—He says that it may be considered as an Appendix to his "La formazione del Tuono."*)

† Congetture sulle tempeste, che sogliono succedere alle aurore boreale. 4to. 5 pp. (*Opusc. Scelti*, i. 191.) *Milano*, 1778

Sopra il Tuono. 8vo. 1779

† Lettera al G. Vivenzio su di una straordinaria Aurora boreali. Londra a 5 Luglio, 1779. 4to. 5 pp. (*Opus. Scelti*, ii. 382.) *Milano*, 1779

Osservazioni fisiche concernenti l'elettricità, il magnetismo e il folgore. 4to. (*Atti dell' Accad. di Napoli*, 1787, pp. 169-195, *on Velocity, pr. in* 1788, 10*th Mem.*) *Napoli*, 1787

Elementi de Fisica. 5 vols. 8vo. *Napoli*, 1802
(*Note.—There is a Venetian edition of* 1819, *and Naples edition of* 1787.)

Saggio sulla calamita e sulle sue virtù medicinali. *Palermo*, 1811

Elementi di Fisica sperimentale. 5 vols. 8vo. *Venezia*, 1824
(*Note.—There are other editions, Naples, &c.*)

Polidori, Luigi Eustachio. *Born at Bientina, near Pisa ; died May* 29, 1830, *at Pisa.*

† Dissertazione sul terremoto. 8vo. 24 pp. (*Ann. di Chim. di Brugnatelli*, v. 30.) *Pavia*, 1794

Poligrafo, Il. Giornale di Scienze, Lettere ed Arti. About 60 vols. 8vo.
Verona, 1830-1847
(*Note.—Many of the Electrical Articles are in the Library.*)

Polinière, Pierre. *Born September* 8, 1671, *at Coulonces, near Vire (Calvados) ; died February* 9, 1734, *at Paris.*

Expériences de Physique. 1st edition. 12mo. 508 pp. 10 plates. *Paris*, 1709

† Expériences de Physique. 4me édition, revue, corrigée et augmentée par l'Auteur. 2 vols. 12mo. *Paris*, 1734

Expériences de Physique. 5me édition, revue, corrigée et augmentée sur les MSS. de l'Auteur. 5th edition. 2 vols. 12mo. *Paris*, 1741
(*Note.—Magnetism*, tom. i. p. 328. *Electricity*, tom. ii. p. 1. *The* 1st *edition is of* 1709 ; 2*nd*, 1718 ; 3*rd* (*last in his life*), 172*ᵔ*. 4*th, revue, &c. par l'Auteur*, 1734.)

† **Polli, G. ed altri.** Esame di un opuscolo di Linati e Coggiati sopra gli effetti dell' elettricità sulle funzioni del gran simpatico. Osservazioni relative di Porta e Magrini. 4to. 2 pp. (*Atti dell' I. R. Istit. Lomb.* i. 294.)
Milano, 1858

Polli. (*Vide* Magrini, L. 1858).

Pollini. Sul passaggio del Fulmine che nel . . 6 Agosto, 1795 . . . scoppiò nel . . Tempio di S. Andrea in Vercelli, e sugli effetti . . 8vo. *Vercelli*, 1796

† **Pollock, T.** An attempt to explain the Phenomena of Heat, Electricity, Galvanism, Magnetism, &c. . . . 8vo. 158 pp. Cuts. *London,* 1832

Pollock. (*Vide* Mackrell, Gann, and Pollock.)

Polo, Marco. Viaggi illustrati e commentati dal Conte B. Baldelli. 4 vols. 4to. with fol. atlas. *Firenze,* 1827
(*There are many other versions and in various languages.*)

† **Polytechnische Bibliothek.** Monatliches Verzeichniss der in Deutschland und dem Ausland neu erschienen en Werke aus den Fächern der Mathematik u. Astronomie, der Physik u. Chemie u.s.w. Mit Inhaltsangebe der wichtigsten Fachzeitschriften. Jeden Monat ersch. eine Nummer. 1866-9

Poma. (*Vide* Porna.)

Poncelet. *Born at Verdun.*

† La nature dans la formation du Tonnerre, &c. 8vo. 172 pp. and 148 pp. 3 plates. *Paris,* 1766

Pongracz, Anton L. B. (*Vide* Bauer, Fulgentius.)

Ponti, A. Cenni storico-critici intorno all' identità di origine delle comete e delle stelle cadenti. 16mo. 21 pp. *Pisa*

† Riassunto generale di Telegrafia elettrica. 8vo. 358 pp. 4 plates. *Milano,* 1854

† **Ponton d'Amécourt.** Exposé du Galvanisme. 8vo. 46 pp. *Paris,* 1803

Pope, F. L. Modern Practice of the Electric Telegraph. 8vo. 1870

† **Pope, W.** The Triumphal Chariot of Friction ; or, a familiar elucidation of the origin of Magnetic Attraction, &c. 4to. 106 pp. 11 plates. *London,* 1829

Poppe, Adolph. *Born January* 11, 1814, *at Frankfurt-a.-M.*

Die Telegraphie von ihrem Ursprunge bis zur neuesten Zeit. 8vo. *Frankfurt-a.-M.* 1848

Chronologische Üebersicht der Erfindungen und Entdeckungen auf dem Gebiete der Physik, Chemie, Astronomie, Mechanik und industriellen Technik von den ältesten Zeiten bis auf unsere Tage. 8vo. 68 pp. *Frankfurt,* 1856

Poppe, Joh. H. M. Gewitterbüchlein zum Schutz und zur Sicherheit gegen d. Gefahren der Gewitter, besond. auch üb. d. Kunst, Blitzableiter auf d. beste Art anzulegen. 8vo. *Tubingen,* 1830

† Die Telegraphen und Eisenbahnen im ganzen Umfange ; ihr Nutzen, ihr verschied. Arten, u. die damit bis auf d. neueste Zeit vorgenomm. neuen Einrichtungen u. Verbessern. 12mo. *Stuttgart,* 1834

Popular Science Review. A Quarterly Miscellany of entertaining and instructive articles on scientific subjects. Edited by James Samuelson. 8vo. *London,* 1862 to —

Porna (or Poma) et **Arnaud.** Medical Electricity? (*Journ. de Méd. de Vandermonde,* 1787.) *Nancy(?),* 1787

Porret, R. On two curious Galvanic Experiments. 8vo. (Voltaic Endosmose, or " Electro-chemical Filtration.") (*Ann. of Phil.* viii. 1816.) *London,* 1816

Porta, Baptista. Magia naturalis, sive de miraculis rerum naturalium, lib. iv. Fol. *Neapolis,* 1558

Posner, Joh. Kaspar. De fulmine Camburgensi. 4to. 1701

† **Postel,** E. Grundzüge der elektrischen Telegraphie. Für angehende Telegraphen Beamte und . . . für Lehrer und reifere Schüler bearbeitet. 8vo. 191 pp. *Langensalza,* 1868

Pott, Johann Heinrich. *Born* 1692, *at Halberstadt ; died March* 29, 1777, *at Berlin.*
Dissertation chimique de Pott, trad. du Latin et Allemand par Damechy. 4 vols. 12mo. 1759

Potter, Richard. *Born June 2, 1799, at Manchester.*

On Difficulties in the Application of some Metals over others. (*Majocchi's Annali, from the Society of Arts ; read April,* 1843.) 1843

Potzsch, C. G. Kurze Darstellung d. Gesch. üb. d. Vorkommen d. gedieg. Eisens, sowohl d. mineral, als auch d. problematisch-meteorischen. 8vo. *Dresden,* 1804

Pouillet, Claude Servais Mathias. *Born February* 16, 1790, *at Cusance, near Baume-les-Dames* (*Doubs*) ; *died at Paris,* 1868.

Sur les phénomènes électro-magnétiques. 8vo. (*Ann. Chim. et Phys.* xxi. p. 77.) *Paris,* 1822

Note sur les phénomènes électro-magnétiques qui se manifestent dans l'acupuncture. (*Jour. de Physiol. Exper.* v. i.) *Paris,* 1825

† Mémoire sur l'électricité des fluides élastiques et sur une des causes de l'électricité de l'atmosphère. Lu à l'Acad. . . le 30 Mai, 1825. 8vo. 20 pp. (*Ext. Ann. de Chimie.*) *Paris,* 1825

† Deuxième mémoire sur l'Electricité qui se développe dans les actions chimiques ; et sur l'origine de l'Electricité de l'atmosphère. Lu à l'Acad. . . le 4 Juil. 1825. 8vo. 15 pp. (*Ext. des Ann. de Chim. et Phys.*) *Paris,* 1825

Eléments de Physique Expérimentale et de Météorologie. 1st edition. 8vo. (The 6th edition in 1853, 2 vols. ; the 7th in 1856.) *Paris,* 1827

† Eléments de Physique Expérimentale et de Météorologie. Tom. i. part ii. 8vo. 31 plates. *Paris,* 1828

† On Electricity of Gases and the Atmosphere. 8vo. 1 p. (*Phil. Mag. or Annals,* iii. 148.) *London,* 1828

† Conducting Power of Metals for Electricity. 8vo. 1 p. (*Phil. Mag. or Annals,* iv. 382.) *London,* 1828

† Eléments de Physique Expérimentale et de Météorologie. Tom. ii. part iii. 8vo. *Paris,* 1829

† Eléments de Physique Expérimentale et de Météorologie. Tom. iii. part iv. 8vo. (*From Ferussac,* xiv. p. 388.) *Paris,* 1830

Eléments de Physique Expérimentale et de Météorologie. 41 planches, 4 vols. 8vo. 1832

Mémoire sur la pile de Volta, et sur la loi générale d'intensité que suivent les courants, soit qu'ils proviennent d'un seul élément, soit qu'ils proviennent d'une pile à petite ou à grande tension. 4to. (*Comptes Rendus,* iv. 267.) *Paris,* 1837

Détermination des basses températures au moyen du pyromètre magnétique et du thermomètre à l'alcohol. 8vo. (*Ann. de Chim. et Phys.* iv. 1837 ; *Comptes Rendus,* iv. 513.) *London,* 1837

Mémoire sur la mesure relative des sources thermo-électriques et hydro-électriques, &c. 8vo. (*Ann. de Chim. et Phys.* iv. 1837 ; *and in Comptes Rendus,* iv. 785.) *Paris,* 1837

Eléments de Physique Expérimentale et de Météorologie. 64 plates. 8vo. 1840

Sur un moyen de mesurer des intervalles de temps extrêmement courts . . etc., et sur un nouveau moyen de comparer les intensités des courants électriques, soit permanents, soit instantanés. 8vo. (*Ann. de Chim. et Phys.* xix. 1844, *and Comptes Rendus,* xix. 1384.) *Paris,* 1844

Note sur l'électro-chimie. 8vo. (*Ann. de Chim. et Phys.* xx. 1845, *and Comptes Rendus,* xx. 1544.) *Paris,* 1845

Elementi di fisica sperimentale e di meteorologia. Quarta edizione voltata in italiano con note e giunte di L. Palmieri e di M. Melloni. *Napoli,* 1846

† Note sur les nouvelles expériences de M. Faraday. 4to. 12 pp. (*Comptes Rendus,* 1846, 1er Semestre, xxii. No. 4, p. 135.) *Paris,* 1846

Pouillet, Claude Servais Mathias.—*continued.*

† Eléments de Physique expérimentale et de Météorologie. 2 vols. plates. 8vo.
The 6th edition in 1853 ; 7th in 1856.   *Paris*, 1847

Note historique sur divers phénomènes d'attraction, de répulsion, et de déviation
qui ont été attribués à des causes singulières, etc. 8vo. (*Ann. de Chim. et
Phys.* xxix. 1849.)   *Paris*, 1849

Rapport sur les Mémoires relatifs aux phénomènes électro-physiologiques
presentés à l'Acad. par M. E. de Bois-Reymond. 4to. (*Comptes Rendus*,
xxxi. 28.)   *Paris*, 1850

† Rapport fait à l'Acad. . . . sur les appareils télégraphiques de M. Siemens
(de Berlin), Regnault, Seguin, et Pouillet rapporteur. 8vo. 16 pp.
(*Edited, with advertisement, for " la Science mise à portée de toutes les intelli-
gences."*)   *Paris*, 1850

Rapport sur les appareils de télégraphie électrique de M. Siemens, de Berlin.
4to. (*Comptes Rendus*, xxx. 500.)   *Paris*, 1850

Supplément à l'Instruction sur les Paratonnerres, presenté par la section de
physique. MM. Becquerel, Babinet, Duhamel, Despretz, Cagniard de La-
tour, Pouillet rapporteur. Séance 18 Déc. 1854. 4to. (*Ext. Comptes Rendus*,
tom. xxxix. 112, and xl.)   *Paris*, 1854-5

Eléments de Physique Expérimentale et de Météorologie. 7e édition ; 2 vols.
in 8vo. et atlas.   *Paris*, 1856

(*Note.—There is a German translation by Müller.*)

† Instructions sur les Paratonnerres des magasins à poudre. Rapport lu 14
Janv. Commissaires Becquerel, Babinet, Duhamel, Vaillant, Pouillet. 4to.
1 plate. 15 pp. (*Ext. des Comptes Rendus*, tom. lxiv. Séance 21 Janv.
1867.)   *Paris*, 1867

(*Vide* also Gay-Lussac and Pouillet.)

Pouillet. (*Vide* Müller-Pouillet.)

† **Poulsen,** C. M. Die Contact-Theorie, vertheidigt gegen Faraday's Abhandlung
über die Quelle der Kraft in der Voltaischen Säule . . 8vo. 52 pp.
Heidelberg, 1845

Poupar. Compte rendu des trav. pend. le 2nd semestre de 1820. 8vo.
Lyon, 1827

† **Pouriau,** A. Observations météorologiques faites à l'Ecole Impériale d'Agriculture de la Saulsaie, Ain, sous la direction de M. A. Pouriau. Année
1854-5. 4to. 151 pp. 1 table. 2 plates.   *Lyon*, 1855

† Etudes sur l'Ozone. 4to. 3 plates. 30 pp. 3 plates. 1 table, 2 plates. *Lyon*, 1855

† **Pouriau,** A. F. Les appareils météorologiques enregistreurs à l'Exposition Universelle de 1867. 8vo. 7 plates.   *Paris*, 1868

(*Note.—This Brochure is composed of several distinct writings.*)

Powell, Baden. On the Communication of Magnetism to Iron in different positions. 8vo. (*Phil. Mag. or Annals*, iii. 1822.)   *London*, 1822

Catalogue of Observations of Luminous Meteors. 6 parts. 8vo. 252 pp.
London, 1850-6

Power, Henry. *Died* 1673.

Experimental Philosophy. . . . (New Magnetical experiments.) 4to. (*From
Watt.*)   *London*, 1664

† **Pownall.** On the Ether suggested by Sir Isaac Newton, compared with the supposed newly-discovered principle of Galvanism. 8vo. 4 pp. (*Phil. Mag.*
xviii. 155.)   *London*, 1804

Pozzi, Cesare Giuseppe. *Died* 1838, *at Milan.*

Lettera sulla turmalina. 12mo.

Prætorius (Richter) Joh. De cometis qui antea visi sunt, et de eo qui novissime
mense Novembri apparuit, etc. 4to.   *Norimberg*, 1579

Prag. Privat Gesellschaft. Abhandl. zur Aufnahme der Mathematik der vaterländischen Geschichte und der Naturgeschichte. Zum Druck befördert von T. Edlen von Born. 8 vols. 8vo. *Prag, 1775-84*

Abhandlungen der königl. böhmischen Gesellschaft der Wissenschaften. 4 vols. 4to. *Prag and Dresden, 1784-9*

Neuere Abhandl. 3 vols. 4to. *Wien and Prag, 1791-8*

Abhandlungen . . . 8vo. 8 vols. *Prag, 1802-23*

Abhandlungen. Neue Folge. 5 vols *Prag, 1824-37*

Beobachtungen magnetische und meteorologische, auf der k.k. Sternwarte zu Prag im Jahre 1868, 29 Jahrg. *Prag, 1869*

Prandi. Sui movimenti del Mercurio. *Bologna, 1826*

Pré. (*Vide* Du Pré.)

Prechtl, Johann Joseph. *Born November 16, 1778, at Bischofsheim of the Rhône, Franken; died October 24, 1854, at Vienna.*

Seven Articles in Gehlen's Journ f. Chem. Phys. u. Min. vol. v. to vol. vii. 1808

Teoria delle meteore elettriche. 8vo. (*Gehlen's Journal,* viii. 1809.) 1809

Six Articles on Electricity, Magnetism, &c. in Gilbert's Annalen, from vol. xxxv. 1810 to vol. lxviii. 1821. (*From Poggendorff,* ii. 520.) 1810-21

Bemerk. zu Configliachi's Prüf. seiner Theorie d. elektr. Meteore. 8vo. (*Schweigg. Journ.* iv. 1812.) *Nürnberg,* 1812

Jahrbücher des k. k. polytechn. Instituts zu Wien. 20 Bde. 8vo. (*From Poggendorff,* ii. 519.) *Wien,* 1819-39

Über d. Transversalmagnetismus, u.s.w. 8vo. (*Schweigg. Jour.* xxxvi. 1822.) *Nürnberg,* 1822

33 Aufsätze in d. Jahrbüchern d. polytechn. Instituts. Vols. iv. 1823 (to qu.) (*Darunter Versuche über d. Bezieh. d. Adharenz d. Metalle zu ihrer elektr. Differenz, u.s.w.* 1829, vol. xiv.) 1823 to —

Erläutr. z. Theorie d. Transversalmagnetismus im galvan. Schliessungsdraht. 8vo. (*In Kastner's Archiv.* ii. 1824.) 1824

Technolog. Encyklopädie, u.s.w. 20 Bde. (*Darin über 90 Artikel von ihm, d. Ency. fortgesetzt von Karmarsch.*) *Stuttgart,* 1830-55

† **Preibsch,** C. Über Blitzstrahlableiter. 32 pp. 1 plate. *Leipzig,* 1825

Über Blizstrahlableiter, deren Nutzbarkeit und Anlegung . . . 2nd edition enlarged &c. (3 B.) 8vo. 46 pp. *Leipzig,* 1830

† **Premoli,** C. P. Nova electricitatis theoria. 8vo. 90 pp. 1 plate. *Mediol.* 1755

Prescott, G. B. History, theory, and practice of the Electric Telegraph. 8vo. (Second edition, 1860 ?) *Boston (U.S.),* 1859

† History, theory, and practice of the Electric Telegraph. 3rd edition, enlarged. 8vo. 508 pp. *Boston,* 1866

Prestel, M. A. F. Die Gewitter des Jahres 1855. Ein Beitrag zur Physiologie der Atmosphäre. 8vo. *Emden,* 1856

† Die geographische Verbreitung der Gewitter in Mittel-Europa im Jahre 1856, so wie über die gegenseitige Beziehung zwischen dem Auftreten der Gewitter, der Temperatur, der Windrichtung und dem Barometerstande. 8vo. 27 pp. 3 plates. (*Aus dem Sitzungsberichten 1858 der mathem.-naturviss. Classe der kais. Akad. der Wissenschaften.*) *Wien,* 1858

Der Sturmwarner und Wetteranzeiger, ein nach wissenschaftlichen Grundsätzen ausgeführtes und durch Beobachtungen und Erfahrungen bewährtes Instrument zur Vorherbestimmung von Sturm und Wetter. 8vo. 55 pp. 3 plates. *Emden,* 1870

Prevost, Jean Louis. Sur l'électricité des muscles. Avec Dumas. (*From Poggendorff,* ii. 527.) 1823

† Sur quelques expériences relatives à l'électricité animale. Lettre à De la Rive. 8vo. 7 pages. (*Extrait des Archives de l'Electricité.*) *Genève,* 1843

Prevost, Jean Louis.—*continued.*

Note sur le développement d'un courant électrique qui accompagne la contraction de la fibre musculaire. 8vo. 5 pp. 1 plate. (*Bibl. Univ. de Genève,* Novembre, 1837.) *Genève,* 1837

Prevost, Pierre. *Born March 3, 1751, at Geneva ; died April 8, 1839, at Geneva.*

† De l'origine des forces magnétiques. 8vo. 228 pp. 2 plates.
Genf (Geneva), 1788

† Vom Ursprung der magnetischen Kraft. aus dem Französischen übers. von D. L. Bourguet mit einer Vorrede von Gren. 8vo. 172 pp. 2 plates. (*Volta's copy.*) *Halle,* 1794

Examen d'une difficulté relative à la théorie de l'Electricité. (*From Poggendorff,* ii. 524. *In Magas. Encycl.* vi. 1796.) 1796

Sur l'époque de la chute d'une pierre. 4to. (*From Poggendorff,* ii. 524. *In Journ. Phys.* lvi. 1803.) *Paris,* 1803

Obs. d'un météore lumineux. 4to. (*Poggendorff,* ii. 524. *In Journ. Phys.* lix. 1804.) *Paris,* 1804

Sur le météore du 15 Mars 1811. 8vo. (*Poggendorff,* ii. 525. *In Bibl. Britann.* 1811.) *Genève,* 1811

Tentative faite dans le but de concilier deux principes fondamentaux de la théorie de l'électricité. 8vo. (*Poggendorff,* ii. 525. *In Bibl. Univ.* 1822.) *Genève,* 1822

Sur l'influence magnétique du soleil. 8vo. (*Poggendorff,* ii. 525. *In Bibl. Univ.* 1826.) *Genève,* 1826

Sur les bolides. 8vo. (*Poggendorff,* ii. 525. *In Ann. Chim. Phys.* xliii. 1830.) *Paris,* 1830

Prevot. Traité du Magnétisme. (*From Exleben,* p. 519.)

Preysinger. Die elektrogalv. Telegraphen. (*Lempertz Cat.* lxxxiv. (1867), p. 122.) *Augsburg,* 1850

Price, James. *Born 1752 ; died August 3, 1783, at London.*

† An Essay on the medical application of Electricity and Galvanism. 8vo. 142 pp. *London,* 1821

Priestley, Joseph. *Born March 13, 1733, at Fieldhead, near Leeds, Yorkshire ; died February 6, 1804, at Northumberland, Pennsylvania.*

† History and present state of Electricity, with original Experiments. 1st ed. 4to. 736 pp. 7 plates. *London,* 1767

† A familiar Introduction to the Study of Electricity. 1st ed. 4to. 51 pp. 4 plates. *London,* 1768

† History and present state of Electricity, with original Experiments, with Additions. 2nd ed. 4to. 712 pp. Index, &c. 8 plates. *London,* 1769

An Introduction to the Study of Electricity. 2nd ed. 8vo. 86 pp. 5 plates. *London,* 1769

Additions to the First Edition of History of Electricity. 4to. *London,* 1770

† Histoire de l'électricité ; traduit de l'Anglois avec des Notes critiques. 3 vols. 12mo. *Paris,* 1771

† Geschichte des gegenwärtigen Zustandes der Elektricitat, nebst eigenthümlichen Versuchen ; nach der zweyten Ausgabe. a. d. Engl. übersetzt von J. G. Krunitz. 4to. 517 pp. 8 plates. *Berlin,* 1772

History of Vision, &c. p. 371. 2 vols. 4to. (*Canton's Composition for Electric Light.*) *London,* 1772

History of the present state of Discoveries on Vision, Light and Colours. 4to. (*From Heath's Cat.* 1866, No. 1, p. 39.) 1772

† An account of a new Electrometer contrived by Mr. Wm. Henley, and of several Electrical Experiments made by him ; in a letter from Dr. Priestley to Dr. Franklin. 4to. 8 pp. 1 plate. (*Phil. Trans.* lxii. 360.) *London,* 1773

Priestley, Joseph.—*continued.*

† History and present state of Electricity, with original Experiments, corrected and enlarged. 3rd ed. 2 vols. 8vo. *London,* 1775

† History and present state of Electricity, with original Experiments, corrected and enlarged. 4th ed. 4to. 691 pp. 8 plates. *London,* 1775

A familiar Introduction to the Study of Electricity. 3rd ed. 8vo.
London, 1777

† Lettera al Volta. 8vo. 2 pp. (*Scelta d'Opuscoli,* xxxii. 107.) *Milano,* 1777

Experiments and Observations relating to various branches of Natural Philosophy, with a continuation of the Observations on Air. 8vo. 3 vols.
London, 1779-86

† A familiar Introduction to the Study of Electricity. 4th ed. 8vo. 85 pp. 5 plates. *London,* 1786

A familiar Introduction to the Study of Electricity. 5th ed. (?) 8vo.
London, 1787

† History and present state of Electricity, with original Experiments. Corrected. 5th ed. 4to. 641 pp. 8 plates. *London,* 1794

Observations and Experiments relating to the Pile of Volta. (*New York Mag.* 1802.) (*Is this from the Phil. Trans. for* 1801, p. 427 ?)
New York, U.S. 1802

† Short account of the Life of the late Dr. Priestley. 8vo. 6 pp. Portr. (*Phil. Mag.* xxii. 166.) *London,* 1805

† Proceedings of the Society on the death of their late eminent associate, Dr. Joseph Priestley. 4to. (*Trans. Amer. Phil. Soc.* Old Series, vol. vi. part i. p. 190.) *Philadelphia,* 1809

Experiments, &c., on Wood Charcoal. (*From Weigel's trans. of Marat,* p. 108.)

Versuche und Beobachtungen über verschiedene Theile der Naturlehre.
(*Weigel refers to* "vol." ii. *Anhang* No. 1, "*Auszug aus einem Briefe des Hrn. Arden* (vom 25 Sep. 1772), S. 313-17.")

An Estimate of the Philosophical Character of Dr. Priestley. By W. Henry 8vo. *York,* 1832

Experiments on Air : iii. Warltire. (*From Dove,* p. 240.)

† **Prieto, A.** Riflessioni sopra le Aurore Boreali. 8vo. 58 pp. *Ferrara* 1794

† **Primo, G.** Su di uno sviluppo di elettricità mediante la stacciatura dello zolfo, osservatosi nell' I. R. Polveriera di Lambrate, presso Milano; e della necessità di raccoglierla e disperderla mediante uno scaricatore. 8vo. 6 pp. (*Bibliot. Ital.* xcv. 250.) *Milano,* 1839
(*Vide* also Moretti and Primo.)

Pringle, Sir John, Bart. *Born April* 10, 1707, *at Stichelhouse, Roxburghshire ; died January* 18, 1782, *at London.*

† A discourse on the Torpedo, delivered at the Anniversary Meeting of the Royal Society, November 30, 1774. 4to. 32 pp. *London,* 1775

† Discorso sulla Torpedine, recitato alla R. Soc. 30 Nov. 1774, with Appendix, by Soave, at p. 60. 12mo. 50 pp. (*Scelta d' Opuscoli* in 12mo. xv. p. 15.)
Milano, 1776
De Torpedine. 2 plates. 4to. *Wittemberg,* 1779

† **Prins, J. H.** Tentamen philosophico-physicum inaugurale Thèses sistens circa questionem an detur analogia inter electricitatem et magnetismum. 4to. 15 pp. *Ludg. Bat.* 1778

† **Privat-Deschanel.** Appareils d'électricité, de magnétisme et de physique mécanique. Exposition univers. de 1867 à Paris. 8vo. 25 pp.
Paris, 1868

Procopius. De bello Vandal. lib. ii. cap. ii. (Stars on spears.) A.D. 560

Profe, Gotfried. Prgrm. Ob die Elektricität des Glasses schon den Alten bekannt gewesen sey. 4to. *Altona,* 1748

† **Programmi.**
Dell' I. R. Istituto Lombardo.
Dell' Accad. dell' Istituto di Bologna. } Per premii sul Galvanismo.
Della Società Medico-Chirurgica di Bologna.
4to. 6 pp. (*Giorn. dell' I. R. Istit. Lomb.* Nuova Serie, viii. 349, 352, 353.)
Milano, 1856

† **Prudhomme.** Rapport . . par M. E. De Fourcy sur un Système d'appareils électriques servant à mettre en communication les voitures d'un train, de M. Prudhomme. 8vo. 15 pp. 1 plate. (*Ext. Ann. des Ponts et Chaussées,* 2e Semestre, 1862.) *Paris,* 1863

Prussian Acad. Mém. de l'Acad. de Prusse. . . . Vols. 12mo. 1768 *to* —

† **Puccinotti, F.** Lezioni sulle malattie nervose date . . in Bologna nel Marzo del 1834. 8vo. 175 pp. *Firenze,* 1834

† **Puccinotti e Pacinotti.** Esperienze sulla esistenza e la legge delle correnti elettro-fisiologiche negli animali a sangue caldo eseguite dai Prof. P. e P. . . nei mesi di Giugno e Luglio, 1839. 8vo. 82 pp. 1 plate. Appendice: Rapporto della Commissione. 8 pp. *Pisa,* 1839

Puget. Tractatus experimentorum magneticorum, sermone Gallico conscriptorum. (*On 4 poles in a Magnet.*)

Puliti, Tito (di Firenze). Account of his Experiments in repetition of Jacobi's (on electro-type), December 19. (*La Gazzetta di Firenze,* No. 152, above date.)
Firenze, 1839
Una Lettera da Firenze (che) ci fa sapere che il dottor Tito Pulito copiò già coll' enunciato metodo (Galvano-plastico) la bella medaglia di Galileo coniata in occasione del Congresso di Pisa," Feb. 29, 1840. (*Rivista Europ.* p. 363, 4 Fas. of the above date.) 1840

Pullen. Red Sea Electric Cable. 8vo. (*From Spon's Cat.* 1871.) 1858

Pulvermacher. Guide pratique pour le traitement électro-médical des maladies rhumatismales . . . etc. au moyen des Chaines Hydro-élect. Pulvermacher. 8vo. 13 pp. &c. *Paris,* 1856

† Médecine physique. L'électricité médicale à l'usage de tout le monde. 8vo. 112 pp. *Paris,* 1859

Puppius, R. De meteoris (Thesis) presid. Bertio, P. 4to. 8 pp. *Lugd. Batav.* 1605

Purchas. Pilgrim, or Collection of Voyages.
(*Note.—From Whiston, "Longitude," &c. who refers to* vol. iii. *for Observations on Inclination by Hudson, who sailed towards the North Pole about* 1608.)

Pusckin. "Expériences sur le Galvanisme par le moyen d'une *colonne tournante.* Extrait d'une lettre de St. Pétersbourg, 29 Sep. 1801 (ou 7 Vendémiaire, an 10). (Experiments made) à la séance extraordinaire de l'Acad. des Sciences de Pétersb.? Voyez le Journal des Débats du 7 Frimaire, an 10, et celui de Paris, du 10 Brumaire, an 9. (*From Sue's Hist.* ii. 257.)
(*Sue says that he has not details of these Experiments, or of others by the same chemist, made on December* 2, 1801 (*chez M. le Comte de Stroganow*), *in presence of the Emperor of Russia, who took great interest in them.*)

† **Putnam, A.** Remarks on L. Baldwin's proposed improvement in Lightning-rods, in a letter to Jed. Morse. Article dated January 12, 1799. 4to. 6 pp.
(*Mems. Amer. Acad.* Old Series, ii. part ii. p. 99.) *Charlestown,* 1804

Q.

† Quarterly Journal of Science and the Arts. W. T. Brande. Edited at the Royal Institution. 8vo. *Commencing* 1815 (*A few vols. are in the Library.*)

Quatrefages. Action de la Foudre sur des êtres organisés. (*In Toulouse Acad.*) *Toulouse*, 1837

Quatremère-Disjouval, D. B. (*Vide Pogg.* ii. 548.) About Spiders and Atmospheric Changes. 8vo. *La Haye*, 1785

Quellmalz, Samuel Theodor. *Born May* 12, 1696, *at Freiberg ; died February* 10, 1758, *at Leipzig.*

Dissertatio de Magnete. 4to. (*From Pogg.* ii. 548.) *Leipzig*, 1722

De homine Electrico. 4to. (*From Pogg.* ii. 548.) *Leipzig*, 1744

Progr. Homo electricus. 4to. (*From Saxtorph*, ii. 516.) *Leipsig*, 1744

Dissertatio de viribus elec. medicis. 4to. (*From Saxtorph*, ii. 516.) *Leipsig*, 1753

Theoria electricitatis. (*From Van Troostwyk.*)

De cadmia fornacum phosphorescente. (*Commerc. Litt. Norimb.* v.)

De succino artificiali. (*Commerc. Litt. Norimb.* vi.)

(*Vide* Anon. Medical Electricity, 1763.)

Querard, J. M. La Littérature Française contemporaine, par F. Bourquelot et A. Maury. 6 vols. 8vo. *Paris*, 1827, &c.

Dictionnaire des ouvrages polyonymes et anonymes de la littérature Française ; 1700 à 1845. 8vo. *Paris*, 1846

Les Ecrivains pseudonymes . . . de la littérature Française, &c. 8vo. *Paris*, 1854-8

Quesneville, G. A. Revue scientifique et industrielle. Series i. 30 vols. 8vo. Series ii. 8vo. *Paris*, 1840-9, 1850 *to —*

Quet, Jean Antoine. *Born October* 18, 1810, *at Nismes.*

† Note relative à l'action des électro-aimants sur l'arc voltaïque. 4to. 4 pp. (*Ext. des Comptes Rendus* . . . xxxiv. 805.) *Paris*, 1852

† Expériences sur le magnétisme du fer doux. 4to. 6 pp. (*Ext. des Comptes Rendus*, . . xxxv. 279.) *Paris*, 1852

† Sur quelques faits relatifs au courant et à la lumière électrique. 4to. 4 pp. (*Ext. Comptes Rendus*, . . xxxv. Séance du 27 Déc. 1852.) *Paris*, 1852

† Des divers phénomènes électriques. 4to. 4 pp. (*Ext. Comptes Rendus* . . . xxxvi. Séance du 6 Juin, 1853.) *Paris*, 1853

† Recueil de Rapports sur les progrès des Lettres et des Sciences en France. De l'Electricité du Magnétisme et de la Capillarité. La. 8vo. 274 pp. *Paris*, 1867

Quetelet, Lambert Adolphe Jacques. *Born February* 22, 1796, *at Ghent.*

Corréspondance mathématique et physique. 8vo. 11 vols. *Bruxelles*, 1825

† Recherches sur l'intensité magnétique de différens lieux de l'Allemagne et des Pays-Bas. 4to. 18 pp. (*Mém. ext. du* vi. vol. des *Mém. de l'Acad. Rle. de Bruxelles.*) (*Arago's copy.*) *Bruxelles*, 1830

† Recherches sur l'intensité magnétique en Suisse et en Italie. 4to. 16 pp. (*Mém. de l'Acad. de Bruxelles*, vol. vi.) (*Arago's copy.*) *Bruxelles*, 1831

Annales de l'Observatoire de Bruxelles. 4to. *Bruxelles*, 1834

Recherches sur les degrés successifs de force magnétique qu'une aiguille d'acier reçoit pendant les frictions multiples qui servent à l'aimanter. 8vo. *Paris*, 1834

416
QUE

Quetelet, Lambert Adolph Jacques.—*continued.*

Table of Contents of his Correspondance Mathématique et Physique de l'Observatoire de Bruxelles, 1re Livraison, 1834, tom. viii. Auroræ, Magnetism Observatory, &c. (*Fusinieri's Annali*, iv. 56, 1834.)

† Sur l'état du Magnétisme terrestre à Bruxelles pendant les douze années de 1827 à 1839. 4to. 40 pp. (*Mém. de l'Acad. de Bruxelles*, tom. xii.)
Bruxelles, 1839

Observations des phénomènes périodiques en Belgique, 1838—1861. 6 vols. 4to. (*From Kohler's Cat.* 1866, p. 35.) *Bruxelles,* 1839

† Catalogue des principales Apparitions d'Etoiles-filantes. 4to. 63 pp. (*Nouv. Mém. de Bruxelles*, xii. 1839.) *Bruxelles,* 1839

† Second Mémoire sur le Magnétisme terrestre en Italie. 4to. 27 pp. 1 plate (map). (*Acad. Rle. de Bruxelles. Ext. du* tom. xiii. *des Mémoires.*)
Bruxelles, 1840

† Nouveau Catalogue des principales apparitions d'Etoiles filantes. Mém. lu ... 6 Nov. 1841, Acad. de Bruxelles. 4to. 60 pp. 1 plate.
Bruxelles, 1842

† Sur l'emploi de la boussole dans les mines. 8vo. 34 pp. *Bruxelles,* 1843

Sur le climat de la Belgique. 2 vols. 4to. *Bruxelles,* 1845 *and* 1853

† Catalogue des livres de la Bibliothèque de l'Observatoire de Bruxelles. 8vo. 80 pp. *Bruxelles,* 1847

† Sur le climat de la Belgique. De l'électricité de l'air. 4to. 73 pp.
Bruxelles, 1849
(*This is part* iii. *of his* "*Sur le Climat de Belgique* 1846 *to* 1857.")

† Instructions sur l'électricité atmosphérique. 4to. 21 pp. (*Ext. de l'Annuaire météorologique de la France Année* 1850, 2e année.) 1850

† On Atmospheric Electricity, especially in 1849. 8vo. 4 pp. (*Phil. Mag.* ser. iv. vol. i. April, 1851, p. 329.) *London,* 1851

Histoire des sciences mathématiques et physiques chez les Belges. 8vo.
Bruxelles

† Sur l'état de l'électricité statique et de l'électricité dynamique pendant plusieurs averses observées à Bruxelles le 14 Juin, 1852. 8vo. 6 pp. (*Ext. Bulletin Acad. Rle. de Bruxelles,* xix. No. 7.) *Bruxelles,* 1852

† Sur l'électricité de l'air d'après les observations de Munich et de Bruxelles. Lettre à M. Lamont. 8vo. 7 pp. (*Ext. Bulletin, Acad. Rle. Bruxelles,* xix. No. 8.) *Bruxelles,* 1852

Sur la physique du globe. 4to. *Bruxelles,* 1861

† Météorologie de la Belgique comparée à celle du Globe. 8vo. 505 pp. 1 plate.
Bruxelles and Paris, 1867
Observations sur la Météorologie, le Magnétisme, etc., faites à l'Observatoire de Bruxelles. 4to. *Bruxelles*

† Résumé des Observations Magnétiques et Météorologiques, faites à des époques déterminées. 4to. 138 pp. 5 plates. *Bruxelles*
(*This forms part of his* "*Observations des Phénomènes Périodiques.*")

† Sur l'Electricité des nuages orageux. 8vo. 12 pp. (*Acad. Royale de Bruxelles* (*Extrait du*), tom. xxi. No. 7 *des Bulletins.*) *Bruxelles*

Many Articles, &c. in Periodicals, &c. &c. some of which may be Electrical, not yet registered in this Catalogue. (*Vide* Poggendorff, ii. 552.)

(*Vide* also Stas and Quetelet.)

(*Vide* also Wheatstone.)

† **Quetelet, De Vaux, Cabry.** Rapport adressé sous la date du 21 Mars, 1850, à M. le Ministre des Travaux Publics par la Commission des Télégraphes électriques. 8vo. 53 pp. 1 map. (*Extrait des Annales des Travaux Publics de Belgique.*) *Bruxelles,* 1850

† Quetelet and Zantedeschi. Sur les courants Electriques telluro-atmosphériques, et leurs rapports avec les perturbations des aimants. Lettre de . Zantedeschi . à . Quetelet, &c. 8vo. 16 pp. (*Bulletins de l'Académie Royale de Belgique*, 2e série, xv. No. 5.) *Bruxelles*

Quincke, G. Über d. Fortführung materieller Theilchen durch strömende Electricität. 8vo. Sammlung von 4 Abhandl. über elektrische Ströme. 8vo. *Berlin and Leipzig*, 1856-61

† Quinet de Certines. Théorie de l'Aimant, appliquée aux déclinaisons et inclinaisons de l'aiguille de boussole, et démontrée par la trigonométrie sphérique. Notice sommaire. 4to. 8 pp. *Paris*, 1809

† Quinet, J. Exposé des variations magnétiques et atmosphériques du globe terrestre ; avec des Tables et des Cartes de la déclinaison et de l'inclinaison de l'aiguille aimantée. Mémoire préliminaire, suivi d'un prospectus. 8vo. 40 pp. *Paris*, 1826

† Second Mémoire sur les Variations magnétiques et atmosphériques. 8vo. 125 pp. *Bourg*, 1826

Quinquet. On the formation of Hail. Electrical experiment in illustration, &c. (*From Arella, Storia*, ii. p. 160.) Observations sur les Paratonnerres. (*Journal de la Société des Pharmaciens de Paris*, tom. i. p. i. 100.)

† Quint. Entwickelung der elektromagnetischen Telegraphen. 4to. 9 pp. (*Programm des K. Progymnasium, in Ste. Wendel* . . 29 August, 1862.) *Wendel*, 1862

Quintine, l'Abbé De La. Dissertation sur le Magnétisme des Corps. 8vo. 42 pp. (*Prize Dissertation of the Bordeaux Academy*, tom. iii.) *Bordeaux*, 1732

R.

R . . . . C. (*Vide* Anon. Lightning, 1846.)

R. R. (*Vide* Anon. Lightning, 1817.)

R . . . . (*Rieu ?*) (*Vide* Anon. Magnetism, 1847, and the four entries following.)

† R., W. The relation of a storm of Thunder, Lightning and Hail, at Oundle, in Northamptonshire, on the 20th March, 1692-3. 4to. 2 pp. (*Phil. Trans.* xvii. for 1693, p. 710.)   *London*, 1694

† Raben, F. Ziveene merkwurdige Wolkenzeuge bey Nysted in Laland. 8vo. 2 pp. (*K. Schwed. Akad.* Abh. xii. 285.)   *Hamburg and Leipzig*, 1750

† Rabiqueau, C. Le spectacle du feu élémentaire, ou cours d'électricité expérimentale . . 8vo. 296 pp. 10 plates.   *Paris*, 1753

Le spectacle de la Nature, etc. 8vo.   *Paris*, 1753

Relation curieuse et intéressante pour le progrès de la physique et de la médecine. 8vo.   *Paris*, 1756

Lettre sur la mort de Richman. 8vo.   *Paris*

Racagni, Giuseppe Maria. *Born January 6, 1741, at Torazza, prov. Voghera ; died March 4, 1822, at Milan.*

† Sopra alcuni conduttori elettrici che sono stati percossi dal fulmine. Memoria. Ricevuta 13 Luglio, 1818. 4to. 14 pp. (*Mem. della Soc. Ital.* xviii. 139.)
                                                                  Modena, 1820

Sopra alcuni edifizii muniti di Parafulmini Frankliniani stati dal Fulmine danneggiati. Memoria. Ricevuta 10 Nov. 1821. 4to. 26 pp. 1 plate. (*Ital. Soc. Mem.* xix. p. 1.)   *Modena*, 1823

† Sopra i Sistemi di Franklin e di Symmer spettanti all' Elettricità. 4to. 31 pp. (*Mem. dell' I. R. Istituto Lombardo-Veneto*, v. 187.)   *Milano*, 1838

Sopra una Memoria di Poisson. Mem. letta il 10 Aprile, 1817. 4to.
                                                                  Milano, 1838

Raccolta d'Opuscoli Scientifici e Filologici. Many (or all) of the Elect. articles are in the Library.

† Rackstrow, B. Miscellaneous Observations, together with a collection of Experiments on Electricity ; to which is annexed (a letter) to the Members of the Academy of Sciences at Bordeaux, on the similarity (*sic*) of Electricity to Thunder and Lightning (at p. 59). 8vo. 72 pp. 1 plate.   *London*, 1748

† Radcliffe, C. B. Dynamics of Nerve and Muscle.   *London*

Radcliffe, John. (*Vide* Duchenne de Boulogne.)

Radouay. Remarques sur la Navigation.   1727

Radoumowski, G., Comte de. Mémoire sur le Phosphorisme des corps du règne minéral, par moyen du frottement. 4to. (*Mém. de Lausanne*, ii. *Hist.* p. 14, *Mém.* pt. i. p. 13.)   *Lausanne*

Rafn, C. G. "M. Rafn a lu, le 8 Mai, 1801," dit le rédacteur du *Magazin Encyclopédique*, " à l'Acad. des Sciences de Copenhague, la continuation de ses Expériences relatives à la Végétation." (*Mag. Encyclopédique*, No. 19, Ventose, an. 10, p. 370 ; *Konig. Dan. Gesels. d. Wissens.*)   *Paris*, 1802

† Ragona, Dom. Risultati delle Osservazioni sull' elettricismo atmosferico instituite nel R. Osservatorio di Modena. Memoria. 4to. 48 pp. 1 plate. (*Inscritta nel t.* xi. *delle Mem. della R. Accad. di Scienz. in Modena.*)   *Modena*, 1870

† Ragona-Scina, D. Un nuovo caso di rotazione dell' ago magnetico. Lettera di Domenico Ragona-Scina. . . di Palermo, al . . A. De la Rive. . . 4to. 3 pp. (*Ann. del Reg. Lomb.-Veneto*, xiv. 123.)   *Vicenza*, 1845

(*Vide* also Scina D. Ragona.)

† **Raillet, M.** Origine de la Grêle. 12mo. 48 pp. *Paris*, 1866

† **Rambert.** Mémoire sur l'utilité des Paragrêles, &c. 8vo. 67 pp. *Paris*, 1826

‡ **Rambosson, J.** Histoire des Météores, et des grands phénomènes de la Nature. 1st edition. 8vo. 408 pp. *Paris*, 1868

Histoire des Météores, et des grands phénomènes de la Nature. 2nd edition. 8vo. 424 pp. 90 gravures. 2 planches. *Paris*, 1869

Ramis. Orologio elettrico a pendolo "fece egli quindi sotto il 18 Marzo, 1815, l'annunzio alla Regia Accad.".. (Monaco). (*Almanacco di Lipsia, nominato Magazzino di tutte le nuove scoperte al* No. 64, art. 11.) *Leipzig*, 1815
(*Note.—For a short account of this clock, vide Gilbert, Neuere Versuche mit Trocknen . Saule, Ann. de Phys.* li. 189, 1815. *The account of my Electric Clock is dated* March 9, 1815.*—F. R.*)

‡ **Rammelsberg, C.** Die chemische Natur der Meteoriten. 4to. 88 pp. (*Abhand. d. k. Akad. d. Wiss.* 1870.) *Berlin*, 1870

Die Zusammensetzung des Turmalins vergleich. mit derj. des Glimmers und Feldspaths.

Ramus, Jochum Friderik. *Born July 3, 1686, at Drontheim ; died January 4, 1769, at Copenhagen.*

† Historico-physica enarratio de stupendis luminis borealis phænomenis, natura et origine. Pars prima. 4to. 78 pp. 15 figs. (*Scriptores Soc. Hafniensis,* pt. i. p. 317 ; *or in Skrifter det Kiobenhavnske Selskab Deel,* i. S. 317.) *Hafniæ*, 1745
(*Note.—The Danish version in Poggendorff,* ii. 567, *is entitled* "*Beskrivelse over Nordlysets forunderlige, Skikkelse Natur og. Oprindelse,* 1845.) (*Not in Heinsius.*)

Historico-physica enarratio de stupendis luminis borealis phænomenis, natura et origine. Pars secunda. 4to. (*Scriptores Soc. Hafniensis,* pt. iii. p. 209 ; *and in Skrifter del Kiöbenhavnske Selskab Deel.* iii. S. 147.) (*Danish in Pog.*) *Hafniæ*

Ranzi. Discovery of the direction of the discharge of the Silurus. (*Vide* Matteucci.)

† **Raoult, F.** Thèse. Etude des forces électromotrices des éléments voltaïques ; soutenue le 13 Mai devant la Commission d'Examen . . 4to. 100 pp. 2 plates. *Paris*, 1863

Rappolt, K. H. De origini succini in littore Sambiensi, &c. 4to. *Regiomont*, 1737

Rapport au Ministre d'Instruction Publique . . Applications de la Pile de Volta. (*Vide* Dumas.)

Raschig. Works in Gilbert's Ann. (about 8).

Ratte, Etienne Hyacinthe de. *Born September* 1, 1722, *at Montpellier ; died August* 15, 1805, *at Montpellier.*

Sur deux Aurores boréales observées en 1726 et en 1730. 4to. 20 pp. (*Montpel. Acad. Hist. &c.* vol. ii. p. 4.) *Montpellier*, 1778

Sur un Tourbillon . . qui fit des ravages à . . Montpellier et aux environs de cette ville. 4to. 9 pp. (*Montpel. Acad. Hist.* ii. p. 24.) *Montpellier*, 1778

Sur l'Aurore boréale du 16 Décembre, 1737 ; et sur quelques Aurores boréales moins considérables. 4to. 4 pp. (*Montpel. Acad. Hist.* ii. 131.) *Montpellier*, 1778

Rauch, C. V. Coup d'œil historique sur l'application de l'électricité à la médecine. Thèse. 8vo. 72 pp. *Strasbourg*, 1851

Raulin, V., et Peslin, H. Quelques vues générales sur les variations séculaire du Magnétisme terrestre. Suivi de : Sur la loi de variation annuelle de la déclinaison de l'aiguille aimantée à Paris. Par M. H. Peslin. 8vo. 92 pp. and plate. *Bordeaux*, 1867

Raven. Account from Carolina of the effects of Lightning on two of the rods affixed to houses for securing them against Lightning.

Raymond. Résumé de tous les faits connus jusqu'en 1824 inclusivement, en faveur des Paragrêles. **1824 ?**

† **Read, I.** Summary view of the Spontaneous Electricity of the Earth and Atmosphere . . 8vo. 160 pp. 1 plate. *London*, **1793**

Experiments and Observations made with the Doubler of Electricity, with a view to determine its real utility in the investigation of the Electricity of atmospheric air in different degrees of purity. 4to. (*Phil. Trans.* an. 1794, p. 266.) *London*, **1794**

† **Reade, J. B.** Observations on the Aurora borealis of the 24th October, 1847. Folio. 1 page. (*Ext. Bucks Herald, Oct.* 30, 1847.) **1847**

Reael. Observatien am den Magneotsteen. **1651**

Réaumur, R. A. F. de. Des effets que produit le poisson appellé en français Torpille. 4to. (*Mém. de Paris*, 1714.) *Paris*, **1714**

Expériences qui montrent avec quelle facilité le fer et l'acier s'aimantent même sans toucher à l'aimant. 4to. (*Mém. de Paris*, 1723.) *Paris*, **1723**

† **Rebold. E.** L'Electricité, moteur de tous les rouages de la vie . . 8vo. 372 pp. 6 planches. *Paris*, **1869**

Récamier. Appareil galvano-électrique portatif du Professeur Récamier, ou cataplasme galvanique. 8vo. 8 pp. *Paris*, **1851**

† **Recupero, G.** Storia naturale e generale dell' Etna. 2 vols. 4to. Portrait and plates. *Catania*, **1815**

† **Recy, C. M.** Télétato-dydaxie, ou Télégraphie électrique. 8vo. 35 pp. 1 plate. *Paris*, **1838**

Rédacteurs du Bulletin des Sciences de la Société Philomatique. Résumé des nouvelles expériences faites sur le Galvanisme, par divers physiciens. (*Bull. de la Soc. Phil.* an. 9, Floreal.) *Paris*, **1801**

Redfield, William C. *Born March* 25, 1789, *at Middletown, Connecticut ; died Feb.* 12, 1857, *at New York.*

On Whirlwind Storms ; with replies to the Objections and Strictures of Dr. Hare. 8vo. 78 pp. Plates. *New York*, **1842**

Observations on the Storm of December 15th, 1839. 4to. (*Trans. Amer. Phil. Soc.* New Series, vol. viii.) *Philadelphia*, **1843**

† On three several Hurricanes in the Atlantic, &c. 8vo. *Newhaven, U.S.* **1846**

On Waterspouts. 8vo. (*Lomb. I. R. Ist. Giorn.* xvi. 1847. *In Silliman's Journal*, xxxvi. 50.)

Redi, Francesco. *Born February* 18, 1626, *at Arezzo ; died March* 1, 1697, *at Pisa.* Experimenta circa res naturales. **1666**

(*Note.—Poggendorff adds* . . "*accessere observationes de viperis* . . *etiamque observationes circa illas guttulus et fila ex vitro quæ rupta* . . 12mo. *Amstelod.* 1675. (*Italiens.* 4to. *Firenze,* 1671 ; *auch Miscell. Acad. Nat. Cur.* 1671.")

† Esperienze intorno a diverse cose naturali, &c. . . in una lettera al Atan. Chircher (Kircher). . 4to. 152 pp. 6 plates. (*On the Torpedo,* pp. 47—54.) *Firenze*, **1671**

Rees, Abraham. *Born* 1743, *in Wales ; died June* 9, 1825, *in London.*

Chambers's Dictionary of Arts. 4 vols. Fol. *London*

The Cyclopædia ; or, Universal Dictionary of Arts, Sciences, and Literature. 39 vols. + 5 vols. of plates. 4to. *London*, **1802-19**

(*The Articles referable to Tensional Electricity are by Cavallo and by Cuthbertson in vol.* xii.; *those on Galvanism, Voltaism, &c., by Sylvester, in vol.* v.; *those on Chemistry, &c., by Aikin (Arthur), Davy (J. and H.), Fletcher, Marcet, Sylvester, and Dalton ; the articles on Meteorology are by Dalton, Howard (Luke), and Dickson ; those on Atmosphere, &c. are by Dr. Rees himself, who wrote other articles more or less connected with the subject of Electricity, &c.; Magnetism by Cavallo.*)

Rees, Richard Van. *Born May* 24, 1797, *at Nimwegen.*

Zwei Meteorsteinfälle in Holland. 8vo. (*Pogg. Ann.* lix. 1843.)
Leipzig, 1843

†   Over de Verdeeling v. het Magnetismus in staalmagneten, en electro-magneten.
4to. 25 pp. (*Nieue Verhandelingen van het. K. Nederl. Instit.* xii. 94.)
Amsterdam, 1846

†   Over de Verdeeling van het Magnetismus in Magneten. 4to. 24 pp. 1 plate.
(*Neu Verhand. von K. Nederl. Instit.* xiii.)   *Amsterdam,* 1847

Über d. elektr. Eigenschaften d. Spitzen u. Flammen. 8vo. (*Pogg. Ann.*
lxxiii. u. lxxiv. 1848 u. 1849.)   *Leipsig,* 1848-9

†   Over de theorie der magnetische Krachtlijnien van Faraday. 4to. (*Verhandl.
d. K. Nederl. Acad. d. Wetensch.* i. 1854.)   1854

Rees, W. Van. Verzameling van Stukken als brydagen tot het Galvanismus, &c.
8vo.   *Arn* (?), 1803-5

Bijdr. tot h. Galvanismus. 2 vols. 8vo. plates.   *Aruth,* 1803

(*Vide also* Moll and others.)

† **Regnard, E.** Mémoire sur la Télégraphie électrique à courants combinés et à
double échappement, et sur l'Horlogerie électrique. 8vo. 46 pp. 3 plates.
Paris, 1855

† **Regnauld, J.** De la production de l'Electricité dans les êtres organisés. Thèse
de concours.   *Paris,* 1847

Una unità elettro-motrice . . . coppia termo-elettrica, bismuto e rame. (*Vide*
Matteucci.)

Thèse de Phys. Recherches sur les forces électromotrices, et sur une nouvelle
méthode propre à les déterminer. 4to. 56 pp. 1 plate.   *Paris,* 1855

Regnault, V. Cours élémentaire de Chimie. 4 vols. 12mo.   *Paris,* 1854

† **Regnault.** Appareils télégraphiques électriques. (*Vide Combes. Rapport fait
. . . sur l'appareil de télégraphie électrique de M. Regnault,* 1855, 4to. Paris,
36 pp. 6 plates. *Extrait du Bulletin de la Société d'Encouragement.*)

Reich, Ferdinand. *Born February* 19, 1799, *at Bernburg.*

Über elektr. Strömungen auf Erzgängen. 8vo. (*Pogg. Ann.* xlviii. 1839.)
Leipzig, 1839

Über d. Wirk. einiger Blitzschläge in Freiberger Gruben. 8vo. (*Pogg. Ann.*
lxv. 1845.)   *Leipzig,* 1845

†   Elektrische Versuche. 4to. 10 pp. (*Abhandl. bei Begründ d. k. Sächs. Gesellsch.
d. Wiss.* 1846.)   1846

†   Versuche über die abstossende Wirkung eines Magnet-poles auf unmagnetische
Körper. 8vo. 5 pp. (*Berichte über die Verhandlungen der König. Sächs.
Gesellsch. . . zu Leipzig,* vii. and *Pogg. Ann.* lxxiii. 1848.)   *Leipzig,* 1847

Beobb. über d. magnet. Polarität d. Pöhlberges. 8vo. (*Pogg. Ann.* lxxvii.
1849.)   *Leipzig,* 1849

Neue Versuche mit der Drehwage. 4to.   *Leipzig,* 1852

†   Leitfaden für den Unterricht über Magnetismus, Elektricität und Licht. 8vo.
101 pp.   *Freiburg,* 1853

Über d. Diamagnet. Wirkung. (*Berichte d. K. Sächs. Gesellsch. d. Wiss.* 1855.)
1855

Über photographische Registrirung d. magnet. Declination. (*Berichte d. K.
Sächs. Gesellsch. d. Wiss.* 1859.)   1859

Beiträge zu d. von Humboldt später von Gauss eingerichteten Beobb. d.
magnetischen Declinations veränderungen.

† **Reichardt, E.** Chemische Verbindung d. anorganischen Chemie, geordnet
nach dem electro-chemischen Verhalten, mit Inbegriff der durch Formeln
aus drückbaren Mineralien. 8vo. 325 pp.   *Erlangen,* 1858

† Reichenbach, K. Freiher Von. Physikal-physiolog. Untersuchungen über die Dynamide des Magnetismus, der Elektricität, der Wärme, des Lichts, in ihren Beziehungen zur Lebenskraft, 2 Aufl. 8vo. 2 vols. 218 and 240 pp. 3 plates. *Braunschweig*, 1850

† Odisch-magnetische Briefe. Erste Reihe. Zweite Ausgebe. 8vo. 201 pp. *Stuttgart*, 1856

† Researches on Magnetism, Electricity, Heat, Light, &c. . . . translated . . . by William Gregory. 1st with preface, notes, &c. Parts i. and ii. 8vo. 455 pp. 2 plates. *London*, 1850

Reichenberger Johann Nepomuk. *Born November* 23, 1737, *at Munich; died* 1805, *at Wemding.*

Directorium magneticum, magneticis quibusdam phænomenis exhibens, experimentis dirigendis, ac observationibus instituendis optatum, ejusque descriptio. (*From Van Swinden.*)

Hydrotica. 8vo. (*From Van Swinden.*) *Ratisbonne*, 1778

Reichenstein. (*Vide* Müller von Reichenstein.)

Reichsanzeiger, Der. A Journal containing articles by Ritter. (*From Augustin*, p. 75.)

† Reid, John. On the relation between Muscular Contractility and the Nervous System. (*From Meyer.*) *Edinburgh*, 1841

† Reid, Sir W. Attempt to develop the Law of Storms. 8vo. *London*, 1838

Progress of development of the Law of Storms. 8vo. *London*, 1849

† An attempt to develop the Law of Storms by means of facts, and to point out a cause for the Variable Winds, with the view to practical use in Navigation. 3rd edition. 8vo. *London*, 1850

† Attempt to develop the Law of Storms. New edition. 8vo. *London*, 1859

Reid, W. Reid's patent method of testing Submarine Cables. 8vo. 11 + 2 pp. (*Extracted from the Engineer, June* 25, 1859. *Idem. Extracted from the Mechanic's Magazine, August* 5, 1859. *Idem. Extracted from the Mining Journal, August* 6, 1889.) *London*, 1853 *and* 1859

† Reid and Bain. Elements of Chemistry and Electricity. (Edited by D. M. Reese, M.D.) 8vo. 411 pp. Cuts. *New York, U.S.* 1849

Reil, J. C. Über thierische Elektricität. (*Gren's Journal*, vi. 1792.) *Leipzig*, 1792

Electricité galvanique, développée par la dissolution du zinc dans l'acide sulfurique. Extrait d'une lettre de . . Reil. 8vo. (*Sue, Hist.* iv. 26. *In Journal de Van Mons,* No. iv. p. 104, *article de correspondance.*) *Bruxelles*

† Reimann. Beiträge zur Lehre von der Wärme und der Elektricität. 4to. 14 pp. *Saalfeld*, 1848

Reimarus, Johann Albert Heinrich. *Born November* 11, 1729, *at Hamburg; died June* 6, 1814, *at Ranzau.*

† Die Ursache des Einschlagens vom Blitze u. dessen natürl. Abwend. von unseren Gebäuden, aus zuverlässigen Erfahrungen von Wetterschlagen vor Augen gelegt. 8vo. 128 pp. *Langensalsa*, 1769

Ursache v. Einschlagen des Blitzes. 8vo. *Leipzig*, 1774

† Vorschriften zur Anlegung einer Bliz-Ableitung an allerley Gebäuden nach zuverlässigen Erfahrungen. 8vo. 24 pp. *Hamburg*, 1778

† Vom Blitze : i. Dessen Bahn u. Wirk. auf versch. Körper, &c. (and two other heads, ii. and iii.) 8vo. 678 pp. *Hamburg*, 1778

† Nachricht von einer Zurüstung, welche die Wirkung der Gewitterwolke sinnlich darstelt. (*Deutsch. Mus.* Oct. 1779, p. 329, f.) 1779

Einige gegen d. Blitzableitung gemachte Einwürfe beantwortet. 8vo. *Frankf.-a.-M.* 1790

† Neuere Bemerkungen vom Blitze dessen Bahn,Wirkung, sicheren und bequemen Ableitung, &c. 8vo. 386 pp. 9 plates. *Hamburg*, 1794

Reimarus, Johann Albert Heinrich.—*continued.*

Ausführliche Vorschriften, &c. 8vo. *Hamburg,* 1794

† Ausführliche Vorschr. z. Blitzableitung an allerley Gebäuden. 8vo. 46 pp.
2 plates. *Hamburg,* 1797

Über Blitzschläge u. Blitzableiter. 8vo. (*Gilb. Ann.* vi. ix. u. xxxvi.)
Leipzig

Lebensbeschreibung von ihm selbst aufgesetzt. 8vo. *Hamburg,* 1814

† Memoriæ Joh. Alb.-Henr. Reimari. 4to. 50 pp. *Hamburg,* 1815

Reinhardt. (*Vide* Œrsted and others.)

Reinhold, Johann Christoph Leopold. *Born* 1769, *at Leipzig ; died November* 28, 1809, *at Leipzig.*

† De Galvanismo. Specimen i. Read 16 December, 1797. 4to. 125 pp. (*Volta's
copy.*) *Lipsiæ,* 1797

† De Galvanismo. Specimen ii. Read 11 March, 1798. 4to. 82 pp.
Lipsiæ, 1798

De Galvanismo. 8vo. *Lipsiæ,* 1798

Ausmittel d. Grundkette d. Voltaschen Säule. 8vo. (*Gilb. Ann.* x. 1802.)
Leipzig, 1802

Natur d. Säule. 8vo. (*Gilb. Ann.* x. u. xii. 1802.) *Leipzig,* 1802

Galvanisch-elektr. Versuche. 8vo. (*Gilb. Ann.* xi. 1802. *Leipzig,* 1802

† Geschichte des Galvanismus. Nach Sue d. a. frey bearbeitet, nebst Zusätzen und
einer Abhandlung über die Anwendung des Galvanismus in der praktischen
Heilkunde. Zwey Abtheil. 8vo. 328 pp. 2 plates. (*Volta's copy.*)
Leipzig, 1803

Über Davy's Versuche. 8vo. (*Gilb. Ann.* xxviii. 1808.) *Leipzig,* 1808

Reinsch, H. Versuch e. neuen Erklärungsweise d. elektrischen Erscheinungen.
8vo. *Nurnberg,* 1841

Ueber den Einfluss tonender Saiten auf die Magnetnadel und eine darauf
gegründete Erklärung der elektrischen und magnetischen Erscheinungen.
8vo. 16 pp. *Speyer,* 1855

Reifer, S. J. W. Beschreibung e. sehr vortheilhaft eingerichteten Elektrisirmaschine, aus e. Schreiben an den Herausgeber. 8vo. 7 pp. (*Voight's Magaz.
f. Phys.* vii. 3 St. p. 73.) *Gotha,* 1791

Reiser. Electrical Plate Machine. 8vo. (*Lichtenberg's Mag.* vii. St. 3, p. 73.)
Gotha

† **Reiser** or **Reusser.** Der elektrische Würfel. 8vo. 3 pp. (*Voight's Magazin,*
vol. vii. 2 Stuck, p. 57.) *Gotha,* 1791

† Schreiben an den Herausgeber. 8vo. 2 pp. (*Voight's Magazin,* B. ix.,
1 Stuck, p. 183.) *Gotha,* 1794

† **Reitlinger,** E. Ueber flüssiger Isolatoren der Elektricität. Vorgelegt. 20 Jan.
1859. 8vo. 32 pp. (*Aus den Sitzungsberichten* 1859, *de mathemat. naturwiss.
Classe der k. Akad. der Wissenchaften abgedruckt,* vol. xxxv. No. 8.)
Wien, 1859

Ueber die Einwirkung der Elektricität auf Springbrunnen. 8vo. 23 pp.
(*Aus den Sitz.* 1860, *der kais. Akad. der Wiss.*) *Wien,* 1860

Sibven Abhandl. (Lichtenberg'sche Figuren, Electro-chem, Untersuch. flüssige
Isolatoren, &c.) 8vo. (*Wien Acad.*) *Wien*

† **Reitlinger,** E., u. **Kuhn.** Über Spectra negativer Elektroden und lange ge-
brauchter Geissler'scher Röhren. 8vo. 9 pp. (*Aus d. Sitzb. Wien Akad.
am* 7 April, 1870, lxi.) *Wien,* 1870

† **Reitlinger** und **Kraus.** Über Brande's electrochemische Untersuchungen. Vorgelegt in der Sitzung, vom 10 Juli, 1862.) 8vo. 26 pp. (*Sitzungsberichte d.
k. Akad. d. Wiss.* xlvi.) *Wien,* 1862

424

REM—REN

† **Remak, R.** Über Methodische Electrisirung gelähmter Muskeln. 8vo. 31 pp.
Berlin, 1855

Über method. Electrisirung gelähmter Muskeln. 2 Aufl. mit Anhg. üb. Galvanisirung motor. Nerven. 8vo. *Berlin,* 1856

† Note additionnelle au Mémoire sur l'action physiologique et thérapeutique du courant galvanique constant sur les nerfs et les muscles de l'homme. 4to. 2 pp. (*Extrait Comptes Rendus,* xliii.) *Paris,* 1856

† Ueber die Heilwirkungen des constanten galvanischen Stromes bei Lähmungen, Schmerzen und Krämpfen. 8vo. 2 pp. (*Der Allg. Med. Central-Zeitung,* 12 Stück, Jahrg. 1857.) *Berlin,* 1857

† Ueber den antiparalytischen Werth inducirter elektrischer Ströme. 8vo. 8 pp. *Berlin,* 1858

† Galvanothérapie der Nerven und Muskelkrankheiten. 8vo. 461 pp. *Berlin,* 1858

Galvanothérapie; ou de l'application du courant galvanique constant au traitement des maladies nerveuses et musculaires. Traduit par A. Morpain, avec les additions de l'auteur. 8vo. 467 pp. *Paris,* 1860

† Ueber Tabes dorsalis. Fol. 8 pp. (*Allgem. Med. Central.*) *Berlin,* 1863

† Ueber vitale Wirkungen des constanten Strömes. 8vo. 5 pp. (*Berliner Klinischen Wochenschrift,* No. 26.) *Berlin,* 1864

† Application du Courant Constant au Traitement des Névroses. 8vo. 41 pp. *Paris,* 1865

Die Verdickung d. Muskeln durch Constante galvan. Ströme.

† Method Electrisirung. 1st edition. 1865

Method Electrisirung. 2nd edition. 1866

Elektrotherapeut. Mittheilgn.

† **Renard, N. A.** Sur la propagation de l'électricité. 8vo. 20 pp. *Nancy,* 1859

† Distribution de l'électricité dans les corps conducteurs en partant de l'hypothèse d'un seul fluide. Premier Mémoire. 8vo. 30 pp. *Nancy,* 1859

† Distribution de l'électricité dans les conducteurs cristallisés, en partant de l'hypothèse d'un seul fluide. 8vo. 53 pp. (*Mém. Acad. Stanislas,* 1861. Zuchold says 1862.) *Nancy,* 1861

† Théorie du Magnétisme Terrestre dans l'hypothèse d'un seul fluide électrique. 8vo. 74 pp. (*Extrait des Méms. de l'Acad. de Stanislas.*) *Nancy,* 1863

† Sur l'établissement des formules fondamentales de l'électro-dynamique dans l'hypothèse d'un seul fluide. 8vo. 42 pp. (*Extrait des Méms. de l'Acad. de Stanislas.*) *Nancy,* 1865

† Théorie d'Induction en partant de l'hypothèse d'un seul fluide. 8vo. 34 pp. (*Extrait des Méms. de l'Acad. de Stanislas.*) *Nancy*

Théorie des actions des aimants dans les corps magnétiques et diamagnétiques dans l'hypothèse d'un seul fluide électrique. 8vo. 38 pp. (*Extrait des Méms. de l'Acad. de Stanislas,* 1868.) *Nancy,* 1868

Renaudot. Dissert. of the Chinese Sciences, &c. (*From Whiston, "Longit. and Lat. found," who refers to pp.* 287, 294 *and* 382, *relative to Renaudot's "reasons against the antiquity of the Compass in the Oriental parts, both as to Arabia and China."*)

† **Renault, B.** Vérification expérimentale de la réciproque de la loi de Faraday. Sur la décomposition des électrolytes. Thèse présentée à la Faculté des Sciences de Paris. 8vo. 59 pp. *Paris,* 1867

Rendorp, W. De magnetismo telluris. 4to. *Lugd. Bat.* 1780

† **Rendu, Abbé.** Influence of Magnetism upon chemical action. 8vo. 1 p. ("*Ann. de Chimie,* xxxviii. 196." *Phil. Mag. or Annals,* iv. 385.) *London,* 1828

Renier, Stefano Andrea. *Born* 1759, *at Chioggia ; died January* 16, 1830, *at Padua.* Intorno alla fosforescenza delle acque de' mari. A Disquisition read on the 14th April, 1814, in the Venetian-Lombardy I. R. Instit. (but not said to be printed). *Vide* vol. ii. p. 9, Notizie delle Dissertationi lette.

† **Renwick.** Told Dr. Alberts (in Bremen) that he had, as amusement, known the action of two metals on the tongue. 8vo. (*Alberts Amerikanisch Ann. d. Artzneykund &c. Bremen*, ii. Heft, 1802.) *Bremen*, 1802

Repertorium der Physik. (*Vide* Dove and Moser.)

Repertorium fur Physikalische Technik. (*Vide* Carl.)

Resal, Amé Henri. *Born January* 27, 1828, *at Plombières, Vosges.*

† Théorie de l'électro-dynamique. 4to. 11 pp. 1 plate. (*Extrait des Méms. de la Société d'Emulation des Doubs.*) *Besançon*

† **Rescalli, P.** (Marq.) Nouvelle méthode pour installer et isoler parfaitement les fils conducteurs des Télégraphes électriques. Fol. 51 pp. 3 plates.
Milan, 1854
(*Note.—Similar to Ronalds' method.*)

Reslhuber, A. Über d. Ozongehalt d. atmosphär. Luft. 8vo. (*Sitzungsberichte d. Wiener. Acad.* v. 1849.) *Wien*, 1849

† Beobachtungen während der Nordlichter am 18 October und 17 November, 1848. 8vo. 16 pp. (*Sitzungsberichte d. Wien Acad.* v. 1849, p. 133.)
Wien, 1849

Über d. Periode, welche die Mittlere Jahresgrösse d. tägl. Schwankung d. magnet. Déclination u. Horizontal-Intensität befolgt. 8vo. (*Pogg. Ann.* lxxxv. 1852.) *Leipzig*, 1852

† Die Constanten von Kremsmünster. 4to. 20 pp. *Linz*, 1853

† Über das magnetische Observatorium in Kremsmünster und die aus den Beobachtungen bis zum Schlusse des Jahres 1850 gewonnenten Resultate. Vorgelegt in der Sitzung. d. Mathem.-Naturw. Classe am 22 Mai, 1852. 4to. 56 pp. 9 plates. *Wien*, 1854

Über das magnetische Observatorium in Kremsmünster und die vom Jahre 1839-50 aus den Beobachtungen abgeleiteten Resultate. Mit 7 Tafeln und 2 Beilagen. Beilage A.—Die täglichen Beobachtungen der magnetischen Declination und Horizontal-Intensität. Beilage B.— Die Beobachtungen an den Terminstagen. (Der Theil " Über das Observatorium und die Resultate " aus dem vi. Bande der Denkschriften der mathematisch-naturwissenschaftlichen Classe der Kaiserl. Akademie der Wissenschaften besonders abgedruckt.) 4to. 56 and 190 pp. 7 plates. (*In der Kaiserl. Akademie der Wissenschaften.*) *Wien*, 1854

Über den Ozongehalt der atmosphärischen Luft. 8vo. 11 pp. (*Aus den Sitzungsberichten* 1854 *der mathematisch-physikalischen Classe der Kaiserlichen Akad. der Wissenschaften.*) *Wien*, 1855

Untersuchungen über d. atmosphärische Ozon. 8vo. 30 pp. (*Sitzungsberichte d. Wiener Acad.* xxi. 1856.) *Wien*, 1856

Über d. Wetterleuchten. 8vo. (*Sitzungsberichte d. Wiener Acad.* xxviii. 1858.) *Wien*, 1858

Mehrjahr stündl. u. monatl. Mittel d. magnet. Beobb. von 1851-54. (*Jahrbücher d. Central-Anstalt f. Meteorologie u. Erdmagnetism. in Wien* v. vi. vii. 1858, 1859, 1860.) *Wien*, 1858, 1859, 1860

Bericht über die am 21 und 29 April, 1859, zu Kremsmünster beobachteten Nordlichter. 8vo. 10 pp. (*Aus den Sitzungsberichten* 1859 *der mathemnaturwiss. Classe der K. Akad. der Wissenschaften.*) *Wien*, 1859

Ressel, Joseph. Biografia publicata per cura di un comitato formatosi all' oggetto di onorare la memoria di lui coll' erigergli un monumento. 8vo. *Trieste*, 1858

† **Restelli, A.** Come evitare alcuni inconvenienti dell' elettro-puntura. Lettera al Prof. Petrequin. 4to. 11 pp. (*Estratto dalla Gazzetta Medica di Milano*, tom. v. No. 32.) *Milano*, 1846

† **Resti-Ferrari, G.** (Review, &c. of) Gerbi, Corso elementare di fisica, Pisa, 1823-25, 5 vols. 8vo.; and of Siena, Elementi di fisica generale ed Elementi di fisica particolare, Palermo, 1828-30, 4 vols. 8vo. 8vo. 18 pp. (*Bibl. Ital.* lxx. 222.) *Milano*, 1833

426   RES—REY

Resti-Ferrari, G.—*continued.*

† Review, &c. of Gerbi's "Corso elementare di fisica," Pisa, 1831-32, 5 vols. Terza edizione. 8vo. 5 pp. (*Bibl. Ital.* lxxi. 336.)   *Milano,* 1833

† Sopra l'azione che le calamite esercitano all' esterno delle spirali. Esperimenti ed Osservazioni. 4to. 12 pp. (*Ann. del Reg. Lomb.-Veneto,* iv. 20.)
                                                                              Padova, 1834

† Intorno all' elettrizzazione del vetro liscio sfregato colla pelle di gatto. Nota. 4to. 1 p. (*Ann. del Reg. Lomb.-Veneto,* iv. 148.)   *Padova,* 1834

† Sull' influenza esercitata dalla forma delle spirali nella produzione dei fenomeni magneto-elettrici. Memoria. 4to. 15 pp. & 11 pp. 1 plate. (*Ann. del Reg. Lomb.-Veneto,* iv. 149 and 185.)   *Padova,* 1834

† Nuova proprietà delle correnti magneto-elettriche. 8vo. 3 pp. (*Bibl. Ital.* lxxiv. 305.)   *Milano,* 1834

† Elettroscopio dinamico universale del Zamboni.
Nuova macchina elettromagnetica del Dal Negro.
Limite delle prevalenze magneto-elettriche. (*Bibl. Ital.* lxxv. 288, 289, 292.)
                                                                              Milano, 1834

Retz. Fragmens sur l'Electricité humaine, 1r et 2d Mémoires. 12mo. 108 pp. (*In Esprit des Journ. &c.*)   *Paris and Amsterdam,* 1785

Retzer. (*Vide* Mako and Retzer.)

Reuss, F. F. Sur un nouvel effet de l'electricité galvanique.   *Wüllig*
Commentationes duæ, altera physica de Electricitates Voltanæ effectu novo quem hydragogum dixit etc. (*From Heinsius.*)   *Moskaw,* 1822

† **Reuss, G. Ch.** Neuere Ergebnisse im Gebiet d. Electricität u. d. Magnetismus, vom Jahr 1840 bis Mitte 1847 zusammengestellt. 8vo. 60 pp.   *Ulm,* 1847

Reuss. (*Vide* Nöggerath and Reuss.)

Reuss, Jeremias David. *Born July 30, 1750, at Rendsburg; died December 15, 1837, at Göttingen.*

Das gelehrte England u.s.w. von 1770-90. 8vo.   *Berlin and Stettin,* 1791

Das gelehrte England, u.s.w. von 1790 bis 1803. 8vo. *Berlin and Stettin,* 1804
Repertorium commentationum a societatibus litterariis editarum secundum disciplinarum ordinem digessit J. D. Reuss. 8vo.

De re electrica. In Vol. iv. p. 78, Nubes, &c. (Elect.); p. 85, Grando; p. 98, Meteora ignita et Phosphorica; p. 100, Globi ardentes; p. 103, Stellæ cadentes, &c.; p. 104, Fulgor et Tonitrus; p. 120, Presteres, &c.; p. 122, Auroræ boreales; p. 139, Meteora varia; p. 344, Electricitas; p. 368, Galvanismus; p. 370, Magnes. In Vol. x. p. 26, Vis galvanica in Corpore hum. offic. ; p. 37, De Elect. animali; p. 36, Combustio human. In Vol. xii. p. 6, Elect. Medica; p. 17, Galvan. (Med. &c.); p. 18, Magnet. (Med.); p. 20, Perkinsismus; &c.   *Göttingen*

Reusser. (*Vide* Reiszer.)

† **Revillas, D.** Observatio Auroræ Borealis Romæ visæ die 16 Dec. 1737. 12mo. 8 pp. (*Calogera, Raccolta,* xvii. 39.)   *Venezia,* 1738

† Lumen Australe Romæ observatum die 27 Jan. an. 1740 a D. de R. 4to. 2 pp. 1 plate. (*Phil. Trans.* xli. 744 for 1739-40-41.)   *London,* 1744

Revue Encyclopédique. Ou Analyse Raisonnée des Productions les plus rémarquables dans la Littérature, les Sciences, et les Arts.   *Paris,* 1819

Rey. Origine française de la Boussole et des cartes à jouer. 8vo. 28 pp. (*Extrait des Nouv. Ann. de Voyages.*)   *Paris,* 1837

† **Reyger, G.** Von einem am 26 August 1750 erschienenen Nordlichte. 4to. 7 pp. (*Versuche d. Naturf. Gesells. in Danzig,* iii. 258.)   *Danzig,* 1756

Reymond. (*Vide* Du Bois-Reymond.)

·† **Reynaud, J. J.** De la Télégraphie électrique. Resumé historique des divers systèmes de télégraphie . . suivi de la Législation. 8vo. 43 pp.
                                                                              Marseille, 1851

† Reynold-Chauvancy, De (le Comte). Télégraphie nautique polyglotte. 8vo.
108 pp. 6 plates.   Paris, 1853
(Note.—There is a 2nd edition, 1855.)

Reynold, Captain. Nautical Telegraph. Code of Signals for all Nations. 2nd
edit. 2 vols. 8vo.   1858

† Reynolds, J. R. Lectures on the Clinical Uses of Electricity. Delivered in
University College Hospital. 8vo.   London, 1872

Rhades, T. H. (?). Dissertatio inaug. de ferro sanguinis humani. ..
                                                          Gottinguæ, 1753

† Rhind, W. G. The Magnet the Life of the Mariner's Compass. 8vo. 24 pp.
Plate.   London, 1853

Rhodes, J. B. Le père de la Nouvelle Télégraphie. 4to. 1 fol. (Giorn. I.
R. Ist. Lomb. Nouv. série, viii. 361, in 4to. 1856.)

Rhostius. (Vide Weidler and Rhostius.)

Rhyzelius, A. O. Enfaldig Lära och sunfardig berattelse om Askedunder,
Blixt och Skott, bestaende af fyra Delar. 4to.   Stockholm, 1721

† Riadore, J. E. On the Remedial Influence of Oxygen, . Nitrous Oxyde, and
other gases, Electricity and Galvanism. 8vo. 177 pp.   London, 1845

† Riatti, V. La pila voltaica in generale, e come la depolarzzazione di una pila ad
un sol liquido possa vantaggiosamente ottenersi col dare al conduttore pas-
sivo la forma circolare ed un continuo movimento di rotazione. 4to. 13 pp.
(Estr. dall' Industriale Romagnolo.)   Forli, 1867

Ribbentrop, Heinrich Gottlieb Friedrich. Born March 31, 1776, at Grassleben,
near Helmstädt ; died April 20, 1834, at Brunswick.

†   Über die Blizröhren oder Fulguriten und besonders über das Vorkommen ders.
am Regenstein bei Blankenburg. 8vo. 46 pp. 1 plate. (Schweigger, Journ.
lvii. 1829.)   Braunschweig, 1830

† Ribright, T. A curious collection of Experiments to be performed on the Electrical Machines of Thomas Ribright. 2nd Ed. enlarged and improved. 8vo.
24 pp. 2 plates.   London, 1788

Ricatti, Jacopo. Sein Opere. 4 vols. 4to.   Lucca, 1765

      (Vol. iv. "Discorsi d'argomento filosofico" contains an eighth Letter addressed
      al " Sig. Conti," etc., "sopra la dissertazione dell' Aurore Boreali." The ninth
      Letter is a "Breve descrizione delle tre Aurore Boreali comparse l'anno 1741.")

Rich, O. Bibliotheca Americana Nova ; or, Catalogue of Books relating to
America since 1700. 3 vols. 8vo.   London, 1835-46

Richard. Hist. nat. de l'air et des Météores. First six volumes. 12mo.
                                                          Paris, 1770
Hist. nat. de l'air et des Météores. Last four volumes. 12mo.   Paris, 1771

Richard. Compte rendu des travaux de l'Acad. 8vo.   Lyon, 1821

Richard, L. Essai sur les instruments et sur les tables de Navigation et d'Astronomie, . . moyens de perfectionner l'Héliomètre, La Boussole, &c. 8vo.
176 pp. 6 plates.   Brest, 1840

Richard, Rudolph. Born July 6, 1763, at Rosenwinkel, near Kyritz Priegnitz ; died
July 29, 1798, at Berlin.

Observations on change of Poles of Iron Bars. . . . (Mag. de Hamburg,
tom. iv. p. 681.)

Richard, or Richardot. (Vide Anon. Lightning, 1846.)

† Richardot, C. Nouveaux appareil contre le danger de la Foudre et le fléau de la
Grêle. 8vo. 44 pp.   Paris, 1825

Nuovo sistema di apparecchi contro i pericoli del Fulmine ed il flagello della
Grandine. Trad. dal Francese. 8vo. 45 pp. Indice e Lettera.
                                                          Milano 1827

428 RIC

Richardson, Sir John. Arctic Searching Expedition. A Journal of a Boat Voyage through Rupert's Land and the Arctic Sea, with an Appendix on Natural History. 2 vols. 8vo. *London,* 1851

Richardson. (*Vide* Lefroy and Richardson.)

Riche, A. Recherches sur l'action du courant électrique sur certains métalloïdes et quelques-uns de leurs composés en dissolution dans l'eau. Thèse presentée à l'Ecole supérieure de pharmacie de Paris. 4to. 24 pp. *Paris,* 1858

Richer, T. Observations astronomiques et physiques faites en l'isle de Cayenne. 4to. *Paris,* 1679

Richerand, A. Nouveaux éléments de physiologie. 8vo. *Paris*
Experiences, observations et résumé sur le galvanisme. Note. 8vo. (*Mém. de la Soc. Méd. d'émulation,* iii. 311.) *Paris, about* 1802

† **Richers, J.** Der Magnetismus, der Galvanismus, und die Electricität. 8vo. 416 pp. *Leipzig,* 1850

Richie. (*Vide* Anon. Electricity, 1830.)

Richman, George William. *Born July* 11, 1711, *at Pernau, Livland ; died July* 26, 1753, *at St. Petersburg.*

† De electricitate in corporibus producenda nova tentamina. 4to. 25 pp. 1 plate. (*Comment. Acad. Scient. Petrop.* xiv. pro. 1744-46, imp. 1751, pp. 23 et 301.) *Petropoli,* 1744-46

† De virtute magnetica absque magnete communicata, experimenta. 4to. 6 pp. plate. (*Novi Comment. Acad. Petrop.* iv. pro 1752-3, imp. 1758, pp. 25 et 235.) *Petropoli,* 1752-3

† De indice electricitatis et de ejus usu in definiendis artificialis phænomenis. 4to. 40 pp. 1 plate. (*Novi Comment. Acad. Petrop.* iv. pro 1752-3, imp. 1758, p. 299.) *Petropoli,* 1752-3

His Electroscope described. 4to. (*Comment. Acad. Petrop.* tom. xiv. p. 302.) *Petropoli*

Richter. Ueber Duchenne's electro-physiologische Arbeiten. (*Schmidt's Jahrbücher,* 1853, B. lxxx. p. 265.) 1853

Richter. Georg Friedrich. *Born October* 26, 1691, *at Schneeberg ; died June* 23, 1742, *at Leipzig.*

† De natalibus fulminum. Tractatus physicus. Accedit appendix qua litteræ et observationes quædam huc pertinentes, Maffei, Lionii, Pagliarini, aliorumque continentur. Sm. 8vo. 12 pp. *Leipzig,* 1725

(*Note.—The letter of Maffei is that to Vallisnieri (from Padua) printed in the " Della Formazione del Fulmine."*)

† **Richter u. Erdman.** IV. Bericht über medicinische Elektricität. 4to. 23 pp. (*Med. Jahrbb.* xciv. 97.) *Dresden*

(*Vide* also Erdman and Richter.)

† **Rico-y-Sinobas,** Don M. Memoria sobre las causas meteorológicofisicas que producen las constantes sequias de Murcia y Almerya. . . . 4to. 391 pp. 9 plates. *Madrid,* 1851

† Noticia sobre las Auroras Boreales observadas en España durante el siglo xviii. y parte del xix. 4to. 15 pp. (*Memorias de la Real Acad. de Ciencias, Madrid,* tom. iii. ?) *Madrid,* 1853

† Estudio del Huracan que pasó sobre una parte de la peninsula Española el dia 29 de Octubre de 1842. 4to. 31 pp. 1 plate (i.e. map). (*Memorias de la Real Acad. de Ciencias, Madrid,* iii. 45 (?). *Madrid,* 1856

(*Note.—At the end " Valladolid,* 27 *de Mayo,* 1853.'')

Resumen de los trabajos meteorologicos correspondientes al año 1854, verificados en el Real Observatorio de Madrid. 4to. or 8vo. *Madrid,* 1857

† Discurso que sobre los fenómenos de la electricidad atmosférica leyò el Sr. D. Manuel Rico y Sinobas. . . . en la Real Acad. de Ciencias. (*Read in the Madrid Academy.*) 4to. 71 pp. *Madrid,* 1858

Rico, Santistéban. Manual de física y elementos de química. 2 edicion. 8vo. 544 pp. *Madrid,* 1858

† **Ricoby, H. de.** Mémoire sur mes travaux electro-chimiques. 8vo. 63 pp.
Paris, 1851

† **Riddel (Lieut.) C. J. B.** Magnetical Instructions for the use of portable instruments . . for Magnetical Surveys and portable Observatories, and for the use of . . small instruments . . for a fixed Magnetical Observatory, with Forms for . . . Observatories. 8vo. 146 pp. *London,* 1844

† Supplement to the Magnetical Instructions, &c. 8vo. 22 pp. *London,* 1846

Riddermarck, And. De lachrymis vitreis. *Lund,* 1698?

† **Ridley, M.** A short Treatise of Magnetical Bodies and Motions. Sm. 4to. 157 pp.
Engraved title-page and portrait. *London,* 1613
Animadversions on a Work, &c. 4to. (*From Watt.*) *London,* 1617

Ridolfi, C. (Marchese). Pensieri intorno ai singolari fenomeni elettro-magnetici.
Firenze, 1821
Lettera intorno al fenomeno elettro-magnetico ecc. *Firenze,* 1821

† Lettera al Taddei intorno di nuovi fenomeni elettro-magnetici, (dated) Firenze 1o Ottob. 1821. 8vo. 5 pp. (*Estratta dall' Antologia.*) *Firenze,* 1821

Description of Novelluci's Machine in the Antologia di Firenze, Agosto, 1824, p. 159. (*From Biblia Ital.* lxiii. 268.

(*Ridolfi's title is "Nuovo miglioramento delle Macchine elettriche a disco." V. Arella.*)

† Lettera . . . al Direttore di questi Annali sopra le scintille elettriche prodotte dal magnetismo. 4to. 2 pp. (*Ann. del Reg. Lomb.-Veneto,* ii. 113.)
Padova, 1832

† Riflessioni sulle Osservazioni e fatti risguardanti i fenomeni elettro-magnetici del Gazzeri. 8vo. 14 pp.

Riecke, Fried. J. P. *Born* 1794.
Über d. Erricht. von Hagelableitern. (*Würtemb. Correspondenzblatt,* vii.)
Würtemburg
Über d. Mittel, die Weinberge gegen Frühlingsfrost zu schützen. (*Würtemb. Correspondenzblatt,* xiv.)
Würtemburg

Riedel, T. G. Gründl. Unterricht vom Gebrauch. d. Boussole in d. pract. Geometrie. 8vo. 12 plates. *Leipzig,* 1795

Rieger, Christian. *Born May* 4, 1714, *at Vienna; died March* 26, 1780, *at Vienna.*
Observaciones fisicas sobre la fuerza electrica, grande y fulmine, confirmada y aumentada con nuevos experimentos. 4to. *Madrid,* 1763

Riess, Peter Theophil. *Born June* 27, 1805, *at Berlin.*
About sixty-seven Articles in Poggendorff's Annalen, vols. xvi. 1829 to vol. cx. 1860, some few of which are also in the Abhandl. d. Berlin Acad., as specified in this Catalogue.

† De telluris magnetismi mutationibus, et diurnis et menstruis. Dissertatio inauguralis. 4to. 62 pp. 1 plate. *Berolini,* 1831

Über gebundene Electricitat (Vertheilung). 8vo. (*Dove's Repert. d. Phys.* ii. 30; *and in Poggend. Ann.* xxxvii. 1836.) *Leipzig,* 1836

† Fortgesetzte Untersuchen über d. Nebenstrom der electrischen Batterie. 8vo. 24 pp. (*Ann. d. Phys. (Pog.)* vol. 1. No. 5.) *Leipzig,* 1840

† Über d. Anordnung d. Elektr. auf Leitern. Gelesen in d. Akad. 22 Feb. 1844. 4to. 47 pp. 1 plate. (*Abhandl. Berlin Acad.* 1844.) *Berlin,* 1844

† Über das Glühen und Schmelzen von Metalldräthen durch Electricität. Read 6th June, 1845. 4to. 52 pp. (*Abhandl. d. K. Akad. d. Wissen. zu Berlin,* pro 1845.) *Berlin,* 1847

Riess, Peter Theophil.—*continued.*

† Über elektrische Figuren und Bilder. Read 5th Feb. 1846. 4to. 50 pp.
(*Abhandl. d. K. Akad. d. Wissen. zu Berlin pro* 1846.) *Berlin, 1846 ?*

† Über die Seitenentladung der elektrischen Batterie. Read 15th Feb. 1849.
4to. 34 pp. (*Abhand. d. K. Akad. zu Berlin, pro* 1849.) *Berlin,* 1849

† Über den elektrischen Entladungsstrom in einem dauernd unterbrochenen
Schliessungsbogen. Read 18th April, 1850. 4to. 36 pp. (*Abhandl. d. K.
Akad. zu Berlin, pro* 1850.) *Berlin,* 1850

† Die Lehre von der Reibungs-Elektricität. 8vo. 2 vols. 9 plates.
Berlin, 1853

† Die Lehre von der Reibungs-Elektricität. 8vo. 2 vols. 12 plates. (*Poggend.*
ii. 642.) *Berlin,* 1853

† Über d. Unterbrechnung des Schliessungsbogens der elektr. Batterie durch
einen Condensator. 8vo. 20 pp. (*Bericht üb. d. Verhandl. d. K. Preuss. Akad.*
Nov. 1853.) *Berlin,* 1853

† On the Generation of Heat by Electricity. 8vo. 2 pp. (*Phil. Mag.* June, 1854,
vol. vii. p. 428.) *London,* 1854

† Note sur l'électricité dissimulée et Mém. sur l'Electrisation par influence, et
la Théorie du Condensateur. Ext. par Verdet. 8vo. 8 pp. (*Ann. de Chim.*)
Paris, 1854 ?

† Über die elektr. Funkenentladung in Flüssigkeiten. (*Poggend. Ann. cii. Aus*
den Monatsbericht. d. Akad. Juli, 1857.) *Leipzig,* 1857

Die Lehre von der Reibungs-Elektricität. 2 vols. 8vo. 12 plates.
Berlin, 1858

† Über die Prüfungsmittel des Stromes d. Leydener Batterie. 8vo. 21 pp.
(*Monats. Bericht d. Berlin Akad.* Jan. 1860.) *Berlin,* 1860

† Sur les Figures Roriques et les Bandes colorées produites par l'Electricité.
Translated by E. Wartmann. 8vo. 7 pp. (*Ext. des Arch. de l'Elect. Suppl.*
à la Bibl. Univer. de Genève.) *Geneva*

† Über d. elektr. Ringfiguren. 4to. 33 pp. (*Abhandl. Berlin Akad.* 1861.)
Berlin, 1861

† Abhandlungen zu der Lehre von der Reibungs-Elektricität. 8vo. 394 pp. 1
plate. *Berlin,* 1867

† **Riess and Moser.** On Magnetising Power of the Solar Rays. 8vo. 1 p. "Ann.
de Chim. Nov. 1829." (*Phil. Mag. or Annals,* viii. 155.) *London,* 1830

Riess, P. T. und Rose. Über d. Pyro-Elektricität d. Minerale mit G. Rose. 4to.
(*Abhandl. d. Berlin Acad.* 1843.) *Berlin,* 1843

Riess. (*Vide* Faraday and Reiss.)

Rieu. (*Vide* Anon. Magnetism, 1847, and the four entries following.)

† **Riffault.** Account of some additional Experiments made by the Galvanic Society
of Paris. Communicated by M. Riffault. 8vo. 5 pp. (*Phil. Mag.* xxviii.
35.) *London,* 1807

Rigaud, S. P. Correspondence of Scientific Men of the 17th Century. 2 vols.
8vo. *London,* 1862

Rijke, Pieter Leonhard. *Born July* 11, 1812, *at Hemmen, Netherlands.*

† Spec. phys. inaugurale de origine electricitatis voltaicæ. Read 30th June,
1836. 4to. 81 pp. *Trajecti ad Mosam,* 1836

Erklärung des durch e. unterbrochn. galvan. Strom unter Umständen verur-
sachten Geräusches. 8vo. (*Pogg. Ann.* lxxxix. 1853.) *Leipzig,* 1853

Über d. Schlagweite d. Ruhmkorffschen Apparats. 8vo. (*Pogg. Ann.* xcvii.
1856.) *Leipzig,* 1856

Über d. Elektricitäts-Erregung, die man beobachtet wenn e. Flüssigk. d.
sphäroi lalen Zustand verlässt. 8vo. (*Pogg. Ann.* xcviii. 1856.)
Leipzig, 1856

Rijke, Pieter Leonhard.—*continued.*

Über d. Extraströme. 8vo. (*Pogg. Ann.* cii. 1857.) *Leipzig,* 1857

Über d. Schlagweite d. elektr. Batterie. 8vo. (*Pogg. Ann.* cvi. u. cvii. 1859, cix. 1860.) *Leipzig,* 1859-60

Über d. Inductionsfunken. 8vo. (*Pogg. Ann.* cxi. 1860.) *Leipzig,* 1860

Rinklake. Beiträge zur Kenntniss des Galvanismus, und Resultate s. Untersuchung. 2 Bde und neue Beitr. i. Heft. *Jena,* 1800-8

Elektrische Versuche an der *Mimosa pudica* (L.) in Paralele mit Versuchen an Fröschen. 4to.

Rinmann, S. *Born June* 12, 1720 (*A. St.*), *at Upsala ; died December* 20, 1792, *at Eskilstuna.*

Anmerkungen vom leuchtenden Spat von Garpenberg. 8vo. 6 pp. (*K. Schwed. Akad.* Abh. ix. 186.) *Hamburg,* 1746

† Anmerkungen eisenhaltige Erd- und Steinarten betreffend. 8vo. 14 pp. (*K. Akad. Schwed.* Abh. xvi. 286.) *Hamburg and Leipzig,* 1754

† Mineralogische Untersuchung vom Tourmalin oder Aschenblaser. 8vo. 12 pp. (*K. Schwed. Akad.* Abh. xxviii. 46.) *Leipzig,* 1766

† Fernere mineralogische Untersuchung der brasilischen Tourmaline. 8vo. 8 pp. 1 plate. (*K. Schwed. Akad.* Abh. xxviii. 114.) *Leipzig,* 1766

Anledning till Stål- och Jernsförädlingen och dess forbattring. 8vo. *Stockholm,* 1772

Forsok till Jernets historia. 2 vols. 4to. *Stockholm,* 1782

Versuche e. Geschichte des Eisens. 2 vols. 8vo. *Berlin,* 1785

Anleit. z. Kentn. d. Eisen- u. Stahlveredlung. 8vo. *Wien,* 1790

Versuche e. Geschichte des Eisens mit Anmerkk. von K. T. B. Karsten. 2 vols. 8vo. *Liegnitz,* 1814

Geschichte des Eisens: übers. von Georgi.

(*Note.—Poggendorff says,* ii. 646, " *Försök till Jernets historia,* 2 vols. 4to. *Stock.* 1782, *and Deutsch,* 2 vols. 8vo. *Berlin,* 1785 ; *mit Anmerkk. von K. T. B. Karsten,* 2 vols. 8vo. *Leignitz,* 1814.)

Rio. (*Vide* Da Rio.)

† **Ritchie,** W. On Electro-Magnetism and Ampère's proposal of Telegraphic Communication by means of this power. 8vo. 1 p. (*Phil. Mag. or Annals,* vii. 212.) *London,* 1830

† Experimental Researches in Electro-Magnetism and Magneto-Electricity. Read March 21, 1833. 4to. 9 pp. 1 plate. (*Phil. Trans.* 1833, p. 313.) *London,* 1833

(*Note.—This paper contains, at p.* 319, *his first actually constructed revolving apparatus. Description and figure.* " *A revolving apparatus of this kind had a power sufficient to raise several ounces over a pulley.*")

† **Rittenhouse,** D. An account of some Experiments on Magnetism, in a Letter to Jno. Page, at Williamsburg. 4to. 1 p. (*Trans. Amer. Phil. Soc.* Old Series, vol. ii. 178.) *Philadelphia,* 1786

(*Vide* also Page and Rittenhouse.)

† **Rittenhouse and Hopkinson.** An account of the effects of a Stroke of Lightning on a house furnished with Two Conductors ; in a letter . . to Mr. R. Patterson. Read Oct. 15, 1790. 4to. 4 pp. (*Trans. Amer. Phil. Soc.* Old Series, vol. iii.) *Philadelphia,* 1793

† **Rittenhouse and Jones.** Account of several Houses in Philadelphia struck by Lightning, June 7th, 1789. Read July 17, 1789. 4to. 4 pp. 1 plate. (*Trans. Amer. Phil. Soc.* Old Series, vol. iii. p. 119.) *Philadelphia,* 1793

Ritter. "Mémoires de Ritter, concernant les effets du Galvanisme. 16mo. (*Almanach ou portefeuille des Chimistes pour l'année,* 1801.) *Weimar,* 1801

Ritter.—*continued.*

Expériences par lesquelles il cherche à prouver l'identité du galvanisme et de l'électricité : détail communiqué par Pfaff. (*Bulletin de la Soc. Philomat.* Thermidor, an. 9, No. 53.) *Paris,* 1801

† Galvanic Experiments from Van Mons' Journal, No. 17. 8vo. 2 pp. (*Phil. Mag.* xxiii. 54.) *London,* 1806

† New Galvanic discoveries by M. Ritter. Extracted from a letter from M. Christ. Bernouilli. " Abridged from Van Mons' Journal, vol. vi." 8vo. 4 pp. (*Phil. Mag.* xxiii. 51.) *London,* 1806

Ritter, Johann Wilhelm. *Born December* 16, 1776, *at Samitz, near Hainau, Schlesien; died January* 23, 1810, *at Munich.*

Expériences sur les Rayons invisibles du spectre solaire. 4to. (*Soc. Philomat.* an. 11, p. 197.) *Paris*

Nouvelles expériences galvaniques. 4to. (*Soc Philomat.* an. 4, tom. ii. p. 181.) *Paris,* 1796

† Beweis, dass ein beständiger Galvanismus den Lebensprocess in dem Thierreich begleitet. Nebst neuen Versuchen und Bemerkungen über den Galvanismus. 8vo. 174 pp. 2 plates. *Wiemar,* 1798

Darstell. d. neuesten Unters. üb. d. Leuchten des Phosphors m Stickstoffgas. 1st Stück. *Jena,* 1800

† Beyträge zu nähern Kenntniss des Galvanismus und der Resultate seiner Untersuchung. 8vo. Erst. Bandes, 1 und 2 Stück, 284 pp. 3 plates. *Jena,* 1800

Expériences par lesquelles il cherche à prouver l'Identité du Galvanisme et de l'Electricité. 4to. (*Soc. Philomat.* an. 9, p. 39.) *Paris,* 1801

Galvan. Versuche über d. chem. Natur. d. Wassers. 8vo. (*Crell's Annal.* 1801.) 1801

Beitr. z. Kenntn. d. Chem. Wirkungen d. Galvanismus. 8vo. (*Gottlings Almanach,* 1801.) 1801

Neue Versuche über d. Galvanismus. (*Reichsanzeiger,* 1802, Bd. i. No. 66, u. Bd. ii. No. 194.) 1802

† Beyträge zur nähern Kenntniss des Galvanismus und der Resultate seiner Untersuchung. 8vo. Erst. Bandes, 3 und 4 Stück, 290 pp. 1 plate. *Jena,* 1802

† Beyträge zur nähern Kenntniss des Galvanismus und der Resultate seiner Untersuchung. 8vo. Zweiten Bandes, 1 Stück, 172 pp. 1 plate. Idem, 2 Stück, 158 pp. 1 plate. *Jena,* 1802

Über verschiedene merkw. Erscheinungen, welche mehre Metalle in d. galvan Kette darbieten. 8vo. (*Gehlen's Journ. f. Chem.* iii. 1804.) *Berlin,* 1804

† Beyträge zur nähern Kenntniss des Galvanismus und der Resultate seiner Untersuchung. Zweiten Bandes, (und litzes) Stück. 8vo. 370 pp. 1 plate. *Jena,* 1805

Twelve Articles (about) on Magnetism, Galvanism, Electricity, Auroræ, Meteorites, &c. in Gilbert's Annalen, vol. ii. 1799, to vol. xix. 1805.

Five Articles (about) in Voigt's Magazine für Naturk. ii. 1800 to ix. 1805, on Galvanism, Electricity, and Magnetism.

† Das elek. System der Körper. Ein Versuch. 8vo. 412 pp. *Leipzig,* 1805

† Physisch. Chemische Abhandlungen in chronologischer Folge. 3 vols. 8vo. 1st Bd. xxx. and 326 pp. 1 plate. 2nd Bd. xiv. and 360 pp. 1 plate. 3rd Bd. viii. and 388 pp. 2 plates. *Leipzig,* 1806

† Extract of a letter to Professor Pictet from a Correspondent at Munich, upon some Galvanico-Magnetic experiments recently made by M. Ritter. Dated Dec. 31, 1805. 8vo. 2 pp. (*From Bibliot. Britan.* xxxi.; *Phil. Mag.* xxv. 368.) *London,* 1806

RIT- ROB 433

Ritter, Johann Wilhelm.—*continued.*

About five Articles on Galvanism and Magnetism in Gehlen's Journ. für Chemie, vol. iii. 1804, to vol. vi. 1806 ; and twelve Articles in Gehlen's Journal für Chemie und Physik, vol. i. 1806, to vol. viii. 1809.

† Der Siderismus. 8vo. 210 pp. *Tubingen,* 1806
(*Note.—This has also the T.-p. Neue Beyträge zur nähern Kenntniss des Galvanismus . . Erstes Bandes, Erstes Stuck.*)

Neue Versuche über den Einfluss des Galvanismus auf die Erregbarkeit thierischer Nerven. 4to. (*Denkschr. der Akad. der Wissench. zu München,* an. 1808, p. 257.) *München,* 1808

Fernere Versuche und Bemerkung üb. Davy's metallähnliche Producte aus Alkalien, etc. 4to. 1808

Abhandl. v. T. W. Ritter. *Berlin,* 1809

Amoretti's phys. u. hist. Untersuchungen über d. Rhabdomantie oder animal Elektrometrie u.s.w. 1 Bd. 1 St. 8vo. *Berlin,* 1809

† Fragmente aus dem Nachlasse eines jungen Physikers. 2 vols. 12mo.
Heidelberg, 1810

Elektr. Versuche an. d. *Mimosa pudica.* 4to. (*Denkschr. d. Münch. Acad. i.* 1814.) *München,* 1814

Beantwortung der Frage; welches Ende der Voltaischen Batterie hat man, Gründen zu Folge, das Zink, welches das Silberende derselben zu nennen? 8vo. (*Gibert's Annalen de Phys.* ix. Bd. 2 St.)

Extrait d'un Mémoire sur le Galvanisme, envoyé par J. G. Ritter à l'Institut National. 4to. (*Soc. Philomat.* an 12, p. 145.) *Paris,* 1804

† **Ritter** e (l'Editore) **Amoretti.** Scoperte galvaniche del Ritter, estratto da una lettera del C. Bernouilli e L'Editore (Amoretti). Nota (sul Romagnosi, Bouvier, &c.). 4to. 2 pp. (*Nuova Scelta d'Opus.* i. 201.) *Milano,* 1804

† Estratto d'una lettera scritta da Monaco al Pictet di Ginevra su alcuni sperimenti galvanico-magnetici, fatti recentemente dal Ritter. Monaco, 31st Dec. 1805. Nota del Traduttore (Amoretti). 4to. 3 pp. (*From Bibliot. Britan.* No. 242 ; *Nuova Scelta d'Opusc.* i. 334.) *Milano,* 1804
(*Note.—The date,* 1804, *in this volume is that of the Title-page. The volume contains works which were printed in Fascicoli subsequent to* 1804.)

Rive, De La. (*Vide* La Rive, De.)

Rivoire, Antoine. *Born March* 13, 1709, *at Lyons ; died about* 1789.

† Traités sur les Aimants artificiels . . traduits de deux ouvrages de Michell et Canton ; avec une Préface historique du traducteur. 12mo. 160 pp. 4 plates. *Paris,* 1752
(*Note.—At the end is a MSS.* (*Anon.*) *entitled " Recueil d'expériences diverses sur l'Aimant, pour servir de suite au traité sur les aimants tant artificiels que naturels. Extrait par un curieux."* Dated 1764, 16 pp. *including the title.*)

Rivot, Louis Edmond. *Born October* 12, 1820, *at Paris.*

† **Rivot** et **Phillips.** Note sur la conductibilité électrique des principales Roches à de hautes températures. 8vo. 11 pp. 1 plate. (*Extrait des Ann. des Mines,* xiv. 1848.) *Paris,* 1848

† **Rizzardi,** G. Proposizioni di Magnetismo, elettro-dinamica, ed elettro-magnetismo. 4to. 15 pp. *Parma,* 1830

Roberts, J. M. On the process of Blasting by Galvanism. 8vo. 36 pp. Plates.
London, 1840

Robertson. Natural History of the Atmosphere. 2 vols. 8vo. *London*

Robertson, Etienne Gaspard. *Born June* 25, 1763, *at Liége ; died July,* 1837, *at Batignolles-Monceaux, near Paris.*

Sur l'électrophore résineux et papiracé. 4to. (*Journ. Phys.* xxxvii. 1790.)
Paris, 1790

Robertson, Etienne Gaspard.—*continued.*

Improvement in Galvanic pile (or couronne de tasses), Galvanometer, &c. (*Journ. de Paris, les numéros du 10 fructidor, du 15 et du 17 du même mois.*)
Paris, 1800

On "Acide Galvanique." (*Journ. de Paris, 2e Jour complémentaire, No. 362.*)
Paris, 1800

Expériences nouvelles sur le fluide galvanique. Lu à l'Institut, 11 Fructidor, an 8. 8vo. (*Ann. de Chim.* xxxvii. 132, 1801.)   Paris, 1801

His Piles métalliques de 2,500 plaques de zinc et autant en cuivre (mentioned by the Editors), and also his new experiment " avec deux charbons ardents," communicating with the poles of a column and producing an intense white light. (*Journ. de Paris du* 22 Ventôse, an 10, 1802.)   Paris, 1802

Magnetic deductions in Jamaica.

Robertson. (*Vide* Volta.) Nouvelle expériences sur le Galvanisme.

Robertus, J. Curationis magneticæ et unguenti amarii magica impostura clare demonstrata . . responsis ad . . Helmont. (*Poggendorff,* ii. 663.)
Luxemb. 1621, Coloniæ, 1622

Robespierre. Un Plaidoyer prononcé dans une cause relative à un Paratonnerre. 8vo. (*From Marget, Etude sur les travaux de Romas,* page 80—*not the exact title. Marget also refers to a mention of the subject in the Journ. des Savans without date. The judgment of the Tribunal d'Arras was on the* 31st *May,* 1783. *I have seen a printed copy in Paris of this " Plaidoyer.*")   1783?

Robida, K. Vibrations-Theorie der Elektricität. 8vo. 37 pp. *Klagenfurt,* 1857

Magnetismus als Fortsetzung und Schluss der Vibrations-Theorie der Elektrizität. 8vo. 60 pp.   *Klagenfurt,* 1858

Erklärung der Lichterscheinungen aus den Grundzügen einer naturgemässen Atomistik. 2 Heft. 8vo. 36 pp. 1 plate.   *Klagenfurt,* 1861

Robillard. (*Vide* Chappe, &c.)

† **Robin, C.** Recherches sur un appareil . . des Raies . . organes élect. "Lu à l'Institut 18 Mai 1846." 8vo. 110 pp. 2 plates.   *Paris,* 1847

Robins, B. A Letter showing that the Electricity of Glass disturbs the Mariner's Compass and also nice balances. 4to. (*Phil. Trans.* 1746, p. 242.)
London, 1746

† **Robinson, T. R.** On the relation between the Temperature of Metallic Conductors and their Resistance to Electric Currents. Read Nov. 30, 1848. 4to. 24 pp. (*Trans. R. Irish Acad.* xxii. pt. i.) (*Faraday's copy.*)   *Dublin,* 1849

† Experimental Researches on the Lifting Power of the Electro-Magnet. Part i. Read June 14, 1852. 4to. 21 pp. (*Trans. R. Irish Acad.* xxii. p. 291.)
Dublin, 1852

† Experimental Researches on the Lifting Power of the Electro-Magnet. Part ii. Read June 26, 1854. 4to. 26 pp. (*Trans. R. Irish Acad.* xxii. p. 499.)
Dublin, 1854

Experimental Researches on the Lifting Power of the Electro-Magnet. Part iii. Read November 30, 1857. 4to. 33 pp. (*Royal Irish Acad.* xxiii.)
Dublin, 1858

† **Robinson, W.** A brief account of the Application of Magnetism to the Manufacture of Wrought Iron. 8vo. 8 pp.   *London,* 1867

Robison, John. *Born* 1739, *at Boghall, near Glasgow ; died January* 30, 1805, *at Edinburgh.*

Encyclopædia Britannica, Supplement, Article " Electricity," vol. i. p. 571, &c. 4to.   1797

Encyclopædia Britannica, Supplement, Article " Thunder." 4to.   1801

Encyclopædia Britannica, Supplement, Article "Magnetism." (*From Young,* p. 442.)

Encyclopædia Britannica, Article "Variation.' (*From Young,* p. 442.)

Elements of Mechanical Philosophy. Vol. i. 8vo.   *London*

† **Robison, J.** and **Brewster.** A System of Mechanical Philosophy, with Notes by D. Brewster. 4 vols. and 1 vol. of plates. 8vo.
London and Edinburgh, 1822

† **Rocaut, De.** Delle variazioni del Barometro. 12mo. 10 pp. (*Scelta d'Opuscoli,* xxxi. 91.) *Milano,* 1777

Roch. Über eine Umgestaltung der Ampère'schen Formel. (*Zeitschr. f. Mathem.* 1859, p. 295.) 1859

Ueber magnetische Momente. (*Zeitschr. f. Mathem.* 1859, p. 374.) 1859

Ueber Magnetismus. (*Zeitschr. f. Mathem.* 1860, p. 415.) 1860

† **Roch, G.** Anwendung d. Potential-Ausdrücke auf d. Theorie d. molekularphysicalischen Fernwirkung u. d. Bewegung d. Electricität in Leitern. 4to. 28 pp. ("*Inaugural Dissertation zur Erlangung der philosophischen Doctorwürde an der Universität Leipzig.*") *Göttingen,* 1862

Rochas. (*Vide* Breton and Beau de Rochas.)

Roche, Edouard Albert. *Born October* 17, 1820, *at Montpellier.*

† Note sur la distribution de l'Electricité à la surface des corps sphériques. 4to. 15 pp. (*Montpell. Acad. Sect. Sciences,* ii. 115.) *Montpellier,* 1852-3

† **Roche, R.** A Letter from Roche, R. to the President, of a fustian frock being set on fire by Electricity. Read May 19, 1747. 4to. 2 pp. (*Phil. Trans.* xlv. 323.) *London,* 1848

Rode, J. Versuch einer fasslichen Darstellung des Elektromagnetischen Telegraphen. 8vo. 28 pp. 1 plate. *Grünberg,* 1852

† **Rodolfi, B.** Sulla frequenza della grandine. Memoria. 8vo. 2 pp. (*Commentarii dell' Ateneo di Brescia per l'anno Accad.* 1827 (*stampata* 1828), p. 104.) *Brescia,* 1828

† **Rodwell, G. F.** Dictionary of Science. (*From the "Athenæum,"* July to December, 1871.) 1871?

† **Roessinger, F.** Fragment sur l'Electricité universelle. 8vo. 187 pp. *Paris et Genève,* 1839

† Résumé de l'ouvrage "Fragment sur l'Electricité Universelle. 8vo. 41 pp. *Genève,* 1840

Roger-Martin. Sur une Foudre ascendante. *Toulouse ?*

Rogers. Electrical Machines. 8vo. (*Libri Sale Cat.* 1861, p. 273.) *London,* 1779

Rogers, James B. *Died June* 15, 1852, *at Philadelphia.*

Telegraphic Dictionary. 8vo. (*From Roorbach's Bibl. Amer. of* 1852.) *United States*

† **Rogers, W.** The variation of the Magnetical Compass, observed by Capt. Rogers, Commander of the ship "Duke," in his passage from Cape St. Lucar, in California, to the Isle of Guam or Guana, one of the Ladrones, with some remarks thereon communicated by the same. 4to. 4 pp. (*Phil. Trans.* xxxi. *for* 1720-21, p. 173.) *London,* 1723

† **Rogers, W. B.** and **Rogers, H. D.** Experimental Enquiry into some of the laws of the elementary Voltaic Battery. 8vo. 23 pp. (*Silliman's Journ.* (*q.*) xxvii. p. 39.) *New York*

Roget, Peter Mark. *Born January* 18, 1779, *at London.*

† Electro-Magnetism in the "Library of Useful Knowledge." 8vo. *London,* 1831

† On the Geometric Properties of the Magnetic Curve, with an account of an instrument for its mechanical description. 8vo. (*Journ. of the R. Instit.* No. 2, p. 311, 1831.) *London,* 1831

† Treatise on Electricity, Galvanism, Magnetism, and Electro-Magnetism (four Treatises comprising a volume of the "Library of Useful Knowledge.") 8vo. *London,* 1832

Roget, Peter Mark.—*continued.*

† Darstellung der Elektrizität, des Magnetismus, des Galvanismus des Electro-Magnetismus, herausgegeben von der Gesellschaft zur Verbreitung nütz-licher Kenntnisse. Aus dem Engl. von Franz Kottenkamp. Small 8vo.
Stuttgart, 1847

† Treatise on Electricity, Galvanism, &c. 8vo. *London,* 1848

On Ampère's Theory of Magnetism. (*Quarterly Review.*) (*From Lamont, Handb.* p. 445.)

Rohault, Jacques. *Born* 1620, *at Amiens; died* 1675, *at Paris.*

Traité de physique. 2 vols. 12mo. *Paris,* 1671-82

Entretiens sur la Philosophie. 12mo. *Paris,* 1671

Œuvres posthumes. 4to. *Paris,* 1682

† Traité de Physique. New edit. 2 vols. 12mo. Figs. (*De l'Aimant,* pp. 216—260, vol. ii. (*and Amber*) *Du Tonnerre, &c.* pp. 310—317, vol. ii.) 1730

Rohde. Système complet de Signaux de Jour et de Nuit &c. 8vo. (*From Bachelier's Cat. of* 1853, p. 85.) 1835

† **Rohr,** J. B. von. Physikalische Bibliothek, herausgegeben von. A. G. Kastner. 8vo. *Leipzig,* 1754

† Physikalische Bibliothek, herausgegeben von Kastner mit vielen Zusätzen und Verbesserungen. 8vo. 694 pp. 1 plate. *Leipzig,*1754

Roi. (*Vide* Du Roi.)

Roiffé. (*Vide* Perrière de Roiffé.)

† **Rollmann,** W. Die Thermoelektricität. 4to. 38 pp. *Stralsund,* 1859

Roloff, J. F. Die Mechanik des Elektromagnetismus, ihren Grundbegriffen nach entwickelt ; so wie mehrere neue elektromagnetische Maschinen und Ap-parate, um den Elektromagnetismus als Triebkraft für Mühlen, Uhren, Telegraphen etc. anzuwenden. 4to. 24 pp. 1 plate. (*From Zuchold's Cat.* 1851, p. 91.) *Berlin,* 1851

Die Mechanik des Elektromagnetismus, ihren Grundbegriffen nach entwickelt. 2 Auflage. 4to. 32 pp. 2 plates. (*From Zuchold's Cat.* 1859, p. 76.) 1859

† Der Elektromagnetismus insbesondere als Triebkraft ; sowie mehrere neue elektromagnetische Maschinen, Wagen und Locomotiven. 8vo. 192 pp. 8 plates. *Berlin,* 1868

(*Note.—The first* (*or short*) *title-page contains the following titles of work by him :* " *Der Electromagnetismus,*" " *Die Mechanik des Electromagnetismus,*" 3 Auflage, " *Die Maschinenlehre des Electromagnetismus.*")

Romagnosi, Gian Domenico. *Born December* 13, 1761, *at Salso Maggiore, near Piacenza ; died June* 8, 1835, *at Milan.*

Articolo sul Galvanismo. (*Gazzetta di Trento,* 3 Aug. 1802.) *Trento,* 1802

(*Note.—Govi gives, at p.* 8, *the Article in full* (*with comments*), *and refers to the probability of it having appeared in the contemporaneous· periodical at Roveredo* (*if this is not quoted in error*). *Govi's remarks, &c. are very good, and contain abundant evidence against magnetic action of the column on the Magnetic Needle. Romagnosi's experiment was in fact a modification of Milner's as performed with static electricity.* (*Vide* " *Milner's Experiments and Observations in Electricity,* 1783. 8vo. p. 35, *and fig.* 3.") *Milner's instrument, an Electrometer, has been reproduced in a different form by Peltier, as a new invention. Poggendorff,* ii. 682, *refers to other particulars concerning Romagnosi's experiment.*)

Account of his Electromagnetic Observations in the Gazzetta di Trento of the 3rd August, 1802. (*Vide Editor's Review of Zantedeschi's* " *Saggio*" *in the Bibl. Ital.* tom. xcviii. p. 60. *They give the account at length, and ascribe* (*not confidently*) *the repulsion of the insulated needle after having been touched by the voltaic galvanic iron, to electricity* (*not to magnetism*). *This remark of theirs was written in* 1840, *and is the true explanation of Romagnosi's Experiment.*)

(*Vide* also Cantu, C.)

Romagnosi, Gian Domenico.—*continued.*

His discovery described in the Gazzetta di Roveredo del 1802, No. 65. (*Vide Amoretti's note attached to an extract from C. Bernouilli's letter in the Nuova Scelta d' Opuscoli,* tom. i. p. 201, 1804.)
Cenni sulla vita di, da De Giorgi.
Biografia di, da Sacchi.

Roman Accad. dei Nuovi Lincei. Atti dell' Accad. Pontificia de' Nuovi Lincei.
compilati dal Segretario. 4to. Figs. *Roma*, 1854

Roman Giornale Arcadico. Giornale Arcadico di Scienze. 4to. *Roma*
(*Note.*—Tom. ix. dated 1821, tom. cxliv. 1856.)

Romas, . . . **de.** *Born beginning of eighteenth century, at Nérac; died* 1776, *at Nérac.*

A Mem. presented to the Acad. de Bordeaux in MS. (*Merget, Etude sur les travaux de Romas, says that the MS. still exists. Romas presented many Mems. to the Acad.*) 1753

Mémoire ou après avoir donné un moyen aisé pour élever fort haut, et à peu de frais, un corps électrisable isolé, un cerf volant, on rapporte des observations frappantes, qui prouvent que plus le corps isolé est élevé au dessus de la terre, plus le feu de l'électricité est abondant. 4to. (*Mém. de Mathém. et de Phys.* ii. p. 393.) *Paris*, 1755

Electrical Experiments made with a Paper Kite, raised to a very considerable height in the air on the 7th June, 1753. 8vo. (*Gentleman's Magazine for* Aug. 1756, p. 378.) *London*, 1756

Sur l'expérience électrique du cerf volant. 4to. (*Mém. de Mathém. et Phys.* iv. 514.) *Paris*, 1763

Account of another Kite. (*Gentleman's Mag. for* Feb. 1763, p. 69.) 1763

† Letter from M. De Romas to the Abbé Nollet containing experiments made with an Electrical Kite. 8vo. (*Gentleman's Mag. for* March, 1764, p. 109.) *London*, 1764

† Mémoire sur les moyens de se garantir de la Foudre dans les maisons; suivi d'une Lettre sur l'invention du Cerf volant électrique, avec les pièces justificatives de cette même lettre. 12mo. 156 pp. 2 plates. *Bordeaux*, 1776

(*The Pièces Justicatives contain testimonials, a certificate of the Bordeaux Acad. &c., which prove that he had invented (imaginé) (but had not used) the Electrical Kite on the 12th July, 1752. Merget, Etude sur les travaux de Romas, imputes to Franklin (by implication) the possibility of having derived the idea from Romas: without foundation, I think.*—F. R.)

Romershausen, E. Apparat zur Beobachtung d. atmosphär. Electricität. 8vo. (*Pogg. Ann.* lxix. and lxxxiii.) *Leipzig*, 1846-58

† Antagonismus d. Electricität und des Magnetismus. Erstes Heft. 12mo. 48 pp. 1 plate. *Halle*, 1846

Die magneto-elektrische Rotationsmaschine u. des Stahlmagnet als Heilmittel u.s.w. *Halle*, 1847

Der einfache galvano-elektr. Bogen als Heil. u. Schutzmittel. *Halle*, 1848

† Galvano-electrischer Apparat zur Förderung der Vegetation und Fruchtbarkeit des Bodens, für Land- und Gartenwirthschaft, Kunstgärtnerei und Blumenculture. 8vo. 16 pp. *Marburg*, 1851

Die Heilkräfte d. Elektricität u. Magnetismus. 1 Aufl. *Marburg*, 1851

† Ein neues Galvanometer zur Ergänzung d. Oersted'schen Fundamental Versuchs, &c. . . 8vo. 16 pp. 1 plate. (*Aus Dingler's Polytech. Journal,* Bd. cxvii. Heft 5, *besond. abgedruckt.*) *Stuttgart*, 1852

Die Heilkraft d. Elektricität u. Magnetismus. 2 Aufl. *Marburg*, 1853

† Die Elektricität in Beziehung auf d. Salubrität unseres Wohnortes, &c. . 8vo. 8 pp. (*Aus Dingler's Polytech. Journal,* cxxxi. Heft 1, *besond. abgedruckt.* *Fortsetzung des Aufsatzes in Polytech. Journal,* Bd. cxxx. S. 193.) *Stuttgart*, 1854

Romershausen, E.—*continued.*

† Beobachtungen d. atmosphärisch und terrestrisch. Electricität. 8vo. 8 pp. (*Dingler's Polytech. Journal*, cxxx. Heft 3.) *Stuttgart*, 1854

Ronalds, (afterwards Sir) Francis. *Born February* 21, 1788, *at London; died August* 8, 1873, *at Battle, Sussex.*

† On Electricity. 8vo. 4 pp. (*Phil. Mag.* xliv. 442.) *London,* 1814
† Experiments on the Variable Action of the Electric Column. 8vo. 6 pp. (*Phil. Mag.* xliii. 414.) *London,* 1814
† On the Electric Column of M. De Luc. 8vo. 2 pp. (*Phil. Mag.* xlv. 466.) *London,* 1815
† On Electro-Galvanic Agency employed as a Moving Power ; with description of a Galvanic Clock. (Dated March 9, 1815.) 8vo. 4 pp. 1 plate. (*Phil. Mag.* xlv. 261.) *London,* 1815
† On correcting the rate of an Electric Clock by a compensation for changes of temperature. 8vo. 2 pp. (*Phil. Mag.* xlvi. 203.) *London,* 1815
† Description of a new mode of Electrical Insulation. 8vo. (*Brande's Quarterly Journal of Science*, ii. 249.) *London,* 1817
† Description of an Electrical Telegraph and of some other Electrical apparatus. 8vo. 83 pp. 8 plates. *London,* 1823

Observations on Atmospheric Electricity, made on Vesuvius in June and July, 1819. 8vo. (*Quarterly Journal Science*, xiv. pp. 332-34.) *London,* 1823
Pendulum Doubler of Electricity. 8vo. (*Edinburgh Phil. Journal*, ix. 1823, pp. 323-25. *Copied from Ronalds' " Descriptions of an Electrical Telegraph,"* &c.) *Edinburgh,* 1823

† Application of the Photographic Camera to Meteorological Registration (described by Collen). 8vo. 3 pp. 1 plate. (*Phil. Mag.* Series iii. vol. xxviii.) *London,* 1846
On the Meteorological Observations at Kew, with an Account of the Photographic Self-registering Apparatus. 8vo. (*British Association Reports*, 1846, pt. ii. pp. 10—11.) *London,* 1846
(*Note.—A short account of the Electric Registering Apparatus is given in the Berlin Fortschrift d. Phys.* vol. iii. p. 586, *by Buys-Ballot,* 1847.)

† On Photographic Self-registering Meteorological and Magnetical Instruments. 4to. 7 pp. 2 plates. (*Phil. Trans.* part i. for 1847.) *London,* 1847
† Experiments made at the Kew Observatory on a new Kite Apparatus for Meteorological Observations. 8vo. 2 pp. (*Phil. Mag.* series iii. xxxi. 191.) *London,* 1847
† Epitome of the Electro-Meteorological and Magnetical Experiments made at the Kew Observatory under the direction of Francis Ronalds. 8vo. 12 pp. *Chiswick,* 1848

† Telegraphie Electricit. (*Förster's Bauzeitung*, 1848, p. 238.)
† Reports concerning the Kew Observatory for 1843-4, printed in 1845. 8vo. 23 pp. 3 plates. *London,* 1845
† Reports concerning the Kew Observatory for 1848-9, printed in 1850. 8vo. 10 pp. 5 plates. *London,* 1850
† Reports concerning the Kew Observatory for 1849-50, printed in 1850. 8vo. 11 pp. 3 plates. *London,* 1850
Reports concerning the Kew Observatory for 1850-51, printed in 1850. 8vo. 35 pp. 6 plates. *London,* 1852
(*British Association Reports for the above-mentioned years.*)
† Report concerning the Kew Observatory for 1851-52. MS. 17 pp. 1852
† Correspondence and Memoranda on various scientific subjects previous to 1853. Folio. MS.
(*Contains improved bifilar Coulomb. Electrometer, drawing only ; Letter to the Astronomer Royal on Electrical Observatory ; Reply of the Astronomer Royal ; Plan of the Leaden Roof of the Greenwich Observatory ; Mr. Weld's Drawing of the Observatory at the Stoneyhurst College, and Francis Ronalds' projected apparatus.*)

Ronalds, Sir Francis—*continued.*

† Description de quelques instruments météorol. et magnét. 8vo. 58 pp. 14 plates. *Paris*, 1855

† Barographe Photographique et Thermographe. 8vo. 9 pp. (*Cosmos par Moigno* tom. viii. p. 541. *Extr. from his " Descriptions," with some variations, &c.*) *Paris*, 1856

† Candélabre électrique. (*Cosmos*, viii. 31, Jan. 1856.) *Paris*, 1856

† Correspondence and Memoranda on Mr. Francis Ronalds' Barograph, Thermograph, and Hygrograph ; and on Mr. Johnson's and Mr. M. Crook's Apparatus at the Radcliffe Observatory, Oxford ; also on Mr. Ronalds' Electro-Atmospherical Apparatus during the period from Feb. 2, 1852, to. June 5, 1856. Folio. MS. 1862

† Correspondence on Mr. Francis Ronalds' Atmospherico-Electrical Apparatus ; and on Magnetic Instruments at the Madrid Observatory, during the period from August 14, 1851, to April 17, 1855. Folio. MS. 1862

† Correspondence and Memoranda on various scientific subjects from March 8, 1853, to September 2, 1862, during a residence in France, Switzerland, and Italy chiefly. Folio. MS. 1862

Vide Silliman, "Principles of Physics," 2nd Edition, 1869, p. 617 : "Ronalds published a volume in 1823 . . detailing . . a moveable disk carrying the letters (of his telegraph), *the type of all Dial Telegraphs.*"

† Correspondence and Memorial on the Electric Telegraph in 1816-52-66-67-70 ; and Extracts relative thereto from Journals in 1866-67-68-70. 4to. MS. &c. *London*, 1870

† Correspondence and Memorial on Self-registering Meteorological Instruments in 1848-49. 4to. MS. *London*, 1870

† Description of an Electric Telegraph. 2nd edition. 8vo. 25 pp. 3 plates. *London*, 1871

(*Vide* also Anon. Magnetism, 1855.)

Ronayne, T. and Henley. Account of some Observations on Atmospherical Electricity in regard of Fogs, Mists, &c., with some Remarks by William Henley. 4to. (*Phil. Trans.* 1772, p. 137.) *London*, 1772

Ronconi, G. B. Delle aeroliti. Discorso inaugurale. 8vo. 40 pp. *Padova*, 1844

Rönnberg, Bernhard Hen. Gedanken von d. Ursachen d. Elektricität. 4to. *Wismar*, 1747

† Ronzoni, C. Intorno ad alcuni fenomeni del magnetismo. 4to. 14 pp. *Padova*, 1855

Roorbach, O. A. Bibliotheca Americana. 8vo. *New York*, 1852

Roret. Manuel de l'Electricité.

Manuel de l'Electricité médicale.

Manuel de l'Electricité atmosphérique.

Manuel de l'Electricité. Galvanoplastrie. 2 vols. 8vo.

Manuel de l'Electricité. Télégraphie élect.

Rose. (*Vide* Riess and Rose.)

Rose, G. Über d. Zusammenhang zwischen d. Form und d. elektr. Polarität d. Krystalle, i. 4to. 2 plates. *Berlin*, 1836?

† Rose, Gustav. Beschreibung u. Eintheilung der Meteoriten, auf Grund der Sammlung im mineralogischen Museum zu Berlin. 4to. 139 pp. 4 plates. (*Aus der Abhandl. d. K. Akad. d. Wissen. zu Berlin*, 1863.) *Berlin*, 1864

(*On the back of the title-page is printed " Gelesen in d. Akad. d. Wissensch. am 7 und 14 Aug. 1862 und am 11 Juni 1863. Anfang des Druckes am 4 Juli 1864, bis wohin einige neue Zusätze hinzugefügt sind. Die Seitenzahl bezeichnet die laufende Pagina des Jahrgangs 1863 in den Abhandl. der physikalischen Klasse der Konigl. Akad. der Wissenschaften.*")

Rose, Heinrich. *Born August 6, 1795, at Berlin.*

On Light from Crystallisation. (*Mars. Annales de Chimie,* 1836.)

† Roseleur, A. Manipulations hydroplastiqnes. Guide pratique du Doreur, de l'Argenteur et du Galvanoplaste. 8vo. 90 figures on Galvanoplastie in text. 308 pp. *Paris,* 1855

† Handbuch der Galvanoplastik oder der hydroelektrischen Metall-Üeberziehung in allen ihren Anwendungsarten. . . Nach dem Französischen. Deutsch bearbeitet nach dem heutigen Standpunkt der Wissenschaft von H. Willich und G. Kaselowsky. 8vo. 228 pp. *Stuttgart,* 1862

Manipulations hydroplastiques. Guide pratique du doreur, de l'argenteur et du galvanoplaste. 2me édition. 8vo. *Paris,* 1866

Galvanic Manipulations. 8vo. (*From the "Athenæum,"* Oct. 12, 1872.) 1872

Rosenberg, Abraham Gottlob. *Died* 1764.

Versuche einer Erklärung von den Ursachen der Elektricität. 8vo. 8½ sheets. *Breslau,* 1745

Rosenschöld. Artis mensor. electr. fundam. 8vo. *Lund ?*

Rosenschöld. (*Vide* Munck.)

Rosenthal, G. E. Über Verhaltungs regeln bei nahen Donnerwettern. 8vo. (*Lichtenberg's Mag. f. Phys.* iv. 1786.) *Gotha,* 1786

Rosenthal, J. (*Vide* Monatsbericht d. K. Preuss. Acad. d. Wiss. zu Berlin, Dec. 1857, p. 640.) *Berlin,* 1857

Ueber den Einfluss höherer Temperaturgrade auf motorische Nerven. Notiz. (*Med. Central-Zeitung,* 1859, No. 96.) 1859

† Electricitätslehre für Mediciner. 8vo. 185 pp. *Berlin,* 1862

† Die Elektrotherapie, ihre Begründung und Anwendung in der Medizin. Für praktische Aerzte bearbeitet. 8vo. 256 pp. *Wien,* 1865

† Von den elektrischen Erscheinungen. Nach drei im Berliner Handwerker Verein im April 1866 gehat. Vorträgen. 8vo. 32 pp. (*Sammlung gemeinverstandl. wissenschaftlicher Vorträge.*) *Berlin,* 1866

† Electricitätslehre für Mediciner. 2 Verm und verbessert Auflage. 8vo. 55 engravings. 216 pp. *Berlin,* 1869

† Rosetti, F. Sull' uso delle coppie termo-elettriche nella misura delle temperature. 8vo. 24 pp. 2 plates. (*Estratta dal Nuovo Cimento,* xxvi. Nov. and Dec. 1867.) *Padovu,* 1868

Rösler, Gottlieb Fried. Diss. de cometis, et arcenda exinde electricitate, ad explicandum systema mundanum a nonnullis advocata. 4to. *Tubingæ,* 1757

Rosignol. Les métaux dans l'antiquité . . suivie d'une Appendice sur les substances appelées Electre. 8vo. *Paris,* 1863

Rosling, Chr. Leberecht. Krit. Prüfungen u. Berichtig. d. bisherig Elektricitätslehre durchgängig auf Experimente gegründet, &c. *Ulm,* 1823

Kritische Prüfungen u. Berichtigungen der bisher Elektricitätslehre. . . 8vo. 1 plate. *Ulm,* 1823

† Der Galvanismus aus d. Dunkel in d. Licht hervorgezogen. 2 vols. 8vo. *Ulm,* 1824

† Rospigliosius, L. De artificiali electricitate phænomenisque ex ea profluentibus. 4to. 22 pp. *Florentiæ,* 1765

Ross, James Clark, Sir. *Born April 15, 1800, at London.*

† On the Position of the North Magnetic Pole. 4to. 6 pp. (*Phil. Trans.* part i. for 1834.) *London,* 1834

Magnetic Observations in the Antarctic Region. 4to. (*Phil. Trans.* 1841.) *London,* 1841

Notice of the Aurora-Borealis. 4to. (*Phil. Trans.* 1842. *London,* 1842

Ross, Sir James Clark.—*continued.*

A Voyage of Discovery and Research in the Southern and Antarctic Regions during the years 1839 to 1843 inclusive. 2 vols. 8vo. *London,* 1847

On the Errors which may be occasioned by disregarding the influence of Solar or Artificial Light on Magnets. 8vo. (*Athenæum,* 1854, p. 1238.)
London, 1854

Ross. (*Vide* Lloyd Sabine and Ross.)

Ross, John, Sir. *Born June 24, 1777, at Balsarroch, Wigtownshire, Scotland; died August 30, 1856, at London.*

An Explanation of Captain Sabine's remarks on the late Voyage of Discovery to Baffin's Bay. 8vo. *London,* 1819

† Letter to Captain Napier from on board the *Isabella* . . . employed in the Northern Expedition, dated off Sugar-Loaf Bay, Davis's Straits, July 12, 1818, lat. 74·2° N. long. 58° W. (*Magnetical Observations at* pp. 307 and 309.) 8vo. 6 pp. (*Phil. Mag.* lii. 305.) *London,* 1818

A Voyage of Discovery made . . 1818, April to November, for the purpose of exploring Baffin's Bay, and inquiring into the probability of a North-West Passage. 2 vols. 8vo. *London,* 1819

Appendix to the Narrative . . of a Second Voyage. 4to. *London,* 1835

Ross, J. and J. C. Narrative of a Second Voyage in search of a North-West Passage, and of a residence in the Arctic Regions during the years 1829 to 1833 inclusive, including the reports of J. C. Ross and the Discovery of the Northern Magnetic Pole. 4to. *London,* 1835

Roffel, Eliz. Paul E. de. Published (edited) Voyage de M. d'Entrecasteaux envoyé à la recherche de Lapérouse. 2 vols. 4to. *Paris,* 1809
(*Note.—He adds his own observations.*)

† **Rossellini,** F. Riflessioni sopra un Articolo del Fusinieri nel 2o Bimestre (1834) degli Annali delle Scienze del Regno Lombardo-Veneto. 8vo. 6 pp.
Firenze, 1834

† Sperienze elettro-fisiologiche. 8vo. 7 pp. *Firenze,* 1834

Rossi, Francesco. *Died December 18, 1841, at Turin.*

Rapports des expériences galvaniques. Le 1er sur les animaux à sang froid, lu . . à Turin le 24 Nivose an 11. Le 2e sur un homme décapité le 23 Nivose et lu le 3 Pluviose. . . Le 3e sur un homme décapité le 2 Pluviose et lu le 3. 4to. (*Bibliot. Ital.* i. 106.) *Turin,* 1803

Rapport des expériences faites sur les animaux à sang chaud et à sang froid; lu à la classe des Sciences exactes, Turin, 24 Nivose, 1803. 4to. *Turin,* 1803

† Extrait d'un Mémoire sur l'application du Galvanisme dans le traitement de quelques maladies. 8vo. 11 pp. (*Bibliot. Ital.* i. 221.) *Turin,* 1803

† Précis de quelques expériences, faites par Julio et Rossi . . . dans le but de découvrir si le fluide galvanique se charge et entraine avec lui, des miasmes putrides. (Publié par Rossi.) 8vo. (*Bibliot. Ital.* ii. 113.) *Turin,* 1803

† Extrait d'observations sur l'usage du Galvanisme dans la cure de l'Hydrophobie. Lu à l'Acad. 3e jour Complim. an 11 (1803). 8vo. 10 pp. (*Bibliot. Ital.* iii. 44.) *Turin,* 1804

† Sur l'électricité animale. Lu le 11 Vent. an 9. 4to. 39 pp. (*Mém. de Turin Ann.* x. et xi. 1802-3.) *Turin,* 1804

† Report of Galvanic Experiments made on Men and Animals. Read to the Academy of Turin by G. Rossi. 8vo. 10 pp. (*Phil. Mag.* xviii. 131.)
London, 1804

Expériences galvaniques. 4to. (*Mém. de Turin,* xiv. lxxxi. xcii. *Hist.*)
Turin, 1805

Observations sur l'électricité et sur le galvanisme. 4to. (*Mém. de Turin,* xvi. viii. *Hist.*) *Turin,* 1809

Sur la guérison des enragés au moyen du galvanisme. 4to. (*Mém. de l'Acad. de Turin,* an. 1805-1808; *Sc. Phys. et Mathém. Hist.* p. xi. cf. tom. viii. *Hist.* p. xci.) *Turin,* 1809?

M M

Rossi, Francesco.— *continued.*

Expériences galvaniques. 4to. (*Mém. de Turin,* xvi. xii. *Hist.*) *Turin,* 1809

† Deuxième Essai sur les Miasmes, avec la description d'un appareil Docimias-
mique. Lu 13 Juin 1824. 4to. 16 pp. 1 plate. (*Mem. di Torino,* tom. xxxi.
p. 95.) *Torino,* 1827

† Rapporto di alcune esperienze sull' Elettricità. 8vo. 11 pp. *Torino,* 1840

† **Rossi et Michelotti.** Expériences sur la décomposition de l'eau par le moyen de
la pile de Volta. 4to. 10 pp. (*Mém. de Turin,* an. 1809-10, p. 57, tom. xviii.)
Turin, 1811

Rossi, Gab. Sulla seconda lettera alla continuazione dello Spallanzani. Lettere
tre. 8vo. *Bologna,* 1824

Rossi. (*Vide* Giulio and Rossi.)

Rossi. (*Vide* Julio, &c.)

Rossi. (*Vide* Vassalli-Eandi.)

† **Rossijn, J.** Tho. Translation of Cavallo. 8vo. 339 pp. *Utrecht,* 1780

(*Vide* also Rossyn, J. Th.)

Rossler, T. F. Progr. de luce primigenia. 4to. *Tubingen,* 1776

(*Note.*—*The light before the creation of the Sun mentioned by Moses was,
according to this author, an electrical light.*)

† **Rossyn, J.** Th. Dissertatio philosophica inauguralis de Tonitru et Fulmine, ex
nova electricitatis theoria deducendis. 4to. 34 pp. *Francqueræ,* 1762

(*Vide* also Rossijn.)

Rost, Joh. Leon. Histor. Beschreibung d. merkwürd. Nordlichts von 1721. 4to.
Nürnberg, 1721

† **Rother, L. F.W.** Der Telegraphenbau, ein Handbuch zum praktischen Gebrauch
für Telegraphen-Techniker und Beamte. Zweite Auflage verb. 8vo. 366
pp. *Berlin,* 1867

† Der Telegraphenbau, ein Handbuch zum praktischen Gebrauch für Telegraphen-Techniker und Beamte. 3te verb. Aufl. 8vo. 306 pp.
Berlin, 1870

Rothmann, Chr. "Beobachtete zuerst (obwohl undeutlich) d. Zodiacallicht."

(*Note.*—*He died long after* 1597, *at Bernberg. Many of his works in MSS.
are preserved at Cassel.*)

Rotterdam Bataafsch Genootschap. Verhandelingen van het Bataafsch
Genootschap der præfondervindelse Wisbegeerde. 4to.
Rotterdam, 1774 *to* —

Roucher Deratte (or de Ratte ?). Mélanges de Physiologie, de Physique, et de
Chimie. Cont. . . Elect. . . Galv. . . Magnét. 2 vols. 8vo. *Paris,* 1803

Traité sur l'élect. le galv. le magnét. &c. 8vo. 1803

Rouelle, G. F. Oxydation des Goldes, durch d. elektr. Funken. (*Journ. de
Médecine, &c. Von Roux. Published probably by his brother. He died* 1770 *at
Passy, near Paris.*)

Rouland, N. Descr. des mach. élect. à taffetas, &c. 8vo. 35 pp. 1 plate.
Amsterdam and Paris, 1785

Sur l'électricité appliquée aux végétaux. 4to. (*Journ. Phys.* xxxv. 1790.)
Paris, 1790

Rouppe. Extrait d'une lettre de M. Rouppe de la Haye, 28 Août, 1801. 8vo.
(*Journ. de Chim. de Van Mons,* i. pp. 106 et 1808. *Bruxelles,* 1801

Rousseau et Martin-Magron. (*Vide* Matteucci.)

Roussier, P. J. Mém. sur le clave cin électrique de M. de la Borde. 4to.
Paris, 1782

† **Roussilhe, L.** Notice sur un nouvel électromoteur. 8vo. 15 pp. 2 plates.
Paris, 1857

† Note sur un perfectionnement de mon Electromoteur, sur un nouveau genre de barrage, et sur une nouvelle Vanne. 8vo. 28 pp. 1 plate. *Paris*, 1857

Routz. (*Vide* Elkington and Routz, De.)

Roux, Augustin. *Born January 26, 1726, at Bordeaux; died June 28, 1776, at Paris.*

Expériences nouvelles sur le platine et sur les différents cobalts soumis à l'étincelle électrique. 8vo. (*Ext. du Journal de Médecine*, Nov. 1793.)
Paris, 1773

† **Roux, F. L.** Conservation des plaques des navires cuirassés et des coques en fer par l'application directe d'un doublage en cuivre. 8vo. 63 pp. 2 plates.
Paris, 1866

† Etude sur la fabrication et la pose des Câbles électriques sous-marins. 8vo. 52 pp.
Paris, 1865

Roux. (*Vide* Le Roux.)

† **Rowell, G. A.** An Essay on the Cause of Rain and the allied phænomena. 8vo. 166 pp. 1 plate. *Oxford*, 1859

† Papers on the cause of Rain, Storms, the Aurora, and Terrestrial Magnetism. Reprinted from the "Edinburgh New Philosophical Journal," &c.; with an Appeal for a Consideration of the Theory advanced. 8vo. 58 pp. 1 map.
London, 1871

† **Rowett, Wm.** The new Submarine Telegraphic Cable; the regulation of its specific gravity, and its true construction and submersion explained, showing its easy adaptation to deep sea as well as to shallow waters, and at a great diminution of expense. 8vo. 43 pp. *London*, 1858

† The Ocean Telegraph Cable; its construction, the regulation of its specific gravity and submersion explained. 8vo. 122 pp. 4 plates. 1 map.
London, 1865

Light Sea Cables. 4to. (*Elect. Teleg. Review, &c.*, September 24, 1870. *From British Association, Liverpool, Section G.*) *London*, 1870

Rowning, John. *Died November*, 1771, *at Lincoln's Inn Fields.*

A Compendious System of Natural Philosophy. 2 vols. 8vo. *London*, 1744

A Compendious System of Natural Philosophy. 2 vols. 8vo. *London*, 1753

Roy. (*Vide* Le Roy.)

† **Royal Geographical Society.** The North Atlantic Telegraph *via* the Færöe Isles, Iceland, and Greenland. Reports of the Surveying Expeditions, &c. 8vo. 94 pp. 1 plate (*i.e.* map). (*Proceedings of the Royal Geographical Society*, Jan. 28 and Feb. 11, 1861, *Lord Ashburton, President*.) *London*, 1861

Royal Institution of Great Britain. Journals of the Royal Institution. Vol. i. 8vo. *London*, 1800

The Journal of Science and the Arts. Edited at the Royal Institution of Great Britain. 8vo. *London*, 1817, &c.

(*The 1st volume is a 2nd edition. At the 7th volume the title is altered to "Quarterly Journal," &c. and the words, "edited at the Royal Institution," are omitted.*)

The Quarterly Journal of Science, Literature, and Art (in continuation of the preceding). 8vo. *London*, 1827, &c.

Index to the First Twenty Volumes of the Quarterly Journal, &c. 8vo.
London, 1826

Proceedings and Abstracts of Discourses delivered at the weekly evening meetings. Issued to the Members in numbers. Sold occasionally. Not published. 8vo. *London*

Royal Polytechnic Society of Cornwall. Containing papers by Snow Harris, &c.

Royal Society of London. Philosophical Transactions, commenced 1665. 4to.
London, 1666

† A General Index to the Phil. Trans. from the 1st to the end of the 70th vol. (1665-1780). 4to. *London,* 1787

A General Index or Alphabetical Table to all the Philosophical Transactions from January, 1677-8, to December, 1693. Also a Catalogue of the books mentioned and abbreviated in these Transactions in an alphabetical order. 4to. 33 pp. (*Phil. Trans.* xvii. for 1693, p. 1005.) *London,* 1694

The Philosophical Transactions and Collections to the end of the year 1700, abridged and disposed under general heads by J. Lowthorp. 3 vols. 4to.
London, 1705

The Philosophical Transactions abridged from 1700 to 1720, by H. Jones. 4to.
London, 1721

The Philosophical Transactions from 1719 to 1733, abridged and disposed under general heads by J. Eames. 4 parts in 2 vols. 4to. *London,* 1734

The Philosophical Transactions from 1743 to 1750, abridged and disposed under general heads (the Latin papers translated) by J. Martyn. 4to.
London, 1756

The History of the Royal Society of London, in which papers not published are inserted as a Supplement to the Transactions. By Thomas Birch. 4 vols. 4to. *London,* 1756-7

Abstracts of the Proceedings of the Royal Society. Vols. i. to vi. 8vo.
London, 1800-30

Philosophical Transactions from the commencement in 1665 to 1800, abridged; with notes and illustrations by Hutton, Shaw, and Pearson. 18 vols. 4to.
London, 1809

Continuation to the Alphabetical Index contained in the *Phil. Trans.* from vol. lxxi. to vol. cx., the year 1781 to 1820. 4to. *London,* 1821

Continuation of the Alphabetical Index of the matter contained in the *Phil. Trans.* from the year 1821 to 1830. 4to. *London,* 1833

† Proceedings of the Royal Society. Vol. vii. to —. 8vo. (*Continuation of Abstracts of Proceedings.*) *London,* 1831
(*Part in Library.*)

Abstract of the Papers printed in the Philosophical Transactions, together with the Proceedings, from 1800 to July 1864. 13 vols. 1832-64

† Catalogue of the Scientific Books in the Library of the Royal Society. 8vo. 776 pp. *London,* 1839

† Report of the President and Council, . . on the Instructions to be prepared for the Expeditions to the Antarctic Regions. 8vo. 79 pp. *London,* 1839

† Revised Instructions for the use of the Magnetic and Meteorological Observatories, &c. by Sir J. F. W. Herschell, Bart. 8vo. 47 pp. *London,* 1842

† Report of the Committee of Physics and Meteorology of the Royal Society, relative to the Observations to be made in the Antarctic Expedition and the Magnetic Observatories. 8vo. 4 plates. 120 pp. *London,* 1840

History of the Royal Society, with Memoirs of the Presidents, compiled from authentic documents by C. R. Welds. 2 vols. 8vo. 1848

Report of the Meteorological Committee of the Royal Society. 8vo.
London, 1868

Rozier François. *Born January* 24, 1734, *at Lyons; died September* 29, 1793, *at Lyons.* (*Poggendorff,* ii. 710.)

† Tableau du Travail annuel de toutes les Acad. de l'Europe. 10 vols. 4to.
Paris, 1773

(*Vide* also Journal de Physique.)

Nouvelle Table. . Académie Royale, 1666 à 1770. 4 vols. 4to. *Paris,* 1776

Rozier. (*Vide* Pilâtre de Rozier.)

† Rudolphi, D. K. A. Über d. elektrisch. Fische und über d. Giftsporn d. männl. Schnabelthiers. Vorgelesen d. 7 Juni 1821. 4to. 14 pp. 3 plates. 1821

Rue, De La. (*Vide* La Rue, De.)

Rugger. On the Leyden Phial.

Ruggero. (*Vide* Fabri-Ruggero.)

Ruhland, Reinhold Ludwig. *Born April* 16, 1786, *at Ulm; died April* 23, 1827, *at Ulm.*

Über d. Gegensatz d. Elektricität u. d. Chemismus. 8vo. (*Gehlen's Journal f. Chemie,* ix. 1810.) *Berlin,* 1810

Über e. neue Art Luft zu galvanisiren u. galvan. Ketten mit plus- u. minus-elektrisch gestalteten Pflanzenwurzeln. 8vo. (*Gehlen's Journ. f. Chim.* ix. 1810.) *Berlin,* 1810

About Five Articles in Schweigger's Journal, vol. i. 1811, to vol. xv. 1815.

System d. allegemein. Chemie. *Berlin & Stettin,* 1828

Ruhmkorff. Appareil d'induction électrique. (Inductorium.) 1850 *or* 1851

(*I can find no account in any periodical of this instrument, or any precise date of its first appearance. Feilitzsch,* p. 416, 1865, *says that it appears to have become known by degrees (nach und nach)* : "*indem eine seine Einführung bezweckende Abhandlung sich nicht vorfindet.*" *Du Moncel describes it, &c. in four or five Editions (first in* 1855 *and last in* 1867 *?*), *entitled,* "*Notice sur l' Appareil d' Induction de Ruhmkorff.*"—F. R.)

(*Vide* also Verdu and Ruhmkorff.)

(*Vide* also Du Moncel.)

† Rundell, Westcott. Author of the Three Reports on the Magnetism of Iron Ships, made by the Compass Committee of Liverpool to the Board of Trade, and presented to Parliament and ordered to be printed.

(*Vide* Liverpool Compass Committee.)

† Rundspaden, A. Über die Electrolyse des Wassers in Berührung mit Silber. Inaugural Dissertation . . an der Universität Göttingen. 8vo. 43 pp. *Göttingen,* 1869

Runeberg, Edward Fredrik. *Born* . . . ; *died May* 27, 1802.

† Versuche, mit Beyhülfe der Elektricität Gewächse zu treiben ; zu Stockholm im Jahr. 1754 angestellt. 8vo. 12 pp. 1 plate. (*Schwedische Acad. Abhandl.* 1757, p. 15, vol. xix.) *Leipzig,* 1757

Forsok att med. electricitetens tillhjelp drifva växter. 8vo. (*Vetensk Acad. Handl.* 1757.) *Stockholm,* 1757

Abhandll. in Vetensk Acad. Handl. 1757, &c.

Electroscope. (*Schwedischen Abhandlungen,* 1759, p. 17.) 1759

† Runnels, J. Specimen phil. inaugurale de causa fulminis ac tonitrus. 4to. 27 pp. *Trajecti ad Rhenum,* 1759

Ruolz, F. A. H. F. De. Nouveau procédé pour dorer et pour argenter les métaux. 4to. (*Compt. Rend.* xiii. 342.) *Paris,* 1841

Mémoire sur les arts insalubres. Remis à l'Acad. le 9 Août, 1841. 4to. (*Ext. du tom.* xi. *du Recueil des Savants Etrangers,* p. 161.) *Paris,* 1841

Several Papers, &c. in the *Comptes Rendus* having respect to his Galvano-Plastic Processes, &c.

(*Many controversial writings of his appear in Christofle's* "*Histoire de la Dorure,*" *&c., without indications of dates, or places of publication.*)

† Ruspini, G. Del modo di preparare esatte ed uniformi carte ozonoscopiche. 4to. 20 pp. 1 plate, coloured. (*Estratto dagli Ann. di Chimica applicata alla Medicina, Luglio,* 1857.) *Milano,* 1857

† **Russegger.** Beiträge zur Ausmittlung der Abweichung der Magnetnadel durch Entgegenhalt der aus alten Karten erhobenen Daten mit den Ergebnissen der gegenwärtig, mit Beibehaltung der gleichen Fixpuncte, erneuert vorgenommenen Vermessung. 8vo. 8 pp. 1 plate. (*Sitzb. d. K. Akad. Wien,* 1849, ix. u. x. Heft, p. 203.)   *Wien,* 1849

† **Russell, W. H.** The Atlantic Telegraph, illustrated by R. Dudley. 4to. 117 pp. 26 plates, coloured.   *London,* 1865?

† **Russian Government,** The. Has made considerable progress towards opening a communication with the northern regions of America by the way of Siberia. 8vo. 1 p. (*Phil. Mag.* xliii. 395.)   *London,* 1814

† **Rutter, J. O. N.** Magnetoid Currents, their forces and directions, with a description of the Magnetoscope, to which is subjoined a Letter from W. King. 8vo. 47 pp.   *London,* 1851

       On Magnetoid Currents, their forces and directions. 8vo.   1851

†   Human Electricity: the means of its development, illustrated by Experiments. With additional notes. 12mo. 182 pp.   *London,* 1854

S.

S., Sir R. (*Vide* Anon. Mag. Variation, 1683.)

Sabatelli, di Napoli. On Violent Fracture (*quasi* spontaneous) of a Glass Electrical machine.

Sabine, Edward. *Born October* 14, 1788, *at Dublin.*

† On Irregularities observed in the direction of the Compass Needles of H.M.ss. "Isabella" and "Alexander," in their late Voyage of Discovery, and caused by the attraction of the iron contained in the ships. 8vo. 7 pp. (*Phil. Mag.* liv. 276.) *London,* 1819

† Remarks on the Account of the late Voyage of Discovery to Baffin's Bay, published by Capt. J. Ross, R.N. (8vo. 40 pp.) 8vo. 11 pp. (*Phil. Mag.* liii. 367.) *London,* 1819

An Account of Experiments . . figure of the Earth, &c. (*From Dove,* p. 278.) *London,* 1825

† On the Dip of the Magnetic Needle, in August, 1828. 4to. 7 pp. (*Phil. Trans.*) *London,* 1829

Observations on the Direction and Intensity of the Terrestrial Magnetic Force in Scotland. 8vo. Map. *London,* 1836

On the Phænomena of Terrestrial Magnetism. 8vo. 30 pp. Fig. (*Brit. Ass. Reports ?*) *London,* 1836

† Report on the Variations of the Magnetic Intensity observed at different points of the Earth's surface. 8vo. 85 pp. 2 plates and 3 charts. (*In 7th Report Brit. Assoc.*) *London,* 1838

† Report on the Magnetic Isodynamic Lines in the British Islands. From observations by Lloyd, Phillips, Fox, James C. Ross, and Sabine. 8vo. 148 pp. 3 plates. Maps. (*Report Brit. Assoc.*) *London,* 1839

Report relative to the Observations to be made in the Antarctic Expedition and in the Magnetic Observatories. 1 vol. 8vo. Map. *London,* 1840

Contributions to Terrestrial Magnetism. 9 parts. 4to. 604 pp. 16 plates. (*Phil. Trans.*) *London,* 1840-49

† Contributions to Terrestrial Magnetism, No. 2. 4to. 25 pp. (*Phil. Trans.* pt. i. *for* 1841.) *London,* 1841

Contributions to Terrestrial Magnetism. 4to. Plates. *London*

Observations made at the Magnetic Observatories of Toronto, in Canada, during a remarkable Magnetic Disturbance of September, 1841. 8vo. Plate. 1841

Observations made at the Magnetic Observatories of Toronto, . . Trevandrum . . and St. Helena. 8vo. 1841

† Magnetic Observations on days of unusual Magnetic Disturbance, made at the British Colonial Observatories. Part i. for 1840-41. 4to. *London,* 1843

† Observations made at the Magnetical and Meteorological Observatory at Toronto. Vol. i. for the years 1840-41-42. 4to. 4 plates. *London,* 1845

† Observations made at the Magnetical and Meteorological Observatory at St. Helena. Vol. i. for 1840-41-42-43. 4to. 6 plates. *London,* 1847

† On the Diurnal Variation of the Magnetic Declination at St. Helena. Read Feb. 18, 1847. 4to. 7 pp. 2 plates. (*Phil. Trans. for* 1847, pt. i. p. 51.) (*Faraday's copy, with marks and notes.*) *London,* 1847

† Contribution to Terrestrial Magnetism, No. ix. "Containing a Map of the Magnetic Declination for 1840 in the Atlantic Ocean between the parallels of 60° N. and 60° S. latitude." 4to. 62 pp. 2 plates. (*Faraday's copy.*) *London,* 1849

Sabine, Edward.—*continued.*

† The means adopted in the British Colonial Magnetic Observatories for determining the absolute values, secular change, and annual variation of the Terrestrial Magnetic Force. 4to. 19 pp. (*Phil. Trans.* pt. i. *for* 1850.)
London, 1850

† Observations . . at Hobart Town, vol. i. commencing with 1841. La. 4to.
London, 1850

† Observations . . during Magnetic Disturbances. Part ii. for 1842-43-44. 4to.
London, 1851

† Observations . . at the Cape of Good Hope in 1841 to 1846, with Abstracts of the Observations from 1841 to 1850 inclusive. La. 4to.   *London,* 1851

† On Periodical Laws discoverable in the mean effects of the larger Magnetic Disturbances. Read Feb. 27, 1851. 4to. 17 pp. (*Phil. Trans. for* 1851, pt. i. p. 123.) (*Faraday's copy.*)   *London,* 1851

† On the Annual Variation of the Magnetic Declination at different periods of the day. Read May 22, 1851; revised October, 1851. 4to. 7 pp. 1 plate. (*Phil. Trans. for* 1851, pt. ii. p. 635.) (*Faraday's copy.*)   *London,* 1851

† On the Periodical Laws discoverable in the mean effects of the Larger Magnetic Disturbances, No. ii. Read May 6, 1852. 4to. 22 pp. (*Phil. Trans. for* 1852, p. 103.) (*Faraday's copy.*)   *London,* 1852

† Observations . . at Hobart Town. Vol. ii. commencing with 1843. La. 4to.
London, 1852

† The Bakerian Lecture. On the Influence of the Moon on the Magnetic Declination at Toronto, St. Helena, and Hobarton. Read Nov. 17, 1853. 4to. 11 pp. (*Phil. Trans. for* 1853.) (*Faraday's copy.*)   *London,* 1853

Magnetical Observations at the Kew Observatory. 4to.   *London,* 1853

† Some of the results obtained at the British Colonial Magnetic Observatories. 8vo. 14 pp. (*Brit. Assoc. Reports.*)   *London,* 1854

† On the Hourly Observations of Magnetic Declination made by Capt. R. Maguire, R.N., and the Officers of H.M.S. "Plover," in 1852-53-54, at Point Barrow, on the shores of the Polar Sea. 4to. 36 pp.   *London,* 1857

Results of the Magnetic Observations at the Kew Observatory, 1857-58. 3 parts. 4to. (*Phil. Trans.* 1863-66.)   *London,* 1863-66

† Address (as President) delivered at the Anniversary Meeting of the Royal Society on Friday, Nov. 30, 1866. 8vo. 18 pp.   *London,* 1866

† Address (as President) delivered at the Anniversary Meeting of the Royal Society on Saturday, Nov. 30, 1867. 8vo. 20 pp.   *London,* 1867

(*Vide* also Lloyd, Sabine, and Ross.)

(*Vide* also Arago.)

✝ **Sabine, Robert.** The Electric Telegraph. 8vo. 428 pp.   *London,* 1867

† The History and Progress of the Electric Telegraph, with descriptions of some of the Apparatus. 2nd edition with addns. 12mo. 280 pp. *London,* 1869

(*Vide* also Clark Latimer, and Sabine.)

Sable. (*Vide* Du Sable.)

*† **Sacharof.** An Account of the Aerial Voyage undertaken at Petersburg on the 30th January, 1804. Read before the Academy of Sciences by Sacharof. 8vo. 8 pp. (*Phil. Mag.* xxi. 193.)   *London,* 1805

Sacher, Wenzel Aloys. Die geographischen, meteorologischen u. erdmagnetischen Constanten Tarnows. 4to.   *Tarnow,* 1851

Die inducirten elektrischen Ströme. 4to. 48 pp.   *Salzberg,* 1855

Sachischen Gesellschaft der Wissensch. zu Leipzig. Abhandlungen der mathematisch-phyischen Clase der Konig. Sächsischen Gesellschaft der Wissenschaften. 4to.   *Leipzig*

Sadeler. (*Vide* Matthæus de Sadeler.)

Sage, Balthazar Georges. *Born May 7, 1740, at Paris; died September 9, 1824, at Paris.*

Observations sur les Paratonnerres. 8vo. *Paris*, 1808

† De la nature et des propriétés de huit espèces d'électricité. 8vo. 36 pp. (*Volta's copy.*) *Paris*, 1809

Description des colonnes électrifères et de leur effets. 8vo. *Paris*, 1814

De l'origine et de la nature des Globes de Feu Météoriques. 8vo. *Paris*, 1815

† De la nature et de la production du Gas Electrifiable. 8vo. 46 pp. *Paris*, 1815

† Probabilité physique sur la cause de l'Intermittence de l'Electroscope. 8vo. 13 pp. *Paris*, 1822

† Recueil historique d'Effets Fulminaires. 8vo. 21 pp. *Paris*, 1822

† Recherches et conjectures sur la formation de l'Electricité Métallique, nommée Galvanisme. 8vo. 14 pp. (*Volta's copy.*) *n. d.*

Sage. (*Vide* Le Sage.)

Saigey. (*Vide* Coulvier Gravier, and Saigey.)

Saget, De. On Aërolithes, &c. *Toulouse*, 1827

Saint-Agobard. Agobardi Sancti, Opera quæ octingentos annos in tenebris delituerant. Nunc e papirii Massoni jurisconsulti bibliotheca proferuntur. Accesserunt binæ Epistolæ Leidradi non ante excusæ. Sm. 8vo. 43 pp. *Paris*, 1605

Agobardi, Sancti, Archiepiscopi Lugdunensis, opera. Item Epistolæ et opuscula Leidradi et Amulonis : Steph. Baluzius emendavit notisque illustravit. 2 vols. 8vo. *Paris*, 1666

Livre " contra insulsam vulgi opinionem de Grandine et Tonitruis." (*Note.—This is contained in the Opera and in the following work.*)

De la Grêle et du Tonnerre. (Anonymous translation of the Liber contra Opinionem.) 8vo. *Lyon*, 1841
(*Note.—This is an improved version of a translation by Péricaud in the " Annuaire de Lyon " for 1837, p. 33.*)

† **Saint-Amans,** De. Account of a Fall of Uranolytes (Aërolites) near Agew. 8vo. 4 pp. ("*Ann. de Chim.* xcii. 25 Oct. 1814.") (*Phil. Mag.* xlv. 23.) *London*, 1815

Saintard. (*Vide* Gillet and Saintard.)

Sainte-Claire-Deville, Ch. Recherches sur les principaux Phénomènes de Météorologie et de physique terrestre aux Antilles. 2 vols. 4to. Plates. *Paris*, 1861

† **Saint-Edme,** E. L'Electricité appliquée aux Arts mécaniques, à la Marine, et au Théâtre. 8vo. 234 pp. *Paris*, 1871

Saint-Lazare. (*Vide* Bertholon de Saint-Lazare.)

† **Saint-Martin** et **Lacoste.** Rapport à M. le Chev. Pullini . . sur l'essai de Paragrélage . . dans les environs de Chambéry. 8vo. 50 pp. Table. *Chambéry*, 1825

Saint-Severe, Prince. Lettres à Nollet. 1re partie. 8vo. 1753

Dissertation sur une Lampe antique, trouvée à Munich, pour servir de suite à la 1re partie de ses lettres à Nollet. 8vo. *Naples*, 1756

Salleron, Jules. Anémographe de. Construit en Mars, 1856. (*From Du Moncel, who gives no indication where first described.*)

Salle. (*Vide* La Salle, De.)

Saltzmann, Joh. Reid. De igne fatuo. (*From Pog.* ii. 743.) *Strasburg*

Salva. Application de l'Electricité à la Télégraphie. (*Gazette de Madrid du 25 Novembre*, 1796.) *Madrid*, 1796

† **Salverte,** E. Conjectures on the Stones which have fallen from the Atmosphere. 8vo. 6 pp. (*From Ann. de Chimie,* No. 133.) (*Phil. Mag.* xv. 354.) *London*, 1803

† Sampson, T. Electrotint . . by means of Voltaic Electricity. 8vo. 26 pp. 3 plates. *London,* 1842

† Samter, J. Die Grenet'sche Batterie und ihre Bedeutung für die operative Heil-Anwendung des Galvanismus. 8vo. 21 pp. *Posen,* 1858

Samuelson, James. The Applications of Photography. 8vo. (*Popular Science Review,* No. xvii. for October.) *London.*

Samuelson and Crookes. Journal of Science, edited by J. Samuelson and W. Crookes. Plates.

Sanctis, B. De. A Letter to Professor Millington, of the Royal Institution, respecting some Frigorific Experiments made on the Magnetic Fluid, and on Sea Water. 8vo. 8 pp. (*Phil. Mag.* lx. 199.) *London,* 1822

Reference, by the Editor of the Magazine, to a letter from Dr. De Sanctis to him, relative to an error into which he had been led relative to Governor Ellis's observations. 8vo. 1 p. (*Phil. Mag.* lxi. 70.) *London,* 1823

Sanden, Heinrich Von. *Born July* 28, 1672, *at Königsberg ; died August* 18, 1728, *at Königsberg.*

Dissert. phys. experim. de succino electricorum principe. 4to. *Regiomons,* 1714
† Dissertatio de Succino electricorum principe . . Edited with Hausenius, Novi profectus, . . by Gottsched. 8vo. *Lipsiæ,* 1746

(*Vide* also Hausenius and De Sanden.)

† Sanderson, W. Observations upon the Variation of the Needle made in the Baltick, Anno 1720. 4to. 1 p. (*Phil. Trans.* xxxviii. *for* 1733-34, p. 120.) *London,* 1735

Sandonnini, G. Relazione sopra un metodo di elettro-doratura e di elettro-argentatura esposta dal . G. Sandonnini all' Accad. di Scienze . . di Modena il 28 Luglio, 1843. 1843

(*Note.—This article does not seem to have been printed separately. It appears in Grimelli's Storia,* p. 90, *as the report of a Commission appointed by the Academy to examine it.*)

Sandwall, T. G. Relation über den Wasser-Wirbel, welcher sich den 10 letz verwichenen August in Mälarzeigte und Langholmstull vorbeyzog. 8vo. (*Neue Schwedische Akad. Abhandl.* an. 1790, p. 198.) *Stockholm,* 1790

† Sangiorgio, P. Analisi delle acque che hanno sentita l'azione della pila di Volta. 8vo. 32 pp. *Milano,* 1805

San-Martino, Giambattista da. *Born* 1739, *at Lupari, near Treviso ; died* 1800, *at Venice.*

Memoria sopra la nebbia de' vegetabili. (" *Coronata dall' Accad. d'Agricult. di Vicenza, il di* 16 Maggio, an. 1785.'') *Vicenza,* 1785

† Riflessioni intorno alla causa d'un fenoḿeno elettrico. 4to. 18 pp. (*Mem. Soc. Ital.* vi. 120.) *Verona,* 1792

† Lettera all' Amoretti intorṇo ad un fenomeno magnetico. 4to. 4 pp. 1 figure. (*Opusc. Scelti,* xvii. 243.) *Milano,* 1794

Lettera intorno agli effetti provenienti dalla varia grossezza de' dischi elettrici di cristallo. 8vo. (*Nuovo Giornale-Enciclop. di Vicenza,* 1794.) *Vicenza,* 1794

† Sanna-Solaro, J. M. Nouvelle théorie de la Grêle. La. 8vo. 18 pp. 2 plates. (*Extr. Ann. Soc. Météorol. de France,* xi. 95, 9 Juin 1863.) *Paris,* 1863

Sans, De. Anweisung, die von einem Schlagflusse gelahmten Kranken durch die Elektrizität zu heilen. (*Journ. de Med.* 1782, *for other work.*) *Augsberg,* 1780

† Sans, l'Abbé. Guérison de Paralysie par l'Electricité. 1re partie. 12mo. 150 pp. Frontispiece and 1 plate. *Paris,* 1772

† Guérison de Paralysie par l'Electricité. 2me partie. 12mo. 234 pp. Frontispiece and 4 plates. *Paris,* 1778

Santanelli, F. Philosophiæ reconditæ ; sive magicæ, magneticæ, mumialis scientiæ explanatio. 2 vols. 4to. *Coloniæ,* 1723

Santi Linari. (*Vide* Linari Santi.)

Santini, Giovanni. *Born January* 30, 1786, *at Caprese, near Borgo of San Sepolcro.*

† Dei diversi metodi per determinare le longitudini geografiche, e dell' applicazione dell' elettro-magnetismo alla loro determinazione. Memoria. 8vo. 36 pp. (*Estratta dalla Rivista periodica dei lavori della I. R. Accad. di Padova Trim.* 3 *e* 4 *dell' anno* 1855-56.)　　　　　　　　　　　　　*Padova,* 1856

† Notizie intorno agli apparati magneto-elettrici per la determinazione delle longitudini geografiche con la descrizione dell' apparato dell' osservatorio di Greenwich. 8vo. 68 pp. 8 plates. (*Mem. letta all' Accad. di Padova,* 10 Giugno, 1866.)　　　　　　　　　　　　　　　　　　　　　　*Padova,* 1867

Santoli. Narrazione dei fenomeni osservati sul suolo Irpino (da V.M.S.) contemporanei all' ultimo incendio del Vesuvio, accaduto in Giugno dell' anno 1794 ; coll' aggiunta di varie importantissime osservazioni dell' istessa classe. 8vo. 156 pp. (*Opuscoli Scelti* (in 4to.), xix. 31.)　　　　*Napoli,* 1795

† **Sarlandiere.** Mémoires sur l'Electro-puncture. 8vo. 150 pp. 2 coloured plates.
　　　　　　　　　　　　　　　　　　　　　　　　　　　　　　　Paris, 1825

Traité du Système Nerveux dans l'état actuel de la science. 2 parts. 8vo. 6 plates on 2 sheets.　　　　　　　　　　　　　　　　　　　　　　　　1840

Sarpi. On the Magnet. (*Vide* Bertelli sopra Pietro Peregrino, p. 23 *et seq.*)

Sarrabat, Nicolas. *Born February* 9, 1698, *at Lyons ; died April* 27, 1737, *at Paris.*

† Nouvelle hypothèse sur les variations de l'Eguille (*sic*) aimantée qui a remporté le Prix . . pour l'an 1727. 12mo. 47 pp. 1 woodcut (sep.). (*In Prix de l'Acad de Bord.* iii.)　　　　　　　　　　　　　　　　　*Bordeaux,* 1727

Sarrau. Journal d'Observations météorologiques depuis 1714 jusqu'en 1770. MS. Fol. 11 vols.　　　　　　　　　　　　　　　　　　　　　　　　*Bordeaux*

Sarti, C. Saggio di congetture su i terremoti.　　　　　　　　　*Lucca,* 1783

Sarton, D. H. Erfand ein neues Echappement . . . ein Elektrometer u. A. ; sämmtlich von ihm in einer Brochure beschrieben.　　　　*Lüttich,* 1822

Sartorius von Waltershausen. Bestimmung d. absolut. Intensität d. Erdmagnetismus zu Waltershausen. 4to. (*Gauss and Weber, Magn. Vereins für* 1837.)　　　　　　　　　　　　　　　　　　　　　　　　　　　　1837

Das Oscillations-Inclinatorium. 4to. (*Gauss and Weber, Magn. Vereins für* 1838.)　　　　　　　　　　　　　　　　　　　　　　　　　　　　1838

† **Sauer,** G. The International Telegraph Treaty, Paris, April 18, 1865. Translated from the French by G. Sauer. 8vo. 28 pp.)　　　　*London,* 1868

† The Telegraph in Europe. A complete Statement of the Rise and Progress of Telegraphy in Europe, showing the cost of construction and working expenses of Telegraphic Communications in the principal countries, &c. Corrected from Official Returns. 8vo. 177 pp. 30 leaves not numbered.
　　　　　　　　　　　　　　　　　　　　　　　　　　　　　Paris, 1869

(To this work is added (and mentioned in the Table of Contents as an Appendix) "The International Telegraph Treaty, Paris, April 18, 1865. Translated from the French by Geo. Sauer." 8vo. 28 pp.)

Saule. (*Vide* Pfaff and Saule.)

† **Saur,** L. Betrachtungen über die Electricität. 8vo. 88 pp.　　*Berlin,* 1832

Saussure, Horace Bénédict De. *Born February* 17, 1740, *at Conches, near Geneva ; died January* 22, 1799, *at Geneva.* (*From Poggendorf,* ii. 755, *who says " Nach ihm becam, durch sienen Sohn, der Saussurit sienen Namen.*")

Dissertatio de Igne. 4to. 36 pp.　　　　　　　　　　　　*Geneve,* 1759

† Dissertatio physica de electricitate, quam, favente Deo, Præside D. D. H. B. De Saussure . . publice tueri sonabitur Amad. Lullin. Die . . Sep. 26. 1766. 8vo. 55 pp. (*This is constantly attributed to Saussure. Vide Lullin.*)
　　　　　　　　　　　　　　　　　　　　　　　　　　Geneva, 1766

Saussure, Horace Bénédict De.—*continued.*

Exposition abrégée de l'utilité des Conducteurs électriques.   4to. 9 pp.
Geneve, 1771
(*Note.—Landriani says " Manifeste ou Exposition."*)

† Manifesto ossia breve Exposizione dell' utilità dei Conduttori.   Tradotto in Italiano da Toaldo.   (*Vide* Toaldo.)   *Venezia,* 1772

Sur l'Electricité de l'Atmosphère au dessus d'une Montagne du Valais.   4to. (*Journ. de Phys.* ii. 1773.)   *Paris,* 1773

Des Effets électriques du Tonnerre observés à Naples.   (*Journ. de Phys.* i. 1773.)   *Paris,* 1773

Sur l'effet de l'Electricité sur les Animaux microscopiques.   4to.   (*Spallanzani's Opusc.* i.)   *Modena,* 1776

† Voyages dans les Alpes. 4 vols. 4to. (*Poggendorff,* ii. 755, *says there is also an edition in* 8vo. 1780—96.)   *Genève,* 1779-96

Sur l'Electricité des Végétaux.   4to.   (*Journ. de Phys.* xxv. 1784.)   *Paris,* 1784

Sur l'électricité naturelle de l'homme, et sur un moyen d'estimer facilement celle de l'atmosphère.   (*Journ. de Paris,* 1784.)   *Paris,* 1784

Découverte des Tourmalines sur le St. Gotthard.   (*Journ. de Paris,* 1784.)
Paris, 1784

Description d'un Electromètre portatif pour l'Electricité de l'Atmosphère. (*Journ. de Paris,* 1785.)   *Paris,* 1785

Observations faites sur le Col de Géant.   4to.   (*Journ. de Phys.* xxxiii. 1788.)
Paris, 1788

Metodo facile e semplice per conoscere, colla calamita, il ferro che è nei minerali.   4to.   (*Opusc. Scelti di Milano,* iii.)   *Milano*

† Mémoire historique sur la vie et les écrits de Horace Bénédict de Saussure. . . par Jean Senebier.   8vo. 219 pp.   *Genève,* 1801

Saussure.   (*Vide* Necker de Saussure.)

Sauvages de la Croix, François. *Born May* 12, 1706, *at Alais ; died February* 19, 1776, *at Montpellier.*

Dissertatio medica ; de hemiplegia per electricitatem curanda.   4to.   (*From Krunitz,* p. 188.)   *Montpellier,* 1749
(*Note.—Printed under the feigned name of Deshais.*)

De usu electricitatis in Rheumatismo.   (*Acta Upsaliensia,* an. 1744-1750, p. 1.)
1744 ?

Nosologia methodica.   8vo.   *Amsterdam,* 1760
(*Vide* also Anon. Medical Electricity, 1763.)

Savary, Félix. *Born October* 4, 1797, *at Paris ; died July* 15, 1841, *at Estagel, East Pyrénées.* (*Poggendorff,* ii. 761.)

† Mémoire sur l'application du calcul aux phénomènes électro-dynamiques.   4to. 26 pp. 1 plate.   *Paris,* 1823

Sur les phénomènes électro-dynamiques.   8vo.   (*Ann. Chim. Phys.* xxii. xxiii.)
Paris, 1823

† Mem. sur l'Aimantation.   Lu à l'Acad. des Sciences le 31 Juillet, 1826.   8vo. 52 pp.   (*Extr. Ann. de Chim.*)   *Paris,* 1826

Savérien, Alexandre. *Born July* 16, 1720, *at Arles ; died May* 28, 1805, *at Paris.*

Histoire des progrès de l'esprit humain dans les sciences naturelles, et dans les arts, qui en dépendent.   *Paris,* 1775

† **Savery,** Servington. Magnetical Observations and Experiments.   4to. 46 pp. (*Phil. Trans.* xxxvi. 295. 1 plate.)   *London,* 1729-30

† **Savi,** P. Etudes anatom. . . Torpille, &c. (*Vide* Matteucci.) 8vo.   *Paris,* 1844

† **Savioli, G.** Dissertatio in causam physicam Auroræ borealis. 8vo. 87 pp.
Bergomi, 1789

† Dissertazione sulla causa fisica dell' Aurora-boreale, tradotta dall' originale latino, con qualche aggiunta dell' Autore. 4to. 40 pp. (*Opus. Scelti,* xiii. 3.)
Milano, 1790

† **Saweljeff, A.** Magnetische Beobachtungen und geographische Orts-Bestimmungen angestelt im Jahre 1841 während einer Reise an den Küsten des Weissen und Eismeers. Gelesen am 31 Mai, 1844. 4to. 32 pp. 1 plate. (*Mem. des Sav. Etrang. de l'Acad. St. Petersb.* vi.) *St. Petersburg,* 1844

Über d. Polarisations-Erscheinungen in d. galvan. Kette. Magister-Diss. (russisch.) *St. Petersburg,* 1845

Über d. galvan. Leitungsfähigkeit d. Flüssigkeiten (russ.) (*In Erman's Archives,* xv. *It received the Demidorf prize.*) *Kasan,* 1853

Royal Saxon Society. Berichte über die Verhandlungen der k. sächsischen Gesellschaft der Wissenschaften zu Leipzig.

† **Saxton.** Magneto-electric Machine. 8vo. (*Exhibited at the Meeting of the British Association,* June, 1833.) *London,* 1833

Improved Magneto-Electric Rotation Machine. 8vo. (*Phil. Mag.* ix. p. 360, 1836.) *London,* 1836

Saxtorph, Friderich. *Born at Roeskilde, Iceland; died* 1806.

Elektricitätsläre. 2 vols. 8vo. *Kiöbnhaven,* 1802-3

† Darstellung der gesammten auf Erfahrung u. Versuchen gegründeten aus dem Dänischen übersetzt von B. Fangel. Erster Theil. 8vo. 552 pp. 6 plates.
Kopenhagen, 1803

† Darstellung der . . Electricitätslehre . . aus dem Dänischen übersetzt von B. Fangel. Zweyter und lezter Theil. 8vo. 521 pp. 2 plates.
Kopenhagen, 1804

Om den bedste Inretning af en Elektrisier-maskine. (*Rafn's Phys. Biblioth.* xii.)

Saxtorph, J. Afhandl. om Galvanism (Anhang zu seine Lärebog i Naturlären. Kiobnh. 1779.) 8vo. *Kiöbnhaven,* 1804

Kort Vejledning til Kundskap om Elektriciteten. 8vo. *Kiöbnhaven,* 1807

Scaguller, G. Di un caso creduto per alcuni stranissimo, e nuovo, cicalata di G. Scaguller. 4to. 1794

Sull' efficacia dei conduttori. 8vo. 27 pp. *Venezia,* 1797

Comento . . alla pagina 102 del Giornale Astro-meteorologico per l'anno 1806. † 8vo. 16 pp. *Venezia?* 1806

Del comune antico proverbio . . applicato al libro "Dubbj sull' efficacia de' † Conduttori elettrici," del Trans. Bragadin . . Cicalata estemporanea seconda. . . 8vo. *Venezia*

Scales, W. Handbook to the Electric Telegraph. 3rd edition. *From Schellen.*
About 1848

† **Scaramella, A.** Le Bussole indeclinabili del MDCCCXVI. ad uso della geodesia e della nautica. 8vo. 36 pp. 1 plate. *Venezia,* 1816

† Le Bussole indeclinabili del 1816 ad uso della Geodesia e della nautica. 2d ed. 8vo. 36 pp. 1 plate. 1820

† La paralizzazione della mobilità dell' ago magnetico alla vicinanza del ferro scoperta l'anno 1816. . . 1st ed. 8vo. 16 pp. 3 plates. *Venezia,* 1820

† **Scaramelli.** Il Paragrandinatore istruito sull' arte e sugli usi dei paragrandini e parafulmini alla Tholard. 8vo. 20 pp. 2 plates. *Venezia,* 1824

Scarella, Giambattista. *Born* 1711, *at Brescia; died February* 26, 1779, *at Brescia.*

De Magnete. Libri quatuor. 2 vols. 4to. 2 plates. *Brixiæ,* 1759

(*Vide also Casali.*)

* **Scarpellina, C.** Termografia Italiana, istituita da Zantedeschi di Padova. Stazione meteorologica sul Campidoglio. Due lettere al Prof. Zantedeschi. 4to. 8 pp. *Padova,* 1865

Scarpellini, la Signora. Ozonometrical Observations made by her, sent regularly to the I. R. Istituto Lombardo. *Rome*

† **Scarso, J.** Disputatio de Magnete. (Adjutore L. Menini.) Thesis. 12mo. 55 pp. *Patavii,* 1813

Scelta di Dissertazioni. Cavate dai più celebri Autori si antichi che moderni intorno ad ogni sorte d'arti e scienze. Editore Savioli A. (libraio.) *Venezia,* 1750 *to* —

Scelta d'Opuscoli interessanti tradotti . . &c. (*Vide* Amoretti.) (*Note.—This series was reprinted in* 3 *vols.* 4to. *Nearly all the Electrical articles are in the Library.*)

Scelta di Opuscoli scientifici e letterari. *Venezia* 1812 *to* —

Schäffer, Jacob Christian. *Born May* 30, 1718, *at Querfurt ; died January* 5, 1790, *at Regensburg.*

† Abbildung und Beschreibung des beständigen Elektricitäts-trägers. . . 4to. 48 pp. 2 plates. (*Volta's copy.*) *Regensburg,* 1776

† Kräfte, Wirkungen, und Bewegungsgesetze des beständigen Elektrophors. . . 4to. 50 pp. 1 plate. (*Volta's copy.*) *Regensburg,* 1776

† Fernere Versuche mit dem beständigen Elektricitäts-trägers, &c. . . 4to. 56 pp. 1 plate. *Regensburg,* 1777

† Abbild. Beschr. d. elekt. Pistole u. e. kl. Elektrophors. 4to. 32 pp. 3 plates. *Regensburg,* 1778

† Versuche mit dem beständigen Electricitästräger 4 Abhand. 4to. 176 pp. 7 plates. *Regensburg,* 1780

Schäffer, Johann Gottlieb. *Born September* 13, 1720, *at Querfurt, Thüring ; died February* 1, 1795, *at Regensburg.*

Die Kraft und Wirkung der Elek. in dem menschlichen Körper. . . 1st ed. 1 vol. 8vo. *Regensburg,* 1752

† Die elektrische Medicin, oder die Kraft und Wirkung der Electricität in dem menschlichen Korper. . . 2nd ed. 4to. 84 pp. 1 frontispiece. *Regensburg,* 1776

† **Schaffrath, L.** De electricitate cœlesti, atque ratione ædificia ab ictu fulminis præservandi. . . 4to. 34 pp. 2 tab. *Pestrini,* 1778

Schaub, Franz. *Born April* 23, 1817, *at Gross-Schweinbarth, Nieder-Oestere.*

Magnetische Beobachtungen im östlichen Theile des Mittelmeers ausgeführt im Jahr 1857 ; auf Befehl seiner K. K. Hoh. *Triest,* 1858

Schaub, J. Historical Memoir on Galvanism. 8vo. (*Archiv. fur Pharmacie und Ärztliche Naturkunde,* iii. Cahir.) *Cassel,* 1802

(*Note.—Contains Bauer's experiment to cure the toothache by zinc and silver rods in* 1772.)

On Galvanism applied for Deafness, &c. (*Schaub und Piepenbring Archive fur Pharmacie u. Medicinische Physik,* Nos. i. and ii.) *Nürnburg or Cassel,* 1802

Untersuch. einer vom Pariser Institut gemachten galvan. Entdeckung. *Cassel,* 1802

" Essai ou Précis complet de l'histoire du Galvanisme, sous les rapports physiques de chimie et de médecine," &c. 8vo. 2 vols. *Nuremberg,* 1805 ?

Schaub. (*Vide* Gmelin and Schaub.)

† **Schaw.** Télégraphie militaire, traduite de l'Anglais par C. Maunior. 8vo. 44 pp. *Paris,* 1863

Scheel, Paul. *Born February* 28, 1773, *at Itzehoe, Holstein ; died June* 17, 1811, *at Copenhagen.*

Scheel, P. Beskrivelse over et apparat til Vandet's Decomposition ved det Voltaiske Batterie. (*Rafn's Nyt Biblioth.* i. 1801.) *Kiobnhaven, 1801*

Scheel, Dr. Paul. Modification of Voltaic pile. (*Nordisches Archiv.* ii. Bd. 2es St. p. 59.) 1801?

Scheffler, Aug. Ch. W. H. Giebt. heraus. : Organ f. d. Fortschritte d. Eisenbahnwesens; Darm, von ihm; Über d. Versenk. unterseeischer Telegraphentaue. (Jahrgang xiii.) *Braunschweig, 1858*

† **Scheffler, J. P. E.** Beyträge zu den Untersuchungen über das Elektrum und den Lyncurm. der Alten. 4to. 13 pp. (*Neue Samml. v. Versuche . . d. Naturf. Gesellsch. in Danzig*, i. 234.) *Danzig, 1778*

Scheibel, Johann Ephraim. *Born September 5, 1736. at Breslau; died May 31, 1809, at Breslau.*

Einige Progr. üb. den a. d. Elisabetkirche zu Breslau erricht. Blitzableiter. . . 1793-4

Scheibel, G. E. D. d. Blitz in Pulverthurm verunglückte Breslau. 4to. (*From Heinsius.*) *Breslau, 1750*

Die Witterungen, ein Gedicht. 8vo. (*From Heinsius.*) *Breslau, 1752*

Scheliha, Von. Torpedo Warfare. A Treatise on Coast Defence : based on the experience gained by Officers of the Corps of Engineers of the Army of the Confederate States, and compiled from Official Reports of Officers of the Navy of the United States, made during the North American War, from 1861 to 1865.

Schellen, Thom.s Joseph Heinrich. *Born March 30, 1818, at Kevelaer, Reg.-Bez. Düsseldorf.*

† Der elektromagnetische Telegraph in den einzelnen Stadien seiner Entwicklung und in seiner gegenwärtig. Ausbildung und Anwendung, &c. . . 8vo. 368 pp. 1st ed. *Braunschweig, 1850*

Beschrijving van den electro-magnetischen Druk-Telegraph van Morse wiens stelsel door het Nederlandsche Gouvernement is aangenomen. Naar het Hoogduitsch; door P. J. M. de Gelder. 12mo. Plates. *Hertogenbosch, 1852*

Der elektromagnetische Telegraph in den Hauptstadien seiner Entwickelung und in seiner gegenwartigen Ausbildung, &c. . . . 2s. ganz umgearbeitete . . . Ausgabe. 2nd edition. 8vo. 259 pp. *Braunschweig, 1854*

† Der elektromagnetische Telegraph in den Hauptstadien seiner gegenwartigen Ausbildung und Anwendung, nebst einem Anhang über den Betrieb der elektrischen Uhren. Ein Handbuch . . . Dritte ganzlich umgearbeitete . . . Auflage. 8vo. 392 pp. *Braunschweig, 1861*

† Der elektromagnetische Telegraph in den Hauptstadien seiner Entwickelung und in seiner gegenwärtigen Ausbildung und Anwendung nebst einem Anhange über den Betrieb der elektrischen Uhren . . . Vierte ganzlich umgearbeitete . . . Auflage. 8vo. 756 pp. *Braunschweig, 1867*

† Das atlantische Kabel, seine Fabrication, seine Legung, und seine Sprechweise. Gemeinfasslich dargestellt. 8vo. *Braunschweig, 1867*

Aufsätze über elektr. Uhren, &c. (*Westermann's Illustrirt. Monatschefte, or in Programmen d. Düsseld. Realschule.*)

† Die Spectralanalyse in ihrer Anwendung auf die Stoffe der Erde und die Natur der Himmels-Körper, gemeinfasslich dargestellt von H. S. 8vo. 452 pp. 2 plates and 4 portraits. *Braunschweig, 1870*

Schellen, Thomas Joseph Heinrich.—*continued.*

† Der elektromagnetische Telegraph in den Hauptstadien seiner Entwicklung, und in seiner gegenwärtigen Ausbildung, und Anwendung ; nebst einem Anhange über den Betrieb der elektrischen Uhren. Ein Handbuch der theoretischen und praktischen Telegraphie . . Fünfte gänzlich umgearbeitete, bedentend erweiterte . . . Auflage. 1st u. 2nd Lieferungen. 8vo. 384 pp. *Braunschweig*, 1870
3rd Lieferung. 191 pp. *Braunschweig*, 1870
4th Lieferung. 290 pp. *Braunschweig*, 1871

Schelling. Zeitschrift für speculative Physik.

† **Schelling,** F. W. J. Von. Über Faradays neueste Entdeckung. (*Zur öffentlichen Sitzung der K. Akad. der Wissen.* am 28 März, 1832.) *Munchen*, 1832
Über Faraday's neueste Entdeckungen. 8vo. 1852

Scherer, Alexander Nicoläus. *Born December* 30, 1771 (*A. St.*), *at St. Petersburg ; died October* 16, 1824 (*A. St.*), *at St. Petersburg.*
Allgem. Journl. d. Chemie. 10 vols. 8vo. *Leipzig & Berlin*, 1798-1803
Archives für d. theoret. Chemie. 4 Hfte. 8vo. *Jena*, 1800-2
Nördlichen Blatter für d. Chemie. 4 Hfte. 8vo. *Halle*, 1817-8
Allgem. nördliche Annal. d. Chemie u.s.w. 8vo. 8 Bde. *St. Petersburg*, 1819-22

Scherer, J. B. A. von. Über d. märischen Meteorsteine u. ihre Incrustirung. 8vo. (*Gilb. Ann.* xxxi. 1809.) *Leipzig*, 1809
Über Problematische d. Meteorsteine u. Eisenmassen. (*Oken's Isis*, 1833.) 1833

Schering, E. C. J. Zur mathematischen Theorie electrischer Ströme. Beweis der allgemein. Lehrsätze der Electro-dynamite insbesondere der Inductionslehre aus dem electrischen Grundgesetze. Eine am 13 Juni 1857. Gekrönte Preisschrift. 4to. 35 pp. *Göttingen*, 1857

† **Scherzer,** K. Narrative of the Circumnavigation of the Globe by the Austrian ship *Novara*, in 1857—1859. 3 vols. 8vo. Maps and woodcuts. 1861-63

Scheuchzer, J. Jb. Historia fulminis d. 23 Mai. (*Ephem. Acad. Nat. Cur.* 1712.) 1710-1712
Meteorologia et Orytographia helvetica, oder Beschreibung d. Luftgeschichten Steine, Metalle, u.s.w. *Zurich*, 1718
De Aurora Boreali d. 2 Oct. 1728. (*Act. Acad. Nat. Cur.* ii. 1730.) 1728-30

† **Schiaparelli,** G. V. Intorno al corso ed all' origine delle stelle meteoriche. Lettere al Secchi. 4to. 33 pp. (*Estr. dal Bullettino Meteorol. dell' Osservat. del Collegio Romano*, vol. v. Nos. 2, 10, 11, 12.) *Roma*, 1866

Schickard, William. Ignis versicolor e cælo sereno delapsus et Tubingæ spectatus anno 1623, die 7 Nov. *Tubingæ*, 1623
De globo seu pila ignea ann. 1624.
(*Note.—Poggendorff*, ii. 794, *says : " Lichtkugel, darin aus Anleitung des neulich erscheinen. Wundalichts nicht allein von demselbigen in specie, sondern zu mal von der gleichen meteoris in genere u.s.w. gchandelt wird.*) *Tubingæ*, 1624

† **Scheick,** Über atmosphärische Elektricität. 8vo. 46 pp. *Oldenburg*, 1870
(*Note.—The Polytechnische Bibliothek for January* 1871, No. i. p. 3, *adds,* " *Prog. der Realschule zu Oldenburg.*")

† **Schiefferdecker,** W. Bericht über die vom Verein für wissenschaftliche Heilkunde in Königsberg in Preussen angestellten Beobachtungen über den Ozongehalt der atmosphärischen Luft, und sein Verhältniss zu den herschenden Krankheiten. Gelesen in d. Vereinssitz. am 30 Mai 1854. 8vo. 47 pp. 15 plates. (*Sitzb. d. Wien Acad.* vol. xvii. 2 Ht. 1855.) *Wien*, 1855

Schilling, von Canstadt, Pawel Lwowitsch. *Born April 5, 1786, at Reval; died July 25, 1837, at St. Petersburg.*

(*Vide* Hamel, J. for some account of his Telegraph, &c.)

(*Vide* also Shaffner, Lenz, Wheatstone, and Kuhn, pp. 834—838 particularly; also Muncke, Phys. Worterb. ix. 111.)

(*Note.—Poggendorff,* vol. ii. 798, *makes a statement relative to Wheatstone and Cooke's experiments, giving the invention, in about 1835, to Schilling, and the introduction of it into England to Cooke, who learnt it from Muncke at Heidelberg. Schilling does not seem to have published. Poggendorff,* ii. 798, *says:* "Durch S. Th. Sömmering für d. Idee d. elektr. Telegraphirens enthusiasmirt, benutzte er schon im Oct. 1812, den galvan. Strom, um quer durch die Newa Minen zu sprengen. Später (wann? sagt Hamel nicht) construite er einen elektromagnet. Telegraphen u. zeigte denselben 1835 in d. Naturforscherversammlung zu Bonn vor. Durch dieses Instrument soll Wheatstone auf die elektr. Telegraphie gebracht worden sien, indem ihn ein junger Engländer William Fothergill Cooke, der dasselbe beim Prof. Muncke in Heidelberg hatte functioniren gesehen, von der Einrichtung desselben in Kenntniss setzte (Bekanntlich haben übrigens Gauss u. W. Weber die Inductions-Elektricität schon 1832 zum Telegraphiren benutzt).*")

Schilling, G. W. Diatribe de morbo in Europa pene ignoto, Jaws dicto. 8vo.   *Trajecti ad Rhenum,* 1770

(*Note.—Bonnefoy, De l'Application,* p.48, *says . . .* "Cui adjuncta est . . . Observatio physica de Torpedine,* 1770. *Mém. Acad. Sciences de Berlin,* 1770." *Ellice (Saggio,* p. 26) *says:* "Il medico Schilling osservò sin dal 1764, ed annunziò nel 1769, che l'ago magnetico si muove quando vi si avvicina l'anguilla tremante;" *and alludes to the* "favole che inserì nell' Accad. Rle. di Berlino per l'anno 1770." *Spallanzani, lèttere al Lucchesini,* p. 89, *refers to* "i pretesi fatti . . . in un' opera . . di questo medico nel 1769, sopra . . . una malattia la javvi.*")

Des écrits au sujet de la sensibilité du Gymnotus electricus au Magnétisme. (*Ingenhauz, Nouv. Exp.* p. 340, *says:* "Publié depuis quelques années," *i.e. before 1785.*)

Schilling, G. W. and **Hahn.** De Lepra commentationes, (avec) Préface de Hahn. 8vo.   *Leidæ,* 1778

(*Note.—Hahn endeavours, in this preface, to explain away Schilling's false statements in his Diatribe . . . relative to the Torpedo.*)

Schilling, Johann Jacob. *Born April 25, 1702, at Cleve; died . . .*

Observationes et Experimenta de vi electrica vitri aliorumque corporum. 4to. (*Miscellanea Berolinensia,* iv. 334, *pro an.* 1734; *et in idem* v. 109, *pro an.* 1737.)   *Berolini,* 1734-37

(*Note.—Waitz, Abhandlung, thought that John Jacob Schilling was the first German experimentalist in Electricity.*)

Schinz, Salomon. *Born January 26, 1734, at Zürich; died May 26, 1784, at Zurich.*

† Specimen physicum de electricitate. . . 4to. 38 pp. (*Read June,* 1776.) (*Volta's copy.*)   *Turici,* 1776

Reflexionen über d. Strahlableiter 4to.   *Zurich,* 1776

† Supplementum speciminis physici de Electricitate.   *Turici,* 1777

Schlagintweit, A. und H. Untersuchungen über die Physikalische Geographie der Alpen in ihren Beziehungen zu den Phænomenen der Gletscher, zur Geologie, Meteorologie und Pflanzengeographie. 4to. 71 engravings. 1850

Neue Untersuchungen über die physicalische Geographie u. die Geologie der Alpen. 4to. Atlas, 22 plates in folio.   *Leipzig,* 1854

Schlagintweit, A. H. and R. Report on the proceedings of the Officers engaged in the Magnetic Survey of India.   *Madras,* 1855

458   SCH

Schlagintweit, Hermann Rudolph Alfred. *Born May* 13, 1826, *at Munich.*
Reports of the Magnetic Survey of India, and the researches connected with it.
Nos. i. to x. (*Indien gedruckt u. auszugsweise in d. Berlin. Zeitschr. f. all-
gem. Erdkunde,* ii. 1857.)   1859

Schlagintweit, Herman, mit Adolph und Robert.   Astronom. Ortbestimm. u.
magnet Beobacht. in Indien u. Hoch Asien.   8vo.   (*From Poggendorff,* ii.
802. *In Poggend. Ann.* cxii. 1861.)   *Leipzig,* 1861
Results of a Scientific Mission to India and High Asia, undertaken between
the years 1854 and 1858, by order of the Court of Directors of the Hon.
East India Company.   *Leipzig and London,* 1861
(*Note. — This is the first* (*of three ?*) *volumes, and contains the Magnetic
Observations.*)

Schlegel, J. De Galvanismo.   8vo.   (*From Muller's Cat. Dissert. No.* 324.)
                                                         Leipzig, 1797

Schlesinger. Die Electricität als Heilmittel; vom physikalischen und experi-
mentell physiologischen Standpunkte erörtert. (*Zeitschrift Wiener Aerzte,*
1852, Juli.)   1852

† **Schlottman,** A. Kritische Geschichte d. Theorieen des Galvanismus. 8vo. 63 pp.
                                                         Breslau, 1856

Schmidt, C. H. Beschreibung aller neuerfund. magneto-elektrischen u. elektromagnet. Apparate u. Maschinen für Gewerbetreibende.   8vo. 3 plates.
                                                         Leipzig, 1841

Die Benutzung d. Galvanoplastik f. technische Zwecke.   8vo. 1 plate.
                                                         Leipzig, 1841

Beschreibung aller neuerfundenen magneto-elektrischen u. elektro-magnetischen Apparate u. Maschinen f. Gewerbtreibende. Als Fortsetzung des
Unterrichts üb. Magnetismus, Electricitäts, u. Elektro-magnetismus.   8vo.
4 plates.   *Leipzig,* 1842

Die Benutzung der Galvanoplastik für technische Zwecke. Enthaltend, aus-
ser e. Bericht über d. neuesten Leistungen der Galvanoplastik, eine sorgfältige
Beschreibung u. Abbild. d. galvanoplast. Apparate zur Herstellung von
Relief-Kupferplatten.   8vo. 1 plate.   *Leipzig,* 1842

Die Kunst des Vergoldens, Versilberns, Verplatinirens, Verzinkens, Ver-
bleiens, Verkupferns, Verkobaltens, und Vernickelns, der Metalle, sowohl
nach den bewährtesten älteren Verfahrungsarten, &c. 8vo.   1843

† Handbuch der Galvanoplastik. Zunächst für Künstler und Gewerbtreibende.
Nach den neuesten Verbess. bearb. 2 Sehr. Verbesserte Auflage. 8vo.
227 pp. 5 plates.   *Quedlinberg,* 1847

Handbuch der Galvanoplastik in allen ihren Anwendungsarten. 3 ganz
umgearbeitete und sehr vermehrte Auflage. 8vo. 238 pp. 8 plates lith.
                                                       Quedlinburg, 1856

Schmidt, Georg Christoph. *Born alout* 1740, *at Gattenhöfen; died July* 29, 1811,
at Jena.

Beschreibung e. Elektrisirmaschine u. deren Gebrauch.   1st edition. 4to.
                                                           Jena, 1774

Erste Sammlung gemeinnützieger Maschinen. Beschreibung einer Elektrisir
Maschine und deren Gebrauch. mit e. Vorrede d. H. Wiedburg. Vermeh.
u. verbesserte Auflage.   *Berlin,* 1778

Samml. gemeinnütz. Maschinen. 2 Hfte. 4to.   *Berlin,* 1778-80

Beschreib. gemeinnütz. Machinen, &c. 4to. 5 plates.   *Jena,* 1784

(*Note.—Ersch enumerates amongst other contents, "electr. Werkzeuge od.
Lampen; Blitzableiter in Bezieh. auf Erdbeben, elektr. Löschbade u. Schiess-
gewehre.*")

Schmidt, Georg Gottlieb. *Born June 17, 1768, at Zwingenburg ; died October 8, 1837, at Giessen.*

Sammlung, &c. . . Phys.-math. Abhandl. 8vo. *Giessen,* 1793

Über d. Verkalken d. Metalle durch d. Elektricität u.s.w. 8vo. *(Gren's New Journ. i.* 1795.) *Leipzig,* 1795

Beschreibung einiger sonderbar scheinenden elektr. Phänomene die eine Walzenmaschine darbot. 8vo. *(Gren's New Journ. i.* 1795.) *Leipzig,* 1795

Eight Articles on Magnetism, Electricity, Electro-Magnetism, Transversal-Magnetism, &c., in Gilbert's Annalen, vol. lxiii. 1820, to vol. lxxiv. 1823.

Schmidt, Karl. *Born November 17, 1762, at Breslau ; died December 4, 1834.*

† Der Zitterstoff (Elektrogen) u. seine Wirkungen in d. Natur. Entdeckt von C. Schmidt. 8vo. 229 pp. *Breslau,* 1803

Der Zitterstoff (Electrogen). . . 8vo. 2 Theil. *Breslau,* 1806

Der Zitterstoff (Electrogen) u. seine Wirkungen in d. Natur. 3 Thle. 8vo. *Breslau,* 1803-6

Schmidt, Johann Friedrich Julius. *Born October 26, 1825, at Entin.*

Über Sternschnuppen-Beobb. 8vo. *(Pogg. Ann.* lxxx.) *Leipzig,* 1850

Über e. Feuermeteor in d. Rheinprovinz, 1850. 8vo. *(Pogg. Ann.* lxxxii. 1851.) *Leipzig,* 1851

† Resultate aus 10 jährigen Beobachtungen über Sternschnuppen. 8vo. 193 pp. *Berlin,* 1852

† Das Zodiacallicht. Übersicht der seitherigen Forchungen, nebst neuen Beobachtungen . . in den Jahren 1843 bis 1855. 8vo. 110 pp. *Braunschweig,* 1856

† Feuermeteor am 18 October 1863 beobachtet von J. T. Julius Schmidt. Ein Sendschreiben ein . . Haidinger. 8vo. 8 and 2 pp. 1 plate. *(Sitzungsb. d. K. Akad. d. Wiss. Wien,* 1863, xlviii. iv. Heft. pp. 551 and 559.) *Wien,* 1863

† Zweiter Bericht über das zu Athen, am 18 October 1863, beobachtete Feuermeteor. Sendschreiben am v. Haidinger ii. Wien. 8vo. 10 pp. *(Sitzungsb. d. K. Akad. d. Wiss Wien,* 1864, xlix. i. Heft. p. 17.) *Wien,* 1864

† Über Feuermeteore ; nach Zahlen, Detonationen, Meteoritenfällen, Schweifen und Farben, verglichen zur Höhe der Atmosphäre. Ein Sendschreiben an Herrn W. Haidinger in Wien. 8vo. 8 pp. *(Aus den Sitzungsberichten der Kaiserl. Akad. der Wissenschaften.)* *Wien,* 1864

Über Feuermeteore 1842 bis 1867 (Schreiben an Hn. K. Hof. Ritter v. Haidinger.) 8vo. 34 pp. *(Ans. d. Sitzungsb. d. K. Akad. d. Wiss.)* *Wien,* 1868

† Astronomische Beobachtungen über Meteorbahnen und deren Ausgangspunkte. 4to. 54 pp. *(Publications de l'Observatoire d'Athènes,* 1e Série, tom. ii. p. 1.) *Athens,* 1869

† Beobachtung der Meteore in der Nacht des 13—14 November, 1866. (Schreiben an Herr K. Hof. W. Haidinger.) 8vo. 14 pp. *(Sitzb. d. Mathem.-naturwiss.* Cl. liv. ii. Abth. p. 775.) *Wien*

Schmidt, Joh. Jac. Bibl. physicus od. Enleit zur bibl. Naturwissenschaft. 8vo. *Zullichau and Jena,* 1748

† **Schmidt,** L. Über d. electr. Ströme u. die Spannungsgesetze bei d. Elektrolyten. 8vo. 19 pp. *(Poggendorff, Annalen,* vol. cix.) *Leipzig*

Schmidt, N. E. A. Vom Magnete. 8vo. *(Hannover Magazin,* 1765.) *Hannover,* 1765

† **Schmidl,** Ad. Notizen über die von ihm aus der Planina-Höle mitgebrachten, und der Classe vorgezeigten Proteen. 8vo. 5 pp. *(Sitzungsb. d. K. Akad. Wien,* 1850, ii. Bd. iii. Heft, p. 228.) *Wien,* 1850

Schmuck, Edmund Joseph. *Born* 1771, *at Heidelberg ; died December* 21, *at Heidelberg.*

On the action of Galvanic Electricity on the Mimosa pudica. (*Seyfer,* p. 406, *refers to " Ludwigii scriptor. nevrolog. minor.* iii. 21.'*)*

† Dissertatio philosophica inauguralis, de electricitate corporum organicorum. 4to. 36 pp. (*Volta's copy.*) *Heidelbergæ,* 1791

† Beiträge zur nähern Kenntniss der thierischen Elektricitat. 8vo. 77 pp.
Manheim, 1792

Schnabel, Karl. De globis igneis et meteorolithis comment. *Marberg,* 1833

† **Schneeberger,** F. J. Der transmundane Telegraph. 8vo. 40 pp. *Leipzig,* 1860

Schneider, E. Das Ungenügende der jetzigen Methode der Tiefenmessung und Vervollkommnung derselben mit Hülfe der Electricität, pp. 265—292. 1 pl. lith. *Pétersbourg, Leipzig, and Riga,* 1861-2

(*Mélanges physiques et chimiques tirés du Bulletin de l'Académie Impériale des sciences de St. Pétersbourg,* tome v. livraison 1—3.)

† Oceanische Tiefenmessung vermittelst der Elektricität. (Lu 28 Fév. 1867.) 4to. 37 pp. 1 plate. (*Bulletin de l'Acad. Imp. de St. Pétersbourg,* xi. 471, 1867.) *Pétersbourg,* 1867

Schneider, F. A. Nachrichten über die Fortschritte der Astrometeorologie. (Als Manuscript gedruckt.) 4to. 97 pp. *Berlin,* 1864

Schneider, Jacob. *Born September* 10, 1818, *at Trier.*

† De figuris electricis. 4to. 10 pp. 1 plate. *Bonnæ,* 1840

Über Phosphorescenz durch mechanische Mittel. 8vo. (*Pogg. Ann.* xcvi. 1855.) *Leipzig,* 1855

† Über elektr. Figuren mit Rucksicht auf verwandte Erscheinungen d. elektr. u. magnet. Gewitters (Progrm.) 4to. *Emmerich,* 1856

Über einige elektr. Meteore. 8vo. (*Pogg. Ann.* xcviii. 1856.) *Leipzig,* 1856

Uber d. elektr. Erscheinungen in d. Vereinigt. Staaten. 8vo. (*Pogg. Ann.* ci. 1857.) *Leipzig,* 1857

Über e. Elektrometeor. 8vo. (*Pogg. Ann.* cx. 1860.) *Leipzig,* 1860

Schnelle, H. Der elektro-magnetische Telegraph in den einzelnen Stadien seiner Entwickelung und in seiner gegenwärtigen Ausbildung und Anwendung, nebst einer Einleitung über die optische und akustische Telegraphie und einem Anhang über die elektrischen Uhren. . 8vo. 368 pp.
Braunschweig, 1851

Schnitzer, A. Über d. rationelle Anwendung des mineral Magnetismus in verschied. Krankheits zuständen nebst e. Anweis. z. Anfertig. v. Stahlmagneten. 8vo. *Berlin,* 1837

† Practische Anleitung zur Anwendung des magneto-elektrischen Rotations Apparates in verschiedenen Krankheiten. 8vo. 121 pp. 1 plate.
Berlin, 1843

Practische Anleitung zur Anwendung des magneto-elektr. Rotations-Apparates in verschied. Krankheiten. 2es, mit einem Nachtrage verm. Ausg. 8vo. 185 pp. 2 plates. *Berlin,* 1850

The Appendix is sold separately under the title, " Der magneto-elektrische Rotations-Apparat und seine Anwendung in verschiedenen Krankheiten. Nach den neuesten Erfahrungen bearb. 8vo. 64 pp. 1 plate. *Berlin,* 1850

Scholefield. (*Vide* Pilkington, James.)

Scholle. (*Vide* Stohrer & Scholle.)

Scholz, B. Anfangsgründe der Physik als Vorbereitung zum Studium der Chemie. *Wien,* 1825

Schomburg. Der Magnetberg auf St. Domingo v. Leonhard u. Bronn. 1855, p. 89. (*Ann. des Voyag.* 1854, ii. 360.) 1855

Schonbein, Christian Friedrich. *Born October 18, 1799, at Metzingen under Urach, Wirtemberg.*

† Nouvelles Observations sur la production et la destruction de la passivité du Fer. 8vo. 12 pp. (*Tiré de la Bibliothèque Universelle de Genève, Juin,* 1837.)
Genève, 1837

Das Verhalten d. Eisens z. Sauerstoff. Ein Beitr. z. Erweiterung elektrochem. Kenntnisse. 8vo. 1 plate. *Basel,* 1837

Über das Verhalten des Eisens zum Sauerstoff. *Basel,* 1837

† Experimental Researches on a peculiar action of Iron upon solutions of some Metallic Salts. 8vo. 10 pp. (*Phil. Mag.* Series iii. vol. x. 1837, 267.)
London, 1837

† Una nuova pila di Volta di straordinaria forza. Lettera del Schonbein al Compilatore della Gazzetta Universale di Augusta. Basilea 28 Decembre (Dalla Gazzetta medesima del 12 Gennajo 1840.) 8vo. 4 pp. (*Bibl. Ital.* xcvi. 125.) *Milano,* 1839

† Beobachtungen über den bei der Elektrolyse des Wassers und dem Ausströmen der gewöhnlichen Elektricität aus Spitzen sich entwickelnden Geruch. Dated end, Basel 10 Ap. 1840. 4to. 22 pp. (*Abhand. der* ii. *Class. d. Akad. d. Wiss.* iii. Bd. Abth. i. p. 225.) *Munchen,* 1840

† An Account of Researches in Electro-Chemistry. 8vo. 12 pp. (*Report of the British Association for the Advancement of Science for* 1840.)
London, 1841

† Beobb. über d. elektr. Wirkungen d. Zitteraals. 8vo. 39 pp. *Basel,* 1841

† Observations sur quelques actions électrolysantes de la Pile simple. Lu 16 Mars 1842 à Bâle. . . Observations sur un état particulier du fer. Lu 13 Avril 1842 à Bâle. . . Notice sur une nouvelle Pile voltaïque. 8vo. 51 pp. (*Extrait des Archives de l'Electricité, Supplément à la Bibliothèque Universelle de Genève.*) *Genève,* 1842?

† Über die Natur des eigenthümlichen Geruchs, welcher sich sowohl am positiven Pole einer Säule während der Wasser Elektrolyse wie auch beim gewöhnlichen Elekt. aus Spitzen entwickelt. 4to. 17 pp. (*In Abhandl. d.* ii. *Cl. d. Akad. d. Wissen,* iii. Bd. Abth. iii.) *München,* 1843

† Über die Häufigkeit der Berührungswirkungen auf dem Gebiete der Chemie. 4to. 20 pp. *Basel,* 1843

Beiträge zur physikal. Chemie. 8vo. *Basel,* 1844

† Über d. Erzeugung d. Ozons auf chem. Wege. 8vo. 159 pp. *Basel,* 1844

† Sulla produzione dell' Ozono per via chimica. 8vo. 136 pp. (*Translation from the original work published in Basle, dated* 13th *May,* 1844.)
Milano, 1845

† Nuove esperienze sulla produzione dell' Ozono. Lettera al De Kramer, cumunicata all' I. R. Istituto 17 Ap. 1845, (dated) Basilea, 23rd Marzo, 1845. 8vo. 7 pp. (*Giorn. I. R. Istit.-Lomb.* xi. 200.) *Milano,* 1845

† Ueber die Natur des Ozons. 8vo. 12 pp. *Basel,* 1845

† L'Ozono non é acido nitroso nitrico. Nota spedita ai trad. Kramer e Belli. 8vo. 9 pp. (*Giorn. I. R. Istit.-Lomb.* x. 258.) *Milano,* 1845

† Beleuchtung der Meinung des Hrn. Fischer betreffend das Ozon. 8vo. 7 pp. (*Annalen der Physik und Chemie,* Bd. lxv.) *Leipzig,* 1845

† Ueber die Einwirkung des Ozons auf organische Substanzen. 8vo. 4 pp. (*Annalen der Physik und Chemie,* Bd. lxv.) *Leipzig,* 1845

† Das Ozon verglichen mit dem Chlor. 8vo. 18 pp. (*Annalen der Physik und Chemie,* Bd. lxv.) *Leipzig,* 1845

† Einige Bemerkungen über die Anwesenheit des Ozons in der atmosphärischen Luft, &c. und die Rolle, welche es bei langsamen Oxidationen spielen dürfte. 8vo. 13 pp. (*Annalen der Physik und Chemie,* Bd. lxv.)
Leipzig, 1845

462

SCH

Schonbein, Christian Friedrich,—*continued.*

† Letter to Faraday, on a new Test for Ozone. 8vo. 2 pp. (*Phil. Mag.* Series iii. xxxi. 176.) *London,* 1847

Denkschr. über d. Ozon. 4to. *Basel,* 1849

† Ueber einige mittelbare physiologischen Wirkungen der atmosphärischen Elektricität. 8vo. 22 pp. (*Vorgetragen in der medicinischen Gesellsch. in Basel im* Oct. 1850.) 1850

† Über d. Einfluss d. Sonnenlichtes auf d. chem. Thätigkeit d. Sauerstoffs u. den Ursprung d. Wolkenelektricität u. d. Gewitters. 4to. 20 pp. *Basel,* 1850

† Sur quelques effets physiologiques indirects de l'Electricité Atmosphérique. Mémoire lu à la Société de Médecine de Bâle, en Octo. 1850. 8vo. 15 pp. (*Tiré de la Bibliothèque Universelle de Genève,* Novembre, 1851.) *Genève,* 1851

Über die Natur des eigenthümlichen Geruches, welcher sich sowohl am positiven Pole einer Säule während der Wasserelektrolyse, wie auch beim Ausströmen der gewohnl. Electricität aus Spitzen entwickelt. 4to. 19 pp. (*Aus. Abhandlungen der 2e Classe der Akad. d. Wissenschaften,* vol. iii. 3 Abthl. *besonders abgedruckt.*) *München,* 1851

† On some Secondary Physiological Effects produced by Atmospheric Electricity. Communicated by M. Faraday. 8vo. 16 pp. *London,* 1851

† Aus einer am 10 Dec. 1851 und 7 Jan. 1852 der naturforschenden Gesellschaft von Basel gemachten Mittheilung. 8vo. 35 pp. (*Journ. f. prakt. Chemie,* lv. 3.) *Basel*

† Über die chemischen Wirkungen der Electricität, der Wärme und des Lichtes 8vo. 52 pp. (*Aus den Verhandlungen der Naturforschenden Gesellschaft in Basel besonders abgedruckt.*) *Basel,* 1854

Darstell. d. Ozonisirt. Sauerstoffs aus Silbersuperoxyd. (*Gelehrt. Anz. d. Munch. Acad.* xli. 1855.) *München,* 1855

† Ueber die Selbstbläuung einiger Pilze, &c. und das Vorkommen von Sauerstofferregern und Sauerstoffträgern in der Pflanzenwelt (Den 24 October und 14 Novembre, 1855.) 8vo. 15 pp. (*Verhandlungen der Naturforschenden Gesellschaft in Basel,* iii. Heft.) *Basel,* 1856

Über d. nächste Ursache d. spontanen Bläuung einiger Pilze. 4to. (*Abhh. d. Münch. Akad.* 1857.) *Munchen,* 1857

† On a peculiar property of Ether and some essential Oils. 8vo. 10 pp. *London*

Many notices, &c., in the "Verhandll. d. naturf. Gesellsch. zu Basel, vol. i. 1857, and vol. ii. 1860;" some of which may relate to Electricity, &c. (*Poggendorff,* ii. 832.)

† Ueber das Verhalten organischer Farbstoffe zur schweflichten Säure. Aus einer der schweizerischen naturforschenden Gesellschaft in Glarus gemachten Mittheilung. 8vo. 11 pp. (*Journ. f. Prakt. Chemie,* liii. 6.)

† Ueber den Zusammenhang zwischen elektrischen und chemischen Thätigkeiten. 8vo. 23 pp. (*Journ. f. Prakt. Chemie,* xx. 3.)

† Ueber verschiedene Zustände des Sauertoffes. 8vo. 44 pp. (*Annalen der Chemie und Pharmacie,* Bd. lxxxix. iii. Heft.)

† Sur la pile à Oxi-Hydrogène. 8vo. 19 pp. *Genève*

† Beobachtungen über die elektrische Polarisation fester und flüssiger Leiter. 8vo. 19 pp.

Forty-one Articles on Passivity of Iron, &c., Voltaic-Electricity, Polarisation, Ozone, Phosphorus, in Poggendorff's Annalen, vol. xxxvii. 1836, to vol. cix. 1860.)

(*Vide* also Faraday and Schönbein.)

Schönberger, Geo. Demonstratio et constructio horologiorum novorum, radio recto, refracto in aqua, reflexo in speculo, solo magnete, horas astronomicas, italicas, babylonicas indicantium. 4to. *Friburg,* 1622

SCH 463

Schoning, Gerh. Om Nordlyets Aelde, &c. (*Danske Vid. Selsk. Skr.* viii. 1760.)
1760

Schott, Gaspar. *Born 1608, at Königshofen, near Würzburg ; died May 22, 1666, at Würzburg.*

Ars Magnetica. (*From Lamont, Handb.* p. 445.)

Magia universali natura, &c. 4 vols. 4to. (*From Watt.*) *Herbipoli,* 1657

De arte mechanica hydraulico-pneumatica, Pt. ii. 4to. *Herbipoli,* 1657-8
(*Note.—Guerick's experiment first published in this work.*)

P. Gasp. Schotti . . Technica curiosa, sive mirabilia artis. Libris xii. 4to.
1044 pp. Numerous plates and Portrait. *Norimburgæ,* 1664
(*Note.—Contains Guerick's experiments, enlarged.*)

Magia universali naturæ et artis. 4to. *Bamberg,* 1677

Schotten und **Herger.** Die Systeme d. Magnet-Curven. Fol. *Leipzig,* 1844

† **Schrader,** C. Dissertatio inauguralis medico-physica de electro-magnetismo . . quam in Acad. Fridericiana . . defendit 8 Sept. 1821. 8vo. 1 plate.
35 pp. *Halæ,* 1821

Schrader, John Gottlieb Friedrich. *Born September 17, 1763, at Salzdahlum, near Wolfenbüttel.*

Versuch einer neuen Theorie der Elekt. *Altona,* 1796

Schreibers, Karl Franz Anton Von. *Born August 15, 1775, at Presburg ; died May 21, 1852, at Vienna.*

Five Articles in Gilbert's Annalen, vol. xxix. 1808, to vol. lv. 1817, on Meteorites, &c., the last of which is on Zamboni's dry Pile.

Proteus anguineus. 4to. *Wien,* 1818

Beiträge zur Geschichte und Kenntnisz meteorischer Stein-und-Metall-Massen, und der Erscheinungen, welche deren Niederfallen zu begleiten pflegen. Als Nachträge zu H. D. Chladni's neuestem Werke über Feuer-Meteore und die mit denselben herabgefallenen Massen. Fol. 92 pp.
10 plates. *Wien,* 1820
(*Note.—The 9th plate is a "Meteor-Eisen Autograph, the 10th a plan. This was published separately from Chladni's work "über Feuer-Meteore," but is included in his title-page. The two are considered as a joint work, therefore, under the names Chladni and Schreibers.*)

Über d. Meteorstein-Niederfall zu Wessely in Mähren. (*Baumgartner's Zeitschrift,* i. 1832.) *Wien,* 1832

Über d. neuerlichst bei Magdeburg zufällig aufgefundene problemat. Metall-masse. (*Baumgartner's Zeitschrift,* ii. 1833.) *Wien,* 1833

(*Vide* also Chladni and Schreibers.)

† **Schroder,** B. G. Von. Versuch einer Abhandlung von den Phosphoris. Erster Abschnit. 4to. 46 pp. (*Neue Sammlung von Versuche u. Abhandl. d. naturf. Gesellsch. in Danzig,* i. 128.) *Danzig,* 1778

Schroder, Chr. F. Ueber Hohenmessungen, zwei entdeckte grosse Magnetfelsen, u.s.w. *Hildesheim,* 1790

† **Schroder,** Van der Kolk H. W. Over het Meten v. d. galvanischen geleidings weerstand enzonderheid bij Metalen. 8vo. 128 pp. 1 plate. *Utrecht,* 1860

Über d. Magnet Störungen im Sept. 1859. 8vo. (*Pogg. Ann.* cxiv. 1861.)
Leipzig, 1861

Schrötter, A. Ueber die Ursache des Leuchtens gewisser Körper beim Erwärmen. 8vo. 12 pp. (*Aus den Sitzungsberichten* 1852 *der K. Akad. der Wissenschaften abgedruckt.*) *Wien,* 1853

Ist die krystallinischa Textur des Eisens von Einfluss auf seine Magnetisirbarkeit ? 8vo. (*Sitzungsb. d. Wien Akad.* xxiii. 1857.) *Wien,* 1857

Schubert. "Kohlencylinder" for galvanoplastic purposes. (*From Martin.*)

Schubler, C. L. Revision d. vorzüglichst. Schwierigkeiten in d. Lehre von d. Elektricität u.s.w. 8vo. *Leipzig*, 1789

Schubler, Gustav. *Born August 17, 1787, at Heilbronn; died September 8, 1834, at Tübingen.*

Diss. inaug. sistens experimenta quædam ad influxum electricitatis in sanguinem et respirationem spectantia. 8vo. *Tübingen*, 1810

Bestimm. d. tägl. Perioden d. atmosphär. Elektricität. 8vo. *Nürnberg*, 1811

About ten Articles in Schweigger's Journal, vol. iii. 1811, to vol. lxix. 1833, on Atmospheric Electricity, De Luc's dry Pile, Zamboni's Pile, Aurora, Magnet-needle, Thunder, &c.

Über Deluc's elektr. Säule, &c. 8vo. *Nürnberg*, 1813

Über Zamboni's trockne Säule. 8vo. *Nürnberg*, 1815-16

Untersuchungen üb. den Einfluss d. Monds auf d. Veränderungen unserer Atmosphäre, mit Nachweisn. der Gesetze nach welchen dieser Einfluss erfolgt. 8vo. *Leipzig*, 1830

Grundsätze d. Meteorologie in näherer Beziehung auf Deutschlands Klima. 8vo. *Leipzig*, 1831

Grundf. d. Meteorologie. *Leipzig*, 1834

† Grundsäze der Meteorologie in näherer Beziehung auf Deutschlands Klima. Neu bearb. von G. A. Jahn. 8vo. 244 pp. 9 plates. *Leipzig*, 1849

Effects of Wind in some cases of Atmospheric Electricity. (*Bibl. Univ.* xlii. 203.)

Schuh. Improved construction of Smee's Battery. (*Vide* Martin.)

Schulthess, Rudolph. *Born 1802, at Zürich; died 1832, at Zürich.*

† Über Elektromagnetismus nebst Angabe e. neuen durch elektromagnet. Kräfte bewegten Maschine. Drey Vorlesungen. 8vo. 106 pp. 1 plate. *Zürich*, 1835

† Part of a Lecture on Electro-magnetism, delivered to the Philosophical Society at Zurich, February 13, 1833, by the late Dr. R. Schulthess. 8vo. 7 pp. 1 plate. (*Translation by E. Solly, jun., from "Schulthess, Über Electromagnetismus," for Taylor's Scientific Memoirs,* vol. i. part iv. p. 534.) *London*, 1837

Schulz. Ueber das Verhalten der Muskeln bei Paralysis N. facialis gegen den inducirten und constanten electrischen Ström. (*Wiener med. Zeitung,* No. 27, 1860.) 1860

Schultz, Ernst Christian. *Born 1740, at Konigsberg, Prussia; died May 31, 1810.*

Über die Elekt. verschiedener Schörle. 8vo. (*Mayer's Sammlung Physik. Aufsätze der Gesellsch Bohmischer Naturf.* tom. i. p. 261.) *Dresden*, 1791

Schultz-Schultzenstein, K. H. Versuche über d. thierische Elektricität u. über d. Elektricität in Krankheiten. (*Froreip's Tagesberichte,* 1851.) *Berlin ?* 1851

† **Schultze, M.** Zur Kentniss der elektrische Organe der Fische. 1 Abth. : Malapterurus Gymnotus. 35 pp. 2 plates. *Halle*, 1858

† Zur Kenntniss der electrischen Organe der Fische. Zweite Abtheil. : Torpedo. La. 4to. 38 pp. 2 plates. *Halle*, 1859

Schulze, J. K. Relation particulière de l'Aurore Boréale observée à Berlin. 4to. (*Mém. Berlin,* 1777.) *Berlin*, 1777

† **Schumann.** Von d. Gewitter u. dem damit verbundenen Erscheinungen. 4to. 45 pp. *Quedlinburg*, 1848

† Eine neue Tangenten Boussole. 4to. 32 pp. *Königsberg*, 1862

Schurer, Jacob Ludwig. Diss. de historia electricitatis. 4to. *Argentoratum*, 1766

Respond. Joh. Jac. Revtlinger, De Historia Electricitatis, Sect. i. *Argentoratum*, 1766

Experimenta circa Electricitatem. 4to. *Argentoratum*, 1767

De electricitate corporum. 8vo. *Argentoratum*, 1775

SCH

465

Schutzenberger, P. Essai sur les substitutions des éléments électro-négatifs aux métaux dans les sels, et sur les combinaisons des acides anhydres entre eux. Propositions de physique données par la Faculté. Thèses présentées à la Faculté des sciences de Paris. 4to. 61 pp. *Strasbourg*, 1863

Schuyler, General. Table of Variations of the Magnetic Needle, copied from one furnished by the late General Schuyler, to S. de Witt, Surgeon-General. 8vo. (*Trans. of the Albany Instit.* i. 4, 1st article.) *Albany, U.S.* 1831

Schwab, J. Prolusio historico-mineralogica de succino, &c. 4to. *Heidelberg ?* 1794

† **Schwarze, C. A.** Ein kleiner Beytrag zu der Geschichte der aus der Luft gefallenen Steine. . . Sm. 4to. 14 pp. *Gorlitz*, 1804

De magnetide Lapide Theophrasti a recentiorum magnete plane diverso. 4to. *Gorlitz*, 1808

Schwedischen Akad. Abhandlungen der Königl. Schwedischen Akad. Translated by Kastner and others. 41 vols. 8vo. *Hambourg*, 1749-1783

Neue Abhandlungen. 12 vols. 8vo. *Hambourg and Leipzig*, 1784 *to* 1792

Schweigger, Johann Salomo Christoph. *Born April 8, 1779, at Erlangen ; died September 6, 1857, at Halle.*

Proposal to simplify Sommering's Conducting System (by employing two instead of thirty-seven wires, &c.) 8vo. (*Schweigger's Journ.* ii. 240.) *Nürnberg, about* 1811

Appendice au Mémoire de Sommering. (*From Moigno.*) (*Schweigger's Polyt. centralblatt.*)
(*Note.—No such journal in Poggendorff by Schweigger.*)

Ueber Benutzung der magnetischen Kraft bei Messung der elektrischen. 8vo. (*Gehlen's Journ. fur Phys. u. Chem.* vii. 1808.) *Berlin*, 1808

About Six Articles on Galvanism, &c., in Gehlen's Journal f. Chem. u. Physik, vol. iv. 1807, to vol. ix. 1810. (*From Poggendorff*, ii. 874.)

Neues Journal für Chemie und Physik. 8vo. 60 vols. *Nürnberg*, 1811-30
(*Note.—This journal follows Gehlen's Journal für die Chemie und Physik.*)
(*Vide Allgemeine Journal der Chemie von Sherer, &c.*)

Journ. f. Chemie u. Physik erst allein Bd. i. xxvii. 1811-19 ; dann mit Meinecke bis Bd. xxxviii. 1823 ; darauf wieder allein bis Bd. xliv. 1825 ; u. nun mit Schweigger-Seidel bis Bd. liv. 1828 ; der es fur sich bis Bd. lxix. 1833, fortsetzte. (*From Poggendorff*, ii. 873.) 1811 *to* 1828

About Twenty Articles in Schweigger's Journal, vol. iii. 1811, to vol. lxiii. 1831, with many remarks, &c. on other writers' articles. (*From Poggendorff*, ii. 874-75.)

Ueber electrische Reizung der Nerven und Muskeln. 8vo. (*Magaz. der Gesellsch. Naturf. Freunde zu Berlin*, an. 6, p. 261. *Also in Schweig. Journ.* x. 1814.) *Berlin*, 1814

† Bemerkungen über Umkehrung der Polarität einer elektrischen Combination. 4to. 13 pp. *Munchen*, 1821

Über naturwiss. Mysterien in ihrem Verhaltn. zur Literatur d. Alterthums. 4to. *Halle*, 1843

† Über das Elektron der Alten und den fortdauernden Einfluss der Mysterien des Alterthums auf die gegenwärtige Zeit, nebst einem Anhange üb. einige neuere Gegenstände angewandter Naturwissenschaft. 8vo. (*Aus Grunert's Archiv. der Mathem. u. Physik*, ix. u. x. Thl. besonders abgedruckt.) *Greifswald*, 1848

† Über d. Umdrehung d. magnet. Erdpole u. ein davon abgeleitetes Gesetz d. Trabanten- u. Planetenlaufs. 4to. 8 pp. (*In Abhandl. d. Naturf. Gesellsch. zu Halle*, i. 1853.) *Halle*, 1853

† Ueber die optische Bedeutsamkeit des am elektromagnetischen Multiplicator sich darstellenden Princips zur Verstärkung des magnetischen Umschwungs. 4to. 38 pp. (*Aus den Abhandlungen der Naturforschenden Gesellschaft zu Halle, besonders abgedruckt.*) *Halle*, 1855

Schweigger, Johann Salomo Christoph.—*continued.*

† Über Magnetismus in akustischer Beziehung, u. damit zusammenhängende weltharmon Gezetze. 4to. 45 pp. (*Abhandl. d. Naturf. Gesellsch. zu Halle,* iii. 1855.)   *Halle,* 1856

† **Schweigger,** J. S. C. und Pfaff, W. Über die Umdrehung d. Magnet-Erdpole, u. ein davon abgeleitetes Gesetz des Trabanten- u. Planeten-Umlaufs; in Briefen an W. Pfaff, nebst einem Schreiben des letstern üb. Kepler's Weltharmonie. 8vo. 90 pp. (*Aus dem Journ. f. Chemie u. Phys.* Bd. x. Heft 1, besonders abgedruckt.)   *Nürnberg,* 1814

Schweigger-Seidel, Franz Wilhelm. *Born October* 16, 1795, *at Weissenfels; died June* 5, 1838, *at Halle.*

Literatur d. Mathematik Natur- u. Gewerbskunde u.s.w. von Ersch. 2 Aufl. (*Vide* Ersch.)   *Leipzig,* 1828

Neues Jahrbuch der Chemie und Physik. 8vo.   *Halle,* 1831

(Edited Schweigger's Journal, partly with J. T. C. Schweigger.)

Schweighardt. Ars magnetica. (*From Young.*)

Schwenkenhardt. Von dem Einfluss der Elekt. auf die Vegetation.

Scientific American. An illustrated Journal. New Series. 1860-62. 5 vols. Fol.

Scientific Memoirs. Selected from the Transactions of Foreign Academies and Learned Societies, and from Foreign Journals. Edited by Richard Taylor. 8vo.   *London,* 1836

Scientific Record. Weekly Journal.

Scientific Review and Journal of "The Inventor's Institute." 8vo.
                                                                        London, 1865

† **Scienziati Italiani.** Atti della Prima Riunione degli . . tenuta in Pisa nell' Ottobre del 1839. (Seconda Ed.) Sezione di Fisica Chimica, Matematica. 4to. 47 pp. Portrait of Galileo.   *Pisa,* 1840
 (*Note.—Contains matter on Electricity by Zantedeschi, Paccinotti, Orioli, Tito Puliti, Majocchi, Cassiani Bolto, Paccinotti e Pucinotti, &c.*)

† Atti della Seconda Riunione, tenuta in Torino, nel Settembre del 1840. Sezione di Fisica, Chimica e Matematica. 54 pp. Sotto-sezione di Chimica, 17 pp; e qualche foglio della sezione di Medicina. 4to. 71 pp. and 28 pp.
                                                                        Torino, 1841

† Atti della Terza Riunione degli . . tenuta in Firenze nel Settembre del 1841 S. Sezione di Fisica e Matematica. 4to. 65 pp.   *Firenze,* 1841

Diario della 3a Riunione degli Sci. Ital. convocati in Firenze nella 2da metà del Settembre, 1841. 4to.   *Firenze,* 1841

† Atti della Quarta Riunione degli . . tenuta in Padova, nel Settembre del 1841. Atti verbali della Sezione di Fisica e Matematica. 4to. 64 pp.
                                                                        Padova, 1843

Atti della Quinta Riunione tenuta in Lucca nel Settembre, 1843. 4to.
                                                                        Lucca, 1844

† Diario del Settimo Congresso degli Scien. Ital. in Napoli dal 20 di Set. al 5 di Ottob. dell' Anno 1845. 4to. 162 pp.   *Napoli,* 1845

† Atti della Sesta Riunione degli . . tenuta in Milano nel Settembre del 1844. Atti verbali della Sezione di Fisica e Matematica. 4to. 98 pp. *Milano,* 1845

† Atti della Settima Riunione tenuta in Napoli nel Set. e Otto. 1845. Atti verbali della Sezione di Fisica e Matematica. 4to. 101 pp. 3 plates.
                                                                        Napoli, 1846

Atti dell' Ottava Riunione tenuta in Genova nell' Ottobre, 1846. Atti verbali della Sezione di Fisica. 4to.   *Genova,* 1847

† Diario del Nono Congresso . . convocato in Venezia nel Settembre, 1847. 4to. 160 pp.   *Venezia,* 1847

Scina, Domenico Ragona. *Born February* 28, 1765, *at Palermo ; died July* 13, 1837, *at Palermo.*

Esperienze e scoperte sull' elettromagnetismo.   (*From Pogg.* ii. 879.)

Introd. alla fisica esperimentale.   *Palermo,* 1803

Elementi di fisica generale.   *Palermo,* 1809

Introduzione alla fisica sperimentale. Terza ed.   (*Solvestre's Biblioteca Scelta d'Opere Ital. antiche e moderne.*)   *Milano,* 1817

Elementi di fisica generale. (*From Pogg.* ii. 879.)   1829

† Lettera del Prof. D. Scina ai Direttori della Biblioteca Italiana. 8vo. 12 pp. (*Bibl. Ital.* lxxiii. 165.)   *Milano,* 1834

† Elementi di fisica particolare. Seconda Edizione. Tomo 1o. 8vo. 357 pp. 5 plates.   *Milano,* 1842

† Elementi di fisica particolare. Seconda Edizione Milanese. Tomo 11o. 8vo. 455 pp. 4 plates.   *Milano,* 1843

† Un nuovo caso di Rotazione dell' Ago magnetico. 4to. 3 pp. (*Ann. delle Scien. Lomb. Ven.* xiv. 125.)   *Vicenza,* 1845

† **Scoffern, John.** Chemistry of Imponderable Agents, Light, Heat, Electricity, and Magnetism, including the Gases, and the Elementary Principles of the Science. 8vo. 277 pp. (*Orr's Circle of the Sciences.*)   *London,* 1855

† **Scoffern and Lowe.** Practical Meteorology : Being a Guide to the Phenomena of the Atmosphere, and the Practical Use of Instruments for Registering and Recording Atmospheric Changes. 8vo.   *London,* 1856

Scoresby, Dr. William. *Born October* 5, 1789, *at Cropton ; died March* 21, 1857, *at Torquay.*

Description of a Magnetimeter, &c. 4to. (*Trans. Edinb. Soc.* ix. 1821.)
                                                      Edinburgh, 1821

Electro-magnetic Experiments. 4to. (*Trans. Edinb. Soc.* ix. 1821.)
                                                      Edinburgh, 1821

Experiments on the Development of Magnetic Properties in Steel and Iron by Percussion. Part i. 4to. (*Phil. Trans.* 1822-24.)   *London,* 1822

Voyage to the Northern Whale Fishery. 8vo.   *London,* 1823

Experiments on the Development of Magnetical Properties in Steel and Iron by Percussion. Part ii. (*Vide* 1822.) 8vo. (*Phil. Trans.* 1824.)
                                                      London, 1824

Magnetical Experiments, &c. 4to. (*Trans. Edinb. Soc.* xi. 1824.)
                                                      Edinburgh, 1824

Tagebuch einer Reise auf dem Wallfischfang, etc. ; aus dem Engl. übers. von F. Kries. 8vo.   *Hornberg,* 1825

On the singular effects of Two Strokes of Lightning upon a Vessel. 8vo. (*Brewster's Journal of Science,* viii. 1828.)   1828

On the Uniform Permeability of all known substances to the Magnetic Influence, &c. 4to. (*Trans. Edinb. Soc.* xii. 1831.)   *Edinburgh,* 1831

On the Uniform Permeability of all known substances to the Magnetic Influence, &c. Continuation. 4to. (*Trans. Edin. Soc.* xiii. 1832.)
                                                      Edinburgh, 1832

An Exposition of some of the Laws and Phænomena of Magnetic Induction, &c. 4to. (*Trans. Edinb. Soc.* xiii. 1832.)   *Edinburgh,* 1832

Observations on the Deviation of the Compass, &c. 4to. (*Trans. Edinb. Soc.* xiv. 1832.)   *Edinburgh,* 1832

Observations on the Deviation of the Compass. Continuation. 4to. (*Trans. Edinb. Soc.* xiv. 1833.)   *Edinburgh,* 1833

† Magnetic Investigations. Part i. 8vo. 92 pp. 2 plates.   *London,* 1839

Magnetical Investigations. Part ii. 8vo.   *London,* 1843

Scoresby, William Dr.— continued.

† Magnetic Investigations. 8vo. 364 pp. 4 plates. *London*, 1844

† Zoistic Magnetism ; being the substance of Two Lectures delivered at Torquay, May, 1849. 8vo. 144 pp. *London*, 1849

Magnetic Investigations. Vol. ii. 8vo. 448 pp. Cuts. *London*, 1852

On the Magnetism of Iron Ships, and its changes. 8vo. (*Rep. Brit. Assoc.* 25th Meeting, 1855 & Abstr. vii. 1854.) *London*, 1855

† Journal of a Voyage to Australia and Round the World for Magnetic Research. Edited by Arch. Smith. 8vo. 315 pp. Portrait and chart. *London*, 1859

Scoresby, W. and Joule. Experiments and Observations on the mechanical powers of Electro-Magnetism, Steam, and Horses. 8vo. (*Phil. Mag.* 3rd Series, xxviii. 1846.) *London*, 1846

† Scoresby, W. Junr.   On the Anomaly in the Variation of the Magnetic Needle as observed on ship-board. 8vo. 9 pp. (*Phil. Mag.* liv. 282.) *London*, 1819

† Observations on the Errors of the Sea-Rates of Chronometers arising from the Magnetism of their balances, with suggestions for removing the source of error. 4to. 12 pp. 1 plate. (*Trans. R. Soc. Edinb.*) *Edinburgh*, 1822

Scott, P. H. (*Vide* Dove, 1862.)

Scots Magazine. (*Vide* vol. xv. p. 73, 1753, for letter signed C. M. (Charles Marshall) on an Electric Telegraph.) 1753

† Scoutetin, H.   L'Ozone ou recherches chimique, météorologiques physiologiques et médicales sur l'oxygène électrisé. 12mo. 284 pp. and 5 tables. 1 plate, coloured. *Paris and Metz*, 1856

† De l'electricité considerée comme cause principale de l'action des Eaux Minérales sur l'organisme. 8vo. 422 pp. *Paris*, 1864

Scudery, D. J.   Fernglas d. Artzney wissenschaft. nebst Abhdl. Schiffe und Häuser v. d. Blitz zu verwahren a. d. Italianen. 8vo. *Münster*, 1774

† Scudery, G. Del folgoreggiante vapore fuocoso. Cannocchiale della medicina. Difesa delle navi, e Palazzi da' fulmini. *Genova*, 1772

Secchi, Angelo.   Born June 29, 1818, at Reggio ; died at Rome, 1878.

† Ricerche di Reometria elettrica. (Memoria.) 8vo. 1 plate. 58 pp. (*Negli Annali di Scienze Matem. e Fis.* tom. i.) *Roma*, 1850

† Researches on Electrical Rheometry. 4to. 59 pp. 3 plates. (*Smithsonian Contributions*, iii. Art. 2.) *Washington*, 1852

† Ricerche sull' attuale valore della Declinazione magnetica in Roma. Memoria. 4to. 19 pp. (*Estr. dagli Atti dell' Accad. Pontif. de' Nuovi Lincei*, an. v. Sessione vi. del 15 Agosto, 1852.) *Roma*, 1854

Memoria sulle variazioni periodiche dell' Ago magnetico. *Roma, about* 1855

Sur l'influence magnétique du Soleil. 8vo. *Bruxelles*, 1855

Descrizione del Nuovo Osservatorio del Collegio Romano. D.C.D.G. e Memoria sui lavori eseguiti dal 1852 a tutto Ap. 1856. 4to. *Roma*, 1856

Sulle variazioni periodiche del magnetismo terrestre. Memoria Seconda, relativa alle perturbazioni straordinarie. 8vo. 23 pp. *Roma*, 1857

Sulle variazioni periodiche del magnetismo terrestre. Memoria Seconda, relativa alle perturbazioni straordinarie. 8vo. 23 pp. *Roma*, 1859

Descrizione di un meteorografo, ossia registratore meteorologico universale. 4to. 2 plates. *Roma*, 1859

Intorno alla corrispondenza che passa tra i fenomeni meteorologici e le variazioni d'intensità del magnetismo terrestre. Memoria. 4to. 16 pp. (*Estr. dagli Atti dell' Accad. Pontif. de' Nuovi Lincei*, an. xiv. Session iii. Feb. 1861.) *Roma*, 1861

† Bullettino Meteorologico dell' Osservatorio del Collegio Romano con corrispondenza e bibliografia per l'avanzamento della fisica terrestre. 4to. *Roma*, 1862 to —

Secchi, Angelo.—*continued.*

† Intorno alla relazione che passa tra i fenomeni meteorologici e le variazioni del magnetismo terrestre. Quattro Memorie. 4to. 40 pp. (*Mem. dell' Osservatorio del Collegio Rom.* Nuova Serie, ii. p. 81.)   *Roma,* 1862
Memorie dell' Osservatorio del Collegio Romano D.C.D.G. Nuova Serie. 4to.
                                                                        Roma, 1863
L'Unità delle forze fisiche, saggio di filosofia naturale. 4to.   *Roma,* 1864

† Ricerche sulla corrente elettrica e sue analogie coi fenomeni idraulici. 4to. 16 pp. (*Atti dell' Accad. Pontif. dei Nuovi Lincei. Est. dalla Sessio.* v. del 3 Ap. 1864, tom. xvii.)   1864

† Sulla relazione di fenomeni meteorologici colle variazioni del magnetismo terrestre. Mem. letta alla Pontificia Accad. Tiberina. 8vo. 23 pp.
                                                                        Roma, 1864
† Riduzioni delle Osservazioni magnetiche fatte all' Osservatorio del Collegio Romano dal 1859 al 1864. Parte Prima. 4to. 36 pp. 1 plate. (*Estratto dagli Atti dell' Accad. Pontif. dei Nuovi Lincei 5 Feb.* 1865, tom. xviii.)
                                                                        Roma, 1865
† Descrizione del Meteorografo dell' Osservatorio del Collegio Romano. 4to. 20 pp. 3 plates. (*Bullettino Meteorologico del Collegio Romano* . . vol. v. No. 4, p. 25.)   *Roma,* 1866

† Riduzione delle Osservazioni magnetiche fatte all' osservatorio del Collegio Romano dal 1859 al 1864. Declinaz. magnet. Parte Seconda. 4to. 5 pp. (*Bullettino Meteorologico dell' Osservatorio del Collegio Romano,* vol. v. No. 1. p. 1.)   *Roma,* 1866
† La météorologie et le météorographe à l'Exposition Universelle. 8vo. 32 pp. 1 plate.   *Paris,* 1867

† Barometro a Bilancia. Modificazione e aggiunta al Meteorografo. 4to. 1 p. (*Bullettino Meteorologico dell' Osservatorio del Collegio Romano,* vol. vi. p. 12.)   *Roma,* 1867
† Notice sur son Météorographe (établi à l'Exposition Universelle). 8vo. 7 pp.
                                                                        Paris, 1867
† Le recenti scoperte astronomiche. Stelle cadenti. Spettri stellari, &c. 8vo. 31 pp. (*Estratta dal Giorn. Arcadico,* tom. liv. della Nuova Serie.) *Roma,* 1868

† Descrizione del meteorografo dell' Osservatorio del Collegio Romano. 8vo. 96 pp. 3 plates.   *Roma,* 1870
† **Secchi, R. P.** (A.?). Le soleil est un Aimant. 8vo. 4 pp. (*The Cosmos,* 453.) *Paris,* 1854

Seckendorf, Ad. V. Appellation an alle Regierungen, Urtheilsverf. und d. gesammte Mensch. wider d. Anstell. galvan. u. and. Versuche an. d. Kopfen d. d. Schwerdt. hinger. Verbrecher. 8vo.   *Leipzig,* 1808

Secondat de Montesquieu, Jean Baptiste, Baron de. *Born* 1716, *at Marthilhac, near Bordeaux; died June* 17, 1796, *at Bordeaux.*

Mémoire sur l'électricité.   *Paris,* 1746

Observations de Physique et Histoire naturelle. Sur les eaux minérales de Dax, &c. . Histoire de l'électricité. . Dissertation sur . l'Aimant. 8vo. 205 pp.   *Paris,* 1750

Seeauer, Beda. Magnetologia. 4to. (*From Pogg.* ii. 889.)   *Salzburg,* 1745

Seebeck, Thomas Johann. *Born April* 9, 1770, *at Reval; died December* 10, 1831, *at Berlin.*

Über e. Magnetnadel aus Kobalt u. den Magnetism d. Kobalts u. Nickels. 8vo.
                                                                        Berlin, 1810
† Über den Magnetismus der galvanischen, Ketten. Gelesen 14 Dec. 1820 und 8 Feb. 1821.) 4to. 58 pp. 3 plates.   1820-21
† On the Magnetic property inherent in all Metals and many Earths, occasioned by difference of temperature. 8vo. 1 p. (*Phil. Mag.* lviii. 462.) *London,* 1821

Über den Magnet. der galvanischen Kette. 4to. (*Abhand. der Berlin Akad.* Bd. *pro* 1821-22.)   *Berlin,* 1822

Seebeck, Thomas Johann.—*continued.*

† Magnetische Polarisation der Metalle und Erze durch Temperateur-Differenz (Auszug) aus vier Vorlesungen, welche in der Akad. der Wissenschaften am 16 Aug. am 18 und 25 Okt. 1821 und am 11 Feb. 1822, gehalten worden. 4to. 2 plates. 109 pp. 2 tables.			*Berlin,* 1822-3

† New Electromagnetic Experiments. 8vo. 1 p. (*Phil. Mag.* lxi. 146.)
				London, 1823

† Von dem in allen Metallen durch Vertheilung zu erregenden Magnetismus. Gelesen 9 Juni 1825. 4to. 22 pp. (*Abhh. d. Berliner Akad.* 1825.)
				Berlin, 1825

† Die magnetische Polarisation verschiedener Metalle, Alliagen u. Oxyde zwischen d. Polen starker Magnetstäbe. Gelesen 11 Juni, 1827. 4to. 10 pp. 1 plate. (*Abhh. d. Berliner Akad.* 1827.)			*Berlin,* 1827

† Über e. von Barlow u. Bonnycastle wahrgenommene anomale Anzieh. d. Magnetnadel durch glühendes Eisen. Gelesen 22 Marz, 1827. 4to. 18 pp. 1 plate. (*Abhh. d. Berliner Akad.* 1827.)			*Berlin,* 1827

Gedächtnissrede auf . . . (*From Pogg.*)			1839

Left unprinted Eight Treatises on Magnetism and Electricity.

Seeger, J. F. Übersetztung d. "Lord Mahon's Principles of Electricity" mit Anmerk. 8vo.			*Leipzig,* 1789

Seegers, C. De motu planetarum secund. leg electrodynam. Weberianam, solem ambient. 4to.			*Göttingen,* 1864

Sefström, Nils. Gab. Resultater af undersökningar om magnetnälens miss visning i atskilliga orter in Sverig.			*Stockholm,* 1845

Seguier, Am. Pierre. Sur le télégraphe électrique. 4to. (*Comptes Rendus,* xxix. 1849.)			*Paris,* 1849

† Seguin, A. Summary ideas on the probabilities of the Origin of Aërolites. 8vo. 4 pp. (*Phil. Mag.* xliv. 212.)			*London,* 1814

† Seguin ainé, M. Mémoire sur les causes et sur les effets de la chaleur, de la lumière et de l'électricité. 8vo. 117 pp.			*Paris,* 1865

Seidel. (*Vide* Schweigger-Seidel.)

Seidmacher, D. Kurze u. allgemein verstnädliche Beschreibung eines höchst einfachen elektromagnet. Telegraphen der von Jedem selbst angefertigt und überall gebraucht werden kann. 8vo. 15 pp. 1 plate.			*Dresden,* 1848

Kurze und allgemein verständliche Beschreibung eines höchst einfachen elektromagnetischen Telegraphen. 8vo. 1 plate.			*Dresden,* 1848

† Die elektrische Sonne. Allgemein verständl. Beschreibung des Apparates mit dem das prachtvollste Licht durch Elektricität hervorgebracht wird.
				Dresden, 1850

Seiferheld, Georg Heinrich. *Born September 12, 1757, at Haberschlacht, Wirtemberg; died July 23, 1818.*

Beschreibung einer sehr wirksamen Elektrisirmaschine als Anwendung d. Weber'schen Luftelektrophors auf Elektrischemach. 8vo.			*Nürnberg,* 1787

Entw. e. elektr. Flinte; u über d. Werth d. Luftmaschinen v. Weber. 8vo. 1 plate.			*Salzburg,* 1787

Sammlung electrischer Spielwerke für junge Electriker. 4 Lieferungen. 8vo.
				Nuremberg and Altdorf, 1787-91

Samml. elek. Spielwerk. 10 vols. or parts. 8vo.			*Nürnberg,* 1787-1808

Elektr. Versuche wodurch Wassertropfen in Hagel verwandelt werden sammt d. Frage : Ist ein Hagelableiter ausfuhrbar u. wie? 8vo.
				Nurnberg & Altdorf, 1790

Elektr. Zauberversuche. 8vo.			*Nürnberg,* 1793

† Seiler, J. De la Galvanisation par l'influence appliquée au traitement des déviations de la Colonne Vertébrale, &c. 8vo. 157 pp.			*Paris,* 1860

Selbiger (Felbiger ?). Eine Abhandlung von der Elektricität. 1734

Selmi, Francesco. *Born April 7, 1817, at Vignola.*

Ricerche sull' elettrodoratura. (*From Pogg.* ii. 900.) *Milano*, 1844
Manuale dell' arte d'indorare e d'inargentare coi metodi elettro-chimici e per
semplice immersione, compilata da Francesco Selmi ; sugli scritti e sui lavori
di Brugnatelli, Boquillon, Boettiger, Bagration, Briant, Dumas, Elkington,
Frankenstein, Graeger, Giorgi e Puccetti, Grimelli, Levol, La Rive, Marchisio, Majocchi, Roulz, Sandonnini, Smee, Wright, ed altri ; ad uso degli
Artefici Italiani. Sm. 8vo. 176 pp. (*Vide Valecourt for French translation
and additions.*) *Reggio*, 1844

Formole di alcuni liquidi da indorare con processo elettro-chimico, ec.
Reggio, 1844
Notizia sulla pila a triplice contatto del Prof. Selmi di Torino, e sugli usi di
essa nella telegrafia elettrica nella elettro-metallurgia, ec. *Torino*, 1857

Selmi and Others. Nouveau Manuel complet de dorure et d'argenture par la
méthode électro-chimique et par simple immersion. Ouvrage dans lequel
on a rassemblé les travaux de tous les chimistes qui se sont occupés de ce
sujet. 18mo. *Paris*, 1856

(*Vide* also Majocchi and Selmi.)

† **Selwyn, J. H.** Explanation of the Floating Cylinders for laying Telegraphic
Submarine Cables. 8vo. 18 pp. 1 plate. *London*, 1862

Sementini, L. Pensieri e sperimenti sui fenomeni della bacchetta divinatoria.
Milano, 1811

Semenzi, A. (*i.e.* G. B. A. S.). Descrizione del Telegrafo elettro-magnetico . .
popolare. 8vo. 20 pp. *Treviso*, 1851

† **Semeyns, M.** Kurze aus der Wirkung des Magnets hergeleitete Abhandlung von
der innern Beschaffenheit der Erdkugel. Aus dem Holländ. übersetz. 4to.
64 pp. 2 plates. *Nurnberg*, 1764

Het nieuw ontdekte magnetische systema. 8vo. *Enkh.* 1767

Semler, Ch. Coniglobium, Globus astronomicus. Methodus triplex inveniendæ
longitudinis maritimæ per acus magneticas verticales, &c. 8vo.
Halæ, 1723

Senarmont, Henri Hureau De. *Born September 6, 1808, at Broué, Département
Eure et Loire.*

† Mémoire sur la conductibilité superficielle des corps cristallisés pour l'électricité
de tension. 8vo. 26 pp. ("*Ce Mém. est imprimé ici tel qu'il a été présenté à l'Acad.* 17 Décembre 1849 *et* 14 Janvier 1850.") *Paris*, 1850 ?

Sendel, N. Electrologia s. varia tentamina historica ac physica de perfectione
succinorum operum. Miss. 1, 2, and 3. 8vo. (*From Heinsius.*)
Elbing. 1725-28

† Electrologiæ per varia tentamina historica ac physica continuandæ. Missus
Secundus, de mollitie succinorum, &c. 4to. 64 pp. *Elbingæ*, 1726

Electrologiæ per varia tentamina historica ac physica, continuandæ. Missus
Tertius, de prosapia succinorum, et eorum variis affectionibus, vi electrica,
colore, odore, sapore. 4to. *Elbinguæ*, 1728

Senebier, Jean. *Born May, 1742, at Geneva ; died July 22, 1809, at Geneva.*

† Articolo di lettera . . al Brugnatelli sopra alcuni fenomeni chimici del Geneva,
17 Agosto, 1802. 8vo. 3 pp. (*Ann. di Chim. di Brugnatelli*, xxi. 90.)
Pavia, 1802

Seneca, Lucius Annæus. *Born* A.D. 2 or 3, . . . *at Corduba, Spain ; died* 65.
Quest. Nat. lib. i. cap. 1. (*From Watson, in Phil. Trans.* xlviii. pt. i. p. 211.)

Senes. Mémoire. Description d'une Aurore boréale, observée à Cuères, en
Provence, le 15 Février 1730. The Mem. dated 30th March, 1730. (*Montpellier Acad. Hist. &c. par De Ratte*, vol. i. p. 329.) *Lyon*, 1766

Senft, Ad. And. Dissertatio Experimenta physico-medica de electricitate et calore animali. 8vo. *Herbipol*, 1778

Sequard. (*Vide* Brown-Sequard.)

† **Serafini, Filippo.** Il telegrafo in relazione alla giurisprudenza civile e commerciale. 8vo. 231 pp. *Pavia*, 1862

† Le Télégraphe dans ses relations avec la Jurisprudence civile et commerciale. Traduit et annoté par Lavialle de Lameillière. 8vo. 175 pp. *Paris*, 1863

† **Serantoni, I. M.** Dialogo intorno alla cagione della celebre Aurora Boreale vedutasi in cielo nella notte susseguente alli 16 Dicembre 1737. 4to. 88 pp. 4 plates. *Lucca*, 1740

Serapion. De simplicibus medicinis.

Séré. (*Vide* Breguet fils et Séré.)

Serinci, J. H. J. Meditationes physicæ de phialis vitreis quæ casu minimi silicis in easdem projecti dissiliunt. (*Act. Acad. Nat. Cur.* x. 1754.) 1754

† **Serra, F. M.** De Boreali Aurora, Genuæ visa 14 Kalendas Januarii, Anni 1737, nuper elapsi. Carmen. 12mo. 27 pp. (*Calogera, Raccolta*, xvii. 69.) *Venezia*, 1738

Serres, Pierre Marcel Toussaint. *Born November 3, 1783, at Montpellier.*

† Observations on the Fall of Stones from the Clouds. 8vo. 16 pp. (*Phil. Mag.* xliv. pp. 217 and 253.) *London*, 1814

Sur l'intensité magnétique des laves. *Marseille*, 1831

Translation with additions, &c. of Œrsted's " Ansicht d. chemischen Naturgesetze."

† **Serrin.** Rapport sur un Régulateur de la Lumière Electrique imaginé par M. Serrin. Commission composée de MM. Becquerel, Despretz, Combes, Pouillet rapporteur. 4to. 7 pp. (*Ext. des Comptes Rendus de l'Acad.* tome liv. *Séance du* 10 Mars 1862.) *Paris*, 1862

† **Servier, E.** Notice sur le moteur électro-magnétique de pression ou de niveau à maxima et à minima. 32mo. 15 pp. *Paris*, 1859

† **Sestier et Méhu.** De la Foudre, de ses formes et de ses effets sur l'homme, les animaux, les végétaux et les corps bruts ; des moyens de s'en préserver, et des paratonnerres par F. Sestier. Rédigé sur les documents laissés . . et complété par C. Méhu. 2 vols. 8vo. *Paris*, 1866

† **Setschenou, S.** Über die elektrische und chemische Reizung der sensiblen Rückenmarksnerven d. Frosches. 8vo. 69 pp. *Graz*, 1868

Seugraff. Un mémoire sur un système de reproduction autographique des signatures dans les communications télégraphiques. 4to. (*Comptes Rendus de l'Acad.* 1855.) *Paris*, 1855

S'Gravesande. (*Vide* Gravesande.)

Severe. (*Vide* St. Severe.)

Sevin. (*Vide* Delalot-Sevin.)

† **Sewall, S.** Magnetical Observations made at Cambridge, by Sewall. 4to. 5 pp. (*Mem. Amer. Acad.* Old Series, i. pt. ii. p. 322.) *Boston*, 1785

Seward, W. H. Der Telegraph um die Erde. Zur Verbind. d. östl. u. westl. Halbkugel. Nach offic. Orig.-Documenten. A. d. Engl. v. Cl. Gerke. 8vo. 1 Weltkarte. *Hamburg*, 1865

Seyffer, Otto Ernst Julius. *Born October 7, 1823, at Stuttgart.*

† Geschichtliche Darstellung des Galvanismus ; erweiterte Ausarbeitung einer, im Jahr 1844, von der philos. Facultät der Univers. zu Heidelb. gekrönten Preisschrift. 4to. 638 pp. *Stuttgart & Tubingen*, 1848

Über Lichtpolarisation . . Versuche über Dampfelektricität, &c. 8vo. (*Pogg. Ann.* xc. 1853.) *Leipzig*, 1853

† **Seyffert, J. H.** Über seine galvan. Batterie. 8vo. (*Gilb. Ann.* xi. 1802, S. 375.)
Leipzig, 1802

Gutachten über Steinhauser's Magnete. (*Anzeig Leipz. Okonom. Soc.* 1809.)
Leipzig, 1809

Sforza. Meteorologicæ lucubrationes. 4to. *Neapoli*, 1570

Sgagnoni, P. Relazione delle osservazioni fatte nel circondario di Borgo San Donnino per verificare un fenomeno meteorologico apparso nelle ville di Cella, di Costamezzana, Pieve di Cusignano, e Varano de' Marchesi. 8vo.
Reggio, 1808

† **Sgagnoni e Guidotti, &c.** (1.) Delle pietre cadute nel Parmigiano; (2) e Notizia su antichi Aeroliti tratta dall' Arabo Kaswini. 4to. 6 pp. (*Nuova Scelta d' Opus.* tom. ii. 327 and 333. *The second in Bibl. Britt.* No. 304.)
Milano, 1807
(*Note.*—This is Amorelli's " Transunto delle osservazioni fisico-chimiche fatte sulle pietre cadute dall' atmosfera nel Parmigiano dai . Scagnoni e Guidotti," and of an article in the Bibliothèque Britannique, No. 304, on an old notice concerning Aërolites, by Kazwini.)

† **Sguario, Eusebi.** Dissertazione sopra le aurore boreali. 4to. 119 pp. 3 plates.
Venezia, 1738

Due dissertazioni della elettricità applicata alla medicina. 8vo. *Venezia*, 1746
(*Note.*—He is said to have been the first writer on Electricity in Italy. Krunitz calls him Squario, Euseb.)

† **Shaffner, T. P.** The Telegraph Manual; a complete history and description of the Semaphoric, Electric, and Magnetic Telegraphs of Europe, Asia, Africa and America, ancient and modern. 8vo. 850 pp. 10 portraits, steel.
New York, 1859

† **Sharpe, Benjamin.** A Treatise on the Construction and Submersion of Deep-sea Electric Telegraph Cables. 8vo. 16 pp. 1 plate. *London*, 1861

Sharpe, R. On his Electric Telegraph, invented in 1813. 8vo. (*Repertory of Arts*, 2nd series, xxix. 23.) *London*, 1816
(*Note.*—No description of this Telegraph appears to have been printed. It was mentioned at the Admiralty after the invention and full description of Sommering's, described fully and with figures in the Denkschriften of the Acad. of Munich for 1809-10, printed in 1811.)

Shaw, George. Born December 10, 1751, at Bierton, Buckinghamshire ; died July 22, 1813, at London.

† Osservazioni sulla Scolopendra elettrica e Scolopendra sotterranea. 8vo. 4 pp. (*Ann di Chim. di Brugnatelli*, ix. 26.) *Pavia*, 1795

† Manual of Electro-Metallurgy. 8vo. 49 pp. *London*, 1842

† Manual of Electro-Metallurgy. 2nd edition, considerably enlarged. 8vo. 202 pp. 2 specimens of glyptography. *London*, 1844

Shebbeare. The Practise of Physic, founded on principles in Physiology hitherto unapplied to physical enquiries. 2 vols. 4to. *London*, 1755

† **Shepard, Ch.** Report on American Meteorites. *New Haven, U.S.* 1848

Shepherd. (*Vide* Vievar and Shepherd.)

† **Shepherd, C.** On the application of Electro-Magnetism as a motor for Clocks. 8vo. 24 pp. Cuts. *London*, 1851

† **Short, James.** An extract from a Letter from J. S. to Rd. Graham. Dated Edinb. Nov. 18, 1736. 4to. 2 pp. (*Phil. Trans.* xli. 368.) *London*, 1739-40-41

† **Short, T.** An account of several Meteors in a letter to the President. Dated Sheffield, March 18, 1740-41. 4to. 6 pp. (*Phil. Trans.* xli. 625.)
London, 1739-40-41

† **Sidney, E.** Electricity : its Phenomena, &c. 16mo. 184 pp. *London*, 1843

Electricity : its Phenomena, Laws, and Results. New edition. 8vo.
London, 1862
(*Note.*—Latimer Clark's Catalogue mentions an edition of 1855.)

O O

Siebold, P. F. Flora Japonica, seu plantæ quas in imperio Japonico collegit, descripsit, ex parte in ips. loc. pingendas curavit. Plantas ornatui vel usui inservientes digess. T. G. Zuccarini. *Lugd. Bat.* 1835

Siebold. Ueber die Kenntniss der Polarität des Magnets und den Gebrauch der Magnetnadel bei den Chinesen in ältester Zeit. (*Verh. d. Naturh. Ver. d. Rheinl.* 1855, vii.) 1855

Siemens, C. W. On the progress of the Electric Telegraph. 8vo. (*Journ. Soc. of Arts,* vi. 1858.) *London,* 1858

† Inaugural Address delivered to the members of the Society of Telegraph Engineers, by C. William Siemens, D.C.L., F.R.S., President, on Feb. 28, 1872. 8vo. 19 pp. *London,* 1872

(*Vide* also Siemens, E. W. and C. W.)

Siemens, Ernst Werner. Anwendung d. elektr. Funkens zur Geschwindigkeitsmessung. 8vo. (*Poggendorff's Ann.* lxvi. 1845.) *Leipzig,* 1845

Eleven Articles on Electric Telegraphs, on Magneto-electric Machines, on Induction, on unusual Electric appearances, upon the Pyramid of Cheops, &c. (*Poggendorff's Annalen,* vol. lxvi. 1845, to vol. cxiii. 1861.)

Verfahren z. Mess. kleiner Zeitintervalle. (*Fortschr. d. Physik.* 1845.)

Histor. Abriss d. Entdeck. u. Vervollkommn. d. elektr. Telegraphie. (*Fortschr. d. Physik.* 1845.) 1845

Bericht über die Fortschritte der Telegraphie im Jahr 1846. (*Berlin Ber.* ii. 540.) *Berlin,* 1846

† Mémoire sur la Télégraphie électrique. Présenté à l'Académie des Sciences le 15 Avril 1850. 8vo. 48 pp. (*Ann. Chim. Phys.* sér. iii. xxix. 1850.) *Paris,* 1851

Mémoire sur la télégraphie électrique, suivi d'un rapport fait sur ce mémoire à l'Académie des Sciences de Paris dans la séance du 29 Avril 1850. 8vo. 64 pp. *Berlin,* 1851

Kurze Darstellung der an den Preussischen Telegraphen-Linien mit untererdischen Leitungen, bis jetzt gemachten Erfahrungen. 8vo. 31 pp. *Berlin,* 1851

Über d. Ladungserscheinungen an untererdischen Telegraphendrähten. (*Zeitschr. d. deutsch-österr. Telegr.-Vereins,* i. 1854.) 1854

Beschreib. e. neuen magneto-elektr. Zeiger-Telegraphen. 8vo. (*Dingler's Polytech. Journ.* cli. 1855. *Stuttgart,* 1855

Über d. transatlant. elektr. Kabel. 8vo. (*Dingler's polytech. Journ.* cli. 1855.) *Stuttgart,* 1855

Über Bonelli's Vorschlag die über-sponnenen Kupferdrähte für Elektromagnete durch Papierbänder mit metallischen Linien zu ersetzen. (*Zeitschr. d. Deutsch-österr. Telegr.-Vereins,* vol. iii. 1856.) 1856

Der Inductions-Schreib-Telegraph von Siemens u. Halske. (*From Poggendorff,* ii. 925. *In Zeitschr. d. Deutsch-österr. Teleg.-Vereins,* vol. iv. 1857.) 1857

Apparate für d. Betrieb langer Untersee-Linien. (*From Poggendorff,* ii. 925. *In Zeitschr. d. Deutsch-österr. Teleg.-Vereins,* vi. 1859.) 1859

Beschreibung e. galvan. Batterie von anhaltend constanter Wirkung. 4to. (*From Remak's Cat.* p. 20. *Abdr. a. d. Zeitschrift d. deutsch-österr. Telegr.-Vereins.)* 1859

Rapport fait à l'Académie de Berlin sur les appareils télégraphiques de M. Siemens (de Berlin).

Vide also Poggendorff, vol. ii. pp. 925-6, who gives a long list of items relative to Siemens, and Siemens and Halsk's operations, &c., but no indications as to where descriptions of them are to be found.

Mémoire sur la Télégraphie électrique. 8vo. (*Comptes Rendus,* xxx. 434.) *Paris,* 1850

Rapport sur le mémoire sur le télégraphe électrique de Siemens, par M. Pouillet. 4to. (*Comptes Rendus,* xxx. 500.) *Paris,* 1850

SIE—SIL 475

Siemens, E. W. and C. W. Outline of the principles and practice involved in dealing with the Electrical conditions of Submarine Electric Telegraphs; with C. W. Siemens. 8vo. *London*, 1860

Siena Accad. Atti dell' Accad. delle Scienze di Siena (detta dei Fisio-Critici). (1760 to . . .) 4to. Plates. *Siena*, 1760-1844 ?
(*Note.—The publication of this was much interrupted.*)

Sigaud de la Fond. *Born 1740, at Dijon; died January 26, 1810, at Bourges.*

Lettre sur l'Electricité médicale. 12mo. *Amsterdam*, 1771

† Traité de l'Electricité; pour servir de suite aux Leçons de physiques. 12mo. 413 pp. 12 plates. *Paris*, 1771
(*Note.—There is a reprint of this Traité with the date 1776. Or there was a new title-page printed for this impression.*)

Sur la fusion de l'Or opérée instantanément par une commotion électrique . . . 4to. (*Journal de Phys.* ii. 1773.) *Paris*, 1773

† Précis historique des phénomènes électriques. 8vo. 742 pp. 9 plates. *Paris*, 1781

† Précis hist. et exp. des phénomènes électriques. 2me Edition, revue et augmentée. 8vo. 624 pp. 10 plates. *Paris*, 1785

Examen de quelques principes erronés en électricité. 8vo. *Paris*, 1795-6

† De l'électricité médicale. 8vo. Plates. *Paris*, 1803

Silberschlag, Johann Esaias. *Born November 16, 1721, at Aschersleben; died November 22, 1791, at Berlin.*

† Theorie d. am. 23 Juli 1762, ersch. Feuerkugel. 4to. 135 pp. 2 plates. *Magdeburg*, 1764

De Aurora boreali. 4to. 2 plates. *Berlin*, 1770

Sendschreiben über das, am 18 des Janners . . 1770, zu Berlin, beobachtete Nordlicht, an seinen Bruder . . . 4to. 23 pp. 2 plates. *Berlin*, 1770

System d. Neigung u. Abweichung d. Magnetnadel. 4to. (*Mém.* Berlin, 1786.) *Berlin*, 1786

† Systema inclinationis et declinationis utriusque acus magneticæ. 4to. 60 pp. 11 plates. (*Mém. Berl.* Août, 1786-87.) *Berlin*, 1786-7

Siljestrom, P. A. Observationer öfver nordskenet. 8vo. (*Vetensk. Acad. Handl.* 1841.) *Stockholm*, 1840

Om magnetiska inklinationens årliga variation. (*Afhandlingar i. fysiska och filosofiska amnen*, No. ii. 1857.) 1857

Silliman, Benjamin. *Born August 8, 1772, at Trumbull, Connecticut.*

† On the Galvanic Deflagrator of Professor Robert Hare. 8vo. 4 pp. (*Phil. Mag.* lix. 113.) *London*, 1822

† Meteor of a green colour, observed by him on crossing the East River between New York and Long Island, on the 11th February, 1828. 8vo. 1 p. (*Phil. Mag. or Annals*, iv. 66.) *London*, 1828

† Silliman and Kingsley. An account of a remarkable shower of Meteoric Stones, at Weston, in America. 8vo. 14 pp. (*Phil. Mag.* xxx. 232.) *London*, 1808

Memoir on the Meteoric Stones which fell from the atmosphere in the State of Connecticut, on the 14th December, 1807. 4to. (*Trans. American Phil. Soc.* Old Series, vol.vi. Part ii.) *Philadelphia*, 1818

Silliman, B. and others. The American Journal of Science and Arts. 71 vols. 8vo. (*Continued.*) *Newhaven*, 1818 *to* 1856

† Silliman, B. (jun.) Principles of Physics, or Natural Philosophy; designed for the use of colleges and schools. 2nd Edition, revised and rewritten. 8vo. 710 pp. *Philadelphia*, 1869

Silvestre, Augustin François de. *Born December 7, 1762, at Versailles; died August 4, 1851, at Paris.*

Silvestre, Augustin François de.—*continued.*

Traité complet d'Electricité par Cavallo. Traduit de l'Anglais sur la 2de
édition de l'auteur enrichie de ses nouvelles expériences. 8vo. 343 pp.
4 plates.   *Paris,* 1785
(*Vide* also Chappe and others.)

Simmons. Essay on the Cause of Lightning, &c. 8vo. 81 pp.   *Rochester,* 1775

Simon, Paul Louis. *Born January* 12, 1767, *at Berlin; died February* 14, 1815, *at
Berlin.*

Beschreib. e. neuen-galvanisch-chem. Vorrichtung u. einiger merkwürd, Versuche damit. 8vo.   *Leipzig,* 1801

Resultate d. neuest. Untersuchungen d. Galvanismus. 8vo.   (*Scherer's Journ.*
vi. 1801.)   *Berlin* (*or Leipzig*), 1801

Versuche über d. Wirkung d. Volta'schen Säule auf d. Wasser. 8vo.   (*Sue
adds a third article,* " *la production d'un acide et d'un alcali par l'action de la
pile de Volta sur l'eau.*")   *Berlin* (*or Leipzig*), 1801

† Observations sur l'attraction. 8vo. 18 pp.   *Paris,* 1819

Simonoff, Iwan Michailowitsch. *Born* 1785, *at Astrakan ; died January* 21, 1855
(*N. St.*), *at Kasan.*

† Recherches sur l'Action magnétique de la Terre. 8vo. 78 pp. 1 plate.
Kazan, 1840

Simons, Gerrit. Electromagnetische proeven. (*Natuur- en Scheikundig Archif,* iii.
1835.)   1835

Simpson. Zymologia physica, &c. 8vo.   *London,* 1675

† **Simpson,** J. Y. Observations regarding the influence of Galvanism upon the
action of the Uterus during Labour. 2nd edition. 8vo. 16 pp. (*Edinb.
Monthly Journ. of Med. Science,* July, 1846, p. 33.)   *Edinburgh,* 1846

† **Sinding,** E. A. H. Magnetiske Undersgeleser foretagne i 1868. Indberetning til
det akademiske Collegium ved det kongelige Frederike Universität. 8vo.
44 pp. 1 plate.   *Christiania,* 1870

Singer, George John. *Born about* 1786 ; *died June* 28, 1817, *at London.*

† Announcement of a new system of Insulation for Electrical Apparatus. 8vo.
1 pp. (*Phil. Mag.* xxxvii. 80.)   *London,* 1811
(*Note.—This method was derived from my suggestion.—*F. R.)

† An Electric Column recently constructed (by him) of twenty thousand pairs
of zinc and silver plates, small ; a series of three thousand larger, &c. ; and
Experiments. 8vo. 1 page. (*Phil. Mag.* xli. 393.)   *London,* 1813

† On Electrical Influence. 8vo. 3 pp. (*Phil. Mag.* xlii. 36.)   *London,* 1813

† On Electricity. 8vo. 4 pp. (*Phil. Mag.* xlii. 261.)   *London,* 1813

† On Electricity. 8vo. 3 pp. (*Phil. Mag.* xliii. 20.)   *London,* 1814

† Elements of Electricity and Electro-Chemistry. 8vo. 480 pp. 4 plates.
London, 1814
(*Note.—This copy has two impressions produced by explosion of wires, in
addition to the four plates.*)

† The Electric Column considered as a Maintaining Power, or first Mover for
mechanical purposes. 8vo. 5 pp. 1 plate. (*Phil. Mag.* xlv. 359.)
London, 1815

† Some Account of the Electrical Experiments of M. De Nelis, of Malines in
the Netherlands, communicated by Mr. Singer. 8vo. 6 pp. 1 plate. (*Phil.
Mag.* xlvi. 259.)   *London,* 1815

† Elements d'Electricité et de Galvanisme par G. Singer, ouv. traduit de l'Anglais et augmenté de Notes par Thyllaye. 8vo. 655 pp. 5 plates.
Paris, 1817

Singer, George John.—*continued.*

†   Elementi di fisica e chimica Elettrica. Traduzione dal Francese. 8vo. 410 pp.
    5 plates.   *Milano,* 1819

†   Elemente der Elektricität und Electro-chemie aus dem Engl. mit Anmerkun-
    gen . . von Muller. 8vo. 502 pp. 4 plates.   *Breslau,* 1819

Singer, G. J. and **Cross,** A. Account of some Electrical Experiments by M. De
    Nelis, of Malines in the Netherlands : with an Extension of them. 8vo. 6 pp.
    1 plate. (*Phil. Mag.* xlvi. 161.)   *London,* 1815

Sinner, Basilius. Beschreibung d. Telegraphen, welchen B. Sinner in d. Bibliothek
    zu Füessen aufgestellt hat.   *Tuessen,* 1795

Sinobas. (*Vide* Rico-y-Sinobas.)

Sjoestén, C. G. Beskrifning på ett nytt sätt magnetisera stalstanger kallad cir-
    kelstrickning. 8vo. (*Vetensk Akad. Handl.* 1802, *Beschreibung einer neuen
    Methode Stahlstangen mittelst des Kreisstriches zu magnetisiren.*) *Stockholm,* 1802

† **Skinner,** J. (Lieut.) Engraving from a drawing by him of a Waterspout, and
    Note by Editor. 8vo. 1 p. 1 plate. (*Phil. Mag.* xx. 374.)   *London,* 1804

Smaasen, Willem. *Born January 5, 1820, at s'Gravenhage; died February 23, 1850
    at Kampen.*

†   Specimen . . de equilibrio dynamico electricitatis in plano et in spatio . .
    submitt. 26 Mart. 1846. 4to. 32 pp. 1 plate. (*Auszug in Pogg. Annal.*
    lxix. 1846 and lxxii. 1847.)   *Trajecti ad Rhenum,* 1846

† **Smaastykker.** Om Galvanismen og Elektro-magnetismen samt deres Anvendelse.
    Meddelt ved Aftenmeder i Tingjellinge Skole af A. Jacobsen. 12mo. 42 pp.
                                             Kjobenhavn, 1871

† **Smeaton,** John. An account of some Improvements of the Mariner's Compass, in
    order to render the Card and Needle proposed by Dr. Knight of general use.
    Presented July 5, 1750, here printed with alterations. 4to. 5 pp. 1 plate.
    (*Phil. Trans.* xlvi. 513.)   *London,* 1749

Smee, Alfred. *Born June 18, 1818, at Camberwell, near London; died 1878.*

†   On the Galvanic Properties of the principal Elementary Bodies, with a de-
    scription of a new Chemico-mechanical Battery. 8vo. 11 pp. (*Lond. and
    Edinb Mag. for* April, 1840.) (*Arago's copy.*)   *London,* 1840

†   Elements of Electro-Metallurgy, or the Art of Working in Metals by the Gal-
    vanic Fluid. 1st edition. 8vo. 163 pp. 1 woodcut.   *London,* 1841

    Nouveau Manuel complet de Galvanoplastie, par Smee. (*Vide* also Valicourt.)
                                                  Paris, 1843

†   Elements of Electro-Metallurgy. 2nd edition, enlarged, &c. 8vo. 338 pp.
    Frontispiece, with electrotypes and woodcuts.   *London,* 1843

†   Nouveau Manuel complet de Galvanoplastie . . par M. Smee, augmenté de
    notes d'après Jacoby, Spencer, Becquerel, De Kobell, De La Rive, Elking-
    ton, De Ruolz, etc.; suivi d'un Traité de Daguerréotypie . . Ouvrage publ.
    par M. P. De Valicourt. Nouvelle édition. 12mo. 566 pp.   *Paris,* 1845

†   Elements of Electro-Biology. 8vo. 164 pp.   *London,* 1849

    Instinct and Reason, deduced from Electro-Biology. 8vo. Plates.   1850

†   Lecture on Electro-Metallurgy, delivered before the Bank of England Library
    and Literary Association, November 25, 1851. 8vo. 23 pp. *London,* 1851

    Elements of Electro-Metallurgy. 3rd edition, revised, corrected, and consi-
    derably enlarged. 8vo. 374 pp. Cuts.   *London,* 1851

†   Elemente der Electro-Metallurgie. Deutsch bearbeitet nach der dritten . .
    engl. orig. Ausgabe. 8vo. 408 pp.   *Leipzig,* 1851

    Elements of Electro-Metallurgy, First American, from the third London
    edition. Revised, corrected, and considerably enlarged by Chilton. 12mo.
    364 pp.   *New York,* 1852

Smee, Alfred.—*continued.*

Nouveau Manuel complet de Galvanoplastie . . Augmenté d'un grand nombre de notes d'après M. M. Jacoby, Spencer, Elsner, Becquerel, De Kobell, De La Rive, Elkington, De Ruolz, etc. Ouvrage publié par E. De Valicourt. Nouvelle édition, traduite sur la 3me édition de l'original anglais, entièrement refondue. 2 vols. 18mo. 791 pp. (*Vide* also Valicourt.)
Paris, 1860

(*Vide* also Anon. Elect. 1852.)

Smith, Archibald. *Born August* 10, 1813, *at Glasgow.*

† Instructions for the computation of a Table of the Deviations of a Ship's Compass; from deviations observed on four, eight, sixteen, or thirty-two points; and for the adjustment of the Table on a change of Magnetic Latitude. Published by order of the . . Admiralty. 8vo. 31 pp. *London,* 1851

On Correcting the Deviations of a Ship's Compass. 8vo. *London,* 1855

Supplement to the practical rules for ascertaining the Deviations of the Compass . . caused by the Ship's Iron, &c. 2nd edition, and a graphic Method of correcting. Published by order of the Lords Commissioners of the Admiralty. 8vo. Figs. *London,* 1855

Smith, J. L., Prof. Manual of Telegraphy. (*From Morse.*) *New York*

Smith, Samuel. Application of Electro-Magnetism. *New York, U.S.* 1850

Smith. (*Vide* Evans and Smith.)

Smithsonian Institution. Smithsonian Contributions to Knowledge. 4to.
Washington. Commencing 1848
Catalogue of Works published by the Institution.

A collection of Meteorological Tables, with other Tables useful in practical Meteorology. Prepared by order of the Smithsonian Institute by Arnot Guyot. 8vo. *Washington,* 1852

Smithsonian Reports. Notices of Public Libraries in the United States, by C. Jewett, Librarian. 8vo. *Washington,* 1851

Tables, Meteorological and Physical, prepared for the Smithsonian Institution, by Guyot. 8vo. *Washington,* 1858

Smithsonian Report. On the Construction of Catalogues of Libraries and their publication, by means of separate stereotyped titles. With Rules and Examples. By C. C. Jewett. 2nd edition. 8vo. *Washington,* 1853

† **Smyth, C. Piazzi.** Contributions to a knowledge of the phænomena of the Zodiacal Light. Read February 7, 1848. 4to. 13 pp. 1 plate. (*Trans. Rl. Soc. of Edinb.* xx. pt. iii.) *Edinburgh,* 1852

Snaith and Gardiner. Magnetism and Electricity. (*From the "Athenæum," Nov.* 18, 1871.) 1871

† **Snart, J.** Physico-mechanical experiments and discussions of the phænomena observable in that casual product of Art, the Hand Grenade, Prince Rupert's Drop, or Glass Tear. 8vo. 6 pp. (*Phil. Mag.* xxii. 334.) *London,* 1805

Snell, C. W. Nachricht von e. starken Luftelektricität. 8vo. (*Lichtenburg's Mag. f. Phys.* v. 1788.) *Gotha,* 1788

Snow, Robert. *Born about* 1806 ; *died August* 4, 1854, *at London.*

† Observations of the Aurora Borealis from September, 1834, to September, 1839. 12mo. 17 pp. (*Also in Proceedings of Astronom. Society.*) *London,* 1842

Snowball, J. C. Cambridge course of Elementary Natural Philosophy. 12mo. 1850

Soave, Francesco. *Born June,* 1743, *at Lugano ; died January* 17, 1806, *at Pavia.*

† Appendice (del Traduttore) ove spiegasi per qual ragione la scossa della Torpedine non sia accompagnata d.i strepito nè da luce. 12mo. 6 pp. (*Scelta di Opuscoli,* xv. p. 60.) *Milano,* 1776

Soave, Francesco.—*continued.*

† Articolo di lettera al C. Amoretti sull' Aurora Boreale dei 28 passato Luglio. 4to. 2 pp. (*Opus. Scelti,* iii. 253.) *Milano,* 1780

† Elogio di Francesco Soave .. scritto da G. Savioli. 4to. 16 pp. (*Nuova Scelta d'Opusc.* i. p. 413.) *Milano,* 1804
(*Note.—He is said in another Elogio to have written "Alcune sue congetture sulla Scossa della Torpedine," in the Scelta d'Opuscoli.*")

† S ... (q. Soave). Descrizione di una nuova macchina elettrica .. (per) amendue le elettricità. 12mo. 9 pp. 1 plate. (*Scelta d'Opuscoli* .. *in* 12mo. viii. 75.) *Milano,* 1775

Società di assicurazione contro il flagello della Grandine. Piano disciplinare e regolamento. 8vo. 16 pp. *Venezia,* 1822

Società Italiana. Memorie di Matematica e Fisica della Società Italiana delle Scienze. 4to. *Verona e Modena,* 1782 *to* —
(*Note.—*Vols. i. *to* vii. *and* xiv. *to* xvii. *printed at Verona; the rest at Modena.* Vol. xxv. *is dated* 1854. *Indices are in vols.* xv. *and* xxi. *Many of the Electrical works are in the Library.*)

Società detta dei XL. *Torino*

Societas Meteorologica Palatina. (*Vide* Manheim.)

Société d'Arcueil. Mémoires de physique et de chimie de la Société d'Arcueil. 8vo. 3 vols. *Paris,* 1807-1810 ?

Société Galvanique de Paris. Nauche President. Was opened on the 24th October, 1802. Fourcroy, Cabanis, and very many naturalists, &c. members. (*From Martens,* p. 334.)

Société Météorologique de France. Bulletin de la. (*From Martin in Montpell. Acad.* iii. 135.)

Société Philomatique. Rédacteurs du Bulletin des Sciences. Résumé des nouvelles expériences faites sur le Galvanisme par divers physiciens. (*Bulletin de la Société Philomatique,* an 9, Floréal.) *Paris,* 1801

Society for the Encouragement of Arts, &c. Transactions of. 8vo. *London,* 1753 *to present time.*

† Society for the Diffusion of Useful Knowledge. Notices of the Electromagnetic Telegraph. 8vo. 20 pp. (*From the Companion to the Almanac for* 1842, *published by the Society.*) *London,* 1842

Society of Arts. Journal of. *London*

† Society of Telegraph Engineers. Journal, including original communications on Telegraphy and Electrical Science. 8vo. *London,* 1872, *to present time.*

Socin, Abel. *Born January* 16, 1729, *at Basle; died October* 20, 1808, *at Basle.*

Tentamina electrica in diversis morborum generibus; quibus accedunt levis electrometri Bernoulliani adumbratio et quorundam experimentorum instituendorum ratio. 4to. (*Acta Helvetica,* tom. iv. pp. 214-230.) *Basil.* 1760

† Anfangsgründe der Elektricität. 8vo. 124 pp. 1 plate. *Hanau und Frankfort,* 1777

† Aufangsgründe der Elektricität. Zweite Auflage. 2nd edition. 8vo. 124 pp. 1 plate. *Hanau,* 1778

Södergen. Specimen academicum graduale, de recentioribus quibusdam in electricitate detectis; præside Klingensterna. 4to. *Upsal,* 1746

† Soderini, T. De electricismo. Dissertatio physica . in ædibus propriis habita. 4to. 13 pp. *Roma,* 1755

Sofka, F. O. Die Kosmisch. Abkühlungen ein meteor. Princip. *Wien,* 1863

† Sohncke, L. A. Bibliotheca mathematica. Catalogue of Books in every Branch of Mathematics, Arithmetic, Higher Analysis, constructive and analytical Geometry, Mechanics, Astronomy, and Geology; which have been published in Germany and other countries, from the year 1830 to the middle of 1854. 8vo. 388 pp. *Leipzig and London,* 1854

Sokoloff, N. Über die Salze der Nitrobenzoesäure und die Einwirkung des Zinks auf eine ammoniakalische Lösung derselben. 8vo. (*Acad. Impériale des Sciences de St. Pétersbourg*, tom. vi. livr. i. pp. 79-84.)
St. Petersburg, Riga, Leipzig, 1864

Solaro. (*Vide* Sanna-Solaro.)

Soldani, Ambrogio. *Born* 1733, *at Poppi, Tuscany ; died July* 14, 1808, *at Siena.*

Sopra una pioggetta di Sassi accaduta nella sera de' 16 Giugno del 1794 in Lusignan d'Asso nel Sanese. 8vo. Plates. *Siena,* 1794

† Storia di quelle Bolidi, che hanno, da se, scagliato Pietre alla terra. 4to. 29 pp. (*Atti dell' Accad. . di Siena.* Tom. ix. p. l.) *Siena,* 1808

† **Solly, E.** On the conducting powers of Iodine, Bromine, and Chlorine, for Electricity . and Further Experiments on conducting power for Electricity. 8vo. 6 pp. *London,* 1836

† On the influence of Electricity on Vegetation. 8vo. 31 pp. (*Journal of the Horticultural Society,* vol. i. part ii.) *London,* 1845

Somerville, Mary. *Born about* 1790, *near Edinburgh.*

On the Magnetizing power of the more refrangible Solar Rays. 4to. (*Phil. Trans.* 1826, p. 132.) *London,* 1826

Connexion of the Physical Sciences. 2nd edition. 12mo. 1834

The connexion of the Physical Sciences. 2nd edition. 12mo. *London,* 1835

Physical Geography. 6th edition. 1851

On the connexion of the Physical Sciences. 9th edition. 8vo. *London,* 1858

The connection of Physical Sciences. New edition. 12mo. *London,* 1858

On Molecular and Microscopic Science. 2 vols. 8vo. *London,* 1869

Sommering, Samuel Thomas. *Born January* 28, 1755, *at Thorn, West Prussia ; died March* 2, 1830, *at Frankfort.*

Über das Organ der Seele. 4to. 2 plates. *Konigsberg,* 1796

† Über einen elektrischen Telegraphen. 4to. 2 plates. 14 pp. (*Denkschr. Münch. Akad.* ii.) *München,* 1811

Über d. Zeichnungen, welche sich bei Auflösung d. Meteoreisens auf demselben bilden. 8vo. (*Schweigg. Journ.* xx. 1817.) *Nürnberg,* 1817

† Samuel Thomas von Sömmering geboren den 28 Jan. 1755, gestorb. d. 2 März, 1830, von Dr. A. W. Otto. 4to. 16 pp.

Dollinger Gedächtnissrede auf Samuel Thomas von Sömmering gehalten in der öffentl. Sitzung der K. bay. Akad. d. Wissen. am 25 Aug. 1830. 4to. *München,* 1830

W. v. Sömmering. Auszüge aus dem Tagebuche von S. Th. v. Sömmering, als Beitrag zur Geschichte des galvanischen Telegraphen. 8vo. (*Pog. Ann.* cvii. 644-647.) *Leipzig*

Der elektrische Telegraph als Deutsche Erfindung Samuel Thomas von Sömmering's aus dessen Tagebüchern nachgewiesen. 23 pp. *Frankfurt,* 1863

On application of Galvanism to ascertain the Reality of Death. (*Note.—Seyffer,* p. 408, *refers to "Ludwig scripter nevrolog.* iii. 23.)

† **Sonntag, A.** Observations on Terrestrial Magnetism in Mexico, conducted under the direction of Baron von Müller ; with Notes and Illustrations of an examination of the Volcanic Popocatepetl and its vicinity. Accepted May, 1859. Folio. 84 pp. 1 plate. (*Smithsonian Contributions.*) *Washington,* 1860

Sorel. His Pile employed for Galvanoplastic and Gilding purposes, &c. (*From Lerebour's Galvanoplastie*)

Soret, J. L. Recherches sur la corrélation de l'électricité dynamique. 4to. *Genève,* 1865

† **Soret, L.** Sur la décomposition des sels de cuivre par la pile et la loi des équivalents électro-chimiques. 8vo. 23 pp. (*Tiré de la Bibliothèque Universelle de Genève, Octobre*, 1854.) *Geneva*, 1854

† Recherches sur la corrélation de l'électricité dynamique et des autres forces physiques. 4to. 88 pp. 2 plates. (*Soc. de Phys. et d'Hist. Nat. Genève.*) *Genève*, 1858

† Recherches sur la corrélation de l'électricité dynamique et des autres forces physiques. 4to. 29 pp. 1 plate. (*Soc. de Phys. et d'Hist. Nat. Genève ?*) *Genève*, 1859

About Seven Articles in the Arch. Scien. Phys. et Nat. Vols. xxv. 1854 to vol. xxxi. 1856 on Décomposition de l'eau, ozone, courants électriques, &c. (*From Pogg*. ii. 961.)

† **Sowerby.** Notice of a plate . the Meteor-Stone . seen to fall in Yorkshire on the 13th December, 1795 ; and engravings of part of the one which fell in Scotland in 1804, and of that which fell in Ireland in 1810 ; all of which are deposited in his museum.) (*Phil. Mag*. xl. 315.) 1812

† **Sowerby, J.** Particulars of the Sword of Meteoric Iron presented by Mr. Sowerby to the Emperor Alexander of Russia. 8vo. 4 pp. (*Phil. Mag*. lv. 49.) *London*, 1820

Spallanzani, Lazaro. *Born January* 12, 1729, *at Scandiano Herz, Modena ; died February* 11, 1799, *at Pavia.*

Opuscoli di fisica animale e vegetabile, aggiuntevi alcune lettere relative ad essi Opuscoli dal celebre Sig. Carlo Bonnet di Ginevra e da altri scritte all' Autore. Tomi due. 4to. *Modena*, 1776

Dissertazioni di fisica animale e vegetabile. Aggiuntevi due lettere relative ad esse dissert. dal. cel. Sig. Bonnet di Genève. 2 vols. 8vo. *Modena*, 1780

† Lettera al Marchese Lucchesini sopra la Torpedine, ed altri argomenti. 4to. 32 pp. (*Opusc. Scelti*, vi. 73.) *Milano*, 1783

Lettera sulla fecondazione artificiale e sulla elettricità della torpedine. *Venezia*

† Lettera relativa a diverse produzioni marine, p. 340. 4to. 53 pp. (*Opusc. Scelti*, vii. 340 and 361.) *Milano*, 1784

Relazione su diversi oggetti fossili e montani scritta nel 1784, 23 Luglio. 4to. (*Soc. Ital*. ii. 11.) *Verona*, 1784

† Lettera relativa a diversi oggetti fossili, e montani. 4to. 34 pp. (*Opusc. Scelti*, viii. 3.) *Milano*, 1785

† Osservazioni sopra alcune Trombe di Mare formatesi su l'Adriatico il di 23 Agosto, 1785. 4to. 7 pp. (*Mem. Soc. Ital*. iv. 476.) *Verona*, 1788

† Lettera al Barletti sopra un fulmine ascendente. 4to. 5 pp. (*Opusc. Scelti*, xiv. 296.) *Milano*, 1791

† Lettera al Fortis sugli sperimenti fatti da Pennet in Pavia. 4to. 9 pp. (*Opusc. Scelti*, xiv. 145.) *Milano*, 1791

† Lettera . al Thouvenel sopra l'elettricità organica, e minerale. Data 7 Marzo 1793. 8vo. (*Ann. di Chim. di Brugnatelli.*) *Pavia*, 1793

† Articolo di risposta del . Spallanzani alla lettera seconda del F. Caldani. 8vo. 7 pp. (*Ann. di Chim. di Brugnatelli*, vii. 201.) *Pavia*, 1795

Chimico esame degli esperimenti del Sig. Gorling di Jena, sopra la luce del Fosforo di Kunkel, &c. 8vo. 171 pp. *Modena*, 1796

† Elogio di Lazaro Spallanzani scritto da P. Pozzetti. 4to. 52 pp. *Parma*, 1800

† Portrait "da un disegno fatto sul vivo da G. B. Busani." C. Piotti-Pirola sculp.

Späth, Johann Leonhard. *Born November* 11, 1759, *at Augsburg ; died March* 31, 1842, *at Munich.*

† Abhandl. üb. Elektrometer. 8vo. 1 plate. 95 pp. *Nürnberg*, 1791

Späth, Johann Leonhard.—*continued.*

Über d. Spannkraft d. Elektricität in d. Leiter e. Elektrisirmaschine u.s.w.
8vo. *Nürnberg,* 1792

† Ueber den natürlichen Magnetismus unserer Erde; über das Nordlicht, Sonnenflecken, Feuerkugeln, Sternschüsze und Cometen. 8vo. 138 pp.
Nürnberg, 1822

Specifications of Patents, Abridgments of. (*Vide* Woodcroft.)

† Spencer, K. Description of a Camp Telegraph invented by him. 8vo. 5 pp.
1 plate. (*Phil. Mag.* xxxvi. 321.) *London,* 1810

† Spencer, Knight. On the Luminous appearance of Sea Water. 8vo. 2 pp.
(*Phil. Mag.* xl. 206.) *London,* 1812

† Spencer, T. An Account of some Experiments for the purpose of ascertaining how far Voltaic Electricity may be usefully applied for Working in Metal, by Thos. Spencer. 8vo. 27 pp. 2 plates. 1840?
(*Note.—This is an account or review of Spencer.*)

† Instructions for the Multiplication of Works of Art in Metal by Voltaic Electricity, with an Introductory chapter. 8vo. 62 pp. (*Griffin's Scientific Miscellany.*) (*Faraday's copy.*) *London,* 1840

Spengler, Lorenz. *Born September 22, 1722, at Schaffhausen, Switzerland; died December 21, 1807, at Copenhagen.*

Briefe . . electrischen Wirkungeni n Krankheiten ; nebst einer . . Beschreibung der elekt. Maschine von Lorenz Spengler. 8vo. 102 pp. 2 plates.
Kopenhagen, 1754

Spidberg, Jens. Christ. Hist. Demonstrat, u. Anmerk. über d. Nordlicht. 8vo.
Halle, 1724

Spiller, P. Gemeinschaftliche Principien für die Erscheinungen des Schalles, des Lichts, der Wärme, des Magnetismus und der Electricität. 8vo.
Posen, 1855

Das Phantom der Imponderabilien in der Physik. Ein Versuch zu einer neuen Theorie des Magnetismus und der Electrizität. 8vo. 56 pp. *Posen,* 1858

† Neue Theorie der Elektrizität und des Magnetismus in ihren Beziehungen auf Schall, Licht und Wärme. Erweiterte Auflage. 8vo. 93 pp. *Berlin,* 1861

Sporschel. Electricität. (*From Lempertz, Cat.* 1867.) *Leipzig,* 1839

Sprenger, J. J. A. Anwendungsart der galvano-voltaisch.-metall-Elekt. zur Abhelfung der Taubheit u. Harthörigheit. 8vo. 2 plates. *Halle,* 1802

Squario. (*Vide* Sguario, E.)

Stack, T. Part of a letter concerning Electricity. Translated from the French by T. Stack. 4to. (*Phil. Trans.* 1748, p. 187.) *London,* 1748

† An account of a Tract intituled, "Jo. Frid. Wiedleri Commentatio de Parheliis . . Accedit de rubore cœli igneo Meuse Dec. 1737, observato corollarium. Vittembergæ, 1738." Drawn up by T. Stack. 4to. 7 pp. (*Phil. Trans.* xli. 459, 1st part.) *London,* 1739-40-41

† Stacquez. Conférences sur l'Electro-Thérapie données a l'hôpital militaire de Liége. 8vo. 238 pp. *Liége,* 1862

† Stadler, D. Magnes, experimenfis, theoriis et problematis explanatus. 8vo.
Dillingen, 1740

† Stadlhofer, I. N. Über d. tödtliche Wirkungsart d. Blitzes. 8vo. *Dresden,* 1791

† Stahelin, C. Die Lehre der Messung von Kräften mittelst der Bifilarsuspension.
4to. 204 pp. 9 plates. *Basel,* 1852

† Stamkart, F. J. Over de Afwijkingen van het Kompas voortgebragt door de Aantrekking van het Scheeps-Ijzer. 4to. 48 pp. (*Uitgegeven door de Koninklijke Akad. van Wettenschappen.*) *Amsterdam,* 1836

† Beschriving van het Model van een werktuigje de afwijk van het Kompas aan te wijzen . . (long title). 8vo. 9 pp. 1 plate. *Amsterdam,* 1857

STA—STE 483

Stamkart, F. J.—*continued.*

† De regeling van compassen aan boord van ijzeren en hooten shepen. 8vo. 344 pp. 5 plates. *Amsterdam,* 1861

† Theorie van het intensiteits-kompas en van zijn gebruik op ijzeren en houten schepen. 4to. 39 pp. 1 plate. (*Naturk. Verhand. der Akad.* vii.) *Amsterdam ?*

Stanhusius, Mich. De Meteoris, libri duo, quorum prior tradit de æthere et elementis, posterior complectitur omnium fere meteorum prolixam explicationem. 12mo. *Vitembergæ,* 1572

De Meteoris (lib. duo). 8vo. *Witteb.* 1578

Stanley, Owen. Made Astronomical and Magnetic observations in Captain Back's Expedition, &c.

† **Stark.** Observation of a Luminous Band on the 7th July, 1817, from the Observatory of Augsburg. 8vo. 1 pp. (*Phil. Mag.* 1. 156.) *London,* 1817

Stark, J. On the Nature of the Nervous Agency, &c. 8vo. (*Edinb. Med. and Surg. Journ.* 1844, lxi. 285 ; 1845, lxiii. 103.) *Edinburgh,* 1844-5

Stas, Jean Servais. *Born September* 20, 1813, *at Löwen.*

Stas et Quetelet. Météorologie nautique. Rapport sur une demande du Gouvernement belge (pp. 8 et 2). Rapporteurs. Ext. du tom. xx. No. 10, des Bulletins de l'Acad. de Belgique. 8vo. (*Giorn. I. R. Ist. Lomb.* Nouv. S. in 4to. 1854.)

Statham. On the Origin of his Electro-magnetic Fuzees for Mining Purposes, &c. 8vo. (*Phil. Mag.* (4) vii. 199.) *London*

† **Steblecki, Albin.** Über d. tellurisch. Magnetismus. 4to. 34 pp. (*Jahres-Bericht. des . . zweiten Lemberger Ober-Gymnasiums fur . .* 1853.) *Lemberg,* 1853

Steege, Van Der. Bericht van de proefnemingen met den door kunst gemaakten magnet. (*Verhandl. van het Bataviaasch. Genootsch.* i. 110.) *Rotterdam, &c.*

† **Steevens, J.** Account of a Meteor seen at London and other places on the night of Monday, March 22, 1813. 8vo. 3 pp. 1 plate. (*Phil. Mag.* xli. 346.) *London,* 1813

Stefan, J. Ueber einige Thermoelemente von grosser elektromotorischer Kraft. 8vo. 3 pp. (*Abd. aus den Sitzungsberichten der Kais. Akad. der Wissenschaften.*) *Wien,* 1865

Über die Grundformeln der Elektrodynamik. 8vo. 77 pp. *Wien,* 1869

† **Stefani, S. e Zantedeschi, F.** Dichiarazione. 4to. 3 pp. (*Ann. del Reg. Lomb. Veneto,* xiv. 84 and 85.) *Vicenza,* 1845

Note.—The above two articles concern priority as to certain experiments in Thermo-Electricity (dynamic).

Steffans, H. (*Vide* Bulmerincq.)

† **Stehr, L.** Der Magnetismus als Urkraft in sein. verschied enen Wirkungen geschildert. 8vo. 160 pp. *Berlin,* 1865

Steiglehner, Cölestin. *Born August* 17, 1738, *at Sindersbühl, near Nuremberg ; died February* 21, 1819, *at Regensburg.*

† Observationes phænomenorum electricorum in Hohen-Gebrachin, etc. 4to. 55 pp. 1 table. *Regensburg,* 1773

Ueber die Analogie der Electricität und des Magnetismus. (*Neue Abhandl. der Baierischen Akad. Philos.* ii. 227.) *München,* 1780

Steinau, C. Katechismus der Elect. u. d. Galv. Nach Biot's Physik u Singer's Electricitatslehre bearb. &c. . . 8vo. *Leipzig,* 1824

† **Steinberg, C.** Die Dynamide, Elect. Magnetis., Licht, Warme. Verwandschaftslehre, &c. und Stochiometrie-Compendium zu Vorlesungen über allgemeine Chemie von Dr. C. Steinberg. 8vo. 81 pp. *Berlin,* 1846

† **Steindachner**, F. Die Gymnotidie des K. K. Hof-Naturallincabinets zu Wien (*Vorgel.* 23 Juli, 1868.) 8vo. 16 pp. 2 plates. (*Sitzb. d. K. Akad. d. Wiss.* Juli Heft, 1868, lviii. Bde.) *Wien*, 1868

Steinhauser, Johann Gottfried. *Born September* 20, 1768, *at Plauen Voigtl; died November* 16-17, 1825, *at Halle.*

Gegen. d. Wasserzersetz. durch magnet. Kraft, u. s. w. 8vo. (*Gilb. Ann.* xiv. 1803.) *Leipzig*, 1803

† De magnetismo telluris commentationes math. phys. Sect. i. Magnetis virtutes in genere proponens. Sm. 4to. 52 pp. 1 plate. *Vittebergæ*, 1806

† De magnetismo telluris commentationes math. phys. Sect. ii. De inclinatione acus magneticæ. . . 4to. 17 pp. 1 plate. *Vittebergæ*, 1810

Nähere Bestimm. d. Bahn d. Magnetism. Innern d. Erde. 8vo. (*Gilb. Ann.* lvii. 1817 u. lxv. 1820.) *Leipzig*, 1817-20

Über d. Magnetism. d. Erde. 8vo. (*Gilb. Ann.* lxv. u. lxvi. 1820.) *Leipzig*, 1820

Über d. Verfertigung von Stahlmagneten. 8vo. (*Schweigg. Journ.* xxxiii. 1821.) *Nürnberg*, 1821

Eight Articles on Magnetism of the Needle, &c., in Voigt's Magazine, viii. 1804 to xi. 1806.

Steinheil, Karl August. *Born October* 12, 1801, *at Rappoltsweiler, Elsass; died October* 14, 1870.

Beschreib. seiner Sternwarte in München. (*Schumacher's Astrom. Nachr.* xi. 1834.) *München*, 1834

Beiträge zu Gauss u. Weber's Resultaten aus Beobb. d. magnet. Vereins. 1836-39

† Letter from Munich dated Dec. 23, 1836 (on his Telegraph). (*Magazine of Popular Sciences*, vol. iii.) 1836?

(*Note.—Shaffner does not say that this letter was written by Steinheil himself; neither does he give the title or date of the article in the Magazine, or the place of publication of the Magazine. He says that the letter was "the first published notice of the invention."*)

"Die erste öffentliche Nachricht über die telegraphischen Arbeiten Steinheils."

("*Augsb. allg. Zeit.* vom 23 Juli, 1837 (ausserord Beil. No. 356-357 p. 1424) und zwar in dem Aufsatze, der über den von Prof. W. Alexander (in Edinburgh) gemachten Vorschlag Mittheilung macht." *Weitere Nachrichten über den Telegraphen findet man in d. Augsb. allg. Zeit.* 31 Jan. 1838, *Beil.* No. 34, p. 245, *und vom* 17 Feb. *ausserord. Beil.* No. 88-94, pp. 350 u. 358.)

Erscheinungen in der galvanischen Telegraphenleitung zwischen Munchen und Augsburg, bei Gewittern, und daraus abgeleitete Einrichtung zur Ableitung des Blitzes von den Stationszimmern. 8vo. (*Dingler's Polytech. Journal*, cix. 350.)

"Hat noch eine andere Einrichtung erdacht, durch welche hörbare Schläge unmittelbar durch die Bewegung des Ankers eines Electromagnetes hervorgebracht werder können." (*Bayr. Kunst- u. Gewerbsblätter*, xxviii. 25, 26.)

Notice sur un Télégraphe électrique de son invention. 4to. (*Comptes Rendus*, vii. 390.) *Paris*, 1838

† Ueber Telegraphie ins besondere durch galvanische Kräfte. Eine öffentliche Vorlesung gehalten in der festlichen Sitzung der Königl. Bayerischen Akad. der Wissensch. am 25 Aug. 1838.

Beilage.—Beschreibung und Abbildung der galvano-magnetischen Telegraphen zwischen München und Bogenhausen errichtet im Jahre 1837 von Prof. Steinheil. 4to. 30 pp. 2 plates. *München*, 1838

Noch ein Wort über galvan. Telegraphen. *Schumacher's Jahrbuch*, 1839.) *München*, 1839

Steinheil, Karl August.—*continued.*

Seine elektromagnet. Rotations-maschine.   (*Kunst- u. Gewerbeblatt.* 1841.)
1841

Teleskopspiegel galvanoplastisch copirt u. galvan. vergoldet u. über Ruolz' Methode zu vergolden. 4to. (*München gelehrt Anzeig.* xv. 1842.)
München, 1842

Vereinfachte Methode Brandstellen bei Nacht zu ermitteln. 4to. (*München gelehrt Anzeig.* xvi. 1843.)
München, 1843

Seine galvan. Uhren.   (*Kunst- u. Gewerbeblatt.* 1843.)
1843

Electromagnetische Zeit Übertrager.   (*Bayer. Kunst- und Gewerbebl.* xxi. 127-142.)

Organised the Neapolitan, Austrian, and Swiss Telegraphs.   1844-49-51

Sein Telegraph für Eisenbahnen.   (*Kunst- u. Gewerbeblatt.* 1846, xxiv. 482.)
1846

† Beschreibung u. Vergleichung der galvanischen Telegraphen Deutchlands nach Besichtigung im April, 1849. 4to. 61 pp. (*Abhandl. der* ii. *Cl. der K. Akad. d. Wiss.* v. Bd. iii. Abth. 4to.)
München, 1849

Über d. Geschwindigk. d. Fortpflanzung d. galvan. Stroms.   (*Schumacher's Astrom. Nachr.* xxix. 1849.)
München, 1849

Beschreibung der Vereinfachung und Verbesserungen an galvanischen Telegraphen. (*Kunst. und Gewerbe-Blatt, herausgegeben v. d. polyt. Verein fur d. Königreich Bayern,* Jahr 1850, Bd. xxviii. pp. 18-20.)
München, 1850

Instruction für d. Telegraphisten d. Schweiz, u.s.w.   8vo.   *Bern,* 1852

† **Steinigeweg,** J. H.   De fluidor. facultate conducendi fluxum electricum. 8vo. 44 pp. 1 plate.
Trajecti ad Rhenun, 1847

Steininger. (*Vide* Chladni and others.)

Steinschneider. Der hängende Sarg Mohammeds.   8vo.   (*Zeitschrift der deutschen Morgenländischen Gesellschaft,* v. 378, 1851.)
Leipzig, 1851

† **Stella,** F. M.   Articolo di lettera al Selva, sopra una nuova maniera di caricare la Pistola elettrica.   4to. 3 pp. (*Opusc. Scelti,* x. 202.)
Milano, 1787

† Lettera al G. Morosini ; nella quale si espongono alcune circostanze, che accompagnarono un fulmine, nell' atto di colpire la casa de' Nobili Sig. Liruti di Udine. 4to. 4 pp. (*Opusc. Scelti,* xii. 329.)
Milano, 1789

† Articolo di lettera all' Amoretti su un fenomeno elettrico. 4to. 1 page. (*Opusc. Scelti,* xiii. 427.)
Milano, 1790

Stella e figli. Bibliografia Italiana delle opere . . stampate in Italia e pubblicate all' estero.   8vo.
Milano, 1835 *to* 1846

Stendardi, Carlo Antonio.   *Born* 1721, *at Siena ; died July* 6, 1764, *at Florence.*

Meteore ed altri fenomeni osservati in Algieri, nel 1753, circa l'Equin. inverno. 12mo. (*Calogera, Nuova Raccolta* . . xiii. No. 19, *Declin. della Bussola verso ponente,* 18mo. 1842.)
Venezia, 1765

† **Stephen,** J.   The Transference of the Telegraphs to the State.   58 pp.
London, 1860

Stepling, Jos.   De pluvia lapidea anni 1763, ad Strkow, pagum Bohemiæ, et ejus causis, meditatio, etc.   8vo.
Pragæ, 1754

Beantwort. verschiedn. Fragen über d. Beschaffenheit d. Lichterscheinung nachts d. 28. Hornnungstage u. über d. Nordlichter. 8vo. *Pragæ,* 1761

Beobachtungen der Magnetnadel in Prag. 8vo. (*Abhandl. d. Privat Gesellschaft Böhmen,* i.)
Pragæ, 1775

Über d. elektr. Ableiter. 8vo. (*Abhandl. d. Privat Gesellsch. Böhmen,* iii. 1777.
Prag, 1777

Sternberg, K. M. Graf von. Galvanische Versuche in manchen Krankheiten.
Regensburg, 1803

Von galvanischen Versuchen in manchen Krankheiten, herausgeg. mit e. Einleit. u. in Bezug auf Erregungstheorie von J. U. G. Schäffer. 8vo.
Regensburg, 1803

Beschreibung and Untersuchung e. merkwürd. Eisengeode, welche zu Radnitz in Böhmen gefunden würde. 8vo. Plate. *Prag,* 1816

Sternberg, Joachim Graf von. *Born August 15, 1755, at Prague; died October 18, 1808, at Przesina.*

Beobachtungen über die Bildung der Donner-Wolken und Entstehung der Donner-Wetter. (*Mayer's Samml. Phys. Aufs. des Böhmischen Naturf.* iii. p. 1.) *Prag,* 1792

† On the Electrogene of Schmidt. By Count Sternberg, Vice-President of the Electoral Regency of Ratisbon. 8vo. 2 pp. (*Phil. Mag.* xxiv. 250.)
London, 1806

Stevens, Hen. Catalogue of American Books in the British Museum. 2 vols. 8vo. (*From Engl. Cat.* p. 734.) *London,* 1862

† **Stevenson, W. F.** Most Important Errors in Chemistry, Electricity, and Magnetism. 2nd edition, revised, &c. 8vo. 68 pp. *London,* 1847

Stevinus, Simon. Portuum investigandorum ratio. *Leyden,* 1599
(*Note.—This may probably be "gesammelt in Wisconstige Gedächtnissen, &c.* 2 vols. fol. *Leyden,* 1605-8; *und Lateinisch von V. Snell und franzos. von A. Girard," (i.e.) Les ouvrages mathém. de Simon Stevin, revus, corrigés, et augmentés,* 1 vol. fol. *Leyden,* 1634.)

† **Stewart, Balfour.** An Account of the Construction of the Self-Recording Magnetographs at present in operation at the Kew Observatory of the British Association. 8vo. 29 pp. 2 plates. (*Rep. Brit. Assoc. for* 1859.)
London, 1860

Makerstoun Magnetical and Meteorological Observations. Appendix. 4to.
Edinb. 1861

† Earth Currents during Magnetic Calms, and their connexion with Magnetic Changes. 4to. 16 pp. (*Trans. R. Soc. Edinb.* xxiii. pt. ii.)
Edinburgh, 1863

Lessons in Elementary Physics. 2nd edition. 12mo. *London,* 1872

Stief, John Ernst. Histor. u. physical. Betracht. über d. Wirkungen des in einem Pulverthurm zu Breslau u.s.w. eingedrungenen Blitzstrahls. 4to.
Breslau, 1749

Das Donnerwetter im Winter, erklärt. (*Fortgesetzte Bemuh. d. Zittauer Gesellsch.* Bd. ii.)

† **Stille, L.** Ueber einen neuen galvanischen Apparat. Inaugural Dissertation. 8vo. 57 pp. *Göttingen,* 1864

Stock, Jno. Chr. Diss. de fulgure, tonitru et fulmine. 4to. *Jenæ,* 1734

Stockholm. Afhandl. i. Fisik, Kemi, och Mineralogi. 1806 *to* —

Stockholm Acad. Kongl. Svenska vetenskaps academiens-handlingar. 8vo.
Stockholm, 1740-1813

Stockholm or Swedish Acad. Abhandlungen der Königl. Schwedisch Acad. Übers. von A. G. Kastner. 41 vols. 8vo. *Hambourg,* 1749 *to* —

Neue Abhandl. 12 vols. 8vo. *Hambourg,* 1784-92

Stohrer, E. Bemerkungen über die Construction magneto-elektr. Maschinen. 8vo. (*Poggend. Ann.* lxi. 1844.) *Leipzig,* 1844

Über d. Benutz. d. Kraft einer elektr. Spirale zu rotirenden Bewegungen. 8vo. (*Pogg. Ann.* lxix. 1846.) *Leipzig,* 1846

Stohrer, E.—*continued.*

Anwendung der electro-magnet Rotations-apparate und constante Säule zu Elektro-telegraphie. 8vo. (*Pogg. Ann.* lxxvii. 1849.) *Leipzig,* 1849

Beitrag zur Vervollkommen. der electro-magnet. Rotations-apparate. 8vo. (*Pogg. Ann.* lxxvii. 1849.) *Leipzig,* 1849

Doppelstift-Apparat mit Relais. 8vo. (*Polyt. Cent. Blatt.* 1852, pp. 66 and 172.) 1852

Verbesserter Inductions-apparat. 8vo. (*Pogg. Ann.* xcviii. 1856.) *Leipzig,* 1856

Five Articles in Poggendorff's Annalen, vol. lxi. 1844, to vol. xcviii. 1856.

Stohrer und Scholle. Die elektromagnetischen Zeitindicatoren. (*Bayer. Kunst- und Gewerbebl.* xxxiii. 449, and *Polyt. C. Bl.* 1855.) 1849?

Stoikowich, Ath. Schutzmittel wider d. Blitz. *Petersburg ?* 1810

Über Blitzableiter. *Petersburg ?* 1826

Stoikowitz. (*Vide* Stoikowich.)

Stoll, J. J. Beleuchtung einiger Vorurtheile in Ansehung der Donnerwetter und Blizableiter. 8vo. *Lindau,* 1790

Stone, G. W. Electro-Biology. 12mo. *London,* 1852

Storia Letteraria d'Italia. (*Vide* Zaccaria, Editor.)

† **Strada, F.** Famiani Stradæ Romani e Societate Jesu. Prolusiones academicæ, nunc demum ab Auctore recognita, atque suis judicibus illustrata. Editio postrema. 16mo. 619 pp. (*Ideal Magnetic Telegraph,* p. 443 *et seq.*) *Mediolani,* 1626

† **Strambio, G.** Relazione delle indagini sperimentali sui bruti, intorno all' azione della stricnina, &c. . . . istituite dai . Restelli e Strambio e degli esperimenti intorno alla formazione di grumi otturanti le arterie per mezzo dell' ago-puntura elettrica, eseguite dai Restelli, &c. Lettere due al . Panizza del Gaet. Strambio. 8vo. 43 pp. 1 plate. *Milano,* 1846

† Galvano-Ago-Puntura dei vasi sanguigni per curare gli Aneurismi e le Varici. Studj storico-critici. 8vo. 195 pp. (*Estratto dalla Gazzetta Medica di Milano,* tom. v. vi.) *Milano,* 1847

† Sperimenti di Galvano-Ago-Puntura istituiti sulle arterie e sulle vene dei Bruti, dai Dottori Strambio, Quaglino, Tizzoni, Restelli. Relazione con Poscritto e note addizionali. 8vo. 86 pp. 1 plate. (*Estratto dalla Gazzetta Medica di Milano,* tom. vi.) *Milano,* 1847

† Su l'Ozono atmosferico durante l'ultima epidemia colerosa in Milano. Sperienze e considerazioni—"Mem. letta all' I. R. Ist. Lomb." 8vo. 53 pp. 2 tables. *Milano,* 1856

Stratico, Simone. *Born* 1733, *at Zara, Dalmatia; died July* 16, 1824, *at Milan.*

† Osservazioni sopra alcuni fenomeni magnetici. 4to. 15 pp. (*Mem. dell' I. R. Istit. del Reg. Lomb.-Veneto,* vol. iii. anni 1816 e 1817. Parte ii. p. 115.) *Milano,* 1824

† Portrait.

Stratingh, Sibrandus. Elektromagnetische Bewegingskracht en anwending daar-van tot een elektromagnetischen wagen. (*Algem. Konst- en Letterbode,* 1835.) *Groningen,* 1835

Beschriving van een verbeterd Faradaysch Magnetisch-elektrisch wirktuig. (*Mulder en Wenckebach's Archeif,* iv. 1836.) 1836

Elektrisch-Archimedische Modelboot. (*Algem. Konst- en Letterbode,* 1840.) *Groningen ?* 1840

Straub, J. C. Uber d. galvan. Wirksamkeit d. Kohle. (*Meissner's Naturwiss. Anzeiger,* iv. 1820.) 1820

Strehlke, F. Anzieh. zwischen gleich- u. ungleichnamig elektrisirten Scheiben. 8vo. (*Pogg. Ann.* xii. 1828.) *Leipzig,* 1828

Einfluss d. Gewitter auf d. Barometerstand. 8vo. (*Pogg. Ann.* xix. 1830.) *Leipzig,* 1830

Einfache Hervorbring. d. magnet. Funkens. 8vo. (*Pogg. Ann.* xxv. 1832.) *Leipzig,* 1832

Über d. Eigenschaften d. von Daguerreschen Lichtbildern erhaltenen galvan. Kupferplatten. 8vo. (*Pogg. Ann.* lx. 1843.) *Leipzig,* 1843

Streizig. " Un Orologio semplicissimo di due ruote, fu annunziato, recando la data di Verona 25 Marzo, 1815." (*Giornale di Venezia,* No. 84, 25 Marzo, 1815.) *Verona,* 1815

Streus, L. La Télégraphie Elect. mise à la portée de tout le monde. 12mo. 80 pp. Cuts. *Bruxelles,* 1855

Stromer, Märten. *Born June* 7, 1707 (*A. St.*), *at Orebro ; died January* 2, 1770, *at Upsala.*

† Untersuchung von der Elektricität. 8vo. 4 pp. (*Schwedische Akad. Abhandl.* an. 1746, p. 154, vol. ix.) *Hamburg,* 1746

(*Note.—Priestley, Hist. 5th Ed.* p. 117, *says Klingensterna and Strömer were the first who properly electrified by the Rubber.*)

Rön angående electricitaten. 8vo. (*Vet. Acad. Handl.* 1847.) *Stockholm,* 1747

Om electricitetens verkan på meniskans kropp, etc. 8vo. (*Vet. Acad. Handl.* 1752.) *Stockholm,* 1782

† Einige Versuche von der Wirkung der elektrischen Kraft auf den menschlichen Körper und ihren Nutzen zur Heilung verschiedener Krankheiten und Zufälle. 8vo. 10 pp. (*Schwedische Akad. Abhandl.* an. 1752, p. 199, vol. xiv.) *Hamburg and Leipzig,* 1752

Specimen de theoria declinationis magneticæ. 4to. *Upsala,* 1755

Stromer. (*Vide* Melander.)

Stromeyer. (*Vide* Anon. Meteorolog. Phen. 1821.)

Strouy, H. Über d. Bestimmung d. Intensität d. Erdmagnetismus. 4to. *Eupen,* 1860

Struve, Christian August. *Born January* 28, 1767, *at Görlitz ; died November* 6, 1807, *at Görlitz.*

Versuch üb. d. Kunst Scheintodte zu beleben und üb. d. Rettung in schnellen Todesgefahren, e. tabellar. Taschenbuch. 8vo. *Hanover,* 1797

System der medicin. Elektats-Lehre. 8vo. *Breslau,* 1802

Galvanodesmos, ein besond. in Krankh. nützl. leicht transportabl. und unverzugl. anwendbar. galvan. Apparat, und beschr. 8vo. 1 plate. *Hanover,* 1804

Der Lebensprüfer oder Anwend. des von mir erfundenen Galvanodesmos " z u Bestimm. d. Nähern v. Scheintodte, um d. Lebendigbegraben zu verhüten.' ' 8vo. 1 plate. *Hanover,* 180 5

Strype. Life of Grindal. Fol. *London,* 171 0
(*Note.—At* p. 481, *in a note of Pilkington, it is said that Strype,* pp. 53-55, *gives a full account of the circumstances of the Fire at St. Paul's, &c.*)

† **Stuart,** Alex. Part of a Letter to the Publisher, concerning some Spouts he observed in the Mediterranean. 4to. 6 pp. 1 plate. (*Phil. Trans.* xxiii. *for* 1702-3, p. 1077.) *London,* 1704

† Experiments to prove the Existence of a Fluid in the Nerves. 4to. 4 pp. (*Phil. Trans.* xxxvii. 327.) *London,* 1731-32

† **Stukely,** W. On the Causes of Earthquakes 4to. (*Phil. Trans.* 1750.) *London,* 1750

Sturgeon, William. *Born* 1783, *at Whittington, near Lancaster ; died December* 8, 1850, *at Prestwich, near Manchester.*

† Improvement upon Ampère's and Marsh's Electro-Magnetic Rotation Apparatus. 8vo. 1 p. (*Phil. Mag.* lxii. 237.) *London,* 1823

† Electro-Magnetical Experiments. 8vo. 6 pp. (*Phil. Mag.* lxiii. 95.) *London,* 1824

A Complete Set of Novel Electro-Magnetic Apparatus. 8vo. (*Trans. Society of Arts,* 1825.) *London,* 1825

† On the Inflammation of Gunpowder and other Substances by Electricity ; with a proposal to employ the term Momentum as expressive of a certain condition of the Electric Fluid. 8vo. 7 pp. (*Phil. Mag. or Annals,* i. 20.) *London,* 1827

Recent Experimental Researches, &c. Views on the Elect. Column. Remarks on the Electricity of the Atmosphere. 8vo. *London,* 1830

† Account of an Aurora Borealis observed at Woolwich on the night of January 7, 1831. 8vo. 5 pp. (*Phil. Mag or Annals,* ix. 127.) *London,* 1831

† On the Thermo-Magnetism of Homogeneous Bodies ; with illustrated experiments. 8vo. 24 and 9 pp. (*Phil. Mag.* x. 1 and 116.) *London,* 1831

† Annals of Electricity, Magnetism, and Chemistry, and Guardian of Experimental Science, conducted by W. Sturgeon. 8vo. Plates. *London, &c.* 1836—1843

(*Note.—The date in the title-page of* 1st *vol. is* 1837 ; *that of the last number is October,* 1843. *There are* 10 *vols. and four monthly numbers. It contains very many papers by himself.*)

† Experimental and Theoretical Researches in Electricity, Magnetism, &c. Fourth Memoir: On Marine Lightning Conductors. 8vo. 31 pp. Plate. (*Ann. of Electricity,* iv. 161.) (*Roget's copy.*) *London,* 1839

On Successions or Series of Electric Currents. Letter to Henry. 8vo. 1 p. (*Ann. of Electricity.*) (*Roget's copy.*) *London,* 1839

† Supplementary Note to Article xvi. (*i.e.* . . on Marine Lightning Conductors). 8vo. 2 pp. (*Ann. of Electricity,* iv. 235.) *London,* 1839

† Fourth Letter to W. S. Harris, Esq., on the subject of Marine Lightning Conductors. 8vo. 6 pp. (*Ann. of Electricity,* iv. 496.) *London,* 1840

† Description of an Electrical Machine made by Watkins and Hill . . for the Royal Victoria Gallery, Manchester. 8vo. 2 pp. Plate. (*Ann. of Electricity,* iv. 504.) (*Roget's copy.*) *London,* 1840

† Letter to W. Snow Harris. 8vo. 4 pp. (*Ann. of Electricity,* v. 220.) *London,* 1840

† Address to the Readers of the Annals of Electricity, Magnetism, and Chemistry, &c. May, 1841. 8vo. 3 pp. (*Ann. of Electricity,* vi. 500.) (*Roget's copy.*) *London,* 1841

† Lectures on Electricity, delivered . . at Manchester. 12mo. 240 pp. Cut. *London,* 1842

Annals of Philosophical Discovery and Monthly Reporter of the Progress of Practical Science. 8vo. *London,* 1843 *to —*

† A course of Twelve Elementary Lectures on Galvanism. 12mo. 231 pp. 100 engravings. *London,* 1843

† Scientific Researches, Experimental and Theoretical, in Electricity, Magnetism, Galvanism, Electro-Magnetism, and Electro-Chemistry. Published by subscription. La. 4to. 563 pp. 18 plates. *Bury,* 1850

Scientific Memoirs ; comprising the Scientific Investigations, Observations, Discoveries, &c. of William Sturgeon, in Electricity, Magnetism, and Electro-Chemistry, within the last twenty-five years. (*Note.—Vide preceding entry.*)

Researches in Electricity, Magnetism, and Chemistry. 4to. *London,* 1852

A Familiar Explication of . . . Electro-Gilding, &c. (*From Dove and Pogg.*)

P P

† **Sturgeon and Backhofner.** On a Massive Iron Core, or better, a Slitted Iron Hollow Cylinder, or still better, a Bundle of separate Varnished Iron Wires enclosed in a Hollow Cylinder. 8vo. (*Sturgeon's Ann.* i. 481.) *London*

(*Vide* also Cartwright and Sturgeon.)

(*Vide* also Barlowe.)

(*Vide* also Faraday and Sturgeon.)

Sturm, Johann Christoph. *Born November 3, 1635, at Hippolstein, Pfalz-Neuberg ; died December 25, 1703, at Altdorf.*

† Collegium experimentale sive curiosum. Pars prima. 4to. 168 pp. and 122 pp. Auctaria. *Norimbergæ*, 1676

† Collegii experimentalis sive curiosi. Partes prima et secunda. 4to. 168, 122, and 256 pp. *Norimbergæ*, 1685

Physica electiva, sive hypothetica. Tom. i. 4to. *Norimbergæ*, 1697

Epistola invitatoria ad observationes magneticæ variationis instituendas. (*Opusc. Act. Erudit. Lyps.* F. i. p. 60.)

† **Sue, J. J.** Recherches physiologiques et expérimentales sur la Vitalité. Lues à l'Institut National de France, le 11 Messidor, an. v. de la République. Suivies d'une nouvelle édition de son opinion sur le Supplice de la Guillotine, ou sur la douleur qui survit à la décollation. 1re édition. 8vo.' 76 pp. 4 plates. *Paris*, 1797

Recherches physiologiques et expérimentales sur la Vitalité. *Paris*, 1798

† Physiolog. Untersuchungen u. Erfahrungen über d. Vitalität, und Abhandl. üb. den Schmerz der Enthauptung, a. d. Franz. von J. C. F. Harles. 8vo. Plate. *Nürnberg*, 1799

† **Sue, J. J.** (and others). Recherches Physiologiques, et expériences sur la Vitalité et le Galvanisme. 3e édition. 8vo. 86 pp. 4 plates. *Paris*, 1803
(*Note.—This edition contains, at p. 77, "Précis des expériences Galvaniques d'Aldini ;" at p. 83, "Note de l'influence Galvanique sur la fibrine du Sang," par C. F. Cireaud; at p. 84, a notice concerning "la Nature du Galvanisme," according to the ideas and experiments of Creve, signed by Witmann. The work of J. J. Sue himself is a mere reprint of his first edition.*)

† **Sue, P.** (aîné). Geschichte des Galvanismus ; aus dem Franzn. . . mit Anmerkungen und Zusätzen von Aug. Clarus. Erster Theil. 8vo. 250 pp. (*Volta's copy.*) *Leipzig*, 1802

† Histoire du Galvanisme, et Analyse des différents ouvrages publiés sur cette découverte, depuis son origine jusqu'à ce jour. Première et seconde parties. 8vo. 1re partie, 335 pp. ; 2e partie, 492 pp. 1 plate. *Paris*, 1802

† Geschichte des Galvanismus ; aus dem Franzn. . . mit Anmerkungen und Zusätzen von Aug. Clarus. Zweiter Theil. 8vo. 208 pp. (*Volta's copy.*) *Leipzig*, 1803

† Geschichte des Galvanismus nach Sue, frey bearbeitet, nebst Zusätzen und eine Abhandlung über die Anwendung des Galvanismus in der praktischen Heilkunde. Zwei Abtheilungen von D. J. C. L. Reinhold. 8vo. 1 Abth. 328 pp. ; 2 Abth. 176 pp. 2 plates. (*Volta's copy.*) *Leipzig*, 1803

† Histoire du Perkinisme. 8vo. *Paris*, 1805

Sue's Histoire du Galvanisme, vol. iii. pp. 62—92, contains indications o many writings on Medical Galvanism which have not been registered.

Histoire du Galvanisme, et Analyse des différents ouvrages publiés sur cette découverte, depuis son origine jusqu'à ce jour. Troisième et quatrième parties. 8vo. 3e pt. 379 pp. ; 4e pt. 362 pp. *Paris*, 1805

Sulzer, Johann Georg. *Born October 16, 1720, at Winterthur, Switzerland ; died February 25, 1779, at Berlin.*

Theorie d. angenehmen u. unangenehmen Empfindungen. 8vo. *Berlin*, 1762

Sulzer, Johann Georg.—*continued.*

† Nouvelle Théorie des Plaisirs, avec des réflexions sur l'origine du Plaisir, par M. Koestner. 8vo. 363 pp. 1767

(*Note.—Various articles collected from the Berlin Academy by an unnamed Editor under the above title. Various changes and corrections. The celebrated note is at p.* 155. *Elice, Saggio,* 2nd edition, p. 32, *says that Sulzer made the experiments in* 1752, *without indicating the source of his information. Seyfer, Geschichte,* p. 9, *says that Sulzer made the discovery in* 1760, *and refers to the "Mém. de l'Acad. de Berlin" (pro)* 1760, *"Theorie der angenehmen und unangenehmen Empfindungen. Berlin, (printed)* 1762, p. 82.*")

Sundelin, K. Anleitung zur medicinischen Anwendung der Elektricität und des Galvanismus. 2 plates. *Berlin,* 1822

Svanberg, Adol. Ferd. Magisterprogram. Försök att förklara orsaken till den dynamiska thermo-electriciteten. *Upsala,* 1851

Försök att Förklara Orsaken till den dynamiska Thermo-Elektriciteten. Inbjudningsskrift. 4to. 27 pp. (*Köngl. Academiska Boktryckeriet.*) *Upsala,* 1851

Svanberg, Gustaf. "Einer d. ersten Theilnehmer an d. Göttinger magnet. Beobachtungen: errichtete schon 1835 ein magnet. Observatorium in Upsala u. stellte darin Beob. an die gedruckt sind in Gauss u. Webers Resultate. . . . (*Vide Poggendorff,* ii. 1052.)

Svanberg, Praes. et Zeipel, Resp. De vi electro-magnetica observationes. 4to. 9 pp. (*Reg. Acad. Typographus.*) *Upsala,* 1851

† **Swaim, J.** Extraits de Rapports et Lettres, sur le système de Télégraphie de Guerre de M. J. Swaim. 8vo. 38 pp. *Paris,* 1859

Swammerdam. Biblia naturæ (à la fin du xvii. siècle). Tom. ii. p. 849.

(*Note.—Contains the account of the Experiments made before the Grand Duke of Tuscany in Galvanism. Vide Du Bois-Reymond, vol.* i. p. 43, *for an account of Swammerdam's experiment, given by Duméril in Annales des Sciences Naturelles,* 2e série, Zoologie, tom. xiii. p. 65 (4 *Février,* 1840.) *Another Experiment is also referred to. Du Bois-Reymond shows that Duméril's account of the Experiments is incorrect. He quotes the book, &c., thus : " Johannis Schammerdamii Amstelodamensis, Biblia Naturæ. Hollandisch und Lateinisch von* Gaubius, tom. ii. Leidæ, 1738, fol. p. 839 ;" *and also " Bibel der Natur. Aus dem Holländischen übersetzt von* Reiske. Leipz. 1752, fol. s. 300." *Matteucci,* p. 3, *refers to the passage in Swammerdam's Biblia Naturæ,* tom. ii. p. 849 (*to which Duméril drew attention*), *and describes the Experiments, and says that the Book appeared toward the middle of the fourteenth century. Figuier's date, end of fourteenth century, wrong. Weber, Cat. Berlin* (1870 ?), p. 79, *has " Swammerdam, J. Biblia Naturæ, sive hist. insectorum. Belgice et Lat. acced. interp.* H. D. Gaubius, 2 vols. *cum multis Tabs. &c. in fol.* 1737-38.*")

Swanberg. (*Vide* Svanberg.)

Swieten, Van. Commentaria in Herm. Boerhaave aphorismos de cognoscendis et curandis morbis. L. B. p. 382. 4to. 1753

(On the course of Lightning.) "Lettera al Bammacari." (*From Toderini,* p. 34.)

Swinden, Jan Hendrik Van. *Born June 8, 1746, at Haag ; died March 9, 1823, at Amsterdam.*

Tentamina theoriæ mathematicæ de phænomenis magneticis, sistens principia genera. . . . 4to. Figures and 7 tables. *Lugd. Bat.* 1772

De paradoxo phænomeno magnetico ; magnetem fortius ferrum purum quam alium magnetem attrahere. 4to. (*Neue Abhandl. der Baierischen Akad. Philos.* i. 351.) *München,* 1778

Swinden, Jan Hendrik Van.—*continued.*

Recherches sur les Aiguilles aimantées, et sur leurs variations singulières—régulières (both terms are given, one at p. 378, the other at p. 90). 4to. (*Mém. de Mathémat. et de Phys.* viii. p. 1.) *Paris*, 1780

(*Note.—In his "Recueil,"* p. xxvi. *of* vol. i. *he calls this his second "ouvrage couronné en* 1777 *par l'Académie de Paris, et publié en* 1730 *dans le* 8e vol. *des Mémoires présentés par des Savants étrangers"* . . *porte le titre de "Recherches sur les Aiguilles aimantées," et contient plus de* 600 *pages.*)

Marche de l'Aiguille magnétique observée pendant l'Aurore-boréale du 26 Février, 1780) à la Haye et à Franeker. 4to. (*Acta Acad. Petropolitanæ,* an. 1780, pt. i. *Hist.* p. 10.) *Petropol.* 1780

De analogia electricitatis et magnetismi. 4to. (*Neue Abhandl. der Baierischen Akad. Philos.* ii. 1.) *Munchen*

(*Note.—His French translation of this, with corrections, is the first Memoir (and* 1st *vol.*) *of his " Recueil de Mémoires sur l'Analogie de l'Electricité et du Magnétisme.*)

† Recueil de Mémoires sur l'Analogie de l'Electricité et du Magnétisme, couronnés et publiés par l'Académie de Bavière. Traduits du latin et de l'allemand, augmentés de notes, et de quelques dissertations nouvelles. 3 vols. all dated 1784. 8vo. *La Haye*, 1784

1. Analogie de l'Electricité et du Magnétisme, ou Recueil de Mémoires . . . 1785 ; 2. Recueil de Mémoires sur l'Analogie de l'Electricité et du Magnétisme . . . 1784.

(*Note.—This copy has two title-pages, as above.*)

Positiones Physicæ. Tom. 1er. (*From Opusc. Scelti,* x. 7.) *Harderovici*, 1786

Sylvester, C. Articles on Galvanism and Voltaism in Rees's Cyclopædia. 4to. *London,* 18—

† **Symes,** R. Fire analysed . . and manner . of making Electricity medicinal and healing. . . 8vo. 87 pp. *Bristol,* 1771

Symmer, Robert. New Experiments and Observations concerning Electricity . . 4to. (*Phil. Trans.* 1759.) *London,* 1759

† **Szafarkiewicz,** W. Beobachtungen über d. Erregung von Wärme oder Kälte beim Uebergang des galvanischen Stromes zwichen heterogenen Leitern. 8vo. 22 pp. 1 plate. *Breslau,* 1853

T.

Tabareau. On the Theories of Heat, Light, . . Magnetism, and Electricity. 8vo.
Lyon, 1825

Tachard, Guy. *Born in Prov. Guienne; died* 1714, *at Bengal.*

Voyage de Siam des Pères Jésuites, envoyés par le Roy aux Indes et à la Chine,
avec leurs observations astronomiques et leurs remarques de physique, de
géographie, d'hydrographie, et d'histoire, 4to. (*Daily Variation of Declination.*)
Paris, 1689

Second Voyage de Siam. (On Declination.) (*From Hartsocker,* p. 172.)

† **Taerg, M.** Letter . . to Mr. Tilloch on the Preparation of Composts for land,
and on the Composition of Potass. 8vo. 2 pp. (*Phil. Mag.* xxv. 358.)
London, 1806

† On the Constituent principles of Potash. 8vo. 1 p. (*Phil. Mag.* xxx. 173.)
London, 1808

Taisnier, Jean. *Born* 1509, *at Ath, Hennegau; died* 1562 (*shortly after*).

† De Natura Magnetis et ejus effectibus. . . Sm. 4to. 84 pp. Portrait (in woodcut. (*Apud Joannem Birckmannum.*)
Coloniæ, 1562
(*This copy contains his name, and signed Dedication. The first printed edition
of Petrus Peregrinus is dated* 1558. *Taisnier's only edition was printed in*
1562.)

De Natura Magnetis . . English translation by Eden. (*From Watt.*)
London, 1579 ?

Tait. (*Vide* Thomson and Tait.)

† **Tanchou.** Enquête sur l'authenticité des phénomènes électriques d'Angélique
Cottin. 8vo. 54 pp.
Paris, 1846

Tan (or Ten) Eyck, Dr. A work on the Application of Electro-Magnetism to
Mechanical purposes.
American

(*Vide* also Henry and Ten Eyck.)

Tarbes. (*Vide* Thollard de Tarbes.)

Tarde, Jn. Les usages du Quadrant à l'aiguille aimantée. Divisé en deux livres. .
4to. 120 pp. Cuts.
Paris, 1621

Targioni-Tozzetti, Giovanni. (*Vide* Tozzetti.)

† **Tassoni.** Pensieri diversi. V. Lib. 5, quesito 6.

† **Tata, D.** Mem. sulla Pioggia di Pietre avvenuta nella Campagna Sanese il 16
Giugno di quest' anno. 8vo.
Napoli, 1794

† **Tatum, J.** On the influence of the Atmosphere on the Electro-Galvanic Column
of M. De Luc. 8vo. 2 pp. (*Phil. Mag.* xlvii. 47.)
London, 1816

On the Electric properties of Metals, and the absolute position and negative
powers of various substances. 8vo. (*Phil. Mag.* li. 438.)
London, 1818

† Electro-Magnetic Experiments. 8vo. 2 pp. 1 figure. (*Phil. Mag.* lvii. 446.)
London, 1821

† On Electro-Magnetism. 8vo. 5 pp. 1 plate. (*Phil. Mag.* lxi. 241.)
London, 1823

† On Electro-Magnetism. 8vo. 3 pp. (*Phil. Mag.* lxii. 107.)
London, 1823

Tauber. Beschreibung der vom H. Tauber erfundenen Kopfbinde (for Galvanic
application). (*From Martens.*)

Tauberi. (*Vide* Teuberi.)

Tavernier, A. de. Blitzableiter, genannt Antijupiter oder Tavernier's gewitter-
ableitende Säule.
Leipzig, 1833

† **Tavignot.** De la méthode Galvano-caustique appliquée . . 2e édition. 8vo. 58 pp.
Paris, 1863

† **Taylor, Brook.** An account of an Experiment made by Dr. Brook Taylor, assisted by Mr. Hawkesbee, in order to discover the law of the Magnetical Attraction. 4to. 2 pp. (*Phil. Trans.* xxix. *for* 1714-15-16, p. 294.)
London, 1717

† Extract of a Letter to Sir Hans Sloane, dated 25th June, 1714. Giving an account of some Experiments relating to Magnetism. 4to. 5 pp. (*Phil. Trans.* xxxi. *for* 1720-21, p. 204.)
London, 1723

Taylor, Richard. *Born May 18, 1781, at Norwich; died December 1, 1858, at London.*

Scientific Memoirs . . . begins. 8vo. *London*, 1836 *to* 18—?

Taylor, Thomas Glanville. *Born November 22, 1804, at Ashburton, Devonshire; died May 4, 1848, at Southampton.*

Taylor. T. G., and Caldecourt. Observations on the direction and intensity of the Terrestrial Magnetic Force in Southern India. 8vo. *Madras*, 1839

Taylor, Dr. W. C. An account of the newly-invented Electro-Magnetic Engine for the propulsion of locomotives, ships, mills, &c. 12mo. *London*, 1841

† **Taylor, W. H.** Patent Electro-Magnetic Engine. 4to. 1 p. (*Extract from the Times*, May 7, 1840. *A handbill headed Colosseum, Regent's Park (for Exhibition).* *London*, 1840

† **Teclu, Nicolæ.** Chemische Untersuchung des Meteoriten v. Goalpara in Asam (Indien). 8vo. 3 pp. (*Sitz. d. K. Akad. Vorgelegt*, 9 Dec. 1870, lxii.)
Wien, 1870

Tecnico, Il. A Periodical containing articles on the Electric Telegraph.
Torino, 1858

† **Tedesch, A.** Grundl. u. auf Erfahr. beruh. Anleit z. Verfert u. Einricht. d. Tholardschen Blitz- u. Hagel-Ableiter u.s.w. nach d. Ital. m. e. Anh. . . 8vo. 30 pp. 1 plate. *Prag*, 1825

† **Telegraphic Cable Committee.** Report of the Joint Committee appointed by the Lords of the Committee of Privy Council for Trade, and the Atlantic Telegraph Company, to inquire into the construction of Submarine Telegraph Cables, &c. with Minutes of Evidence and Appendix presented to both Houses. 519 pp. Folio. The Committee were: J. Stuart Wortley, Latimer Clark, Capt. D. Galton, C. Wheatstone, C. F. Varley, and George Saward. *London*, 1861

† **Télégraphie Electrique.** Documents relatifs à l'établissement de lignes télégraphiques en Belgique. 8vo. 53 pp. 1 map. (*Extrait des Annales des Travaux Publiqes de Belgique.* *Vide Quetelet, De Vaux, and Labry.*) *Brussels*, 1850

Telegraph Engineers. (*Vide* Society of Telegraph Engineers.)

Telegraph Journal, The. A weekly record of Electrical and Scientific Progress. 4to. *London*, 1864 *to* 18—

Télégraphique Conférence de Vienne. (*Vide* Conférence.)

Teller, C. G. Dissertatio de Therapia per electrum. *Leipzig*, 1785

Ten Eyck. (*Vide* Tan Eyck.)

Tentzelius. Medicina diostatica (Magnet. ?) from the Latin by Parkhurst. 8vo. (*From Watt. Not in Pogg.*) 1653

Tenzel, F. B. K. Samml. verschiedn. Merkmale, welche Seeleute im Adriat. u. Mittelländ. Meere von d. bevorstehenden Wetter haben, nebst Beobb. d. Neigung d. Magnetnadel auf einer Seereise im J. 1807. 8vo. *Erlang*, 1821

† **Termeyer, R. M. De.** Esperienze e Rifiessioni sulla Torpedine. 4to. 48 pp.
(*Raccolta Ferrarese di Opuscoli Scientifici . . di Autori Ital.* tom. viii. p. 23.)
Venezia, 1781
(*Note.*—*This work " Sulla Torpedine," is the same as that quoted, &c. in the
Opuscoli Scelti,* tom. iv. p. 324, 1781, *under the title Esperienze del . .
Termeyer su l'Anguilla tremante. Vide* p. 26 *of the former, and* p. 334 *of
the latter, &c. &c. It seems to have been printed in* 1774 *" incirca" in*
vol. viii. *although that volume is dated* 1781. *Vide* p. 156 *of his Opusc.* xi.
Intorno ad un' Anguilla, &c. . .)

† Esperienze su l'Anguilla tremante. 4to. 12 pp. (*Opusc. Scelti,* iv. 324.)
Milano, 1781

† Intorno ad un'Anguilla, ossia Ginnoto Americano. Conghietture della cagione
dei mirabili effetti risultanti dal mediato, ed immediato contatto del
medesimo. 4to. 69 pp.
(*Note.*—*This work contains the same matter (and nearly the same wording) as
the work headed " Esperienze Riflessioni sulla Torpedine," or as that
quoted and described in the Opuscoli Scelti,* tom. iv. p. 324, 1781, *under the
heading " Su l'Anguilla Tremante.*")

Ternaux Compans, H. Bibliothèque Américaine. 8vo. *Paris,* 1837

Terquem, Paul Aug. Observations de la déclinaison de l'Aiguille aimantée en
1860-61. (*Mém. Soc. Dunkerque pour l'encouragement des Sciences, &c.* 1861.)
Dunkerque, 1861

Terzagi, P. M. Musæum Septalianum descriptum, cum logocentrionibus de
natura Crystalli . . Succini, Ambr et Magnetis, &c. 4to. *Dertonæ,* 1664
(*Note.*—*" At* p. 43, *account of a monk at Milan having been killed by the fall
of an Aerolite in the presence of Septala himself.*")

† **Tesini, D.** Descrizione del Telegrafo elettrico del Prof. D. Tessini di Cremona.
4to. 8 pp. 1 plate. (*Giornale dell' Ingegnere Architetto,* No. 13.)
Milano, 1854

Teske, Jno. Gottfd. Abhandl. von d. Elektricität nebst zwei anderen Abhandl.
gleichen Inhalts (siehe Unger) von d. Acad. in Berlin zum Druck befördert.
4to. *Berlin,* 1745

Newe Entdeck. verschiedener bisher noch unbekannter Wirkkn. u. Eigenschaften der Elektricität. 4to. *Königsberg,* 1746

Dissertatio de phialis Vitreis ab illabente minimo silice dissilientibus. 4to.
Königsberg, 1751

Vom Nutzen . Elektricität in Abwend. d. Ungewitter. (*Wochentl. Königsb.
Nachr.* 1753.) *Königsberg,* 1753

Abhandl. von d. nöthigen Sorgfalt bei Einricht. d. Elektrisirkugeln, in
Absicht auf d. darin befindliche Luft. (*Hamb. fr. Urtheile u. Nachr.* 1757.)
1757

Neue Versuche in Curirung d. Zahnschmerzen vermittelst d. magnet Stahls.
8vo. *Königsberg,* 1765

† **Testelin, A.** Essai de théorie sur la formation des images photographiques rap-
portée à une cause électrique. . . 8vo. 80 pp. *Paris and Gand,* 1860

† De l'électricité et du magnétisme relativement à la théorie de la pile de Volta,
des aimants, et des moteurs électriques. 4to. 79 pp. (*Reinwald's Cat. for*
1861.) *Paris,* 1860

Testu. On Hail (viz.) "nel suo viaggio aerostatico fatto in Parigi, 18 Giugno,
1786." (*Bibliot. Ital.* xcix. 89.) 1786

Tetens, Johann Nicolaus. *Born September* 16, 1736, *at Tetenbüll; died August* 15,
1807, *at Copenhagen.*

Üb. d. beste Sicherung einer Person bey einem Gewitter. 8vo.
Bützow and Wismar, 1774

Schreiben eines Naturforschers über d. Magnet-curen. 8vo.
Bützow and Wismar, 1775

Teuberi, M. C. Instrumentum novum variationem magnetis exhibens. M. C. Tauberi (*sic*.) (*Opusc. Act. Erudit. Lips.* tom. i. p. 401.)

Teyler van der Hulst, Pieter. *Born March* 25, 1702, *at Harlem ; died April* 8, 1778, *at Harlem.*

Teyler's Tweede Genootschap.   *Haarlem,* 1781—1859

† **Thackeray.** Electricity and the Electric Telegraph. 8vo. 14 pp. (*Cornhill Magazine,* ii. No. 7, 61.)   *London,* 1860

Thalen, T. R. La longitude terrestre déterminée au moyen de signaux galvaniques.   *Upsala,* 1856

Recherches sur les propriétés magnétiques du fer. 4to. 43 pp. (*Nova Acta Reg. Soc. Upsala,* iii. Série.)   *Upsala,* 1862

Thales. *orn* 639 (B.C.), *in Ionia ; died* 548 (B.C.).

On Amber. (*From Pogg.* ii. 1088.)   B.C. 600

Thenard, L. J. (*Vide* Gay-Lussac et Thenard.) (*Vide* Vanquelin and others.)

Theophrastus. *Born* 371 (B.C.) *at Ereson, in Lesbos ; died* 286 (B.C.), *at Athens.*

De lapidibus (sect. 53.) (Amber.) (*From Priestley, Martin, and others.*) *Lugd. Bat.* 1613?

History of Stones. Translated by Sir John Hill 2nd edition. 8vo. (*Greek text. English translation and notes.*)   1774

† **Theorell,** A. G. Printing Meteorograph. 4to. 4 pp. *Stockholm,* 1871

Thiers. (*Vide* Laccassagne and Thiers.)

Thillaye, Antoine. *Born October* 2, 1782, *at Rouen ; died March* 24, 1806, *at Paris.*

Bulletin du traitement des malades dans les cabinets de l'Ecole de Médecine de Paris qu'a tenu le Cit. Thillaye fils. 8vo.   *Paris,* 1801

Essai sur l'emploi méd. de l'Elect. et du Galv. (*Sue, Hist.* iii. 14, *says : " Thèse présentée . . à l'Ecole de Méd. le* 15 Floréal, an. xj.)   *Paris,* 1803

† Eléments de l'Elect. et Galvan. par G. Singer, trad. de l'Anglais, et augmentée de Notes. 8vo. 655 pp. 5 plates.   *Paris,* 1817

† Table . . du Dict. de Médecine de l'Encyclop. méthodique. 4to. 174 pp. et Table des princip. Rédacteurs.   *Paris, n. d.*

Thilorier, A. Sur une inscription de Memphis relative à l'observation d'un phénomène céleste. 4to.   *Paris,* 1840

Thilorier, J. C. Genèse philosophique, précédée d'une diss. sur les pierres tombées du ciel. 8vo.   *Paris,* 1803

Thollard de Tarbes. Moyens préservatifs de la Foudre et de la Grêle.

Thom, A. Nature and Course of Storms in the Indian Ocean, South of the Equator. 8vo. With chart.   1845

† **Thomsen, J.** Den elektromotoriske Kraft udrykt i Varmeenheder. 4to. 23 pp. (*Særskilt aftrykt af det Kongl. Danske Videnskabernes Selskabs Skrifter,* 5te *Række, naturhistorisk ogmathematisk Afdeling,* 5te *Band.*) *Kjöbenhaven,* 1858

Die Polarisations Batterie. (*Repertorium für physikalische Technik für mathematische u. astronom. Instrumentkunde von Carl.* i. 171. *Aus der Tidsskrift for Physik og Chemie,* 1864, iii. 195.)   *Kopenhagen,* 1864

Die Polarisations-Batterie, ein neuer Apparat zur Hervorbringung eines kontinuirlichen elektrischen Stromes von hoher Spannung und Konstanter Stärke, mit Hülfe eines einzelnen galvanischen Elementes. 8vo. 32 pp. (*From Zuchold's Cat.* 1865, p. 138. *Aus der Zeitschrift für Physik und Chemie.*) *Hamburg,* 1865

Thomsen, H. P. J. J. Om den elektromotoriske Kraft bestant i. Arbeidsmangder. (*Vidensk. Selsk. Skr.* Bd. v. Hft. i.)   *Kopenhagen,* 1859

Die constante Kupferkohlenkette. 8vo. (*Pogg. Ann.* cxi. 1860.) *Leipzig,* 1860

Thomson, M. System of General Night Signals. Sm. 4to. *London,* 1850

Thomson, Dr. Thomas. *Born April 12, 1773, at Crieff, Perthshire; died July 2, 1852, at Kilmun, Argyllshire.*

† An Outline of the Sciences of Heat and Electricity. 1st edition. 8vo. 583 pp. 1 plate. *London,* 1830

† An Outline of the Sciences of Heat and Electricity. 2nd edition, remodelled, much enlarged, and illustrated with woodcuts and maps. 8vo. 585 pp. 1 plate. 2 maps. *London, Paris, and Leipzig,* 1840

(*Vide* Annals of Philosophy.)

Thomson, (now Sir) William. *Born June 25, 1824, at Belfast, Ireland.*

† On a Mechanical Representation of Electric, Magnetic, and Galvanic Forces. 8vo. 4 pp. (*Extracted from the Cambridge and Dublin Mathematical Journal,* Dec. 1847.) *Cambridge,* 1847

† On the Forces experienced by Small Spheres under Magnetic Influence, and on some of the Phenomena presented by Diamagnetic Substances. 8vo. 6 pp. *Extracted from Cambridge and Dublin Mathematical Journal,* May, 1847.) *Cambridge,* 1847

† On the Mathematical Theory of Electricity in Equilibrium. 8vo. 43 pp. (*Extracted from the Cambridge and Dublin Mathematical Journal,* March, 1848, Nov. 1848; Nov. 1849; Feb. 1850.) *Cambridge,* 1850?

† Applications of the Principle of Mechanical Effect to the Measurement of Electro-Motive Forces, and of Galvanic Resistances in Absolute Units. 8vo. 12 pp. (*From the Phil. Mag. for Dec.* 1851.) *London,* 1851

† A Mathematical Theory of Magnetism. Read June 21, 1849. 8vo. 43 pp. (*Phil. Trans. for* 1851, pt. i. p. 243.) (*Faraday's copy.*) *London,* 1851

† Magne-crystallic Property of Calcareous Spar. 8vo. 1 p. (*Phil. Mag. for Dec.* 1851.) *London,* 1851

† On the Theory of Magnetic Induction in Crystalline and Non-Crystalline Substances. Read before the British Association, August, 1850. 8vo. 10 pp. (*From the Phil. Mag. for March,* 1851.) *London,* 1851

† Note on Induced Magnetism in a Plate. 8vo. 4 pp. (*Extracted from Cambridge and Dublin Mathematical Journal,* i. p. 34, New Series.)

† On the Mechanical Theory of Electrolysis. 8vo. 16 pp. (*From the Phil. Mag. for Dec.* 1851.) *London,* 1851

† On the Mutual Attraction or Repulsion between two Electrified Spherical Conductors. 8vo. 12 pp. (*From the Phil. Mag. for April and August,* 1853.) *London,* 1853

† On Transient Electric Currents. Read at the Glasgow Philosophical Society on the 19th Jan. 1853. 8vo. 13 pp. (*From the Phil. Mag. for June,* 1853.) *London,* 1853

† Reprint of Papers on Electrostatics and Magnetism. 8vo. 592 pp. *London,* 1872

† **Thomson and Tait.** Elements of Natural Philosophy. 8vo. Part i. *London,* 1872

Thompson, F. B. Fire: Its Causes considered and explained on the basis of Chemical and Electrical Science, including the Theory of "Spontaneous Combustion," together with Scientific Facts. 8vo. 32 pp. *London,* 1857

† **Thoresby, Ralph.** An Account of a Young Man slain with Thunder and Lightning, Dec. 22, 1698. 4to. 2 pp. (*Phil. Trans.* xxi. *for* 1699, p. 51.) *London,* 1700

Thorp, Robert William Disney. *Born about 1767; died July 4, 1849, at Kemerton.*

† Dissertatio physico-medica inauguralis de Electricitate. Subm. 12 Mart, 1790. 8vo. 69 pp. *Lugduni Bat.* 1790

Thouret. Extrait de la Correspondance de la Société Royale de Médecine relativement au magnétisme animal. *Hist. Mém. de la Soc. Rl. de Méd.* an. 1782 et 1783, *Hist.* p. 217.) *Paris,* 1783?

Thouret. (*Vide* Andry and Thouret.)

Thouri, De. De l'influence de l'élect. sur le corps humain. Mém. . . ("*Prix Acad. de Lyon.*" *And in Journ. de Phys.* Juin, 1777.) *Paris,* 1777

Thouvenel, Pierre. *Born 1747, in Lothringen; died February 28, 1815, at Paris.* Méms. i. et ii. montrant les rapports évidents entre les phénomènes de la baguette divinatoire, du magnétisme et de l'électricité, etc. *Londres et Faris,* 1781-84

Mém. sur l'électricité organique et minérographie. 8vo. *Brescia,* 1790

† Transunto d'una Memoria del . Thouvenel su l'Elettrometria organica. 4to. 19 pp. (*Nuova Scelt. Opusc.* xv. 397.) *Milano,* 1792

La guerra di dieci anni raccolta polemico-fisica sull' elettrometria galvanico-organica. 8vo. *Verona,* 1802

† **Thouvenel** (Th* * *). Mém. sur l'aérologie et l'électrologie. 3 vols. 8vo. 3 plates. (*Forms part of "Mélanges d'Hist. Naturelle, de Phys. et de Chimie."*) *Paris,* 1806

(*Note.—There are other works of Thouvenel not noted in the Catalogue.*)

Thurn. (*Vide* Zallinger.)

Thury. (*Vide* Héricart de Thury.)

Thyllaye. (*Vide* Singer, 1817.)

† **Tiburtius, T.** Von einem merkwürdigen Wolkenzuge beym Wreta Kloster. 8vo. 3 pp. (*K. Schwed. Akad. Abh.* xx. 39.) *Hamburg and Leipzig,* 1758

Tieenk, J. Bericht wegens de miswyzing van het compas, door den donders. (*Verhandel. van het Genootschte Vlissingen,* iii. 615.)

† **Tietz, J.** Die Erfindung und erste Verbreitung d. Blitzableiters. 4to. 17 pp. (*Jahrsbericht über das Kön. Kath. Gymnasium zu Braunsberg,* 1850-59.) *Braunsberg,* 1859

† **Tilas, D.** Von einem Donnerschlage in Oesterwahla. Kirchspiele und Waszmannlands Hauptmannschaft, im J. 1740. 8vo. 6 pp. (*K. Schwed. Akad. Abh.* iv. 43.) *Hamburg,* 1742

† **Tilden, J.** An Account of a Singular Property of Lamprey Eels. 4to. (*Mem. Amer. Acad.* Old Series, iii. part ii. p. 335.) *Boston, U.S.* 1815

† **Tilesius von Tilenau,** Wm. Gott. Die Wirkung des Blitzes auf den menschlichen Körper durch einen merkwürdigen Fall erläutert. 8vo. 13 pp. 1 plate. (*Journ. f. Chem.* N.R. ix. 129.)

Tilloch, Alexander. *Born February 28, 1759, at Glasgow; died January 26, 1825, at Islington, London.*

Letter to Pictet. Answer to a Letter from Paris on Volta's Experiments. 8vo. (*Bibliothèque Britan.* No. 144, p. 390.) *Genève,* 1801

(*Vide* Philosophical Magazine.)

† **Timbs, J.** Year Book of Facts in Science and Art. 12mo. *London,* 1839-66?

Curiosities of Science Past and Present. 3rd edition. First Series. 8vo. 248 pp. *London,* 1862

† Curiosities of Science. Second Series. 2nd edition. Fourth thousand. 8vo. 248 pp. *London,* 1865

Timæus, Louis. De mundi anima. . . (Amber.) (*From Dutens,* p. 150.) B.C. 500

"**Times**" (Journal). Notices, Letters, &c., concerning the Atlantic Telegraph.

"**Times**"—Editor. On Export of Telegraphic Wire. "Telegraphic Wire.—Telegraphic wire has become rather an important article of commerce. During the last ten years it has been exported to the following values :— 1853, £72,584 ; 1854, £81,566 ; 1855, £163,737 ; 1856, £80,076 ; 1857, £302,246 ; 1858, £224,708 ; 1859, £742,306 ; 1860, £251,712 ; 1861, £214,441 ; and 1862, £321,044. We have thus in the ten years an aggregate export of the value of £2,474,410. *London, August* 22, 1863

† **Tinelli, L.** Stralcio di Lettere di L. Tinelli . . di Nuova York, al di lui fratello Carlo in Milano. (Data 30 Luglio, 1839). 8vo. 3 pp. (*Bibl. Ital.* xcvii. 131.) *Milano*, 1840

Tingry, P. F. Sur la phosphoréscence des corps et particulièrement sur celle des eaux de la mer. 4to. (*Journ. Phys.* xlvii. 1798.) *Paris*, 1798

Sur la nature du fluide électrique. 4to. (*Journ. Phys.* xlvii. 1798.) *Paris*, 1798

† **Tip, Joh.** Over de electrische warmte-en lichtverschijnselen die door de rigting des strooms gwijzigd word. 8vo. 72 pp. 1 plate. *Utrecht*, 1854

Tipaldo. Biografia degli Italiani illustri nelle scienze. 8vo. (*Milano Stella*, "*Saranno circa* 8 *Volumi* . . *ne son pubbl.* 7.") *Venezia*, 1834

Tiraboschi. Storia della letteratura Italiana di Girol. Tiraboschi. 8vo. (In tom. iv. p. 285, and viii. p. 384.) *Milano*, 1823

(*Note.—Many quotations on the history of the Mariner's Compass, &c.*)

Tissier. Mém. pour établir la Surphosphoréscence des Corps. 8vo. *Lyon*, 1807

Notice sur l'Aurore boréale qui a paru à Lyon le Vendredi, 7 Janv. 1831. Lu à la Société d'Agriculture 14 Jan. 1831. 8vo. *Lyon*, 1831

Tissot. Tract. de Variolis, Apoplexia et Hydrope. 12mo. *Lausanne*, 1761

Epistol. ad Hallerum. (*From Bertholon.*)

Tith, J. D. Progr. de electrici experimenti lugdunensis inventore primo. 4to. (*Vide* Titius.) *Witteberg*, 1771

Titius (Tietz), Johann Daniel. *Born January* 2, 1729, *at Konitz, West Prussia; died December* 16, 1796, *at Wittenberg.*

De electrici experimenti lugdunensis inventore primo. 4to. *Viteberg*, 1771

Tittlmann. Ideen über die Natur des Galvanismus. 8vo. (*Allgem. Medizin Annalen*, Sep. 1802, p. 641.) 1802

Tittman. On Difference in the Action of the Two Poles of a Column. 8vo. (*Dresdner gelehrte Anzeigen*, 22 Stück, 1802.) *Dresden*, 1802

Toaldo, Giuseppe. *Born July* 11, 1719, *at Pianezza, near Vicenza; died November* 11, 1797, *at Padua.*

Trattato della vera Influenza degli Astri . . ossia Saggio meteorologico. 4to. 1st edition. 1769

(*Note.—This Treatise was often reprinted.*)

† Saggio Meteorologico. Della vera influenza degli Astri . . Si aggiungono i Pronostici di Arato tradotti dall' A. L. Bricci, &c. . . (1a editione. 4to. 222 pp. 5 tables. 1 plate. *Padova*, 1770

Giornali Astro-meteorologici. 8vo. *Padova*, 1773-98

† Dell' uso dei Conduttori metallici. . . Apologia colla Descrizione del Conduttore . . di Padova. 4to. 32 pp. 1 plate. *Venezia*, 1774

† Del Conduttore elettrico posto nel Campanile di S. Marco in Venezia. . . (1a ed.) 4to. 37 pp. 1 plate. *Venezia*, 1776

Relazione del Fulmine caduto nel Conduttore della Specola di Padova. (1st ed.) 4to. *Padova*, 1777

Toaldo, Giuseppe.—*continued.*

† Dei Conduttori per preservare gli edifizj . . Memorie, in questa nuova ed. ritoccate ed accresciute di un' Appendice. . . 4to. 104 pp. 2 plates.

Venezia, 1778

(*Note.*—*This work contains his " Informazione al popolo" of* 1772, *including his translation of Saussure's " Manifesto." His " Dell' uso dei Conduttori . . Apologia . . of* 1774, *colla Descriz. del Cond. di Padova." His "Del Condutt. . . di S. Marco," &c. of* 1776. *His " Relazione del Fulmine caduto nel Condutt. della Specola, Padova," of* 1777. *His " Notizia del Fulmine . . nella Torre dell' università, Padova." His "Appendice sui fatti . . recenti,"* 1778 : *new matter. It also contains an Italian translation of Barbier's " Considérations en général,"* . . *which is a memoir appended to Barbier's French translation of this work of Toaldo. This Italian translation is by a printer, and not dated.*)

† Saggio meteorologico. Della vera influenza degli Astri sulle Stagioni e mutazioni di tempo. 2da ed. 4to. 230 pp. 4 tables. *Padova,* 1781

Osservazioni meteorologiche del mese di Giugno, 1783. Con un Discorsetto sulla Nebbia straordinaria ed influenza de' Fulmini nella corrente stagione. Sm. 8vo. 13 pp. *Padua ?*

† Osservazioni meteorologiche del mese di Giugno, 1783, con un discorsetto sulla Nebbia straordinaria ed influenza de' Fulmini nella corrente stagione. 4to. 9 pp. (*Opusc. Scelti,* vi. 265.) *Milano,* 1783

In his " Giornale Astro-Meteorologico," for or of 1784, " Dei principali accidenti dell' anno 1783." The first division is headed "Della Nebbia, e della Influenza de' Fulmini," and in which he refers to much writing on these subjects by himself and others in the " Giornale enciclopedico di Vicenza." 1784

† Fenomeno singolare d'un Fulmine descritto, e proposto all' esame de' fisici. 4to. 4 pp. (*Opus. Scelti,* vii. 35.) *Milano,* 1784

Essai météorologique . . . Nouvelle éd. beaucoup augmentée trad. de l'Ital. par Daquin on y a joint . . les Pronostics d'Aratus . . trad. par Brieri. New edition. 4to. 317 pp Text. *Chambéry,* 1784

† Descrizione d'un' Aurora Boreale osservata in Padova il di 29 Feb. 1780. 4to. 17 pp. (*Saggi . . dell' Accad. di Padova,* tom. i. p. 178.) *Padova,* 1786

La Meteorologia applicata all' Agricoltura. 8vo. *Vicenza,* 1786

Della Fiamma Volante, ossia del Globo di fuoco degli 11 Settembre, 1784. 4to. 22 pp. (*Padua Acad. Saggi,* iii. p. cv.) *Padova,* 1794

† Appendice : Riflessioni sopra i colpi di fulmine (alla Memoria del Marzari, "Descrizione d'una tempesta di fulmini.") . . . Letta 8 Feb. 1787. 4to. (*Vide* Marzari.) (*Saggi dell' Accad. di Padova,* iii. 212, pt. i. *Padova,* 1794

† Dei moti del Barometro nei Temporali. Ricevuta 28 Decemb. 1796. 4to. 5 pp. (*Mem. Soc. Ital.* viii. pt. i. p. 11.) *Modena,* 1799

† Elogio di Giuseppe Toaldo scritto da A. Fabbroni. Ricevuto 22 Ottob. 1798. 4to. 19 pp. (*Mem. Soc. Ital.* tom. viii. pt. i. p. 29.) *Modena,* 1799

† Completa Raccolta di Opuscoli, osservazioni e notizie diverse contenute nei Giornali Astro-Meteorologici dall' anno 1773 sino all' anno 1798 del fu Sig. G. Toaldo . . coll' aggiunta di alcune altre sue produzioni meteorologiche e pubblicate ed inedite. 8vo. 4 volumes. Vol. i. 36 and 248 pp. Portrait and 1 plate (at p. 17). Vol. ii. 264 pp. Vol. iii. 306 pp. Vol. iv. 296 pp.

Venezia, 1802

(*Note.*—*The first three vols. are dated* 1802, *the fourth* 1803. *A fifth was promised* (*at p.* 289 *of the fourth*), *but it never appeared. The editor is Tiato. An " Elogio di G. Toaldo scritto da A. Fabbroni," is at p.* 7 *of vol.* i. *Another plate is referred to, but was not printed. Tiato was assisted by Chiminello.*)

Some MSS. on Electrical Matters, &c., in the Seminario, Padova.

(*Vide* also Lorgna and Toaldo.)

(*Vide* also Marzari C. and Toaldo.)

† **Toaldo, G.** (or Anonym.) and **Saussure.** Della maniera di preservare gli edifizi dal Fulmine : Informazione al popolo. 4to. 19 pp. 1st edition.

Venezia, 1772

(*Note.—Annexed is his translation of Saussure's Exposition under the title "Manifesto ossia Breve esposizione ;" the paging being continued from 20 to 38. The date of Saussure's work is Geneva,* 1771.)

† **Tocchi.** Essai de statique électrique. 8vo. 106 pp. 1 plate. *Marseille,* 1828

Todd. Clinical Lectures on Paralysis, &c. *London,* 1856

† **Todd, J. T.** Some Observations and Experiments made on the Torpedo of the Cape of Good Hope in the year 1812. From Phil. Trans. for 1816, part i. 8vo. 5 pp. (*Phil. Mag.* xlviii. 14.) *London,* 1816

An Account of some Experiments on the Torpedo electricus at Rochelle. 4to. (*Phil. Trans.* 1817.) *London,* 1817

† **Toderini, G.** Filosofia Frankliniana. . , 8vo. 65 pp. *Modena,* 1771

Tomcsányi, Adam. Diss. de theoria phænomenorum electricitatis galvanianæ. 8vo. *Budæ,* 1809

† **Tomlinson, Charles.** Cyclopædia of Useful Arts. 2 vols. 8vo. *London,* 1854

The Thunder Storm. An Account of the Properties of Lightning, and of Atmospheric Electricity in various parts of the World. 12mo. 348 pp. *London,* 1859

Useful Arts and Manufactures. New edition. 2 vols. 12mo. *London,* 1860

† **Topler.** An Electrical Machine formed on the principles of Influence. 1864

Torino Società Agraria. Memorie della Soc. Agraria. 8vo. (*From Opuscoli Scelti* (in 4to.) xii. 11, 1789.) *Torino*

Torre, Giovanni Maria della. *Born* 1713, *at Rome ; died March* 7, 1782, *at Naples.*

Scienze della natura particolare. Parte ii. cap. vi. No. 416.

(*Note.—On Magnetism, Phosphorescence, Igneous Meteors, &c., as effects of the sun's rays.*)

Scienze della Natura. . . 2 vols. 4to. 1750

(*Note.—This is a reprint by Ricurti, with new matter on Magnetism and Electricity added by Della Torre to his other accounts.*)

Tortolini, Barnaba. Annali di scienze matematiche e fisiche. 8 vols. 8vo. *Roma,* 1850-7

Annali di matematica pura e applicata. 4to. *Roma,* 1856-61

† **Toscanelli, C. M.** Descrizione del Telegrafo. 8vo. 16 pp. 4 plates. *Torino,* 1806

Tosetti, G. B. Nuova macchina atta al raccoglimento delle due elettricità. 3 pp. 1 plate (of 4 figures). (*In Biblioteca Marciana, Venice.*)

Toulouse Acad. des Sciences, &c. Hist. et Mém. de l'Acad. Rle. des Sciences, Inscriptions et Belles Lettres de Toulouse.

† Tables des Matières contenues dans les Vols. i. ii. iii. iv. v. de la 4me Série des Mém. de l'Acad. 8vo. *Toulouse,* 1851-55

Table alphabétique des matières contenues dans les 16 premiers Tomes des Mém. de l'Acad. Impériale. 8vo. 83 pp. *Toulouse,* 1854

Tour. (*Vide* Du Tour.)

† **Tourdes G.** Relation médicale de l'accident occasionné par la foudre, le 13 Juillet, 1869, au pont du Rhin, près de Strasbourg. 8vo. 32 pp. *Paris,* 1869

Tourdes, J. Lettre au Prof. Volta sur l'élect. animale. An. x. (*Décade philosophique* No. 3, de l'an x. p. 118. *On the motion of the fibrous part of the Blood, galvanised Discovery of.*) 1802

† **Toutain.** Electricité Médicale. Nouv. méthode d'application de l'électricité pour la guérison des maladies. 12mo. 352 pp. *Paris,* 1870

† **Towson, J. T.** Practical information on the Deviation of the Compass, for the use of Masters and Mates of Iron Ships. Published by order of the Committee of Privy Council for Trade. 8vo. 122 pp. *London*, 1863

A work published and circulated by the Board of Trade on Compass Deviations and Corrections, &c. (*Title not given, nor date. From Correspondence* .. *in Proceedings of the Royal Society*, Nov. 1865, p. 539.)

Tozzetti, Targioni. *Born September* 11, 1712, *at Florence; died January* 7, 1783, *at Florence.*

Atti e Memorie inedite dell'Accad. del Cimento e Notizie, Aneddoti dei progressi della scienza in Toscana... Elett. tom ii. p. 421; Calamita, p. 539; Ambra, &c. 546. 3 vols. 4to. *Firenze*, 1780

Notizie degli aggrandimenti delle scienze fisiche accaduti in Toscana nel corso di anni 60 nel secolo 17. 3 vols. (in 4). 4to. *Firenze*, 1780

Notizie sulla storia delle scienze fisiche in Toscana cavate da un manoscritto inedito di Giov. Targioni Tozzetti. 4to. *Firenze*, 1852

† **Traill** (of Liverpool). Expériences et Observations sur le Thermo-Magnétisme. 4to. 26 pp. *Genève*, 1828

(*Note.—The first part is an extract only of the researches communicated to the Geneva Society in* 1827, *and printed in their Mém.* tom. iv. p. 94, 1828. *The second part translated entire.*)

Tralles, Johann Georg. *Born October* 15, 1763, *at Hamburg ; died November* 18-19, 1822, *at London.*

Beyträge zur Lehre von der Elektricität. 4to. 14 pp. (*Volta's copy.*)
Berne, 1786

(*Note.—A copy of a MS. letter of Tralles to Volta, dated 2nd November,* 1788, *on the Electricity of Waterfalls, made for me by the Count Zanino Volta, is bound with this work.*)

Über d. Elektricität d. Staubbachs, u.s.w. 8vo. (*Gren's Journ.* i. 1790.)
Leipzig, 1790

Volta's Eudiometer, u.s.w. 8vo. (*Gilb. Ann.* xxviii. 1808. *Leipzig*, 1808

† **Tredwey, Robert.** Part of a Letter to Dr. Leonard Plukenet, dated Jamaica, February 12, 1696-7, giving an account of a great piece of Ambergriese, thrown on that Island ; with the opinion of some there about the way of its production. 4to. 2 pp. (*Phil. Trans.* xix. *for* 1695-96-97, p. 711.)
London, 1698

Trembley, A. On the Light caused by Quicksilver shaken in a Glass Tube, proceeding from Electricity. 4to. (*Phil. Trans. for* 1746, p. 58.) *London*, 1746

Tremery, J. L. Observations sur les aimants elliptiques. (*Journ. des Mines*, vi. 1797.) 1797

Tremeschini. Télégraphe controleur. (*From* Du Moncel.)

Tressan, Louis Elisabeth de la Vergne. *Born November* 4, 1705, *at Le Mans; died October* 31, 1783, *at St. Lew.*

† Essai sur le Fluide Electrique considéré comme agent universel. 2 vols. 8vo. Vol. i. xl. and 396 pp. Vol. ii. 487 pp. *Paris*, 1786

Tressan, De. Partie physique du Voyage autour du Monde de la Frégate la *Venus.* Observations Météorologiques et Magnétiques. 5 vols. 8vo. Carte et Table. 1842-44

(*Vide* Du Petit Thouars.)

Treviranus, A. C. Über gewisse in Westpreussen u. Schlesien angebl. mit einem Gewitterregen gefallene Saamen Körner. 8vo. *Breslau*, 1823

Treviranus, Gottfried Reinhold. *Born February* 4, 1776, *at Bremen ; died February* 16, 1837, *at Bremen.* (*Poggendorff,* ii. 1132.)

Einfluss d. galvan. Agens u. einiger chem. Mittel auf d. vegetative Leben. (*Nordisches Archiv f. Natur- und Arzneiw.* 1st Bd. 2tes Stück; *and in Gilbert's Ann.* vii. 1801.) *Kiel,* 1800

Einfluss d. Galvanismus auf d. thier. Reizbarkeit. 8vo. (*Gilb. Ann.* viii. 1801.) *Leipzig,* 1801

Galvanisch-meteorolog. Ideen. 8vo. (*Gilb. Ann.* viii. 1801.) *Leipzig,* 1801

De protei anguinei encephalo et organis sensuum disquisitiones zootomicæ. 4to. 2 plates. *Göttingen,* 1820

Treviso Athenæum. Memorie scientifiche e letterarie dell' Ateneo di Treviso. 4to. *Treviso, Venezia, Padova,* 1817-47

Treviso Giornale. Giornale sulle Scienze e Lettere delle Provincie Venete. Treviso. 18 vols. 8vo. *Treviso,* 1821-30 (*Note.—Was suspended after the number for June,* 1829, *and was renewed in November,* 1829 ; *ceased in October,* 1830.)

Trew, Abdias. De meteoris. (*From Pogg.* ii. 1134.) *Argentorat.* 1654

† **Triewald,** S. Von. Versuch eines künstlichen Nordlichtes. 8vo. 2 pp. 1 plate. (*K. Schwed. Akad. Abh.* vi. 103.) *Hamburg,* 1744

† **Tripier,** A. Manuel d'Electrothérapie. Exposé pratique et critique des applications médicales et chirurgicales de l'électricité. 8vo. 624 pp. *Paris,* 1861

† La Galvanocaustique chimique. 8vo. 19 pp. (*Ext. des Archives générales de Médecine de* Janvier, 1866.) *Paris,* 1866

† **Troili,** Domenico. *Born April* 11, 1722, *at Macerata ; died February* 14, 1792, *at Macerata.*

† Della caduta di un Sasso dall' aria. 4to. 120 pp. *Modena,* 1766

† Il Ragionamento della caduta di un Sasso dall' aria difeso da una lettera apologetica. 4to. 71 pp. *Modena,* 1767

† Della elettricità. Lezioni di fisica sperimentale fatte nella Università di Modena il primo anno del suo rinnovamento. 8vo. 351 pp. (*Note.—Grimelli,* "*Storia* . . " *calls the date* 1774, *and the place Modena.*)

† **Trolliet.** Rapport sur les Paragrêles présenté à la Société d'Agriculture du Rhône. 8vo. 18 pp. 1 plate. *Lyon,* 1825

Notice sur les Paragrêles, par M. le Dr. Troillet, au nom d'une Commission. 8vo. (*Extrait du Précurseur,* 15 *et* 16 Mai, 1827.) *Lyon,* 1827

† **Trombelli,** Giangrisostomo (or J. C.). *Born* 1697, *near Nonantola ; died January* 24, 1784, *at Bologna.*

De Acus nauticæ inventore ad F. M. Zanottum. 4to. 39 pp. (*Comment. de Bonon. Scient. Instit.* tom. ii. pars iii. p. 333.) *Bononiæ,* 1747

Trommsdorf, Johann Bartholomäus. *Born May* 8, 1770, *at Erfurt ; died March* 8, 1837, *at Erfurt.*

Expériences Galvaniques. 8vo. (*Crell's Chemisch. Annalen a.* 1801. 4to. Cahir, p. 337. *Extrait du Journ. de Chim. de Van Mons,* i. 41, and p. 98.) *Helmstadt,* 1801

† Geschichte des Galvanismus ; oder der Galvanischen Elekt. vorzüglich in chemischer Hinsicht. Besonders abgedruckt aus J. B. Tromsdorff, Chemie im Felde der Erfahrung. 8vo. 260 pp. *Erfurt,* 1803

Geschichte des Galvanismus. 8vo. (*From Ersch,* p. 119.) *Erfurt,* 1808

† Grundritz der Physik als Vorbereitung zum Studium der Chemie. 8vo. 488 pp. 2 plates. *Gotha,* 1817

Troostwijk, Adriaan Paets Van. *Born March* 1, 1752, *at Utrecht ; died April* 3, 1837, *at Breukelen.*

Troostwijk, A. P. and Deiman. Verhandl. over de geneeskonstige electriciteit mit J. R. Deimann. 4to. (*Verhandl. Genootsch te Rotterdam*, viii. 1787.)
Rotterdam, 1787
La Réponse à la question proposée par la Soc. de Phys. Expérimentale à Rotterdam, " Quelle influence l'Electricité naturelle et ses différents états dans notre atmosphère ont elles sur le corps humain ?" (*Mém. de la Soc. de Phys. expér. Rotterdam*, tom. viii.)
1787
Antwoord op de Vraage, welken invloed heeft de natürlyke electriciteit en derzelver verschillende verdeeling in onzen dampkring op gezonde en ziekelüke ligchaamen? In welke ongesteldheben en ziekten is de konstige electriciteit dienstig tot geneezing of verligting? Op wat wüze werkt zü tot det einde? En welke is de beste manier om'er zig mit dat ogmerk van te bedienen? (*Verhandl. van het Genootsch te Rotterdam*, viii. p. 63.)
Rotterdam

† **Troostwijk, Van, et Krayenhoff.** De l'Application de l'Electricité à la Physique et à la Médecine. Ouvrage Couronné. 4to. 319 pp. 4 plates.
Amsterdam, 1788
Verhandeling over zekere onderschiedene Figuuren, welcken door beide soorten van Electriciteit worden voordgebragt, (1s Band *des Algem. Magazyn Nat. und übers. in den Leipziger Samml. zur Phys. u. Naturg.* iv. Band, 4 Stuck.)

Troostwijk, A. P. and Van Marum. Expériences sur la cause de l'électricité des substances fondues et refroidies avec mit Van Marum. 4to. (*Journ. de Phys.* xxxiii. 1788.)
Paris, 1788
(*Vide* also Marum, Van.)
(*Vide* also Carradori.)

Trousseau et Pidoux. Traité de thérapeutique et de matière médicale. Tom. i.
Paris, 1847

† **Trubner, N.** Bibliographical Guide to American Literature. A classed list of Books published in the U.S. during the last forty years, with Biographical Introduction, Notes, and Index. 8vo. 554 pp.
London, 1859

Trullard. Sur une nouvelle manière de faire les Aimants artificiels d'une très grande force, sans . . l'aimant naturel. 8vo. (*Mém. de l'Acad. de Dijon*, tom. i. p. 66.)
Dijon, 1769

† **Tschermak, G.** Über den Meteorstein von Golpara und über die leuchtende Spur der Meteore. 8vo. 11 pp. (*Sitzb. d. K. Akad. d. Wissen.* lxii. 1870.)
Wien, 1870

† Nachrichten über den Meteoritenfall bei Murzuk im Dec. 1869. Mitgetheilt von G. Tschermak. 8vo. 3 pp. (*Sitzb. d. K. Akad.* Juni 1870, lxii.)
Wien, 1870

† Der Meteorit von Lodran. 8vo. 11 pp. 1 Taf. (*Sitzb. d. K. Akad. d. Wiss.* Apr. 1870, lxi.)
Wien, 1870

† **Tschudi, Von J. J.** Beobachtungen über Irrlichter (Mitgetheilt von Fenzl.) dated Barbacina 5 Jän. 1857. 8vo. 2 pp. (*Sitzungsberichten d. Wien Acad.*) *Wien*

† Über einige electr. Erscheinungen in den Cordilleras der Westküste Süd-Amerikas. 8vo. 16 pp. (*Wien Acad. Sitzb.* vol xxxvii. No. 20.) *Wien*

Turchini. Sopra due elettro-calamite a rocchetto di grandi dimensioni. 8vo. 1 p. (*Rendiconto* . . *R. Accad. dei Georgofili*, p. 180, Gennajo, 1855.) *Firenze*, 1855

Turin Acad. Miscellanea philosophica Societatis privatæ. Tomus primus (ligna edit.). 4to. 146 pp. Conspectus and 4 plates. *Augustæ Taurinensis*, 1759

1st Series.—Miscellanea Taurinensia. 4 vols. Tom. ii. al tom. v. sono per gli anni 1760 al 1773. Questi 4 vol. hanno anche il titolo " Mélanges de Philosophie," &c. (Tom. i. *is the above Miscellanea Philosophica.*)

2nd Series.—Mémoires de l'Acad. Royale des Sciences. 6 vols. per gli anni 1784 al 1799 (no printed dates).

3rd Series.— Mémoires de l'Acad. des Sciences, Littérature, et B. Arts de Turin. 10 vols. Per gli anni 1802-3 al 1812-13.

Turin Acad., Miscellanea philosophica.—*continued.*

4th Series.—Mémoires de l'Acad. Royale des Sciences de Turin. 11 vols.

(*Note.—The vol. for* 1813 *and* 1814 *was printed in* 1816, *and is called* vol. xxii. *The ten following vols. are numbered* xxiii. *to* xxxii. *and were printed in various years between* 1818 *and* 1828 *inclusive, and continued to* 1835.)

Another Series begins in 1839. Vols. i. and ii. are dated 1839 and 1840.

† **Turini, P.** Considerazioni intorno all' Elettricità delle Nubi. 4to. 68 pp.
Venezia, 1780

Turkeim, W. Über Electricität Gym. Prgm. Breslau, 1834 ?

Turnbull, L. Lectures on the Electro-Magnetic Telegraph. 8vo.
Philadelphia, 1852

† Electro-Magnetic Telegraph ; with an historical account of its rise, progress, and present condition. 2nd edition, revised, &c. 8vo. 264 pp. 2 plates.
Philadephia, 1853

† **Turnbull and McRea.** Railway Accidents, and the Means by which they may be prevented, by the use of the Electro-Magnetic Safety Apparatus. 8vo. 64 pp. Philadelphia, 1854

Turner, Ed. Elements of Chemistry by Liebig and Gregory. 8vo.
London, 1847

† **Tuxen, J. C.** Die Deviation der Compassnadel, &c. 8vo. 66 pp. Stettin, 1856

† **Tyer, Ed.** Compagnie des Signaux électriques. Système Tyer. 8vo. 22 pp. (*Ext. Journal l'Ingénieur.*) Paris, 1855

Tylee, J. P. Observations on Galvanism. 12mo. London, 1848

† **Tyndall, John.** On Diamagnetism and Magne-crystallic Action. 8vo. 24 pp. (*From Phil. Mag. for* September, 1851.) London, 1851

† On the Laws of Magnetism. 8vo. 31 pp. (*Phil. Mag. for* April, 1851.)
London, 1851

On the Reduction of Temperatures by Electricity. 8vo. (*From Lieber's Cat.* 1865.) London, 1852

The Polymagnet. 8vo. Plate. (*From Lieber's Cat.* 1865.) London, 1855

Further Researches on the Polarity of the Diamagnetic Force. 4to. (*From Lieber's Cat.* 1865.) London, 1855

On the Nature of the Force by which Bodies are repelled from the Poles of a Magnet. 4to. (*From Lieber's Cat.* 1865.) London, 1855

† On the Relation of Diamagnetic Polarity to Magne-crystallic Action. 8vo. 13 pp. (*Phil. Mag. for* February, 1856.) London, 1856

† Researches on Diamagnetism and Magne-crystallic Action ; including the Question of Diamagnetic Polarity. 8vo. 361 pp. London, 1870

† Notes of a Course of Seven Lectures on Electrical Phenomena and Theories, delivered at the Royal Institution of Great Britain, April-June, 1870. 8vo. 40 pp. London, 1870

† **Tyndall and Knoblauch.** On the Deportment of Crystalline Bodies between the Poles of a Magnet. 8vo. 6 pp. (*Phil. Mag. for* March, 1850.)
London, 1850

(*Vide* also Knoblauch and Tyndall.)

U.

Udine Acad. Atti dell' Accad. di Udine . . 1844-5 dal Zambra. 8vo. (*From Giorn. I. R. Istit. Lomb.* xii. 378.) 1845

Unger, F. Mikroskopische Untersuchung des atmosphärischen Staubes von Gratz. 8vo. 7 pp. 5 plates. (*Sitsb. d. K. Akad. d. Wissens. Wien*, 1849, ix. u. x. Heft. p. 230.) *Wien*, 1849

Über Krystallbildungen in dem Pflanzensetzling. 4to. 60 pp. 7 plates. (*Ann. des Wien. Museum*, ii. Bd.) *Wien*

Unger, Johann Friedrich von. *Born* 1716, *at Brunswick ; died February* 8, 1781, *at Brunswick.*

Abhandlung von der Natur der Electricität. 4to. 1745
(*Note.—Poggend. says Zwei Abhandl. von d. Natur d. Elect. ; von d. Acad. in Berlin zum Druck befördert.* 4to. *Berlin*, 1745. *Die eine Abhandl. von T. G. Teske.*)

Von d. Elektricität. 8vo. (*Gelehrt Hanover Anzeig*, 1750.) *Hanover*, 1750

Vom Uebergang d. elektr. Materie aus e. Körper in d. anderen. 8vo. (*Hamb. Magaz.* viii. 1751.) *Hamburg*, 1751

Entwurf einer Maschine, wodurch alles was auf d. Clavier gespielt wird sich von selber in Noten setzt u.s.w. 4to. *Braunschweig*, 1744

Ungerer. (*Vide* Miege and Ungerer.)

† **Unsgaard,** T. J. Reglement for afbenyttelsen af de danske Statstelegraphlenier. 4to. 15 pp. *Kjobenhavn*, 1857

Unterberger, Leopold Freiherr von. *Born October* 12, 1734, *at Strengberg, Nieder-Oesterr ; died February* 9, 1818, *at Vienna.*

Nützl. Begriffe von d. Gewittermaterie, nebst Beobachtungen üb. die beste Art, Blitzableiter anzulegen. 8vo. (*See next.*) *Wien*, 1811

Nützliche Anmerkungen von den Wirkungen der Electricität und Gewittermaterie. 8vo. *Wien*, 1811

† **Unverdorben,** F. X. Über das Verhalten des Magnetismus zur Wärme. (Inauguralschrift). 8vo. 78 pp. 1 plate. *München*, 1866

† **Upington,** H. Account of an Electrical Increaser for the unerring manifestation of small portions of the Electric Fluid. 8vo. 6 pp. 1 plate. (*Phil. Mag.* lii. 47.) *London*, 1818

Upsal. Acta nova regiæ societatis scientiarum Upsaliensis. Ser. iii. 4to. *Upsaliæ*

Urban, C. G. Widerlegung gewisser Vorurtheile, welche noch b. Gewittern herrschen. 8vo. *Eisenach*, 1792

† **Urbanski,** A. Theorie des Potenzials u. dessen Anwendung auf Elektricität. 8vo. 142 pp. ("*Zweite Ausgabe der Vorträge über höhere Physik, erste Abtheilung.*") *Berlin*, 1864

† **Ure.** Experiments on the Body of a Criminal executed at Glasgow on the 4th Nov. 1818. An account read at the Glasgow Literary Society on Dec. 10, 1818. 8vo. 6 pp. *Phil. Mag.* liii. 56.) *London*, 1819
(*Note.—This is not the whole af the paper read. It seems to have been copied from the Journal of Science and the Arts*, No. xii.)

Ussher, Henry. Aurora Borealis seen in Full Sunshine. 4to. (*Trans. Inst. Acad.* 1788.) *Dublin*, 1788

† **Usiglio,** C. (Estratto). Alquante osservazioni sul Galvanismo per curare alcune malattie degli occhi. 8vo. 1 page. (*Giorn. I. R. Istit. Lomb.* ix. 12.) *Milano*, 1844

† **Utting,** J. Aurora Borealis? Particulars of the Meteor visible on the 29th September, 1828. Extracted from a Note to the Editor of the *Norwich Mercury*. Dated Lynn Regis, October 1, 1828. 8vo. 2 pp. (*Phil. Mag. or Annals*, iv. 393.) *London*, 1828

V.

Vacca Berlinghieri. (*Vide* Berlinghieri.)

† **Vail, Alfred.** The American Electro-Magnetic Telegraph; with the Reports of Congress. (1st issue, 1845.) 8vo. 208 pp. Cuts. *Philadelphia, 1847*

Le Télégraphe électro-magnétique Américain avec le rapport du Congrès et la description de tous les télégraphes connus, où sont mis en usage l'électricité et le galvanisme; trad. de l'Anglais par H. Watteman. 8vo. 263 pp.

† Gründliche Darstellung d. elektr.-magnet. Telegraphen, nach dem System des Prof. Morse . . Aus dem Englischen übersetzt von C. Gerke. 8vo. 1 plate. 27 pp. *Hamburg, 1848*

† Le meraviglie della Telegrafia elettrica ossia Telegrafo elettro-magnetico Americano . . Opera di A. Vail . . Voltata in Italiano da Lorenzo Polettini. Con note ed aggiunte. 4to. 169 pp. 5 plates. 1850

Vaillant. Voyage . . sur la Bonite pend. 1836 et 1837. 8vo. *Paris, 1840-46*

† **Valentin, G. G.** Beiträge zur Anatomie des Zitteraales (Gymnotus elect.). 4to. 74 pp. 3 plates. (*Aus dem 6 Bande d. Neuen Denkschriften d. allgem. Gesellsch. für d. gesamm Naturwissenschaft, besonders abgedruckt.*) *Neuchatel, 1841*

Ueber d. Möglichkeit d. Stimmungsrichtung eines galvan. Froschpräparates wilkuhrlich umzukehren. (*Vierordt's Archiv.* 1853.) 1853

Neue Untersuchungen über d. Polarisations Ercheinungen d. Krystall-linsen d. Menschen u. d. Thiere. *Arch. f. Ophthalm.* iv. 1859.) 1859

Valentini, Ch. B. Dissertatio inaug. de lapide ceraunio, vulgo von der Donner-Axt, et de fulmine tactis. 4to. *Gissæ, 1717*

Valerius, H. Mémoire sur l'emploi de l'électricité en médecine. 8vo. 83 pp. (*Ann. de la Soc. de Méd. de Gand,* xxix. 115.) *Gand, 1852*

Valerius Maximus. Quoted by Anobius (or Numa.) (*From Dutens,* p. 150.) A.D. 32

† **Valicourt.** Nouveau Manuel complet de Dorure et d'Argenture par la méthode électro-chimique et par simple immersion. . . Trad. de l'Italien et augmenté. 18mo. 173 pp. (*The translator quotes a Manuel complet de Galvano-plastie, nouvelle éd.* 1845.) *Paris, 1845*

(*Note.—This is a translation of the Manuel published at Reggio by Selmi.*)

† Nouveau Manuel complet de Galvano-plastie. 18mo. 2 vols. 2 plates. *Paris, 1854*

Valicourt et Smee. Manuel complet de Galvano-plastie. Nouvelle éd. 18mo. *Paris, 1854*

Vallemont, Pierre le Lorrain. *Born September 10, 1649, at Pont Audemer (Eure); died December 30, 1721, at Pont Audemer.*

La physique occulte ou traité de la baguette divinatoire. *Paris, 1693*

† Description de l'aimant qui s'est formé à la pointe du clocher neuf de Notre Dame de Chartres. 12mo. 215 pp. *Paris, 1692*

Valli, E. Procès-verbal des expériences de M. Valli sur l'électricité animale. (*La Médecine éclairée par les sciences physiques,* iv. 66. "*L'Acad. a chargé MM. Leroi, Vicq d'Azyn, et Coulor de répéter ces Exp. avec M. Valli. Les principales ont été faites . . le 12 Juillet,* 1792.) *Paris, 1792*

9 Lettres sur l'électricité animale. 4to. (*Journ. de Phys.* tom. xli. pp. 66, 72, 185, 189, 193, 197, 200, 435; et tom. xlii. p. 74.) *Paris, 1792-3*

Discours sur le sang, considéré dans l'état de santé, et de maladie, avec quelques expériences relatives. (*From Sue, Hist. du Mag.* vol. i. p. 45.) 1792

Lettera sull' Elettricità animale ad un suo amico. 4to. 15 pp. (*Volta's copy.*) *Pavia, 1792*

Valli, E.—*continued.*

† Lettere sull' Elettricità animale ad un suo amico. (Pavia, 5 Ap. 1792.) 8vo. 25 pp. *Torino,* 1792

† Experiments on Animal Electricity, with their application to Physiology. 8vo. 323 pp. *London,* 1793

† Lettera al Brugnatelli sull' Elettricità Animale. (Data) Mantova, 8 Settem. 1794. 8vo. 5 pp. (*Ann. di Chim. di Brugnatelli,* vii. 40.) *Pavia,* 1795

† Lettera all' Aldini. (Movimenti del cuore, &c.) 8vo. 5 pp. (*Giorn. Fis. Med. di Brugnatelli,* anno 8, tom. i. p. 264.) *Pavia,* 1795

† Lettera xi. sull' Elettricità Animale. Data 15 Ottobre, 1794, Mantova. 8vo. 15 pp. (*Ann. di Chim. di Brugnatelli,* vii. 213.) *Pavia,* 1795

† Lettera xii. indirizzata al . . Brugnatelli, sull' Elettricità animale. (Senza data.) 8vo. 11 pp. (*Ann. di Chim. di Brugnatelli,* vii. 228.) *Pavia,* 1795

Lettres sur le Galvanisme ou électricité animale. (*Soc. Philomat.* tom. i. pp. 27, 31, 43.) *Paris*

His Experiments tried by a Commission of the Paris Academy, viz. Leroi, Vicq d'Azyr, Coulomb, and Fourcroy. *Paris*

Vallisnieri, Antonio. *Born May 3, 1661, at Trasilico, in Modenesischen; died Jan. 18, 1730, at Padua.*

† Due Lettere . . sopra alcune cose di Storia-naturale. 12mo. 49 pp. (*Calogera, Raccolta d'Opuscoli,* tom. ii. p. 1. *Fuoco lambente, Luce degli Animali e fuochi Fatui. Fuoco volante incend. &c.*) *Venezia,* 1729

De pluvia lapidea. (*Ephemer. Acad. Nat. Curios-Cent.* 5 e 6, p. 195.)

Vallot. Note sur un mém. de M. D'Agout constatant la périodicité des Etoiles filantes. 8vo. (*Bord. Acad.* Actes 2me an. p. 689.) *Bordeaux,* 1840

Rapport qui existe entre la pierre nommée par Pline Calcédoine, et la Tourmaline. *Bordeaux, Bourges*

Valvasor, J. W. Account of the Zirknitz Sea. 4to. (*Phil. Trans.* 1687.) *London,* 1687

Lacus cirkniensis potiora phænomena ex principiis physicis et mathematicis explanata. (*Acta Erudit.* 1689.) *Lypsiæ,* 1689

Van Barneveld. (*Vide* Barneveld.)

Van Charante. (*Vide* Charante.)

Van der Kolk. (*Vide* Kolk.)

Vanderlot (Cayenne). A little work on the Surinam Eel. (*Humboldt, Voyage Zoologie,* p. 88, says, " un mémoire sur les propriétés médicales des Gymnotes électriques.*")

Van Marum. (*Vide* Marum.)

Van Musschenbroek. (*Vide* Musschenbroek.)

Note.—All names having the prefix VAN (OR VAN DER) *will be found entered under the first letter of the second portion of the name as above.*

† **Varley, C.** On Atmospheric Phænomena : particularly the Formation of Clouds ; their Permanence ; their Precipitation in Rain, Snow, and Hail ; and the consequent rise of the Barometer. 8vo. 7 pp. (*Phil. Mag.* xxvii. 115.) *London,* 1807

† Meteorological Observations on a Thunder Storm ; with some remarks on Medical Electricity. 8vo. 3 pp. 1 plate. (*Phil. Mag.* xxxiv. 161.) *London,* 1809

† Further Remarks on Thunder Storms. 8vo. 2 pp. (*Phil. Mag.* xxxiv. 201.) *London,* 1809

† **Varley,** Cromwell Fleetwood. Description of the Translating Apparatus and Universal Galvanometer invented by C. F. Varley. 8vo. 8 pp. 5 separate woodcuts. *London,* 1863

† Instructions for the Use of C. F. Varley's Universal Testing Apparatus. 8vo. 12 pp. *London,* 1863 ?

† The Phenomena of the Atlantic and the other Long Cables. Fol. 2 pp. (*Engineer,* Feb. 22, 1867.) *London,* 1867

† **Varley.** S. A. On the Electrical qualifications requisite in Long Submarine Telegraphic Cables. 8vo. (*Proceedings of the Institute of Civil Engineers,* xvii. 1857-58.) *London,* 1858

Varro, Terentius. De lingua Latina. (*From Dutens,* p. 150.) (*Numa's religious ceremonies : Jupiter Elicias.*) B.C. 115

Vasari, G. Opere. 8vo. Plates. 2 vols. *Firenze,* 1832-8

Vasquez-y-Morales. Ensayo sobre la Electricidad . . iscrito en idioma francès por . . Nollet . . traducido en Castellano por D. Jos. Vasquez-y-Morales : añadida la Historia de la Elect. 4to. 1 plate. 131 pp. *Madrid,* 1747

Vassalli-Eandi, Antonio Maria. *Born January 30, 1761, at Turin ; died July 5, 1825, at Turin.*

† Memoria sopra il Bolide degli xi. Sett. 1784 e sopra i bolidi in generale. 12mo. 113 pp. *Torino,* 1786

Dell' influsso dell' elettricità nella vegetazione, e dell' azione della vegetazione sopra l'aria. 8vo. (*Mem. della Soc. Agraria di Torino,* tom. i.) *Torino,* 1786

Spiegazione delle esperienze contra l'influsso dell' elettricità nella vegetazione dei Sig. Ingenhousz e Schwankhardt. 8vo. (*Mem. della Soc. Agraria,* tom. i. 1786 ?) *Torino,* 1786 ?

Esame della "Elettricità delle Meteore" del Bertholou. (*Biblioteca Oltremontana,* tom. ix. (*sic :* q. error.) *Torino,* 1787

Esame della teoria sull' elettricità e sopra il magnetismo dell' Ab. Hauy. (*Biblioteca Oltremontana,* tom. ii. e iii. ?) *Torino,* 1788

† Sperienze elettriche sopra l'acqua e sopra il ghiaccio. 4to. 14 pp. (*Mem. Soc. Ital.* iv. 263.) *Verona,* 1788

Aghi calamitati, che non soffrono alcuna declinazione. (*Opusc. Scelti,* xix. 216.) *Torino,* 1788

† Memorie fisiche. 8vo. 143 pp. *Torino,* 1789

Osservazioni sull' agghiacciamento dell' acqua elettrizzata. (*Giornale Scientif. Letter. d'una Soc. Filosofica di Torino,* tom. i.) *Torino,* 1789

Sperienze sopra l'influsso dell' elettricità nel colore dei vegetabili. (*Giornale Scientifico d'una Soc. Filosofica di Torino,* tom. iii.) *Torino,* 1789

† Lettere fisico-meteorologiche dei . fisici Senebier, De Saussure, e Toaldo con le risposte di A. M. Vassalli. 8vo. 223 pp. *Torino,* 1789

† Lettera sopra diversi argomenti di fisica diretta al Sig. Brugnatelli. 8vo. 3 pp. (*Bibl. Fis. d'Europa,* tom. xvii. p. 144.) *Pavia,* 1790

† Articolo di Lettera del . Vassalli al Brugnatelli sopra l'elettricità di diversi corpi, ed altri argomenti di Chimica. 8vo. 5 pp. (*Ann. di Chim. di Brugnatelli,* i. 53.) *Pavia,* 1790

Sperienze elettriche . acqua e ghiaccio. (*In Mem. della Soc. Ital.* tom. iii.) 1790

Theses ex universa philosophia selectæ. 4to. *Derthonæ,* 1790

† Saggio sopra i principali fenomeni della Meteorologia del Sig. Monge, colle riflessioni del Sig. Ab. A. M. Vassalli. 8vo. *Torino,* 1791

† Conghietture sopra l'arte di tirare i Fulmini appo gli Antichi. 8vo. (*Opuscoli Scelti di Milano in* 4to. tom. xiv.) 1791 ?

Vassalli-Eandi, Antonio Maria.—*continued.*

Sul colore dei Vegetabili. Lettera di P. F. con la risposta di A. M. Vassalli,
1791. 8vo. *Torino,* 1791
(*Note.—Vassalli published his opinion that colour in vegetables was, in some
measure, due to Electricity.*)

† Expériences électrométriques. 4to. 36 pp. 1 plate. (*Mem. Accad. Torino,* x. 57.
Lu le 15 Déc. 1790.) *Turin,* 1792

† Articolo di lettera al Brugnatelli sopra diversi argomenti. 8vo. 5 pp. (*Giorn.
Fis. Med.* ii. 110.) *Pavia,* 1792

Physica experimentalis lineamenta. (*Vide* Eandi.) 1793-4

Lettre à J. Buniva . . sur l'électricité animale. 8vo. (*Recueil périodique de
Littérature médicale étrangère, de Sedillot,* ii. 266.) *Paris ?*

† Lettera al C. Amoretti (del 30 Nov. 1796, Torino) I metodi di preparare
aghi calamitati che non soffrano alcuna declinazione, &c. 4to. 2 pp.
(*Opuscoli Scelti,* xix. 215, &c.) *Milano,* 1796

Lettera allo Spallanzani sopra i suoi viaggi alle due Sicilie. 4to. (*Bibliot.
Italiana.*) *Milano,* 1797

† Sopra alcuni istrumenti meteorologici che segnano, per se stessi, le variazioni
atmosferiche per 24 ore o più. Ricevuta 3 Mag. 1799. 4to. 5 pp. 1 plate.
(*Ital. Soc. Mem.* viii. pt. ii. p. 516.) *Modena,* 1799

Lettre sur le Galvanisme et l'origine de l'électricité animale. 4to. (*Journal de
Phys.* an. 7, p. 336.) *Paris,* 1799

Sur les phénomènes de la Torpille. 4to. (*Journal de Phys.* xlix. 69.) *Paris,* 1799

Sur le Vitalomètre. 4to. (*Journal de Phys.* l.) *Paris,* 1800

Essai sur l'utilité des conducteurs électriques, prouvée par la foudre tombée . .
sur l'hôtel de M. le Marquis Graneri. 4to. (*Mem. dell' Accad. di Torino,* vi.
part i. p. 57.) *Torino,* 1800

† Expériences galvaniques sur les décapités. 4to. *Turin,* 1802

† Rapport présenté à la Classe des Sciences exactes de l'Acad. de Turin . 2 Ne-
vôse, an. xl. (1803), sur l'action du Galvanisme, et sur l'application de ce
fluide et de l'électricité à l'art de guérir. 4to. 12 pp *Turin,* 1803 ?

† Rapporto fatto alla Classe delle Scienze esatte dell' Accad. di Torino nella
Sessione de' 2 Nevôso, an. xi. (Dec. 23, 1802) sull' azione del Galvanismo, e
sull' applicazione del fluido galvanico e dell' elettricità all' arte di guarire.
4to. 8 pp. (*Opusc. Scelti,* xxii. 76.) *Milano,* 1803

† Report presented to the Class of the Exact Sciences of the Academy of Turin,
January 12, 1803, on the action of Galvanism, and the application of this
fluid and of Electricity to Medicine. 8vo. 7 pp. (*Phil. Mag.* xv. 319.)
London, 1803

. Saggio sopra il fluido galvanico. Ricevuto 21 Marzo, 1803. 4to. 33 pp. Carta.
(*Mem. Soc. Ital.* x. part ii. p. 733.) *Modena,* 1803

† Lettera sopra la natura del fluido Galvanico al Senebier. Ricevuta 1 Luglio,
1803. 4to. 3 pp. (*Ital. Soc.* x. 802.) *Modena,* 1803

† Expériences galvaniques, ou Notice de la dernière séance du cours public des
expériences physiques, faites dans le théâtre physique et anatomique des
Ecoles Spéciales de Turin. 8vo. 36 pp. (*Bibliothèque Italienne par Giulio
Gioberti, &c.* tom. ii. p. 25.) *Turin,* 1803

† Notice d'un météorographe. Lu le 25 Brumaire, an 12. 4to. 19 pp. 2 plates.
(*Mém. de Turin,* an x. *et* xi. 1802-3 (*printed* 1804), p. 426. *At* p. 443, *Action
du Céraunographe et de l'Electromètre.*) *Turin,* 1804

† Ricerche sulla natura del fluido-galvanico. Letta nell' Accad. di Torino ai
30 Settem. 1804, " tradotto poco meno che letteralmente." 4to. 13 pp.
(*Nuova Scelta d'Opuscoli,* i. 167.) *Milano,* 1804

† Sur la vitesse du fluide galvanique. 8vo. (*Bibl. Italien. par Giulio, Gioberti, &c.*
vol. i. p. 128.) *Torino,* 1804

Vassalli-Eandi, Antonio Maria.—*continued.*

Notice sur la vie et les ouvrages d'Eandi (T. A. F. J.), par A. M. Vassalli-Eandi. (*Vide* Eandi, 1804.) (*Mém. de Turin,* xii. 1.) 1804

† Expériences et observations sur le fluide de l'électromoteur de Volta. Lu le 30 Floréal, an 9.) 4to. 34 pp. (*Mém de Turin,* ann. x. *et* xi. 1802-3, *printed in* 1804.) *Turin,* 1804

Recherches sur la nature du fluide galvanique. Lu le 3me jour complémentaire an 11. 4to. 31 pp. (*Méms. de Turin,* Ann. 12 *et* 13, p. 144, tom. xiv.) *Turin,* 1805

† Lettera sopra la costruzione del cervovolante, e la maniera di servirsene. (*Giornale di Torino.*) 1805

† Saggio di un trattato di Meteorologia Ricevuto 26 Nov. 1805. Proemio Storico. 4to. 15 pp. (*Ital. Soc. Mem.* xiii. part ii. p. 85.) *Modena,* 1807

Sur le tremblement de terre, 1808. 8vo. *Pavia ?* 1808

Résultat des ebservations météorol. faites à Turin de 1787 à 1807. 1809

Nota sopra un mezzo facile di preservare le case rustiche dal Fulmine. (*Calend. Georg.* 1814.) 1810

*† Saggio di un Trattato di Meteorologia. Memoria. Ricevuta li 19 Dec. 1814. 4to. (*Ital. Soc. Mem.* xvii. *Parte Fisica,* p. 230.) *Verona,* 1815

*† Indice degli Autori e delle Materie dei volumi della R. Accad. delle Scienze dal 1759 al 1815. 4to. (*Mem. dell'Accad. di Torino,* tom. xxii.) *Torino,* 1816

† Notizia sopra la Vita, &c. di Beccaria. (*Vide* Beccaria.) 1816

*† La Meteorologia Torinese : ossia Risultamento delle Osservazioni fatte dal 1757 al 1817. (*Mem. dell' Accad. di Torino,* tom. xxiv.)

*† Nota sulla virtù igrometrica dei capelli delle Mummie. '(*Mem. dell' Accad. di Torino,* tom. xxix.)

*† Nota sopra le straordinarie variazioni del Barometro, &c. . nel 1821. (*Mem. dell' Accad. di Torino,* tom. xxvii.)

† Memorie istoriche intorno alla Vita ed agli studii di Gianfrancesco Cigna. 4to. (*Mem. dell' Accad. di Torino,* tom. xxvi. p. xiii.) *Torino,* 1821

*† Nota sopra lo straordinarissimo abbassamento del Barometro . 2 Feb. 1823 ; e sopra un fenomeno in alcuni Pozzi d'acqua viva. (*Mem. dell' Accad. di Torino,* tom. xxvii.)

† Descrizione di un nuovo Atmidometro per misurare l'evaporazione dell' acqua, del ghiaccio, e di altri corpi a varie temperature. Ricevuta Aprile 29, 1823. 4to. 7 pp. 1 plate. (*Ital. Soc. Mem.* xix. 347.) *Modena,* 1823

† Saggio sulla vita e sugli scritti del Prof. A. M. Vassalli-Eandi scritto dal di lui Nipote Secondo Berrutti. 8vo. 198 pp. *Torino,* 1825

† Notizie biografiche del Professore Abate Vassalli-Eandi, Membro e Segretario perpetuo della Reale Accademia delle Scienze di Torino: Raccolte dal . . Carena. Lette 4 Dicem. 1825. 4to. 23 pp. (*Mem. di Torino,* tom. xxx. p. xix.) *Torino,* 1826

† Elogio : scritto dal Berrutti. Ricevuto 11 Dec. 1837. 4to. 32 pp. (*Ital. Soc. Mem.* xxii. p. liv.) *Modena,* 1832
(*Vide* also Eandi.)

† **Vassalli-Eandi, Giulio,** e **Rossi.** Rapport présenté à la Classe des Sciences exactes de l'Acad. de Turin, le 27 Thermidor, sur les expériences galvaniques faites, les 22 et 26 du même mois, sur la tête et le tronc de trois hommes, peu de temps après leur décapitation. 4to. 19 pp. (*Volta's copy.*) *Turin,* 1802

† Transunto del Rapporto fatto all' Accad. delle Scienze di Torino ai 27 Termidoro anno x. sulle sperienze Galvaniche fatte ai 22 e 26 dello stesso mese sul tronco di tre uomini poco dopo la loro decapitazione. 4to. 6 pp. (*Opusc. Scelti,* xxii. 51.) *Milano,* 1803

(*Vide* also Giulio, &c.)

Vassalli-Eandi, Giulio e Rossi.—*continued.*

† Report presented to the . Acad. of Turin, August 15, 1802, in regard to the Galvanic Experiments made by them on the 10th and 14th of the same month, on the head and trunk of three men a short time after their Decapitation. 8vo. 8 pp. (*Phil. Mag.* xv. 38.)   *London,* 1803

† Vassalli-Eandi, Rossi, et Michelotti. Précis de nouvelles expériences galvaniques. Lu 28 Juillet, 1805. 4to. 7 pp. (*Mém. de Turin,* ann. 1805-1808, p. 160.)   *Turin,* 1809

Vatlon (Watson?). Essai sur les causes de l'électricité. (*From Mangin.*)

† Vauquelin, C. On Stones supposed to have fallen from the Clouds, (and discussion thereon) in the French National Institut. (Read by Lacroix.) 8vo. 2 pp. (*Phil. Mag.* xv. 187.)   *London,* 1803

† Memoir on the Stones said to have fallen from the Heavens. Read in the French National Institute. 8vo. 8 pp. (*Phil. Mag.* xv. 346.)   *London,* 1803

Vauquelin, L. N. Mémoire sur les pierres dites tombées du ciel. 8vo. (*Journ. des Mines,* xiii. 1802-3.)   *Paris,* 1802-3

Vauquelin, Fourcroy, et Thenard. Nouvelles expériences galvaniques. 4to. (*Mém. des Soc. Savantes et Litter.* i. 204.)   *Paris,* 1801

Vedova. (*Vide* Kreil and Vedova.)

Vegelin, J. H. De materia magnetica, ejusque actione in ferrum et magnetem. 4to.   *Frankfort,* 1763

† Veladini, G. (Rivista) Misura assoluta . di Gauss tradotto . da Frisiani e Osservazioni sull' intensità, &c. della forza magnetica di Kreil e Vedova. 8vo. 26 pp. (*Giorn. I. R. Istit.* vi. 207.)   *Milano,* 1843

*† Sulla prima applicazione del pendolo. (*Giorn. I. R. Istit.* Lomb. Nuova Ser. 4to. vi. 191.)

† Vene, A. Essai sur une nouvelle théorie de l'électricité. 8vo. 118 pp. 1 plate. *Arras,* 1820

Venetian Athenæum. Esercitazioni scientifiche e letterarie dell' Ateneo di Venezia. 4to. Tom. i. 1827; tom. ii. 1838; tom. iii. 1839; tom. iv. 1841; tom. v. 1846; tom. vi. 1847; tom. vii. 1855, 1860, &c. *Venezia,* 1827 *to —*
    (*Note.—This collection contains many notices, &c. relative to Electricity and Magnetism, &c. in "Relazioni" by various members. There are only two entire Memoirs in the collection, viz. those of Marianini in the first volume.*)

Venetian Imperial Royal Institution. Atti delle Adunanze dell' I. R. Istituto Veneto di Scienze Lettere ed Arti. 8vo. 1st Series, 1841—1848, 7 vols. ; 2nd Series, 1850—1855, 6 vols. ; 3rd Series, 1856—1859, 4 vols. ; and continued after 1859. The vols. have no designating number in the title-pages.
                                                        Venezia, 1841 *to —*
    (*Note.—This collection contains many Articles relative to Electricity and Magnetism, &c., some few of which are original, entire memoirs, &c. ; but the principal number consist of accounts, more or less abridged, of memoirs, &c. read, and of conversations held at Meetings, and many of these have been printed elsewhere.*)

Memorie dell' I. R. Istituto Veneto di Scienze, Lettere, ed Arti. 4to. Vol. i. 1843 ; vol. ii. 1845 ; vol. iii. 1847 ; vol. iv. 1852 ; vol. v. 1855 ; vol. vi. 1856 ; vol. vii. 1858 ; vol. ix. 1860.   *Venezia,* 1843 *to —*

Venetian Lombardy Imperial Royal Institution. Memorie dell' Imperiale Regio Istituto del Regno Lombardo-Veneto. 5 vols.? 4to.
                                                        Milano, 1819 *to* 1838 ?

Venturoli, Gius. Sull' elettricità atmosferica. (*Opusc. Scientif. di Bologna,* ii. 1818.)   *Bologna,* 1818

Vérard de Sainte Anne, M. Télégraphie Electrique, lignes d'Europe, d'Asie, d'Afrique, d'Océanie, d'Amérique, &c. Section de Mossoul à Haïderabad, de Calcutta à Bangkok et Singapour. Mémoire présenté à l'Institut Impérial de France, aux gouvernements des cinq parties du monde. 4e édition, augmentée et précédée d'une préface par L. Giraudau. 8vo. 45 pp.
                                                        Paris, 1862

Verati. Storia del Magnetismo animale. 4 vols. 8vo. 1845-6

Veratti, Giuseppe. *Born January 30,* 1707, *at Bologna ; died March* 24, 1793, *at Bologna.*

† De Aurora boreali anni 1732. 4to. 3 pp. (*Comment. de Bonon. Scient. Instit.* tom. ii. pars iii. p. 493.) *Bononiæ,* 1747

† Osservazioni fisico-mediche intorno alla elettricità. 8vo. 143 pp. *Bologna,* 1748

† Observations physico-médicales sur l'électricité ; auxquelles on a joint des expériences faits à Montpellier, pour guérir les paralytiques. Trad. 12mo. 151 pp. *La Haye,* 1750

Osservazione fatta in Bologna l'anno 1753 dei fenomeni elettrici nuovamente scoperti in America e confermati in Parigi. 4to. 4 pp. *Bologna* (*Note.—Troili,* p. 39, *refers to Experiments on Elettricità celeste made by Veratti in* 1752, *and described in the Storia Letteraria,* vol. vi. lib. iii. cap. iii. p. 686. *In* vol. vi. p. 686, 1854, *is noticed " un' Osservazione di Veratti fatta in Bologna l'anno* 1752, *dei fenomeni nuovamente scoperti in America, e confermati in Parigi."*)

† De electricitate cælesti. 4to. 5 pp. (*Comment. de Bonon. Scient. Instit.* tom. iii. p. 200.) *Bononiæ,* 1755

† De electricitate medica. 4to. 25 pp. (*Comment. de Bonon. Scient. Instit.* tom. iii. p. 454.) *Bononiæ,* 1755

† Experimenta magnetica. 4to. 14 pp. (*Comment. de Bonon. Scient. Instit.* tom. vi. p. 31. *Bononiæ,* 1782

De animalibus electrico ictu percussis. 4to. (*Comment. de Bonon. Scient. Instit.* vii. c. p. 41.) *Bononiæ,* 1791 (*Vide* also Anon. Med. Elect. 1763.)

Verdeil, François. *Born* 1747, *at Berlin ; died February* 21, 1832, *at Lausanne.*

Mémoire sur les brouillards électriques, vus en Juin et Juillet, 1783, et sur le tremblement de terre arrivé à Lausanne le 6 Juillet de la même année. 4to. (*Mém. de Lausanne,* tom. i. p. 110.) *Lausanne,* 1784

Observations et expériences faites à l'occasion d'un coup de foudre tombé sur l'église cathédrale de Lausanne. 4to. (*Mém. de Lausanne,* tom. i. p. 158.) *Lausanne,* 1784

Verdet, Marcel Emile. *Born March* 13, 1824, *at Nîmes, Département du Gard.*

Note sur le Mémoire de Masson relatif à la constitution des courants induits de divers ordres. 8vo. (*Ann. de Chim. et Phys.* liii. 1858.) *Paris,* 1858

† Thèse . . Recherches sur les Phénomènes d'Induction produits par les décharges électriques. 4to. 42 pp. *Paris,* 1848

Note sur les courants induits d'ordre supérieur. Communiqué à la Soc. Philom. 8 Déc. 1849. 8vo. 6 pp. (*Ann. de Chim.* 3e série, tom. xxx.) (*Arago's copy.*) *Paris,* 1852

Sur les courants d'ordre supérieur. 8vo. (*Ann. de Chim.* 3e série, xxix. 501.) *Paris*

† Récherches sur les Phénomènes d'Induction produits par le mouvement des métaux magnétiques ou non magnétiques. Présenté à l'Acad. 1850. 8vo. 31 pp. 1 plate. (*Ann. de Chim.* 3e série, xxxi. 187.) *Paris,* 1851

† Récherches sur les propriétés optiques développées dans les corps transparents par l'action du magnetisme. 8vo. 1re partie. 44 pp. 1 plate. 2e partie. 8 pp. 1 plate. 3e partie. 35 pp. (*Ext. des Ann. de Chim. et Phys.* 3e série, vols. xli. xliii. *et* lii.) *Paris,* 1854-5-8

Verdu, G. Nouvelles expériences pour mettre le feu aux fourneaux des mines au moyen de l'électricité. Rapport de M. le Maréchal Vaillant. 4to. (*Comptes Rendus,* xxxviii. 801-4.) *Paris,* 1854

Note relative à de nouvelles expériences sur l'application de l'électricité à l'explosion des mines militaires. 4to. (*Comptes Rendus,* xxxviii. 1024-26.) *Paris,* 1854

Verdu, G.—*continued.*

Nouvelles mines de Guerre appliquées à la défense, suivant un nouveau procédé pour mettre le feu aux fourneaux de poudre, à l'aide de l'électricité. Traduit de l'Espagnol. 8vo. *Paris et Bruxelles,* 1855

Verdu and Ruhmkorff. Mémoire sur de nouvelles expériences pour mettre le feu aux fourneaux des mines au moyen de l'électricité. 4to. (*Comptes Rendus,* xxxvi. 649-652.) *Paris,* 1853

Verney, De. "Irritando, col coltello anatomico, i nervi d'una rana morta, osservò, sino dall' anno 1700, delle convulsioni simili alle surriferite" (viz. those obtained by Galvani in 1791). 4to. (*Hist. de l'Acad. Paris ; et Gerbi, Corso Elementare di Fis.* tom. iii. p. 314, Pisa 1828.) *Paris,* 1823

† **Vernier,** De la distribution de l'électricité à la surface des corps conducteurs. 4to. 22 pp. *Paris,* 1824

Verona Accad. di Agricultura. Memorie dell' Accad. di Agricultura, Commercio, ed Arti di Verona. 8vo. *Verona,* 1807 *to* 1859 ?

Verona e Modena Soc. Italiana. Memorie di Matem. e Fisica. (*Vide* Italian Society.) 1782

† **Verona Poligrafo.** Poligrafo. Giornale di Scienze, Lettere, ed Arti. G. G. Orte, Direttore. 60 vols. 8vo. (*Many of the Electrical articles, &c. are in the Library.*) *Verona,* 1830—47

Verrier. (*Vide* Le Verrier.)

Vershour Forsten, W. Dissertatio de arteriorum et venarum vi irritabili. *About* 1750

Verschuir, Gualth. De arteriorum et venarum vi irritabili ejusq. in vasis excessu et inde oriunda sanguinis direct. abnormi. 4to. *Grœningen,* 1766

Vesci. (*Vide* Meyer.)

† **Viacinna, C.** Del fulmine e della sicura maniera di evitarne gli effetti. Dialoghi Tre. 8vo. 156 pp. *Milano,* 1766

† **Vianelli, G.** Nuove scoperte intorno la luce notturna dell' acqua marina, spettanti alla naturale Storia; fatte da G. V. 8vo. 28 pp. 1 plate. *Venezia,* 1749

† **Viano, G. C.** Meccanismo e natura della Elettricità, Arie fattizie, &c. Dissertazioni. 8vo. 30 pp. *Asti,* 1805

Viard, Henri Stanislas. *Born October* 28, 1821, *at Rouen.*

† Thèses de Phys. et de Chim. 1re. Du Rôle de l'oxygène libre dans les Piles. 2e. Programme d'une Thèse De la Corrosion des métaux au contact de l'air. 4to. 58 pp. 2 plates. 19 figs. *Paris,* 1850

† Du Rôle Electro-chimique de l'Oxygène. 8vo. 23 pp. (*Annales de Chim. et de Phys.*) *Paris,* 1854

Vicenza Giornale Enciclopedico. "Giornale Enciclopedico di Vicenza." 8vo. *Vicenza,* 1779 *to* 1784
(*Note.—This Journal contains, says Toaldo, much writing, by himself and others, on* "*Nebbia and della Influenza de' Fulmini.*")

Vidal. On the Magnet, and Experiments. *Toulouse,* 1827

Vidal, J. Mémoire sur l'Aimant. (*Rec. des ouvr. lus dans le Lycée de Toulouse,* an. 8, p. 42.) 1800 ?

Videt, F. F. Quelques considérations sur l'électricité médicale. *Montpellier,* 1853

Viechelmann, C. Elemente der unterseeischen Télégraphie. (*Vide* Delamarche.) *Berlin,* 1859

Viegeron, P. D. Mémoire sur la force des pointes pour soutirer le fluide électrique.

Vienna Acad. Jahrbücher der K. K. Central-Anstalt für Meteorologie und Erd-magnetismus von Karl Kreil. 8vo. *Wien,* 1848

Sitzungsberichte der Kais. Akad. der Wissenschaften. Mathemat-naturwis-senschaftliche Classe. 8vo. *Wien,* 1848 *to* 18—

Vienna Acad.—*continued.*

Denkschriften der K. Akad. der Wissenschaften. Mathemat-naturwissenschaft-liche Classe. 4to. *Wien,* 1850 *to* —

Vienna Polytech. Institut. Jahrbucher des Kaiserl. Königl. Polytechnischen Instituts in Wien herausgeg. von J. T. Precthl. 8vo. *Wien,* 1819 *to* 18—

† **Vievar and Shepherd.** Two Observations of Explosions in the air; one heard at Halsted, in Essex, by A. Vievar, the other by S. Shepherd, of Springfield, in the same county. 4to. 3 pp. (*Phil. Trans.* xli. 288.)
London, 1739-40-41

Vigan, Paul De. Appareil d'induction. (*From Du Moncel.*)

† **Viglione,** G. F. Nuova Discussione della Teoria Frankliniana. 4to. 499 pp. 1 plate. (*The* 7 pp. *contain a letter from and a letter to Volta.*) *Novara,* 1784

† Lettera di V. al Sig. Alte Canonica. 8vo. 32 pp. *Novara,* 1785 —?

† **Vignotti.** Recherches et résultats d'expériences relatifs à la mise en service des chronoscopes électro-balistiques. 8vo. 215 pp. 4 pp. *Paris,* 1859

† **Villain,** B. Analyse chimique de la lumière, et nouvelle théorie des phénomènes magnétiques électriques et galvaniques. 8vo. 136 pp. 1 plates. *Paris,* 1810

Villallongue. Rapport sur le Télégraphe . . inventé par Villallongue. 4to.
Perpignan, 1841

Villeneuve, Olivier de. Essai de Dissertation médico-physique sur les ex-périences de l'électricité. 8vo. 22 pp. *Paris,* 1748

Villette, F. Made a paper-Electrophorus and Electrical Experiments, &c. (*Vide Nollet, as to Experiments, Letters,* vol. iii. p. 209, &c.)

Vimercati, G. (Editor). Rivista Scientifico-Industriale.
(*Note.—Signor Gherardi in an article,* April, 1871, *shows how the first idea of the Magnetic Telegraph dates from at least fifty years earlier than* A.D. 1636, *the date given by some recent writers on the subject : Signor Gherardi maintains that the idea originated and was developed in Italy earlier than in any other country.*)

† **Vinall,** J. Experiments in Electricity. Communication dated May 23, 1790. 4to. 2 pp. (*Mem. Amer. Acad.* Old Series, ii. pt. i. p. 144.)
Boston, U.S. 1793

Vince, Samuel. On a very remarkable Waterspout. 4to. (*Trans. Irish Acad.* xii. 1811.) *Dublin,* 1811

Vincendon Doumoulin. Voyage sur l'Astrolabe et la Zélée.

Viola, N. Maravigliosa et inestimabil Ruina fatta dalla Saetta nella torre della monitione del castello della Città di Tortona. 8vo. 1609

† **Vismara,** G. Dei fulmini che hanno colpito il torrazzo di Cremona. Memoria. 8vo. 24 pp. (*Extr. del fascicolo di* Feb. 1841, *degli Ann. di Fisica, &c.*)
Milano, 1841

† **Vitalis.** Letter to Mr. Bouillon Lagrange on the Amalgam of Mercury and Silver, called Arbor Dianæ. (*Ann. de Chim.* lxxii. 93.) 8vo. 2 pp. (*Phil. Mag.* xxxvi. 143.) *London,* 1810

Vitalis, Hieronymus. De magnetica vulnerum curatione. 8vo. *Parisiis,* 1668

Vitry, Jacques de. *Died* 1244, *at Rome.*

Histoire Hyérosolymitaine. Cap. lxxxix. (*Tiraboschi Storia Letteraria,* iv. 290.) *About* 1200
(*Note.—Poggendorf,* ii. 1184, *says his " Historia Orientalis," written between* 1215 *and* 1220, *contains the oldest European document on the use of the compass, and refers to Klaproth.*)

† **Vivenzio,** G. Teoria e pratica della elettricità medica del Sig. T. Cavallo . . e della forza dell' Elettricità nella cura delle supp. di menst. del Chirurgo, G. Birch. Tradotto dall' Inglese, di alcune annotazioni corredate, e dall' istoria dell' elettricità medica precedute. 4to. 157 pp. 4 plates.
Napoli, 1784

Viviani, Domenico. *Born* 1772, *at Legnaro Riviera di Levante; died February* 15, 1840, *at Genoa.*

† Phosphorescentia maris, 14 lucentium animalculorum novis speciebus illustrata, &c. 4to. 17 pp. 5 plates. (*Pogg.* ii. 1213, *says in Mem. della Sos. Med. d'emulazione,* iv.)   *Genuæ,* 1805

Vlach, F. Generalia quædam de electricitate. Diss. inauguralis. 8vo. 32 pp.
Patavii, 1838

Vlacovich, Nicc. Sulla scarica istantanea della bottiglia di Leyda. 8vo. 41 pp.
(*Aus den Sitzungsberichten der Kaiserl. Akad. der Wissenschaften.*)
Wien, 1863
Abhandl. z. Theorie d. Electricität u. d. Magnetismus. 8vo. (*Wien Acad.*)
Wien

Vliet, Aug. Fred. Van. Over de electrische verhonding van eene grostire, tegen over eene kleinere zinkplat, beide in een zuur gedompeld (mit P. J. Haax-man). (*Mulder's Nat. en Scheik Archief,* i. 1833.)   *Rotterdam?* 1833

Action de l'acide sulfurique étendu sur le zinc distillé placé dans des vaisseaux isolants et non isolants. (*Bull. des Scienc. Phys. et Nat. en Neerlande,* 1838.)
1838

Vogel, Fr. Chr. Max. Vereinfach. d. Volta'schen Eudiometers. 8vo. (*Schweigger's Journ.* v. 1812.)   1812

† **Vogt,** C. Dei Salmonidi. A volume in 8vo. of 326 pp. and atlas in fol. of 7 plates (double).   *Neuchatel,* 1842

Voight. (*Vide* Lichtenberg, L. C. and Voight.)

Voight, Fred. Wil. Das Toposcope oder d. sogen. Pyrotelegraph. 8vo.
Leipzig? 1803

Voigt, Johann Heinrich. *Born June* 27, 1751, *at Gotha; died September* 6, 1823, *at Jena.*

† Versuche einer neuen Theorie des Feuers, d.Verbrennung, und künstl. Luftarten, d. Athmens, d. Gärung, Elektricität u. d. Magnetismus. 8vo. 408 pp. 1 plate.   *Jena,* 1793

Übersicht d. vornehmsten meteorolog. Instrumente. (*Gothaischen Hof- u. Ta-schenkalenders,* Jahrg. 1796.)   1796

Mag. f. d. neuesten Zustand d. Naturkunde mit Rücksicht auf d. dazu gehör. Hülfswissenschaft. 12 vols. 8vo.   *Jena and Weimar,* 1797 *to* 1806

Allgem. Witterungslehre, &c. 8vo.   *Rudolstadt,* 1808

Construction nouvelle et commode de la pile de Volta. 8vo. (*Journ. de Chim. de Van Mons,* No. ix. p. 326.)   *Bruxelles*

Voith, Ignaz Edler Von. Über d. Einfluss elektrischer Siegellackstangen auf magnetisirte und unmagnisirte Nadeln in einer Boussole, und über ander magnetische Erscheinungen. 8vo. (*Voight's Mag. f. Naturk.* xi. 1806, p. 46.)   *Jena und Weimar,* 1806

Volckamer, J. G., **Wurtzelbau,** J. P. et **Eimmart,** G. C. Acus magneticæ variationis, quæ Norimbergæ paucis abhinc annis deprehensa fuit, observatio, anno currente 1685, ibidem repetita. 4to. (*Phil. Trans.* year 1685, p. 1253.)   *London,* 1685

† **Volpicelli,** P. Descrizione della Lampada elettro-dinamica del Sig. Duboscq-Soleil, e Indicazioni delle principali sperienze ottiche da eseguirsi colla medesima. Nota del Prof. P. Volpicelli. 4to. 15 pp. 1 plate. (*Estratta dagli Atti dell' Accad. Pontif. de' Nuovi Lincei,* anno iv. sessione de' 23 Marzo, 1851.) (*Faraday's copy.*)   *Roma,* 1851

† Fisica. Sopra un principio elettro-statico, riconosciuto dal Sig. Dr. Palagi. Nota. 4to. 4 pp. (*Atti dell' Accad. Pontif. dei Nuovi Lincei,* ann. v. sess. iv. del 23 Mag. 1852.)   *Roma,* 1852

Volpicelli, P.—continued.

† Sur un principe d' Electrostatique reconnu par Palagi. Lettre de Volpicelli à Arago. 8vo. 8 pp. (*Estratta dagli Ann. di Scienze Matemat. e fisiche pubblicati in Roma*, Giugno, 1853, tom. iv.) *Roma*, 1853

† Di alcune nuove esperienze fisiche. 4to. 2 pp. (*Giorn. dell' I. R. Istit. Lomb.* Nuova Serie, v. 146.) *Milano*, 1853

Sopra una nuova proprietà elettro-statica. Nota di P. V. . . letta nell' Accad. Pontificia de' Nuovi Lincei, 22 Gennajo, 1854. 8vo. (*Estratto dagli Annali di Scienze Matem. e Fis. pubblicati in Roma*, Febbrajo . . .) 1853 ?

Sulla polarità elettro-statica. Seconda Nota (del V.) . . 8vo. (*Estratto dagli Annali di Scienze Matemat. e Fisiche, pubblicati in Roma*, Giugno, 1854.) *Roma*, 1854

Delle due Memorie sul magnetismo delle Rocce del Cav. M. Melloni, estratto del Prof. P. Volpicelli. Letto 22 Genn. 1854, nell' Accad. Pontif. de' Nuovi Lincei. 4to. (*Estratto dagli Atti dell' Accad.* suddetta, anno v. sessione vi. del 15 Agosto, 1852.) *Roma*, 1854

† Sur l'induction électrostatique. Lettre de Volpicelli à . Regnault. 4to. 6 pp. (*Ext. des Comptes Rendus* . . tom. xl. *Séance du* 29 Janvier, 1855.) *Paris*, 1855

† Sur l'induction électrostatique. Seconde lettre . . à M. V. Regnault. 4to. 5 pp. (*Ext. des Comptes Rendus*, tom. xli. *Séance du* 8 Octobre, 1855.) *Paris*, 1855

† Sull' associazione di piu condensatori fra loro per l'aumento della elettrostatica tensione. Memoria. 4to. 36 pp. (*Estratta dagli Atti dell' Accad. Pontif. de' Nuovi Lincei*, anno vi. sessione ii. e iii.) *Roma*, 1855

† Sur l'induction électrostatique. Seconde lettre à M. V. Regnault. 8vo. 5 pp. (*Ext. des Comptes Rendus des Séances de l'Académie des Sciences*, tome xli. *Séance de Lundi*, 8 Octobre, 1855.) *Paris*, 1855

† Sur l'induction électrostatique. Troisième lettre à M. Regnault. 8vo. 7 pp. (*Ext. des Comptes Rendus* . . tom. xliii. 13 Octobre, 1856.) *Paris*, 1856

† Sur l'induction électrostatique. Quatrième lettre à M. V. Regnault. 8vo. 7 pp. (*Ext. des Comptes Rendus des Séances de l'Académie des Sciences*, tom xliv. *Séance du* 4 Mai, 1857, p. 917.) *Paris*, 1857

† Sulle immagini elettrografiche prodotte mediante la induzione statica. Comunicazione. 4to. 3 pp. (*Atti dell' Accad. Pontif. de' Nuovi Lincei*, an. 1857.) *Roma*, 1857

† Sulla elettro-statica induzione. Quarta comunicazione. 4to. 33 pp. 1 plate. (*Atti dell' Accad. Pontif. de' Nuovi Lincei*, tom. x. p. 280, 1857.) *Roma*, 1857

† Sull' elettro-statica induzione. Quinta comunicazione. 4to. 14 pp. (*Atti dell' Accad. de' Nuovi Lincei*, tom. xi. p. 411, 1857.) *Roma*, 1858

† Sur quelques observations électrométriques et électroscopiques. Lettre à M. Despretz. 8vo. 6 pp. (*Comptes Rendus*, xlvi. p. 533, 1858.) *Paris*, 1858

Sugli elettromotori. (*Atti Pont. Accad. de' Nuovi Lincei*, pp. 37, 114, 253, 311, and 453, anno 1858.) *Roma*, 1858

† Terza comunicazione sulla polarità elettro-statica e Appendice alla terza comunicazione sulla polarità elettro-statica. 4to. 5 pp. (*Atti Pontif. Accad. de' Nuovi Lincei*, p. 143 and 270, sess. iii. del 7 Feb. e sess. v. dell' Ap. 1858.) *Roma*, 1858

† Sulla polarità elettro-statica. Quarta comunicazione. 4to. 22 pp. (*Estratta dagli Atti dell' Accad. de' Nuovi Lincei*, sessione iii. del 6 Feb. 1859.) *Roma*, 1859

† Sul cognito fenomeno elettro-statico di Libes. Nota. 4to. 4 pp. (*Estratta dagli Atti dell' Accad. Pontif. dei Nuovi Lincei*, tom. xii. sess. vi. dell' 8 Maggio, 1859, p. 375.) *Roma*, 1859

† Descrizione di un nuovo Anemometrografo, e sua Teorica. Memoria. 4to. 16 pp. (*Estratta dagli Atti della Accad. de' Nuovi Lincei*, sess. iv. del 13 Marzo, 1859.) *Roma*, 1859

Volpicelli, P.—*continued*.

† Formules électromètriques. Lettre de. 4to. 2 pp. (*Comptes Rendus*, 1859.)
Paris, 1859

† Osservazioni sul magnetismo. Nota. (*Estratta dagli Atti dell' Accad. Pontif. de' Nuovi Lincei*, anno xiv. sessione iii. del 3 Feb. 1861.) *Roma*, 1861
Sulla elettricità dell' atmosfera. Seconda nota. 4to. (*Atti dell' Accad. de'Nuovi Lincei*, xiv. 270.) 1861

† Alcune osservabili formule che si ottengono da un integrale definito relativo alla Elettrostatica. Nota. 4to. 14 pp. (*Estratta dagli Atti dell' Accad. Pontif. de' Nuovi Lincei*, an. xv. del 1 Giugno, 1862.) *Roma*, 1862
(*Vide* Pianciani, Necrologico cenno intorno al Pianciani.)

† Rapporti fra le accumulazioni elettriche sopra due sfere conduttrici di raggio cognito assegnati generalmente in termini finiti. Nota. Con appendice. 4to. 12 pp. (*Estratta dagli Atti dell' Accad. de' Nuovi Lincei* . . dell' an. xvi. del 1862.) *Roma*, 1863

† Formule per determinare mediante il condensatore la elettricità terrestre, o qualunque altra indeficente, senza bisogno di uno stato elettrico assoluto. Nota. 4to. 7 pp. (*Estratta dagli Atti della Accad. Pont. de' Nuovi Lincei*, xvii. p. 164.) *Roma*, 1864
(*Vide* also Plana, C. A. A. 1864.)

† Sulla Elettro-statica induzione. Decima comunicazione. 4to. 7 pp. (*Estratta dagli Atti della Accad. Pontif. de' Nuovi Lincei*, sess. i. del 4 Dicemb. 1864, tom. xviii. p. 59.) *Roma*, 1865

† Considerazioni sulla tensione, tanto in elettrostatica, quanto in elettro-dinamica, e sulla elettrica influenza. Undecima comunicazione. 4to. 11 pp. (*Estr. della Sessione* i. *del 3 Dicem.* 1865, tom. xix. pp. 11—21. *Atti dell' Acad. Pont. de' Nuovi Lincei.*) *Roma*, 1865

† Ricerche analitiche sul Bifilare tanto magnetometro, quanto elettrometro, sulla curva bifilare e sulla misura del Magnetismo terrestre. Memoria. 4to. 82 pp. 1 plate. (*Estr. d. Atti dell' Accad. Pontificia de' Nuovi Lincei*, xvii.) *Roma*, 1865

† Sulla necessità di proteggere dal fulmine le masse metalliche, stabilite nella cima degli edifici. Nota. 4to. 5 pp. (*Atti dell' Accad. Pontif. dei Nuovi Lincei*, sess. i. del 3 Dicem. 1865, tom. xix. pp. 22—26.) *Roma*, 1865

† Sulle osservazioni meteorologiche e magnetiche, nell' osservatorio dell' Infante D. Luigi, a Lisbona. Cenno. 4to. 4 pp. (*Atti dell' Accad. Pontif. dei Nuovi Lincei*, sess. iv. del 5 Marzo, 1865, tom. xviii.) *Roma*, 1865

† Observations sur la Tension tant en Electrostatique qu'en Electrodynamique et sur l'influence électrique. 8vo. 8 pp. (*Extrait par l'auteur . Comptes Rendus*, lxi. p. 548, 1865.) *Paris*, 1865

† Analisi e rettificazioni di alcuni concetti, e di alcune sperienze che apparten-gono alla elettrostatica. Memoria Prima. 4to. 180 pp. (*Atti dell' Accad. Pontif. de' Nuovi Lincei*, xix. p. 312, 1865.) *Roma*, 1866

† Intorno alle prime scoperte delle proprietà che appartengono al Magnete. Cenno istorico compilato del Prof. Volpicelli. 4to. 16 pp. (*Estratti dagli Atti dell' Acad. Pont. de' Nuovi Lincei*, tom. xix. Marzo, 1866, pp. 205, 218.) *Roma*, 1866

† Pubblicazioni del Prof. P. Volpicelli sino a tutto l'anno 1866. 4to. 12 pp. *Roma*, 1866

† Introduzione alle formule per la teoria dell' elettromotore voltaico. Ragiona-mento. 4to. 29 pp. *Roma*, 1867

† Sugli elettrofori a rotazione continua, e spiegazione di una sperienza eseguita recentemente coll' elettroforo di Holtz. Nota. 4to. 7 pp. (*Accad. Pontif. de' Nuovi Lincei*, sess. iv. del 19 Aprile, 1868.) *Roma*, 1868

† Sopra il telegrafo di Gloesener. Comunicazione del Volpicelli. 4to. 2 pp. (*Accad. Pontif. de' Nuovi Lincei*, sess. iv. del 19 Aprile, 1868.) *Roma*, 1868

Volpicelli, P.—*continued.*

† Sulla causa dell' inversione delle cariche di elettricità nei coibenti armati; e sull' influenza elettrica nei gas rarefatti. 4to. 16 pp. (*Estratta della sessione* vii. *del* 6 *Giugno*, 1869, *dell' Accad. Pontificia de' Nuovi Lincei.*)

Roma, 1869

† Di un Barometro fotografico, e formule per compensare automaticamente gli effetti della temperatura in un barometro qualunque. Memoria. 4to. 42 pp. 1 plate. Roma, 1870
(*Note.*—*This is Sir F. Ronalds' Barograph.*)

† Della distribuzione elettrica sui conduttori isolati. (*Mem. d. Roma.*) 4to. 20 pp. (*Estr. dagli Ann. di Matemat. pure ed applicate,* Serie ii. tom. iii. Fasc. iii. da p. 249 e p. 268.) Roma, 1870

† Esposizione del modo col quale per la prima volta, fu applicato il calcolo alla elettrostatica e ne fu concluso che la elettricità indotta non tende. Memoria del Volpicelli per servire alla storia della elettricità. 4to. 51 pp. (*Estratto dagli Atti della reale Accad. de' Lincei della sessione* i. *del* 4 *Dicembre,* 1870.)

Roma, 1870

Volta, Alessandro. *Born February* 19, 1745, *at Como ; died March* 5, 1827, *at Como.*

† De vi attractiva ignis electrici, ac phænomenis inde pendentibus . ad Joannem Bapt. Beccariam. Dissertatio epistolaris. 4to. 72 pp. (*Presented by Count Zanino Volta.*) Nova-Comi, 1769
(*Note.*—*His first Electrical work, much esteemed.*)

† Novus ac simplicissimus electricorum tentaminum apparatus : seu de corporibus eteroelectricis quæ fiunt idioelectrica experimenta, atque observationes Alexandri de Volta. Sm. 4to. 38 pp. (*Volta's copy.*) Novo-Comi, 1771

† Squarci di due lettere scritte dal Volta. al P. C. G. Campi. lma, 13 Giugno ; 2da, 22 Giugno, 1775. 12mo. 4 pp. (*Scelta d'Opuscoli* in 12mo. viii. 127.)

Milano, 1775

† Articolo di una lettera del . Volta al Dot. Priestley. 12mo. 17 pp. (*Scelta di Opuscoli* in 12mo. ix. 91.) Milano, 1775

† Seguito della lettera al Priestley. 27 pp. 1 plate. (*Scelta di Opuscoli,* x. 87. Milano, 1775

† Articolo di lettera . al Fromond. 12mo. 12 pp. (*Scelta d'Opuscoli,* xii. 94.)

Milano, 1775

† Lettre . à l'Auteur de ce Recueil,sur l'Electrophore perpétuel de son invention. 4to. (*Rozier Obser.* vii. Juill. 1776, pp. 21-24.) Paris, 1776

† Lettere sull' Aria infiammabile nativa delle Paludi. Al Padre C. G. Campi. 8vo. 16 pp. Milano, 1776

† Articolo di lettera al Fromond Como 21 Dec. 1775. 12mo. 14 pp. (*Scelta d'Opuscoli,* xiv. 84.) Milano, 1776

† Lettera al G. Klinkosch, dated May, 1776. 12mo. 36 pp. Milano, 1776

Schreiben an den Hern Jos. Klinkosh den beständigen Elektricitäts träger betreffend. Prag, 1777

Beschreib. ein. neuen elektr. Geräthschaft, Elektrophor genammt ; a. d. Ital. 8vo. Prag, 1777

† Lettera del A. Volta al Dot. Priestley. Eudiometer. 12mo. 19 pp. 1 plate. (*Scelta d'Opuscoli* in 12mo. xxxiv. 65.) Milano, 1777

Lettere sull' aria infiammabile nativa delle paludi. 8vo. 147 pp. (7 Letters.)

Milano, 1777

† Lettere del Volta sull' Aria infiammabile. 12mo. 36 pp. (*Scelta d'Opuscoli* in 12mo. xxviii. 43.) Milano, 1777
(*Note.*—*This is the Editor's account of the book.*)

† Lettera i. al F.Castelli sopra la costruzione di un moschetto e d'una Pistola ad aria infiammabile. 11 pp. 1 plate. Lettera ii. sul medesimo soggetto. 14 pp. Lettera iii. sul medesimo soggetto. 22 pp. 12mo. 47 pp. 1 plate. (*Scelta d'Opuscoli. The* 2 *first in* tom. xxx. *the* 3rd *in* tom. xxxi.)

Milano, 1777

Volta, Alessandro.—*continued.*

† Lettre sur l'air inflammable des Marais. 8vo.   *Strasbourg*, 1778
† Briefe über die entzündbare Luft. der Sümpfe. Nebst Drey andern Briefen.
  die aus dem Maylandischen Journal genommen sind. Aus d.Ital. übers. von
  C. H. Kostlin. 8vo. 226 pp. 1 plate.   *Strasbourg*, 1778
† Osservazioni sulla capacità dei conduttori elettrici, e sulla commozione che anche
  un simplice conduttore è atto a dare eguale a quella della boccia di Leyden ;
  in una Lettera al De Saussure. 4to. 32 pp. 1 plate.   (*Opusc. Scelti*, i. 273
  and 289.)   *Milano*, 1778
Of the method of rendering very sensible the weakest Natural or Artificial
  Electricity. 4to. (*Phil. Trans*.1782, p. 237, and *Append.* p. 7, vol. lxxii. pt. i.
  *This is Cavallo's translation of the Italian version, which is printed with it, and
  the title of which is "Del modo di rendere sensibilissima la più debole elettricità,
  sia artificiale, sia naturale.*)   *London*, 1782
Mémoire sur les grands avantages d'une espèce d'isolement trés imparfait. trad.
  de l'Ital. par M. . Sur la capacité des Conducteurs conjugués. Prem. Mém.
  dans lequel on démontre les avantages très-considerables d'une sorte d'isolement si imparfait, qu'on peut à peine lui donner ce nom, sur l'isolement le
  plus parfait. 4to. (*Rozier, Observ.*. xxii. *Mai* 1783, pp. 325-50.) *Paris*, 1783
Seconde partie du mémoire sur les isolements imparfaits. Expériences qui
  demonstrent un autre avantage, trés-considérable, attaché à la même espèce
  d'isolement imparfait, consistant; en ce qu'il rend un conducteur propre à
  recevoir l'électricité plus aisément, et en dose beaucoup plus grande, que s'il
  étoit parfaitement isolé. 4to. (*Rozier, Observ.* xxiii. *Juil.* 1783, pp. 3-16.)
                                                                 Paris, 1783
Suite du Mémoire sur les conducteurs électriques. 4to. (*Rozier, Observ.* xxiii.
  Août, 1783.)   *Paris*, 1783
Sur les avantages d'un isolement imparfait. 4to. (*Rozier, Observ.* tom. i. p. 323.
  et tom. ii. p. 3.)   *Paris*, 1783
  (*Note.*—*These form the French version of the Paper in the Phil. Trans.* 1782,
  *entitled "Of the method of rendering very sensible the weakest Electricity,"
  and of the Memoir in the Opuscoli Scelti,* tom. vii. *entitled "Del modo di
  rendere sensibilissima la più debole elettricità,"* &c.)
† (1) Del modo di rendere sensibilissima la più debole Elettricità sia naturale, sia
  artificiale. (Parte Prima) p. 128. (2) In qual modo un Conduttore accostandosi a un altro sotto certe condizioni acquisti una straordinaria capacità di
  ricevere, e contenere l'Elettricità. Parte Seconda, p. 145. 4to. 35 pp.
  (*Opusc. Scelti*, vii. 128 e 145.)   *Milano*, 1784
Lettera al Viglione, data 29 Feb. 1776. 4to. 2 pp.   *Novara*, 1784
† Mem. sopra i fuochi de' terreni e delle fontane ardenti in generale, e sopra quelli
  di Pietra-Mala in particolare. 4to. 14 pp. and 8 pp. (*Mem. Soc. Ital.* ii.
  part ii. pp. 662 and 900.)   *Verona*, 1784
  (*Note.*—*He refers at* p. 675 *to his theory on the influence of inflammable air
  (thus produced) in conjunction with Electricity on the production of some
  kinds of Igneous Meteors.*)
† Estratto d'una Lettera . al Sig. L. Brugnatelli. Como 20 Agos. 1788. 8vo.
  3 pp. (*Bibliot. Fisica d'Europa*, tom. iv. p. 133, *Pavia*.) *Pavia*, 1788
Osservazioni sull' Elettricità dei vapori dell' acqua. 8vo. (*Biblioteca Fisica
  d'Europa* (*di Brugnatelli*, i. 149.)   *Pavia*, 1788
Osservazioni sull' elettricità del ghiaccio. 8vo. (*Biblioteca Fisica d'Europa
  di Brugnatelli*, vi. p. 164.)   *Pavia*, 1788
† Lettere del Volta . sulla Meteorologia elettrica. No. ix. 8vo. 2 plates.
  (*Biblioteca Fisica d'Europa di Brugnatelli.*)   *Pavia*, 1788-89-90
† Della maniera di far servire l'elettrometro atmosferico portatile all' uso di un
  igrometro sensibilissimo. Memoria in cui si rischiarano molte cose intorno
  al trascorrimento del fluido elettrico ne' conduttori imperfetti. 4to. 30 pp.
  (*Mem. Soc. Ital.* v. 551.)   *Verona*, 1790

Volta, Alessandro.—*continued.*

† Descrizione dell' Eudiometro ad aria infiammabile, il quale serve inoltre di
apparato universale per l'accensione al chiuso delle arie infiammabili di ogni
sorta mescolate in diverse proporzioni con aria respirabile più o meno pura;
e per l'analisi di quelle e di questa, inventato e perfezionato dal Sig. D.
Ales. Volta. 8vo. 61 pp. 1 plate. *Pavia,* 1790

† Seguito della descrizione dell' Eudiometro ad aria infiammabile. 8vo. 26 pp.
(*Ann. di Chim. di Brugnatelli,* ii. 161.) *Pavia,* 1791

† Seguito della descrizione dell' Eudiometro ad aria infiammabile. 8vo. 10 pp.
(*Ann. di Chim. di Brugnatelli,* iii. 36.) *Pavia,* 1791

Observationum circa electricitatem animalem, specimen. 8vo. (*Commentarii
de rebus in Scientia naturali et in medicina gestis,* vol. xxxiv. pars ii. p. 685.)
Lipsiæ

The early Experiments of Volta relative to Galvanic Phenomena are described.
(*Commentarii de rebus in Scientia Naturali et in Medicina gestis,* xxxiv. 684,
Article xiv. des *nouvelles littéraires, Journal de Leipzig.*) *Leipzig,* 1792

† Memoria Prima, sull' Elettricità Animale. 8vo. 42 pp. (*Giorn. Fis.-Med.* ii. 146.
"*Discors. recitato 5 Maggio,* 1792.") *Pavia,* 1792

† Memoria Seconda sull' Elettricità Animale. 8vo. 30 pp. (*Giorn. Fis.-Med.*
ii. 241.) *Pavia,* 1792

† Continuazione della Seconda Memoria sopra l Elettricità Animale. 8vo. 39 pp.
(*Giorn. Fis.-Med.* iii. 35, Luglio.) *Pavia,* 1792

† Nuove osservazioni sull' Elettricità Animale. Comunicate dal Volta. 8vo.
5 pp. (*Giorn. Fis.-Med.* iv. 192, Novemb.) *Pavia,* 1792

† Transunto di Osservazioni sull' Elettricità Animale, ed alcune nuove proprietà
del fluido elettrico. 4to. 3 pp. (*Opusc. Scelti,* xv. 213.) *Milano,* 1792

(*Note.—This is the Editor's account of Volta's early Experiments, &c.*)

† Transunto delle nuove osservazioni sull' Elettricità Animale del Volta. Articolo
tratto dal Giornale Fisico-Medico del Brugnatelli, Novembre. 4to. 3 pp.
(*Opusc. Scelti,* xv. 425.) *Milano,* 1792

† Lettera all' Autore (Bondioli) dell' Opera (sopra l'Aurora Boreale). 8vo. 14 pp.
(*Giorn. Fis.-Med.* i. 66.) *Pavia,* 1792

† Lettera del 3 Aprile 1792 al Baronio sull' Elettricità Animale. 8vo. 9 pp.
(*Giorn. Fis.-Med.* ii. 122.) *Pavia,* 1792

† (Editori) Sull' Elettricità Animale ed alcune nuove proprietà del fluido elettrico,
del Volta. 8vo. 4 pp. (*Giorn. Fis.-Med.* ii. 287.) *Pavia,* 1792

† Memoria Terza sull' Elettricità Animale compresa in alcune lettere. Lettera
Prima, al Prof. Aldini. 8vo. 19 pp. (*Giorn. Fis.-Med.* v. 63.) *Pavia,* 1793

Schriften über die thierische Elektricität ; aus dem Ital. von Joh. Mayr. 8vo.
Prag, 1793

Account of some discoveries made by Mr. Galvani, with Experiments and Observations on them. In two letters to Cavallo : 1st dated September 13, 1792 ;
2nd, October 25, 1792. 4to. (*Phil. Trans.* 1793, part i.) *London,* 1793

Meteorologische Beobachtungen, besond. über die atmosphär. Electricität, aus
d. Ital. mit Anmerk.(auch unter d. Titel Meteorol. Briefe, nebst d. Beschreib.
s. Eudiometers). Aus d. Ital. mit Anmerk. 8vo. *Leipzig,* 1793

Meteorologische Briefe a. d. Ital. 8vo. (*From Dove,* p. 228.) *Leipzig,* 1793

Volta, Alessandro.—*continued.*

† Nuova Memoria sull' Elettricità Animale, in alcune lettere al Vassalli. Lettera prima. 8vo. 10 pp. (*Ann. di Chim. di Brugnatelli*, v. 132.) *Pavia*, 1794

† Lettera seconda. . al . Vassalli sull' Elettricità Animale. 8vo. 25 pp. *Pavia*, 1794

† Lettera terza. al . Vassalli sull' Elettricità Animale. 24 Ottob. 1795. 8vo. 45 pp. (*Ann. di Chim. di Brugnatelli*, xi. 84.) *Pavia*, 1796 (*Vide* also Anon. Electricity, 1796.)

Fortgesetz Schriften über thierische Elekt. od Schreiben an I. M. Vassalli üb. d. thier. Elect. aus d. Ital. v. J. Mayer. 8vo. *Prag*, 1796

Letters 1st, 2nd, and 3rd to Gren on Galvanism or on Electricity excited by contact of dissimilar Conductors, in German. 8vo. (*Gren's Neue Journ. de Phys.* iii. 1796 and iv. 1797.) *Leipzig*, 1796-7

Expériences sur l'électricité dite animale. 8vo. (*Ann. de Chimie*, xxx. p. 276.) *Paris*, 1797

† Estratto di una lettera . . al Gren . . sul Galvanismo, ossia sull' elettricità eccitata dal contatto dei conduttori dissimili. 1 Agosto, 1796. 8vo. 49 pp. 1 plate. (*Ann. di Chim. di Brugnatelli*, xiii. 226.) *Pavia*, 1797

† Lettera ii. sul Galvanismo, in continuazione della precedente . . al . . Gren. Agosto, 1796. 8vo. 37 pp. (*Ann. di Chim. di Brugnatelli*, xiv. 3.) *Pavia*, 1797

† Lettera iii. sul Galvanismo, al Gren. (Not dated ; written in 1797 ; *vide* p. 40 of this 3rd letter.) 8vo. 35 pp. (*Ann. di Chim. di Brugnatelli*, xiv. 40.) *Pavia*, 1797

(*Note.*—*Configliachi says* (*in reference to the three letters*) : "*Vedi inoltre les Ann. de Chim.* vol. xxix. p. 91 *et Paris* (an. 1799) *e le Journ. de Phys.* . . t. xlviii. p. 336, 1799.")

† Estratto di lettere del Cit. N. N. Comasco al Cit. Aldini . . intorno alla pretesa elettricità animale nelle sperienze del Galvanismo. Lettera i. (Data Como Aprile, 1798. 8vo. 25 pp. (*Ann. di Chim. di Brugnatelli*, xvi. 3.) *Pavia*, 1798

† Appendice (alla Lettera i.). Articolo tratto dagli elementi di Fisica di Gren intorno alla così detta elettricità animale. 8vo. 15 pp. (*Ann. di Chim. di Brugnatelli*, xvi. 27.) *Pavia*, 1798

† Lettera ii. Estratto di lettere del Cit. N. N. (Comasco). . . (*Senza data : the second*). 8vo. 47 pp. (*Ann. di Chim. di Brugnatelli*, xvi. 42.)

(*Note.*—*These two letters are known to be by Volta from Configliachi's statement at* p. 17 *of the* "*Identità del fluido elettrico.*")

Lettera al Carradori. ("*Memorie per servire alla Storia letteraria e civile* anno 1798." (*Venezia*.) *Scrittura non compresa nella* "*Collezione.*") *Venezia*, 1798

Meteorologische Beobachtungen, besonders über die atmosphärische Elektricität. Aus d. Ital. mit Anmerkungen des Herausgebers (Herausgeg. von Lichtenberg, übers. von Schäffer.) 8vo. *Leipzig*, 1799

On the Electricity excited by the mere contact of Conducting Substances of different kinds. In a letter to Sir Joseph Banks, F.R.S. Read June 26, 1800. (*Dated Como*, 20 *March*, 1800 ; *written in French*.) 4to. (*Phil. Trans.* 1800, part ii. p. 403.) *London*, 1800

(*Note.*—*Configliachi* "*Questa Mem. interessantissima è scritta in francese. Veggasi anche la Description du nouvel appareil electrico galvanique, etc. nel J. de Phys. etc. Paris*, 1800, t. li. p. 334, *e nella Bibliothèque Britannique* . . . 1800, vol. xv. p. 3, *e Journ. de Nicholson.*" (*Account, &c.* . . *nel Luglio*, 1800.) (*Contains the first account of his Pile and couronne de tasses.*)

Volta, Alessandro.—*continued.*

† Lettera . . al . . Brugnatelli sopra alcuni fenomeni chimici ottenuti col nuovo apparecchio elettrico. 8vo. 7 pp. 1 plate. (*Ann. di Chim. di Brugnatelli*, xviii. 3.) *Pavia*, 1800

† Lettera . . al . . Mas. Landriani sopra l'elettrico. Como, 22 Sep. 1800. 8vo. 17 pp. (*Ann. di Chim. di Brugnatelli*, xviii. p. 7.) *Pavia*, 1800

Extrait d'une lettre au C. Dolomieu sur quelques tentatives pour rendre l'appareil galvanique encore plus commode. (*Bulletin . . Soc. Philomat.* No. 54, p. 48, an. 9.) *Paris*, 1801

Lettre à Delametherie sur les phénomènes galvaniques. 4to. ("*Journal de Phys. et Chim.* t. liii. p. 309. *La medesima in Italiano col titolo Sopra gli Elettromotori negli Opusc. Scelti*, . . t. xxi. p. 373, 1801.") *Paris*, 1801-2

(*Note.—Configliachi says: "Questa Mem. è interessantissima ed è come la prima parte della seguente stata letta dal Volta all' Istituto Nazionale di Francia nel Novembre, 1801.*")

Réponse aux observations de Nicholson sur ma théorie. Aux Rédacteurs de la Bibliothèque Britannique. 8vo. (*Bibl. Britan.* tom. xix. p. 274; *Sciences et Arts*, part i. p. 274; part ii. p. 339, an. x.) *Genève*, 1801-2

† Lettera a Delametherie sopra gli Elettro-motori. (Dal) Giornale di Fisica e Chim. (*i.e.* Journal de Phys.) tom. liii. 309, an. 10 (1801-2). Parigi vendemmiale anno 10. 4to. 7 pp. (*Opusc. Scelti*, xxi. 373. *Translation with a note by the translator, i.e. Soave.*) *Milano*, 1801

Description of the Apparatus which Beron used for his Demonstrations of Volta's Experiments, &c., and which he (Beron), presented to the "Société de Médecine et à l'Institut." *Paris*, 1801

Sur les phénomènes galvaniques. 4to. *Paris*, 1801

Paris Institute Commission Report . . Sur les expériences du Cit. Volta, viz., Laplace, Coulomb, Hallé, Monge, Fourcroy, Vauquelin, Pelletan, Charles, Brisson, Sabathier, Guyton de Morveau, and Biot. *Paris*, 1801

† Exposition abrégée des principales expériences répétées par lui, en présence des commissaires de l'Institut National par lesquelles il a rendu évidente l'identité de principe entre les phénomènes du galvanisme et ceux de l'électricité. 4to. (*Soc. Philomat.* an. 10, p. 74.) *Paris*, 1802

De l'électricité galvanique (ossia) le fluide galvanique ne diffère point du fluide électrique. 8vo. (*Ann. de Chim. e Phys.* vol. xi. p. 225. "*La medesima Mem. col titolo ' Sull' identità del fluido elettrico col fluido galvanico' ritrovasi negli Annali di Chimica di Pavia,* tom. xix. p. 38 (1802), *colla Continuazione, la quale manca negli Annali di Chim. di Parigi.* (*La continuazione (Ital.) è inscritta nel* tom. xxi. (*degli Annali di Pavia*), p. 163.") *Paris*, 1802

† Memoria . . sull' identità del fluido elettrico col fluido Galvanico, letta nell' Istituto Naz. di Francia. 8vo. 37 pp. (*Ann. di Chim. di Brugnatelli*, xix. 38.) *Pavia*, 1802

(*Note.—In this memoir Volta announces his laws of tension in reference to the electricity of his Pile.*)

† Continuazione della Memoria . . sopra l'identità del fluido elettrico col fluido galvanico. 8vo. 97 pp. (*Ann. di Chim. di Brugnatelli*, xxi. 163.) *Pavia*, 1802

Observation sur l'identité du fluide galvanique avec le fluide électrique. (*Sedillot, Rec. Périod. de la Soc. de Méd. de Paris*, ix. 97 et 231.) *Paris*

† Articolo di lettera del. . . Volta sopra alcuni fenomeni elettrici (29 Sep. 1802.) 8vo. 3 pp. (*Ann. di Chim. di Brugnatelli*, xxi. 79.) *Pavia*, 1802

Volta, Alessandro.—continued.

† Lettera . . al Brugnatelli sopra l'applicazione dell' elettricità ai sordomuti dalla nascita. 8vo. 5 pp. (*Ann. di Chim. di Brugnatelli*, xxi. 100.)

Pavia, 1802

"Lettre de Volta sur l'identité du fluide électrique avec le prétendu fluide galvanique. Pavie, 29 Floréal, an. ix. Elle a été communiquée par le professeur Brugnatelli." 8vo. (*Journ. de Chim. de Van Mons,* No. ii. p. 167.)

Bruxelles, 1803 ?

Schreiben über Electricität und Galvanismus ; herausgegeben von C. F. Nasze. i. Th. 8vo. *Halle,* 1803

Neuest. Versuche über Galvanismus Beschreib. ein. neuen Galvanometers, und and. kleine Abhandl. üb. diesen Gegenstand. 8vo. *Wien,* 1805

† Estratto di un MSS., sull' insussistenza della genesi del Clorino, e dell' Alcali nell' acqua sottoposta all' azione degli Elettromotori. (" *Il presente Estratto trovasi pubblicato nel Saggio di naturali Osservazioni sull' Elettricità Voltiana del . . Baronio a* p. 102.) 1806

† Sopra la Grandine. Ricevuta 1 Agosto, 1804. 4to. 66 pp. (*Mem. dell' I.tit. Nazion. Ital.* tom. i. pt. ii. p. 125.) *Bologna,* 1806

(*Note.—It is also in the Giorn. d. Fis. Chim. . . di Brugnatelli,* lo Bimestre, p. 31 ; 2o, p. 129 ; *e* 3o, p. 179 (*Pavia,* 1808) ; *and in the Journ. Phys.* pp. 289 and 333 of vol. lxix.)

† Sopra la Grandine. Memoria inserita nella p. ii. del tom. i. dell' Istituto nazionale Italiano. 8vo. 19 pp. (*Giorn. dell' Ital. Lettera del Dal Rio,* xviii. p. 249.) *Padova,* 1807

(*Note.—This is an extract and notice merely by the editor of the* " *Giornale.*")

Lettera al . . Zuccagni, A., responsiva ad altra di esso, sopra un Ignivomo. (Data) Milano, 16 Feb. 1807. (*Pubblicata nel* tom. vi. p. 87, *del Giornale Pisano.*) 1807

Estratto d'una lettera relativa alla Memoria del Chimico Sig. Porati, *Sulla possibilità d'un' accensione spontanea.* 8vo. *Milano,* 1810

† L'identità del fluido elettrico col cosidetto fluido galvanico vittoriosamente dimostrata, con nuove esperienze ed osservazioni. Memoria comunicata al Sig. P. Configliachi . . e da lui pubblicata con alcune note. Fol. 145 pp. Portrait. *Pavia,* 1814

† Collezione dell' Opere del Cav. Conte Alessandro Volta Patrizio Comasco, &c. . . (da V. Antinori). 3 vols. 8vo. (in 5 parts). 7 plates, and portrait. *Firenze,* 1816

Trattato meteorologico sopra la natura e la formazione dei Bolidi e delle Stelle-cadenti. Letto 31 Dec. 1812. 4to. (*Mem. dell' I. R. Istit. Reg. Lomb.- Veneto,* i. 24.) *Milano,* 1819

† Risposta del Volta al Marzari relativa ai cosi detti Paragrandini. 8vo. 3 pp. (*Volta's copy.*) *Como,* 1823

† Mémoire sur la formation de la Grêle, traduit par Veau-Delaunay. Première Partie. 4to. 14 pp. (*In Journ. de Phys. ?*) *Paris*

† Relazione di A. Volta di un suo viaggio letterario nella Svizzera ; ora, per la prima volta, pubblicata. . . 8vo. 47 pp. *Milano,* 1827

† In morte del Conte Aless. Volta. Cantica di Giorn. Fogliani. 4to. 12 pp. *Como,* 1827

† Elogio morale del Conte Alessandro Volta di G. Zuccala. 8vo. 42 pp. *Bergamo,* 1827

Volta, Alessandro.—*continued.*

† Della vita del Conte Aless. Volta, Patrizio Comasco. 8vo. 138 pp. Portrait and 1 plate.
Como, 1829

† Elogio del Conte Alessandro Volta, Patrizio Comasco . . dal F. Mocchetti. 4to. 82 pp. ("*Letta* 7 Dec. 1832 . . *per l'inaugurazione del Busto di Aless. Volta.*")
Como, 1833

† Elogio scientifico di Al. Volta, scritto dal . . Configliachi. 4to. 56 pp. Portrait by Garavaglia.
Como, 1834

† Lettere inedite di Aless. Volta (G. I. Montanari, editore.) 8vo. 212 pp. 1 plate. (*Configliachi's* " *Elogio scientifico di Volta* " *is added at* p. 170.)
Pesaro, 1834

Notice biographique de, par Arago. (*From Arago's Sale Cat.* p. 179.)
1835?

† Elogio storico di Alessandro Volta del Sig. Arago . . estratto dagli Annali di Chimica e di Fisica di Parigi. 8vo. 80 pp. (*This translation from the French is taken from the Journal* " *l'Indicatore Lombardo,*" *and appeared in the* " *Raccolta pratica,*" *a periodical edited by Luigi and Zanino Volta. Luigi adds many footnotes, some of which are important and interesting.*)
Como, 1835

† Elogio storico di Alessandro Volta del Sig. Arago. . . Estratto dagli Annali di Chimica e di Fisica di Parigi. 8vo. 80 pp. (*This translation from the French is copied from the Indicatore Lombardo, a journal printed in Milan. Menini is the translator.*)
Como, 1835

(*Note.—The translation is furnished with many good footnotes, signed L. V* (*i.e. Luigi Volta, his second son*). *Printed in the Raccolta Pratica di Scienze, &c.* Nos. xxxiii. *e* xxxiv. March and April, 1835 (*Como*).

† Erigendosi in Como il Monumento di Aless. Volta, Patrizio Comasco. Discorsi due del Prof. F. Mocchetti. 8vo. 13 pp.
Milano, 1838

† Biografia di Alessandro Volta (di G. Chiappa). 8vo. 12 pp. (*Estr. dalla Gazzetta Provinciale di Pavia.*)
Pavia, 1844

† Alessandro Volta, per L. A. Girardi. 16mo. 78 pp. Portrait. ("*I contemporarii Italiani,*" *Galleria Nazionale del secolo* xix.)
Torino, 1861

† Notizie biografiche, &c. su Aless. Volta esposte dal Magrini. . . 4to. 53 pp. 1 plate. (*Estratto dal* vol. ii. *degli Atti del Rle. Istituto di Scienze.*)
Milano, 1861

Un Manoscritto autografo in data di Como 15 Aprile, 1777 . . diretto, probabilmente, al Barletti. "Contiene varie sperienze sulle sue pistole : . . e la pro-posta di trasmettere segnali col mezzo della elettricità ordinaria." 4to.
Milano, 1861

(*Note.—This is from a short paragraph in a* "*Comunicazione sui manoscritti di Volta fatta* 28 *Giugno,* 1860," *by Magrini to the R. Istituto Lombardo di Scienze . . in varie tornate del* 1861 *di esso Ist.*)

† Sulla importanza dei Cimelj e dei Manoscritti di A. Volta. Discorso del Cav. Prof. L. Magrini. Letto 7 Agosto, 1864. 8vo. 43 pp.
Milano, 1864

Sul periodo de' Temporali. Lettera al . . Configliachi. 4to. (*Estratta dal* tom. x. p. 17, *del Giornale di Fisica.*)
Pavia

(*Note.— "Deve riguardarsi come una Continuazione delle* 10 (*Lettere*) *sulla Meteorologia elettrica . . Le prime 9 sono dirette al . . Lichtenberg . . la decima poi è quella sulla formazione della Grandine.*)

Articolo "Eudiometro" nella traduzione di Scopolo del Dizionario di Chimica di Macquer.

Poscritto di una lettera diretta al Vassalli. "Separato dalla Terza Lettera al Vassalli per essere estraneo alla materia . . di quella lettera" (viz.) dated Como, 24 Oct. 1795.

Volta, Alessandro.—*continued.*

Nouvelles expériences sur le galvanisme, répétées à Paris par Robertson. (*Rec. des Actes de la Soc. de Santé de Lyon,* tom. ii. p. 376.) *Lyon*

Description abrégée do la pile électrique de Volta. 8vo. (*Journ. de Van Mons,* p. 129, No. 2.) *Bruxelles*

Über Volta's Kondensator der Elektricitat. (*Leipsiger Samml. zur Physik und Naturgesch.* iii.)

The *Times* of the 26th Jan., 1860, contains an article referring to the manuscripts and the Electric Machine of Prof. Volta. *London,* 1860

† Portrait by Rades inc. Focosi des.

† Portrait by Garavaglia des. et inc.

(*Vide* also Anon. Miscellaneous, 1833.)

(*Vide* also Cuthbertson.)

† Volta e Bellani. Sulla formazione della Grandine Memoria del Volta. Con un Articolo sul medesimo argomento del Bellani. 8vo. 170 pp. *Milano,* 1824
(*Note.—This is a reprint of Volta's Mem. in* tom. i. *dell' Istit. Nat. Ital.; and of Bellani's Mem. in* tom. ii. *degli Opuscoli Matem. e Fisica.*)

† Volta e Configliachi. Lettera del . . Volta al Configliachi . . sopra esperienze ed osservazioni da intraprendersi sulle Torpedini. Como 15 Luglio, 1805.

Risposta del Configliachi. Porto Veneto, 6 Agosto, 1805. 8vo. 34 pp. (*Ann. di Chim. di Brugnatelli,* xxii. 223-249.) *Pavia,* 1805

Volta, A. e Mazari. Giudizio definitivo del Prof. Conte Volta sulla questione dei Paragrandini. 8vo. *Venezia,* 1823
(*Note.—This is a letter to Mazari . . from Volta, signed at Como 9th July, 1823, together with a few lines prefaced by Mazari. The letter (without MS. preface) was printed at Como.*)

Voltolini, Rudolph. Die Anwendung der Galvanokaustik im Innern des Kehlkopfes und Schlundkopfes nebst einer Rurzen Anleitung zur Laryngoskopie und Rhinoskopie. 8vo. *Wien,* 1867

Von Dalberg. (*Vide* Dalberg.)

Von Guericke. (*Vide* Guericke.)

Von Kolke. (*Vide* Kolke.)

Von Weber. (*Vide* Weber.)
(*Note—All names having the prefix Von will be found entered under the first letter of next portion of the name, as above.*)

Vorsselmann de Heer, Pieter Otto Coenraad. *Born September 20, 1809, at Valburg, Geldern; died December 26, 1841, at Utrecht.*

Note sur le calcul de l'inclinaison magnétique. 8vo. (*Bibl. Univ.* lx. 1835.) *Genève,* 1835

Waarneming van vallende sterren in d. nacht v. 12-13 Nov. (*Algem. Konst- en Letterb.* 1836.) 1836

Electro-magnet. Proefningen met gegoten ijzer. (*Algem. Konst- en Letterb.* 1836.) 1836

Middel om van de Magneto-Electriciteit tot chemische ontledingen zich te bedienen. (*Algem. Konst- en Letterb.* 1837.) 1837

Over de Thermo-Electriciteit van het Kwikzilver. 2 Aufsätze. (*Algem. Konsten Letterb.* 1838.) 1838

Jets over eene proef van de la Rive. 8vo. (*Algem. Konst- en Letterb.* 1838.) 1838

Vorsselmann de Heer, Pieter Otto Coenraad.—*continued.*

Over het Electro-magnetismus als bewegende Kracht. (*Meijlink's Nieuwe Schei-Artsenijmeng-en Naturkundige Bibl.* iii. 1838; *auch Pogg. Ann.* xlvii.) 1838

† Théorie de la Télégraphie électrique, avec la description d'un nouveau Télégraphe fondé sur les actions physiologiques de l'Électricité. 8vo. 30 pp. 1 plate.   *Deventer,* 1839

Über d. thermische Wirkung elektrischer Entladungen. 8vo. (*Pogg. Ann.* xlviii. 1839.)   *Leipzig,* 1839

Recherches sur quelques points de l'électricité voltaïque. 2 prt. 8vo. 78 pp. 2 plates. (*Bull. de Sciences Phys. en Néerlande, and from Freidlander's Cat. of* 1867, p. 43.)   *Rotterdam,* 1839-40

Über d. Übergangswiderstand. 8vo. (*Pogg. Ann.* liii. 1841.) *Leipzig,* 1841

Vruggink, W. Telegrafie gegrond op de natuurkunde. Eenvondig en bevattelijk voorgesteld. 8vo. 70 pp.   *Rotterdam,* 1856

W.

W. (*Vide* Anor. Elect. 1817.)

W., F. (*Vide* Anon. Elect. 1745.)

W., J. (*Vide* Anon. Meteorolog. Phen. Aurora, 1721.)

† **Waddel, J.** and **Knight, G.** A Letter from Captain Waddel to N. Franks concerning the effects of Lightning in destroying the polarity of a Mariner's Compass; to which are subjoined some Remarks thereon by G. Knight. Read April 13, 1749. 4to. 7 pp. (*Phil. Trans.* xlvi. 111.) *London*, **1849-50**

Wagner. Untersuchungen über die Contractilität der Milz mittelst des electromagnetischen Rotations-Apparates. (*Jena'sche Annalen*, 1849, Heft i.) **1849**

Wagner, Johann Georg. *Born* . . . *at Breslau; died October* 31, 1756, *at Liegnitz.* Erforschung der Ursachen der electrischen Wirkungen. 8vo. *Liegnitz*, **1747**

Wagner, J. P. Invented, in 1836, an Electro-Magnetical Rotation Apparatus; in 1838, a little Electro-Magnetical Carriage; and in 1840, an Apparatus for producing sound in unmagnetical metals by means of the Interrupted Galvanic Current. (*Poggendorff*, ii. 1241, says: "*Auf der Naturforscher-Versamml. zu Erlangen durch Prof. R. Böttger vorgezeigt.*")

Erfinder des Electro-magnet Hammers. Apparatus for the automatic closing and opening of a Voltaic arrangement, exhibited 25th February, 1837. (*Vorzeigte im physikal Verein zu Frankfurt a. M.* 27th Feb. 1837. *Siehe auch Neeff in Pogg. Ann.* xlvi. S. 107.)

† Erfolge der Bestrebung, den Elektromagnetismus als Triebkraft nutzbar zu machen. 8vo. 18 pp. 1 plate. (*Besonder-Abdruck aus dem* liii. Bde. *der Sitzungsb. der K. Akad. d. Wissens.*) *Wien*, **1866**

Wagner, R. C. Erzehlung derer zu Helmstadt am 17ten Martii . . gesehnen Meteorum igneorum welche bestanden in einer starken Helle-und-Luft Erleuchtung, &c. 4to. 44 pp. *Helmstadt*, **1716**

Wagner, Rud. Über den feineren Bau des elektrischen Organs im Litterrochen. 4to. 28 pp. 1 plate. (*Abhandl. K. Gesellsch. d. Wissen. zu Göttingen besonders abgedruckt.* iii. Bande.) *Göttingen*, **1847**

Waitz, Jacob Siegismund von. *Born May* 16, 1698, *at Gotha; died November* 7, 1777, *at Berlin.* (*From Pogg.* ii. 1243.)

(*Note.—This seems to be an erroneous name; his work has the initials T. H.*)

† Abhandlung : Von der Electricität und deren Ursachen : welche . . den Preis erhalten hat.

† Anonymous (*i.e.* Waitz). Zweite Abhand.: Von der Natur der Elect. . . des Drucks würdig geschätzt worden.

† Anonymous (*i.e.* Waitz). Dritte Abhand.: Von den Eigenschaften, Wirkungen und Ursachen der Elect. . . des Drucks würdig geschätz worden.

† Anonymous (*i.e.* Waitz). Dissertation sur la cause de l'Electricité des corps, et des phénomènes qui en dépendent. 4to. (*These four treatises contained in* 1 vol. *of* 237 pp. *and* 5 plates.) *Berlin*, **1745**

Waldschmidt, J. J. De magnete. 4to. (*From Pogg.* ii. 1246; *Lamont's Handb. says* 1682.) *Marperg*, **1683**

Diss. Meteori ignei in aere nuper conspecti considerationem physicam sist. 4to (*From Pogg.* ii. 1246.) *Marburg*, **1683**

Waldung (Baldung) **Wolfgang.** Disput. de meteororum causis in genere et de meteoris ignibus puris in specie. (*From Pogg.* ii. 124.) *Altdorf*, **1605**

Walker. Account of Experiments with a Constant Voltaic Battery. 4to. (*Phil. Trans. ?*) 1839

Walker, C. V. Electrotype Manipulation. 12mo. *London*, 1837

Electrotype Manipulation. Third edition. 12mo. *London*, 1841

† The effects of a Lightning-Flash on the Steeple of Brixton Church, and observations on Lightning Conductors generally. Large 8vo. 18 pp. 1 plate. (*Proceed. Lond. Elect. Soc.*) *London*, 1842

† On the action of Lightning Conductors. La. 8vo. 15 pp. 1 plate. (*Proceed. Lond. Elect. Soc.*) *London*, 1842

† Memoir on the difference between Leyden Discharges and Lightning Flashes, &c. La. 8vo. 42 pp. (*Proceed. Lond. Elect. Soc.*) *London*, 1842

Electrical Magazine, begun 1843. 8vo. *London*, 1843

† Electrotype Manipulation. 2 parts. 1st part, 1844, 14th edition. 12mo. 60 pp. 2nd part, 1843, 7th edition. 60 pp. (*Forms part of his "Manipulations in the Scientific Arts."*) *London*, 1843

† Die Galvanoplastik für Künstler, Gewerbetreibende und Freunde der Numismatik von C. Walker. Nach der 10ten Auflage mit Anmerk. von Dr. C. H. Schmidt. 177 pp. 4 plates. *Weimar*, 1843

† On the Electricity of Paper-Mills, and an Analysis of Friction. 8vo. 3 pp. (*Elect. Magaz. for* Oct. 1845.) *London*, 1845

† Replies to certain questions relative to the Electric Telegraphs in England. 8vo. 14 pp. 2 tables. *London*, 1850

† Galvanoplastik für Künstler u. Freunde d. Numismatik. Deustch V. Schmidt 2 Aufl. vermehrte nach der 18 Aufl. des Engl. Werks; nebst Zusätz. des Uberset. 1850. 8vo. 194 pp. 5 plates. *Weimar*, 1850

† Electric Telegraph Manipulation. 2nd Thous. 16mo. 107 pp. Cuts. (*Forms Part v. of Manipulations in the Scientific Arts. Vide Magnier, for French translation.*) *London*, 1850

Gemeinfasslicher Unterricht über die elektrischen Telegraphen mit ihren neuesten Einrichtungen und Vervollkomnungen. Bearbeitet und durch Zusätze vermehrt von Chr. Heinr. Schmidt. 8vo. 157 pp. 6 plates. *Quedlinburg*, 1851

Nouveau Manuel complet de la Télégraphie électrique . . traduit de l'anglais par M. D. Magnier. *Paris*, 1851

† Telegraphic Train-Signals. Electro-Magnetic Telegraph Semaphore. 4to. 2 pp. *London*, 1865

† Train Signalling in Theory and in Practice. Reprinted from the "Popular Science Review" for April, 1865. 8vo. 19 pp. 1 woodcut. *London*, 1865

† Manipulations électrotypiques; ou, Traité de Galvanoplastie . . traduit de l'anglais sur la 18e édition. 7e édition, par M. J. Fau. 12mo. 184 pp. *Paris*, 1866

† A Few Hours in the Signal-Box of the South-Eastern Company (London-Bridge Station-yard) on Easter Monday, April 2, 1866. Reprinted from the "Railway News" of April 7. 8vo. 8 pp. *London*, 1866

(*Vide* also Lardner and Walker.)

(*Vide* also London Electrical Society.)

(*Vide* La Rive, De.)

† **Walker, E.** On Mr. Bennet's Electrometer. 8vo. 2 pp. (*Phil. Mag.* xli. 415.) *London*, 1813

† On Electricity. 8vo. 3 pp. 1 plate. (*Phil. Mag.* xlii. 161.) *London*, 1813

† On Electricity by position or induction. 8vo. 3 pp. (*Phil. Mag.* xlii. 215.) *London*, 1813

† Electrical Phænomena. 8vo. 1 p. (*Phil. Mag.* xlii. 485.) *London*, 1813

† On Electricity; in answer to Mr. Singer's remarks 8vo. 3 pp. (*Phil. Mag.* xlii. 364.) *London*, 1814

† **Walker, Ed.** Terrestrial and Cosmical Magnetism. The Adam Prize Essay for 1865. 8vo. 336 pp. 10 plates. *Cambridge, 1866*

† **Walker, R.** A Treatise on Magnetism, &c. 8vo. 226 pp. 7 plates. *London, 1794*
A Treatise on the Magnet. 8vo. *London, 1798*

Walker, S. C. Researches concerning the periodical Meteors of August and November. 4to. (*Trans. Amer. Phil. Soc.* New Series, vol. viii.)
Philadelphia, 1843

† **Walker, W.** (Captain). The Magnetism of Ships and the Mariner's Compass; being a rudimentary exposition of the Induced Magnetism of Iron in seagoing vessels. 8vo. 207 pp. 1 plate. *London, 1853*
Het magnetismus van schepen en het scheepskompas, zijnde eene grondige verklaring van het opgewekte magnetismus van het scheepsijzer, en zijnen, invloed op de Kompasnaald, op verschillende breedten en in verschillende omstandigheden Naar het Engelsch, door J. M. Heijbroek. 12mo. Plates.
Amsterdam, 1854

Walker, W. Memoirs of Distinguished Men of Science in 1807-8. 8vo.
London, 1862

† **Walker and Hunt.** Memoirs of the Distinguished Men of Science of Great Britain, living A.D. 1807-8, by W. Walker, jun., with an Introduction by Robert Hunt, F.R.S. 2nd edition. 160 pp. *London, 1864*

† **Walker and Mitchel.** Moigno says, "Plusieurs Mémoires de . . Walker et Michel ont été publiés (on the subject of the Retardation of Signals) in the "Journal Astronomique de Cambridge" de 1848 (U.S.).

Walkiers. Nachricht von einer neuen Elektrisirmaschine des Herrn Walkiers von St. Amand. 8vo. 4 pp. (*Lichtenberg's Mag.* iii. 1 St. p. 118.)
Gotha, 1785

Wall. Experiments of the Luminous Qualities of Amber, Diamonds, and Gumlac. In a letter to Sloane. 4to. (*Phil. Trans.* 1708, xxvi. 69, 76.)
London, 1708
Luminous Qualities of Amber, &c. (*Phil. Trans.* Ab. iv. 275.)

Wall, Arth. Improvement of Metals by Voltaic Electricity. 8vo. *London, 1846*

† **Waller, R.** Essays of Natural Experiments made in the Academy del Cimento . . Englished by R. Waller. 4to. 160 and 10 pp. 19 plates. (*Latimer Clark* says 2nd edition *in Firenze*, 1691.) *London, 1684*

† **Wallerius, G.** Versuch von der Vegetation des Quecksilbers ohne Beymischung anderer Metalle. 8vo. 8 pp. (*K. Schwed. Akad. Abh.* xvi. 257, *and in Vetensk Acad. Handl.* 1754.) *Hamburg and Leipzig, 1754*

Wallerius, Martin Joh. Dissert. Meteorologia generalis. *Upsal, 1736*

Wallerius, N. Observationes nonnullæ circa lumen nocturnum boreale Upsaliæ, etc. habitæ. (*Act. Litt. et Scient. Sueciæ,* 1737.) *Stockholm, 1737*

Wallis, John. Of an unusual Meteor. 4to. (*Phil. Trans.* 1677.) *London, 1677*
On the Production and Effects of Hail, Thunder, and Lightning. 4to. (*Phil. Trans.* 1697-8.) *London, 1697-8*

† A Letter to Sloan, Sec. R. S. concerning some supposed alteration of the Meridian Line; which may affect the Declination of the Needle, and the Poles' elevation, Oxford, June 21, 1699. 4to. 2 pp. (*Phil. Trans.* xxi. *for* 1699, p. 285.) *London, 1699*

On the Invention and Improvement of the Mariner's Compass. 4to. (*Phil. Trans.* 1701.) *London, 1701*

Letter relating to John Sommer's Treatise of Chartham News, and some Magnetic Affairs. 4to. (*Phil. Trans.* 1701.) *London, 1701*

Letter concerning Captain Edmund Halley's Map of Magnetic Variations, &c. 4to. (*Phil. Trans.* 1702.) *London, 1702*

Wallis, John.—*continued.*

† A Letter to Capt. Edmund Halley, concerning the Captain's Map of Magnetick Variations, and some other things relating to the Magnet. 4to. 7 pp. (*Phil. Trans.* xxiii. *for* 1702-3, p. 1106.) *London*, 1704

Wallmark, Lars Johan. Om d. S. K. swarta diamanteus lednings förmaga för electriciteten. 8vo. (*Pogg. Ann.* lxxxiii. 1849.) *Lulea*, 1849

Walsh, John. On the Electric Property of the Torpedo. 4to. (*Phil. Trans.* 1773.) *London*, 1773

On the Torpedo found on the Coast of England. 4to. (*Phil. Trans.* 1774.) *London*, 1774

Waltenhofen, A. K. Edler von. Über d. Kohlenzink-Kette bei Anwend. verschiedn. Ladungsflüssigk. 8vo. (*Dingler's Journ.* clxiv. 1862.) *Stuttgart*, 1862

† Über das elektro-magnetische Verhalten des Stahles. 8vo. 30 pp. 1 plate lith. (*Aus den Sitzungsberichten der Kaiserl. Akad. der Wissenschaften*, xlviii. iv. Heft, p. 518.) *Wien*, 1863

† Über eine anomale Magnetisirung des Eisens. 8vo. 5 pp. (*Sitzungsb. d. K. Akad. d. Wiss. Wien*, 1863, xlviii. iv. Heft, p. 564.) *Wien*, 1863

Einige Beobachtungen über das elektrische Licht in höchst verdünnten Gasen. 8vo. 11 pp. (*Abdr. aus den Sitzungsberichten der Kaiserl. Akademie der Wissenschaften.*) *Wien*, 1865

Elektro-magnetische Untersuchungen mit besonderer Rücksicht auf die Anwendbarkeit der Müller'schen Formel. 1 Abhandl. enthaltend die Versuche mit massiven Cylindern. 8vo. 28 pp. (*Abdruck aus den Sitzungsberichten der Kaiserl. Akademie der Wissenschaften.*) *Wien*, 1865

Über den Lullin'schen Versuch und die Lichtenberg'schen Figuren. 8vo. 18 pp. (*Aus den Sitzungsberichten der Kaiserl. Akademie der Wissenschaften.*) *Wien*, 1866

† Über d. Grenzen d. Magnetisirbarkeit des Eisens und des Stahles. 8vo. 19 pp. (*Siszungsb. K. Akad. d. Wissensch.* ii. Abth. April Heft, 1869, *Vorgel.* 29th April, 1869.) *Wien*, 1869

† Elektromagnetische Untersuchungen mit besonderer Rücksicht auf die Anwendbarkeit der Müller'schen Formel 2 Abhandlung, enthaltend die Versuche mit discontinuirlichen Eisenmassen, nebst einem Anhang über die Grenzen der Giltigkeit des Lenz-Jacobischen Gesetzes. 8vo. 26 pp. 1 plate. (*Sitzb. d. K. Akad. Vorgelegt*, 19 Mai, 1870, lxi.) *Wien*, 1870

† Über die Anziehung, welche eine Magnetiserungs-Spirale auf einen beweglichen Eisenkern ausübt. 8vo. 16 pp. 1 plate. (*Abdr. a. d. Sitzb. d. Akad.* lxii. 21 Juli, 1870.) *Wien*, 1870

† Über einen einfachen Apparat zur Nachweisung des magnetischen Verhaltens eiserner Röhren. 8vo. 3 pp. 1 plate. (*Sitzb. d. K. Akad. d. Wiss.* lxii. 14 Juli, 1870.) *Wien*, 1870

Über elektromagnetische Tragkraft. 8vo. 16 pp. 2 plates. (*Sitzb. d. K. Akad. d. Wiss.* lxi. 12 Mai, 1870.) *Wien*, 1870

Walter, A. H. Beschouwing d. elektro-magnet. telegrafs. 8vo. 1 plate. *Amsterdam*, 1849

Waltershausen. (*Vide* Sartorius Von Waltershausen.)

Walther, A. F. Systematisches Repertorium über die Schriften sämmtlicher historischer Gesellschaften Deutschlands. 8vo. *Darmstadt*, 1845

Walther, P. F. Ueber die therapeutische Anwendung und Technicismus der galvanischen Operationen. 8vo. 2 plates. *Wien*, 1803

† **Ward.** The Ocean Marine Telegraph. 8vo. 4 pp. 4 plates. *London*, 1860

Ward's International Prize Medal Signal Telegraph. 8vo. *London*, 1863

† **Ward, S.** Magnetis reductorium theologicum tropologicum, in quo ejus novus verus et supremus usus indicatur. 12mo. 166 pp. *Londini*, 1639

† **Warden,** D. B. Description and Analysis of the Meteoric Stone which fell at Weston, in North America, the 4th December, 1807. 8vo. 3 pp. (*Phil. Mag.* xxxvi. 32, *and Ann. de Chim.* March, 1810.) *London,* 1810

† **Ware.** In his Remarks on Diseases of the Eye confirms the utility of Electricity in these diseases, by an example. (*Kuhn, Hist.* ii. 183.)

Wargentin, Pehr Vilhelm. *Born September* 22, 1717, *at Sunne Prestgard, Jemtland; died December* 13, 1783, *at Stockholm.*

† Beobachtungen an der Magnetnadel. 8vo. 8 pp. (*K. Schwed. Akad. Abh.* xii. 54.) *Hamburg and Leipzig,* 1750

Observationer på magnet-nalen. 8vo. (*Vetensk Acad. Handl. Abhandl.* an. 1750, p. 53 ; *and in Swedische Akad. Abhandl.* 1750, p. 54.) *Stockholm,* 1750

On the Variation of the Magnetic Needle. 4to. (*Phil. Trans.* year 1751, p. 126.) *London,* 1751 ?

† Geschichte der Wissenschaften : Vom Nordscheine. 8vo. 10 pp. (*K. Schwed. Akad. Abh.* xiv. 169.) *Hamburg and Leipzig,* 1752

† Fortsetzung der Gesch. vom Nordscheine. 8vo. 11 pp. (*K. Schwed. Akad. Abh.* xv. 85.) *Hamburg and Leipzig,* 1753

† Om nord-skenet. 8vo. (*Schwed. Akad. Abhandl.* an. 1752, p. 169, 1753, p. 85.) *Stockholm,* 1752-3

Wartmann, Elie François. *Born November* 7, 1817, *at Geneva.*

† Essai historique sur les phénomènes et les doctrines de l'électro-chimie. 8vo. 167 pp. *Genève,* 1838

† Note sur de nouvelles expériences sur la production des sons musicaux, par M. Delezenne . . Communiqué par M. E. Wartmann, et Observations du Rédacteur (*i.e.* De la Rive). 8vo. 2 pp. (*Bibliot. Univ.* xvi. 199, Juillet, 1838.) *Genève,* 1838

† Des travaux et des opinions des Allemands sur la Pile voltaïque. 8vo. 45 pp. (*Arch. de l'Elect.* tom. i. p. 31.) (*Arago's copy.*) *Genève,* 1841

† Mém. sur la Diathermansie électrique des couples métalliques. (Lu 18 Juin, 1840.) 4to. 23 pp. 1 plate. (*Soc. de Phys. de Geneva,* ix. 119.) *Genève,* 1841

† Expériences sur la non-caloricité propre de l'électricité. 8vo. 6 pp. (*Archives de l'Elect.* ii. 1842.) *Genève,* 1842

† Sur les relations qui lient la lumière à l'électricité, lorsque l'un des deux fluides produit une action chimique. 8vo. 7 pp. (*Arch. de l'Elect.* tcm. ii. p. 596.) (*Arago's copy.*) *Genève,* 1842

† Sur le refroidissement des corps électrisés. 8vo. (*Arch. de l'Elect.* tom. iii. 429. (*Arago's copy.*) *Genève,* 1843

† Notice sur divers travaux de Wheatstone. 8vo. 16 pp. 1 plate. (*Arch. de l'Elect.* tom. iii. p. 468.) (*Arago's copy.*) *Genève,* 1843

† Recherches historiques sur les courbes magnétiques. 8vo. (*Arch. de l'Elect.* iii. 1843.) *Genève,* 1843

† Mém. sur divers phénomènes d'induction. 8vo. 26 pp. (*Acad. de Bruxelles,* Ext. x. No. 10, *Bulletins.*) (*Arch. de l'Elect.* iv. 34.) *Bruxelles*

† Deuxième Mém. sur l'induction. 8vo. 11 pp. (*Acad. de Bruxelles. Ext.* xii. No. 10, *Bulletins. Commun. le* 19 *Mars* 1845, *à la Société Vaudoise des Sciences, and to others.*) *Bruxelles,* 1845

† De la méthode dans l'électricité et le magnétisme ; à propos du "Trattato del magnetismo e dell' elettricità del Sig. Zantedeschi." 8vo. 15 pp. (*Arch. de l'Elect.* tom. v. p. 320.) (*Arago's copy.*) *Genève,* 1845

† Troisième Mém. sur l'induction. 8vo. 18 pp. 1 plate. (*Acad. de Bruxelles. Ext.* xiv. No. 3, *Bulletins. Commun. le* 20 *Mai,* 1846, *à la Société Vaudoise des Sciences, and to others.*) *Bruxelles,* 1846

† Mém. sur deux Balances à réflexion. 4to. 31 pp. Plates. (*Soc. de Genève,* xi. 115, 1846.) *Genève,* 1846

Wartmann, Elie François.—*continued.*

† Quatrième Mém. sur l'induction. 8vo. 16 pp. 1 plate. (*Archives de l'Elect.*)
(*Faraday's copy.*) *Genève*

† Cinquième Mém. sur l'induction. 8vo. 9 pp. 1 plate. (*Acad. de Belgique, Ext.*
xv. No. 4, *Bulletins.*) *Bruxelles*

Sixième Mém. sur l'induction. 8vo. (*Bibl. Univ. de Genève*, viii. 45, 1848.)
Genève, 1848

† Septième Mém. sur l'induction. 8vo. 15 pp. (*Bibl. Univ. de Genève*, Juillet,
1848.) *Genève*, 1848

† Sixième et Septième Mémoires sur l'Induction. 8vo. 18 pp. (*Ext. Acad. Rle.*
de Belgique, xv. No. 7 *des Bulletins.*) (*Faraday's copy.*) *Bruxelles*
(*Note.*—*The seventh Memoir was printed separately* (*see* 1848), *and somewhat*
differently.)

Nouv. recherches relatives à l'action du magnétisme sur différents corps. 8vo.
(*Arch. d. Scienc. Phys. et Nat.* viii.) *Genève*, 1848

Huitième Mém. sur l'induction. 8vo. (*Bibl. Univ. de Genève*, Jan. 1850, viii.
35.) *Genève*, 1850 ?

† Recherches sur l'électricité animale par M. E. Du Bois Reymond (Analysis).
8vo. 8 pp. (*Bibl. Univ. de Genève*, Juin, 1850.) *Genève*, 1850

† Note sur les courants électriques qui existent dans les végétaux. 8vo. 8 pp.
(*Bibl. Univ. de Genève*, Dec. 1850.) *Genève*, 1850

Note sur la polarisation de la chaleur atmosphérique. Lu 19 Juil. 1849. 4to.
(*Soc. de Phys. de Genève*, xii. 349.) *Genève*, 1850

† Note sur quelques expériences faites avec le fixateur électrique. 8vo. 8 pp.
(*Bibl. Univ. de Genève*, Aug. 1852.) *Genève*, 1852

Tentative d'éclairage élect. pour l'éclairage public. 8vo. (*Archiv. de Sciences*
Phys. et Natur. xxi. *et* xxxvi.) *Genève*, 1852 and 1857

† Sur l'éclairage électrique. 8vo. 12 pp. 1 plate. (*Bibl. Univ. de Genève*, Dec.
1857.) *Genève*, 1857

† Description d'appareils destinés à établir une correspondance immédiate entre
deux quelconques des stations situées sur une même ligne télégraphique.
8vo. 22 pp. 1 plate. (*Bibl. Univ. de Genève*, Mai, 1853.) *Genève*, 1853

† Recherches sur la conductibilité des minéraux pour l'électricité voltaïque. Lu
Nov. 20, 1851. 4to. 12 pp. and 8 pp. of tables. (*Soc. de Phys. de Genève*,
xiii. 199.) *Genève*, 1854

† Sur la transmission simultanée de dépêches électriques entre deux·stations
télégraphiques jointes par un seul fil de ligne. 8vo. 7 pp. 1 plate. (*Bibl.*
Univ. de Genève, Mars, 1856.) *Genève*, 1856

De l'influence de la pression sur la conductibilité électrique des métaux. 8vo.
(*Archiv. des Scien. Phys. et Nat.* Nouv. Période, iv. 1859.) *Genève*, 1859

† Description du compensateur voltaïque destiné à maintenir l'intensité d'une
pile. Lu à la Classe d'Industrie (Acad. de Genève) le 19 Déc. 1853. 8vo.
7 pp. 1 plate. (*Archives des Sciences* . . Janv. 1858. Dated 1 Sep. 1853.)
Genève, 1858 ?

† Mémoire sur l'échange simultané de plusieurs dépêches télégraphiques entre
deux stations qui ne communiquent que par un fil de ligne. 4to. 15 pp.
(*Lu à la Soc. de Phys. et Hist. Nat. dans les Séances du* 16 Avril 1857, *et du*
4 Mars 1860.) *Genève*, 1860

Appareil électro-magnétique destiné à remplacer les compteurs. 8vo. (*Archiv.*
de Scien. Phys. et Nat. Nouv. Période, xii. 1861.) *Genève*, 1861

† Sur les vibrations qu'un courant élect. discontinu fait naître dans le fer doux ;
et sur la non-existence d'un courant électr. dans les nerfs des animaux
vivants. 8vo. 5 pp. (*Ext. Acad. de Belgique*, xiii. No. 5, *Bulletins. Ext.*
d'une lettre de M. E. Wartmann . . *M. Quetelet.*) *Bruxelles*

Wartmann, Elie François.—*continued.*

Descrip. du Télégraphe militaire de M. Hipp. 8vo. (*Bibl. Univ. de Genève,* xxxiii. 109.) *Genève*

† Sur les nouveaux Rapports entre la Chaleur, l'Elect. et le Magnétisme. 8vo. (*Bibl. Univ. de Genève,* Nouv. Série, i. 457.) *Genève*

Wartmann, Louis François. *Born January* 7, 1793, *at Geneva.*

† Notice sur l'Aurore boréale observée à Genève le 18 Oct. 1836. Lu à la Soc. de Genève . . 3 Nov. 1836. 8vo. 11 pp. (*Tiré de la Bibl. Univ. de Genève,* Oct. 1836.) *Genève,* 1836?

† Notice sur les Météores périodiques du 13 Nov. Lu à la Soc. de Genève, 15 Déc. 1836. 8vo. 13 pp. (*Tiré de la Bibl. Univ. de Genève,* Juin, 1837.) *Genève,* 1837

† Notice sur les Etoiles filantes. Lu à la Soc. de Genève, 6 Sep. 1838. 8vo. 7 pp. (*Tiré de la Bibl. Univ. de Genève,* Août, 1838.) *Genève,* 1838

† Mèm. sur les Etoiles filantes observées à Genève . . 10-11 Août, 1838. Lu à la Soc. de Genève, 21 Mars 1839. 8vo. 77 pp. 1 plate. (*Correspondance* . . *de Quetelet,* tom. xi.) *Bruxelles,* 1839

† Note sur l'apparition remarquable d'étoiles filantes accompagnée d'une aurore boréale et d'une perturb. . . magn. 21 Sep. 8vo. 7 pp. (*Correspondance* . . *de Quetelet,* tom. xi., *and in Bibl. Univ. de Genève,* Nov. 1840.) *Genève,* 1840

† Note sur divers phénomènes météorologiques. 8vo. 8 pp. (*Acad. Rle. de Bruxelles, Bulletins,* tom. vii.) *Bruxelles*

† Nouveau cas d'une apparition d'aurore boréale . . (Lettre) à la Redaction. 8vo. 2 pp. (*Tiré de la Bibl. Univ. de Genève,* Déc. 1843.) *Genève,* 1843

Waszmuth, Ant. Über die Abhängigkeit des erregten Magnetismus von den Dimensionen der Magnetisirungs spirale. 8vo. 6 pp. *Wien,* 1868

Über die Ströme in Nebenschliessungen zusammengesetzter Ketten. 8vo. 9 pp. (*Aus den Sitzungsb. d. K. Akad. d. Wiss.*) *Wien,* 1868

† Über ein neues Verfahren, den Reductionsfactor einer Tangentboussole zu bestimmen. 8vo. 7 pp. (*Sitzb. d. K. Akad. d. Wissensch.* ii. Abth. Jan. Heft, 1870.) *Wien,* 1870

† **Watkins.** A popular sketch of Electro-Magnetism; or Electro-Dynamics. 8vo. 83 pp. 3 plates. *London,* 1828

A popular sketch of Electro-Magnetism. 8vo. (*From Dove,* p.153.) *London,* 1832

† **Watkins,** F. On the Magnetic powers of Soft Iron. (Read April 25, 1833.) 4to. 10 pp. (*Phil. Trans.* 1833, p. 333.) *London,* 1833

† On Thermo-Electricity. 8vo. 4 pp. (*Phil. Mag.* September, 1837.) *London,* 1837

† On the Thermo-Electric Spark, as obtained from a single pair of Metallic Elements. 8vo. 1 page. (*Phil. Mag.*) *London*

Electro-Magnetism. 12mo. *London,* 1856

Watkins, Fr. A particular Account of the Experiments published to this time on Electricity. 8vo. *London,* 1747

Watson, William. *Born* 1715, *at London; died May* 10, 1787, *at London.*

† Experiments and Observations tending to illustrate the Nature and Properties of Electricity. Read at several Meetings of the Royal Society between March 28 and October 24, 1745, here printed with alterations. 4to. 21 pp. (*Phil. Trans.* xliii. 481.) *London,* 1744-5

Further Experiments and Observations, &c. 4to. (*Phil. Trans. Ab.* x. 290.) *London,* 1745-6

† Experiments and Observations tending to illustrate the Nature and Properties of Electricity; in one letter to Martin Folks, President, and two to the Royal Society. With Continuation and Preface. 8vo. 59 pp. *London,* 1746

Watson, William.—*continued.*

† Sequel to the Experiments and Observations tending to illustrate the Nature and Properties of Electricity . . Addressed to the Royal Society. 8vo. 80 pp.
London, 1746

† Observations upon so much of Le Monnier the younger's Memoir, lately presented to the Royal Society, as relates to the communicating the Electric Virtue to Non-Electrics. Read January 29, 1746-7.) 4to. 8 pp. (*Phil. Trans.* xliv. 388.)
London, 1746-7

† A continuation of a Paper concerning Electricity by W. Watson, printed in these Trans. N. 477, Article i. ending p. 501. Read February 6, 1745-6. 4to. 10 pp. (*Phil. Trans.* xliv. 695.)
London, 1746-7

† A Sequel to the Experiments and Observations tending to illustrate the Nature and Properties of Electricity; in a letter to the Royal Society. Read October 30, 1746. 4to. 46 pp. 1 plate. (*Phil. Trans.* xliv. 704.)
London, 1746-7

† An Account of the Experiments made by some gentlemen of the Royal Society in order to measure the absolute Velocity of Electricity. Read October 27, 1748. 4to. 6 pp. (*Phil. Trans.* xliv. 491.)
London, 1748

† An Account of the Experiments made by several gentlemen of the Royal Society in order to discover whether or no the Electric power, when the conductors thereof were not supported by Electrics *per se,* would be sensible at great distances. With an Inquiry concerning the respective Velocities of Electricity and Sound : to which is added an Appendix containing some further Inquiries into the Nature and Properties of Electricity. 4to. 72 pp. (*Phil. Trans.* xlv. 49 and 93.)
London, 1748

(*Vide* Anon. Electricity, 1748.)

† A Letter from W. Watson, declaring that he, as well as many others, have not been able to make Odours pass through Glass by means of Electricity ; and giving a Particular Account of Bose at Wittemberg his Experiment of Beatification, or causing a Glory to appear round a Man's Head by Electricity. Read March 1, 1749-50. 4to. 9 pp. (*Phil. Trans.* xlvi. 348.)
London, 1749-50

Some Observations relating to the Lyncurium of the Ancients. 4to. (*Phil. Trans.* 1759, li.)
London, 1759

Observations upon the effects of Electricity applied to a Tetanus, or Muscular Rigidity, of four months' continuance. 4to. (*Phil. Trans.* 1763, p. 10.)
London, 1763

Watson, Walter. General Telegraphic List of Ships' Names. 18mo. *London,* 1840

† **Watson,** T. T. W. A few Remarks on Electrical Illumination, &c. 8vo. 31 pp.
London, 1853

† Remarques sur l'état actuel . de l'éclairage électrique et sur la production sans frais de l'électricité, avec une description des inventions brevetées de l'auteur relatives aux batteries galvaniques et aux lampes électriques. *Paris,* 1855

† **Watt,** A. Electro-Metallurgy practically treated. 8vo. 116 pp. *London,* 1860
Electro-Metallurgy. New edition, enlarged.

† Bibliotheca Britannica. 4 vols. 4to. *London,* 1824

† **Watts.** H. A Dictionary of Chemistry and the allied Branches of other Sciences. 5 vols. 8vo. *London,* 1872 ?

† Supplement to his Dictionary of Chemistry, bringing the Record of Chemical Discovery down to the end of the year 1869; including also several Additions to, and corrections of, former Results which have appeared in 1870 and 1871. 5 vols. 8vo. *London,* 1872

(*Vide* also Gmelin.)

Weare, R.   Electro-Motor.   (*From Du Moncel.*)

† **Webb,** F. C.   On the Practical Operations connected with paying out and repairing Submarine Telegraphic Cables.   8vo.   (*Proc. Inst. of Civil Engineers,* xvii. 1857-8.)   *London,* 1858

† A Treatise on the Principles of Electric Accumulation and Conduction, in two parts.   Part i.   8vo.   156 pp.   *London,* 1862

  (*Note.—Part* ii. *was never published, but a portion of it appeared in the* "*Electrician.*"*—*Ed.)

† On Inductive Circuits, or the Application of Ohm's Law to Problems of Electrostatics.   8vo.   9 pp.   (*Phil. Mag.* May.)   *London,* 1868

Weber, E.   Questiones physiologicæ de phænom. galvano-magneticis in corpore humano observatis.   Commentatio, 1836.   (*From Meyer.*)

Questiones physiologicæ de phænomenis galvano-magneticis in corpore humano observatis.   (*From Lawrence,* 1858.)   *Lipsiæ,* 1836

Weber, Ed. und E. H.   Wirkung des magnet-electr. Stromes auf die Blutgefässe. 8vo.   (*Muller's Archiv.* 1847, Heft ii. *und* iii.   1847

Weber, F. A.   Abhandlung von Gewittern u. Gewitterableitern.   8vo.
                                           Zurich und Leipzig, 1792

† **Weber,** H.   Ueber die Bestimmung des galvanischen Widerstandes der Metalldrähte, aus ihrer Erwärmung durch den galvanischen Strom nach absolutem Maase.   Inaug. Diss.   4to.   33 pp.   2 plates.   *Leipzig,* 1863

Weber, Joseph.   *Born September* 23, 1753, *at Rain, Bayern*!; *died February* 14, 1831.

Vom Luftelektrophor.   4to.   (*Neue Abhandl. d. Bairisch Akad. Philos.* i. 169.)
                                        München, 1778 or 9

Abhandlung von dem Luftelektrophor.   Zweite Auflage mit neuen Erfahrungen. neuen Instrumenten und mit einem Unterrichte von Zubereitung der brennbaren Luft, vermerht und bereichert.   8vo.   *Ulm,* 1779

Beschreibung des Luftelektrophors.   Nebst angehängten neuen Erfahrungen, neuen Instrumenten, einem Unterrichte von Zubereitung der brennbaren Luft, und verschiedener Versuche mit derselben.   Neueste mit der Beschreibung der elektrischen Lampe vermehrte Auflage.   8vo.   *Augsberg,* 1779

† Neue Erfahrungen, idioelektrische Körper ohne einiges Reiben zu elektrisiren. 8vo.   118 pp.   3 plates.   *Augsburg,* 1781

† Positiver Luftelektrophor; samt der Anwendung desselben auf einer Elektrisirmaschine.   Sm. 8vo.   118 pp.   2 plates.   *Augsberg,* 1782

Die Theorie der Elektricität.   (*Schriften der Berliner Gesellsch. Naturf-Freunde,* iv. 330.)   *Berlin,* 1783

† Theorie der Elektricität.   Sm. 8vo.   64 pp.   (*Ausgetheilt bei der Gradverleihung im* August 1784.)   (*Volta's copy.*)   *Dilingen,* 1784

Unterricht vom Verwahrungsmitteln gegen d. Gewitter f. d. Landmann.   8vo.   (*Pogg. says Dilling,* 1784.)   *Salzbourg,* 1784

† Theorie der Elektricität nebst Halfenreider's Vorschlag die Blizableiter zu verbessern.   8vo.   76 pp.   *Salzburg,* 1785

† Neue elektrische Versuche.   8vo.   24 pp.   *Salzburg,* 1786

Abhandlung vom Feuer.   8vo.   *Regensbourg,* 1788

† Vollständige Lehre von den Gesetzen der Electricität.   8vo.   368 pp.   2 plates.   (*This belongs to his* "*Vorlesungen aus der Naturlehre.*"   *Vide Ersch,* pp. 115 and 73.)   *München,* 1791

Physikal Chemie.   2 Aufl.   8vo.   *Landshut,* 1791

Über die Unwirksamkeit des Schietzens auf Gewitter.   8vo.   *Landshut,* 1791

† Der Galvanismus ; einer Zeitschrift.   1, 2, 3 Heft, 1802, 4 Heft, 1803.   8vo.
                                         Landshut, 1802-3

Lehrbuch der Naturwissenschaft.   4 Hefte.   8vo   *Landshut,* 1805

Weber, Joseph.—*continued.*

† Theorie der Elektricität. 8vo. 132 pp. *Landshut*, 1808

† Begriff und Construction des Doppel-Elektrophors aus Harz u. Glass. 8vo. 5 pp. (*Gilbert's Ann. d. Phys.* li. 198.) *Leipzig*, 1815

Neuer Versuch. den Galvanism. zu erklären. 8vo. (*Gilb. Ann.* li. 1815.) *Leipzig*, 1815

Der Galvanismus u. dessen Theorie. 8vo. *München*, 1816

Der thierische Magnetism aus dynamisch-chem. Kräften verständlich gemacht. 8vo. (*Gilb. Ann.* liv. 1816.) *Leipzig*, 1816

† Vom dynam. Leben der Natur überhaupt und vom elektrischen Leben im Doppelelektrophor insbesondere. 8vo. 151 pp. *Landshut*, 1816

Der Elektrophantes, u.s.w. 8vo. (*Gilb. Ann.* lv. 1817.) *Leipzig*, 1817

† Das Wesen der Electricität durch neue Versuche, mit seidenen Bandern, darges-telt. sammt Beschreibung und Theorie des Elektrophantes. 8vo. 88 pp. *Sulzbach*, 1819

Von den Meteorsteinen u. ihrem Entstehen. 8vo. *Landshut*, 1820

Vom Verhältniss d. Elektricität zum Magnetismus. 8vo. *München*, 1821

† Die Sicherung unserer Gebäude durch Blizstralableiter theoretisch und prak-tisch beleuchtet und bewahrt, samt einer Beurtheilung der Ableiter aus Stroh von Lapostolle. Eine Vorlesung. 8vo. 46 pp. *Landshut*, 1822

Vorlesungen aus der Naturlehre, 1—9 Abhandlungen. 8vo. *Landshut.*

Luftelektrophor in seiner Vervollständigung u. Zuruckführung seiner Erschei-nungen auf bestimmte Gesetze. 8vo. 1 plate. (*Sendschreiben an d. Kgl. Bayr. Hrn. Geh. Rath. Frh. C. E. v. Moll, etc. etc.* La. 8vo.) *München*, 1831

Weber, M. L. Über das Potential von Kreis und Spirale, sowie S. Verwendung in der Theorie inducirter elektr. Ströme. 4to. (*From Kohler's Cat.* No. 217 1871, p. 18.) *Leipzig*, 1868

† **Weber**, M. M. (Freih.) von. Das Telegraphen- und Signalwesen der Eisenbahnen. Geschichte und Technik desselben. 8vo. 319 pp. 1 plate. *Weimar*, 1867

Weber, R. Beskrivning öfwer elektro-magnetiska Telegraphen. Populär Fram-ställning. 1 Häft. 8vo. *Stockholm*, 1852

Weber, Wilhelm Eduard. *Born October* 24, 1804, *at Wittenberg.*

Ueber die Elasticität der Seidenfäden. (*Gott. gel. Anz.* 1835, St. 8.) *Göttingen*, 1835

Comm. de fili bombycini vi elastica. (*Comment. recent. Soc. Gott. a.* 1832—37, vii. 1841.) *Göttingen*, 1841

De natura chalybdis magnetica. 4to. *Lipsiæ*, 1843

† Elektro-dynam. Maasbestimmungen, Abh. i. 4to. 170 pp. (*Abh. bei Begrund d. K. Sächs. Gesellsch. d. Wiss.* 1846.) *Leipzig*, 1846

Electro-dynamische Maasbestimmungen insbesondere Widerstands messungen. 4to. 186 pp. (*Abhandl. der K. Sachs. Gesellschaft des Wissens. zu Leipzig.*) *Leipzig*, 1850

Elektro-dynamische Maasbestimmungen. Widerstandsmessungen. La. 8vo. (*Abhandl. d. Kon. Sächs. Gesellsch.* tom i. 1852.) *Leipzig*, 1852

Nine articles in Poggendorff's Annalen, vol. xlii. 1838 to vol. lxxxvii. 1852.

† Elektro-dynamische Maasbestimmungen insbesondere über Diamagnetismus. 4to. 93 pp. 1 plate. (*Abhandl. d. K. S. Gesells. d. Wiss. zu Leipzig,* i.) (*Faraday's copy.*) *Leipzig*, 1852
(*Note.—This is the third Abhandlung under the heading "Elektro-dyn. Maasbestim."*)

Weber, Wilhelm Eduard.—*continued.*

† Ueber die Anwendung der magnetischen Induction auf Messung der Inclination mit dem Magnetometer. 4to. 58 pp. 1 plate. (*Abhandl. d. Gesellsch. d. Wiss. in Göttingen*, 1853, p. 17.)   *Göttingen*, 1853

† Bestimmung der rechtwinkeligen Componenten der erdmagnetischen Kraft in Göttingen in dem Zeitraume von 1834-1853. Vorgelegt am 27 Nov. 1854. 4to. 46 pp. (*Abhandl. Gött. Gesellsch.* vi. 1856.)   *Göttingen*, 1854-6

Elektrodynam. Maassbestimmungen Abh. iv. : Zurückfuhr d. Stromintensitäts-stress auf mechan. Maass. mit R. Kohlrausch. 4to. (*Abh. d. K. Sächs. Gesellsch. d. Wiss.* v. 1857.)   *Leipzig*, 1857

Bestimm. d. galvan. Widerstands d. Dräthe aus d. Erwärmung. 4to.
                                                                   Leipzig, 1862

Zur Galvanometrie. (*Abh. d. Gött. Gesellsch. d. Wiss.* ix. 1862.)
                                                                   Göttingen, 1862

† Elektrodynamische Maassbestimmungen insbesondere Widerstandsmessungen. 2 Abdruck. Abhandl. ii. 4to. 186 pp. (*Aus den Abhandl. der Königl. Sächsischen Gesellschaft der Wissenschaften*, vol. i.)   *Leipzig*, 1863

† Elektrodynamische Maassbestimmungen, insbesondere über Elektrische Schwingungen. 5te Abhandl. 4to. 146 pp. (*Abhandl. d. Kön. Sächs. Gesellsch.* tom. ix. 1864.)   *Leipzig*, 1864

Elektrodynamische Maassbestimmungen insbesondere über Diamagnetismus. 2 Abdruck. 4to. 96 pp. (*Abhandl. d. K. Sächsischen Gesells. d. Wissen.*)
                                                                   Leipzig, 1867

† Elektrodynamische Maassbestimmungen insbesondere über das Princip der Erhaltung der Energie. 6 Abhandl. La. 8vo. 61 pp. (*Des* x. *Bandes der Abhandl. d. Mathemat-Phys. d. Classe d. K. Sächischen Gesellsch. d. Wissensch.* No. 1.)   *Leipzig*, 1871

  (*N.B.—The printer's mark at the feet of the sheets is vol.* xv.)

Weber. (*Vide* Gauss and Weber.)

  (*Vide* Kohlrausch and Weber.)

† **Webster, J.** On the Agency of Electricity in constituting the peculiar properties of Bodies, and producing Combustion. 8vo. 4 pp. (*Phil. Mag.* xliii. 17.)
                                                                   London, 1814

† **Webster, J. W.** Chemical Examination of a Fragment of a Meteor which fell in Maine, August, 1823. 8vo. 4 pp. (*Phil. Mag.* lxiii. 16.)   *London*, 1824

† **Webster, W. H. B.** The recurring monthly Periods and Periodic System of Atmospheric Actions, with evidences of the Transfer of Heat and Electricity. 8vo. 286 pp. 1 plate.   *London*, 1857

† **Webster, R.** On the Causes of the Variation of the Magnetic Pole. 8vo. 3 pp. (*Phil. Mag.* lxi. 165.)   *London*, 1823

† **Wedelstaedt, L. von.** Elektricität, Wärme, Licht, Versuch d. Lösung d. Problems der Weltbildung, Weltbewegg. u. Welterhaltung. 8vo. 112 pp.
                                                                   Berlin, 1870

† **Weeks, W. H.** On Atmospherical Electric Apparatus and Experiments, described in a Letter to the Editor. 8vo. 6 pp. 1 plate. (*Ann. of Elect.* vi. 446.)   *London*, 1841

Weidler, Johann Friedrich. *Born* 1692, *at Gross-Neuhausen, Thüring ; died November* 30, 1755, *at Wittenberg.*

Exercitatio de phosphoro mercuriali. . . . 4to. 71 and 16 pp.
  (*From Kruniz*, p. 158.)   *Vittembergæ*, 1715
  (*Note.—The* 16 pp. (*about*) *are a* "*Schediasma*," *and not numbered.*)

† An Account of a Book intituled, Jo. Fried. Weidleri observationes Meteorologicæ et Astronomicæ, Annorum 1728, 1729, &c. Wittembergæ, anno 1729. 4to. 7 pp. (*Phil. Trans.* xxxvi. 250.)   *London*, 1729-30

† Commentatio de Aurora boreali. An. 1729 Dei 16 Novembris. 4to. 72 pp.
                                                                   Wittembergæ, 1730

† Descriptio Luminum Borealium Vitembergæ. Anno 1732 conspectorum. 4to. 3 pp. (*Phil. Trans.* xxxviii. *for* 1733-34, p. 291.)   *London*, 1735

† **Weidlerus, T. F. and Rhostius.** De meteoro lucido singulari, a 1730 in Octobri conspecto, dissertatio qua observat. Madritensis et Vittembergensis inter se comparantur . . publice proposita d. 23 Juni a 1731. 4to. 32 pp.
Vitembergæ, 1731

Weigel, Christian Ehrenfried. *Born May* 24, 1748, *at Stralsund; died August* 8, 1831, *at Greifswald.*

Grundriss der reinen u. angewandt. Chemie. 8vo. 2 vols. *Greifs,* 1777
(*Note.—He very frequently refers to this work in his notes in his translation of Marat.*) (*Vide* Marat.)

Weinhold, Karl August. *Born October* 6, 1782, *at Meissen; died September* 29, 1829, *at Halle.*

† Physikalische Versuche über den Magnetismus als scheinbaren Gegensatz des elektro-chemischen Processes in der Natur. 8vo. *Meissen,* 1812

Physikal Vorlesung über d. Magnetismus. 8vo. *Meissen,* 1819

Weinlig, Christian Albert. *Born* 1812, *at Dresden.*

Examen theoriæ electrochimatomist. 8vo. (*From Weigel's Cat.* p. 171.)
Lipsiæ, 184

† **Weisens,** Chr. Curiöse Gedanken von Wolkenbrüchen. Aus dem Latinschen übersetzet durch M. M. 12mo. 29 pp. *Dresden,* 1701

† **Weiske,** H. A. Dr. Die Überführung des Chlor bei der Electrolyse seiner Verbindung mit den Metallen die Alkalien und alkalischen Erden. 8vo. 46 pp.
Leipzig, 1857

† **Weiss,** A. Die galvanischen Grundversuche, mathematisch erklärt und die Theorie des Condensators. 4to. 70 pp. *Ansbach,* 1851

Weiss, E. Beiträge zur Kentniss der Sternschnuppen. 8vo. 62 pp. (*Den Sitzungsb. d. K. Akad. d. Wissen.*)
Wien

† Beiträge zur Kenntniss der Sternschnuppen. 2 Abhandlung. 8vo. 68 pp. (*Abhand. d. K. Akad. d. Wiss.* lxii. Bd. Juli-Heft. Jahrg. 1870. Vorgelegt 19 Mai.)
Wien, 1870

† **Weisse,** M. Variationen der Declination der Magnetnadel beobachtet in Krakau. 4to. 38 pp. (*Aus den Denkschriften der Kais. Akad. der Wissenschaften,* vol. xviii. Vorgelegt 15 Juli, 1858.)
Wien, 1859

Weitling, W. Der bewegende Urstoff in seinem kosmoelectro-magnetischen Wirkungen. 8vo. *New York,* 1856

†† **Welch,** J. Report to F. Ronalds on the performance of his three Magnetographs during the Experimental Trial at the Kew Observatory, April 1 till Oct. 1, 1851. 8vo. 8 pp. *Rep. Brit. Assoc. for* 1851.) *London,* 1852

† **Weld,** A. Account of the Aurora Borealis as seen at Stoneyhurst Observatory, Oct. 1848. 8vo. 3 pp. (*Phil. Mag,* Nov. 1848.) *London,* 1848

Welds, C. R. (*Vide* Royal Society.)

Wells, D. A. Annual of Scientific Discovery; or, Year Book of Facts in Science and Art; exhibiting the most important discoveries in Mechanics, Useful Arts, Natural Philosophy, Chemistry. Astronomy, Meteorology, &c. . . together with a list of recent scientific publications, a classified list of Patents, obituaries of eminent scientific men, an Index of important Papers in Scientific Journals, Reports, &c. 12mo. *Boston, U.S.* 1850 *to* 18—

Wells, William Charles. *Born May,* 1757, *at Charlestown, South Carolina; died September* 18, 1817, *at London.*

Observations on the Influence which incites the Muscles of Animals to contract in Mr. Galvani's experiments. 4to. (*Phil. Trans.* an. 1795, p. 246.)
London, 1795

† **Wenckebach,** E. De magneto-elektrische telegraaf van Gauss en Steinheil. 8vo. (*From Muller's Cat.* Jan. 1865.)
1838

† **Wenkebach, W.** Sur Petrus Adsigerius et les plus anciennes observations de la déclinaison de l'aiguille animantée . . traduit de l'hollandais par T. Hooiberg. 4to. (*Extrait des Annali di Matematica Pura ed Applicata*, tom. vii. No. 3.) (*Vide Bertelli on Petrus Peregrinus.*) *Rome*, 1865

† **Wenckh.** Notæ unguenti magnetici et ejusdem act. 12mo. (*From Friedlander's Cat.* 1861.) *Dilingæ*, 1626

Wenzel, C. A. W. Abhandlung über die Blitzableiter aus d. Franz. 8vo. *Wesel*, 1818

Abhandlung über die Blitzableiter ; aus d. Franz, frei übers.f.angeh.Ingenieur-Officire. 2 Abtheil. 8vo. *Berlin*, 1823-4

Werenberg, Johann Georg. *Born* 1702, *at Lüneburg ; died* 1780, *at Hamburg.* Gedanken von der Elek. (*Altonaer Zeitung*, 1845.) 1745

Werner. Referred the Magnetism of the Loadstone to Electricity. (*Vide* Hoppman, Handbuch der Mineralogie, fortgesetzt von Breithaupt, 4 Bd.1811-1818.)

Werner, Abraham Gottlob. *Born September* 25, 1750, *at Wehrau Oberlausitz ; died June* 30, 1817, *at Dresden.*

Cronstedt's Mineralogie. Uebersetzung. 1 Th. 8vo. *Freiberg*, 1780

In Cronstedt's Mineralogy. 1782

Beschrieb. d. drei Arten d. Strahlsteins. (*Bergman's Journal*, 1789, i.) 1789

† **Werner, F.** Die Galvanoplastik in ihrer technischen Anwendung. 8vo. 84 pp. 12 plates. *Petersburg*, 1844

Wertheim, Wilhelm. *Born February* 22, 1815, *at Vienna : died January* 20, 1861, *at Tours.*

Note sur les vibrations qu'un courant galvanique fait naître dans le fer doux. 4to. (*Comptes Rendus*, Fév. 1846.) *Paris*, 1846

Réponse aux remarques de M. De la Rive sur la note précédente. 4to. (*Comptes Rendus, Mars*, 1846.) *Paris*, 1846

Note sur l'influence du courant galvanique, et de l'électro-magnétisme sur l'élasticité des métaux. 8vo. (*Ann. de Chim.* 3me série, xii. 610.) *Paris*

Sur les courants d'induction produits par la torsion. 4to. (*Comptes Rendus,*xxxv.) *Paris*, 1852

† Mémoire sur les sons produits par le courant électrique. Présenté à l'Acad. 1er Mai, 1848. 8vo. 26 pp. 1 plate. *Paris*, 1848 ?

Wesermann, H. M. Der Magnetismus u. die allgem. Weltsprache. 8vo. *Erefeld*, 1822

Wesley, John. Desideratum ; or, Electricity made plain. 12mo. *London*, 1760

The Desideratum ; or, Electricity made plain and useful. 3rd edition. 12mo. 72 pp. *London*, 1790

† The Desideratum ; or, Electricity made plain and useful, by a Lover of mankind and common sense. *London*, 1871 (*From the Athenæum*, July 15, 1871.)

West, F. (*Vide* Peschel.)

† **Westenholz, R.** Reglement for den indenlandfke Afbenyttelse af de danske Statstelegraphlinier. 4to. 15 pp. *Kiobenhavn*, 1859

Westrumb. Description of New Galvanic Apparatus. (*Crell's Chimisch Annalen*, x. Cahir, No. iv.) (*From Sue, Hist.* iv. 241, *not correct title.*) *Helmstadt*

Wetzler, J. E. Beobachtungen über den Nutzen des Keilsehen magneto-elektrischen Rotations Apparats, in Krankheiten. 8vo. *Leipzig*, 1842

Weyr, Emil. Ein Beitrag zur Theorie transversal-magnetischer Flächen. 8vo. 13 pp. *Wien*, 1868

Wheatstone (afterwards Sir), Charles. *Born* 1802, *at Gloucester.; died* 1875, *at Paris.*

† An Account of some Experiments to measure the Velocity of Electricity, and the duration of Electric Light. 4to. 9 pp. 2 plates. (*Phil. Trans.* part ii. for 1834.)   *London,* 1834

Experiments on the Velocity of Electricity, &c. in June, 1836, in a course of Lectures at King's College. 8vo. (*The Magazine of Popular Science,* vol. iii.)
                                                                    London, 1837

On the Thermo-Electric Spark. 8vo. (*Phil. Mag.* Series iii. x. 1837.)
                                                                    London, 1837

Essais relatifs à un télégraphe électrique qui doit être établi entre Londres et Liverpool. 4to. (*Comptes Rendus,* vi. §1.)   *Paris,* 1838

Note sur un appareil de télégraphie électrique qui doit être établi entre Londres et Liverpool, par M. Wheatstone; Lettre de M. le docteur Buckland. 4to. (*Comptes Rendus,* vi. 51.)   *Paris,* 1838

Some matter relative to Cooke and Wheatstone's First Telegraph, and, Quere, Schilling's. (*Morning Chronicle, Jan.* 25, 1838.)   *London,* 1838

M. Quetelet lut . une note . sur le procédé que M. Wheatstone se propose de suivre. 4to. (*Acad. de Bruxelles, Séance du* 10 Fév. 1838.)   *Bruxelles,* 1838

A Pamphlet, probably a reprint of the Specification of Cook and Wheatstone's first Patent.   *London,* 1839

Evidence before the Select Committee on Railways, February, 1840.
                                                                    London, 1840

Expériences que M. Wheatstone venait de faire, à l'Observatoire Royal, au moyen du nouveau Télégraphe électrique de son invention. Note de M. Quételet. 4to. (*Comptes Rendus de l'Acad. de Bruxelles,* tom. vii. 2me part. p. 131 *et* 132.)   *Bruxelles,* 1840

Description of the Electro-Magnetic Clock. 8vo. (*Proc. Roy. Soc.* 1840.)
                                                                    London, 1840

Chronoscope. . Extension du mécanisme de son télégraphe électrique. 8vo. (*Bulletin de l'Acad. de Brux.* Oct. 1840.)   *Bruxelles,* 1840

Electro-magnetic Registerer of Meteorological Observations. 8vo. (*Brit. Assoc. Reports,* 1842, p. 9.)   *London,* 1842

(*Note.—This instrument, said by Moigno, Kuhn, Du Moncel, &c. to exist at the Kew Observatory, was* never there, *although ordered for that establishment by the British Association, who paid Wheatstone for it.*—FRANCIS RONALDS, *Director of the Observatory.*)

Thermomètre télégraphique, etc. Note. 8vo. (*Bulletin de l'Acad. de Bruxelles,* Mai, 1843.)   *Bruxelles,* 1843

† An Account of several new Instruments and Processes for determining the constants of a Voltaic circuit. 4to. 25 pp. 4 plates. (*Phil. Trans. for* 1843, part ii. p. 303.)   *London,* 1843

Note sur le chronoscope électro-magnétique. 4to. (*Comptes Rendus,* xx.1554.)
                                                                    Paris, 1845

Note sur le télégraphe électrique qu'il vient d'établir entre Paris et Versailles. 4to. (*Comptes Rendus,* xx. 1703.)   *Paris,* 1845

† On Professor Quetelet's Investigations relating to the Electricity of the Atmosphere, made with Peltier's Electrometer. Communicated by Professor Wheatstone. 8vo. 4 pp. (*British Association Report for* 1849.) *London,* 1850

† The Universal Private Telegraphic Company, incorporated 24 and 25 Vic. (Royal Assent June 7, 1861.) 8vo. 11 pp.   *Glasgow*

† *Vide* an Article in the Quarterly Review.   *June,* 1854

† Reply to Mr. Cooke's pamphlet "The Electric Telegraph; was it invented by Prof. Wheatstone?" 8vo. 74 pp.   *London,* 1855

An Account of some Experiments made with the Submarine Cable of the Mediterranean Electric Telegraph. 8vo. (*Proc. Roy. Soc.*)   *London,* 1855

† The Universal Private Telegraphic Company. Incorporated by Act of Parliament, 24 and 25 Vic. (Royal Assent June 7, 1861. Wheatstone's Patents.) 8vo. 16 pp.   *Glasgow*

(*Vide* also Cooke, W. F. and Wheatstone.)

Wheeler, Granville. *Died May* 16, 1770.

† Some Electrical Experiments, chiefly regarding the repulsive force of Electrical bodies. Communicated in a Letter to C. Mortimer, Sec. R. S. (dated January 17, 1737-8). 4to. (*Phil. Trans.* xli. 98 and 112.) *London,* 1739-41

† A Letter from G. W. to C. Mortimer (Sec. R. S.), containing some Remarks on the late Stephen Gray his Electrical Circular Experiment. Dated Otterden Place, February 20, 1737-8. 4to. 8 pp. (*Phil. Trans.* xli. 118.)
London, 1739-41

† Two Letters from G. W. to the President concerning a rotatory motion of Glass Tubes about their axes. 4to. 8 pp. (*Phil. Trans.* xliii. 341.)
London, 1744-5

Whewell, William. *Born May* 24, 1794, *at Lancaster ; died* 1867.

† Report on the recent progress of the Mathematical theories of Electricity and Magnetism and Heat. 8vo. 34 pp. *London,* 1836

History of the Inductive Sciences, from the Earliest to the Present Time. 3 vols. 8vo. 1837

Astronomy and Physics. (Bridgewater Treatise.) New edition. 12mo.
London, 1847

History of the Inductive Sciences. New edition. 3 vols. 8vo. *London,* 1857

(*Vide* also Challis.)

† **Whipple,** Blodget, &c. Astronomical, Magnetical, and Climatological Observations made on a Route from the Mississippi to the Pacific. 4to. 290 pp.
Washington, 1856

Whiston, William. *Born December* 9, 1667, *at Norton, Leicestershire ; died August* 22, 1752, *at London.*

An account of a surprising Meteor seen in the air, &c. 8vo. *London,* 1716

An account of another surprising Meteor. 8vo. *London,* 1719

† The longitude and latitude found by the Inclinatory : . . to which is subjoined Mr. Robert Norman's new attractive. 8vo. 115 pp. 2 plates of Whiston. 43 pp. of Norman (reprint dated 1720). (*Vide Norman.*) *London,* 1721

† **White,** W. H. On the variation of the Needle, as observed during a voyage to and from India. 8vo. 1 p. (*Phil. Mag. or Annals,* vi. 153.) *London,* 1829

† Observations on Auroræ Boreales, witnessed at Bedford at various times from April 19th, 1830, to January 11th, 1831. 8vo. 3 pp. (*Phil. Mag. or Annals,* ix. 393.) *London,* 1831

Whitehouse, E. O. W. The Atlantic Telegraph. 8vo. *London,* 1858

Wibel. Versuche über die Reduction von Kupferoxyd-Salzen durch Eisenoxydul-Salze zu metallischen Kupfer oder Kupferoxyd. 47 pp. *Hamburg,* 1864

Wiedeburg, Johann Ernst Basilius. *Born June* 24, 1733, *at Jena ; died January* 1, 1789, *at Jena.*

Beobachtungen und Muthmassungen über d. Nordlichter. 8vo. *Jena,* 1771

Von d. Sternbildern, Nordlichtern, etc. Nebst Erklärungen. *Jena,* 1771 ?

† **Wiedemann,** G. Über die Bewegung von Flüssigkeiten im Kreise der geschlossenen galvanischen Säule. 8vo. 32 pp. (*Poggend. Ann.* lxxxvii.) *Leipzig,* 1852 ?

† Die Lehre vom Galvanismus und Elektro-magnetismus. 2 vols. 8vo. 680 pp. and 1183 pp. *Braunschweig,* 1863

Die Lehre vom Galvanismus u. Electro-magnetismus. Vol. i. Galvanismus; ii. Verm. Auflage. 8vo. *Braunschweig ?* 1872

Wiedemann, G. M. About twelve Articles in Poggendorff's Annalen on Galvanism, Electro-galvanism, Magnetism, Heat, &c. Vol. lxxiv. 1848, to vol. cxvii. 1862. N.B.—Some of which are also in the Verhandl. d. Naturf. Gesellsch. zu Basel.

Wienholt, A. Abhandl. über Magnetismus, herausgeb. von J. C. F. Scherf. 8vo.
Bremen, 1807

(*Note.—He wrote two other works, both on Animal Magnetism.*)

Wies. Thesis. (Galvanism.) (*Not in Pogg.*) *Strasbourg*, 1804

† **Wikstrom, A.** Erfahrung wegen der Störung der Magnetnadel, durch die Elektricität. 8vo. 2 pp. (*K. Schwed. Akad. Abh.* xx. 157.)
Hamburg and Leipzig, 1758

† **Wilbrand, J. B.** Das Gesetz des polaren Verhaltens in der Natur. "Dargestelt in den magnetischen, elektrischen u. chemischen Natur . . " 8vo. 353 pp.
Giessen, 1819

Wilcke, Johann Karl. *Born September* 6, 1732, *at Wismar, Mecklenbourg; died April* 18, 1796.

† Anmerkung zu. Eliz. Chr. Linnæa; Vom Blizen der indianischen Kresse.
Disputatio inauguralis physica experimentalis de electricitatibus contrariis. 4to.
Rostock, 1757

† Franklin's Briefe . .; nebst Anmerkungen. 12mo. *Leipzig*, 1758

† Electriska rön och-försök om den electriska laddningens och stötens ästad kommande vid flera kroppa än glas oth porcellain. 8vo. (*Vetensk. Acad. Handl.* an. 1758, p. 250.) *Stockholm*, 1758

† Elektrische Versuche und Untersuchungen, wie die elektrische Ladung und Schlag durch mehrere Körper als Glas und Porzellan erhalten werden können. 8vo. (*Schwedische Akad. Abhandl.* an. 1758, p. 241, vol. xx.)
Stockholm, 1758

† De natuurkunnigas meningar om orsakerna til äske-dundret. 8vo. (*Vetensk. Acad. Handl.* an. 1759, pp. 79 and 159.) *Stockholm*, 1759

† Die Meynungen der Naturforscher von den Ursachen des Donners. 8vo. 19 pp. (*Schwedische Akad. Abhandl.* an. 1759, pp. 81 and 155.)
Hamburg and Leipzig, 1759

† Rön och tankar om snofigurens skiljaktighet; zuei Aufsätze. 8vo. (*Vetensk. Acad. Handl.* 1761.) *Stockholm*, 1761

† Ytterligare rön och försök om contraire electricitaterne vid laddningen och där til horande de lar. 8vo. (*Vetensk. Acad. Handl.* an. 1762, pp. 206 and 245.) *Stockholm*, 1762

† Fernere Untersuchungen von den entgegengesetzten Elektricitäten bey der Ladung und den dazu gehörenden Theilen. 8vo. 46 pp. and 22 pp. 1 plate. (*Schwedische Akad. Abhandl.* an. 1762, pp. 213 and 253, vol. xxiv.)
Hamburg and Leipzig, 1762

† Beskrifning pa en ny declinations-compass hvarmed magnet-nalens afvikande ifrän Norrstreket finnas kan, utan middags-linia. 8vo. (*Vetensk. Acad. Handl.* an. 1763, p. 143.) *Stockholm*, 1763

† Beschreibung eines neuen Abweichungs-Compasses womit die Abweichung der Magnet-Nadel von Norden ohne Mittags-Linie zu finden ist. 8vo. 11 pp. 1 plate. (*Schwedische Akad. Abhandl.* an. 1763, p. 154, vol. xxv.)
Hamb. u. Leipz. 1763

† Electriska försök med Phosphorus. 8vo. (*Vetensk. Acad. Handl.* an. 1763, p. 195.) *Stockholm*, 1763

† Elektrische Versuche mit Phosphorus. 8vo. 20 pp. 1 plate. (*Schwedische Akad. Abhandl.* an. 1763, p. 207, vol. xxv.) *Hamburg and Leipzig*, 1763

† Tal om Magneter. 8vo. (*From Young*, p. 436.) *Stockholm*, 1764

† Om magnetiska kraftens upväkande genom electricitet. 8vo. (*Vetensk. Acad. Handl.* an. 1766, p. 294.) *Stockholm*, 1766

† Von Erregung der magnetischen Kraft durch die Electricität. 8vo. 22 pp. 1 plate. (*Schwedische Akad. Abhandl.* an. 1766, p. 306, vol. xxviii.)
Leipzig, 1766

† Historien om Turmalinen. 8vo. (*Vetensk. Akad. Handl.* 1766 and 1768.)
Stockholm, 1766 *and* 1768

† Geschichte des Tourmalins. 8vo. 19, 24, and 24 pp. (*K. Schwed. Akad. Abh.* vol. xxviii. 95 pp. and xxx. 3 pp. and 105 pp.) *Leipzig*, 1868

Wilcke, Johann Karl.—*continued.*

† Anmürkning vid föregäende rön. 8vo. (*Vetensk. Acad. Handl.* an. 1767, p. 319, *and in Schewedische Akad. Abhandl.* an. 1767, p. 333, vol. xxix. 2 pp.)
Stockholm, 1767

† Försök til en magnetisk inclinations-charta. 8vo. (*Vetensk. Akad. Handl.* an. 1768, p. 193.)
Stockholm, 1768

† Versuch einer magnetischen Neigungscharte. 8vo. 29 pp. Chart. (*K. Schwed. Akad. Abh.* xxx. 209.)
Leipzig, 1768

† Electriska försök pä här och smalta metaller. 8vo. (*Vetensk. Acad. Handl.* an. 1769, p. 319.)
Stockholm, 1769

† Elektrische Versuche mit Haaren und geschmelzten Metallen. 8vo. 6 pp. (*Schwedische Akad. Abl. andl.* an. 1769, p. 317, vol. xxxi.)
Leipzig, 1769

Nya Rön om vattnetts frysning till snölika is-figurcr. 8vo. (*Vetensk. Acad. Handl.* 1769.)
Stockholm, 1769

Electriska försök pår hår och smälta metaller. 8vo. (*Vetensk. Acad. Handl.* 1769.)
Stockholm, 1769

† Anmärkningar vid et d. 30 Maji, 1769, här i Staden timadt äske-slag. 8vo. (*Vetensk. Acad. Handl.* an. 1770, p. 112.)
Stockholm, 1770

† Bemerkungen bey einem d. 30 May in Stockholm geschehenen Donner-Schlage. 8vo. 11 pp. 1 plate. (*Schwedische Akad. Abhandl.* an. 1770, p. 115, vol. xxxii.)
Leipzig, 1770

Om magnetiska inclinationen med beskrifning pä twänne inclinations-compes er. 8vo. (*Vetensk. Acad. Handl.* an. 1772, p. 287.)
Slockholm, 1772

† Neigung der Magnetnadel, nebst Beschreibung zweener Neigungs compasse. 8vo. 13 pp. 2 plates. (*K. Schwed. Akad. Abh.* xxxiv. 285.)
Leipzig, 1772

† Anmarkningar vid C. G. Ekeberg's ingifne observationer öfver magnetisker inclinationen. 8vo. (*Vetensk. Acad. Handl.* an. 1775, p. 298.)
Stockholm, 1775

† Anmerkungen zu H. Capt. Ekeberg's eingegebenen Beobachtungen über die magnetische Nutzung. 8vo. 8 pp. (*K. Schwed. Akad. Abh.* xxxvii. 298.)
Leipzig, 1775

† Undersokning om de vid Volta's nya elettrophoro perpetuo forekommande electrische phenomener. 8vo. (*Vetensk. Acad. Handl.* an. 1777, pp. 56, 128, 216.)
Stockholm, 1775

† Untersuchung der bey Volta's neuen electrophoro perpetuo vorkommenden elektrischen Erscheinungen. 8vo. 25, 15, 17 pp. 1 plate. (*Schwedische Akad. Abhandl.* an. 1777, pp. 54, 116, 200, vol. xxxix.)
Leipzig, 1777

† Rön om magnet-nalens ärliga och dugeliga andringa i Stockholm. 8vo. (*Ve!ensk. Acad. Handl.* an. 1777, p. 273.)
Stockholm, 1777

† Über der Magnet-Nadel jährliche und tägliche Aenderungen zu Stockholm. 8vo. 26 pp. (*Schwedische Akah. Abhand.* an. 1777, p. 259, vol. xxxix.)
Leipzig, 1777

Tal om de nyeste förklaringar af norrskenet. 8vo.
Stockholm, 1778

† Försök til uplysning om luft-hujrflar och sky-drag. 8vo. (*Vetensk. Acad. Nya Handl.* an. 1780, pp. 1 and 83 ; an. 1782, p. 3 ; an. 1785, p. 290 ; an. 1786, p. 3.)
Stockholm, 1780-2-5-6

Versuche zur Erläuterung der Luft-Wirbel und Wasser-Hosen. 8vo. (*Neue Schwedische Akad. Abhcndl.* an. 1780, pp. 3 and 81 ; 1782, p. 3 ; 1785, p. 271 ; 1786, p. 3.
Leipzig, 1780-2-5-6

Von den ncuesten Erklärungen des Nordlichts. 8vo. (*Schwedisches Musæum,* tom. i. p. 31.)
Wismar, 1783

Anmärkn. vid ett af Gerdes beskrifvet skydrag. (*Vitcnsk. Acad. Handl.* 1790.)
1790

Wilcke, Johann Karl.—*continued.*

† Anmerkung über vorerwähnten Wolken-zug (*i.e.*) der Wasser-Wirbel des 10 August, 1790, in Malar. (*Vide* Sandwall, Relation . .) (*Neue Schwedische Akad. Abhandl.* an. 1790, p. 199.) *Stockholm,* 1790

† Über den Magneten. Germ. by Gröning. "Vorgelesen, 1 Feb. 1764." 8vo. 40 pp. (*Volta's copy.*) *Leipzig,* 1794

Über d. Magneten. Vorlesung gehalten in der Schwedischen Akademie der Wissench. zu Stockholm a. d. Schwed. übers. u. herausg, v. Gröning. 8vo. *Leipzig,* 1794-5

Wild, Franz Samuel. *Born* 1743, *at B₁rn; died April* 16, 1802.

Expériences sur l'Electricité des Cascades. (*Mém. de Lausanne,* iii. *Hist.* p. 13, 1790.) *Lausanne,* 1790

† Ragguaglio di due fenomeni straordinari. Tratto da una lettera . . all' . . Amoretti. 4to. 1 p. *Milan,* 1792

† **Wild, H.** Theorie d. Nobilischen Farbenringe. 4to. 42 pp. 1 plate. *Zürich,* 1857

† Die Selbstregistrirenden Meteorologischen Instrumente der Sternwarte in Bern. 8vo. 41 pp. 9 plates. (*Extraabdruck aus dem* ii. *Bande von Carl's Repertorium.*) *München,* 1866

Wild, M. F. Weitere Beschreibung der in des vii. Bandes, 3ten Stück, Seite 73 vorgekommenen Elektrisirmaschinen. 8vo. 27 pp. 4 plates. (*Voight's Mag.* vii. 4th Stuck, p. 77.) *Gotha,* 1792

Wilde. New Magneto-electric Machine. *London,* 1866-7

Wilhelm, H. M. Semicenturia observationum electricarum. *Wirceburg,* 1774

Wilkens. (*Vide* Linnæa, E. C.)

Wilkes, C. (Commander of the U.S. Exploring Expedition). Meteorology. Journal of Meteorological Observations, vol. xi. (of the collection of 16 vols.) 4to. 726 pp. 25 illustrations, map, and 24 plates. *Philadelphia,* 1851

† Theory of the Zodiacal Light. (Read at the American Association for Advt. of Science, at Montreal, August, 1857.) 4to. 16 pp. 1 plate. *Philadelphia,* 1857

Wilkins, John (Bishop of Chester; First Secretary of the Royal Society). *Born* 1614, *at Fawsley, near Daventry; died November* 19, 1672, *in London.* (*From Poggendorff,* ii. 1328.)

Mercury; or, The secret and swift Messenger, showing how a man may with privacy and speed communicate his thoughts to a friend at any distance. 8vo. (*Also in his Mathematical and Philosophical Works.* 8vo. London, 1708.) *London,* 1641

Wilkinson, C. H. Tentamen philosophico-medicum de Electricitate. 8vo. *Edinburgh,* 1783

An Essay on the Leyden Phial ("with a view of explaining this remarkable phenomenon on pure mechanical principles.") 8vo. 220 pp. Plates. *London,* 1798

† An Analysis of a course of Lectures on . . Natural Philosophy. 8vo. 141 pp. *London,* 1799

(*Note.—This is annexed to his "Effects of Electricity in Paralytic, &c. affections.*)

† The Effects of Electricity in Paralytic . . affections, &c. 8vo. 76 pp. *London,* 1799

† Elements of Galvanism in Theory and Practice . . 2 vols. 8vo. *London,* 1804

† Description of an improved Galvanic Trough 8vo. 2 pp. 1 plate. (*Phil. Mag.* xxix. 243.) *London,* 1808

† Explanation of his Mechanical Theory of Electricity, 8vo. 2 pp. (*Phil. Mag.* xlix. 299.) *London,* 1817

Wilkinson, George. Medical Facts . iii. p. 52. 1792

† Willard, J.　Observations made at Beverley, lat. 42° 36′ N., long. 70° 45′ W., to determine the variation of the Magnetical Needle.　4to.　4 pp.　(*Mem. Amer. Acad.* Old Series, i. 318, pt. ii.)　　　　　　　　　　*Boston, U.S.* 1785

† Wille, C. F.　Om Kompassets deviation vaesentligst paa jernskibe, samt kortfatlet udsigt over laeren om magnetisme med 23 tegninger og 3 karter.　Udgivet med bidrag ef det k. marine- og post-department.　8vo.　84 pp.
　　　　　　　　　　　　　　　　　　　　　　　　　　　Kristiania, 1869

Willemet.　Sur l'usage du Fluide Electrique dans l'économie animale.

† Williams, Samuel.　A Memoir on the Latitude of the University of Cambridge : with Observations of the Variation and Dip of the Magnetic Needle. , 4to.　8 pp.　(*Mem. Amer. Acad. of Arts, &c.* i. 62.)　　　　　　*Boston, U.S.* 1785

Magnetic Observations made at the University of Cambridge . in the year 1785.　4to.　1 page.　(*Trans. Amer. Phil. Soc.* Old Series, vol. iii. p. 115.)
　　　　　　　　　　　　　　　　　　　　　　　　　　　Philadelphia, 1793

Williams, W.　Compass Deviation Recorder for iron Ships.　Fol.　*London*, 1859

Williams, Z.　An attempt . . Longitude . . Magnetic Needle.　(In Italian and English.)　4to.　(*From Watt.*)　　　　　　　　　　　　*London*, 1755.

Willigen, van der V. S. M.　Proeven betreffende den Galvanischen Lichtboog 8ve.　　　　　　　　　　　　　　　　　　　　*Deventer*, 1854

† Wilmer, B.　Lettera a Guglielmo Sharpe sullo strano abbruciamento di una Donna seguito a Coventry.　4to.　2 pp.　(*Scelta d'Opusc.* Nuova ed. iii. 36.　*Trans.* by Soave, with his Note.　12mo. edition, printed in 1777.)　　　*Milano*

Wilson, Benjamin.　*Born about* 1708 ; *died June* 6, 1788, *at London.*

† Essay towards an Explanation of the Phænomena of Electricity. . 8vo.　95 pp.　1 plate.　　　　　　　　　　　　　　　　　　　　*London*, 1746

Experiments on the Law of Accumulation of Electricity in the Leyden Phial.　(Read, as per minutes of Royal Society, Oct. 23, 1746, " though the original was lost or mislaid.")　　　　　　　　　　　　　　　　　1746

Letter to Mr. Ellicot on the issue of the Electric Fluid from the Earth, not from the globe (of the Electric Machine), when rubbed.　*About* 1746

Short View of Electricity.　8vo.　　　　　　　　　　　*London*, 1750

† Treatise on Electricity, by B. W.　1st edition.　8vo.　5 plates.　223 pp.
　　　　　　　　　　　　　　　　　　　　　　　　　　　London, 1750

† Treatise on Electricity.　2nd edition.　8vo.　224 pp.　Plates.　*London*, 1752

Observations on a series of Electrical Experiments by Dr. Hoadly and Mr. Wilson.　2nd edition.　With alterations, and the addition of some experiments, letters, and explanatory notes, by B. Wilson.　4to.　*London*, 1759
　　(*Note.—There is a German translation of* 1763.)

† Auszug aus einem Briefe . an Thorbern Bergman . . einige neue Versuche, die Elektricität betreffend.　8vo.　3 pp.　(*K. Schwed. Akad. Abh.* xxiii. 323.)
　　　　　　　　　　　　　　　　　　　　　Hamburg and Leipzig, 1761

Observations on Lightning, and the method of securing Buildings from its effects.　In a letter to Sir Charles Frederick.　4to.　68 pp.　(*Phil. Trans.* an. 1773, p. 49.)　　　　　　　　　　　　　　　　*London*, 1773

† Further Observations on Lightning.　4to.　26 pp.　　　　*London*, 1774

† A Series of Experiments relating to Phosphori . (92 pp.), together with a translation of two Memoirs from the Bologna Acts upon the same subject, by T. B. Beccari, Prof. of Bologna.　1st edition.　4to.　96 pp.
　　　　　　　　　　　　　　　　　　　　　　　　　　　London, 1775

† A Series of Experiments on the subject of Phosphori and their prismatic colours ; in which are discovered some new properties of Light.　Also a translation of two Memoirs of the late T. B. Beccari, Professor of Philosophy at Bologna, taken from the Bologna Acts.　2nd edition, with additions.)　4to.　117 pp.　　　　　　　　　　　　　　　　*London*, 1776

New Experiments and Observations on the nature and use of Conductors.　4to.　*Phil. Trans.* pt. i. p. 245.)　　　　　　　　　　　　*London*, 1777

Wilson, Benjamin.—*continued.*

† An account of Experiments made at the Pantheon, on the nature and use of Conductors ; to which are added some new Experiments with the Leyden Phial. Read at the meetings of the Royal Society. 4to. 100 pp. 4 plates.

London, 1778

(*Note.—First printed in Phil. Trans.* vol. lxviii.)

† Short View of Electricity. 4to. 37 pp. 1 plate. *London,* 1780

Short View, &c. 4to. (*From Watt, error in date ?*) *London,* 1781

Letter to Hoadly on a Chain, &c. (*From Priestley.*)

(*Vide* also Hoadly and Wilson.)

Wilson, George. *Born February* 21, 1818, *at Edinburgh ; died November* 22, 1859, *at Edinburgh.*

† Electricity and the Electric Telegraph, &c. 12mo. 77 pp. *London,* 1852

† The Progress of the Telegraph. 12mo. 60 pp. *Cambridge,* 1859

Electricity and Electric Telegraph ; to which is added The Chemistry of the Stars. New edition. 12mo. 50 pp. *London,* 1859

† **Wilson, W.** Hints respecting a speedy decomposition of Water by means of Galvanism. 8vo. 3 pp. (*Phil. Mag.* xxii. 260.) *London,* 1805

Windham, Colonel. Observations on the Dipping Needle, made about 1672.

Window, F. R. On the Electric Telegraph, and the principal improvements in its construction. 1852

Wingfield, John. (*Vide* Cuthbertson.)

Winkler, G. Gott u. d. Christ im Gewitter, nebst e. physikal. Anhange v. Gewitter. *Leipzig,* 1784

Winkler, Johann Heinrich. *Born March* 12, 1703, *at Wingendorf, Oberlausitz ; died May* 18, 1770, *at Leipzig.*

Gedanken von den Eigenschaften, Wirkungen und Ursachen der Electricität, nebst einer Beschreibung zwo neuer elektrischen Maschinen. 8vo. 3 pls.

Leipzig, 1744

Abstract from his book just published at Leipsig, 1744. 4to. (*Phil. Trans. Ab.* x. 270.) *London,* 1744

† Abstract of what is contained in a book concerning Electricity just published at Leipzic, 1744, by J. H. Winkler. Read Nov. 22, 1744. 4to. 4 pp. (*Phil. Trans.* xliii. 166.) *London,* 1744-5

New Observations on Electricity. 4to. (*Phil. Trans. Ab.* x. 273.)

London, 1744-5

† Regiæ Societati Anglicanæ Scient. quædam electricitatis recens observata exhibet J. H. Winkler. Presented March 21, 1744-5. 4to. 8 pp. 1 plate. (*Phil. Trans.* xliii. 307.) *London,* 1744-5

Die Eigenschaften der elektrischen Materie u. des elektrischen Feuers, aus verschiedenen neuen Versuchen erkläret, und nebst etlichen neuen Maschinen zum Electisiren beschrieben. 8vo. 164 pp. 4 plates. *Leipzig,* 1745

Quædam electricitatis recens observata. 4to. (*Phil. Trans.* an. 1745, p. 307.) *London,* 1745

Die Stärke der elektrischen Kraft des Wassers in gläsernen Gefässen, welche durch den Musschenbrockischen Versuch bekannt geworden, erklärt von J. H. Winkler. 8vo. 164 pp. 2 plates. *Leipzig,* 1746

Letter on Effects of Electricity on himself and his wife. 4to. (*Phil. Trans. Ab.* x. 327.) *London,* 1746

Abhandlung von der Ursprung des Wetterleuchtens. 1746

† Epistola Winkleri J. H. ad Societatem Regalem Londinensem data, quæ continet descriptionem et Figuras Pyrorgani sui Electrici. Read May 7, 1747. 4to. 6 pp. 1 plate. (*Phil. Trans.* xliv. 497.) *London,* 1746-7

Winkler, Johann Heinrich.—*continued.*

Epistola quæ continet descriptionem et figuras pyrorgani sui electrici. 4to. (*Phil. Trans.* an. 1747, p. 497.) *London,* 1747

† Novum reique medicæ utile Electricitatis inventum exponit. Read March 31, 1748. 4to. 9 pp. (*Phil. Trans.* xlv. 262.) *London,* 1748

† Novum rei medicæ utile electricitatis inventum. 4to. (*Phil. Trans.* an. 1748, p. 262.) *London,* 1748

† Essai sur la nature, les effets, et les causes de l'électricité, avec úne description de deux nouvelles machines à électricité. Traduit de l'Allemand. 12mo. 156 pp. 2 plates. *Paris,* 1748

Grundriss zu einer ausführl. Abhandl. von d. Elektricität. 1750?

† De imagine motuum cælestium viribus electricitatis effecta. 4to. 20 pp. 1 plate. *Lipsiæ,* 1750

An Account of his Experiments relating to Odours passing through electrised Globes and Tubes, &c. 4to. (*Phil. Trans.* 1751.) *London,* 1751

De avertendi fulminis artificio secundum electricitatis doctrinam Commentatio. 4to. 1 plate. *Leipzig,* 1753

Account of two Electrical Experiments. 4to. (*Phil. Trans.* an. 1754, p. 772.) *London,* 1754

Elements of Natural Philosophy, translated from the 2nd German edition. 2 vols. 8vo. *London,* 1757

Conjectura de vi vaporum solarum in lumine boreali. 4to. (*From Kruniz.*) 1763

Progr. de commercio luminis borealis cum acu magnetica. (*From Kruniz.*) *Leipzig,* 1767

Disquisitio qua ratione ignis et materia electrica inter se differant. Pro-gramme. (*From Kuhn.*) 1767

Prog. de vi luminis borealis in commovenda acu magnetica. (*From Kruniz.*) *Leipsig,* 1768

Tentamina quæstiones et conjecturæ circa electricitatem animantium. 4to. (*From Pogg.* ii. 1338.) *Leipzig,* 1770

Winter. His plate Electrical Machine with a Ring Conductor.

Winterl, Jacob Joseph. *Born April* 15, 1732, *at Eisenerz Steyerm; died November* 23, 1809, *at Pesth.*

Prolusiones in chemium seculi decimi novi. 8vo. (*From Œrsted.*) *Budæ,* 1800

Materialen für eine chemie des nennzehorten Jahrhunderts. (*From Œrsted.*) 1803

Accessiones novæ ad prolusionem suam primam et secundam. 8vo. (*From Pogg.*) *Budæ,* 1803

Darstellung d. 4 Bestandtheile d. anorganischen Natur, e. Umarb. d. 1sten The. f. Prolusionen u. Accessionen d. d. Verfasser, a. d. Latein. von J. J. Schuster. 8vo. *Jena,* 1804

Über Ritter's Pendelversuche. 8vo. (*Gehlen's Journ.* iii. 1807.) *Berlin,* 1807

Über seine angebl. Entdeckung. 8vo. (*Gehlen's Journ.* ii. 1810.) *Berlin,* 1810

Wirdig. Nova medicina spirituum. 12mo. *Hamburgi,* 1673

† **Wittenbach.** Articolo di lettera in cui s'annunziano alcuni nuovi esperimenti elettrici. 4to. 1 page. (*Opusc. Scelti,* v. 375.) *Milano,* 1782

† **Wittiber.** Über atmosphär. Electricität und Gewitter, insbesondere die Gewitter der Grafschaft. 4to. 23 pp. *Glatz,* 1860

† **Wittmütz,** C. R. A. Zur Theorie magnetischer Reflexions apparate, insbesondere d. Gaussischen Magnetometers. (Thesis.) 4to. 40 pp. *Hamburg,* 1844

Witry de Abt. On preparation of Mosaic Gold for Electric Machines. 8vo. (*Lichtenberg's Mag.* iv. St. 3, pp. 58-61.) *Gotha*

Wleugel, P. J. Forsog om Magnetnaalen Kan sikkres mot Jernets Paavirkning, &c. 4to. *Kiobenhaven,* 1828

† **Wohler.** The Black Deposit on the sides of the tubes, which Becquerel supposed to be Carbon, . is Sulphuret of Copper. 8vo. 1 page. (*Phil. Mag.* 151.)
London, 1831

Ueber ein magnetisches Chromoxyd. *Göttingen,* 1859
(*Vide* also Partsch and Wöhler.)

Wolf, Christian. *Born January* 24, 1679, *at Breslau; died April* 9, 1754, *at Halle.*
Relatio de phænomeno luminoso d. 17 Martii, etc. 4to. (*Act. Erudit.* 1716.)
Leipzig, 1716
Gedanken über das ungewöhnliche Phænomenon vom 17 Martii new, 1716.)
4to. *Halle,* 1716
Nützliche Versuchen. Part ii. cap. 10, § 173, von den elect. Lichte. (*From Waitz.*)
1722
Gesammelte kleine philos. Schriften meistentheils aus d. Lateinischen von G. F. H. 6 Thle. 8vo. *Halle,* 1736-40
Übrige, theils noch gefundene kleine Schriften v. einzelne Betrachtungen. 4to.
Halle, 1755
Schreiben von der Electricität, von dem sel. Hrn. Kanzler von Wolff, an Hrn. Prodechant Wolshoffer zu Rostall abgelassen; zum Druck befördert von T. . . 4to. *Frankf.-u.-Leips.* 1755

† **Wolf und Bina.** Physica experimentalis Chr. Wolfii . . nunc primum ex Germanico idiomate in Latinum translata opera et studio D. And. Bina. . . . Accedit Dissertatio interpretis de Electricitate. (Ad vol. ii. partem ii.) 2 vols. in 4 parts. 8vo. *Venetiis,* 1753-56
(*Note.—Bina's " Dissertatio " . . is a much enlarged edition of his " Electricorum effectuum Explicatio.*")

† **Wolf, R.** Nachrichten von der Sternwarte in Bern. Beobachtungen der Sternschnuppen im Winterhalbjahre 1852 auf 1853. (Vorgelegt am 7 Mai, 1853.) 8vo. 9 pp. *Bern,* 1853?

† Über den Jährlichen Gang der magnetischen Declinations-Variation. (Vorgetragen den 4 Juni, 1853.) 8vo. 7 pp. *Bern,* 1853?

† Über d. Ozongehalt der Luft und seinen Zusammenhang mit d. Mortalität. Vorträge gehalten in d. bernischen naturforsch. Gesellschaft. 8vo. 21 pp. 1 plate. *Bern,* 1855

Wolfram, Erdmann. *Born September* 7, 1760, *at Förbau; died December* 28, 1828, *at Gross-Tinz, Schles.*

His Electrical Machine : a receiver rubbed inside and outside. (*Ferussac, Bulletin* 1824.)

† **Wolke, C. H.** Nachricht von den zu Jever durch die Galvani-Voltaische Kunst beglükten Taubstummen und von Sprenger's Methode sie durch die Voltaische Electricität auszuüben. 8vo. 224 pp. 1 plate. (*Volta's copy.*) *Oldenberg,* 1802

Wollaston, W. H. *Born August* 6, 1766; *died December* 22, 1828.

† On the Agency of Electricity on Animal Secretions. 8vo. 3 pp. (*Phil. Mag.* xxxiii. 488.) *London,* 1809

† On the Apparent Magnetism of Metallic Titanium. Phil. Trans. for 1823, part ii. 8vo. 2 pp. (*Phil. Mag.* lxiii. 15.) *London,* 1824

† Magnetism of Titanium, Cobalt, and Nickel. 8vo. 1 p. (*Edinb. Phil. Journal,* x. 183.) *Edinburgh,* 1824

† Death of, announced, and Eulogy, by Dr. Fitton, Pres. of the Geological Society. 8vo. 2 pp. (*Phil. Mag. or Annals,* v. 444.) *London,* 1829

† **Woodbury.** The History of the Electric Telegraph, as detailed in the opinion of Judge Woodbury, of the Supreme Court of the United States.

† **Woodcroft, B.** Patents for Inventions. Abridgments of Specifications relating to Electricity and Magnetism; their Generation and Applications. Printed by Order of the Commissioners of Patents. 8vo. 769 pp. *London,* 1859

Woodcroft, B.—*continued.*

Patents for Inventions. Abridgments of Specifications relating to Electricity and Magnetism ; their Generation and Applications. Part ii. A.D. 1858-1866. Printed by Order of the Commissioners of Patents. 8vo. 863 pp.
London, 1870

† **Woods, S.** Essay on the Franklinian Theory of Electricity. Read before the Askesian Society in the Session 1802-3. 8vo. 15 pp. 1 fig. (*Phil. Mag.* xvii. 97.)   *London,* 1803

† Essay on the Phænomena of the Electrophorus; with an Attempt to Reconcile them with the principles of the Franklinian Theory. Read before the Askesian Society in the Session 1803-4. 8vo. 15 pp. (*Phil. Mag.* xxi. 289.)
London, 1805

Wrangel, Ferd. Lud. Physikal Beobb. während seiner Reisen auf. d. Eismeere in d. J. 1821-23 ; herausgegeb. von G. F. Parrot. 8vo.   *Berlin,* 1827

Wrede, K. F. Versuche u. Beantwort d. Preisfrage d. Berlin Acad. : Wirkt d. Elektricität auf Stoffe die gähren ? 8vo.   *Berlin,* 1804

Wright, Edward. *Born about* 1560, *at Graveston, Norfolkshire ; died* 1615, *at London.*

The Haven-finding Art. Translation of Simon Stevenus, Portuum investigandorum ratio.   *London,* 1599

† **Wright and Bain.** A few remarks on Wright and Bain's Patent Electro-Magnetic Printing Telegraph. 8vo. 8 pp.   *London, about* 1841

(*Note.—Including* 4th *Report of a Committee of the House of Commons, and a letter signed* " *G. A.*" *to the* " *Railway Magazine.*")

Wucherer, Wm. Fred. Von Anlegung d. Blitzableiter auf Kirchen u s.w.
Carlsruhe, 1839

Wüllner, A. Uber d. Elektricität-Entwickl. biem Lösen von Salzen. 8vo. (*Pogg. Ann.* cvi. 1859.)   *Leipzig,* 1859

Über d. Electricität-Entwickl. durch chem. Processe. 8vo. (*Pogg. Ann.* cix. u. cxi.)   *Leipzig,* 1860

† Lehrbuch der Experimentalphysik mit theilweiser Benutzung von Jamin's Cours de Physique de l'Ecole Polytechnique. Zweite Bande, Zweite Abtheilung. Vierter Thiel, Die Lehre von Magnetismus und der Elektricität. 8vo. 1352 pp.   *Leipzig,* 1865

Wünsch, C. Ern. Lucifer, oder Nachtrag zu d. bisher angestellt. Untersuchungen d. Erdatmosphere. 8vo. 2 vols.   *Lipsiæ,* 1802

Wurtzelbau. (*Vide* Volckamer.)

Wurzelbau, J. Phil. Acus magneticæ variationis observatio. (Mit T. G. Volckamer u. G. Ch. Ermmart.)

Wurzer, F. On a Sensation of Cold in the Hand in contact with the Copper Pole of a Pile. (*Allgem. Mediz. Ann.* Feb. 1802, p. 26.)   1802

Nouveaux phénomènes de l'électricité galvanique. 8vo. (*Van Mons, Journ. de Chim.* v. 1805, *and Gehlen, Journ.* v. 1805.)   *Bruxelles,* 1805

Activité de la pile, considérablement accrue par l'interposition de l'acide nitrique affaibli. Ext. d'une lettre du Prof. Wurzer. 8vo. (*Van Mons, Journ. de Chim.* No. vi. p. 326,—*Gehlen's Journ.* v. 1805.)   *Bruxelles,* 1805

Wynne, James. Lives of Literary and Scientific Men of America. 8vo.
London, 1851

551

X.

Ximenes, Leonardo. *Born December 27, 1716, at Trapani, Sicily; died May 3, 1786, at Florence.*

Osservazione dell' Aurora-boreale comparsa la sera del di 19 Agosto, 1751.
8vo. 2 pp. (*Zaccaria's Storia*, iii. 656.) *Venezia*, 1752

Osservazione dell' Aurora-boreale del di 3 Feb. 1750, &c. . . Osserv. dell' Aur. bor. dal 26 Ag. 1750. 20 pp. (*Symbolæ litterariæ Opuscula varia Florentiæ*, 1753, vol. x. p. 73.) *Florentiæ*, 1753

Y.

Yates. (*Vide* Liverpool Compass Committee.)

† **Yatman,** M. A familiar Analysis of the Fluid capable of producing the Phenomena of Electricity and Galvanism, or Combustion, &c. 8vo. 73 pp.
London, 1810

A Letter, &c., on Davy's Galvanic Girdle. 8vo. *London*, 1811

† Davy's Enquiries concerning the relation of Galvanism to Living Action. Illustrated, &c. 8vo. *London*, 1814

† **Yeates,** T. Theory of the Magnetical Variation. 8vo. 5 pp. (*Phil. Mag.* lii. 295.)
London, 1818

A new Variation Chart of the Navigable Globe from 60 deg. North to 60 deg. South Latitude. With account. (*Phil. Mag.* lii. 222.) *London*
(*Note.—The account only is in the Library.*)

Yelin, Julius Konrad von. *Born October 22, 1771, at Wassertrüdingen, Rezatkr, Bayern; died April 20, 1826, at Edinburgh.*

Lehrbuch d. Experimental-Naturlehre. 1 Th. 8vo. 1796

† Über Magnetismus und Elektricität als identische und Urkräfte . . Akademische Rede. 4to. 76 pp. *München*, 1818

† Versuche und Beobachtungen zur nähern Kenntniss der Zambonischen trockenen Säule. 4to. 69 pp. *München*, 1820

Über d. Zusammenhang d. Elektricität u. d. Magnetismus. 8vo. (*Gilb. Ann.* lxvi. 1820 u. lxviii. 1821.) *Leipzig*, 1820-21

Die Akademie der Wissenschaften und ihre Gegner. 8vo. *München*, 1822
(*Note.—On the discovery of Thermo-Magnetism*, p. 31.)

† Der Thermo-Magnetismus in einer Reihe neuer elektro-magnetischer Versuche dargestellt. Nach Zusät in den Sitzungen der K. Akad. vom 12 u. 26 April, d. J. gehalten mit Versuchen begleiteten Vorlesungen. 4to. 12 pp. 1 plate.
München, 1823

Le thermo-magn. exposé dans une série de nouvelles exp. électro-magnétiques.
4to. (*From Zantedeschi's Trattato.*) *Monaco (or München)*, 1823

† Neue electro-magnetische Versuche. Die magnetomotorische Wirkung der flüssigen Säuren, Basen und Salze mittelst einfacher metallischer Leiter, &c. . . dargestellt von Dr. Julius von Yelin. 4to. 15 pp. *München*, 1823

Über d. Elektricität d. Papiers. 8vo. (*Gilb. Ann.* lxxv. 1823.) *Leipzig*, 1823

Über d. Blitzableiter aus Messingstricken u. üb. d. am 30 Ap. 1822, erfolgt. merkwürd. Blitzschlag auf d. Kirchthurm zu Rosstall. 8vo.
München, 1823

Über die Blitzableiter aus Messingdrahtstricken. 2e Aufl. 8vo. *München*, 1824

Young, Arthur. *Born September 7, 1741, in London ; died April 12, 1820, in London.*

Travels in France during 1787-8-9. 4to.   *Bury St. Edmunds,* **1792**

Relation des Voyages de A. Young, en France. (*Transmission of Signs by Electricity.*)

† **Young,** James. An account of a new Voltaic Battery, being a modification of the construction recommended by Mr. Faraday. 8vo. 4 pp. 1 plate. (*Phil. Mag. for* April, 1837.) (*Faraday's copy.*)   *London,* **1837**

Young, Thomas. *Born June 13, 1773, at Milverton, Somersetshire ; died May 10, 1829, at London.*

†   A course of Lectures, &c. 1st ed. 2 vols. 4to.   *London,* **1807**

†   Arago, Eloge historique du Dr. Thòmas Young. Read 26 Nov. 1832. 4to. 50 pp. (*Acad. des Scien.* 1831, p. lvii. *Hist.*)   *Paris,* **1835**

A Course of Lectures on Natural Philosophy by Kelland. New ed. in 8vo.
London, **1845**

Miscellaneous Works, including his Scientific Memoirs, Hieroglyphical Essays and Correspondence, &c., with Life by Dean Peacock. Portrait and engravings. 4 vols. 8vo.   *London,* **1855**

† **Younghusband,** C. W. On periodical Laws in the larger Magnetic Disturbances. Read Feb. 24, 1853. 4to. 13 pp. (*Phil. Trans. for* 1853, p. 165.) (*Faraday's copy.*)   *London,* **1853**

Z.

Zaccaria, F. A. Storia letteraria d'Italia. 14 vols. + others connected with it. 8vo. (*From Libri's Sale Cat.* p. 795.) 1750-59

Annali letterari d'Italia dal 1756 al 1758. 3 vols. 8vo. *Modena,* 1762-4

Storia della Elettricità. (*Zaccaria, Editor.*)

Zach, Fr. Baron de. L'attraction des montagnes et ses effets sur les fils à plomb, etc. 2 Theile. 8vo. *Marseille,* 1815

Zaddach, C. Beobachtungen über die magnetische Polarität des Basalts und der trachytischen Gesteine. *Bonn,* 1851

† Über natürliche Magnete. 8vo. 22 pp. (*Unterhaltungen Königsberger Naturwissenschaftliche,* Heft iii. Bd. ii.) *Königsberg,* 1852

† **Zaliwski, J.** Attraction universelle des corps au point de vue de l'électricité. 12mo. 53 pp. (*Reinwald's Cat. for* 1859.) *St. Denis,* 1855

Discours sur l'attraction universelle des corps par l'électricité. 8vo. 15 pp. *Paris,* 1857

La gravitation c'est l'électricité. Nouvelle édition. 18mo. 35 pp. *Paris,* 1858

† **Zaliwski-Mikorski.** La gravitation par l'électricité. 8vo. 43 pp. *Paris,* 1865

Zallinger zum Thurn, Franz Seraphim. *Born February* 14, 1743, *at Botzen; died October* 2, 1828, *at Innsbruck.*

" Von d. Elekt. des in Tyrol gefundenen Turmalins, 8vo. Innsbruck, 1779." (*From Pogg.* ii. 1391.) *Vienna,* 1779

Von d. elektr. Grundsätzen. 8vo. (*From Pogg.* ii. 1391.) *Innsbruck,* 1779

Abhandl. ü. d. Grundsätze d. Electricität. 2nd ed. 8vo. *Innsbruck,* 1801

Zamboni, Giuseppe. *Born June* 1, 1776, *at Verona; died July* 25, 1846, *at Verona.*

† Della pila elettrica a secco. 8vo. 55 pp. 3 plates. *Verona,* 1812

† An Instrument of his own Construction, presented to the Royal Society . . . attempt . . . at Perpetual Motion. 8vo. 1 p. (*Phil. Mag.* xlv. 67.) *London,* 1815

† Lettera all' Accad. di Monaco, sopra i miglioramenti da lui fatti alla sua pila elettrica. 8vo. 39 pp. *Verona,* 1816

† L'Elettromotore perpetuo. Trattato. 8vo. Parte Prima. 298 pp. *Verona,* 1820

† L'Elettromotore perpetuo. Trattato. 8vo. 361 pp. Parte Seconda. *Verona,* 1822

An article headed " All' Accademia Reale delle Scienze di Parigi," on his Electrical Clock. 8vo. 6 pp. 1 plate. (*Verona Poligrafo,* v. 1831, p. 87.) *Verona,* 1831

† Lettera al Direttore . . sopra un micrometro magneto-elettrico. Veronà li 16 Agosto, 1832. 4to. 1 page (*Ann. del Reg. Lomb.-Veneto,* ii. 229.) *Padova,* 1832

† Descrizione di un nuovo Galvanometro ossia elettroscopio dinamico universale. Memoria. 4to. 7 pp. 1 plate. (*Inserita nel Bimestre* 5 e 6 *degli Annali . . del Reg. Lomb.-Veneto,* 1833, tom. iii. p. 290.) *Padova,* 1833

† Elettroscopio dinamico universale. Memoria. 8vo. 4 pp. (*Commentarii dell' Ateneo di Brescia, per l'anno* 1832 (stamp. 1833), p. 38.) *Brescia,* 1833

† Sulla teoria elettro-chimica delle pile Voltiane al Sig. A. Fusinieri. Lettera ii. 4to. 7 pp. (*Estratta dagli Ann. del Reg. Lomb.-Veneto,* tom. vi. Bim. i. di Gennajo e Feb. 1836.) *Padova,* 1836

† Sull' argomento delle Pile secche contro la teoria elettro-chimica. Risposta . . ad una Nota del Fusinieri negli Annali delle Scienze del Regno Lombardo-Veneto, an. 1836, p. 143. 8vo. 35 pp. *Verona,* 1836

Zamboni, Giuseppe.—*continued.*

† Difesa degli argomenti tratti dalle pile secche per la teoria Voltiana contro le obbiezioni del Sig. de la Rive. Memoria. Ricevuta 5 Luglio, 1837. 4to. 19 pp. (*Mem. Soc. Ital.* xxi. 368.) *Modena,* 1837

Sulla durata della tensione elettrica nelle pile secche. " Letta nel quarto Congresso . degli Italiani a Padova." 8vo. 6 pp. (*Majocchi, Ann. di Fisica* . . viii. 1842, p. 14.) *Milano,* 1842

Sulla elettricità statica. *Verona,* 1842

† Sull' Elettromotore perpetuo istruzione teoretico-pratica. 8vo. 93 pp. 2 plates. *Verona,* 1843

Sulla teoria dell' elettroforo. 4to. (*Mem. Soc. Ital.* xxiii. 1844.) *Verona,* 1844

† Esame della memoria del . Peclet sullo sviluppo della elettricità statica nel contatto de' corpi. 4to. 10 pp. (*Mem. dell' Istit. Veneto,* tom. ii. p. 239.) *Venezia,* 1845

† Esame di una memoria del . Buff intorno all' elettroforo, e sulla migliore costruzione di questa macchina. 4to. 10 pp. (*Mem. dell' Istit. Veneto,* tom. ii. p. 251.) *Venezia,* 1845

† Elogio dell' Abate Giuseppe Zamboni dell' Ab. Ant. Rivato. 8vo. 47 pp. *Verona,* 1847

(*Note.*—*A letter from the Secretary of the " I. R. Istituto di Scienze . . . Venezia " to the members, announcing Zamboni's death, and adding some lines to his memory, dated 30 July, 1846, and signed Passini, G.*)

Elogio di, da Maggi, P. 8vo. *Verona,* 1851

(*Vide* also Ampère.)

† Portrait.

† **Zamboni, G. e Fusinieri, A.** Sulla teoria elettro chimica delle pile Voltiane. Lettera del Zamboni al Fusinieri. 4to. (*Ann. . Reg. Lomb.-Veneto,* tom. iv. pp. 128 and 132.) *Padova,* 1834

Articolo di Lettera al Fusinieri. Verona, 26 Mar. 1836. 4to. (*Ann. del Reg. Lomb.-Veneto,* vi. 142 and 143.) *Padova,* 1836

Zambra, Bernardino. *Born about* 1813 ; *died January* 7, 1859, *at Treviso.*

† Dell' arte galvanoplastica. 8vo. 14 pp. (*Giorn. I. R. Istit. Lomb.* i. 402.) *Milano,* 1841

† I principj e gli elementi della fisica esposti. 8vo. 473 pp. *Milano,* 1851

(*Note.*—*This is called* tom. i., *but not so printed in the title-page.*)

† I principj e gli elementi di fisica esposti. Tomo ii. 8vo. 415 pp. *Milano,* 1854

(*Vide* also Baumgartner and Zambra.)

Zamminer, F. G. K. Seine Magnetisirungs-Versuche (gemeinschaftlich mit Buff). (*From Lamont, Handb.*)

Zanetti. (*Vide* Larcher and others.)

Zangen, K. G. Über d. Läuten beym Gewitter. 8vo. *Marburg,* 1791

Zanon, Bartolomeo. *Born January* 21, 1792, *at Chias di Alpago, near Belluno.* Intorno un punto della nuova dottrina del Sig. G. Pelletier relativa all' influenza elettro-chimica delle varie terre sulla vegetazione. Osservazioni. 8vo. 12 pp. *Belluno,* 1840

Zanotti, Eustachio. *Born November* 27, 1709, *at Bologna ; died May* 15, 1782, *at Bologna.*

De Boreali Aurora anni 1737. 4to. (*Commentarii Bononienses,* ii. p. 476.) *Bononiæ*

Descrizione di una Aurora Boreale osservata nella specola dell' Istituto delle Scienze di Bologna dei 16 Decembre, 1737. 4to. 8 pp. 1 plate. *Bologna,* 1738

† Descrizione di una Aurora Boreale osservata . a Bologna 16 Dec. 1737. 12mo. 16 pp. *Venezia,* 1738

† De quibusdam luminibus septentrionalibus anno MDCCXXX. mense Martii observatis. 4to. 4 pp. *Bononiæ,* 1747

Zanotti e Matteucci, P. De borealibus auroris quibusdam anni 1739. 4to.
(*Comment. Bononienses*, ii. pt. i. p. 479.) *Bononia*

Zanotti. (*Vide* Pivati.)

Zanotti, Francesco Maria. *Born January* 6, 1692, *at Bologna ; died December* 25, 1777, *at Bologna.*
De globi cujusdam ignei trajectione. 4to. (*Comment. Bononienses*, tom. ii. p. 464.)
Bononiæ

Zantedeschi, Francesco. *Born August* 18, 1797, *at Dolcé, Prov. Verona.*
Sullo stato attuale dell' elettro-magnetismo in Italia. 1829
Dell' origine del magneto-elettricismo. 8vo. (*Bibl. Ital.* 1829 *and Bibl. Univ.* xlviii. 1830.) *Milano*, 1829

† Nota sopra l'azione della calamita e di alcuni fenomeni chimici. 8vo. 5 pp.
(*Bibl. Ital.* liii. 398.) *Milano*, 1829

Estratto di una lettera di A. De la Rive a lui diretta ; pubblicata nell' Ateneo Italiano che stampasi in Parigi (u. 7 Aprile, 1854, p. 6 e seq.)

(Analysis of) Esperienze elettro-magnetiche del . Moll. G. 8vo. 4 pp. (*Verona Poligrafo*, vii. 1831, p. 45.) *Verona*, 1831

Dell' influenza delle Calamite nella produzione de' fenomeni chimici. 8vo. 5 pp.
(*Verona Poligrafo*, viii. 1831, p. 3.) *Verona*, 1831

† Relazione intorno alla forza elettro-motrice del magnetismo. 8vo. 13 pp.
(*Poligrafo di Verona*, ix. 338.) *Verona* 1832

† Relazione dell' influenza, che esercita il contatto dei metalli eterogenei nelle loro proprietà chimiche. 8vo. 7 pp. (*Poligrafo di Verona*, x. 333.)
Verona, 1832

† Nota sopra qualche fenomeno elettrometrico. 8vo. 2 pp. (*Poligrafo di Verona*, x. 340.) *Verona*, 1832

† Riflessioni intorno alla Memoria de' Sig. Zantedeschi e Meyer risguardante la scossa della rana. 8vo. 4 pp. (*Poligrafo di Verona*, xiv. 61.) *Verona*, 1833

† Nuove esperienze sull' elettricità terrestre. Memoria (Account of). 8vo. 3 pp.
Brescia, 1833

† Della priorità della scoperta nella scienza elettro-magnetica. Memoria. 8vo. 6 pp. *Brescia*, 1833

† Sopra le scosse delle rane col mezzo di calamite. Articolo di lettera al direttore. 4to. 3 pp. (*Ann. Reg. Lomb.-Veneto*, iii. 57.) *Padova*, 1833

† Relazione delle principali scoperte magneto-elettriche. Presentata all' Ateneo di Brescia il 30 Gennajo, 1834. 8vo. 16 pp. (*Estratto dal Poligrafo.*)
Verona, 1834

† Lettera . sopra il passaggio pei liquidi delle correnti termo-elettriche. 4to.
1 page. (*Ann. . Reg.-Lomb. Veneto*, iv. 31.) *Padova*, 1834

Lettera al . G. G. Orti, Direttore del Poligrafo di Verona, &c. 8vo. (*Verona Poligrafo*, iii. 1834, p. 205.) *Verona*, 1834
(*Note.—This letter concerns writings by Ellice and Botto.*)

Progressi della scienza elettro-magnetica estratti da due memorie del S. Dal Negro. . 8vo. 14 pp. (*Verona Poligrafo.*) *Verona*, 1834

Progressi della Scienza Magneto-elettrica del Zantedeschi. 8vo. 9 pp.
(*Verona Poligrafo*, viii. 1835, p. 16.) *Verona*, 1835

† Effetti fisiologici ottenuti colle correnti magneto elettriche. Del Prof. al F. Zantedeschi. 4to. 1 page. (*Ann. del Reg. Lomb-Veneto*, v. 44.)
Padova, 1835

† Esperienze risguardanti la direzione e intensità delle correnti magneto-elettriche 8vo. 10 pp. *Brescia*, 1835

† Della dinamica e statica Magneto-Elettrica. Memoria. Presentata con lettura all' Ateneo di Brescia, 8 Mar. 1836. 8vo. 11 pp. (*Bibl. Ital.* lxxx. 399.) *Milano*, 1836

Zantedeschi, Francesco.—*continued.*

† Della polarizzazione dei conduttori isolati diretti a determinati punti del globo e di un nuovo apparecchio per esplorare l'elettricità atmosferica chiamato elettro-magnetometro. 8vo. 13 pp. *Milano,* 1837

† Della dinamica e statica magneto-elettrica. Memoria presentata con lettura all' Ateneo di Brescia 8 Marzo, 1836. 4to. 8 pp. *Padova,* 1837

†̇ Dell' influenza reciproca dell' elettro-magnetismo dei corpi ; nota. 4to. 4 pp. *Padua,* 1837

† Sulla priorità dell' azione elettrica dell' atmosfera nell' alterare la virtù delle magnete. 8vo. 2 pp. (*Bibl. Ital.* lxxxv. 135.) *Milano,* 1837

Della natura delle calamite e degli scandagli magnetici. Memoria. 8vo. 8 pp. (*Bibl. Ital.* lxxxvi. 134.) *Milano,* 1837

† Ricerche sul termo-elettricismo dinamico e luce-magnetico ed elettrico. 72 pp. 1 plate. *Milano,* 1837.

† Esperimenti d'induzione e polarizzazione termo-elettrica. 8vo. 4 pp. (*Bibl. Ital.* lxxxix. 123.) *Milano,* 1838

† Saggi dell' elettro-magnetico e magneto-elettrico. 8vo. 169 pp. 3 plates. *Venezia,* 1839

† Brano di lettera del . . F. Zantedeschi al . prof. Carlini. 8vo. 3 pp. (*Bibl. Ital.* xcvii. 124.) *Milano,* 1840

† Risposta all' articolo della Bibl. Ital. sui Saggi elettro-magnetico, etc. . 8vo. 4 pp. (*Bibl. Ital.* tom. c. pp. 269 and 272.) *Milano,* 1840

† Memoria sulle Leggi fondamentali che governano l'elettro-magnetismo. 8vo. 22 pp. (*Poligrafo di Verona,* tom. i. p. 200.) *Verona,* 1840

† Relazione storico-critica-sperimentale sull' elett.-magn. 8vo. 56 pp. 1 plate. *Venezia,* 1840

(*Note.—Called* (*at* p. 10) *Mem. prima, but no second was printed.*)

† Della Elettrotipia. Memorie. 4to. 52 pp. 6 plates. *Venezia,* 1841

† Risultamenti . . di nuove esperienze fatte sull' induzioni dinamiche. (Presentati all' I. R. Istituto, 9 Marzo, 1841.) Nota sull' induzione termo-elettrica . . letta alla riunione scientifica di Torino tenuta nel Settembre del 1840. 4to. 3 pp. (*Annali . del Reg. Lomb.-Veneto,* xi. 35 and 37.) *Vicenza,* 1841

Delle correnti elettro-vitali. *Atti Istit. Veneto,* Ser. i. vol. i. 1841.) *Venezia,* 1841

Delle correnti elettriche delle Torpedini. (*Fauna Ital. del Principe di Canino, all' Articolo Torpedo Narce ; Bull. Acad. Brux.* viii. 1841.) 1841

† Estratto di alcune esperienze dalla memoria inedita del Prof. Zantedeschi che ha per titolo, "Dei nodi termo-elettrici dell' apparato Voltiano, letto all' I. R. Istituto in Venezia . . 12 Luglio, 1841." 4to. 4 pp. 1 woodcut. (*Ann. del Reg. Lomb.-Veneto,* xi. 135.) *Vicenza,* 1841

† Lettera al Fusinieri sulla induzione dinamica attraverso involucri e diaframmi di ferro. 4to. (*Ann. . del Reg. Lomb.-Veneto,* xi. 223.) *Vicenza,* 1841

† Lettera al Fusinieri sull' indizionometro dinamico differenziale, &c. . . 4to. 1 p. (*Annali Sci. Reg. Lomb.-Veneto,* xi. 226 and 238.) *Vicenza,* 1841

† Sui conduttori bipolari e unipolari termo-elettrici. 4to. 2 pp. (*Ann. Scien. Reg. Lomb.-Veneto,* xi. 261.) *Vicenza,* 1841

† Osservazioni ed Esperienze sulle condizioni, e sulle leggi, dei fenomeni elettro-termici dell' apparato Voltiano ; e sulle cause che sono loro assegnate dai fisici. Memoria. 4to. 20 pp.

† Continuazione e fine della Memoria. . . 4to. 6 pp.

† Tabella elettro-termica relativa alla Memoria del Zantedeschi al precedente pag. 23. 4to. 1 p. (*Ann. . del Reg. Lomb.-Venet ,* xii. 13, 65, and page not numbered.) *Vicenza,* 1842

ZANTEDESCHI 557

Zantedeschi, Francesco.—continued.

† Di alcune modificazioni fatte alla macchina magneto-elettrica di Newman e dei speciali esperimenti eseguiti con essa. Memoria. Presentata all' . . I. R. Istituto Veneto. 18 Dec. 1840, e letta . . 27 Dec. 1840. 4to. 10 pp. 1 plate. (*Ann. Scien. Lomb.-Veneto*, xii. 73.) *Vicenza*, 1842

† Risposta all' accuse date sulla priorità di alcune scoperte dal . . Majocchi al . Zantedeschi (Annali di Fisica, Chimica e Matematiche di Milano, Aprile, 1842, pag. 89.) 4to. 14 pp.

† Seguito della Risposta . . e Nota di Fusinieri. 4to. 2 pp. (*Ann. . del Reg. Lomb.-Veneto*, xii. pp. 193 and 276.) *Vicenza*, 1842

Sopra alcuni fenomeni che presentano i poli di un Elettromotore Voltiano, e, precisamente, sopra la facoltà calorifica e combustiva. Memoria. 8vo. (*Ital. Soc. Mem.* xxiii. 10.) *Venezia*, 1842

† Dei conduttori bipolari e unipolari termo-elettrici. 8vo. 2 pp. (*Giorn. I. R. Istit. Lomb.* iii. 441.) *Milano*, 1842

† Le leggi del Magnetismo nel filo congiuntivo percorso dalla corrente voltiana. 8vo. 7 pp. 1 plate. *Venezia*, 1843

Dell' influsso dei raggi solari sulla vegetazione delle piante. 4to. (*Mem. dell' Istit. Veneto*, i. 269.) *Venezia*, 1843

Le leggi elettro-magnetiche. 8vo. *Verona*, 1843

† Risposta . . all' articolo del Prof. Majocchi intitolato, "Alcune osservazioni risguardanti le correnti magneto-elettriche." 4to. 3 pp. (*Ann. . del Reg. Lomb.-Veneto*, xiii. 10.) *Vicenza*, 1844

† Dell' esistenza delle due opposte correnti di materia attenuata nell' elettromotore Voltiano. Esperienze del . F. Zantedeschi (comunicata il giorno 8 Gennajo, 1844, all' Ateneo Veneto.) (Con piccola Nota di Fusinieri.) 4to. 4 pp. (*Ann. . del Reg. Lomb.-Veneto*, xiii. 13.) *Vicenza*, 1844

Delle induzioni dinamiche leido-elettriche. (*Atti Istit. Veneto*, Ser. i. vol. iii. 1844.) *Venezia*, 1844

† Del movimento vorticoso o a spirale della luce voltiana, e di altri fenomeni osservati ai due poli dell' elettromotore del Volta. 4to. 2 pp. (*Ann. . del Reg. Lomb.-Veneto*, xiii. 107.) *Vicenza*, 1844

† Del trasporto della materia pesante nelle due opposte correnti dell' apparato voltiano, della loro Natura, e del moto vorticoso, o a spirale, dell' arco luminoso. Memoria. 4to. 11 pp. (*Ann. . del Reg. Lomb.-Veneto*, xiii. 169.) *Vicenza*, 1844

† Memoria sul Termo-Elettricismo Dinamico nei circuiti formati di un solo metallo. 4to. 13 pp. 3 figures. (*Ann. . del Reg. Lomb.-Veneto*, xiii. 191.) *Vicenza*, 1844

† Trattato del Magnetismo e della Elettricità. Parte i. 8vo. 389 pp. 3 plates. *Venezia*, 1844

† Trattato del Magnetismo e della Elettricità. Parte ii. 8vo. 546 pp. 4 plates. *Venezia*, 1845

Lettera al Principe di Canino. (*Trattato*, ii. 341.) *Venezia*, 1844

† Descrizione di una Macchina a disco per la doppia elettricità, e delle esperienze eseguite con essa comparativamente a quelle dell' elettromotore Voltiano. Letta il 16 Luglio, 1843. 4to. 11 pp. 1 plate. (*Mem. dell' Istit. Veneto*, tom. ii. p. 171.) *Venezia*, 1845

† Osservazioni . . alla descrizione della batteria magneto-elettrico-tellurica ed alla continuazione delle ricerche intorno ai fenomeni d'induzione del magnetismo terrestre di Luigi Palmieri. 4to. 9 pp. (*Ann. Reg. Lomb.-Veneto*, xiv. 45.) *Vicenza*, 1845
   Note.—"*Articoli estratti dal No.* 17 *del Rendiconto della Rle. Accad. delle Scienze di Napoli.*")

T T 2

Zantedeschi, Francesco.—*continued.*

† Memoria sugli effetti fisici chimici e fisiologici prodotti dalle alternative dell correnti d'induzione della macchina elettro-magnetica di Callan. Letta ii. Agosto, 1844, all' I. R. Istituto Veneto. 4to. 8 pp. (*Ann. Scien. Reg. Lomb.-Veneto*, xiv. 110.) *Vicenza*, 1845

† Della teoria fisica delle Macchine magneto-elettriche ed elettro-magnetiche. Letta all' Ist. Veneto, 19 Genn. 1845. 4to. 12 pp. (*Ann. Scien. Reg. Lomb.-Veneto*, xiv. 231.) *Vicenza*, 1845

† Della struttura dell' organo elettrico della Torpedine. Osservazioni. 4to. 4 pp. (*Inserito nei Bim.* v. *e* vi. 1845, *degli Ann. delle Scien. del Reg. Lomb.-Veneto.*) *Vicenza*, 1845

† Raccolta fisico-chimica Italiana, ossia Collezione di Memorie originali edite ed inedite di Fisici, Chimici, e Naturalisti Italiani. Tom. i. 4to. 573 pp. 3 plates. (*Vide* 1847 and 1848.) *Venezia*, 1846

Sulla virtù illuminante del polo negativo e calorifico del polo positivo dell' elettromotore Voltiano. 8vo. *Venezia*, 1846

Osserv. ed esperim. sulla termo-elettricità della piroscellina di Schönbein. 8vo. *Venezia*, 1846

† Giudizii di elettricisti oltramontani sul Trattato del Magnetismo e della Elettricità del . Zantedeschi. 4to. 6 pp. *Venezia*, 1846

Mention by Minotto in "Relazioni" of three works of Zantedeschi (communications, &c.) One of them is " sull' uffizio elettrico della midolla allungata nella Torpedine of 18 Mag. 1824." 8vo. (*Venetian Athenæum Esercitazioni*, v. 323 *et seq.*) *Venezia*, 1846

Ricerche fisico-chim. . . sulla luce. La. 4to. (*From Arago's Sale Cat.*) *Venezia*, 1846

† Sulle vibrazioni dei corpi sottoposti all' influenza del magnetismo e della elettricità, &c. 8vo. 16 pp. (*Contained in his* "*Raccolta,*" tom. ii. p. 467.) *Venezia*, 1847

† Raccolta fisico-chimica Italiana. Tom. ii. 4to. 574 pp. 5 plates. *Venezia*, 1847

† Sulla universalita dell influenza elettro-magnetica nei corpi. 8vo. 3 pp. (" *Dalla Gazz. Piem. del* 16 *Otto* 1847," No. 246.) 1847

† Dei movimenti che presenta la fiamma sottoposta all' influenza elettro-magnetica. 8vo. 7 pp. (*Dalla Gazz. Piem. del* 12 *Otto* 1847, No. 242.) 1847

Illustrazione di alcuni fenomeni di elettro-magnetismo. 8vo. *Venezia*, 1848

Dell' influenza elettro-magnetica nei corpi, coll' analisi di una nota del Prof. Bancalari. 8vo. *Venezia*, 1848

† Della condizione magnetica e diamagnetica proprie del regno inorganico ; e della condizione diamagnetica generale ai composti dei regni organici. 4to. 4 pp. (*Estratto dal fasc.* viii. *del* t. iii. *della Raccolta fisico-chim. Ital.* p. 391.) *Venezia*, 1848

† Raccoltà fisico-chimica Italiana. . . Tomo iii. 4to. 558 pp. 7 plates. *Venezia*, 1848

† Dello sviluppo della elettricità nell' atto della contrazione muscolare. Nuove esperienze del . Z. 4to. 7 pp. *Roma*, 1849
(*Note.*—" *Lettera estratta dalla Corrispondenza Scientifica in Roma. Bullettino Universale*, anno ii. *di sua fondazione*, No. 15, 12 Dec. 1849.")

Della bipolarità chimico-elettrica delle irradiazioni prismatiche, &c. 8vo. *Venezia*, 1847

† Annali di fisica. 8vo. 400 pp. 1 plate. *Padova*, 1849-50

Di una nuova esperienza di Alfredo Smee, &c. &c. (*Comptes Rendus*, xxxii. 1851.) *Venezia*, 1850

Dell' influenza del magnetismo sull' arco voltiano. *Venezia*, 1850

Delle variazioni di temperatura immediatamente prodotte dal magnetismo. *Venezia*, 1850

Dell' azione calorifica ai poli dell' elettromotore voltiano. *Venezia*, 1850

Zantedeschi, Francesco.—*continued*.

Dei fenomeni luminosi studiati ai due poli dell' elettromotore voltiano.
Venezia, 1850

† Descrizione di un nuovo Dinamoscopio atomico. 8vo. 3 pp. (*Estr. dalla Puntata* ii. p. 29 *del Giorn. Fis. Chim. Ital.* 1851.) *Venezia*, 1851

Risposta alle osservazioni e nuove sperienze sopra un fenomeno avvertito da Dubois-Reymond, di Luigi Magrini. 8vo. (*Akad. Wien*, 1851, vii. Bd. i. Heft, p. 220.) *Venezia*, 1851

" Giornale Fisico-chimico Italiano." 2 vols. *Venezia e Padova*, 1851-52

† Memorie di Fisica del F. Zantedeschi. 4to. 28 p. (*Estratte dalle Puntate* ii. iii. e iv. *del Giornale Fis. Chim. Ital. del* 1852.) *Padova*, 1852

Nouvelles expériences d'électricité animale. 4to. (*Comptes Rendus*, xxxv. 1852.) *Paris*, 1852

† De la différence du pouvoir des deux Electricités. Note communiquée par M. Arago. 4to. 1 p. (*Comptes Rendus . de l'Acad.* xxxv. *Séance* Sept. 1852.) (*Faraday's copy.*) *Paris*, 1852

Dei fenomeni elettrici della macchina di Armstrong, &c. (" Letta il 9 Agosto, 1846." 4to. 28 pp. (*Mem. dell' I. R. Istit. Veneto*, . . vol. iv. p. 45.) *Venezia*, 1852

† Della elettricità degli stami e pistilli delle piante esplorata all' atto della fecondazione e di una nuova classificazione delle linfe o succhi vegetabili fondata sul numero e sulla direzione delle correnti elettriche longitudinali e trasversali. Memorie. 4to. 56 pp. 1 plate. (*Atti dell' Accad. dei Nuovi Lincei*, 1853.) *Padova*, 1853

Dell' esistenza e della natura delle correnti elettrici nei fili telegrafici. (*Abhandl. d. K. Akad. d. Wissensch.* Nov. 1853.) *Wien*, 1853

La termocrosi di Melloni dimostrata insussistente ; e l'autore in opposizione con se stesso. Ricerche del Z. 4to. 7 pp. *Padua ?* 1853

Nuovi esperimenti risguardanti l'origine della elettricità atmosferica e dell' induzione elettrostatica dei conduttori solidi isolati. 8vo. *Venezia*, 1854

† All' insig. Chimico Dumas . . dell' Istituto. Dell' azione reciproca di due correnti elettriche dirette nel medesimo senso e in senso opposto nello stesso filo ; e dell' azione induttiva laterale delle medesime in fili isolati paralleli vicinissimi. 4to. 3 pp. (*Faraday's copy.*) *Padova*, 1854

† Sur le Principe électrostatique de Palagi, et ses expériences. Lettre . . à M. Quetelet. 4to. 2 pp. (*Extrait du* tom. xxi. No. 2 *des Bulletins. Acad. Rle. de Belgique, Séance du* 4 Février, 1874.) (*Faraday's copy.*) *Padova*, 1854

Nota intorno uno scaricatore elettro-telegrafico delle stazioni. (*Atti Istit. Venet.* Ser. ii. vol. v. 1854.) *Venezia*, 1854

Della interferenza luminosa che presenta il filo metallico comune a due circuiti chiusi. (*Sitzungsb. Wien Acad.* xvi. 1855.) *Wien*, 1855

Ricerche sulla contemporaneità del passaggio delle opposte correnti elettriche in un filo metallico. (*Wien Acad.* xvii. 1855.) *Wien*, 1855

Mem. sugli argomenti comprovanti il simultaneo passaggio delle opposte correnti sullo stesso filo conduttore comune a due circuiti chiusi ed isolati. (*Atti Istit. Venet.* Ser. ii. vol. vi. 1855.) *Venezia*, 1855

Mem. sul simultaneo passaggio delle correnti elettriche opposte, etc. 4to. (*Atti Istit. Venet.* Ser. ii. vi. 1855.) *Venezia*, 1855

Documents à l'appui de sa réclamation de priorité concernant les variations de température produits par le magnétisme. 4to. (*Comptes Rendus*, xli. 55.) *Paris*, 1855

Etudes d'Electro-physiologie. Expériences sur le système ganglionnaire, et sur le système cérébro spinal. 8vo. (*La Science, Journal du Progrès des Sciences* . . 8 Sept. 1855.) *Paris*, 1855

Nuovo elettroscopio par le due elettricità d'influenza. 4to. (*Sitzungsb. Wien Acad.* 1855.) *Wien*, 1855

Zantedeschi, Francesco.—*continued.*

Telegrafo elettro-magnetico delle Stazioni e delle Locomotive delle Strade
ferrate, di Zantedeschi.
(*Note.*—*This is dated Padova il di* 27 Gennajo 1855. *No printing date* (*Tip.
Sicca*). 1 p. *only.*)

Risposta .. ai Cenni della Relazione del Sig. Dott. Gintl. intorno al contem-
poraneo passaggio delle correnti opposte in un solo filo. 4to. 2 pp.
                                                              Padova, 1855
Elettricità dinamica. Di alcune nuove proprietà delle correnti elettriche in
circuiti comunicanti fra di loro. (Dated Padova, il 14 Feb. 1855.) 8vo.
1 p. (*Estratta dalla Gazzetta ufficiale di Venezia,* No. 38 and 16, Feb. 1855.)
                                                              Padova, 1855
Della contemporaneità e sincronismo delle opposte correnti attraverso un conduttore comune a due circuiti chiusi ; e degli effetti, non che delle applicazioni
che ne derivano. 4to. 3 pp.   *Padova,* 1855
Telegrafo a correnti dirette successive e derivate e contemporanee per la doppia
simultanea corrispondenza sopra un solo filo comunicante colla terra. 4to.
2 pp.   *Padova,* 1855

† An article headed "Physique Appliquée," in the journal *La Science, Journal
du Progrès des Sciences* .. quoting his note "sur les courants électriques
dirigés en sens opposé sur le même fil" .. presented to the French Acad.
16 and 17 August is the date of the number of the journal. Folio. 148 pp.
                                                                        1855

† A Notice headed Electricity, concerning two little works presented to the
Academy. 4to.   *Paris,* 1855

† Sur les Courants électriques . en sens opposé .. en relation avec la Télé
graphie. 4to. 3 pp. (*Comptes Rendus,* xli. 1855.)   *Paris,* 1855

† De l'influence lumineuse que présente le fil . commun à deux circuits formés,
&c. . 4to. 3 columns of *La Science.*   *Paris,* 1855

A large plate and figure, headed, "Disposizione degli apparati magneto-elettrici co' quali furono eseguiti gli esperimenti del passaggio simultaneo di due
correnti opposte sul medesimo filo telegrafico, nel giorno 27 Ottobre nell' I. R.
Università di Padova."
(*Note.*—*This plate is lettered. It is not described in the article headed Telegrafo*·
a correnti dirette, &c.)

Del moto rotatorio dell' arco luminoso dell' elettromotore voltiano. (*Sitzungsb.
d. Wien Acad.* xxi. 1856.)   *Wien,* 1856

Risultamenti ottenuti da un giroscopio. (*Sitzungsb. Wien Acad.* xvii. 1397.)
                                                              Wien, 1857
La non simultanea esistenza di due correnti opposte sul medesimo filo conduttore. (*Sitzungsb. Wien Acad.* xxii. u. xxvii.)   *Wien,* 1857

† De mutationibus quæ contingunt in spectro solari fixo Elucubratio. 4to.
11 pp. 1 plate. (*Akad. d. Wissensch.* viii. 1 Abth. (ii. Classe.)
                                                              Monachii, 1857
Della misura dei limiti della sensibilità nervo-muscolare dell' uomo, studiata
comparativamente alla forza dello stesso : Esperienze. 4to. (*Atti dell' I.
R. Ist. Veneto,* p. 58, 1858.)   *Venezia,* 1858

† Ueber die physikalischen Studien und Entdeckungen der Italiener im Jahre
1858. 8vo.   *Padua,* 1858

† Osservazioni ai nuovi sforzi fatti dal Belli, a difesa dei due esperimenti addotti
dal Matteucci, e dal Petrina, contro la simultanea esistenza di due' opposte
correnti elettriche sul medesimo filo conduttore. 8vo. 8 pp. *Wien,* 1858
(*Nota II.*—*Dal fascicolo di Decembre dell' anno* 1857, *della Classe di Matematica
e Scienze Naturali dell' Accademia Imp. delle Scienze ; specialmente
stampato.*)

Di alcuni nuovi esperimenti, co' quali si è creduto di comprovare la non simultanea esistenza di due correnti opposte sul medesimo filo conduttore. 8vo.
7 pp.   *Vienna,* 1858
(*Nota.*—*Dal fascicolo d'Ottobre dell' anno* 1856, *della Classe di Matematica e
Scienze naturali dell' Accademia Imp. delle Scienze ; specialmente stampato.*)

Zantedeschi, Francesco.—*continued.*

† Intorno alla influenza dell' elettrico nella formazione della gragnuola, e dei mezzi economici a preservare le campagne dai danni della grandine, &c. 8vo. 24 pp. *Padova*, 1860

† Dei presagi delle burrasche e della dottrina della rugiada e della brina. Illustrazioni con un' Appendice . . 8vo. 19 pp. *Padova*, 1865

† Della natura elettrica dell' Ozono ed Antozono, della loro propagazione ed effetti. 8vo. 12 pp. (*Atti dell' Istituto*, Ser. iii. vol. xi.) *Venezia*, 1866

† Gli allarmi magnetici delle burrasche, e i presagi della telegrafia meteorologica. Documenti storici. 8vo. 28 pp. *Padova*, 1866

† Risposta documentata . . all' articolo del P. A. Secchi nel Bollet. meteorol. dell' Osservat. del Collegio romano ; vol. v. p. 107. N. di Ottob. 1866, intorno ai presagi delle meteore e delle burrasche, con documenti. 8vo. 20 pp. *Padova*, 1866

† Elenco generale dei principali capitoli di Memorie o note Fisiche pubblicati dal Prof. F. Zantedeschi dal 1820 al 1867. 8vo. 8 pp. *Padova*, 1867

† Intorno all' elettricità indotta e d'influenza, negli strati aerei dell' atmosfera. 8vo. 7 pp. 1 plate. (*Atti dell' Istituto*, ser. iii. vol. xii.) *Venezia*, 1867

Meteorology in Italy. Report on Temperature in Italy, showing the hourly, daily, monthly, and annual oscillations of Heat, for 1867. 4th Report.

Publiche date del magneto-elettrico ed elettrico-magnetico. 8vo. 28 pp. *Padova*, 1868

Intorno al Magnetismo transversale alla direzione della corrente elettrica. 8vo. 12 pp. *Padova*, 1869

† Documenti raccolti intorno alle date di alcune moderne scoperte di elettricità applicata. 8vo. 17 pp. (*Estr. dal* vol. xiv. serie iii. *degli Atti del R. Istituto Veneto.*) *Venezia*, 1869

Intorno all' elettro-chimica applicata all' industria ed alle belle arti ; lettera all' ill. Sig. Besso. 8vo. 7 pp. *Padova*, 1870

Delle oscillazioni calorifiche orarie, diurne, mensili ed annuali del 1867, con alcune indicazioni di meteore, uragani, terremoti e fulmini accaduti nel 1867, e della loro connessione colla elettricità atmosferica e coi perturbamenti di magneti e dei fenomeni astronomici. 8vo. 105 pp. (*Estratto dal* vol. xv. serie iii. *degli Atti dell' Istituto Veneto di Scienze ed Arti.*) *Venezia*, 1870

Breve riassunto di tremoti, di vittime e di fulmini e di grandini desolatrici. 4to. *Venezia*, 1870

† Risposta . . all' Articolo del Cosmos. 19 livraison, 4 Nov. 1859. Vol. xv. page 524. 8vo. 4 pp. (*Estratto dalla Gazzetta di Trento*, No. 273.)

Reclamo contro un Articolo del Sig. D. Al. Donnè sulle scoperte magneto-elettriche. 8vo. *Milano*

† Sur la Télégraphie électrique. Extrait d'une lettre à Quetelet. 8vo. 2 pp. *Bruxelles*

Trattato del Magnetismo e dell' Elettricità. 2e edition. 16mo. *Milano*

(*Vide* also Anon. Elect. 1870.)

† Portrait. Lithog. 1856.

⊹ **Zantedeschi** e **Mayer.** Esperienze intorno alle alterazioni della virtù magnetica per l'azione del calorico e di qualche altro fenomeno relativo. Memoria. 8vo. 42 pp. 2 plates. (*Estr. dal Poligrafo*, Agosto, 1831.) *Verona*, 1831

† Sopra le osservazioni ed esperienze risguardanti la scossa della rana sottomessa all' influenza degli elettromotori voltaici ed i conduttori che fanno arco di comunicazione. Poligrafo, Luglio, 1832. Articolo comunicato. 4to. 2 pp. (*Ann. Reg. Lomb.-Veneto*, ii. p. 357.) *Padova*, 1832

† Nuove esperienze intorno all' origine dell' elettricità terrestre. Memoria. 8vo. 9 pp. (*Poligrafo di Verona*, ix. 8,) *Verona*, 1832

Zantedeschi e Mayer.—*continued.*

† Osservazioni ed Esperienzo risguardanti la scossa della rana, sottomessa all influenza degli elettromotori voltaici ed i conduttori che fanno arco di comunicazione. Memoria. 8vo. 46 pp. (*Poligrafo di Verona*, x. 3, xi. 321, xii. 161.) *Verona*, 1832

(*Vide* also Fario and Zantedeschi.)

(*Vide* also Stefani and Zantedeschi.)

(*Vide* also Barlocci and Zantedeschi.)

(*Vide* also Quetelet and Zantedeschi.)

† **Zecchinello, L.** Assertiones tres physico-mathematicæ quas . . in publicum certamen exponit Laurentius Zecchinello . . Adiutore A. Zuccoli. 8vo. 137 pp. 1 plate. *Patavii*, 1809

Zegollström. Theoria declinationis magn. *Upsala*, 1755

† **Zehfuss, G.** Beiträge zur Theorie der statistischen Electricität. 8vo. 36 pp. *Frankfurt*, 1865

† Die kosmische Bedeutung der Aerolithen namentlich gegenüber der Sonne, den Eiszeiten und dem Magnetismus der Himmels Körper. In gedrängter Darstellung. 8vo. 18 pp. *Frankfurt*, 1869

† **Zeihero, J. E.** Acus novæ declinatoriæ descriptio. 4to. 4 pp. 1 plate. (*Novi Comment. Acad. Petrop.* vii. *pro* 1758 *et* 1759 ; *imp.* 1761, p. 309.) *Petrop.* 1758-9

Zeipel, E. V. E. De vi electro-magnetica observationes. 4to. 1851

(*Vide* also Şwanberg and Zeipel.)

Zeitschrift für Physik und Mathematik. 8vo. *Wien*, 1806-31-2

(*Vide* Baumgartner and Ettingshausen.)

Zeitschrift d. deutsch österreichischen Telegraphen-Vereins. 4to. *Berlin*, 1854 *to* 1870 ?

Zendrini, Bernardino. *Born April* 7, 1679, *at Saviore ; died May* 18, 1747, *at Venice.*

Discorso fisico matematico sopra il Turbine accaduto in Venezia, l'anno 1708. (*Galleria di Minerva*, iv. 1708, *auch Act. Erudit.* 1708.) 1708

† Osservazione dell' Aurora boreale fatta in Venezia. 16th Dec. 1737. 12mo. 6 pp. (*Calogera, Raccolta*, xvii. 15.) *Venezia*, 1738

Zengen, C. G. Von. Über das Läuten bei Gewittern besonders in Hinsicht der deshalb zu treffenden Polizeyverfügungen. 8vo. *Giessen*, 1791

† **Zenger, W.** Über eine indirecte Methode z. Bestimmung d. Inclination. Vorgelegt am 14 Dec. 1854. 8vo. 14 pp. (*Sitzungen. Wien Acad.* vol. xv. 1 Heft.) *Wien*, 1854 ?

† Über die Messung der Strom-Intensität mit der Tangenten Boussole. Vorgelegt in d. Sitzung vom 19 April, 1855. 8vo. 14 pp. (*Sitzungsbericht d. Wien Acad.*) *Wien*, 1855 ?

Zenneck, L. H. Elektr. Verprüfung-Instrument. (*Erdman's Journ. f. Pract. Chem.* x. 1837.) *Tubingen*, 1837

† **Zetzell, P.** Anmerkung von der Lahmheit. 8vo. 5 pp. (*K. Schwed. Akad. Abh.* xvii. 59.) *Hamburg*, 1755

Nouvelles expériences sur les effets de l'électricité dans plusieurs maladies. 8vo. (*Journ. d. Médecine*, Octobre 1756.) *Paris*, 1856

Versuch über die Wirkungen der Elekt. in verschiedenen Krankheiten. (*From Kuhn, Hist.* ii. 379.) 1756

Zetzell sub presidio C. Linnæi. Consectaria electro-medica. 4to. *Upsalæ* 1754

(*Vide* also Anon. Medical Elect. 1763.)

† **Zetzsche, E.** Die Elektricitätslehre vom Standpuncte der Undulations theorie. Ein Versuch. 8vo. 13 pp. (*Zeitschrift für Mathem. u. Phys.* iii. 365, iv. 131.) *Wein*

Zetzsche, Karl Eduard. Die Copirtelegraphen, die Typendruck-telegraphen und die Doppeltelegraphie. Ein Beitrag zur Geschichte der elektrischen Telegraphie. 8vo. 199 pp. *Leipzig,* 1865

Zeune, Aug. Ueber Basalt polarität. (*Allg. Lit. Zeit. Inst.* 1805, 169.)
Berlin, 1805-9

Zeuner, Gust. Anton. Die Weisbachschen Versuche über d. Stoss d. isolert. Wasserstrahls gegen ruhende u. bewegte Flächen. (*Civil-Ing.* i. 1854.) 1854

Ziemssen, H. Die Elektricität in d. Medicin. 1st edition. 8vo. 4 plates.
Berlin, 1857

† Die Elektricität in der Medecin. Studien. 2nd edition. 8vo. 169 pp. 1 plate.
Berlin, 1864

Die Elektricität in der Medecin. Studien. 3 umgearbeitete Auflage. 8vo.
Berlin, 1866

Ueber Lähmung von Gehirnnerven durch Affectionen an der Basis cerebri. (*Virchow's Archiv,* Bd. xiii. Heft ii. *u.* iii.)

Zimmermann, Wilhelm Ludwig. *Born October* 7, 1780, *at Bickenbache, Hesse Darmstadt; died July* 19, 1825, *at Giessen.* (*Poggendorff,* ii. 1412.)

† Einige merkwürdige die Metallvegetation begleitende Phenomena. 4to. 24 pp. *Giessen,* 1811

Über eine neue Entstehungs art mehrerer Metallothion-und-Hydrothionmetall-Arten. 4to. *Giessen,* 1816

† New Facts respecting the Atmosphere. 8vo. 1 p. (*Phil. Mag.* lxii. 154.)
London, 1823

Ueber magnetischen Serpentin vom Frankensteiner Schloss, bei Darmstadt. (*Gilbert's Ann.* xxviii. 483.)

Zimmermann, W. F. A. Naturkräfte u. Naturgesetze, Popul. Handbuch der Physik. 1 Bd. Elektricität, Magnetismus, Galvanismus. 8vo.
Berlin, 1856

† Elektricität. Magnetismus. Galvanismus. Was wissen wir bis jetzt darüber? und welche praktische Anwendung haben wit gefunden? Eine populäre Darstellung. 8vo. 632 pp. *Berlin,* 1856

† **Zimpel.** C. F. Die Reibungs-elektricität in Verbindung mit Imponderabilien als Heilmittel nach dem System von C. Beckensteiner. 8vo. 207 pp.
Stuttgart, 1859

† **Zollinger,** H. Ueber die Gewitter und andere damit verwandte meteorologische Erscheinungen im Indischen Archipel. 8vo. 166 pp. (*Aus der Vierteljahrschrift der Naturforschenden Gesellschaft in Zürich.* Bd. iii. Heft iii. u. iv. abgedrt.) *Zürich,* 1858

Zöllner, J. K. Fried. Photometrische Untersuchungen, insbesonderer über d. Lichtenwickl. galvan. glühender Platindrähte (Diss.). 4to. (*Auszug in Pogg. Ann.* cix. 1860.) *Basel,* 1859

† **Zorn** von **Plobsheim,** F. A. Beschreibung der auf d. Tab. abgebildeten siberischen gediegenen Eisenstufen. 4to. 3 pp. 1 plate. (*Neue Samml. von Versuche* . . *d. Naturf. Gesells. in Danzig,* i. 288.) *Danzig,* 1778

† **Zornlin.** What is a Voltaic Battery? 16mo. *London,* 1842 (?)

Zuccala, G. (*Vide* Volta, 1827.)

Zucchi, Nicolò. *Born December* 6, 1586, *at Parma; died May* 21, 1670, *at Rome.* (*Poggendorff,* ii. 1421.)

Nova de machinis philosophia. 4to. *Romæ,* 1649

Zucconi, Ludovico. *Born about 1706, at Venice; died June 30, 1783, at Venice.* (*Poggendorff,* ii. 1421.)

† L'elettrometro o sia la misura di forza elettrica . . da D. L. Z. in 4 Lettere. 8vo. 28 pp. 1 plate. (*Volta's copy.*)   *Venezia,* 1756

† Descrizione ed esame di un grazioso fenomeno osservato nel Dec. del 1765 e nell' anno corrente 1767 da D. L. Z. Sm. 8vo. 14 pp. ("*Disegno a ghiaccio finaturale . . formato in alcune delle mie finestre.*")   1767

Zuchold, E. A. Bibliotheca Historico-Naturalis, Physica, Chimica, et Mathematica; oder systematisch geordnete Übersicht der in Deutschland und dem Auslande auf dem Gebiete der gesammten Naturwissenschaften und der Mathematik neuerscheinenden Bücher. 2 vols. per ann. 8vo.
Göttingen, 1851 *to* —

Zurcher. (*Vide* Margollé.)

Zurla, P. Di Marco Polo e degli altri Viaggiatori Veneziani più illustri. Dissertazioni; con Appendice sulle antiche Mappe idro-geografiche lavorate in Venezia. 2 vols. 4to.   *Venezia,* 1818

† **Zuylen Van Nyevelt,** Van. Notice respecting some new Electro-Magnetic Phænomena. Abridged and translated from the Bibliot. Univ. Aug. 1823, p. 274. 8vo. 2 pp. (*Edinb. Phil. Journal,* x. 130.)   *Edinburgh,* 1824

Zwinger, Theodor. *Born August 26, 1658, at Basle; died April 22, 1724, at Basle.* Scrutinum magnetis physico-medicum. 8vo. (*From Pogg.* ii. 1424.)   *Basel,* 1697

Specimen Physicæ electrico-experimentalis e compendio phisico H. Suiceri aliisq. admotum. 2 vols. 12mo.   *Basel,* 1707

THE END.

Printed in the United States
By Bookmasters